Nanotechnologies and Nanomaterials: Selected Papers from CCMR—Volume I

Nanotechnologies and Nanomaterials: Selected Papers from CCMR—Volume I

Editors

Jihoon Lee
Ming-Yu Li

MDPI

Basel • Beijing • Wuhan • Barcelona • Belgrade • Novi Sad • Cluj • Manchester

Editors

Jihoon Lee
Department of Electronic Engineering, College of Electronics and Information, Kwangwoon University
Seoul
Korea

Ming-Yu Li
School of Science, Wuhan University of Technology
Wuhan
China

Editorial Office
MDPI
St. Alban-Anlage 66
4052 Basel, Switzerland

This is a reprint of articles from the Special Issue published online in the open access journal *Nanomaterials* (ISSN 2079-4991) (available at: https://www.mdpi.com/journal/nanomaterials/special_issues/nano_CCMR_Series).

For citation purposes, cite each article independently as indicated on the article page online and as indicated below:

Lastname, A.A.; Lastname, B.B. Article Title. *Journal Name* **Year**, *Volume Number*, Page Range.

Volume I
ISBN 978-3-7258-0879-3 (Hbk)
ISBN 978-3-7258-0880-9 (PDF)
doi.org/10.3390/books978-3-7258-0880-9

Volume I-II
ISBN 978-3-7258-0807-6 (Hbk)
ISBN 978-3-7258-0808-3 (PDF)

Contents

About the Editors

Jihoon Lee

Jihoon Lee received his Ph.D. degree in Microelectronics and Photonics from the University of Arkansas, Fayetteville, AR, USA, in 2008, and a Management of Business Administration degree from Sam M. Walton College of Business, University of Arkansas, Fayetteville, AR, USA, in 2006. In 2009, he joined the faculty of Kwangwoon University, Seoul, South Korea, where he is currently a professor in the Department of Electronics Engineering, College of Electronics and Information. His current research interests include the fabrication and characterization of various hybrid metals and semiconductors, quantum- and nanostructures, and their applications in biosensors, detectors, and energy.

Ming-Yu Li

Ming-Yu Li is a faculty member at Wuhan University of Technology (WUT). He received his Ph.D. in Electronic Engineering from Kwangwoon University, South Korea, in 2016. He served as a postdoctoral fellow for 3 years at Huazhong University of Science and Technology (HUST). His research interests focus on nanomaterial synthesis, nanostructure fabrication and characterization, functional materials, and optoelectronic devices. He has published more than 92 research papers in international journals and has authorized more than 16 patents.

nanomaterials

MDPI

ditorial

Nanotechnologies and Nanomaterials: Selected Papers from CCMR

Ming-Yu Li [1,*] and Jihoon Lee [2]

1 School of Science, Wuhan University of Technology, Wuhan 430070, China
2 Department of Electronic Engineering, College of Electronics and Information, Kwangwoon University, Nowon-gu, Seoul 01897, Republic of Korea; jihoonleenano@gmail.com
* Correspondence: mingyuli.oliver@gmail.com

Nanomaterial technology for the synthesis, processing, and fabrication of low-dimensional materials is where disciplines merge into a remarkable range of applications, from optoelectronics to health care (contribution 1–7), which affect the lives of millions. With the rapid development of science and technology, people have gradually realized that the variety of materials currently available cannot entirely meet the ever-increasing demand for minimalization and integration for devices. Nanomaterials, endowed with a facile controllability of directionally optical and electrical properties, have great potential to overcome the huge challenges related to the restricted area and balanced performance of multifunctional devices (contribution 8–10). Thus, the swift development of nanomaterial technology especially benefits burgeoning fields, such as Internet of Things [1], energy conversion [2], intelligent sensing [3], and even biomimetics [4].

Establishing how to efficiently control topology of nanomaterials is an essential issue in order to tailor their optical and electrical properties, which can be achieved by means of controllable growth in certain orientations, as well as stoichiometry. For chemical synthesis, the reaction of precursors is generally associated with catalyst and thermal supply, and isotropically spherical nanostructures can be obtained through individual nucleation and growth processes [5]. Meanwhile, anisotropic nanostructures (i.e., nanowires, nanorods, nanoflowers, etc.) are commonly synthesized via preferential growth in certain directions via selective passivation [6], and the obtained nanostructures can be uniformly dispersed into various solutions after ligand exchange. However, the long-term stability and well-aligned patterns of these nanomaterials still remain major problems for the practical applications of chemical synthesized nanostructures, and thus abundant research has focused on the synthesis methods [7], ligand exchange and passivation [8], and coating approaches [9], which will be briefly described in this Special Issue. For example, Henrieta Markevičiūtė et al. chemically deposited Ag–Se nanostructure films on a–Se/nylon, and the structural and optical properties of the composite nanostructures were systematically investigated (contribution 11). On the other hand, the nanostructures can also be directly fabricated via deposition and lithography, and the stoichiometry during deposition and the manufacturing cost are the two main concerns for the technique. In another study in this Special Issue, the fabrication approach for In-Sn-Zn oxide (ITZO) nanocomposite films via high-power impulse magnetron sputtering at room temperature is systematically investigated, and variations in carrier mobility can be correspondingly optimized through the a control of pulse off-time (contribution 12). To lower the expense for the lithography process, the nanopillar and nanohole arrays fabricated using the roll-to-roll nanoimprint lithography are systematically discussed in this Special Issue (contribution 13), which can uniformly realize large-scale manufacturing for well-defined nanostructures (contribution 14). In addition, various methods have been successfully realized to systematically engineer the surface morphology and stoichiometry for nanomaterials, i.e., electrostatic nano-assembly (contribution 15), pulsed-flow-induced fluidization (contribution 16), high-power-impulse magnetron sputtering (contribution 17), ink-jet printing

itation: Li, M.-Y.; Lee, J. Nanotechnologies and Nanomaterials: Selected Papers from CMR. *Nanomaterials* **2024**, *14*, 36. ttps://doi.org/10.3390/ano14010036

eceived: 10 October 2023
evised: 17 November 2023
ccepted: 13 December 2023
ublished: 22 December 2023

(contribution 18), optimized e-beam-lithography (contribution 19), the simple template-fre hydrothermal method (contribution 20), femtosecond laser-assisted fabrication (contribu tion 21), magnesiothermic catalysis (contribution 22), and even a simple grinding metho (contribution 23). To date, the relatively facile approaches to achieve adjustable optica and electrical properties include variations to temperature (contribution 24,25), annealin durations (contribution 26), light radiation (contribution 27,28), and background gas (co tribution 29,30). To accurately predict the effect of growth conditions, numerous simulatio models and characterization systems have been established based on first principles stu ies (contribution 31–33) and micro-nano characterization techniques (contribution 34–36 respectively.

Photodetection can convert the message carried by light irradiation (i.e., wavelengt polarization, intensity) into readable electronic signals, and the performance of photodete tors strongly depends on photoactive layers. Nanomaterials with tunable optical propertie depending on the topology provide a feasible approach for achieving an accurate respons to light irradiation, triggering enthusiastic interest in developing sensitive low-dimensiona photoactive materials. However, the disconnected morphology induced by many crack during deposition is commonly believed to hinder carrier extraction and transfer withi photoactive layers fabricated by nanomaterials, which urgently requires a plausible stra egy to address. In another study in this Special Issue, ZnO quantum dots in a toluen solution are used as antisolvents for the formation of continuous films with 2D $(PEA)_2PbI$ nanosheets as a result of accelerated crystallization, and the carrier transfer highway i spontaneously established with the existence of ZnO quantum dots owing to the wel matched bandgaps (contribution 37). Collective oscillation of surface free electrons on meta nanoparticles can effectively concentrate incident light into photoactive layers, resulting i radically improved light absorption. Thus, silver triangular nanoprism arrays on WSe_2 ar proposed for fabricating spectra and polarization dual-sensitive photoreactors, which ar realized by directionally boosted light absorption due to the silver triangular nanopris arrays (contribution 38).

Additionally, the emission efficiency has been noticeably boosted by intensified hol injection (contribution 39) and spatial light modulation (contribution 40,41) with the exis tence of nanomaterials for light emitting and terahertz devices. In addition to light-emittin devices, nanomaterials have been widely witnessed in energy conversion devices an the synthesis of nanocomposites (contribution 42–44), and elaborately designed (contribu tion 45) nanostructures provide a facile approach for the enhancement of the conversio efficiency for Li-ion batteries and solar cells. Nanocomposites have been successfull introduced to electrochemical catalysts or photocatalysts during the hydrogen evolutio (contribution 46–51) and oxygen evolution reactions (contribution 52–55), facilitating th carrier transition with semiconductor nanocomposites and even metallic materials. Nan otechnology has also emerged in the fabrication of transistors (contribution 56–59) an artificial synapses (contribution 60) for enhancing the performance with a controllabl resistance for the barriers, and the conductivity of electrodes (contribution 61) and capacit (contribution 62) for supercapacitors can be also effectively optimized via nanotechnologie

This Special Issue also includes studies on the novel strategies for distinctive sensin techniques, including chemical sensors, gas monitoring, biosensors, acoustic sensors, an even special detection for different irradiations (contribution 63–7 6). We sincerely hop readers will find useful information for their research, and that all of these works can spar more interesting ideas in various research fields.

Conflicts of Interest: The authors declare no conflict of interest.

List of Contributions

1. Balasooriya, I.L.; Chen, J.; Korale Gedara, S.M.; Han, Y.; Wickramaratne, M.N. Applications of Nano Hydroxyapatite as Adsorbents: A Review. *Nanomaterials* **2022**, *12*, 2324. https://doi.org/10.3390/nano12142324.
2. Li, X.; Peng, Y.; Deng, Y.; Ye, F.; Zhang, C.; Hu, X.; Liu, Y.; Zhang, D. Recycling and Reutilizing Polymer Waste via Electrospun Micro/Nanofibers: A Review. *Nanomaterials* **2022**, *12*, 1663. https://doi.org/10.3390/nano12101663.
3. Dzulkharnien, N.S.F.; Rohani, R. A Review on Current Designation of Metallic Nanocomposite Hydrogel in Biomedical Applications. *Nanomaterials* **2022**, *12*, 1629. https://doi.org/10.3390/nano12101629.
4. Li, X.; Peng, Y.; He, Y.; Zhang, C.; Zhang, D.; Liu, Y. Research Progress on Sound Absorption of Electrospun Fibrous Composite Materials. *Nanomaterials* **2022**, *12*, 1123. https://doi.org/10.3390/nano12071123.
5. Ali Dheyab, M.; Abdul Aziz, A.; Jameel, M.S.; Moradi Khaniabadi, P. Recent Advances in Synthesis, Medical Applications and Challenges for Gold-Coated Iron Oxide: Comprehensive Study. *Nanomaterials* **2021**, *11*, 2147. https://doi.org/10.3390/nano11082147.
6. Yang, J.; Yue, L.; Yang, Z.; Miao, Y.; Ouyang, R.; Hu, Y. Metal-Based Nanomaterials: Work as Drugs and Carriers against Viral Infections. *Nanomaterials* **2021**, *11*, 2129. https://doi.org/10.3390/nano11082129.
7. Yang, S.; Lei, G.; Xu, H.; Lan, Z.; Wang, Z.; Gu, H. Metal Oxide Based Heterojunctions for Gas Sensors: A Review. *Nanomaterials* **2021**, *11*, 1026. https://doi.org/10.3390/nano11041026.
8. Lozovoy, K.A.; Izhnin, I.I.; Kokhanenko, A.P.; Dirko, V.V.; Vinarskiy, V.P.; Voitsekhovskii, A.V.; Fitsych, O.I.; Akimenko, N.Y. Single-Element 2D Materials beyond Graphene: Methods of Epitaxial Synthesis. *Nanomaterials* **2022**, *12*, 2221. https://doi.org/10.3390/nano12132221.
9. Douloudi, M.; Nikoli, E.; Katsika, T.; Vardavoulias, M.; Arkas, M. Dendritic Polymers as Promising Additives for the Manufacturing of Hybrid Organoceramic Nanocomposites with Ameliorated Properties Suitable for an Extensive Diversity of Applications. *Nanomaterials* **2021**, *11*, 19. https://doi.org/10.3390/nano11010019.
10. Chen, H.; Liu, Z.-G.; Geng, M.-Y.; Meng, X.-Y.; Fu, W.-L.; Ju, L.; Yu, B.-Y.; Yang, W.; Dai, Y.-Q.; Lu, W.-B. A Study on the Dynamic Tunning Range of CVD Graphene at Microwave Frequency: Determination, Prediction and Application. *Nanomaterials* **2022**, *12*, 4424. https://doi.org/10.3390/nano12244424.
11. Krylova, V.; Dukštienė, N.; Markevičiūtė, H. Ag–Se/Nylon Nanocomposites Grown by Template-Engaged Reaction: Microstructures, Composition, and Optical Properties. *Nanomaterials* **2022**, *12*, 2584. https://doi.org/10.3390/nano12152584.
12. Sun, H.; Li, Z.-Y.; Chen, S.-C.; Liao, M.-H.; Gong, J.-H.; Bai, Z.; Wang, W.-X. In-Sn-Zn Oxide Nanocomposite Films with Enhanced Electrical Properties Deposited by High-Power Impulse Magnetron Sputtering. *Nanomaterials* **2021**, *11*, 2016. https://doi.org/10.3390/nano11082016.
13. Jang, H.-I.; Yoon, H.-S.; Lee, T.-I.; Lee, S.; Kim, T.-S.; Shim, J.; Park, J.H. Creation of Curved Nanostructures Using Soft-Materials-Derived Lithography. *Nanomaterials* **2020**, *10*, 2414. https://doi.org/10.3390/nano10122414.
14. Tahir, U.; Kim, J.I.; Javeed, S.; Khaliq, A.; Kim, J.-H.; Kim, D.-I.; Jeong, M.Y. Process Optimization for Manufacturing Functional Nanosurfaces by Roll-To-Roll Nanoimprint Lithography. *Nanomaterials* **2022**, *12*, 480. https://doi.org/10.3390/nano12030480.
15. Yokoi, A.; Tan, W.K.; Kuroda, T.; Kawamura, G.; Matsuda, A.; Muto, H. Design of Heat-Conductive hBN–PMMA Composites by Electrostatic Nano-Assembly. *Nanomaterials* **2020**, *10*, 134. https://doi.org/10.3390/nano10010134.
16. Asif, M.; Al-Ghurabi, E.H.; Fatehmulla, A. Pulsed Fluidization of Nanosilica: Rigorous Evaluation of the Efficacy of Pulsation Frequency. *Nanomaterials* **2022**, *12*, 2158. https://doi.org/10.3390/nano12132158.

17. Zhao, M.-J.; Zhang, J.-F.; Huang, J.; Chen, Z.-Z.; Xie, A.; Wu, W.-Y.; Huang, C.-J.; Wuu, D.-S.; Lien, S.-Y.; Zhu, W.-Z. Role of Ambient Hydrogen in HiPIMS-ITO Film during Annealing Process in a Large Temperature Range. *Nanomaterials* **2022**, *12*, 1995. https://doi.org/10.3390/nano12121995.
18. Feng, F.; Hong, H.; Gao, X.; Ren, T.; Ma, Y.; Feng, P. Effectiveness of Oxygen during Sintering of Silver Thin Films Derived by Nanoparticle Ink. *Nanomaterials* **2022**, *12*, 1908. https://doi.org/10.3390/nano12111908.
19. Lee, Y.; Lee, S.H.; Son, H.S.; Lee, S. Reduced Electron Temperature in Silicon Multi-Quantum-Dot Single-Electron Tunneling Devices. *Nanomaterials* **2022**, *12*, 603. https://doi.org/10.3390/nano12040603.
20. Man, L.-F.; Kwong, T.-L.; Wong, W.-T.; Yung, K.-F. Mesoporous Zn/MgO Hexagonal Nano-Plates as a Catalyst for Camelina Oil Biodiesel Synthesis. *Nanomaterials* **2021**, *11*, 2690. https://doi.org/10.3390/nano11102690.
21. Hwang, T.-Y.; Kim, Y.-d.; Cho, J.; Lee, H.-J.; Lee, H.-S.; Lee, B. Multi-Angular Colorimetric Responses of Uni- and Omni-Directional Femtosecond Laser-Induced Periodic Surface Structures on Metals. *Nanomaterials* **2021**, *11*, 2010. https://doi.org/10.3390/nano11082010.
22. Niu, T.; Zhou, B.; Zhang, Z.; Ji, X.; Yang, J.; Xie, Y.; Wang, H.; Du, A. Low-Temperature Synthesis of Monolithic Titanium Carbide/Carbon Composite Aerogel. *Nanomaterials* **2020**, *10*, 2527. https://doi.org/10.3390/nano10122527.
23. Wang, H.; Sun, Z.; Wei, Z.; Wu, Y. A Simple Grinding Method for Preparing Ultra-Thin Boron Nanosheets. *Nanomaterials* **2022**, *12*, 1784. https://doi.org/10.3390/nano12111784.
24. Lee, S.; Kim, J.; Lee, S.; Cha, H.-J.; Son, C.-S.; Son, Y.-G.; Hwang, D. Variations in the Physical Properties of RF-Sputtered CdS Thin Films Observed at Substrate Temperatures Ranging from 25 °C to 500 °C. *Nanomaterials* **2022**, *12*, 1618. https://doi.org/10.3390/nano12101618.
25. Huang, P.-H.; Zhang, Z.-X.; Hsu, C.-H.; Wu, W.-Y.; Ou, S.-L.; Huang, C.-J.; Wuu, D.-S.; Lien, S.-Y.; Zhu, W.-Z. Deposition Mechanism and Characterization of Plasma-Enhanced Atomic Layer-Deposited SnOx Films at Different Substrate Temperatures. *Nanomaterials* **2022**, *12*, 2859. https://doi.org/10.3390/nano12162859.
26. Abujabal, M.; Abunahla, H.; Mohammad, B.; Alazzam, A. Tunable Switching Behavior of GO-Based Memristors Using Thermal Reduction. *Nanomaterials* **2022**, *12*, 1812. https://doi.org/10.3390/nano12111812.
27. Liu, Y.; Li, C.; Fan, S.; Song, X.; Wan, Z. The Effect of Annealing and Optical Radiation Treatment on Graphene Resonators. *Nanomaterials* **2022**, *12*, 2725. https://doi.org/10.3390/nano12152725.
28. Lee, H.Y.; Haidari, M.M.; Kee, E.H.; Choi, J.S.; Park, B.H.; Campbell, E.E.B.; Jhang, S.H. Charge Transport in UV-Oxidized Graphene and Its Dependence on the Extent of Oxidation. *Nanomaterials* **2022**, *12*, 2845. https://doi.org/10.3390/nano12162845.
29. Wu, G.; Sahoo, A.K. Influence of Oxygen Flow Rate on Channel Width Dependent Electrical Properties of Indium Gallium Zinc Oxide Thin-Film Transistors. *Nanomaterials* **2020**, *10*, 2357. https://doi.org/10.3390/nano10122357.
30. Yu, H.; Jiang, K.; Kang, S.G.; Men, Y.; Shin, E.W. Hexagonal and Monoclinic Phases of $La_2O_2CO_3$ Nanoparticles and Their Phase-Related CO_2 Behavior. *Nanomaterials* **2020**, *10*, 2061. https://doi.org/10.3390/nano10102061.
31. Li, W.; Wang, X.; Feng, X.; Du, Y.; Zhang, X.; Xie, Y.; Chen, X.; Lu, Y.; Wang, W. Deformation Mechanism of Depositing Amorphous Cu-Ta Alloy Film via Nanoindentation Test. *Nanomaterials* **2022**, *12*, 1022. https://doi.org/10.3390/nano12061022.
32. Zhang, L.; Cui, Z. Theoretical Study on Electronic, Magnetic and Optical Properties of Non-Metal Atoms Adsorbed onto Germanium Carbide. *Nanomaterials* **2022**, *12*, 1712. https://doi.org/10.3390/nano12101712.

33. Chung, Y.J.; Lee, G.H.; Beom, H.G. Atomistic Insights into the Phase Transformation of Single-Crystal Silicon during Nanoindentation. *Nanomaterials* **2022**, *12*, 2071. https://doi.org/10.3390/nano12122071.
34. Choi, H.-K.; Cha, J.; Choi, C.-G.; Kim, J.; Hong, S. Effect of Point Defects on Electronic Structure of Monolayer GeS. *Nanomaterials* **2021**, *11*, 2960. https://doi.org/10.3390/nano11112960.
35. Ali, S.S.; Arsad, A.; Roberts, K.L.; Asif, M. Effect of Voidage on the Collapsing Bed Dynamics of Fine Particles: A Detailed Region-Wise Study. *Nanomaterials* **2022**, *12*, 2019. https://doi.org/10.3390/nano12122019.
36. Kovalev, A.I.; Vakhrushev, V.O.; Beake, B.D.; Konovalov, E.P.; Wainstein, D.L.; Dmitrievskii, S.A.; Fox-Rabinovich, G.S.; Veldhuis, S. Damage Accumulation Phenomena in Multilayer (TiAlCrSiY)N/(TiAlCr)N, Monolayer (TiAlCrSiY)N Coatings and Silicon upon Deformation by Cyclic Nanoindentation. *Nanomaterials* **2022**, *12*, 1312. https://doi.org/10.3390/nano12081312.
37. Liu, S.; Li, H.; Lu, H.; Wang, Y.; Wen, X.; Deng, S.; Li, M.; Liu, S.; Wang, C.; Li, X. High Performance 0D ZnO Quantum Dot/2D (PEA)2PbI4 Nanosheet Hybrid Photodetectors Fabricated via a Facile Antisolvent Method. *Nanomaterials* **2022**, *12*, 4217–4217. https://doi.org/10.3390/nano12234217.
38. Guskov, A.; Lavrov, S.; Galiev, R. Polarization Sensitive Photodetectors Based on Two-Dimensional WSe_2. *Nanomaterials* **2022**, *12*, 1854. https://doi.org/10.3390/nano12111854.
39. Park, C.Y.; Choi, B. Enhanced Hole Injection Characteristics of a Top Emission Organic Light-Emitting Diode with Pure Aluminum Anode. *Nanomaterials* **2021**, *11*, 2869. https://doi.org/10.3390/nano11112869.
40. Peng, K.; Zhang, N.; Zhang, J.; Chen, P.; Yan, J.; Zheng, C.; Jiang, Z.; Zhong, Z. Extensive Broadband Near-Infrared Emissions from GexSi1−x Alloys on Micro-Hole Patterned Si(001) Substrates. *Nanomaterials* **2021**, *11*, 2545. https://doi.org/10.3390/nano11102545.
41. Yin, S.; Zeng, D.; Chen, Y.; Huang, W.; Zhang, C.; Zhang, W.; E, Y. Optically Controlled Terahertz Dynamic Beam Splitter with Adjustable Split Ratio. *Nanomaterials* **2022**, *12*, 1169. https://doi.org/10.3390/nano12071169.
42. Xu, J.; Wang, Q.; Li, B.; Yao, W.; He, M. Ti3SiC.75Al0.25C2 Nanosheets as Promising Anode Material for Li-Ion Batteries. *Nanomaterials* **2021**, *11*, 3449. https://doi.org/10.3390/nano11123449.
43. Dasarathan, S.; Ali, M.; Jung, T.-J.; Sung, J.; Ha, Y.-C.; Park, J.-W.; Kim, D. Vertically Aligned Binder-Free TiO_2 Nanotube Arrays Doped with Fe, S and Fe-S for Li-ion Batteries. *Nanomaterials* **2021**, *11*, 2924. https://doi.org/10.3390/nano11112924.
44. Sim, G.S.; Shaji, N.; Santhoshkumar, P.; Park, J.W.; Ho, C.W.; Nanthagopal, M.; Kim, H.K.; Lee, C.W. Silkworm Protein-Derived Nitrogen-Doped Carbon-Coated Li[Ni0.8Co0.15Al0.05]O_2 for Lithium-Ion Batteries. *Nanomaterials* **2022**, *12*, 1166. https://doi.org/10.3390/nano12071166.
45. Su, D.; Lv, L.; Yang, Y.; Zhou, H.-L.; Iqbal, S.; Zhang, T. Simple Self-Assembly Strategy of Nanospheres on 3D Substrate and Its Application for Enhanced Textured Silicon Solar Cell. *Nanomaterials* **2021**, *11*, 2581. https://doi.org/10.3390/nano11102581.
46. Kaewmeesri, R.; Nonkumwong, J.; Witoon, T.; Laosiripojana, N.; Faungnawakij, K. Effect of Water and Glycerol in Deoxygenation of Coconut Oil over Bimetallic NiCo/SAPO-11 Nanocatalyst under N_2 Atmosphere. *Nanomaterials* **2020**, *10*, 2548. https://doi.org/10.3390/nano10122548.
47. Cui, Y.; Song, H.; Shi, Y.; Ge, P.; Chen, M.; Xu, L. Enhancing the Low-Temperature CO Oxidation over CuO-Based α-MnO_2 Nanowire Catalysts. *Nanomaterials* **2022**, *12*, 2083. https://doi.org/10.3390/nano12122083.

48. Nimal, R.; Yahya, R.; Shah, A.; Khan, M.A.; Zia, M.A.; Shah, I. Development o Electrolyzer Using NiCo(OH)2 Layered Double Hydroxide Catalyst for Efficien Water Oxidation Reaction. *Nanomaterials* **2022**, *12*, 1819. https://doi.org/10.3390/nano12111819.
49. Su, J.-C.; Hsieh, T.-L.; Yang, S.-M.; Chao, S.-C.; Lu, K.-C. Fabrication and Photoca alytic Properties of Zinc Tin Oxide Nanowires Decorated with Silver Nanoparticle *Nanomaterials* **2022**, *12*, 1201. https://doi.org/10.3390/nano12071201.
50. Li, C.; Cheng, J.; Ye, Q.; Meng, F.; Wang, X.; Dai, H. The Deactivation Mechanism o the Mo-Ce/Zr-PILC Catalyst Induced by Pb for the Selective Catalytic Reduction o NO with NH3. *Nanomaterials* **2021**, *11*, 2641. https://doi.org/10.3390/nano1110264
51. Burse, S.; Kulkarni, R.; Mandavkar, R.; Habib, M.A.; Lin, S.; Chung, Y.-U.; Jeong J.-H.; Lee, J. Vanadium-Doped FeBP Microsphere Croissant for Significantly Enhance Bi-Functional HER and OER Electrocatalyst. *Nanomaterials* **2022**, *12*, 3283. https://doi.org/10.3390/nano12193283.
52. Saber, O.; Osama, A.; Alshoaibi, A.; Shaalan, N.M.; Osama, D. New Approach for De signing Zinc Oxide Nanohybrids to Be Effective Photocatalysts for Water Purificatio in Sunlight. *Nanomaterials* **2022**, *12*, 2005. https://doi.org/10.3390/nano12122005.
53. Yuan, G.; Zhang, G.; Li, K.; Li, F.; Cao, Y.; He, J.; Huang, Z.; Jia, Q.; Zhang, S.; Zhang H. Preparation and Photocatalytic Performance for Degradation of Rhodamine B o $AgPt/Bi_4Ti_3O_{12}$ Composites. *Nanomaterials* **2020**, *10*, 2206. https://doi.org/10.3390 nano10112206.
54. Kobkeatthawin, T.; Chaveanghong, S.; Trakulmututa, J.; Amornsakchai, T.; Ka jitvichyanukul, P.; Smith, S.M. Photocatalytic Activity of $TiO_2/g\text{-}C_3N_4$ Nanocom posites for Removal of Monochlorophenols from Water. *Nanomaterials* **2022**, *12*, 285 https://doi.org/10.3390/nano12162852.
55. Huang, R.; Zhang, M.; Zheng, Z.; Wang, K.; Liu, X.; Chen, Q.; Luo, D. Photocatalyti Degradation of Tobacco Tar Using $CsPbBr_3$ Quantum Dots Modified Bi_2WO_6 Com posite Photocatalyst. *Nanomaterials* **2021**, *11*, 2422. https://doi.org/10.3390/nano11092422.
56. Shin, D.-H.; You, Y.G.; Jo, S.I.; Jeong, G.-H.; Campbell, E.E.B.; Chung, H.-J.; Jhang S.H. Low-Power Complementary Inverter Based on Graphene/Carbon-Nanotub and Graphene/MoS_2 Barristors. *Nanomaterials* **2022**, *12*, 3820. https://doi.org/10.3390/nano12213820.
57. Lee, J.-H.; Choi, I.; Jeong, N.B.; Kim, M.; Yu, J.; Jhang, S.H.; Chung, H.-J. Simulation o Figures of Merit for Barristor Based on Graphene/Insulator Junction. *Nanomaterial* **2022**, *12*, 3029. https://doi.org/10.3390/nano12173029.
58. Sánchez, F.; Sánchez, V.; Wang, C. Independent Dual-Channel Approach to Meso scopic Graphene Transistors. *Nanomaterials* **2022**, *12*, 3223. https://doi.org/10.3390/nano12183223.
59. Na, J.; Park, C.; Lee, C.H.; Choi, W.R.; Choi, S.; Lee, J.-U.; Yang, W.; Cheong, H.; Camp bell, E.E.B.; Jhang, S.H. Indirect Band Gap in Scrolled MoS_2 Monolayers. *Nanomaterial* **2022**, *12*, 3353. https://doi.org/10.3390/nano12193353.
60. Ryu, H.; Kim, S. Improved Pulse-Controlled Conductance Adjustment in Trilaye Resistors by Suppressing Current Overshoot. *Nanomaterials* **2020**, *10*, 2462. https://doi.org/10.3390/nano10122462.
61. Wang, P.; Ding, X.; Zhe, R.; Zhu, T.; Qing, C.; Liu, Y.; Wang, H.-E. Synchronous De fect and Interface Engineering of $NiMoO_4$ Nanowire Arrays for High-Performanc Supercapacitors. *Nanomaterials* **2022**, *12*, 1094. https://doi.org/10.3390/nano12071094.
62. Saber, O.; Ansari, S.A.; Osama, A.; Osama, M. One-Dimensional Nanoscale Si/Co Based on Layered Double Hydroxides towards Electrochemical Supercapacitor Elec trodes. *Nanomaterials* **2022**, *12*, 1404. https://doi.org/10.3390/nano12091404.
63. Vázquez-Vélez, E.; Martínez, H.; Castillo, F. Degradation of Acid Red 1 Catalyzed by Peroxidase Activity of Iron Oxide Nanoparticles and Detected by SERS. *Nanomaterial* **2021**, *11*, 3044. https://doi.org/10.3390/nano11113044.

64. Kulkarni, R.; Kunwar, S.; Mandavkar, R.; Jeong, J.-H.; Lee, J. Hydrogen Peroxide Detection by Super-Porous Hybrid CuO/Pt NP Platform: Improved Sensitivity and Selectivity. *Nanomaterials* **2020**, *10*, 2034. https://doi.org/10.3390/nano10102034.
65. Abd-Elsabour, M.; Alsoghier, H.M.; Alhamzani, A.G.; Abou-Krisha, M.M.; Yousef, T.A.; Assaf, H.F. A Novel Electrochemical Sensor for Detection of Nicotine in Tobacco Products Based on Graphene Oxide Nanosheets Conjugated with (1,2-Naphthoquinone-4-Sulphonic Acid) Modified Glassy Carbon Electrode. *Nanomaterials* **2022**, *12*, 2354. https://doi.org/10.3390/nano12142354.
66. Zhang, H.; Du, W.; Zhang, J.; Ahuja, R.; Qian, Z. Nitrogen-Containing Gas Sensing Properties of 2-D Ti_2N and Its Derivative Nanosheets: Electronic Structures Insight. *Nanomaterials* **2021**, *11*, 2459. https://doi.org/10.3390/nano11092459.
67. Chang, H.-W.; Dong, C.-L.; Chen, Y.-H.; Xu, Y.-Z.; Huang, T.-C.; Chen, S.-C.; Liu, F.-J.; Lai, Y.-H.; Tsai, Y.-C. Extended Graphite Supported Flower-like MnO_2 as Bifunctional Materials for Supercapacitors and Glucose Sensing. *Nanomaterials* **2021**, *11*, 2881. https://doi.org/10.3390/nano11112881.
68. Cai, H.; Wang, M.; Wu, Z.; Liu, J.; Wang, X. Performance Enhancement of SPR Biosensor Using Graphene–MoS_2 Hybrid Structure. *Nanomaterials* **2022**, *12*, 2219. https://doi.org/10.3390/nano12132219.
69. Li, C.; Xiao, X.; Liu, Y.; Song, X. Evaluating a Human Ear-Inspired Sound Pressure Amplification Structure with Fabry–Perot Acoustic Sensor Using Graphene Diaphragm. *Nanomaterials* **2021**, *11*, 2284. https://doi.org/10.3390/nano11092284.
70. Kallas, P.; Haugen, H.J.; Gadegaard, N.; Stormonth-Darling, J.; Hulander, M.; Andersson, M.; Valen, H. Adhesion of Escherichia Coli to Nanostructured Surfaces and the Role of Type 1 Fimbriae. *Nanomaterials* **2020**, *10*, 2247. https://doi.org/10.3390/nano10112247.
71. Kumar, R.; Kehr, N.S. 3D-Printable Oxygen- and Drug-Carrying Nanocomposite Hydrogels for Enhanced Cell Viability. *Nanomaterials* **2022**, *12*, 1304. https://doi.org/10.3390/nano12081304.
72. Singh, J.P.; Park, J.Y.; Chae, K.H.; Ahn, D.; Lee, S. Soft X-ray Absorption Spectroscopic Investigation of Li(Ni0.8Co0.1Mn0.1)O_2 Cathode Materials. *Nanomaterials* **2020**, *10*, 759. https://doi.org/10.3390/nano10040759.
73. Islam, K.; Haque, M.; Kumar, A.; Hoq, A.; Hyder, F.; Hoque, S.M. Manganese Ferrite Nanoparticles ($MnFe_2O_4$): Size Dependence for Hyperthermia and Negative/Positive Contrast Enhancement in MRI. *Nanomaterials* **2020**, *10*, 2297. https://doi.org/10.3390/nano10112297.
74. Aguila, M.A.C.; Esmenda, J.C.; Wang, J.-Y.; Chen, Y.-C.; Lee, T.-H.; Yang, C.-Y.; Lin, K.-H.; Chang-Liao, K.-S.; Kafanov, S.; Pashkin, Y.A.; et al. Photothermal Responsivity of van der Waals Material-Based Nanomechanical Resonators. *Nanomaterials* **2022**, *12*, 2675. https://doi.org/10.3390/nano12152675.
75. Huang, C.-H.; Lu, Y.-J.; Pan, Y.-C.; Liu, H.-L.; Chang, J.-Y.; Sie, J.-L.; Pijanowska, D.G.; Yang, C.-M. Nanohollow Titanium Oxide Structures on Ti/FTO Glass Formed by Step-Bias Anodic Oxidation for Photoelectrochemical Enhancement. *Nanomaterials* **2022**, *12*, 1925. https://doi.org/10.3390/nano12111925.
76. Ali, S.S.; Arsad, A.; Hossain, S.S.; Asif, M. A Detailed Insight into Acoustic Attenuation in a Static Bed of Hydrophilic Nanosilica. *Nanomaterials* **2022**, *12*, 1509. https://doi.org/10.3390/nano12091509.

References

1. Ghder Soliman, W.; Swathi, C.; Yasasvi, T.; Keerthi Priya, B.; Akhila Reddy, D. Review on Poly(Ethylene Oxide)-Based Electrolyte and Anode Nanomaterials for the Internet of Things Node-Level Lithium-Ion Batteries. *Mater. Today Proc.* **2021**, *42*, 429–435. [CrossRef]
2. Ge, R.; Huo, J.; Sun, M.; Zhu, M.; Liu, Y.; Chou, S.; Li, W. Surface and Interface Engineering: Molybdenum Carbide–Based Nanomaterials for Electrochemical Energy Conversion. *Small* **2019**, *17*, 1903380. [CrossRef] [PubMed]

3. Zhu, X.; Lin, L.; Wu, R.; Zhu, Y.; Sheng, Y.; Nie, P.; Liu, P.; Xu, L.; Wen, Y. Portable Wireless Intelligent Sensing of Ultra-Trace Phytoregulator α-Naphthalene Acetic Acid Using Self-Assembled Phosphorene/Ti_3C_s-MXene Nanohybrid with High Ambient Stability on Laser Induced Porous Graphene as Nanozyme Flexible Electrode. *Biosens. Bioelectron.* **2021**, *179*, 113062. [CrossRef] [PubMed]
4. Jackman, J.A.; Bo Kyeong Yoon; Ouyang, L. ; Wang, N.; Abdul Rahim Ferhan; Kim, J.; Tetsuro Majima; Cho, N.-J. Biomimetic Nanomaterial Strategies for Virus Targeting: Antiviral Therapies and Vaccines. *Adv. Funct. Mater.* **2020**, *31*, 2008352. [CrossRef]
5. Luo, D.; Zheng, L.; Zhang, Z.; Li, M.; Chen, Z.; Cui, R.; Shen, Y.; Li, G.; Feng, R.; Zhang, S.; et al. Constructing Multifunctional Solid Electrolyte Interface via In-Situ Polymerization for Dendrite-Free and Low N/P Ratio Lithium Metal Batteries. *Nat. Commun.* **2021**, *12*, 186. [CrossRef] [PubMed]
6. Velázquez-Salazar, J.J.; Bazán-Díaz, L.; Zhang, Q.; Mendoza-Cruz, R.; Montano-Priede, L.; Guisbiers, G.; Large, N.; Link, S.; José-Yacamán, M. Controlled Overgrowth of Five-Fold Concave Nanoparticles into Plasmonic Nanostars and Their Single-Particle Scattering Properties. *ACS Nano* **2019**, *13*, 10113–10128. [CrossRef] [PubMed]
7. Shah, A.H.; Rather, M.A. Effect of Calcination Temperature on the Crystallite Size, Particle Size and Zeta Potential of TiO_2 Nanoparticles Synthesized via Polyol-Mediated Method. *Mater. Today Proc.* **2020**, *44*, 482–488. [CrossRef]
8. Gualdrón-Reyes, A.F.; Roser Fernández-Climent; Masi, S. ; Mesa, C.A.; Echeverría-Arrondo, C.; Aiello, F.; Balzano, F.; Uccello-Barretta, G.; Rodriguez-Pereira, J.; Gimenez, S.; et al. Efficient Ligand Passivation Enables Ultrastable CsPbX3 Perovskite Nanocrystals in Fully Alcohol Environments. *Adv. Opt. Mater.* **2023**, *11*, 2203096. [CrossRef]
9. Braim, F.S.; Ab Razak NN, A.N.; Aziz, A.A.; Dheyab, M.A.; Ismael, L.Q. Layla Qasim Ismael Optimization of Ultrasonic-Assisted Approach for Synthesizing a Highly Stable Biocompatible Bismuth-Coated Iron Oxide Nanoparticles Using a Face-Centered Central Composite Design. *Ultrason. Sonochem.* **2023**, *95*, 106371. [CrossRef] [PubMed]

nanomaterials

MDPI

Communication

Design of Heat-Conductive hBN–PMMA Composites by Electrostatic Nano-Assembly

Atsushi Yokoi [1], Wai Kian Tan [1,*], Taichi Kuroda [2], Go Kawamura [2], Atsunori Matsuda [2] and Hiroyuki Muto [1,2,*]

1 Institute of Liberal Arts and Sciences, Toyohashi University of Technology, Toyohashi 441-8580, Japan; yokoi@ion.ee.tut.ac.jp

2 Department of Electrical & Electronics Information Engineering, Toyohashi University of Technology, Toyohashi 441-8580, Japan; t123231@edu.tut.ac.jp (T.K.); kawamura.go.km@tut.jp (G.K.); matsuda@ee.tut.ac.jp (A.M.)

* Correspondence: tan@las.tut.ac.jp (W.K.T.); muto@ee.tut.ac.jp (H.M.); Tel.: +81-532-44-6800 (W.K.T.)

Received: 23 December 2019; Accepted: 9 January 2020; Published: 12 January 2020

Abstract: Micro/nanoscale design of composite materials enables alteration of their properties for advanced functional materials. One of the biggest challenges in material design is the controlled decoration of composite materials with the desired functional additives. This study reports on and demonstrates the homogeneous decoration of hexagonal boron nitride (hBN) on poly(methylmethacrylate) (PMMA) and vice versa. The formation of the composite materials was conducted via a low environmental load and a low-energy-consuming, electrostatic nano-assembly method which also enabled the efficient usage of nano-sized additives. The hBN/PMMA and PMMA/hBN composites were fabricated in various size combinations that exhibited percolated and layer-oriented structures, respectively. The thermal conductivity behaviors of hBN/PMMA and PMMA/hBN composites that exhibited good microstructure were compared. The results showed that microstructural design of the composites enabled the modification of their heat-conducting property. This novel work demonstrated the feasibility of fabricating heat-conductive PMMA matrix composites with controlled decoration of hBN sheets, which may provide a platform for further development of heat-conductive polymeric materials.

Keywords: electrostatic adsorption; electrostatic assembly; composite; heat-conductive; hBN; PMMA

1. Introduction

Polymer matrix composites with the desired properties attracted the interest of many researchers due to the significant potential for low-cost fabrication of high-performance functional polymeric materials. The composite's properties can be changed by altering the composition of its organic or inorganic materials. Poly(methylmethacrylate) (PMMA) is a widely used polymeric material due to its unique properties such as good optical clarity, high mechanical strength, and good thermal stability. PMMA also exhibits an amorphous polymer processing dimensional stability that allows it to be used as a host in composite material fabrication [1]. Various PMMA composite materials with desired properties were reported such as functional transparent composites [2–4] and infrared (IR) shielding composites [5].

In the rapid development of wearable and high-power electronic devices, the requirement of effective heat-dissipating composite materials is more important than ever [6–8]. High-performance electronic devices tend to generate heat rapidly, making heat dissipation a key factor in a device's design, as heat entrapment will not only affect the performance of the device but also reduce the device's lifetime. In the electronics industry, synthetic polymers are commonly used in encapsulation, as well as for packaging materials. However, the synthetic polymeric materials used possess low thermal conductivities of 0.1–0.5 W·m^{-1}·K^{-1} [9]. Therefore, the development and improvement of heat-conductive polymeric composite materials are crucial to improve the heat-dissipation of electronic devices. Recently, two-dimensional (2D) boron nitride (BN) sheets which possess good thermal conductivity, oxidation resistance, high elastic modulus, and electrical insulation were widely used as 2D micro/nano thermal conductive fillers [9–11]. The unique honeycomb-configured sp^2-bonded boron and nitrogen promotes anisotropic thermal conductivity with in-plane and out-of-plane thermal conductivity of 600 and 30 W·m^{-1}·K^{-1}, respectively [12]. Due to the strong structural correlation of 2D hexagonal boron nitride (hBN) micro and nanosheets with their in-plane thermal conduction, it is imperative to maintain its structure for effective heat conduction. This makes conventional mechanical mixing methods such as mechanical mixing inappropriate as they would destroy the sheet-like structure and hamper the thermal conductivity of the composite, in addition to there being issues when mixing them [1,13]. Furthermore, a high loading amount of more than 50 vol.% is required to generate good thermal conductivity, but this compromises the composite's mechanical property [13]. In order to preserve the sheet-like structure of hBN sheets, electrostatic nano-assembly (EA) was used in this study. Previously, our group demonstrated the feasibility of an EA method for fabricating various composite materials [14]. The composites were used for applications such as IR shielding [5], optical property-controlled ceramic composite films [15], selective laser sintering [16], mechanical property control of carbon-based alumina composite [17], and fabrication of the negative electrode for rechargeable Fe–air batteries [18]. The EA method not only allows for the homogeneous decoration of additives onto desired primary particles, but also the preservation of the shape and size of the starting materials. Moreover, compared to most reported methods for fabrication of hBN polymeric composite materials such as the chemical vapor deposition method [10], which involve high cost and energy consumption, the EA method is a facile and smart material processing method that involves low environmental load, as well as low energy consumption, which aligns with recently established sustainable development goals. As the fabrication process is carried out in an aqueous medium that is environmentally friendly, chemical contamination could be avoided. Moreover, the mixing is conducted at room temperature with minimal energy consumption compared to high-energy-consuming equipment such as mechanical ball-milling, heated roller mixing [6], and chemical vapor deposition [10]. The EA method also allows the efficient usage of the raw materials, which helps to reduce wastage due to the good recovery rate of the composites while maintaining the shape and structure of the raw materials [14].

In previous reported studies, the effects of the hBN sheet orientation, applied voltage waveform, and composite molding temperature on the electrical breakdown strength and thermal conductivities of only layer-oriented PMMA/hBN composites using hBN with large diameters of 10 and 45 μm as the primary (core) particles were investigated [19,20]. In this study, the feasibility of controlling the microstructure of the pellets using a wider size range combination of PMMA particles and hBN sheets was further demonstrated. A systematic investigation was carried out by adjusting the size of the starting materials (hBN and PMMA), and the microstructural morphologies obtained by decoration of hBN sheets on PMMA particles, and vice versa, were also tabulated. Decoration of hBN sheets onto PMMA particles led to the formation of a percolate-structured microstructure, while decoration of PMMA onto hBN sheets resulted in the formation of a layer-oriented microstructure. The correlation of composite particle assembly, microstructural formation, and heat conductivity of the aforementioned composites was evaluated using IR thermography. The hBN/PMMA composite materials possessed good potential for application as heat-dissipation materials for electronic industries. The results obtained in this work on the controlled design of hBN/PMMA composite materials could be beneficial for further development of heat-conductive polymeric materials.

2. Materials and Methods

The experiments were carried out using commercially available poly(methylmethacrylate) (PMMA) particles (average particle diameter of 0.3, 5, and 12 μm, Sekisui Chemical, Tokyo, Japan) and hexagonal boron nitride sheets (average particle size of 0.5 μm, Showa Denko 5, 18 μm Tokyo, Japan, Denki Kagaku, Tokyo, Japan). The polycation and polyanion used were polydiallyldimethyl ammoniumchloride (PDDA) (average molecular weight 100,000 to 200,000, Sigma-Aldrich, Missouri, United States) and polysodium styrenesulfonate (PSS) (average molecular weight 70,000, Sigma-Aldrich, Missouri, MO, United States), respectively. The surfactant used for the initial coating onto PMMA was sodium deoxycholate (SDC). After that, an alternative layer of PDDA/PSS/PDDA was coated onto PMMA to induce positive zeta-potential-modified PMMA primary particles. Similarly, for the hBN sheets, the surface charge was firstly modified using SDC, followed by PDDA and PSS. to obtain a negative zeta-potential surface. Then, the oppositely charged PMMA particles (positive) and hBN nanosheets (negative) in aqueous solution were mixed and stirred to obtain the electrostatically assembled hBN/PMMA composite particles. The coverage amount of additive particles (hBN or PMMA) on intended primary particles (PMMA or hBN) was adjusted in vol.% according to previous reported work [14].The suspension was then dried to obtain the composite powder. For the heat distribution evaluation of the hBN/PMMA composite obtained, the composite powders were pressed into pellets using a hot press with a pressure of 100 MPa at 200 °C for 15 min. Prior to pressing, 50 μL of methyl methacrylate (MMA) was added to promote the bonding of the main chain and to improve the mold-ability. The pressed pellets were then annealed at 80 °C for 15 min. Six different combinations of hBN with PMMA consisting of different sizes, as mentioned above, were fabricated for comparison. The morphological structures of the composites obtained were observed using an S-4800 field-emission scanning electron microscope (FE-SEM, Hitachi S-4800, Tokyo, Japan). For the cross-sectional observation, the composite pellets were cut into half and polished using a surface grinder prior to SEM observation. The heat distribution properties of the nanocomposite pellets were evaluated using infrared thermography (Testo corporation, Testo 881-1, Lenzkirch, Germany).

3. Results and Discussion

The SEM images of the starting materials used for this study are shown in Figure 1. The spherical PMMA particles, with average sizes of 0.3, 5, and 12 μm, are shown in Figure 1a–c, respectively. On the other hand, the SEM images of the sheet-like hBN, with average sizes of 0.5, 5, and 18 μm, are shown in Figure 1d–f respectively. Depending on the size of the primary particle (either PMMA or hBN), smaller-sized secondary/additive particles (hBN or PMMA) were used for electrostatic decoration on the surface of the primary particle. For example, smaller-sized hBN sheets (0.5 μm) were used for electrostatic assembly onto the surface of larger spherical PMMA particles (12 μm).

In this study, two different composites with either spherical PMMA particles or two-dimensional hBN sheets were used as the primary particle (core). Different size combinations of PMMA and hBN were investigated systematically. Firstly, decoration of sheet-like hBN with an average diameter of 0.5 or 5 μm was carried out on PMMA with an average diameter of 5 or 12 μm. The hBN/PMMA composite particles obtained, with the combinations of (0.5 μm) hBN/PMMA (5 μm), (0.5 μm) hBN/PMMA (12 μm), and (5 μm) hBN/PMMA (12 μm) are shown in Figure 2. Regardless of the hBN sheet size, homogeneous decoration of hBN sheets on PMMA particles was obtained. A good surface coverage of 0.5-μm hBN on both 5- and 12-μm PMMA was observed, as shown in Figure 2a,b, respectively. Figure 2c further demonstrates the feasibility of decorating 12-μm PMMA particle with larger hBN having an average diameter of 5 μm.

Subsequently, hBN sheets with different sizes, 5 and 18 μm, were used as the primary matrix (core) while PMMA with different average particle sizes of 0.3 and 5 μm was used as the additives in the electrostatic nano-assembly. PMMA/hBN composite particles with the combinations of (0.3 μm) PMMA/hBN (5 μm), (0.3 μm) PMMA/hBN (18 μm), and (5 μm) PMMA/hBN (18 μm) are shown in Figure 3. PMMA particles were observed to be homogeneously distributed on the surface of hBN sheets. The morphological observations of the composites obtained in Figures 2 and 3 indicated that an electrostatic nano-assembly method could be used for the homogeneous assembly of hBN sheets on PMMA particles or vice versa.

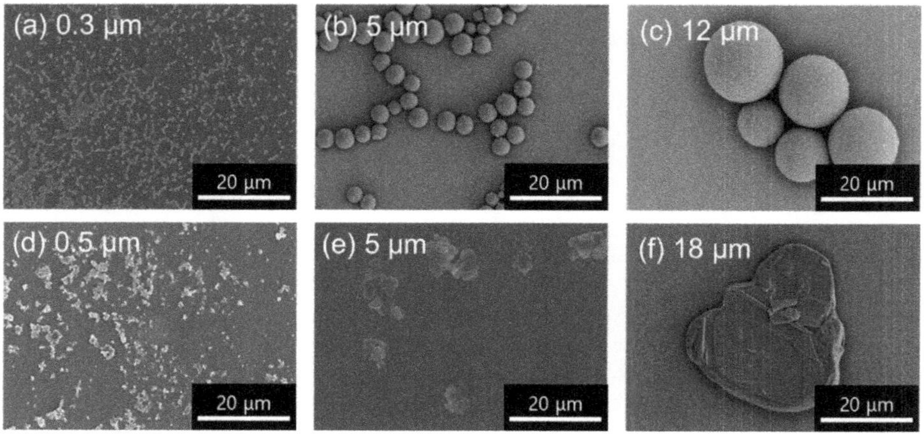

Figure 1. SEM images of the starting materials used for hexagonal boron nitride (hBN)/poly (methylmethacrylate) (PMMA) composite formation. PMMA particles with different sizes of (**a**) 0.3, (**b**) 5, and (**c**) 12 μm, and hBN sheets with different sizes of (**d**) 0.5, (**e**) 5, and (**f**) 18 μm were used.

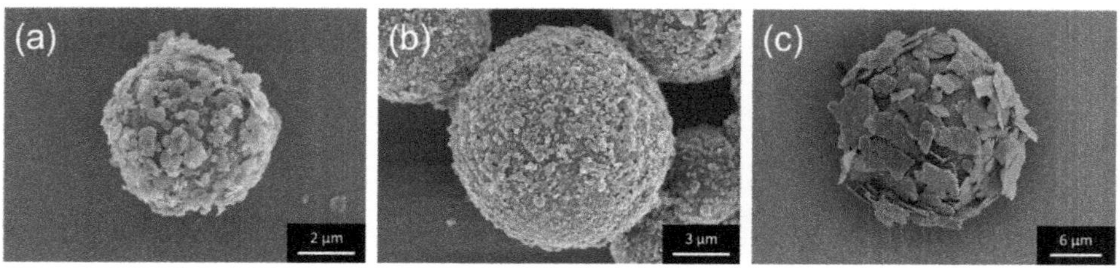

Figure 2. SEM images of the hBN/PMMA (PMMA core) composite particles obtained after electrostatic nano-assembly method using various sizes combination of (**a**) 0.5-μm hBN on 5-μm PMMA, (**b**) 0.5-μm hBN on 12-μm PMMA, and (**c**) 5-μm hBN on 12-μm PMMA.

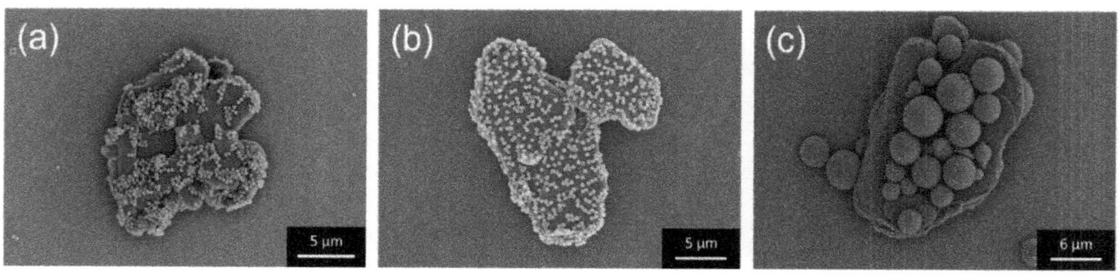

Figure 3. SEM images of the PMMA/hBN (hBN core) composite particles obtained after electrostatic assembly, using various combinations: (**a**) 0.3-μm PMMA on 5-μm hBN sheets, (**b**) 0.3-μm PMMA on 18-μm hBN sheets, and (**c**) 5-μm PMMA on 18-μm hBN sheets.

Subsequently, the hBN/PMMA and PMMA/hBN composite powders shown in Figures 2 and 3 were hot-pressed into pellet and grinded for cross-sectional SEM observation. The cross-sectional SEM images of the composite pellets obtained are shown in Figure 4. Interestingly, hBN/PMMA and PMMA/hBN composite pellets exhibited different microstructural morphologies of percolate structure or layer-oriented structure, respectively. The percolation of hBN on the surface of the PMMA grain boundaries was clearly observed as shown in Figure 4a–c. The degree of percolation of hBN within the PMMA matrix increased with the size of PMMA particles used. This was due to the lower overall surface area available for the electrostatic adsorption of hBN sheets when larger PMMA particles were used. This promoted a denser decoration of hBN on the PMMA surface, generating an interconnected percolation structure. The difference can be observed by comparing Figure 4a,b, where similarly sized 0.5-μm hBN sheets were decorated on PMMA particles with different diameter sizes of 5 and 12 μm, respectively. Similar phenomena were also observed in our previous work involving the decoration of carbon nanoparticles on alumina granules [17]. The highest percolation degree was observed using the composite powders with 5-μm hBN sheets as additives with 12-μm PMMA as the primary particles. On the other hand, disconnected, sheet-like layered microstructures were observed using PMMA-decorated hBN sheets as shown in Figure 4d–f. The interlayer distance between hBN layers was observed to increase when larger PMMA particles were used, as shown in the SEM images of Figure 4e,f, where similar hBN sheets with an average diameter of 18 μm were used. Meanwhile, the hBN sheets were observed to exhibit a more random orientation within the matrix when composites that consisted of smaller PMMA particles decorated on smaller hBN sheets were used compared to those on larger hBN sheets as shown in Figure 4d,e. In previous studies, the orientation probability of layer-oriented PMMA/hBN was altered by changing the size of PMMA particles; decoration of larger PMMA particles on hBN sheets led to irregular hBN sheet direction (larger angle difference) in the composite matrix as compared to when smaller PMMA particles were used [19,20]. Thus, this study shows that the orientation could also be affected by changing the size of the hBN sheets. These results demonstrated that the interlayer distance, as well as the orientation of the layer-structured PMMA/hBN composites, could be controlled by altering the size of PMMA particles decorated on hBN sheets during the assembly step. The schematic illustration in Figure 5 shows the inter-correlation of the hBN/PMMA and PMMA/hBN composite morphologies, with the corresponding microstructures obtained after hot-pressing.

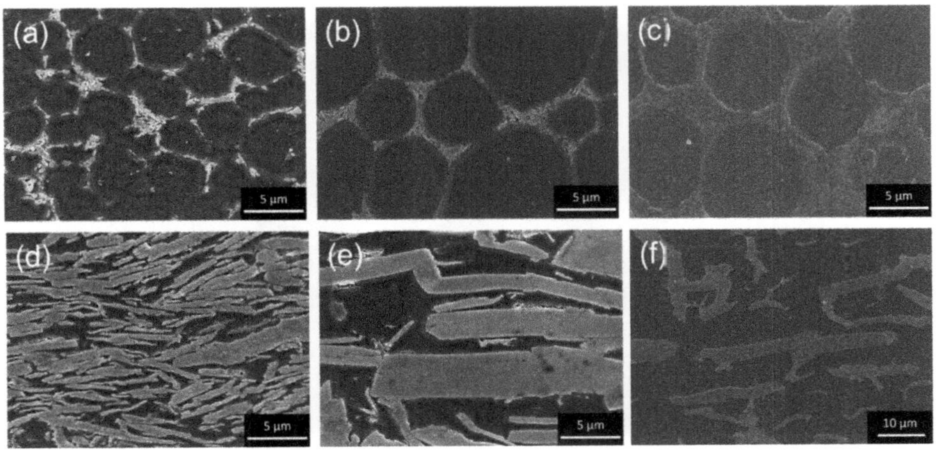

Figure 4. Cross-sectional SEM images of the respective hBN/PMMA composite pellets obtained after hot-pressing, using the following composite powders: (**a**) 0.5-μm hBN on 5-μm PMMA, (**b**) 0.5-μm hBN on 12-μm PMMA, (**c**) 5-μm hBN on 12-μm PMMA, (**d**) 0.3-μm PMMA on 5-μm hBN sheets, (**e**) 0.3-μm PMMA on 18-μm hBN sheets, and (**f**) 5-μm PMMA on 18-μm hBN sheets.

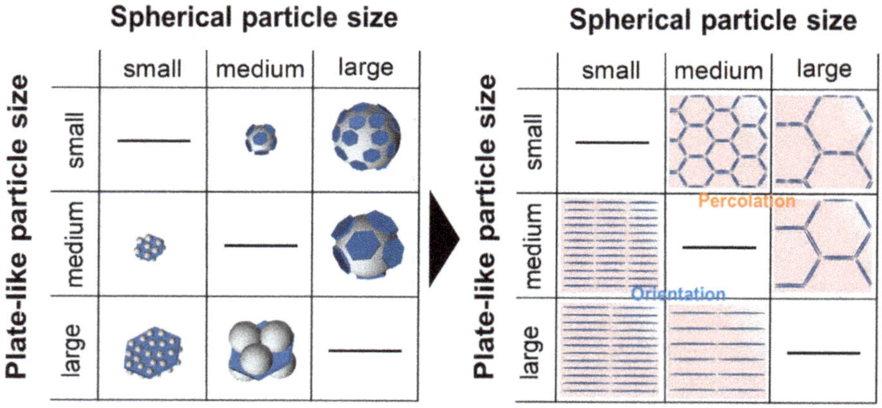

Figure 5. Schematic illustrations showing the morphologies of hBN/PMMA and PMMA/hBN composite particles with the corresponding microstructures obtained.

To gain insight into the thermal conduction properties, infrared thermography comparison was carried out using the pellets fabricated using only PMMA (12 μm), (0.3 μm) PMMA/hBN (18 μm), and (5 μm) hBN/PMMA (12 μm). The composite pellets of (0.3 μm) PMMA/hBN (18 μm) and (5 μm) hBN/PMMA (12 μm) were chosen due to the ordered microstructure obtained compared to others. The thermograph images obtained are shown in Figure 6. From the thermograph image comparison, the heat signatures of the composite pellets, (0.3 μm) PMMA/hBN (18 μm) and (5 μm) hBN/PMMA (18 μm), were observed to be higher than the PMMA pellets. After 16 s, the temperature exhibited by the (0.3 μm) PMMA/hBN (18 μm) and (5 μm) hBN/PMMA (18 μm) pellets was 83.3 and 84.9 °C, respectively, while that of the PMMA pellet was 62.6 °C. These results demonstrated that hBN incorporation into a PMMA matrix enhanced the heat conductance of the composite pellets.

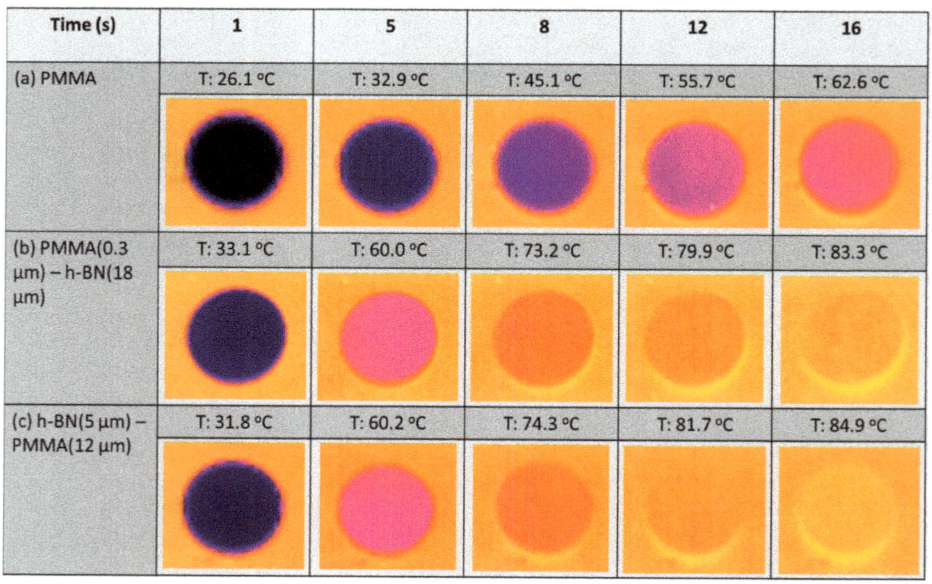

Figure 6. Thermograph images of (**a**) PMMA, (**b**) PMMA (0.3 μm)/hBN (18 μm). and (**c**) hBN (5 μm)/ PMMA (18 μm) composite pellets after infrared thermography irradiation.

A graph showing plots of temperature change against time for PMMA (18 µm), (0.3 µm) PMMA/hBN (18 µm). and (5 µm) hBN/PMMA (18 µm) is shown in Figure 7. The results show that heat conduction reached a plateau after 20 s for all the samples, with the composite pellets achieving temperature higher than 80 °C. Although both composite pellets demonstrated almost comparable heat conductivity, it is important to note that the amount of hBN added into the composites differed. The amount of hBN present in (0.3 µm) PMMA/hBN (18 µm) and (5 µm) hBN/PMMA (12 µm) was 57 vol.% and 26 vol.%, respectively. This indicates that, despite having half the volume of hBN incorporated, the (5 µm) hBN/PMMA (18 µm) composite pellet that had a percolated structure exhibited heat-conducting properties comparable to the layer-oriented structure of the (0.3 µm) PMMA/hBN (18 µm) composite pellet. Due to the higher heat conductivity of the hBN sheets in the *a*- and *b*-axis, the percolated microstructure allowed better heat conductivity in the PMMA matrix [21]. The calculated heat conductivity for (0.3 µm) PMMA/hBN (18 µm) and (5 µm) hBN/PMMA (12 µm) was 1.27 and 1.42 W/m·K, respectively. Zhi et al. reported an approximately three-fold increase in thermal conductivity from 0.17 to 0.50 W/m·K using a 10 vol.% BN nanotube/PMMA composite film [22]. In comparison with the work reported by Pullanchiyodan et al., using Ag-decorated BN nanosheets in PMMA, they reported a thermal conductivity of 1.48 W/m·K by incorporating 35 vol.% of this hybrid filler into PMMA matrix [23]. Therefore, the thermal conductivity of 1.42 W/m·K obtained in this study using percolate-structured hBN/PMMA (26 vol.%) demonstrates promising potential, given the simplicity and good reproducibility of this method. The interconnectivity and interaction of hBN sheets were the determining factors in this heat-conducting property, and the percolated structure of hBN/PMMA composite provided a thermal conductive pathway compared to the layer-structured PMMA/hBN composite. Similar findings were also reported by Mosanenzadeh et al. in their study on the thermal behavior of ordered and random hBN networks using different types of polymers [24]. In addition, the higher content of hBN within the PMMA matrix reduced the mechanical properties drastically, as hBN acted as a defect in the polymer matrix [13]. A better approach would be to precisely design the composite powders to promote a particulate interaction, which not only promotes the efficient use of hBN but also does not compromise its mechanical properties [24].

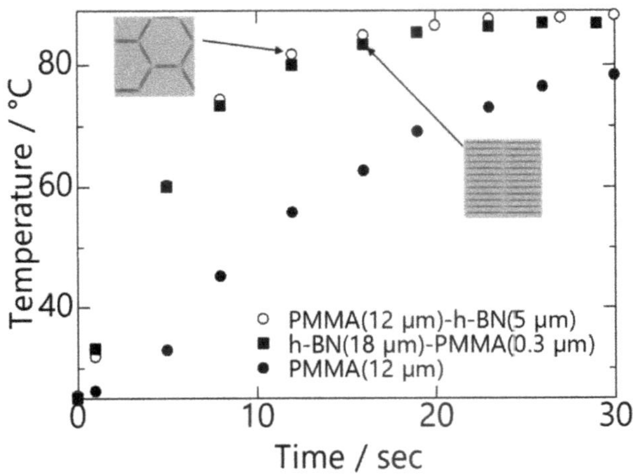

Figure 7. Comparison of temperature change against time for PMMA, layer-oriented PMMA/hBN, and percolation-oriented hBN/PMMA composites.

This study demonstrated that the microstructure of an hBN/PMMA composite is crucial to its property of heat conductivity. Therefore, controlled design of composite formation is crucial using a bottom-up fabrication approach, which is not achievable using the conventional mixing method.

4. Conclusions

The controlled microstructure formation of hBN/PMMA and PMMA/hBN composite pellets was systematically investigated using an electrostatic nano-assembly method. The composite pellets of both hBN/PMMA (PMMA core) and PMMA/hBN (hBN core) composites exhibited promising heat-conducting properties. By varying the sizes of the starting materials in the composite powder design, percolate-structured hBN/PMMA and layer-structured PMMA/hBN composite pellets were fabricated from the composite powders obtained. With hBN composition at 26 vol.% for the percolate-structured hBN/PMMA composite compared to 57 vol.% for the layer-structured PMMA/hBN composite, both composites demonstrated almost comparable thermal conductivity behavior during infrared thermography characterization. The thermal conductivity obtained for (0.3 μm) PMMA/hBN (18 μm) and (5 μm) hBN /PMMA (12 μm) was 1.27 and 1.42 W/m·K, respectively. The results indicated that the percolate-structured hBN/PMMA composite pellet (half the hBN content in comparison to PMMA/hBN composite) exhibited higher thermal conductivity compared to the layer-structured PMMA/hBN composite pellet due to a better heat-conducting route in the percolated structure. This study demonstrated the importance and feasibility of microscale structural design of composite materials in achieving the desired properties such as improved thermal conductivity. The results obtained in this work demonstrate the feasibility of a controlled fabrication of heat-conductive PMMA matrix composites by controlled decoration of hBN sheets and provide a platform for further development of heat-conductive polymeric materials. This electrostatic nano-assembly method not only enables controlled design of the composite particles, which is not achievable using conventional mixing methods, but it also has a low environmental load and low energy consumption. Smart fabrication processes with a low environmental load and low-energy-consumption methods are indispensable as we move toward a sustainable society, and electrostatic nano-assembly is one of the feasible methods in heat-conductive material design.

Author Contributions: H.M., W.K.T., and A.Y. designed the study and contributed to the manuscript writing. T.K. and A.Y. carried out the experimental work. G.K. and A.M. provided technical and scientific insight for discussion. All authors have read and agreed to the published version of the manuscript.

Funding: This work was supported by the Japan Society for Promotion of Science (JSPS) KAKENHI JP18H01706, Early-Career Scientist JP18K14013, and the Kazuchika Okura Memorial Foundation.

Acknowledgments: The authors would like to acknowledge the TUT Research Promotional Research Fund for supporting the publication fee.

Conflicts of Interest: The authors have no conflict of interest to declare.

References

1. Mittal, G.; Rhee, K.Y.; Park, S.J. Processing and characterization of PMMA/PI composites reinforced with surface functionalized hexagonal boron nitride. *Appl. Surf. Sci.* **2017**, *415*, 49–54. [CrossRef]
2. Sugumaran, S.; Bellan, C.S. Transparent nano composite PVA–TiO2 and PMMA–TiO2 thin films: Optical and dielectric properties. *Optik* **2014**, *125*, 5128–5133. [CrossRef]
3. Cierech, M.; Osica, I.; Kolenda, A.; Wojnarowicz, J.; Szmigiel, D.; Lojkowski, W.; Kurzydlowski, K.; Ariga, K.; Mierzwinska-Nastalska, E. Mechanical and Physicochemical Properties of Newly Formed ZnO-PMMA Nanocomposites for Denture Bases. *Nanomaterials* **2018**, *8*, 305. [CrossRef] [PubMed]
4. Arlindo, E.P.S.; Lucindo, J.A.; Bastos, C.M.O.; Emmel, P.D.; Orlandi, M.O. Electrical and Optical Properties of Conductive and Transparent ITO@PMMA Nanocomposites. *J. Phys. Chem. C* **2012**, *116*, 12946–12952. [CrossRef]
5. Tan, W.K.; Yokoi, A.; Kawamura, G.; Matsuda, A.; Muto, H. PMMA-ITO Composite Formation via Electrostatic Assembly Method for Infra-Red Filtering. *Nanomaterials* **2019**, *9*, 886. [CrossRef]
6. Muratov, D.S.; Stepashkin, A.A.; Anshin, S.M.; Kuznetsov, D.V. Controlling thermal conductivity of high density polyethylene filled with modified hexagonal boron nitride (hBN). *J. Alloy. Compd.* **2018**, *735*, 1200–1205. [CrossRef]
7. Xu, M.; Lei, Y.; Ren, D.; Chen, S.; Chen, L.; Liu, X. Synergistic Effects of Functional CNTs and h-BN on Enhanced Thermal Conductivity of Epoxy/Cyanate Matrix Composites. *Nanomaterials* **2018**, *8*, 997. [CrossRef]

8. Liu, J.; Li, W.; Guo, Y.; Zhang, H.; Zhang, Z. Improved thermal conductivity of thermoplastic polyurethane via aligned boron nitride platelets assisted by 3D printing. *Compos. Part A Appl. Sci. Manuf.* **2019**, *120*, 140–146. [CrossRef]
9. Wang, X.; Yu, Z.; Jiao, L.; Bian, H.; Yang, W.; Wu, W.; Xiao, H.; Dai, H. Aerogel Perfusion-Prepared h-BN/CNF Composite Film with Multiple Thermally Conductive Pathways and High Thermal Conductivity. *Nanomaterials* **2019**, *9*, 1051. [CrossRef]
10. Ashton, T.S.; Moore, A.L. Foam-like hierarchical hexagonal boron nitride as a non-traditional thermal conductivity enhancer for polymer-based composite materials. *Int. J. Heat Mass Transf.* **2017**, *115*, 273–281. [CrossRef]
11. Pan, C.; Kou, K.; Jia, Q.; Zhang, Y.; Wu, G.; Ji, T. Improved thermal conductivity and dielectric properties of hBN/PTFE composites via surface treatment by silane coupling agent. *Compos. Part B Eng.* **2017**, *111*, 83–90. [CrossRef]
12. Liang, W.; Ge, X.; Ge, J.; Li, T.; Zhao, T.; Chen, X.; Zhang, M.; Ji, J.; Pang, X.; Liu, R. Three-Dimensional Heterostructured Reduced Graphene Oxide-Hexagonal Boron Nitride-Stacking Material for Silicone Thermal Grease with Enhanced Thermally Conductive Properties. *Nanomaterials* **2019**, *9*, 938. [CrossRef] [PubMed]
13. Kim, K.; Ju, H.; Kim, J. Vertical particle alignment of boron nitride and silicon carbide binary filler system for thermal conductivity enhancement. *Compos. Sci. Technol.* **2016**, *123*, 99–105. [CrossRef]
14. Tan, W.K.; Araki, Y.; Yokoi, A.; Kawamura, G.; Matsuda, A.; Muto, H. Micro- and Nano-assembly of Composite Particles by Electrostatic Adsorption. *Nanoscale Res. Lett.* **2019**, *14*, 1–9. [CrossRef] [PubMed]
15. Tan, W.K.; Shigeta, Y.; Yokoi, A.; Kawamura, G.; Matsuda, A.; Muto, H. Investigation of the anchor layer formation on different substrates and its feasibility for optical properties control by aerosol deposition. *Appl. Surf. Sci.* **2019**, *483*, 212–218. [CrossRef]
16. Kuwana, T.; Tan, W.K.; Yokoi, A.; Kawamura, G.; Matsuda, A.; Muto, H. Fabrication of Carbon-decorated Al_2O_3 Composite Powders using Cellulose Nanofiber for Selective Laser Sintering. *J. Jpn. Soc. Powder Powder Metall.* **2019**, *66*, 168–173. [CrossRef]
17. Tan, W.K.; Hakiri, N.; Yokoi, A.; Kawamura, G.; Matsuda, A.; Muto, H. Controlled microstructure and mechanical properties of Al_2O_3-based nanocarbon composites fabricated by electrostatic assembly method. *Nanoscale Res. Lett.* **2019**, *14*, 245. [CrossRef]
18. Tan, W.K.; Asami, K.; Maeda, Y.; Hayashi, K.; Kawamura, G.; Muto, H.; Matsuda, A. Facile formation of Fe3O4-particles decorated carbon paper and its application for all-solid-state rechargeable Fe-air battery. *Appl. Surf. Sci.* **2019**, *486*, 257–264. [CrossRef]
19. Hamasaki, N.; Yamaguchi, S.; Use, S.; Kawashima, T.; Muto, H.; Nagao, M.; Hozumi, N.; Murakami, Y. Electrical and Thermal Properties of PMMA/h-BN Composite Material Produced by Electrostatic Adsorption Method. *IEEJ Trans. Fundam. Mater.* **2019**, *139*, 60–65. [CrossRef]
20. Yamaguchi, S.; Hamasaki, N.; Use, S.; Kawashima, T.; Muto, H.; Nagao, M.; Hozumi, N.; Murakami, Y. Influence of PMMA and h-BN particles sizes on electrical and thermal properties of PMMA/h-BN composite materials produced by electrostatic adsorption method. In Proceedings of the 2017 IEEE Conference on Electrical Insulation and Dielectric Phenomenon (CEIDP), Fort Worth, TX, USA, 22–25 October 2017; pp. 245–248.
21. Tanimoto, M.; Ando, S. Effects of chain rigidity/flexibility of polyimides on morphological structures and thermal diffusivity of hBN-filled composites. *Compos. Sci. Technol.* **2014**, *99*, 103–108. [CrossRef]
22. Zhi, C.Y.; Bando, Y.; Wang, W.L.; Tang, C.C.; Kuwahara, H.; Golberg, D. Mechanical and Thermal Properties of Polymethyl Methacrylate-BN Nanotube Composites. *J. Nanomater.* **2008**, *2008*, 1–5. [CrossRef]
23. Pullanchiyodan, A.; S. Nair, K.; Surendran, K.P. Silver-Decorated Boron Nitride Nanosheets as an Effective Hybrid Filler in PMMA for High-Thermal-Conductivity Electronic Substrates. *Acs Omega* **2017**, *2*, 8825–8835. [CrossRef] [PubMed]
24. Ghaffari Mosanenzadeh, S.; Naguib, H.E. Effect of filler arrangement and networking of hexagonal boron nitride on the conductivity of new thermal management polymeric composites. *Compos. Part B Eng.* **2016**, *85*, 24–30. [CrossRef]

nanomaterials

MDPI

Article

Soft X-ray Absorption Spectroscopic Investigation of $Li(Ni_{0.8}Co_{0.1}Mn_{0.1})O_2$ Cathode Materials

Jitendra Pal Singh [1], Jae Yeon Park [2], Keun Hwa Chae [3], Docheon Ahn [1,*] and Sangsul Lee [1,*]

1 Pohang Accelerator Laboratory, Pohang University of Science and Technology, Pohang 37673, Korea; jitendra2029@postech.ac.kr

2 Radiation Equipment Research Division, Korea Atomic Energy Research Institute, Jeongup 56212, Korea; jaeyeon@kaeri.re.kr

3 Advanced Analysis Center, Korea Institute of Science and Technology, Seoul 02792, Korea; khchae@kist.re.kr

* Correspondence: adc4055@postech.ac.kr (D.A.); sangsul@postech.ac.kr (S.L.); Tel.: +82-54-279-1569 (S.L.)

Received: 30 March 2020; Accepted: 12 April 2020; Published: 15 April 2020

Abstract: Herein, we report the soft X-ray absorption spectroscopic investigation for $Li(Ni_{0.8}Co_{0.1}Mn_{0.1})O_2$ cathode material during charging and discharging. These measurements were carried out at the Mn *L*-, Co *L*-, and Ni *L*-edges during various stages of charging and discharging. Both the Mn and Co *L*-edge spectroscopic measurements reflect the invariance in the oxidation states of Mn and Co ions. The Ni *L*-edge measurements show the modification of the oxidation state of Ni ions during the charging and discharging process. These studies show that e_g states are affected dominantly in the case of Ni ions during the charging and discharging process. The O *K*-edge measurements reflect modulation of metal–oxygen hybridization as envisaged from the area-ratio variation of spectral features corresponding to t_{2g} and e_g states.

Keywords: cathode materials; orbital symmetry states; metal–oxygen hybridization

1. Introduction

Rechargeable lithium batteries with layered oxide cathode materials have rapidly risen to prominence as fundamental devices for green and sustainable energy [1–3]. Thus, efforts to understand electrochemistry to improve these batteries' performances in terms of commercial development for electric vehicles [4] and large-scale grid storage applications [5] are underway. Charging capacity, thermal stability, and capacity fading are certain factors that determine a battery's performance [6,7], leaving ample scope for researchers to investigate. These efforts are reflected in the recent structure and electronic/atomic structural investigation of cathode materials by numerous techniques in the context of battery performance [8–10]. The metal ions of these materials undergo transformations in terms of oxidation state and local structural order during typical battery operation [11,12]. This affects the nature of metal–oxygen hybridization. It has been reported that these factors are influenced by the orbital symmetry states of materials' constituent ions [13,14]. Thus, investigations related to these orbital energy levels using an appropriate technique can provide atomic level insights during the charging and discharging of batteries.

X-ray absorption spectroscopic (XAS) measurements based on soft X-rays can probe the *L*-edge of transition metals, which are important constituents of layered oxide cathode materials. These measurements, termed near-edge X-ray absorption fine structures (NEXAFS), depict orbital symmetry states associated with the oxidation of metal ions [15,16]. The O *K*-edge of these layered oxide materials can also be successfully investigated with soft X-rays. The O *K*-edge NEXAFS measurements infer the symmetry states associated with metal–oxygen interactions in these materials [17,18]. Thus, soft XAS measurements can give a complete account of underlying phenomena due to the orbital symmetry states in the oxide cathode material during charging and discharging. To depict these phenomena, the Ni-rich cathode material, $Li(Ni_{0.8}Co_{0.1}Mn_{0.1})O_2$, is selected for the present investigation and denoted as NCM811. NCM811 is a well-known layered oxide cathode material and preferred for Li rechargeable batteries due to its high capacity [19,20]. Thus, this work investigates NCM811 cathode materials during charging and discharging using soft XAS.

2. Experimental Details

X-ray diffraction (XRD) experiments prior to electrochemical testing of cathode material were performed with the 9B high-resolution powder diffraction (HRPD) beamline of the Pohang accelerator laboratory (PAL) [21]. Pristine cathode material was also investigated using X-ray absorption near-edge spectroscopy (XANES) imaging measurements to reveal the local electronic structure. These measurements at the Mn *K*-, Co *K*-, and Ni *K*-edge were performed with the 7C X-ray nano imaging (XNI) beamline of the PAL. This beamline utilizes zone plate-based transmission X-ray nanoscopy, which is a kind of transmission X-ray microscopy (TXM), for image formation. A zone plate of diameter 150 μm and 40 nm outermost zone width was used for measurements. This arrangement gave a spatial resolution of 40 nm and a field of view (FOV) of around 50 μm. The X-ray absorption images based on TXM were captured at various energies within a 1 eV interval in the range −20 to 80 eV from the main edge energies of each respective element. The XANES spectrum was extracted from these X-ray absorption images [22].

Soft XAS measurements of this cathode material at various charging and discharging states were performed with the 10D XAS-KIST beamline of the same laboratory in total electron yield (TEY) mode. The grating with 1100 grooves/mm was used to measure spectra at the Mn *L*-, Co *L*-, Ni *L*-, and O *K*-edges. The obtained spectra were background-subtracted and normalized with respect to the post-edge height [23].

3. Results and Discussion

Figure 1 shows the Rietveld refinement of the synchrotron high-resolution powder diffraction patterns of the NMC811 cathode material. The diffraction peaks in the XRD patterns are associated with rhombohedral R3m space group with Li-ion on the 3a site, transition metal ions on the 3b site, and oxygen ion on the 6c site [24]. The Rietveld refinement procedure is adopted to estimate the structural parameter for this material. Table 1 collates the structural and refined parameters. The refined values of the lattice parameter "a" and "c" are 2.871 and 14.195 Å, respectively. The unit cell volume is 101.34 Å^3. The transition metal oxide (TM–O) and Li–O bond distances are 1.978 and 2.0978 Å, respectively. These parameters are similar to those reported for this composition of NMC cathode material [25].

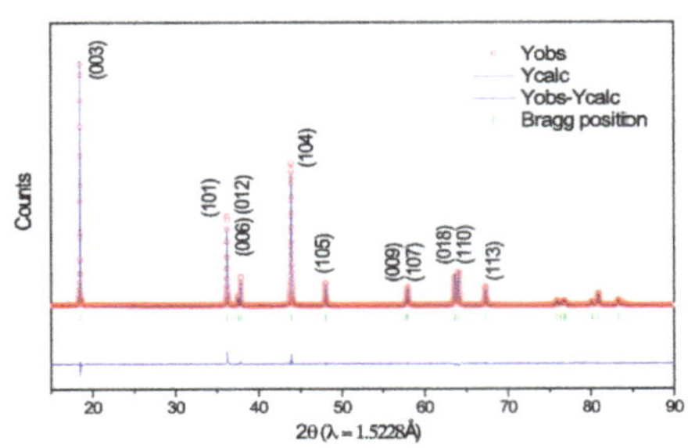

Figure 1. Rietveld refinement of the synchrotron high-resolution powder diffraction patterns of NCM811 cathode material prior to electrochemical testing.

Table 1. Structural parameters and reliability factors estimated from Rietveld refinement of the high-resolution powder diffraction (HRPD) pattern of the NMC811 cathode material prior to electrochemical testing.

a **(Å)/*c*** **(Å)**		**2.87402 (2)/14.2018 (1)**			
Atom	Site	Wyckoff positions			Occupancy *,$
Li	3a	0	0	0	0.0812 (2)
Ni2	3a	0	0	0	0.0021 (2)
Ni1	3b	0	0	0.5	0.0703 †
Co	3b	0	0	0.5	0.0075 †
Mn	3b	0	0	0.5	0.0032 †
O	6c	0	0	0.2419 (1)	0.166 †
Reliability factors		R_p = 6.67%, R_{wp} = 9.22%, R_{exp} = 5.99%, S = 1.54			

* The normalized site occupation numbers in % are the following: Li_1:Ni_2(97.52:2.48), Ni_1:Co:Mn (84.48:8.99:3.88), O (100) $ Fixed parameter. † Occupancy was achieved using the constraints. $Li_{3a} + Ni_{3a} = 0.08333$.

In this structure, almost 2.52% of Ni ions occupy the Li-ions site (Table 1). This kind of effect is common in NCM cathode materials and associated with the almost similar size of Li and Ni ions [26].

The chemical state of the constituent ions of the NCM811 cathode material was investigated using XANES-imaging measurements. Figure 2 shows the XANES spectra of the NCM811 cathode material extracted from the TXM images. In Figure 2, these spectra are shown at the Ni K-edge, Mn K-edge, and Co K-edge extracted from the original TXM images. These spectra exhibit that the main edge energy of Mn, Co, and Ni ions exist at the energy of 6.552 ± 1, 7.721 ± 1, and 8.342 ± 1 eV, respectively, in this cathode material. These values are higher than those corresponding to metal edges for each ion showing a higher oxidation state of Mn (6.539 eV), Co (7.709 eV), and Ni (8.333 eV) ions in cathode material [27]. The value of the Mn K-edge energy is similar to that of MnO_2, revealing the 4+ oxidation state of Mn ions in the NCM811 cathode material [28,29]. The Mn L-edge NEXAFS spectrum of this material also supports this oxidation state (Figure S1). The main energy of the Co ions occurs at 7.721 ± 1 eV, as estimated by the XANES spectrum that coincides with the spectrum of materials with a 3+ oxidation state [30,31]. The Co L-edge NEXAFS spectrum further favors this (Figure S2). The main edge energy value for Ni ions is associated with the 3+ oxidation state, which is characteristic of the NCM811 cathode material and reported by numerous authors [32–34]. This oxidation state of Ni ions is also evident from the Ni L-edge NEXAFS spectrum (Figure S3).

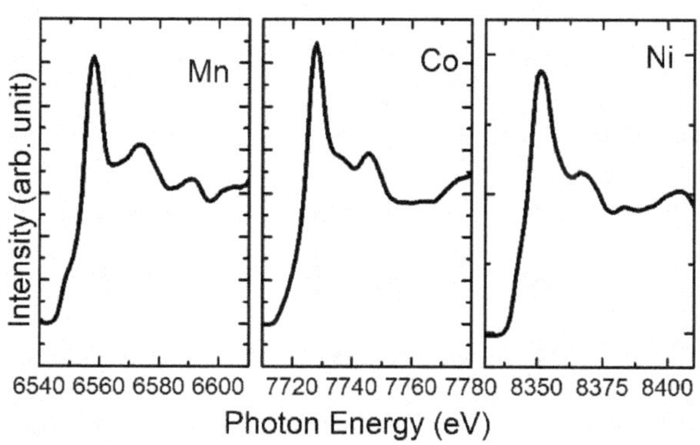

Figure 2. X-ray absorption near-edge spectroscopy (XANES) spectra at the Mn, Co, and Ni *K*-edges extracted (bulk mode) from transmission X-ray microscopy (TXM) images for NMC811 cathode material.

The secondary particles of cathode materials are generally 10–25 μm: thus the XANES spectrum of the selected region equivalent to the size of the secondary particles is extracted from the specific region of interest (ROI). Figure 3a shows the extracted Mn K-edge spectra of the selected ROI. These spectra exhibit spectral features a_1 and b_1 centered at 6.558 ± 1 and 6.573 ± 1 eV, respectively (Figure 3a). These spectral features are clearly visible in the spectra extracted from region R_1, R_2, and R_3 (Figure 3a). This envisages that Mn ions exist at almost the same oxidation state in each region of the cathode materials. Similarly, the Co K-edge spectra for each region R_1, R_2, and R_3 (Figure 3b) exhibit spectral features a_2, b_2, and c_2 centered at 7.728 ± 1, 7.737 ± 1, and 7.746 ± 1 eV, respectively (Figure 3b). The oxidation state of Ni ions in NMC cathode materials plays an important role during the charging and discharging process. Thus, the Ni K-edge spectra were also measured using XANES-imaging and shown in Figure 3c. The Ni K-edge spectra in the different regions R_1, R_2, and R_3, (Figure 3c) exhibit spectral features a_3, b_3, c_3, and d_3. These spectral features are centered at 8.351 ± 1, 8.368 ± 1, 8.384 ± 1, and 8.404 ± 1 eV, respectively (Figure 3c). Thus, these measurements show that all metal ions exist in an almost matching oxidation state despite the different ROIs.

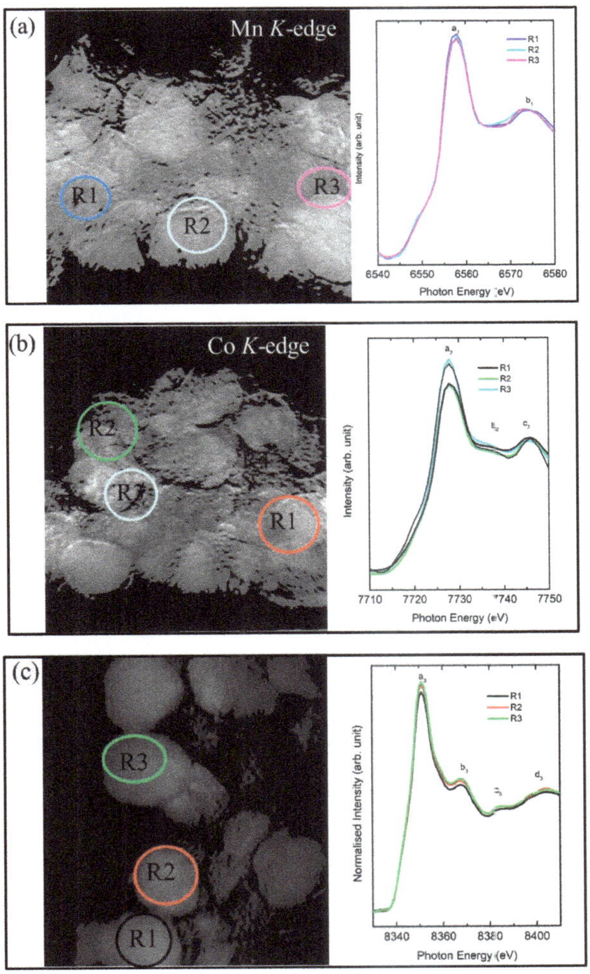

Figure 3. (**a**) Mn *K*-, (**b**) Co *K*-, and (**c**) Ni *K*-edge XANES spectra extracted from each region of interest (ROI), R_1, R_2, and R_3, equivalent to the dimensions of secondary particles.

The NCM811 cathode material exhibits structure and local electronic structure concurrent with previous reports; therefore, electrochemical testing was performed on this material. Figure 4 shows the charging and discharging curve of the NCM811 cathode material. Various states of charging (SOC) are shown by the numerals, 1, 2, 3, 4, and 5. These numerals denote the charging capacity of 0 (pristine), 40, 100, 160, and 215 mAhg^{-1}, respectively. The values at the capacity of 247, 302, 357, and 409 mAhg^{-1} represent the depth of discharging and are represented by the numerals 6, 7, 8, and 9, respectively.

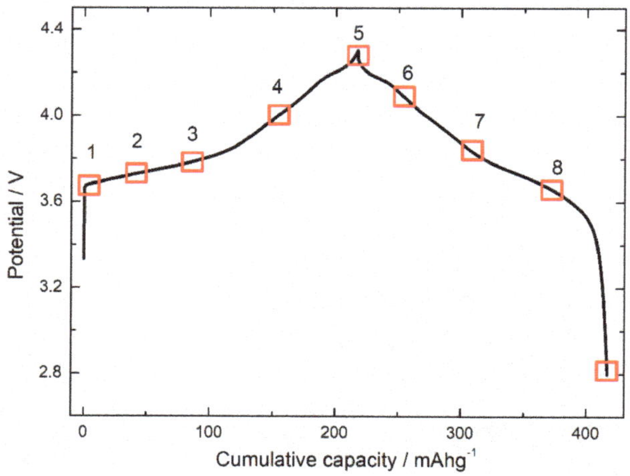

Figure 4. Various states of charging and discharging for NCM811 cathode materials.

Figure 5a shows the Mn *L*-edge spectra of NCM811 during charging and discharging. During the charging and discharging process, the spectral features A_1, B_1, and C_1 present in the spectra at various stages of charging and discharging (Figure 5a). The spectral features A_1, B_1, and C_1 are centered around 640.4 ± 0.4, 642.9 ± 0.4, and 653.3 ± 0.4 eV, respectively, in all states of charging and discharging (i.e., 1–9). The spectral features A_1 and B_1 are associated with L_3; however, C_1 is associated with the L_2-edge. The presence of spectral features that correspond to the L_3 and L_2-edges in the Mn *L*-edge spectra are due to electronic transitions to the 3d levels from the $2p_{3/2}$ and $2p_{1/2}$ levels, respectively. The presence of spectral features at an almost identical position (Figure S4) and the lack of change of shape of spectra reveal that Mn ions exist in the same oxidation state during the charging and discharging process. These observations were also obtained from the XANES study of cathode materials during battery operation [32,33].

Figure 5b shows the Co *L*-edge spectra of NCM811 during charging and discharging. During these processes (i.e., 1–9), the spectral features A_2 and B_2 appeared in the spectra (Figure 5b). These spectral features were assigned to the L_3- and L_2-edges. Further splitting into t_{2g} and e_g symmetry state was absent in these spectra. This kind of behavior is observed in Co-based layered oxide cathode materials [35,36]. Spectral features A_2 centered at 779.1 ± 0.4 and 793.3 ± 0.4 eV remain at almost the same position during charging and discharging (Figure S5). This demonstrates that during the charging and discharging process, the oxidation state of Co ions does not alter in NCM811 cathode materials. Figure 6a shows the Ni *L*-edge spectra of cathode material during charging (stages 1–5) and discharging (stages 6–9). The spectral features A_3, B_3, C_3, and D_3 appear in the spectra for charging and discharging (Figure 6a). These spectral features are centered at 851.4 ± 0.4 (A_3), 853.3 ± 0.4 (B_3), 868.4 ± 0.4 (C_3), and 869.5 ± 0.4 (D_3) eV in the spectra of the pristine NCM811 cathode. The presence of these spectral features is associated with the presence of Ni ions in an octahedral crystal field. These spectral features represent t_{2g} (L_3), e_g (L_3), t_{2g} (L_3), and e_g (L_3) symmetry states. The positions of spectral feature A_3 remain the same during charging and discharging. The position of spectral feature B_3 is slightly modified during charging and discharging (Figure 6b); this evinces that the e_g symmetry states are affected dominantly during the charging and discharging state. No role is played by the t_{2g} states.

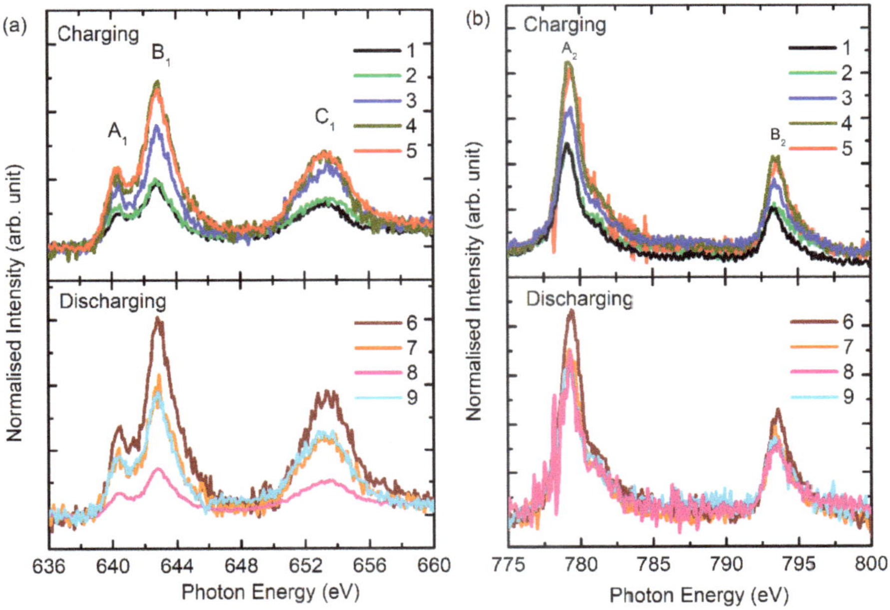

Figure 5. (**a**) Mn *L*-edge and (**b**) Co *L*-edge near-edge X-ray absorption fine structures (NEXAFS) spectra of NCM811 cathode material during charging and discharging.

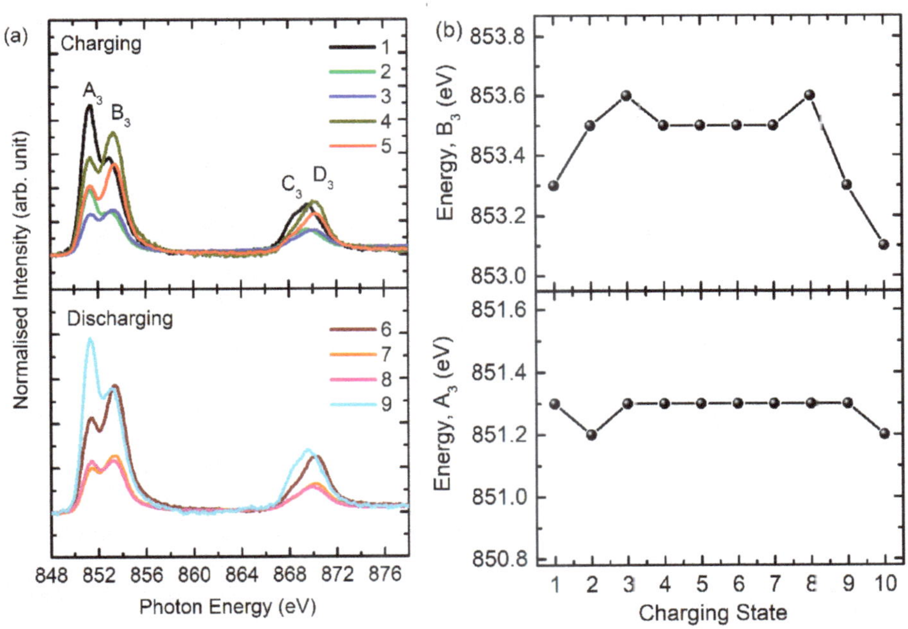

Figure 6. (**a**) Ni *L*-edge spectra and (**b**) energy of spectral features A_3 and B_3 of NCM811 cathode materials at stages 1–5 (charging) and 6–9 (discharging).

Another interesting change observed in the Ni *L*-edge spectra is the relative change in intensity of spectral features A_3 and B_3 associated with the t_{2g} and e_g symmetry states. The changes are associated with the modification of the oxidation state of Ni ions in the cathode material during charging and discharging [37]. To get deeper insights into this behavior, the area ratio of the spectral features A_3 and B_3 was determined (Figure 7a) by de-convoluting the L_3-edge region using the Gaussian function. The de-convolution spectra for states 1, 5, and 9 are shown in Figure 7b–d. The ratio of these spectral features changes alongside the charging and discharging states. The ratio is reduced with charging, indicating transformation to a higher oxidation state. After discharging, the ratio equivalent to the pristine state is observed, indicating that the oxidation state of Ni ions is reversible.

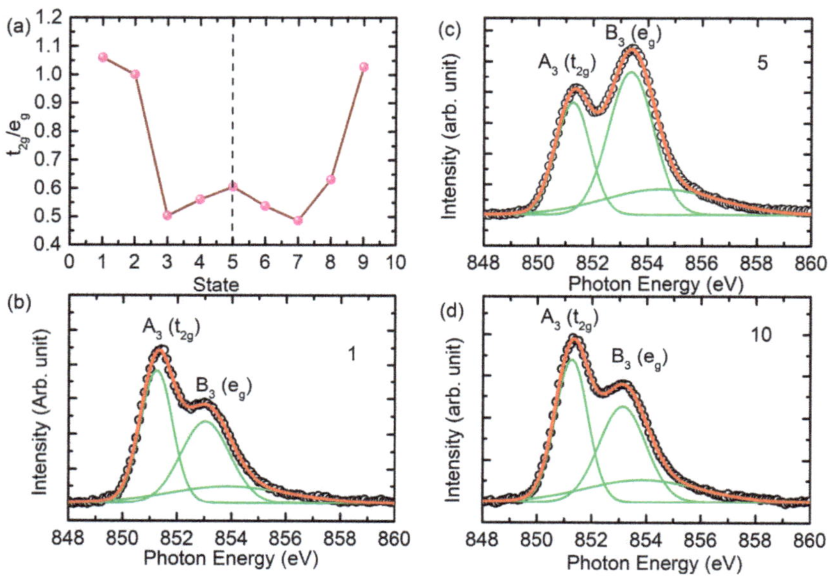

Figure 7. (**a**) The t_{2g}/e_g ratio estimated from the Ni *L*-edge spectra at various stages of charging and discharging. Here, (**b–d**) show the representative de-convoluted spectra for states 1, 5, and 9.

Figure 8 shows the O *K*-edge spectra of the NCM811 cathode during the charging and discharging process. Both the pre-edge and post-edge region of the O *K*-edge spectra reflect the systematic changes during charging and discharging. The pre-edge region of the O *K*-edge spectra exhibits spectral features A_4 and B_4 centered around 528.5 and 531.5 eV, respectively, for state 1 (Figure 8b). These spectral features also appear in the spectra for the charging/discharging states 2–9 at almost the same position. The intensities of these features change during charging and discharging.

Features C_4 and D_4 appear in the post-edge region of the O *K*-edge spectra (Figure 8a). These spectral features occur at 537.8 ± 0.4 and 541.9 ± 0.4 eV, respectively, for various charging (1–5) and discharging (6–9) states.

Because the pre-edge region of the O *K*-edge spectrum is effective to gather information of metal (3d)–O(2p) hybridization in transition metal oxides [17,18,38]; hence, the intensity ratio of spectral features A_4 and B_4 corresponding to t_{2g} and e_g symmetry state is determined and shown in Figure 9a. Figure 9b–d shows the de-convoluted spectra for states 1, 5, and 9. The intensity ratio is at its maximum for the fully charged cathode and reduces thereafter. Similar behavior is also observed by Tian et al. [39]. This shows the modulation of metal–oxygen hybridization during charging and discharging. The ratio is different from that obtained from the Ni *L*-edge spectra (Figure 7a); this may be due to the influence of these symmetry states from other metal ions such as Mn and Co [39] as well as to the creation of oxygen vacancies during the charging and discharging process [18].

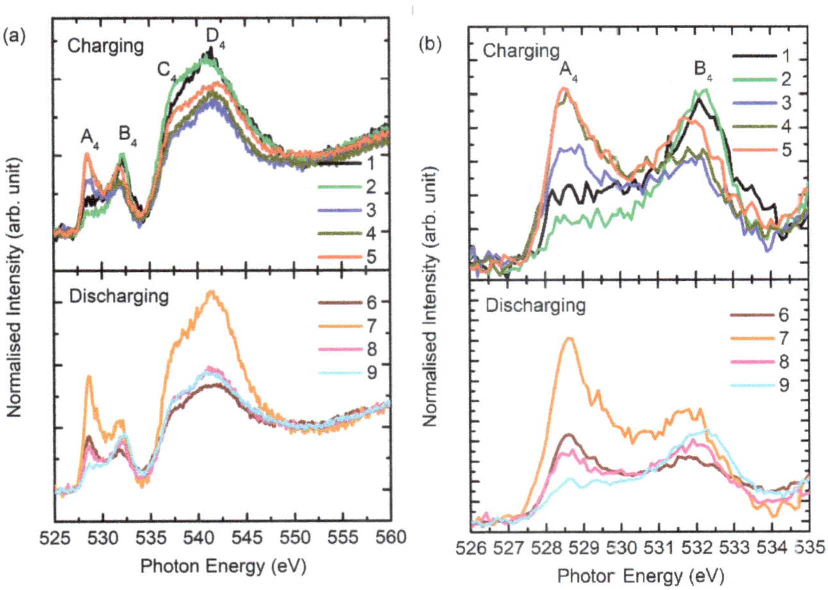

Figure 8. (**a**) O *K*-edge spectra and its (**b**) pre-edge region for the NCM811 cathode material during charging and discharging.

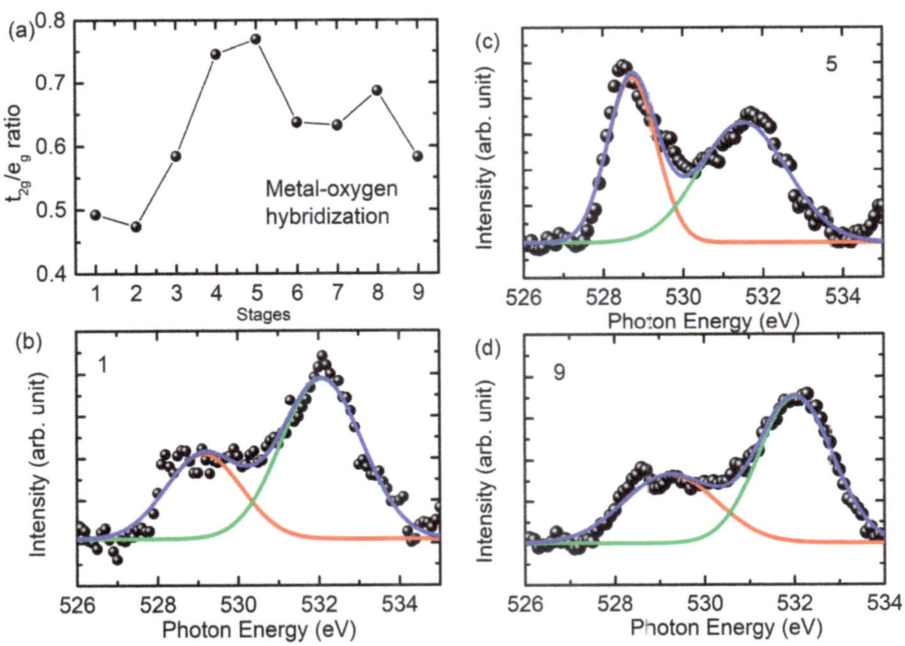

Figure 9. (**a**) The t_{2g}/e_g ratio estimated from the pre-edge region of the O *K*-edge spectra at various charging and discharging stages. Here, (**b**), (**c**), and (**d**) show the representative de-convoluted spectra for the 1, 5, and 9 states.

4. Conclusions

In conclusion, we have successfully performed soft X-ray spectroscopic measurements during charging and discharging on a well-characterized NCM811 cathode material. These results envisage that oxidation state of Mn and Co ions do not influenced by the charging and discharging process. The oxidation state of Ni ions is affected by charging and discharging; however, in the case of Ni ions, e_g states are dominantly influenced. Although the oxidation state of Ni ions changes, metal–oxygen hybridization during charging and discharging is affected by the presence of other metal ions.

Supplementary Materials: The following are available online at http://www.mdpi.com/2079-4991/10/4/759/s1, Figure S1: Mn *L*-edge NEXAFS spectra of NCM811 with reference oxide; Figure S2: Co *L*-edge NEXAFS spectra of NCM811 with reference oxide; Figure S3: Ni *L*-edge NEXAFS spectra of NCM811 with NiO; Figure S4: Positions of spectral features A_1 and B_1 of Mn *L*-edge spectra for various states; Figure S5: Positions of spectral features A_2 and B_2 of Co *L*-edge spectra for various states.

Author Contributions: Conceptualization, J.P.S., D.A., and S.L.; methodology, J.Y.P., K.H.C., and J.P.S; software, J.Y.P.; validation, S.L., D.A., and K.H.C.; formal analysis, J.P.S.; investigation, J.P.S.; resources, S.L.; data curation, D.A.; writing—original draft preparation, J.P.S.; writing—review and editing, S.L. and D.A.; visualization, K.H.C.; supervision, S.L.; project administration, S.L. and D.A.; funding acquisition, S.L. All authors have read and agreed to the published version of the manuscript.

Funding: This research was supported by the Basic Science Research Program (grant number: 2017R1C1B1011518) and EEMAP (grant number: 2017M3A7B4049173) through the National Research Foundation of Korea (NRF) funded by the Ministry of Education.

Conflicts of Interest: The authors declare no conflicts of interest.

References

1. Goodenough, J.B.; Park, K.S. The Li-Ion Rechargeable Battery: A Perspective. *J. Am. Chem. Soc.* **2013**, *135*, 1167–1176. [CrossRef]
2. Nitta, N.; Wu, F.; Lee, J.T.; Yushin, G. Li-ion Battery Materials: Present and Future. *Mater. Today* **2015**, *18*, 1369–7021. [CrossRef]
3. Li, M.; Lu, J.; Chen, Z.; Amine, K. 30 Years of Lithium-Ion Batteries. *Adv. Mater.* **2018**, *30*, 1800561. [CrossRef] [PubMed]
4. Kim, T.; Song, W.; Son, D.-Y.; Ono, L.K.; Qi, Y. Lithium-ion Batteries: Outlook on Present, Future, and Hybridized Technologies. *J. Mater. Chem. A* **2019**, *7*, 2942–2964. [CrossRef]
5. Güra, T.M. Review of Electrical Energy Storage Technologies, Materials and Systems: Challenges and Prospects for Large-Scale Grid Storage. *Energy Environ. Sci.* **2018**, *11*, 2696–2767. [CrossRef]
6. Tarascon, J.M.; Armand, M. Issues and Challenges Facing Rechargeable Lithium Batteries. *Nature* **2001**, *414*, 359–367. [CrossRef] [PubMed]
7. Biasi, L.D.; Schwarz, B.; Brezesinski, T.; Hartmann, P.; Janek, J.; Ehrenberg, H. Chemical, Structural, and Electronic Aspects of Formation and Degradation Behavior on Different Length Scales of Ni-Rich NCM and Li-Rich HE-NCM Cathode Materials in Li-Ion Batteries. *Adv. Mater.* **2019**, *31*, 1900985. [CrossRef]
8. Li, Y.; Pei, A.; Yan, K.; Sun, Y.; Wu, C.-L.; Joubert, L.M.; Chin, R.; Koh, A.L.; Yu, Y.; Perrino, J.; et al. Atomic Structure of Sensitive Battery Materials and Interfaces Revealed by Cryo–Electron Microscopy. *Science* **2017**, *358*, 506–510. [CrossRef]
9. Yang, Z.; Gu, L.; Hu, Y.S.; Li, H. Atomic-Scale Structure-Property Relationships in Lithium Ion Battery Electrode Materials. *Ann. Rev. Mater. Res.* **2017**, *47*, 175–198. [CrossRef]
10. Çelik, G.; Aktaş, S.; Ateş, Ş.; Özkendir, O.M.; Klysubun, W. Crystal and Electronic Structure Study of the $Li_2Mn_{1-x}Nd_xO_3$ Battery Cathode. *Prog. Nat. Sci. Mater. Int.* **2019**, *29*, 119–123. [CrossRef]
11. Alemu, T.; Wang, F.M. In situ Electrochemical Synchrotron Radiation for Li-ion Batteries. *J. Synchrotron Rad.* **2018**, *25*, 151–165. [CrossRef] [PubMed]
12. Hwang, B.J.; Tsai, Y.W.; Santhanam, R.; Wu, Y.W.; Hu, S.G.; Lee, J.F.; Liu, D.G. Evolution of Local Electronic and Atomic Structure of Co-Doped $LiMn_2O_4$ Cathode Material for Lithium Rechargeable Batteries. *J. Power Sources* **2003**, *123*, 206–215. [CrossRef]
13. Ragavendran, K.R.; Mandal, P.; Yarlagadda, S. Correlation between Battery Material Performance and Cooperative Electron-Phonon Interaction in $LiCo_yMn_{2-y}O_4$. *Appl. Phys. Lett.* **2017**, *110*, 143901. [CrossRef]

14. Liu, C.; Neale, Z.G.; Cao, G. Understanding Electrochemical Potentials of Cathode Materials in Rechargeable Batteries. *Mater. Today* **2016**, *19*, 109–123. [CrossRef]
15. Tokura, Y.; Nagaosa, N. Orbital Physics in Transition-Metal Oxides. *Science* **2000**, *288*, 462–468. [CrossRef]
16. Singh, V.; Major, D.T. Electronic Structure and Bonding in Co-Based Single and Mixed Valence Oxides: A Quantum Chemical Perspective. *Inorg. Chem.* **2016**, *55*, 3307–3315. [CrossRef]
17. Schuster, M.E.; Teschner, D.; Popovic, J.; Ohmer, N.; Girgsdies, F.; Tornow, J.; Willinger, M.G.; Samuelis, D.; Titirici, M.M.; Maier, J.; et al. Charging and Discharging Behavior of Solvothermal $LiFePO_4$ Cathode Material Investigated by Combined EELS/NEXAFS Study. *Chem. Mater.* **2014**, *26*, 1040–1047. [CrossRef]
18. House, R.A.; Maitra, U.; Jin, L.; Lozano, J.G.; Somerville, J.W.; Rees, N.H.; Naylor, A.J.; Duda, L.C.; Massel, F.; Chadwick, A.V.; et al. What Triggers Oxygen Loss in Oxygen Redox Cathode Materials? *Chem. Mater.* **2019**, *31*, 3293–3300. [CrossRef]
19. Xia, Y.; Zheng, J.; Wang, C.; Gu, M. Designing Principle for Ni-Rich Cathode Materials with High Energy Density for Practical Applications. *Nano Energy* **2018**, *49*, 434–452. [CrossRef]
20. Schipper, F.; Erickson, E.M.; Erk, C.; Shin, J.Y.; Chesneau, F.F.; Aurbach, D. Recent Advances and Remaining Challenges for Lithium Ion Battery Cathodes. *J. ElectroChem. Soc.* **2017**, *164*, A6220. [CrossRef]
21. Kim, J.; Ahn, D.; Kulshreshtha, C.; Sohn, K.-S.; Shin, N. Lithium Barium Silicate, Li_2BaSiO_4, from Synchrotron Powder Data. *Acta Cryst. C* **2009**, *65*, i14–i16. [CrossRef] [PubMed]
22. Park, J.Y.; Singh, J.P.; Lim, J.; Lee, S. Development of XANES Nanoscopy on BL7C at PLS-II. *J. Synchrotron Rad.* **2020**, *27*, 545–550. [CrossRef] [PubMed]
23. Singh, J.P.; Kim, S.H.; Won, S.O.; Lee, I.J.; Chae, K.H. Atomic-Scale Investigation of MgO Growth on Fused Quartz Using Angle-Dependent NEXAFS Measurements. *RSC Adv.* **2018**, *8*, 31275–31286. [CrossRef]
24. Lee, W.; Muhammad, S.; Kim, T.; Kim, H.; Lee, E.; Jeong, M.; Son, S.; Ryou, J.-H.; Yoon, W.-S. New Insight into Ni-Rich Layered Structure for Next-Generation Li Rechargeable Batteries. *Adv. Energy Mater.* **2018**, *8*, 1701788. [CrossRef]
25. Weigel, T.; Schipper, F.; Erickson, E.M.; Susai, F.A.; Markovsky, B.; Aurbach, D. Structural and Electrochemical Aspects of $LiNi_{0.8}Co_{0.1}Mn_{0.1}O_2$ Cathode Materials Doped by Various Cations. *ACS Energy Lett.* **2019**, *4*, 508–516. [CrossRef]
26. Yoshio, M.; Noguchi, H.; Itoh, J.; Okada, M.; Mouri, T. Preparation and Properties of $LiCo_yMn_xNi_{1-x-y}O_2$ as a Cathode for Lithium Ion Batteries. *J. Power Sources* **2000**, *90*, 176–181. [CrossRef]
27. Ravel, B.; Newville, M. ATHENA, ARTEMIS, HEPHAESTUS: Data analysis for X-ray absorption spectroscopy using IFEFFIT. *J. Synchrotron Rad.* **2005**, *12*, 537–541. [CrossRef]
28. Nam, K.W.; Kim, M.G.; Kim, K.B. In Situ Mn K-edge X-ray Absorption Spectroscopy Studies of Electrodeposited Manganese Oxide Films for Electrochemical Capacitors. *J. Phys. Chem. C* **2007**, *111*, 749–758. [CrossRef]
29. Manceau, A.; Marcus, M.A.; Grangeon, S. Determination of Mn valence states in mixed-valent manganates by XANES spectroscopy. *Am. Mineral.* **2012**, *97*, 816–827. [CrossRef]
30. Okumura, T.; Yamaguchi, Y.; Shikano, M.; Kobayashia, H. Correlation of Lithium Ion Distribution and X-ray Absorption Near-Edge Structure in O_3- and O_2-Lithium Cobalt Oxides from First-Principle Calculation. *J. Mater. Chem.* **2012**, *22*, 17340–17348. [CrossRef]
31. Ekwongsa, C.; Rujirawat, S.; Butnoi, P.; Vittayakorn, N.; Suttapun, M.; Yimnirun, R.; Kidkhunthod, P. Temperature Dependent Local Structure of $LiCoO_2$ Determined by in-situ Co *K*-edge X-ray Absorption Fine Structure (EXAFS). *Rad. Phys. Chem.* **2019**, 108545. [CrossRef]
32. Bak, S.M.; Hu, E.; Zhou, Y.; Yu, X.; Senanayake, S.D.; Cho, S.J.; Kim, K.B.; Chung, K.Y.; Yang, X.Q.; Nam, K.W. Structural Changes and Thermal Stability of Charged $LiNi_xMn_yCo_zO_2$ Cathode Materials Studied by Combined In Situ Time-Resolved XRD and Mass Spectroscopy. *ACS Appl. Mater. Interfaces* **2014**, *6*, 22594–22601. [CrossRef] [PubMed]
33. Sun, H.; Zhao, K. Electronic Structure and Comparative Properties of $LiNi_xMn_yCo_zO_2$ Cathode Materials. *J. Phys. Chem. C* **2017**, *121*, 6002–6010. [CrossRef]
34. Boesenberg, U.; Marcus, M.A.; Shukla, A.K.; Yi, T.; McDermott, E.; Teh, P.F.; Srinivasan, M.; Moewes, A.; Cabana, J. Asymmetric Pathways in the Electrochemical Conversion Reaction of NiO as Battery Electrode with High Storage Capacity. *Sci. Rep.* **2014**, *4*, 7133. [CrossRef] [PubMed]

35. Lin, F.; Markus, I.M.; Nordlund, D.; Weng, T.C.; Asta, M.D.; Xin, H.L.; Doeff, M.M. Surface Reconstruction and Chemical Evolution of Stoichiometric Layered Cathode Materials for Lithium-Ion Batteries. *Nat. Commun.* **2014**, *5*, 3529. [CrossRef] [PubMed]
36. Zhang, N.; Long, X.; Wang, Z.; Yu, P.; Han, F.; Fu, J.; Ren, G.X.; Wu, Y.; Zheng, S.; Huang, W.; et al. Mechanism Study on the Interfacial Stability of a Lithium Garnet-Type Oxide Electrolyte against Cathode Materials. *ACS Appl. Energy Mater.* **2018**, *11*, 5968–5976. [CrossRef]
37. Sallis, S.; Pereira, N.; Mukherjee, P.; Quackenbush, N.F.; Faenza, N.; Schlueter, C.; Lee, T.-L.; Yang, W.L.; Cosandey, F.; Amatucci, G.G.; et al. Surface Degradation of $Li_{1-x}Ni_{0.80}Co_{0.15}Al_{0.05}O_2$ Cathodes: Correlating Charge Transfer Impedance with Surface Phase Transformations. *Appl. Phys. Lett.* **2016**, *108*, 263902. [CrossRef]
38. Singh, J.P.; Kuang, H.; Lee, S.; Chae, K.H. Investigation of Metal–Oxygen Hybridization Process during the Growth of $ZnFe_2O_4$ Films on MgO (100) Substrates. (To be communicated).
39. Tian, C.; Nordlund, D.; Xin, H.L.; Xu, Y.; Liu, Y.; Sokaras, D.; Lin, F.; Doeff, M.M. Depth-Dependent Redox Behavior of $LiNi_{0.6}Mn_{0.2}Co_{0.2}O_2$. *J. Electrochem. Soc.* **2018**, *165*, A696. [CrossRef]

nanomaterials

MDPI

Article

Hydrogen Peroxide Detection by Super-Porous Hybrid CuO/Pt NP Platform: Improved Sensitivity and Selectivity

Rakesh Kulkarni, Sundar Kunwar, Rutuja Mandavkar, Jae-Hun Jeong * and Jihoon Lee *

Department of Electronic Engineering, College of Electronics and Information, Kwangwoon University, Nowon-gu, Seoul 01897, Korea; rkulkarninanotech@gmail.com (R.K.); kunwarankees23@gmail.com (S.K.); rutuja.27rrm@gmail.com (R.M.)

* Correspondence: myloveofjh@gmail.com (J.-H.J.); jihoonlee@kw.ac.kr (J.L.)

Received: 9 September 2020; Accepted: 13 October 2020; Published: 15 October 2020

Abstract: A super-porous hybrid platform can offer significantly increased number of reaction sites for the analytes and thus can offer advantages in the biosensor applications. In this work, a significantly improved sensitivity and selectivity of hydrogen peroxide (H_2O_2) detection is demonstrated by a super-porous hybrid CuO/Pt nanoparticle (NP) platform on Si substrate as the first demonstration. The super-porous hybrid platform is fabricated by a physiochemical approach combining the physical vapor deposition of Pt NPs and electrochemical deposition of super-porous CuO structures by adopting a dynamic hydrogen bubble technique. Under an optimized condition, the hybrid CuO/Pt biosensor demonstrates a very high sensitivity of 2205 μA/mM·cm^2 and a low limit of detection (LOD) of 140 nM with a wide detection range of H_2O_2. This is meaningfully improved performance as compared to the previously reported CuO-based H_2O_2 sensors as well as to the other metal oxide-based H_2O_2 sensors. The hybrid CuO/Pt platform exhibits an excellent selectivity against other interfering molecules such as glucose, fructose, dopamine, sodium chloride and ascorbic acid. Due to the synergetic effect of highly porous CuO structures and underlying Pt NPs, the CuO/Pt architecture offers extremely abundant active sites for the H_2O_2 reduction and electron transfer pathways.

Keywords: H_2O_2 detection; super-porous CuO/Pt electrode; dynamic hydrogen bubble technique; biosensor kit

1. Introduction

Hydrogen peroxide (H_2O_2) is one of the most important elements in the field of biomedical, environmental analysis, textile and food manufacturing industries due to its strong oxidizing property [1]. Also, it plays a crucial role as a signaling molecule in regulating various biological processes [2]. Furthermore, H_2O_2 has emerged as a key byproduct for many enzymatic reactions for the biosensing and diverse commercial industries [2]. The significant importance of H_2O_2 detection in various fields attracts many research groups to develop efficient sensors. Here, what is required is high sensitivity with a low limit of detection (LOD) along with good selectivity and a fast response.

Various techniques have been developed for the H_2O_2 detection including the colorimetric assay [3,4], fluorescence detection [5,6], electrochemical luminescence [7,8], surface-enhanced Raman spectroscopy (SERS) [9,10] etc. Among these, electrochemical sensing [11] can offer one of the handiest approaches for the detection of H_2O_2 due to its high sensitivity, fast response, precision and simple operation. The working principle of electrochemical H_2O_2 detection is based on the reduction of H_2O_2 into H_2O by the active electrode materials [12]. The sensitivity, selectivity and LOD of H_2O_2 sensing directly depend upon the morphological, electrical and catalytic properties of the sensing electrodes [13]. In terms of the electrode materials, recently, noble metallic nanoparticles (NPs) have gained much attention in H_2O_2 detection due to the small particle size, high surface area and high electrocatalytic activity [14–16]. At the same time, metal oxides such as CuO, NiO, MnO_2 and Ag_2O nanostructures have emerged as a new class of materials for non-enzymatic electrochemical sensors [17–21]. Among various metal oxides, the CuO as a p-type semiconductor having 1.2 eV bandgap can offer high stability, better electrochemical properties and low manufacturing cost [19–21]. On the other hand, the Pt NPs are well known for their excellent catalytic activity and higher stability over other metals as well as high electrical conductivity [15,21]. Thus, the combination of CuO structures such as highly porous CuO nanostructures on a Pt NP template could provide a significantly increased number of active sites and improved catalytic activity for the H_2O_2 reduction and efficient electron transfer pathways for the electrochemical detection. The fabrication of highly porous CuO nanostructures is enabled by the dynamic hydrogen bubble technique. This platform including the porous CuO nanostructures and Pt NPs on Si substrate has been demonstrated for the first time in this work. Often, the CuO layers were utilized for the H_2O_2 detection but the porous nanostructures of CuO by the dynamic hydrogen bubble technique has been demonstrated first time in this work. Figure 1c shows the energy-dispersive X-ray spectroscopy (EDS) spectra of CuO-5A sample with the corresponding maps. The super-porous CuO/Pt hybrid platform demonstrates a high sensitivity of 2205 $\mu A/mM{\cdot}cm^2$ for the H_2O_2 detection as seen in Figure 1d and an excellent selectivity against glucose, fructose, dopamine, sodium chloride, citric acid and ascorbic acid as clearly seen in Figure 1e. It also shows a low LOD of 140 nM with a wide detection range. This is the first demonstration of super-porous CuO nanostructures and of the hybrid architecture with the Pt NPs for the H_2O_2 sensing.

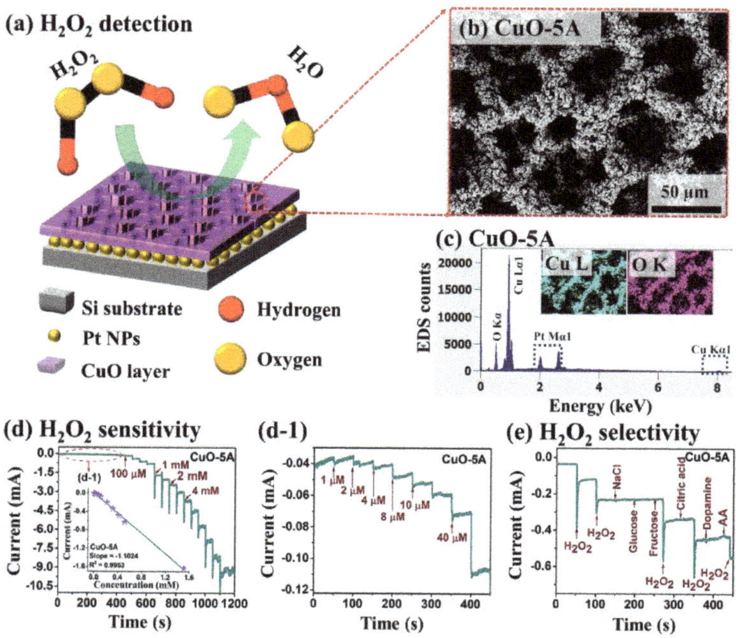

Figure 1. (**a**) Schematic representation of H_2O_2 detection by super-porous hybrid platform, made of CuO nanostructures on Pt nanoparticles (NPs) (CuO/Pt). (**b**) Scanning electron microscope (SEM) image of typical porous CuO nanostructures. (**c**) Corresponding energy-dispersive X-ray spectroscopy (EDS) spectra and elemental maps. (**d**,**d-1**) Amperometric response of CuO-5A sample upon the dropwise addition of H_2O_2 concentration from 1 μM–4 mM in a 0.1 M phosphate-buffered saline (PBS) solution of a pH 7.4 at −0.4 V potential. (**e**) Selectivity response of CuO-5A upon the successive addition of 0.1 mM H_2O_2, NaCl, glucose, fructose, citric acid, dopamine and ascorbic acid (AA) to 0.1 M PBS (pH 7.4).

In this work, a novel H_2O_2 sensing platform is demonstrated by the super-porous hybrid CuO nanostructures on Pt NPs, i.e., CuO/Pt, on Si substrate. The schematic representation of H_2O_2 molecule detection is represented in Figure 1a. The super-porous CuO/Pt hybrid electrode is demonstrated by a physical vapor deposition of Pt NPs on Si substrate followed by the electrochemical deposition of porous CuO nanostructures as shown in Figure 1b.

2. Materials and Methods

2.1. Materials

Copper sulfate ($CuSO_4$), sulfuric acid (H_2SO_4), glucose ($C_6H_{12}O_6$), fructose ($C_6H_{12}O_6$), dopamine ($C_8H_{11}NO_2$), ascorbic acid (AA, $C_6H_8O_6$), sodium chloride (NaCl), hydrogen peroxide (H_2O_2), citric acid ($C_6H_8O_7$) and phosphate-buffered saline tablets (PBS) were purchased from Sigma–Aldrich (St. Louis, Mo, United States). All the reagents were of analytical grade and used without further purification. Deionized (DI) water was used as a solvent throughout the experiment.

2.2. Fabrication of Pt Nanoparticles (NPs)

Initially, the Si substrate was degassed in the pulsed laser deposition (PLD) (DaDa TG, Daegu, South Korea) chamber under the 1.0×10^{-4} Torr at 725 °C for 30 min to remove the trapped gases, water vapors and contaminants. After degassing, the substrate was transferred to a plasma-assisted sputtering chamber for the deposition of 50 nm Pt film with the ionization current of 7 mA under 1.0×10^{-1} Torr. Subsequently, the Pt deposited Si sample was annealed in the PLD chamber at 425 °C for 30 min to form the Pt NPs and enhance the adhesion of Pt on Si. During the annealing process, the chamber pressure was kept constant at 1.0×10^{-4} Torr and the temperature was increased at 4 °C/s to reach the target temperature. To finish the sample growth, the heating system was turned off and the sample was kept under the same vacuum until the temperature was dropped to an ambient over time. The surface morphology of the Pt/Si sample after the annealing process is shown in Figure S1, which shows the root mean squared (RMS) roughness (Rq) and surface area ratio (SAR) were much increased with the formation of Pt NPs.

2.3. Fabrication of CuO Nanostructures

The electrochemical deposition of Cu was carried out in a three-electrode system comprising the Pt/Si substrate, platinum (Pt) electrode and Ag/AgCl electrode, as working (WE), counter (CE) and reference (RE) electrodes, respectively. The working electrode size was 0.5×1 cm^2. For the deposition of porous Cu film on the Pt/Si, the precursor solution of 0.1 M $CuSO_4$ and 0.05 M H_2SO_4 was prepared in 20 mL DI water [22]. Then, various Cu layers were deposited by varying the deposition time such as 5, 10, 20, 30 and 50 s at a fixed current density of 2 A/cm^2 and also at the deposition current density of 0.5, 1, 3 and 5 A/cm^2 with the applied potential of 1 V (vs Ag/AgCl). Then, the samples were transferred to the PLD chamber for the oxidation of Cu. A stepwise annealing at 300 and 500 °C for 30 min each was equally adapted with the continuous O_2 (20 CC) flow. The PLD chamber vacuum was 1.7×10^{-1} Torr during the annealing under the O_2 supply. After the completion of the annealing process, the pure Cu metal layer was converted into the CuO through the oxidation [23]. The CuO samples are named as CuO-10 s, CuO-20 s, CuO-30 s, CuO-50 s and CuO-0.5A, CuO-1A, CuO-2A, CuO-3A and CuO-5A, respectively, based on the variation of deposition time and current density.

2.4. Physical Characterizations

The surface morphology of prepared CuO/Pt/Si samples was characterized by a scanning electron microscope (SEM, COXEM CX-200, Daejeon, Korea) and atomic force microscopy (AFM, Park Systems Corp. XE-70, Gyeonggi-do, South Korea). For an elemental characterization of samples, an energy-dispersive X-ray spectroscope (EDS, Noran System 7, Thermo Fisher, Waltham, MA, USA) was used under the spectral and imaging modes. For the Raman measurement, a NOST system (Nostoptiks, Gyeonggi-do, Korea) was utilized, which is integrated with the 532 nm laser, spectrograph (ANDOR sr-500), charge-coupled device (CCD) and various optics. All the electrochemical measurements were carried out with the Wizmac-1200Premium system (Wizmac, Daejeon, Korea).

2.5. Electrochemical Measurement

All electrochemical performances of as-prepared electrodes were measured on the Wizmac-1200Premium system (Wizmac, Daejeon, Korea). The as-prepared hybrid electrode was used as the working electrode, and the Ag/AgCl and Pt plate were used as reference and counter electrodes respectively. The working electrode, the CuO/Pt/Si sample, size was 0.5×1 cm^2. 0.1 M PBS (pH~7.4) was used as electrolytes for H_2O_2 detection, and a specified concentration of H_2O_2 was added continuously to 0.1 M PBS under a stirring condition. In order to maintain the O_2-free environment, N_2 was purged into the electrolyte solution before the electrochemical measurements. All sensing performances were examined at ambient conditions.

3. Results and Discussion

Figure 2 shows the physical properties of super-porous CuO/Pt hybrid electrodes by the variation of deposition-duration during the electrochemical deposition of Cu at 2 A/cm^2 in a solution containing 0.1 M $CuSO_4$ and 0.05 M H_2SO_4. Initially, the Pt nanoparticle (NP) templates were fabricated by the sputtering of 50 nm Pt film on Si substrate and subsequent annealing. The tiny Pt NPs were fabricated after annealing as clearly seen in the AFM top-views and line-profile in Figure 2a,b-1. Indeed, the Rq and SAR were significantly increased as shown in Figure 2c after the formation of Pt NPs. The average height and diameter of Pt NPs were estimated to be around 5 and 40 nm from the cross-sectional line-profiles in Figure 2b-1. With the annealing of Pt layers under the high temperature and vacuum, the strong adhesion between Pt and Si can be achieved as well, which could further facilitate the adsorption of Cu atoms during the electrochemical deposition. Figure 2d–g show the SEM images of highly porous CuO. The zoom-in SEM images are shown in Figure 2d-1–g-1. All the samples fabricated in this work clearly depicted the formation of highly porous structures as displayed the SEM images. Generally, the CuO nanostructures exhibited large surface pores as well as numerous small pores on the vertical side walls. Thus, the term "super-porous" is used to indicate such porous structures, which is achieved by the electrodeposition of Cu film along with the hydrogen bubbling. It can be observed that the partially connected CuO dendrites were formed on the surface of the Pt NP template with the deposition time (T_d) of 10s in Figure 2d. As the T_d was increased, the gradual growth of interconnected porous structures was observed due to the additional deposition of Cu atoms in Figure 2e–g. The porous nature of CuO nanostructures was due to the simultaneous deposition of Cu and generation of hydrogen bubbles as described in Figure S2 [22]. Due to the large overpotential, the co-reduction process occurs, in which the Cu ions are reduced simultaneously with the H^+ as described by Equations (1) and (2) [18].

$$2H^+_{(aq)} + 2e^- \rightarrow H_2 \quad (1)$$

$$Cu^{2+}_{(aq)} + 2e^- \rightarrow Cu_s \quad (2)$$

The generation of hydrogen bubbles functions as a dynamic bubble template for the porous Cu deposition. The pore size was found to be increased with the deposition time due to the coalescence of Cu nanostructures. Meanwhile, the intensive dendrites and corn-like agglomerates were grown towards the interior of the pores, resulting in the formation of highly porous 3D Cu nanostructures. The complete mechanism of CuO fabrication on Pt/Si substrate is shown in Figure S2 and additional SEM images are provided in Figure S3. The Cu particle size was not significantly affected as seen in Figure 2d-1–g-1.

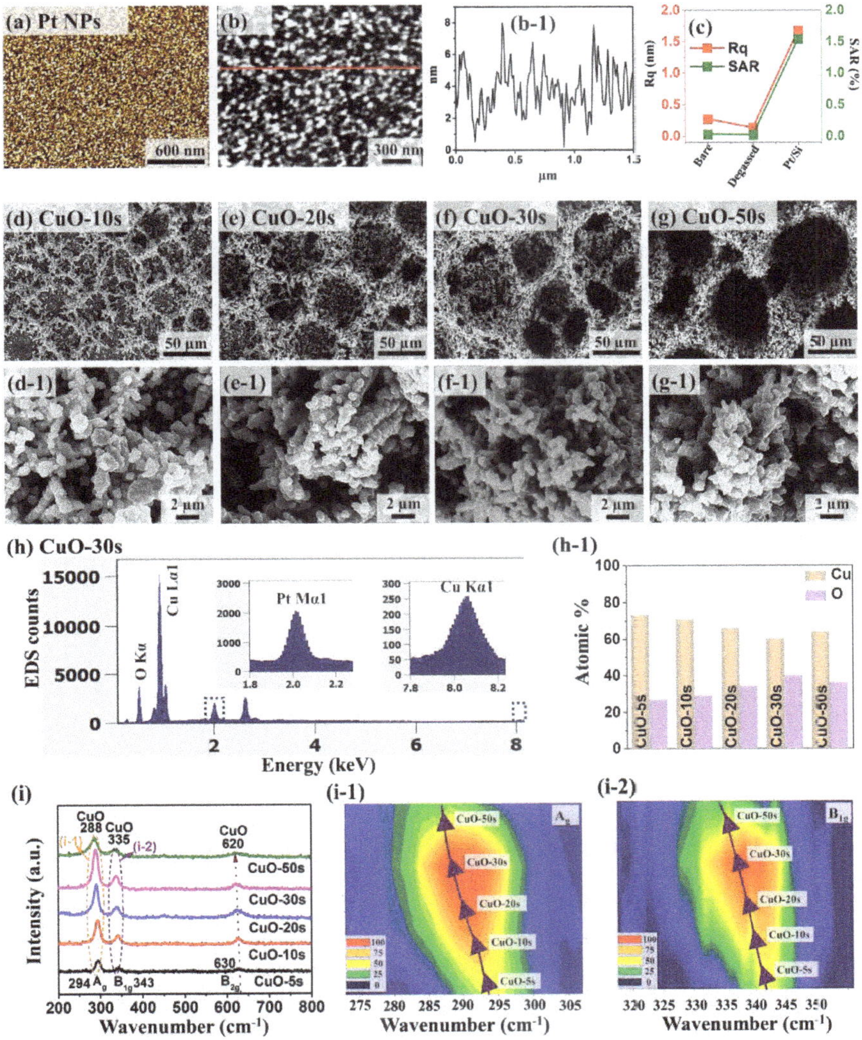

Figure 2. Structural analysis of porous CuO/Pt hybrid detectors fabricated at different deposition duration by an electrochemical deposition at 2 A/cm^2 cathodic current density in a solution of 0.1 M $CuSO_4$ and 0.05 M H_2SO_4. (**a**) Atomic force microscopy (AFM) top-view of Pt NPs on Si substrate. (**b**,**b-1**) Magnified AFM top-view and corresponding cross-sectional line-profile. (**c**) Rq and surface area ratio (SAR) plots at different conditions. (**d**–**g**) SEM images of the porous CuO/Pt hybrid nanostructures for different deposition durations from 10 to 50 s. The CuO-10 s stands for the ten-second deposition duration. (**d-1**–**g-1**) Corresponding zoom-in SEM images. (**h**) EDS spectra of CuO-30 s. (**h-1**) Summary of atomic percentage of Cu and O from different samples as a function of deposition time. (**i**) Raman spectra of the porous CuO/Pt hybrid samples. (**i-1**,**i-2**) Contour plots of the Raman peaks of CuO corresponding to the A_g and $B1_g$ modes.

After the fabrication of highly porous Cu nanostructures, each sample was annealed at 500 °C for 30 min under the continuous O_2 (20 CC) flow, which converts the Cu into CuO by the oxidation without much difference in the morphology. Furthermore, the elemental characterization was carried out as shown in Figure 2h,h-1. The EDS spectra of other samples are provided in Figure S4. The EDS spectra reveal the presence of Cu, O and Pt elemental peaks, indicating the formation of CuO on the

Pt NP template. In addition, the corresponding atomic percentage of Cu and O for each sample is summarized in Figure 2h-1. This indicates that the oxygen amount was gradually increased up to 30 s of Cu deposition and the oxygen amount was decreased at 50 s likely due to the deposition of a thicker Cu layer, which can prevent the exposure of Cu atoms during the annealing under O_2. From the EDS analysis, it was also observed that the atomic percentage ratio of Cu to O is ~2:1. This could be due to the unreacted or unoxidized Cu deep in the structures. In addition, Raman scattering spectra were obtained from the CuO samples for the crystal phase characterization of the deposited materials as shown in Figure 2i. All samples exhibited three Raman peaks at 294, 343 and 630 cm^{-1}, corresponding to the A_g, B_{1g} and B_{2g} phonon modes of CuO [24]. The Cu_2O phase showed the Raman peaks around 220, 415, 520 and 630 cm^{-1}. This clearly indicates that the oxidation status of our samples is mostly CuO with the strong Raman peaks at 294 and 343 cm^{-1}. It was observed that the intensity of Raman peaks was gradually increased up to the CuO-30 s, which may be due to the gradually increased size. It can also be observed from the previous results that the Raman peak position can slightly vary depending upon the annealing temperature and crystallinity of the samples [25]. However, with the 50 s, a lower intensity of Raman peak was observed, which could be due to the poor crystallinity by an inefficient oxidation with the thick structure formation as discussed with the EDS spectra [26]. The counterplots of A_g and B_{1g} peaks are shown in Figure 2i-1,i-2, which demonstrate the blue shift of the A_g and B_{1g} Raman peaks and broadening. The gradual peak broadening and shift in the Raman peaks at ~288, 335 and 620 cm^{-1} can be due to the gradually increased size effect [27,28]. When the deposition time was increased, the grain size was gradually increased, and the formation of thick CuO walls and formation of large dendrite structures were observed.

Figure 3 shows the electrochemical characterizations of the deposition-duration variation set via cyclic voltammetry (CV) and chronoamperometry (CA). The CV and amperometric response of Si and fabricated Si/Pt substrate with the addition of 0.1 M H_2O_2 in 0.1 mM PBS (pH 7.4) electrolyte is shown in Figure S5. First, the CV plots of CuO/Pt hybrid electrodes was measured in a stirred 0.1 M PBS (pH 7.4) containing 0.4 mM H_2O_2 at a scan rate of 50 mV/s as shown in Figure 3a. From the CV results, all the CuO samples showed a gradual increase in the oxidation and reduction peaks along with the deposition time up to 30 s. With the increase in the deposition time, the CuO thickness was increased in both vertical and lateral directions, which resulted in the evolution of porous CuO nanostructures. Due to the much-increased thickness of Cu for the CuO-50 s, the conversion rate of Cu to CuO was diminished, resulting in the increased atomic percentage of Cu as clearly demonstrated by the EDS and Raman spectra analyses in the previous section. The highly porous nature of the CuO-30 s is not only effective for the electron pathways but also provides a significantly increased number of active sites, which is helpful for the enhancement of electrochemical detection of H_2O_2. The overall mechanism of H_2O_2 reduction can be expressed by the relation [29]:

$$H_2O_2 + 2CuO \rightarrow Cu_2O + O_2 + H_2O \tag{3}$$

The electrocatalytic reduction of H_2O_2 by CuO can be described in two steps: (i) electrochemical reduction of Cu(II) to Cu(I) and (ii) electron transmission and O_2 generation, reducing the H_2O_2 into H_2O. From the CV measurement, the CuO-30 s showed the two high intensity reduction peaks at around −0.2 and −0.4 V and two oxidation peaks at −0.1 and 0 V in Figure 3a. The two reduction peaks can be corresponded to the stepwise one-electron reduction of Cu(II) to Cu(I) and Cu(I) to Cu(0), whereas the two oxidation peaks can likely correspond to the one-electron oxidation of Cu(0) to Cu(I) and of Cu(I) to Cu(II) [30]. Since the CuO-30 s demonstrated the highest oxidation and reduction peaks, it was further studied for the amperometric response at different applied potentials upon the drop-wise addition of 0.1 mM H_2O_2 solution as shown in Figure 3b. The maximum and stable current response was obtained at −0.4 V. It is well known that the applied potential in the CA has a great influence on the sensitivity, stability and selectivity of the sensor [31]. Figure 3c displays the CV response of the CuO-30 s by varying the scan rate in the range of 20–200 mV/s in a 0.1 M PBS (pH 7.4) containing 0.1 mM H_2O_2. With the higher applied scan rate, the peak potential was increased, which consequently indicates that the electrocatalytic activities are enhanced by the absorbed analytes at a higher scan rate [32]. The corresponding graph in Figure 3d demonstrates the linear plot of capacitive current ($\Delta j_{-0.2}$) versus the square root of the scan rate, in which the slope of the plot corresponds to the double layer capacitance (C_{dl}) of the electrode [33]. The C_{dl} can be utilized to qualitatively evaluate the electrochemical-active surface area (ECSA) of an electrode. Obviously, the higher C_{dl} can indicate the increased ECSA when the geometrical area is fixed. Here, the C_{dl} for the CuO-30 s sample was 0.04581 mF, which is a decently good value as compared with other conventional electrodes. Furthermore, the current versus potential relationship with the variation of H_2O_2 concentration was studied at the scan rate of 50 mV/s as shown in Figure 3e. With the increased concentration of H_2O_2 from 0.5 to 3.5 mM, the current intensities of reduction and oxidation peaks were gradually increased due to the strong electrolyte reaction of H_2O_2. The increased reduction current can be ascribed to the increased Cu(II) species from Cu(I) by means of simultaneous reduction of H_2O_2 [18]. Figure 3f shows the relation between the peak current at −0.5 V with respect to the H_2O_2 concentration. As the peak current increases progressively at −0.5 V, it can indicate the high electrochemical activity of H_2O_2 reduction around this voltage.

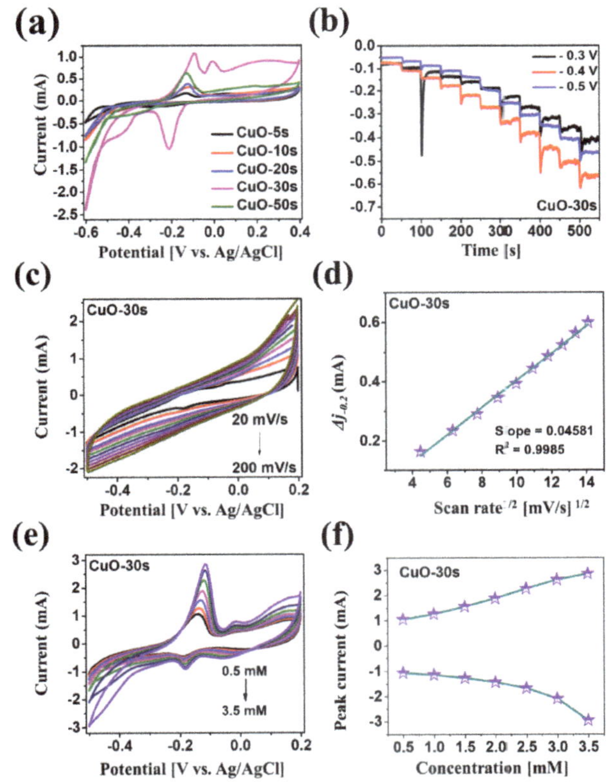

Figure 3. (**a**) Cyclic voltammetry (CV) response of various porous CuO/Pt hybrid detectors in 0.1 M PBS (pH 7.4) containing 0.4 mM H_2O_2 at a scan rate of 50 mV/s. (**b**) Amperometric response of CuO-30 s sample with the dropwise addition of 0.1 mM H_2O_2 at different applied potential. (**c**) CVs response of the CuO-30 s sample at different scan rates from 20 to 200 mV/s in 0.1 M PBS (pH 7.4) containing 0.1 mM H_2O_2. (**d**) Capacitive current Vs square root of scan rate calibration plots for the CuO-30 s at −0.2 V ($\Delta j_{-0.2} = (j_a - j_c)/2$). (**e**) CV of CuO-30 s sample in 0.1 M PBS (pH 7.4) containing different concentrations of H_2O_2 ranging from 0.5 to 3.5 mM at the scan rate of 50 mV/s. (**f**) Relation between peak current with respect to H_2O_2 concentration at −0.5 V.

Figure 4 presents the cyclic voltammetry (CV) response for the electrochemical detection of H_2O_2 by the CuO samples fabricated at different deposition time. The CA response was measured by a dropwise addition of H_2O_2 concentrations in the N_2-saturated 0.1 M PBS (pH 7.4) under a stirring condition at a potential of −0.4 V. The electrochemical CA response was performed using a three-electrode system comprising the Pt/Si substrate, platinum (Pt) electrode and Ag/AgCl electrode, as working (WE), counter (CE) and reference (RE) electrodes, respectively. Figure 4a shows the steady-state amperometric current response for the deposition time variation set upon the successive addition of 0.2 mM H_2O_2. The reduction current was sharply increased and stabilized approximately within 2 s after the ingestion of H_2O_2 drop. Specifically, the CuO-30 s demonstrated the highest current with a rapid electrochemical response among all samples. This can be due to the large active surface area of porous CuO nanostructures that allows the diffusion of H_2O_2 molecules with the best crystalline quality and efficient electron transfer at the interface during the reduction process. Figure 4b presents the current versus concentration relationship, in which the linear current response was observed as a function of concentration for all the samples and the CuO-30 s exhibited the highest current difference. Moreover, various concentrations of H_2O_2 were detected by the CuO-30 s as displayed in Figure 4c, which demonstrates that the current response was sharply increased upon the addition of each drop of H_2O_2 from 1 µM to 4 mM. To gain a clear understanding of low concentration H_2O_2 detection, the concentration regime from 1 µM–1.5 mM is separately plotted in Figure 4c-1, which displays a stepwise steady-state current after the successive addition of H_2O_2. In addition, Figure 4d shows the two linear region calibration curves for the low- and high-concentration ranges. The two linear regions can be observed due to the different activation and adsorption behavior of the hybrid CuO/Pt biosensor along with the increased H_2O_2 concentration [34]. Figure 4d-1 displays the linear regression at low concentration where the electrocatalytic mechanism is dominant, which is described by Equation (4) [9].

$$y = ax + b\left(R^2\right) \tag{4}$$

$y = -0.7492x - 6.4111 \times 10^{-4}$ ($R^2 = 0.9976$) for the first linear range. As the geometric surface area exposed to the electrolyte is 0.5×1 cm^2, the sensitivity (S) is determined to be 1498 µA/mM·cm^2. The definition of sensitivity (S) is shown in the Equation (5). The LOD is calculated to be 325 nM at a signal to noise (S/N) ratio of 3 based on the Equation (6) for the H_2O_2, where the σ represents the standard deviation [35]. Here, the σ was obtained to be 0.08114 µA by averaging 25 blank readings of CuO-30 s in the 0.1 M PBS electrolyte. Each reading was taken for 25 s.

$$S = \frac{Slope}{Geometicla\ area\ of\ working\ electorde} \tag{5}$$

$$LOD = \frac{3 \times \sigma}{slope} \tag{6}$$

Figure 5 shows the physical characterization of porous CuO/Pt hybrid electrodes fabricated by the variation of current density between 0.5 and 5 A/cm^2 on Pt/Si substrate. The SEM images of CuO/Pt hybrid electrodes at different current densities are shown in Figure 5. At a low current density of 0.5 and 1 A/cm^2, the discrete and irregular vertical growth of dendrite-like structures can be observed on the surface in Figure 5a,b. As the current density was increased further, more porous structures and large vertical dendrites were formed due to the abundant hydrogen bubble formation and clustering of electrodeposited metal atoms in Figure 5c,d. The morphology of porous structures directly depends upon the current density such that higher density and small pore size were obtained with higher current density, which can be due to the faster generation and desorption of hydrogen bubbles [36]. The CV and CA electrochemical response of these various CuO/Pt hybrid samples based on current density variation are presented in Figure S6. Specifically, the CuO-5 A demonstrated the C_{dl} of 0.1005 mF as seen in Figure S6d, which is a higher value than the CuO-30 s as discussed in Figure 3d. The increased C_{dl} further confirms that the CuO-5A sample effectively enhances the catalytic active sites with the increased electrochemical active surface area (ECSA) of the electrode and improves the electron transfer rate for the catalytic performance and demonstrates high sensitivity for H_2O_2 reduction. From the CV and CA measurements, the electrochemical performance of CuO-5A for the H_2O_2 detection was found to be maximum in this set. Thus, the CuO-5A sample was further explored in terms of physical properties and electrochemical performance for the H_2O_2 detection. Figure 6a–d show the SEM and EDS elemental maps of the CuO-5A respectively. The EDS elemental maps demonstrate the presence of Cu L and O K peaks in the CuO-5A and matches well with the SEM morphology. This clearly shows the co-existence of Cu and O elements in the porous nanostructures. In addition, the corresponding EDS line-profile analysis of the selected area is shown in Figure 6e–e-2, which confirms the uniform distribution of Cu and O in the nanostructures. The EDS spectrum of the porous CuO-5A sample in Figure 6f also confirms the presence of Cu, O and Pt elements in the sample. From the atomic percentage plot in Figure 6f-1, the atomic percentage of Cu was gradually reduced while the atomic percentage of O was increased with the increased current density. This indicates that the amount of oxidized Cu is greater in the high current density samples likely due to the higher porosity and pore density. The EDS elemental spectra of other samples in this set are provided in Figure S7. To gain structural insights, Raman spectra analysis was performed as shown in Figure 6g. Generally, the Raman vibration peaks were observed at 295, 345 and 629 cm^{-1} for all CuO samples, which corresponds to A_g, B_{1g} and B_{2g}. In comparison to the Raman peak of the single crystal of CuO, the Raman peaks showed a gradual blue shift with a broadening as shown in Figure 6g-1,g-2. Again, this can be related to the increased grain size along with the increased current [26,37,38].

Figure 7 shows the CA response of porous CuO/Pt hybrid electrodes fabricated at the controlled current density. The CA response was measured by the dropwise addition of H_2O_2 concentrations in the N_2-saturated 0.1 M PBS (pH 7.4) under the stirring condition at a potential of −0.4 V. Figure 7a shows the steady-state amperometric current response of samples upon the dropwise addition of 0.2 mM H_2O_2. Specifically, the CuO-5A showed the highest current under the same concentration of H_2O_2, indicating the superior diffusion of analyte due to the high porosity and crystallinity of CuO nanostructures. Figure 7b presents the current versus concentration relationship, in which the linear current response was observed as a function of concentration for all the samples and the CuO-5A exhibited the highest current slope. Based on the high current response of the CuO-5A sample, this was further examined for the electrochemical H_2O_2 detection. The amperometric response of CuO-5A upon the dropwise addition of H_2O_2 from 1 μM to 4 mM is presented in Figure 7c, which clearly showed the current increment upon the addition of H_2O_2. The low concentration H_2O_2 detection between 1 and 40 μM was further plotted as shown in Figure 7c-1, which presents the steady and stable stepwise current state after the addition of H_2O_2. Furthermore, Figure 7d shows the linear calibration curve of current versus H_2O_2 concentration ranging from 1 μM to 4 mM. Figure 7d-1 displays the corresponding linear curve of current versus H_2O_2 concentration, and the linear equation is given by Equation (4), $y = -1.1024x - 0.339$ ($R^2 = 0.9953$) for the first linear range, and the sensitivity is calculated to be about 2205 $\mu A/mM{\cdot}cm^2$ and LOD is 140 nM [35]. In addition, Table 1 summarizes the performance parameters of previously reported CuO-based devices and our biosensors. Our device showed relatively good performance factors compared to other devices in terms of the sensitivity and LOD with a quite wide linear range. Figure 7e shows the selectivity characterization of the CuO-5A sample by varying the various organic molecules of 0.1 mM concentration, i.e., NaCl, glucose, fructose, citric acid, dopamine and ascorbic acid (AA). Interestingly, no current response was observed with the addition of other organic molecules than H_2O_2. The alternated molecules dropping sequence showed that the current was sharply increased only for the addition of H_2O_2. This result clearly confirms that the CuO-5A is highly selective for the H_2O_2 detection at −0.4 V applied bias against typical organic molecules. One step ahead, we measured the selectivity response for the mixture of organic molecules with and without the addition of H_2O_2 as shown in Figure S8. Figure S8a,b clearly indicate the sensor very selectively responds to H_2O_2 the reduction. The CA response was sharply changed with each drop of 0.1 mM solution with the H_2O_2 in Figure S8a. Meanwhile, there was no response for the mixture solution with the H_2O_2 in Figure S8b. Thus, it clearly shown that our fabricated biosensor is highly selective for the H_2O_2 at −0.4 V. Furthermore, the stability and reproductivity tests were conducted with the working electrode CuO-5A as shown in Figure 7f. During the 250 s test period, all four CuO-5A electrodes exhibited similar and stable current upon the injection of the same concentration of H_2O_2. Thus, the porous CuO/Pt hybrid electrodes can be a promising platform for the fabrication of H_2O_2 detectors with the high sensitivity, selectivity and reproductivity. This can be attributed to the effective absorption of the H_2O_2 on the super-porous CuO interface and fast charge transport pathways through the underlying Pt NPs during the electrolysis process [39].

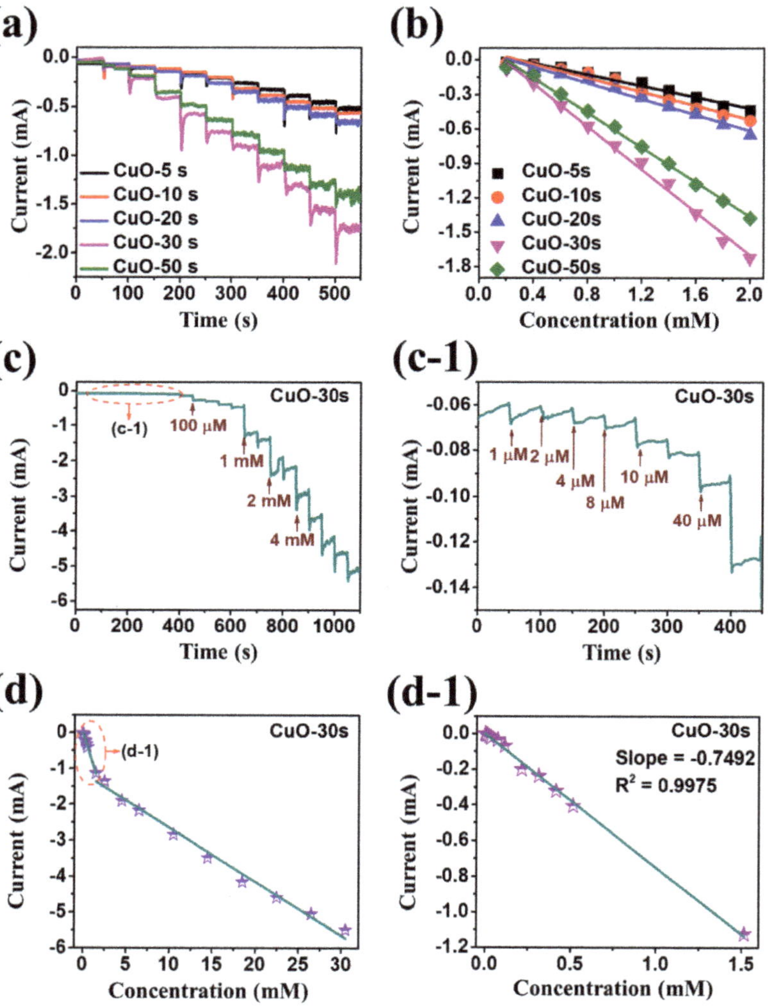

Figure 4. (**a**) Steady-state current time response of porous CuO/Pt hybrid detectors fabricated at a different duration between 5 and 50 s upon the successive addition of 0.2 mM H_2O_2 in N_2- saturated 0.1 M PBS (pH 7.4) at an applied potential of −0.4 V. (**b**) Corresponding calibration curves of various CuO samples for H_2O_2 detection. (**c**,**c-1**) Amperometric response of CuO-30 s to dropwise addition of H_2O_2 from low (1 μM) to high (4 mM) concentrations. (**d**,**d-1**) Linear calibration curve of CuO-30 s based on the current versus H_2O_2 concentration at high concentration range from 0.1–4 mM and low concentration range from 1 μM–1.5 mM.

Figure 5. SEM images of porous CuO nanostructures based on the control of current density between 0.5 and 5 A (CuO-0.5A ~ CuO-5A) for the fixed time of 30 s. (**a**–**d**) Large scale SEM images of the CuO nanostructures as labeled. (**a-1**–**d-1**) High-magnification SEM images.

Figure 8 presents the typical amperometric responses of CuO-5A for the detection of different organic molecules at different measurement conditions such as (a) glucose in 0.1 M NaOH at 0.6 V (b) fructose in 0.1 M NaOH at 0.4 V (c) dopamine in 0.1 M PBS at 0.6 V and (d) ascorbic acid in 0.1 M NaOH at 0.5 V. All the measurements were carried out the under a N_2 saturation and stirring condition. Since, the applied potential has a great impact on the sensitivity, stability and selectivity of sensors, the applied potential for different molecules was determined based on the CV and CA measurement as shown in Figure S9 [48]. Figure 8a–d show the amperometric response of the CuO-5A sample upon the dropwise addition of different molecules as labeled with various concentrations ranging from 1 μM to 2 mM. Figure 8a-1–d-1 present the magnified section of Figure 8a–d at low concentration ranging from 1–10 μM. It was found that the CuO-5A sample exhibited a decent current response with the consequent step change for each drop of different molecules at different potentials. These results confirm that the porous CuO/Pt hybrid electrodes can exhibit excellent sensitivity with different organic molecules, which can be ascribed to high electroconductivity and good electrocatalytic activity. The corresponding linear calibration curves of current versus concentration of each molecules at high concentration ranging from 10 μM to 2 mM is shown in Figure 8e–h. Similarly, the low concentration 1–10 μM linear calibration curves are presented in Figure 8e-1–h-1 for different molecules as labeled.

From these results, it can be concluded that the porous CuO/Pt hybrid electrodes can be applied for the detection of various organic molecules at different applied potentials as well.

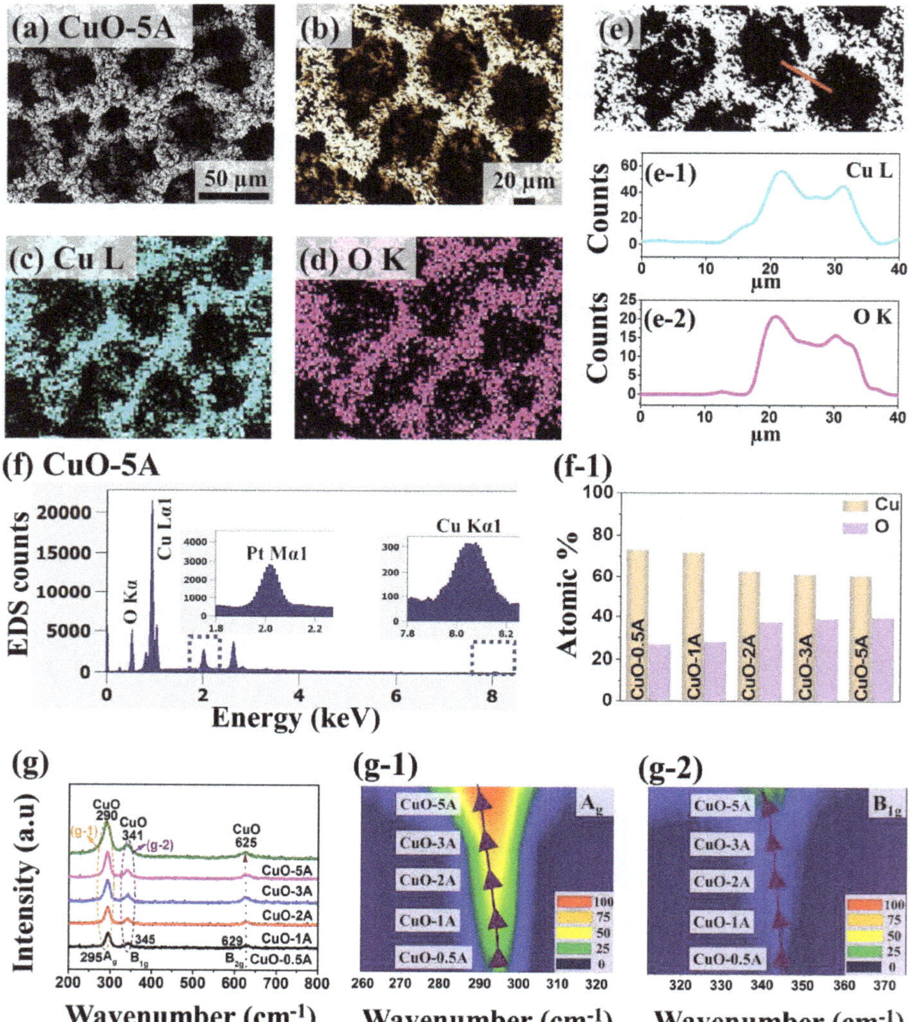

Figure 6. (**a**) SEM image of porous CuO/Pt hybrid detector fabricated at 5 A/cm^2 for 30 s and denoted as CuO-5A. (**b–d**) Enlarged SEM image and elemental maps of Cu and O for the CuO-5A. (**e–e-2**) Elemental line-profiles of Cu L and O K. (**f,f-1**) EDS spectra of CuO-5A and summary of atomic percentage of Cu and O as a function of deposition current. (**g**) Raman spectra for the porous CuO samples fabricated at different currents. (**g-1,g-2**) Contour maps of the Raman peaks of CuO corresponding to the A_g and $B1_g$ modes.

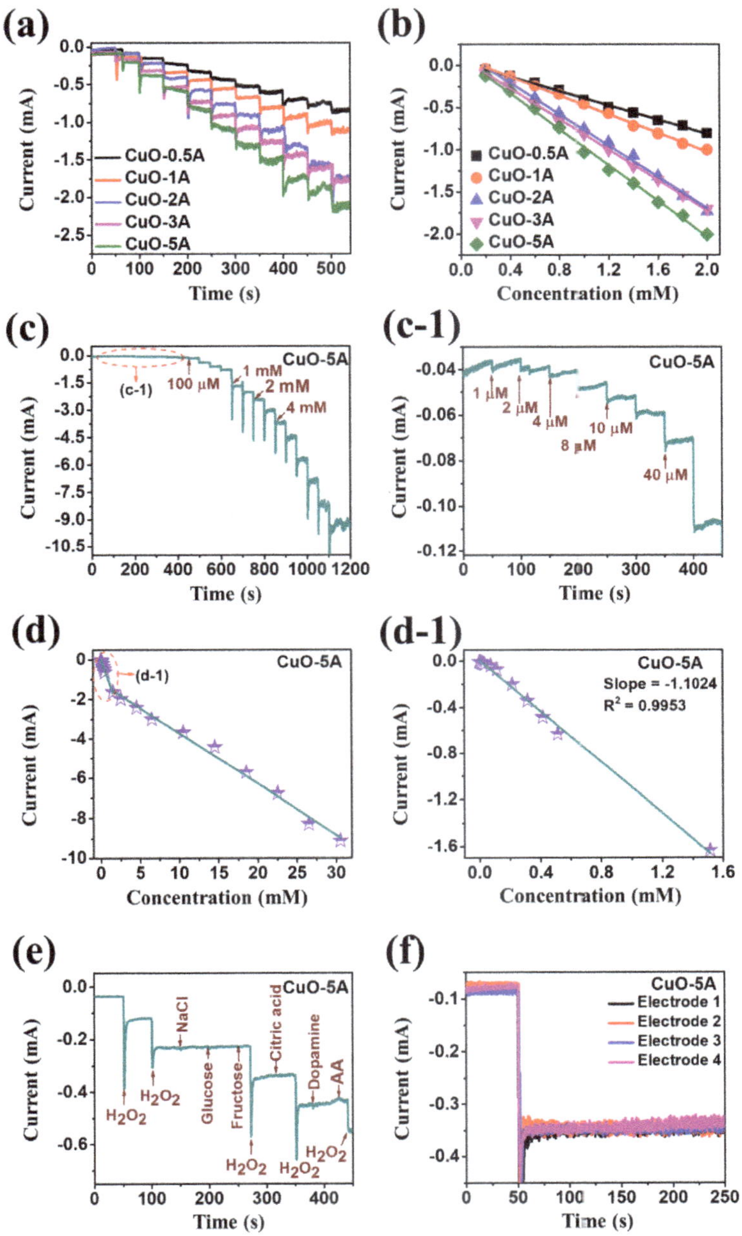

Figure 7. (**a**) Steady-state relation between current and time of current variation set upon successive addition of 0.2 mM H_2O_2 in the N_2-saturated 0.1 M PBS (pH 7.4) at an applied potential of −0.4 V. (**b**) The corresponding calibration curves for H_2O_2 detection. (**c**,**c-1**) Amperometric response of CuO-5A upon dropwise addition of various H_2O_2 concentrations from 1 µM to 4 mM. (**d**,**d-1**) Calibration curve of current versus H_2O_2 concentration at different H_2O_2 concentration range. (**e**) Selectivity response of CuO-5A sample upon addition of 0.1 mM H_2O_2, NaCl, glucose, fructose, citric acid, dopamine and ascorbic acid (AA) to 0.1 M PBS (pH 7.4). (**f**) Reproductivity test of CuO-5A samples 1–4 in the 0.1 M PBS (pH 7.4) at the potential of −0.4 V.

Table 1. Comparison of electrochemical H_2O_2 sensing performance for the CuO-based materials.

Electrode Material	Limit of Detection	Linear Range	Electrolyte Solution	Sensitivity [$\mu A\ mM^{-1}\ cm^{-2}$]	Reference
CuO-5A	140 nM	1 μm–1.5 mM	0.1 M PBS	2205	Present Work
CuO-30 s	325 nM	1 μm–1.5 mM	0.1 M PBS	1498	
NP-PdCu	1.9 μM	0.1–30 mM	PBS + 1 mM H_2O_2	1.6	[9]
Cu_2O PLNWs/Cu foam	1.05 μM	5–1770 μM	0.1 M NaOH	1.4773	[11]
CuO-NP	1.6 μM	0.01–13.18 mM	0.1 M PBS	22.27	[19]
CuO nanorods	-	0.25–18.75 mM	0.1 M NaOH	84.89	[40]
CuO nanosheet	10 μM	10–20,000 μM	1 M NaOH	25.5	[18]
CuO@Cu_2O-NWs/PVA/GC	0.35 μM	1 μM–3 mM	0.1 M NaOH	39.5	[41]
CuO/rGO/Cu_2O	0.05 μM	0.5 μM–9.7 mM	0.1 M NaOH	366.2	[42]
ZnO_3-CuO_7/CPE	2.4 μM	0.003–0.53 mM	0.1 M KCl	1.11	[43]
CuOx/NiOy	90 nM	0.03 μM–9.0 mM	0.10 M NaOH	271.1	[44]
3D CuO/Cu	2 μM	2 μM–19.4 mM	0.1 M NaOH	103	[45]
CuO nanostructures	43 nM	250 nm–2 mM	0.1 M PB	2015.7	[46]
3 DOI Au/NiO@CuO	3.7 nM	20 nM–20 μM	0.1 M PBS	650.2	[47]

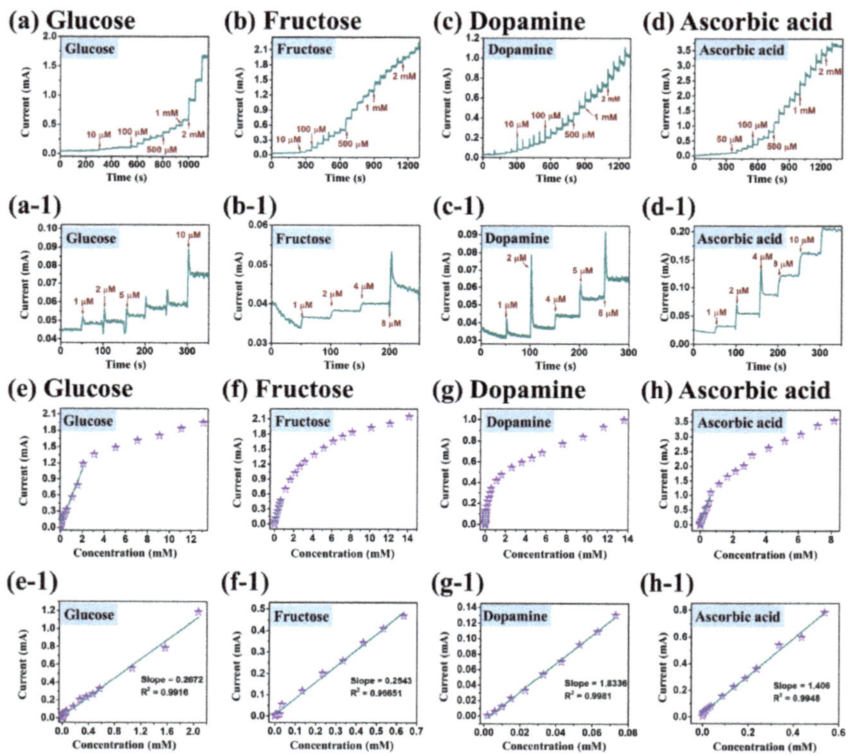

Figure 8. Amperometric response of CuO-5A sample upon dropwise addition of (**a**) glucose into 0.1 M NaOH at a potential of 0.6 V, (**b**) fructose into 0.1 M NaOH at a potential of 0.4 V, (**c**) dopamine into 0.1 M PBS at a potential of 0.6 V and (**d**) ascorbic acid into 0.1 M NaOH at a potential of 0.5 V, respectively. (**a-1–d-1**) Amperometric response of glucose, fructose, dopamine and ascorbic acid at lower concentrations. (**e–h**) Calibration curves CuO-5A based on the current versus high concentration of glucose, fructose, dopamine and ascorbic acid. (**e-1–h-1**) Linear calibration curve based on the current versus low concentration of glucose, fructose, dopamine and ascorbic acid.

4. Conclusions

In summary, a significantly improved H_2O_2 detection performance has been demonstrated by the uniquely designed super-porous hybrid nanostructures of CuO/Pt on Si fabricated by the combined physicochemical approach. In particular, the physical vapor deposition of Pt NPs on the Si substrate was followed by the electrochemical deposition of highly porous nanostructures of CuO by the dynamic hydrogen bubbles approach at different deposition times and current densities. It has been found that the performance of the H_2O_2 sensor is highly dependent on deposition conditions. The highly porous structure of CuO deposited at a current density of 5 A/cm^2 for 30 s (CuO-5A) showed the best performance with the highest sensitivity, wide linear range and selectivity towards the H_2O_2. Under the optimized conditions, the CuO-5A demonstrated the sensitivity of 2205 μA/mM·cm^2 with wide detection range. It also demonstrated an excellent selectivity against other organic molecules like glucose, fructose, dopamine and ascorbic acid along with the limit of detection of 140 nM. The enhanced sensing performance was due to the increased active sites and improved H_2O_2 adsorption and interfacial electron transport, which was achieved by a unique manufacturing method of the dynamic hydrogen bubble technique. Furthermore, the CuO-5A showed the detection of other organic materials at different applied potential. This work demonstrates that the highly porous CuO/Pt platform could be a promising candidate to develop an efficient electrochemical biosensor for the H_2O_2 sensor application.

Supplementary Materials: The following are available online at http://www.mdpi.com/2079-4991/10/10/2034/s1, Figure S1. (a–c) Atomic force microscope (AFM) image of Si substrate before degassing, after degassing and after 50 nm Pt layer deposition and annealing at 425 °C for 30 min. (a-1–c-1) Cross-sectional line profiles from the corresponding AFM images. (d) Summary of RMS roughness (Rq) and surface area ratio (SAR) under different conditions. (d-1) Rq and SAR summary table, Figure S2. (a–d) Schematics of porous CuO fabrication procedure on Pt/Si, Figure S3. (a–e) Large scale SEM images of porous CuO samples by the electrochemical deposition time variation at 2 A/cm^2 cathodic current density. (a-1–e-1) High magnification SEM images of the porous CuO, Figure S4. EDS spectra of porous CuO based on the variation of deposition time (a) CuO-5 s, (b) CuO-10 s, (c) CuO-20 s, and (d) CuO-50 s. Insets show the enlarged Pt Mα1 and Cu Kα1 peaks in each sample. The elemental composition of Cu and O is summarized in corresponding tables, Figure S5. (a,b) CV responses of Si substrate and Pt/Si substrate with and with the addition of the H_2O_2 in 0.1 mM PBS (pH 7.4) at a scan rate of 50 mV/s. (c) Amperometric response of Si substrate and Pt/Si substrate and (d) linear calibration curve of current versus concentration of H_2O_2 with Pt/Si substrate, Figure S6. (a) Cyclic voltammetry (CV) response of various CuO samples in 0.1 M PBS (pH 7.4) containing 0.4 mM H_2O_2 at a scan rate of 50 mV/s. (b) Amperometric response of CuO-5A sample with dropwise addition of 0.1 mM H_2O_2 at different applied potential. (c) CVs response of the CuO-5A sample at different scan rates from 20 to 200 mV/s in 0.1 M PBS (pH 7.4) containing 0.1 mM H_2O_2. (d) Corresponding capacitive plot current Vs square root of scan rate for CuO-5A at −0 2 V ($\Delta j_{-0.2} = (j_a - j_c)/2$). (c,e) CV of CuO-5A sample in 0.1 M PBS (pH 7.4) containing different concentrations of H_2O_2 ranging from 0.5 to 3.5 mM at the scan rate of 50 mV/s. (f) Relation between peak current and H_2O_2 concentration at −0.2 V. Figure S7. (a–d) EDS spectra of CuO samples at different deposition current density CuO-0.5A–CuO-3A. Insets show the enlarged Pt Mα1 and Cu Kα1 peaks. The elemental composition of Cu and O is summarized in corresponding tables, Figure S8. Amperometric response of CuO-5A sample upon the successive addition of mixture solution of 0.1 mM dopamine (DA), ascorbic acid (AA) and and Uric acid (UA) (a) with H_2O_2 and (b) without H_2O_2 to 0.1 M PBS (pH 7.4) at applied potential of −0.4 V. Figure S9. (a–g) CV response of CuO-5A with and without the addition of 0.1 mM organic molecules such as glucose, fructose, and ascorbic acid in 0.1 M NaOH and, dopamine in 0.1 M PBS at a scan rate of 50 mV/s. (b–d) Amperometric current response of corresponding organic molecules at different applied potentials.

Author Contributions: R.K., S.K., R.M., J.-H.J. and J.L. participated in the experiment design and carried out the experiments. R.K., S.K., R.M., J.-H.J. and J.L. participated in the characterizations and analysis of data. J.-H.J. and J.L. designed the experiments and testing methods. R.K., S.K. and J.L. carried out the writing. All authors helped in drafting and read and approved the final manuscript.

Funding: Financial support from the National Research Foundation of Korea (NRF) Grant funded by the Korean Government (MSIP) (no. NRF-2019R1A2C4069438 and NRF-2018R1A6A1A03025242) and in part by the research grant of Kwangwoon University in 2020 is gratefully acknowledged.

Conflicts of Interest: The authors declare no conflict of interest. The funders had no role in the design of the study; in the collection, analyses, or interpretation of data; in the writing of the manuscript; or in the decision to publish the results.

References

1. Hsu, Y.-K.; Chen, Y.-C.; Lin, Y.-G. Spontaneous formation of CuO nanosheets on Cu foil for H_2O_2 detection. *Appl. Surf. Sci.* **2015**, *354*, 85–89. [CrossRef]
2. Jia, W.; Guo, M.; Zheng, Z.; Yu, T.; Rodriguez, E.G.; Wang, Y.; Lei, Y. Electrocatalytic oxidation and reduction of H_2O_2 on vertically aligned Co_3O_4 nanowalls electrode: Toward H_2O_2 detection. *J. Electroanal. Chem.* **2009**, *625*, 27–32. [CrossRef]
3. Masud, M.K.; Kim, J.; Billah, M.M.; Wood, K.; Shiddiky, M.J.A.; Nguyen, N.T.; Parsapur, R.K.; Kaneti, Y.V.; Alshehri, A.A.; Alghamidi, Y.G.; et al. Nanoarchitectured peroxidase-mimetic nanozymes: Mesoporous nanocrystalline α-or γ-iron oxide? *J. Mater. Chem. B* **2019**, *7*, 5412–5422. [CrossRef] [PubMed]
4. Song, C.; Ding, W.; Zhao, W.; Liu, H.; Wang, J.; Yao, Y.; Yao, C. High peroxidase-like activity realized by facile synthesis of FeS_2 nanoparticles for sensitive colorimetric detection of H_2O_2 and glutathione. *Biosens. Bioelectron.* **2020**, *151*, 111983. [CrossRef]
5. Choi, Y.J.; Takahashi, K.; Misawa, N.; Hizawa, T.; Iwata, T.; Sawada, K. Multi-wavelength fluorescence detection of submicromolar concentrations using a filter-free fluorescence sensor. *Sens. Actuators B Chem.* **2018**, *256*, 38–47. [CrossRef]
6. Nie, F.; Ga, L.; Ai, J.; Wang, Y. Trimetallic PdCuAu nanoparticles for temperature sensing and fluorescence detection of H_2O_2 and glucose. *Front. Chem.* **2020**, *8*, 244. [CrossRef] [PubMed]
7. Babamiri, B.; Bahari, D.; Salimi, A. Highly sensitive bioaffinity electrochemiluminescence sensors: Recent advances and future directions. *Biosens. Bioelectron.* **2019**, *142*, 111530. [CrossRef]
8. Tian, K.; Li, D.; Tang, T.; Nie, F.; Zhou, Y.; Du, J.; Zheng, J. A novel electrochemiluminescence resonance energy transfer system of luminol-graphene quantum dot composite and its application in H_2O_2 detection. *Talanta* **2018**, *185*, 446–452. [CrossRef]
9. Yang, H.; Wang, Z.; Li, C.; Xu, C. Nanoporous PdCu alloy as an excellent electrochemical sensor for H_2O_2 and glucose detection. *J. Colloid Interface Sci.* **2017**, *491*, 321–328. [CrossRef]
10. Wang, L.; Qi, H.; Chen, L.; Sun, Y.; Li, Z. Self-assembled Ag-Cu_2O nanocomposite films at air-liquid interfaces for surface-enhanced raman scattering and electrochemical detection of H_2O_2. *Nanomaterials* **2018**, *8*, 332. [CrossRef]
11. Lu, W.; Sun, Y.; Dai, H.; Ni, P.; Jiang, S.; Wang, Y.; Li, Z.; Li, Z. Direct growth of pod-like Cu_2O nanowire arrays on copper foam: Highly sensitive and efficient nonenzymatic glucose and H_2O_2 biosensor. *Sens. Actuators B Chem.* **2016**, *231*, 860–866. [CrossRef]
12. Zhao, J.; Yan, Y.; Zhu, L.; Li, X.; Li, G. An amperometric biosensor for the detection of hydrogen peroxide released from human breast cancer cells. *Biosens. Bioelectron.* **2013**, *41*, 815–819. [CrossRef] [PubMed]
13. Xu, S.; Zhang, X.; Wan, T.; Zhang, C. A third-generation hydrogen peroxide biosensor based on horseradish peroxidase cross-linked to multi-wall carbon nanotubes. *Microchim. Acta* **2011**, *172*, 199–205. [CrossRef]
14. Wilson, N.M.; Flaherty, D.W. Mechanism for the Direct Synthesis of H_2O_2 on Pd Clusters: Heterolytic Reaction Pathways at the Liquid-Solid Interface. *J. Am. Chem. Soc.* **2016**, *138*, 574–586. [CrossRef]
15. Huang, X.; Zhu, Y.; Yang, W.; Jiang, A.; Jin, X.; Zhang, Y.; Yan, L.; Zhang, G.; Liu, Z. A self-supported CuO/Cu nanowire electrode as highly efficient sensor for COD measurement. *Molecules* **2019**, *24*, 3132. [CrossRef]
16. Huang, Y.; Ferhan, A.R.; Dandapat, A.; Yoon, C.S.; Song, J.E.; Cho, E.C.; Kim, D.H. A strategy for the formation of gold-palladium supra-nanoparticles from gold nanoparticles of various shapes and their application to high-performance H_2O_2 sensing. *J. Phys. Chem. C* **2015**, *119*, 26164–26170. [CrossRef]
17. Chen, S.; Yuan, R.; Chai, Y.; Hu, F. Electrochemical sensing of hydrogen peroxide using metal nanoparticles: A review. *Microchim. Acta* **2013**, *180*, 15–32. [CrossRef]
18. Liu, W.B.; Chai, G.C.; Zhang, J.; Wang, M.G.; Dai, Y.X.; Yang, Q. Preparation of Cu_2O nanocubes with different sizes and rough surfaces by a seed-mediated self-assembly process and their application as a non-enzymatic glucose sensor. *New J. Chem.* **2020**, *44*, 15662–15670. [CrossRef]
19. Huang, J.; Zhu, Y.; Zhong, H.; Yang, X.; Li, C. Dispersed CuO nanoparticles on a silicon nanowire for improved performance of nonenzymatic H_2O_2 detection. *ACS Appl. Mater. Interfaces* **2014**, *6*, 7055–7062. [CrossRef]
20. Wang, M.; Ma, J.; Guan, X.; Peng, W.; Fan, X.; Zhang, G.; Zhang, F.; Li, Y. A novel H_2O_2 electrochemical sensor based on $NiCo_2S_4$ functionalized reduced graphene oxide. *J. Alloys Compd.* **2019**, *784*, 827–833. [CrossRef]

21. Ma, M.; Zhang, Y.; Gu, N. Peroxidase-like catalytic activity of cubic Pt nanocrystals. *Colloids Surf. A Physicochem. Eng. Asp.* **2011**, *373*, 6–10. [CrossRef]
22. Li, Y.; Jia, W.Z.; Song, Y.Y.; Xia, X.H. Superhydrophobicity of 3D porous copper films prepared using the hydrogen bubble dynamic template. *Chem. Mater.* **2007**, *19*, 5758–5764. [CrossRef]
23. Choudhary, S.; Sarma, J.V.N.; Pande, S.; Ababou-Girard, S.; Turban, P.; Lepine, B.; Gangopadhyay, S. Oxidation mechanism of thin Cu films: A gateway towards the formation of single oxide phase. *AIP Adv.* **2018**, *8*. [CrossRef]
24. Sahoo, R.K.; Das, A.; Samantaray, K.; Singh, S.K.; Mane, R.S.; Shin, H.C.; Yun, J.M.; Kim, K.H. Electrochemical glucose sensing characteristics of two-dimensional faceted and non-faceted CuO nanoribbons. *Cryst. Eng. Comm.* **2019**, *21*, 1607–1616. [CrossRef]
25. Deng, Y.; Handoko, A.D.; Du, Y.; Xi, S.; Yeo, B.S. In situ raman spectroscopy of copper and copper oxide surfaces during electrochemical oxygen evolution reaction: Identification of CuIII oxides as catalytically active species. *ACS Catal.* **2016**, *6*, 2473–2481. [CrossRef]
26. Walid, A.; Hocine, G.; Beniaiche, A. Effect of deposition time on the properties of zinc oxide thin films produced via electrochemical method. *Appl. Surf. Sci.* **2016**, *01001*, 1–6.
27. Xu, J.F.; Ji, W.; Shen, Z.X.; Li, W.S.; Tang, S.H.; Ye, X.R.; Jia, D.Z.; Xin, X.Q. Raman spectra of CuO nanocrystals. *J. Raman Spectrosc.* **1999**, *30*, 413–415. [CrossRef]
28. Yu, T.; Zhao, X.; Shen, Z.X.; Wu, Y.H.; Su, W.H. Investigation of individual CuO nanorods by polarized micro-Raman scattering. *J. Cryst. Growth* **2004**, *268*, 590–595. [CrossRef]
29. Ni, Y.; Sun, Z.; Zeng, Z.; Liu, F.; Qin, J. Hydrothermal fabrication of hierarchical CuO nanoflowers for dual-function amperometric sensing of hydrogen peroxide and glucose. *New J. Chem.* **2019**, *43*, 18629–18636. [CrossRef]
30. Chen, L.; Wang, Y.; Hasebe, Y.; Yang, X.; Zhang, D.; Zhang, Z.; Hu, Z. Copper(II) ion-doped polyimide composite for nonenzymatic electrochemical hydrogen peroxide sensing. *Int. J. Electrochem. Sci.* **2019**, *14*, 4891–4902. [CrossRef]
31. Wang, Q.; Yang, Y.; Gao, F.; Ni, J.; Zhang, Y.; Lin, Z. Graphene oxide directed one-step synthesis of flowerlike graphene@HKUST-1 for enzyme-free detection of hydrogen peroxide in biological samples. *ACS Appl. Mater. Interfaces* **2016**, *8*, 32477–32487. [CrossRef] [PubMed]
32. Zhang, W.; Fan, G.; Yi, H.; Jia, G.; Li, Z.; Yuan, C.; Bai, Y.; Fu, D. Interfacial engineering of hierarchical transition metal oxide heterostructures for highly sensitive sensing of hydrogen peroxide. *Small* **2018**, *14*, 1703713. [CrossRef] [PubMed]
33. Chen, H.; Fan, G.; Zhao, J.; Qiu, M.; Sun, P. A portable micro glucose sensor based on copper based nanocomposite structure. *New J. Chem.* **2019**, *43*, 7806. [CrossRef]
34. Bai, Z.; Dong, W.; Ren, Y.; Zhang, C.; Chen, Q. Preparation of nano Au and Pt alloy microspheres decorated with reduced graphene oxide for nonenzymatic hydrogen peroxide sensing. *Langmuir* **2018**, *34*, 2235–2244. [CrossRef]
35. Huang, Z.; Zhang, A.; Zhang, Q.; Pan, S.; Cui, D. Electrochemical biosensor based on dewdrop-like platinum nanoparticles-decorated silver nanoflowers nanocomposites for H_2O_2 and glucose detection. *J. Electrochem. Soc.* **2019**, *166*, B1138–B1145. [CrossRef]
36. Lange, G.A.; Eugénio, S.; Duarte, R.G.; Silva, T.M.; Carmezim, M.J.; Montemor, M.F. Characterisation and electrochemical behaviour of electrodeposited Cu-Fe foams applied as pseudocapacitor electrodes. *J. Electroanal. Chem.* **2015**, *737*, 85–92. [CrossRef]
37. Tran, T.H.; Nguyen, V.T. Phase transition of Cu_2O to CuO nanocrystals by selective laser heating. *Mater. Sci. Semicond. Process.* **2016**, *46*, 6–9. [CrossRef]
38. Fang, J.; Xuan, Y. Investigation of optical absorption and photothermal conversion characteristics of binary CuO/ZnO nanofluids. *RSC Adv.* **2017**, *7*, 56023–56033. [CrossRef]
39. Luo, B.; Li, X.; Yang, J.; Li, X.; Xue, L.; Li, X.; Gu, J.; Wang, M.; Jiang, L. Non-enzymatic electrochemical sensors for the detection of hydrogen peroxide based on Cu_2O/Cu nanocomposites. *Anal. Methods* **2014**, *6*, 1114–1120. [CrossRef]
40. Chakraborty, P.; Dhar, S.; Debnath, K.; Mondal, S.P. Glucose and hydrogen peroxide dual-mode electrochemical sensing using hydrothermally grown CuO nanorods. *J Electroanal. Chem.* **2019**, *833*, 213–220. [CrossRef]

41. Chirizzi, D.; Guascito, M.R.; Filippo, E.; Malitesta, C.; Tepore, A. A novel nonenzymatic amperometric hydrogen peroxide sensor based on CuO@Cu_2O nanowires embedded into poly(vinyl alcohol). *Talanta* **2016**, *147*, 124–131. [CrossRef]
42. Zhao, C.; Wu, X.; Li, P.; Zhao, C.; Qian, X. Hydrothermal deposition of CuO/rGO/Cu_2O nanocomposite on copper foil for sensitive nonenzymatic voltammetric determination of glucose and hydrogen peroxide. *Microchim. Acta* **2017**, *184*, 2341–2348. [CrossRef]
43. Daemi, S.; Ghasemi, S.; Akbar Ashkarran, A. Electrospun CuO-ZnO nanohybrid: Tuning the nanostructure for improved amperometric detection of hydrogen peroxide as a non-enzymatic sensor. *J. Colloid Interface Sci.* **2019**, *550*, 180–189. [CrossRef]
44. Long, L.; Liu, X.; Chen, L.; Li, D.; Jia, J. A hollow CuOx/NiOy nanocomposite for amperometric and non-enzymatic sensing of glucose and hydrogen peroxide. *Microchim. Acta* **2019**, *186*, 5–10. [CrossRef] [PubMed]
45. Yuan, R.; Li, H.; Yin, X.; Zhang, L.; Lu, J. Stable controlled growth of 3D CuO/Cu nanoflowers by surfactant-free method for non-enzymatic hydrogen peroxide detection. *J. Mater. Sci. Technol.* **2018**, *34*, 1692–1698. [CrossRef]
46. Gowthaman, N.S.K.; Arul, P.; Lim, H.N.; John, S.A. Negative potential-induced growth of surfactant-free CuO nanostructures on an Al-C substrate: A dual in-line sensor for biomarkers of diabetes and oxidative stress. *ACS Sustain. Chem. Eng.* **2020**, *8*, 2640–2651. [CrossRef]
47. Zhou, X.; Xu, L.; Lv, J.; Yang, S.; Zhu, S.; Chen, X.; Sun, X.; Dong, B.; Bai, X.; Lu, G.; et al. Au anchored three-dimensional macroporous NiO@CuO inverse opals for in-situ sensing of hydrogen peroxide secretion from living cells. *Sens. Actuators B Chem.* **2019**, *297*, 126729. [CrossRef]
48. Yang, Y.J.; Li, W.; Chen, X. Highly enhanced electrocatalytic oxidation of glucose on $Cu(OH)_2$/CuO nanotube arrays modified copper electrode. *J. Solid State Electrochem.* **2012**, *16*, 2877–2881. [CrossRef]

Publisher's Note: MDPI stays neutral with regard to jurisdictional claims in published maps and institutional affiliations.

nanomaterials

MDPI

Article

Hexagonal and Monoclinic Phases of $La_2O_2CO_3$ Nanoparticles and Their Phase-Related CO_2 Behavior

Hongyan Yu [1,2], Kaiming Jiang [1,2], Sung Gu Kang [1], Yong Men [2] and Eun Woo Shin [1,*]

[1] School of Chemical Engineering, University of Ulsan, Daehakro 93, Nam-gu, Ulsan 44610, Korea; yuyubobo0908@163.com (H.Y.); hg041111138@163.com (K.J.); sgkang@ulsan.ac.kr (S.G.K.)

[2] College of Chemistry and Chemical Engineering, Shanghai University of Engineering Science, Shanghai 201620, China; men@sues.edu.cn

* Correspondence: ewshin@ulsan.ac.kr

Received: 28 September 2020; Accepted: 15 October 2020; Published: 19 October 2020

Abstract: In this study, we prepared hexagonal and monoclinic phases of $La_2O_2CO_3$ nanoparticles by different wet preparation methods and investigated their phase-related CO_2 behavior through field-emission scanning microscopy, high-resolution transmission electron microscopy, Fourier transform infrared, thermogravimetric analysis, CO_2-temperature programmed desorption, and linear sweeping voltammetry of CO_2 electrochemical reduction. The monoclinic $La_2O_2CO_3$ phase was synthesized by a conventional precipitation method via $La(OH)CO_3$ when the precipitation time was longer than 12 h. In contrast, the hydrothermal method produced only the hexagonal $La_2O_2CO_3$ phase, irrespective of the hydrothermal reaction time. The $La(OH)_3$ phase was determined to be the initial phase in both preparation methods. During the precipitation, the $La(OH)_3$ phase was transformed into $La(OH)CO_3$ owing to the continuous supply of CO_2 from air whereas the hydrothermal method of a closed system crystallized only the $La(OH)_3$ phase. Based on the CO_2-temperature programmed desorption and thermogravimetric analysis, the hexagonal $La_2O_2CO_3$ nanoparticles (HL-12h) showed a higher surface CO_2 adsorption and thermal stability than those of the monoclinic $La_2O_2CO_3$ (PL-12h). The crystalline structures of both $La_2O_2CO_3$ phases predicted by the density functional theory calculation explained the difference in the CO_2 behavior on each phase. Consequently, HL-12h showed a higher current density and a more positive onset potential than PL-12h in CO_2 electrochemical reduction.

Keywords: monoclinic; hexagonal; $La_2O_2O_3$ phase; CO_2 behavior; precipitation method; hydrothermal method

1. Introduction

Recently, the synthesis of nanomaterials with controllable morphologies and phases has attracted considerable attention in the fields of materials science and inorganic chemistry because the physicochemical and structural properties of the nanomaterials strongly correlate with the types of crystal structures as well as the morphologies of nanoparticles [1–6]. The unique properties of nanomaterials can be properly tuned by controlling various factors, which results in potential applications of nanomaterials in catalysis, biological labeling, sensing, and optics [1,7–9]. Among the methods for synthesizing nanomaterials, wet chemical processes have been considered as the most effective and convenient approaches for the controllable phases of ceramic materials [10].

Lanthana (La_2O_3) has been widely used as a promoter or support in heterogeneous catalysis [11–13]. The basicity of La_2O_3 readily induces the adsorption of CO_2 to form the lanthanum oxycarbonate ($La_2O_2CO_3$) phase, which is an important species in the La_2O_3-containing catalytic reaction [4,13–16]. The crystalline structures of $La_2O_2CO_3$ can be divided into three types of different polymorphs: a tetragonal $La_2O_2CO_3$ (type I), a monoclinic $La_2O_2CO_3$ (type Ia), and a hexagonal $La_2O_2CO_3$

(type II) [16–18]. The hexagonal type II $La_2O_2CO_3$ has a higher chemical stability to water and carbon dioxide than the monoclinic type Ia [4,19,20]. In addition, the different crystalline structures of the $La_2O_2CO_3$ phases affect the interaction between $La_2O_2CO_3$ and ZnO in the $La_2O_2CO_3$/ZnO composite materials as well as the catalytic behavior of the composite materials on glycerol carbonation with CO_2 [4,21]. Meanwhile, the monoclinic type Ia $La_2O_2CO_3$ phase closely resembles the crystalline structure of lanthanum (La) oxysalts (e.g., oxysilicates, oxyhalides, and oxysulfates), whereas the hexagonal type II one is similar to A-type La sesquioxides. Thus, the type Ia $La_2O_2CO_3$ phase has been readily prepared by the thermal decomposition of La compounds (e.g., oxalates and acetates); however, it is difficult to prepare type II $La_2O_2CO_3$ in a single phase by the conventional wet preparation methods [20]. Accordingly, it is necessary to investigate i) the preparation conditions used to form type Ia and type II $La_2O_2CO_3$ phases in the conventional methods and ii) the CO_2 behavior on the $La_2O_2CO_3$ structures, which is an essential step in the CO_2-involving catalytic reactions, as well as the formation of the different $La_2O_2CO_3$ phases.

In this study, we prepared the nanoparticles with type Ia and type II $La_2O_2CO_3$ crystal structures by conventional wet preparation methods and investigated the formation of different $La_2O_2CO_3$ phases with Fourier transform infrared (FT-IR), X-ray diffraction (XRD), field-emission scanning electron microscopy (FE-SEM), and high-resolution transmission electron microscopy (HR-TEM). Furthermore, the CO_2 behavior on the different $La_2O_2CO_3$ crystal structures was observed by CO_2-temperature programmed desorption (TPD), thermogravimetric analysis (TGA), and linear sweeping voltammetry (LSV) of CO_2 electrochemical reduction. The superior CO_2 behavior of the hexagonal $La_2O_2CO_3$ phase to the monoclinic phase was additionally explained by the crystalline structures of both $La_2O_2CO_3$ phases, which was predicted by the density functional theory (DFT) calculation.

2. Materials and Methods

2.1. Materials

A total of 1.00 g of $La(NO_3)_3{\cdot}6H_2O$ was added to 50.0 mL of deionized water, and the resultant solution was vigorously stirred to ensure complete dissolution. The pH of the solution was adjusted to 12 with a 10 wt% NaOH solution, which yielded a white precipitate after the mixture was stirred for approximately 10 min. The sample was continuously stirred for another 6, 12, or 24 h, and the obtained product was centrifuged. The separated precipitate was washed with distilled water and ethanol and then dried at 80 °C for 12 h, followed by the calcination step at 500 °C for 2 h. Depending on the precipitation time, the solid samples prepared by the precipitation method were denoted as PL-xh (x = 6, 12, or 24), where x represents the precipitation time.

For the hydrothermal method, the procedure was almost the same as that in the precipitation method, except using an autoclave for the hydrothermal treatment. The pH-adjusted solution containing the La precursor was transferred to an autoclave (200 mL), heated to 160 °C, and maintained at this temperature for 6, 12, or 24 h. The obtained product was centrifuged, and the remained steps were also the same as those in the precipitation method. The $La_2O_2CO_3$ samples synthesized by the hydrothermal method were designated as HL-yh (y = 6, 12, or 24), where y represents the hydrothermal treatment time.

2.2. Characterizations

The morphologies of the samples were observed by a field-emission scanning electron microscope (JEOL, JSM-600F, Tokyo, Japan) instrument equipped with an energy-dispersive spectrometer. HR-TEM images were obtained using a JEOL JEM-2100F instrument (JEOL Ltd., Tokyo, Japan). The samples were prepared by suspending and grinding in an ethanol solution whose drops were placed on a carbon-film-coated copper grid. XRD patterns were measured at room temperature on a Rigaku D/MAX-2200 powder X-ray diffractometer (Rigaku Corporation, Tokyo, Japan) using a Cu Kα radiation source (λ = 0.15418 nm). The X-ray tube was operated at 35 kV and 20 mA, and the 2θ angle was scanned from 10° to 90° (with a step of 0.02°) at a speed of 2°/min. The FT-IR spectra of the samples were collected for the KBr powder-pressed pellets on a Nicolet 380 FT-IR spectrophotometer (Thermo Fisher Scientific, Waltham, MA, USA) under ambient conditions.

The CO_2-TPD experiments were conducted in a quartz flow reactor. The calcined samples were preheated from room temperature up to 600 °C (with a ramping rate of 15 °C/min) for 1 h under He flow (100 mL/min). The CO_2 gas (10 vol.% CO_2/He) was fed into the reactor with a flow rate of 30 mL/min at 50 °C for CO_2 adsorption before conducting the CO_2-TPD measurements. Finally, the temperature was increased from 50 to 600 °C at the ramping rate of 1.5 °C/min in He flow (30 mL/min). The weight loss in the samples was determined by a thermogravimetric analyzer (TA Instruments Q50, New Castle, DE, USA). A total of 20 mg of the samples was charged into the sample pan and heated to 1000 °C at a rate of 5 °C/min in air flow. The CO_2 electrochemical reduction was carried out via the LSV measurement with an Ag/AgCl electrode as a reference electrode and Pt wire as a counter electrode. The working electrode was prepared by dispersing 10 mg of the samples in a mixture of 2 mL of alcohol and 100 μL of 5% Nafion and then pipetting 10 μL of suspension on the GCE (0.07065 cm^2). The working electrode was tested 20 times at a scan rate of 20 mV/s. The electrolyte was 0.1 M $NaHCO_3$ saturated with CO_2. Before each experiment, high-purity CO_2 gas was bubbled at a flow rate of 30 mL/min for 30 min to remove all oxygen from the electrolyte. The gases in the measurement were analyzed by a GC instrument.

Using the Vienna Ab initio Simulation Package (VASP) [22,23], DFT calculations were conducted along with the GGA–PBE (Perdew–Burke–Ernzerhof) functional [24]. The cutoff energy of 600 eV was chosen in our calculations. The criteria of convergence of energies and forces for geometry optimization were 10^{-4} eV and 10^{-2} eV/Å, respectively. For the calculation of disordered hexagonal $La_2O_2CO_3$, the lowest energy configuration among the other randomly selected 50 structures was used. The Monkhorst-Pack *k*-point meshes of 3 × 5 × 2 and 9 × 9 × 3 were used for the geometry optimization of monoclinic and hexagonal phase of $La_2O_2CO_3$, respectively [25].

3. Results and Discussion

3.1. Synthesis of Monoclinic and Hexagonal $La_2O_2CO_3$ Nanoparticles

Figure 1 shows the XRD patterns of $La_2O_2CO_3$ nanoparticle materials prepared at each reaction time. The two types of $La_2O_2CO_3$ phases are primarily detected in the PL samples: the monoclinic type Ia and hexagonal type II $La_2O_2CO_3$ phases. For 6 h of precipitation (PL-6h), the characteristic XRD peaks in the hexagonal $La_2O_2CO_3$ crystal phase are clearly observed at 2θ = 25.7, 30.2, 47.2, and 56.6° (JCPDS 37-0804) (Figure 1(Aa)) [1,4,20,21,26,27]. However, when the precipitation time is increased to 12 and 24 h, the characteristic XRD peaks in the monoclinic $La_2O_2CO_3$ phase clearly appear at 2θ = 22.8, 29.3, 31.0, 39.9, and 44.4° with a C12/c1 space group (JCPDS 48-1113) (Figure 1(Ab,Ac)), which indicates the prevalence of the hexagonal $La_2O_2CO_3$ phase during the initial precipitation time, followed by the transformation into the monoclinic $La_2O_2CO_3$ phase after 12 h of precipitation. In contrast, the HL samples show the XRD patterns that contain the characteristic peaks in only the hexagonal type II $La_2O_2CO_3$ phase, regardless of the reaction time during the hydrothermal preparation, which demonstrates that there is no change in the $La_2O_2CO_3$ phase during the preparation process (Figure 1(Ba–Bc)).

The FT-IR spectra of the PL and HL samples also confirm the formation of each $La_2O_2CO_3$ crystal phase depending on the preparation methods, as shown in Figure 2. According to the assignments of typical FT-IR bands for carbonates in the $La_2O_2CO_3$ phases, the bands at 745, 855, 1066, and 1518 cm^{-1} are interpreted as CO_3^{2-} vibrations related to the $La_2O_2CO_3$ phase [4,6,21,27]. The three-fold splitting bands at approximately 845 cm^{-1} (υ_2) and a strong band at 1367 cm^{-1} (υ_3) are assigned to the unique carbonate vibrational mode for the monoclinic type Ia $La_2O_2CO_3$ phase. The FT-IR spectra in Figure 2b,c of only the PL-12h and PL-24h samples show the characteristic bands (υ_2 and υ_3) of type-Ia $La_2O_2CO_3$, whereas the FT-IR spectra of the other samples show the typical bands of the $La_2O_2CO_3$

phase, which further confirms that the formation of the type Ia and II $La_2O_2CO_3$ phases depends on the preparation conditions. In the precipitation method, the monoclinic type Ia $La_2O_2CO_3$ phase is mainly formed when the precipitation time is longer than 12 h, whereas the hydrothermal method produces only the hexagonal type II $La_2O_2CO_3$ phase. This is consistent with the XRD results in this study.

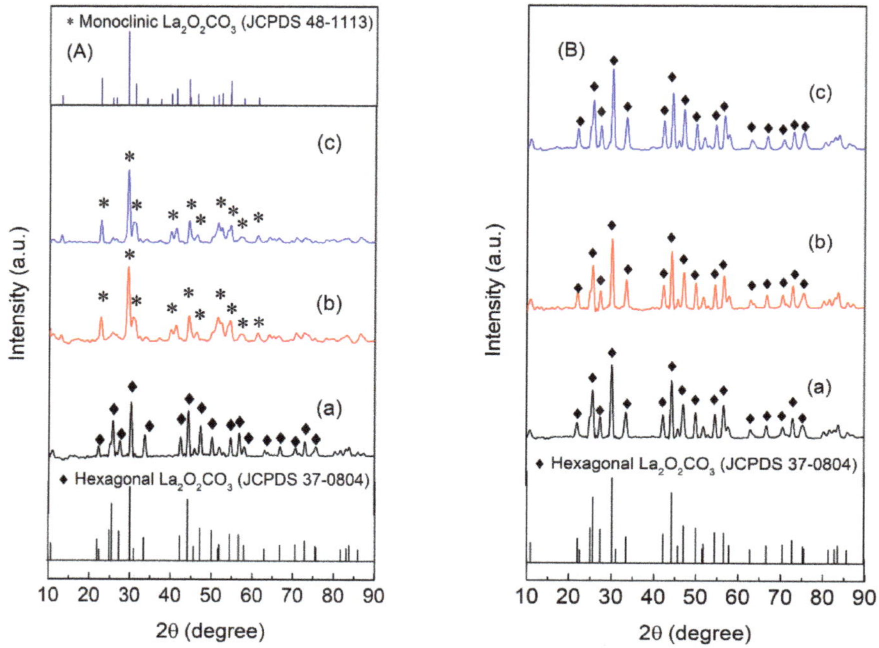

Figure 1. XRD patterns of **A**(**a–c**) PL-6h, PL-12h, and PL-24h and **B**(**a–c**) HL-6h, HL-12h, and HL-24h.

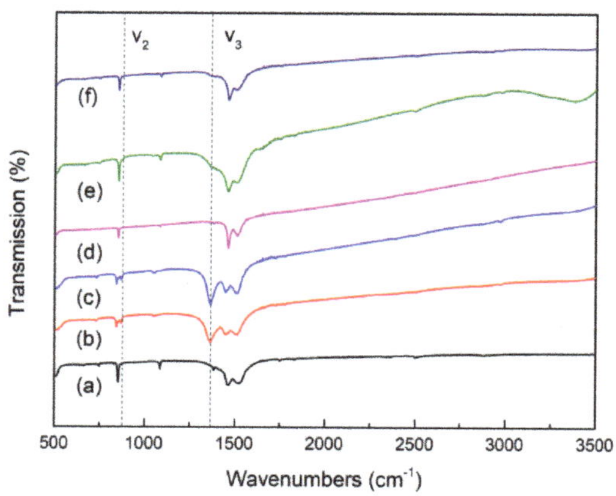

Figure 2. FT-IR spectra of (**a**) PL-6h, (**b**) PL-12h, (**c**) PL-24h, (**d**) HL-6h, (**e**) HL-12h, and (**f**) HL-24h.

Moreover, TEM measurements also provide additional evidence for the existence of the monoclinic and hexagonal $La_2O_2CO_3$ phases in the samples. Figure 3a–c shows the TEM images and fast Fourier transform patterns of PL-6h, PL-12h, and HL-12h. The (207) plane of the monoclinic type Ia $La_2O_2CO_3$ phase is detected in the PL-12h sample, whereas the (260) plane of the hexagonal type II $La_2O_2CO_3$ phase is observed in the HL-12h sample. Similarly, the PL-6h sample shows the (004) plane of the type II $La_2O_2CO_3$ phase, which is in good agreement with the XRD and FT-IR data. However, the morphological structures of PL-12h, HL-12h, and PL-6h samples are similar, as shown by the FE-SEM images; the aggregates of nanoparticles have different sizes: smaller than 10 nm for PL-12h, 10–30 nm for HL-12h, and 30–60 nm for PL-6h (Figure 3d–f).

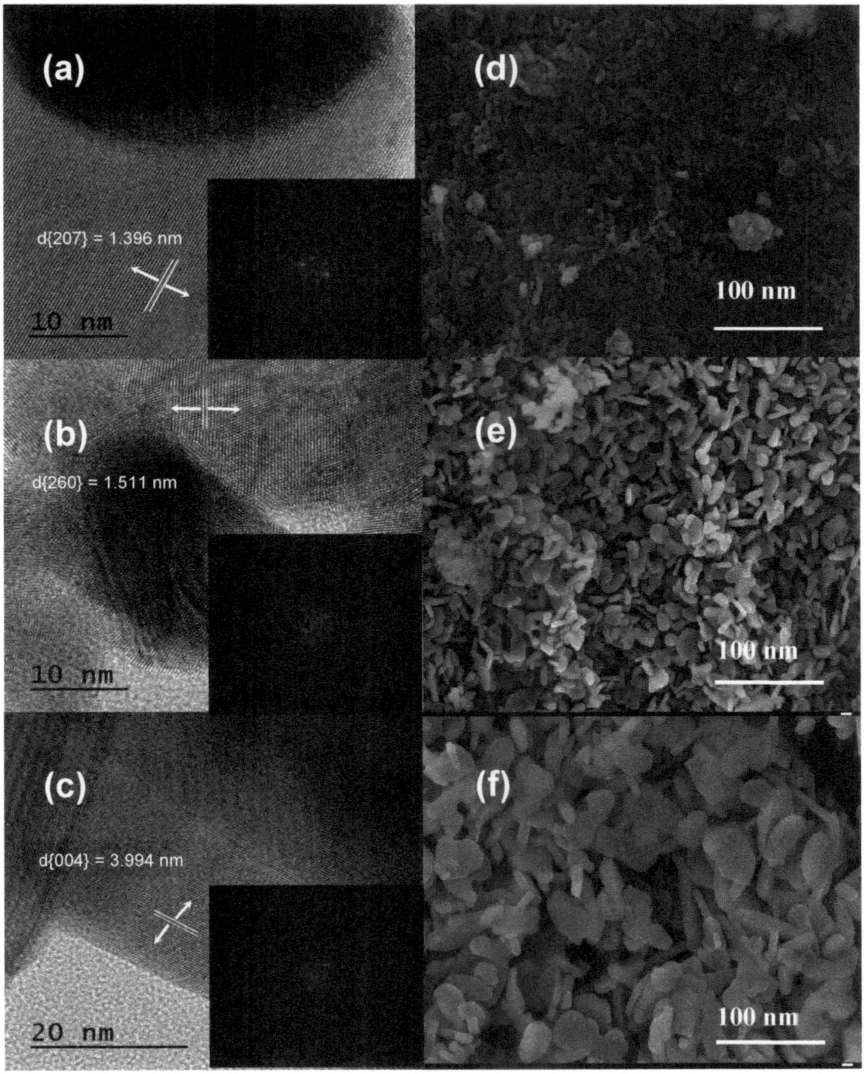

Figure 3. High-resolution transmission electon microscopy (HR-TEM) images of (**a**–**c**) PL-12h, HL-12h, and PL-6h, and FE-SEM images of (**d**–**f**) PL-12h, HL-12h, and PL-6h. The insets of (**a**–**c**) show their fast Fourier transform patterns.

To further understand the formation mechanism of the monoclinic and hexagonal $La_2O_2CO_3$ phases, uncalcined samples after precipitation or hydrothermal treatment were investigated. The XRD and FT-IR measurements indicate that different chemical products are also produced depending on the preparation conditions (Figure 4). The XRD peaks in Figure 4(Aa–Ac), shown as circles, are indexed to the pure hexagonal phase of $La(OH)_3$ with a P63/m(176) space group (JCPDS 36-1481) [1,6,12,21,26–28], which clearly shows that the initial $La(OH)_3$ phase remains unchanged in the hydrothermal method. With an increase in the preparation time during the hydrothermal method, the crystallinity of the $La(OH)_3$ structure becomes stronger with sharper XRD characteristic peaks. Meanwhile, in the precipitation method, the $La(OH)_3$ phase is produced with a very low crystallinity for PL-6h (weak and broad characteristic XRD peaks in Figure 4(Ac)). However, when the precipitation time is increased up to 12 h, the characteristic XRD peaks assigned to the orthorhombic $La(OH)CO_3$ structure (JCPDS 49-0981) appear with the disappearance of the XRD peaks in the $La(OH)_3$ structure (Figure 4(Ad)) [9,29]. Therefore, in the precipitation method, the dominant phase evolves from $La(OH)_3$ to $La(OH)CO_3$, with an increase in the precipitation time. However, the initial $La(OH)_3$ phase in the hydrothermal method is more crystallized during the hydrothermal treatment.

The FT-IR spectra of the uncalcined samples are monitored to confirm the existence of $La(OH)_3$ and $La(OH)CO_3$. First, the strong bands at 1438 and 1491 cm^{-1} shown in Figure 4(Bd) can be assigned to the bending vibrations of CO_3^{2-}, which confirms the presence of carbonate species in the intermediate [9]. A band at 3616 cm^{-1} and a broad band at 3410 cm^{-1} represent the O–H stretching mode in La–OH [6,9,27]. The bands at 850 and 1052 cm^{-1} correspond to the vibrational modes of carbon-related bonds, such as CH and CO, which remain before the calcination step. Thus, the FT-IR spectrum in Figure 4(Bd) clearly confirms the existence of $La(OH)CO_3$ as an intermediate in the PL-12h sample, which is consistent with the XRD data shown in Figure 4A. For $La(OH)_3$, the characteristic FT-IR bands for the O–H stretching and bending modes in La–OH are clearly observed at 3616, 3410, and 1640 cm^{-1}, as shown in Figure 4(Ba–Bc) [6,9,27]. Other bands at approximately 2800–3000, 850, and 1052 cm^{-1} can also be assigned to the vibrational modes of carbon-related bonds. Interestingly, for the samples in the precipitation method, the characteristic IR bands for CO_3^{2-} at approximately 1350–1500 cm^{-1} become sharp and strong with an increase in the reaction time (Figure 4(Bc,Bd)), whereas the characteristic IR band for OH at 3616 cm^{-1} is strongly intensified during the hydrothermal method (Figure 4(Ba,Bb)). Therefore, the precipitation method induces the transformation from $La(OH)_3$ into $La(OH)CO_3$ through the reaction with CO_2. In the hydrothermal method, the crystallization of $La(OH)_3$ goes further, which results in the high crystallinity of $La(OH)_3$.

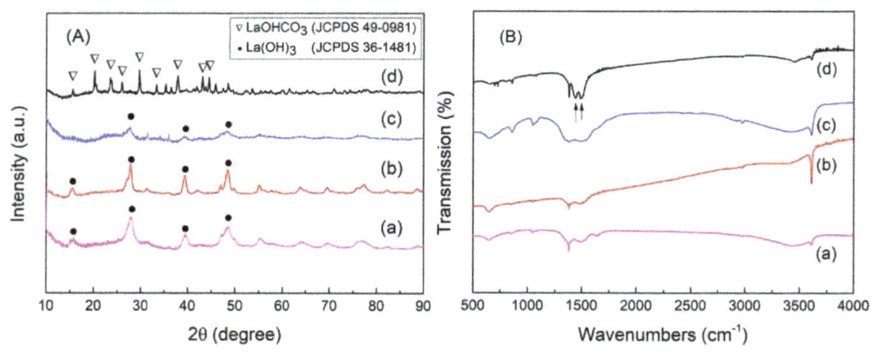

Figure 4. (**A**) XRD patterns and (**B**) FT-IR spectra of uncalcined samples. (**a**) HL-6h, (**b**) HL-12h, (**c**) PL-6h, and (**d**) PL-12h.

A critical difference between the two preparations is an open or closed reaction system, which is related to the supply of carbonate sources. For either the precipitation or hydrothermal method, the La precursor in the aqueous solution is dissociated into La cations and is then readily crystallized into the $La(OH)_3$ phase, because the initial pH conditions are strongly basic (i.e., pH = 12). In the hydrothermal method, a Teflon-lined autoclave reactor is used as a closed reaction system. Because it is a closed system, there is no further transformation of the La intermediate, which only results in the strong crystallization of the $La(OH)_3$ phase for the HL-12h and HL-24h samples. However, in the precipitation method, the precipitation is carried out in an open beaker; thus, the carbonate source (i.e., CO_2 from the air) can be continuously dissolved into the aqueous solution. Therefore, the initial phase, $La(OH)_3$, can be converted into the $La(OH)CO_3$ phase by the reaction with CO_2 at a time longer than 12 h, even though the 6-h precipitation produces only a weakly crystallized $La(OH)_3$. Under the continuous CO_2 supply condition, there is a transformation from $La(OH)_3$ into $La(OH)CO_3$. In the literature, it was reported that $La(OH)_3$ changed into an La carbonate when it was exposed to air [6,27,28]. More importantly, the $La(OH)CO_3$ phase is finally converted into the monoclinic type Ia $La_2O_2CO_3$ phase in the precipitation, while $La(OH)_3$ is transformed into the hexagonal type II structure in the hydrothermal method. The sufficient supply of CO_2 into the aqueous solution produces the $La(OH)CO_3$ that can be changed into the monoclinic $La_2O_2CO_3$ phase.

3.2. CO_2 Behavior on $La_2O_2CO_3$ Nanoparticles

To investigate the CO_2 behavior on each $La_2O_2CO_3$ phase, TGA, CO_2-TPD and CV of CO_2 electrochemical reduction for PL-12h (monoclinic type Ia $La_2O_2CO_3$ phase) and HL-12h (hexagonal type II $La_2O_2CO_3$ phase) were conducted in this study. Figure 5A shows the derivative TGA (DTGA) profiles of PL-12h and HL-12h, where the decomposition peaks correspond to CO_2 gases that leave from the $La_2O_2CO_3$ phases. The weight loss due to the thermal decomposition occurs at 326 °C and in the temperature range of 770–800 °C. According to previous studies [4,30], the CO_2 peak, which appears during the decomposition of $La_2O_2CO_3$ above 600 °C, can be assigned to CO_2 gases leaving from the bulk structure of the $La_2O_2CO_3$ phases, which is then transformed into the La_2O_3 phase. The CO_2 decomposition from the bulk structure of the hexagonal $La_2O_2CO_3$ phase occurs at approximately 800 °C, which is higher than the temperature of CO_2 production during the decomposition of the bulk structure of the monoclinic $La_2O_2CO_3$ phase. This result shows that the thermal stability of the hexagonal $La_2O_2CO_3$ phase is higher than that of the monoclinic phase [21]. The weight loss at approximately 326 °C is assumed to be due to the release of CO_2 gas that is adsorbed on the surface of the $La_2O_2CO_3$ phase. The decomposition peak at approximately 326 °C has a much smaller intensity than that at 650 °C, which indicates that a much lower amount of CO_2 is adsorbed onto the surfaces of the $La_2O_2CO_3$ phase than that released from the bulk structure. Furthermore, based on each peak's intensity, shown in Figure 5A, the hexagonal type II $La_2O_2CO_3$ phase contains more CO_2 on the surface than that on the monoclinic type Ia phase.

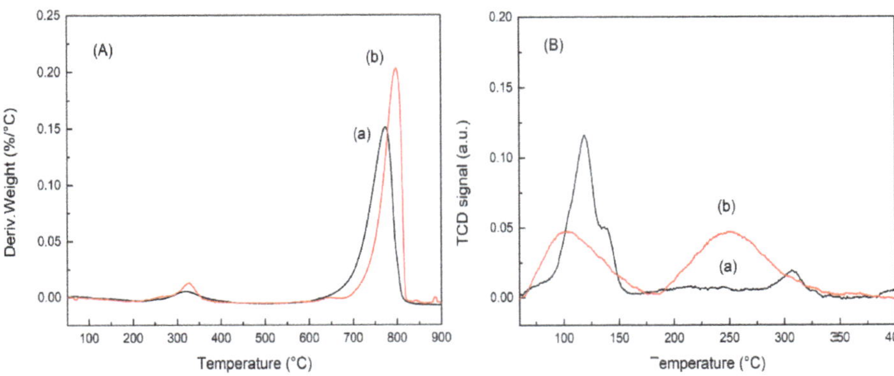

Figure 5. (**A**) Derivative thermogravimetric analysis (TGA) profiles and (**B**) CO_2-tempurate-programmed desorption (TPD) patterns of (**a**) PL-12h and (**b**) HL-12h.

To better understand the CO_2 adsorption ability on the surface of each $La_2O_2CO_3$ phase, the CO_2-TPD profiles of PL-12h and HL-12h were acquired. Before conducting the CO_2-TPD experiments, both samples were thermally treated at 600 °C for 1 h in He gas, and then CO_2 was introduced into the reactor at 50 °C to perform the CO_2 adsorption. Therefore, CO_2 can be assumed to adsorb on the surface of $La_2O_2CO_3$ phases and then desorb from the adsorption surface sites, which demonstrates the CO_2 adsorption behavior on the monoclinic and hexagonal $La_2O_2CO_3$ phases. In Figure 5B, the CO_2 desorption peaks can be approximately categorized into three types. The peak at approximately 100 °C is related to a weak basic site, and the peaks in the range of 200–400 °C correspond to medium and strong basic sites [2,12,31,32]. The CO_2 adsorption modes on each basic site have been studied by a combination of FT-IR spectroscopy and CO_2-TPD measurements [31,32]. Manoilova et al. [30] investigated the CO_2 adsorption onto La_2O_3 by IR spectroscopy, TPD, and DFT calculations. The DFT calculation for the CO_2 adsorption on La_2O_3 predicted that CO_2 gas adsorbed on the surface in the form of polydentate and monodentate species as a starting structure, and then La_2O_3 made a stable connection with polydentate and asymmetric CO_2 adsorptions at the saturated coverage. The CO_2 desorption peak at approximately 290 °C in the CO_2-TPD profile of LaOCl was assigned to the decomposition of coupled bridged CO_2 adsorbate species [31]. On the basis of the results from the FT-IR and CO_2-TPD measurements of Mg–Al basic oxides, Di Cosimo et al. [32] suggested that the three types of CO_2 adsorption modes (e.g., bicarbonate, bidentate carbonate, and unidentate carbonate) were low-strength, medium-strength, and high-strength basic sites, respectively. It was determined that bidentate and unidentate carbonates remained on the surface at approximately 300 °C; only unidentate carbonate was detected at 350 °C [32]. Therefore, in this study, the peak at 110 °C, peaks at approximately 240 °C, and shoulders at approximately 310 °C can be assigned to the desorption of CO_2 species adsorbed on weak, medium and strong basic sites, respectively. Figure 5B and Table 1 shows that the HL-12h sample has a higher combined intensity of medium and strong basic sites than PL-12h, which suggests that the hexagonal type II $La_2O_2CO_3$ phase provides more CO_2 adsorption sites on the surface. This observation is in good agreement with the TGA results shown in Figure 5A.

Table 1. The peak intensities quantified in the CO_2-TPD patterns.

Samples	Temperature at Maximum (°C)	Quantity (cm^3/g STP)
PL-12h	119 306	31.7 3.38
HL-12h	109 241	24.6 29.0

A DFT calculation was performed to optimize the bulk structures of both $La_2O_2CO_3$ phases (Figure 6). The lattice constant of $La_2O_2CO_3$ in the disordered hexagonal structure was predicted by considering the ratio (c/a) of lattice parameters (a and c) of the hexagonal structure [33]. Our DFT calculated lattice constants of $La_2O_2CO_3$ nanoparticles in both monoclinic and hexagonal structures, similar to the available experimental data from the literature, which are shown in Table 2 [34,35]. On the basis of the DFT calculation, we can optimize the hexagonal type II and monoclinic type Ia $La_2O_2CO_3$ nanopartilces, as shown in Figure 7. From the optimized structure of each phase, the La atom is determined to have seven and eight oxygen atoms as nearest neighbors in monoclinic and hexagonal structures, respectively. The eight coordination numbers of the La atom in the hexagonal type II $La_2O_2CO_3$ nanoparticles can produce stronger bonding with carbonate species, which results in the higher stability of the hexagonal type II structure compared to that of the monoclinic type Ia $La_2O_2CO_3$.

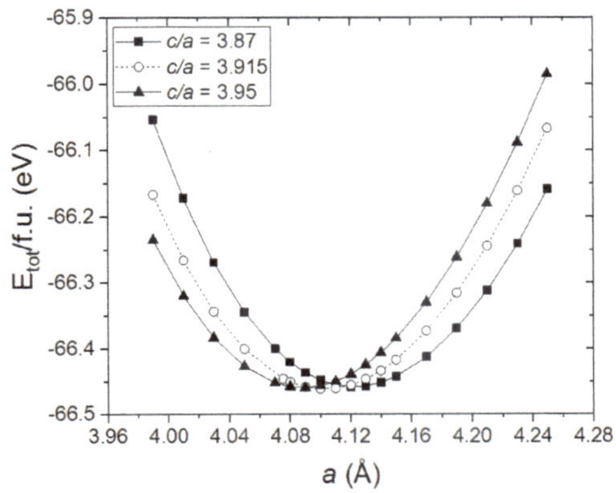

Figure 6. Density functional theory (DFT) total energy (per formula unit) of $La_2O_2CO_3$ in the hexagonal structure for three c/a values.

Table 2. Lattice constants of $La_2O_2CO_3$ in monoclinic and hexagonal structures. The experimental lattice data of monoclinic [34] and hexagonal [35] structures are available from the literature.

$La_2O_2CO_3$	DFT Calculated Data	Experimental Data [33,34]
Monoclinic	a = 12.286 Å b = 7.097 Å c = 16.531 Å β = 75.677	a = 12.239 Å b = 7.067 Å c = 16.465 Å β = 75.690
Hexagonal	a = 4.100 Å b = 4.100 Å c = 16.053 Å γ = 120	a = 4.076 Å b = 4.076 Å c = 16.465 Å γ = 120

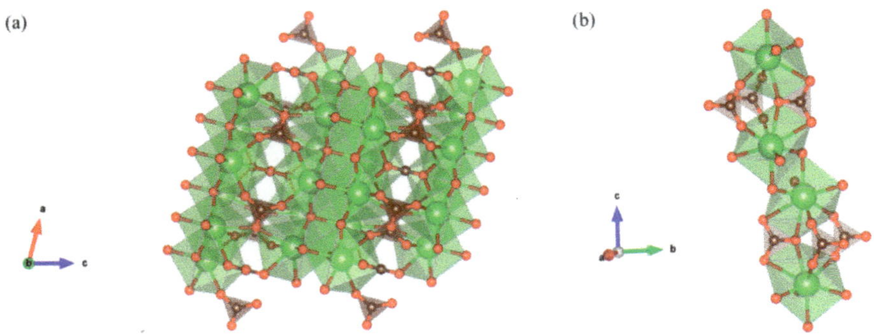

Figure 7. DFT-optimized structure of $La_2O_2CO_3$ in (**a**) monoclinic and (**b**) hexagonal phases.

3.3. CO_2 Electrochemical Reduction

Figure 8 shows LSV curves ranging from 0 to −0.6 V vs. Ag/AgCl for PL-12h and HL-12h in CO_2-saturated 0.1 M $NaHCO_3$ electrolyte. HL-12h exhibits a maximum total current density of −25.2 mA/cm^2 at −1.26 V vs. Ag/AgCl ,whereas a maximum current density of −17.97 mA/cm^2 for PL-12h is achieved at −1.438 V vs. Ag/AgCl. In addition, HL-12h shows a more positive onset potential toward CO_2 electrochemical reduction than PL-12h in Figure 8. Both the higher current density and more positive onset potential apparently indicate a higher activity toward the CO_2 electrochemical reduction in HL-12h compared to that of PL-12h.

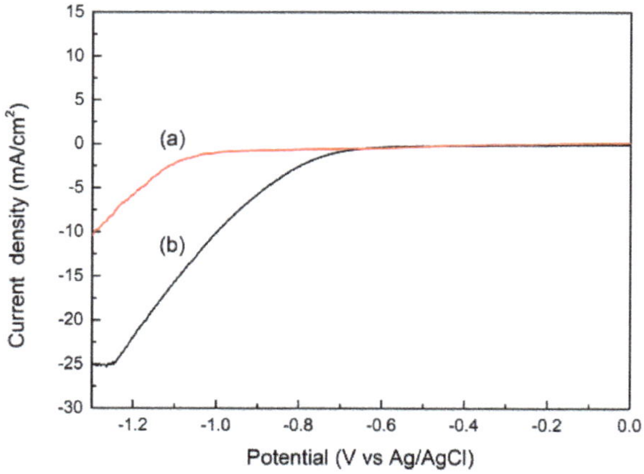

Figure 8. Linear sweeping voltammetry (LSV) curves of electrodes at various reaction times in a 0.1M $NaHCO_3$ solution at a scan rate of 20 mV/s: (**a**) PL-12h; (**b**) HL-12h.

The chronoamperometry (CA) experiments were performed at different potentials for each 10 min, and gaseous products were determined by GC. For both PL-12h and HL-12h, the main gaseous products are CH_4, C_2H_4, C_2H_6 and H_2. Figure 9 shows the Faraday efficiency (FE) of carbon-containing products for PL-12h and HL-12h, resulting in a much higher FE for HL-12h than those for PL-12h. C_2H_4 is a dominant carbonaceous product at lower potential. A maximum of the ethene FE (9.4%) for HL-12h is achieved at −0.6 V (vs. Ag/AgCl), while that for PL-12h is lower than 5%. Interestingly, CO gas was not detected in the potential range, even for the two $La_2O_2CO_3$ samples. This indicates that $La_2O_2CO_3$ catalysts are efficient for C-C coupling rather than desorption to form CO gas, since CO is an intermediate for CO_2 transformation to ethene during CO_2 reduction [36]. The superior electrocatalytic activity of HL-12h to PL-12h would result from the better CO_2 adsorption ability which can optimize the first step involving electron and proton transfer to form a *COOH intermediate, which is then converted to other carbonaceous products [37]. The higher electronegativity of hexagonal $La_2O_2CO_3$ of HL-12h leads to the better CO_2 adsorption ability [38].

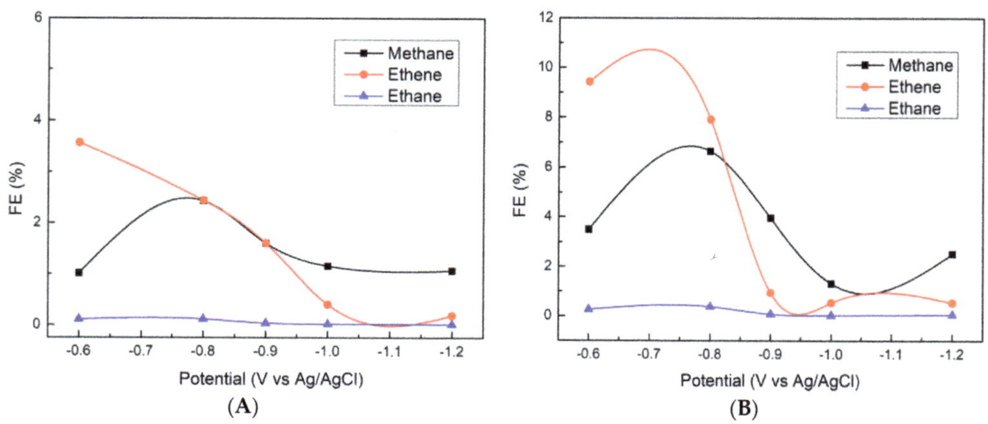

Figure 9. FE values for the (**A**) PL-12h and (**B**) HL-12h as a function of the potential.

4. Conclusions

In this study, $La_2O_2CO_3$ nanoparticles with hexagonal and monoclinic phases were prepared by different preparation methods, and the CO_2 behavior on each crystalline structure was investigated by CO_2-TPD, TGA measurements, and CO_2 electrochemical reduction. The hydrothermal method produced the hexagonal type II $La_2O_2CO_3$ phase, whereas the monoclinic type Ia phase was synthesized by the precipitation method (PL-12h and PL-24h). The initial $La(OH)_3$ phase was transformed into the $La(OH)CO_3$ phase by the reaction with CO_2 supplied from air in the precipitation method. The hexagonal $La_2O_2CO_3$ phase showed a higher CO_2 adsorption ability on the surface and a higher stability in the bulk structure than the monoclinic phase, owing to the differences in optimized crystalline structures predicted by the DFT calculation. Consequently, the hexagonal $La_2O_2CO_3$ phase of HL-12h had a higher current density and a more positive onset potential than the monoclinic $La_2O_2CO_3$ of PL-12h in CO_2 electrochemical reduction.

Author Contributions: H.Y. performed the experiments and wrote the paper; K.J., S.G.K. and Y.M. contributed data analysis; E.W.S. supervised the work and polished the paper. All authors have read and agreed to the published version of the manuscript.

Funding: This research was supported by the National Research Foundation of Korea (NRF) grant funded by the Korea government (MSIT) (No. 2018R1A2B6004219 and No. 2020R1A4A4079954).

Conflicts of Interest: The authors declare no conflict of interest.

References

1. Ding, D.; Lu, W.; Xiong, Y.; Pan, X.; Zhang, J.; Ling, C.; Du, Y.; Xue, Q. Facile synthesis of $La_2O_2CO_3$ nanoparticle films and its CO_2 sensing properties and mechanisms. *Appl. Surf. Sci.* **2017**, *426*, 725–733. [CrossRef]
2. Wang, F.; Zhang, Z.; Wei, X.; Fang, Q.; Jiang, X. The shape effect of $La_2O_2CO_3$ in Pd/$La_2O_2CO_3$ catalyst for selective hydrogenation of cinnamaldehyde. *Appl. Catal. A Gen.* **2017**, *543*, 196–200. [CrossRef]
3. Li, X.; Li, D.; Tian, H.; Zeng, L.; Zhao, J.W.; Gong, J. Dry reforming of methane over Ni/La_2O_3 nanorod catalysts with stabilized Ni nanoparticles. *Appl. Catal. B Environ.* **2017**, *202*, 683–694. [CrossRef]
4. Park, C.Y.; Nguyen, P.H.; Shin, E.W. Glycerol carbonation with CO_2 and $La_2O_2CO_3$/ZnO catalysts prepared by two different methods: Preferred reaction route depending on crystalline structure. *Mol. Catal.* **2017**, *435*, 99–109. [CrossRef]
5. Bosch, C.E.; Copley, M.P.; Eralp, T.; Bilbé, E.; Thybaut, J.W.; Marin, G.B.; Collier, P. Tailoring the physical and catalytic properties of lanthanum oxycarbonate nanoparticles. *Appl. Catal. A Gen.* **2017**, *536*, 104–112. [CrossRef]
6. Mu, Q.; Wang, Y. Synthesis, characterization, shape-preserved transformation, and optical properties of $La(OH)_3$, $La_2O_2CO_3$, and La_2O_3 nanorods. *J. Alloys Compd.* **2011**, *509*, 396–401. [CrossRef]
7. Sun, C.; Sun, J.; Xiao, G.; Zhang, H.; Qiu, X.; Li, H.; Chen, L. Mesoscale organization of nearly monodisperse flowerlike ceria microspheres. *J. Phys. Chem. B* **2006**, *110*, 13445–13452. [CrossRef] [PubMed]
8. Yi, G.; Lu, H.; Zhao, S.; Ge, Y.; Yang, W. Synthesis, characterization, and biological application of size-controlled nanocrystalline NaYF 4: Yb, Er phosphors. *Nano Lett.* **2004**, *4*, 2191–2196. [CrossRef]
9. Li, G.; Peng, C.; Zhang, C.; Xu, Z.; Shang, M.; Yang, D.; Kang, X.; Wang, W.; Li, C.; Cheng, Z.; et al. Eu^{3+}/Tb^{3+}-Doped $La_2O_2CO_3$/La_2O_3 nano/microcrystals with multiform morphologies: Facile synthesis, growth mechanism, and luminescence properties. *Inorg. Chem.* **2010**, *49*, 10522–10535. [CrossRef]
10. Lin, J. Multiform oxide optical materials via the versatile pechini-type sol-gel process. *Proc. Int. Meet. Inf. Disp.* **2008**, *8*, 1247–1250. [CrossRef]
11. Van Le, T.; Che, M.; Kermarec, M.; Louis, C.; Tatibouët, J.M. Structure sensitivity of the catalytic oxidative coupling of methane on lanthanum oxide. *Catal. Lett.* **1990**, *6*, 395–400. [CrossRef]
12. Wang, F.; Ta, N.; Li, Y.; Shen, W. $La(OH)_3$ and $La_2O_2CO_3$ nanorod catalysts for Claisen-Schmidt condensation. *Chin. J. Catal.* **2014**, *35*, 437–443. [CrossRef]
13. Lacombe, S.; Geantet, C.; Mirodatos, C. Oxidative coupling of methane over lanthana catalysts. I. Identification and role of specific active-sites. *J. Catal.* **1995**, *151*, 439–452. [CrossRef]

14. Chen, H.; Yu, H.; Peng, F.; Wang, H.; Yang, J.; Pan, M. Efficient and stable oxidative steam reforming of ethanol for hydrogen production: Effect of in situ dispersion of Ir over Ir/La_2O_3. *J. Catal.* **2010**, *269*, 281–290. [CrossRef]
15. Wang, H.; Fang, Y.; Liu, Y.; Bai, X. Perovskite $LaFeO_3$ supported bi-metal catalyst for syngas methanation. *J. Nat. Gas Chem.* **2012**, *21*, 745–752. [CrossRef]
16. Turcotte, R.P.; Sawyer, J.O.; Eyring, L. On the rare earth dioxymonocarbonates and their decomposition. *Inorg. Chem.* **1969**, *8*, 238–246. [CrossRef]
17. Olafscn, A.; Fjcllvåg, H. Synthesis of rare earth oxide carbonates and thermal stability of $Nd_2O_2CO_3$ II. *J. Mater. Chem.* **1999**, *9*, 2697–2702. [CrossRef]
18. Hirsch, O.; Kvashnina, K.O.; Luo, L.; Süess, M.J.; Glatzel, P.; Koziej, D. High-energy resolution X-ray absorption and emission spectroscopy reveals insight into unique selectivity of La-based nanoparticles for CO_2. *Proc. Natl. Acad. Sci. USA* **2015**, *112*, 15803–15808. [CrossRef]
19. Hölsä, J.; Turkki, T. Preparation, thermal stability and luminescence properties of selected rare earth oxycarbonates. *Thermochim. Acta* **1991**, *190*, 335–343. [CrossRef]
20. Koyabu, K.; Masui, T.; Tamura, S.; Imanaka, N. Synthesis of a new phosphor based on rare earth oxycarbonate. *J. Alloys Compd.* **2006**, *408–412*, 867–870. [CrossRef]
21. Yu, H.; Men, Y.; Shin, E.W. Structural properties of disordered macroporous $La_2O_2CO_3$/ZnO materials prepared by a solution combustion method. *Korean J. Chem. Eng.* **2019**, *36*, 522–528. [CrossRef]
22. Kresse, G.; Furthmüller, J. Efficient iterative schemes for ab initio total-energy calculations using a plane-wave basis set. *Phys. Rev. B* **1996**, *54*, 11169–11186. [CrossRef] [PubMed]
23. Kresse, G.; Furthmüller, J. Efficiency of ab-initio total energy calculations for metals and semiconductors using a plane-wave basis set. *Comput. Mater. Sci.* **1996**, *6*, 15–50. [CrossRef]
24. Perdew, J.P.; Burke, K.; Ernzerhof, M. Generalized gradient approximation made simple. *Phys. Rev. Lett.* **1996**, *77*, 3865–3868. [CrossRef] [PubMed]
25. Monkhorst, H.J.; Pack, J.D. Special points for Brillouin-zone integrations. *Phys. Rev. B* **1976**, *13*, 5188–5192. [CrossRef]
26. Hou, Y.H.; Han, W.C.; Xia, W.S.; Wan, H.L. Structure sensitivity of $La_2O_2CO_3$ catalysts in the oxidative coupling of methane. *ACS Catal.* **2015**, *5*, 1663–1674. [CrossRef]
27. Fleming, P.; Farrell, R.A.; Holmes, J.D.; Morris, M.A. The rapid formation of $La(OH)_3$ from La_2O_3 powders on exposure to water vapor. *J. Am. Ceram. Soc.* **2010**, *93*, 1187–1194. [CrossRef]
28. Fedorov, P.P.; Nazarkin, M.V.; Zakalyukin, R.M. On polymorphism and morphotropism of rare earth sesquioxides. *Crystallogr. Rep.* **2002**, *47*, 281–286. [CrossRef]
29. Bakiz, B.; Guinneton, F.; Arab, M.; Benlhachenu, A.; Gavarria, J.-R. Elaboration, characterization of $LaOHCO_3$, $La_2O_2CO_3$ and La_2O_3 phases and their gas solid interactions with CH_4 and CO gases. *Moroc. J. Condens. Matter Soc.* **2010**, *12*, 60–67.
30. Rémiás, R.; Kukovecz, Á.; Darányi, M.; Kozma, G.; Varga, S.; Kónya, Z.; Kiricsi, I. Synthesis of zinc glycerolate microstacks from a ZnO nanorod sacrificial template. *Eur. J. Inorg. Chem.* **2009**, *2009*, 3622–3627. [CrossRef]
31. Manoilova, O.V.; Podkolzin, S.G.; Tope, B.; Lercher, J.; Stangland, E.E.; Goupil, J.M.; Weckhuysen, B.M. Surface acidity and basicity of La_2O_3, LaOCl, and $LaCl_3$ characterized by IR spectroscopy, TPD, and DFT calculations. *J. Phys. Chem. B* **2004**, *108*, 15770–15781. [CrossRef]
32. Di Cosimo, J.I.; Díez, V.K.; Xu, M.; Iglesia, E.; Apesteguía, C.R. Structure and surface and catalytic properties of Mg-Al basic oxides. *J. Catal.* **1998**, *178*, 499–510. [CrossRef]
33. Sholl, D.S.; Steckel, J.A. *Density Functional Theory: A Practical Introduction*; John Wiley & Sons, Inc.: Hoboken, NJ, USA, 2009.
34. Olafsen, A.; Larsson, A.K.; Fjellvåg, H.; Hauback, B.C. On the crystal structure of $Ln_2O_2CO_3$ II (Ln = La and Nd). *J. Solid State Chem.* **2001**, *158*, 14–24. [CrossRef]
35. Attfield, J.P. Structure determinations of $La_2O_2CO_3$-II and the unusual disordered phase $La_2O_{2.52}(CO_3)_{0.74}Li_{0.52}$ using powder diffraction. *J. Solid State Chem.* **1989**, *82*, 132–138. [CrossRef]
36. Zhang, B.; Zhang, J.; Hua, M.; Wan, Q.; Su, Z.; Tan, X.; Liu, L.; Zhang, F.; Chen, G.; Tan, D.; et al. Highly electrocatalytic ethylene production from CO_2 on nanodefective Cu nanosheets. *J. Am. Chem. Soc.* **2020**, *142*, 13606–13613. [CrossRef] [PubMed]

37. Zhang, J.; Yin, R.; Shao, Q.; Zhu, T.; Huang, X. Oxygen vacancies in amorphous InO_x nanoribbons enhance CO_2 adsorption and activation for CO_2 electroreduction. *Angew. Chem. Int. Ed.* **2019**, *58*, 5609–5613. [CrossRef] [PubMed]
38. Horiuchi, T.; Hidaka, H.; Fukui, T.; Kubo, Y.; Horio, M.; Suzuki, K.; Mori, T. Effect of added basic metal oxides on CO_2 adsorption on alumina at elevated temperatures. *Appl. Catal. A Gen.* **1998**, *167*, 195–202. [CrossRef]

Publisher's Note: MDPI stays neutral with regard to jurisdictional claims in published maps and institutional affiliations.

nanomaterials

MDPI

Article

Preparation and Photocatalytic Performance for Degradation of Rhodamine B of AgPt/$Bi_4Ti_3O_{12}$ Composites

Gaoqian Yuan [1], Gen Zhang [1], Kezhuo Li [1], Faliang Li [1,2,*], Yunbo Cao [1], Jiangfeng He [1], Zhong Huang [1], Quanli Jia [3], Shaowei Zhang [4] and Haijun Zhang [1,*]

1. The State Key Laboratory of Refractories and Metallurgy, Wuhan University of Science and Technology, Wuhan 430081, China; yuangaoqian@126.com (G.Y.); zhang0812m@163.com (G.Z.); likezhuo0206@126.com (K.L.); 15623242376@163.com (Y.C.); hejiangfeng823102@163.com (J.H.); huangzhong@wust.edu.cn (Z.H.)
2. Jiangxi Engineering Research Center of Industrial Ceramics, Pingxiang 337022, China
3. Henan Key Laboratory of High Temperature Functional Ceramics, Zhengzhou University, 75 Daxue Road, Zhengzhou 450052, China; jiaquanli@zzu.edu.cn
4. College of Engineering, Mathematics and Physical Sciences, University of Exeter, Exeter EX4 4QF, UK; s.zhang@exeter.ac.uk

* Correspondence: lfliang@wust.edu.cn (F.L.); zhanghaijun@wust.edu.cn (H.Z.); Tel.: +86-133-6727-5129 (F.L.); +86-27-6886-2829 (H.Z)

Received: 13 October 2020; Accepted: 3 November 2020; Published: 5 November 2020

Abstract: Loading a noble metal on $Bi_4Ti_3O_{12}$ could enable the formation of the Schottky barrier at the interface between the former and the latter, which causes electrons to be trapped and inhibits the recombination of photoelectrons and photoholes. In this paper, AgPt/$Bi_4Ti_3O_{12}$ composite photocatalysts were prepared using the photoreduction method, and the effects of the type and content of noble metal on the photocatalytic performance of the catalysts were investigated. The photocatalytic degradation of rhodamine B (RhB) showed that the loading of AgPt bimetallic nanoparticles significantly improved the catalytic performance of $Bi_4Ti_3O_{12}$. When 0.10 wt% noble metal was loaded, the degradation rate for RhB of $Ag_{0.7}Pt_{0.3}$/$Bi_4Ti_3O_{12}$ was 0.027 min^{-1}, which was respectively about 2, 1.7 and 3.7 times as that of Ag/$Bi_4Ti_3O_{12}$, Pt/$Bi_3Ti_4O_{12}$ and $Bi_4Ti_3O_{12}$. The reasons may be attributed as follows: (i) the utilization of visible light was enhanced due to the surface plasmon resonance effect of Ag and Pt in the visible region; (ii) Ag nanoparticles mainly acted as electron acceptors to restrain the recombination of photogenerated electron-hole pairs under visible light irradiation; and (iii) Pt nanoparticles acted as electron cocatalysts to further suppress the recombination of photogenerated electron-hole pairs. The photocatalytic performance of $Ag_{0.7}Pt_{0.3}$/$Bi_4Ti_3O_{12}$ was superior to that of Ag/$Bi_4Ti_3O_{12}$ and Pt/$Bi_3Ti_4O_{12}$ owing to the synergistic effect between Ag and Pt nanoparticles.

Keywords: photocatalytic; Rhodamine B; photoreduction; $Bi_4Ti_3O_{12}$; AgPt/$Bi_4Ti_3O_{12}$

1. Introduction

The rapid growth of the chemical industry has led to a large volume of organic dye wastewater. Most of the organic pollutants have carcinogenic effects, posing a huge threat to organisms and human health [1–3]. Therefore, organic pollutants must be detoxified before they enter aquatic ecosystems. Some traditional wastewater treatment processes, such as physical adsorption, chemical oxidation and microbial treatment [4–6], have been proposed to deal with organic dyestuff sewage. However, these methods have suffered from several shortcomings including low efficiency, secondary pollution, and mild degree of harmlessness of pollutant [7–9]. Photocatalysis technology, which can non-selectively oxidize and degrade all kinds of organic matter with the ability of deep oxidation, is considered as an acceptable process for the treatment of organic contaminants since only water and non-toxic inorganic substances are produced. In the process of photocatalysis, photocatalysts are always required for achieving excellent catalytic effect. However, for the most widely used semiconductor photocatalysts (such as TiO_2 [10], $SrTiO_3$ [11], ZnS [12] and ZnO [13]), the utilization of solar energy is usually very low since their bandgap width E_g is always larger than 3.0 eV and can only just absorb ultraviolet light (only 4% of the total sunlight). Thus, how to improve the usage of solar energy in the photocatalytic removal of organic pollutants has become an important issue.

$Bi_4Ti_3O_{12}$ can respond to visible light since its bandgap is about 2.9 eV, and shows strong photocatalytic activity for the removal of organic pollutants [14–16]. However, the usage of $Bi_4Ti_3O_{12}$ is still limited because the photon efficiency is low and the photogenerated electrons and holes are easy to recombine. Noble metal nanoparticles (NPs) can be used as the accumulation ground of photogenerated electrons to facilitate the catalytic reactions involving electrons, and thus are widely employed as modifiers to enhance the photocatalytic performance of semiconductors [17–19]. When precious metals and semiconductors contact together, electrons on the surface of the latter will migrate to the surface of the former until their Fermi energy levels equalize. Since charges on the metal surface and holes on the semiconductor surface are both excessive, a Schottky barrier can be formed on the metal-semiconductor interface. As a result, the separation of photogenerated electrons and holes is promoted, and an improved photocatalytic performance is obtained [20–22].

In the present paper, $Bi_4Ti_3O_{12}$ nanosheets were firstly prepared by a molten salt method, and then AgPt bimetallic NPs were assembled on the prepared $Bi_4Ti_3O_{12}$ nanosheets via an in situ photoreduction method to prepare AgPt/$Bi_4Ti_3O_{12}$ composites, finally the photocatalytic performance of as-prepared AgPt/$Bi_4Ti_3O_{12}$ composites on the degradation of rhodamine B (RhB) under visible light was investigated. As far as we know, there has been no study on the decoration of AgPt bimetallic NPs on $Bi_4Ti_3O_{12}$ photocatalyst.

2. Materials and Methods

2.1. Materials

Titanium oxide (TiO_2) and bismuth oxide (Bi_2O_3) were purchased from Shanghai Makclin Biochemical Co., Ltd. (Shanghai, China). Silver nitrate ($AgNO_3$) was purchased from Tianjin Kaitong Chemical Reagent Co., Ltd. (Tianjin, China). Chloroplatinic acid hexahydrate ($H_2PtCl_6{\cdot}6H_2O$) was purchased from Shanghai Aladdin Bio-chem Technology Co., Ltd. (Shanghai, China). Potassium chloride (KCl) and sodium chloride (NaCl) were purchased from Tianjin Bodi Chemical Reagent Co., Ltd. (Tianjin, China) All chemicals were used as purchased without further purification.

2.2. Preparation of $Bi_4Ti_3O_{12}$ Powders

Preparation of $Bi_4Ti_3O_{12}$ powders via the molten salt method was similar to our previously published paper [23–27]. Typically, stoichiometric amounts of Bi_2O_3, TiO_2, NaCl and KCl were weighed firstly according to a predetermined ratio shown in Table 1. After that, the raw materials and molten salt medium were mixed in a planetary ball mill for 3 h under a rotating speed of 300 r/min with ethanol as milling medium. Subsequently, the mixed powders were dried and then subjected to 2 h heating treatment at 700, 800 and 900 °C in a muffle furnace. Finally, $Bi_4Ti_3O_{12}$ powders were obtained after washing, filtration and drying, and the samples were labeled as $Bi_4Ti_3O_{12}$-T-M (T is the heat treatment temperature, M is the mass ratio of molten salt medium and raw material).

2.3. Assembly of Ag and Pt Nanoparticles (NPs) on $Bi_4Ti_3O_{12}$ Nanosheets

Ag/$Bi_4Ti_3O_{12}$, Pt/$Bi_4Ti_3O_{12}$ and AgPt/$Bi_4Ti_3O_{12}$ were prepared by a photoreduction method. The whole preparation procedure involved four steps as follows: (i) 1 g as-prepared $Bi_4Ti_3O_{12}$ powders were dispersed in 200 mL deionized water; (ii) 0.1 mol/L $AgNO_3$ solution, 7.72 mmol/L $H_2PtCl_6{\cdot}6H_2O$ solution and the mixed solution of $AgNO_3$ and $H_2PtCl_6{\cdot}6H_2O$ was separately added to the as-prepared

$Bi_4Ti_3O_{12}$ suspension with the predetermined proportions shown in Table 2; (iii) The as-prepared suspensions were irradiated under a 300 W xenon lamp for 60 min (a 400 nm filter was used to block ultraviolet (UV) light (λ < 400 nm)). In this process, Ag^+ and Pt^{4+} was respectively reduced to Ag NPs and Pt NPs; and iv) The Ag/$Bi_4Ti_3O_{12}$, Pt/$Bi_4Ti_3O_{12}$ and AgPt/$Bi_4Ti_3O_{12}$ powders were obtained after the mixture was filtered, washed and dried (drying conditions: 80 °C for 12 h in an electric drying oven).

Table 1. Synthesis conditions and batch of $Bi_4Ti_3O_{12}$.

Firing Conditions	Salt Medium Composition (Molar Ratio)	Mass Ratio of Salt to Reactant
700 °C/2 h		
800 °C/2 h	NaCl:KCl (1:1)	1:1
900 °C/2 h		
700 °C/2 h		
800 °C/2 h	NaCl:KCl (1:1)	2:1
900 °C/2 h		
700 °C/2 h		
800 °C/2 h	NaCl:KCl (1:1)	3:1
900 °C/2 h		

Table 2. Batch compositions of Ag/$Bi_4Ti_3O_{12}$, Pt/$Bi_4Ti_3O_{12}$ and AgPt/$Bi_4Ti_3O_{12}$ photocatalysts.

Metal Composition (Mass Ratio)	Loading Capacity	Illumination Condition
	0.1 wt%	
	0.2 wt%	
Ag	0.5 wt%	
	1.0 wt%	
	3.0 wt%	300 W xenon lamp (λ > 400 nm) 60 min
	5.0 wt%	
Pt	0.1 wt%	
Ag:Pt (1:1)		
Ag:Pt (7:3)		
Ag:Pt (3:7)	0.1 wt%	
Ag:Pt (9:1)		
Ag:Pt (1:9)		

2.4. Characterization

X-ray diffraction (XRD) was performed on MiniFlex 600 with Cu Kα radiation (λ = 1.54178 Å) to investigate the crystal structure of the as-prepared powders. The field-emission scanning electron microscopy (FE-SEM) images, transmission electron microscopy (TEM) images, selected area electron diffraction (SAED) images and energy dispersive spectroscopy (EDS) elemental mapping images were respectively taken on a JEOL JSM-6700F SEM and JEM-2100 HR TEM to observe the microstructure of the as-prepared catalysts. Fourier transform infrared (FT-IR) spectra ranged from 3000–450 cm^{-1} were recorded on a Nicolet iS50 spectrometer in air at room temperature to differentiate the functional groups formed on the surface of as-prepared catalysts. Chemical composition of as-prepared catalysts was analyzed by the inductively coupled plasma mass spectroscopy (ICP-MS, Spectro Flame, Spectro

Analytical Instrument, Kleve, Germany). X-ray photoelectron spectroscopy (XPS) measurements were performed using an AMICUS ESCA 3400 XPS with Al Kα radiation. In order to investigate the elemental information (Bi, Ti, O, Ag and Pt), all XPS spectra were calibrated by shifting the detected adventitious carbon C 1s peak to 284.8 eV. A RF-6000 fluorescence spectrophotometer was used to measure the photoluminescence (PL) spectra of the samples (excitation wavelength: 320 nm). Ultraviolet-visible (UV-Vis) absorption spectra were recorded using a UV-Vis spectrophotometer (Shimadzu UV-3600, Kyoto, Japan) to study the responsive behavior under UV and visible light irradiation of the as-prepared catalysts.

2.5. Photocatalytic Activity

RhB aqueous solution was chosen to simulate wastewater and the variation of its concentration under the irradiation of xenon lamp (λ > 400 nm) was measured at λ = 554 nm by a UV-Vis spectrophotometer. Four steps were performed to investigate the performance of the as-prepared catalysts: (i) 3.2 mg/L RhB solution was prepared firstly and then its absorption spectrum was determined by a UV-Vis spectrophotometer; (ii) 200 mg as-prepared photocatalysts were put into 200 mL RhB solution with an initial concentration of 3.2 mg/L, and then, in order to achieve adsorption-desorption equilibrium, the obtained mixture was stirred in the dark for 1 h. After that, 3 mL solution was taken out to be centrifuged and the supernatant was taken out for absorption spectrum measurement, from which the adsorbed amount of RhB can be calculated; (iii) The residual mixture was exposed to a xenon lamp irradiation (λ > 400 nm) for 30 min and then 3 mL solution was taken out to be centrifuged, and the supernatant was taken out to measure its absorption spectrum; and (iv) Step 3 was repeated until the identified absorption spectra were almost unchanged. In order to eliminate the influence of temperature on the photocatalytic behavior, the whole experimental process was carried out in a cooling cycle device.

2.6. Detection of Reactive Species

In a typical reactive species trapping experiment, three reactions were carried out to distinguish the reactive species of hole (h^+), hydroxy radical (•OH) and superoxide radical (•O^{2-}) for the photocatalytic degradation process. At first, 200 mL RhB solution and 0.2 g as-prepared catalyst were respectively mixed with 20 mL ethanol (EtOH), 0.2 mmol benzoquinone (BQ) and 2 mmol triethanolamine (TEOA). Then the mixtures were stirred for 1 h in dark for achieving adsorption-desorption equilibrium. Finally, the catalytic process as described in Section 2.5 was carried out.

3. Results

Figure 1 showed the XRD patterns of $Bi_4Ti_3O_{12}$ powders ($Bi_4Ti_3O_{12}$-800-1) prepared by molten salt method and $(AgPt)_{0.001}/Bi_4Ti_3O_{12}$ composites prepared by a photoreduction method. It can be clearly seen that the diffraction peaks of the as-prepared $Bi_4Ti_3O_{12}$ matched well with the standard card JCPDS-01-080-2143, indicating the crystallinity of synthesized $Bi_4Ti_3O_{12}$ was high. The diffraction peaks of Ag and Pt NPs were not observed, and the reason may be ascribed to the low amount of loaded AgPt bimetallic NPs (0.1 wt%).

Figure 2 presents the SEM images of as-prepared $Bi_4Ti_3O_{12}$ and $(AgPt)_{0.001}/Bi_4Ti_3O_{12}$ composite photocatalysts prepared by the photoreduction method. It revealed that all the samples showed a lamellar structure with a grain size of 1–5 μm, and that Ag and Pt NPs were too small to be observed in the SEM images.

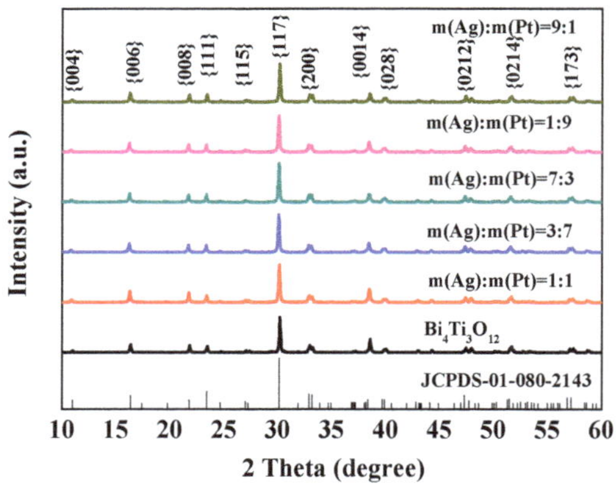

Figure 1. X-ray diffraction (XRD) patterns of $Bi_4Ti_3O_{12}$ and $(AgPt)_{0.001}/Bi_4Ti_3O_{12}$ composite photocatalysts prepared by photoreduction method.

Figure 2. Scanning electron microscopy (SEM) images of pristine $Bi_4Ti_3O_{12}$ and $(AgPt)_{0.001}/Bi_4Ti_3O_{12}$ composite photocatalysts prepared by photoreduction method: (**a**) $Bi_4Ti_3O_{12}$; (**b**)m(Ag):m(Pt) = 1:1; (**c**) m(Ag):m(Pt) = 7:3; (**d**) m(Ag):m(Pt) = 3:7; (**e**) m(Ag):m(Pt) = 9:1 and (**f**) m(Ag):m(Pt) = 1:9.

The representative TEM image of the as-prepared $(Ag_{0.7}Pt_{0.3})_{0.001}/Bi_4Ti_3O_{12}$ photocatalyst is shown in Figure 3a, confirming that $(Ag_{0.7}Pt_{0.3})_{0.001}/Bi_4Ti_3O_{12}$ photocatalyst exhibited a nano-sheet structure with proper thickness. As can be seen from the SAED pattern (Figure 3b), single crystalline $Bi_4Ti_3O_{12}$ was prepared. Figure 3c obviously showed that Ag and Pt NPs were modified on the surface of the $(Ag_{0.7}Pt_{0.3})_{0.001}/Bi_4Ti_3O_{12}$ sample (dashed area). Furthermore, two different d-spacing values of 0.226 nm and 0.239 nm were calculated from the lattice fringes of loaded particles (Figure 3d), which respectively correspond to the (111) crystal plane of cubic Pt and the (111) crystal plane of metallic Ag [28,29]. Figures 4–6 respectively presented the TEM images and the corresponding EDS elemental mapping images of $(Ag_{0.7}Pt_{0.3})_{0.001}/Bi_4Ti_3O_{12}$, $Ag_{0.001}/Bi_4Ti_3O_{12}$, and $Pt_{0.001}/Bi_4Ti_3O_{12}$ photocatalysts prepared by photoreduction method. It can be clearly seen from Figure 4a that a large amount of spherical particles with an average size of 9 nm were randomly distributed over the surface of as-prepared $Bi_4Ti_3O_{12}$. EDS mapping results presented in Figure 4b–f indicated that the spherical particles were Ag and Pt NPs, confirming $(Ag_{0.7}Pt_{0.3})_{0.001}/Bi_4Ti_3O_{12}$ composite photocatalysts were successfully prepared. Similarly, Figures 5 and 6 separately proved that $Ag_{0.001}/Bi_4Ti_3O_{12}$ and $Pt_{0.001}/Bi_4Ti_3O_{12}$ were successfully synthesized.

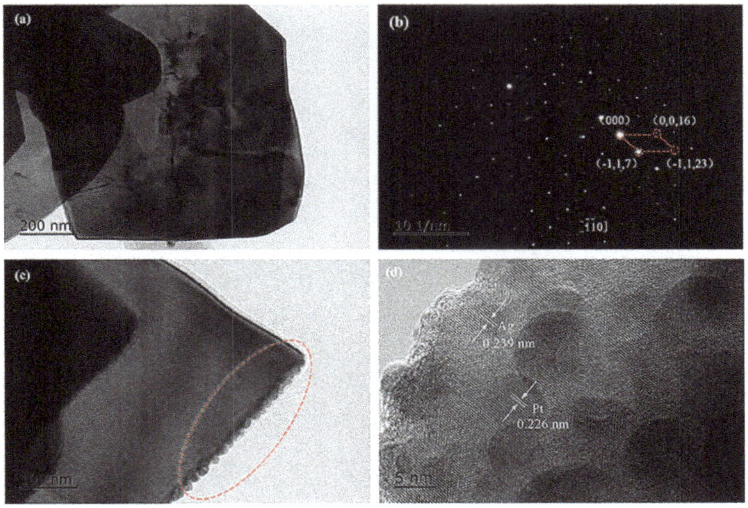

Figure 3. Transmission electron microscope (TEM) images (**a**,**c**), SAED image (**b**) and high-resolution TEM (HRTEM) image (**d**) of $(Ag_{0.7}Pt_{0.3})_{0.001}/Bi_4Ti_3O_{12}$.

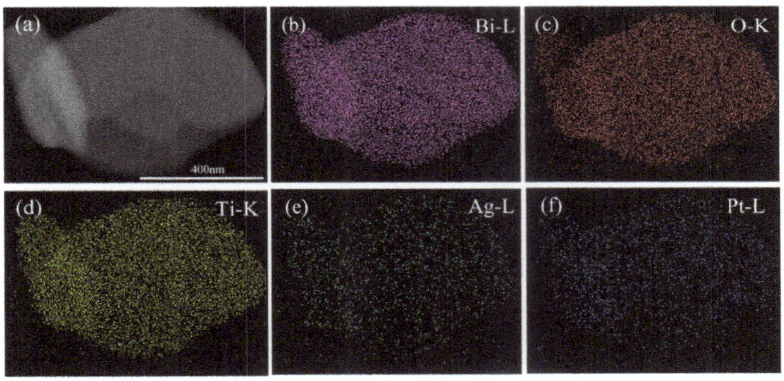

Figure 4. TEM image (**a**) and EDS maps of Bi, O, Ti, Ag and Pt (**b**–**f**) of as-prepared $(Ag_{0.7}Pt_{0.3})_{0.001}/Bi_4Ti_3O_{12}$.

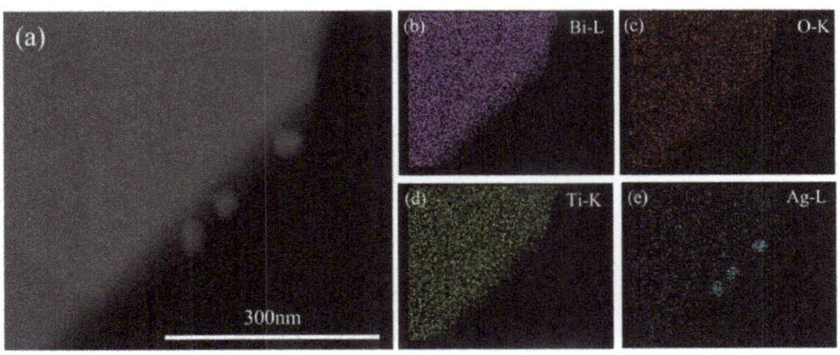

Figure 5. TEM image (**a**) and EDS maps of Bi, O, Ti and Ag (**b**–**e**) of as-prepared $Ag_{0.001}/Bi_4Ti_3O_{12}$.

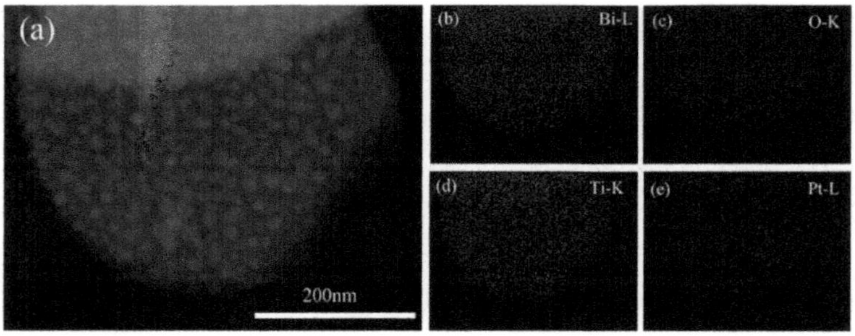

Figure 6. TEM image (**a**) and EDS maps of Bi, O, Ti and Pt (**b–e**) of as-prepared $Pt_{0.001}/Bi_4Ti_3O_{12}$.

ICP-MS results presented in Table 3 revealed that the content of Ag in $Ag_{0.001}/Bi_4Ti_3O_{12}$ and Pt in $Pt_{0.001}/Bi_4Ti_3O_{12}$ photocatalysts was respectively 0.0347 wt% and 0.0782 wt%, while the content of Ag and Pt in $(Ag_{0.7}Pt_{0.3})_{0.001}/Bi_4Ti_3O_{12}$ was respectively 0.0122 wt% and 0.0144 wt%, demonstrating that Ag and Pt have been successfully loaded on $Bi_4Ti_3O_{12}$ nanosheets. However, the actual amount was lower than that theoretical value, indicating that the Ag^+ and Pt^{4+} were not completely photo-reduced and loaded on $Bi_4Ti_3O_{12}$ under present irradiation of Xe lamp.

Table 3. The content of Ag and Pt in the as-prepared photocatalysts.

As-Prepared Photocatalyst	Element	Content (wt%)
$Ag_{0.001}/Bi_4Ti_3O_{12}$	Ag	0.0347
$Pt_{0.001}/Bi_4Ti_3O_{12}$	Pt	0.0782
$(Ag_{0.7}Pt_{0.3})_{0.001}/Bi_4Ti_3O_{12}$	Ag	0.0122
	Pt	0.0144

XPS was performed to investigate the elemental information of as-prepared $(Ag_{0.7}Pt_{0.3})_{0.001}/Bi_4Ti_3O_{12}$ photocatalyst. As shown in Figure 7a, Bi, Ti, O, Ag and Pt elements were identified, demonstrating the successful preparation of the $(Ag_{0.7}Pt_{0.3})_{0.001}/Bi_4Ti_3O_{12}$ composite photocatalyst. The high-resolution XPS spectrum of Bi-4f, Ti-2p, O-1s, Ag-3d and Pt-4f were investigated (Figure 7b–f). Two peaks at 159.2 and 164.5 eV were respectively ascribed to Bi $4f_{7/2}$ and Bi $4f_{5/2}$ (Figure 7b), indicating that only Bi^{3+} existed in the $(Ag_{0.7}Pt_{0.3})_{0.001}/Bi_4Ti_3O_{12}$ composite [30–32]. The high resolution XPS spectrum of Ti 2p can be deconvoluted into three peaks at 458.1 eV, 463.8 eV and 466.2 eV (Figure 7c), which were assigned to Ti $2P_{3/2}$, Ti $2p_{1/2}$ and Bi $4d_{3/2}$, respectively [30–32], demonstrating that only Ti^{4+} specie existed in the $(Ag_{0.7}Pt_{0.3})_{0.001}/Bi_4Ti_3O_{12}$ composite. From the high-resolution XPS spectrum of O 1s (Figure 7d), the peak at 529.9 eV was assigned to lattice oxygen in $Bi_4Ti_3O_{12}$ and the peak at 532.2 eV was ascribed to the oxygen adsorbed on the surface of as-prepared $(Ag_{0.7}Pt_{0.3})_{0.001}/Bi_4Ti_3O_{12}$ [30]. As shown in Figure 7e, peaks at 368.1 and 374.1 eV were, respectively, assigned to Ag $3d_{5/2}$ and Ag $3d_{3/2}$, providing conclusive evidence for Ag metal in the as-prepared $(Ag_{0.7}Pt_{0.3})_{0.001}/Bi_4Ti_3O_{12}$ photocatalyst. As presented in Figure 7f, an asymmetric peak at 74.35 eV assigned to Pt $4f_{5/2}$ was observed, illustrating the existence of Pt metal in the as-prepared $(Ag_{0.7}Pt_{0.3})_{0.001}/Bi_4Ti_3O_{12}$ sample [33,34]. Since no additional peaks were observed in the Ag 3d and Pt 4f spectra, it can be determined that Ag and Pt do not exist in the form of oxidation states.

Fourier transform infrared (FTIR) spectra further offered the functional group information of the $Bi_4Ti_3O_{12}$ and $(Ag_{0.7}Pt_{0.3})_{0.001}/Bi_4Ti_3O_{12}$ (Figure 8). The peak at 832 cm^{-1} belonged to the Bi-O characteristic stretching vibration, the strong peak at 573 cm^{-1} and the weak peak at 472 cm^{-1} belonged to the stretching vibration of Ti-O [35,36]. These results indicated that the structure of orthorhombic phase $Bi_4Ti_3O_{12}$ was not damaged by loading of Ag and Pt NPs. Besides, no characteristic peaks of Pt and Ag oxides were observed, indicating that both Pt and Ag existed in metallic state in $(Ag_{0.7}Pt_{0.3})_{0.001}/Bi_4Ti_3O_{12}$, which was in line with the XPS results presented the Figure 7.

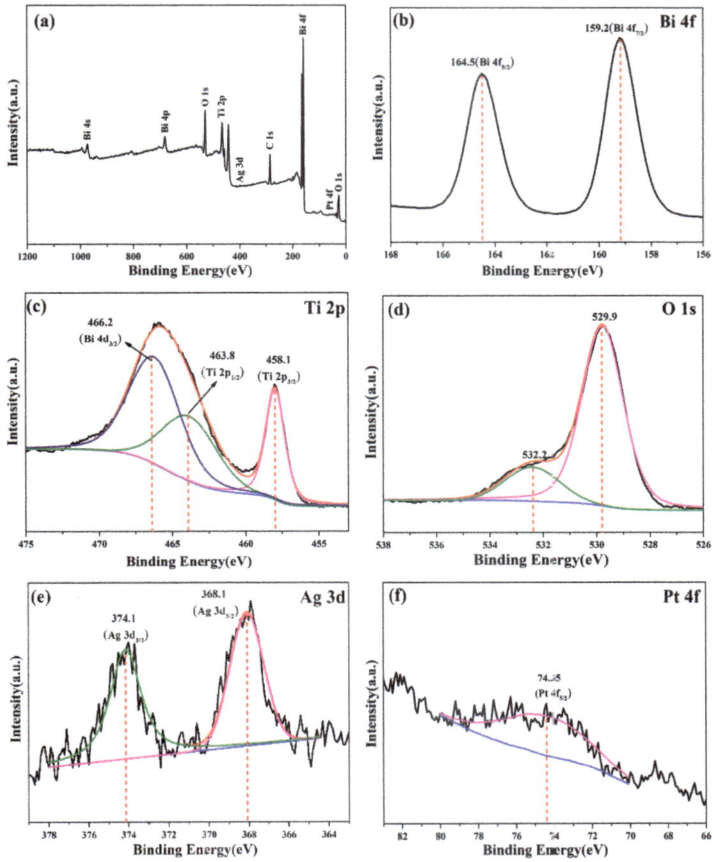

Figure 7. (**a**) X-ray photoelectron spectroscopy (XPS) survey scan spectrum and (**b**–**f**) high resolution XPS spectra of Bi 4f, Ti 2p, O 1s, Ag 3d and Pt 4f of $(Ag_{0.7}Pt_{0.3})_{0.001}/Bi_4Ti_3O_{12}$.

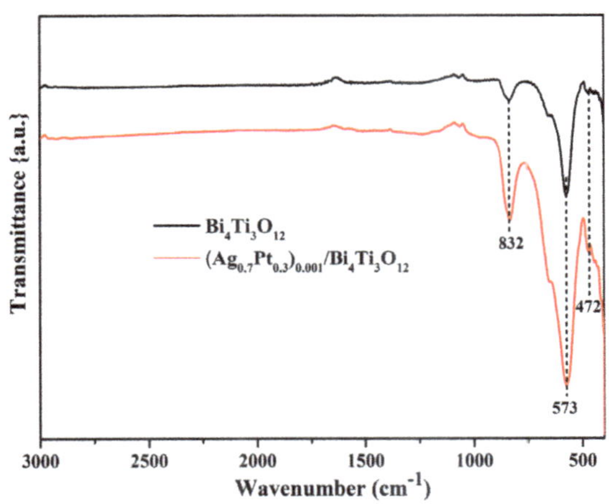

Figure 8. Fourier transform infrared (FTIR) spectra of $Bi_4Ti_3O_{12}$ and $(Ag_{0.7}Pt_{0.3})_{0.001}/Bi_4Ti_3O_{12}$.

Figure 9 showed the UV-Vis absorption spectra of pristine $Bi_4Ti_3O_{12}$ and as-prepared $(AgPt)_{0.001}/Bi_4Ti_3O_{12}$ composites. Obviously, $(AgPt)_{0.001}/Bi_4Ti_3O_{12}$ composites exhibited stronger visible light absorption than pristine $Bi_4Ti_3O_{12}$. As reported in previous works, the enhancement in visible light absorption may be dominated by the surface plasma resonance effect of Ag and Pt NPs [37,38]. When the internal electron oscillation frequency of the AgPt bimetallic NPs was equal to the frequency of the irradiated light, local surface plasmon resonance was induced and then the visible light absorption of $(AgPt)_{0.001}/Bi_4Ti_3O_{12}$ was improved. Meanwhile, no obvious change was observed in the absorption edge for all $(AgPt)_{0.001}/Bi_4Ti_3O_{12}$ samples, suggesting that the bandgap of $Bi_4Ti_3O_{12}$ and $(AgPt)_{0.001}/Bi_4Ti_3O_{12}$ photocatalysts was almost the same.

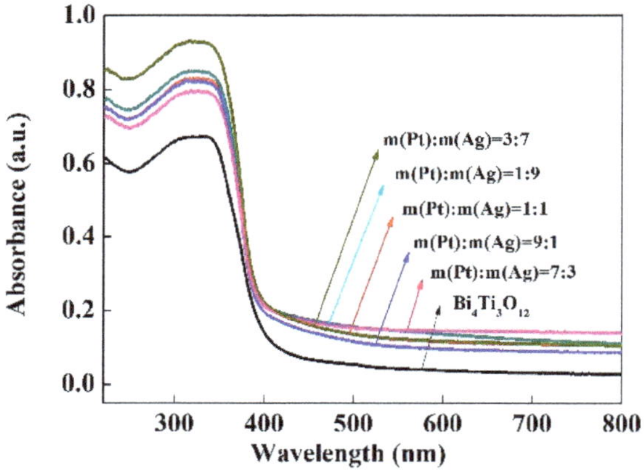

Figure 9. Ultraviolet-visible (UV-Vis) absorption spectra of starting $Bi_4Ti_3O_{12}$ and $(AgPt)_{0.001}/Bi_3Ti_4O_{12}$ composite photocatalysts prepared by photoreduction method.

Figure 10a showed the dependence of C/C_0 (C_0 and C are respectively the initial and instantaneous concentration of the RhB solution) on irradiation time during the RhB degradation process when $Bi_4Ti_3O_{12}$ and $(AgPt)_{0.001}/Bi_4Ti_3O_{12}$ were used as photocatalysts. It can be seen that the degradation rate ($(C_0 - C)/C_0 \times 100\%$) of RhB increased from about 69.7% to about 91.5% after irradiation of 90 min when 0.1 wt% $Ag_{0.7}Pt_{0.3}$ bimetallic NPs were loaded on $Bi_4Ti_3O_{12}$. Additionally, the photocatalytic activity of $(AgPt)_{0.001}/Bi_4Ti_3O_{12}$ was monitored by varying the mass ratio of Ag to Pt, and the $(Ag_{0.7}Pt_{0.3})_{0.001}/Bi_4Ti_3O_{12}$ catalysts displayed the best photocatalytic activity for RhB degradation.

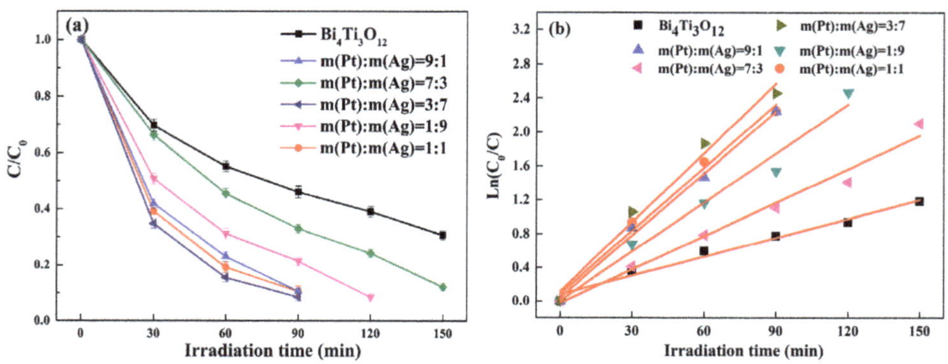

Figure 10. (**a**) Photocatalytic activity of pristine $Bi_4Ti_3O_{12}$ and as-prepared $(AgPt)_{0.001}/Bi_4Ti_3O_{12}$ composite photocatalysts prepared by photocatalytic reduction method on the degradation of rhodamine B (RhB); (**b**) plot of $Ln(C_0/C)$ versus irradiation time for the photodegradation of RhB.

In order to investigate the reaction kinetics of $Bi_4Ti_3O_{12}$ and $(AgPt)_{0.001}/Bi_4Ti_3O_{12}$ photocatalysts for RhB degradation, relationship between $Ln(C_0/C)$ and irradiation time was measured and the results were shown in Figure 10b. All the photocatalytic reactions can be fitted well by pseudo-first-order kinetics,

$$Ln\ (C/C_0) = -kt \quad (1)$$

When $Ln(C/C_0)$ was plotted versus irradiation time t, the apparent reaction rate k can be obtained by calculating the slope of the fitted curves. The results shown in Figure 10b demonstrate that the k value relied heavily on the loading of AgPt bimetallic particles, and that the highest k value of RhB degradation catalyzed by $(Ag_{0.7}Pt_{0.3})_{0.001}/Bi_4Ti_3O_{12}$ was calculated as 0.027 min^{-1}, which was 3.7 times of that by $Bi_4Ti_3O_{12}$.

The photocatalytic activities of $Ag_{0.001}/Bi_3Ti_4O_{12}$ and $Pt_{0.001}/Bi_3Ti_4O_{12}$ composite photocatalysts prepared by the photocatalytic reduction method on the degradation of RhB after a 150 min illumination were also investigated (Figure 11). Figure 11a revealed that the degradation rate of RhB catalyzed by $Pt_{0.001}/Bi_4Ti_3O_{12}$ reached as high as 92.4%, which was higher than that by $Ag_{0.001}/Bi_4Ti_3O_{12}$ (87.7%) and by $Bi_4Ti_3O_{12}$ (69.3%). Meanwhile, the apparent reaction rate k in the $Pt_{0.001}/Bi_4Ti_3O_{12}$ and $Ag_{0.001}/Bi_4Ti_3O_{12}$ catalyzed system was separately calculated as 0.01561 min^{-1} and 0.01366 min^{-1} which was, respectively, 2.1 and 1.8 times that in $Bi_4Ti_3O_{12}$ case (Figure 11b). And excitingly, $(AgPt)_{0.001}/Bi_4Ti_3O_{12}$ composites displayed higher photocatalytic activity towards RhB degradation than $Ag_{0.001}/Bi_4Ti_3O_{12}$ and $Pt_{0.001}/Bi_4Ti_3O_{12}$. The reasons may be as follows: (i) the surface plasma resonance effect of Ag and Pt NPs enhanced the visible light absorption (Figure 8) [34]; (ii) Ag NPs can be used as electronic acceptors with excellent electrical conductivity to promote the separation of electron-hole pairs; and (iii) Pt NPs can be applied as electron cocatalyst to accelerate the capture of photogenerated electron, and then facilitated the proton reduction reaction [39,40].

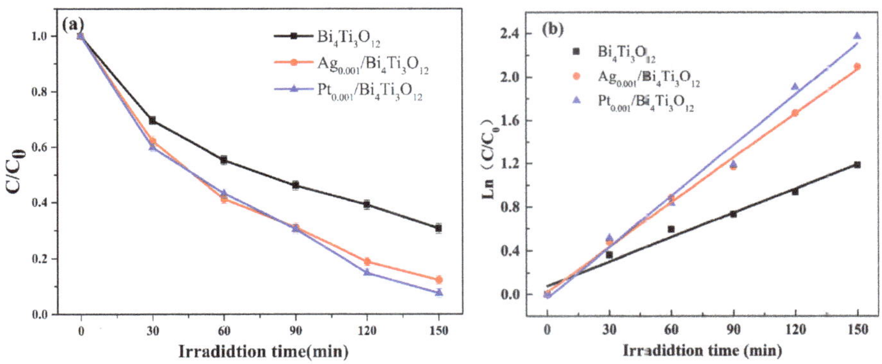

Figure 11. (**a**) Photocatalytic activity of pristine $Bi_4Ti_3O_{12}$, $Ag_{0.001}/Bi_3Ti_4O_{12}$ and $Pt_{0.001}/Bi_3Ti_4O_{12}$ composite photocatalysts prepared by photocatalytic reduction method on the degradation of RhB; (**b**) plot of $Ln(C_0/C)$ versus irradiation time for the photodegradation of RhB.

Hole and radical trapping experiments were carried out to investigate the effective reactants in the photocatalytic RhB degradation process. TEOA, EtOH and BQ were respectively added to RhB solution as capture agents for h^+, •OH, and $•O^{2-}$ scavengers [41,42]. The main active species were determined by the change of degradation effect after the photocatalysis experiment. As can be seen from Figure 12a, the photocatalytic performance of RhB removal was almost unchanged by introducing EtOH, implying that •OH played a tiny effect on the degradation of RhB. By contrast, after adding TEOA or BQ, the photocatalytic RhB degradation activity of $(Ag_{0.7}Pt_{0.3})_{0.001}/Bi_4Ti_3O_{12}$ was remarkably suppressed and the corresponding efficiencies for photodegradation of RhB were calculated as low as 9.4% or 10.8%, respectively (Figure 12b). It can thus be reasonably concluded that h^+ and $•O^{2-}$ were the active groups in the present $(Ag_{0.7}Pt_{0.3})_{0.001}/Bi_4Ti_3O_{12}$ photocatalyzed process.

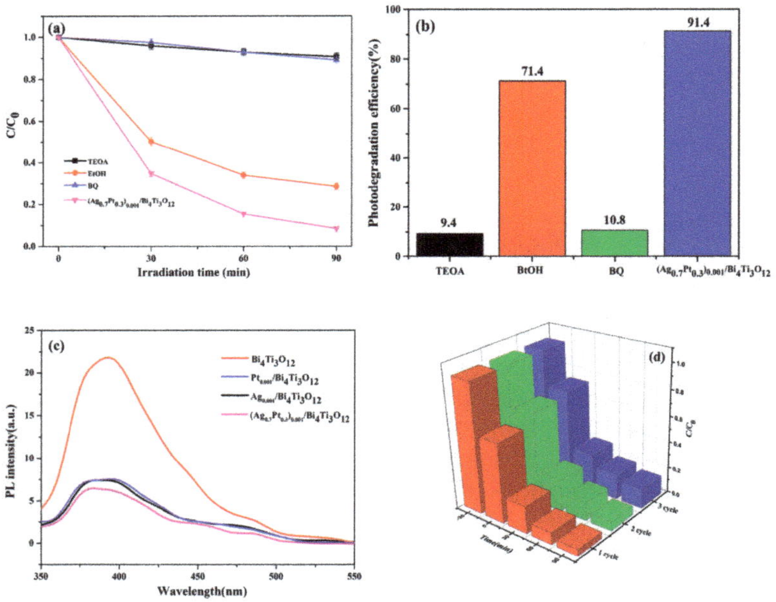

Figure 12. (**a**,**b**) Effects of various scavengers on the photocatalytic efficiency of $(Ag_{0.7}Pt_{0.3})_{0.001}/Bi_4Ti_3O_{12}$; (**c**) photoluminescence spectra of $Bi_4Ti_3O_{12}$ and $AgPt/Bi_4Ti_3O_{12}$; (**d**) Cyclic photodegradation of RhB by $(Ag_{0.7}Pt_{0.3})_{0.001}/Bi_4Ti_3O_{12}$ photocatalyst.

In order to study the separation behavior of electron hole pairs, photoluminescence (PL) spectroscopy was implemented. As shown in Figure 12c, phase pure $Bi_4Ti_3O_{12}$, $Pt_{0.001}/Bi_4Ti_3O_{12}$, $Ag_{0.001}/Bi_4Ti_3O_{12}$ and $(Ag_{0.7}Pt_{0.3})_{0.001}/Bi_4Ti_3O_{12}$ samples exhibit a similar PL emission curve when they were excited by a 320 nm light. However, even though the characteristic peaks of the four PL curves were all observed at about 400 nm, the PL spectroscopy of $(Ag_{0.7}Pt_{0.3})_{0.001}/Bi_4Ti_3O_{12}$ composite exhibited the lowest emission intensity, illustrating that the separation rate of photogenerated electron hole pairs in the $(Ag_{0.7}Pt_{0.3})_{0.001}/Bi_4Ti_3O_{12}$ sample was the highest [43,44]. As a result, more active species were formed to participate in the degradation of RhB in the case.

The stability of the photocatalyst was very important for its practical application. Therefore, the cycle experiment of RhB degradation was implemented (Figure 12d). To our amazement, the RhB degradation efficiency excited by $(Ag_{0.7}Pt_{0.3})_{0.001}/Bi_4Ti_3O_{12}$ photocatalyst decreased slightly after three consecutive cycles. This demonstrated that the as-prepared $(Ag_{0.7}Pt_{0.3})_{0.001}/Bi_4Ti_3O_{12}$ catalyst had good photocatalytic stability for degradation of RhB.

Based on the results obtained, a possible degradation mechanism of RhB degradation over $AgPt/Bi_4Ti_3O_{12}$ was proposed (Figure 13) and described as follows: (i) under the irradiation of light, photogenerated electrons e^- and holes h^+ were generated on the surface of $Bi_4Ti_3O_{12}$ [45]. Generally, h^+ can react with H_2O or OH^- to form active hydroxyl •OH, e^- can combine with O_2 to produce superoxide radical $\bullet O^{2-}$. Since h^+, •OH and $\bullet O^{2-}$ all have strong oxidability, they can react directly with RhB to generate water, CO_2 and inorganic small molecules [46–49]. However, the photogenerated h^+ and e^- will recombine rapidly on the surface of $Bi_4Ti_3O_{12}$ and then the photocatalytic activity was reduced; (ii) when Ag NPs were modified on the surface of $Bi_4Ti_3O_{12}$, the photogenerated h^+ and e^- will be redistributed since the Fermi level of Ag was lower than that of $Bi_4Ti_3O_{12}$ [50]. Due to the higher Fermi level of $Bi_4Ti_3O_{12}$, photogenerated e^- transferred from the surface of $Bi_4Ti_3O_{12}$ to Ag NPs with lower Fermi level and thus Schottky barrier was formed on their interface [51]. Therefore, the photogenerated electrons e^- were trapped and the recombination of photogenerated electron-hole pairs was inhibited [34,38,52–54]. As a result, more photogenerated h^+ was left to oxidize the RhB

(Equations (2)–(10)); (iii) Pt NPs cocatalysts can provide adsorption sites for protons as electron sinks, and thus the number of h^+ was further improved [21] and (iv) due to the surface plasma resonance effect of Ag and Pt NPs, the absorption of visible light by $Bi_4Ti_3O_{12}$ was greatly improved [18,55], and more electrons and holes were generated. The reaction formulas were as follows:

$$Bi_4Ti_3O_{12} + h\nu \rightarrow Bi_4Ti_3O_{12}(e^- + h^+) \quad (2)$$

$$Bi_4Ti_3O_{12}(e^-) + Ag/Pt \rightarrow Bi_4Ti_3O_{12} + Ag/Pt(e^-) \quad (3)$$

$$Bi_4Ti_3O_{12}(h^+) + H_2O \rightarrow \bullet OH + H^+ \quad (4)$$

$$Bi_4Ti_3O_{12}(h^+) + OH^- \rightarrow \bullet OH \quad (5)$$

$$Bi_4Ti_3O_{12}(e^-) + O_2 \rightarrow Bi_4Ti_3O_{12} + \bullet O^{2-} \quad (6)$$

$$Ag/Pt\ (e^-) + O_2 \rightarrow Ag/Pt + \bullet O^{2-} \quad (7)$$

$$\bullet OH + RhB \rightarrow CO_2 + H_2O + \quad (8)$$

$$\bullet O^{2-} + RhB \rightarrow CO_2 + H_2O + .. \quad (9)$$

$$Bi_4Ti_3O_{12}(h^+) + RhB \rightarrow CO_2 + H_2O + \quad (10)$$

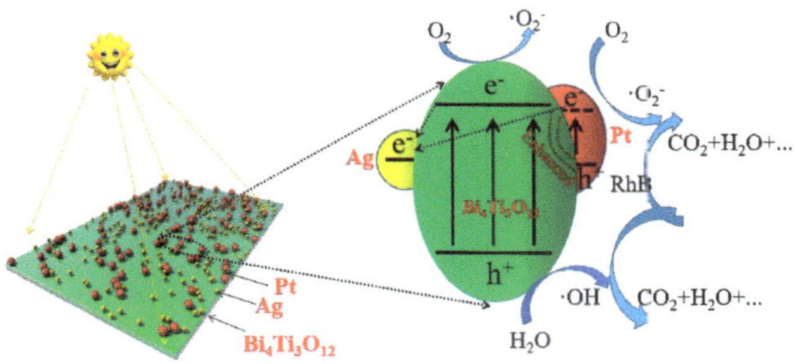

Figure 13. Proposed photocatalytic mechanism for degradation of RhB using as-prepared AgPt/$Bi_4Ti_3O_{12}$ composite photocatalysts.

4. Conclusions

$Bi_4Ti_3O_{12}$ particles with lamellar structure were prepared firstly by a molten salt method as precursor, and then Ag, Pt and AgPt bimetallic NPs were loaded on the as-prepared sheet-like $Bi_4Ti_3O_{12}$ to synthesize AgPt/$Bi_4Ti_3O_{12}$ composites through the photoreduction procedure. The results revealed that the crystal structure and morphology of the as-prepared $Bi_4Ti_3O_{12}$ were almost unchanged by loading Ag, Pt and AgPt bimetallic NPs.

It is exciting to note that the photocatalytic activity of $Bi_4Ti_3O_{12}$ for RhB degradation was sharply enhanced via loading Ag, Pt or AgPt bimetallic NPs. When the loading amount was fixed at 0.1 wt%, the apparent reaction rate of the as-prepared Ag/$Bi_4Ti_3O_{12}$ and Pt/$Bi_4Ti_3O_{12}$ photocatalysts and $Ag_{0.7}Pt_{0.3}$/$Bi_4Ti_3O_{12}$ composite photocatalysts was respectively calculated as 0.01366 min^{-1}, 0.01561 min^{-1} and 0.027 min^{-1}, which was respectively 1.8 times, 2.1 times and 3.7 times that of pristine $Bi_4Ti_3O_{12}$ (0.00744 min^{-1}). The enhanced activity may be ascribed to the surface plasma resonance effect of noble metal and Schottky barrier formed on the interface between noble metal NPs and $Bi_4Ti_3O_{12}$. The reactive species trapping experiment demonstrated that h^+ and $\bullet O^{2-}$ played the main role in present $(Ag_{0.7}Pt_{0.3})_{0.001}$/$Bi_4Ti_3O_{12}$ photocatalyzed process, and the $(Ag_{0.7}Pt_{0.3})_{0.001}$/$Bi_4Ti_3O_{12}$ catalyst exhibited good photocatalytic stability for RhB degradation.

Author Contributions: G.Y. and G.Z. did experimental work and wrote the paper; H.Z., F.L., S.Z., Q.J. and Z.H. designed and reviewed the manuscript and provided some suggestions; K.L., Y.C. and J.H. reviewed the manuscript assisted in the experimental work. All authors contributed to and critically reviewed the manuscript. H.Z. and F.L. made particularly major contributions to the writing and editing. All authors have read and agreed to the published version of the manuscript.

Funding: This work was financially supported by Natural Science Foundation of Jiangxi Province, China (Contract No. 20181BAB216006); National Natural Science Foundation of China (Grant No. 51702241, and 51872210); Program of Hubei Province, China (Contract No. 2017CFA004 and T201602).

Conflicts of Interest: The authors declare no conflict of interest.

References

1. Alansi, A.M.; Al-Qunaibit, M.; Alade, I.O.; Qahtan, T.F.; Saleh, T.A. Visible-light responsive BiOBr nanoparticles loaded on reduced graphene oxide for photocatalytic degradation of dye. *J. Mol. Liq.* **2018**, *253*, 297–304. [CrossRef]
2. Lam, S.M.; Sin, J.C.; Mohamed, A.R. A review on photocatalytic application of g-C_3N_4/semiconductor (CNS) nanocomposites towards the erasure of dyeing wastewater. *Mater. Sci. Semicond. Process.* **2016**, *47*, 62. [CrossRef]
3. Zangeneh, H.; Zinatizadeh, A.; Habibi, M.; Akia, M.; Isa, M.H. Photocatalytic oxidation of organic dyes and pollutants in wastewater using different modified titanium dioxides: A comparative review. *J. Ind. Eng. Chem.* **2015**, *26*, 1–36. [CrossRef]
4. Chávez, A.; Gimeno, O.; Rey, A.; Pliego, G.; Oropesa, A.; Álvarez, P.; Beltrán, F. Treatment of highly polluted industrial wastewater by means of sequential aerobic biological oxidation-ozone based AOPs. *Chem. Eng. J.* **2019**, *361*, 89–98. [CrossRef]
5. Do, M.H.; Phan, N.H.; Nguyen, T.K.P.; Pham, T.T.S.; Nguyen, V.K.; Vu, T.T.T. Activated carbon/Fe_3O_4 nanoparticle composite: Fabrication, methyl orange removal and regeneration by hydrogen peroxide. *Chemosphere* **2011**, *85*, 1269–1276. [CrossRef] [PubMed]
6. Forgacs, E.; Cserháti, T.; Oros, G. Removal of synthetic dyes from wastewaters: A review. *Environ. Int.* **2004**, *30*, 953–971. [CrossRef] [PubMed]
7. Sivakumar, P.; Murugesan, B.; Loganathan, A. A review on decolourisation of dyes by photodegradation using various bismuth catalysts. *J. Taiwan Inst. Chem. Eng.* **2014**, *45*, 2300–2306. [CrossRef]
8. Mahmood, T.; Wang, X.S.; Chen, C.C.; Ma, W.H. Photocatalytic degradation of persistent and toxic organic pollutants. *Chin. J. Catal.* **2007**, *21*, 1117. [CrossRef]
9. Xie, J.; Guo, N.; Liu, A.; Cao, Y.; Hu, J.; Jia, D. Simple solid-state synthesis of BiOCl/$Bi_2O_2CO_3$ heterojunction and its excellent photocatalytic degradation of RhB. *J. Alloys Compd.* **2019**, *784*, 377–385. [CrossRef]
10. Ji, Y.C.; Yang, R.Q.; Wang, L.W.; Song, G.X.; Wang, A.Z.; Lv, Y.W.; Gao, M.M.; Zhang, J.; Yu, X. Visible light active and noble metal free Nb_4N_5/TiO_2 nanobelt surface heterostructure for plasmonic enhanced solar water splitting. *Chem. Eng. J.* **2020**, *402*, 126226. [CrossRef]
11. Xian, T.; Yang, H.; Di, L.; Ma, J.; Zhang, H.; Dai, J. Photocatalytic reduction synthesis of $SrTiO_3$-graphene nanocomposites and their enhanced photocatalytic activity. *Nanoscale Res. Lett.* **2014**, *9*, 327. [CrossRef] [PubMed]
12. Peng, H.; Liu, D.; Zheng, X.; Fu, X. N-Doped Carbon-Coated ZnS with Sulfur-Vacancy Defect for Enhanced Photocatalytic Activity in the Visible Light Region. *Nanomaterials* **2019**, *9*, 1657. [CrossRef]
13. Shekofteh-Gohari, M.; Habibi-Yangjeh, A.; Abitorabi, M.; Rouhi, A. Magnetically separable nanocomposites based on ZnO and their applications in photocatalytic processes: A review. *Crit. Rev. Environ. Sci. Technol.* **2018**, *48*, 806–857. [CrossRef]
14. Liu, Y.; Zhang, M.Y.; Li, L.; Zhang, X.T. In situ ion exchange synthesis of the $Bi_4Ti_3O_{12}$/Bi_2S_3 heterostructure with enhanced photocatalytic activity. *Catal. Commun.* **2015**, *60*, 23. [CrossRef]
15. Shen, G.D.; Pu, Y.P.; Sun, R.J.; Shi, Y.; Cui, Y.F.; Jing, P.P. Enhanced visible light photocatalytic performance of a novel heterostructured $Bi_4Ti_3O_{12}$/BiOBr photocatalyst. *New J. Chem.* **2019**, *43*, 1. [CrossRef]
16. Zheng, C.X.; Yang, H.; Cui, Z.; Zhang, H.M.; Wang, X.X. A novel $Bi_4Ti_3O_{12}$/Ag_3PO_4 heterojunction photocatalyst with enhanced photocatalytic performance. *Nanoscale Res. Lett.* **2017**, *12*, 608. [CrossRef]
17. Liu, Y.; Wang, Z.; Fan, W.; Geng, Z.; Feng, L. Enhancement of the photocatalytic performance of Ni-loaded TiO_2 photocatalyst under sunlight. *Ceram. Int.* **2014**, *40*, 3887–3893. [CrossRef]

18. Zhang, F.; Liu, W.; Liu, Y.; Wang, J.; Ji, G. Fabrication and enhanced photocatalytic properties of Pt@SiO$_2$@TiO$_2$ composites by surface plasma resonance from Pt nanoparticles. *J. Nanopart. Res.* **2015**, *17*, 62. [CrossRef]
19. Fujishima, A.; Rao, T.N.; Tryk, D.A. Titanium dioxide photocatalysis. *J. Photochem. Photobiol. C Photochem. Rev.* **2000**, *1*, 1–21. [CrossRef]
20. Khan, M.R.; Chuan, T.W.; Yousuf, A.; Chowdhury, N.K.; Cheng, C.K. Schottky barrier and surface plasmonic resonance phenomena towards the photocatalytic reaction: Study of their mechanisms to enhance photocatalytic activity. *Catal. Sci. Technol.* **2015**, *5*, 2522–2531. [CrossRef]
21. Wang, W.; Lai, M.; Fang, J.J.; Lu, C.H. Au and Pt selectively deposited on {001}-faceted TiO$_2$ toward SPR enhanced photocatalytic Cr(VI) reduction: The influence of excitation wavelength. *Appl. Surf. Sci.* **2018**, *439*, 430–438. [CrossRef]
22. Chen, D.; Chen, Q.; Ge, L.; Yin, L.; Fan, B.; Wang, H.; Lu, H.; Xu, H.; Zhang, R.; Shao, G. Synthesis and Ag-loading-density-dependent photocatalytic activity of Ag@TiO$_2$ hybrid nanocrystals. *Appl. Surf. Sci.* **2013**, *284*, 921–929. [CrossRef]
23. Tian, L.; Li, J.; Liang, F.; Wang, J.; Li, S.; Zhang, H.; Zhang, S. Molten salt synthesis of tetragonal carbon nitride hollow tubes and their application for removal of pollutants from wastewater. *Appl. Catal. B Environ.* **2018**, *225*, 307–313. [CrossRef]
24. Li, J.Y.; Tian, L.; Liang, F.; Wang, J.K.; Han, L.; Zhang, J.; Ge, S.T.; Dong, L.H.; Zhang, H.J.; Zhang, S.W. Molten salt synthesis of hierarchical porous N-doped carbon submicrospheres for multifunctional applications: High performance supercapacitor, dye removal and CO$_2$ capture. *Carbon* **2019**, *141*, 739. [CrossRef]
25. Tian, L.; Li, J.; Liang, F.; Chang, S.; Zhang, H.; Zhang, M.; Zhang, S. Facile molten salt synthesis of atomically thin boron nitride nanosheets and their co-catalytic effect on the performance of carbon nitride photocatalyst. *J. Colloid Interface Sci.* **2019**, *536*, 664–672. [CrossRef]
26. Huang, Z.; Liu, J.; Huang, L.; Tian, L.; Wang, S.; Zhang, G.; Li, J.; Liang, F.; Zhang, H.; Jia, Q.; et al. One-step synthesis of dandelion-like lanthanum titanate nanostructures for enhanced photocatalytic performance. *NPG Asia Mater.* **2020**, *12*, 1–12. [CrossRef]
27. Huang, Z.; Duan, H.; Liu, J.; Zhang, H. Preparation of lanthanum cerate powders via a simple molten salt route. *Ceram. Int.* **2016**, *42*, 10482–10486. [CrossRef]
28. Sun, Y.; Zhao, Y.-F.; Sun, H.; Jia, F.-C.; Kumar, P.; Liu, B. Synthesis and room-temperature H$_2$S sensing of Pt nanoparticle-functionalized SnO$_2$ mesoporous nanoflowers. *J. Alloys Compd.* **2020**, *842*, 155813. [CrossRef]
29. Lin, C.; Ma, C.Q.; Yi, F.T.; Zhang, H.N.; Qian, Y.X.; Zhang, K.F. Ag NPs modified plasmonic Z-scheme photocatalyst Bi$_4$Ti$_3$O$_{12}$/Ag/Ag$_3$PO$_4$ with improved performance for pollutants removal under visible light irradiation. *Ceram. Int.* **2020**, *46*, 14650–14661. [CrossRef]
30. Zhao, X.; Yang, H.; Li, S.; Cui, Z.; Zhang, C. Synthesis and theoretical study of large-sized Bi$_4$Ti$_3$O$_{12}$ square nanosheets with high photocatalytic activity. *Mater. Res. Bull.* **2018**, *107*, 180–188. [CrossRef]
31. Pooladi, M.; Shokrollahi, H.; Lavasani, S.; Yang, H. Investigation of the structural, magnetic and dielectric properties of Mn-doped Bi$_2$Fe$_4$O$_9$ produced by reverse chemical co-precipitation. *Mater. Chem. Phys.* **2019**, *229*, 39–48. [CrossRef]
32. Du, C.; Li, D.H.; He, Q.Y.; Liu, J.M.; Li, W.; He, G.N.; Wang, Y.Z. Design and simple synthesis of composite Bi$_{12}$TiO$_{20}$/Bi$_4$Ti$_3$O$_{12}$ with a good photocatalytic quantum efficiency and high production of photo-generated hydroxyl radicals. *Phys. Chem. Chem. Phys. Pccp.* **2016**, *18*, 26530. [CrossRef]
33. Li, Z.; Chen, M.; Hu, H.; Zhang, Q.; Tao, D. Mechanochemical synthesis of novel Pt modified ZnAl-LDH for effective ciprofloxacin photodegradation. *J. Solid State Chem.* **2020**, *290*, 121594. [CrossRef]
34. Zhao, X.; Yang, H.; Cui, Z.; Yi, Z.; Yu, H. Synergistically enhanced photocatalytic performance of Bi$_4$Ti$_3$O$_{12}$ nanosheets by Au and Ag nanoparticles. *J. Mater. Sci. Mater. Electron.* **2019**, *30*, 13785–13796. [CrossRef]
35. Ganesh, R.S.; Sharma, S.K.; Abinnas, N.; Durgadevi, E.; Raji, P.; Muthamizhchelvan, C.; Ponnusamy, S.; Hayakawa, Y.; Kim, D.Y. Fabrication of the flexible nanogenerator from BTO nanopowders on graphene coated PMMA substrates by sol-gel method. *Mater. Chem. Phys.* **2017**, *192*, 274–281. [CrossRef]
36. Lin, X.; Guan, Q.F.; Liu, T.T.; Zhang, Y.; Zou, C.J. Controllable synthesis and photocatalytic activity of Bi$_4$Ti$_3$O$_{12}$ particles with different morphologies. *Acta Phys. Chim. Sin.* **2013**, *29*, 411–417.
37. Liu, H.; Wu, R.; Tian, L.; Kong, Y.; Sun, Y. Synergetic photocatalytic effect between 1 T@2H-MoS$_2$ and plasmon resonance induced by Ag quantum dots. *Nanotechnology* **2018**, *29*, 285402. [CrossRef]
38. Zhang, X.; Wang, Y.; Hou, F.; Li, H.; Yang, Y.; Zhang, X.; Yang, Y.; Wang, Y. Effects of Ag loading on structural and photocatalytic properties of flower-like ZnO microspheres. *Appl. Surf. Sci.* **2017**, *391*, 476–483. [CrossRef]

39. Sun, K.; Shen, J.; Liu, Q.; Tang, H.; Zhang, M.; Zulfiqar, S.; Lei, C. Synergistic effect of Co(II)-hole and Pt-electron cocatalysts for enhanced photocatalytic hydrogen evolution performance of P-doped g-C_3N_4. *Chin. J. Catal.* **2020**, *41*, 72–81. [CrossRef]
40. Liu, J.; Li, Y.; Zhou, X.; Jiang, H.; Yang, H.G.; Li, C. Positively charged Pt-based cocatalysts: An orientation for achieving efficient photocatalytic water splitting. *J. Mater. Chem. A* **2020**, *8*, 17–26. [CrossRef]
41. Lin, D.; Gao, M.; You, L. Fabrication of novel Ag/$AgVO_3$/WO_3 homojunction/heterojunction nanomaterials with highly enhanced photocatalytic activity-Investigation on type plus Z-scheme mechanism. *J. Alloys Compd.* **2020**, *846*, 156274. [CrossRef]
42. Shen, X.; Yang, J.; Zheng, T.; Wang, Q.; Zhuang, H.; Zheng, R.; Shan, S.; Li, S. Plasmonic p-n heterojunction of Ag/Ag_2S/Ag_2MoO_4 with enhanced Vis-NIR photocatalytic activity for purifying wastewater. *Sep. Purif. Technol.* **2020**, *251*, 117347. [CrossRef]
43. Chang, M.J.; Cui, W.N.; Liu, J.; Wang, K.; Luo, Z.M. Construction of novel TiO_2/$Bi_4Ti_3O_{12}$/MoS_2 core/shell nanofibers for enhanced visible light photocatalysis. *J. Mater. Sci. Technol.* **2020**, *36*, 97–105. [CrossRef]
44. Liu, Y.; Zhu, G.; Gao, J.; Hojamberdiev, M.; Zhu, R.; Wei, X.; Guo, Q.; Liu, P. Enhanced photocatalytic activity of $Bi_4Ti_3O_{12}$ nanosheets by Fe^{3+}-doping and the addition of Au nanoparticles: Photodegradation of Phenol and bisphenol A. *Appl. Catal. B Environ.* **2017**, *200*, 72–82. [CrossRef]
45. Chen, Z.; Jiang, X.; Zhu, C.; Shi, C. Chromium-modified $Bi_4Ti_3O_{12}$ photocatalyst: Application for hydrogen evolution and pollutant degradation. *Appl. Catal. B Environ.* **2016**, *199*, 241–251. [CrossRef]
46. Zhao, X.; Yang, H.; Cui, Z.; Li, R.; Feng, W. Enhanced photocatalytic performance of Ag–$Bi_4Ti_3O_{12}$ nanocomposites prepared by a photocatalytic reduction method. *Mater. Technol.* **2017**, *32*, 870–880. [CrossRef]
47. Chen, Z.W.; Jiang, H.; Jin, W.L.; Shi, C.K. Enhanced photocatalytic performance over $Bi_4Ti_3O_{12}$ nanosheets with controllable size and exposed {001} facets for Rhodamine B degradation. *Appl. Catal. B* **2016**, *180*, 698. [CrossRef]
48. Dutta, D.P.; Tyagi, A. Facile sonochemical synthesis of Ag modified $Bi_4Ti_3O_{12}$ nanoparticles with enhanced photocatalytic activity under visible light. *Mater. Res. Bull.* **2016**, *74*, 397–407. [CrossRef]
49. Guo, Y.; Li, J.H.; Gao, Z.Q.; Zhu, X.; Liu, Y.; Wei, Z.B.; Zhao, W.; Sun, C. A simple and effective method for fabricating novel p-n heterojunction photocatalyst g-C_3N_4/$Bi_4Ti_3O_{12}$ and its photocatalytic performances. *Appl. Catal. B* **2016**, *192*, 57. [CrossRef]
50. Kamat, P.V. Meeting the Clean Energy Demand: Nanostructure Architectures for Solar Energy Conversion. *J. Phys. Chem. C* **2007**, *111*, 2834–2860. [CrossRef]
51. Zhu, J.; Zhou, Y.; Wu, W.; Deng, Y.; Xiang, Y. A Novel Rose-Like CuS/Bi_2WO_6 Composite for Rhodamine B Degradation. *ChemistrySelect* **2019**, *4*, 11853–11861. [CrossRef]
52. Herrmann, J.-M.; Tahiri, H.; Ait-Ichou, Y.; Lassaletta, G.; González-Elipe, A.R.; Fernández, A. Characterization and photocatalytic activity in aqueous medium of TiO_2 and Ag-TiO_2 coatings on quartz. *Appl. Catal. B Environ.* **1997**, *13*, 219–228. [CrossRef]
53. Low, J.; Qiu, S.; Xu, D.; Jiang, C.; Cheng, B. Direct evidence and enhancement of surface plasmon resonance effect on Ag-loaded TiO_2 nanotube arrays for photocatalytic CO_2 reduction. *Appl. Surf. Sci.* **2018**, *434*, 423–432. [CrossRef]
54. Yan, T.; Tian, J.; Guan, W.; Qiao, Z.; Li, W.; You, J.; Huang, B. Ultra-low loading of Ag_3PO_4 on hierarchical In_2S_3 microspheres to improve the photocatalytic performance: The cocatalytic effect of Ag and Ag_3PO_4. *Appl. Catal. B Environ.* **2017**, *202*, 84–94. [CrossRef]
55. Gao, J.; Ren, X.; Chen, D.; Tang, F.; Ren, J. Bimetallic Ag–Pt hollow nanoparticles: Synthesis and tunable surface plasmon resonance. *Scr. Mater.* **2007**, *57*, 687–690. [CrossRef]

nanomaterials MDPI

Article

Adhesion of *Escherichia Coli* to Nanostructured Surfaces and the Role of Type 1 Fimbriae

Pawel Kallas [1], Håvard J Haugen [1,*], Nikolaj Gadegaard [2], John Stormonth-Darling [2], Mats Hulander [3], Martin Andersson [3] and Håkon Valen [4]

1 Department of Biomaterials, Institute of Clinical Dentistry, University of Oslo, 0455 Oslo, Norway; pawel.kallas@odont.uio.no

2 School of Engineering, University of Glasgow, G12 8QQ Glasgow, UK; Nikolaj.Gadegaard@glasgow.ac.uk (N.G.); John.Stormonth-Darling@glasgow.ac.uk (J.S.D.)

3 Department of Chemistry and Chemical Engineering, Chalmers University of Technology, 412 58 Göteborg, Sweden; mats.hulander@chalmers.se (M.H.); martin.andersson@chalmers.se (M.A.)

4 Nordic Institute of Dental Materials, 0855 Oslo, Norway; hakon.valen@niom.no

* Correspondence: h.j.haugen@odont.uio.no; Tel.: +47-2285-2170

Received: 9 October 2020; Accepted: 10 November 2020; Published: 12 November 2020

Abstract: Bacterial fimbriae are an important virulence factor mediating adhesion to both biotic and abiotic surfaces and facilitating biofilm formation. The expression of type 1 fimbriae of *Escherichia coli* is a key virulence factor for urinary tract infections and catheter-associated urinary tract infections, which represent the most common nosocomial infections. New strategies to reduce adhesion of bacteria to surfaces is therefore warranted. The aim of the present study was to investigate how surfaces with different nanotopography-influenced fimbriae-mediated adhesion. Surfaces with three different nanopattern surface coverages made in polycarbonate were fabricated by injection molding from electron beam lithography nanopatterned templates. The surfaces were constructed with features of approximately 40 nm width and 25 nm height with 100 nm, 250 nm, and 500 nm interspace distance, respectively. The role of fimbriae type 1-mediated adhesion was investigated using the *E. coli wild type BW25113* and Δ*fimA* (with a knockout of major pilus protein FimA) and Δ*fimH* (with a knockout of minor protein FimH) mutants. For the surfaces with nanotopography, all strains adhered least to areas with the largest interpillar distance (500 nm). For the *E. coli wild type*, no difference in adhesion between surfaces without pillars and the largest interpillar distance was observed. For the deletion mutants, increased adhesion was observed for surfaces without pillars compared to surfaces with the largest interpillar distance. The presence of a fully functional type 1 fimbria decreased the bacterial adhesion to the nanopatterned surfaces in comparison to the mutants.

Keywords: nanostructured surface; injection molding; anti-adhesive; *E. coli*; fimbriae; anti-bacterial; biomaterial-associated infections (BAI)

1. Introduction

Bacterial adhesion, colonization, and biofilm formation on medical devices are responsible for a large proportions of nosocomial infections, and biomaterial-associated infections are common post-operative [1–4]. Catheter-associated urinary tract infections (CAUTI) are the most common hospital acquired bacterial infection and are associated with increased mortality and morbidity [5]. The gram negative bacterium *Escherichia coli* (*E. coli*) is the most frequent cause of CAUTIs [6].

E. coli is naturally found in the intestinal flora of humans and animals [7] and includes both non-pathogenic and pathogenic strains [8–10]. Some *E. coli* strains express type 1 fimbria, a bacterial adhesion that has been reported to facilitate adhesion to other bacteria, host cells, and surfaces of medical devices [11]. Type 1 fimbria have a polymeric protein structure, made of FimA, FimF, FimG, and FimH monomers, with a tubular structure of 7 nm in width and 1–2 μm in length [12–14]. Expression of fimbriae has been reported to be an important virulence factor for uropathogenic *E. coli*, important for biofilm formation on abiotic surfaces [15]. *E. coli* expressing type 1 fimbriae was observed to contribute to catheter-associated urinary tract infections in a dynamic catheterized bladder model [16]. The biofilm may act as a natural barrier, protecting bacteria from antimicrobial treatment [17], and biofilms are associated with reduced susceptibility towards antimicrobial treatment [18].

Fimbrial adhesins can also be present in other Gram-negative species like *Serratia marcescens* or *Klebsiella pneumoniae*, where they a play crucial role in biofilm formation [19,20]. Fimbria is an important virulence factor for urinary tract infections [16], where *E. coli* has been reported to be the most prevalent pathogen [21–25], and the presence of fimbriae type 1 enhances the survival of *E. coli* in the urinary tract [26].

Finding new solutions in the fight against bacteria and, therefore, the need to develop medical devices that prevent adhesion of bacteria is highly desired. Controlling bacterial colonization can be done by modifying surface properties of medical devices, such as adhesion properties and coatings, surface roughness, and chemistry [27,28]. Changing surface topography, by altering the surface roughness and its complexity, can affect bacterial adhesion and therefore lower the risk of a potential infection [29–33]. Previous studies have shown that bacterial adhesion can also be influenced by the presence of metallic nanoparticles that show antimicrobial properties, such as such as silver (Ag), gold (Au), and zinc oxide (ZnO) [34]. Another type of nanoparticles that presents antibacterial and biocompatible features is silicon nanoparticles [35–38].

An important obstacle for nanomaterials is the step from scientific idea to a product on the market. Manufacturers need to have high scale, high quality production, and preservation of the desired properties of the nanomaterials [39]. Electron beam lithography (EBL) is a production technique that may enable high production with reproducible quality of materials with surface modifications in the nano-range [40].

The aim of our study was to investigate whether surfaces with different nanoscale topographies affected fimbriae-mediated adhesion of *E. coli*. The hypothesis was that presence of fimbriae has an effect on bacterial attachment to nanostructured surfaces. Electron beam lithography was used to produce controlled nanoscale surface topographies on a template for fabricating injection molded polycarbonate substrates as a model system with varying surface coverages. The higher the interspace distance between nanopillars, the lower the contact area between bacteria and nanopillars. If were to increase that distance, the individual bacteria would only have contact with a few pillars. Therefore, we decided to fabricate surfaces with interspace distances between nanopillars smaller than the actual size of the tested bacteria, namely 100, 250, and 500 nm.

2. Materials and Methods

2.1. Preparation of Nanostructured Surfaces

Nanostructured polymer surfaces were prepared by injection molding of patterns mastered by electron beam lithography on silicon masters. Silicon samples were solvent and oxygen plasma cleaned, dehydrated, and oxygen plasma treated again (this time for adhesion promotion) before immediately being spun on hydrogen silsesquioxane (HSQ) resist (FOx 12) at a thickness corresponding to the desired nanofeature height. Patterns were written in a Vistec VB6 UHR EWF EBL tool (Vistec Electron Beam GmbH, Jena, Germany) operating at 100 kV and developed in 25% tetramethylammonium hydroxide (TMAH) solution at 23 °C followed by rinsing in water and isopropyl alcohol (IPA). As it was not possible to spin HSQ films thinner than ~40 nm, the 20 nm features had to be defined by dry etching with the HSQ pattern, and the mask was subsequently removed by hydrofluoric acid (HF). Dry etching was performed in an STS ICP silicon etch tool using a mixed process and always preceded by a dilute HF dip (2000:1, 48% HF:RO water, by vol.) to prevent micromasking by native oxide. All chemicals were provided by Sigma-Aldrich, Saint-Louis, MO, USA.

Nanopatterns were then replicated using an UV-based nanoimprint lithography (UV-NIL) (EV Group, Sankt Florian am Inn, Austria) into the working stamp material. Silicon masters were first spin coated with the EVG anti-stick material, followed immediately by the working stamp material, which was UV cured against the master onto a carried foil by way of the NIL tool's standard working stamp recipe. UV-NIL was performed using a custom-built pocket imprinter. Each working stamp replica was laser cut to the appropriate dimensions of the injection molding tool, as described in [41].

Injection molding was performed in an Engel Victory 28 hydraulic injection molding machine (Engel Austria GmbH, Schwertberg, Austria) to produce multiple polystyrene samples for biological experiments. Each surface consisted of 9 × 4 repetitions of pattern divided into three sections (total size 1 × 3 mm) with different surface coverages, namely 2.5, 3.5, and 20%. The total number of surface coverages produced was n = 972, although due to production process, some gradients were not included in the image analysis. Therefore, the total number of nanostructured surfaces with different coverages measured was 191.

2.2. *Surface Characterization*

The heights of the pillars were characterized with an atomic force microscope (Veeco Dimension 3100, Santa Barbara, CA, USA). AFM images were processed using Gwyddion software version 2.51 (Free Software Foundation, Boston, MA, USA) to obtain depth data. Water contact angles measurements were performed on the experimental surfaces using a model 100-00-230 NRL contact angle goniometer (Ramé-Hart Inc. Mountain Lakes, NJ, USA) in order to assess surface wettability. A small 5 µL MQ water droplet was applied on the surface. The average contact angle was measured based on seven measurements at 30 s time point.

2.3. *Bacterial Preparation and Growth*

E. coli strains BW25113 7636 (later referred as *E. coli-WT*), JW4277-1 11065 (ΔfimA, *E. coli–ΔfimA*), and JW4283-3 11068 (ΔfimH, *E. coli–ΔfimH*) (The Coli Genetic Stock Center, Yale University, New Haven, CT, USA) were grown in tryptic soy broth (TSB, Sigma-Aldrich, Oslo, Norway) medium overnight at 37 °C in centrifuge tubes and 5% CO_2 atmosphere. *E. coli–ΔfimA* and *E. coli–ΔfimH* are mutant strains that lack the fimA and fimH genes, respectively. Such genes play a role in the production of fimbriae, and their parent is the *E. coli–WT* strain that has type 1 fimbriae [42]. The overnight culture was diluted 10 times the morning after and left to grow again in the same conditions until optical density reached OD_{600} = 1 (Thermo Scientific Spectronic 200E, Waltham, MA, USA). After that, samples were centrifuged at 5000 rpm at 21 °C to obtain a pellet. Medium residues were discarded and exchanged for PBS in order to obtain the same OD_{600} = 1. Afterwards, the pellet, in phosphate buffered saline (PBS, Lonza, Verviers, Belgium), was shaken, and the obtained solution was used later for injection. Average colony-forming units (CFUs) were measured by culturing bacteria overnight on TSB agar plates (overnight at 37 °C and 5% CO_2 atmosphere). There was no observed significant difference in the number of colony forming units (CFU) at OD_{600} = 1 between the strains investigated.

2.4. *Bacterial Adhesion*

After placing the sample in the flow chamber, the system was flushed with distilled water for about 1 min at a constant flow of 20 mL/min to remove any air bubbles trapped in the system. Then, 10 mL of bacteria in PBS was injected in the system manually using a syringe. The valves were then closed, and bacteria were let to adhere under static conditions for 5 min at room temperature. This procedure was followed by manually injecting 10 mL of 0.01% acridine orange (AO) to stain the cells for later viewing with fluorescence microscopy. After 3 min staining, valves were open again and sample was flushed for 5 min with distilled water at the same flow rate as before (20 mL/min). The setup is presented in Figure 1. Each strain was tested three times, and the total number of tested surface coverages was 191.

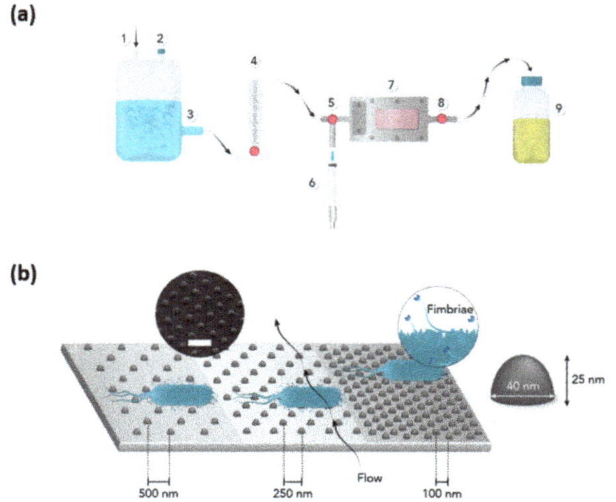

Figure 1. Flow chamber setup (**a**): 1—air supply, 2—air outlet, 3—distilled water tank, 4—flow meter, 5—right valve, 6—syringe input, 7—flow chamber with nanopatterned sample ($n = 191$), 8—left valve, 9—waste container; (**b**) graphical overview of tested surface (not in scale) with pillar height and width. Blue is bacteria. Arrows show direction of the flow. Interspace between pillars: 1–100 nm, 2–250 nm, 3–500 nm. Scanning electron microscope (SEM) image shows nanopattern (30° tilt) with scale bar = 200 nm.

2.5. Fluorescence Microscopy

For the counting of bacteria on the surfaces, the closed flow chamber, including glass cover, was transferred to a fluorescence light microscope (Olympus BX51, Olympus Optical, Tokyo, Japan). Pictures were taken at magnification of 90× by using a 10× magnification objective with U-MNB2 filter (excitation BP 470–490 nm and emission LP 520 nm), and one image was taken for each surface coverage for all replicas ($n = 191$).

2.6. Scanning Electron Microscopy

Samples were stored overnight in 2.5% glutaraldehyde buffered with 0.1 M Sørensen phosphate buffer, and afterwards were flushed with ethanol and PBS [43]. A field emission scanning electron microscope (SEM) was used to characterize the surfaces (Hitachi S-4800, Toyko, Japan). A thin layer of platinum (3 nm) was sputtered onto surfaces (Cressington 308R Coating System, Watford, UK), prior to SEM imaging. Pictures were taken at a magnification of 15k with the working distance (WD) set at 1.5 mm and acceleration voltage of 2.0 kV. The tilted image showing the nanopattern was taken at a magnification of 20k with the working distance set at 13.3 mm, acceleration voltage of 10 kV, and tilt set at 30°.

2.7. ImageJ Analysis

Image analysis was performed using ImageJ software version 1.53a (NIH, Bethesda, MD, USA). To calculate the coverage of the nanopattern, each picture was set to 8-bit, and the level of threshold was set to obtain a visible contrast between nanopattern and surface, which were later measured using the "Analyze particles" plugin. To calculate the number of bacteria, we used an already established macro plugin [44], which allowed us to crop the original image in order to avoid artefacts from vignetting during the automated counting. The number of bacteria was then counted using the "find maxima" feature in the program, using a noise threshold of 12.

2.8. Statistical Analysis

Statistical analysis was performed using GraphPad Prism version 6.07 (GraphPad software, La Jolla, CA, USA). The effect of the different surface coverages for adhesion of the individual bacteria was analyzed using one-way ANOVA with repeated measurements followed by the Tukey's multiple comparisons test. For analysis of the main effect of bacteria and surface coverage, a two-way ANOVA with repeated measurements (two factors in the experiment: surface coverage and bacteria strain) was performed followed by the Tukey's multiple comparisons test for the simple effects within each column, and each bacterial strain compared to each other for each surface coverage was performed. For a subset of surfaces ($n = 10$) Students T-test was used to analyze differences in adhesion between surfaces without pillars compared to surfaces with the largest interpillar distance. Results were considered significant with a p-value ≤ 0.05. SPSS, version 25 (IBM Corp., New York, NY, USA), was used for calculating marginal means of bacteria number for all tested strains with significance at p-value ≤ 0.05. Total number of analyzed surface coverages was $n = 191$. Power analysis was conducted using G*Power version 3.1 (The G*Power Team, Düsseldorf, Germany). The effect size and total sample size were calculated with an alpha probability of 0.05 and power of 0.95 [45,46].

3. Results

3.1. Surface Characteristics

Surfaces with injection molded 40 nm diameter features in polycarbonate were fabricated. Each pattern was divided into three sections with different surface coverages. The AFM measurements provided the interspace (100, 250, and 500 nm), pillar diameter of 40 nm, and pillar height of 25 ± 5 nm (Figure 2).

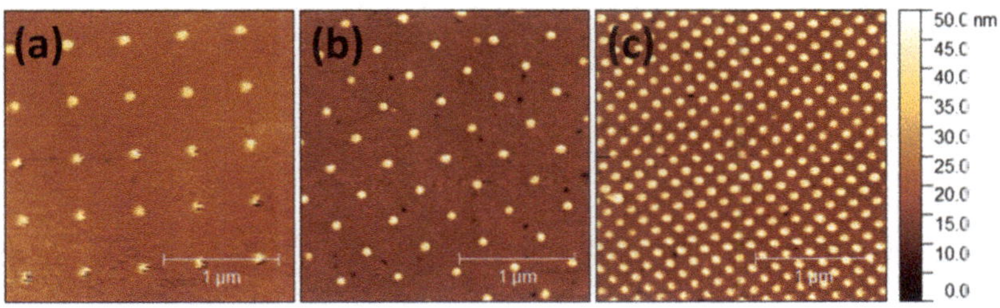

SURFACE COVERAGE

Surface	Coverage (%)	Interspace (nm)	Diameter (nm)	Height (nm)	Contact angle (°)
(**a**) – Low Coverage	2.5	500			
(**b**) – Medium Coverage	3.5	250	40	25 ± 5	67 ± 4
(**c**) – High Coverage	20	100			

Figure 2. AFM images of surfaces with scale presented height (nm): (**a**) low coverage (LC), (**b**) medium coverage (MC), (**c**) high coverage (HC), and table below showing the characterization of the nanopillars with interspace, diameter, and height ($n = 5$).

Wettability analysis along the surface showed that the characterized contact angle was smaller than 90° (after 30 s of measurement, an average contact angle was 67°), which means that surfaces used in our study were hydrophilic. Due to the small size of the nanopatterns (1 × 3 mm), we were limited to individual measurements of each nanopattern and present here only the overall surface contact angle.

3.2. Bacterial Adhesion

Fluorescent microscopy images of adhered bacteria were used to quantify bacterial adhesion to the surfaces. An example of low coverage (Figure 3a), medium coverage (Figure 3b), and high coverage (Figure 3c) under fluorescence microscope is presented. Based on the power analysis, the calculated effect size was 0.29. Therefore, the calculated sample size was 186. In our study we used 191 surface coverages, which is higher than the calculated one. For each of the *E. coli–WT*, *E. coli–ΔfimA*, and *E. coli–ΔfimH*, increased adhesion with increasing surface coverage of nanopatterns was found (Figure 4a). The two-way ANOVA showed a significant main effect for both surface area coverage and the presence of type 1 fimbriae or not on adhesion. There was a significant increased adhesion for the mutants not expressing type 1 fimbriae compared to the wild type for each of the different surface coverages (low coverage, medium coverage, high coverage, Figure 4b). In order to compare an overall bacteria adhesion to different nanopattern surface coverages, the marginal means for all tested bacteria were calculated (Figure 4a). A graphical presentation of bacterial behavior on tested surface coverages is presented on Figure 4c.

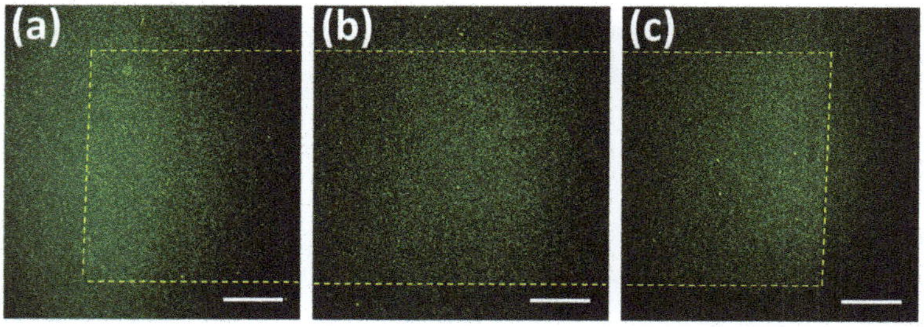

Figure 3. Fluorescent images of *E. coli-WT* after 5 min adhesion to different surface coverages at 10× magnification: (**a**) Low Coverage, (**b**) Medium Coverage, (**c**) High Coverage. The yellow dash marks the edges of the nanopattern. Green is bacteria. Scale bars = 250 μm.

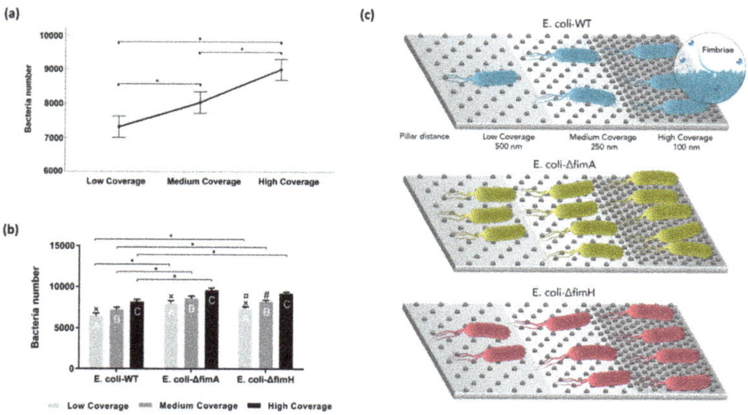

Figure 4. Bacteria adhesion to different surface coverages: (**a**) Bacteria number (calculated marginal means) for all tested strains. Significant at * p-value < 0.05. Median values with error bars represent the 95% confidence interval ($n = 191$, * $p < 0.05$); (**b**) average bacteria number to different surface coverages with the standard error of the mean. Comparison within tested strains (significant at p-value < 0.05, $n = 191$): × low coverage *vs.* medium coverage, # low coverage *vs.* high coverage, ¤ medium coverage *vs.* high coverage; (**c**) graphical presentation of tested strains attaching to different surface coverages.

Bacteria were observed to adhere to the nanopattern surfaces without losing their physiological morphology, keeping the characteristic rod shapes (Figure 5). No damaged bacteria cells were spotted during our study. The number of nanopillars in contact with one bacterium for the different surface coverages was calculated. Depending on the position of a single bacterium, the number of contact points for low coverage surfaces were 1–4 pillars, for medium coverage surfaces 8–10 pillars, and for high coverage surfaces approximately 45 pillars.

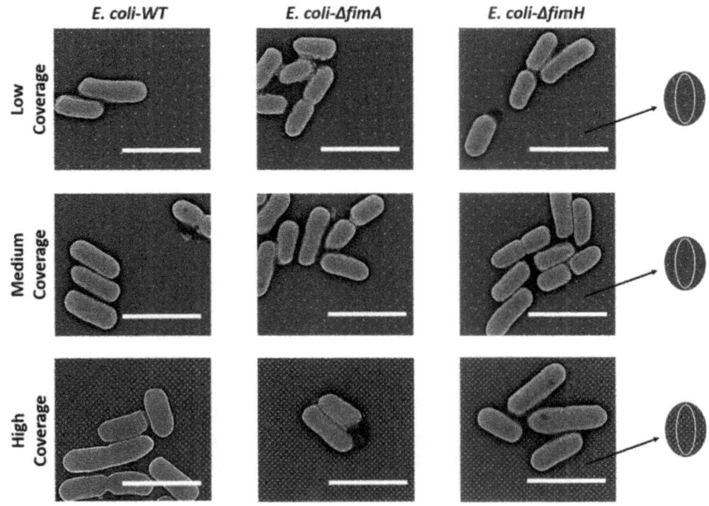

Figure 5. SEM images of tested bacteria. Fimbriae were not visible in all cases. Black arrows point at the theoretical position of bacteria (white oval) on the nanopattern. Scale bars = 2 μm.

The effect of fimbriae on adhesion was observed by calculating and comparing marginal means (Figure 6a), where higher attachment was observed for *E. coli* without fimbriae. To analyze the effect of the presence of pillars, adhesion to areas without pillars, that is, smooth surfaces, was compared to areas with low coverage. Increased adhesion for the *fimA* and fim*H* mutants to the low coverage surfaces was observed compared to areas without pillars, whereas no difference was observed for the *E. coli–WT* (Figure 6b).

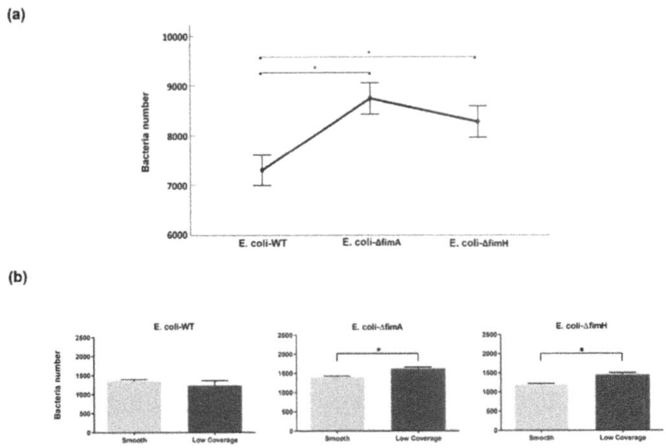

Figure 6. Effect of fimbriae on adhesion for all tested *E. coli* strains. (**a**) Bacteria number (calculated marginal means) significant at * *p*-value < 0.05. Median values with error bars represent the 95% confidence interval (n = 191, * $p < 0.05$). (**b**) Comparison within tested strains between smooth and low coverage. Average bacteria number to different surface coverages with the standard error of the mean. Significant at * *p*-value < 0.05, n = 10.

4. Discussion

Our work focused on investigating how nanopatterned surfaces affected the adhesion of *E. coli* and whether presence of fimbria would affect adhesion. Using a replication-based fabrication process, which combines an electron beam lithography and injection molding, we were able to create and examine a large number of surfaces with identical patterns and features. This process, as described in the previous work of Stormonth-Darling et al. [47], realizes the possibility of manufacturing almost any desired pattern and is a method considered to be a low cost and high speed alternative to other fabricating methods, such as poly dimethyl siloxane (PDMS) casting [48]. The electron beam lithography combined with injection molding method allows one to fabricate a large number of identical surfaces in a short time and at low cost. In comparison to other studies, often performed using few samples (typically with sample size of three) to investigate the different nano-patterned features [49–51], we used a larger sample size than the optimal sample size (n = 186, calculated by power analysis). When studying nano-patterned feature surface effects on bacteria, it is therefore important to always have the appropriate sample size. Unfortunately, when studying bacteria, optimal sample size is larger and it may be difficult to produce nano-patterned feature surfaces in large enough quantities.

The findings in the present study show that all strains investigated were affected by the presence of nanopillars on the surface. Although type 1 fimbriae is an important virulence factor for medical device infections, and expression is associated with increased biofilm formation and catheter-associated urinary tract infections [15,17,18,26,52,53], we observed reduced adhesion of the wild type *E. coli* compared to deletion mutants for the type 1 fimbriae monomers fimA or fimH for surfaces with nanopillars.

Our experimental findings showed an overall hydrophilic nature of the tested surfaces. Due to the size of the surfaces (1 × 3 mm), separate surface angle measurements for the different surfaces coverages were not performed. The hydrophilicity of surfaces has been described to affect bacterial adhesion. Otto et al. [54] used microscope slides in order to obtain hydrophilic (contact angle determined as <10°) and hydrophobic (contact angle determined as >85°) surfaces. They found that the expression of type 1 fimbriae by *E. coli* reduced initial adhesion compared to mutants not expressing fimbriae, in accordance with the present findings. In addition, the authors reported increased adhesion strength for *E. coli* expressing type 1 fimbriae to hydrophobic surfaces compared to hydrophilic surfaces, whereas no difference was observed for the *E. coli* not expressing type 1 fimbriae.

Several studies have shown that adhesion of *E. coli* is affected by changes in surface topography. Widyaratih et al. [55] used electron beam induced deposition (EBID) in order to create silicon surfaces with four different nanopattern types. Their surfaces resembled osteogenic nanopatterns, and the adhesion of the *E. coli* K-12 strain was investigated. Topographies with tall nanopillars (over 100 nm) showed bactericidal effects and changes in the physiological morphology of cells. Among the changes were structure deformations and lack of ability to divide. For surfaces with nanopillars with lower heights (15–20 nm), comparable to the present study, a reduced adhesion was suggested compared to non-patterned surfaces, and no changes in morphology were observed (bacteria kept their rod shape). Another study by Bandara et al. [56] showed how natural nanotopography of dragonfly wing would affect adhesion of *E. coli* (NCTC 10418). Dragonfly wings are covered with nanopillars consisting of two populations with different heights, namely 189 ± 67 nm and 311 ± 52 nm, respectively. The presence of nanopillars led to damage of bacterial membrane as a result of strong adhesion between nanopillars and extracellular polymeric substances (EPS) from bacteria, as well as shear force which appeared during bacterial attempts to move on the nanopillars. In comparison, the nanopillars were approximately 4.7 and 7.5 times higher than the nanopillars in the present study. However, the height of the nanopillars is not the only factor affecting bacterial adhesion and survival on surfaces. Hasan et al. [57] fabricated surfaces inspired by the dragonfly wings based on deep reactive ion etching of a silicon wafer. The surfaces were highly hydrophobic with nanopillars of 4 nm height and 220 nm in diameter that exerted bactericidal activity against both *E. coli* and *Staphylococcus aureus*.

In addition to the morphology of the nanopillars on the surfaces, the density of the nanopillars has been shown to affect both adhesive and bactericidal properties of surfaces. Dickson et al. [58] showed that by increasing the density of nanopillars on the surfaces of poly (methyl methacrylate) (PMMA) films, an increase in the bactericidal activity and decreased adhesion of *E. coli* was observed.

The present study investigated surfaces with interspace distances varying between 100 and 500 nm. The higher the distance, the lower number of pillars that bacteria may adhere to. Depending on the cell size, the highest interspace distance would allow a single *E. coli* cell to establish contact with only one to four pillars. In contrast to the study by Dickson et al. [58], we observed increased initial adhesion to the surfaces with higher nanopillar density. This observation may be explained by, for example, a different contact time used to investigate adhesion and difference in chemistry of the surfaces

5. Conclusions

We have presented a method of producing a large number of identical nanostructured surfaces. *E. coli* wild type and deletion mutants of the type 1 fimbriae monomers FimA or FimH were tested for their ability to adhere to such surfaces with different interpillar distance of 100, 250, and 500 nm, respectively. An increase in interpillar distance (reduced surface coverage) was associated with reduced adhesion of *E. coli* wild type and the deletion mutants. In addition, the presence of a functional type 1 fimbria decreased adhesion to the nanopatterned surfaces in comparison to deletion mutants, leading us to the conclusion that there is a relationship between presence of a functional fimbriae and adhesion towards tested surfaces. The hypothesis that the presence of fimbriae had an effect on bacterial attachment to nanostructured surfaces was verified. These current results could provide insight into development of new nano-patterned structures with anti-adhesion bacterial properties.

Author Contributions: Conceptualization, H.J.H., H.V., and N.G. methodology, M.H., H.J.H, H.V., and N.G.; formal analysis, data curation investigation P.K. and J.S.-D.; writing—original draft preparation, PK and H.V.; writing—review and editing, M.A., M.H., H.J.H., H.V., and N.G.; supervision, M.A., P.K. and H.V.; project administration, P.K. and H.V. All authors have read and agreed to the published version of the manuscript

Funding: This research received no external funding

Acknowledgments: Antje Hofgaard at the Department of Biosciences, UiO, is acknowledged for her help and expertise in SEM measurements.

Conflicts of Interest: The authors declare no conflict of interest

References

1. Khan, H.A.; Baig, F.K.; Mehboob, R. Nosocomial infections: Epidemiology, prevention, control and surveillance. *Asian. Pac. J. Trop. Bio.* **2017**, *7*, 478–482. [CrossRef]
2. Inweregbu, K.; Dave, J.; Pittard, A. Nosocomial infections. *Cont. Care Pain* **2005**, *5*, 14–17. [CrossRef]
3. Darouiche, R.O. Device-associated infections: A macroproblem that starts with microadherence. *Clin. Infect. Dis.* **2001**, *33*, 1567–1572. [CrossRef] [PubMed]
4. Afle, F.C.D.; Agbankpe, A.J.; Johnson, R.C.; Houngbegnon, O.; Houssou, S.C.; Bankole, H.S. Healthcare-associated infections: Bacteriological characterization of the hospital surfaces in the University Hospital of Abomey-Calavi/so-ava in South Benin (West Africa). *BMC. Infect. Dis.* **2019**, *19*, 28. [CrossRef] [PubMed]
5. Öztürk, R.; Murt, A. Epidemiology of urological infections: A global burden. *World J. of Uro.* **2020**. [CrossRef]
6. Friedman, C.; Newsom, W. *IFIC Basic Concepts of Infection Control*; International Federation of Infection Control: Portadown, UK, 2011; p. 259.
7. Campbell, N.A.; Reece, J.B. *Biology: International Edition*, 6th ed.; Pearson Education, Inc.: London, UK, 2002; pp. 595–615, 6th ed.
8. Salyers, A.A.; Whitt, D.D.; Whitt, D.D. *Bacterial Pathogenesis: A Molecular Approach*, 1st ed.; ASM Press: Washington, DC, USA, 1994; p. 190, 1st ed.
9. Stecher, B.; Hardt, W.D. The role of microbiota in infectious disease. *Trends Microbiol.* **2008**, *16*, 107–114. [CrossRef] [PubMed]
10. Johnson, J.R.; Moseley, S.L.; Roberts, P.L.; Stamm, W.E. Aerobactin and other virulence factor genes among strains of Escherichia coli causing urosepsis: Association with patient characteristics. *Infect. Immun.* **1988**, *56*, 405–412. [CrossRef] [PubMed]

11. Martinez, J.J.; Mulvey, M.A.; Schilling, J.D.; Pinkner, J.S.; Hultgren, S.J. Type 1 pilus-mediated bacterial invasion of bladder epithelial cells. *EMBO. J.* **2000**, *19*, 2803–2812. [CrossRef]
12. Klemm, P.; Krogfelt, K.A.; Hedegaard, L.; Christiansen, G. The major subunit of Escherichia coli type 1 fimbriae is not required for D-mannose-specific adhesion. *Mol. Microbiol.* **1990**, *4*, 553–559. [CrossRef]
13. Russell, P.W.; Orndorff, P.E. Lesions in two Escherichia coli type 1 pilus genes alter pilus number and length without affecting receptor binding. *J. Bacteriol.* **1992**, *174*, 5923–5935. [CrossRef]
14. Jones, C.H.; Pinkner, J.S.; Roth, R.; Heuser, J.; Nicholes, A.V.; Abraham, S.N.; Hultgren, S.J. FimH adhesin of type 1 pili is assembled into a fibrillar tip structure in the Enterobacteriaceae. *Proc. Natl. Acad. Sci. USA* **1995**, *92*, 2081–2085. [CrossRef] [PubMed]
15. Blumer, C.; Kleefeld, A.; Lehnen, D.; Heintz, M.; Dobrindt, U.; Nagy, G.; Michaelis, K.; Emödy, L.; Polen, T.; Rachel, R.; et al. Regulation of type 1 fimbriae synthesis and biofilm formation by the transcriptional regulator LrhA of Escherichia coli. *Microbiology (Reading)* **2005**, *151*, 3287–3298. [CrossRef] [PubMed]
16. Reisner, A.; Maierl, M.; Jorger, M.; Krause, R.; Berger, D.; Haid, A.; Tesic, D.; Zechner, E.L. Type 1 fimbriae contribute to catheter-associated urinary tract infections caused by Escherichia coli. *J. Bacteriol.* **2014**, *196*, 931–939. [CrossRef] [PubMed]
17. Donlan, R.M. Biofilms and device-associated infections. *Emerg. Infect. Dis.* **2001**, *7*, 277–281. [CrossRef] [PubMed]
18. Valen, H.; Scheie, A.A. Biofilms and their properties. *Eur. J. Oral Sci.* **2018**, *126*, 13–18. [CrossRef]
19. Kalivoda, E.J.; Stella, N.A.; O'Dee, D.M.; Nau, G.J.; Shanks, R.M. The cyclic AMP-dependent catabolite repression system of Serratia marcescens mediates biofilm formation through regulation of type 1 fimbriae. *Appl. Environ. Microbiol.* **2008**, *74*, 3461–3470. [CrossRef]
20. Stahlhut, S.G.; Struve, C.; Krogfelt, K.A.; Reisner, A. Biofilm formation of Klebsiella pneumoniae on urethral catheters requires either type 1 or type 3 fimbriae. *FEMS Immunol. Med. Microbiol.* **2012**, *65*, 350–359. [CrossRef]
21. Platt, R.; Polk, B.F.; Murdock, B.; Rosner, B. Mortality associated with nosocomial urinary-tract infection. *N. Engl. J. Med.* **1982**, *307*, 637–642. [CrossRef]
22. Lee, D.S.; Lee, S.J.; Choe, H.S. Community-Acquired Urinary Tract Infection by Escherichia coli in the Era of Antibiotic Resistance. *Biomed. Res. Int.* **2018**, *2018*, 7656752. [CrossRef]
23. Alanazi, M.Q.; Alqahtani, F.Y.; Aleanizy, F.S. An evaluation of E. coli in urinary tract infection in emergency department at KAMC in Riyadh, Saudi Arabia: Retrospective study. *Ann. Clin. Microbiol. Antimicrob.* **2018**, *17*, 3. [CrossRef]
24. Garofalo, C.K.; Hooton, T.M.; Martin, S.M.; Stamm, W.E.; Palermo, J.J.; Gordon, J.I.; Hultgren, S.J. Escherichia coli from urine of female patients with urinary tract infections is competent for intracellular bacterial community formation. *Infect Immun.* **2007**, *75*, 52–60. [CrossRef] [PubMed]
25. Marrs, C.F.; Zhang, L.; Foxman, B. Escherichia coli mediated urinary tract infections: Are there distinct uropathogenic E. coli (UPEC) pathotypes? *FEMS Microbiol. Lett.* **2005**, *252*, 183–190. [CrossRef] [PubMed]
26. Connell, I.; Agace, W.; Klemm, P.; Schembri, M.; Marild, S.; Svanborg, C. Type 1 fimbrial expression enhances Escherichia coli virulence for the urinary tract. *Proc. Natl. Acad. Sci. USA* **1996**, *93*, 9827–9832. [CrossRef] [PubMed]
27. Mozetic, M. Surface Modification to Improve Properties of Materials. *Materials (Basel)* **2019**, *12*, 441. [CrossRef]
28. Rahmati, M.; Silva, E.A.; Reseland, J.E.; C, A.H.; Haugen, H.J. Biological responses to physicochemical properties of biomaterial surface. *Chem. Soc. Rev.* **2020**, *49*, 5178–5224. [CrossRef]
29. Tripathy, A.; Sen, P.; Su, B.; Briscoe, W.H. Natural and bioinspired nanostructured bactericidal surfaces. *Adv. Colloid Interface Sci.* **2017**, *248*, 85–104. [CrossRef]
30. Øilo, M.; Bakken, V. Biofilm and Dental Biomaterials. *Materials* **2015**, *8*, 2887–2900. [CrossRef]
31. Wang, L.; Hu, C.; Shao, L. The antimicrobial activity of nanoparticles: Present situation and prospects for the future. *Int. J. Nanomed.* **2017**, *12*, 1227–1249. [CrossRef]
32. Xing, R.; Lyngstadaas, S.P.; Ellingsen, J.E.; Taxt-Lamolle, S.; Haugen, H.J. The influence of surface nanoroughness, texture and chemistry of TiZr implant abutment on oral biofilm accumulation. *Clin. Oral Implants Res.* **2015**, *26*, 649–656. [CrossRef]
33. Haugen, H.; Lyngstadaas, S. Antibacterial effects of titanium dioxide in wounds. In *Wound Healing Biomaterials-Volume 2: Functional Biomaterials*; Ågren, M.S., Ed.; Elsevier: Amsterdam, The Netherlands, 2016; p. 439.

34. Dhull, N.; Nidhi; Gupta, V.; Tomar, M. Antimicrobial properties of metallic nanoparticles: A qualitative analysis. *Mater. Today: Proc.* **2019**, *17*, 155–160. [CrossRef]
35. Smirnov, N.A.; Kudryashov, S.I.; Nastulyavichus, A.A.; Rudenko, A.A.; Saraeva, I.N.; Tolordava, E.R.; Gonchukov, S.A.; Romanova, Y.M.; Ionin, A.A.; Zayarny, D.A. Antibacterial properties of silicon nanoparticles. *Laser Phys. Lett.* **2018**, *15*, 105602. [CrossRef]
36. Asefa, T.; Tao, Z. Biocompatibility of mesoporous silica nanoparticles. *Chem. Res. Toxicol.* **2012**, *25*, 2265–2284. [CrossRef] [PubMed]
37. Sviridov, A.P.; Osminkina, L.A.; Kharin, A.Y.; Gongalsky, M.B.; Kargina, J.V.; Kudryavtsev, A.A.; Bezsudnova, Y.I.; Perova, T.S.; Geloen, A.; Lysenko, V.; et al. Cytotoxicity control of silicon nanoparticles by biopolymer coating and ultrasound irradiation for cancer theranostic applications. *Nanotechnology* **2017**, *28*, 105102. [CrossRef] [PubMed]
38. Tamarov, K.P.; Osminkina, L.A.; Zinovyev, S.V.; Maximova, K.A.; Kargina, J.V.; Gongalsky, M.B.; Ryabchikov, Y.; Al-Kattan, A.; Sviridov, A.P.; Sentis, M.; et al. Radio frequency radiation-induced hyperthermia using Si nanoparticle-based sensitizers for mild cancer therapy. *Sci. Rep.* **2014**, *4*, 7034. [CrossRef]
39. Charitidis, C.A.; Georgiou, P.; Koklioti, M.A.; Trompeta, A.-F.; Markakis, V. Manufacturing nanomaterials: From research to industry. *Manufacturing Rev.* **2014**, *1*, 11. [CrossRef]
40. Altissimo, M. E-beam lithography for micro-nanofabrication. *Biomicrofluidics* **2010**, *4*, 026503. [CrossRef]
41. Stormonth-Darling, J.M.; Gadegaard, N. Injection Moulding Difficult Nanopatterns with Hybrid Polymer Inlays. *Macromol. Mater. Eng.* **2012**, *297*, 1075–1080. [CrossRef]
42. Chao, Y.; Zhang, T. Probing Roles of Lipopolysaccharide, Type 1 Fimbria, and Colanic Acid in the Attachment of Escherichia coli Strains on Inert Surfaces. *Langmuir* **2011**, *27*, 11545–11553. [CrossRef]
43. Lönn-Stensrud, J.; Benneche, T.; Scheie, A.A. Furanones and Thiophenones in Control of Staphylococcus epidermidis Biofilm Infections? In *Science and Technology Against Microbial Pathogens*; World Scientific, Toh Tunk Link: Singapore, 2011; pp. 155–159. [CrossRef]
44. Hulander, M.; Valen-Rukke, H.; Sundell, G.; Andersson, M. Influence of Fibrinogen on Staphylococcus epidermidis Adhesion Can Be Reversed by Tuning Surface Nanotopography. *ACS Biomater. Sci. Eng.* **2019**, *5*, 4323–4330. [CrossRef]
45. Faul, F.; Erdfelder, E.; Lang, A.G.; Buchner, A. G*Power 3: A flexible statistical power analysis program for the social, behavioral, and biomedical sciences. *Behav. Res. Methods* **2007**, *39*, 175–191. [CrossRef]
46. Faul, F.; Erdfelder, E.; Buchner, A.; Lang, A.G. Statistical power analyses using G*Power 3.1: Tests for correlation and regression analyses. *Behav. Res. Methods* **2009**, *41*, 1149–1160. [CrossRef] [PubMed]
47. Stormonth-Darling, J.M.; Pedersen, R.H.; How, C.; Gadegaard, N. Injection moulding of ultra high aspect ratio nanostructures using coated polymer tooling. *J. Micromech. Microeng.* **2014**, *24*, 12. [CrossRef]
48. Darling, J.M.; Saeed, A.; Reynolds, P.M.; Gadegaard, N. Injection Molding Micro- and Nanostructures in Thermoplastic Elastomers. *Macromol. Mater. Eng.* **2016**, *301*, 964–971. [CrossRef] [PubMed]
49. Yu, Q.; Cho, J.; Shivapooja, P.; Ista, L.K.; Lopez, G.P. Nanopatterned smart polymer surfaces for controlled attachment, killing, and release of bacteria. *ACS Appl. Mater. Interfaces* **2013**, *5*, 9295–9304. [CrossRef] [PubMed]
50. Liu, L.T.; Ercan, B.; Sun, L.L.; Ziemer, K.S.; Webster, T.J. Understanding the Role of Polymer Surface Nanoscale Topography on Inhibiting Bacteria Adhesion and Growth. *ACS Biomater Sci. Eng.* **2016**, *2*, 122–130. [CrossRef]
51. Linklater, D.P.; Baulin, V.A.; Juodkazis, S.; Crawford, R.J.; Stoodley, P.; Ivanova, E.P. Mechano-bactericidal actions of nanostructured surfaces. *Nat. Rev. Microbiol* **2020**, 1–15, Epub ahead of print. [CrossRef] [PubMed]
52. Melican, K.; Sandoval, R.M.; Kader, A.; Josefsson, L.; Tanner, G.A.; Molitoris, B.A.; Richter-Dahlfors, A. Uropathogenic Escherichia coli P and Type 1 fimbriae act in synergy in a living host to facilitate renal colonization leading to nephron obstruction. *PLoS. Pathog.* **2011**, *7*, e1001298. [CrossRef]
53. Klemm, P. Fimbrial adhesions of Escherichia coli. *Rev. Infect. Dis.* **1985**, *7*, 321–340. [CrossRef]
54. Otto, K.; Elwing, H.; Hermansson, M. The role of type 1 fimbriae in adhesion of Escherichia coli to hydrophilic and hydrophobic surfaces. *Colloids Surf. B-Biointerfaces* **1999**, *15*, 99–111. [CrossRef]
55. Widyaratih, D.S.; Hagedoorn, P.L.; Otten, L.G.; Ganjian, M.; Tumer, N.; Apachitei, I.; Hagen, C.W.; Fratila-Apachitei, L.E.; Zadpoor, A.A. Towards osteogenic and bactericidal nanopatterns? *Nanotechnology* **2019**, *30*, 20LT01. [CrossRef]

56. Bandara, C.D.; Singh, S.; Afara, I.O.; Wolff, A.; Tesfamichael, T.; Ostrikov, K.; Oloyede, A. Bactericidal Effects of Natural Nanotopography of Dragonfly Wing on Escherichia coli. *ACS Appl. Mater. Interfaces* **2017**, *9*, 6746–6760. [CrossRef] [PubMed]
57. Hasan, J.; Raj, S.; Yadav, L.; Chatterjee, K. Engineering a nanostructured "super surface" with superhydrophobic and superkilling properties. *RSC Adv.* **2015**, *5*, 44953–44959. [CrossRef] [PubMed]
58. Dickson, M.N.; Liang, E.I.; Rodriguez, L.A.; Vollereaux, N.; Yee, A.F. Nanopatterned polymer surfaces with bactericidal properties. *Biointerphases* **2015**, *10*, 021010. [CrossRef] [PubMed]

Publisher's Note: MDPI stays neutral with regard to jurisdictional claims in published maps and institutional affiliations.

nanomaterials

MDPI

Article

Manganese Ferrite Nanoparticles ($MnFe_2O_4$): Size Dependence for Hyperthermia and Negative/Positive Contrast Enhancement in MRI

Khairul Islam [1,2], Manjurul Haque [2], Arup Kumar [1], Amitra Hoq [1,3], Fahmeed Hyder [3] and Sheikh Manjura Hoque [1,*]

[1] Materials Science Division, Atomic Energy Centre, Dhaka 1000, Bangladesh; khairul.aeceiu@gmail.com (K.I.); arup.mb@gmail.com (A.K.); anhoq120@gmail.com (A.H.)

[2] Department of Electrical and Electronic Engineering, Islamic University, Kushtia 7003, Bangladesh; manju_iu@yahoo.com

[3] Department of Radiology & Biomedical Imaging, Magnetic Resonance Research Center (MRRC), Yale University, New Haven, CT 06519, USA; fahmeed.hyder@yale.edu

* Correspondence: manjura_hoque@baec.gov.bd; Tel.: +88-02-9611247

Received: 15 October 2020; Accepted: 6 November 2020; Published: 20 November 2020

Abstract: We synthesized manganese ferrite ($MnFe_2O_4$) nanoparticles of different sizes by varying pH during chemical co-precipitation procedure and modified their surfaces with polysaccharide chitosan (CS) to investigate characteristics of hyperthermia and magnetic resonance imaging (MRI). Structural features were analyzed by X-ray diffraction (XRD), high-resolution transmission electron microscopy (TEM), selected area diffraction (SAED) patterns, and Mössbauer spectroscopy to confirm the formation of superparamagnetic $MnFe_2O_4$ nanoparticles with a size range of 5–15 nm for pH of 9–12. The hydrodynamic sizes of nanoparticles were less than 250 nm with a polydispersity index of 0.3, whereas the zeta potentials were higher than 30 mV to ensure electrostatic repulsion for stable colloidal suspension. MRI properties at 7T demonstrated that transverse relaxation (T_2) doubled as the size of CS-coated $MnFe_2O_4$ nanoparticles tripled in vitro. However, longitudinal relaxation (T_1) was strongest for the smallest CS-coated $MnFe_2O_4$ nanoparticles, as revealed by in vivo positive contrast MRI angiography. Cytotoxicity assay on HeLa cells showed CS-coated $MnFe_2O_4$ nanoparticles is viable regardless of ambient pH, whereas hyperthermia studies revealed that both the maximum temperature and specific loss power obtained by alternating magnetic field exposure depended on nanoparticle size and concentration. Overall, these results reveal the exciting potential of CS-coated $MnFe_2O_4$ nanoparticles in MRI and hyperthermia studies for biomedical research.

Keywords: manganese ferrite; X-ray diffraction; nanomaterials; specific loss power; magnetic resonance angiography

1. Introduction

The applications of nanomaterials in the biomedical field allows solving many issues such as targeted drug delivery [1,2], contrast-enhancing dye in magnetic resonance imaging (MRI) [3–8], mediators for hyperthermia applications [9–14], cell labeling and tracking [15], angiography with MRI [16–18], cellular transfection using magnetic fields [19], cerebral blood volume (CBV) experiments of functional MRI (fMRI) [20], drug distribution in the brain [21], and antimicrobial activity agent [22]. Surface functionalized/modified spinel ferrite nanoparticles such as $MnFe_2O_4$, $MgFe_2O_4$, $CoFe_2O_4$, $ZnFe_2O_4$, Fe_3O_4 are excellent mediators for cancer thermotherapy and MRI contrast agents [23–27]. These nanoparticles are biocompatible, biodegradable, possess high transition temperatures, and have excellent chemical stability. Moreover, nanomagnetism of ferrite nanoparticles provides the opportunity for several biomedical applications because these possess higher magnetic susceptibility than normal superparamagnetic materials and negligible coercivity (i.e., field needed to demagnetize) and retentivity (i.e., residual magnetism after field removal).

Properties of ferrites depend on their composition and microstructure, which in turn depend on their synthesis processes. There are various chemical and physical methods [28–41] to synthesize ferrite nanoparticles, such as chemical co-precipitation, sol-gel auto combustion (i.e., combustion of solution of metal salts and organic fuel forms a gel), reverse micelle, microwave hydrothermal, sonochemical, forced hydrolysis, one-step, high energy ball milling, solvothermal, and microemulsion method. Chemical co-precipitation has several advantages over others, such as (i) uniform and homogeneous nanoparticles of semi-spherical sizes, (ii) control of particle size by varying the reaction parameters such as reaction temperature and pH of the solution, (iii) composition flexibility, (iv) facile, and (v) large scale preparation technique. The co-precipitation method involves the simultaneous occurrence of nucleation at several locations and inhibits the growth mechanism. Since this process requires less heat, only about 80 °C for ferritization reaction, particle sizes are relatively smaller than any other synthesis method.

$MnFe_2O_4$ nanoparticles are of great interest for their remarkable inherent biocompatibility because of the presence of Mn^{2+} ions, tunable magnetic properties, higher transition temperature, and excellent chemical stability for room temperature applications. There are recent reports on $MnFe_2O_4$ nanoparticles as one novel agent for magnetic hyperthermia and MRI contrast [42–45]. Manganese based ferrites have added advantages than other cations because manganese can be consumed in the amount of 0.67–4.99 mg with a mean value of 2.21 mg per day [46].

Magnetic particle hyperthermia is a very efficient technique for localized destruction of cancer cells, targeted drug delivery, and synergistic use of hyperthermia with chemotherapy and radiotherapy [47]. Specific loss power (SLP), i.e., the ease with which nanoparticles heat the surrounding media, depends on Néel and Brownian relaxation mechanisms and the hysteresis loss [48]. Recent studies suggest that the contributions of hysteresis losses related to the areas of the hysteresis loops are vital for SLP. The SLP also depends strongly on the magnetization and anisotropy of the nanomaterials. In the nanoparticulate systems, where the conventional size law breaks down, magnetization and Curie temperatures are no more intrinsic properties of materials; rather, they depend strongly on the nanoparticle size and surface functionalization/modification. $MnFe_2O_4$ possessing divalent Mn^{2+} cations have five unpaired electrons in the d-orbitals resulting in a magnetic moment of 5. Since Fe^{3+} ions are antiparallel in the tetrahedral and octahedral sites, the uncompensated magnetic moments of Mn^{2+} provide reasonable magnetic moments for higher SLP, which can further be tuned by modifying nanoparticle size. Further, nanoparticle size change also incurs shape anisotropy that has a strong influence on SLP [11,48,49]. In the monodomain range, an increase of nanoparticle size would reduce anisotropy since anisotropy is inversely proportional to the nanoparticle size. Limited value of anisotropy enhances SLP, beyond which it is detrimental because higher anisotropy inhibits Néel and Brownian relaxation mechanisms.

Ferrites are ideal MRI contrast agents for higher transverse relaxation time (T_2), and $MnFe_2O_4$ is one among them [45]. However, recently, applications of $MnFe_2O_4$ are explored as the contrast agent for magnetic resonance angiography (MRA). The MRA sequences, which use low repetition and echo time, create images that are weighted by longitudinal relaxation time (T_1). Mn^{2+} cation has five unpaired d electrons that give rise to the magnetic moment, which increases T_1 relaxation, and therefore, possesses higher T_1 relaxation [50]. Mn^{2+} based compounds such as MnO, Mn_3O_4, Mn_3O_4@SiO_2, hollow MnO are some of the recent developments of Mn-based T_1 agents. Though $MnFe_2O_4$ is a T_2 contrast agent, Zhang et al. [51] demonstrated that $MnFe_2O_4$ enhances the T_1 relaxation when the nanoparticle sizes are exceedingly small and monodisperse.

In this work, we synthesized $MnFe_2O_4$ nanoparticle using a chemical co-precipitation technique. We found in a previous study that the use of NaOH as the co-precipitating agent results in larger particle size for which it is impossible to form stable colloidal suspensions (not published). Therefore, we used

NH_4OH as the co-precipitating agent. Since Mn^{2+} possesses an unstable valence state, we synthesized $MnFe_2O_4$ nanoparticle at room temperature and varied the nanoparticle sizes with pH variations without applying any heat. Thus, we examined the size-dependence of SLP and efficacy of $MnFe_2O_4$ nanoparticle as a negative/positive MRI contrast agent in the rat model.

2. Materials and Methods

2.1. Sample Preparation

We synthesized Manganese ferrites ($MnFe_2O_4$) nanoparticles using a chemical co-precipitation method for which we used the starting materials ($MnCl_2{\cdot}4H_2O$ and $FeCl_3$) of analytical grade and NH_4OH as the co-precipitating agent. The $MnCl_2{\cdot}4H_2O$ and $FeCl_3$ salts were dissolved in distilled water in the required molar ratio of 1:2 and underwent thorough mixing. Then 8M of NH_4OH solution were added drop-wise by micropipette (H17662, VWR, Radnor, PA, USA) into the above salts' solutions under continuous stirring by magnetic stirrer (SP250, Lab Depot, Dawsonville, GA, USA). Extra NH_4OH (6M) was added to maintain the pH to the desired level of 9–12 that plays a determining role in controlling the precipitation and the precipitated particles' size. The precipitates collected through centrifugation (2–16 P, Sigma, Harz, Germany) at 13,000 rpm for 20 min were washed ten times by centrifugations. The silver nitrate test confirmed that the sample was free from NH_4OH. The product was then dried in an oven at 80 °C for 72 h for perfect ferritization. The as-dried powder was ground with an agate mortar and pestle to obtain the as-dried $MnFe_2O_4$ nanoparticles. The precipitates of the $MnFe_2O_4$ nanoparticles were obtained according to the following reaction,

$$MnCl_2{\cdot}4H_2O + 2FeCl_3 + 8NH_4OH \rightarrow MnFe_2O_4 + 8NH_4Cl + 8H_2O$$

The aqueous chitosan solution (Sigma-Aldrich, St. Louis, MI, USA) with low molecular weight, 75–80% deacetylated, and viscosity 20 cps (1% solution in 1% acetic acid; Brookfield, Middleboro, MA, USA) was prepared using acetic acid. We mixed 0.40 g chitosan into 40 mL distilled water with a magnetic stirrer at 500 rpm. To form a homogeneous solution, we added 1 mL (2 N) acetic acid drop-wise and stirred until chitosan was fully dissolved in water. Then, the chitosan solution was centrifuged twice at 13,000 rpm for 20 min. The chitosan solution was formulated for coating and surface modification of nanoparticles.

2.2. Characterization

Structural characterization of $MnFe_2O_4$ nanoparticles at different particle sizes was performed by a powder X-ray diffractometer (XRD) (PW3040, X'Pert Pro, Philips, Amsterdam, The Netherland) using Rigaku CuK_α radiation source in the 2θ ranges from 15° to 75° at 40 kV, 30 mA. The crystalline nature and coating condition were observed by Fourier transform infrared spectroscopy (FTIR) (L1600300 Spectrum TWO UTA ETHERNET, Perkinelmer, Shelton, CT, USA). We analyzed the shape and microstructure of samples using a high-resolution transmission electron microscope (TEM) (F200X Talos, Thermo Fisher, Waltham, MA, USA) at the operating voltage of 200 kV and energy-dispersive X-ray spectroscopy (EDX) (S50 QLD9111, FEI, Amsterdam, The Netherlands). The magnetic state of the samples was determined by the physical properties measurement system (PPMS) (D235, Quantum Design, San Diego, CA, USA) at 5 and 300K using a 5 Tesla magnetic field and also using Mössbauer spectroscopy (MS4, Vincent, Belgrade, Serbia). The hydrodynamic size, polydispersity index (PDI), and zeta potential of chitosan-coated samples at different particle sizes were investigated using dynamic light scattering (DLS) (ZEN3600 Zeta Potential Instrument, Malvern, UK) and zeta potential or electrophoretic mobility technique. The time-dependent temperature profiles of the chitosan-coated samples at different particle sizes were carried out at different concentrations of 1, 2, 3, and 4 mg/mL using the hyperthermia set-up (EASYHEAT 5060LI, Ambrell, Rochester, NY, USA) with a radio-frequency (RF) induction coil of 4 mm diameter and 8 turns with an alternating current (AC) magnetic field of 20 mT at a resonance frequency (342 kHz). The temperature rise of the samples was measured using an optical fiber thermometer. The cytotoxicity assay of chitosan-coated

$MnFe_2O_4$ nanoparticles at 2 mg/mL was carried out using live/dead cell assay. HeLa cell line was cultured using a BioSafety cabinet (NU-400-E, NuAire, Plymouth, MN, USA) and a 37 °C + 5% CO_2 incubator (NuAire, Plymouth, MN, USA), trinocular microscope with a camera of Hemocytometer (Optika, Ponteranica, Italy). Nanoparticle size-dependent of T_2 relaxivities were determined on phantoms composed of a small tube that contained chitosan-coated $MnFe_2O_4$ at five different concentrations (e.g., 0.17, 0.34, 0.51, 0.68, and 1.03 mM) in four sets of a larger tube of four different nanoparticle size. MRI in vivo studies were carried out in a rat model by a horizontal-bore 7T MRI scanner (MRS7017, MR Solution, Guildford, UK).

2.3. Animal Handling and In-Vivo MRI

We performed animal handling following The Guide for the Care and Use of Laboratory Animals (1996) and based on the ARRIVE Guidelines for reporting animal research [52]. We minimized the sufferings of the experimental animal according to the requirement of the Ethical Review Committee of Animal Experiments of Atomic Energy Centre Dhaka, which approved the protocol with Memo No: AECD/ROD/EC/20/201.

Albino Wister rats (male, age 11–12 weeks, weight 190–200 g) received an intraperitoneal injection of ketamine/xylazine (0.5 mg/kg and 10 mg/kg) to sedate (i.e., complete loss of reflexes) them before the MRI scan. Rats received a single dose of CS-coated $MnFe_2O_4$ (10 mg/kg) using a 26G needle in the tail vein. The CS-coated $MnFe_2O_4$ nanoparticles suspended in phosphate buffered solution had a concentration of 2 mg/mL.

The rat was placed on an imaging bed of 66 mm diameter and 426 mm in length that was graduated along its axis to ensure positioning reproducibility. The bedding was devised with a tubular structure allowing warmed up air circulation and a tooth bar to immobilize the animal. There was provision for physiological monitoring and gating. The rat's head was put inside a transmit/receive RF head coil, which is a quadrature birdcage of inner diameter 65 mm and length 70 mm. The conveyor mechanism transported the entire assembly inside the homogeneous region of the magnetic field of the 7T MRI scanner.

The Carr–Purcell–Meiboon–Gill (CPMG) pulse sequence was used to determine nanoparticles size-dependent T_2 relaxivity of CS-coated $MnFe_2O_4$ as a negative contrast dye. This experiment using phantoms represented the relaxivity of water protons (r_2) in presence of per mM of CS-coated nanoparticles as a contrast agent. MRA experiments were carried out to examine the performance of CS-coated $MnFe_2O_4$ nanoparticles as MRA or blood pool/positive contrast agent of sizes 6, 10, and 15 nm in rat brain with and without contrast agents using the time-of-flight (TOF) three-dimensional (3D) sequence and Maximum Intensity Projection (MIP).

3. Results and Discussion

3.1. X-ray Diffraction (XRD) Analysis

XRD analysis reveals the structural characterization of the $MnFe_2O_4$ nanoparticles for the determination of average particle size and phase of the nanoparticles. The XRD patterns of as-dried $MnFe_2O_4$ at different pH presented in Figure 1a provides clear evidence of the formation of the ferrite phase. Bragg's reflections indexed as (111), (220), (311), (420), (511), (440), and (620) confirmed the formation of a well-defined single phase cubic spinel structure without any detectable impurity phase belonging to the space group Fd3m shown in Figure 1a. A significant broadening of the XRD peaks indicates that the ferrite particles are of nanometric size. The crystallite size of the samples was determined using the maximum intensity peak (311) by Scherrer's formula [53]: $d = 0.9\lambda/\beta cos\theta$, where β is the full-width half maxima measured in radians, θ is the Bragg's angle, and λ = 1.5418 Å is the wavelength of Cu Kα energy. The interplanar spacing (d_{ip}) obtained from the XRD pattern, yielded the lattice parameter using the formula, $a = d_{ip}\sqrt{(h^2 + k^2 + l^2)}$, where h, k and l are the Miller indices. The d_{ip} values and intensities of diffraction peaks matched with the single crystalline $MnFe_2O_4$

(JCPDS Card No. 074–2403). The XRD results of $MnFe_2O_4$ at different pH shown in Figure 1 are comparable with the previously reported results [54,55].

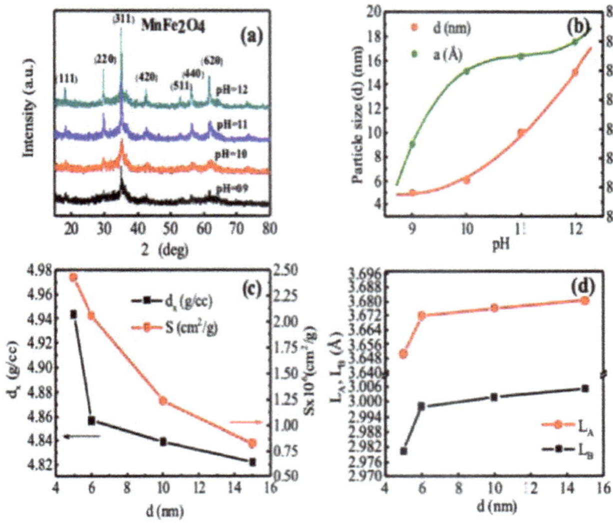

Figure 1. XRD studies of as-dried $MnFe_2O_4$ nanoparticles at different pH. (**a**) XRD patterns of the samples synthesized at pH of 9, 10, 11, and 12, (**b**) pH dependence particle size (d) and lattice parameter (a), (**c**) nanoparticle size dependence of X-ray density (d_X) and specific surface area (S), (**d**) nanoparticle size dependence of hopping length (L) for tetrahedral (A) and octahedral (B) site.

Figure 1b shows the pH dependence of nanoparticle size (d) and lattice parameters (a) extracted from the XRD patterns comparable with the previously reported values [54,55].

We see in Figure 1b that the particle size increases monotonically with the increase of pH, which is 5, 6, 10, and 15 nm at the pH of 9, 10, 11, and 12, respectively, synthesized in four separate batches. Thus, we varied the nanoparticle size by controlling the pH. There was a steady but non-linear increase of lattice parameters with increased pH, which was 8.43, 8.48, 8.49, and 8.50 Å, respectively. From the experimental lattice parameter and grain size, we determined the X-ray density, d_x, the hopping length L_A and L_B (the distance between the magnetic ions on A and B-sites), and the specific surface area of the particles S [56] using the relations $d_x = 8M_w/Na^3$, $L_A = a_{exp}\sqrt{3}/4$, $L_B = a_{exp}\sqrt{2}/4$, and $S = 6/(d \times d_x)$. Figure 1c,d show the variations of d_x, L_A, L_B, and S with the crystallite size d. Both d_x and S decreased with the increase of nanoparticle size d. The hopping length L_B was lower than the L_A, while both L_A and L_B slightly increased with d because of the change of cation distributions.

In the bulk condition, $MnFe_2O_4$ is a normal spinel with 80% Mn^{2+} occupying tetrahedral (A) sites, while 20% of Mn^{2+} occupying octahedral (B) sites [57]. However, an inversion factor of x changes for the $MnFe_2O_4$ nanoparticles. Barik et al. [58] demonstrated that the change in the degree of inversions (x) with the lattice constants and listed theoretical and experimental lattice parameters of several spinel ferrites of different compositions in which the theoretical and experimental lattice parameters of $MnFe_2O_4$ are quite close. Further, O'Neill et al. [59] demonstrated through a histogram the difference of experimental and calculated lattice constants Δa on 66 simple oxide spinels where Δa was small for most of the oxide spinels. Considering $a_{exp} \approx a_{th}$ we would get an estimation of the inversion parameter x considering other interactions negligible from the relation of $a_{th} = 8/3\sqrt{3}\left[\left((r_A + R_O) + \sqrt{3}(r_B + R_O)\right)\right]$. In this relation, r_A, r_B, and R_o are the ionic radii of Mn^{2+}, Fe^{3+}, and O^{2-}, respectively. Considering the random cation distribution $(Mn_{1-x}Fe_x)\left[Mn_{x/2}Fe_{1-x/2}\right]_2O_4$, we determined the inversion parameter x as 0.79, 0.44, 0.37, and 0.30 for the nanoparticle size d of 5, 6, 10, and 15, respectively. The cation distributions became $(Mn_{0.21}Fe_{0.79})$ $[Mn_{0.79}Fe_{1.21}]$ O_4, $(Mn_{0.56}Fe_{0.44})$ $[Mn_{0.44}Fe_{1.56}]$ O_4,, $(Mn_{0.63}Fe_{0.37})$

$[Mn_{0.37}Fe_{1.63}]\ O_4$, $(Mn_{0.70}Fe_{0.30})\ [Mn_{0.30}Fe_{1.70}]\ O_4$ for the particle size of 5, 6, 10 and 15 nm, respectively. This was in line with the fact that with the increase of particle size, the occupancy of Mn^{2+} on A-site increased, which was 80% in the bulk state. Figure 2A shows nanoparticle size (d) dependence of ionic radii on tetrahedral and octahedral sites (r_A and r_B) (left axis) and bond length (d_{AL} and d_{BL}) (right axis), oxygen parameters $u^{\bar{4}3m}$ and $u^{\bar{3}m}$ and shared tetrahedral edge (d_{AE}), shared octahedral edge (d_{BE}), unshared octahedral edge (d_{BEU}). In Figure 2A, the variations of the ionic radii in tetrahedral and octahedral sites r_A and r_B demonstrate that $r_A < r_B$, but with the increase of the nanoparticle size r_A increases as 0.525, 0.585, 0.597, and 0.609Å, while r_B decreases as 0.707, 0.679, 0.674, and 0.668 Å. This was because with the increase of nanoparticle size, the value of x decreased, which led to the transfer of Mn^{2+} from B site to A site progressively. Similar behavior was observed for tetrahedral and octahedral bond length with the change in the nanoparticle size. We determined the tetrahedral bond length and octahedral bond length, $d_{AL} = a_{exp}\sqrt{3}\left(u^{\bar{4}3m} - 0.25\right)$ and $d_{BL} = \sqrt{a_{exp}\left(3\left(u^{\bar{4}3m}\right)^2 - 11/4u^{\bar{4}3m} + 43/64\right)}$. O'Neill et al. [59] suggested that since the cations in ferrites have different sizes, therefore, any change in x would define the lattice parameter, a and oxygen parameter u, which have a unique relation, $u^{\bar{4}3m} = (r_A + R_o)/\sqrt{3}a_{exp}$. There are two ways of representation for the oxygen parameter: $u^{\bar{4}3m}$, when the origin of the unit cell is considered on an A-site cation $\bar{4}3m$ and $u^{\bar{3}m}$ when the origin of the unit cell is at octahedral vacancy $\bar{3}m$, where, $u^{\bar{3}m} = u^{\bar{4}3m} - 1/8$. The oxygen parameters of $u^{\bar{4}3m}$ and $u^{\bar{3}m}$ plotted with the variations of particle size in Figure 2A, are in the same order but different in values. For a perfect fcc structure for which the $u^{\bar{4}3m}_{ideal}$ and $u^{\bar{3}m}_{ideal}$ are 0.375 and 0.250, respectively [56]. O'Neill et al. [59] demonstrated a relationship between the Madelung constant M and the oxygen parameter u. There exists a crossover for $u = 0.2555$ between normal and inverse spinel. The structure is normal above 0.2555 and inversely smaller than this value. We obtained $u^{\bar{3}m}$ as 0.2514, 0.2547, 0.2553, and 0.2560 for 5, 6, 10, and 15 nm, respectively. Thus, we saw that for 5 nm, the structure was mostly inverse with $x = 0.79$. For the size 6 and 10 nm, the structures were mostly disordered with $x = 0.44$ and 0.37, and for the size of 15 nm, the structure was mostly normal with $x = 0.30$. We determined shared tetrahedral edge, $d_{AE} = a_{exp}\sqrt{2}\left(2u^{\bar{4}3m} - 0.5\right)$, shared octahedral edge, $d_{BE} = a_{exp}\sqrt{2}\left(1 - 2u^{\bar{4}3m}\right)$ and unshared octahedral edge $d_{BEU} = a_{exp}\sqrt{\left(4u^2 + 3u^{\bar{4}3m} + 11/16\right)}$ and presented this in Figure 2A. The shared tetrahedral edge increased with the increase in nanocrystallite size, and the shared octahedral edge decreased while the unshared octahedral edge slightly increasesd. This was because of the transfer of larger cation Mn^{2+} on the A-site and the smaller cation Fe^{3+} on the B-site with increased nanocrystallite sizes.

Figure 2b shows the nanoparticle size dependence of interionic distances between cations (Me-Me) (b, c, d, e, f), between cations and anions (p, q, r, s) and bond angles ($\theta_1, \theta_2, \theta_3, \theta_4, \theta_5$). The interionic distances between cations b, c, d, e, and f were calculated using the relations $b = \left(a_{exp}/4\right)\sqrt{2}$, $c = \left(a_{exp}/8\right)\sqrt{11}$, $d = \left(a_{exp}/4\right)\sqrt{3}$, $e = \left(3a_{exp}/8\right)\sqrt{3}$, $f = \left(a_{exp}/4\right)\sqrt{3}$ and cations and anions p, q, r, and s were calculated using the relations $p = a_{exp}\left(1/2 - u^{\bar{3}m}\right)$, $q = a_{exp}\left(u^{\bar{3}m} - 1/8\right)\sqrt{3}$, $r = a_{exp}\left(u^{\bar{3}m} - 1/8\right)\sqrt{11}$, $s = a_{exp}/3\left(u^{\bar{3}m} - 1/2\right)\sqrt{3}$. The bond angle was calculated using the relations $\theta_1 = cos^{-1}\left(\left(p^2 + q^2 - c^2\right)/2pq\right)$, $\theta_2 = cos^{-1}\left(\left(p^2 + r^2 - e^2\right)/2pr\right)$, $\theta_3 = cos^{-1}\left(\left(2p^2 - b^2\right)/2pr\right)$, $\theta_4 = cos^{-1}\left(\left(p^2 + s^2 - f^2\right)/2ps\right)$, $\theta_5 = cos^{-1}\left(\left(r^2 + q^2 - d^2\right)/2rq\right)$. We found that when nanoparticle size increased, the value of x decreased, and therefore, Mn^{2+} was transferred to the A-site replacing Fe^{3+} to B-site. From Figure 2b, it can be assumed that the magnetization should increase with the increase in nanocrystallite size because of the increase of B-B interactions as the bond length B-O, p decreases and the bond angle B-O-B, θ_3 and θ_4 increase because of the migration of Mn^{2+} to A-site and Fe^{3+} to B-site with the increase of crystallite size.

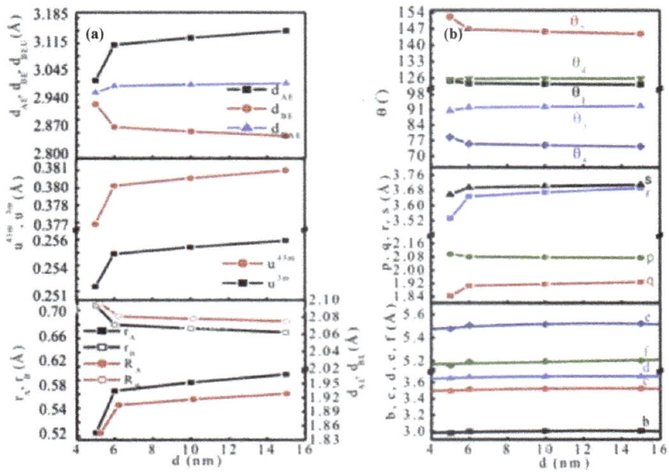

Figure 2. XRD studies of as-dried $MnFe_2O_4$ nanoparticles synthesized at different pH. (**a**) nanoparticle size dependence of ionic radii on tetrahedral and octahedral sites (r_A and r_B) (left axis) and bond length (d_{AL} and d_{BL}) (right axis), oxygen parameters (u) and shared tetrahedral edge (d_{AE}), shared octahedral edge (d_{BE}), unshared octahedral edge (d_{BEU}), (**b**) nanoparticle size dependence of interionic distances between cations (Me-Me) (b, c, d, e, f), between cations and anions (*p*, *q*, *r*, *s*) and bond angles ($\theta_1, \theta_2, \theta_3, \theta_4, \theta_5$).

3.2. TEM and EDX Analysis

Figure 3a–d shows the TEM bright field, dark field, selected area diffraction pattern, and high-resolution TEM images of the CS-coated nanoparticles synthesized at pH = 10 with the crystallite size of 6 nm obtained from XRD. Figure 3a shows bright-field images acquired from the transmission beam, while Figure 3b shows dark field images acquired by a diffracted beam. Selected area diffraction (SAED) patterns in Figure 3c provide information on the nanoparticles' structure and the high-resolutoin TEM image presented in Figure 3d demonstrates crystallinity. The d-values of the diffraction rings in Figure 3c were determined using Velox software and yielded values that exactly match the literature, and the diffraction pattern was indexed accordingly. The SAED pattern was consistent with the XRD pattern showing the cubic spinel structure of the $MnFe_2O_4$ nanoparticles belonging to the Fd3m space group.

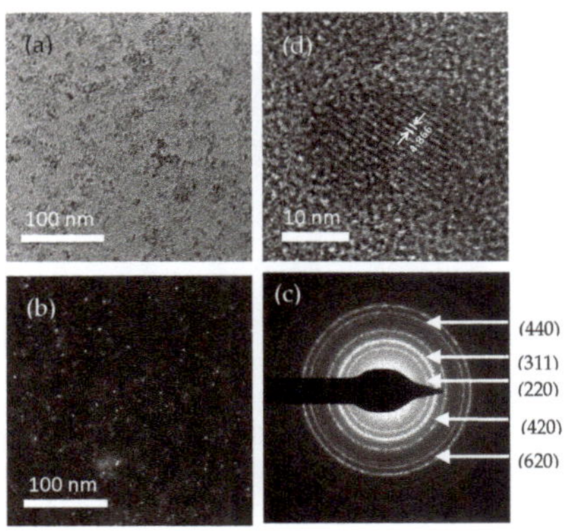

Figure 3. TEM images of the CS-coated $MnFe_2O_4$ nanoparticles at pH 11. (**a**) bright field (BF), (**b**) dark-field (DF), (**c**) selected area diffraction (SAED) pattern, and (**d**) high-resolution TEM image.

From the diffraction circle, conical dark-field images acquired using Velox software. Both the bright field and dark field images show the dispersion of the coated nanoparticles. The high-resolutoin TEM image in Figure 3d along the zone axis indicates good crystallinity. The d-spacing depicted from the high-resolutoin TEM image was d_{311} = 251.7 pm.

The EDX technique provides an effective atomic concentration of the sample on top surface layers of the solids under investigations. The EDX spectrum of $MnFe_2O_4$ nanoparticle in as-dried condition at room temperature shows the peaks of Mn, Fe, and O along with the C substrate peak. The microanalysis of EDX data indicates the constituent elements as Mn (at% 12.33) Fe (at% 31.53), and O (at% 36.65), respectively.

3.3. Magnetic Measurements

Figure 4a–d shows M-H hysteresis loops of the bare $MnFe_2O_4$ nanoparticle of sizes of 5, 6, 10, and 15 nm measured at 5 and 300 K with a maximum applied magnetic field, H_{max} = 5 Tesla. The negligible coercivity indicated that the $MnFe_2O_4$ nanoparticles of different particle sizes exhibited a typical superparamagnetic nature [56] with small interactions between the particles. A small volume fraction of ferromagnetic phases was embedded in the superparamagnetic matrix. The increase of all the parameters with the nanoparticle size, such as maximum magnetization (M_{max}), coercivity (H_c), and remnant ratio (M_r/M_{max}) indicated that inter-particle interactions increased with the increase of nanoparticle size. Two types of interactions exist in the magnetic nanoparticle, which are exchange and dipolar interactions. For sufficiently small particle size, thermal energy dominates over exchange energy, which reduces M_{max}, H_c, M_r/M_{max}.

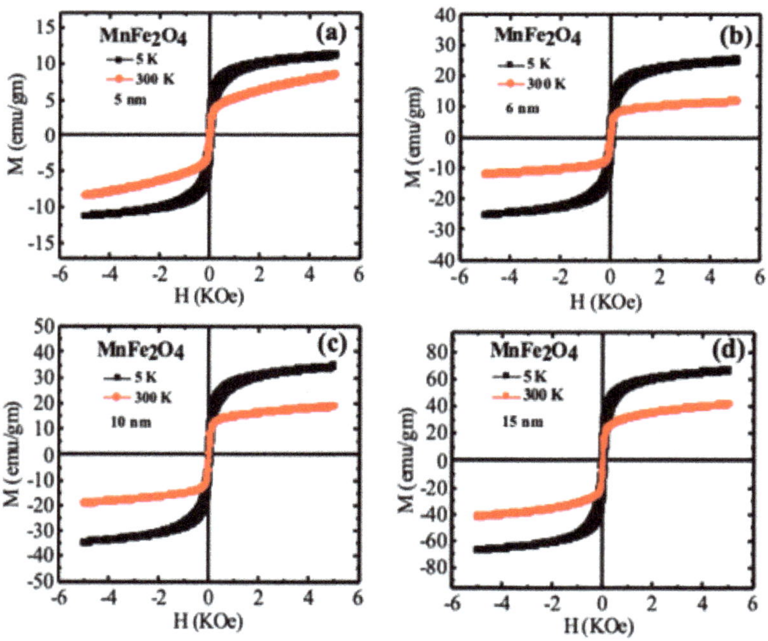

Figure 4. M-H hysteresis loops of $MnFe_2O_4$ nanoparticles in the as-dried condition measured at 5 and 300 K with a maximum applied field of 5 Tesla. The M-H loops of the $MnFe_2O_4$ nanoparticles at different pH are presented, (**a**) 5 nm, (**b**) 6 nm, (**c**) 10 nm, and (**d**) 15 nm.

Further demagnetization occurs in superparamagnetic nanoparticles by the dipolar interactions, which reduces all above parameters [60]. At 5 K, however, exchange energy dominates over thermal energy for which exchange energy overcomes the barrier of thermal energy. The dipolar interactions weaken as a result of which M_{max}, H_c, and M_r/M_{max} increase, and the samples show ferrimagnetic nature. M-H curves indicate that at 300 K with an applied magnetic field of 5 T, the magnetization is not saturated. Maximum magnetization recorded from the M-H curve was 11, 25, 34, and 66 emu/g with an applied field of 5 Tesla for the particle sizes of 5, 6, 10, and 15 nm, respectively at 5 K, whereas maximum magnetization values were 8, 12, 19 and 41 emu/g at 300 K. These values were lower than that of the bulk manganese ferrite, which was 80 emu/g, reported in the literature [51,62]. The smaller quantity of magnetization may be due to the higher percentages of atoms located on the nanoparticles' surface, producing a magnetically inactive layer or disordered layer on the surface.

Figure 5 shows the variations of M_{max}, magnetic moment (n_B), H_c, and M_r/M_{max} with particle sizes at 5 and 300 K acquired from M-H loops. The magnetic phase transition from ferromagnetic to superparamagnetic state occurred between these two temperatures of 5 and 300 K. In Figure 5a, there is an increase of M_{max} from 11 to 66 emu/gm at 5 K and 8 to 41 emu/gm at 300 K with the increase of particle size from 5 to 15 nm. Both M_r/M_{max} and H_c decreased significantly from 5 K to 300 K because of the ferrimagnetic to superparamagnetic transition. Using the formula, $n_B^e = \frac{M_s \times M_w}{5585}$ *Bohr magneton/f.u* gives an experimental magnetic moment n_B^e. Figure 5b shows the variations of n_B^e with particle size. The theoretical magnetic moment of $MnFe_2O_4$ using Néel's two sublattice model was $5\mu_B$ irrespective of cation distribution for $MnFe_2O_4$ because of the similar magnetic moment of $5\mu_B$ for both Mn^{2+} and Fe^{3+}. The values of n_B^e for the particle sizes of 5, 6, 10, and 15 nm were 0.46, 1.05, 1.43, and 2.77μ_B at 5 K, whereas, 0.34, 0.50, 0.79, and 1.72 μ_B at 300 K. The differences between the theoretical and experimental magnetic moment showed that we need to invoke Yafet–Kittel's three sublattice model [63]. We determined the Canting angles using the relation, $\alpha_{YK} = cos^{-1}\left(\frac{n_B^e + M_A}{M_B}\right)$, where, M_A and M_B are the magnetic moments on A and B sites. Figure 5c presents the nanoparticle size-dependence of canting angles, which shows almost a linear relationship. The α_{YK} at both 5 and 300 K decreases with the increase of nanoparticle size from 5 to 15 nm.

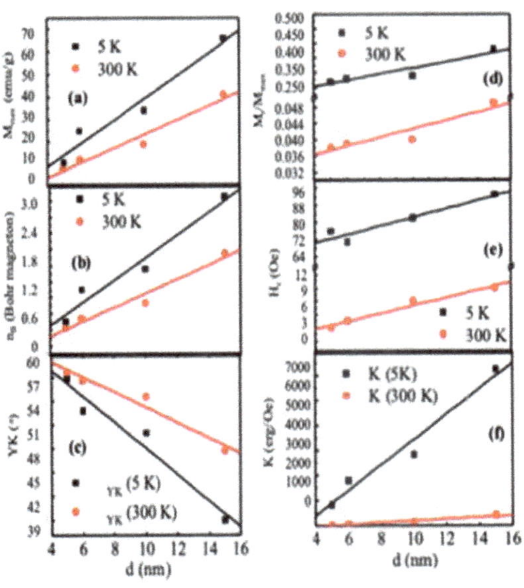

Figure 5. (**a**) Maximum magnetization (**b**) Bohr magneton, (**c**) Canting angles, (**d**) squareness ratio, (**e**) coercivity, and (**f**) anisotropy with size variations of $MnFe_2O_4$ nanoparticles.

Consequently, we see a linear increase of magnetic moment and magnetization with an increase in nanoparticle diameter. Figure 5d–f presents the variations of effective anisotropy, squareness ratio, and coercivity with the size of nanoparticles. A large enhancement in anisotropy at 5 K than 300 K demonstrates that the nanoparticles are in the blocked state at 5 K, whereas at 300 K the nanoparticles are in the superparamagnetic state.

This change manifests in the variations of M_r/M_{max} and H_c. The M_r/M_{max} changed from 0.27 to 0.38 at 5 K, and 0.038 to 0.045 at 300 K with the increase of the nanoparticle size, and the H_c increased from 76 to 93 Oe at 5 K and from 4 to 10 Oe at 300 K.

We analyzed magnetization data with an inversion parameter x, oxygen parameter $u^{4\bar{3}m}$, bond length p, and q and the bond angle θ_2 and θ_3. The results presented in Figure 6a show that $u^{4\bar{3}m}$ decreases linearly with the increase of x, i.e., with the increase of Mn^{2+} on the A-site as a result of which lattice parameter increases and also increases the oxygen parameter. $MnFe_2O_4$ is essentially a normal spinel in the bulk state with 20% Mn^{2+} on the B-site. We see that with the increase of nanoparticle size, the concentration of Mn^{2+} ions on the B-site decreases, which reduces the distortion of the $MnFe_2O_4$ spinel structure. In Figure 6a, magnetization increases with the decrease in the inversion parameter, x, i.e., with the decrease in Mn^{2+} ions on B-site and with subsequent increase in Fe^{3+} ions on B-site shown in Figure 6b. The increase in Fe^{3+} ions on the B-site oxygen parameter increases as a result of which magnetization increases, as shown in Figure 6c. We already found from Figure 2B that the nanoparticle size dependence of bond length, p decreases, and the bond angles θ_3 and θ_4 increase. Magnetization increases because of an increase in the B-O-B bond angles, θ_3 and θ_4, and reduction of B-O bond length, p, because of the increase in B-B interactions.

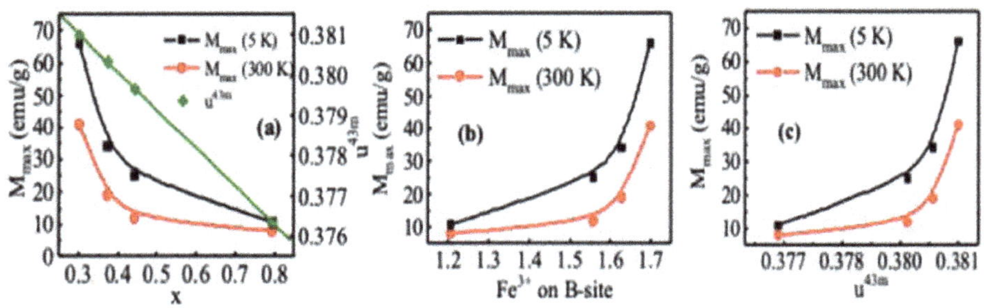

Figure 6. Maximum magnetization with (**a**) inversion parameter, (**b**) with cation distribution in the octahedral site, and (**c**) with oxygen parameter.

3.4. *Mössbauer Spectroscopy Analysis*

Mössbauer spectroscopy is a tool to probe the hyperfine parameter of magnetic nanoparticles. Figure 7a–d represents the Mössbauer spectroscopy of $MnFe_2O_4$ nanoparticles for nanoparticle size of 5, 6, 10, and 15 nm. Table 1 presents the hyperfine parameters such as chemical shift, quadrupole splitting, hyperfine magnetic field, and relative area extracted from the experimental and theoretical fitting of the Mössbauer spectra. We observed from Figure 7 that the samples of different sizes consist of a central doublet region, which confirmed that the major magnetic phase of the samples for all the nanoparticle sizes was largely superparamagnetic. From Table 1, we observe that three subspecies were required to fit the experimental data. Two of them exhibited fast relaxation and represented superparamagnetic phases, and the other one showed slow relaxation that represents ferrimagnetic phases. The area of the ferrimagnetic phase increased with the increase of nanoparticle size, which was 10, 15, 25, and 25% for the size of 5, 6, 10, and 15 nm. Thus, the volume fraction of the ferromagnetic phase increased with the increase of nanoparticle size. The quadrupole splitting, ΔE_q of Fe^{2+} was about 3 mm/s, which was much larger than Fe^{3+}, and the ΔE_q values of a low spin of Fe^{3+} were smaller than 0.8 mm/s. The ΔE_q values presented in Table 1 indicate that iron is present in the form of Fe^{3+} and low spin [64]. It is critical to estimate cation distributions because most of the Fe^{3+} are magnetically isolated with the nonmagnetic coordination atoms, and they do not contribute to the long-range magnetic order [64].

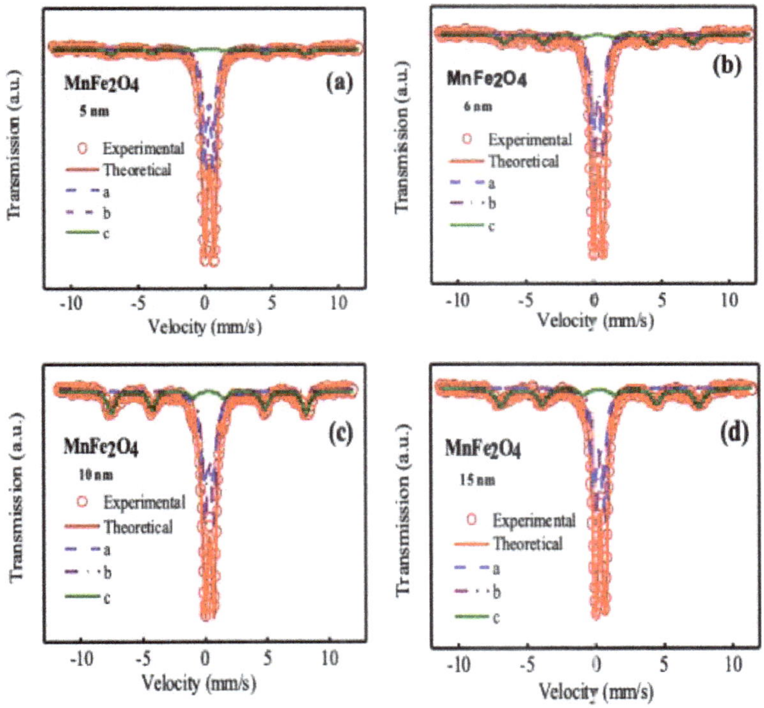

Figure 7. Mössbauer spectra of $MnFe_2O_4$ nanoparticles in the as-dried condition for different nanoparticle sizes measured at room temperature and without any applied magnetic field (**a**) 5 nm, (**b**) 6 nm, (**c**) 10 nm, and (**d**) 15 nm. In the spectrum, hollow red circles represent experimental data, the solid red line represents theoretical fitting, and the other three lines represent sub spectra of three subspecies named a, b, and c to fit the data.

Table 1. Hyperfine parameters of $MnFe_2O_4$ nanoparticles with different pH without magnetic field and at room temperature.

Particle Size (nm)	FWHM	Isomer Shift mm/s	Quadruple Splitting mm/s	Hyperfine Field kG	Area
	0.390	0.258	0.700	0	0.450
5	0.373	0.428	0.700	0	0.450
	0.800	0.319	0.137	453	0.100
	0.447	0.277	0.700	0	0.420
6	0.413	0.412	0.700	0	0.430
	0.800	0.396	0.264	432	0.150
	0.410	0.260	0.700	0	0.370
10	0.380	0.420	0.700	0	0.380
	0.800	0.330	0.295	450	0.250
	0.708	0.319	0.700	0	0.370
15	0.426	0.370	0.700	0	0.380
	0.718	0.336	0.470	488	0.250

3.5. Fourier Transform Infrared Spectroscopy (FTIR) Analysis

Figure 8a–d presents the FTIR spectra of uncoated, CS-coated $MnFe_2O_4$ nanoparticles and pure CS having different nanoparticle sizes. For the uncoated sample, two peaks were found near the wave number 400 cm^{-1} and 600 cm^{-1}, due to metal- oxide stretching bonds at tetrahedral sites and octahedral sites. Another peak was found near the wave number 3500 cm^{-1}, which was due to the O-H stretching band of associated water bound with the sample in the as-dried condition. For pure CS, peaks were found at waves number 1090, 1420, 1610, and 2850 cm^{-1} due to the stretching vibration of C-O-C- in a glycosidic linkage, CH_3 in amide group, NH_2 in the amino group, and CH_2 stretching vibration to a pyranose ring which was a characteristic peak of chitosan. CS-coated $MnFe_2O_4$ samples have similar peaks of CS, and the peaks are shifted, which means the samples coated well with CS [65]. We observed the similarity in behaviors for other particle sizes, i.e., for the nanoparticle sizes of 6, 10 and 15 nm.

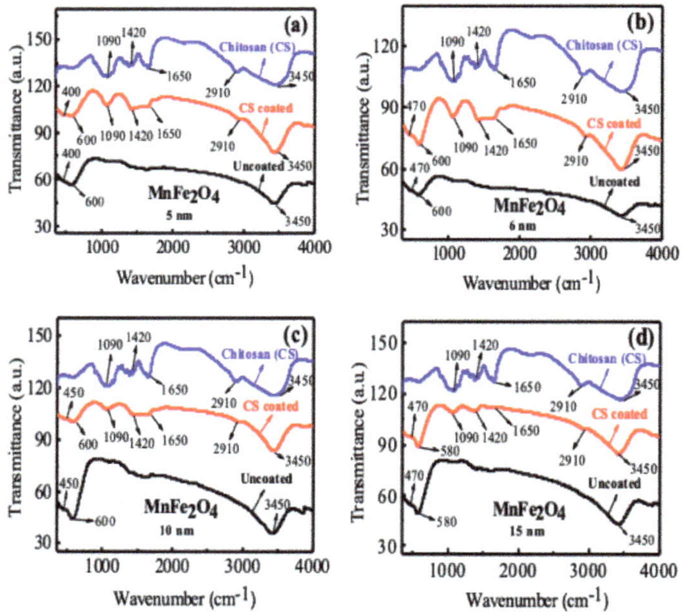

Figure 8. FTIR spectra of uncoated $MnFe_2O_4$, CS-coated $MnFe_2O_4$, and CS nanoparticles for sizes of (**a**) 5 nm, (**b**) 6 nm, (**c**) 10 nm, and (**d**) 15 nm.

3.6. Dynamic Light Scattering (DLS) Measurements

We carried out DLS measures to determine the hydrodynamic size (H_d) and polydispersity index (PDI). The hydrodynamic size (H_d) is the size of the MNPs in association with the hydration layer around the nanoparticle, and the polydispersity index (PDI) indicates the degree of dispersion of nanoparticle in a colloidal suspension. Figure 9a,b represents the concentration and size dependence of the hydrodynamic diameter distribution (H_d). Figure 9c,d presents the mean H_d and PDI of CS-coated $MnFe_2O_4$ nanoparticles with concentration and nanoparticle sizes.

In the DLS measurement, laser beam incident on the nanoparticle moves with dragging force to modify their surfaces due to coating elements and hydration layer, which causes their sizes to increase from 5 to 100 nm [66]. The average H_d was found to vary from 86 to 149 nm for different concentrations of nanoparticles and from 94 to 150 nm for different nanoparticle sizes, whereas the average core size was in the range of 5–15 nm from XRD. Therefore, H_d size distributions are greater than the core diameter observed by TEM. Demirci et al. [67] found the particle agglomerations in size range ~60–300 nm for $MnFe_2O_4$ nanoparticles. For biomedical applications, hydrodynamic diameter and the PDI are critical parameters. The PDI determines the extent of aggregation, and a lower PDI value is a prerequisite for biomedical applications. The range of hydrodynamic diameter for the biomedical applications should be less than 250 nm with a PDI value of less than 0.300 [68]. The H_d of the CS-coated $MnFe_2O_4$ nanoparticles in this study was less than 250 nm, and the PDI was nearly 0.300, which were satisfactory for biomedical applications.

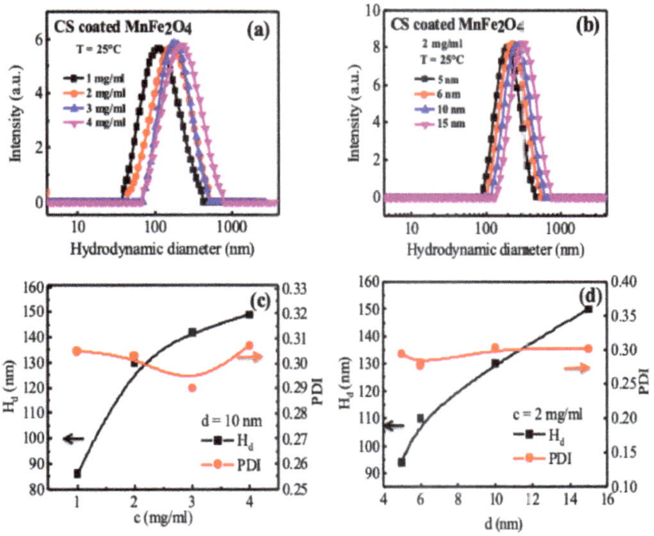

Figure 9. Hydrodynamic diameter (H_d) of CS-coated $MnFe_2O_4$ nanoparticles measured at 25 °C. In the figure, we present (**a**) distribution of H_d with concentration and (**b**) with nanoparticle size at 25 °C. Analyzing the data in (**a**) and (**b**), we presented (**c**) concentration and (**d**) nanoparticle size dependence of hydrodynamic diameter (H_d) and the polydispersity index (PDI).

3.7. Zeta Potential

We studied zeta potentials of CS-coated $MnFe_2O_4$ nanoparticles across sizes. Nanoparticles have a surface charge that attracts a thin layer of ions of opposite charge to the nanoparticle surface. The double layers of ions travel with the nanoparticle as it diffuses throughout the solution. The electric potential at the boundary of the double layer is known as the zeta potential of the particles and has values that typically range from +100 mV to −100 mV. Zeta potential is a tool for understanding the stability and coating condition of the nanoparticle in the colloidal suspension. The surface charge may be positive or negative, depending on the solution's nature by generating ionizable functional groups of nanoparticles [69].

Zeta potential increases with rising electrophoretic, electrostatic, hydrophilic material, and hydrophobic organic surface charge mobility by changing the (+ve), (−ve) charge of nanoparticles in disperse solution. If the suspension is stable, it means that the suspension possesses a high zeta potential value. The magnitude of the zeta potential is a determining factor of colloidal stability, which nominally holds values greater than +30 mV or less than −30 mV for stable colloidal suspension [70,71]. Dispersions with a low zeta potential will eventually aggregate due to Van der Waals's interparticle attractions. Figure 10 presents the distribution of the zeta potential for CS-coated $MnFe_2O_4$ of the concentration of 2 mg/mL. We see from Figure 10 that the value of zeta potential for CS-coated $MnFe_2O_4$ nanoparticles at 2 mg/mL concentration is +47 mV, +46 mV, +44 mV, and +41 mV for the nanoparticle sizes of 5, 6, 10, and 15 nm respectively, which are higher than +30 mV and is satisfactory for stable colloidal suspension.

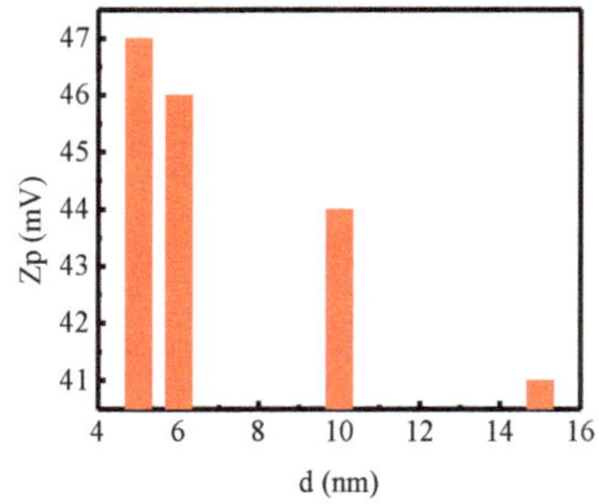

Figure 10. Zeta potential of CS-coated $MnFe_2O_4$ nanoparticles for 5 nm, 6 nm, 10 nm, and 15 nm at room temperature during the DLS measurement.

3.8. *Cytotoxicity Analysis*

Cytotoxicity is crucial for biomedical applications. We examined the cytotoxicity of CS-coated $MnFe_2O_4$ nanoparticles in HeLa cells cultured to a confluent state in DMEM (Dulbecco's Modified Eagles' Medium) containing 1% penicillin-streptomycin, gentamycin, and 10% fetal bovine serum (FBS). Cells (4×10^5/200 μL) were seeded onto 48 well plates and incubated at 37 °C + 5% CO_2. After 24 h, we added 25 μL of the sample (filtered) in each well. We examined cell mortality under an inverted light microscope after 24 h of incubation. After 24 h of incubation, insoluble samples washed out with fresh media. We then examined the cytotoxicity using a Hemocytometer and inverted light trinocular microscope. We repeated this examination twice. Figure 11a,b shows the images of the HeLa cells, which reflect the cytotoxic effect of CS-coated $MnFe_2O_4$ samples with the particle sizes of 5, 6, 10, and 15 nm at the concentration of 2 mg/mL, as well as the medium using a solvent and without solvent as a control. By this figure, it is clear that the cells' survival was 100% and 95% in the absence and presence of the solvent, respectively.

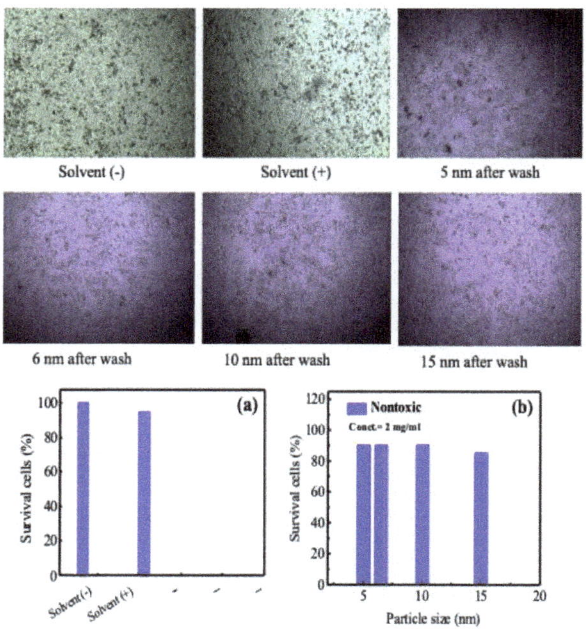

Figure 11. Cytotoxicity results assessed by survival of Hela cells (**a**) with and without solvent and (**b**) CS-coated $MnFe_2O_4$ nanoparticles at different particle sizes, where the corresponding cell culures are pictured on the top panel.

On the other hand, to observe the effect of toxicity of the coated particles with biocompatible polymer materials on Hela cell lines, four different solutions with different particle sizes were prepared, and the toxicity effect was observed using a hemocytometer. The survival cells were 90%, 90%, 90%, and 85% observed for the particle sizes of 5, 6, 10, and 15 nm, respectively, as shown in Figure 11b. Therefore, no cytotoxic effect was observed for the chitosan-coated $MnFe_2O_4$. Thus, the prepared manganese ferrite nanoparticles were nontoxic. Almost every cell survived when incubated with the sample, which meant the sample itself was nontoxic. The mortality of the cells will not occur when this sample is applied to the body for localized hyperthermia and as an MRI contrast dye. These nanoparticles are, therefore, noncytotoxic and viable for cell lines. These culture studies should be taken in the context of the doses used in the in vivo MRI analysis, but future in vivo cytotoxicity tests would further establish the lethal doses for these nanomaterials.

3.9. Magnetic Hyperthermia with Specific Loss Power (SLP)

Hyperthermia is a minimally invasive therapeutic technique for the selective heat treatment approach in cancer therapy, justified by the cancerous cell's vulnerability to high temperatures. Cancer cells have a high potential to be destroyed at about 42 °C while normal cells can survive at temperatures higher than 46 °C, which offers a window of hyperthermia therapy [72]. Magnetic particle hyperthermia is based on the magnetic nanoparticles as heat mediators when subjected to an alternating magnetic field.

Figure 12a–d shows the time-dependent temperature rise of chitosan-coated $MnFe_2O_4$ nanoparticles of different concentrations with different particle sizes. The heat was applied with a radio frequency magnetic field with an amplitude of 20 mT and a resonance frequency of 342 kHz. The sample was placed inside an induction coil attached to a heating station, and the power controlled by the power supply. The RF current switched on and off for different holding times. An optical fiber thermometer recorded the temperature. From Figure 12a–d, it is clear that the temperature increases with the increase of particle sizes and the concentrations of $MnFe_2O_4$ nanoparticles. The synthesized nanoparticles with particle sizes of 6, 10, and 15 nm for different concentrations reached threshold temperatures (42 °C). Figure 12d represents the maximum temperature (T_{max}) rise of chitosan-coated $MnFe_2O_4$ nanoparticle solutions with the nanoparticle size, d. The rise of temperature of the chitosan-coated $MnFe_2O_4$ nanoparticle in an AC magnetic field could be attributed to the different processes of magnetization reversal such as magnetic hysteresis, Néel and Brownian relaxation, and eddy current losses of the magnetic nanoparticles activated by the AC magnetic field [9,13,73–76]. Eddy current losses are negligible due to the high resistivity of ferrites. In the current study, $MnFe_2O_4$ nanoparticles showed superparamagnetic behavior, and therefore, heat generation due to hysteresis loss partially contributed to the total heat. Néel rotations caused by the spins of the nanoparticles and Brownian rotations resulted from the particle movement to align itself with the magnetic field contribute to the total amount of heat generation. The required therapeutic temperature of the samples having particle sizes of 6, 10, and 15 nm was reached within 20 min for 2, 3, and 4 mg/mL sample concentrations. For 4 mg/mL concentration, the time required to reach the desired hyperthermia temperature was less than the 2 and 3 mg/mL concentrations. This is because there are more magnetic particles for 4 mg/mL concentration resulting in an increased particle-particle interaction, which increases the exchange coupling energy and affects the heating characteristics[33]. The plateau temperature of the time-dependent temperature curve over a prolonged period opens up the possibility of using nanoparticle heating for drug delivery, hyperthermia treatment of cancer, and other targeted therapy. The T_{max} for all samples with different concentrations and different particle sizes shown in Figure 12e shows that the temperature increases with particle sizes. The efficiency of the heating capacity of a magnetic material is quantified through SLP, defined as the amount of heat generated per unit gram of magnetic material per unit time. The specific loss power from the time dependence temperature profile of each curve was determined from the slope of the linear rise of temperature with time using the formula: $SLP = C/m \times \Delta T/\Delta t$, where C is the specific heat of water, $\Delta T/\Delta t$ is the linear rise of temperature with time, and m is the mass of the nanoparticles of solution. The SLP value should be as high as possible to minimize the number of magnetic nanoparticles applied for hyperthermia. It is interesting to note that though there is a temperature rise in the case of higher sample concentration, the values of SLP are higher for lower sample concentration and particle sizes shown in Figure 12f. This may be attributed to high particle concentration results in particle agglomeration, suggesting stronger dipole–dipole interactions, which decrease magnetic heating efficiency [77]. Similar results were observed by Urtizberea et al. [78], Presa et al. [79], and Martinez-Boubeta et al. [80] who found that SLP decreases with increasing concentration of iron-oxide nanoparticles. Smaller nanoparticle size gives rise to higher Néel and Brownian motion. Further, smaller particle size gives rise to a smaller hysteresis area. Again, at lower concentrations, Néel and Brownian motion and the hysteresis area are smaller, giving rise to a higher specific loss. Concentration optimization of the nanoparticle is crucial for biomedical applications.

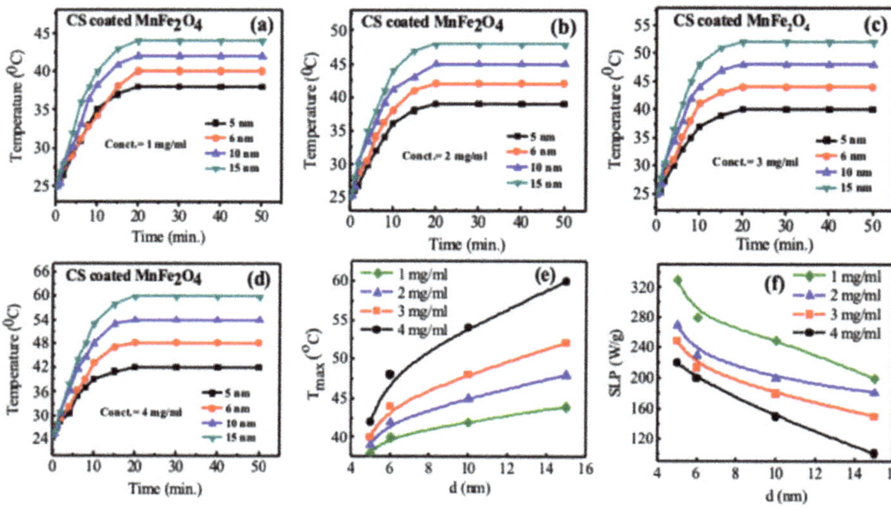

Figure 12. Time dependence temperature curves of CS-coated $MnFe_2O_4$ nanoparticles with a RF magnetic field of an amplitude of 26 mT and a frequency of 342 kHz. The curve presents time dependence of temperature curves of (**a**) 1 mg/mL, (**b**) 2 mg/mL, (**c**) 3 mg/mL, (**d**) 4 mg/mL. Subsequently, nanoparticle size dependence of (**e**) maximum temperature (T_{max}), and (**f**) specific loss power (SLP) are presented.

3.10. MRI Analysis

3.10.1. $MnFe_2O_4$ as Negative Contrast Agent

MRI is a diagnostic technique widely employed due to its ability to distinguish between healthy and pathological tissues. Manganese ferrites are good T_2 contrast agents of MRI though, at exceedingly small nanoparticle size, $MnFe_2O_4$ nanoparticles exhibit good positive blood pool contrast [81,82]. As a T_2 contrast agent, the parameter which determines the quality of contrast is known as relaxivity (r_2). To determine r_2 relaxivity in this study, a series of phantom images were acquired at a different echo time. Phantom images were acquired for $MnFe_2O_4$ nanoparticle solutions at different particle sizes. We developed phantoms by filling small Eppendorf tubes with five different concentrations for each particle size. The Eppendorf tubes were then inserted inside a 50 mL falcon tube, which was placed inside a mouse body coil. The CMPG pulse sequence was used to determine T_2 relaxation in which the repetition time T_R was 4000 ms. The series of images procured at echo times (TEs) are 7, 14, 21, 28, 35, 42, 49, 56, 63, 70, 77, 84, 91, 98, 105, and 112 ms.

Figure 13a represents a slice at TE of 14 ms for the particle sizes of 5, 6, 10, and 15 nm. The degree of darkening increases with the increase of the concentration of the nanoparticle. We recorded each voxel's intensity at different echo times, which dropped exponentially with the increase of echo time.

The relaxation time of each voxel determined from the exponential relation of intensity drop with the relaxation time: $I = I_o \, exp((-\tau)/T_2)$, where I_0 is the maximum intensity, T_2 is echo time, and τ is relaxation time. The inverse of relaxation time $1/T_2$ is also known as relaxation. In Figure 13b, concentration dependence of relaxation at different particle size = 5, 6, 10, 15 nm are presented.

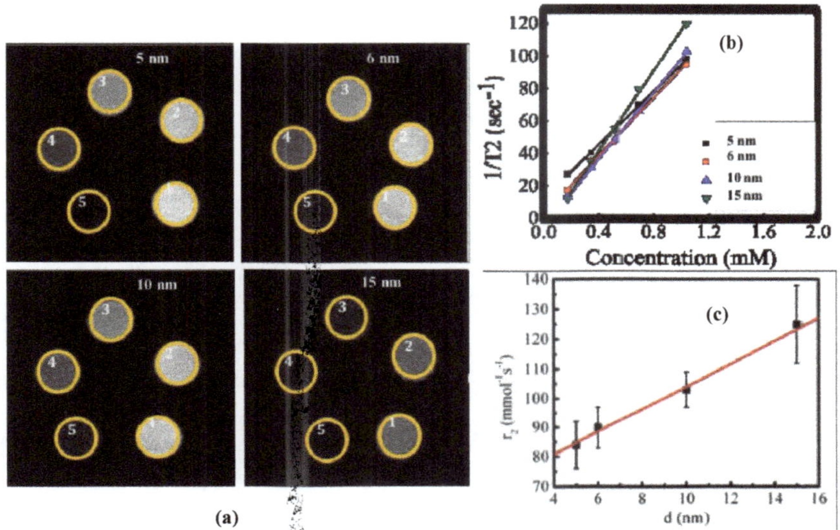

Figure 13. MRI data of CS-coated $MnFe_2O_4$ spinel ferrites nanoparticles. (**a**) Images acquired at TE of 14 ms (particle sizes of 5, 6, 10, and 15 nm) for CS-coated $MnFe_2O_4$ spinel ferrites nanoparticles with different values of concentrations (0.17, 0.34, 0.51, 0.68 and 1.03 mM) inside the five tubes in each image *1* to *5* represents lower to higher concentration of the nanoparticles in solution demonstrating contrast agents at different particle sizes. (**b**) Absolute R_2 (or $1/T_2$) mapping images for particle sizes of 5, 6, 10, and 15 nm at different concentrations. (**c**) Nanoparticle size dependence of relaxivity (r_2) of CS-coated $MnFe_2O_4$ nanoparticles.

The variation of relaxation with concentration is fully linear. From the slope of the linear fitting, we determined relaxivity r_2 for each particle size. The r_2 relaxivities were 84 (±08) $mM^{-1}s^{-1}$, 90 (±07) $mM^{-1}s^{-1}$, 103 (±06) $mM^{-1}s^{-1}$, and 125 (±13) $mM^{-1}s^{-1}$ for the nanoparticle sizes of 5, 6, 10, and 15 nm, respectively. The relaxivity (r_2) or relaxation (R_2) is directly related to the magnetic moment and particle sizes of nanoparticles [83]. Therefore, particles of higher magnetic moments, smaller particle sizes produce shorter relaxation time, which is essential for good negative contrast enhancement. Figure 13c shows linear dependence of nanoparticle size dependence of r_2 relaxivity.

3.10.2. Magnetic Resonance Angiography (MRA) with Time-of-Flight (TOF)

We tested the contrast efficacy of CS-coated $MnFe_2O_4$ nanoparticles for the blood pool imaging or MRA experiment using the TOF 3D sequence. Figure 14a shows image slices with and without contrast agents for 5 nm, while Figure 14b represents images for 10 nm, and Figure 14c shows the image slices for the size of 15 nm. To acquire the angiography images, we used the repetition time (TR) of 22 milliseconds, echo time (TE) of 3 ms, with a 30° flip angle. The total numbers of scans were 168, with a thickness of 2 mm and FOV of 36 × 36. The reconstruction of the images was achieved using Maximum Intensity Projection (MIP) post-processing software. For the nanoparticle size of 6 and 10 nm in Figure 14a,b the contrast agents enhanced the contrast of the rat veins than the veins of pre-injection control, while for the nanoparticle size of 15 nm, the contrast of the image deteriorated in the post-injection image than the pre-injection control in Figure 14c. Previous studies [82,83] demonstrated that Mn-based contrast agents possess similar characteristics as Gd-based contrast agents that enhance T_1 or positive contrast of MRI. $MnFe_2O_4$ is the most nontoxic form among other Mn-based compounds. Previous studies [84,85] demonstrated that to act as a positive contrast agent, r_2/r_1 should be minimum where $r_{1,2}$ are the relaxivities of contrast agents for which nanoparticle size should be exceedingly small, preferably less than 5 nm. In this study, we see in Figure 14a,b the pre- (A and B) and post-contrast agent injecting images (A′ and B′) that contrast-enhancing efficacy of chitosan-coated $MnFe_2O_4$ as the MRA contrast agent is higher when the nanoparticle size is 5 and 10 nm. With the increase of nanoparticle size as 15 nm in Figure 14c, the contrast-enhancing efficacy of the $MnFe_2O_4$ nanoparticles degrades as evidenced by comparing the images of pre- (A and B) and post-contrast agent (A′ and B′) injecting images.

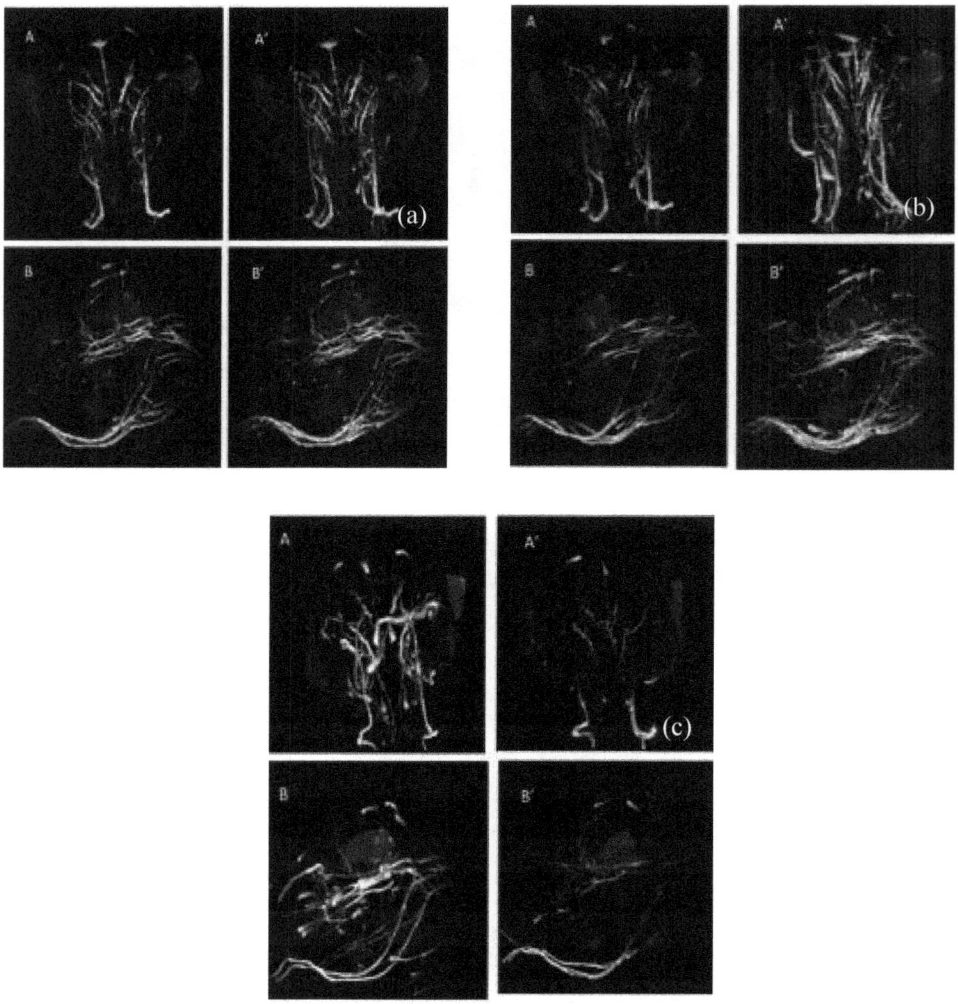

Figure 14. MRA with TOF 3D in vivo demsontrated in rat brain. Images from maximum intensity projection (MIP) are shown, where **A** and **A′** represents a slice of MIP with and without contrast agents at 3° from the horizontal position, **B** and **B′** represents with and without contrast agents at the sagittal position for the nanoparticle size of (**a**) 6 nm, (**b**) 10 nm, and (**c**) 15 nm.

4. Conclusions

Different characterization techniques yielded good coordination to demonstrate that CS-coated $MnFe_2O_4$ nanoparticles of different sizes undergo successful surface modifications. From the hyperthermia studies, we see that with a RF magnetic field of amplitude 26 mT, the SLP in the range of 100–330 depending on the particle sizes and concentrations. The results suggested that to obtain a trade-off between higher specific loss power and T_{max}, $MnFe_2O_4$ nanoparticles sizes should be in the range of 6–10 nm and the concentration in the range of 2–3 mg/mL. In vitro and in vivo imaging performed by MRI at 7T shows promising results. T_2 calculated from the concentration dependence of relaxation and were 84 (±08) $mM^{-1}s^{-1}$, 90 (±07) $mM^{-1}s^{-1}$, 103 (±06) $mM^{-1}s^{-1}$ and 125 (±13) $mM^{-1}s^{-1}$ for the nanoparticle sizes of 5, 6, 10, and 15 nm, respectively. The MRA studies are in line with previous studies, which confirm that to use nontoxic, biocompatible, and biodegradable $MnFe_2O_4$ as MRA contrast agents, nanoparticle sizes should be exceedingly small, preferably less than 5 nm. Thus, CS-coated $MnFe_2O_4$ has the promise of using both positive and negative contrast agents of MRI and hyperthermia with nanoparticle size variations.

Author Contributions: Conceptualization, K.I.; methodology, K.I., A.K.; formal analysis, K.I., A.K., A.H.; investigation, K.I., A.K.; resources S.M.H.; data curation, K.I.; writing–original draft preparation, K.I.; writing–review and editing, S.M.H., M.H., A.H. and F.H.; visualization, S.M.H.; supervision, S.M.H. and M.H.; project administration, S.M.H. and M.H.; funding acquisition, S.M.H. and F.H. All authors have read and agreed to the published version of the manuscript.

Funding: The authors would like to acknowledge the ADP project ID: Nano 5003 of Planning Commission, Bangladesh. The authors highly appreciate the contribution from the grant of R01-EB023366 of NIH. The support of the International Science Program (ISP), Uppsala University, Sweden to BAN−02/2 programme is highly acknowledged. The authors are grateful to the Bangladesh Atomic Energy Commission for the support to perform this research work. We recognize continual support of Islamic University, Kushtia−7003, Bangladesh. The authors are also grateful to the Ministry of Science and Technology, Government of the People's Republic of Bangladesh for providing the NST fellowship to M. Khairul Islam.

Conflicts of Interest: The authors declare no conflict of interest.

References

1. Arruebo, M.; Fernández-Pacheco, R.; Ibarra, M.R.; Santamaría, J. Magnetic nanoparticles for drug delivery. *Nano Today* **2007**, *2*, 22–32. [CrossRef]
2. Kumar, C.S.S.R.; Mohammad, F. Magnetic nanomaterials for hyperthermia-based therapy and controlled drug delivery. *Adv. Drug Deliv. Rev.* **2011**, *63*, 789–808. [CrossRef] [PubMed]
3. Ahmad, T.; Bae, H.; Iqbal, Y.; Rhee, I.; Hong, S.; Chang, Y.; Lee, J.; Sohn, D. Chitosan-coated nickel-ferrite nanoparticles as contrast agents in magnetic resonance imaging. *J. Magn. Magn. Mater.* **2015**, *381*, 151–157. [CrossRef]
4. Wang, Y.; Cao, X.; Liu, G.; Hong, R.; Chen, Y.; Chen, X.; Li, H.; Xu, B.; Wei, D. Synthesis of Fe3O4 magnetic fluid used for magnetic resonance imaging and hyperthermia. *J. Magn. Magn. Mater.* **2011**, *323*, 2953–2959. [CrossRef]
5. Qiao, Y.; Gumin, J.; MacLellan, C.J.; Gao, F.; Bouchard, R.; Lang, F.F.; Stafford, R.J.; Melancon, M.P. Magnetic resonance and photoacoustic imaging of brain tumor mediated by mesenchymal stem cell labeled with multifunctional nanoparticle introduced via carotid artery injection. *Nanotechnology* **2018**, *29*, 165101. [CrossRef] [PubMed]
6. Chen, W.-H.; Luo, G.-F.; Lei, Q.; Cao, F.-Y.; Fan, J.-X.; Qiu, W.-X.; Jia, H.; Hong, S.; Fang, F.; Zeng, X.; et al. Rational design of multifunctional magnetic mesoporous silica nanoparticle for tumor-targeted magnetic resonance imaging and precise therapy. *Biomaterials* **2016**, *76*, 87–101. [CrossRef]
7. Hoque, S.M.; Hossain, S.; Choudhury, S.; Akhter, S.H.; Hyder, F. Synthesis and characterization of ZnFe2O4 nanoparticles and its biomedical applications. *Mater. Lett.* **2016**, *162*, 60–63. [CrossRef]
8. Mohapatra, J.; Mitra, A.; Tyagi, H.; Bahadur, D.; Aslam, M. Iron oxide nanorods as high-performance magnetic resonance imaging contrast agents. *Nanoscale* **2015**, *7*, 9174–9184. [CrossRef]

9. Jordan, A.; Wust, P.; Fählin, H.; John, W.; Hinz, A.; Felix, R. Inductive heating of ferrimagnetic particles and magnetic fluids: Physical evaluation of their potential for hyperthermia. *Int. J. Hyperth.* **1993**, *9*, 51–68. [CrossRef]
10. Jadhav, S.; Kim, B.; Lee, H.; Im, I.; Rokade, A.; Park, S.; Chun, B.-S.; Kim, G.; Yu, Y.S.; Lee, S. Induction heating and in vitro cytotoxicity studies of MnZnFe2O4 nanoparticles for self-controlled magnetic particle hyperthermia. *J. Alloy. Compd.* **2018**, *745*, 282–291. [CrossRef]
11. Lee, J.-H.; Jang, J.-T.; Choi, J.-S.; Moon, S.H.; Noh, S.-H.; Kim, J.-W.; Kim, J.-G.; Kim, I.-S.; Park, K.I.; Cheon, J. Exchange-coupled magnetic nanoparticles for efficient heat induction. *Nat. Nanotechnol.* **2011**, *6*, 418–422. [CrossRef] [PubMed]
12. Du, Y.; Liu, X.; Liang, Q.; Liang, X.-J.; Tian, J. Optimization and Design of Magnetic Ferrite Nanoparticles with Uniform Tumor Distribution for Highly Sensitive MRI/MPI Performance and Improved Magnetic Hyperthermia Therapy. *Nano Lett.* **2019**, *19*, 3618–3626. [CrossRef] [PubMed]
13. Hergt, R.; Dutz, S.; Müller, R.; Zeisberger, M. Magnetic particle hyperthermia: Nanoparticle magnetism and materials development for cancer therapy. *J. Phys. Condens. Matter* **2006**, *18*, S2919–S2934. [CrossRef]
14. Hadadian, Y.; Ramos, A.P.; Pavan, T.Z. Role of zinc substitution in magnetic hyperthermia properties of magnetite nanoparticles: Interplay between intrinsic properties and dipolar interactions. *Sci. Rep.* **2019**, *9*, 1–14. [CrossRef]
15. Wang, Y.-X.J.; Leung, K.C.-F.; Xuan, S.; Chak, C.-P.; Sheng, H.; Zhang, G.; Qin, L. Durable Mesenchymal Stem Cell Labelling by Using Polyhedral Superparamagnetic Iron Oxide Nanoparticles. *Chem. A Eur. J.* **2009**, *15*, 12417–12425. [CrossRef]
16. Knollmann, F.D.; Böck, J.C.; Rautenberg, K.; Beier, J.; Ebert, W.; Felix, R. Differences in Predominant Enhancement Mechanisms of Superparamagnetic Iron Oxide and Ultrasmall Superparamagnetic Iron Oxide for Contrast-Enhanced Portal Magnetic Resonance Angiography. *Investig. Radiol.* **1998**, 33, 637–643. [CrossRef]
17. Mohs, A.M.; Lu, Z.-R. Gadolinium(III)-based blood-pool contrast agents for magnetic resonance imaging: Status and clinical potential. *Expert Opin. Drug Deliv.* **2007**, *4*, 149–164. [CrossRef]
18. Schnorr, J.; Wagner, S.; Abramjuk, C.; Wojner, I.; Schink, T.; Kröncke, T.; Schellenberger, E.; Hamm, B.; Pilgrimm, H.; Taupitz, M. Comparison of the Iron Oxide-Based Blood-Pool Contrast Medium VSOP-C184 With Gadopentetate Dimeglumine for First-Pass Magnetic Resonance Angiography of the Aorta and Renal Arteries in Pigs. *Investig. Radiol.* **2004**, *39*, 546–553. [CrossRef]
19. Plank, C.; Zelphati, O.; Mykhaylyk, O. Magnetically enhanced nucleic acid delivery. Ten years of magnetofection—Progress and prospects. *Adv. Drug Deliv. Rev.* **2011**, *63*, 1300–1331. [CrossRef]
20. Mandeville, J.B.; Marota, J.J.A.; Kosofsky, B.E.; Keltner, J.R.; Weissleder, R.; Rosen, B.R.; Weisskoff, R.M. Dynamic functional imaging of relative cerebral blood volume during rat forepaw stimulation. *Magn. Reson. Med.* **1998**, *39*, 615–624. [CrossRef]
21. Maritim, S.; Coman, D.; Huang, Y.; Rao, J.U.; Walsh, J.J.; Hyder, F. Mapping Extracellular pH of Gliomas in Presence of Superparamagnetic Nanoparticles: Towards Imaging the Distribution of Drug-Containing Nanoparticles and Their Curative Effect on the Tumor Microenvironment. *Contrast Media Mol. Imaging* **2017**, *2017*, 1–15. [CrossRef] [PubMed]
22. Franco, O.L.; Lopez-Abarrategui, C.; Figueroa-Espi, V.; Lugo-Alvarez, M.B.; Pereira, C.D.; Garay, H.; Barbosa, J.A.R.G.; Jimenez-Hernandez, L.; Estevez-Hernandez, O.; Reguera-Ruiz, E.; et al. The intrinsic antimicrobial activity of citric acid-coated manganese ferrite nanoparticles is enhanced after conjugation with the antifungal peptide Cm-p5. *Int. J. Nanomed.* **2016**, *11*, 3849–3857. [CrossRef] [PubMed]
23. Albino, M.; Fantechi, E.; Innocenti, C.; López-Ortega, A.; Bonanni, V.; Campo, G.; Pineider, F.; Gurioli, M.; Arosio, P.; Orlando, T.; et al. Role of Zn2+ Substitution on the Magnetic, Hyperthermic, and Relaxometric Properties of Cobalt Ferrite Nanoparticles. *J. Phys. Chem. C* **2019**, *123*, 6148–6157. [CrossRef]
24. Kim, D.-H.; Nikles, D.E.; Johnson, D.T.; Brazel, C.S. Heat generation of aqueously dispersed CoFe2O4 nanoparticles as heating agents for magnetically activated drug delivery and hyperthermia. *J. Magn. Magn. Mater.* **2008**, *320*, 2390–2396. [CrossRef]
25. Lu, J.; Ma, S.; Sun, J.; Xia, C.; Liu, C.; Wang, Z.; Zhao, X.; Gao, F.; Gong, Q.; Song, B. Manganese ferrite nanoparticle micellar nanocomposites as MRI contrast agent for liver imaging. *Biomaterials* **2009**, *30*, 2919–2928. [CrossRef] [PubMed]

26. Hoque, S.M.; Tariq, M.; Liba, S.I.; Salehin, F.; Mahmood, Z.H.; Khan, M.N.; Chattopadhayay, K.; Islam, R.; Akhter, S. Thermo-therapeutic applications of chitosan- and PEG-coated NiFe2O4nanoparticles. *Nanotechnology* **2016**, *27*, 285702. [CrossRef] [PubMed]
27. Arulmurugan, R.; Jeyadevan, B.; Vaidyanathan, G.; Sendhilnathan, S. Effect of zinc substitution on Co–Zn and Mn–Zn ferrite nanoparticles prepared by co-precipitation. *J. Magn. Magn. Mater.* **2005**, *288*, 470–477. [CrossRef]
28. Amighian, J.; Mozaffari, M.; Nasr, B. Preparation of nano-sized manganese ferrite (MnFe2O4) via coprecipitation method. *Phys. Status Solidi C* **2006**, *3*, 3188–3192. [CrossRef]
29. Hoque, S.M.; Huang, Y.; Cocco, E.; Maritim, S.; Santin, A.D.; Shapiro, E.M.; Coman, D.; Hyder, F. Improved specific loss power on cancer cells by hyperthermia and MRI contrast of hydrophilic FexCo1-xFe2O4nanoensembles. *Contrast Media Mol. Imaging* **2016**, *11*, 514–526. [CrossRef]
30. Khot, V.; Salunkhe, A.; Thorat, N.; Phadatare, M.R.; Pawar, S. Induction heating studies of combustion synthesized MgFe2O4 nanoparticles for hyperthermia applications. *J. Magn. Magn. Mater.* **2013**, *332*, 48–51. [CrossRef]
31. Desai, H.B.; Hathiya, L.J.; Joshi, H.H.; Tanna, A.R. Synthesis and Characterization of Photocatalytic MnFe2O4 Nanoparticles. *Mater. Today: Proc.* **2020**, *21*, 1905–1910. [CrossRef]
32. Jasso-Terán, R.A.; Cortés-Hernández, D.A.; Sánchez-Fuentes, H.J.; Reyes-Rodríguez, P.Y.; De-León-Prado, L.E.; Escobedo-Bocardo, J.; Almanza-Robles, J.M. Synthesis, characterization and hemolysis studies of Zn(1−x)CaxFe2O4 ferrites synthesized by sol-gel for hyperthermia treatment applications. *J. Magn. Magn. Mater.* **2017**, *427*, 241–244. [CrossRef]
33. Vestal, C.R.; Zhang, Z.J. Synthesis and Magnetic Characterization of Mn and Co Spinel Ferrite-Silica Nanoparticles with Tunable Magnetic Core. *Nano Lett.* **2003**, *3*, 1739–1743. [CrossRef]
34. Yoo, P.S.; Lee, B.W.; Liu, C. Effects of pH Value, Reaction Time, and Filling Pressure on the Hydrothermal Synthesis of ZnFe2O4 Nanoparticles. *IEEE Trans. Magn.* **2015**, *51*, 1–4. [CrossRef]
35. Shafi, K.V.P.M.; Gedanken, A.; Prozorov, R.; Balogh, J. Sonochemical preparation and siz-dependent propertis of nanostructured CoFe2O4 particles. *Chem. Mater.* **1998**, *10*, 3445–3450. [CrossRef]
36. Duong, G.; Hanh, N.; Linh, D.; Groessinger, R.; Weinberger, P.; Schafler, E.; Zehetbauer, M. Monodispersed nanocrystalline Co1−xZnxFe2O4 particles by forced hydrolysis: Synthesis and characterization. *J. Magn. Magn. Mater.* **2007**, *311*, 46–50. [CrossRef]
37. Bohara, R.A.; Thorat, N.D.; Yadav, H.M.; Pawar, S.H. One-step synthesis of uniform and biocompatible amine functionalized cobalt ferrite nanoparticles: A potential carrier for biomedical applications. *New J. Chem.* **2014**, *38*, 2979–2986. [CrossRef]
38. Sharma, S.K.; Kumar, R.; Kumar, S.; Knobel, M.; Meneses, C.; Kumar, V.V.S.; Reddy, V.R.; Singh, M.; Lee, C.G. Role of interparticle interactions on the magnetic behavior of Mg0.95Mn0.05Fe2O4ferrite nanoparticles. *J. Phys. Condens. Matter* **2008**, *20*, 235214. [CrossRef]
39. Hou, C.; Yu, H.; Zhang, Q.; Li, Y.; Wang, H. Preparation and magnetic property analysis of monodisperse Co–Zn ferrite nanospheres. *J. Alloy. Compd.* **2010**, *491*, 431–435. [CrossRef]
40. Feltin, N.; Pileni, M.P. New Technique for Synthesizing Iron Ferrite Magnetic Nanosized Particles. *Langmuir* **1997**, *13*, 3927–3933. [CrossRef]
41. Carta, D.; Casula, M.F.; Floris, P.; Falqui, A.; Mountjoy, G.; Boni, A.; Sangregorio, C.; Corrias, A. Synthesis and microstructure of manganese ferrite colloidal nanocrystals. *Phys. Chem. Chem. Phys.* **2010**, *12*, 5074–5083. [CrossRef] [PubMed]
42. Liu, X.L.; Wang, Y.T.; Ng, C.T.; Wang, R.; Jing, G.; Yi, J.B.; Yang, J.; Bay, B.H.; Yung, L.-Y.L.; Di Fan, D.; et al. Coating Engineering of MnFe2O4Nanoparticles with SuperhighT2Relaxivity and Efficient Cellular Uptake for Highly Sensitive Magnetic Resonance Imaging. *Adv. Mater. Interfaces* **2014**, *1*, 1300069. [CrossRef]
43. Kim, D.-H.; Nikles, D.E.; Brazel, C.S. Synthesis and Characterization of Multifunctional Chitosan-MnFe2O4 Nanoparticles for Magnetic Hyperthermia and Drug Delivery. *Materials* **2010**, *3*, 4051–4065. [CrossRef] [PubMed]
44. Kim, D.-H.; Zeng, H.; Ng, T.C.; Brazel, C.S. T1 and T2 relaxivities of succimer-coated MFe23+O4 (M=Mn2+, Fe2+ and Co2+) inverse spinel ferrites for potential use as phase-contrast agents in medical MRI. *J. Magn. Magn. Mater.* **2009**, *321*, 3899–3904. [CrossRef]

45. De Castro, C.H.; Nunes, A.D.C.; Ramalho, L.S.; Colugnati, D.B.; Mendes, E.P.; Sousa, M.H.; Bakuzis, A.F.; Souza, A.P.S.; Zufelato, N. Manganese ferrite-based nanoparticles induce ex vivo, but not in vivo, cardiovascular effects. *Int. J. Nanomed.* **2014**, *9*, 3299–3312. [CrossRef]
46. Tripathi, R.; Mahapatra, S.; Raghunath, R.; Sastry, V.; Krishnamoorthy, T. Daily intake of manganese by the adult population of Mumbai. *Sci. Total. Environ.* **2000**, *250*, 43–50. [CrossRef]
47. Rao, W.; Deng, Z.-S.; Liu, J. A Review of Hyperthermia Combined with Radiotherapy/Chemotherapy on Malignant Tumors. *Crit. Rev. Biomed. Eng.* **2010**, *38*, 101–116. [CrossRef]
48. Dennis, C.L.; Ivkov, R. Physics of heat generation using magnetic nanoparticles for hyperthermia. *Int. J. Hyperth.* **2013**, *29*, 715–729. [CrossRef]
49. Guardia, P.; Di Corato, R.; Lartigue, L.; Wilhelm, C.; Espinosa, A.; Garcia-Hernandez, M.; Gazeau, F.; Manna, L.; Pellegrino, T. Water-Soluble Iron Oxide Nanocubes with High Values of Specific Absorption Rate for Cancer Cell Hyperthermia Treatment. *ACS Nano* **2012**, *6*, 3080–3091. [CrossRef]
50. Zhu, D.; Liu, F.; Ma, L.; Liu, D.; Wang, Z. Nanoparticle-Based Systems for T1-Weighted Magnetic Resonance Imaging Contrast Agents. *Int. J. Mol. Sci.* **2013**, *14*, 10591–10607. [CrossRef]
51. Zhang, H.; Li, L.; Liu, X.L.; Jiao, J.; Ng, C.-T.; Yi, J.B.; Luo, Y.E.; Bay, B.-H.; Zhao, L.Y.; Peng, M.L.; et al. Ultrasmall Ferrite Nanoparticles Synthesized via Dynamic Simultaneous Thermal Decomposition for High-Performance and Multifunctional T1 Magnetic Resonance Imaging Contrast Agent. *ACS Nano* **2017**, *11*, 3614–3631. [CrossRef] [PubMed]
52. Kilkenny, C.; Browne, W.J.; Cuthill, I.C.; Emerson, M.; Altman, D.G. Improving Bioscience Research Reporting: The ARRIVE Guidelines for Reporting Animal Research. *PLoS Biol.* **2010**, *8*, e1000412. [CrossRef] [PubMed]
53. Cullity, B.D.; Weymouth, J.W. Elements of X-Ray Diffraction. *Am. J. Phys.* **1957**, *25*, 394–395. [CrossRef]
54. Kim, J.-G.; Seo, J.-W.; Cheon, J.; Kim, Y.-J. Rietveld Analysis of Nano-crystalline MnFe2O4 with Electron Powder Diffraction. *Bull. Korean Chem. Soc.* **2009**, *30*, 183–187.
55. Sugimoto, M. The Past, Present, and Future of Ferrites. *J. Am. Ceram. Soc.* **2004**, *82*, 269–280. [CrossRef]
56. Satalkar, M.; Kane, S.N. On the study of Structural properties and Cation distribution of Zn0.75-xNixMg0.15Cu0.1Fe2O4nano ferrite: Effect of Ni addition. *J. Phys. Conf. Ser.* **2016**, *755*, 012050. [CrossRef]
57. Szotek, Z.; Temmerman, W.M.; Ködderitzsch, D.; Svane, A.; Petit, L.; Winter, H. Electronic structures of normal and inverse spinel ferrites from first principles. *Phys. Rev. B* **2006**, *74*, 174431. [CrossRef]
58. Brik, M.; Suchocki, A.; Kamińska, A. Lattice Parameters and Stability of the Spinel Compounds in Relation to the Ionic Radii and Electronegativities of Constituting Chemical Elements. *Inorg. Chem.* **2014**, *53*, 5088–5099. [CrossRef]
59. O'Neill, H.S.C.; Navrotsky, A. Simple spinels: Crystallographic parameters, cation radii, lattice energies, and cation distribution. *Am. Mineral.* **1983**, *68*, 181–194.
60. Modak, S.; Karan, S.; Roy, S.; Mukherjee, S.; Das, D.; Chakrabarti, P. Preparation and characterizations of SiO2-coated nanoparticles of Mn0.4Zn0.6Fe2O4. *J. Magn. Magn. Mater.* **2009**, *321*, 169–174. [CrossRef]
61. Li, Q.; Conde, C.F.; Conde, A.; Kiss, L.F. Relationship between coercivity and magnetic moment of superparamagnetic particles with dipolar interaction. *Phys. Rev. B* **2005**, *72*, 174424. [CrossRef]
62. Suominen, T.; Raittila, J.; Salminen, T.; Schlesier, K.; Linden, J.; Paturi, P. Magnetic properties of fine SFMO particles: Superparamagnetism. *J. Magn. Magn. Mater.* **2007**, *309*, 278–284. [CrossRef]
63. Yafet, Y.; Kittel, C. Antiferromagnetic Arrangements in Ferrites. *Phys. Rev.* **1952**, *87*, 290–294. [CrossRef]
64. Sharifi, I.; Shokrollahi, H. Nanostructural, magnetic and Mössbauer studies of nanosized Co1−xZnxFe2O4 synthesized by co-precipitation. *J. Magn. Magn. Mater.* **2012**, *324*, 2397–2403. [CrossRef]
65. Pawlak, A.; Mucha, M. Thermogravimetric and FTIR studies of chitosan blends. *Thermochim. Acta* **2003**, *396*, 153–166. [CrossRef]
66. Lim, J.; Yeap, S.P.; Che, H.X.; Low, S.C. Characterization of magnetic nanoparticle by dynamic light scattering. *Nanoscale Res. Lett.* **2013**, *8*, 381. [CrossRef]
67. Demirci, Ç.E.; Manna, P.K.; Wroczynskyj, Y.; Aktürk, S.; Van Lierop, J. A comparison of the magnetism of cobalt-, manganese-, and nickel-ferrite nanoparticles. *J. Phys. D Appl. Phys.* **2017**, *51*, 025003. [CrossRef]
68. Danaei, M.; Dehghankhold, M.; Ataei, S.; Davarani, F.H.; Javanmard, R.; Dokhani, A.; Khorasani, S.; Mozafari, M.R. Impact of Particle Size and Polydispersity Index on the Clinical Applications of Lipidic Nanocarrier Systems. *Pharmaceutics* **2018**, *10*, 57. [CrossRef]
69. Doymus, K. The Effect of Ionic Electrolytes and pH on the Zeta Potential of Fine Coal Particles. *Turkish J. Chem.* **2007**, *31*, 589–597.

70. Wang, P.; Keller, A.A. Natural and Engineered Nano and Colloidal Transport: Role of Zeta Potential in Prediction of Particle Deposition. *Langmuir* **2009**, *25*, 6856–6862. [CrossRef]
71. Cruz, E.F.D.L.; Zheng, Y.D.; Torres, E.; Li, W.; Song, W.H.; Burugapalli, K. Zeta Potential of modified multi-walled carbon nanotubes in presence of poly (vinyl alcohol) hydrogel. *Int. J. Electrochem. Sci.* **2012**, *7*, 3577–3590.
72. Hejase, H.; Hayek, S.S.; Qadri, S.M.; Haik, Y. MnZnFe nanoparticles for self-controlled magnetic hyperthermia. *J. Magn. Magn. Mater.* **2012**, *324*, 3620–3628. [CrossRef]
73. Bae, S.; Lee, S.W. Applications of NiFe2O4 nanoparticles or a hyperthermia agent in biomedicine. *Appl. Phys. Lett.* **2006**, *89*, 252503. [CrossRef]
74. Rosensweig, R. Heating magnetic fluid with alternating magnetic field. *J. Magn. Magn. Mater.* **2002**, *252*, 370–374. [CrossRef]
75. Fortin, J.-P.; Wilhelm, C.; Servais, J.; Ménager, C.; Bacri, J.-C.; Gazeau, F. Size-Sorted Anionic Iron Oxide Nanomagnets as Colloidal Mediators for Magnetic Hyperthermia. *J. Am. Chem. Soc.* **2007**, *129*, 2628–2635. [CrossRef] [PubMed]
76. Muela, A.; Muñoz, D.; Martín-Rodríguez, R.; Orue, I.; Garaio, E.; De Cerio, A.A.D.; Alonso, J.; García, J.Á.; Fdez-Gubieda, M.L. Optimal Parameters for Hyperthermia Treatment Using Biomineralized Magnetite Nanoparticles: Theoretical and Experimental Approach. *J. Phys. Chem. C* **2016**, *120*, 24437–24448. [CrossRef]
77. Jadhav, S.V.; Nikam, D.S.; Khot, V.M.; Thorat, N.D.; Phadatare, M.R.; Ningthoujam, R.S.; Salunkhe, A.B.; Pawar, S.H. Studies on colloidal stability of PVP-coated LSMO nanoparticles for magnetic fluid hyperthermia. *New J. Chem.* **2013**, *37*, 3121–3130. [CrossRef]
78. Urtizberea, A.; Natividad, E.; Arizaga, A.; Castro, M.; Mediano, A. Specific Absorption Rates and Magnetic Properties of Ferrofluids with Interaction Effects at Low Concentrations. *J. Phys. Chem. C* **2010**, *114*, 4916–4922. [CrossRef]
79. De La Presa, P.; Luengo, Y.; Multigner, M.; Costo, R.; Morales, M.P.; Rivero, G.; Hernando, A. Study of Heating Efficiency as a Function of Concentration, Size, and Applied Field in γ-Fe2O3 Nanoparticles. *J. Phys. Chem. C* **2012**, *116*, 25602–25610. [CrossRef]
80. Martínez, F.P.; Simeonidis, K.; Makridis, A.; Angelakeris, M.; Iglesias, O.; Guardia, P.; Cabot, A.; Yedra, L.; Estradé, S.; Peiró, F.; et al. Learning from Nature to Improve the Heat Generation of Iron-Oxide Nanoparticles for Magnetic Hyperthermia Applications. *Sci. Rep.* **2013**, *3*, srep01652. [CrossRef]
81. Dendrinou-Samara, C.; Katsikini, M.; Sakellari, D.; Paloura, E.C.; Kalogirou, O.; Dendrinou-Samara, C. Reducing the inversion degree of MnFe2O4 nanoparticles through synthesis to enhance magnetization: Evaluation of their 1H NMR relaxation and heating efficiency. *Dalton Trans.* **2014**, *43*, 12754–12765. [CrossRef]
82. Fernández-Barahona, I.; Muñoz-Hernando, M.; Ruiz-Cabello, J.; Herranz, F.; Pellico, J. Iron Oxide Nanoparticles: An Alternative for Positive Contrast in Magnetic Resonance Imaging. *Inorganics* **2020**, *8*, 28. [CrossRef]
83. Bin Na, H.; Song, I.C.; Hyeon, T. Inorganic Nanoparticles for MRI Contrast Agents. *Adv. Mater.* **2009**, *21*, 2133–2148. [CrossRef]
84. Cai, Y.; Wang, Y.; Xu, H.; Cao, C.; Zhu, R.; Tang, X.; Zhang, T.; Pan, Y. Positive magnetic resonance angiography using ultrafine ferritin-based iron oxide nanoparticles. *Nanoscale* **2019**, *11*, 2644–2654. [CrossRef]
85. Wei, H.; Bruns, O.T.; Kaul, M.G.; Hansen, E.C.; Barch, M.; Wisniowska, A.E.; Chen, O.; Chen, Y.; Li, N.; Okada, S.; et al. Exceedingly small iron oxide nanoparticles as positive MRI contrast agents. *Proc. Natl. Acad. Sci. USA* **2017**, *114*, 2325–2330. [CrossRef]

nanomaterials

MDPI

Article

Influence of Oxygen Flow Rate on Channel Width Dependent Electrical Properties of Indium Gallium Zinc Oxide Thin-Film Transistors

Gwomei Wu * and Anup K. Sahoo

Institute of Electro-Optical Engineering, Chang Gung University, Chang Gung Memorial Hospital, Taoyuan 333, Taiwan; anup140387@gmail.com

* Correspondence: wu@mail.cgu.edu.tw; Tel.: +886-3-211-8800

Received: 3 October 2020; Accepted: 25 November 2020; Published: 27 November 2020

Abstract: The effects of various oxygen flows on indium gallium zinc oxide (IGZO) based thin-film transistors (TFTs) with different channel width sizes have been investigated. The IGZO nano-films exhibited amorphous phase while the bandgap energy and sheet resistance increased with increasing oxygen flow rate. The electrical characteristics were evaluated with different sizes in channel width using fixed channel length. The distributions in terms of threshold voltage and current on–off level along the different channel width sizes have been discussed thoroughly. The minimum distribution of threshold voltage was observed at an oxygen flow rate of 1 sccm. The TFT electrical properties have been achieved, using an oxygen flow rate of 1 sccm with 500 μm channel width, the threshold voltage, ratio of on-current to off-current, sub-threshold swing voltage and field effect mobility to be 0.54 V, 10^6, 0.15 V/decade and 12.3 $cm^2/V{\cdot}s$, respectively. On the other hand, a larger channel width of 2000 μm could further improve the ratio of on-current to off-current and sub-threshold swing voltage to 10^7 and 0.11 V/decade. The optimized combination of oxygen flow and channel width showed improved electrical characteristics for TFT applications.

Keywords: IGZO; TFT; oxygen flow; channel width; electrical property

1. Introduction

The flat panel display applications such as active-matrix liquid crystal displays (AMLCDs), active-matrix organic light-emitting diode (AMOLED) displays and electrophoretic displays are widely dependent on transparent conducting oxide (TCO) based thin-film transistor (TFT) as a switching element [1–3]. The TCOs are also useful for the next generation optoelectronic devices for the potential to produce high/low conductivity with high visual transparency at the same instant. The amorphous indium gallium zinc oxide (*a*-IGZO) has become one of the emerging TCO materials for TFT applications due to its high transmittance, low-cost processing at low temperature, and excellent surface quality [4,5]. It exhibits better performance than those based on hydrogenated amorphous silicon (*a*-Si:H), poly-Si and other TCO materials such as ZnO and IZO, in particular by electrical properties including high field-effect mobility, sharp sub-threshold swing, and high on/off current ratio characteristics [6–8]. Besides that, the *a*-IGZO film has been used in other applications such as an alignment layer for LCD and bio-sensor technology [9,10]. In addition, it contemplates an encouraging material for higher field effect mobility in amorphous phase which can be easily grown at lower temperature on Si, glass or flexible substrates using several deposition methods [11–13]. Among all, radio frequency (RF) sputtering has been used widely to achieve high-quality *a*-IGZO films. The other important issue is a suitable gate dielectric material for TFT devices. A simple, low-cost e-beam deposited SiO_2 has been established as an alternative promising gate dielectric material to obtain high-performance *a*-IGZO TFT applications [14].

Inevitably, it is still desirable to optimize the process parameters such as argon flow, oxygen flow, deposition radio frequency power, substrate temperature, channel layer thickness and chamber pressure, etc. There have been several reports on the performances of *a*-IGZO TFTs to consider those types of process parameters during film formation [15–17]. Especially, the electrical properties of *a*-IGZO TFTs were influenced by the oxygen flow rate (OFR) during film growth immensely [18]. The physical properties of film could be affected by the absorption of oxygen molecules during film formation [19]. As an n-type material for doping, each oxygen vacancy generates two free electrons in the conductor band. The adsorbed oxygen at the surface can further influence the stability of IGZO TFTs [20]. The electron concentration decreased and would increase the resistivity in the oxide film with increased oxygen content, which influenced the electrical characteristics of the *a*-IGZO TFT [21]. On the other hand, the TFT performances depend not only on the process parameters but are also influenced by channel dimension with contact source-drain of *a*-IGZO TFT [22–24]. Generally, the TFT performances are not influenced too much by the variation of the channel width. The infant change of threshold voltage with a width of TFTs is considered as an infringe effect due to the instability of devices from the poor fabrication process. However, several researchers reported the noticeable variation of electrical properties with changing channel width of TFTs based on tunneling effect, heating effect and trap charge effect. For example, Lee et al. focused on the various channel width at a fixed length of TFTs using high-k dielectric material ZrO_2 and obtained good performance and showed better stability of TFTs using argon ambient deposited *a*-IGZO film [25]. They also showed that the electrical performance changed slightly with a change in the size of the channel width of *a*-IGZO TFTs. In addition, Liu et al. investigated channel width-dependent threshold voltage variation in *a*-IGZO TFTs with different drain voltages on lower size to higher size channel width and have shown the noticeable variation of threshold voltage with channel width at low drain to source voltage [26]. Kwon et al. studied the performance of organic TFTs with variation in channel width and reported the large variation of threshold voltage and mobility with increased channel width [27]. Hatzopoulos et al. investigated the channel width-dependent electrical characteristics of bilayer amorphous and nano-crystalline silicon TFTs [28]. However, the conjugate effect of O_2 flow rate and channel width on the electrical performances of *a*-IGZO TFTs has not been established. This type of study would be important to develop TFTs with different channel widths controlled by a common bottom gate structure deposited on the glass substrate at room temperature fabrication process. It is scientifically interesting to understand the basic device operation phenomena without a complicated high-temperature annealing process and the passivation layer. It is desirable to investigate the *a*-IGZO TFT performances at different channel dimensions using low-cost, easy processing e-beam deposited gate dielectrics fabricated on glass substrates for flat-panel display applications. It can be figured out with different O_2 flow rates along the different sizes of channel width in *a*-IGZO TFTs with a possible mechanism. The variation of threshold voltage along the different size of channel width could be minimized with an optimized O_2 flow rate. The different flow rates of O_2 create more variation of electrical performances TFTs in threshold voltage, field mobility, the ratio of on-current to off-current and sub-threshold swing voltage along with changes of channel resistance, parasitic source-to-drain resistance, surface roughness and trap charge density, etc.

In this report, we focused on the influence of O_2 flow rate on the electrical properties of different sizes of channel width *a*-IGZO TFTs. This study attempted to find the suitable oxygen flow during film growth for achieving uniform electrical characteristics from the low size to higher size width of TFTs with a possible mechanism. Thus, the fabrication of TFTs with different sizes of channel width would be more favorable to be employed in the display application. The variation of other performance parameters such as field effect mobility, the ratio of on-current to off-current, sub-threshold swing voltage, trap charge density was investigated in addition to a threshold voltage at different oxygen contents as well as different sizes of the channel width. It is therefore possible to obtain high-performance *a*-IGZO TFT with optimized O_2 flow rate, the dimension of channel width, and argon flow rate using low-cost e-beam deposited SiO_2 gate dielectric.

2. Materials and Methods

The indium-tin oxide (ITO)-coated glass substrates had a sheet resistance of 15 Ω/sq. They were washed in acetone, alcohol and DI water, subsequently, before being dried by N_2 gun and a hotplate. A vacuum tape was used to provide the common bottom gate of ITO. The experimental samples were deposited with gate dielectric SiO_2 ~200 nm by an e-beam evaporator system. The e-beam chamber's base vacuum pressure was 8×10^{-6} Torr. During the evaporation, it increased to about 10^{-4}–10^{-5} Torr due to the evaporation of the depositing materials [29]. The chamber temperature was also slightly increased during the deposition to 23–30 °C due to the heating effect from the e-beam sources. The *a*-IGZO ~40 nm films were deposited by RF sputtering using the power of 70 W at 3 mTorr pressure under a fixed Ar flow rate of 30 sccm and different O_2 flow rates, including 0, 1, 3 and 6 sccm. Then, 100 nm Al metallic film was deposited by a thermal evaporator system and was patterned to form the different sizes of TFT source/drain following photolithography and lift-off. The channel width was varied at 500 µm, 800 µm, 1500 µm and 2000 µm. The channel length was fixed at 200 µm. On the other hand, SiO_2 was deposited on p + Si at the same time for capacitance measurement and *a*-IGZO film grown on clean bare glass for the investigation of physical characteristics of films. A schematic illustration of the device structure is shown in Figure 1. A micrograph of the un-patterned common gate with different channel width based *a*-IGZO TFTs is displayed in Figure 2. The optical transmittance characteristics (Jasco, ISN-723, Tokyo, Japan), X-ray diffraction (XRD, Bruker, D2 Phaser, Billerica, MA, USA) analysis, and X-ray photoelectron spectroscopy (XPS, VG Scientific Microlab 350, Waltham, MA, USA) were carried out. The sheet resistance of *a*-IGZO film was measured using a Hall-effect automatic measurement system (KeithLink Tech, Taipei, Taiwan). For quality assurance, the films were grown on cleaned bare glass at the same time when preparing for the TFT samples. In addition, the TFT devices' electrical properties were evaluated by a semiconductor parameter analyzer (Agilent, B1500A, Santa Clara, CA, USA).

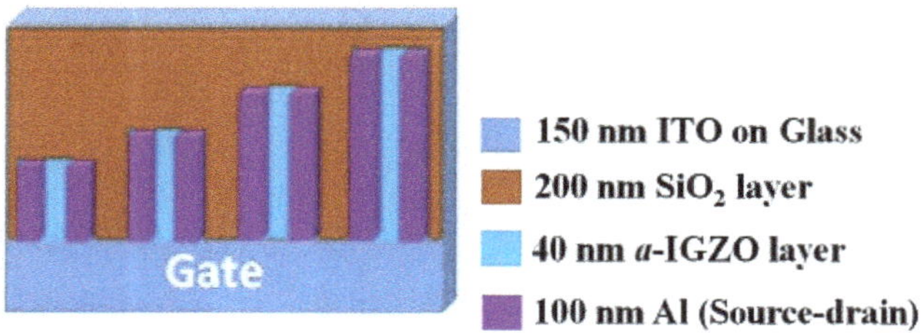

Figure 1. Schematic illustration of the device structure of amorphous indium gallium zinc oxide (*a*-IGZO) thin-film transistor (TFT) fabricated on a glass substrate.

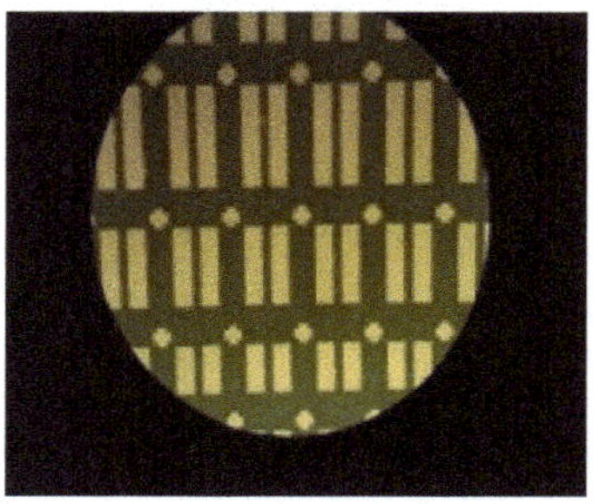

Figure 2. A micrograph of un-patterned *a*-IGZO TFT devices with different channel sizes.

3. Results and Discussion

To investigate the effects of O_2 flow rate during sputtering of IGZO film, the film microstructure was characterized by X-ray diffraction analysis. The amorphous structure was confirmed by XRD analysis at different O_2 flow rates at constant argon flow. The amorphous structure remained unchanged with the increased oxygen flow during film growth, as shown in Figure 3. The notable halo crest occurs at 23° from SiO_2 of the glass substrate but there was no individual sharp diffraction peak for crystalline structure because of the irregular arrangement of atoms inside of layer. The physical structure of *a*-IGZO films at different oxygen contents has not been well-justified because of the amorphous phase. Nevertheless, the crystalline properties can be achieved at higher O_2 flow with high-temperature annealing in vacuum ambient [30].

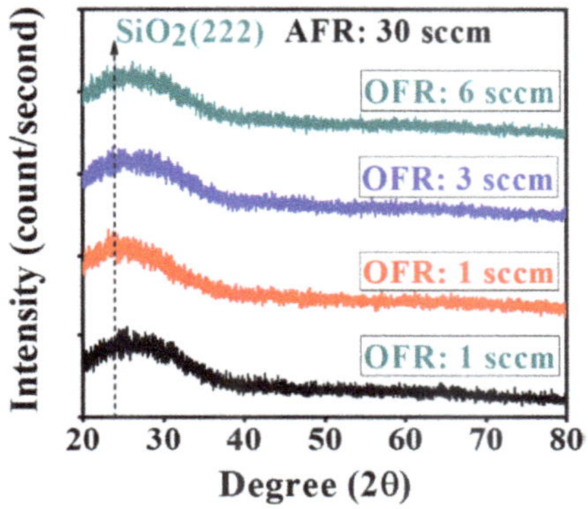

Figure 3. The amorphous phase of the as-deposited IGZO films at different O_2 flow rates was confirmed by XRD analysis.

The optical transmittance spectra at different oxygen flow rates of the as-deposited *a*-IGZO films on the clean bare glass are shown in Figure 4. It has appeared that the transmittance is higher than 90% in the visible and near-infrared range (wavelength 450 nm to 1000 nm) for the RF sputtered films under the various O_2 flow rates. The transmittance increased with an increased O_2 flow rate from 0 sccm to 3 sccm at the longer wavelength region due to improved internal film quality with a decreased defect. The oxygen ion could easily escape during the deposition. However, the transmittance at the highest O_2 flow rate of 6 sccm dropped slightly, likely caused by the defects in the nanostructure of film due to the high density of oxygen content. A sharp absorption occurred near 350 nm only for the oxygen-based samples because of the strong electronic band transitions of carriers. It is also shown that at lower than 370 nm wavelength region the transmittance drops sharply for all the IGZO films (inset (a)). This suggests improved film quality using oxygen flow in conjunction with the argon flow process. The bandgap energy of each film can be calculated using the plot method by the equation in [31]:

$$\alpha hv = k\,(hv - E_g)^{1/2}, \tag{1}$$

where α, h and v represent the absorption coefficient of *a*-IGZO film, Planck constant and the radiation frequency, and k is a constant. The absorption coefficient is determined using the following equation [32]

$$\alpha = [\ln(1/T)]/d, \tag{2}$$

where T is the measured transmittance and d is the film thickness. The optical band gap (Eg) can be obtained by extrapolating the straight-line portion of $(\alpha hv)^2$ vs. hv plots to the energy axis. The bandgap energy extends from 3.47 eV to 3.55 eV with the increase in O_2 flow rate from 0 to 6 sccm. This phenomenon is likely caused by a change in carrier concentration [33,34].

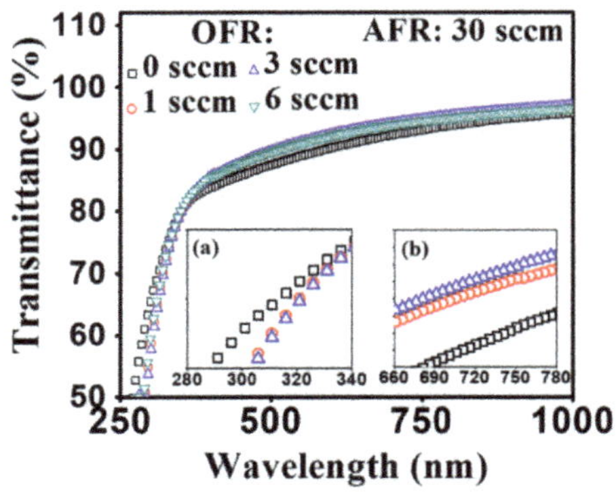

Figure 4. Optical transmission spectra of radio frequency (RF) sputtered *a*-IGZO films on clean bare glass with different O_2 flow rates during film growth. Subfigures are shown for the variation of transmittance at (**a**) shorter and (**b**) longer wavelength range.

The XPS analysis was carried out for further physical characteristics of the as-deposited *a*-IGZO films, and the results are shown in Figure 5. The variation of the binding energy of In $3d_{5/2}$, Ga $2p_{3/2}$ and Zn $2p_{3/2}$ are not so much affected as compared with the O1s state by the change in oxygen flow rate. The chemical shift of O1s is more prominent than the other electronic states because of the insensitive nature of In, Ga and Zn at the normal environment. The binding energy values at the peak intensity of O1s state with an oxygen flow rate of 0 sccm, 1 sccm, 3 sccm and 6 sccm are 530.2 eV, 530.4 eV, 530.5 eV and 530.7 eV, respectively. The binding energy values at peak intensity for all the elements with different OFR are shown in Table 1. The enhancement in binding energy suggested that the carrier concentration of film was altered when more oxygen content was introduced in the film. The drop in intensity at a high O_2 flow rate of 6 sccm agrees well with the earlier results in enhanced defect properties with decreased transmittance. Yao et al. revealed that the electron concentration decreased with increased oxygen flow rate while the atom ratios of ZnO, Ga_2O_3, and In_2O_3 remained almost the same [35]. Lee et al. found that the value of Ga/(In + Ga + Zn) and Zn/(In + Ga + Zn) ratio was increased and In/(In + Ga + Zn) decreased with increasing oxygen flow rate [36]. The noticeable variation of the binding energy of In 3d5/2, Ga 2p3/2 and Zn 2p3/2 could be changed with various RF power during a-IGZO film growth [37].

Table 1. The binding energy values at peak intensity of In $3d_{3/2}$, Ga $2p_{3/2}$, Zn $2p_{3/2}$, and O 1s electronic state with different oxygen flow rate (OFR), along with the corresponding standard values.

Element		$In3d_{3/2}$	$Ga2p_{3/2}$	Zn $2p_{3/2}$	O1s
Standard value		451.3	1117.4	1021.5	531.4
OFR	0	451.8	1117.5	1021.7	530.2
	1	451.8	1117.5	1021.7	530.4
	3	451.8	1117.5	1021.7	530.5
	6	451.8	1117.5	1021.6	530.7

In order to change the parasitic resistance of source to drain with different width channel of *a*-IGZO TFTs, the surface sheet resistance was measured to understand the surface resistance properties of the as-deposited *a*-IGZO films at different O_2 flow rates. The sheet resistance values of the *a*-IGZO surface are 9.64×10^{11} Ω/sq, 3.54×10^{12} Ω/sq, 8.53×10^{12} Ω/sq and 9.54×10^{13} Ω/sq for the various O_2 flow rate of 0 sccm, 1 sccm, 3 sccm and 6 sccm, respectively. The sheet resistance increasing tendency revealed the high surface roughness as well as decreased carrier concentration of film due to the higher density of oxygen content plasma. The sheet resistances were evaluated to set up the possible correlation with different sizes of channel width at each O_2 flow rate. The variation of sheet resistance and bandgap energy with different oxygen flow are plotted together in Figure 6.

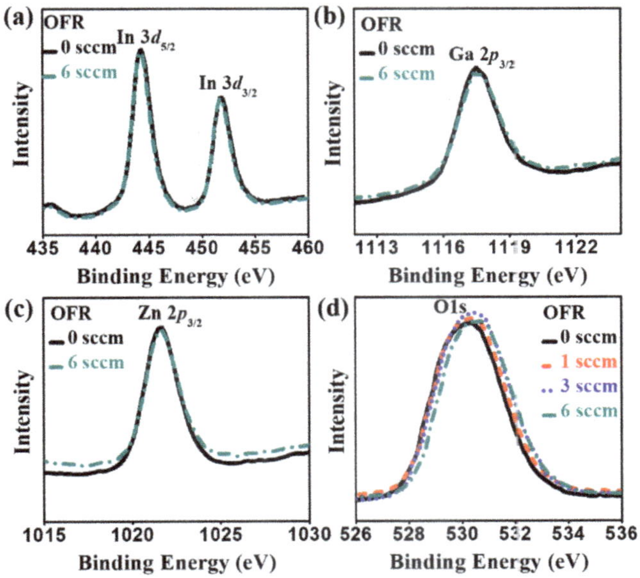

Figure 5. The XPS analysis of the as-deposited *a*-IGZO films with the different oxygen flow rates: (**a**) In $3d_{5/2}$ (**b**) Ga $2p_{3/2}$ (**c**) Zn $2p_{3/2}$ and (**d**) O1s (solid, dash, dot and dash–dot lines are indicated for OFR at 0 sccm, 1 sccm, 3 sccm and 6 sccm, respectively).

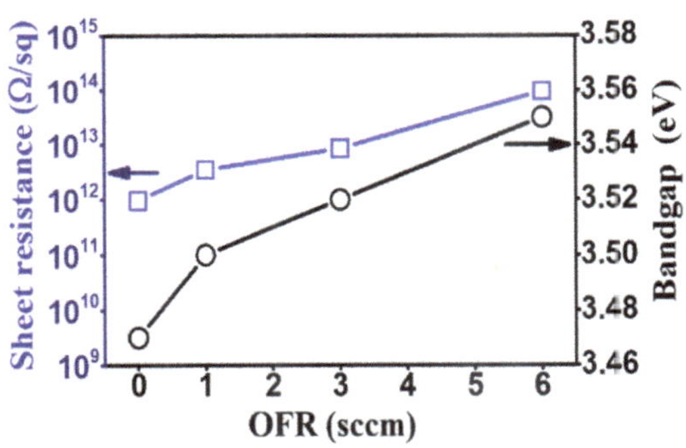

Figure 6. The variation of sheet resistance and bandgap energy of the as-deposited *a*-IGZO films with various oxygen flow rates.

The influence of O_2 flow rate is substantial by channel width variation. Figure 7 displays the transfer characteristics of *a*-IGZO TFT with the different O_2 flow rate at 0 sccm, 1 sccm, 3 sccm and 6 sccm. The drain voltage was fixed at 1 V, and the swing gate voltage was varied from −1 V to +3 V. In general, the transfer characteristics showed that a minimum distribution in terms of threshold voltage variation, on-current and off-current level along the increasing channel width, was observed from Figure 7b using the oxygen flow rate of 1 sccm. On the other hand, the maximum distribution was found from Figure 7d with the highest O_2 flow rate at 6 sccm. It should be noted that the off-current level increased up at the beginning of negative voltage through the lower size to higher size devices, especially at the −1 to 0 V negative bias. Also, there has been smoother characteristics respect to off-current level, and on-voltage from sweep gate voltage −1 V to 3 V achieved only at oxygen flow rate of 1 sccm rather than the other flow rates. In order to understand the impact on the electrical prosperities by the process parameter of various O_2 flow rate and physical dimension of channel width, we further characterized each transfer characteristic curve to obtain field-effect mobility (μ_{fet}), threshold voltage (V_{th}), sub-threshold swing voltage (*SS*) and the ratio of on current to off current (I_{on}/I_{off}). Consequently, the reason for the dissimilarity of each parameter would be discussed on the mechanism due to oxygen flow treatment and channel width. After that, both effects would be combined to show how they influence together on electrical characteristics of *a*-IGZO TFTs. Furthermore, the effects of OFR and channel width on gate leakage current are shown in Figure 7e,f. The leakage current increased substantially with increasing OFR for the channel width of 500 μm, which could impact the device's performances. On the other hand, the effect of channel width on gate leakage current at OFR of 1 sccm could be considerably smaller.

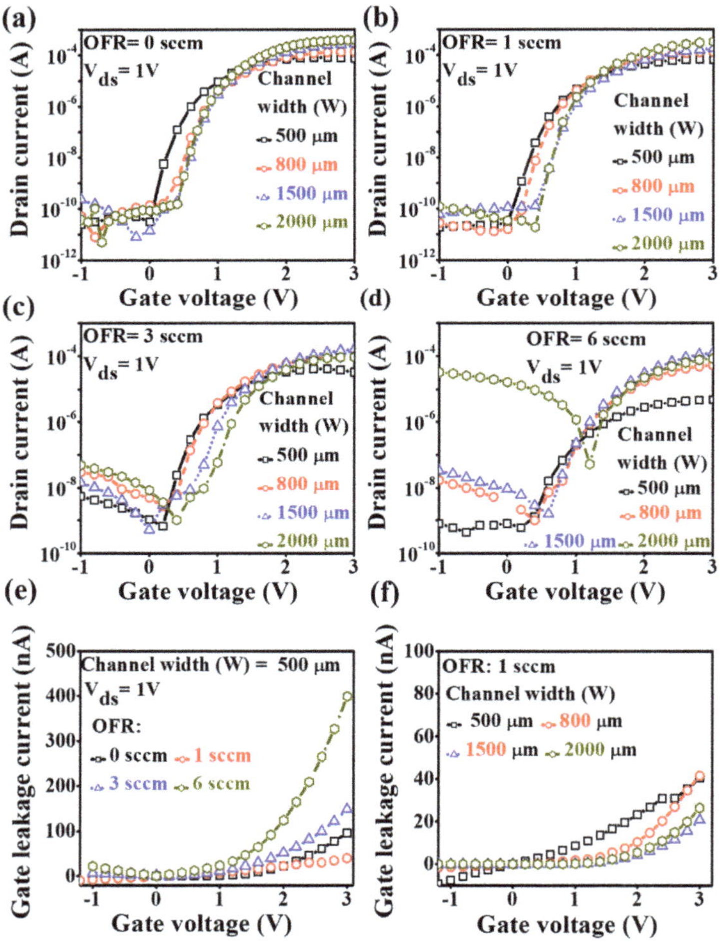

Figure 7. Typical transfer characteristics of *a*-IGZO based TFT with oxygen flow rate of (**a**) 0 sccm, (**b**) 1 sccm, (**c**) 3 sccm and (**d**) 6 sccm. The fixed drain voltage is 1 V, gate voltage sweeping from −1 V to +3 V (the open square with solid line, open circle with dash line, open triangle with dot and open hexagon with dash–dot line represent the transfer characteristics for channel width 500 μm, 800 μm, 1500 μm and 2000 μm, respectively). The effects of (**e**) OFR and (**f**) channel width (W) on gate leakage current are presented.

The field-effect mobility can be extracted using the following equation:

$$\mu_{\text{fet}} = \frac{g_{\text{m}} \times L}{C_{\text{ox}} \times W \times V_{\text{ds}}}, \tag{3}$$

where g_m is the transconductance that is defined by the derivative of drain current with respect to gate voltage, L is the length of the channel, W is the width of the channel and C_{ox} is gate oxide capacitance per unit area which was obtained at 4 nF/cm^2 by C-V measurement of MOS capacitor using the same thickness and maintained same deposition condition of SiO_2. The obtained field effect mobility without O_2 flow-based sample was 11.9 cm^2/V·s, 9.4 cm^2/V·s, 8.3 cm^2/V·s and 7.4 cm^2/V·s for the corresponding channel widths of 500 µm, 800 µm, 1500 µm and 2000 µm, respectively. The mobility was slightly improved for each size of channel width based TFTs when introduced with small amount of oxygen flow rate at 1 sccm and the values became 12.3 cm^2/V·s, 10.8 cm^2/V·s, 9.4 cm^2/V·s and 9.1 cm^2/V·s. However, the mobility has become lower at the higher oxygen flow rates at 3 sccm and 6 sccm, with a maintained tendency of decreased associates along with increased channel width size. The calculated mobility results for all sizes of channel width at different oxygen flow rates are presented in Tables 2–5. The field effect mobility decreased with increasing O_2 flow rate for each size of channel width because of the increased defects. This trend is in good agreement with the reported results based on the oxygen flow rate study [38]. It is also effectively degraded with increasing channel width at each fixed oxygen flow rate due to the increased surface to volume ratio of the channel surface with different contact size of channel width-based source-drain. However, the mobility could be increased with increasing channel width due to the decrease of the source-drain contact resistance at a fixed ratio of W/L [39]. The mobility could be improved using the little oxygen content of plasma during the deposition. The fluctuations of mobility along the channel width were also suppressed at the lower oxygen flow rate. These changes of mobility may arise from the dimension effect of TFTs rather than just oxygen flow rate. The high oxygen flow rate at 6 sccm greatly influenced the mobility of TFTs and resulted in very low mobility for all types of channel width because of not only higher defects and decreased carrier concentration but also the increased parasitic effect due to higher sheet resistance with increasing sizes of source-drain. These results also suggest that field effect mobility with the various channel width can be controlled using the appropriate oxygen flow rate in plasma during film growth.

Table 2. The electrical performance parameters of *a*-IGZO TFT with different sizes of channel width at an oxygen flow rate of 0 sccm.

Channel Width (µm)	V_{th} (V)	I_{on}/I_{off}	SS (V/decade)	μ_{fet} (cm^2/V·s)
500	0.41	2.2×10^6	0.18	11.9
800	0.65	1.6×10^6	0.22	9.4
1500	0.89	3.0×10^7	0.16	8.3
2000	0.90	1.1×10^7	0.13	7.4

Table 3. The electrical performance parameters of *a*-IGZO TFT with different sizes of channel width at an oxygen flow rate of 1 sccm.

Channel Width (µm)	V_{th} (V)	I_{on}/I_{off}	SS (V/decade)	μ_{fet} (cm^2/V·s)
500	0.54	2.6×10^6	0.15	12.3
800	0.66	1.0×10^7	0.17	10.8
1500	0.77	1.2×10^6	0.13	9.4
2000	0.86	1.5×10^7	0.11	9.1

Table 4. The electrical performance parameters of *a*-IGZO TFT with different sizes of channel width at an oxygen flow rate of 3 sccm.

Channel Width (μm)	V_{th} (V)	I_{on}/I_{off}	SS (V/decade)	μ_{fet} (cm²/V·s)
500	0.58	5.9 × 10⁵	0.23	10.2
800	0.69	3.4 × 10⁵	0.27	8.8
1500	0.98	2.7 × 10⁵	0.21	7.6
2000	1.0	8.2 × 10⁴	0.20	5.8

Table 5. The electrical performance parameters of *a*-IGZO TFT with different sizes of channel width at an oxygen flow rate of 6 sccm.

Channel Width (μm)	V_{th} (V)	I_{on}/I_{off}	SS (V/decade)	μ_{fet} (cm²/V·s)
500	0.64	5.9 × 10³	0.35	5.5
800	0.94	4.2 × 10⁴	0.26	4.6
1500	0.98	7.1 × 10⁴	0.25	3.9
2000	1.16	1.2 × 10³	0.44	-

The threshold voltage (V_{th}) is calculated from each transfer characteristic curve by extra-plotting of $I_{ds}{}^{1/2}$ vs. V_{gs}, and the calculated values are also shown in Tables 2–5. The threshold voltages for the pure argon based samples are 0.41 V, 0.65 V, 0.89 V and 0.90 V, corresponding to the channel width of 500 μm, 800 μm, 1500 μm and 2000 μm, respectively. The threshold voltage differs by 0.49 V from the lower size to the higher size channel width. On the other hand, the threshold voltage for the oxygen flow rate at 1 sccm based samples are 0.54 V, 0.66 V, 0.77 V and 0.86 V, with a corresponding channel width of 500 μm, 800 μm, 1500 μm and 2000 μm, respectively. It has been noted that a hysteresis analysis by taking a double sweep could help to extract information on gate insulator and interface quality [40,41]. The variation of the threshold voltage is about only 0.32 V from the lower to higher size channel width. The threshold voltage was increased with increasing channel width, while the variation values from lower to higher size channel width at an oxygen flow rate of 3 sccm and 6 sccm are 0.43 V and 0.52 V. It has been suggested that the threshold voltage is small for the smaller channel width of the device and the threshold voltage is increased with increasing channel width at each O_2 flow rate. It increased as well with an increased O_2 flow rate at a fixed channel width. The threshold voltage change with channel width has been well-maintained with respect to the O_2 flow rate fixed at 1 sccm. The effects of both oxygen content and dimension of channel width played some important roles. It became more irregular when the oxygen flow increased to a higher rate at 6 sccm. The higher size devices associated with 1500 μm and 2000 μm squeeze the threshold voltage at an oxygen flow rate of 1 sccm even more than the samples with no oxygen flow. The results indicate that the proper amount of oxygen flow can be helpful to improve the uniformity of threshold voltage and to avoid the sort of channel effect of TFT from low order to high order device size for TFTs with different sizes of the channel width. The distribution of threshold voltage would be also affected by the contact of source-drain on channel surface due to sheet resistance. The total volume to surface ratio changes with channel width as the length remains constant.

Sub-threshold swing voltage (SS) is one of the key factors to characterize transistor performance for high speed and low power operation in low voltage portable devices. It is determined from the slope of log I_{ds} versus V_{gs} curve and the corresponding equation is

$$SS = \frac{d\,V_{gs}}{d(\log I_{ds})}. \tag{4}$$

The evaluated values of SS at pure argon based samples are 0.18 V/decade, 0.22 V/decade, 0.16 V/decade and 0.13 V/decade, and at an oxygen flow rate of 1 sccm they are 0.15 V/decade,

0.17 V/decade, 0.13 V/decade and 0.11 V/decade with the corresponding channel width of 500 μm, 800 μm, 1500 μm and 2000 μm, respectively. The very small *SS* is obtained at about 0.11 V/decade from the biggest size of the channel width of 2000 μm at an O_2 flow rate of 1 sccm and all the other data are also shown in Tables 2–5. The results have indicated that the uniformity of *SS* can be obtained with a small amount of oxygen flow during film deposition. The *SS* values are somewhat increased when much more oxygen is introduced to the channel deposition. The SS values are shown lower along with the sizes of channel width with the low oxygen flow rate of 1 sccm because of the smoother surface of channel material with respect to the other samples. It is well known that small SS values can lead to higher mobility and good stability of TFTs. Therefore, the optimized O_2 flow rate and channel width would be associated with 1 sccm and 2000 μm, respectively. The small *SS* values are also attributed to the lower number of interfacial trap charge densities in the interfacial surface of the channel and gate dielectric. It would be expected with a lower number of interfacial charges produced at an O_2 flow rate of 1 sccm because of lower values of *SS*. The higher number of interfacial charge densities may play a more important role to increase the off-current level due to higher leakage current with increased O_2 flow rate, and also increased channel width.

The TFT performances are very much dependent on the ratio of on-current to the off-current (I_{on}/I_{Off}) and all the evaluated values with channel width and O_2 flow rate are shown in Tables 2–5. It is significantly changed when increased O_2 flow rate on the *a*-IGZO film deposition. Even without oxygen inlet, the I_{on}/I_{Off} achieved higher than 10^7 for the larger width of the channel but remarkably decreased at the higher O_2 flow rates of 3 and 6 sccm. It is shown about 10^3–10^4 for the higher channel width and even so for the smaller sizes of channel width-based *a*-IGZO TFTs. The on-current level was increased with increasing channel width at an oxygen flow rate of 0 sccm and 1 sccm because of the increased current in the channel. The phenomena are in good agreement with the reported results by Liu et al. [26]. The distribution in both on- and off-current levels was achieved along the different channel width sizes when more oxygen was introduced during the film formation. The off-current level at the initial negative voltage is increased with the increasing channel width from lower sizes to higher sizes at each fixed O_2 flow rate. The higher channel width thus produced more leakage current that increased the off-current level. This phenomenon disturbed strongly at the higher oxygen flow rate at 6 sccm. The off-current level reached above 10^{-5} order for the largest channel width of 2000 μm based TFTs. This disturbance might be caused by the increased self-heating effect from the higher surface roughness of the channel [42]. The corresponding threshold voltage was also increased to 1.16 V. On the other hand, the on-current level for the higher size devices behaved normally like the higher on-current level corresponding to the higher size devices at an oxygen flow of 0 and 1 sccm. This effect became prohibited for the higher oxygen flow rate samples.

In the discussion, the *a*-IGZO TFTs electrical performance parameters are dependent upon the O_2 flow rate during sputtering. The transfer characteristics from the lower to higher size channel width at different O_2 flow rate showed lots of dissimilarities, such as distribution of on- and off-current level. The variation of the off-current level along O_2 flow rate followed a more-or-less linear fashion. The worst distribution was observed at the high O_2 flow rate at 6 sccm. The leakage current increased above 10^{-6} at higher sizes of channel width-based *a*-IGZO TFTs. It should be noted that the leakage current variation at negative gate voltage 1 V along the channel width depends on the oxygen flow rate. It is varied only by one order at O_2 flow rate of 1 sccm and is increased the order at 0 sccm as well as the higher flow rates of 3 sccm and 6 sccm. The variations are shown in the same manner for each channel width-based TFTs. In this report, the threshold voltage was also increased with increasing O_2 flow rate due to the higher resistivity of the film. The subthreshold swing voltages increased with increasing O_2 flow rate because of the increased defects. The off-current level is increased with increasing oxygen flow rate due to the increased defect in the interfacial region and the lower density of oxygen vacancy suppresses the carrier concentration in the film. In addition, the channel width dependent field effect mobility, threshold voltage, subthreshold swing voltage and ratio of on-current to the off-current at a fixed O_2 flow rate of 1 sccm are plotted together, as shown in

Figure 8. The variation of threshold voltage with channel width is short of uniformity as compared with conventional devices. However, the increased tendency of mobility and on-current with increasing channel width has been reported by Lee et al. without considering the oxygen flow effect on the *a*-IGZO surface [24]. Nevertheless, the process parameter of O_2 flow rate changed the sheet resistance of the channel surface which could increase the parasitic effect of source-drain on channel material. This effect gradually increased as the channel width of the *a*-IGZO TFT was increased [22]. Therefore, the O_2 flow rate effect on the TFTs with different sizes of channel width lead to large variation without considering the optimization of O_2 flow rate during the deposition of *a*-IGZO films. The source to drain resistance could influence the electrical performances of *a*-IGZO TFT and as a result to have achieved the short channel effect due to the lack of ohmic contact. This affected the voltage drop at the source and drain contacts. In addition, the water and oxygen molecules in the atmosphere can also affect the device performance of the oxide semiconductor-based TFTs, due to the lack of a passivation layer or cap layer [20]. A suitable passivation layer could be thus employed to improve the uniformity of the *a*-IGZO TFTs. The variation (Δ) of each of the electrical parameters from low value to high value with the various O_2 flow rates are plotted together, as shown in Figure 9. In addition, the e-beam deposited SiO_2 is a promising gate dielectric at a low drain voltage of 1 V.

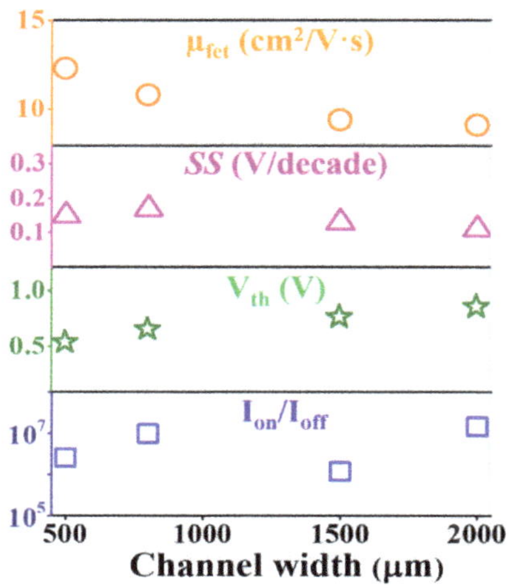

Figure 8. The field effect mobility (μ_{fet}), sub-threshold swing voltage (*SS*), threshold voltage (V_{th}), and ratio of on-current to off-current (I_{on}/I_{off}) with different sizes of channel width at an oxygen flow rate of 1 sccm.

The phenomena regarding changing performance parameters of *a*-IGZO TFTs at the higher fixed drain voltage of 2~5 V exhibited a similar trend. However, at drain voltage higher than 6 V, the performances were degraded due to the intrinsic characteristics of the as-deposited gate dielectric SiO_2. Further annealing and/or plasma treatment can be developed to improve the quality of the SiO_2 dielectric film.

The output characteristics of the *a*-IGZO TFT are shown in Figure 10 for the devices from O_2 flow rate of 0 sccm, 1 sccm and 6 sccm for the lower size to higher size channel width. All the output characteristics exhibited positive drain current at zero gate voltage because of carrier transportation of channel by the drain to source voltage. Again, the output characteristics supported the better ohmic performance of *a*-IGZO TFT at an oxygen flow rate of 1 sccm than the other OFR samples. Three different output characteristics were shown with saturation region, saturation current and pinch of voltage. On the other hand, the hump effect increased with increasing O_2 flow rate. The effect might be originated from structural defects which act as traps for the charge carriers under bias. Thus, higher OFR introduced higher structural defects and decreased the ohmic contact properties of metal on the *a*-IGZO surface, especially for large channel width. The stress-induced hump phenomenon was also observed by Choi and Han for wide channel width (W > 100 μm) [43]. This problem could be avoided by employing the post-fabrication annealing process. However, the characteristics have been shown with excellent n-type enhancement mode with low drain current at zero gate voltage. In addition, the output characteristics revealed that the ohmic contact could be affected by sheet resistance with unsuitable O_2 content in the a-IGZO film. The stability and better ohmic contact could be further improved with a surface treatment such as annealing, plasma process, or even by replacing the metal of source-drain contact.

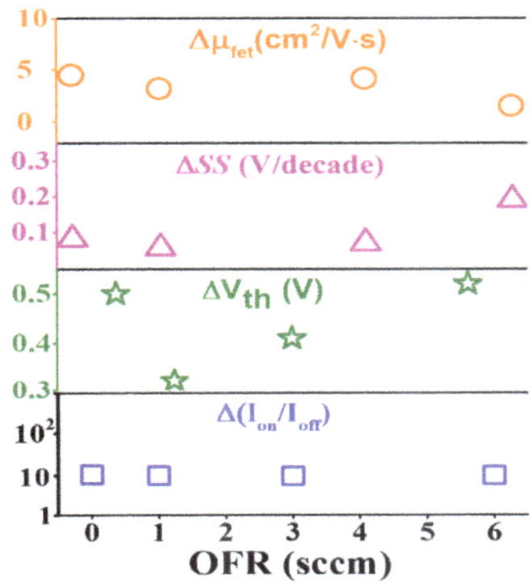

Figure 9. The variation (Δ) of each electrical parameter from the lower to higher sizes channel width with the various O_2 flow rates.

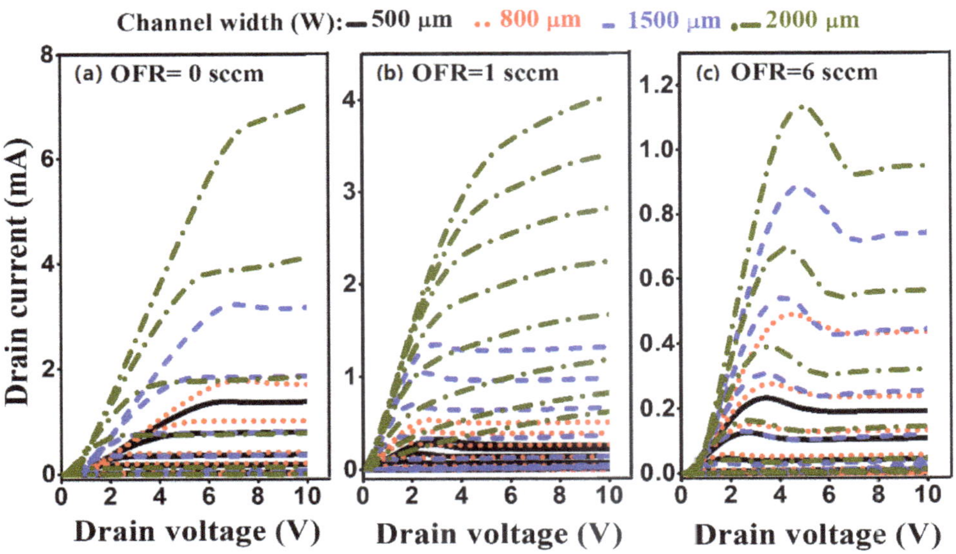

Figure 10. The output characteristics of *a*-IGZO TFT at an oxygen flow rate of (**a**) 0 sccm, (**b**) 1 sccm and (**c**) 6 sccm. The gate voltage varied −1 V to 3 V with an increased step voltage of 0.5 V. (Solid, dot, dash and dot–dash lines represented for channel width 500 μm, 800 μm, 1500 μm and 2000 μm, respectively).

4. Conclusions

In conclusion, O_2 flow rate is one of the key factors for obtaining better performances, especially for TFTs with different sizes of the channel width. The results showed that the suitable channel width at an optimized oxygen flow rate could produce high-performance *a*-IGZO TFTs. However, the very high O_2 flow rate could degrade the performance of TFT, in particular I_{on}/I_{off} ratio as compared with the small amount of O_2 in the irrespective channel width. The small amount O_2 flow rate would be sufficient to obtain high performances and also to exhibit less variation with channel width in TFTs. This study has paved the way to minimize the variation of threshold along the different dimension TFTs usingthe single optimized oxygen flow rate. The optimized TFT electrical properties were achieved at the oxygen flow rate of 1 sccm with 500 μm channel width. The threshold voltage, the ratio of on-current to off-current, sub-threshold swing voltage and field effect mobility were 0.54 V, 10^6, 0.15 V/decade and 12.3 cm^2/V·s, respectively. The mobility decreased with increasing oxygen flow rate from 1 to 6 sccm by ~55%. The threshold voltage increased by ~16% at the constant channel width of 500 μm. The better ohmic contact behavior and high performance TFTs can be achieved while considering both the effects of O_2 flow rate and the channel width design for the future TFT applications.

Author Contributions: Methodology, A.K.S.; investigation, G.W. and A.K.S.; writing—original draft preparation, A.K.S.; writing—review and editing, G.W.; supervision, G.W. All authors have read and agreed to the published version of the manuscript.

Funding: This research was funded in part by the Ministry of Science and Technology under research grant MOST108-2221-E182-020-MY3. The APC was funded by CGMH financial support of CMRP246 and CMRPD3G0063.

Conflicts of Interest: The authors declare no conflict of interest.

References

1. Park, J.S.; Maeng, W.J.; Kim, H.S.; Park, J.S. Review of recent developments in amorphous oxide semiconductor thin-film transistor devices. *Thin Solid Films* **2012**, *520*, 1679–1693. [CrossRef]
2. Kim, H.; Jung, C.; Kwon, J.Y.; Kim, S. Analysis of key factors for high yield AMOLED display. *J. Display Technol.* **2015**, *11*, 783–787. [CrossRef]
3. Hu, S.; Ning, H.; Lu, K.; Fang, Z.; Li, Y.; Yao, R.; Xu, M.; Wang, L.; Peng, J.; Lu, X. Mobility enhancement in amorphous In-Ga-Zn-O thin-film transistor by induced metallic in nanoparticles and Cu electrodes. *Nanomaterials* **2018**, *8*, 197. [CrossRef] [PubMed]
4. Wang, J.L.; Yang, P.Y.; Juang, M.H.; Hsieh, T.Y.; Hwang, C.C.; Juan, C.P.; Lee, I.C. Zinc oxide thin-film transistors fabricated via low-temperature hydrothermal method. *Surf. Coat. Technol.* **2013**, *231*, 428–432. [CrossRef]
5. Moreira, M.; Carlos, E.; Dias, C.; Deuermeier, J.; Pereira, M.; Barquinha, P.; Branquinho, R.; Martins, R.; Fortunato, E. Tailoring IGZO composition for enhanced fully solution-based thin-film transistors. *Nanomaterials* **2019**, *9*, 1273. [CrossRef]
6. Xu, F.; Liu, A.; Liu, G.; Shin, B.; Shan, F. Solution-processed yttrium oxide dielectric for high-performance IZO thin-film transistors. *Ceram. Int.* **2015**, *41*, S337–S343. [CrossRef]
7. Jeong, H.; Nam, S.; Park, K.; Yoon, S.; Park, C.; Jang, J. Significant performance and stability improvements of low-temperature IGZO TFTs by the formation of In-F nanoparticles on an SiO_2 buffer layer. *Nanomaterials* **2020**, *10*, 1165. [CrossRef]
8. Cho, S.; Cho, W. Performance enhancement of electrospun IGZO-nanofiber-based field-effect transistors with high-k gate dielectrics through microwave annealing and postcalcination oxygen plasma treatment. *Nanomaterials* **2020**, *10*, 1804. [CrossRef]
9. Wu, G.M.; Liu, C.Y.; Sahoo, A.K. RF sputtering deposited a-IGZO films for LCD alignment layer application. *Appl. Surf. Sci.* **2015**, *354*, 48–54. [CrossRef]
10. Shen, Y.C.; Yang, C.H.; Chen, S.W.; Wu, S.H.; Yang, T.L.; Huang, J.J. IGZO thin film transistor biosensors functionalized with ZnO nanorods and antibodies. *Biosens. Bioelectron.* **2014**, *54*, 306–310. [CrossRef]

11. Huang, S.Y.; Chang, T.C.; Chen, M.C.; Chen, T.C.; Jian, F.Y.; Chen, Y.C.; Huang, H.C.; Gan, D.S. Improvement in the bias stability of amorphous InGaZnO TFTs using an Al_2O_3 passivation layer. *Surf. Coat. Technol.* **2013**, *231*, 117–121. [CrossRef]
12. Wu, C.H.; Chang, K.M.; Huang, S.H.; Deng, I.C.; Wu, C.J.; Chiang, W.H.; Chang, C.C. Characteristics of IGZO TFT prepared by atmospheric pressure plasma jet using PE-ALD Al_2O_3 gate dielectric *IEEE Electr. Device Lett.* **2012**, *33*, 552–554. [CrossRef]
13. Choi, J.H.; Shim, J.H.; Hwang, S.M.; Joo, J.; Park, K.; Kim, H. Effect of sintering time at low temperature on the properties of IGZO TFTs fabricated by using the sol-gel process. *J. Korean Phys. Soc.* **2010**, *57*, 1836–1841.
14. Sahoo, A.K.; Wu, G.M. Effects of argon flow rate on electrical properties of amorphous indium gallium zinc oxide thin-film transistors. *Thin Solid Films* **2016**, *605*, 129–135. [CrossRef]
15. Tak, Y.J.; Ahn, B.D.; Park, S.Y.; Kim, S.J.; Song, A.R.; Chung, K.B.; Kin, H.J. Activation of sputter-processed indium-gallium-zinc oxide films by simultaneous ultraviolet and thermal treatments. *Sci. Rep.* **2016**, *6*, 21869. [CrossRef]
16. Hino, A.; Takanashi, Y.; Tao, H.; Morita, S.; Ochi, M.; Goto, H.; Hayashi, K.; Kugimiya, T. Effects of thermal annealing on variations of electron traps in the channel region of amorphous In-Ga-Zn-O thin film transistor. *J. Vac. Sci. Technol. B* **2014**, *32*, 031210. [CrossRef]
17. Kim, J.; Park, J.; Yoon, G.; Khushabu, A.; Kim, J.S.; Pae, S.; Cho, E.C.; Yi, J. Effect of IGZO thin films fabricated by pulsed-DC and RF sputtering on TFT characteristics. *Mater. Sci. Semicond. Process.* **2020**, *120*, 105264. [CrossRef]
18. Chen, T.C.; Chang, T.C.; Hsieh, T.Y.; Tsai, M.Y.; Tsai, C.T.; Chen, S.C.; Lin, C S.; Jian, F.Y. Analyzing the effects of ambient dependence for InGaZnO TFTs under illuminated bias stress. *Surf. Coat. Technol.* **2013**, *231*, 465–470. [CrossRef]
19. Chong, E.; Chun, Y.S.; Kim, S.H.; Lee, S.Y. Effect of oxygen on the threshold voltage of a-IGZO TFT. *J. Electr. Eng. Technol.* **2011**, *6*, 539–542. [CrossRef]
20. Fuh, C.S.; Liu, P.T.; Chou, Y.T.; Teng, L.F.; Sze, S.M. Role of oxygen in amorphous In-Ga-Zn-O thin film transistor for ambient stability. *ECS J. Solid State Sci. Technol.* **2013**, *2*, Q1–Q5. [CrossRef]
21. Liu, P.; Chen, T.P.; Li, X.D.; Liu, Z.; Wong, J.I.; Liu, Y.; Leong, K.C. Effect of exposure to ultraviolet-activated oxygen on the electrical characteristics of amorphous indium gallium zinc oxide thin film transistors. *ECS Solid State Lett.* **2013**, *2*, Q21–Q24. [CrossRef]
22. Heo, Y.W.; Cho, K.M.; Sun, S.Y.; Kim, S.Y.; Lee, J.H.; Kim, J.J.; Norton, D.P.; Pearton, S.J. Effects of channel dimensions on performance of a-$InGaZnO_4$ thin-film transistors. *J. Vac. Sci. Technol.* **2011**, *29*, 021203. [CrossRef]
23. Kim, H.W.; Kim, E.S.; Park, J.S.; Lim, J.H.; Kim, B.S. Influence of effective cannel length in self-aligned coplanar amorphous-indium-gallium-zinc-oxide thin-film transistors with different annealing temperatures. *Appl. Phys. Lett.* **2018**, *113*, 022104. [CrossRef]
24. Wu, G.; Sahoo, A.K.; Chen, D.W.; Chang, J.W. A comparative study of e-beam deposited gate dielectrics on channel width-dependent performance and reliability of *a*-IGZO thin-film transistors. *Materials* **2018**, *11*, 2502. [CrossRef]
25. Lee, J.S.; Chang, S.; Koo, S.M.; Lee, S.Y. High performance a-IGZO TFT with ZrO_2 gate dielectric fabricated at room temperature. *IEEE Electr. Device Lett.* **2010**, *31*, 225–227.
26. Liu, K.H.; Chang, T.C.; Wu, M.S.; Hung, Y.S.; Hung, P.H.; Hsieh, T.Y.; Chou, W.C.; Chu, A.K.; Sze, S.M.; Yeh, B.L. Investigation of channel width-dependent threshold voltage variation in a-InGaZnO thin-film transistors. *Appl. Phys. Lett.* **2014**, *104*, 133503. [CrossRef]
27. Kwon, J.H.; Shih, S.I.; Choi, J.; Chung, M.H.; Oh, T.Y.; Kim, K.H.; Cho, M.J.; Kim, K.N.; Choi, D.H.; Ju, B.K. Channel width effect for organic thin film transistors using TIPS-pentacene employed as a dopant of poly-triarylamine. *Org. Electron.* **2009**, *10*, 729–734. [CrossRef]
28. Hatzopoulos, A.T.; Arpatzanis, N.; Tassis, D.H.; Dimitriadis, C.A.; Templier, F.; Oudwan, M.; Kamarinos, G. Effect of channel width on the electrical characteristics of amorphous/nanocrystalline silicon bilayer thin-film transistors. *IEEE Electr. Device Lett.* **2007**, *54*, 1265–1269. [CrossRef]
29. Brezna, W.; Fischer, M.; Wanzenboeck, H.D.; Bertagnolli, E.; Smoliner, J. Electron-beam deposited SiO_2 investigated by scanning capacitance microscopy. *Appl. Phys. Lett.* **2006**, *83*, 122116. [CrossRef]

30. Thakur, A.; Yoo, H.; Kang, S.J.; Baik, J.Y.; Lee, I.J.; Lee, H.K.; Kim, K.; Kim, B.; Jung, S.; Park, J.; et al. Effects of substrate temperature on structural, electrical and optical properties of amorphous In-Ga-Zn-O thin films. *ECS J. Solid State Sci. Technol.* **2012**, *1*, Q11–Q15. [CrossRef]
31. Chen, H.; Shen, W.Z. Optical characterization of boron-doped nanocrystalline Si:H thin films. *Surf. Coat. Technol.* **2005**, *198*, 98–103. [CrossRef]
32. Aqili, A.K.S.; Maqsood, A. Determination of thickness, refractive index, and thickness irregularity for semiconductor thin films from transmission spectra. *Appl. Opt.* **2002**, *41*, 218–224. [CrossRef] [PubMed]
33. Gobbiner, C.R.; Veedu, M.A.A.; Kekuda, D. Influence of oxygen flow rate on the structural, optical and electrical properties of ZnO films grown by DC magnetron sputtering. *Appl. Phys. A* **2016**, *122*, 272. [CrossRef]
34. Lee, D.U.; Kim, S.P.; Lee, K.S.; Pak, S.W.; Kim, E.K. Band gap modulation of $ZnTe_{1-x}O_x$ alloy film by control of oxygen gas flow rate during reactive magnetron sputtering. *Appl. Phys. Lett.* **2013**, *103*, 263901. [CrossRef]
35. Yao, J.; Xu, N.; Deng, S.; Chen, J.; She, J.; Shieh, H.P.D.; Liu, P.T.; Huang, Y.P. Electrical and photosensitive characteristics of a-IGZO TFTs related to oxygen vacancy. *IEEE Electr. Device Lett.* **2011**, *58*, 1121–1126.
36. Lee, Y.S.; Yen, T.W.; Lin, C.I.; Lin, H.C.; Yeh, Y. Electrical characteristics of amorphous In-Ga-Zn-O thin-film transistors prepared by radio frequency magnetron sputtering with varying oxygen flows. *Displays* **2014**, *35*, 165–170. [CrossRef]
37. Wu, G.M.; Sahoo, A.K.; Liu, C.Y. Influence of RF power on performance of sputtered a-IGZO based liquid crystal cells. *Thin Solid Films* **2015**, *596*, 56–62. [CrossRef]
38. Lee, Y.S.; Yu, E.K.H.; Shim, D.H.; Kong, H.S.; Bie, L.; Kanicki, J. Oxygen flow effects on electrical properties, stability, and density of states of amorphous In-Ga-Zn-O thin-film transistors. *Jpn. J. Appl. Phys.* **2014**, *53*, 121101. [CrossRef]
39. Lee, J.S.; Chang, S.; Bouzid, H.; Koo, S.M.; Lee, S.Y. Systematic investigation on the effect of contact resistance on the performance of a-IGZO thin-film transistors with various geometries of electrodes. *Phys. Status Solidi A* **2010**, *207*, 1694–1697. [CrossRef]
40. Choi, S.; Kim, J.Y.; Kang, H.; Ko, D.; Rhee, J.; Choi, S.J.; Kim, D.M.; Kim, D.H. Effect of oxygen content on current stress-induced instability in bottom-gate amorphous InGaZnO thin-film transistors. *Materials* **2019**, *12*, 3149. [CrossRef]
41. Kim, S.; Jeon, Y.W.; Kim, Y.; Kong, D.; Jung, H.K.; Bae, M.K.; Lee, J.H.; Ahn, B.D.; Park, S.Y.; Park, J.H.; et al. Impact of oxygen flow rate on the instability under positive bias stresses in DC-sputtered amorphous InGaZnO thin-film transistors. *IEEE Electron. Dev. Lett.* **2012**, *33*, 6264. [CrossRef]
42. Hsieh, T.Y.; Chang, T.C.; Chen, T.C.; Chen, Y.T.; Tsai, M.Y.; Chu, A.K.; Chung, Y.C.; Ting, H.C.; Chen, C.Y. Self-heating-effect-induced degradation behaviors in a-InGaZnO thin-film transistors. *IEEE Electr. Device Lett.* **2013**, *37*, 63–65. [CrossRef]
43. Choi, S.H.; Han, M.K. Effect of channel widths on negative shift of threshold voltage, including stress-induced hump phenomenon in InGaZnO thin-film transistors under high-gate and drain bias stress. *Appl. Phys. Lett.* **2012**, *100*, 043503. [CrossRef]

nanomaterials

MDPI

Article

Creation of Curved Nanostructures Using Soft-Materials-Derived Lithography

Hyun-Ik Jang [1,2], Hae-Su Yoon [1,2], Tae-Ik Lee [3], Sangmin Lee [4], Taek-Soo Kim [4], Jaesool Shim [5,*] and Jae Hong Park [1,2,*]

1 National NanoFab Center, 291 Daehak-ro, Yuseong-gu, Deajeon 34141, Korea; hijang2018@nate.com (H.-I.J.); hsyoon2018@nate.com (H.-S.Y.)
2 Nanoin Inc., 291 Daehak-ro, Yuseong-gu, Deajeon 34141, Korea
3 Joining R&D Group, KITECH, 156 Gaetbeol-ro, Yeonsu-gu, Incheon 21999, Korea; tilee@kitech.re.kr
4 Department of Mechanical Engineering, KAIST, 291 Daehak-ro, Yuseong-gu, Deajeon 34141, Korea; lsm417@kaist.ac.kr (S.L.); tskim1@kaist.ac.kr (T.-S.K.)
5 School of Mechanical Engineering, Yeungnam University, 280 Daehak-ro, Gyeongsan-si, Gyeongbuk 38541, Korea
* Correspondence: jshim@ynu.ac.kr (J.S.); jhpark@nnfc.re.kr (J.H.P.)

Received: 3 November 2020; Accepted: 29 November 2020; Published: 3 December 2020

Abstract: In this study, curved nanostructures, which are difficult to obtain, were created on an Si substrate through the bonding, swelling, and breaking processes of the polymer and silicone substrate. This method can be utilized to obtain convex nanostructures over large areas. The method is simpler than typical semiconductor processing with photolithography or compared to wet- or vacuum-based dry etching processes. The polymer bonding, swelling (or no swelling), and breaking processes that are performed in this process were theoretically analyzed through a numerical analysis of permeability and modeling. Through this process, we designed a convex nanostructure that can be produced experimentally in an accurate manner.

Keywords: nanostructure; nanopatterning; soft lithography; soft material; swelling

1. Introduction

Nanoscale structures are recognized for their diverse application and versatility in a variety of fields such as electronics and communications technology, optical technology, biotechnology, environment and energy technology, and national defense technology [1–4]. There has been an increasing interest in nanostructure processing with curved surfaces. This is because nanostructures used in essential parts such as light guide panels, lenses, labs-on-chips, among others required in displays, optics, and nano-biotechnology are decreasing in size but increasing in precision. There is a large demand for highly reproducible nanostructures. Previously, methods such as the lithography thermal reflow technique, gray-tone photolithography, diamond milling, and beam direct writing have been suggested as methods for the creation of curved micro- and nano-scale structures [5–9]. However, there are still some unresolved issues regarding the use of the methods for the fabrication of high-quality curved structures. In addition, because these methods are heavily equipment-dependent and require high processing costs, and taking into account processing time and energy efficiency, they are not adequate for mass-production. Such methods show low structural reproducibility and have a limitation in scaling down to smaller structures. Park et al. reported a method that can implement nano-lenses in an area of 1 μm or less us organic vapor deposition. Although the above method does not require the use of expensive semiconductor processes, the dimensions and shapes of the standardized nano-lenses are not uniform because they are crystallized on a specific surface-treated substrate by controlling organic vapor in a vacuum. Additionally, lenses can be aggregated together [10].

There is an increasing interest in soft lithography approaches, which are capable of mass replicating nanostructures [2,4,9–17]. Kim et al. demonstrated that line-type nanostructures can be transferred uniformly onto SiO_2 substrate using the PDMS (polydimethylsiloxane) swelling method/applied soft lithography technique [18]. These technologies semi-permanently conserve the nanostructure's Si master mold, which requires high processing costs. They are advantageous for mass production as they enable infinitely reversible and repeatable processes. Additionally, since the processing precision of the semiconductor process is maintained, the error range of the fabricated nanostructures is significantly small. With such advantages, several functional types of soft lithography have been developed, including technology for replication between different types of material, 3D replication technology, etc. [14,19].

In this study, convex structures were uniformly formed on a silicon substrate using the bonding, swelling, and breaking process of modified polydimethylsiloxane (MPDMS), which was replicated from an Si master mold. The bonding, swelling (or lack thereof), and breaking processes of the PDMS nanostructures on the silicon substrate were analyzed theoretically with permeability modeling. Through our experimental results and theoretical analysis, we have developed a novel fabrication method for convex nanostructures. Our method allows for easy and accurate fabrication, suggesting the method's applicability.

2. Materials and Methods

2.1. Fabrication of SPDMS (Soft PDMS)

The soft polydimethylsiloxane (SPDMS) (Dow Corning) mixture of Sylgard 184 elastomer and curing agent was fabricated in a mass ratio of 10:1 and stirred using a glass rod for approximately 5 min in a disposable container. Air bubbles in the mixed SPDMS were removed in a vacuum oven (OV-12, JEIO Tech, Deajeon, Korea) for 30 min, and the SPDMS mixture was poured into a thermally cured hard PDMS (HPDMS) substrate. Subsequently, it was heated in an oven at 70 °C for 1 h. For reference, the Young modulus of SPDMS is about 2 MPa [19].

2.2. Fabrication of HPDMS (Hard PDMS)

The HPDMS mixture was prepared by mixing VDT-731 (vinylmethylsiloxane-dimethylsiloxne copolymer, trimethylsiloxy terminated, Gelest Inc., Morrisville, PA, USA), a Pt catalyst (platinum1,3-divinyl-1,1,3,3-tetramethyl-disiloxane-complex, Karstedt catalyst in Xylene, JSI Silicon Corp, Seongnam, Republic of Korea), and a monomer (2,4,6,8-tetramethyl-2,4,6,8-tetravinylcycloterasiloxane, Sigma Aldrich Corp, Saint Louis, MO, USA) at a ratio of 3.4 g:2 drops:1 drop. The mixture was placed in a disposable container and mixed for 1 min using a glass rod. Air bubbles in the mixed HPDMS were removed in the vacuum oven at room temperature for approximately 5 min. Next, 1 g of HMS-301 (methylhydosiloxane-dimethylsiloxane copolymer, trimethylsiloxane terminated, Gelest Inc., Morrisville, PA, USA) was mixed into the HPDMS mixture. It was then stirred slowly with a glass rod to prevent air bubbles. The HPDMS mixture was poured into the Si master with nanostructures and cured by heating in an oven at 70 °C for 20 min. The tensile modulus of the HPDMS was 6.32 MPa in air, and significantly decreased to 0.60 MPa when swelled by the THF solvent.

2.3. Fabrication and Treatment of MPDMS (Modified PDMS)

The modified PDMS (MPDMS) was composed of SPDMS (soft polydimethylsiloxane) on HPDMS (hard polydimethylsiloxane), which is shown in Figure 1. The thermally curable MPDMS stamp was fabricated by the replica molding process (one of the lithography methods) using an Si substrate with a nano-scale structure as the master mold [14]. Subsequently, the Si substrate was cleaned using ultrasonication for 1 min in each of acetone, ethanol, and deionized water, sequentially. The Si substrate was completely dried in N_2 gas. Next, the prepared MPDMS stamp and Si substrate were treated with plasma on each surface for 2 min 30 s using oxygen at a flow rate of 800 cc per minute and

85 watts of energy using a plasma cleaner (CUTE, FEMTO SIENCE) for the irreversible bonding process. Immediately after the Si substrate and MPDMS surfaces were exposed to the plasma, the MPDMS stamp was carefully placed on the Si substrate. Subsequently, the MPDMS stamp placed on the Si substrate was heated in an oven at 70 °C for 5 min. As a result, the Si substrate and the MPDMS stamp were bonded.

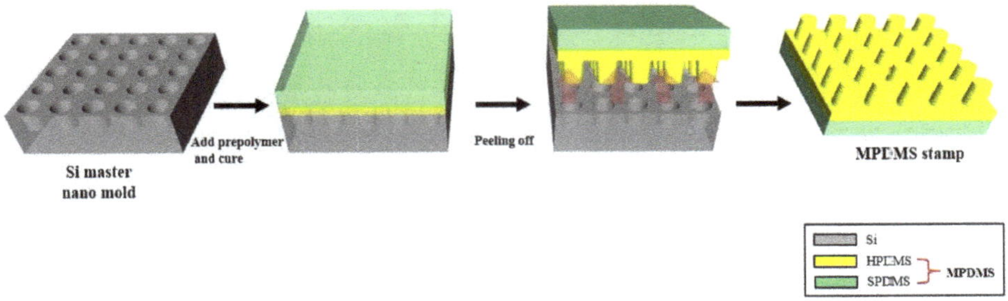

Figure 1. Schematic illustrations of the replica molding process for fabricating the modified PDMS (MPDMS) stamp from the Si master nano-mold.

2.4. Fabrication of Fully Swelled MPDMS

The bonded MPDMS stamp and Si substrate were dipped for 5 min in swelling organic solvents such as tetrahydrofuran (THF, swelling ratio: 1.38), hexane (swelling ratio: 1.34), 1,2-dichlorobenzene (1,2-DCB, swelling ratio: 1.28), all of which have high penetration properties for PDMS. The bonded MPDMS stamp and Si substrate were completely dried in N_2 gas. Subsequently, the MPDMS stamp and the Si substrate were separated from the edge of the bonded plane.

2.5. Fabrication of Convex-Type Nanostructures on Si

The polymer transferred onto the Si substrate was etched using a deep silicon etcher (TCP-9400DFM, LAM) in 17 mTorr (2.2644 Pa) of $Cl_2/HBr/O_2$, with a main power of 300 watts and a bottom bias of −225 watts. Subsequently, the remaining polymer was completely removed in a polymer remover solution (Dynasolve 218, DYNALOY) at 45 °C with an agitation of 150 rpm for 2 h. As a result, convex nanostructures were uniformly formed on the Si substrate.

3. Results and Discussion

3.1. Curved Nanostructure Creation via the Bonding, Swelling (or Lack Thereof), and Breaking Process from Replica Molding

In this study, we performed convex nanostructure fabrication using PDMS nanopattern transfer technology. PDMS nanopattern transfer technology is divided into two steps. The first step is the replica molding step [13]. Figure 1 illustrates the fabricating process of replicating MPDMS stamps through molding and curing processes using an Si-based master mold with micro- and nano-scale structures [14]. We defined the nanostructured HPDMS (hard polydimethylsiloxane) on the SPDMS backing substrate as modified PDMS (MPDMS).

Figure 2a is an optical image of an Si master nano-mold with a 1 inch × 1 inch size. Figure 2b shows the nanostructure of Figure 2a. Nanostructures of circular holes with a diameter of 400 nm, an interval of 200 nm, and a depth of 520 nm were fabricated on an Si wafer using the photolithography and etching process with an inductively coupled plasma reactive ion etcher (ICP-RIE). Figure 2c shows an optical image of the MPDMS stamp, which was replicated from the Si master nano-mold in Figure 2a and whose structure is reversed through the replica molding method. Figure 2d shows the nanostructure of Figure 2c.

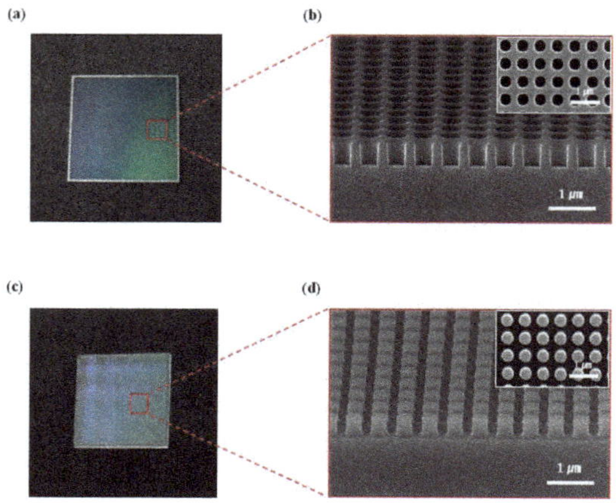

Figure 2. (**a**) Optical image of the uniformly etched Si master nano-mold. (**b**) Cross-sectional and top SEM images of arrayed circular nano-holes with a relief depth of 520 nm. (**c**) Optical image of the replicated MPDMS stamp from the Si master nano-mold. (**d**) Cross-sectional and top SEM images of the replicated MPDMS stamp with a circular pillar structure.

Figure 3a is a schematic diagram showing the bonded state of the oxygen-plasma-treated MPDMS stamp and Si substrate. Figure 3b is a schematic diagram of a state where the MPDMS stamp is dipped into THF (tetrahydrofuran) and THF has completely penetrated into the entire MPDMS stamp. If enough time is given for THF to penetrate into the MPDMS stamp, the concentration of THF in the MPDMS stamp remains constant regardless of the penetration path and the penetration area. In other words, the degree of breakdown of the bonds between HPDMS's molecular chains is the same throughout the entire polymer. If a relatively low critical tensile stress is applied in this state, then the probability of a fracture occurring at the bottom of the HPDMS pillar with the smallest cross-sectional area is high [20]. As above, if the MPDMS stamp does not swell at all, the state of the bonds between HPDMS's molecular chains is the same throughout. If a relatively high critical tensile stress is applied in this state, then the probability of a fracture occurring at the bottom of the HPDMS pillar with the smallest cross-sectional area is high. Figure 3c shows the broken MPDMS structure when critical tensile force is applied to Figure 3b. Figure 3d shows the MPDMS convex nanostructure transferred to a Si substrate.

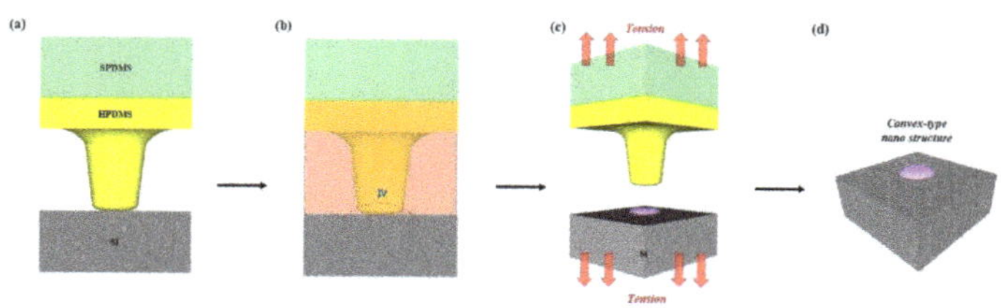

Figure 3. Schematic process flow to fabricate the convex nanostructure from the MPDMS stamp: (**a**) The irreversible bonding step; (**b**) the fully swelling step; (**c**) the breaking step on applied tension; (**d**) the resulting convex nanostructure transferred to the Si substrate.

3.2. Curved Nanostructure Creation via the Bonding, Swelling (or Lack Thereof), and Breaking Process from Replica Molding

Figure 4a,b shows the top and cross-sectional views, respectively, of convex patterns of MPDMS stamps (diameter of 300 nm, spacing of 300 nm, and height of 40 nm) uniformly transferred onto the Si substrate through the bonding, no swelling (the swelling was skipped) and breaking process. Figure 4c,d shows the top and cross-sectional views, respectively, of the convex patterns of the MPDMS stamp uniformly transferred to the Si substrate through the bonding, full swelling, and breaking process. As the transferred MPDMS convex nanopatterns are used as hard masks for the etching process, convex Si nanostructures (diameter of 250 nm, spacing of 350 nm, and height of 160 nm) that are difficult to form by conventional photo lithography were easily formed. An important parameter that affects the change of curvature is determined by the degree of swelling. Swelling is determined by the concentration of THF, the time of swelling, and the shape and size of the nanostructure. For example, in the case of full swelling, in which HPDMS nanostructures are completely penetrated by THF, the effect is the same as no swelling. However, in the case of the full swelling step, less energy is required for the breaking step than in the case of no swelling step (i.e., swelling is skipped). Therefore, the uniformity and accuracy of the nanostructure formation step are improved even with a small force. Figure 4e,f shows the Si nano lens structures, which were formed after deep RIE using the convex patterns in Figure 4c,d as hard masks for the etching process. The convex nanostructures fabricated through this process are used as a functional structure in biological and medical fields as well as in the development of optical lenses and sensors [21]. In this study, experiments were conducted with one type of nano-structure. However, by using the bonding, full swelling, and breaking process, curvature is not controllable, but is determined naturally via swelling and breaking mechanics. Since the diameter and height of the fabricated curved nano-structure depends on the diameter of the initial nano-structure of the Si master nano-mold, it is thought that if a smaller Si nano-mold structure is used, a smaller-diameter curved structure can be fabricated.

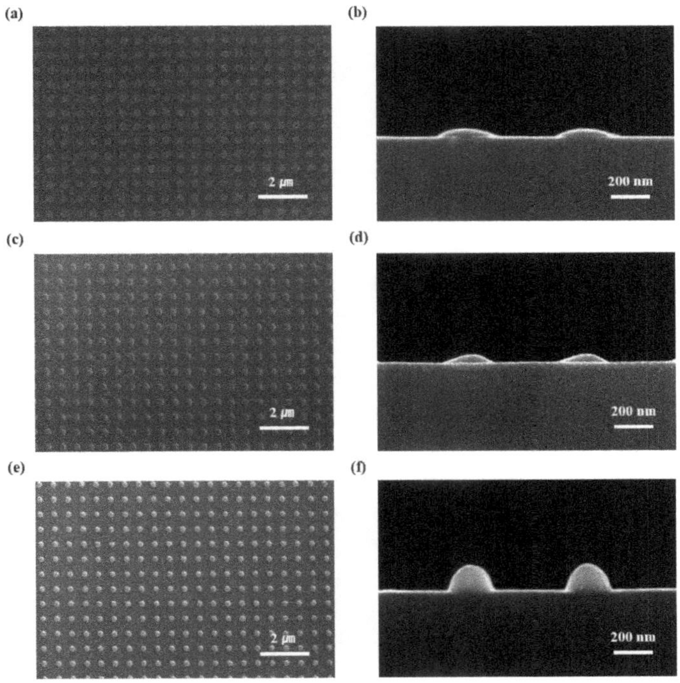

Figure 4. Fabricated convex nanostructures after transfer to the Si substrate by the MPDMS bonding, swelling (i.e., no swelling or full swelling), and breaking process (used as a hard mask against dry etching): (**a**,**b**) are SEM images of the top and cross-sectional views, respectively, of transferred convex patterns (diameter of 300 nm, spacing of 300 nm, and height of 40 nm) onto the Si substrate by the bonding and breaking process (no swelling). (**c**,**d**) are SEM images of the top and cross-sectional views of fully swelled and transferred convex patterns, respectively. (**e**,**f**) present convex Si nanostructures formed by deep reactive ion etching (RIE) through the MPDMS hard masks with convex patterns as shown in (**c**,**d**).

3.3. *Mechanical Modulus Measurement of the HPDMS*

A dynamic mechanical analyzer (DMA) (DMA 8000, Perkin Elmer, Waltham, MA, USA) was used to measure the tensile modulus of the HPDMS. In order to investigate the effect of THF swelling on the mechanical properties of HPDMS, a ceramic-coated fluid bath of the DMA machine was utilized. Among DMA testing configurations, the film tension mode was adopted because the tensile elastic response is important when analyzing stress concentration in the nanostructures during the initial stage of the transfer process. The HPDMS specimens were fabricated with dimensions of 20 mm length, 5 nm width, and 1mm thickness using an aluminum mold. The HPDMS solution was thoroughly mixed and then poured into rectangular grooves. After degassing trapped air in the mixed solution under vacuum, thermal curing was conducted at 90 °C for 30 min. With the gauge length of 10 mm, the dynamic actuation was conducted both in the air and in the THF solvent. For consistency, the measurement in the THF solvent started after 2 min of immersion. The measured tensile modulus values were averaged with results from at least five specimens.

Figure 5 shows the modulus measurement result of the HPDMS specimens under ambient conditions and under the THF-swelled condition. The tensile modulus of the HPDMS was 6.32 MPa in the air and significantly decreased to 0.60 MPa when swelled by the THF solvent. It is verified that the THF treatment greatly weakens the intermolecular attraction among the polymeric chains. Note that the modulus value during 3 min of the cyclic measurement was kept constant, meaning that the mechanical stiffness was saturated to the modulus level of 0.60 MPa.

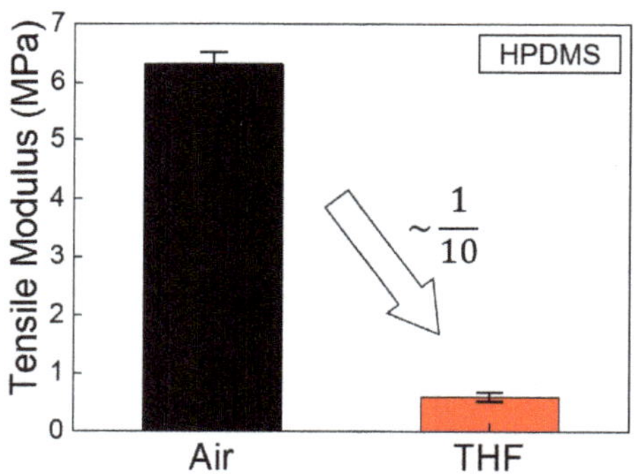

Figure 5. Tensile modulus of HPDMS in air and in THF.

3.4. *Numerial Analysis of Permeability and Modeling in the Bonding, Controlled Swelling, and Breaking Process*

A stationary stress analysis was conducted to validate the failure process of convex nanostructures. Interpenetrating swelling properties of the HPDMS and polymer networks were obtained from the experiment (Figure 5) and were utilized to carry out a numerical simulation by using Young's modulus (tensile modulus) of the fully swelled HPDMS. We assume that Young's modulus is the tensile modulus in an environment where tensile force is applied. The failure process was studied by examining two main tensors: diffusion of the chemical solvent and stationary stress under the specific swelled conditions.

3.4.1. Concentration Diffusion Analysis

The concentration equation is defined as:

$$\frac{\partial C}{\partial t} + \vec{U} \cdot \nabla C = \nabla \cdot (D \nabla C) + R \tag{1}$$

where C is the concentration (mole/m^3) and D is the diffusion coefficient (m^2/s). In addition, $\vec{U}$ is the fluid velocity and R is the net rate at which C is produced in a chemical reaction. In this simulation, convection is suppressed to obtain a pure diffusion profile with time. The reaction of the PDMS polymer does not take place during swelling.

3.4.2. Stationary Stress Analysis

The stationary stress–strain equation is expressed as:

$$\nabla \cdot \sigma + F_v = 0 \tag{2}$$

$$\sigma = \sigma_0 + C : (\varepsilon - \varepsilon_0 - \alpha\theta) \tag{3}$$

$$\varepsilon = \frac{1}{2}\left(\nabla u + (\nabla u)^T\right) \tag{4}$$

where σ is the stress tensor, F_v is the body force, ε is the strain tensor, σ_0 and ε_0 are the initial stress and strain tensors, respectively, α is the thermal expansion, $\theta = T - T_{ref}$, u is the displacement field, and C is the 4th-order elasticity tensor.

This displacement field gradient can be defined as:

$$\nabla u = \begin{bmatrix} \frac{\partial u}{\partial X} & \frac{\partial u}{\partial Y} & \frac{\partial u}{\partial Z} \\ \frac{\partial v}{\partial X} & \frac{\partial v}{\partial Y} & \frac{\partial v}{\partial Z} \\ \frac{\partial w}{\partial X} & \frac{\partial w}{\partial Y} & \frac{\partial w}{\partial Z} \end{bmatrix} \tag{5}$$

where (X, Y, Z) are material coordinates and (x, y, z) are spatial coordinates.

3.4.3. Numerical Simulation and Modeling of Permeability

In this study, two failure modes, such as fracture at upper regions and fracture at lower regions of pillar type nanostructures, were investigated by using experimental data as numerical input data. We studied how to propagate the concentration of chemical solvent with time before the stationary stress analysis was carried out to study the failure process.

We used the diffusion coefficient 1E-16 m^2/s for the HPDMS and chemical solvents [21]. Figure 6 shows the schematic of a cylindrical bead nanostructure. For the boundary conditions of concentration diffusion, the axial symmetry condition is used at the axial line, whereas the bottom regions and the top-right are set as the insulation boundary. Only the constant concentration is provided at the blue regions as C = 12.3 M.

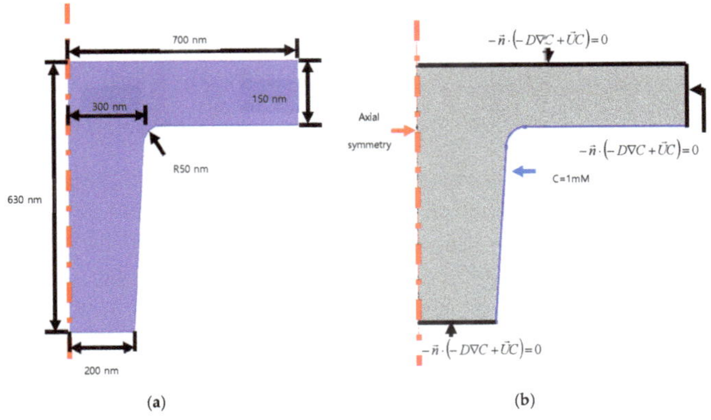

Figure 6. Schematic of cylindrical bead nanostructure: (**a**) geometry; (**b**) boundary conditions of concentration simulation.

The concentration propagation with time is revealed in Figure 7. The PDMS swelling due to chemical solvents starts from the outside before moving into the PDMS (as shown in Figure 7a,b) and the concentration still does not reach the top of the center of the nanostructure in Figure 7c. Finally, the PDMS is fully swelled in Figure 7d. We also conducted a stationary stress analysis by using the numerical results of concentration diffusion. The result at 10 s (Figure 7c) was used as the partially swelled condition and 120 s (Figure 7d) was used as the fully swelled condition.

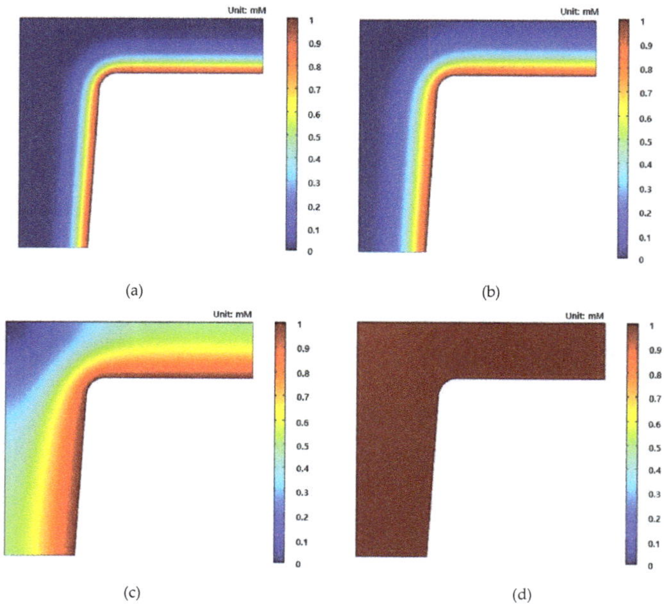

Figure 7. Concentration profile of a cylindrical bead nanostructure with time: (**a**) 1 s; (**b**) 5 s; (**c**) 10 s; (**d**) 120 s.

In Figure 8, the partially swelled regions are divided into region I (non-swelled region) and region II (fully swelled region). The classified curve was obtained from the Bezier polygon, which is based on {r, z} = {[−1E−7m, 2E−7m], [1E−7, 6E−7], [5E−7, 8E−7]}. For boundary conditions, a 2D axial symmetric boundary was applied to the axisymmetric line. A fixed boundary condition was applied to the bottom of the pillar nanostructure and the prescribed displacement of 80 nm for the partially swelled condition and 170 nm were set to the top of nanostructure. All other boundaries were used as free boundary conditions. In Figure 8, the schematic of the partially swelled nanostructure is shown for simulation of the swelling condition. Area I of the HPDMS is dried, as E = 6.32 MPa, and area II of the HPDMS is swollen, as E = 0.6. The Poisson ratios are v = 0.45 and 0.47 for the dried condition and the swollen condition, respectively.

Figure 9 shows the simulation results for the partially swelled PDMS (Figure 9a) and for the fully swelled PDMS (Figure 9b). The maximum stress of the partially swelled PDMS is 0.762 MPa at the edge of the bottom, whereas the maximum stress of the fully swelled PDMS is 0.475 MPa. It was reported that the maximum strengths of dry PDMS and swollen PDMS are 0.76 and 0.45 MPa, respectively [22]. As compared to the literature, the maximum stresses for both the partially swelled PDMS and the fully swelled PDMS are greater than the material strength. This means that the partially swelled PDMS breaks near the top of the nanostructure and the fully swelled HPDMS fractures at the bottom of the nanostructure.

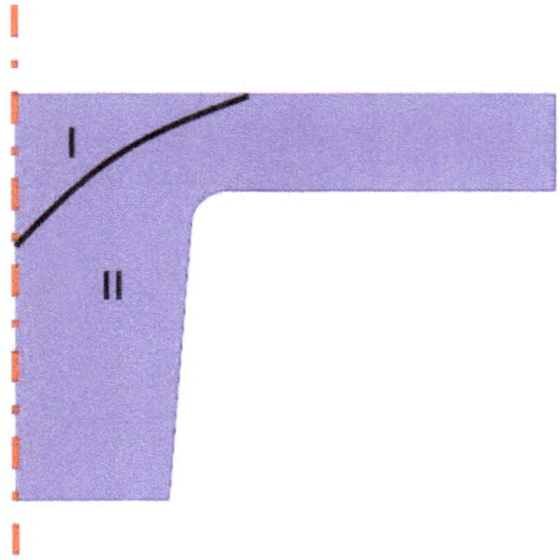

Figure 8. The schematic of the partially swelled nanostructure.

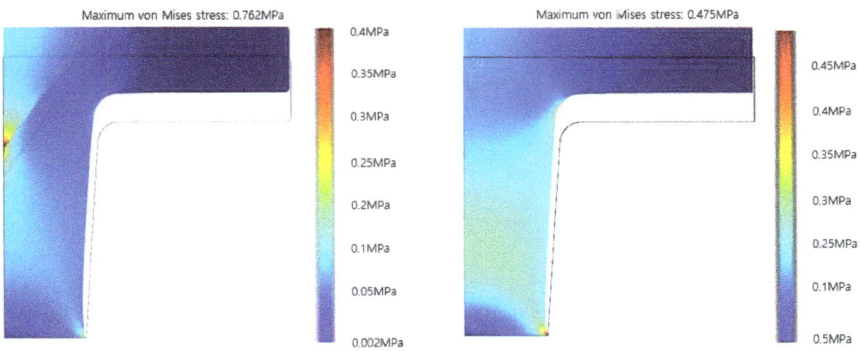

(a) (b)

Figure 9. Stress analysis results for (**a**) the partially swelled condition and (**b**) the fully swelled condition.

4. Conclusions

In this study, convex nanostructures over large areas were formed on a silicon substrate by using the bonding, swelling, and breaking of MPDMS that was replicated from an Si master mold. In addition, the bonding, swelling (or lack thereof), and breaking process of the PDMS nano-mold on the silicon substrate was analyzed theoretically through numerical analysis of permeability and modeling. Through this process, we have developed a novel fabrication method for convex nanostructures that can easily and accurately be fabricated, which usually represents a difficult engineering problem. Easily fabricated, curved nanostructures can be essential parts required in the fields of displays, optics, and nano-bio diagnosis, and they can play an important role in light guide panels, lenses, and labs-on-chips.

Author Contributions: Conceptualization, J.H.P. and H.-I.J.; methodology, J.H.P., H.-I.J., T.-I.L., S.L., T.-S.K., and J.S.; experiment H.-I.J. and H.-S.Y.; experimental analysis, J.H.P and H.-I.J.; mechanical experiment and analysis, T.-I.L., S.L., and T.-S.K.; numerical analysis, J.S.; writing—original draft preparation, J.H.P., H.-I.J., T.-I.L., S.L., T.-S.K., and J.S.; writing—review and editing, J.H.P. and J.S.; visualization, H.-I.J.; supervision, J.H.P. and J.S.; funding acquisition, J.H.P. All authors have read and agreed to the published version of the manuscript.

Funding: This research was supported by the Ministry of Trade, Industry & Energy (MOTIE), Korea Institute for Advancement of Technology (KIAT) and Ulsan Institute for Regional Program Evaluation (IRPE) through the Encouragement Program for The Industries of Economic Cooperation Region (P0002278). This work was supported by the Technology Development Program (S2640831) funded by the Ministry of SMEs and Startups

(MSS, Korea). This research was supported by the Technology Development Program (2020-DD-RD-0044) funded by the Korea Innovation Foundation (INNOPOLS, Korea).

Conflicts of Interest: The authors declare no conflict of interest. The funders had no role in the design of the study; in the collection, analysis, or interpretation of data; in the writing of the manuscript; or in the decision to publish the results.

References

1. Geissler, M.; Xia, Y.N. Patterning: Principles and Some New Developments. *Adv. Mater.* **2004**, *14*, 1249–1269. [CrossRef]
2. Park, W.I.; Kim, K.H.; Jang, H.I.; Jeong, J.W.; Kim, J.M.; Choi, J.S.; Park, J.H.; Jung, Y.S. Directed Self-Assembly with Sub-100 Degrees Celsius Processing Temperature, Sub-10 Nanometer Resolution, and Sub-1 Minute Assembly Time. *Small* **2012**, *8*, 3762–3768. [CrossRef] [PubMed]
3. Pang, C.H.; Lee, G.Y.; Kim, T.I.; Kim, S.M.; Kim, H.N.; Ahn, S.H.; Suh, K.Y. A flexible and highly sensitive strain-gauge sensor using reversible interlocking of nanofibers. *Nat. Mater.* **2012**, *11*, 795–801. [CrossRef] [PubMed]
4. Jeong, J.W.; Yang, S.R.; Hur, Y.H.; Kim, S.W.; Baek, K.M.; Yim, S.M.; Jang, H.I.; Park, J.H.; Lee, S.Y.; Park, C.O.; et al. High-resolution nanotransfer printing applicable to diverse surfaces via interface-targeted adhesion switching. *Nat. Commun.* **2014**, *5*, 5387–5398. [CrossRef]
5. Cheng, H.C.; Huang, C.F.; Lin, Y.; Shen, Y.K. Brightness field distributions of microlens arrays using micro molding. *Opt. Express* **2010**, *18*, 26887–26904. [CrossRef]
6. Yao, Y.; Su, J.Q.; Du, J.L.; Zhang, Y.X.; Gao, F.H.; Gao, F.; Guo, Y.K.; Cui, Z. Coding gray-tone mask for refractive microlens fabrication. *Microelectron. Eng.* **2000**, *53*, 531–534. [CrossRef]
7. Yi, A.Y.; Li, L. Design and fabrication of a microlens array by use of a slow tool servo. *Opt. Lett.* **2005**, *30*, 1707–1709. [CrossRef]
8. Mathis, A.; Courvoisier, F.; Froehly, L.; Furfaro, L.; Jacqot, M.; Lacourt, P.A.; Dudley, J.M. Micromachining along a curve: Femtosecond laser micromachining of curved profiles in diamond and silicon suing accelerating beams. *Appl. Phys. Lett.* **2012**, *101*, 071110. [CrossRef]
9. Xia, Y.N.; Whitesides, G.M. Soft Lithography. *Angew. Chem.* **1998**, *37*, 550–575. [CrossRef]
10. Park, Y.S.; Han, K.H.; Kim, J.H.; Cho, D.H.; Lee, J.H.; Han, Y.J.; Lim, J.T.; Cho, N.S.; Yu, B.G.; Lee, J.I.; et al. Crystallization-assisted nano-lens array fabrication for highly efficient and color stable organic light emitting diodes. *Nanoscale* **2017**, *9*, 230–236. [CrossRef]
11. Odom, T.W.; Christopher Love, J.; Wolfe, D.B.; Paul, K.E.; Whitesides, G.M. Improved Pattern Transfer in Soft Lithography Using Composite Stamps. *Langmuir.* **2002**, *18*, 5314–5320. [CrossRef]
12. Rogers, J.A.; Nuzzo, R.G. Recent progress in soft lithography. *Mater. Today Sci.* **2005**, *8*, 50–56. [CrossRef]
13. Qin, D.; Xia, Y.N.; Whitesides, G.M. Soft lithography for micro- and nanoscale patterning. *Nat. Protoc.* **2010**, *5*, 491–502. [CrossRef] [PubMed]
14. Jang, H.I.; Ko, S.H.; Park, J.Y.; Lee, D.E.; Jeon, S.W.; Ahn, C.W.; Yoo, K.S.; Park, J.H. Reversible creation of nanostructures between identical or different species of materials. *Appl. Phys. A* **2012**, *108*, 41–52. [CrossRef]
15. Yu, S.H.; Han, H.J.; Kim, J.M.; Yim, S.M.; Sim, D.M.; Lim, H.H.; Lee, J.H.; Park, W.I.; Park, J.H.; Kim, K.H.; et al. Area-Selective Lift-Off Mechanism Based on Dual-Triggered Interfacial Adhesion Switching: Highly Facile Fabrication of Flexible Nanomesh Electrode. *ACS Nano* **2017**, *11*, 3506–3516. [CrossRef] [PubMed]
16. Han, H.J.; Jeong, J.W.; Yang, S.R.; Kim, C.G.; Yoo, H.G.; Yoon, J.B.; Park, J.H.; Lee, K.J.; Kim, T.S.; Kim, S.W.; et al. Nanotransplantation Printing of Crystallographic-Orientation-Controlled Single-Crystalline Nanowire Arrays on Diverse Surfaces. *ACS Nano* **2017**, *11*, 11642–11652. [CrossRef]
17. Lee, J.K.; Kim, B.O.; Park, J.C.; Kim, J.B.; Kang, I.S.; Sim, G.S.; Park, J.H.; Jang, H.I. A bilayer Al nanowire-grid polarizer integrated with an IR-cut filter. *Opt. Mater.* **2019**, *98*, 109409. [CrossRef]
18. Kim, P.N.; Kwak, R.K.; Lee, S.H.; Suh, K.Y. Solvent-Assisted Decal Transfer Lithography by Oxygen-Plasma Bonding and Anisotropic Swelling. *Adv. Mater.* **2010**, *22*, 2426–2429. [CrossRef]
19. Park, J.Y.; Park, J.H.; Kim, E.H.; Ahn, C.W.; Jang, H.I.; Rogers, J.A.; Jeon, S.W. Conformable Solid-Index Phase Masks Composed of High-Aspect-Ratio Micropillar Arrays and Their Application to 3D Nanopatterning. *Adv. Mater.* **2011**, *23*, 860–864. [CrossRef]

20. Chan, P.N. Adhesion of Patterned Polymer Interfaces. Ph.D. Dissertation, University of Massachusetts Amherst, Amherst, MA, USA, February 2014. Available online: https://scholarworks.umass.edu/dissertations_1/1104/ (accessed on 1 January 2007).
21. Liong, M.; Hoang, A.N.; Chung, J.H.; Gural, N.; Ford, C.B.; Min, C.W.; Shah, R.R.; Ahmad, R.; Fernandez-Suarez, M.; Fortune, S.M.; et al. Magnetic barcode assay for genetic detection of pathogens. *Nat. Commun.* **2013**, *4*, 1752–1772. [CrossRef]
22. Yoo, S.H.; Cohen, C.; Hui, C.Y. Mechanical and swelling properties of PDMS interpenetrating polymer networks. *Polymer* **2006**, *47*, 6226–6235. [CrossRef]

nanomaterials

MDPI

Article

Improved Pulse-Controlled Conductance Adjustment in Trilayer Resistors by Suppressing Current Overshoot

Hojeong Ryu and Sungjun Kim *

Division of Electronics and Electrical Engineering, Dongguk University, Seoul 04620, Korea; hojeong.ryu95@gmail.com

* Correspondence: sungjun@dongguk.edu

Received: 10 November 2020; Accepted: 8 December 2020; Published: 9 December 2020

Abstract: In this work, we demonstrate the enhanced synaptic behaviors in trilayer dielectrics ($HfO_2/Si_3N_4/SiO_2$) on highly doped n-type silicon substrate. First, the three dielectric layers were subjected to material and chemical analyses and thoroughly investigated via transmission electron microscopy and X-ray photoelectron spectroscopy. The resistive switching and synaptic behaviors were improved by inserting a Si_3N_4 layer between the HfO_2 and SiO_2 layers. The electric field within SiO_2 was mitigated, thus reducing the current overshoot in the trilayer device. The reset current was considerably reduced in the trilayer device compared to the bilayer device without a Si_3N_4 layer. Moreover, the nonlinear characteristics in the low-resistance state are helpful for implementing high-density memory. The higher array size in the trilayer device was verified by cross-point array simulation. Finally, the multiple conductance adjustment was demonstrated in the trilayer device by controlling the gradual set and reset switching behavior.

Keywords: resistive switching; X-ray photoelectron spectroscopy; synaptic device; metal oxide; current overshoot

1. Introduction

Resistive switching memory is very attractive for a wide range of applications due to its various resistive switching characteristics stemming from a number of resistive switching materials by easily tunable resistive switching parameters such as on-resistance, off-resistance, and operation voltage [1–5]. Moreover, its simple structure, such as metal–insulator–metal (MIM) with $4F^2$ (F is feature size), can be scaled down via a lithography process [1–6]. Further, the multiple resistance states triggered by electrical pulses can be used for high-density memory. Finally, the good retention properties, such as NAND flash and high endurance, provide an edge over other competing memory products. The types of resistive switching should be characterized depending on the possible applications, such as in storage class memory, neuromorphic devices, and logic devices. Among them, a neuromorphic device using resistive switching memory is attracting considerable attention [7–10]. To meet the needs of efficient data processing in the era of big data, neuromorphic computing provides a major breakthrough that can replace the existing Von Neumann computing. In particular, neuromorphic systems are specialized in data processing, such as complex pattern recognition. Moreover, they take an energy-efficient approach by carrying out data processing in a parallel manner. In a neuromorphic system, the conductance of the resistive switching memory cell placed on a cross-point array has multiple states and can be updated and controlled by the input pulse from the neuron circuit. The conductance control of resistive switching is similar to the synaptic weight adjustment in biological synapses in the human nervous system.

Resistive switching and artificial synaptic behaviors are observed in many insulators, such as oxide [11–13], nitride [13–15], organic materials [16], and 2D materials [17]. Among them, metal–oxide-based resistive switching memories such as HfO_2 have proven to have the best resistive switching performances in terms of, e.g., endurance, retention, and variability. Excellent resistive switching has been reported when using metal bottom electrodes [18,19]. HfO_2-based resistive switching memory with ITO electrode can also be used for flexible and transparent electronic devices [20]. On the other hand, the HfO_2-based resistive memory with a silicon bottom electrode has not yet been reported as superior to the metal bottom electrode. However, the metal–oxide–semiconductor structure has other advantages, such as self-rectification and low-power operation. The most effective way to enhance resistive switching is to design multiple dielectric stacks [21]. The use of an oxygen reservoir, such as a TiO_x layer, is popular in metal–oxide-based resistive switching memory. Abundant oxygen vacancies are created in the main resistor for resistive switching. A tunnel barrier, such as SiO_2 and Al_2O_3, with a large band gap, can enhance the resistive switching properties by reducing the operation current and increasing the nonlinearity of the I–V curve in the low-resistance state (LRS) [22,23]. The SiO_2 layer can be easily formed when using silicon substrate as the bottom electrode and different methods such as native oxide, thermal oxide, and chemical vapor deposition (CVD). Another advantage of inserting the tunnel barrier with a high band gap is a reduction in the LRS current [21,22]. To employ the advantageous tunnel barrier in resistive switching, the insulating property of the SiO_2 layer is maintained after the forming and set processes. If the excess electric field is applied on the tunnel barrier with a high compliance current, breakdown of the tunnel barrier can occur. Therefore, the use of a careful device stack design is necessary to ensure a stable tunnel barrier layer; for example, the thicknesses of the tunnel barrier and the main resistor are important. In addition, the dielectric constant should be considered to properly distribute the electric field throughout multiple dielectric layers.

In this work, we fabricated a trilayer (HfO_2/Si_3N_4/SiO_2) resistive switching memory device and demonstrated low current switching by suppressing the current overshoot and nonlinear I–V curves in the LRS. The trilayered dielectric stacks were confirmed via high-resolution transmission electron microscopy (TEM) and X-ray photoelectron spectroscopy (XPS) before electrical characterization. Through a comparative study with a control group without a Si_3N_4 layer, it was verified that the Si_3N_4 layer can relieve the concentration of the electric field in SiO_2. Finally, we demonstrated the improved synaptic behaviors by achieving gradual conductance control in the trilayer structure compared to the device without a Si_3N_4 layer.

2. Materials and Methods

The Ni/HfO_2/Si_3N_4/SiO_2/Si device was prepared as follows: The ion implantation was conducted in the Si substrate to increase the conductivity on the single crystalline Si surface as the bottom electrode. Phosphorus (P) as an impurity was used to form an n-type Si bottom electrode, where the dose and energy were 5×10^{15} cm^{-2} and 40 keV, respectively. The Si lattice damage caused by ion implantation was cured by the annealing process. Next, a 2.5-nm-thick SiO_2 film was deposited via low-pressure chemical vapor deposition (LPCVD) by reacting SiH_2Cl_2 (40 sccm) and N_2O (160 sccm) at 785 °C after removing native oxide through HF cleaning. Then, a 3.5-nm-thick Si_3N_4 layer was deposited via LPCVD by reacting SiH_2Cl_2 (30 sccm) and NH_3 (100 sccm) at 785 °C. After that, a 3.5-nm-thick HfO_2 layer was deposited by atomic layer deposition (ALD) system by reacting tetrakis (ethylmethylamino) hafnium (TEMAH) and ozone (O_3) at 300 °C. Finally, a 100-nm-thick Ni top electrode was deposited by a thermal evaporator and patterned by a shadow mask containing circular patterns with a diameter of 100 µm. A Ni/HfO_2/SiO_2/Si device was prepared as a control device in the same way, except for the Si_3N_4 layer.

The electrical properties were characterized both in DC mode using a Keithley 4200-SCS semiconductor parameter analyzer (Keithley Instrumnets, Cleveland, OH, USA) and in pulse mode using a 4225-PMU ultrafast module (Keithley Instrumnets, Cleveland, OH, USA) During the measurements, a bias voltage and pulse were applied to the Ni top electrode, while the Si bottom electrode was grounded. XPS depth analysis was conducted with a Nexsa (ThermoFisher Scientific,

Waltham, MA, USA) with a Microfocus monochromatic X-ray source (Al-Kα (1486.6 eV)), a sputter source (Ar+), an ion energy of 1 kV, and a beam size of 100 μm × 100 μm.

3. Results and Discussion

Figure 1a,b respectively shows the schematics and a TEM image of the Ni/HfO_2/Si_3N_4/SiO_2/Si device. Single crystalline Si substrate and amorphous HfO_2, Si_3N_4, and SiO_2 layers could be observed in the TEM image. In addition, the TEM image provides information about the exact film thicknesses of HfO_2 (3.5 nm), Si_3N_4 (3.5 nm), and SiO_2 (2.5 nm). The energy-dispersive X-ray spectra (EDS) line scan was obtained through scanning transmission electron microscopy (STEM) and is shown in Figure S1. Next, the XPS depth profile of HfO_2/Si_3N_4/SiO_2/Si was investigated to determine the elements in each layer. Figure 1c shows the XPS spectra Hf 4f of HfO_2 as the first dielectric layer [24]; Hf 4f is typically composed of a 4f 5/2 and 4f 7/2 spin–orbit doublet, which are respectively centered at 20 and 18.5 eV. This result is consistent with existing literature about HfO_2 on a Si substrate [24]. Figure 1d shows the Si 2p spectra for the Si_3N_4 layer, SiO_2 layer, and Si substrate. The peak intensity that is located at about 99.5 eV is higher at the deeper etching level (level 11) than it is at level 7. This indicates that the Si substrate is more exposed by X-ray beams at the deeper etching level (level 11) [21]. Moreover, the peak point at etch level 11 is shifted to the right compared to that at etch level 7, indicating that the Si–O bond located at 103.5 eV is increased at level 11 [25]. Figure 1e shows the N 1s spectra at level 7 and level 11, where the peak is centered at about 398 eV [26]. The peak intensity at level 11 is much weaker than that at level 7. This result is consistent with the Si 2p result shown in Figure 1d.

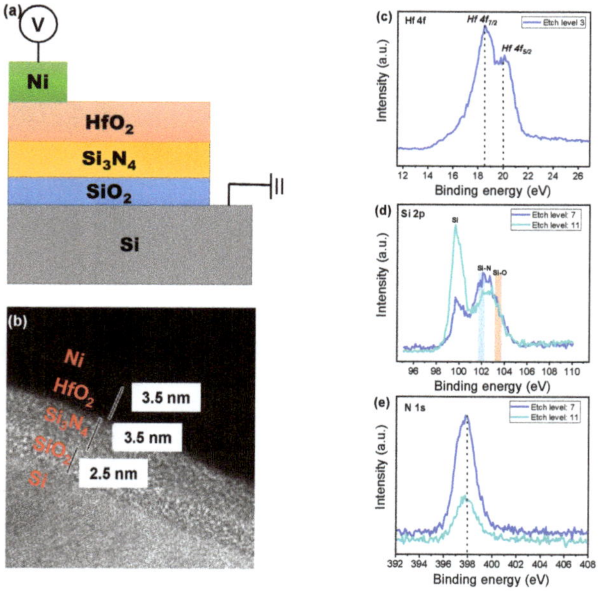

Figure 1. Device configuration and material analysis of the Ni/HfO_2/Si_3N_4/SiO_2/Si device. (**a**) Schematic drawing of the device stack; (**b**) TEM image; (**c**) XPS Hf 4f spectra at etch level 3; (**d**) XPS Si 2p spectra at etch levels 7 and 11; and (**e**) XPS N 1s spectra at etch levels 7 and 11.

Figure 2a,b shows the I–V characteristics of the Ni/HfO_2/SiO_2/Si and Ni/HfO_2/Si_3N_4/SiO_2/Si devices. For a fair comparison, the compliance current (CC) of 5 μA is applied to both devices. The initial cells are activated with the positive bias DC sweep. The current is significantly increased during the reverse sweep, which indicates that soft breakdown occurs within the dielectrics. The CC can protect the device from permanent breakdown. Subsequently, the reset process is conducted by the negative bias sweep, causing the state of the device to be changed to the high-resistance state (HRS).

This process can be explained by the rupture of the conducting path in dielectrics. Then, the set process follows to make the state of device be the LRS again. The HRS and LRS of the device can be repeatedly switched in the repetitive set and reset process. It should be noted that high current (~10 mA) flows within the Ni/HfO_2/SiO_2/Si device in the LRS. The current is very high despite the fact that 5 μA is applied on the Ni/HfO_2/SiO_2/Si device during the forward and reverse sweep in the LRS under the positive bias. Subsequently, the high LRS current without CC in a negative bias is the real current level. This suggests that current overshoot occurs during the set process, meaning that the LRS current cannot be tightly controlled by CC. An abrupt transition is observed during the reset process, indicating that the conducting path is ruptured at once. The I–V characteristics of the Ni/HfO_2/Si_3N_4/SiO_2/Si device are substantially different from those of the Ni/HfO_2/SiO_2/Si device, as shown in Figure 2b. The LRS current in a negative region is lower than the CC of 5 μA. This implies that the current overshoot is suppressed during the set process at a positive region. The bipolar resistive switching is driven by the temperature and electric field [27–30]. The Ni/HfO_2/SiO_2/Si device shows abrupt reset with high current, indicating that Joule heating is the dominant mechanism of the reset process. On the other hand, the electric field may be more important for the Ni/HfO_2/Si_3N_4/SiO_2/Si device considering the switching at low current.

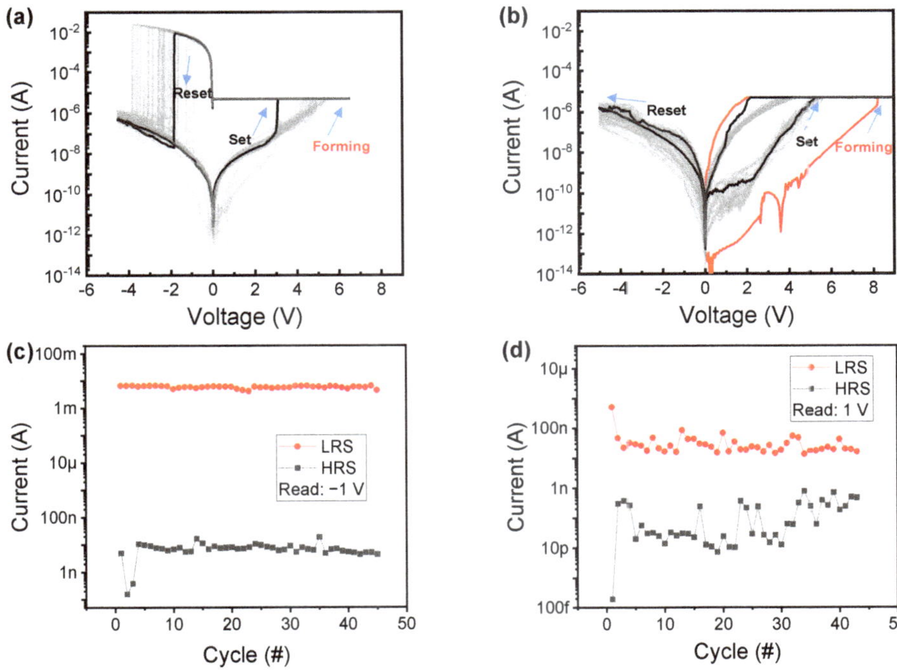

Figure 2. I–V characteristics of the (**a**) Ni/HfO_2/SiO_2/Si device and (**b**) Ni/HfO_2/Si_3N_4/SiO_2/Si device including forming, set, and reset processes; cycling data of I–V characteristics of (**c**) Ni/HfO_2/SiO_2/Si device and (**d**) Ni/HfO_2/Si_3N_4/SiO_2/Si device.

Figure 2c,d shows the cycling trend of the Ni/HfO_2/SiO_2/Si and Ni/HfO_2/Si_3N_4/SiO_2/Si devices, respectively. Time series statistical analysis could provide the indirect information of filament evolution [31,32]. Both HRS and LRS are stable except for the initial few points during the cycling for the Ni/HfO_2/SiO_2/Si device. This indicates that the large size of conducting filament could be uniformly formed and ruptured. On the other hand, the read current of the Ni/HfO_2/Si_3N_4/SiO_2/Si device has larger variation during the cycling, and the read current in the HRS is increased. The larger variation is probably due to the fact that the filaments are formed and ruptured in multiple layers

($HfO_2/Si_3N_4/SiO_2$), and these formations and ruptures would be quite random processes spatially inside the insulators.

Figure 3a shows the statistical distribution of the Ni/HfO_2/SiO_2/Si and Ni/HfO_2/Si_3N_4/SiO_2/Si devices in the LRS and HRS. The LRS resistance of the Ni/HfO_2/Si_3N_4/SiO_2/Si device is much higher than that of the Ni/HfO_2/SiO_2/Si device. However, the variations of the LRS and HRS of the Ni/HfO_2/Si_3N_4/SiO_2/Si device are worsened. Figure 3b shows the ratio between reset current (I_{RESET}) and CC. From this ratio, we can obtain information on how much CC suppresses the overshoot current during the set process. I_{RESET} is rather smaller than CC in the Ni/HfO_2/Si_3N_4/SiO_2/Si device. However, the I_{RESET}/I_{CC} ratio of the Ni/HfO_2/Si_3N_4/SiO_2/Si device is more than 1000. Other advantages of the Ni/HfO_2/Si_3N_4/SiO_2/Si device are its high nonlinear I–V characteristic and its low-current operation. The nonlinearity is defined as the ratio between the current at read voltage (V_{READ}) and the current at half read voltage ($1/2 \cdot V_{READ}$) for the half bias scheme in the cross-point array (Figure S2). Further, the nonlinearity is defined as the ratio between the current at V_{READ} and the current at $1/3 \cdot V_{READ}$ for the 1/3 read scheme. The LRS resistance is the main leakage path in the cross-point array structure when reading the target cell with the HRS. Therefore, the current at $1/2 \cdot V_{READ}$ or the current at $1/3 \cdot V_{READ}$ should be suppressed to reduce crosstalk among the cells. Figure 3c shows the nonlinearity of both devices when applying the 1/2 read scheme and the 1/3 read scheme. The nonlinearities of the Ni/HfO_2/SiO_2/Si device in the LRS are about 2 and 3 for the 1/2 read scheme and the 1/3 read scheme, respectively. This indicates that the LRS follows Ohmic conduction with a slope of 2. The nonlinearity of the Ni/HfO_2/Si_3N_4/SiO_2/Si device in the LRS is substantially higher due to its nonlinear I–V characteristics.

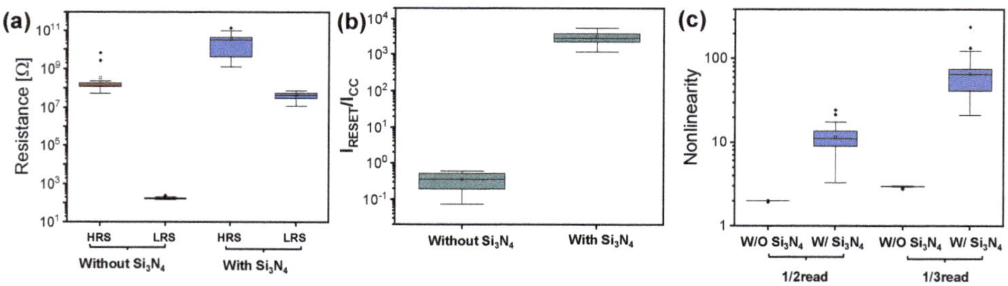

Figure 3. Statistical distributions of (**a**) high-resistance state (HRS) and low-resistance state (LRS) resistance, (**b**) I_{RESET}/I_{CC}, and (**c**) nonlinearity for Ni/HfO_2/SiO_2/Si and Ni/HfO_2/Si_3N_4/SiO_2/Si devices.

Figure 4 shows the read margin as a function of the number of word lines for the Ni/HfO_2/SiO_2/Si and Ni/HfO_2/Si_3N_4/SiO_2/Si devices. Here, the 1/2 read scheme and the 1/3 read scheme are applied to a virtual cross-point array without line resistance. The detailed array read schemes are well known in the literature, and we discuss in detail equation and the scheme in Figure S3. The Ni/HfO_2/Si_3N_4/SiO_2/Si device shows higher read margin compared to the Ni/HfO_2/SiO_2/Si device. This is due to the fact that Ni/HfO_2/Si_3N_4/SiO_2/Si has higher LRS resistance and nonlinear I–V curves in the LRS. The read margin at the 1/3 read scheme is also higher than that of the 1/2 read scheme.

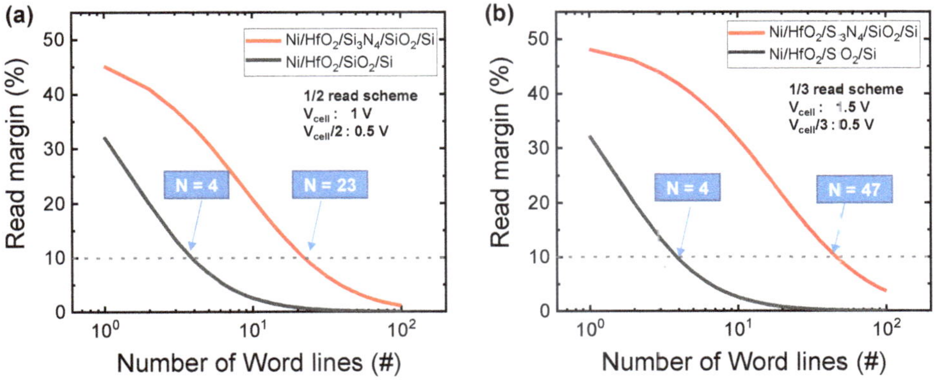

Figure 4. Read margin of $Ni/HfO_2/SiO_2/Si$ and $Ni/HfO_2/Si_3N_4/SiO_2/Si$ devices when applying (**a**) 1/2 read scheme and (**b**) 1/3 read scheme in cross-point array.

The conducting path would be formed in the SiO_2 layer during the set process for the $Ni/HfO_2/Si_3N_4/SiO_2/Si$ device. The electric field is concentrated within the SiO_2 layer with consideration of the dielectric constants (HfO_2: ~20 and SiO_2: ~4). Therefore, high current cannot be avoided in the LRS after the set process. On the other hand, the overshoot current was mitigated in the $Ni/HfO_2/Si_3N_4/SiO_2/Si$ device during the set process. This can be explained by the dispersion of the focused electric field of the SiO_2 layer due to the Si_3N_4 layer. The dielectric constant of the Si_3N_4 layer (Si_3N_4: ~7) is slightly higher than that of SiO_2 and lower than that of HfO_2. Therefore, a Si_3N_4 layer between the HfO_2 layer and the SiO_2 layer is a good buffer layer to reduce the current overshoot.

Next, we compared the tendency of conductance change as a function of identical pulse during the set and reset process. Figure 5a shows the conductance changes of the $Ni/HfO_2/SiO_2/Si$ device for potentiation (set process) and depression (reset process), respectively. The pulse amplitude voltages with a pulse width of 450 μs are 6 V and −3.5 V for potentiation and depression, respectively. The conductance values are extracted from the middle point of read pulse (1 V and 450 μs). For potentiation, the conductance value increases abruptly in response to the 18th pulse. The depression curve shows several fluctuations after the first decrease in conductance. Such randomness and abrupt conductance change are not suitable for a hardware-based neuromorphic synaptic device. On the other hand, the conductance values in the $Ni/HfO_2/Si_3N_4/SiO_2/Si$ device are gradually controlled by the potentiation and depression pulses (Figure 5b). The voltages of the set pulse and reset pulse are 7 and −4.5 V, respectively, and the pulse width is 450 μs. Further, the read pulse (1 V and 450 μs) is inserted between set pulses or reset pulses to obtain the conductance value. It should be noted that a gradual conductance update is possible when the same pulse is repeatedly applied on the device for potentiation and depression. Moreover, the conductance value of the $Ni/HfO_2/Si_3N_4/SiO_2/Si$ device is substantially lower than that of the $Ni/HfO_2/SiO_2/Si$ device. Therefore, the improved synaptic properties, such as the low energy and multiple conductance, of the $Ni/HfO_2/Si_3N_4/SiO_2/Si$ device are beneficial for synaptic applications. The conductance update method of the $Ni/HfO_2/Si_3N_4/SiO_2/Si$ device is suitable for offline learning. To apply it to online learning that provides information by reading conductance values in real time, improvement in variation will be required [33].

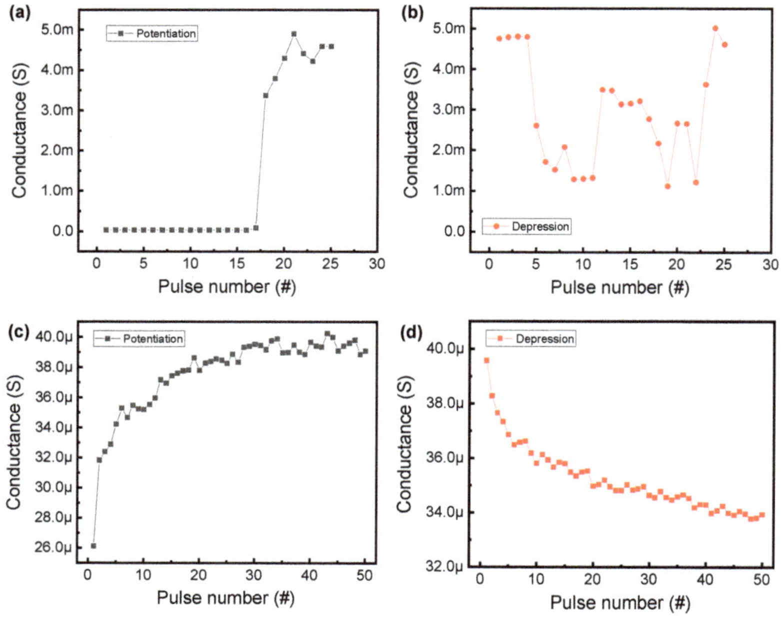

Figure 5. Potentiation and depression characteristics for (**a**,**b**) Ni/HfO_2/SiO_2/Si device and (**c**,**d**) Ni/HfO_2/Si_3N_4/SiO_2/Si device.

4. Conclusions

In summary, we fabricated a CMOS-compatible trilayer device (Ni/HfO_2/Si_3N_4/SiO_2/Si) and characterized its resistive and synaptic characteristics. The TEM and XPS provide the exact dielectric thickness and chemical information of the trilayer device. The Si_3N_4 layer could alleviate the concentrated electric field into the SiO_2 layer in the trilayer design, so the conducting paths are not formed in all dielectrics in the LRS. This property can reduce reset current and provide a nonlinear I–V curve in the LRS. The high nonlinearity in the trilayer device can enlarge the array size in the cross-point array architecture. Finally, we demonstrated that gradual set and reset switching in a trilayer device can be highly suitable for emulating the synaptic behavior of a biological synapse in the human nervous system by controlling multiple conductance.

Supplementary Materials: The following are available online at http://www.mdpi.com/2079-4991/10/12/2462/s1, Figure S1: The STEM image and EDS line scan of Ni/HfO_2/Si_3N_4/SiO_2/Si. Figure S2: Definition of nonlinearity of I–V in the LRS. Figure S3: Read operation scheme in virtual cross-point array structure.

Author Contributions: H.R. conducted the electrical measurements and wrote the manuscript. S.K. designed the experiment concept and supervised the study. All authors have read and agreed to the published version of the manuscript.

Funding: This work was supported in part by the National Research Foundation of Korea (NRF), grant funded by the Korean government (MSIP) under Grant 2018R1C1B5046454.

Conflicts of Interest: The authors declare no conflict of interest.

References

1. Lanza, M.; Wong, H.-S.P.; Pop, E.; Ielmini, D.; Strukov, D.; Regan, B.C.; Larcher, L.; Villena, M.A.; Yang, J.J.; Goux, L.; et al. Recommended Methods to Study Resistive Switching Devices. *Adv. Electron. Mater.* **2018**, *5*, 1800143. [CrossRef]
2. Linn, E.; Rosezin, R.; Kügeler, C.; Waser, R. Complementary resistive switches for passive nanocrossbar memories. *Nat. Mater.* **2010**, *9*, 403–406. [CrossRef]

3. Kim, S.; Jung, S.; Kim, M.-H.; Chen, Y.-C.; Chang, T.-C.; Ryoo, K.-C.; Cho, S.; Lee, J.-H.; Park, B.-G. Scaling Effect on Silicon Nitride Memristor with Highly Doped Si Substrate. *Small* **2018**, *14*, 1704062. [CrossRef]
4. Zhou, F.; Chang, Y.F.; Fowler, B.; Byun, K.; Lee, J.C. Stabilization of multiple resistance levels by current-sweep in SiOx-based resistive switching memory. *Appl. Phys. Lett.* **2015**, *106*, 063508. [CrossRef]
5. Pan, F.; Gao, S.; Chen, C.; Song, C.; Zeng, F. Recent progress in resistive random access memories: Materials, switching mechanisms, and performance. *Mater. Sci. Eng. R Rep.* **2014**, *83*, 1–59. [CrossRef]
6. Maikap, S.; Banergee, W. In Quest of Nonfilamentary Switching: A Synergistic Approach of Dual Nanostructure Engineering to Improve the Variability and Reliability of Resistive Random-Access-Memory Devices. *Adv. Electron. Mater.* **2020**, *6*, 2000209. [CrossRef]
7. Mikhaylov, A.; Pimashkin, A.; Pigareva, Y.; Gerasimova, S.; Gryaznov, E.; Shchanikov, S.; Zuev, A.; Talanov, M.; Lavrov, I.; Demin, V.; et al. Neurohybrid Memristive CMOS-Integrated Systems for Biosensors and Neuroprosthetics. *Front. Mol. Neurosci.* **2020**, *14*, 358. [CrossRef] [PubMed]
8. Kuzum, D.; Yu, S.; Wong, H.-S.P. Synaptic electronics: Materials, devices and applications. *Nanotechnology* **2013**, *24*, 382001. [CrossRef] [PubMed]
9. Graves, C.E.; Li, C.; Sheng, X.; Miller, D.; Ignowski, J.; Kiyama, L.; Strachan, J.P. In-Memory Computing with Memristor Content Addressable Memories for Pattern Matching. *Adv. Mater.* **2020**, *32*, 2003437. [CrossRef] [PubMed]
10. Xia, Q.; Yang, J.J. Memristive crossbar arrays for brain-inspired computing. *Nat. Mater.* **2019**, *18*, 309–323. [CrossRef]
11. Ryu, H.; Kim, S. Pseudo-Interface Switching of a Two-Terminal TaO_x/HfO_2 Synaptic Device for Neuromorphic Applications. *Nanomaterials* **2020**, *10*, 1550. [CrossRef] [PubMed]
12. Choi, J.; Kim, S. Nonlinear Characteristics of Complementary Resistive Switching in $HfAlO_x$-Based Memristor for High-Density Cross-Point Array Structure. *Coatings* **2020**, *10*, 765. [CrossRef]
13. Ryu, H.; Kim, S. Voltage Amplitude-Controlled Synaptic Plasticity from Complementary Resistive Switching in Alloying HfO_x with AlO_x-Based RRAM. *Metals* **2020**, *11*, 1410. [CrossRef]
14. Cho, S.; Kim, S. Emulation of Biological Synapse Characteristics from Cu/AlN/TiN Conductive Bridge Random Access Memory. *Nanomaterials* **2020**, *10*, 1709. [CrossRef] [PubMed]
15. Kim, S.; Kim, H.; Jung, S.; Kim, M.H.; Lee, S.; Cho, S.; Park, B.G. Tuning resistive switching parameters in Si_3N_4-based RRAM for three-dimensional vertical resistive memory applications. *J. Alloy. Compd.* **2016**, *663*, 419–423. [CrossRef]
16. Emelyanov, A.V.; Nikiruy, K.E.; Serenko, A.V.; Sitnikov, A.V.; Presnyakov, M.Y.; Rybka, R.B.; Sboev, A.G.; Rylkov, V.V.; Kashkarov, P.K.; Kovalchuk, M.V.; et al. Self-adaptive STDP-based learning of a spiking neuron with nanocomposite memristive weights. *Nanotechnology* **2020**, *31*, 045201. [CrossRef]
17. Li, D.; Wu, B.; Zhu, X.; Wang, J.; Ryu, B.; Lu, W.D.; Liang, X. MoS_2 Memristors Exhibiting Variable Switching Characteristics toward Biorealistic Synaptic Emulation. *ACS Nano* **2018**, *12*, 9240–9252. [CrossRef]
18. Chand, U.; Alawein, M.; Fariborzi, H. Enhancement of Endurance in HfO_2-Based CBRAM Device by Introduction of a TaN diffusion Blocking Layer. *ECS Trans.* **2017**, *77*, 1971–1976. [CrossRef]
19. Chand, U.; Huang, K.C.; Huang, C.Y.; Ho, C.H.; Lin, C.H.; Tseng, T.Y. Investigation of thermal stability and reliability of HfO_2 based resistive random access memory devices with cross-bar structure. *J. Appl. Phys.* **2015**, *117*, 184105. [CrossRef]
20. Mahata, C.; Lee, C.; An, Y.; Kim, M.H.; Bang, S.; Kim, C.S.; Ryu, J.H.; Kim, S.; Kim, H.; Park, B.G. Resistive switching and synaptic behaviors of an HfO_2/Al_2O_3 stack on ITO for neuromorphic systems. *J. Alloy. Compd.* **2020**, *826*, 154434. [CrossRef]
21. Mikhaylov, A.; Belov, A.; Korolev, D.; Antonov, I.; Kotomina, V.; Kotina, A.; Gryaznov, E.; Sharapov, A.; Koryazhkina, M.; Kryukov, R.; et al. Multilayer Metal-Oxide Memristive Device with Stabilized Resistive Switching. *Adv. Mater. Technol.* **2020**, *5*, 1900607. [CrossRef]
22. Kim, S.; Park, B.G. Nonlinear and multilevel resistive switching memory in $Ni/Si_3N_4/Al_2O_3/TiN$ structures. *Appl. Phys. Lett.* **2016**, *108*, 212103. [CrossRef]
23. Kim, M.H.; Kim, S.; Bang, S.; Kim, T.H.; Lee, D.K.; Cho, S.; Park, B.G. Uniformity Improvement of SiN_x-Based Resistive Switching Memory by Suppressed Internal Overshoot Current. *IEEE Trans. Nanotechnol.* **2018**, *17*, 824–828. [CrossRef]
24. Jung, R. Fermi-Level Pinning at the Poly-Si/HfO_2 Interface. *J. Korean Phys. Soc.* **2009**, *55*, 2501–2504. [CrossRef]

25. Bommali, R.K.; Singh, S.P.; Prakash, G.V.; Ghosh, S.; Srivastava, P. Growth and tailoring of physical properties of Si quantum dots in a-SiN_x:H matrix. *Energy Procedia* **2013**, *41*, 50–56. [CrossRef]
26. Kim, T.; Koka, S.; Surthi, S.; Zhuang, K. Direct Impact of Chemical Bonding of Oxynitride on Boron Penetration and Electrical Oxide Hardening for Nanoscale Flash Memory. *IEEE Electron. Dev. Lett.* **2013**, *34*, 405–407. [CrossRef]
27. Ielemini, D. Modeling the Universal Set/Reset Characteristics of Bipolar RRAM by Field- and Temperature-Driven Filament Growth. *IEEE Trans. Electron Devices* **2011**, *58*, 4309. [CrossRef]
28. Aldana, S.; García-Fernández, P.; Rodríguez-Fernández, A.; Romero-Zaliz, R.; González, M.B.; Jiménez-Molinos, F.; Campabadal, F.; Gómez-Campos, F.; Roldán, J.B. A 3D kinetic Monte Carlo simulation study of resistive switching processes in Ni/HfO_2/Si-n^+-based RRAMs. *J. Phys. D Appl. Phys.* **2017**, *50*, 335103. [CrossRef]
29. Vandelli, L.; Padovani, A.; Larcher, L.; Bersuker, G. Microscopic Modeling of Electrical Stress-Induced Breakdown in Poly-Crystalline Hafnium Oxide Dielectrics. *IEEE Trans. Electron Devices* **2013**, *60*, 1754. [CrossRef]
30. Guy, J.; Molas, G.; Blaise, P.; Bernard, M.; Roule, A.; Carval, G.L.; Delaye, V.; Toffoli, A.; Ghibaudo, G.; Clermidy, F.; et al. Investigation of Forming, SET, and Data Retention of Conductive-Bridge Random-Access Memory for Stack Optimization. *IEEE Trans. Electron Devices* **2015**, *62*, 3482. [CrossRef]
31. Roldán, J.B.; Alonso, F.J.; Aguilera, A.M.; Maldonado, D.; Lanza, M. Time series statistical analysis: A powerful tool to evaluate the variability of resistive switching memories. *J. Appl. Phys.* **2019**, *125*, 174504. [CrossRef]
32. Miranda, E.; Mehonic, A.; Ng, W.H.; Kenyon, A.J. Simulation of Cycle-to-Cycle Instabilities in SiO_x-Based ReRAM Devices Using a Self-Correlated Process With Long-Term Variation. *IEEE Electron Device Lett.* **2019**, *40*, 28. [CrossRef]
33. Kim, C.-H.; Lim, S.; Woo, S.Y.; Kang, W.M.; Seo, Y.-T.; Lee, S.T.; Lee, S.; Kwon, D.; Oh, S.; Noh, Y. Emerging memory technologies for neuromorphic computing. *Nanotechnology* **2018**, *30*, 032001. [CrossRef] [PubMed]

nanomaterials

MDPI

Article

Low-Temperature Synthesis of Monolithic Titanium Carbide/Carbon Composite Aerogel

Tingting Niu [1,2], Bin Zhou [1,2,*], Zehui Zhang [1,2], Xiujie Ji [1,2], Jianming Yang [1,2], Yuhan Xie [1,2], Hongqiang Wang [1,2] and Ai Du [1,2,*]

1 School of Physics Science and Engineering, Tongji University, Shanghai 200092, China; 94niutingting@tongji.edu.cn (T.N.); 1910105@tongji.edu.cn (Z.Z.); 1710867@tongji.edu.cn (X.J.); 1810908@tongji.edu.cn (J.Y.); luwietse@tongji.edu.cn (Y.X.); 1910759@tongjiedu.cn (H.W.)

2 Shanghai Key Laboratory of Special Artificial Microstructure Materials and Technology, Tongji University, Shanghai 200092, China

* Correspondence: zhoubin863@tongji.edu.cn (B.Z.); duai@tongji.edu.cn (A.D.)

Received: 20 November 2020; Accepted: 12 December 2020; Published: 16 December 2020

Abstract: Resorcinol-formaldehyde/titanium dioxide composite (RF/TiO_2) gel was prepared simultaneously by acid catalysis and then dried to aerogel with supercritical fluid CO_2. The carbon/titanium dioxide aerogel was obtained by carbonization and then converted to nanoporous titanium carbide/carbon composite aerogel via 800 °C magnesiothermic catalysis. Meanwhile, the evolution of the samples in different stages was characterized by X-ray diffraction (XRD), an energy-dispersive X-ray (EDX) spectrometer, a scanning electron microscope (SEM), a transmission electron microscope (TEM) and specific surface area analysis (BET). The results showed that the final product was nanoporous TiC/C composite aerogel with a low apparent density of 339.5 mg/cm^3 and a high specific surface area of 459.5 m^2/g. Comparing to C aerogel, it could also be considered as one type of highly potential material with efficient photothermal conversion. The idea of converting oxide–carbon composite into titanium carbide via the confining template and low-temperature magnesiothermic catalysis may provide new sight to the synthesis of novel nanoscale carbide materials.

Keywords: titanium carbide; sol-gel; confining template; magnesiothermic catalysis

---

1. Introduction

Carbide aerogel is a type of foamed material with three-dimensional nanoporous structure. It combines the advantages of traditional aerogels such as low density, large specific surface area, opening-pore structure and the intrinsic properties of carbides. The crystal structure of typical carbide consists of hybrid bonds including ionic bonds, covalent bonds and metallic bonds, leading to the fundamental characteristics of high hardness, high melting point, wear resistance and conductivity. Therefore, carbide aerogel could be prospectively applied not only in heterogeneous catalysis, heat-resisting materials, superhard additives, fuel cells and sensor elements, etc., but also in high technology fields such as machinery, the chemical industry, carburization and microelectronics, etc. [1–5]. Traditional synthesis methods of carbide include the direct carbonization method [6–8], the carbothermic method [9–15], self-propagation high-temperature synthesis [16–20], the mechanical alloying method [21–24], the microwave method [25,26], the molten salt method [27–30] and the plasma method [31,32]. The carbothermic method is widely promoted because it possesses the advantages of low cost, simple equipment, high production efficiency and mass production capacity. However, this method needs high temperature (1200–1500 °C for SiC and 1300–2000 °C for TiC) in an inert or reducing atmosphere, usually leading to the agglomeration and growth of the carbide [9,10,12–15]. Thus, it is difficult to prepare the ultrafine and nanoscale carbide materials with monolithic appearance.

Only a few works on nanoporous monolithic carbide aerogel have been reported. For example, in 2012 our group developed a facile method for the low-temperature magnesiothermic conversion from C/SiO_2 aerogel to SiC aerogel via the confining template reaction method at 700 °C [33]. In 2015, a novel low-temperature pseudomorphic transformation of TiC and NbC aerogels was achieved via a solid–gas–solid reaction between the metal and carbon aerogels in aid of iodine catalysts [34]. Kong et al. carried out a series of works about the preparation, control and properties of silicon-carbide-based aerogels, including pure SiC aerogel, mainly by using carbothermal reduction at 1500 °C [35]. These works provide exciting ideas to develop novel nanoporous carbide aerogels.

Previous studies have successfully demonstrated the facile route to carbide-based aerogel. Overall, however, it is difficult to find a delicate balance between the ultrafine structure (high porosity, nanoscale skeleton, high specific surface area, etc.) and rigorous preparation conditions (high temperature and long treating time). By comparison with silicon carbide, titanium carbide normally needs higher treating temperature, indicating the difficulty to maintain both its nanoscale structure and its monolithic appearance. By comparison with the pseudomorphic transformation method, the solid-state reaction is relatively easy to control and scale up. Carbon aerogel is an excellent template because of its low density and high porosity [36–41], so it was a good choice for us to increase the strength of the framework structure. Thus, in this paper, we tried to apply the simple low-temperature magnesiothermic conversion method to synthesize titanium carbide in the TiO_2/C composite aerogel to maintain its nanoporous microstructure and overcome the disadvantage of high temperature and high energy consumption in the preparation process. The main idea was to develop the methods of (1) preparing low-density C/TiO_2 composite aerogel via a co-gelled process and (2) achieving the magnesium-catalyzed conversion with relatively low temperature and short reaction time. We finally obtained a self-supported monolithic nanoporous TiC/C composite aerogel with low density (339.5 mg/cm^3) and large specific surface area (459.5 m^2/g) via a magnesium-catalyzed reaction of only 800 °C and 4 h. This method achieved the goal of controllable structure and adjustable property for the carbide-based aerogels, which may promote the studies of the other novel carbide aerogels with high performances.

2. Experimental Setup

2.1. Materials

The reagents, including resorcinol (AR, 99.5 wt.%), formaldehyde (AR, 37.0 wt.% in water), acetonitrile (AR, 99.0 wt.%), acetic acid (AR, 99.5 wt.%), nitric acid (AR, 65.0 wt.%), tetra-n-butyl titanate (CP, 98.0 wt.%), ethanol, concentrated hydrochloric acid (AR, 37.0 wt.%) were purchased from Sinopharm Chemical Reagent Co., Ltd. (Shanghai, China). Magnesium powders (1.74 g/mL at 25 °C, 99%) were obtained from Sigma-Aldrich (St. Louis, MO, USA). All reagents were used without further purification.

2.2. Preparation of RF/TiO_2 Composite Gel and Aerogel

The synthesis process of preparation is detailed in Scheme 1, displaying the mixture of two parts named as RF system and TiO_2 system. The key of RF/TiO_2 composite gel preparation was to successfully compound two monomer gels into one co-gel, so it was very important to control the proportion and concentration of precursors and the reaction condition based on the reaction mechanisms. Firstly, 1.1 g of resorcinol and 2.5 mL of formaldehyde solution (37.0 wt.%) were uniformly mixed with the molar ratio 3:1 in 10 mL of acetonitrile at 30 °C to achieve colorless transparent solution A for the RF system, while 10 mL of acetonitrile, 2 mL of acetic acid, 150 uL of 65.0 wt.% nitric acid and 4 mL of tetra-n-butyl titanate were uniformly mixed at 0 °C in an ice bath to achieve pale yellow semitransparent solution B for another part. Then, solution A was poured slowly into solution B to achieve a red uniform solution. The pH value was in the range of 1~3. Here we determined the molar ratio of the carbon element and the titanium element at the appropriate value 7.6 according to the initial experiments. It must be noted that the mixed solution was stirred at 0 °C because of the temperature sensitivity of tetra-n-butyl titanate. Four milliliters of distilled water could be added to promote the hydrolysis of tetra-n-butyl titanate. After stirring for 30 min, the mixed solution was transferred to a proper sealed container (such as plastic tubes) and gelled at 50 °C within 2 h. The color turned opaque dark red. The wet hybrid gel was placed at room temperature (RT) and aged for 24 h, then removed and repeatedly substituted by ethanol every 6 h for 6 times in order to remove water and residual chemicals. The gel was cut into small slices with a thickness of less than 2 mm, which was beneficial for the magnesiothermic catalysis. At last, they were dried by CO_2 supercritical fluid to obtain red-brown RF/TiO_2 aerogels.

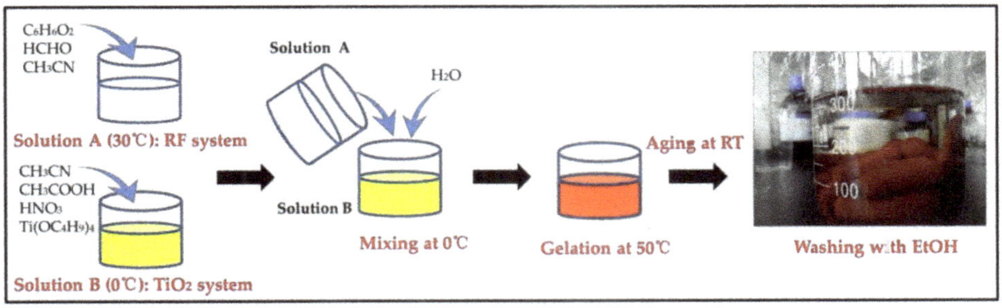

Scheme 1. Scheme of the preparation of RF/TiO_2 hybrid gel.

2.3. *Conversion of RF/TiO_2 Aerogel into TiC/C Composite Aerogel*

The RF/TiO_2 aerogel was placed in a high-temperature carbide furnace under flowing nitrogen (250 mL/min) at 800 °C for 2 h with a heating rate of 3 °C/min to completely carbonize it, so as to obtain black C/TiO_2 aerogel. Subsequently, C/TiO_2 aerogel was subject to low-temperature magnesiothermic reduction and acid treatment to be converted into TiC/C composite aerogel. In a typical procedure, 100 mg C/TiO_2 aerogel and 110 mg magnesium powders were sealed in a steel ampoule with an inner volume of 80 cm^3 in a glove box full of argon. The magnesium was separately dispersed next to the samples. Then, the ampoule was pushed into a tube furnace under flowing argon (120 mL/min). The temperature was increased to 800 °C and maintained for 4 h. It should be noted that the thickness of the samples was controlled less than 1 mm due to the penetration depth of magnesium vapor. After it cooled to ambient temperature, the ampoule was opened under an inert atmosphere to avoid oxidation. The collected samples were immersed in 10 mL ethyl alcohol. To remove the redundant Mg and the by-product MgO, the corrosion treatment was carried out for 6 h by adding the mixed solution of 10 mL ethyl alcohol, 1.6 mL distilled water and 2 mL concentrated hydrochloric acid (HCl). Then, the purified samples were repeatedly washed by ethyl alcohol every 2 h for more than 4 times and dried by CO_2 supercritical fluid.

2.4. *Characterizations*

The phase structure of the sample was analyzed by a Model Rigaku D/Max-RB powder X-ray diffractometer (Rigaku, Tokyo, Japan) to analyze its diffraction pattern and obtain the information such as the composition and the structure of the atoms or molecules inside the material. Cu target K_{α}radiation ($\lambda = 0.15406$ nm) was adopted in the test, with the working voltage of 40 kV and current of 0.04 A. The scanning step was adopted with the step length of 0.08° and the scanning range of the diffraction angle of 10°~80°. Raman spectra were detected at 514 nm excitation using a HORIBA Jobin-Yvon HR800 Raman system (HORIBA Jobin-Yvon, Paris, France) at room temperature. X-ray photoelectron spectroscopy (XPS) was examined by Thermo ESCALAB 250XI (Thermo Fisher Scientific, Waltham, MA, USA). The morphology of the sample was collected by a Philips XL30FEG scanning electron microscope (Royal Philips Electronics, Amsterdam, The Netherlands) with the acceleration voltage of 10 kV; the sample was subject to gilding before observation. Composition analysis was carried out on the samples by an Oxford energy-dispersive X-ray (EDX) spectrometer

(Oxford Instruments NanoAnalysis, Concord, MA, USA). Transmission electron microscopy (TEM) was conducted with a Model JEM-2100 electron microscope (JEOL Corp, Tokyo, Japan) made by Japanese JEOL Corp. operating at 200 keV. Nitrogen adsorption–desorption isotherms were measured at liquid nitrogen temperature (77 K) by a Quantachrome Autosorb-1MP analyzer (Quantachrome, Boynton Beach, FL, USA) after the samples were degassed in a vacuum at 150 °C for at least 12 h. The specific surface area was calculated by Brunaur–Emmett–Teller (BET) method.

3. Results and Discussion

3.1. Appearance and Structural Characterization

Figure 1 shows the appearances of the three samples of RF/TiO_2 aerogel, C/TiO_2 aerogel and the objective TiC/C composite aerogel. Because all of the samples we obtained had monolithic appearances of regular cylinders, we measured the mass, the diameter and the thickness directly to calculate the density (weighting method). Table 1 lists their average values of densities, diameters and linear shrinkage ratios after repeated measurements. According to Figure 1, the volume of C/TiO_2 aerogel shrunk slightly after carbonization, while the TiC/C composite aerogel had no obvious shrinkage and maintained a monolithic appearance similar to its original template.

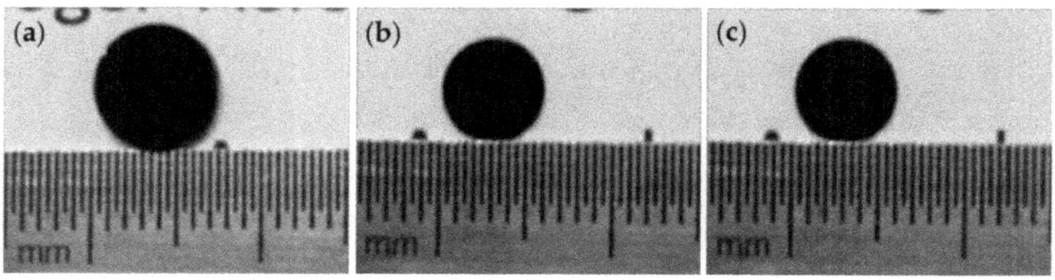

Figure 1. Appearances of samples of (**a**) RF/TiO_2 aerogel, (**b**) C/TiO_2 aerogel and (**c**) TiC/C composite aerogel subject to magnesiothermy catalysis, HCl treatment and CO_2 supercritical treatment.

Table 1. Densities and diameters of RF/TiO_2, C/TiO_2 and TiC/C composite aerogels.

Samples	Density (mg/cm^3)	Diameter (cm)	Linear Shrinkage Ratio (%)
RF/TiO_2	234.8 ± 7.6	0.723	–
C/TiO_2	355.1 ± 10.4	0.582	19.5%
TiC/C	339.5 ± 7.9	0.569	2.2%

The XRD analysis results of RF/TiO_2, C/TiO_2 and TiC aerogels are shown in Figure 2. Only one broad peak in the range of 15–30° was found in the XRD spectrum of RF/TiO_2 aerogel, which indicated the existence of an amorphous substrate. Several obvious diffraction peaks of C/TiO_2 aerogel were detected and indexed as anatase-TiO_2 phase (PDF No. 21-1272). However, the diffraction peaks, located at $2\theta = 35.906°, 41.710°, 60.448°, 72.369°$ and $76.139°$ in the sample of objective product TiC/C aerogel, corresponded to the typical (111), (200), (220), (311) and (222) diffraction peaks of cubic-TiC phase, respectively (PDF No. 32-1383). Meanwhile, the phase of TiO_2 disappeared. This indicated that TiC had been generated from the reaction between TiO_2 and C through the magnesiothermic catalysis reaction. The final sample contained crystalline titanium carbide. The successful conversion of C/TiO_2 aerogel into TiC/C aerogel could also be further verified in the following SAED (selected area electron diffraction) patterns. In addition, a broad peak of amorphous carbon was also displayed in TiC/C composite aerogel.

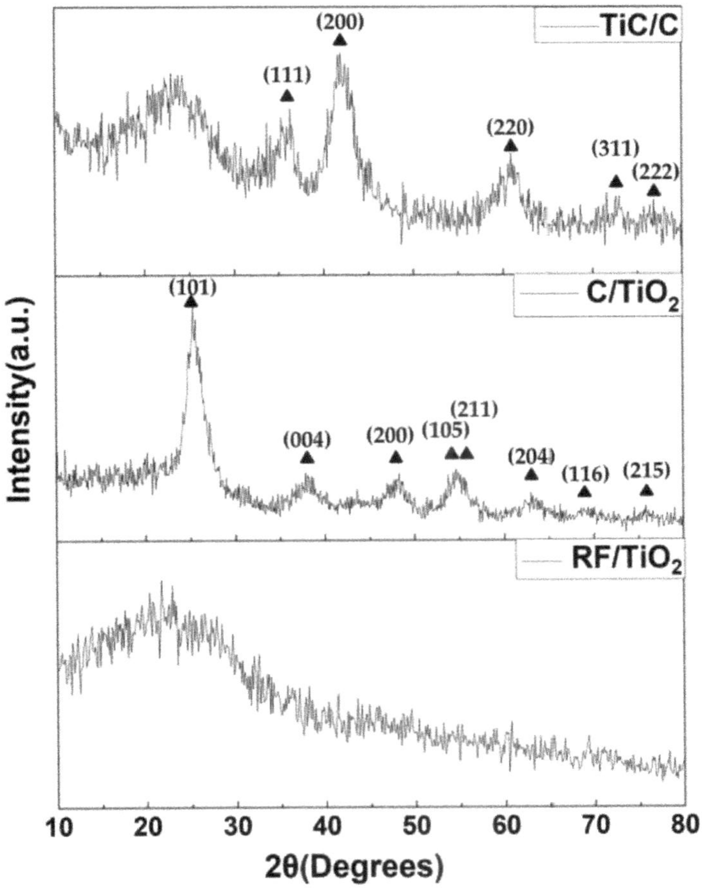

Figure 2. XRD spectra of RF/TiO_2 aerogel, C/TiO_2 aerogel and the objective product TiC/C aerogel.

Raman spectra of C/TiO_2 and TiC/C aerogels are shown in Figure 3. For C/TiO_2 aerogel, the high-frequency branch of the E_g mode of anatase-TiO_2 was detected at 152.1 cm^{-1}, as well as B_{1g}, A_{1g} + B_{1g} and E_g peaks at 411.4 cm^{-1}, 518.6 cm^{-1} and 616.7 cm^{-1}. However, there was just one weak peak at 155.3 cm^{-1} for TiC/C aerogel assigned to the strongest mode E_g, which was in accordance with the XRD analysis. The D band and G band of carbon were observed obviously at about 1353.3 cm^{-1} and 1586.6 cm^{-1} for both C/TiO_2 aerogel and TiC/C aerogel. X-ray photoelectron spectroscopy (XPS) emissions of carbon (1 s) and titanium (2 p) of these two samples are shown in Figure 4 to obtain more detailed information regarding the chemical environment of elements. In both aerogels, it was found that the binding energies (BEs) of C 1 s belonged to C-C, C-O, C=O and O-C=O. Two strong peaks centered at about 457.9 eV and 463.6 eV were ascribed to Ti $2p_{3/2}$ and $2p_{1/2}$. As for the TiC/C aerogel, another weak but non-negligible peak could be found at 281.35 eV in the C (1 s) region, which further confirmed the presence of titanium carbide [42,43]. The binding energies at 456.06 eV and 460.81 eV belonging to Ti(III), at 455.12 eV and 459.84 eV belonging to Ti(II), as well as at 454.47 eV and 458.6 eV belonging to Ti(0) were probably caused by the partial reduction of TiO_2 [44,45].

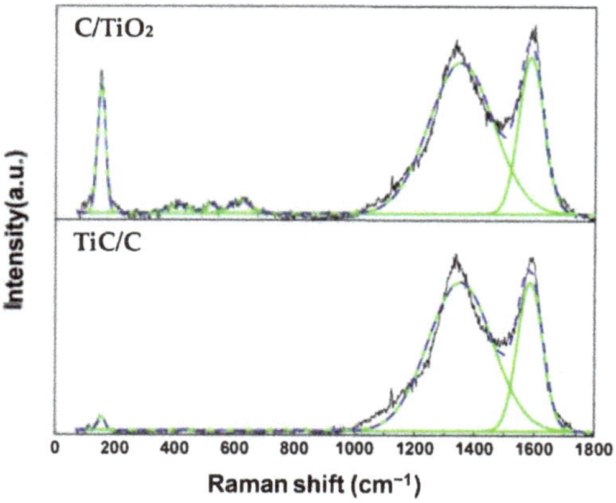

Figure 3. Raman spectra of C/TiO_2 and TiC/C aerogels.

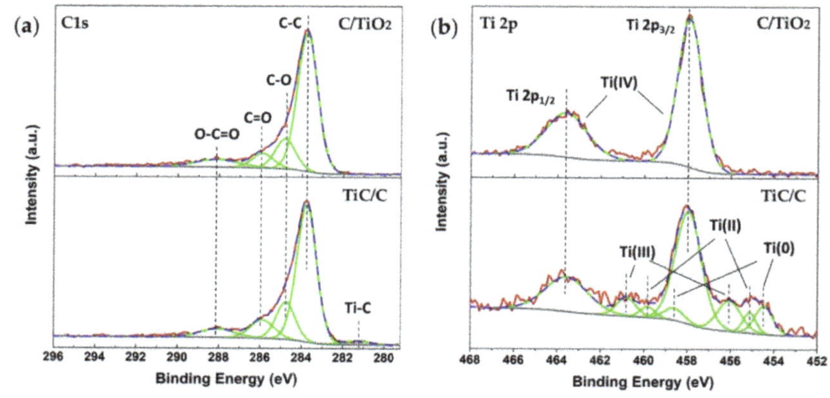

Figure 4. XPS spectra of (**a**) C/TiO_2 and (**b**) TiC/C aerogels.

3.2. Morphology and Composition Analysis

Figure 5 shows the SEM diagrams (a–c) of RF/TiO_2, C/TiO_2 and the resulting TiC/C aerogels, respectively. According to the SEM diagrams, these three aerogels all exhibited random nanoporous network structures. The nanoparticles on the framework were distributed uniformly in a globular manner. TiC/C aerogel (Figure 5c) acquired through magnesiothermic catalysis of C/TiO_2 aerogel (Figure 5b) maintained the porous structure of the original template, which was consistent with the appearances in Figure 1. SEM images indicate that the objective product TiC/C aerogel had no obvious deformation or agglomeration, with a similar spherical particles structure. The results showed that the growth of TiC particles was confined in the microstructure of the original template. Hence, the structure of TiC/C aerogel could be facilely adjusted by controlling the microstructure and shape of the C/TiO_2 aerogel template.

According to the EDX spectra, relative oxygen content decreased significantly after carbonization and magnesiothermic reduction, which indicated the success of conversion. A small amount of oxygen (Figure 5f) was related to oxidation at the surface of TiC nanoparticles exposed to etching solution or the oxygen absorption at the surface of C nanoparticles. In spite of this, the weight ratio of oxygen and titanium fell from 1.18 to 0.23 after magnesiothermic reduction. Additionally, in order to confirm the content of the Ti element in the resulting TiC/C aerogel, we measured the mass loss after air calcination at 600 °C for 1 h and calculated that the weight percent of Ti was about 30%, which was close to the identification of the EDX spectra.

Transmission electron microscopy (TEM) was investigated in order to further understand the microstructure and nanocrystalline structure of TiC/C composite aerogel. Figure 6a,b indicates that the mutually supported framework structures mainly consisted of regular spherical or near-spherical clusters and irregular pores. According to Figure 6c, all the elements had a uniform distribution, while the oxygen content was obviously limited, which was identical to the EDX spectra analysis, indicating a homogeneous TiC distribution. The SAED patterns (insets in Figure 6a,b) show two kinds of different concentric diffraction rings corresponding to the Miller indices of the tetragonal lattice of the anatase-TiO_2 phase and the cubic lattice of the cubic TiC phase, respectively, which is consistent with the relevant XRD results. The lattice fringes of both aerogels were also easily observed in the high-resolution transmission electron images (HRTEM, Figure 7).

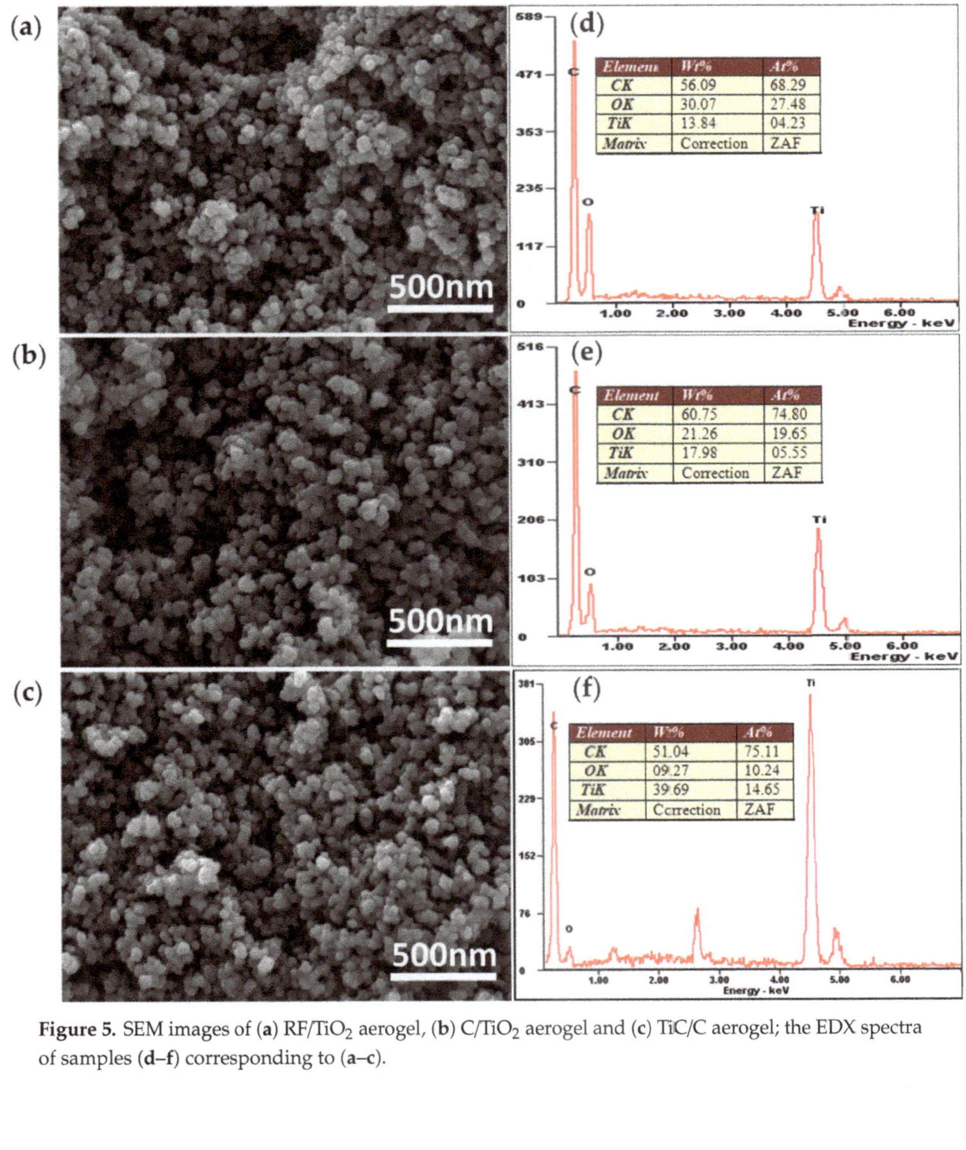

Figure 5. SEM images of (**a**) RF/TiO_2 aerogel, (**b**) C/TiO_2 aerogel and (**c**) TiC/C aerogel; the EDX spectra of samples (**d**–**f**) corresponding to (**a**–**c**).

Figure 6. TEM images with SAED inset patterns of (**a**) C/TiO_2 aerogel and (**b**) TiC/C aerogel; (**c**) mapping results of TiC/C aerogel.

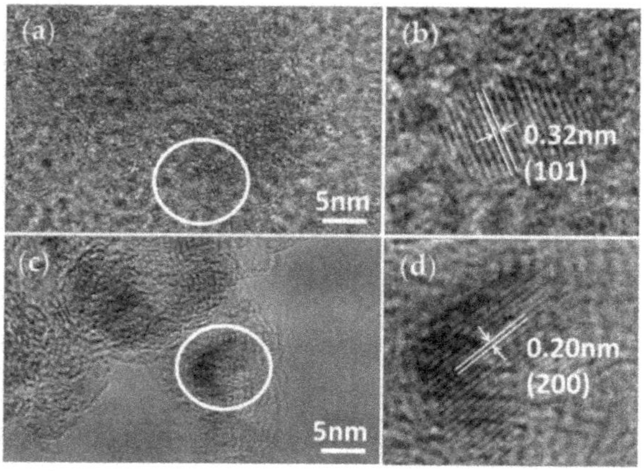

Figure 7. HRTEM images of (**a**) C/TiO_2 aerogel and (**c**) TiC/C aerogel with (**b**,**d**) the magnified images of the corresponding selected areas in the white circles.

3.3. Specific Surface Area Analysis

The nitrogen adsorption/desorption isotherm (Figure 8a,b) of C/TiO_2 aerogel and TiC/C aerogel showed Type IV curves, which suggests that they were typical mesoporous materials. They exhibited mixing H1 and H3 hysteresis loops, which are deemed mesoporous systems with cylindrical pores and wedge-shaped pores stacked by nanoparticles. Based on the BJH (Barrett–Joyner–Halenda model) analysis on the desorption curves (shown as insets in Figure 8a,b), pore-size distributions were similar with a single peak around 36 nm. Specific surface area analysis is listed in detail in Table 2. The specific surface area of TiC/C aerogel after magnesiothermic reduction arrived at 459.5 m^2/g, which was obviously reduced compared to C/TiO_2 aerogel (780.6 m^2/g). This may be because the aerogels of the two frameworks reacted to form TiC/C aerogel during magnesiothermic catalysis, which made the original crosslinking framework stack. Another reason may be the dissolution of some impurities containing magnesium during the corrosion process after magnesiothermic catalysis. Due to the reduction in micropore contents, the specific surface area reduced and the average pore size increased.

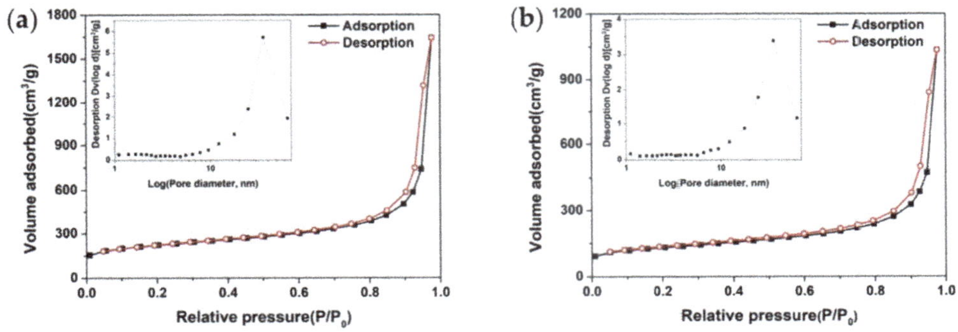

Figure 8. Adsorption and desorption curves and pore size distribution curves of (**a**) C/TiO_2 aerogel and (**b**) TiC/C aerogel.

Table 2. Data related to the pore structures of RF/TiO_2 aerogel, C/TiO_2 aerogel and TiC/C aerogel.

Sample	Specific Surface Area (m^2/g)	Average Pore Diameter (nm)	Pore Size of Maximum Distribution (nm)	Total Pore Volume (cm^3/g)
C/TiO_2	780.6	13.05	36.1	2.547
TiC/C	459.5	13.97	35.7	1.605

3.4. Photothermal Conversion

A simple experiment was conducted to compare the photothermal conversion properties of three different samples (C aerogel, C/TiO_2 aerogel and TiC/C aerogel). Here, C aerogel was obtained after the corrosion of C/TiO_2 aerogel with HF. The results of SEM, EDX spectrum and Raman are shown in Figure S1. These three samples were placed under illumination at different light intensities (600, 800, 1000, 1200 W/m^2) within 8 min. The whole process, including the temperature rise during the illumination and the temperature reduction after the shut-off light source, was recorded using a thermal camera connected to the application of AnalyzIR (Video S1). Figure 9d shows all the temperature variation curves (the average value of a selected area, as shown in Figure 9a) of these three aerogels at four different light intensities. It indicates that C aerogel delivered the fastest response both at the beginning and the ending of illumination. The first derivative curves at the beginning and the ending of illumination within 1 s were compared in Figure S2a,b to further understand the response speeds of different aerogels. We found that C aerogel had the highest sensibility. Nevertheless, the temperature variations of TiC/C aerogel reached the maximum, which exceeded that of C aerogel after a period of illumination (Figure 9b with the highest temperature of the whole area marked). Figure 9e shows the average temperature variation within 54 s before the ending of illumination at different intensities. Meanwhile, the comparisons of temperature variation with C aerogel within 54 s before the ending of illumination are shown amplified in Figure S2c,d. It was obvious that TiC/C aerogel arrived at the highest equilibrium temperature regardless of the light intensity. The opposite results may be due to the difference of thermal conductivity coefficients between C and TiC. According to the cooling part of these curves, TiC/C aerogel dissipated heat slower than C aerogel (obviously observed in Figure 9c), which indicated its coefficient of thermal conductivity was lower. The equilibrium temperature was determined by the combined effect of heat absorption and heat dissipation. Under illumination, these three aerogels were able to convert the energy of illumination into thermal energy. Some thermal energy was absorbed, leading to the rise in temperature, while another part was dissipated. Because of the lower thermal conductivity, TiC/C aerogel had no obvious heat dissipation effect and thus could reach a higher equilibrium temperature. This showed that the TiC/C aerogel was also a highly potential material of efficient photothermal conversion.

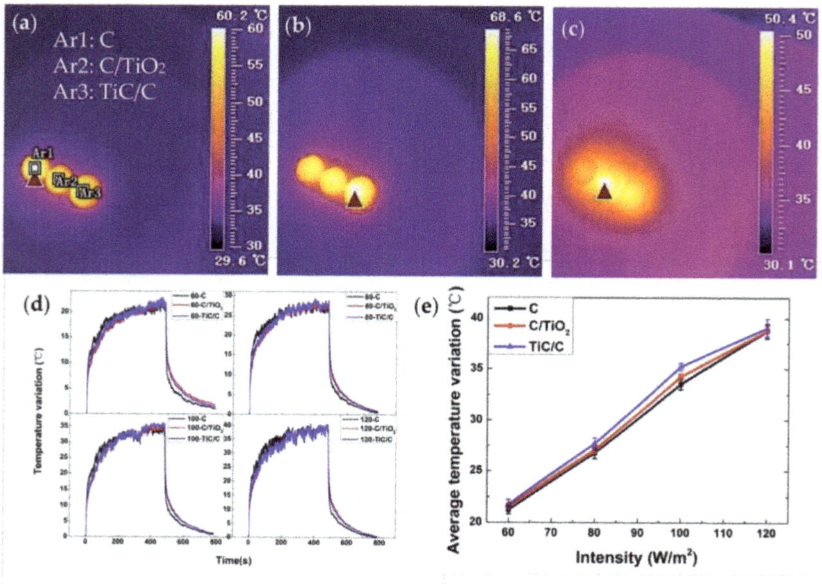

Figure 9. Thermal camera image of three samples after illuminating for 90 s: (**a**, **b**) in the equilibrium state; (**c**) within 6 s after turning off the illumination; (**d**) temperature variation curves of C, C/TiO_2 and TiC/C aerogels at four different light intensities; (**e**) average temperature variation within 54 s before the ending of illumination at different intensities of three samples.

4. Conclusions

With RF/TiO_2 aerogel as the template, a novel nanoporous TiC/C aerogel material was prepared by means of a confining template of magnesiothermic catalysis under low temperature (800 °C) and short reaction time (4 h). Firstly, RF/TiO_2 aerogel was acquired via sol-gel of a compound acid catalysis. After carbonization and subsequent magnesiothermic reaction, the objective product TiC/C aerogel was successfully obtained. It mainly maintained the nanoporous structure of the original material with very integral appearance and ultrafine morphology. Moreover, it had low density (339.5 mg/cm^3) and comparatively large specific surface area (459.5 m^2/g). XRD and TEM analyses indicated the formation of cubic TiC phase. The TiC/C aerogel was discovered as a potential material of highly efficient photothermal conversion. This preparation method can be expected to solve the bottleneck problems of preparing nanoscale titanium carbide materials with controllable structure, adjustable composition and density as well as integrated functional structure at low temperature. In addition, this confining reaction method with controllable template structure can expand the study of conversion of carbon/transition metallic oxides to carbides at low temperature, and it is expected to further expand to the study of transition metals such as Hf, Ta and W.

Supplementary Materials: The following are available online at http://www.mdpi.com/2079-4991/10/12/2527/s1, Figure S1: (**a**) The corrosion of C/TiO_2 aerogel with HF; (**b**) Raman spectrum, (**c**) SEM image, (**d**) EDX spectrum of C aerogel, Figure S2. The 1st derivative curves at the beginning (**a**) and the ending (**b**) of illumination within 1 s; (**c**) the comparisons of temperature variation within 54 s before the ending of illumination; (**d**) the difference in temperature variation compared with C aerogel within 54 s before the ending of illumination, Video S1: The whole process of the temperature rise and the temperature reduction.

Author Contributions: Conceptualization, B.Z., A.D. and T.N.; methodology, A.D., T.N., H.W. and X.J.; investigation, T.N., Z.Z. and J.Y.; formal analysis, Y.X., X.J.; data curation, T.N. and Z.Z.; writing—original draft preparation, T.N.; writing—review and editing, all the authors; supervision, B.Z. and A.D. All authors have read and agreed to the published version of the manuscript.

Funding: This research is supported by the National Key Research and Development Program of China (2017YFA0204600) and National Natural Science Foundation of China (11874284).

Acknowledgments: We would like to appreciate all the fellows from our research group.

Conflicts of Interest: The authors declare no conflict of interest.

References

1. Wang, X.; Chen, L.; Wang, L.; Fan, Q.; Pan, D.; Li, J.; Chi, F.; Xie, Y.; Yu, S.; Xiao, C.; et al. Synthesis of novel nanomaterials and their application in efficient removal of radionuclides. *Sci. China Chem.* **2019**, *62*, 933–967. [CrossRef]
2. Duong-Viet, C.; Ba, H.; Liu, Y.; Truong-Phuoc, L.; Nhut, J.-M.; Pham-Huu, C. Nitrogen-doped carbon nanotubes on silicon carbide as a metal-free catalyst. *Chin. J. Catal.* **2014**, *35*, 906–913. [CrossRef]
3. Gupta, S.; Yadav, A.; Singh, S.; Verma, N. Synthesis of Silicon Carbide-Derived Carbon as an Electrode of a Microbial Fuel Cell and an Adsorbent of Aqueous Cr(VI). *Ind. Eng. Chem. Res.* **2017**, *56*, 1233–1244. [CrossRef]
4. Yang, P.; Zhao, H.; Yang, Y.; Zhao, P.; Zhao, X.; Yang, L. Fabrication of N, P-Codoped Mo_2C/Carbon Nanofibers Via Electrospinning as Electrocatalyst for Hydrogen Evolution Reaction. *ES Mater. Manuf.* **2020**, *7*, 34–39. [CrossRef]
5. Han, D.; Mei, H.; Xiao, S.; Cheng, L. Macroscopic carbon nanotube assembly/silicon carbide matrix composites produced by gas phase route. *Adv. Compos. Hybrid Mater.* **2019**, *2*, 142–150. [CrossRef]
6. Kumari, S.; Kumar, R.; Agrawal, P.R.; Prakash, S.; Mondal, D.P.; Dhakate, S.R. Fabrication of lightweight and porous silicon carbide foams as excellent microwave susceptor for heat generation. *Mater. Chem. Phys.* **2020**, *253*, 123211. [CrossRef]
7. Wu, H.; Wang, Q.; Qin, M.; Yin, R.; Zhang, Z.; Jia, B.; Qu, X. Synthesis of tungsten carbide nanopowders by direct carbonization of tungsten oxide and carbon: Effects of tungsten oxide source on phase structure and morphology evolution. *Ceram. Int.* **2020**, *46*, 8787–8795. [CrossRef]
8. Yi, Z.; Shao, G.; Duan, X.; Sun, P.; Shi, X.; Xiong, Z.; Guo, J. Preparation of WC-Co powder by direct reduction and carbonization. *China Particuology* **2005**, *3*, 286–288. [CrossRef]
9. Feng, L.; Fahrenholtz, W.G.; Hilmas, G.E.; Zhou, Y. Synthesis of single-phase high-entropy carbide powders. *Scr. Mater.* **2019**, *162*, 90–93. [CrossRef]
10. Mojaki, S.C.; Mishra, S.B.; Mishra, A.K. Synthesis, characterization and material properties of titanium carbide nanocomposite derived from biochar. *Mater. Lett.* **2020**, *264*, 127317. [CrossRef]
11. Wang, K.-F.; Tang, X.-L.; Jiao, S.; Chou, K.-C.; Zhang, G.-H. A short and facile process to synthesize WC-Co cemented carbides. *Int. J. Refract. Met. Hard Mater.* **2020**, *92*, 105288. [CrossRef]
12. Rambo, C.R.; Cao, J.; Rusina, O.; Sieber, H. Manufacturing of biomorphic (Si, Ti, Zr)-carbide ceramics by sol–gel processing. *Carbon* **2005**, *43*, 1174–1183. [CrossRef]
13. Worsley, M.A.; Kuntz, J.D.; Pauzauskie, P.J.; Cervantes, O.; Zaug, J.M.; Gash, A.E.; Satcher, J.H.; Baumann, T.F. High surface area carbon nanotube-supported titanium carbonitride aerogels. *J. Mater. Chem.* **2009**, *19*, 5503–5506. [CrossRef]

14. An, Z.; Zhang, R.; Fang, D. Synthesis of monolithic SiC aerogels with high mechanical strength and low thermal conductivity. *Ceram. Int.* **2019**, *45*, 11368–11374. [CrossRef]
15. Chen, X.; Fan, J.; Lu, Q. Synthesis and characterization of TiC nanopowders via sol-gel and subsequent carbothermal reduction process. *J. Solid State Chem.* **2018**, *262*, 44–52. [CrossRef]
16. Alsawat, M.; Altalhi, T.; Alotaibi, N.F.; Zaki, Z.I. Titanium carbide—Titanium boride composites by self propagating high temperature synthesis approach: Influence of zirconia additives on the mechanical properties. *Results Phys.* **2019**, *13*, 102292. [CrossRef]
17. Jin, S.; Shen, P.; Lin, Q.; Zhan, L.; Jiang, Q. Growth Mechanism of TiCxduring Self-Propagating High-Temperature Synthesis in an Al−Ti−C System. *Cryst. Growth Des.* **2010**, *10*, 1590–1597. [CrossRef]
18. Jin, S.; Shen, P.; Zhou, D.; Jiang, Q. Self-propagating high-temperature synthesis of nano-TiCx particles with different shapes by using carbon nano-tube as C source. *Nanoscale Res. Lett.* **2011**, *6*, 515. [CrossRef]
19. Kurbatkina, V.V.; Patsera, E.I.; Levashov, E.A.; Timofeev, A.N. Self-propagating high-temperature synthesis of single-phase binary tantalum-hafnium carbide (Ta,Hf)C and its consolidation by hot pressing and spark plasma sintering. *Ceram. Int.* **2018**, *44*, 4320–4329. [CrossRef]
20. Patsera, E.I.; Levashov, E.A.; Kurbatkina, V.V.; Kovalev, D.Y. Production of ultra-high temperature carbide (Ta,Zr)C by self-propagating high-temperature synthesis of mechanically activated mixtures. *Ceram. Int.* **2015**, *41*, 8885–8893. [CrossRef]
21. Rahaei, M.B.; Yazdani rad, R.; Kazemzadeh, A.; Ebadzadeh, T. Mechanochemical synthesis of nano TiC powder by mechanical milling of titanium and graphite powders. *Powder Technol.* **2012**, *217*, 369–376. [CrossRef]
22. Jia, H.; Zhang, Z.; Qi, Z.; Liu, G.; Bian, X. Formation of nanocrystalline TiC from titanium and different carbon sources by mechanical alloying. *J. Alloys Compd.* **2009**, *472*, 97–103. [CrossRef]
23. Stanciu, V.I.; Vitry, V.; Delaunois, F. Study of the milling parameters optimization in the direct carburization of WO_3 by mechanical alloying. *Int. J. Refract. Met. Hard Mater.* **2020**, *87*, 105160. [CrossRef]
24. Zhao, Z.; Hu, W. Synthesis and characterization of chromium carbide nanopowders processed by mechanical alloying assisted microwave heating route. *Int. J. Refract. Met. Hard Mater.* **2016**, *58*, 206–210. [CrossRef]
25. Lee, C.C.; Kahar, S.M.; Voon, C.H. Microwave synthesis of silicon carbide nanowhiskers: Effect of molar ratio. *Mater. Today Proc.* **2020**, *4*, 571.
26. Moshtaghioun, B.M.; Poyato, R.; Cumbrera, F.L.; de Bernardi-Martin, S.; Monshi, A.; Abbasi, M.H.; Karimzadeh, F.; Dominguez-Rodriguez, A. Rapid carbothermic synthesis of silicon carbide nano powders by using microwave heating. *J. Eur. Ceram. Soc.* **2012**, *32*, 1787–1794. [CrossRef]
27. Ding, J.; Zhu, H.; Deng, C.; Li, G.; Wang, K.; Liu, J. Preparation and characterisation of porous biomorphic SiC/C ceramic from molten salt. *Ceram. Int.* **2015**, *41*, 11539–11545. [CrossRef]
28. Kan, X.; Ding, J.; Yu, C.; Zhu, H.; Deng, C.; Li, G. Low-temperature fabrication of porous ZrC/C composite material from molten salts. *Ceram. Int.* **2017**, *43*, 6377–6384. [CrossRef]
29. Nadimi, H.; Soltanieh, M.; Sarpoolaky, H. The formation mechanism of nanocrystalline TiC from KCl–LiCl molten salt medium. *Ceram. Int.* **2020**, *46*, 18725–18733. [CrossRef]
30. Yang, L.; Wang, Y.; Liu, R.; Liu, H.; Zhang, X.; Zeng, C.; Fu, C. In-situ synthesis of nanocrystalline TiC powders, nanorods, and nanosheets in molten salt by disproportionation reaction of Ti(II) species. *J. Mater. Sci. Technol.* **2020**, *37*, 173–180. [CrossRef]
31. Tian, X.; Zhang, X.; Hu, Y.; Liu, B.; Yuan, Y.; Yang, L.; Chen, Q.; Liu, Z. Fabrication of iron carbide by plasma-enhanced atomic layer deposition. *J. Mater. Res.* **2019**, *35*, 813–821. [CrossRef]
32. Scandurra, R.; Scotto d'Abusco, A.; Longo, G. A Review of the Effect of a Nanostructured Thin Film Formed by Titanium Carbide and Titanium Oxides Clustered around Carbon in Graphitic Form on Osseointegration. *Nanomaterials* **2020**, *10*, 1233. [CrossRef] [PubMed]
33. Chen, K.; Bao, Z.; Du, A.; Zhu, X.; Wu, G.; Shen, J.; Zhou, B. Synthesis of resorcinol–formaldehyde/silica composite aerogels and their low-temperature conversion to mesoporous silicon carbide. *Microporous Mesoporous Mater.* **2012**, *149*, 16–24. [CrossRef]
34. Chen, K.; Huang, X.; Zhang, Z.; Du, A.; Zhou, B.; Xu, Y.; Zhou, Z.; Wang, Y. Low temperature pseudomorphic synthesis of nanocrystalline carbide aerogels for electrocatalysis. *J. Mater. Chem. A* **2015**, *3*, 11745–11749. [CrossRef]
35. Kong, Y.; Shen, X.; Cui, S.; Fan, M. Preparation of monolith SiC aerogel with high surface area and large pore volume and the structural evolution during the preparation. *Ceram. Int.* **2014**, *40*, 8265–8271. [CrossRef]

36. Schuchardt, A.; Braniste, T.; Mishra, Y.K.; Deng, M.; Mecklenburg, M.; Stevens-Kalceff, M.A.; Raevschi, S.; Schulte, K.; Kienle, L.; Adelung, R.; et al. Three-dimensional Aerographite-GaN hybrid networks: Single step fabrication of porous and mechanically flexible materials for multifunctional applications. *Sci. Rep.* **2015**, *5*, 8839. [CrossRef]
37. Yuksel, R.; Buyukcakir, O.; Panda, P.K.; Lee, S.H.; Jiang, Y.; Singh, D.; Hansen, S.; Adelung, R.; Mishra, Y.K.; Ahuja, R.; et al. Necklace-like Nitrogen-Doped Tubular Carbon 3D Frameworks for Electrochemical Energy Storage. *Adv. Funct. Mater.* **2020**, *30*, 1909725. [CrossRef]
38. Chen, Y.; Wang, Y.; Su, T.; Chen, J.; Zhang, C.; Lai, X.; Jiang, D.; Wu, Z.; Sun, C.; Li, B.; et al. Self-Healing Polymer Composites Based on Hydrogen Bond Reinforced with Graphene Oxide. *ES Mater. Manuf.* **2019**, *4*, 31–37. [CrossRef]
39. Liu, C.; Fang, Q.; Wang, D.; Yan, C.; Liu, F.; Wang, N.; Guo, Z.; Jiang, Q. Carbon and Boron Nitride Nanotubes: Structure, Property and Fabrication. *ES Mater. Manuf.* **2019**, *3*, 2–15. [CrossRef]
40. Prehal, C.; Fitzek, H.; Kothleitner, G.; Presser, V.; Gollas, B.; Freunberger, S.A.; Abbas, Q. Persistent and reversible solid iodine electrodeposition in nanoporous carbons. *Nat. Commun.* **2020**, *11*, 4838. [CrossRef]
41. Ran, F.; Yang, X.; Shao, L. Recent progress in carbon-based nanoarchitectures for advanced supercapacitors. *Adv. Compos. Hybrid Mater.* **2018**, *1*, 32–55. [CrossRef]
42. Lewin, E.; Persson, P.O.Å.; Lattemann, M.; Stüber, M.; Gorgoi, M.; Sandell, A.; Ziebert, C.; Schäfers, F.; Braun, W.; Halbritter, J.; et al. On the origin of a third spectral component of C1s XPS-spectra for nc-TiC/a-C nanocomposite thin films. *Surf. Coat. Technol.* **2008**, *202*, 3563–3570. [CrossRef]
43. Samuelsson, M.; Sarakinos, K.; Högberg, H.; Lewin, E.; Jansson, U.; Wälivaara, B.; Ljungcrantz, H.; Helmersson, U. Growth of Ti-C nanocomposite films by reactive high power impulse magnetron sputtering under industrial conditions. *Surf. Coat. Technol.* **2012**, *206*, 2396–2402. [CrossRef]
44. Biesinger, M.C.; Lau, L.W.M.; Gerson, A.R.; Smart, R.S.C. Resolving surface chemical states in XPS analysis of first row transition metals, oxides and hydroxides: Sc, Ti, V, Cu and Zn. *Appl. Surf. Sci.* **2010**, *257*, 887–898. [CrossRef]
45. Godfroid, T.; Gouttebaron, R.; Dauchot, J.P.; Leclère, P.; Lazzaroni, R.; Hecq, M. Growth of ultrathin Ti films deposited on SnO_2 by magnetron sputtering. *Thin Solid Films* **2003**, *437*, 57–62. [CrossRef]

Publisher's Note: MDPI stays neutral with regard to jurisdictional claims in published maps and institutional affiliations.

nanomaterials

Article

Effect of Water and Glycerol in Deoxygenation of Coconut Oil over Bimetallic NiCo/SAPO-11 Nanocatalyst under N_2 Atmosphere

Rungnapa Kaewmeesri [1,2], Jeeranan Nonkumwong [2], Thongthai Witoon [3,4], Navadol Laosiripojana [1] and Kajornsak Faungnawakij [2,4,*]

1 The Joint Graduate School of Energy and Environment (JGSEE), King Mongkut's University of Technology Thonburi, Bangkok 10140, Thailand; rungnapa@nanotec.or.th (R.K.); navadol@jgsee.kmutt.ac.th (N.L.)
2 National Nanotechnology Center (NANOTEC), National Science and Technology Development Agency (NSTDA), Pathum Thani 12120, Thailand; jeeranan.non@nanotec.or.th
3 Center of Excellence on Petrochemical and Materials Technology, Department of Chemical Engineering, Kasetsart University, Bangkok 10900, Thailand; fengttwi@ku.ac.th
4 Research Network of NANOTEC-KU on NanoCatalysts and NanoMaterials for Sustainable Energy and Environment, Kasetsart University, Bangkok 10900, Thailand
* Correspondence: kajornsak@nanotec.or.th; Tel.: +66-2-564-7100 (ext. 6638); Fax: +66-2-564-6981

Received: 18 November 2020; Accepted: 14 December 2020; Published: 18 December 2020

Abstract: The catalytic deoxygenation of coconut oil was performed in a continuous-flow reactor over bimetallic NiCo/silicoaluminophosphate-11 (SAPO-11) nanocatalysts for hydrocarbon fuel production. The conversion and product distribution were investigated over NiCo/SAPO-11 with different applied co-reactants, i.e., water (H_2O) or glycerol solution, performed under nitrogen (N_2) atmosphere. The hydrogen-containing co-reactants were proposed here as in-situ hydrogen sources for the deoxygenation, while the reaction tests under hydrogen (H_2) atmosphere were also applied as a reference set of experiments. The results showed that applying co-reactants to the reaction enhanced the oil conversion as the following order: N_2 (no co-reactant) < N_2 (H_2O) < N_2 (aqueous glycerol) < H_2 (reference). The main products formed under the existence of H_2O or glycerol solution were free fatty acids (FFAs) and their corresponding C_{n-1} alkanes. The addition of H_2O aids the triglyceride breakdown into FFAs, whereas the glycerol acts as hydrogen donor which is favourable to initiate hydrogenolysis of triglycerides, causing higher amount of FFAs than the former case. Consequently, those FFAs can be deoxygenated via decarbonylation/decarboxylation to their corresponding C_{n-1} alkanes, showing the promising capability of the NiCo/SAPO-11 to produce hydrocarbon fuels even in the absence of external H_2 source.

Keywords: deoxygenation; coconut oil; in-situ hydrogen; nickel–cobalt bimetallic nanoparticles; biorefinery

1. Introduction

Hydrotreating is a typical process in the production of petroleum fuels from fossil crude oils using active sulfided catalysts, e.g., supported NiMo, CoMo, and NiW, operated under hydrogen atmosphere [1]. However, in order to secure economic feasibility as well as the environmental sustainability, many studies have been carried out to improve the process by finding new kinds of alternative resources, applying an inert ambient to the process, and/or using non-sulfided catalysts [2–4]. Nowadays, renewable oils, animal fats, and biodiesel from low-grade oils/fats are considered as alternative sources for hydrocarbon fuel production [5–7]. Lately, an upgrading of hydrocarbon fuels, which mostly consists of linear alkanes, to higher fraction of branched alkanes for the use as jet fuel-like hydrocarbons has received considerable attention because the demand of fuel for the aviation sectors has been increasing.

Principally, jet fuel-like hydrocarbon is produced from renewable oils through well-known processes where the oxygen in the oil's molecules is removed by hydrodeoxygenation followed by hydro-isomerization/cracking, and distillation as the last step. To this procedure, a huge amount of hydrogen is required. For example, one molecule of triglyceride (TG) in vegetable oil reacts with twelve molecules of hydrogen gas (H_2) for hydrodeoxygenation (HDO), six molecules of H_2 for decarbonylation (DCO), or three molecules of H_2 for decarboxylation (DCO_2) route to form three molecules of alkanes [8,9]. The H_2 consumption for the catalytic deoxygenation via those three major pathways is in the order of HDO > DCO > DCO_2 [10]. Therefore, the DCO/DCO_2 pathway has been more widely studied due mainly to its less hydrogen consumption. Specifically, three molecules of H_2 are initially used for breaking a TG molecule into propane and free fatty acids (FFAs). Then, FFAs will further convert to hydrocarbons via DCO_2 reaction without the need of hydrogen source. To promote the deoxygenation rate, well-designed non-sulfided catalysts are essential.

Hollak et al. [11] studied the catalytic activity of Pd/C catalyst for HDO of triglycerides (TGs) and FFAs. The results showed that even though the hydrogen produced in-situ from glycerol reforming can help for hydrocarbon product formation, the main products were still long-chain unsaturated hydrocarbons. This could be because the hydrogen production from glycerol using Pd/C catalyst was rather low. Similarly, Chiapper et al. [12] found that unsaturated hydrocarbon products were also the most selective in the deoxygenation of coconut oil and palm kernel oil over $PtSnK/SiO_2$, even though mild hydrogenation treatment was performed to saturate mono- and polyunsaturated bonds in TGs prior to deoxygenation. A comparison of three catalysts, i.e., 1 wt% Pt/C, 5 wt% Pd/C, and 20 wt% Ni/C, was studied by Morgan and co-workers [13] for the deoxygenation of TGs (tristearin, triolein, or soybean oil) under N_2 atmosphere. The supported non-noble Ni metal was found to provide significantly higher TGs conversion and fraction of C_8–C_{17} compounds than the noble Pd and Pt metals. Miao et al. [14] also attempted to convert palmitic acid through non-noble Ni metal supported on ZrO_2 with low-pressure or without external supply of H_2 in the presence of water. The conversion of palmitic acid was significantly improved, and the results showed that hydrogenolysis promoted palmitic conversion and DCO occurred as the major reaction. Although non-noble Ni metal can be efficient to drive the reaction under H_2 atmosphere, about 20 wt% of Ni loading was needed in their research. Apart from the widely used Ni-based catalysts, Co-based catalysts also have been reported providing high yield of hydrocarbons from deoxygenation of TGs (triolein, ceiba oil, and sterculia oil) under inert N_2 flow condition via DCO_2 pathway [15]. Moreover, it is well-documented that lower acidity of Co than Ni may assist lowering of catalyst coking and subsequent deactivation. Thus, Co shows potential to be addressed as an alternative or a component in Ni-based bimetallic catalysts.

The H_2 produced from glycerol aqueous-phase reforming (APR) is of interesting to utilize for the in-situ deoxygenation of TG molecules. In 2002, the pioneer work [16] applied glycerol into the system to generate in-situ H_2 for the reaction. However, the hydrogen production is always accompanied by several side reactions, such as methanol decomposition, water gas shift reaction, and methanation [17–20]. Therefore, the hydrogen yield is highly related to the performance of the catalysts which actively involves for the C–C, C–H, and C–O bonds scission, especially the cleavage of the C–H bond. The development of bimetallic catalysts such as Ni-Cu [17,18] or Ni-Co [19] is an efficient way to control the selectivity as the second metal can affect metal particle sizes or strengthen metal-support interaction to improve water gas shift reaction or inhibit methanation reaction.

Previously, Zhong and co-worker [21] studied the in-situ hydrogen production from high temperature water with Ni catalysts. In the system, high temperature water, with weak hydrogen bonds, was oxidized by active reductant metals as following; $M + H_2O \rightarrow H_2 + M_xO_y$. Those results suggested that co-feeding with glycerol and water is clearly proved to be a facile hydrogen source for the deoxygenation of TGs to hydrocarbon fuels [21,22].

In this work, we studied the deoxygenation of TG to hydrocarbon alkanes via catalytic deoxygenation (under N_2 ambient) and in-situ HDO (by H_2O or glycerol addition) in comparison with conventional HDO (under H_2 ambient) to yield oxygen-free hydrocarbons. The NiCo (about 10 wt% metal loading) supported on commercially available silicoaluminophosphate-11 (SAPO-11) was used as the deoxygenation catalyst. SAPO-11 is a molecular sieve material with 10-ring porous structure that has been extensively utilized due to its stability and potential activity in many catalytic processes [23,24]. The catalytic hydro/deoxygenation reactions of medium-chain (mostly C_{12}) TGs in coconut oil, without hydrogenation treatment prior to deoxygenation, were examined to describe the reaction pathway via the degree of oxygen removal and product distribution, depending on the variable feed molecules. To the best of our knowledge, it is reported for the first time that SAPO-11-supported NiCo nanoparticles could directly convert coconut oil to jet fuel-like hydrocarbons with the in-situ hydrogen donors.

2. Materials and Methods

2.1. Materials

Commercial coconut oils were obtained from local companies in Thailand. Analytical grade glycerol ($C_3H_8O_3$, 99% purity) was obtained from Fisher Scientific, Fair Lawn, NJ, USA. Hydrogen gas (H_2, 99.99% purity) and nitrogen gas (N_2, 99.5% purity) were obtained from S.I. Technology Co., Ltd., Bangkok, Thailand. SAPO-11 powder, used as the catalyst support, was purchased from ACS Material, Pasadena, CA, USA. Deionized (D.I.) water (18.2 MΩ·cm, at 25 °C) was produced by water purification machine in laboratory. Hydrocarbon standards (C_8–C_{18}) for calibration were purchased from Sigma-Aldrich, Buchs, Switzerland. Analytical reagent grade of nitrate salts of nickel ($Ni(NO_3)_2{\cdot}6H_2O$) and cobalt ($Co(NO_3)_2{\cdot}6H_2O$) were purchased from Ajax FineChem, Taren Point, NSW, Australia.

2.2. Catalyst Preparation

The catalyst was prepared by co-impregnation technique by dissolving nitrate salts of Ni and Co in D.I. water. Then, SAPO-11 powder was further added to obtain 5 wt% loading of each metal. Mixing as well as drying were employed using a magnetic stirrer hot plate at 85 °C for 3 h and transferred to vacuum oven at the same temperature for overnight. The resultant powder was calcined in static air at 500 °C for 5 h. The final sample was denoted as NiCo/SAPO-11. Before catalytic testing, the catalyst powder was pelleted, sieved into 0.5–1.0 mm. in size.

2.3. Catalyst Characterization

Crystalline phase of the prepared catalyst was analysed by X-ray diffraction (XRD) measurement using an X-ray diffractometer (D8 ADVANCE, Bruker, Ltd., Karlsruhe, Germany) with Cu Kα radiation (wavelength of 1.5406 Å). The diffraction angle was scanned from 2θ of 5° to 80°, with steps of 0.5°/min and counting at 0.02 s/step. The specific surface area and pore structures of the samples were calculated by the Brunauer–Emmett–Teller (BET) and the density functional theory (DFT)/Barrett–Joyner–Halenda (BJH) methods, respectively, via nitrogen adsorption–desorption technique (Quantachrome Autosorb-iQ3, Boynton Beach, FL, USA). The total pore volume was measured at the relative pressure (P/P0) of 0.98. Before measurement, the sample of approximately 0.2 g was outgassed at 120 °C for 3 h under helium flow. Total acidity and strength of acid sites of catalyst were evaluated by ammonia (NH_3)-temperature programmed desorption (NH_3-TPD) carried out using Quantachrome Chemisorption Analyzer ChemStar TPX Series (Boynton Beach, FL, USA). Scanning electron microscopy (SEM) analysis was performed with a S-3400N (Hitachi-Science & Technology, Berkshire, UK), operated at 5 kV. Transmission electron microscopy (TEM) images were acquired by JEOL-JEM-2100Plus instrument (Akishima, Tokyo, Japan) equipped with an energy dispersive X-ray spectrometer (EDS). The TEM machine was operated at an accelerating voltage of

200 kV. Before measurement, the sample was reduced at 700 °C in a flow of H_2 for 2 h. The obtained powder was grounded then dispersed in ethanol, followed by sonication for 30 min to make the particle well-dispersed, and dropped onto a gold grid.

2.4. Catalytic Deoxygenation Testing in Continuous-Flow Fixed-Bed Reactor

The deoxygenation reaction was operated in a custom-made continuous-flow trickle-bed reactor made of the stainless steel 316, with an external diameter of 1.28 cm, internal diameter of 1 cm, and length of 70 cm. The catalyst bed was located at the middle of the fixed-bed reactor and held in position with quartz wool plugs at uniform temperature position (confirmed by inside thermocouples inserted in the catalyst bed). The reaction in the continuous system consists of a feed tank, a reactor unit, and a gas–liquid product separation unit, as shown in Figure 1. The temperature was controlled by using electrical furnace with two positions of temperature controller. A K-type thermocouple was inserted to contact the catalyst bed for temperature monitoring. The pressure of the reactor was manually controlled by using back pressure regulator (BPG) for two-phase flow. High-performance liquid chromatography (HPLC) pump and mass flow controller (MFC) were used to control the flow rate of liquid and gas feed, respectively.

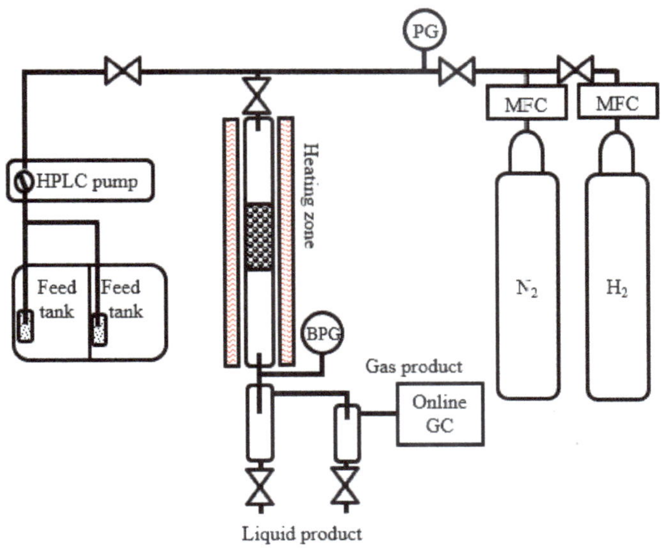

Figure 1. Schematic diagram of continuous-flow reactor system for catalytic testing.

In a typical run, 4 mL of catalyst was loaded into a reactor and reduced with 50 cm^3/min of pure H_2 at 650 °C for 2 h. After the catalyst reduction step, the continuous reaction system was immediately flushed with reaction gas (N_2 or H_2) and set up to the desired temperature under pressurized condition. Liquid feeds—coconut oil, water, and glycerol solution—were separately introduced by two-suction line in the HPLC pump.

Four different experimental practices were performed as follows:

Practice #1 was designed as a reference experiment for the purpose of investigating the catalytic performance on HDO of feed oil over NiCo/SAPO-11 catalyst. The reaction was performed at 330 °C under 50 bar of H_2 atmosphere, and liquid feed (i.e., coconut oil) was introduced into the reactor with liquid hourly space velocity (LHSV) of 1 h^{-1}.

Practice #2 aimed to investigate the catalytic performance on the deoxygenation with absent H_2. The procedure was similar to that described in practice #1, except the operating gas which was N_2.

Practice #3 aimed to investigate the catalytic performance on the coupling deoxygenation of feed oil with in-situ H_2 production by glycerol-APR. In short, the feed was introduced into the reactor by two-suction lines in HPLC pump, one for oil feed (80% by vol.) and one for glycerol solution feed (20% by vol.). The glycerol solution of 36 wt% glycerol in water (140 g of glycerol in 250 g of water) which was 0.1 mole ratio of glycerol: water) was used at a total LHSV of 1 h^{-1}.

Practice #4 aimed to investigate the catalytic performance on the solvent-free HDO by adding 5% by volume of water into the reaction.

2.5. Liquid and Gaseous Product Analysis

Two-phase back pressure regulator was used to reduce a mixed-phase flow streams from high pressure into ambient before they went to a liquid–gas phase separator placed downstream. The liquid products were analysed by gas chromatography–mass spectrometry (GC–MS, Shimadzu GCMS-QP2020, Kyoto, Japan) equipped with a capillary column (DB-1HT, 30 m × 0.32 mm × 0.1 μm) to inform the quantitative/qualitative data of FFAs, hydrocarbons, and other intermediate products. Detailed information of this method is applied elsewhere [25,26]. Briefly, 10 μL of sample was injected into the injection port in a split mode with the inlet temperature of 340 °C. In a GC oven, the oven temperature was kept at 40 °C for 5 min then increased to 240 °C with a rate of 15 °C/min, then ramped by 8 °C/min to 370 °C, which was maintained for another 15 min. The total analysis run time was about 50 min. In the MS zone, the temperatures of ion source and interface were held at 250 °C during the analysis. The m/z was scanned from 2 to 500. The calibration curves of alkane standards (n-C_8 to n-C_{18}) were used to quantify their weight in the liquid products, and the response factor for isomer products was assumed to be equal to its corresponding n-alkanes. For other oxygenated intermediates, the calculation was referenced by the same type of known standards.

Gas phase was analysed using an online GC equipped with two columns (molecular sieve 5A and Porapak Q), thermal conductivity detector (TCD) and flame ionization detector (FID). The GC was calibrated for all of the gaseous products obtained, including CO, CO_2, CH_4, C_2H_6, and C_3H_8.

Feed conversion and product selectivity were calculated based on mass balance. The catalytic performance was evaluated according to oil conversion (Equation (1)), liquid hydrocarbon yield (Equation (2)), selectivity to liquid hydrocarbon product (Equation (3)), and selectivity to gas species (Equation (4)):

$$\text{Oil conversion (\%)} = ((\text{mass of oil fed} - \text{mass of oil remaining})/(\text{mass of oil fed})) \times 100, \quad (1)$$

$$\text{Liquid hydrocarbon yield (\%)} = (\text{mass of liquid hydrocarbon produced/mass of oil fed}) \times 100, \quad (2)$$

$$\text{Selectivity to liquid hydrocarbon product (\%)} = (\text{mass of each liquid hydrocarbon product/mass of liquid hydrocarbon product in total}) \times 100, \quad (3)$$

$$\text{Selectivity to gas species (\%)} = (\text{mass of each gas species/mass of gas in total}) \times 100, \quad (4)$$

Contribution percentages of the HDO (Equation (5)) and DCO/DCO_2 (Equation (6)) pathways were calculated based on the total moles of n-alkanes with even numbers or odd numbers of carbon atoms in the liquid product as follows:

$$\text{HDO (\%)} = (\text{mass of } C_8, C_{10}, C_{12}, C_{14}, C_{16}, \text{ and } C_{18} \text{ in product/mass of feed}) \times 100, \quad (5)$$

$$\text{DCO+DCO}_2 \text{ (\%)} = (\text{mass of } C_9, C_{11}, C_{13}, C_{15}, \text{ and } C_{17} \text{ in product/mass of feed}) \times 100, \quad (6)$$

3. Results and Discussion

3.1. Catalyst Characterization

The XRD pattern of NiCo/SAPO-11 (Figure 2a) suggested an alloy formation between Ni and Co, matched with PDF 01-074-5694, in accordance with Huynh and co-workers [27] who prepared bimetallic Ni-Co supported on different acidic materials used in Phenol HDO. Figure 2b,c show porous characteristics of the NiCo/SAPO-11 and SAPO-11 support. The obtained isotherms for all samples could be classified as type IV with hysteresis loops at $0.4 < P/P_0 < 0.8$ of H4 type [28]. These results showed that samples possess mesopore structures [29]. High adsorption of N_2 appeared in the low relative pressure range, and the hysteresis loops were detected, suggesting the existence of both micropores and mesopores. The pore size distribution plots calculated by DFT method revealed two types of peaks with narrow micropores (high intensity at pore diameter < 2 nm) and mesopores (broad peak at pore diameter ≥ 2 nm) at approximately 0.98 and 7.78 nm, respectively. Upon NiCo loading, the BET surface area and pore volume of NiCo/SAPO-11 were ~134.4 m^2/g and ~0.140 cm^3/g, respectively, which were slightly lower than those of pristine SAPO-11 (~134.7 m^2/g and ~0.139 cm^3/g, respectively).

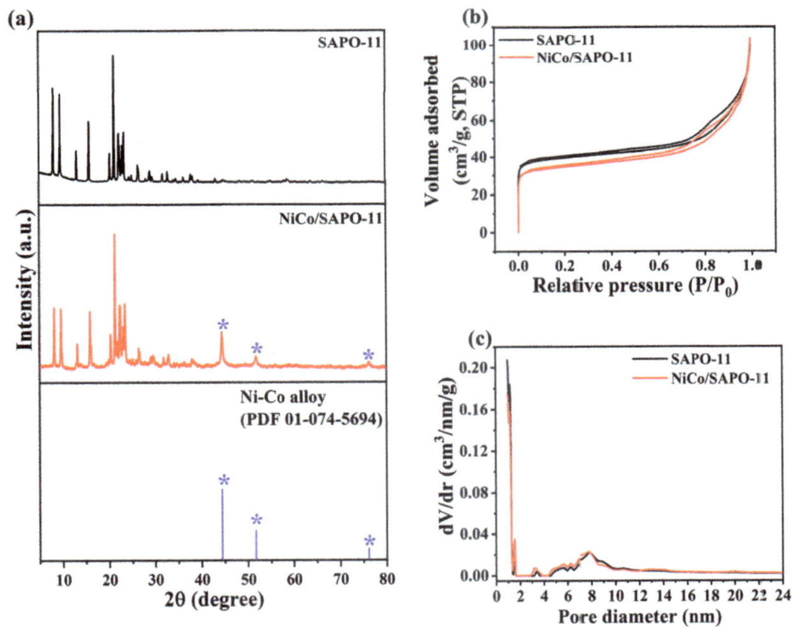

Figure 2. (**a**) XRD patterns, * 2θ = 44.41°, 51.78°, and 76.09° (powder diffraction file (PDF) for Ni-Co alloy (01-074-5694)); (**b**) N_2 adsorption-desorption isotherms; and (**c**) pore-size distributions of SAPO-11 and the reduced NiCo/SAPO-11 catalyst.

The SEM image (Figure 3a) indicated that 10 wt% metal loading does not change the microstructure of SAPO-11 support, which is normally rod-like and has non-smooth surface. The bright field TEM image (Figure 3b) of the catalyst clearly shows Ni-Co nanoparticles, presented in black spots, in a spherical shape dispersed on the support, presented in the dark grey area. The metal particle size, estimated by measuring at least 100 particles, was about 9–18 nm. This range is consistent with San-Jose-Alonse and co-workers [30], who varied the proportion of Ni:Co. They showed that the particle size of samples prepared by impregnation was ~14.1 nm. (for Co), ~8.6 nm. (for NiCo), and ~7.0 nm. (for Ni). It is evidenced by EDS elemental mappings (Figure 3c,d) of the corresponding

TEM image (Figure 3b) that Ni and Co species appeared at identical areas of the bimetallic particles, possibly due to the alloy structure formation, in good agreement with previous XRD results (Figure 2). Figure 3e,f display the high-resolution TEM (HRTEM) image of a single NiCo nanoparticle and its corresponding EDS line profile, respectively. These results confirm the co-existence of Ni and Co with comparable content of each metal, which was consistent with the analysis result from an inductively coupled plasma–mass spectrometry (ICP-MS) (results not shown). Additionally, lattice spacings of around 1.28 and 2.42 nm which were rotated by 53° were evidenced [31]. Evaluated by the NH_3-TPD technique, total acidity of the SAPO-11 was 0.456 mmol/g, whereas NiCo/SAPO-11 catalyst showed lower values at 0.338 mmol/g (results not shown), coincided with the lower surface area of the supported catalyst.

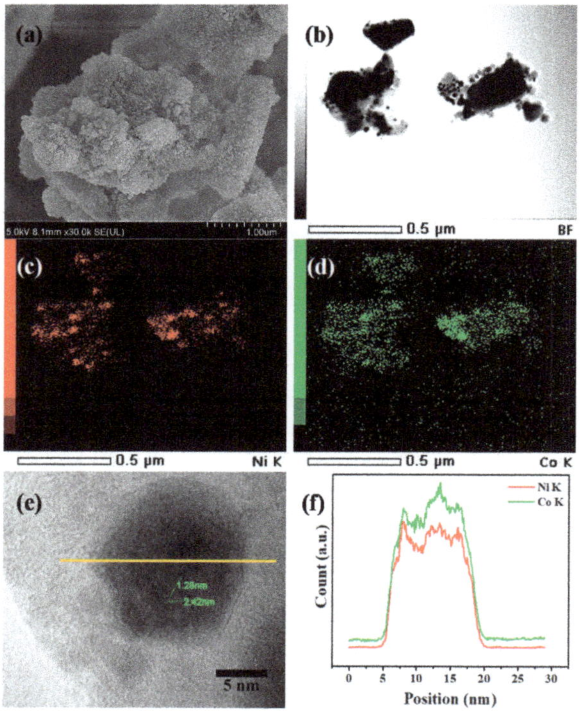

Figure 3. (**a**) Scanning electron microscopy (SEM) image; (**b**) transmission electron microscopy (TEM) image; (**c**,**d**) corresponding energy dispersive X-ray spectrometer (EDS) elemental mappings of Ni and Co; (**e**,**f**) EDS line profile of the NiCo/SAPO-11 catalyst.

3.2. Feed Compositions

Table 1 shows fatty acid compositions directly evaluated by GC–MS. It is seen that C_{12}, C_{16}, and C_{18} fatty acids are major components in coconut oil, with oxygen content ~13 wt%, the ratio of saturated/unsaturated FFAs ~3 times, which might induce different product distributions. The molecular weights for further calculation were 779.7 g/mol for coconut oil.

Table 1. Fatty acid composition of coconut oil.

Fatty Acid	Content (wt%)
Octanoic acid (C8:0)	0.9
Decanoic acid (C10:0)	3.3
Dodecanoic acid (C12:0)	30.1
Myristic acid (C14:0)	21.5
Palmitic acid (C16:0)	17.9
Stearic acid (C18:0)	5.2
Oleic acid (C18:1)	15.9
Linoleic acid (C18:2)	4.8
Eicosanoic acid (C20:0)	0.2
Eicosenoic acid (C20:1)	0.2
Saturated FFAs	79.1
Unsaturated FFAs	20.9
Oxygen	12.9

3.3. Hydrogen Production from Glycerol Aqueous-Phase Reforming (Gly-APR) over NiCo/SAPO-11

As of the complexity during the glycerol-APR process, the reactions of dehydration, dehydrogenation and hydrogenolysis could possibly take place on NiCo/SAPO-11 since the fact that this bifunctional catalyst comprises two main catalytic sites; (1) metal sites prefer to dehydrogenation and (2) acid sites prefer dehydration. Table 2 summarizes the main products such as propylene glycol ($C_3H_8O_2$), ethylene glycol ($C_2H_6O_2$), acetaldehyde (C_2H_4O), 1-propanol, 2-methyl- ($C_4H_{10}O$), 2-propanone, 1-hydroxy- ($C_3H_6O_2$), and ethanol (C_2H_5OH) obtained in the condensable phase, which can be confirmed by the fact that side reactions occurred. The most abundant liquid product was ethylene glycol (~17.8%), followed by propylene glycol (~11.0%), and 2-propanone, 1-hydroxy- (~10.5%). In addition, trace amounts of 1-propanol, 2-methyl-, acetaldehyde, and ethanol could be detected (~7.1%). These results show a good agreement with the literature in that the same intermediates were reported [32,33]. The comparatively higher selectivity to ethylene glycol indicated that the primary hydroxyl group of glycerol is eliminated by the dehydrogenation pathway. However, the formation of propylene glycol and 2-propanone, 1-hydroxy- indicated that both pathways of the dehydration/hydrogenolysis were presented on Lewis sites of the catalyst. It also points out the high C–C, C–H cleavage activity of catalyst, accompanied by an increased production of hydrogen. As NiCo/SAPO-11 catalyst was identified to be of weak and medium strength acid sites (i.e., Lewis sites), this would lead to the formation of oxygenated hydrocarbons in the liquid phase (i.e., acetaldehyde, 1-propanol, 2-methyl-, 2-propanone, 1-hydroxy-, and ethanol), suggesting that Lewis sites are involved in C–O cleavage. The result also signifies that dehydrogenation, which competes with dehydration/hydrogenolysis, was significantly favoured over bifunctional NiCo/SAPO-11 catalyst.

Table 2. The list of investigated intermediates and their selectivity in liquid product during the glycerol aqueous-phase reforming over NiCo/SAPO-11 catalyst according to the GC–MS analysis.

Liquid Product		Retention Time (min)	Selectivity (%)
Acetaldehyde	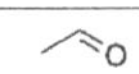	1.050	1.13
Ethanol	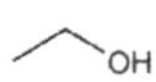	1.259	1.97
1-propanol, 2-methyl-	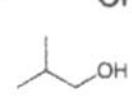	4.843	30.47
2-propanone, 1-hydroxy-	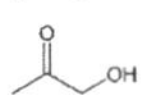	10.585	56.75
Propylene glycol	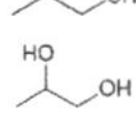	11.035	7.89
Ethylene glycol	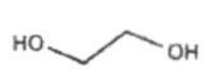	17.858	1.80

Reaction conditions: glycerol solution 36 wt%, and 4 mL catalyst, 230 °C, 30 bar N_2.

Overall, the reaction pathways could be proposed that the reaction involves the C–O bond scission, followed by hydrogenation which consumes H_2. Therefore, a part of H_2 could be consumed and resulted in limited hydrogen production. The formation of ethylene glycol generated from dehydrogenation would support this assumption.

Table 3 shows the main components of the gaseous products: $CH_4 > CO > CO_2 > H_2$, respectively. As mentioned above, H_2 yield was hardly maintained because H_2 was consumed by side reaction (in liquid phase). Notably, the formation of CH_4 also confirmed that H_2 was used in CO/CO_2 methanation ($CO + 3H_2 \leftrightarrow CH_4 + H_2O$, $\Delta H_{298} = -206$ kJ/mol and $CO_2 + 4H_2 \leftrightarrow CH_4 + 2H_2O$, $\Delta H_{298} = -164.94$ kJ/mol) [33]. Generally, Ni species prefer to generate more CO. However, the amount of CO formed was higher than that of CO_2, suggesting that bimetallic NiCo/SAPO-11 catalyst was favourable to the decomposition of methanol.

Table 3. The yield of gas products during the glycerol aqueous-phase reforming over NiCo/SAPO-11 catalyst according to the GC–MS analysis.

Gas Product	Retention Time (min)	Yield ($\times 10^{-9}$ mole)	Yield ($\times 10^{-5}$ mol/mol Glycerol)
H_2	1.78	0.0000084	0.00000024
CO	3.919	5.58528	1.5926
CH_4	4.393	6.85314	1.95412
CO_2	8.862	2.8679	0.81775

Reaction conditions: glycerol solution 36 wt%, and 4 mL catalyst, 230 °C, 30 bar N_2.

Co-based catalysts have been reported as less effective materials for the reforming of methanol [34,35]. However, Xue and co-worker [36] promoted Co into NiCu/Al catalyst for methanol reforming. The increased H_2 amount showed that Co could promote the reaction by enhancement of water gas shift reaction. Papadopoulou and co-worker [37] investigated the adsorption of methanol on Co-based catalysts. TPD results indicated that the number of active sites provided the catalyst to be able to adsorb or decompose methanol molecules. In Table 3, methanol was not detected among the condensable phase products suggesting the possible reactions of methanol reforming ($CH_3OH + H_2O \leftrightarrow CO_2 + 3H_2$) or methanol decomposition ($CH_3OH \leftrightarrow CO + 2H_2$) over NiCo/SAPO-11 catalyst. This behaviour coincided with the synergistic effects of Ni-CO alloy in a NiCo/SAPO-11 catalyst in a positive way for H_2 production.

3.4. *Catalytic Performance of NiCo/SAPO-11 over In-Situ HDO of Coconut Oil*

Figure 4 displays the results of coconut oil HDO according to the change of co-reactant feed at 330 °C, LHSV 1 h^{-1}, and 50 bar of reaction gas over 4 mL of NiCo/SAPO-11 catalyst. Under H_2 ambient, the catalyst exhibited high activity with 100% conversion, while more than 85% of coconut oil conversion could yield the liquid product. Under an inert N_2 atmosphere, the conversion was 30% and liquid product accounted for 15%. With the help of co-reactants, H_2O and aqueous glycerol, the conversion increased to 70% and 90%, whereas liquid product yield also increased to 47% and 87%, respectively. Compared to the conventional H_2 ambient, the results verified that the presence of H_2O and glycerol helped to promote both coconut oil conversion and product yield. In addition, the experimental practices applying H_2 below 50 bar (i.e., 30 bar) with and without glycerol solution were also performed. Even though 100% conversion was also achieved, hydrocarbon liquid yields were dropped to 65–67%, compared with using either 50 bar H_2 without glycerol (Practice #1) or 50 bar N_2 with glycerol solution (Practice #3). This could be because the existence of H_2O under H_2 ambient caused side reaction. Thus, the glycerol did not clearly selective to glycerol reforming.

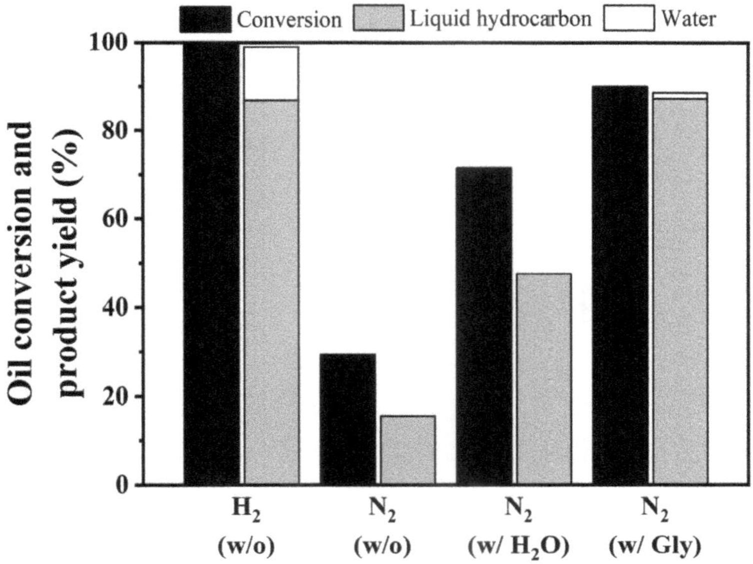

Figure 4. Oil conversion and product yield of coconut oil deoxygenation under different atmospheres with (w/) and without (w/o) co-reactant feeds. (Condition: Coconut oil, LHSV of 1 h^{-1}, 330 °C, and 50 bar of reaction gas, Gly = glycerol solution).

Figure 5 shows the liquid product distribution over NiCo/SAPC-11 catalyst. With the help of H_2, the catalyst could provide excellent selectivity to alkanes at 97%. Several intermediates (i.e., ketones, aldehydes, alcohols, cycloalkanes esters, FFAs, and alkenes) were detected in the liquid phase, while the C_3H_8, H_2, CO, and CO_2 were mainly formed in the gas phase. However, the product distribution was altered (see Figure 6). Under N_2 atmosphere, main products are alkanes (C_{11}–C_{18}) and a small amount of FFAs, alkenes, and some alcohols. This finding is consistent with Morgan and co-workers [13], who established the thermal deoxygenation of TGs without H_2 over 1 wt% Pt/C, 5 wt% Pd/C, and 20 wt% Ni/C catalysts. However, the existence of fatty acid composition was an important intermediate for alkane formation. Long chains of alkenes and alcohols were produced via hydrocracking. It can be implied that H_2 generated from cracking affected the hydrocarbon production [38]. In the presence of H_2O, CO and CO_2 were mainly formed, while propane was not found in the bulk of gas product (see Figure 6). It can be considered that FFAs are successfully produced by the TG hydrolysis reaction route rather than deoxygenation which requires H_2 molecules. However, the produced FFAs will either undergo DCO_2 or DCO without H_2 gas.

As also shown in Figure 6, the existence of methane (CH_4) showed the consumption of H_2. As a result, in-situ H_2 produced via the water gas shift reaction ($CO + H_2O \leftrightarrow CO_2 + H_2$, $\Delta H^0_{298} = -41$ kJ/mol) [37] circulated in the system. Interestingly, two major side products were detected: high molecular weight esters or fatty esters (such as dodecanoic acid, dodecyl ester) and acetate compounds (such as octacosyl acetate). The formation of fatty esters might be due to the dehydration reaction of FFAs and alkyl alcohols on acid sites. It is believed that large molecular weight hydrocarbons lead to the deactivation of the catalyst by blocking the catalyst pores and depositing as coke. In addition, with an acidic functionality, acetate can be obtained from dehydrogenation of alcohol couples with aldehyde under an inert atmosphere to yield an acetate [38].

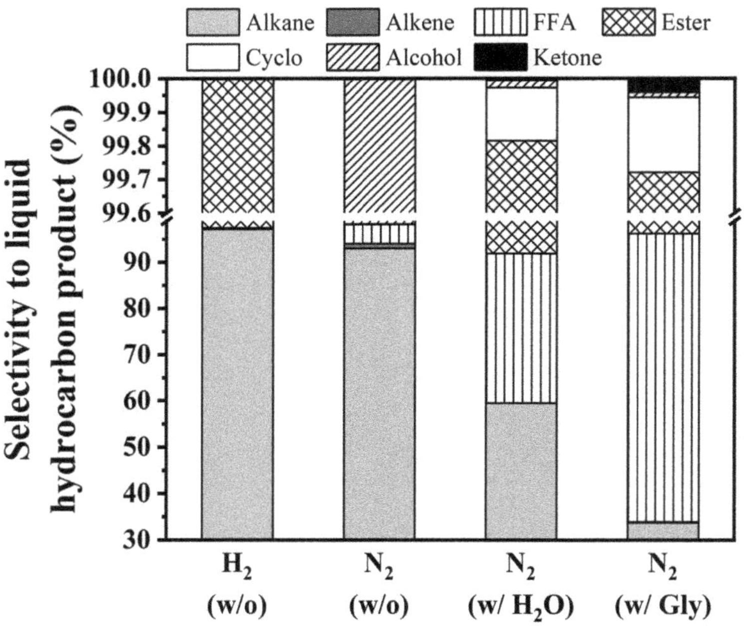

Figure 5. Distribution of liquid hydrocarbon products of coconut oil deoxygenation under different atmospheres with (w/) and without (w/o) co-reactant feeds. (Condition: Coconut oil, LHSV of 1 h^{-1}, 330 °C, and 50 bar of reaction gas, Gly = glycerol solution).

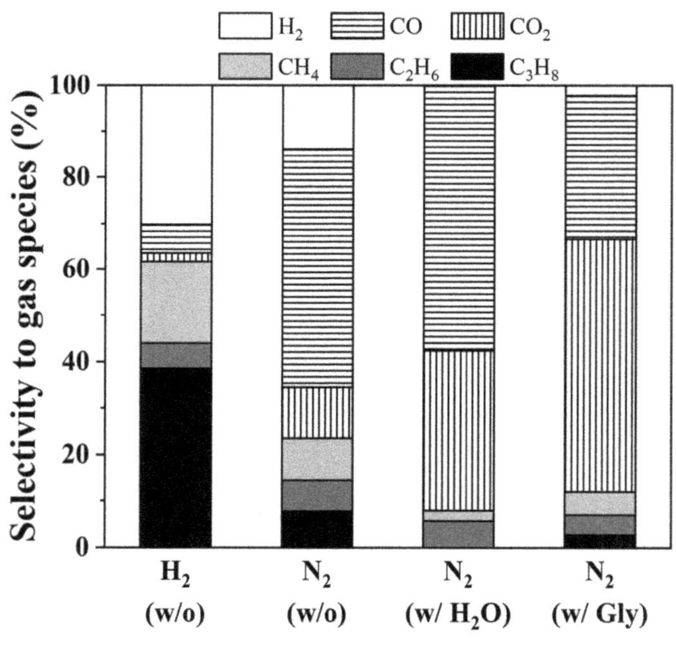

Figure 6. Distribution of gas species of coconut oil deoxygenation under different atmospheres with (w/) and without (w/o) co-reactant feeds. (Condition: Coconut oil, LHSV of 1 h^{-1}, 330 °C, and 50 bar of reaction gas, Gly = glycerol solution).

In H_2O, the amount of FFAs, originating from hydrolysis of TGs, was much higher. This is in accordance with the literature [26,39,40], in which the disappearing of C=O of esters (in the TGs molecules) and the obtaining of C=O of carboxylic acids (in the FFAs molecules) in the reaction by using H_2O were reported. This result confirmed that hydrolysis was the dominant pathway. The presence of H_2O accelerated and promoted the conversion of TG molecules to FFAs. There was no consumption of hydrogen for the hydrolysis of the TGs. However, the H_2 could be further utilized to promote deoxygenation of FFAs (DCO or DCO_2). Therefore, the formation of the desired hydrocarbons was very low. It is known that when there was a lack of H_2, which inhibits the catalytic performance, the reaction underwent cracking, for example: $C_{10}H_{22} \rightarrow C_3H_6 + C_7H_{16}$. Initially, hydrogen is present in the system because of some hydrocarbon molecules rearranging itself into a cycloalkane or an aromatic hydrocarbon with the loss of H_2—for example: $C_6H_{14} \rightarrow C_6H_{12} + H_2$—and thus after prolonged reaction times, when this reservoir was consumed, an enhanced formation of aromatic C_{n-1} hydrocarbons was observed. Those of FFAs will either undergo DCO_2 or DCO. This interesting phenomenon showed that alkenes were formed using either N_2 or H_2O. In the latter case, there was enough hydrogen and the formation hydrocarbons was almost constant.

The initial in-situ H_2 can be generated from glycerol reforming ($C_3H_8O_3 + 3H_2O \leftrightarrow 3CO_2 + 7H_2$, ΔH^0_{298} = +128 kJ/mol) [41] and glycerol decomposition ($C_3H_8O_3 \leftrightarrow 3CO + 4H_2$, ΔH^0_{298} = +250 kJ/mol) [42,43]. The formation of H_2, CO, and CO_2 (Figure 6) confirmed the utilization of glycerol co-feeding to generate in-situ H_2. These H_2 initially aid the triglyceride breakdown to form propane and FFAs. Unfortunately, the abundant FFAs still remained in the liquid product. It exhibited the insufficiency of in-situ H_2 so that alkanes could not be considerably produced. However, this in-situ H_2 helps to decrease the level of high molecular weight hydrocarbon compounds.

Table 4 shows the proportion of DCOx/HDO and CO_2/CO. The ratio of DCOx/HDO declined when applying co-reactant as follows: N_2 > aqueous glycerol > H_2O. The results indicated that generated H_2 had a significant impact on reaction pathways. At the initial of reaction with glycerol co-reactant, an abundance of generated H_2 can break C–O bonds in TG molecules through the HDO reaction to produce propane and FFAs. However, considering the final gas product, the reaction prefers to promote via DCO_2 which does not require H_2. It can be said that the rate of breaking of the C–O bonds in FFA should be slower than the C–O bonds in TGs and C–C bonds in FFA itself. On the other hand, the reaction seems to contribute to DCO when using H_2O co-reactant. These results indicated that the insufficient H_2 at the beginning suppressed HDO. Once FFA was generated, H_2 from water gas shift reaction promoted DCO pathway, consistent with other reports [10,14,42–46]. In all conditions, the selectivity to jet fuel range (C_8–C_{14}) was higher than that to diesel fuel range (C_{15}–C_{18}).

Table 4. Summary reaction pathway and selectivity toward jet and diesel range fuels of the alkane products.

Feed	Co-Reactant	Selectivity (%)		DCO_x/HDO [3]	CO_2/CO
		Jet Fuel [1]	Diesel Fuel [2]		
Coconut oil	H_2	64.09	35.91	1.08	0.30
	N_2	28.73	71.27	2.90	0.21
	H_2O	62.58	37.42	1.05	0.59
	Glycerol	58.74	41.26	1.18	1.75

[1] Jet fuel range = C_8–C_{14} alkanes, [2] Diesel fuel range = C_{15}–C_{18} alkanes, [3] DCO_x = DCO + DCO_2.

4. Conclusions

The catalytic deoxygenation of coconut oil was performed on NiCo/SAPO-11 nanocatalyst in a continuous-flow reactor for hydrocarbon production. Reaction condition in assistance with in-situ H_2 produced from glycerol (50 wt% aqueous solution), without any external H_2 source (under N_2 ambient) and addition of H_2O (5% by volume) was investigated on the deoxygenation reaction and product distribution. The results showed that under N_2 atmosphere applying co-reactants H_2O and glycerol significantly increased the oil conversion. The main products were FFAs and their corresponding C_{n-1} alkanes, whereas a small fraction of the intermediates was found. Considering FFA products, the addition of H_2O aids the triglyceride breakdown into FFAs, and the addition of glycerol solution as the hydrogen donor is favourable for the initial hydrogenolysis into FFAs. Then, those FFAs can be either decarbonylated or decarboxylated to their corresponding C_{n-1} alkanes, in the fraction of jet and diesel fuels. The NiCo/SAPO-11 showed great potential in the production of advanced biofuels in the absence of an external H_2 source.

Author Contributions: Conceptualization, K.F., N.L., T.W. and R.K.; formal analysis, R.K. and J.N.; investigation, R.K.; methodology, R.K.; resources, K.F.; supervision, K.F., N.L. and T.W.; writing—original draft, R.K. and J.N.; writing—review and editing, K.F., N.L., R.K., and J.N. All authors have read and agreed to the published version of the manuscript.

Funding: The financial supports from the NANOTEC platform project (P1750381), the CAS-NSTDA project (P1952712), and the Royal Golden Jubilee Ph.D. (RGJ-PHD) Program (Grant No. PHD/0239/2558) are acknowledged.

Conflicts of Interest: The authors declare no conflict of interests.

References

1. Yeletsky, P.M.; Kukushkin, R.G.; Yakovlev, V.A.; Chen, B.H. Recent advances in one-stage conversion of lipid-based biomass-derived oils into fuel components—aromatics and isomerized alkanes. *Fuel* **2020**, *278*, 118255. [CrossRef]
2. The European Parliament; The Council of the European Union. *Directive 2009/28/EC of the European Parliament and of the Council of 23 April 2009, on the Promotion of the Use of Energy from Renewable Sources and Amending and Subsequently Repealing Directives 2001/77/EC and 2003/30/EC*; The European Parliament: Brussels, Belgium, 2009. Available online: https://eur-lex.europa.eu/eli/dir/2009/30/oj (accessed on 18 November 2020).
3. Abdullah, A.Z.; Razali, N.; Mootabadi, H.; Salamatinia, B. Critical technical areas for future improvement in biodiesel technologies. *Environ. Res. Lett.* **2007**, *2*, 34001. [CrossRef]
4. Ogunkunle, O.; Ahmed, N.A. A review of global current scenario of biodiesel adoption and combustion in vehicular diesel engines. *Energy Rep.* **2019**, *5*, 1560–1579. [CrossRef]
5. Huang, D.; Zhou, H.; Lin, L. Biodiesel: An alternative to conventional fuel. *Energy Procedia* **2012**, *16*, 1874–1885. [CrossRef]
6. Phumpradit, S.; Reubroycharoen, P.; Kuchonthara, P.; Ngamcharussrivichai, C.; Hinchiranan, N. Partial hydrogenation of palm oil-derived biodiesel over Ni/Electrospun silica fiber catalysts. *Catalysts* **2020**, *10*, 993. [CrossRef]
7. Li, X.; Luo, X.; Jin, Y.; OchoaLi, J.; Zhang, H.; Zhang, A.; Xie, J. Heterogeneous sulfur-free hydrodeoxygenation catalysts for selectively upgrading the renewable bio-oils to second generation biofuels. *Renew. Sustain. Energy Rev.* **2018**, *82*, 3762–3797. [CrossRef]
8. Kordulis, C.; Bourikas, K.; Gousi, M.; Kordouli, E.; Lycourghiotis, A. Development of nickel based catalysts for the transformation of natural triglycerides and related compounds into green diesel: A critical review. *Appl. Catal. B Environ.* **2016**, *181*, 156–196. [CrossRef]
9. Jin, W.; Pastor-Pérez, L.; Villora-Picó, J.J.; Sepúlveda-Escribano, A.; Gu, S.; Reina, T.R. Investigating new routes for biomass upgrading: "H_2-free" hydrodeoxygenation using Ni-based catalysts. ACS Sustain. *Chem. Eng.* **2019**, *7*, 16041–16049. [CrossRef]
10. Choo, M.-Y.; Juan, J.C.; Oi, L.E.; Ling, T.C.; Ng, E.-P.; Noorsaadah, A.R.; Centi, G.; Lee, K.T. The role of nanosized zeolite Y in the H_2-free catalytic deoxygenation of triolein. *Catal. Sci. Technol.* **2019**, *9*, 772–782. [CrossRef]
11. Hollak, S. Catalytic Deoxygenation of Fatty Acids and Triglycerides for Production of Fuels and Chemicals. Ph.D. Thesis, Utrecht University, Utrecht, The Netherlands, 2014.
12. Chiappero, M.; Do, P.T.M.; Crossley, S.; Lobban, L.L.; Resasco, D.E. Direct conversion of triglycerides to olefins and paraffins over noble metal supported catalysts. *Fuel* **2011**, *90*, 1155–1165. [CrossRef]
13. Morgan, T.; Grubb, D.; Santillan-Jimenez, E.; Crocker, M. Conversion of triglycerides to hydrocarbons over supported metal catalysts. *Top. Catal.* **2010**, *53*, 820–829. [CrossRef]

14. Miao, C.; Marin-Flores, O.; Davidson, S.D.; Li, T.; Dong, T.; Gao, D.; Wang, Y.; Garcia-Pérez, M.; Chen, S. Hydrothermal catalytic deoxygenation of palmitic acid over nickel catalyst. *Fuel* **2016**, *166*, 302–308. [CrossRef]
15. Asikin-Mijan, N.; Lee, H.V.; Juan, J.C.; Noorsaadah, A.R.; Ong, H.C.; Razali, S.M.; Taufiq-Yap, Y.H. Promoting deoxygenation of triglycerides via Co-Ca loaded SiO_2-Al_2O_3 catalyst. *Appl. Catal. Gen.* **2018**, *552*, 38–48. [CrossRef]
16. Cortright, R.D.; Davda, R.R.; Dumesic, J.A. Hydrogen from catalytic reforming of biomass-derived hydrocarbons in liquid water. *Nature* **2002**, *418*, 964–967. [CrossRef]
17. Polychronopoulou, K.; Charisiou, N.; Papageridis, K.; Sebastian, V.; Hinder, S.; Dabbawala, A.; AlKhoori, A.; Baker, M.; Goula, M. The effect of Ni addition onto a Cu-based ternary support on the H_2 production over glycerol steam reforming reaction. *Nanomaterials* **2018**, *8*, 931. [CrossRef]
18. Chen, Z.; Kukushkin, R.G.; Yeletsky, P.M.; Saraev, A.A.; Bulavchenko, O.A.; Millan, M. Coupling hydrogenation of guaiacol with in-situ hydrogen production by glycerol aqueous reforming over Ni/Al_2O_3 and Ni-X/Al_2O_3 (X = Cu, Mo, P) Catalysts. *Nanomaterials* **2020**, *10*, 1420. [CrossRef]
19. Aman, D.; Radwan, D.; Ebaid, M.; Mikhail, S.; van Steen, E. Comparing nickel and cobalt perovskites for steam reforming of glycerol. *Mol. Catal.* **2018**, *452*, 60–67. [CrossRef]
20. Li, K.T.; Yen, R.H. Aqueous-phase hydrogenolysis of glycerol over Re promoted Ru catalysts encapuslated in porous silica nanoparticles. *Nanomaterials* **2018**, *8*, 153. [CrossRef]
21. Zhong, H.; Yao, G.; Cui, X.; Yan, P.; Wang, X.; Jin, F. Selective conversion of carbon dioxide into methane with a 98% yield on an in-situ formed Ni nanoparticle catalyst in water. *Chem. Eng. J.* **2019**, *357*, 421–427. [CrossRef]
22. Hu, X.; Li, P.; Zhang, X.; Yu, B.; Lv, C.; Zeng, N.; Luo, J.; Zhang, Z.; Song, J.; Liu, Y. Ni-based catalyst derived from NiAl layered double hydroxide for vapor phase catalytic exchange between hydrogen and water. *Nanomaterials* **2019**, *9*, 1688. [CrossRef]
23. Ma, Z.; Liu, Z.; Song, H.; Bai, P.; Xing, W.; Yan, Z.; Zhao, L.; Zhang, Z.; Gao, X. Synthesis of hierarchical SAPO-11 for hydroisomerization reaction in refinery processes. *Appl. Petrochem. Res.* **2014**, *4*, 351–358. [CrossRef]
24. Zuo, H.; Liu, Q.; Wang, T.; Ma, L.; Zhang, Q.; Zhang, Q. Hydrodeoxygenation of methyl palmitate over supported Ni catalysts for diesel-like fuel production. *Energy Fuels* **2012**, *26*, 3747–3755. [CrossRef]
25. Srifa, A.; Faungnawakij, K.; Itthibenchapong, V.; Assabumrungrat, S. Roles of monometallic catalysts in hydrodeoxygenation of palm oil to green diesel. *Chem. Eng. J.* **2015**, *278*, 249–258. [CrossRef]
26. Kaewmeesri, R.; Srifa, A.; Itthibenchapong, V.; Faungnawakij, K. Deoxygenation of waste chicken fats to green diesel over Ni/Al_2O_3: Effect of water and free fatty acid content. *Energy Fuels* **2015**, *29*, 833–840. [CrossRef]
27. Huynh, T.M.; Armbruster, U.; Phan, B.M.Q.; Nguyen, D.A.; Martin, A. The influence of cobalt in bimetallic Ni-Co catalyst supported on H-Beta for phenol hydrodeoxygenation. *Chim. Oggi. Chem. Today* **2014**, *32*, 40–44.
28. Sotomayor, F.J.; Cychosz, K.A.; Thommes, M. Characterization of micro/mesoporous materials by physisorption: Concepts and case studies. *Acc. Mater. Surf. Res.* **2018**, *3*, 36–37.
29. Said, S.; Zaky, M.T. Pt/SAPO-11 catalysts: Effect of platinum loading method on the hydroisomerization of n-hexadecane. *Catal. Lett.* **2019**, *149*, 2119–2131. [CrossRef]
30. San-José-Alonso, D.; Juan-Juan, J.; Illán-Gómez, M.J.; Román-Martínez, M.C. Ni, Co and bimetallic Ni–Co catalysts for the dry reforming of methane. *Appl. Catal. A Gen.* **2009**, *371*, 54–59. [CrossRef]
31. Gao, X.; Tan, Z.; Hidajat, K.; Kawi, S. Highly reactive Ni-Co/SiO_2 bimetallic catalyst via complexation with oleylamine/oleic acid organic pair for dry reforming of methane. *Catal. Today* **2017**, *281*, 250–258. [CrossRef]
32. Reynoso, A.J.; Iriarte-Velasco, U.; Gutiérrez-Ortiz, M.A.; Ayastuy, J.L. Highly stable Pt/$CoAl_2O_4$ catalysts in aqueous-phase reforming of glycerol. *Catal. Today* **2020**, in press. [CrossRef]
33. Reynoso, A.J.; Ayastuy, J.L.; Iriarte-Velasco, U.; Gutiérrez-Ortiz, M.A. Cobalt aluminate spinel-derived catalysts for glycerol aqueous phase reforming. *Appl. Catal. B Environ.* **2018**, *239*, 86–101. [CrossRef]
34. Wei, Y.; Li, S.; Jing, J. Synthesis of Cu–Co catalysts for methanol decomposition to hydrogen production via deposition–precipitation with urea method. *Catal. Lett.* **2019**, *149*, 2671–2682. [CrossRef]
35. Wang, L.; Guan, E.; Wang, Y. Silica accelerates the selective hydrogenation of CO_2 to methanol on cobalt catalysts. *Nat. Commun.* **2020**, *11*, 1–9. [CrossRef] [PubMed]

36. Xue, H.; Xu, J.; Gong, X.; Hu, R. Performance of a Ni-Cu-Co/Al_2O_3 catalyst on in-situ hydrodeoxygenation of bio-derived phenol. *Catalysts* **2019**, *9*, 952. [CrossRef]
37. Papadopoulou, E.; Ioannides, T. Methanol reforming over cobalt catalysts prepared from fumarate precursors: TPD investigation. *Catalysts* **2016**, *6*, 33. [CrossRef]
38. Krobkrong, N.; Itthibenchapong, V.; Khongpracha, P.; Faungnawakij, K. Deoxygenation of oleic acid under an inert atmosphere using molybdenum oxide-based catalysts. *Energy Convers. Manag.* **2018**, *167*, 1–8. [CrossRef]
39. Authayanun, S.; Arpornwichanop, A.; Paengjuntuek, W.; Assabumrungrat, S. Thermodynamic study of hydrogen production from crude glycerol autothermal reforming for fuel cell applications. *Int. J. Hydrogen Energy* **2010**, *35*, 6617–6623. [CrossRef]
40. Li, Y.; Zhang, C.; Liu, Y.; Tang, S.; Chen, G.; Zhang, R.; Tang, X. Coke formation on the surface of Ni/HZSM-5 and Ni-Cu/HZSM-5 catalysts during bio-oil hydrodeoxygenation. *Fuel* **2017**, *189*, 23–31. [CrossRef]
41. Gosselink, R.W.; Hollak, S.A.; Chang, S.W.; van Haveren, J.; de Jong, K.P.; Bitter, J.H.; van Es, D.S. Reaction pathways for the deoxygenation of vegetable oils and related model compounds. *ChemSusChem* **2013**, *6*, 1576–1594. [CrossRef]
42. Avasthi, K.S.; Reddy, R.N.; Patel, S. Challenges in the production of hydrogen from glycerol—A biodiesel byproduct via steam reforming process. *Procedia Eng.* **2013**, *51*, 423–429. [CrossRef]
43. Wang, M.; Wang, Z.; Gong, X.; Guo, Z. The intensification technologies to water electrolysis for hydrogen production—A review. *Renew. Sustain. Energy Rev.* **2014**, *29*, 573–588. [CrossRef]
44. Hollak, S.A.; Ariëns, M.A.; de Jong, K.P.; van Es, D.S. Hydrothermal deoxygenation of triglycerides over Pd/C aided by in-situ hydrogen production from glycerol reforming. *ChemSusChem* **2014**, *7*, 1057–1062. [CrossRef]
45. Hu, W.; Wang, H.; Lin, H.; Zheng, Y.; Ng, S.; Shi, M.; Zhao, Y.; Xu, R. Catalytic decomposition of oleic acid to fuels and chemicals: Roles of catalyst acidity and basicity on product distribution and reaction pathways. *Catalysts* **2019**, *9*, 1063. [CrossRef]
46. Sousa, F.P.; Silva, L.N.; de Rezende, D.B.; de Oliveira, L.C.A.; Pasa, V.M.D. Simultaneous deoxygenation, cracking and isomerization of palm kernel oil and palm olein over beta zeolite to produce biogasoline, green diesel and biojet-fuel. *Fuel* **2018**, *223*, 149–156. [CrossRef]

nanomaterials

MDPI

Review

Dendritic Polymers as Promising Additives for the Manufacturing of Hybrid Organoceramic Nanocomposites with Ameliorated Properties Suitable for an Extensive Diversity of Applications

Marilina Douloudi [1,*], Eleni Nikoli [1], Theodora Katsika [1], Michalis Vardavoulias [2] and Michael Arkas [1,*]

1 Institute of Nanoscience Nanotechnology, NCSR "Demokritos", Patriarchou Gregoriou Street, 15310 Athens, Greece; h.nikoli@inn.demokritos.gr (E.N.); theodorakatsika@gmail.com (T.K.)
2 PYROGENESIS S.A., Technological Park 1, Athinon Avenue, 19500 Attica, Greece; mvardavoulias@pyrogenesis-sa.gr
* Correspondence: marilina.douloudi@gmail.com (M.D.); m.arkas@inn.demokritos.com (M.A.); Tel.: +30-210-650-3669 (M.A.)

Abstract: As the field of nanoscience is rapidly evolving, interest in novel, upgraded nanomaterials with combinatory features is also inevitably increasing. Hybrid composites offer simple, budget-conscious and environmental-friendly solutions that can cater multiple needs at the same time and be applicable in many nanotechnology-related and interdisciplinary studies. The physicochemical idiocrasies of dendritic polymers have inspired their implementation as sorbents, active ingredient carriers and templates for complex composites. Ceramics are distinguished for their mechanical superiority and absorption potential that render them ideal substrates for separation and catalysis technologies. The integration of dendritic compounds to these inorganic hosts can be achieved through chemical attachment of the organic moiety onto functionalized surfaces, impregnation and absorption inside the pores, conventional sol-gel reactions or via biomimetic mediation of dendritic matrices, inducing the formation of usually spherical hybrid nanoparticles. Alternatively, dendritic polymers can propagate from ceramic scaffolds. All these variants are covered in detail. Optimization techniques as well as established and prospected applications are also presented.

Keywords: dendritic polymers; ceramic compounds; biomimetic; nanomaterials; dendrimers; hyper-branched polymers; silica composites; hybrid materials

Citation: Douloudi, M.; Nikoli, E.; Katsika, T.; Vardavoulias, M.; Arkas, M. Dendritic Polymers as Promising Additives for the Manufacturing of Hybrid Organoceramic Nanocomposites with Ameliorated Properties Suitable for an Extensive Diversity of Applications. *Nanomaterials* **2021**, *11*, 19. https://dx.doi.org/10.3390/nano11010019

Received: 9 November 2020
Accepted: 20 December 2020
Published: 24 December 2020

1. Introduction

With the advent of nanotechnology and the introduction of nanoscale matter, revolutionary progress was observed in the field of materials science. The synthesis of nanoscale analogues to conventional implements with enhanced properties and a broad spectrum of applications became a routine process. Nanocomposites confirmed the hypothesis that size and shape play a pivotal role in the determination of their exquisite physical and chemical profile. The establishment of superiority over conventional microcomposites was an expected consequence. The exceptionality of these materials originates from their multilayer, core-shell-surface architecture. Depending on their core substance, they can be classified into three key categories: ceramic nanocomposites, metal nanocomposites and polymeric nanocomposites [1–3].

Ceramic materials are considered as an emerging trend of recent decades due to their mixed and flexible features, their porous, small-size structure and their broad spectrum of applications. They can appear in various forms, with the most prominent ones being ceramic membranes, ceramic nanostructures and biogenic ceramics, each form possessing exceptional uses and serving different purposes. Their capability to adopt controlled and diverse combined conformations through easily accessible, low-cost, ambient temperature preparation methods render them excellent candidates for the production of hybrids with optimized properties such as core-shell nanoparticles. They also constitute excellent alternatives in order to avoid the undesirable defects of their elementary equivalents, for instance toxicity, instability and high solubility [4–7].

Among the numerous properties of ceramic membranes, the most distinguished are mechanical tolerance, chemical inertia, thermal stability, damage impedance, energy conservation, high hydrophobicity, satisfactory flux at low pressures and an excessive working lifetime [6–8]. All these characteristics contribute to the suitability and expanded applicability of these membranes in various fields such as gas separation [9–11] and filtration technology [7,8], water decontamination processes, water and wastewater treatment, desalination [12–16] and catalysis [11,17]. Another advantage of ceramic membranes is their employment in both microfiltration (MF) and ultrafiltration (UF) processes and less frequently in nanofiltration (NF) and reverse-osmosis (RO) [18]. Specifically, for water and wastewater purification, depending on the targeted configuration (flat-sheet or tubular) and the overall properties of the fabricated membrane, the preferred ceramic materials used are alumina, titania, zirconia and silica. Still, the non-specialized nature resulting primarily from their inability to selectively extract ultra-low-sized, "mobile" contaminants and to achieve accurately desired pore size distribution, remains one of their major drawbacks [6,19].

Common problems of unprocessed ceramics such as brittle fracture, poor toughness and strength degeneration highlight the need to introduce secondary components to these materials in order to reach nanoscale ceramic products with upgraded properties. Alumina, titania, zirconia, silica and hydroxyapatite are again useful candidates for the construction of ceramic nanoparticles [20]. These nanomaterials are unique in terms of biocompatibility. Thanks to their surface modification, encapsulation and interaction with biological molecules, such as proteins, lipids and antibodies, they serve as novel diagnostic and therapeutic tools. Some examples include drug-delivery vehicles, biomedical imaging and photodynamic and photothermal therapy [21–25]. In addition, advantages linked to biocompatibility, such as increased dissolution [26], resistance to microbes, stable porosity and high selectivity, along with the facility of synthesis procedures without requirement of toxic substances and extreme conditions, make nano-ceramics the epicenter of pharmaceutical research. They combine safety with the ease of administration and controlled drug availability [20,27,28]. For example, the photooxidative and cytotoxic features of titania can be efficiently manipulated for the fabrication of thermally stable anatase nanoparticles. Therefore, they can be employed as excipients and in the context of photodynamic therapy for the selective treatment of several types of tumors. On the other hand, silica mesoporous nanoparticles can be readily functionalized with proteins and radioisotopes (e.g., radioiodine) to facilitate cancer detection and subsequent radiotherapy [2,3]. Similar forms can also be combined with titania analogues and yield core-shell nanomaterials with advanced catalytic and absorbing performance, as well as anti-pollutant properties [29].

Another category, bioceramics, is considered a convenient type of biomaterial in terms of applicability in biomedical engineering fields. Their biological importance stems from their surface similarities with living tissues and molecules with osseous origin, facilitating therefore their interaction with bones and teeth and thus their involvement in orthopedic therapy and tissue engineering [30–34]. To date, modern bioceramic products replace first-generation inert predecessors claiming the overthrow of conventional tissue damage treatment regimes, since they provide novel therapeutic approaches that rely on tissue repair and replacement in parallel and less on simple substitution. Second generation bioceramics include varieties of different crystallization degrees. They are characterized by high bioactivity, biodegradability and resorbability and are ideal for repairing purposes where a strong and controlled bond with the implant is required. These materials may be based on apatite sources (hydroxyapatite, calcium phosphate mixtures, bone cements based on calcium salts) that function as bone imitators against bone loss or as surface coatings for better bone deposition rates. Alternatively, they may contain a glass source (bioglasses, glass-ceramics) that stimulate bone growth processes by interacting with body fluids or by filling bone cavities. Likewise, third generation counterparts introduce a porous framework for the construction of scaffolds with an adjusted size and shape to accommodate specific molecular components [31,34]. Silica-based mesoporous structures and organic/inorganic hybrid composites intervene in differentiation pathways of mesenchymal cells and regulate osteogenesis and thus promote bone regeneration by exposing active molecules, such as growth factors, hormones and cells at the site of damage [30,31,34].

There is a constantly increasing interest towards the construction of new embodiments that are inspired by living systems, mimicking natural molecules and recapitulating key cellular and molecular processes that have prompted nanotechnologists to incorporate molecules bearing a dendroid architecture into their constructs. Dendritic polymers are ubiquitous materials, graced by exceptional physicochemical properties, well-defined highly branched discoid or spherical shapes and an adjustable surface area. They constitute the fourth major class of polymeric architecture and are classified as (a) monodispersed symmetric dendrimers named after the Greek word δένδρο for tree because of their structural resemblance; (b) asymmetric species, called hyperbranched polymers; (c) fragments of both categories defined as dendrons; and (d) dendrigrafts, which are combinations of conventional (usually linear) polymers with dendritic patterns [35,36]. Their uniqueness emanates from their separate structural units (Figure 1): the core which is the substrate of the radial functionalization. The interior, where the "branches" reside, create a well-suited environment for the encapsulation of guest elements. The exterior part, forming a peripheral surface, functions as an active nano-scaffold and interacts with the nearby functional groups and environmental stimuli. Some of the most established dendrimers are diaminobutane poly(propylene imine) DAB/PPI and poly(amidoamine) PAMAM (Figure 2), while their common non-symmetric hyperbranched relatives include polyglycerol PG and poly(ethylene imine) PEI (Figure 3) [37–40]. Dendritic polymer's biomimetic behavior favors their applicability in clinical fields [41,42] as drug carriers, targeted and controlled delivery vehicles [43–47] for anticancer therapeutics [48], in photodynamic therapy [49], imaging [50], diagnostics [51,52] as sensors, gene transfer agents or as artificial molecules per se [30,35]. Additional applications are encountered in the fields of separation systems [36,37], chromatography, membranes, hydrogels [38–44], solvent extraction [45] and color chemistry such as dyes [46].

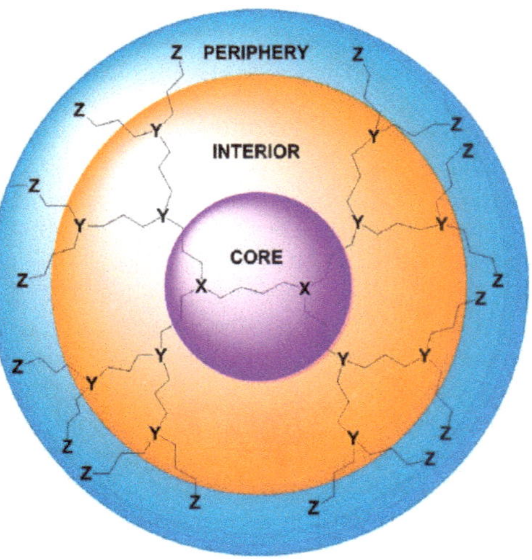

Figure 1. Schematic representation of the three structural parts of a dendritic polymer.

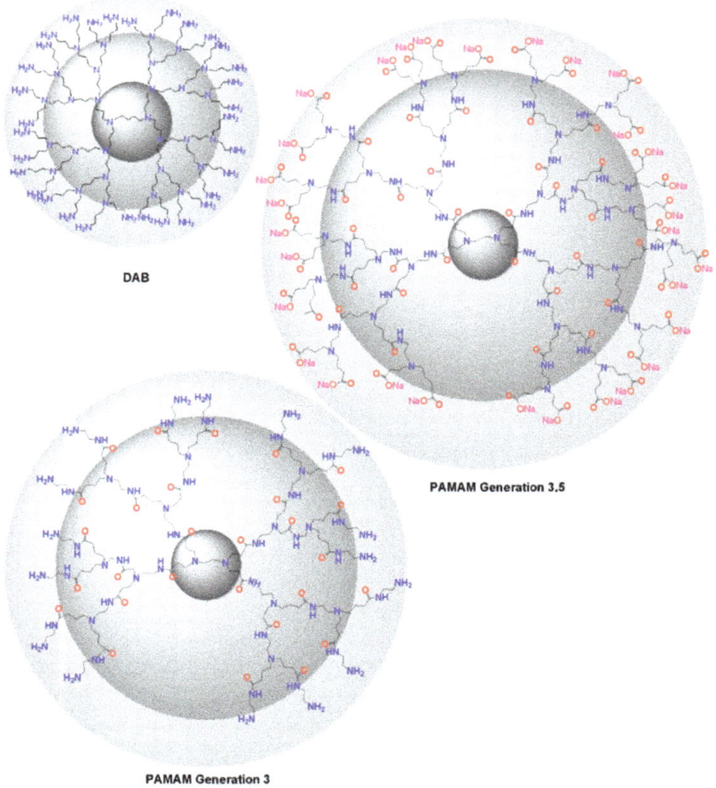

Figure 2. Chemical structures of Diaminobutanepoly(propylene imine) DAB (PPI), and Poly(amido amine) PAMAM, 3rd and 3.5rd generations.

Undoubtedly, the inauguration of interventions in the nano-scale gave the opportunity to access new conformations, synthetic or semi-synthetic, of augmented functionality, tractable structure and enhanced multi-functionality, compared to their raw counterparts. Still, the margins for improvement of nanosynthetic constructs will remain large, as long as there are unexplored aspects of different combinations of distinctive precursors. The fabrication of novel, tailor-made hybrid products profiting from combinatorial properties of suitably functionalized dendritic polymers and ceramic compounds represent powerful tool of the utmost importance. The resulting blends have the potential to substitute their unilateral sources, by considerably eliminating most of their inherent defects and benefiting from their individual assets. An inorganic/organic (ceramic/dendrimer) combination for instance can incorporate stabilizing and guest-host characteristics of the polymer matrices, in parallel with the mechanical superiority provided by the ceramic substrate [53]. The yielding organoceramics possess a confirmed potential for a multiplicity of applications, among them the most outstanding include catalysis [29,54,55], water treatment [56–59] and magnetoceramics [60]. An overview of the prevailing methods used for the synthesis of organoceramic materials will be presented in the framework of this review, along with a summary of potential barriers encountered and the proposed solutions for each given case. The most important applications of these materials will be also thoroughly outlined with a final note concerning the outlooks for the future.

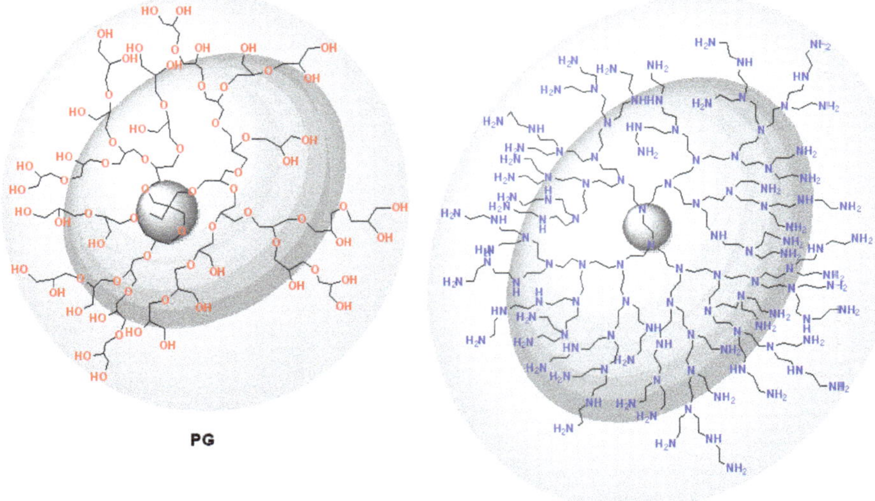

PG

PEI

Figure 3. Chemical structures of hyperbranched polyglycerol (PG), and hyperbranched poly(ethylene imine) (PEI).

2. Methods of Preparation

2.1. Simple Coating by Absorption of Dendritic Additives into the Pores of the Ceramic

One of the easiest and fastest methods to obtain hybrids that follow a dendritic/ceramic pattern, is the direct immersion of the inorganic support possessing suitable porosity into a solution of the selected polymer [61]. In this manner, complex organic/inorganic formulations are attained that incorporate the functionalities of both constituents: (a) elevated and selective encapsulation properties, due to the highly accessible empty cavities formed by the branches of the dendritic compound; (b) perfect mechanical properties and absorption/filtering capacity that ceramic supports of variant porosity sizes and shapes can furnish. The most astonishing characteristic of this specific technique is the high polymer impregnation. Loading efficacy depends mostly on the structural components and the resting free surface area of the ceramic substrate, as well as the functionalized polymer compatibility, in terms of reactivity, with the given porous ceramic.

A first typical immersion of a silica monolith to a PAMAM G4 solution yielded a solid anion exchange polyelectrolyte gel for Prussian Blue (PB), iron(III) hexacyanoferrate(II), and cobalt(II) hexacyanoferrate(II) [62]. G2, G4 and G6 PAMAM with silica substrates have been employed in order to take advantage of the electrostatic attractions of positive charged amino groups and mount a second layer of DNA molecules for gene therapy applications [63]. Alternate layering of PAMAM and chondroitin sulphate onto mesoporous silica nanoparticles of an intensified negative charge due to the presence of a primary substrate of carboxylated PAMAM G1 led to ideal carriers for protracted release of doxorubicin and curcumin [64].

After initial capping of the silanol groups of the external surface by $(CH_3)_2SiCl_2$, G1-G3 PPI dendrimers peripherally modified by amidoferrocenes were incorporated by intermolecular hydrogen bonds to the pores of mesoporous MCM-41 silica affording hybrid redox-active ceramics [65]. Another distinguishing example that employs this impregnation approach concerns alumina filters of distinctive porosity directly immersed into an alkylated dendritic polymer solution under mild heating conditions. Octyl functionalized poly(ethyleneimine) (PEI), the proposed hyperbranched polymeric additive, serves as a cheap and facile-to-synthesize alternative to symmetric dendrimer "nanosponges", exhibiting equivalently excellent pollutant inclusion constants ($k_{inclusion}$) [66]. It creates a highly homogenous water-insoluble coating, a thoroughly efficient depollution film. The hallmark of this technique is the attainment of a polymer impregnation percentage of up to 22%. Interestingly, there is an analogy between the concentration of the dendritic polymer and its impregnation percentage into the same alumina filter, meaning that an increasing concentration (0.2–30% *w*/*w*) of the alkylated PEI allows better impregnation percentages until a certain critical point. When this threshold is surpassed, small pores clog up and no additional polymer can be integrated. This critical point that marks the loading capacity of the specific ceramic compound depends primarily on the compositional characteristic of its pores. The efficient coverage of the ceramic pores is confirmed by the linearity of the plots correlating with the impregnation percentage/concentration ratio as a function of the surface area of the given alumina tube filter.

Another instance that adopts the elementary impregnation approach involves the fabrication of titania processed by PAMAM dendrimers adsorbed into its surface pores through a slurry mixture preparation [67]. The yielding nanocomposites benefit from the non-toxicity of the organic layer and the metal chelation properties introduced by functionalization of the periphery with hydroxy-terminal groups. Herein, only a partial, localized coating of the pores is achieved, driven by the electrostatic attractions developed between the hydroxyl groups of the dendrimer and the surface of titania (Figure 4). These intermolecular forces favor the immobilization process during which the dendritic molecules occupy a rather limited space compared to the total porous volume of pure titania. Specifically, this method can reach an impregnation level of 1% organic content by weight to the inorganic substrate. The composites withstood 2 h at 300 °C, exhibiting minimal mass loss of their active layer (0.16%), proving their high thermal stability.

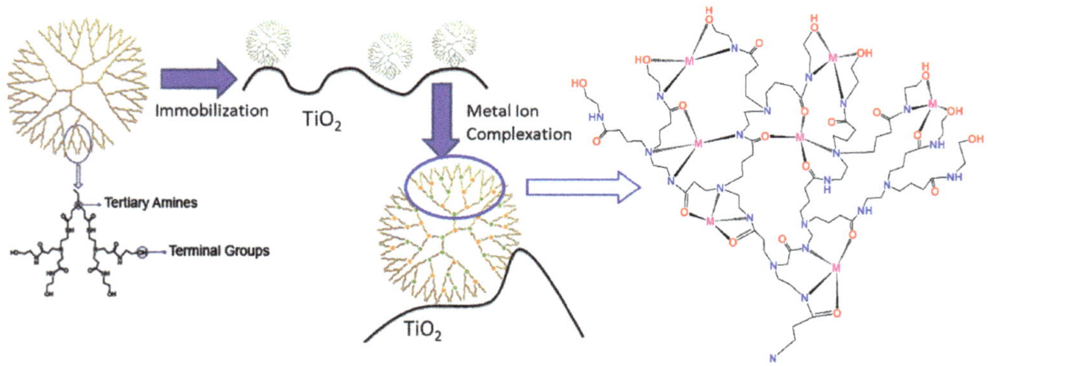

Figure 4. Schematic representation of the immobilization process of G4 dendrimers onto the surface of titania. (Reproduced with permission from [67]; Copyright Elsevier, 2013).

As expected, the impregnation rate does not depend only on the polymer concentration. The chemical nature of the selected dendrimer is a crucial factor too. The molecules are attracted to the inorganic surfaces due to intermolecular, particularly electrostatic, forces [68]. Typically, accumulation increases with increasing dendrimer generation while it is considerably larger for those with negatively charged groups (e.g., carboxylic anions) on positively charged surfaces (alumina) than that of analogues containing positive charged groups (e.g., ammonium cations) on negatively charged surfaces (silica) [69]. For this reason, the pH and the ionic strength of the solution exert an important influence in as much as they affect the charge of the molecules and the ceramic support [70]. Organic loading is also governed by the pore parameters of each particular ceramic [71]. This was verified by a study that evaluated the impregnation degree of silylated dendrimers and hyperbranched polymers into three ceramic filters of different origins and compared them with the homologous β-cyclodextrin derivative [58]. The resulting data demonstrated that ceramic membranes manufactured by Al_2O_3, TiO_2 or SiC, when immersed into a solution of triethoxy silyl dendritic polymers of the same concentration (30%), yielded hybrid organoceramics of different polymer contents (Figure 5). All of them, however, exhibited higher saturation limits compared to the β-cyclodextrin derivative. As was anticipated, ceramic filters with the highest porosity values (Al_2O_3 and SiC) scored the best inclusion values, while the ones with the lowest porosity values (TiO_2) presented considerably inferior impregnation properties. Pore size also regulates polymer incorporation degree considering that smaller pores allow a wider pore surface area. Therefore, the impregnation percentage of ceramics having the same porosity is determined according to the pore size. The smallest pore size, such as in alumina, in that case leads to a higher content in additives. The type of guest molecules also plays a crucial role to the extent of their incorporation, as each specific solution is characterized by different viscosity. This apparently explains the high impregnation percentages obtained by ethoxysilated polyglycerol compared to the other polymers employed in the context of this study.

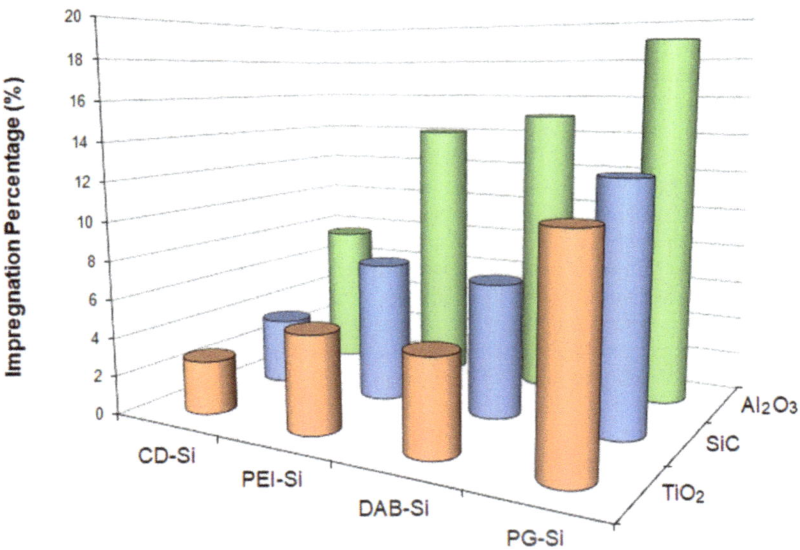

	CD-Si	PEI-Si	DAB-Si	PG-Si
TiO_2	2.75	5.01	4.84	11.13
SiC	3.5	7.25	6.9	12.51
Al_2O_3	7.63	14	15	18.93

Figure 5. Loading percentage for three ceramic filters of different origins (TiO_2, SiC, Al_2O_3) after their immersion into 30% solutions of triethoxy silated dendritic and cyclodextrin polymers (Reproduced with permission from [59]; Copyright Elsevier, 2006).

2.2. Attachment of Polymers with Covalent Bonds to the Surface of the Ceramic Substrate

The performance of the nanocomposites is largely dictated by their surface features. A well-defined and fully characterized external surface leads to better prediction and management of the resulting material properties, such as active ingredient solubility/release and reactivity with the other active factors present. Chemical reactions with active groups induced on the ceramic, for instance by etching with strong acid or base are the conventional, versatile method to obtain highly specialized and flexible modified surfaces. Dendritic polymers, known for the high local concentration of exposed external surface groups, branching diversity and hence susceptibility to chemical interventions, represent unique options to afford hybrid materials with desired chemical profiles. The functionalization of dendritic polymers with groups based on silicon, has gained considerable interest. They offer active sites, commonly silanols, which can subsequently react with chemically relative ceramic substrates, forming stable siloxane bridges through curing processes [56,72]. Some of the most prominent advantages of these Si-heteroatom dendrimers over the conventional ones are as follows: (a) unrestrained shaping and tailoring of the dendrimer structure density with Si atoms as the branching centers of the last 2 or 3 generations; (b) efficient diffusion into the ceramic filler or membrane and improved impregnation rate. This results in lower activation energy requirements and secure reduced curing temperatures and cost effectiveness; and (c) a variety of synthetic pathways.

Production of organosilicon-based ceramics can be implemented through methods such as cross-linking, sol-gel processing, pyrolysis and ceramization. During the fabrication procedure, emphasis should be given to the specifications of the selected dendrimer, and more specifically in its molecular weight, which should be as high as possible, in its solubility into the solution of the ceramic precursor and in its stereochemical convenience to perform cross-linking. Chemical stability of the resulting hybrid material depends on the type of formed bonds. Covalent bonds are the most suitable for sustainability. By being stable they provide the option of regeneration without substantial organic material loss and therefore allow repeated use. Formation of chemical bonds between the dendritic polymer and the ceramic substrate can be established through the sol-gel reaction. The effectiveness of this method in producing organoceramics is based on the employment of appropriate functionalized silyl-reactants, particularly alkoxylates, established for their capability to undergo hydrolysis under acidic conditions [59]. Full conversion is crucial for the shaping of the final product, as heterogeneity affects negatively both the compatibility and the structure of the hybrid ceramic. Triethoxylated derivatives of dendritic molecules, for instance DAB, PEI and PG, are susceptible to rapid hydrolysis, delivering silanols, which in turn can react intramolecularly or intermolecularly with adjacent hydroxyl groups via a condensation mechanism (Figure 6a). Dendritic silanols can condensate with the hydroxyl groups of the targeted ceramic surface as well, enabling the formation of a chemically bound film (Figure 6b). Activation of the ceramic surface and formation of hydroxyl groups is a prerequisite for the reaction and can be performed by using several techniques. An interesting activation example proposed by Ottenbrite is the etching through immersion into the 1 M NaOH solution for some hours, followed by extensive washing with methanol and water and drying under vacuum conditions [73]. This process combined with curing at slightly increased temperatures, successfully promoting the development of nanostructured hyperbranched PEI covalently attached to an activated ceramic filter for absorption of mono and polycyclic aromatic hydrocarbons, pesticides, trihalogen alkanes and methyl *tert*-butyl ethers [59]. Implementation of this exact synthetic pathway with G3 PPI dendrimer loaded with ibuprofen or levofloxacin on mesoporous silica nanoparticles afforded antimicrobial agents to be defined as nanoantibiotics [74,75]. Interference of disulfide bridges between sulfur-containing G1–G5 PPI and accordingly modified mesoporous silica nanoparticles loaded with fluorescent dyes (fluorescein disodium salt and carboxyfluorescein) generated redox-responsive stimuli-triggered release carriers [76]. The transformation of disulfides to sulfhydryl groups in the presence of a reductive reagent removes the dendrimers and permits the discharge of the active ingredients.

Antibacterial PAMAM and carbosilane dendrimers and dendrons have been grafted to silica surface too [77], either by mediation of triethoxysilylpropyl succinic anhydride silane (TESPSA) or by direct grafting, biocompatible G2 and G3 carbosilane dendrons bearing ammonium or tertiary amine groups were attached to mesoporous silica nanoparticles in order to serve as nonviral gene delivery systems [78]. Alternative carbosilanes with Si-Cl at the focal point and peripherally modified zirconocene were used as catalysts [79]. Co-condensation of PAMAM dendron (G1.0) containing a triethoxysilyl coupling agent and tetraethoxy silane (TEOS) in different molar ratios with the addition of cetyltrimethylammonium bromide (CTAB) as pore-directing agents, yielded Hg^{2+} absorbing silica-gel sol-gel [80]. Non-activated mesoporous silica required reflux for 12 h in DMF under nitrogen in order to form bonds with triethoxy-sillylated PAPAM dendrons and provides composite nanoparticles for drug delivery (curcumin) and fluorescence imaging [81].

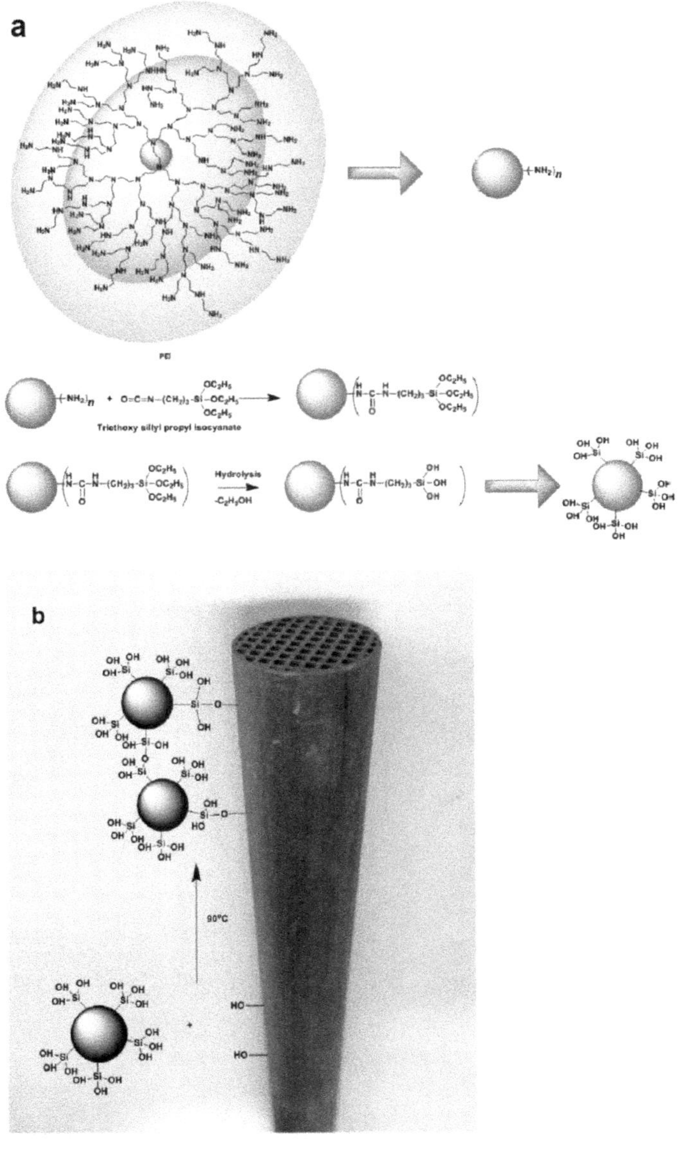

Figure 6. (**a**) Hydrolysis of triethoxysilyl derivatives forming silanols susceptible to polycondensation with neighboring hydroxy groups; (**b**) attachment of functionalized dendritic polymers on the surface of silica ceramic filters by formation of siloxane bridges (Reproduced with permission from [59]; Copyright Elsevier, 2006).

The adjustable external surface and internal cavities of dendrimers renders them extremely selective and prone to interventions aiming to enhance their chemical affinity to specific targets. The application of their derivatives that are prone for sol-gel reactions as sorbents for undesirable substances removal and analytical extractions, such as solid phase microextraction (SPME) and capillary microextraction, is common [82,83]. Deficient immobilization of the polymeric coating on the fused silica capillary is a typical shortcoming encountered in conventional SPME methods, delaying or even blocking full extraction of a wide range of analytes. Sol-gel processing, however, guarantees firm attachment, allowing for improved extraction performance and maximum use of the overall analytical potential. One of the first studies that investigated adsorptive capacity of dendritic molecules in solventless microextraction and polycondensation, introduced benzyl-terminated dendrons as a chemically stable coating on the inner walls of fused silica capillaries. Principles of sol-gel processing were exploited in order to form and maintain a strong chemical connection between the organic coating and the inorganic substrate and achieve the highest possible stability towards thermal and solvent distortions. The attachment of benzyl-terminated dendrons was accomplished by the mediation initially of 3-(triethoxysilyl)ethylamine and then methyltrimethoxysilane. The introduction of ethoxysil groups into the roots of the dendron enables sol-gel hydrolysis and polycondensation, creating thus a dendritic stationary phase bonded to the column walls, characterized by remarkable uniformity and a roughened, porous appearance (Figure 7). Various types of bifunctional poly(phosphorhydrazone) dendrons and dendrimers were likewise modified by triethoxysilyl moieties for grafting to silica and by amines for trapping CO_2 [84].

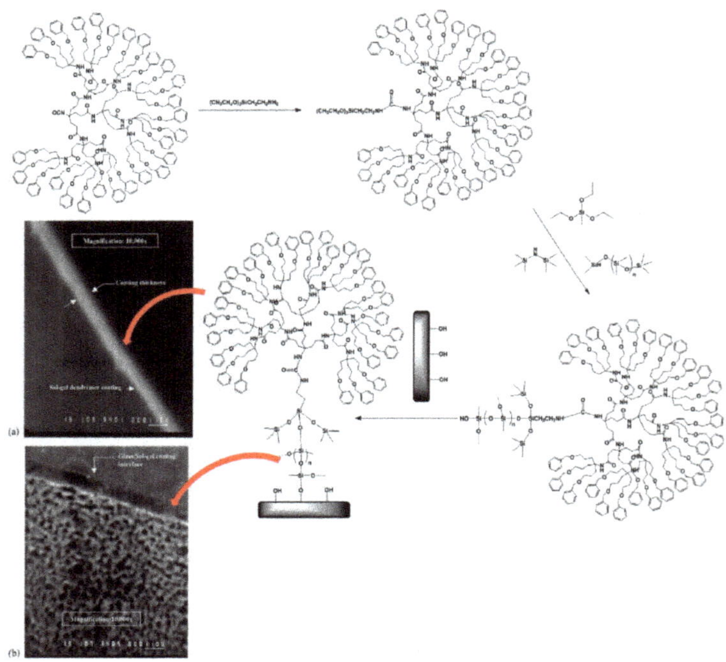

Figure 7. Synthesis of Phenyl-terminated dendrimer with triethoxysilyl root and subsequent immobilization to a fused silica capillary. SEM micrographs of coating thickness (**a**) (0.5 µm) and roughened porous texture (**b**). (Reproduced with permission from [83]; Copyright Elsevier, 2004).

Another variation that involves reaction with activated rough surfaces and subsequent sol-gel siloxane bond formation, exploits the selectivity potential of hyperbranched PEI by grafting onto the surface of a magnetic porous ceramic [85]. The selected ceramic powder was composed of Fe_3O_4 and SiO_2 in order to combine and inherit the magnetism of magnetite with surface adaptability and chemical stability of silica. This powder was efficiently prepared through dispersion polymerization that involves the coupling of γ-chloropropyltrimethoxysilane. The production of SiO_2@Fe_3O_4-PEI nanocomposites follows a simple two-step process. The magnetic porous ceramic substrate is first activated with methanesulfonic acid, promoting the creation of silanol groups. The latter then underwent a reaction with the corresponding analogues of hydrolized γ-chloropropyltrimethoxysilane at 80 °C to yield a chlorine functionalization reaction. The final step includes the synthesis of the anticipating adsorptive composite by reaction of the porous magnetic ceramic compound with 10% PEI solution at 90 °C for 6 h. In this way, magnetic porous adsorbents with an excessive separation performance and a faster adsorption equilibrium are produced (Figure 8). Another variant implementing magnetic silica (SiO_2-Fe_3O_4) and Pyridylphenylene dendrons adopted two approaches with 3-(iodopropyl)trimethoxysilane or (3-aminopropyl)triethoxysilane (Figure 9) in order to get efficient stabilizing hosts of Pd catalysts [86].

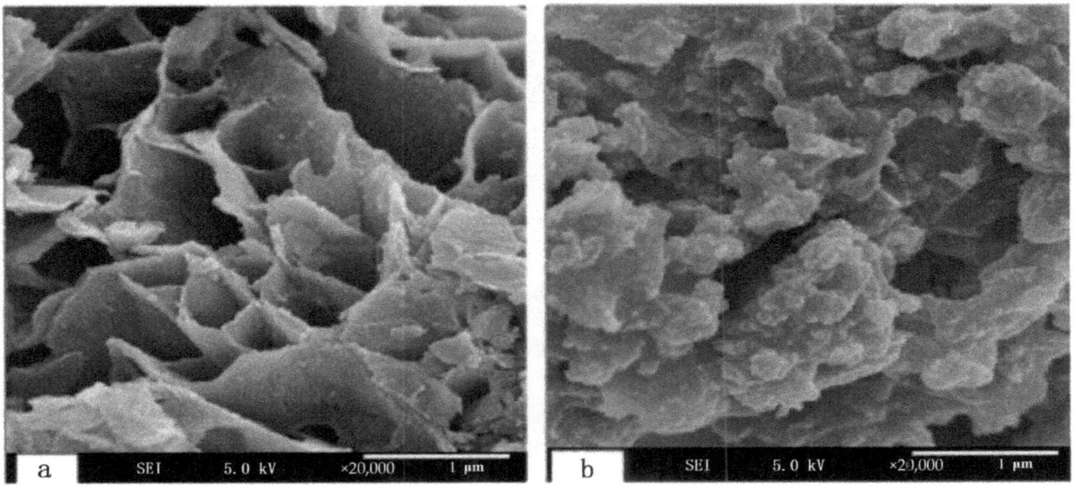

Figure 8. SEM micrographs of Fe_3O_4 and SiO_2 composite magnetic powder (**a**) and PEI-grafted adsorbent (**b**). (Reproduced with permission from [85]; Copyright Elsevier, 2011).

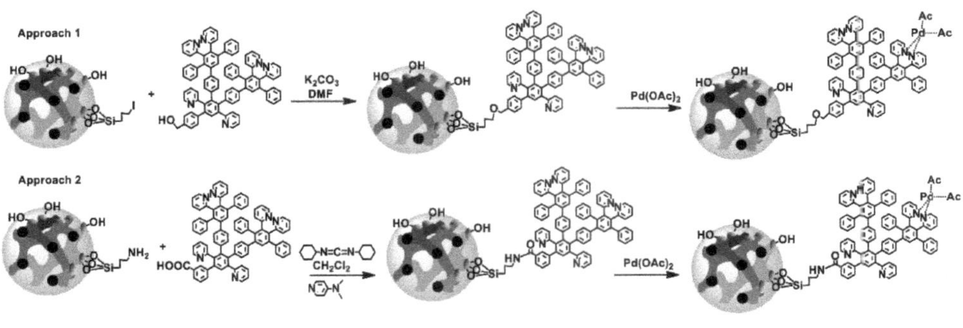

Figure 9. Attachment of Pyridylphenylene dendrons to magnetic silica with the aid of iodine (Approach 1) or amino (Approach 2) groups succeeded by the complexation with Pd acetate. (Reproduced with permission from [86]; Copyright Elsevier, 2019).

While symmetric dendrimers have been intensively studied for their catalytic properties in homogeneous catalysis, limited information is available for their contribution to the field of heterogeneous or the so-called immobilized catalysts on porous media. A very hopeful area that addresses the insufficiently selective and enantioselective behavior of heterogeneous inorganic supports in asymmetric reactions. It also incorporates the prospect of chiral dendrimers on inorganic supports as catalytic alternatives for enantiomeric reactions [87]. Specifically, the study conducted by Chung et al. describes the application of 3-glycidoxypropyl-trimethoxysilane in the development of silica-supported chiral catalysts based on PAMAM dendrimers (Figure 10) for the enantioselective addition of diethylzinc to benzaldehyde. The chemical grafting reaction of the *n*th PAMAM generation dendrimer derivative onto the silica substrate proceeds under the presence of methanol and reflux for 4 h. The resulting composite can then be further treated to achieve the desired peripheral chiral functionality by reaction with (1*R*, 2*R*)-(+)-1-phenylpropylene oxide and the optional presence of a long alkyl chain spacer.

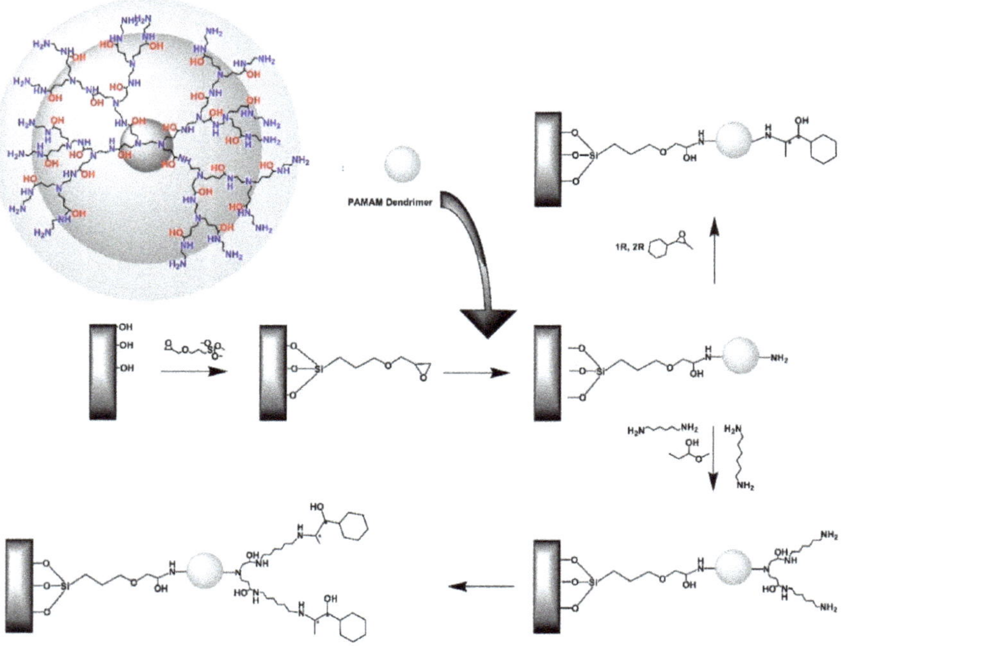

Figure 10. Synthetic path for the immobilization of PAMAM dendrimers decorated by chiral functionality onto silica substrate. (Reproduced with permission from [87]; Copyright Elsevier, 2003).

Implementation of the same concept by the aid of chloropropyl trichloro silane and polyallylamine permitted the coupling of G3 PPI to an amorphous silica gel and the posterior immobilization of palladium nanoparticles for selective hydrogenation catalysis [88]. On the other hand, aminopropyl triethoxy silane (APTES) was required in order to bind polyphosphorhydrazone (PPH) dendrimers on thermally activated silica nanoparticles [89]. Further derivatization of the remaining aldehyde groups with amine-functionalized Poly(Ethylene (Glycol) (Figure 11a) afforded hybrid nanocarriers for silver and silver oxide colloids that exhibit antibacterial activity. Carboxy derivatized glutamic acid-based chiral dendrimers with peptide linkages coupled with APTES gave silica chiral stationary phases for use in HPLC [90], whereas the same reagent in conjunction with 1,10-carbonyldiimidazole was used for binding G1, G3, G4, G5 PAMAM. Their subsequent loading with fluorescein isothiocyanate or rhodamine B isothiocyanate led to photoluminescence sensors for cyanide and copper ions, respectively [91]. Dendritic fragments bearing specialized sol-gel bonds forming groups at their focal point demonstrate another beneficial feature. They serve as convenient media to secure high local concentrations of active ingredients by protracting, for instance, the lifetime of hydrophobic species in aqueous solutions. This stabilizing aspect is confirmed in a study where PAMAM dendrons displayed a protecting effect over water-dispersible TiO_2 nanocomposites. The establishment of a strong chemical Si-O-Ti bond created a screening polar shell to the solubilized metal oxide from the aqueous environment and thus hampered flocculation [92]. The formation of TiO_2 nanoparticles, and the chemical attachment of PAMAM dendrons (generation 1–3) carrying siloxy focal points and long alkyl chains (hexyl, C6), was realized through hydrolysis of tetraisopropylorthotitanate and subsequent formation of Ti-O-Si bonds. Both reactions proceeded simultaneously in the same pot, mixing various ratios of the titania precursor and the functionalized dendron in 1-propanol solution at 0 °C and then pouring the mixture in water. The final configuration of dendron-protected titania (Figure 11b) was defined by the metal ion/dendron ratio and the generation of the latter, while its size was only generation-dependent. This functionalization of titania nanoparticles enhanced their photocatalytic activity towards 2,4-dichlorophenoxyacetic acid degradation.

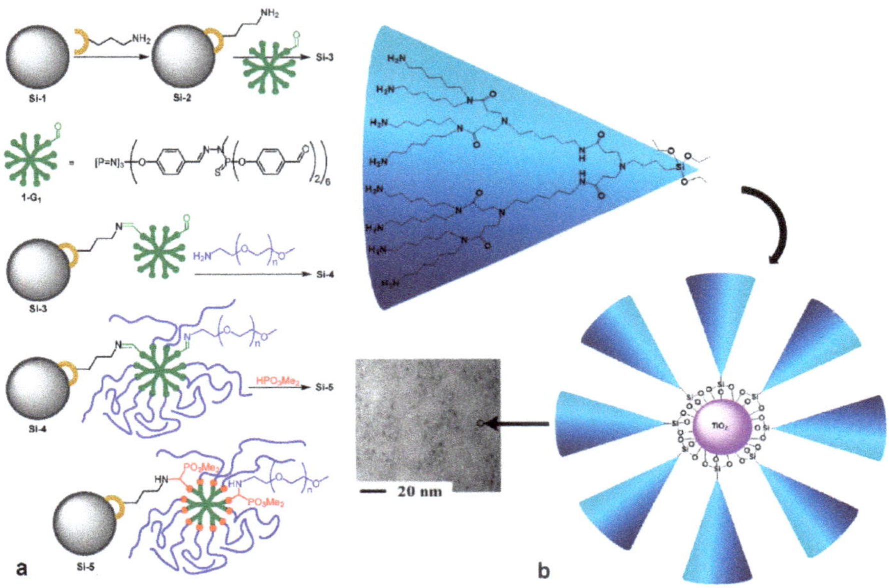

Figure 11. (**a**) Synthetic path for grafting polyphosphorhydrazone dendrimers onto functionalized silica and following modification of its periphery with Poly(EthyleneGlycol) chains. (Reproduced with permission from [89]; Copyright Royal Society of Chemistry, 2013). (**b**) Pathway for the clipping of the PAMAM dendron protective layer to titania photocatalytic nanoparticles and SEM micrograph of the complexes (Reproduced with permission from [92]; Copyright Elsevier, 2005).

Some other reactive groups may be introduced to ceramic surfaces. Such organosilicates are even commercially available. Silica gels bearing isocyanate or maleimide groups were successfully used for binding PAMAM dendrimers and derivatives thereof [93] for Cu^{2+} absorption studies [94], whereas 3-(1-Piperazino)propyl-functionalized gels presented an ideal substrate to immobilize melamine-based dendrons and were implemented for pesticide (atrazine) containment [95].

2.3. *Direct Growth of Dendritic Polymers on the Surface of Ceramics*

Another strategy for the decoration of the ceramic surface by dendritic sequences is the direct propagation of dendrons on their surface through conventional divergent pathways. This solid-phase synthesis technique is a quick and simple alternative for the preparation of hybrid "dendro-ceramers". Most of the studies employing this direct growth method are based on silica gels. Nevertheless, the main stages formulating the synthesis schemes that ultimately lead to the product of interest remain the same, both for gels and ceramics: (a) surface activation reaction; (b) dendron core binding; and (c) outspread of the branches.

The grafting of PAMAM dendrimers into the silica surface can be easily achieved by performing three successive reaction steps [85,96]. At the outset, the amino core of PAMAM polymers is introduced to the activated silica surface through the interaction between the silanol surface groups and 3-aminopropyltriethoxylsilane (APES). This reaction is carried out at 110 °C in toluene for 48 h. PAMAM sprouts then proliferate by repetitive additions of a branching unit. Evolution of dendrons on silica surface can be accomplished through a series of successive alternate Michael addition and amidation reactions. Michael addition of methyl acrylate (MA) to the terminal amine generates amino propionate esters. Subsequent amidation with ethylene diamine (EDA) results in the first generation dendron. Repetitive Michael additions and amidations afford higher generations [97]. This primary PAMAM coating of silica exhibited high complexation affinity for Zn (Figure 12a) and equally excellent metal ion absorption efficiency for Zn^{2+} [98], Mn^{2+} [99], Ni^{2+} [100], Co^{2+} [98,101], Cd^{2+} [102,103], Fe^{3+} [103,104] (Figure 12b), Pb^{2+} [105] and U(IV) [106]. Employing exactly the same pattern, Alper et al. grew G4 PAMAM dendrons inside the channels of pore-expanded periodic mesoporous silica [107]. Mesoporous silica particles incorporating PAPAM were used as well for generation-dependent size exclusion chromatography columns [108]. Further treatment of PAMAM dendron external groups with salicylaldehyde [109] enables the effective removal of Hg^{2+} from water, whereas processing by methyl isothiocyanate permits absorption of both Hg^{2+} [110], (Figure 13a,b) and Ag^{+} [111] (Figure 13c) and also from ethanol [112].

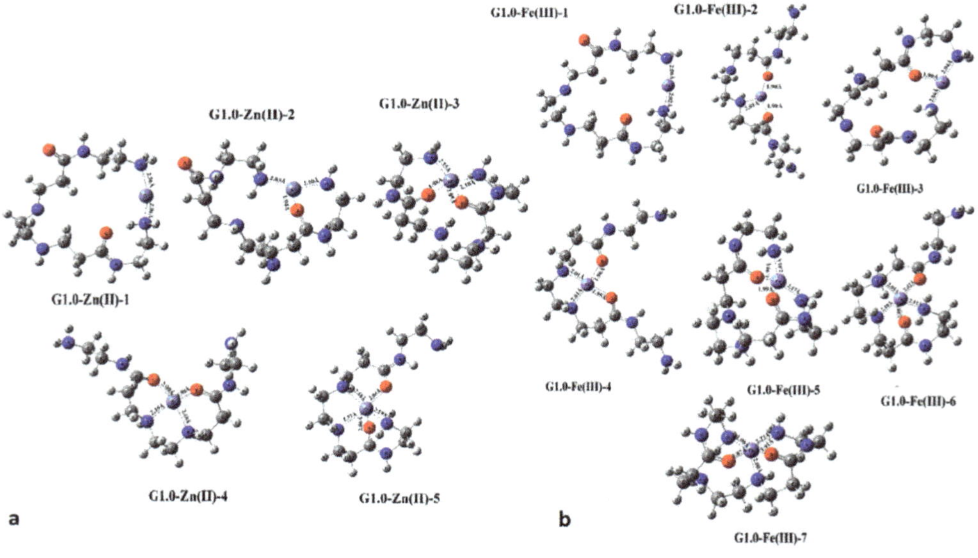

Figure 12. Optimized coordination structures of PAMAM G1.0 with (**a**) Zn(II) (Reproduced with permission from [98]; Copyright Elsevier, 2020) and (**b**) Fe(III) (Reproduced with permission from [103]; Copyright Elsevier, 2018).

Addition of n-octadecylisocyanate [97] leads to the introduction of long aliphatic chains to the periphery of the dendrons (Figure 14). In this way, production of amphiphilic silica bearing a hydrophilic dendritic core combined with a hydrophobic surface is obtained and is designed to deal with lipophilic pollutant encapsulation issues.

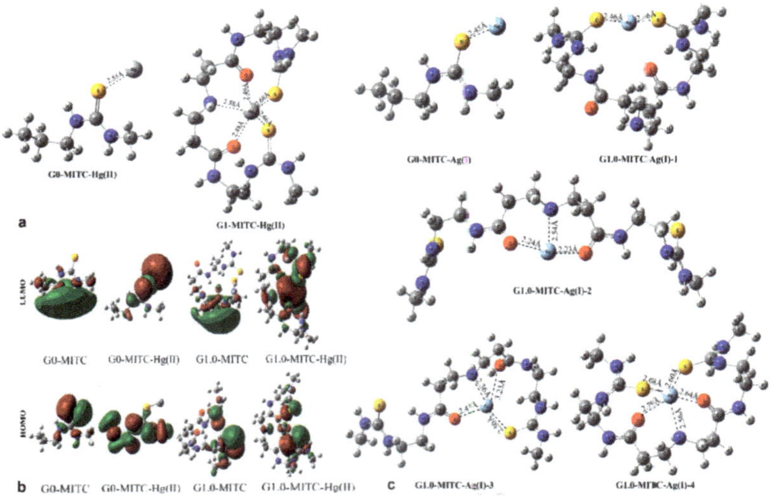

Figure 13. Optimized geometries (**a**) and contour plots (**b**) of the highest occupied molecular orbitals HOMOs (red) and lowest unoccupied molecular orbitals LUMOs (green), of Hg^{2+} complexes with methylisothiocyanated (MITC) G0-PAMAM and G1.0-PAMAM. (Reproduced with permission from [110]; Copyright American Chemical Society, 2016); (**c**) Optimized geometries of the respective silver complexes (Reproduced with permission from [111]; Copyright Elsevier, 2018).

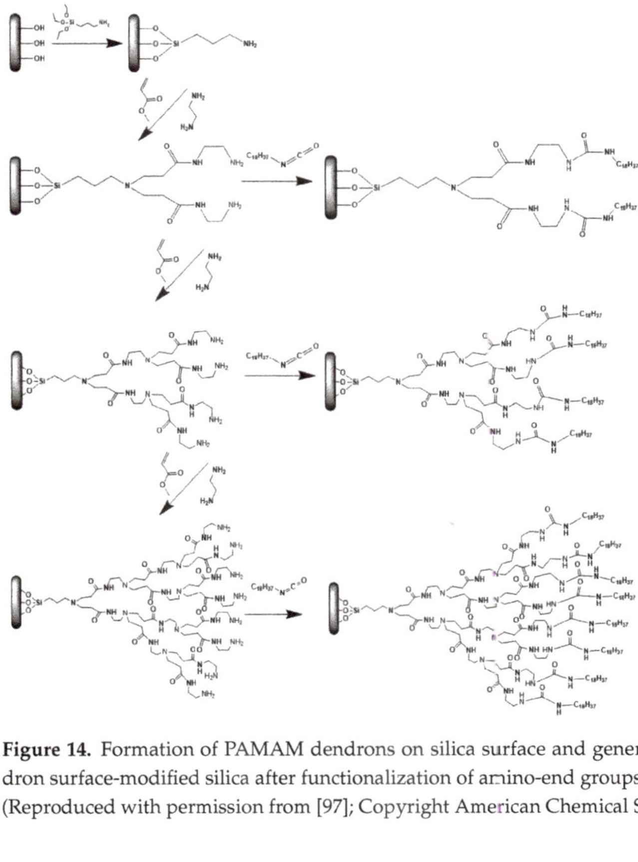

Figure 14. Formation of PAMAM dendrons on silica surface and generation of amphiphilic dendron surface-modified silica after functionalization of amino-end groups with octadecylisocyanate (Reproduced with permission from [97]; Copyright American Chemical Society, 2008).

In another context, usually the third or the fourth generation of the dendron is required prior to the inclusion of the metal-catalyst in order to amplify its catalytic activity as is the case for recyclable gold nanoparticles that promote oxidation of alcohols [113]. In most instances, an additional stage is required: introduction of complexation sites. Specifically, PAMAM segments are phosphonated in order to establish attractive coordination sites for typical metal catalysts such as rhodium and palladium. Peripherally restructured derivatives can then readily complex with chloro(dicarbonyl)rhodium(I) or dichlorobis-(benzonitrile)palladium(II). Resulting composites are applied in catalytic hydroformylation of olefins [114–116], carbonylation [117–119], hydroesterification [120] and Heck reaction of aryl bromides with butyl acrylate and styrene [121]. Another modification of the PAMAM periphery by salicylaldehyde affords a Schiff base. The reaction of the latter with $Mn(OAc)_2 \cdot 4H_2O$ yields an Mn(II) complex, immobilizes Mn(II) onto PAMAM–SiO_2 hybrids and leads to olefin epoxidation catalysts with increased activity (Figure 15a) [122]. Absorption and crosslinking of external amino groups with cellulase stabilizes and enhances enzymolysis efficiency [123], whereas propagation of up to third generation PAMAM dendrons from silica-coated gold nanorods produces trifunctional composites for anticancer photothermal-chemo-gene therapy and co-delivery of anticancer drugs (doxorubicin) and B-cell lymphoma 2-(Bcl-2)-targeted siRNA [124].

Aziridine polymerization is the prevailing method in order to evolve hyperbranched poly(ethyleneimine) dendrons. Pre-attachment of a reactive intermedia on the ceramic is not required [125,126]. Of particular interest is an unconventional, environmentally friendly technique implementing the dual functionality of compressed CO_2 as reaction medium and catalyst (Figure 15b) [127].

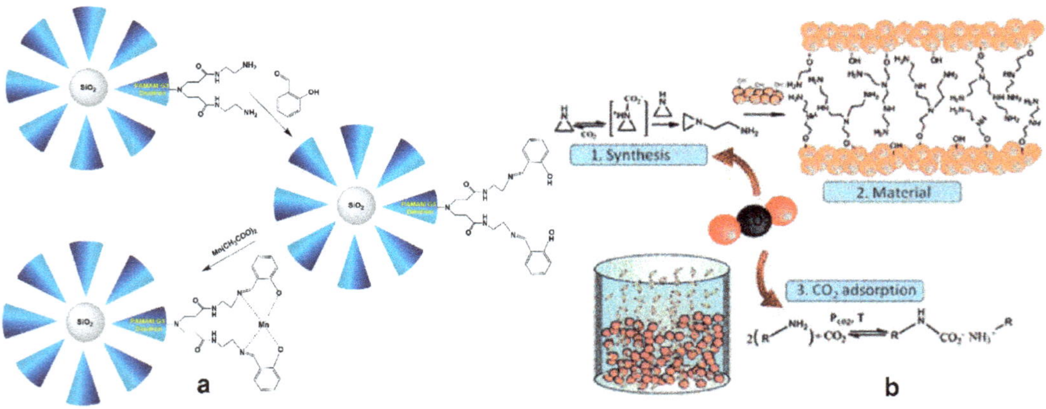

Figure 15. (**a**) Procedure for assembling Silica-PAMAM Dendron-Manganese Composite Catalysts. (**b**) Schematic representation of PEI dendron growth onto silica using CO_2, and the mechanism of posterior CO_2 adsorption (intended application) (Reproduced with permission from [127]; Copyright Royal Society of Chemistry, 2013).

Another compelling example of this method involves the synthetic pathway of various generations, melamine-based dendrimers directly sprouted on the surface of mesoporous silica. Initiation of this synthesis is performed by a functionalization reaction of the silica support with APES and is followed by succeeding reactions with triazinetrichloride and 4-aminomethylpiperidin (Figure 16). Silica derivatives containing the primary amino groups are alternatively incubated with the solutions of the growth-promoting reagents at 4 °C for 24 h [128]. Replacement of 4-aminomethylpiperidin by tris(2-aminoethyl) amine (TREN) evolves dendrons suitable for CO_2 absorption [129]. In a related study, combined γ-alumina membranes comprised of a mesoporous γ-alumina layer, grown into a macroporous alumina solid support, undergo a specialized activation process. Direct evolution of the dendrons off the alumina surface is once more triggered by the attachment of APES on the support, and is differentiated by the consecutive treatment by diisopropylethylamine and piperazine instead of 4-aminomethylpiperidin. The obtained modified membrane was further functionalized with dodecylamine in order to augment its potential in polarity-based separation applications [130].

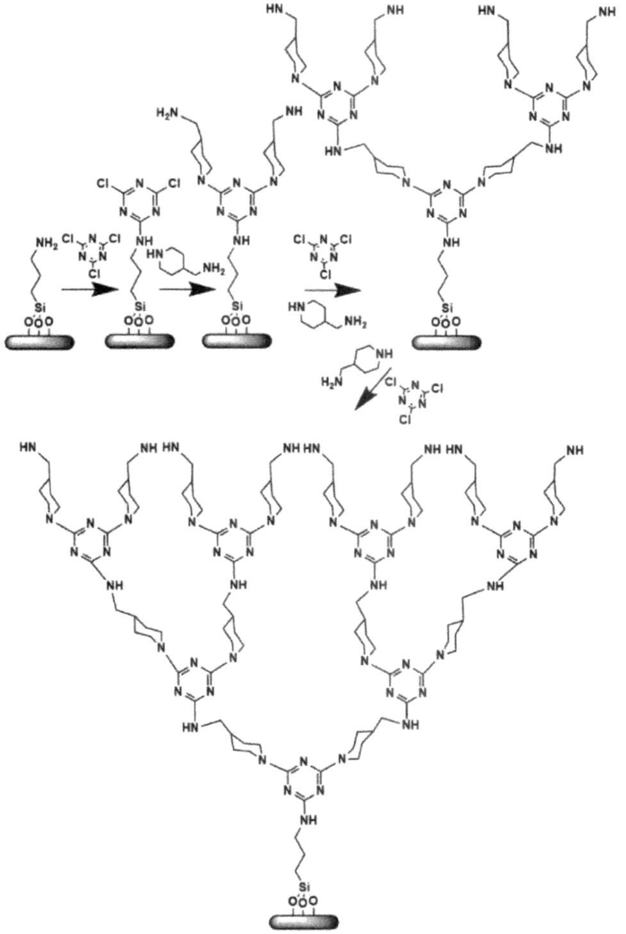

Figure 16. Evolution stages of melamine-based dendrons (Reproduced with permission from [128]; Copyright WILEY-VCH Verlag GmbH & Co., 2004).

There is no reason for the restriction of the dendritic evolution to conventional forms only. Dendrigrafts are a distinctive and rather exotic class, and the building blocks of the internal branches are replaced by polymeric side chains may equally emanate from a suitable focal point. Tsubokawa et al. produced such a type of coating, starting by aminopropyl triethoxy silane root and "postgraft polymerization" of vinyl groups [131,132]. Although entirely symmetrical branching cannot be attained through solid-phase strategies, still, it furnishes a convenient and inexpensive alternative in order to immobilize dendritic functionalities through a single sol-gel reaction onto diverse ceramic frameworks such as silica, alumina, titania, ceria, lanthania, hafnia and zirconia.

2.4. Sol-Gel Cross-Linking of Silyl Dendritic Polymers

Silicon-containing dendritic polymers are a promising subcategory, distinguished for the mixed properties mostly originated from the radially multilayered copolymeric morphology and the distinctive characteristics that each constituent layer impose (Figure 17) [133]. Poly(amidoamine-organosilicon) (PAMAMOS) dendrimers present a hydrophilic poly (amidoamine) (PAMAM) molecular core and a hydrophobic organosilicon surface. They provide different structures as building blocks or active sites for various synthetic pathways depending on the chemical reactivity of their corresponding end-groups. Among them, alkoxysilyl-functionalized groups are susceptible to crosslinking chemistry, giving rise to PAMAMOS highly defined three-dimensional networks (Figure 18) [133,134]. PAMAMOS "honeycomb like" compositions are prepared through conventional cross-linking methods, which enable the formation of organosilicon structures with nanoscale dimensions, tunable sizes and shapes and high applicability in the areas of films, sheets and coatings. These sol-gel reactions follow the same two-step synthetic pattern as previously described for the dendritic moieties on the surface of ceramics: hydrolysis of alkoxysilyl (Si-O-R) end-groups to the corresponding silanols (Si-OH) and condensation of resulting silanol intermediates into final siloxane (Si-O-Si) bridges. The rate of the crosslinking reaction is greatly determined by several external factors and reagents, such as additional catalysts and reaction conditions [133]. For example, the optional presence of a catalyst, such as bis(2-ethylhexanoate)tin and variations to curing periods and temperatures of the solvent-free PAMAMOS blocks, lead to a multiplicity of organic-inorganic nanohybrids with completely different properties depending each time on the chosen conditions. Multifunctional crosslinking agents such as tetramethoxysilane or tetraethoxysilane may be employed to increase crosslink density of the dendrimers and silica content. Moreover, controlled thermal degradation of PAMAM molecules allows the formation of porous nanocomposites, where the final pore size is dictated by the diameter of the selected PAMAM core [135].

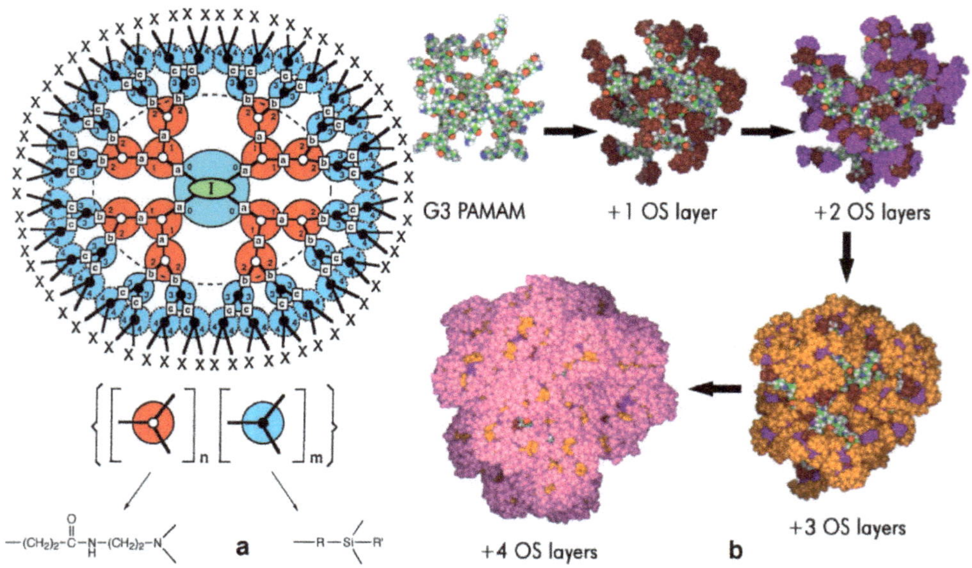

Figure 17. (**a**) Generalized representation of PAMAMOS dendrimer structure I: core; X: end-groups; 1, 2, 3, 4, denote generations; letters: a, PAMAM-PAMAM bonds b, PAMAMOS bonds c, OS-OS bonds. (**b**) Molecular models of Generation 3 PAMAM interior (green–red–white–blue) with increasing number of organosilicon exterior layers. First layer brown; 2nd layer purple; 3rd layer yellow; 4th layer magenta (Reproduced with permission from [133]; Copyright WILEY-VCH Verlag GmbH & Co., 2006).

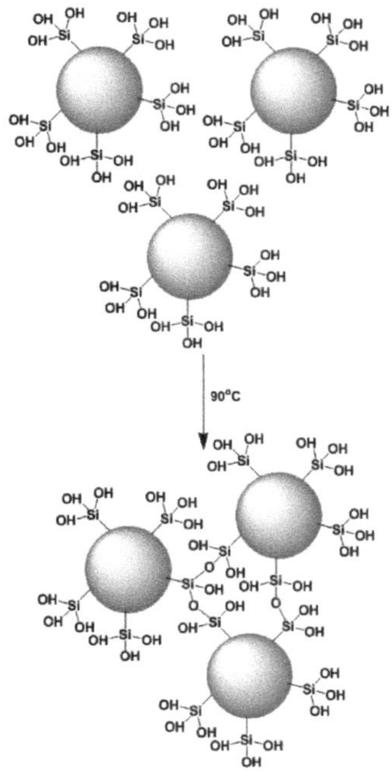

Figure 18. Formation of honeycomb-like PAMAMOS networks from silyl dendrimer precursors (Reproduced with permission from [133,134]; Copyright WILEY-VCH Verlag GmbH & Co., 2006; American Chemical Society, 2002).

Hybrid, silica-containing hyperbranched polymers can be also be produced by the aid of PEIs (MW 5000 or 25,000) and 3-(triethoxysilyl)-propylisocyanate as the silica precursor in a *N*,*N*-Dimethylformamide (DMF) or chloroform solution. The crosslinking density remarkably relies on ethoxysilylated polymer end groups reactivity; hydrolysis and concomitant polycondensation rates, that ultimately drive the architecture of the silica-PEI network. Hybrid products can be exploited, either as they are or impregnated/covalently bound to another appropriate ceramic substrate. They also offer the option of pyrolytic dendritic template removal, affording nanoporous silica powders with advanced textural characteristics. Nanomaterials produced by sol/gel processes display similar morphology characteristics, such as surface areas varying between 300 and 500 $m^2 \cdot g^{-1}$ and a pore diameters of around 2 nm [136]. Cohydrolysis and polycondensation is not limited to dendritic polymers containing solely silicon heteroatoms. Phosphorus dendrons of first, second, and third generation bearing ethoxy silane groups at their apex cross-link with tetraethoxy silane (TEOS) leads to silica xerogels with adjustable textures too [137]. Very recently it has been discovered that even "traditional" (i.e., not containing silicon) hyperbranched PEIs form nanocomplexes via hydrogen bonding with orthosilicic acid produced by the acid hydrolysis of TEOS. These nanostructures grow faster at neutral pH (7.5) and substantially slower at a lower pH (pH 5) and transform to hydrogels. When dried they became organoceramic xerogels. Active ingredients may be incorporated into the dendritic cavities at any step of the process. In the framework of a multinational European Project (Novel Marine Biomolecules Against Biofilm Application to Medical Devices: NoMorFilm) this procedure was tested on titanium microporous coating of a stainless-steel model orthopaedical implants with excellent results [138].

2.5. Biomimetic Reactions

Evolution has equipped living organisms with advanced self-constructing capabilities in order to secure better survival possibilities. Self-assemblage biogenic mechanisms are introduced for the development of complex organizations. The process where biological molecules activate and control independently a cellular pathway leading to the construction of metal nanoparticles is called biomineralization [139]. Biosilicification, the analogous procedure for biogenic formation for silica-based materials, occurs naturally in many microorganisms too, either terrestrial (higher plants) or aquatic ones (diatoms, sponges). It proceeds rapidly, under mild aqueous conditions, such as ambient temperature, pressure and neutral pH [140]. Biogenic silica occurs in a great diversity of nanopatterned frameworks, with varying sizes, shapes and organizations that are species-specific and malleable in nature, all deriving from the single reaction of silicic acid condensation [141]. This "manipulating siloxane bridge formation" advantage is attributed to the mediation of several biopolymers [142,143]. Specifically, it is established that post-translationally modified polycationic peptides, such as silaffins and R5 peptides, both isolated, from diatom cell walls, precipitate silica nanospheres almost instantly and in a peptide concentration-dependent fashion. There are numerous subsequent studies that successfully used conventional synthetic substitutes of biomolecules to imitate the evolution of biocomposites. For instance, a variety of homopeptides such as poly(arginine) and poly(lysine) [144] and a combination of triethanolamine and cetyltrimethylammonium chloride [145] can serve as biomimetic templates, forming silica nanoparticles with controlled specifications.

Amine-terminated dendrimers of various generations constitute exceptional templates that stand out for their similarity to proteins and tunable design that is primarily dictated by the high concentration of well-localized nitrogen functionalities, which resemble the units found in their biological counterparts and are responsible for silica nanospheres production [142]. For instance, poly(amidoamine) PAMAM and poly(propylene imine) PPI represent popular matrices for the formulation of nanopatterned silica (Figure 19) and alumina (Figure 20) [146]. The silica ratio of the final aggregated nanoproduct depends highly on the number of amino groups of the selected dendrimer. Knecht et al. described biomimetic sol-gel reactions as a powerful method to produce PAMAM/PPI-Si nanospheres. The sol-gel reaction proceeds at room temperature through the addition of metastable silicic acid into the dendrimer solution. Use of a phosphate buffer is required to obtain a pH of 7.5. Organic (dendritic) scaffolds are encapsulated into the gradually forming silica nanospheres and then the hybrids precipitate from a water solution in a way similar to that described before for the case of bioactive peptides. The outcome of the reaction and the production of nanospheres with distinct and specific dimensions hangs on the appropriate selection of dendritic reagents, with proper amine concentration suitable conditions and the presence of inorganic salts [143]. Given the fact that hydroxyl-terminated PAMAM analogues do not cause silica precipitation, it is ascribed exclusively to the presence of amino groups.

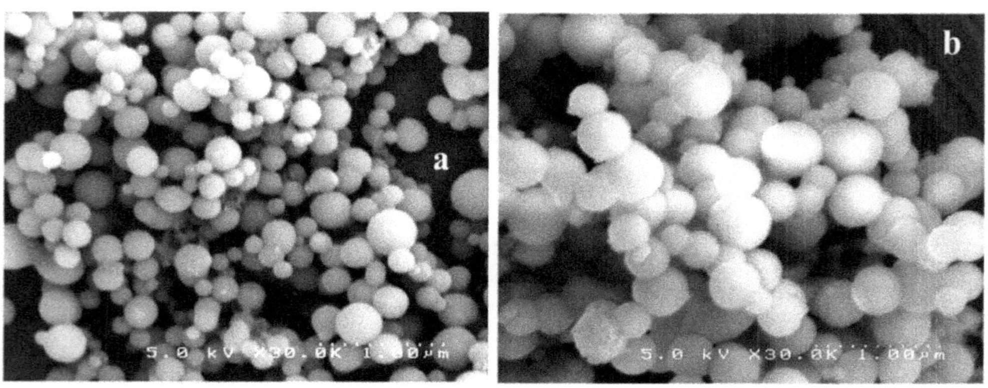

Figure 19. SEM micrographs of silica nanospheres, biomimetically templated by Amine-Terminated Dendrimers (**a**) G4 PPI and (**b**) G4 PAMAM (Reproduced with permission from [142]; Copyright American Chemical Society, 2004).

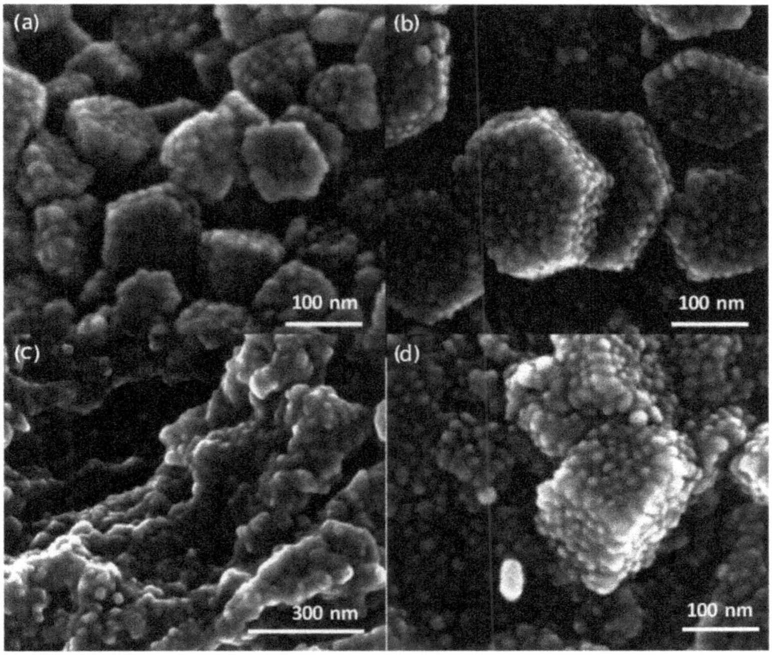

Figure 20. SEM micrographs of alumina nanospheres templated by hyperbranched PEI for different quantities of nano-dispersed boehmite ((**a**) 0.02, (**b**) 0.04, (**c**) 0.06, and (**d**) 0.08 g/mL) (Reproduced with permission from [146]; Copyright The American Ceramic Society, 2018).

Poly(amidoamine), PAMAM (G5.5) anionic starburst dendrimers functionalized with surface carboxylate groups have also been used as templates for the synthesis of hybrid nanocrystalline ZrO_2, CeO_2 and Y_2O_3 [147]. These ceramic powders were obtained through hydrolysis of the methylester-terminated dendrimers in the presence of stoichiometric NaOH volumes. The selection of the highest possible dendrimer generation is essential for the inhibition of extensive crystallization and the control of the final morphology of the nanoparticles. Thermal treatment at 600 °C for at least 2 h was a necessary step for the shaping of the amorphous nanoproduct into highly ordered spherical nanocrystalline powders. As long as the encapsulated dendritic polymers and their optional active guests in their internal cavities retain their properties, the resulting nanomaterials may be applied in various processes, including environmental remediation and catalysis. Nanospheres obtained through non-symmetrical analogues serve as inexpensive alternatives for both organic and inorganic pollutant sorption and water decontamination. Hyperbranched Poly(ethylene-imines) were proven ideal matrices as they caused silica nanosphere precipitation even in the absence of phosphate buffer, albeit to a noticeably lesser degree. The resulting hybrids removed toxic heavy metal ions (Pb^{2+}, Cd^{2+}, Hg^{2+}, $Cr_2O_7^{2-}$) and polycyclic aromatic hydrocarbons (pyrene, phenanthrene) from water [148]. For the attainment of catalytic activity, templated metal nanoparticle nucleation occurs by reduction of trapped ions prior to inorganic shell organization. Composites containing Au and Ag successfully catalyzed 4-nitrophenol reduction oxidation of methylene blue and selective oxidative transformation of benzyl alcohol to benzaldehyde [149].

An evolution of the above procedure proposes the fusion of two different biomimetic processes, biosilicification and biomineralization in one pot synthesis procedure. Both syntheses are regulated by the same hyperbranched polymer in the role of protein substitute [150]. In this way, silica-PEI-silver nanoparticle composites have been assembled, retaining antibacterial activity, catalytic properties and protracted release profile of the respective metal-dendritic polymer complexes. The synthesis of these highly compartmentalized ceramic-organic metal nanocomposites is attained via two successive steps that use the same functional scaffold: PEI Mr = 25,000 (Figure 21). At first, silver ions absorbed into the PEI pockets undergo reduction to metallic silver (core), then silica precipitation occurs from the silicification in the periphery of the dendritic polymer (shell). The bactericide activity of these hybrid nanoparticles is of particular interest. *E. coli, P. aeruginosa* and *S. aureus* assays established that Si-PEI-Ag nanoparticles efficiently prevented their proliferation, consistent with the concept of conservation of the complexes features despite their enclosure into silica capsules.

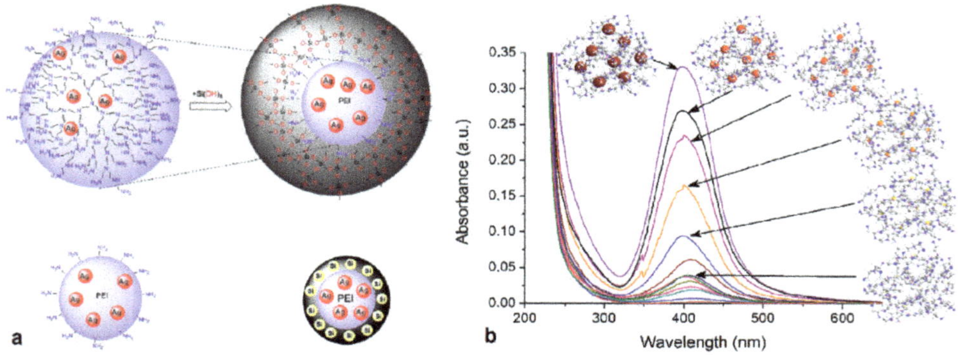

Figure 21. (**a**) Schematic representation of hybrid silica-PEI-Ag nanoparticles; (**b**) UV-Vis spectra of silica-PEI-Ag nanoparticle suspension in different time periods (Reproduced with permission from [150]; Copyright Elsevier, 2018).

Bearing in mind the large impact of electrostatic forces between dendrimers and silica precursors, a similar influence in the size and shape of the biomimetically produced silica particles is rationally expected. Ionic strength plays a pivotal role. In the case of PAMAM, for example, for all generations (G1–G6), it has been established that silica corpuscles get bigger with augmenting phosphate concentrations (Figure 22a) [143]. On top of that, at pH 7.5, when the phosphate buffer was replaced with individual alkali metal-chlorine solutions, SEM analysis unveiled a similar linear increase in particle diameter with increasing concentrations up to 100 mM (Figure 22b). Smaller cations up to K^+ generated typically larger composites, most probably due to preferential interaction with a single silanol group. In contrast, larger cations tend to affect multiple silanols, provoking imperfect surface coverage and limited charge neutralization. These results highlight further the role of neutralizing agents on the growth of silica nanospheres, i.e., securing as minimal as possible electrostatic repulsions and as much as possible free space for unhindered agglomeration.

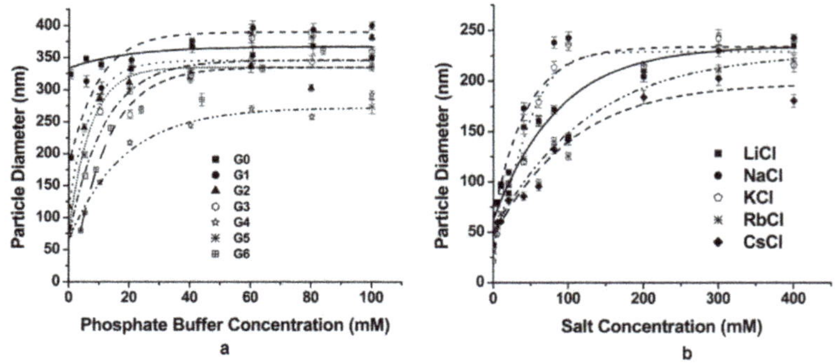

Figure 22. (**a**) Particle size of nanospheres produced by different PAMAM generations as a function of phosphate buffer concentration (**b**) as a function of salt concentration for different alkali metal chlorides (Reproduced with permission from [143]; Copyright American Chemical Society, 2005).

Dendritic scaffolds are likely to deliver many roles in the fulfillment of the silica condensation reaction. Generally, polypeptides, polyamines and polyamidoamines are partially protonated (NH_3^+ groups) at neutral pH, suggesting that an equilibrium is achieved between the opposing charges of the template and the surface of developing silica [151–153]. This neutralization results in the creation of an extensive network with NH_3^+, Si–O^- interacting groups and solubilizing water molecules. These interactions also drive the incorporation procedure of the matrices within the silica nanospheres. Overall, silica PEI-nanospheres display a negative potential at low pH that reaches the value of nearly −80 mV at pH 11, as indicated by zeta-potential measurements. In addition, the isoelectric point (IEP) after pyrolysis is nearly 2, similar to the IEPs typically encountered in the case of amorphous silica. The presence of the dendritic core in the nanospheres shifts the IEP towards higher pH values. Indeed, the hybrid organic-inorganic nanospheres show an IEP of roughly 8, persistent with the presence of positively charged amino groups. These hybrid nanocomposites present a positive surface potential of nearly +50 mV at pH 2, that decreases to about −70 mV at pH 11.

Besides creating novel ceramic nanostructures, dendritic polymers can serve as crystal modifiers for the biomimetically controlled production of nanoscale biomaterials. The development of these nanoproducts has attracted increasing interest recently, especially in the field of drug delivery systems and biomolecule expression. Aquasomes, for instance, are nanocrystalline ceramics with flexible surfaces that can be appropriately modified with carbohydrates in order to interact with specialized proteins and pharmaceuticals and to transfer them to tissue-specific areas [154,155]. An interesting relevant study attempts to fabricate spherical hydroxyapatite nanocores (Figure 23) by employing carboxylic acid-terminated PAMAM of 3.5 and 4.5 generations as starting materials for the production of hemoglobin-carrying aquasomes [156]. Noteworthily, hydrothermal treatment with generation 5.5 PAMAM drives the production of highly crystalline hydroxyapatite nanorods [157]. In general, dendritic PAMAMs, due to amido groups in their external surface, induce hydroxyapatite nanoparticles with an adjustable size and shape according to their generation through a hydrothermal crystallization. Increasing PAMAM generation number from G1.0 to G4.0 causes a decrease in the particle size from 82 to 38 nm respectively. In parallel, a transition occurs in the resulting shape from rod-like to ellipsoid-like [158]. Analogous effect is observed with increased concentrations of PAMAM G4.0. On the other hand, their polyhydroxy-substituted analogs in most cases formulate elliptical particles with an average grain size of about 20 nm and rarely short nano-rods with an average length of about 30 nm and an average width of about 25 nm [159].

Cationic fourth generation diaminobutane poly(propylene imine) dendrimers (DAB) strongly favor the formation of hydroxyapatite over all the other possible calcium phosphate crystalline phases [160]. The crystallization process may be regulated by introducing different thermal conditions and dendrimer:calcium ratios (Figure 24). The resulting nanoparticles display a characteristic elongated hexagonal rod-like configuration, with dimensions that are largely dictated by the particular reaction conditions. For example, increasing the temperature of hydrothermal processing from 80 to 130 °C leads to a noticeable lengthening, while maintaining room temperature conditions induces a fine nanostructure with distinctive mean length and width values (i.e., 11 and 5 nm). The presence of DAB dendrimers plays a vital role in the homogeneity of the developing hydroxyapatite nanocrystals, as well as in energy conservation and cost reduction of the process. Conventional hydroxyapatite crystal generation in the absence of dendritic mediators requires high temperatures (i.e., 130 °C) and exhibits a highly inhomogeneous morphology with broad particle size distribution.

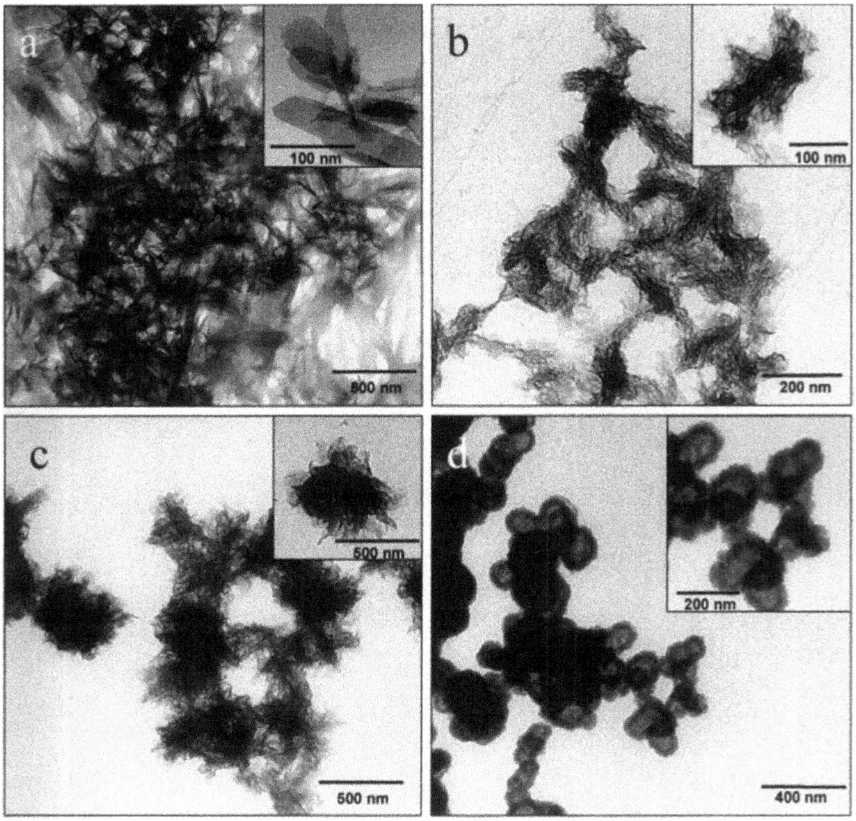

Figure 23. TEM image of hydroxyapatite prepared by self-precipitation from SBF. (**a**) Direct precipitation from SBF at pH 9.0 in absence of dendrimer. (**b**) Self-precipitation in the presence of 4.5 generation PAMAM dendrimer from SBF at pH 7.4. (**c**) Self-precipitation in the presence of 3.5 generation PAMAM dendrimer from SBF at pH 8.5. (**d**) Self-precipitation in the presence of 3.5 generation PAMAM dendrimer from SBF at pH 8.0. Insets in all figures show typical hydroxyapatite particle shape and morphology under magnification (Reproduced with permission from [146]; Copyright The American Ceramic Society, 2018).

Hyperbranched PEI bearing long alkyl chains, when mixed with single-chain surfactants such as octadecylamine, haxedecyl trimethylammonium bromide and sodium dodecyl sulfate afford a well-defined dendrimer/HAP biocomposite that mimics principal structural and mechanical properties of bones, such as plastic deformation and high stiffness and strength. The selected surfactant clearly affects not only the size and shape of the resulting hydroxyapatite composite, but also its mechanical idiosyncrasy, with octadecylamine and haxedecyl trimethylammonium bromide yielding composites exhibiting the highest toughening [161].

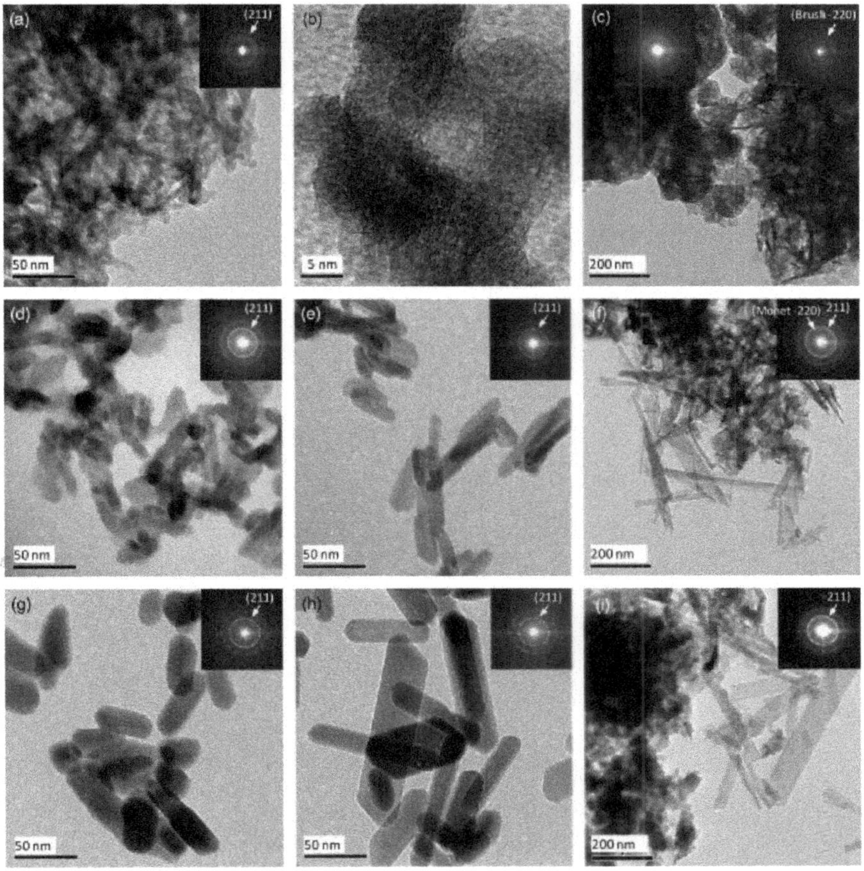

Figure 24. TEM images and respective selected area electron diffraction patterns of nanoparticles obtained under different conditions. Upper row for samples prepared at room temperature: (**a**) HAP dendrimer:calcium 1:1, (**b**) high resolution image, and (**c**) control (without DAB); middle row for samples hydrothermally treated at 80 °C: (**d**) HAP dendrimer:calcium 2:1, (**e**) HAP dendrimer:calcium 1:2, and (**f**) control; lower row for samples hydrothermally treated at 130 °C: (**g**) HAP dendrimer:calcium 2:1, (**h**) HAP dendrimer:calcium 1:2, and (**i**) control (Reproduced with permission from [160]; Copyright The American Ceramic Society, 2011).

3. Improvement Strategies for Performance Optimization

As discussed above, the performance of filters in applications such as water purification is directly proportional to the impregnation percentage of the dendritic polymer. The latter is governed by the particular pore architecture of the ceramic by the pH and the ionic strength of the immersion solution but mainly by its concentration [87,162,163]. A common emerging problem during the impregnation process is the rapid pore blocking which is aggravated abruptly with viscous dendrimeric derivatives as concentration of the coating solution raises [61], due to the intrinsic spongy labyrinthine configurations of the porous structures. Another issue encountered at denser immersion solutions is the structural breaks/cracks that mark the final composite product, and can undermine the filter absorption performance. These deformations originate from the stress applied during the drying process. Smoother, flawless, ceramic surfaces with optimum absorption properties (Figure 25) can be attained through a simple multiple immersion technique to solutions of gradually increasing concentrations. This method combines both adequate polymer loading with controllable porosity decrease, and avoids pore blocking and all the other disadvantages of immersion in concentrated media.

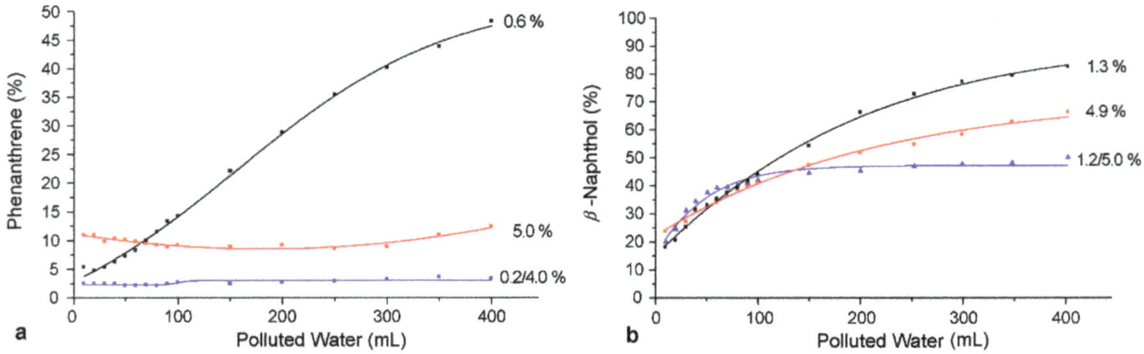

Figure 25. Percentage of phenanthrene (**a**) and β-naphthol (**b**) remaining in water in continuous filtration through filters impregnated by different percentages of octyl-substituted hyperbranched PEI employing a single or a two-step immersion stage (Reproduced with permission from [61]; Copyright Elsevier, 2008).

The involvement of dendritic polymers in the evolution of original nanoscale species with optimized characteristics is raised continuously. Their full biomimetic potential is exploited in parallel. Knecht et al. were the first to introduce pH regulating agents, such as phosphates or HCl, as a means to achieve optimal ionic strength and pH stabilization, as well as to control the final nanosphere dimensions [142]. The idea of double biomimicry gave rise to the synthesis of more complex inorganic/organic/inorganic nanostructures that generally exhibit combinatorial, enhanced properties [150]. These hybrid materials usually carry a dendrimeric matrix stuffed by metal nanoparticles and surrounded by a ceramic shell. Each single biomimetic reaction is successively and independently carried out in order to retain the full properties of the two constituents. It is possible to micromanage the specifications of the final hybrid product, such as the size and shape of both inner metal and outer ceramic particles, the overall composition and the porosity and pore dimensions of the inorganic segment, by adjusting either the biomineralization incubation time or the ratios of the silicification precursor towards the dendritic matrix or the molecular weight of the latter. For example, higher molecular weight of the organic core causes faster formation of metal nanoparticles and optimal catalytic performance of the final product. Thus, biomimetic process optimization can offer unique hybrid solutions with both combinatorial and manageable structural characteristics, evolving nanomaterial synthesis to a whole new level.

Numerous factors can perturb the perfectly symmetrical growth process of the dendritic polymer onto the ceramic surface, affecting the overall properties of the final nanoproduct. For example, side reactions usually co-occur with the propagation pathway of PAMAM on the silica surface, inducing thus major structural diversifications such as dimer and intramolecular cyclization (Figure 26a) [96]. Specifically, a high density of trihydroxy aminopropyl silane molecules bond to the ceramic surface leads to an equally high concentration of precursor- "dendron core" amino groups, which in turn cause serious formational defects. Lower density of dendron-generating amino groups leads to a decrease in steric hindrance, as well as an increase in the in-between distance of secondary amino groups of the dendritic branches, thus preventing the formation of undesired inter-dendron and intra-dendron cross-linked structures in higher generations (Figure 26b). This effect can be successfully avoided through the co-immobilization of 3-(triethoxysilyl) propionitrile on silica, which serves the role of the "spacer", reduces the "dendron-core" amino group density and thereby improves the whole evolution procedure.

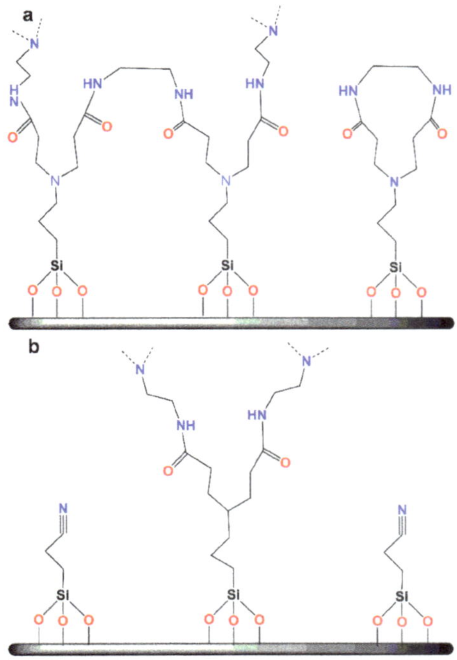

Figure 26. (**a**) Common structural defects occurring during the dendron propagation reaction; (**b**) introduction of triethoxysilyl propylonitrile as a spacer for the elimination of cross-linking structural defects (Reproduced with permission from [96]; Copyright American Chemical Society, 2004).

The non-uniform shape and irregular size of the pores are two other factors contributing to the steric/crowding effect. A prominent solution includes the exploitation of silica bearing highly ordered, hierarchical channels (e.g., hexagonally packed) that shows narrow size distribution, such as MCM-41, and SBA-15 [164,165]. These afford dendron development up to fourth or higher generations with negligible structural defects.

4. Concluding Remarks—Perspectives

Hybrid dendritic/ceramics materials offer a multitude of perspectives in a variety of interdisciplinary fields replacing traditional, non-functionalized materials that usually suffer from structural inhomogeneity, absorption instability and thus limited applicability [166,167]. The use of dendritic polymers as additives is now widely accepted. In fact, there are several studies substantiating their superiority as stabilizing agents that do not undermine the physicochemical properties of the guest molecules. Some applications covered herein are summarized in Table 1. A plethora of perspectives are envisaged. For instance, PAMAM polymers have been used as stabilizers of semiconductor quantum dots, CdS synthesized at the nanometer-scale [168]. The stabilization effect lies in the effective solvation of these highly insoluble nanoclusters, avoiding precipitation, which occurs in the absence of PAMAM. Dendritic polymers also stabilize colloidal suspensions of gold by in situ reduction of its ions encapsulated into their cavities [169]. Silver nanoparticles may be obtained too through a reduction of cations encapsulated into dendrigrafts, even without a reductant [170]. Optical and structural characteristics of these nanoparticles, such as UV-Vis absorption spectrum, eventual light emission, size and shape are determined by the chemical nature of the dendritic matrix and the host/guest concentrations ratio. Dendritic polymers are furthermore well-known for their ability to deliver an immense variety of colored metal complexes. The composites with ceramics retain this characteristic, along with luminescence properties according to the metal particle conformation. These hybrid sensors can be particularly useful in the detection of explosive chemical warfare agents, as well as other substances, even at the smallest concentrations.

Table 1. General applications of ceramics/dendritic polymer composites.

Dendritic Polymer-Ceramic	Application	Reference
PPI-SiO_2	Electrochemical (Redox-Active Materials)	[65]
PAPAM-PDMS	High-Resolution Contact Printing	[68]
Benzyl-Terminated Dendrons-SiO_2	High-Resolution Capillary Gas Chromatography	[82]
Benzyl-Terminated Dendrons-SiO_2	Capillary microextraction	[83]
L-glutamic acid-Chiral-Dendrimers-SiO_2	HPLC	[90]
PAMAM-SiO_2	Photoluminescence Sensors	[91]
PAMAM-TiO_2	TiO_2 Nanoparticle Stabilization	[92]
PAMAM-SiO_2	Size Exclusion Chromatography	[108]
Melamine Dendrons-Al_2O_3	Polarity-Based Separation	[130]
PAMAMOS	Electronics, Photonics, Magnetics, Sensors, Coatings	[133–135]

Moreover, these metal organoceramic nanocomposites are noteworthy candidates for catalysis reactions [171–173] (Table 2). They are particularly useful when gas phases are involved in reactions such as low molecular hydrocarbon hydrogenation and epoxidation. Ceramics may act not only as solid supports but also local reactant concentration enhancers, mostly through electrostatic attractions. Moreover, when both components exhibit photocatalytic properties, an enhanced performance is achieved. In some cases, the incorporation of the dendritic polymers increases the porosity of the inorganic supports to the point that they approximate the kinetics of homogenous catalysis. The same phenomenon occurs when inorganic nanoparticles are coated by dendrons. Concurrently, they allow easy and rapid retrieval regeneration and reuse.

Table 2. Applications of ceramics/dendritic polymer composites in catalysis.

Dendritic Polymer-Ceramic	Application	Reference
PEI-SiO_2-TiO_2	Nitro-compounds Reduction	[29]
CBS-SiO_2	Ethylene Polymerization	[79]
Pyridylphenylene dendrons-SiO_2	Suzuki cross-coupling	[86]
Chiral-PAMAM-SiO_2	Enantiomeric Reactions	[87]
DAB-SiO_2	Selective Hydrogenation	[88]
PAMAM-SiO_2	Alcohol Oxidation	[113]
PAMAM-SiO_2	Olefin Hydroformylation	[114–116]
PAMAM-SiO_2	Carbonylation	[117–119]
PAMAM-SiO_2	Hydroesterification	[120]
PAMAM-SiO_2	Heck reaction	[121]
PAMAM-SiO_2	Olefin Epoxidation	[122]
PAMAM-SiO_2	Enzymolysis	[123]
PAMAM-SiO_2	Benzyl Alcohol, Methylene Blue Oxidation, Nitrophenol Reduction	[149]
DAB-SiO_2	Phenols Hydrogenation	[171]
PEI-SiO_2	CO, NO, CH_4, C_3H_6, C_3H_8 Oxidation	[172]
PEI-SiO_2-CeO_2	Nitrophenol Reduction	[173]

Besides, due to their antimicrobial attributes against a wide range of bacteria, viruses, fungi parasites and other microorganisms, they may serve as leather or textile disinfection purposes [150]. Decontamination devices may also benefit from the implementation of these ceramic-organic-metal ternary systems. Thanks to their combined adsorption (Table 3) and microbicide capacity, they can be used for water treatment processes. In contrast to other absorbents, they are immune to bacterial infections and, apart from absorbing toxic pollutants, they may provide in parallel a certain level of sterilization.

Table 3. Adsorption properties of ceramics/dendritic polymer composites.

Dendritic Polymer-Ceramic	Application	Reference
DAB, PEI, PG-Al_2O_3, TiO_2 SiC	PAH, BTX, THM, MTBE, Pesticides	[58,59]
PEI-Al_2O_3	PAH	[61]
PAMAM-SiO_2	$Fe_4[Fe(CN)_6]_3$, $Co_2[Fe(CN)_6]$	[62]
PAMAM-TiO_2	Cu(II), Ni(II), Cr(III)	[67]
PAMAM-SiO_2	Hg(II)	[80,109,110,112]
PPH-SiO_2	CO_2	[84]
PEI-SiO_2	Cu(II), Zn(II), Cd(II)	[85]
PAMAM-SiO_2	Cu(II)	[94]
Melamine Dendrons-SiO_2	Atrazine	[95]
PAMAM-Dendrons-SiO_2	Organic Dyes, Amphiphilic Surfactants	[97]
PAMAM-SiO_2	Zn(II), Co(II)	[98]
PAMAM-SiO_2	Mn(II)	[99]
PAMAM-SiO_2	Ni(II)	[100]
PAMAM-SiO_2	Co(II)	[101]
PAMAM-SiO_2	Cd(II)	[102,103]
PAMAM-SiO_2	Fe(III)	[103,104]
PAMAM-SiO_2	Pb(II)	[105]
PAMAM-SiO_2	U(IV)	[106]
PAMAM-SiO_2	Ag(I)	[111,112]
PEI-SiO_2	CO_2	[125,126]
TREN-SiO_2	CO_2	[129]
PEI-SiO_2	PAH, Pb(II), Hg(II), Cd(II), Cr(VI)	[148]

Another compelling aspect of dendro-ceramics is that they can be employed in nanomedicine (Table 4) as controlled DNA expression vehicles, emulating the modularity model that viruses follow in order to target distinct regions of the cell [174,175]. Silica nanoparticles modified by aminopropyltriethoxysilane (APES) and methyl triethoxysilane (MTES) can generate ternary complexes with DNA and dendrimers. The dense configuration of silica nanoparticles accelerates DNA accumulation into the surface of the cells and subsequent DNA endosomal-lysomal uptake, leading to better cell transfection levels. A biogenic silica system produced by a complex of DNA and a dendrimer bearing an amino group could simplify the formation of the silica nanoparticles and optimize the overall transfection process, as long as this ternary complex remains non-toxic.

Table 4. Ceramics/dendritic polymer composites applications in nanomedicine.

Dendritic Polymer-Ceramic	Application	Reference
PEI-HA	Hydroxy apatite formation	[30]
PAMAM-SiO_2	Gene therapy	[63]
PAMAM, CBS-SiO_2	Antibacterials	[77]
CBS-SiO_2	Oligonucleotide Delivery Carriers	[78]
PAMAM-SiO_2	Fluorescence Imaging	[81]
PPH-SiO_2	Antibacterials	[89]
PAMAM-SiO_2	Anticancer Photothermal Therapy	[124]
PEI-SiO_2	Orthopedic Implants	[138]
PEI-SiO_2	Antibacterials	[150]
PAMAM-HA	Hemoglobin Aquasomes	[156]
PAMAM-SiO_2	Gene transfection	[175]

Last but not least, drug delivery systems based on hybrid ceramics bearing dendritic polymers is a hot topic as well (Table 5). There is an established potential for controlled protracted and stimuli triggered release as well as targeted therapy [176]. Shortcomings of dendritic polymer carriers are mainly an undesired response by the immune system [177], rapid clearance from blood circulation due to small size [178] and toxicity of cationic high Mw species [179]. They may be addressed by charge neutralization and magnification, resulting in the combination with silica. On the other side, there are biocompatibility issues on silica nanoparticle forms, such as toxicity due to aggregation or hemolytic activity. These are alleviated either by the formation of the composites or by additional coating by a biocompatible polymer [64].

Table 5. Applications of ceramics/dendritic polymer composites in drug delivery.

Dendritic Polymer-Ceramic	Application	Reference
PAMAM-SiO_2	Doxorubicin, Curcumin	[64]
DAB-SiO_2	Ibuprofen	[74]
DAB-SiO_2	Levofloxacin	[75]
DAB-SiO_2	Redox-responsive release	[76]
PAMAM-SiO_2	Curcumin	[81]
PAMAM-SiO_2	Doxorubicin and Bcl-2 targeted siRNA	[124]

Author Contributions: Conceptualization, M.A.; Data Curation, M.D., E.N., and T.K.; Formal analysis, M.D., E.N., and T.K.; Scientific Literature Investigation, M.D., E.N., and T.K.; Methodology, M.A. and M.D.; Project administration, M.D. and M.V.; Resources, M.A., M.D., E.N. and T.K.; Supervision, M.A.; Visualization, M.D., E.N., and T.K.; Writing—original draft, M.D.; Writing—review and editing, M.A., M.D., and M.V.; Funding Acquisition, M.A., E.N., and M.V. All authors have read and agreed to the published version of the manuscript.

Funding: This work was co-financed by Greece/Greek General Secretariat for Research and Technology and European Union under the frame of EPAnEK 2014–2020 Operational Programme Competitiveness, Entrepreneurship Innovation, project "MEDNANOLEAT" grant number T6YBΠ-00081.

Conflicts of Interest: The authors declare no conflict of interest.

References

1. Camargo, P.H.C.; Satyanarayana, K.G.; Wypych, F. Nanocomposites: Synthesis, structure, properties and new application opportunities. *Mater. Res.* **2009**, *12*, 1–39. [CrossRef]
2. Parveen, S.; Misra, R.; Sahoo, S.K. Nanoparticles: A boon to drug delivery, therapeutics, diagnostics and imaging. *Nanomed. Nanotechnol. Biol. Med.* **2012**, *8*, 147–166. [CrossRef] [PubMed]
3. Khan, I.; Saeed, K.; Khan, I. Nanoparticles: Properties, applications and toxicities. *Arab. J. Chem.* **2019**, *12*, 908–931. [CrossRef]
4. Niihara, K. New design concept of structural ceramics. Ceramic nanocomposites. *J. Ceram. Soc. Jpn. Int. Ed.* **1991**, *99*, 945–952. [CrossRef]
5. Kiani, A.; Rahmani, M.; Sivakumar, M.; Tan, B. Nanoceramics: Synthesis, characterization, and applications. *J. Nanomater.* **2014**, *2014*. [CrossRef]
6. Thomas, S.; Harshita, B.S.P.; Mishra, P.; Talegaonkar, S. Ceramic Nanoparticles: Fabrication Methods and Applications in Drug Delivery. *Curr. Pharm. Des.* **2015**, *21*, 6165–6188. [CrossRef]
7. Meulenberg, W.A.; Schulze-Küppers, F.; Deibert, W.; Gestel, T.V.; Baumann, S. Ceramic Membranes: Materials–Components–Potential Applications. *ChemBioEng Rev.* **2019**, *6*, 198–208. [CrossRef]
8. Benfer, S.; Árki, P.; Tomandl, G. Ceramic membranes for filtration applications—Preparation and characterization. *Adv. Eng. Mater.* **2004**, *6*, 495–500. [CrossRef]
9. Acharya, M.; Foley, H.C. Spray-coating of nanoporous carbon membranes for air separation. *J. Membr. Sci.* **1999**, *161*, 1–5. [CrossRef]
10. de Lange, R.S.A.; Keizer, K.; Burggraaf, A.J. Analysis and theory of gas transport in microporous sol-gel derived ceramic membranes. *J. Membr. Sci.* **1995**, *104*, 81–100. [CrossRef]
11. Montanaro, L.; Saracco, G. Influence of some precursors on the physico-chemical characteristics of transition aluminas for the preparation of ceramic catalytic filters. *Ceram. Int.* **1995**, *21*, 43–49. [CrossRef]
12. Bodzek, M.; Konieczny, K. Comparison of ceramic and capillary membranes in the treatment of natural water by means of ultrafiltration and microfiltration. *Desalination* **1998**, *119*, 191–197. [CrossRef]
13. Bodzek, M.; Konieczny, K.; Rajca, M. Membranes in water and wastewater disinfection—Review. *Arch. Environ. Prot.* **2019**, *45*, 3–18. [CrossRef]
14. Russo, C. A new membrane process for the selective fractionation and total recovery of polyphenols, water and organic substances from vegetation waters (VW). *J. Membr. Sci.* **2007**, *288*, 239–246. [CrossRef]
15. Wang, P.; Xu, N.; Shi, J. A pilot study of the treatment of waste rolling emulsion using zirconia microfiltration membranes. *J. Membr. Sci.* **2000**, *173*, 159–166. [CrossRef]

16. Yang, C.; Zhang, G.; Xu, N.; Shi, J. Preparation and application in oil-water separation of ZrO_2/α-Al_2O_3 MF membrane. *J. Membr. Sci.* **1998**, *142*, 235–243. [CrossRef]
17. Nair, P.; Mizukami, F.; Okubo, T.; Nair, J.; Keizer, K.; Burggraaf, A.J. High-temperature catalyst supports and ceramic membranes: Metastability and particle packing. *AIChE J.* **1997**, *43*, 2710–2714. [CrossRef]
18. Komolikov, Y.I.; Blaginina, L.A. Technology of ceramic ultrafiltration membranes. *Ogneup. Tekhnicheskaya Keram.* **2002**, *43*, 20–28.
19. Koo, O.M.; Rubinstein, I.; Onyuksel, H. Role of nanotechnology in targeted drug delivery and imaging: A concise review. *Nanomed. Nanotechnol. Biol. Med.* **2005**, *1*, 193–212. [CrossRef]
20. Moreno-Vega, A.I.; Gómez-Quintero, T.; Nuñez-Anita, R.E.; Acosta-Torres, L.S.; Castaño, V. Polymeric and ceramic nanoparticles in biomedical applications. *J. Nanotechnol.* **2012**, *2012*. [CrossRef]
21. Manzano, M.; Vallet-Regí, M. Mesoporous Silica Nanoparticles for Drug Delivery. *Adv. Funct. Mater.* **2020**, *30*, 3–5. [CrossRef]
22. Croissant, J.G.; Fatieiev, Y.; Almalik, A.; Khashab, N.M. Mesoporous Silica and Organosilica Nanoparticles: Physical Chemistry, Biosafety, Delivery Strategies, and Biomedical Applications. *Adv. Healthc. Mater.* **2018**, *7*, 1700831. [CrossRef] [PubMed]
23. Castillo, R.R.; Vallet-Regí, M. Functional mesoporous silica nanocomposites: Biomedical applications and biosafety. *Int. J. Mol. Sci.* **2019**, *20*, 929. [CrossRef] [PubMed]
24. Li, Z.; Mu, Y.; Peng, C.; Lavin, M.F.; Shao, H.; Du, Z. Understanding the mechanisms of silica nanoparticles for nanomedicine. *Wiley Interdiscip. Rev. Nanomed. Nanobiotechnol.* **2020**, *13*, e1658. [CrossRef]
25. Tang, F.; Li, L.; Chen, D. Mesoporous silica nanoparticles: Synthesis, biocompatibility and drug delivery. *Adv. Mater.* **2012**, *24*, 1504–1534. [CrossRef]
26. Croissant, J.G.; Fatieiev, Y.; Khashab, N.M. Degradability and Clearance of Silicon, Organosilica, Silsesquioxane, Silica Mixed Oxide, and Mesoporous Silica Nanoparticles. *Adv. Mater.* **2017**, *29*. [CrossRef]
27. Singh, D.; Singh, S.; Sahu, J.; Srivastava, S.; Singh, M.R. Ceramic nanoparticles: Recompense, cellular uptake and toxicity concerns. *Artif. Cells Nanomed. Biotechnol.* **2016**, *44*, 401–409. [CrossRef]
28. Vakifahmetoglu, C.; Zeydanli, D.; Colombo, P. Porous polymer derived ceramics. *Mater. Sci. Eng. R* **2016**, *106*, 1–30. [CrossRef]
29. Kitsou, I.; Panagopoulos, P.; Maggos, T.; Arkas, M.; Tsetsekou, A. Development of SiO_2@TiO_2 core-shell nanospheres for catalytic applications. *Appl. Surf. Sci.* **2018**, *441*, 223–231. [CrossRef]
30. Tsiourvas, D.; Tsetsekou, A.; Arkas, M.; Diplas, S.; Mastrogianni, E. Covalent attachment of a bioactive hyperbranched polymeric layer to titanium surface for the biomimetic growth of calcium phosphates. *J. Mater. Sci. Mater. Med.* **2011**, *22*, 85–96. [CrossRef]
31. Vallet-Regí, M.; Ruiz-Hernández, E. Bioceramics: From bone regeneration to cancer nanomedicine. *Adv. Mater.* **2011**, *23*, 5177–5218. [CrossRef] [PubMed]
32. El-Ghannam, A. Bone reconstruction: From bioceramics to tissue engineering. *Expert Rev. Med. Devices* **2005**, *2*, 87–101. [CrossRef] [PubMed]
33. Habraken, W.J.E.M.; Wolke, J.G.C.; Jansen, J.A. Ceramic composites as matrices and scaffolds for drug delivery in tissue engineering. *Adv. Drug Deliv. Rev.* **2007**, *59*, 234–248. [CrossRef] [PubMed]
34. Vallet-Regí, M.; Salinas, A.J. *Ceramics as Bone Repair Materials*, 2nd ed.; Elsevier Ltd.: Amsterdam, The Netherlands, 2019; ISBN 9780081024515.
35. Tomalia, D.A.; Fréchet, J.M.J. Discovery of dendrimers and dendritic polymers: A brief historical perspective. *J. Polym. Sci. Part A Polym. Chem.* **2002**, *40*, 2719–2728. [CrossRef]
36. Fréchet, J.M.J.; Tomalia, D.A. (Eds.) *Dendrimers and Other Dendritic Polymers*; John Wiley & Sons, Ltd.: Chichester, UK, 2001; ISBN 9780470845820.
37. Jaymand, M.; Lotfi, M.; Lotfi, R. Functional dendritic compounds: Potential prospective candidates for dental restorative materials and: In situ re-mineralization of human tooth enamel. *RSC Adv.* **2016**, *6*, 43127–43146. [CrossRef]
38. Svenson, S.; Tomalia, D.A. Dendrimers in biomedical applications-reflections on the field. *Adv. Drug Deliv. Rev.* **2012**, *64*, 102–115. [CrossRef]
39. Sheikhpour, M.; Barani, L.; Kasaeian, A. Biomimetics in drug delivery systems: A critical review. *J. Control. Release* **2017**, *253*, 97–109. [CrossRef]
40. Leiro, V.; Moreno, P.M.; Sarmento, B.; Durão, J.; Gales, L.; Pêgo, A.P.; Barrias, C.C. *Design and Preparation of Biomimetic and Bioinspired Materials*; Elsevier Ltd.: Amsterdam, The Netherlands, 2017; ISBN 9780081007464.
41. Fruchon, S.; Poupot, R. Pro-inflammatory versus anti-inflammatory effects of dendrimers: The two faces of immuno-modulatory nanoparticles. *Nanomaterials* **2017**, *7*, 251. [CrossRef]
42. Fan, Y.; Sun, W.; Shi, X. Design and Biomedical Applications of Poly(amidoamine)-Dendrimer-Based Hybrid Nanoarchitectures. *Small Methods* **2017**, *1*, 1700224. [CrossRef]
43. Yetisgin, A.A.; Cetinel, S.; Zuvin, M.; Kosar, A.; Kutlu, O. Therapeutic Nanoparticles and Their Targeted Delivery Applications. *Molecules* **2020**, *25*, 2193. [CrossRef]
44. Le, N.T.T.; Nguyen, T.N.Q.; Cao, V.D.; Hoang, D.T.; Ngo, V.C.; Thi, T.T.H. Recent progress and advances of multi-stimuli-responsive dendrimers in drug delivery for cancer treatment. *Pharmaceutics* **2019**, *11*, 591. [CrossRef] [PubMed]
45. Devnarain, N.; Osman, N.; Fasiku, V.O.; Makhathini, S.; Salih, M.; Ibrahim, U.H.; Govender, T. Intrinsic stimuli-responsive nanocarriers for smart drug delivery of antibacterial agents—An in-depth review of the last two decades. *Wiley Interdiscip. Rev. Nanomed. Nanobiotechnol.* **2020**, *13*, e1664. [CrossRef] [PubMed]

46. Majumder, J.; Minko, T. Multifunctional and stimuli-responsive nanocarriers for targeted therapeutic delivery. *Expert Opin. Drug Deliv.* **2020**, 1–23. [CrossRef]
47. Saravanakumar, K.; Hu, X.; Ali, D.M.; Wang, M.-H. Emerging Strategies in Stimuli-Responsive Nanocarriers as the Drug Delivery System for Enhanced Cancer Therapy. *Curr. Pharm. Des.* **2019**, *25*, 2609–2625. [CrossRef] [PubMed]
48. Shukla, R.; Singh, A.; Pardhi, V.; Sunil Dubey, P.K. Dendrimer (polyamidoamine, polypropylene imine, poly-L-lysine, carbosilane dendrimers, triazine dendrimers) as promising tool for anticancer therapeutics. In *Polymeric Nanoparticles as a Promising Tool for Anti-Cancer Therapeutics*; Wolff, A.G., Ed.; Elsevier Ltd.: Amsterdam, The Netherlands, 2019; pp. 233–255, ISBN 9780128169636.
49. Wang, B.Y.; Liao, M.L.; Hong, G.C.; Chang, W.W.; Chu, C.C. Near-Infrared-Triggered photodynamic therapy toward breast cancer cells using Dendrimer-Functionalized upconversion nanoparticles. *Nanomaterials* **2017**, *7*, 269. [CrossRef] [PubMed]
50. Shahbazi-Gahrouei, D.; Khaniabadi, P.; Khaniabadi, B.; Shahbazi-Gahrouei, S. Medical imaging modalities using nanoprobes for cancer diagnosis: A literature review on recent findings. *J. Res. Med. Sci.* **2019**, *24*, 38. [CrossRef] [PubMed]
51. Shi, X.; Wang, S.H.; Van Antwerp, M.E.; Chen, X.; Baker, J.R. Targeting and detecting cancer cells using spontaneously formed multifunctional dendrimer-stabilized gold nanoparticles. *Analyst* **2009**, *134*, 1373–1379. [CrossRef] [PubMed]
52. Li, H.; Sun, J.; Zhu, H.; Wu, H.; Zhang, H.; Gu, Z.; Luo, K. Recent advances in development of dendriticpolymer-based nanomedicines for cancer diagnosis. *Wiley Interdiscip. Rev. Nanomed. Nanobiotechnol.* **2020**, e1670. [CrossRef]
53. El Kadib, A.; Katir, N.; Bousmina, M.; Majoral, J.P. Dendrimer-silica hybrid mesoporous materials. *New J. Chem.* **2012**, *36*, 241–255. [CrossRef]
54. Gaikwad, A.V.; Boffa, V.; ten Elshof, J.E.; Rothenberg, G. Cat-in-a-Cup: Facile Separation of Large Homogeneous Catalysts. *Angew. Chem.* **2008**, *120*, 5487–5490. [CrossRef]
55. Safari, J.; Zarnegar, Z.; Sadeghi, M.; Enayati-Najafabadi, A. Dendritic macromolecules supported Ag nanoparticles as efficient catalyst for the reduction of 4-nitrophenol. *J. Mol. Struct.* **2016**, *1125*, 772–776. [CrossRef]
56. Arkas, M.; Tsiourvas, D.; Paleos, C.M. Organosilicon dendritic networks in porous ceramics for water purification. *Chem. Mater.* **2005**, *17*, 3439–3444. [CrossRef]
57. Arkas, M.; Tsiourvas, D.; Paleos, C.M. Functional dendritic polymers for the development of hybrid materials for water purification. *Macromol. Mater. Eng.* **2010**, *295*, 883–898. [CrossRef]
58. Arkas, M.; Allabashi, R.; Tsiourvas, D.; Mattausch, E.M.; Perfler, R. Organic/inorganic hybrid filters based on dendritic and cyclodextrin "nanosponges" for the removal of organic pollutants from water. *Environ. Sci. Technol.* **2006**, *40*, 2771–2777. [CrossRef]
59. Allabashi, R.; Arkas, M.; Hörmann, G.; Tsiourvas, D. Removal of some organic pollutants in water employing ceramic membranes impregnated with cross-linked silylated dendritic and cyclodextrin polymers. *Water Res.* **2007**, *41*, 476–486. [CrossRef] [PubMed]
60. Zhang, G.; Qin, Y.; Zhang, H.; Shang, Y.; Sun, M.; Liu, B.; Li, Z. Electronic structure-transport property relationships of polyferrocenylene, polyferrocenylacetylene, and polyferrocenylsilane. *J. Phys. Chem. C* **2010**, *114*, 9469–9477. [CrossRef]
61. Tsetsekou, A.; Arkas, M.; Kritikaki, A.; Simonetis, S.; Tsiourvas, D. Optimization of hybrid hyperbranched polymer/ceramic filters for the efficient absorption of polyaromatic hydrocarbons from water. *J. Membr. Sci.* **2008**, *311*, 128–135. [CrossRef]
62. Zamponi, S.; Kijak, A.M.; Sommer, A.J.; Marassi, R.; Kulesza, P.J.; Cox, J.A. Electrochemistry of Prussian Blue in silica sol-gel electrolytes doped with polyamidoamine dendrimers. *J. Solid State Electrochem.* **2002**, *6*, 528–533. [CrossRef]
63. Ainalem, M.L.; Campbell, R.A.; Nylander, T. Interactions between DNA and poly(amido amine) dendrimers on Silica surfaces. *Langmuir* **2010**, *26*, 8625–8635. [CrossRef]
64. Xu, X.; Lü, S.; Gao, C.; Bai, X.; Feng, C.; Gao, N.; Liu, M. Multifunctional drug carriers comprised of mesoporous silica nanoparticles and polyamidoamine dendrimers based on layer-by-layer assembly. *Mater. Des.* **2015**, *88*, 1127–1133. [CrossRef]
65. Díaz, I.; García, B.; Alonso, B.; Casado, C.M.; Morán, M.; Losada, J.; Pérez-Pariente, J. Ferrocenyl dendrimers incorporated into mesoporous silica: New hybrid redox-active materials. *Chem. Mater.* **2003**, *15*, 1073–1079. [CrossRef]
66. Arkas, M.; Tsiourvas, D.; Paleos, C.M. Functional dendrimeric "Nanosponges" for the removal of polycyclic aromatic hydrocarbons from water. *Chem. Mater.* **2003**, *15*, 2844–2847. [CrossRef]
67. Barakat, M.A.; Ramadan, M.H.; Alghamdi, M.A.; Algarny, S.S.; Woodcock, H.L.; Kuhn, J.N. Remediation of Cu(II), Ni(II), and Cr(III) ions from simulated wastewater by dendrimer/titania composites. *J. Environ. Manag.* **2013**, *117*, 50–57. [CrossRef] [PubMed]
68. Li, H.; Kang, D.J.; Blamire, M.G.; Huck, W.T.S. High-Resolution Contact Printing with Dendrimers. *Nano Lett.* **2002**, *2*, 347–349. [CrossRef]
69. Esumi, K.; Goino, M. Adsorption of poly(amidoamine) dendrimers on alumina/water and silica/water interfaces. *Langmuir* **1998**, *14*, 4466–4470. [CrossRef]
70. Tokarczyk, K.; Jachimska, B. Quantitative interpretation of PAMAM dendrimers adsorption on silica surface. *J. Colloid Interface Sci.* **2017**, *503*, 86–94. [CrossRef]
71. Ottaviani, M.F.; Turro, N.J.; Jockusch, S.; Tomalia, D.A. EPR investigation of the adsorption of dendrimers on porous surfaces. *J. Phys. Chem. B* **2003**, *107*, 2046–2053. [CrossRef]
72. Fu, S.; Zhu, M.; Zhu, Y. Organosilicon polymer-derived ceramics: An overview. *J. Adv. Ceram.* **2019**, *8*, 457–478. [CrossRef]
73. Ottenbrite, R.M.; Yin, R.; Zengin, H.; Suzuki, K.; Siddiqui, J.A. Surface Modification of Silica Particles and Silica Glass Beads Polymer Surface Modification of Particles. *Spec. Monomers Polym.* **2000**, *13*, 170–183.

74. González, B.; Colilla, M.; De Laorden, C.L.; Vallet-Regí, M. A novel synthetic strategy for covalently bonding dendrimers to ordered mesoporous silica: Potential drug delivery applications. *J. Mater. Chem.* **2009**, *19*, 9012–9024. [CrossRef]
75. González, B.; Colilla, M.; Díez, J.; Pedraza, D.; Guembe, M.; Izquierdo-Barba, I.; Vallet-Regí, M. Mesoporous silica nanoparticles decorated with polycationic dendrimers for infection treatment. *Acta Biomater.* **2018**, *68*, 261–271. [CrossRef] [PubMed]
76. Nadrah, P.; Porta, F.; Planinšek, O.; Kros, A.; Gaberšček, M. Poly(propylene imine) dendrimer caps on mesoporous silica nanoparticles for redox-responsive release: Smaller is better. *Phys. Chem. Chem. Phys.* **2013**, *15*, 10740–10748. [CrossRef] [PubMed]
77. Sánchez-Milla, M.; Gómez, R.; Pérez-Serrano, J.; Sánchez-Nieves, J.; de la Mata, F.J. Functionalization of silica with amine and ammonium alkyl chains, dendrons and dendrimers: Synthesis and antibacterial properties. *Mater. Sci. Eng. C* **2020**, *109*, 110526. [CrossRef] [PubMed]
78. Martínez, Á.; Fuentes-Paniagua, E.; Baeza, A.; Sánchez-Nieves, J.; Cicuéndez, M.; Gómez, R.; De La Mata, F.J.; González, B.; Vallet-Regí, M. Mesoporous Silica Nanoparticles Decorated with Carbosilane Dendrons as New Non-viral Oligonucleotide Delivery Carriers. *Chem.-A Eur. J.* **2015**, *21*, 15651–15666. [CrossRef]
79. Andrés, R.; Jesús, E.D.; Fierro, J.L.G.; Terreros, P. Bifunctional carbosilane dendrons for the immobilization of zirconocene catalysts on silica. *New, J. Chem.* **2011**, *35*, 2203–2211. [CrossRef]
80. Sun, X.; Qu, R.; Sun, C.; Zhang, Y.; Sun, S.; Ji, C.; Yin, P. Sol-gel preparation and Hg(II) adsorption properties of silica-gel supported low generation polyamidoamine dendrimers polymer adsorbents. *Ind. Eng Chem. Res.* **2014**, *53*, 2878–2888. [CrossRef]
81. Xu, X.; Lü, S.; Gao, C.; Wang, X.; Bai, X.; Gao, N.; Liu, M. Facile preparation of pH-sensitive and self-fluorescent mesoporous silica nanoparticles modified with PAMAM dendrimers for label-free imaging and drug delivery. *Chem. Eng J.* **2015**, *266*, 171–178. [CrossRef]
82. Newkome, G.R.; Yoo, K.S.; Kabir, A.; Malik, A. Synthesis of benzyl-terminated dendrons for use in high-resolution capillary gas chromatography. *Tetrahedron Lett.* **2001**, *42*, 7537–7541. [CrossRef]
83. Kabir, A.; Hamlet, C.; Yoo, K.S.; Newkome, G.R.; Malik, A. Capillary microextraction on sol-gel dendrimer coatings. *J. Chromatogr. A* **2004**, *1034*, 1–11. [CrossRef]
84. Riegert, D.; Bareille, L.; Laurent, R.; Majoral, J.P.; Caminade, A.M.; Chaumonnot, A. Silica Functionalized by Bifunctional Dendrimers: Hybrid Nanomaterials for Trapping CO_2. *Eur. J. Inorg. Chem.* **2016**, *2016*, 3103–3110. [CrossRef]
85. Pang, Y.; Zeng, G.; Tang, L.; Zhang, Y.; Liu, Y.; Lei, X.; Li, Z.; Zhang, J.; Xie, G. PEI-grafted magnetic porous powder for highly effective adsorption of heavy metal ions. *Desalination* **2011**, *281*, 278–284. [CrossRef]
86. Sorokina, S.A.; Kuchkina, N.V.; Lawson, B.P.; Krasnova, I.Y.; Nemygina, N.A.; Nikoshvili, L.Z.; Talanova, V.N.; Stein, B.D.; Pink, M.; Morgan, D.G.; et al. Pyridylphenylene dendrons immobilized on the surface of chemically modified magnetic silica as efficient stabilizing molecules of Pd species. *Appl. Surf. Sci.* **2019**, *488*, 865–873. [CrossRef]
87. Chung, Y.M.; Rhee, H.K. Silica-supported dendritic chiral auxiliaries for enantioselective addition of diethylzinc to benzaldehyde. *Comptes Rendus Chim.* **2003**, *6*, 695–705. [CrossRef]
88. Karakhanov, E.; Maximov, A.; Kardasheva, Y.; Semernina, V.; Zolotukhina, A.; Ivanov, A.; Abbott, G.; Rosenberg, E.; Vinokurov, V. Pd nanoparticles in dendrimers immobilized on silica-polyamine composites as catalysts for selective hydrogenation. *ACS Appl. Mater. Interfaces* **2014**, *6*, 8807–8816. [CrossRef]
89. Hameau, A.; Collière, V.; Grimoud, J.; Fau, P.; Roques, C.; Caminade, A.-M.; Turrin, C.-O. PPH dendrimers grafted on silica nanoparticles: Surface chemistry, characterization, silver colloids hosting and antibacterial activity. *RSC Adv.* **2013**, *3*, 19015. [CrossRef]
90. Mathews, B.T.; Beezer, A.E.; Snowden, M.J.; Hardy, M.J.; Mitchell, J.C. The synthesis of immobilised chiral dendrimers. *New J. Chem.* **2001**, *25*, 807–818. [CrossRef]
91. Gerrans, K.; Luhrs, A.; Feider, C.; Margerum, L.D. Silica nanoparticles functionalized with polyamidoamine (PAMAM) dendrimers as platforms for photoluminescence (PL) sensing of copper and cyanide ions. *J. Colloid Interface Sci.* **2016**, *470*, 276–283. [CrossRef]
92. Nakanishi, Y.; Imae, T. Preparation of siloxy focal dendron-protected TiO_2 nanoparticles and their photocatalysis. *J. Colloid Interface Sci.* **2006**, *297*, 122–129. [CrossRef]
93. Pawlaczyk, M.; Pasieczna-Patkowska, S.; Ryczkowski, J.; Schroeder, G. Photoacoustic infrared spectroscopic studies of silica surface functionalized by dendrimers. *Vib. Spectrosc.* **2019**, *103*, 102943. [CrossRef]
94. Pawlaczyk, M.; Schroeder, G. Adsorption studies of Cu(II) ions on dendrimer-grafted silica-based materials. *J. Mol. Liq.* **2019**, *281*, 176–185. [CrossRef]
95. Acosta, E.J.; Gonzalez, S.O.; Simanek, E.E. Synthesis, characterization, and application of melamine-based dendrimers supported on silica gel. *J. Polym. Sci. Part A Polym. Chem.* **2005**, *43*, 168–177. [CrossRef]
96. Bu, J.; Li, R.; Quah, C.W.; Carpenter, K.J. Propagation of PAMAM dendrons on silica gel: A study on the reaction kinetics. *Macromolecules* **2004**, *37*, 6687–6694. [CrossRef]
97. Chu, C.C.; Ueno, N.; Imae, T. Solid-phase synthesis of amphiphilic dendron-surface-modified silica particles and their application toward water purification. *Chem. Mater.* **2008**, *20*, 2669–2676. [CrossRef]
98. Wu, K.; Luan, L.; Xing, J.X.; Ma, S.; Xue, Z.; Xu, W.; Niu, Y. Removal of Zn(II) and Co(II) from N,N-dimethylformamide by polyamidoamine dendrimers decorated silica: Performance and mechanism. *J. Mol. Liq.* **2020**, *308*, 113073. [CrossRef]
99. Fu, T.; Niu, Y.; Zhou, Y.; Wang, K.; Mu, Q.; Qu, R.; Chen, H.; Yuan, B.; Yang, H. Adsorption of Mn(II) from aqueous solution by silica-gel supported polyamidoamine dendrimers: Experimental and DFT study. *J. Taiwan Inst. Chem. Eng.* **2019**, *97*, 189–199. [CrossRef]

100. Qiu, Z.; Niu, Y.; Fu, T.; Wang, K.; Mu, Q.; Wang, F. Removal of Ni(II) from fuel ethanol by PAMAM dendrimers/silica hybrid materials: Combined experimental and theoretical study. *Chem. Eng. Res. Des.* **2019**, *144*, 174–184. [CrossRef]
101. Song, X.; Niu, Y.; Zhang, P.; Zhang, C.; Zhang, Z.; Zhu, Y.; Qu, R. Removal of Co(II) from fuel ethanol by silica-gel supported PAMAM dendrimers: Combined experimental and theoretical study. *Fuel* **2017**, *199*, 91–101. [CrossRef]
102. Ren, B.; Wang, K.; Zhang, B.; Li, H.; Niu, Y.; Chen, H.; Yang, Z.; Li, X.; Zhang, H. Adsorption behavior of PAMAM dendrimers functionalized silica for Cd(II) from aqueous solution: Experimental and theoretical calculation. *J. Taiwan Inst. Chem. Eng.* **2019**, *101*, 80–91. [CrossRef]
103. Zhu, Y.; Niu, Y.; Li, H.; Ren, B.; Qu, R.; Chen, H.; Zhang, Y. Removal of Cd(II) and Fe(III) from DMSO by silica gel supported PAMAM dendrimers: Equilibrium, thermodynamics, kinetics and mechanism. *Ecotoxicol. Environ. Saf.* **2018**, *162*, 253–260. [CrossRef]
104. Zhang, S.; Niu, Y.; Chen, Z.; Chen, H.; Yang, Z.; Bai, L.; Yuan, B. Removal of Fe(III) from ethanol by silica-gel supported ester-terminated pamam dendrimers: Experimental and dft calculation. *Desalin. Water Treat.* **2019**, *164*, 310–318. [CrossRef]
105. Niu, Y.; Qu, R.; Sun, C.; Wang, C.; Chen, H.; Ji, C.; Zhang, Y.; Shao, X.; Bu, F. Adsorption of Pb(II) from aqueous solution by silica-gel supported hyperbranched polyamidoamine dendrimers. *J. Hazard. Mater.* **2013**, *244–245*, 276–286. [CrossRef]
106. Shaaban, A.F.; Khalil, A.A.; Lasheen, T.A.; Nouh, E.S.A.; Ammar, H. Polyamidoamine dendrimers modified silica gel for uranium(VI) removal from aqueous solution using batch and fixed-bed column methods. *Desalin. Water Treat.* **2018**, *102*, 197–210. [CrossRef]
107. Reynhardt, J.P.K.; Yang, Y.; Sayari, A.; Alper, H. Polyamidoamine dendrimers prepared Inside the channels of pore-expanded periodic mesoporous silica. *Adv. Funct. Mater.* **2005**, *15*, 1641–1646. [CrossRef]
108. Sakai, K.; Kunitake, M.; Teng, T.C.; Katada, A.; Harada, T.; Yoshida, K.; Yamanaka, K.; Asami, Y.; Sakata, M.; Hirayama, C. Designable Size Exclusion Chromatography Columns Based on Dendritic Polymer-Modified Porous Silica Particles. *Chem. Mater.* **2003**, *15*, 4091–4097. [CrossRef]
109. Niu, Y.; Qu, R.; Chen, H.; Mu, L.; Liu, X.; Wang, T.; Zhang, Y.; Sun, C. Synthesis of silica gel supported salicylaldehyde modified PAMAM dendrimers for the effective removal of Hg(II) from aqueous solution. *J. Hazard. Mater.* **2014**, *278*, 267–278. [CrossRef]
110. Niu, Y.; Yang, J.; Qu, R.; Gao, Y.; Du, N.; Chen, H.; Sun, C.; Wang, W. Synthesis of Silica-Gel-Supported Sulfur-Capped PAMAM Dendrimers for Efficient Hg(II) Adsorption: Experimental and DFT Study. *Ind. Eng. Chem. Res.* **2016**, *55*, 3679–3688. [CrossRef]
111. Zhang, P.; Niu, Y.; Qiao, W.; Xue, Z.; Bai, L.; Chen, H. Experimental and DFT investigation on the adsorption mechanism of silica gel supported sulfur-capped PAMAM dendrimers for Ag(I). *J. Mol. Liq.* **2018**, *263*, 390–398. [CrossRef]
112. Song, X.; Niu, Y.; Qiu, Z.; Zhang, Z.; Zhou, Y.; Zhao, J.; Chen, H. Adsorption of Hg(II) and Ag(I) from fuel ethanol by silica gel supported sulfur-containing PAMAM dendrimers: Kinetics, equilibrium and thermodynamics. *Fuel* **2017**, *206*, 80–88. [CrossRef]
113. Li, H.; Zheng, Z.; Cao, M.; Cao, R. Stable gold nanoparticle encapsulated in silica-dendrimers organic-inorganic hybrid composite as recyclable catalyst for oxidation of alcohol. *Microporous Mesoporous Mater.* **2010**, *136*, 42–49. [CrossRef]
114. Reynhardt, J.P.K.; Yang, Y.; Sayari, A.; Alper, H. Rhodium complexed C2-PAMAM dendrimers supported on large pore Davisil silica as catalysts for the hydroformylation of olefins. *Adv. Synth. Catal.* **2005**, *347*, 1379–1388. [CrossRef]
115. Bourque, S.C.; Alper, H.; Manzer, L.E.; Arya, P. Hydroformylation reactions using recyclable rhodium-complexed dendrimers on silica [4]. *J. Am. Chem. Soc.* **2000**, *122*, 956–957. [CrossRef]
116. Reynhardt, J.P.K.; Yang, Y.; Sayari, A.; Alper, H. Periodic mesoporous silica-supported recyclable rhodium-complexed dendrimer catalysts. *Chem. Mater.* **2004**, *16*, 4095–4102. [CrossRef]
117. Lu, S.M.; Alper, H. Intramolecular carbonylation reactions with recyclable palladium-complexed dendrimers on silica: Synthesis of oxygen, nitrogen, or sulfur-containing medium ring fused heterocycles. *J. Am. Chem. Soc.* **2005**, *127*, 14776–14784. [CrossRef] [PubMed]
118. Lu, S.M.; Alper, H. Synthesis of large ring macrocycles (12-18) by recyclable palladiumComplexed dendrimers on silica gel catalyzed intramolecular cyclocarbonylation reactions. *Chem.-A Eur. J.* **2007**, *13*, 5908–5916. [CrossRef] [PubMed]
119. Antebi, S.; Arya, P.; Manzer, L.E.; Alper, H. Carbonylation reactions of iodoarenes with PAMAM dendrimer-palladium catalysts immobilized on silica. *J. Org. Chem.* **2002**, *67*, 6623–6631. [CrossRef]
120. Reynhardt, J.P.K.; Alper, H. Hydroesterification Reactions with Palladium-Complexed PAMAM Dendrimers Immobilized on Silica. *J. Org. Chem.* **2003**, *68*, 8353–8360. [CrossRef]
121. Alper, H.; Arya, P.; Bourque, S.C.; Jefferson, G.R.; Manzer, L.E. Heck reaction using palladium complexed to dendrimers on silica. *Can. J. Chem.* **2000**, *78*, 920–924. [CrossRef]
122. Bu, J.; Judeh, Z.M.A.; Ching, C.B.; Kawi, S. Epoxidation of olefins catalyzed by Mn(II) salen complex anchored on PAMAM-SiO2 dendrimer. *Catal. Lett.* **2003**, *85*, 183–187. [CrossRef]
123. Wang, S.; Su, P.; Ding, F.; Yang, Y. Immobilization of cellulase on polyamidoamine dendrimer-grafted silica. *J. Mol. Catal. B Enzym.* **2013**, *89*, 35–40. [CrossRef]
124. Zhang, Q.; Wang, L.; Jiang, Y.; Gao, W.; Wang, Y.; Yang, X.; Yang, X.; Liu, Z. Gold Nanorods with Silica Shell and PAMAM Dendrimers for Efficient Photothermal Therapy and Low Toxic Codelivery of Anticancer Drug and siRNA. *Adv. Mater. Interfaces* **2017**, *4*, 1–9. [CrossRef]
125. Hicks, J.C.; Drese, J.H.; Fauth, D.J.; Gray, M.L.; Qi, G.; Jones, C.W. Designing adsorbents for CO_2 capture from flue gas-hyperbranched aminosilicas capable of capturing CO_2 reversibly. *J. Am. Chem. Soc.* **2008**, *130*, 2902–2903. [CrossRef] [PubMed]

126. Drese, J.H.; Choi, S.; Lively, R.P.; Koros, W.J.; Fauth, D.J.; Gray, M.L.; Jones, C.W. Synthesis-structure-property relationships for Hyperbranched aminosilica CO_2 adsorbents. *Adv. Funct. Mater.* **2009**, *19*, 3821–3832. [CrossRef]
127. López-Aranguren, P.; Vega, L.F.; Domingo, C. A new method using compressed CO_2 for the in situ functionalization of mesoporous silica with hyperbranched polymers. *Chem. Commun.* **2013**, *49*, 11776–11778. [CrossRef]
128. Acosta, E.J.; Carr, C.S.; Simanek, E.E.; Shantz, D.F. Engineering nanospaces: Iterative synthesis of melamine-based dendrimers on amine-functionalized SBA-15 leading to complex hybrids with controllable chemistry and porosity. *Adv. Mater.* **2004**, *16*, 985–989. [CrossRef]
129. Bhagiyalakshmi, M.; Park, S.D.; Cha, W.S.; Jang, H.T. Development of TREN dendrimers over mesoporous SBA-15 for CO_2 adsorption. *Appl. Surf. Sci.* **2010**, *256*, 6660–6666. [CrossRef]
130. Yoo, S.; Yeu, S.; Sherman, R.L.; Simanek, E.E.; Shantz, D.F.; Ford, D.M. Reverse-selective membranes formed by dendrimers on mesoporous ceramic supports. *J. Membr. Sci.* **2009**, *334*, 16–22. [CrossRef]
131. Tsubokawa, N.; Hayashi, S.; Nishimura, J. Grafting of hyperbranched polymers onto ultrafine silica: Postgraft polymerization of vinyl monomers initiated by pendant azo groups of grafted polymer chains on the surface. *Prog. Org. Coat.* **2002**, *44*, 69–74. [CrossRef]
132. Hayashi, S.; Fujiki, K.; Tsubokawa, N. Grafting of hyperbranched polymers onto ultrafine silica: Postgraft polymerization of vinyl monomers initiated by pendant initiating groups of polymer chains grafted onto the surface. *React. Funct. Polym.* **2000**, *46*, 193–201. [CrossRef]
133. Dvornic, P.R. PAMAMOS: The first commercial silicon-containing dendrimers and their applications. *J. Polym. Sci. Part A Polym. Chem.* **2006**, *44*, 2755–2773. [CrossRef]
134. Dvornic, P.R.; Li, J.; De Leuze-Jallouli, A.M.; Reeves, S.D.; Owen, M.J. Nanostructured dendrimer-based networks with hydrophilic polyamidoamine and hydrophobic organosilicon domains. *Macromolecules* **2002**, *35*, 9323–9333. [CrossRef]
135. Chujo, Y.; Matsuki, H.; Kure, S.; Saegusa, T.; Yazawa, T. Control of pore size of porous silica by means of pyrolysis of an organic-inorganic polymer hybrid. *J. Chem. Soc. Chem. Commun.* **1994**, 635–636. [CrossRef]
136. Tsiourvas, D.; Tsetsekou, A.; Papavasiliou, A.; Arkas, M.; Boukos, N. A novel hybrid sol-gel method for the synthesis of highly porous silica employing hyperbranched poly(ethyleneimine) as a reactive template. *Microporous Mesoporous Mater.* **2013**, *175*, 59–66. [CrossRef]
137. Turrin, C.O.; Maraval, V.; Caminade, A.M.; Majoral, J.P.; Mehdi, A.; Reyé, C. Organic-Inorganic hybrid materials incorporating phosphorus-containing dendrimers. *Chem. Mater.* **2000**, *12*, 3848–3856. [CrossRef]
138. Vardavoulias, M.; Gkomoza, E.; Tsetsekou, A.; Kitsou, I.; Arkas, M.; Papageorgiou, M.; González, S.M.S.; Cubillos, Y.L.; Cepas, V.; Jensen, H.E.; et al. Hydrogel and Xerogel Active Ingredient Carriers Made from Dendritic Polymers and Silica for Solid Substrate Coating Applications. Greek Patent Application Number 2019010000585, 30 December 2019.
139. Talham, D.R. Biomineralization: Principles and Concepts in Bioinorganic Materials Chemistry Stephen Mann. Oxford University Press, New York, 2001. *Cryst. Growth Des.* **2002**, *2*, 675. [CrossRef]
140. Sumper, M.; Brunner, E. Learning from diatoms: Nature's tools for the production of nanostructured silica. *Adv. Funct. Mater.* **2006**, *16*, 17–26. [CrossRef]
141. Lechner, C.C.; Becker, C.F.W. Silaffins in silica biomineralization and biomimetic silica precipitation. *Mar. Drugs* **2015**, *13*, 5297–5333. [CrossRef] [PubMed]
142. Knecht, M.R.; Wright, D.W. Amine-terminated dendrimers as biomimetic templates for silica nanosphere formation. *Langmuir* **2004**, *20*, 4728–4732. [CrossRef] [PubMed]
143. Knecht, M.R.; Sewell, S.L.; Wright, D.W. Size control of dendrimer-templated silica. *Langmuir* **2005**, *21*, 2058–2061. [CrossRef]
144. Coradin, T.; Durupthy, O.; Livage, J. Interactions of amino-containing peptides with sodium silicate and colloidal silica: A biomimetic approach of silicification. *Langmuir* **2002**, *18*, 2331–2336. [CrossRef]
145. Kobler, J.; Möller, K.; Bein, T. Colloidal Suspensions of Functionalized Mesoporous Silica Nanoparticles. *ACS Nano* **2008**, *2*, 791–799. [CrossRef] [PubMed]
146. Petrakli, F.; Arkas, M.; Tsetsekou, A. α-Alumina nanospheres from nano-dispersed boehmite synthesized by a wet chemical route. *J. Am. Ceram. Soc.* **2018**, *101*, 3508–3519. [CrossRef]
147. Zhang, F.; Yang, S.P.; Wang, W.M.; Chen, H.M.; Wang, Z.H.; Yu, X. Bin Preparation of nanocrystalline ceramic oxide powders in the presence of anionic starburst dendrimer. *Mater. Lett.* **2004**, *58*, 3285–3289. [CrossRef]
148. Arkas, M.; Tsiourvas, D. Organic/inorganic hybrid nanospheres based on hyperbranched poly(ethylene imine) encapsulated into silica for the sorption of toxic metal ions and polycyclic aromatic hydrocarbons from water. *J. Hazard. Mater.* **2009**, *170*, 35–42. [CrossRef] [PubMed]
149. Nemanashi-Maumela, M.; Nongwe, I.; Motene, R.C.; Davids, B.L.; Meijboom, R. Au and Ag nanoparticles encapsulated within silica nanospheres using dendrimers as dual templating agent and their catalytic activity. *Mol. Catal.* **2017**, *438*, 184–196. [CrossRef]
150. Arkas, M.; Kithreoti, G.; Boukos, N.; Kitsou, I.; Petrakli, F.; Panagiotaki, K. Two completely different biomimetic reactions mediated by the same matrix producing inorganic/organic/inorganic hybrid nanoparticles. *Nano-Struct. Nano-Objects* **2018**, *14*, 138–148. [CrossRef]
151. Kröger, N.; Deutzmann, R.; Bergsdorf, C.; Sumper, M. Species-specific polyamines from diatoms control silica morphology. *Proc. Natl. Acad. Sci. USA* **2000**, *97*, 14133–14138. [CrossRef]

152. Mizutani, T.; Nagase, H.; Fujiwara, N.; Ogoshi, H. Silicic Acid Polymerization Catalyzed by Amines and Polyamines. *Bull. Chem. Soc. Jpn.* **1998**, *71*, 2017–2022. [CrossRef]
153. Knecht, M.R.; Wright, D.W. Functional analysis of the biomimetic silica precipitating activity of the R5 peptide from Cylindrotheca fusiformis. *Chem. Commun.* **2003**, *3*, 3038–3039. [CrossRef]
154. Kossovsky, N.; Gelman, A.; Sponsler, E.E.; Hnatyszyn, H.J.; Rajguru, S.; Torres, M.; Pham, M.; Crowder, J.; Zemanovich, J.; Chung, A.; et al. Surface-modified nanocrystalline ceramics for drug delivery applications. *Biomaterials* **1994**, *15*, 1201–1207. [CrossRef]
155. Kossovsky, N.; Gelman, A.; Rajguru, S.; Nguyen, R.; Sponsler, E.; Hnatyszyn, H.J.; Chow, K.; Chung, A.; Torres, M.; Zemanovich, J.; et al. Control of molecular polymorphisms by a structured carbohydrate/ceramic delivery vehicle—Aquasomes. *J. Control. Release* **1996**, *39*, 383–388. [CrossRef]
156. Khopade, A.J.; Khopade, S.; Jain, N.K. Development of hemoglobin aquasomes from spherical hydroxyapatite cores precipitated in the presence of half-generation poly(amidoamine) dendrimer. *Int. J. Pharm.* **2002**, *241*, 145–154. [CrossRef]
157. Zhang, F.; Zhou, Z.H.; Yang, S.P.; Mao, L.H.; Chen, H.M.; Yu, X. Bin Hydrothermal synthesis of hydroxyapatite nanorods in the presence of anionic starburst dendrimer. *Mater. Lett.* **2005**, *59*, 1422–1425. [CrossRef]
158. Yan, S.J.; Zhou, Z.H.; Zhang, F.; Yang, S.P.; Yang, L.Z.; Yu, X. Bin Effect of anionic PAMAM with amido groups starburst dendrimers on the crystallization of $Ca_{10}(PO_4)_6(OH)_2$ by hydrothermal method. *Mater. Chem. Phys.* **2006**, *99*, 164–169. [CrossRef]
159. Zhou, Z.H.; Zhou, P.L.; Yang, S.P.; Yu, X.B.; Yang, L.Z. Controllable synthesis of hydroxyapatite nanocrystals via a dendrimer-assisted hydrothermal process. *Mater. Res. Bull.* **2007**, *42*, 1611–1618. [CrossRef]
160. Tsiourvas, D.; Tsetsekou, A.; Kammenou, M.I.; Boukos, N. Controlling the formation of hydroxyapatite nanorods with dendrimers. *J. Am. Ceram. Soc.* **2011**, *94*, 2023–2029. [CrossRef]
161. Donners, J.J.J.M.; Nolte, R.J.M.; Sommerdijk, N.A.J.M. Dendrimer-based hydroxyapatite composites with remarkable materials properties. *Adv. Mater.* **2003**, *15*, 313–316. [CrossRef]
162. Cahill, B.P.; Papastavrou, G.; Koper, G.J.M.; Borkovec, M. Adsorption of poly(amido amine) (PAMAM) dendrimers on silica: Importance of electrostatic three-body attraction. *Langmuir* **2008**, *24*, 465–473. [CrossRef]
163. Longtin, R.; Maroni, P.; Borkovec, M. Transition from completely reversible to irreversible adsorption of poly(amido amine) dendrimers on silica. *Langmuir* **2009**, *25*, 2928–2934. [CrossRef]
164. Kresge, C.T.; Leonowicz, M.E.; Roth, W.J.; Vartuli, J.C.; Beck, J.S. Ordered mesoporous molecular sieves synthesized by a liquid-crystal template mechanism. *Nature* **1992**, *359*, 710–712. [CrossRef]
165. Kruk, M.; Jaroniec, M. Relations between pore structure parameters and their implications for characterization of MCM-41 using gas adsorption and x-ray diffraction. *Chem. Mater.* **1999**, *11*, 492–500. [CrossRef]
166. Arkas, M. Hybrid organoceramics deriving from dendritic polymers, methods of preparation, optimization techniques and prospected applications. In *Recent Advances in Ceramic Materials Research (Materials Science and Technologies)*; Nova Science Publishers: Hauppauge, NY, USA, 2013; pp. 1–30, ISBN 9781624177293.
167. Arkas, M. Hybrid Organic/Inorganic Materials Based on Functionalized Dendritic Polymers: Methods of Preparation, Applications and Future Prospects. *J. Mater. Sci. Eng.* **2012**, *1*, 1000e103. [CrossRef]
168. Sooklal, K.; Hanus, L.H.; Ploehn, H.J.; Murphy, C.J. A Blue-Emitting CdS/Dendrimer Nanocomposite. *Adv. Mater.* **1998**, *10*, 1083–1087. [CrossRef]
169. Garcia, M.E.; Baker, L.A.; Crooks, R.M. Preparation and characterization of dendrimer-gold colloid nanocomposites. *Anal. Chem.* **1999**, *71*, 256–258. [CrossRef] [PubMed]
170. Manoth, M.; Manzoor, K.; Patra, M.K.; Pandey, P.; Vadera, S.R.; Kumar, N. Dendrigraft polymer-based synthesis of silver nanoparticles showing bright blue fluorescence. *Mater. Res. Bull.* **2009**, *44*, 714–717. [CrossRef]
171. Karakhanov, E.; Maximov, A.; Zolotukhina, A.; Mamadli, A.; Vutolkina, A.; Ivanov, A. Dendrimer-stabilized Ru nanoparticles immobilized in organo-silica materials for hydrogenation of phenols. *Catalysts* **2017**, *7*, 86. [CrossRef]
172. Tsiourvas, D.; Papavasiliou, A.; Deze, E.G.; Papageorgiou, S.K.; Katsaros, F.K.; Romanos, G.E.; Poulakis, E.; Philippopoulos, C.J.; Xin, Q.; Cool, P. A green route to copper loaded silica nanoparticles using hyperbranched poly(Ethylene imine) as a biomimetic template: Application in heterogeneous catalysis. *Catalysts* **2017**, *7*, 390. [CrossRef]
173. Kitsou, I.; Arkas, M.; Tsetsekou, A. Synthesis and characterization of ceria-coated silica nanospheres: Their application in heterogeneous catalysis of organic pollutants. *SN Appl. Sci.* **2019**, *1*. [CrossRef]
174. Gemeinhart, R.A.; Luo, D.; Saltzman, W.M. Cellular fate of a modular DNA delivery system mediated by silica nanoparticles. *Biotechnol. Prog.* **2005**, *21*, 532–537. [CrossRef]
175. Radu, D.R.; Lai, C.Y.; Jeftinija, K.; Rowe, E.W.; Jeftinija, S.; Lin, V.S.Y. A polyamidoamine dendrimer-capped mesoporous silica nanosphere-based gene transfection reagent. *J. Am. Chem. Soc.* **2004**, *126*, 13216–13217. [CrossRef]
176. Zang, S.; Chang, S.; Shahzad, M.B.; Sun, X.; Jiang, X.; Yang, H. Ceramics-based Drug Delivery System: A Review and Outlook. *Rev. Adv. Mater. Sci.* **2019**, *58*, 82–97. [CrossRef]
177. Janaszewska, A.; Lazniewska, J.; Trzepiński, P.; Marcinkowska, M.; Klajnert-Maculewicz, B. Cytotoxicity of dendrimers. *Biomolecules* **2019**, *9*, 330. [CrossRef] [PubMed]
178. Patri, A.K.; Majoros, I.J.; Baker, J.R. Dendritic polymer macromolecular carriers for drug delivery. *Curr. Opin. Chem. Biol.* **2002**, *6*, 466–471. [CrossRef]
179. Yousefi, M.; Narmani, A.; Jafari, S.M. Dendrimers as efficient nanocarriers for the protection and delivery of bioactive phytochemicals. *Adv. Colloid Interface Sci.* **2020**, *278*. [CrossRef] [PubMed]

nanomaterials

Review

Metal Oxide Based Heterojunctions for Gas Sensors: A Review

Shulin Yang [1,2], Gui Lei [1,2], Huoxi Xu [1,*], Zhigao Lan [1], Zhao Wang [2] and Haoshuang Gu [1,2,*]

1 Hubei Key Laboratory for Processing and Application of Catalytic Materials, School of Physics and Electronic Information, Huanggang Normal University, Huanggang 438000, China; yangsl@hgnu.edu.cn (S.Y.); leig@hgnu.edu.cn (G.L.); lanzhg@hgnu.edu.cn (Z.L.)
2 Hubei Key Laboratory of Ferro & Piezoelectric Materials and Devices, Faculty of Physics and Electronic Sciences, Hubei University, Wuhan 430062, China; wangzhao@hubu.edu.cn
* Correspondence: xuhuoxi@hgnu.edu.cn (H.X.); guhsh@hubu.edu.cn (H.G.)

Abstract: The construction of heterojunctions has been widely applied to improve the gas sensing performance of composites composed of nanostructured metal oxides. This review summarises the recent progress on assembly methods and gas sensing behaviours of sensors based on nanostructured metal oxide heterojunctions. Various methods, including the hydrothermal method, electrospinning and chemical vapour deposition, have been successfully employed to establish metal oxide heterojunctions in the sensing materials. The sensors composed with the built nanostructured heterojunctions were found to show enhanced gas sensing performance with higher sensor responses and shorter response times to the targeted reducing or oxidising gases compare with those of the pure metal oxides. Moreover, the enhanced gas sensing mechanisms of the metal oxide-based heterojunctions to the reducing or oxidising gases are also discussed, with the main emphasis on the important role of the potential barrier on the accumulation layer.

Keywords: metal oxide; heterojunctions; gas sensor; sensing mechanism; review

Citation: Yang, S.; Lei, G.; Xu, H.; Lan, Z.; Wang, Z.; Gu, H. Metal Oxide Based Heterojunctions for Gas Sensors: A Review. *Nanomaterials* **2021**, *11*, 1026. https://doi.org/10.3390/nano11041026

Academic Editors: Jihoon Lee and Ming-Yu Li

Received: 18 March 2021
Accepted: 16 April 2021
Published: 17 April 2021

1. Introduction

Gas sensors based on nanostructured metal oxides have attracted significant interest over the last few decades due to their advantages of low cost, ease of fabrication, high sensor response and short response/recovery times [1–4]. Various metal oxides have been successfully assembled as gas sensors since Seiyama et al. reported their research on the gas sensing performance of the ZnO thin film in the 1960s [5,6]. According to the sensing behaviours of the metal oxides, sensing metal oxides are typically divided into two main groups: n-type metal oxides and p-type metal oxides. Normally, the resistances of the n-type metal oxides decrease (or increase) towards reducing gases such as H_2, H_2S, CO, CH_4, NH_3 and other volatile organic compounds (or oxidising gases such as NO_2, NO, O_3, SO_2, etc.), while p-type metal oxides exhibit the opposite behaviour [7,8]. The n-type metal oxides of SnO_2, TiO_2, WO_3, MoO_3, Nb_2O_5, ZnO, etc., and the p-type metal oxides of CuO, Co_3O_4, Cr_2O_3, NiO, PdO, etc., have been widely studied for their gas sensing behaviours towards both reducing and oxidizing gases [9–12].

In recent years, the advancement of new technologies and methods has induced a boom in nanomaterials. Nanostructured metal oxides with various morphologies, such as nanoparticles, nanosheets, nanowires, nanorods, nanoribbons, nanofibres, nanoflowers and nanocages, have been successfully prepared through the routes of hydrothermal processing, thermal oxidation, sol-gel processing, atomic layer deposition, etc. Nanoscale metal oxides have been reported to exhibit promising gas sensing performances, benefiting from their high specific surface areas [13] and active surface states [14]. Specifically, the sensor response of the pure ZnO nanowires was ~15 towards 0.5 ppm NO_2 at a working temperature of 225 °C [15]. Other sensors based on SnO_2 nanowires [16], TiO_2 nanotubes [17], WO_3 nanoparticles [18] and In_2O_3 nanofibres [19] have also been found to respond to the gases of NO_2, formaldehyde, H_2S and CO, respectively. The gas sensing performances of the gas sensors mentioned above could be further improved to better meet the demands of practical applications via compositing the metal oxide with another (different) metal oxide to form a heterojunction between them. It is noteworthy that when the main phase of a metal oxide was decorated or composited with the second phase of a different semiconductor, the interface between them was known as the structure of a heterojunction in the sensing material [20]. Reports show the specific surface area of the composite is higher than that of the pure metal oxide [21], and the modulation of the potential barrier or accumulation layer in the composite effectively improves the gas sensing behaviour [22,23]. More and more researchers have focused their attention on the studies of the high-performance sensors based on nanostructured metal oxide heterojunctions, as shown in Figure 1. For example, the ordered mesoporous WO_3/ZnO nanocomposites synthesised with a hydrothermal method displayed an enhanced sensor response of 168.7 to 1 ppm NO_2 at a working temperature of 150 °C, over 10 times higher than that of the pure WO_3 [24]. Moreover, the sensor response of CeO_2 nanostructures modified with NiO was reported to be ~1570, much higher than that of the pure CeO_2 (139). The response time and the recovery time of the composite was 15 s and 19 s, respectively, which is also shorter than that of the pure CeO_2 (96 s/118 s) [25]. Therefore, the construction of heterojunctions could be a successful method to improve the gas sensing performances of sensors based on metal oxides.

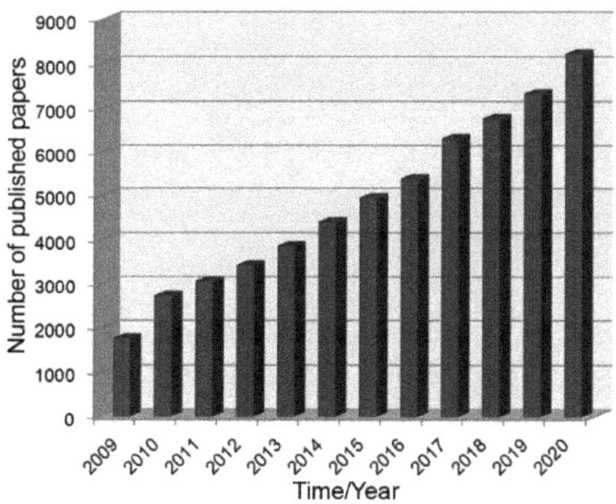

Figure 1. The number of the published papers on nanostructured metal oxide heterojunctions for high-performance gas sensors during 2009–2020 as obtained from the Web of Science. The words "nanostructured metal oxide heterojunctions" or "high-performance gas sensors" were keyed into the "topic" search box.

As reported, there have been different kinds of heterojunctions assembled to improve the gas sensing performances of metal oxides, such as n-n, n-p, p-n or p-p heterojunctions [26–29]. Note that, in this paper, the type of heterojunction is defined according to the dominant material or the main phase in the composite [30,31]. Accordingly, an n-p heterojunction is formed when the main phase of an n-type metal oxide is modified with a second phase of a p-type semiconductor. Similarly, a p-n heterojunction is established through compositing a p-type metal oxide with an n-type semiconductor. An n-n (or p-p) heterojunction would also be constructed if an n-type (p-type) metal oxide is decorated with a different n-type (p-type) semiconductor in the composite. For example, CuO-decorated ZnO or ZnO-decorated WO_3 are the typical n-p or n-n heterojunctions. The ZnO-decorated CuO or NiO-decorated CuO are defined as the p-n or p-p heterojunctions. With the development in the techniques to synthesise nanomaterials, it is facile to construct heterojunctions in composites composed of metal oxides. The metal oxide-based heterojunctions have been successfully established through various combined technologies, such as the thermal oxidation [28], hydrothermal method [32], electrospinning [33,34], chemical vapour deposition (CVD) [35], pulsed laser deposition (PLD) [36], the co-precipitation method [37] and the solvothermal method [38]. For example, the hydrothermal method was reported to effectively prepare the ZnO/SnO_2 [39] and the NiO/SnO_2 composites [32]. The difference in the Fermi levels of the two metal oxides in the obtained composite would lead to the formation of the potential barrier at their interfaces, an important factor for improving sensing property of the gas sensor based on heterojunctions. The hydrothermal method combined with CVD was also used by Li et al. to establish heterojunctions composed of vertically aligned MoS_2/ZnO nanowires [40]. SnO_2-CuO heterojunctions were successfully constructed via electrospinning [41]. Their results clearly indicated that the sensor response of the sensor based on the MoS_2/ZnO nanowires or the SnO_2-CuO heterojunctions was highly improved compared with that of the bare ZnO or SnO_2. Though there have been a number of references reviewing the developments in the gas sensing performances of given metal oxides, only a few articles provide a comprehensive review of the effects of the heterojunctions on the enhanced gas sensing performances of composites based on nanostructured metal oxides or separately discuss heterojunctions with n-n, n-p, p-n or p-p structures. Moreover, the enhanced gas sensing mechanism of a given type of heterojunction to a reducing or an oxidising gas should also be studied and summarised. The synthesised methods and the gas sensing performances of the normally studied metal oxides as well as the important roles of the heterojunctions need to be systematically summarised and compared. This will allow us to fully understand the improved gas sensing properties of metal oxide heterojunctions.

In this review, the typical synthetic routes of n-n, n-p, p-n and p-p heterojunctions based on metal oxides are introduced. The gas sensing behaviours of the n-n/n-p heterojunctions (or p-n/p-p heterojunctions) are based on SnO_2 and TiO_2. ZnO, WO_3, MoO_3, In_2O_3, CuO, Cr_2O_3, NiO and Co_3O_4, etc., semiconductors are reviewed and compared to show the effects of the heterojunctions on the gas sensing performances of the metal oxides. The enhanced gas sensing mechanisms of the composites towards reducing and oxidising gases are also discussed in detail to systematically understand the role of the built heterojunctions in improving the gas sensing properties of the composites.

2. Nanostructured Metal Oxide Heterojunctions for High-Performance Gas Sensors

As reported, the formation of heterojunctions could be a positive effective strategy to improve the gas sensing performance of the metal oxides. Various methods such as hydrothermal [42], PLD [36], vapour-liquid-solid (VSL) [43], anodic oxidation [44], solvothermal treatment [45], sputtering [46], thermal evaporation [47], electrospinning [33], sol-gel [48] and spin-coating [49] have been successfully applied to assemble the heterostructures in the sensors, and are generally combined to form various heterojunctions (n-n, n-p, p-n or p-p types), as displayed in Figure 2. Other methods to assemble nanostructured metal oxide heterojunctions are listed in Tables S1–S4 (see Supporting Materials), along with the improved gas sensing performances of sensors based on n-p, n-n, p-n and p-p heterojunctions. Some of the typical nanostructured heterojunctions with the n-type (or p-type) metal oxides as the main phases are discussed in the following sections.

2.1. Enhanced Gas Sensing Performances of n-n Junctions or n-p Junctions

Heterojunctions with the n-p or the n-n structure in the sensing materials have been reported to be successful strategies to enhance their gas sensing properties. When the p-type metal oxide (acting as the second phase) is attached to an n-type metal oxide (acting as the main phase), an n-p heterojunction is formed between the two sensing metal oxides. Additionally, n-n heterojunctions can also be assembled in a similar way. One of the common routes to establish the n-n (or n-p) heterojunctions is to prepare the main n-type metal oxides and then decorate the prepared n-type metal oxides with the n-type (or p-type) metal oxides [50]. The modulation of the built potential barrier in the n-n or n-p heterojunction can effectively modify the resistance of the sensing material, and thus greatly improve the gas sensing properties of the sensor composed with the n-n (or n-p) heterojunctions. Meanwhile, it is also noticed that the majority of heterojunctions are assembled as a decorated structure, core-shell structure or mixed structure (one metal oxide mixed with another metal oxide), discussed in the following subsections.

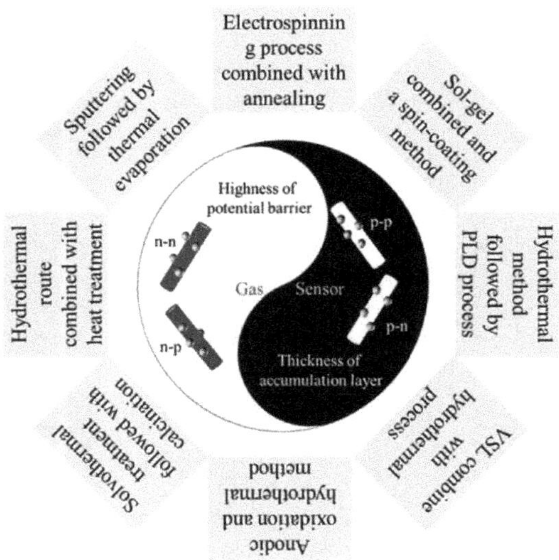

Figure 2. Typical types of heterojunctions for high-performance gas sensors and assembly strategies.

2.1.1. Gas Sensors Based on n-n Junctions

Many references report the improved gas sensing performances of sensors based on n-n heterojunctions. For example, Lu et al. reported the gas sensing properties of the ZnO-decorated SnO_2 hollow spheres towards the ethanol synthesised via a two-step hydrothermal method [39]. Hollow spheres of SnO_2 of ~100 nm thickness were synthesised via a facile template-free hydrothermal route (see Figure 3a,b) with ZnO nanoparticles of 10–30 nm diameter (see Figure 3c,d) uniformly decorated on its surface via a solution route. The sensor response of the ZnO-decorated SnO_2 hollow spheres was calculated to be 34.8 towards 30 ppm ethanol at their optimised operating temperature of 225 °C (see Figure 3e), much higher than that of the bare SnO_2 (~5.7 times). Their further research indicated that the composite also exhibited promising selectivity to acetone compared with methanol (Figure 3f). The recovery time of the composite towards 30 ppm ethanol (50 s) was also much shorter than that of acetone (120 s) at the same concentrations as shown in Figure 3g. In addition, the SnO_2 compositing with Co_3O_4 and SiO_2 have been reported to be promising gas sensing materials. The SnO_2/SiO_2 heterojunctions were synthesised via a facile method of a magnetron sputtering process and exhibited promising H_2 sensing performance at room temperature [51]. Hybrid Co_3O_4/SnO_2 core-shell nanospheres prepared with a one-step hydrothermal method demonstrated a measured response of 13.6 to 100 ppm NH_3 at 200 °C, a value two times higher than that of the solid nanospheres [52]. CeO_2-decorated ZnO nanosheets were prepared by a hydrothermal process in combination with the wet impregnation method, exhibited an enhanced sensor response of 90 to 100 ppm ethanol at 310 °C [53]. Additionally, Kim et al. have fabricated ZnO-SnO_2 nanofibres through an electrospinning process to effectively detect CO [54].

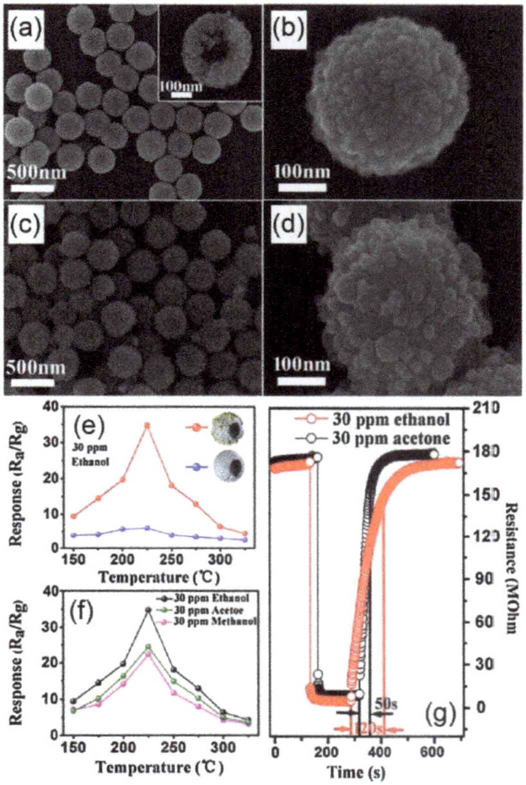

Figure 3. SEM images of the pure SnO_2 hollow spheres (**a**,**b**) and ZnO-decorated SnO_2 (n-n junctions) (**c**,**d**), sensor response of pure SnO_2 hollow spheres and the ZnO-decorated SnO_2 hollow spheres to 30 ppm ethanol at different operating temperatures (**e**), the sensor response of ZnO-decorated SnO_2 hollow spheres to 30 ppm ethanol, acetone and methanol at different operating temperatures (**f**), and the dynamic sensing performance of the decorated SnO_2 hollow spheres towards 30 ppm ethanol or acetone at 225 °C (**g**). Copied with permission from reference [39]. Copyright 2017, Elsevier.

The α-MoO_3/TiO_2 core/shell nanorods have been synthesised through a hydrothermal process combined with the following annealing process in air atmosphere [55]. Uniform α-MoO_3 nanorods were first prepared and then coated with a shell of TiO_2 via a modified wet-chemical method. It was found that the core/shell nanorods exhibited an improved gas sensing performance to 10 ppm ethanol at 180 °C with a short response time of less than 40 s. Meanwhile, the SnO_2-core/ZnO-shell nanowires [56] and the Ga_2O_3-core/ZnO-shell nanorods [57] were successfully synthesised through a plasma-enhanced CVD and atomic layer deposition (ALD), respectively, which also exhibited promising gas sensing performances. The MoO_3 nanorods decorated with the ZnO nanoparticles were also reported to be a promising material to detect 100 ppm ethanol with a sensor response of ~30 at the working temperature of 250 °C [58]. Besides the nanocomposites discussed above, it was reported that α-MoO_3 compositing with WO_3 through a sol-gel method [59] or with Fe_2O_3 nanoparticles via a hydrothermal method [60] also showed improved gas sensing performances towards O_2 or xylene, respectively.

In addition, ZnO nanorods/TiO_2 nanoparticles [61] and the ZnO/$La_{0.8}Sr_{0.2}Co_{0.5}Ni_{0.5}O_3$ heterojunction structure [62] were successfully constructed to research their improved gas sensing performances to NO_2 and CO, respectively. WO_3 compositing with SnO_2 was reported to be a potential material to detect acetone [63], while WO_3-modified ZnO nanoplates synthesised via the hydrothermal route were assembled for the detection of

NH_3 [26]. Other effective methods to synthesise n-n heterojunctions and their gas sensing performances are listed in Table S1.

2.1.2. Gas Sensors Based on n-p Junctions

The sensors based on the n-p heterojunctions have been found to show promising gas sensing properties towards various gases. For example, PdO nanoparticles-decorated flower-like ZnO structures (see Figure 4) were prepared by Zhang et al. through a surfactant-free hydrothermal process combined with a further heat treatment [42]. The $Zn(AC)_2{\cdot}2H_2O$ was used in the study to synthesise the flower-like ZnO structures, a certain amount of which was dissolved in a solution of NaOH, ethanol and deionized water. The obtained precursor was kept at 150 °C for 24 h. Before decorating with PdO nanoparticles, the flower-like ZnO structures were treated by an annealing process. The annealed ZnO nanoflowers were then dispersed in methanol solvent dissolving $PdCl_2$, and the collected products were calcined at 350 °C for 1 h to obtain the PdO-modified ZnO structures. The decorated flower-like ZnO was reported to show a gas sensor response of 35.4 to 100 ppm ethanol at 320 °C (see Figure 4e), which was much higher than that of the pure ZnO (~10 as shown in Figure 4f). Moreover, the composite presented a shorter recovery time of 7 s than that of the ZnO (14 s). ZnO/Co_3O_4 composite nanoparticles [64] and Al-doped ZnO/CuO nanocomposites [65] were reported to be sensitive to NO_2 and ammonia, respectively.

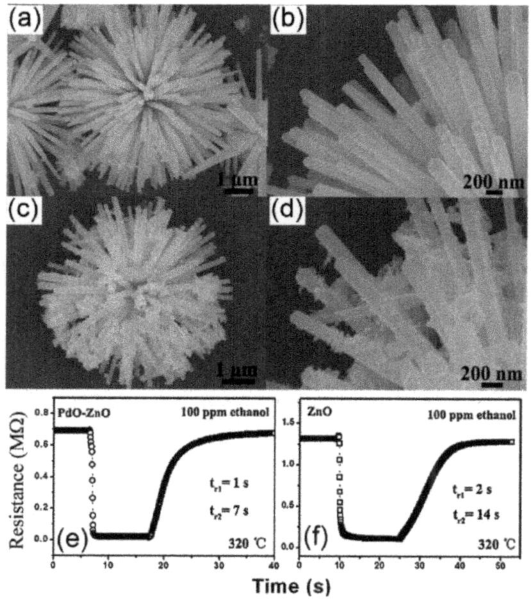

Figure 4. SEM images of pure flower-like ZnO (**a**,**b**) and the PdO nanoparticle-decorated ZnO (n-p junctions) (**c**,**d**), the dynamic sensing performance of the sensor based on the PdO nanoparticle-decorated ZnO (**e**) or pure ZnO (**f**) to 100 ppm ethanol at 320 °C. Copied with permission [42]. Copyright 2013, Elsevier.

Gao et al. synthesised CuO nanoparticles-decorated MoO_3 nanorods through a hydrothermal process combined with an annealing process [66]. In the first step, MoO_3 nanorods were prepared with the raw material $(NH_4)_6Mo_7O_{24}{\cdot}4H_2O$. Then, the obtained MoO_3 nanorods were dispersed in a solution of anhydrous ethanol and copper nitrate under high intensity ultrasonication. The final collected samples were annealed at 550 °C for 2 h. The CuO nanoparticles-decorated MoO_3 nanorods showed a higher H_2S sensor response of 272 at 270 °C compared with that of pure MoO_3, which was mainly attributed to the formation of n-p heterojunctions in the sensing material as reported in their article.

Nano-coaxial Co_3O_4/TiO_2 heterojunctions were successfully assembled through a typical two-step process by Yang et al. [44]. The authors firstly synthesised uniform TiO_2 nanotubular arrays via anodic oxidation of a Ti plate which were then decorated with Co_3O_4 nanoparticles by a hydrothermal process at 120 °C for 5 h. The sensor based on the nano-coaxial Co_3O_4/TiO_2 heterojunctions showed an enhanced sensor response of 40 to 100 ppm ethanol at 260 °C with a short response/recovery time of 1.4 s/7.2 s. The SnO_2-Co_3O_4 composite nanofibres were prepared through electrospinning combined with annealing with the working voltage of 15 kV [67]. The PdO nanoparticle-decorated WO_3 nanorods were also reported to be synthesised via a modified precipitation process combined with annealing at 300 °C for 2 h [68]. The SnO_2-Co_3O_4 composite nanofibres and the PdO nanoparticle-decorated WO_3 nanorods were found to be sensitive to 10 ppm C_6H_6 at 350 °C and 3.0 vol% of H_2 at 25 °C with enhanced sensor responses of 20 and 80.4, respectively.

CuO/ZnO heterostructural nanorods were prepared by Cao et al. via a combination of hydrothermal and wet-chemical processes [69]. The CuO nanoparticles-decorated ZnO nanorods array showed a sensor response of ~8 to 50 ppm towards triethylamine at a relatively low working temperature of 40 °C (higher than that of the bare ZnO nanorods of ~2.4), and a response time of 5 s (significantly shorter than that of the pure ZnO nanorods of ~11 s). Improved gas sensing performances were also observed in sensing materials composed of CuO-decorated SnO_2 nanowires [70], CuO nanoparticles-decorated ZnO flowers [71], flower-like p-CuO/n-ZnO nanorods [72], NiO@ZnO heterostructured nanotubes [73], n-ZnO/p-NiO heterostructured nanofibres [74] and Co_3O_4 decorated flower-like SnO_2 nanorods [75]. The various methods used to assemble the n-p heterojunctions and their gas sensing properties are provided in Table S2.

Besides the nanocomposites discussed above, sensors based on the n-n or n-p heterojunctions have also been assembled to enhance the gas sensing performances of metal oxides. TiO_2 composited with ZnO, MoS_2, MoO_3, V_2O_5 and WO_3 have been designed and successfully established, exhibiting improved gas sensing performances towards ethanol, NO_2, alcohol and ammonia [33,55,76–78]. For example, the ZnO-decorated TiO_2 nanotube layer (prepared by anodic oxidation combined with atomic layer deposition) [76], TiO_2/V_2O_5 branched nanoheterostructures (synthesised by an electrospinning process followed by an annealing treatment) [33] and a TiO_2-WO_3 composite (obtained via plasma spraying technology using mixed feedstock suspensions) [77] have each exhibited promising gas sensing performances to 1170 ppm ethanol, 100 ppm ethanol and 100 ppm NO_2, respectively. α-Fe_2O_3 composited with SnO_2, In_2O_3 and CdO have also been successfully synthesised through hydrothermal, carbon sphere template and co-precipitating processes, enabling excellent gas sensitivity towards acetone, TMA and CO, respectively [37,79,80]. In_2O_3 composited with WO_3, Fe_2O_3, TiO_2 and SnO_2 were also synthesised to assemble the high-performance gas sensors [81–84]. A series of In_2O_3-WO_3 nanofibres were prepared via an electrostatic spinning technology, which was reported to show an enhanced gas sensing performance to acetone with the n-n semiconductor heterojunctions formed at the interface between WO_3 and the In_2O_3 [81]. The sensor based on mixed Fe_2O_3-In_2O_3 nanotubes was also reported to show a high gas sensor response of ~33 towards 100 ppm of formaldehyde at 250 °C [82]. TiO_2 nanoparticle-functionalised In_2O_3 nanowires [83], SnO_2/In_2O_3 composite hetero-nanofibres [84], an octahedral-like ZnO/CuO composite [85] and a nanoporous SnO_2@TiO_2 heterostructure [86] were reported to show enhanced gas sensing properties towards acetone, formaldehyde and H_2S, respectively.

Based on the research discussed above, it is clear that the establishment of n-n or n-p heterojunctions can effectively improve the gas sensing properties of n-type metal oxides. Typically, n-type metal oxides are decorated with zero-dimensional nanoparticles and two-dimensional nanosheets, the concentrations of which have significant effects on the performance of the main n-type phase [67,87–89]. More specifically, the gas sensing performance of the main phase in a sensing material improves with increasing concentration of the second phase up to an optimal value, which can be attributed to the increase in the specific surface area of the composite. However, the sensing property of the main phase always degenerates when the content of the second phase is further elevated. The interconnection of the second phase and decrease in the effective surface area was reported to be the two main factors causing a weakened sensor response of the composite. However, most of the reported sensors based on the n-n or n-p heterojunctions always worked at temperatures above 100 °C. We also found there are no clear strategies to indicate which material should be chosen to enhance the gas response of a certain n-type metal oxide, which may be paid more attention in the future by researchers.

2.2. Improved Gas Sensing Properties of p-n or p-p Junctions

Sensors based on the heterojunctions with p-n or p-p structures were also reported to exhibit promising gas sensing properties with high sensor responses and short response/recovery times. Similar to the formation of n-n or n-p heterojunctions, p-n and p-p junctions are also built with p-type metal oxides (as the main phase) decorated or coated with n-type or p-type metal oxides (as the second phase). The different Fermi levels of the metal oxides in the constructed p-n or p-p junctions induce the formation of a thick accumulation layer and thin depletion layer in the sensing composite. The modulation of the thickness of the accumulation layer (acting as the conductive channel of the carriers) significantly influences the conductivity of the sensors, further resulting in improved gas sensing properties of the p-n or p-p junctions.

2.2.1. Gas Sensors Based on p-n Junctions

The p-n heterojunctions in sensing materials have been reported to be effective in improving the gas sensing properties of metal oxides. For example, the SnO-SnO_2 composite (p-n heterojunction) was successfully prepared by a facile two-step method with the raw materials of $SnCl_2{\cdot}2H_2O$, NaOH and CTAB at 140 °C for 5 h. Black SnO nanopowders were synthesised via a hydrothermal method at 140 °C for 5 h, and the obtained sample was then treated with an annealing process at a high temperature of 300–500 °C in air atmosphere to obtain the SnO–SnO_2 composite. The sensor response of the SnO-SnO_2 composite was 2.5 towards 200 ppm NO_2 at room temperature, significantly higher than that of pure SnO_2 (1.27) or bare SnO (1.1) [90]. The hydrothermal method was also applied to synthesise SnO_2-decorated NiO nanostructures (see Figure 5) with the raw sources of $NiCl_2{\cdot}6H_2O$ and $SnCl_4{\cdot}5H_2O$ at 160 °C for 12 h [91]. It is worth noting that NiO was modified with SnO_2 nanoparticles through a one-step process without any catalysts. The SnO_2-decorated NiO nanostructure was reported to show enhanced gas sensor responses to 1–200 ppm toluene (see Figure 5e,f). The calculated sensor response of the composite was measured to be 66.2 to 100 ppm toluene at 250 °C, more than 50 times higher than that of the pure NiO nanospheres (1.3). Moreover, the detection limit of this sensor was reported to be as low as 10 ppb toluene with a promising sensor response of 1.2.

Novel TiO_2-decorated Co_3O_4 acicular nanowire arrays were also successfully synthesised by Li et al. with a hydrogen thermal method combined with pulsed laser deposition. The s acicular nanowire arrays modified with TiO_2 nanoparticles were found to present a high sensor response of 65 to 100 ppm ethanol at 160 °C, much higher than that of the pure Co_3O_4 nanowires (~25) [92]. In_2O_3-decorated CuO nanowires were also prepared through thermal oxidation of Cu meshes followed by the deposition of amorphous indium hydroxide from $In(AC)_3$ solution in ammonia [93]. The decorated CuO nanowires showed a shorter response time of 12 s to CO than that of the pure CuO nanowires (25 s). The novel rod-like α-Fe_2O_3/NiO heterojunction nanocomposites were synthesised with a one-step hydrothermal method, exhibiting an enhanced sensor response of 290 to 100 ppm acetone at 280 °C with a response time or a recovery time being 28 s or 40 s, respectively [94].

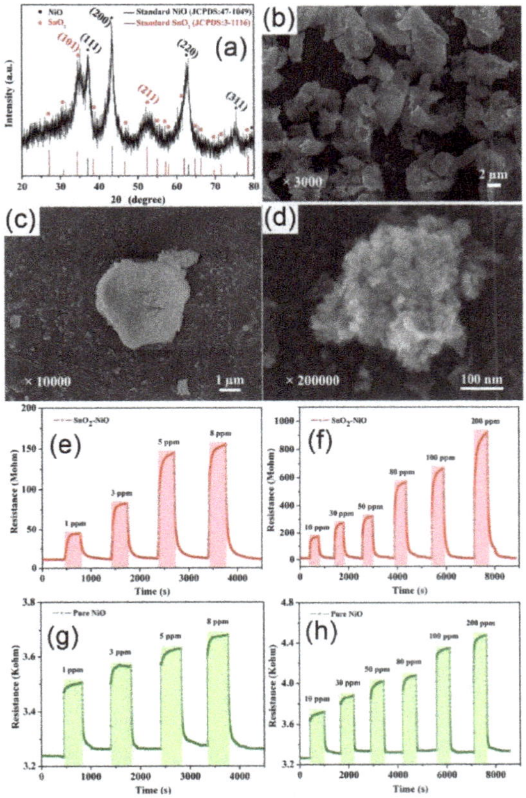

Figure 5. The XRD pattern (**a**) and the SEM images (**b–d**) of SnO_2 nanoparticles-modified NiO nanostructure (p-n junctions), the dynamic sensing performances of the pure NiO and the SnO_2 nanoparticles-modified NiO nanostructures towards 1–200 ppm toluene at 250 °C (**c**,**e–h**). Copied with permission [91]. Copyright 2018, Elsevier.

Additionally, CuO composited with TiO_2 [95] and SnO_2 [35] were constructed to investigate their improved gas-sensing properties. The nanofibres composed of SnO_2-CuO heterojunctions have been reported to be successfully synthesised by an electrospinning process and exhibited an improved sensor response of ~95 compared with that of the pure CuO (<10) [35]. Co_3O_4 composited with In_2O_3 [96], SiO_2 [97] and TiO_2 [98] were also successfully prepared via hydrothermal, thermal conversion and facile nanoscale coordination polymer routes, respectively, which showed better gas sensing properties than those of pure Co_3O_4. The reported sensors based on p-n heterojunctions and their gas sensing performances are listed in Table S3.

2.2.2. Gas Sensors Based on p-p Junctions

The p-p heterojunctions have been found to enhance the gas sensing performance of metal oxides. Li et al. prepared NiO@CuO nanocomposites (a p-p junction) via a facile reflux and hydrothermal process [99]. In their work, the $Ni(OH)_2$ was firstly synthesised with the raw material of nickel nitrate hexahydrate through a hydrothermal method at 140 °C for 5 h. Then, the obtained $Ni(OH)_2$ and the $Cu(CH_3CO)_2{\cdot}H_2O$ compounds were added in a solution separately with a certain amount of NaOH added during a reflux process to obtain the $Ni(OH)_2$@$Cu_2(OH)_3NO_3$. The synthesised products were finally treated by a calcination process in air atmosphere at 450 °C for 2 h. The prepared hierarchical flower-like nanostructured NiO-CuO composite exhibited an enhanced gas sensing performance to NO_2 at room temperature with a higher gas sensor response compared to pure NiO. The response time of the composite to the 100 ppm NO_2 was measured to be as low as 2 s, much shorter than that of the pure NiO. Moreover, the NiO/$NiCr_2O_4$ nanocomposite was also found to be more effective at detecting xylene than the pure NiO nanoparticles [100].

Co_3O_4 hollow nanocages (HNCs) decorated with PdO nanoparticles (see Figure 6a–d) were successfully assembled by the infiltration of metal precursors combined with a subsequent reduction process [101]. The gas sensing performance of the pure Co_3O_4 hollow nanocages was significantly improved when composited with PdO nanoparticles (PdO-Co_3O_4 HNCs), with the sensor response measured to be 2.51 towards 5 ppm acetone at 350 °C (see Figure 6e), which was higher than that of the Co_3O_4 powders (1.96), Co_3O_4 HNCs (1.45) or PdO-Co_3O_4 powders (1.98). Moreover, the PdO-Co_3O_4 HNCs also exhibited outstanding stability to 1 ppm acetone, which is shown in Figure 6f.

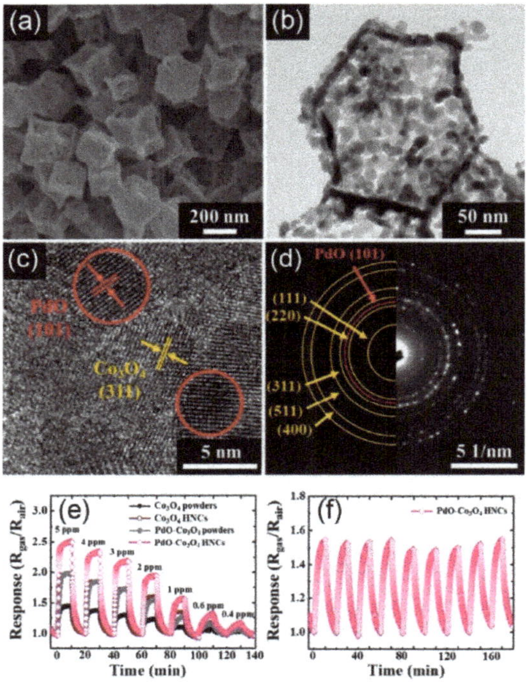

Figure 6. SEM image (**a**), TEM image (**b**), HRTEM (**c**) and SEAD patterns (**d**) of the Co_3O_4 hollow nanocages (HNCs) decorated with PdO nanoparticles (p-p junctions), the dynamic sensing behaviours of the sensors based on Co_3O_4 powders, pure Co_3O_4 hollow nanocages, PdO-Co_3O_4 powders and PdO-Co_3O_4 HNCs towards 0.4–5 ppm acetone at 350 °C (**e**), the stability of the sensing performance of PdO-Co_3O_4 HNCs towards 1 ppm acetone (**f**). Copied with permission [101]. Copyright 2017, American Chemical Society.

Lee et al. prepared TeO_2/CuO core-shell nanorods by a combined method of thermal evaporation and sputter deposition [102]. In the reported study, the Te powders were used as the raw material to synthesise TeO_2. The TeO_2 nanorods were prepared on a substrate of p-type Si (100) by thermal evaporation of Te powders at 400 °C in air in a quartz tube furnace. Then, a thin layer of CuO was directly sputtered on the surface of the obtained TeO_2 nanorods through a radio frequency magnetron sputtering process with a target of CuO. The sensor response of TeO_2-core/CuO-shell nanorods was found to be 4.25 to 10 ppm NO_2 at 150 °C, which was over two times higher than that of the pure TeO_2. However, the relatively low sensor response of the TeO_2/CuO core-shell nanorods is a drawback that limits their application. Further studies are required to further improve the gas sensing performance of the TeO_2/CuO core-shell nanorods.

Meanwhile, p-$NiFe_2O_4$ nanoparticle-decorated p-NiO nanosheets were also synthesised with a solvothermal method [103]. The NiO precursor was firstly synthesised after which $FeCl_3 \cdot 6H_2O$ was added to prepare NiO nanosheets decorated with $NiFe_2O_4$ nanoparticles. The ratio of Fe/Ni was found to have a significant effect on the gas sensing performance of the decorated NiO nanosheets. The composite with the Fe/Ni-24.9 exhibited the optimal sensing performance to 50 ppm acetone at 280 °C, with a high response of ~23.0. The release of captured electrons back to the sensing material breaks the dynamic carrier balance between p-NiO and p-$NiFe_2O_4$. This resulted in a reduced potential barrier near the surfaces of the heterojunctions and yielded a large variation in resistance, improving the sensor response of the Fe/Ni-24.9 at%. The in situ formation of a second phase (p-$NiFe_2O_4$) on the first phase (p-NiO) was a novel and effective strategy to improve the interaction between the targeted gas and the sensing composite. Similar improvements in CuO-NiO nanotubes with controllable element content of Cu/N developed by a one-pot synthesis was also found, with a sensing capability towards 100 ppm glycol at 110 °C [104]. Based on the studies listed above, the in situ preparation of the second phase required further attention to improve the gas sensing performance of the sensor based on the metal oxide. Other sensors composed with p-p heterojunctions and their gas sensing performances are listed in Table S4.

Other types of heterojunctions based on metal oxides that improve gas sensing performances also exist. Duy et al. assembled n-p-n heterojunctions with the structure of SnO_2-carbon nanotube-SnO_2 by the method of CVD combined with spray coating process [105]. The obtained n-p-n heterojunctions showed a high response of 17.9 to 100 ppm NO_2 at 100 °C. The n-p-n heterostructure of the ZnO-branched SnO_2 nanowires decorated with Cr_2O_3 nanoparticles [106] or the p-n-p heterojunctions of PANI coated CuO-TiO_2 nanofibres [107] were also reported to exhibit improved gas sensing performance towards hydrogen and ammonia, respectively. However, only a few references report the study of the sensor based on n-p-n or p-n-p heterojunctions. More research should be conducted to systematically investigate the gas sensing properties of metal oxide heterojunctions comprising the n-p-n or the p-n-p structures.

Based on the discussions above, many kinds of metal oxides heterojunctions have been successfully assembled to enhance the gas sensing performance towards various gases. The sensor response of sensors based on heterojunctions was much higher than that of the pure metal oxides and the response time was improved. The n-n, n-p, p-n or p-p (even the n-p-n or p-n-p) heterojunctions can be chosen to be constructed to assemble gas sensors with outstanding properties. We should point out that the enhanced gas sensing mechanisms of certain heterojunctions towards the oxidising or reducing gases need to be clearly discussed and compared to fully understand the role of the heterojunctions. Therefore, in the next section, we review the mechanisms of the improvements in the gas-sensing properties of the metal oxide heterojunctions.

3. Enhanced Gas Sensing Mechanisms of the Metal Oxide Heterojunctions

Compared with the pure metal oxides, sensors based on metal oxide heterojunctions show improved gas sensing performances towards the targeted gases. When in contact with each other, the transfer of carriers between the two semiconductor materials is induced due to inconsistent Fermi levels at their interfaces. In the n-n or n-p heterojunctions, the Fermi levels of the two metal oxides will move up or down to an equilibrium state, resulting in the bending of their energy bands and the formation of a potential barrier between them [28]. The gas sensing performances of the studied metal oxides are reported to be mainly attributed to the redox reactions of the adsorbed targeted gases on the surfaces of the sensing materials, which has been widely reported by researchers to explain the gas sensing mechanisms of the assembled sensors [12,108]. The variation in the concentration of the carriers induced by the redox reactions on the surfaces of the composites could be of importance to affect the height of the built-in potential barrier. This process further influences the resistance or conductivity of the sensor based on n-n or n-p heterojunctions according to Equation (1):

$$\Delta R \propto \exp\{-e\Delta V_b/k_B T\} \tag{1}$$

where the ΔR is the change of the resistance of the sensor, ΔV_b is the reduction of the height of the potential barrier, k_B is the Boltzmann constant and T is the temperature [109]. Therefore, little change in the height of a potential barrier would make the resistance of the investigated sensor vary greatly, leading to an improved gas sensing property of the heterojunction [110]. In the case of the p-n or p-p heterojunctions, the interaction between the targeted gases and the sensing materials would also modify the carriers (especially holes) in the sensors, which would further result in the variation of the thickness of the accumulating layer in the heterojunctions, making a more effective modulation in the width of the conduction channel for the carriers. As a result, sensors based on p-n or p-p heterojunctions also show improved gas sensing properties to the reducing or oxidising gases [108]. Moreover, the composites composed of metal oxide heterojunctions always show higher specific surface areas than the pure metal oxides, which was confirmed by BET measurements of the composites. The higher specific surface area enables gas molecules to diffuse more smoothly to the surface and more easily interact with the composite as well as provide more active sites. The size of the pore volume can be increased with a higher specific surface area, facilitating the diffusion of gas molecules into the sensing material and increasing the active surface in internal parts of the composite for gas molecule adsorption. The absorption and the desorption of the gas molecules can also be accelerated during the response and recovery process of the sensor based on metal oxide heterojunctions. Therefore, the high specific surface area forms another positive factor contributing to the comprehensive improvements in the gas sensing performance of the composite [111–115].

Compared with the effect of the specific surface area, it is more complex to study the enhanced gas sensing mechanisms of the heterojunctions in the sensing materials. The role of the heterojunctions in enhanced gas sensing performances should be analysed in detail to fully understand their direct and significant effects on the enhancement of the gas sensing properties of the sensors based on the composites. In the following section, the gas sensing mechanisms of the metal oxides to the common reducing and oxidising gases are discussed, and the effects of various commonly studied heterojunctions on the improved gas sensing properties of the composites are systematically investigated. In order to make the discussions clear, H_2 (a typical reducing gas) and NO_2 (a typical oxidising gas) were selected for the discussion of the enhanced gas sensing mechanisms of the metal oxides due to their immense studies in the area of gas sensors.

3.1. Enhanced Gas Sensing Mechanisms to Reducing Gases

Gas sensors based on n-n or n-p heterojunctions always exhibit typical n-type sensing performances at relatively low working temperatures towards reducing gases such as H_2, H_2S, CO, NH_3 and ethanol. The widely studied ZnO-based material is taken as an example to more clearly illustrate the sensing mechanism of the n-type metal oxide to a reducing gas. The resistance of ZnO-based sensors has been reported to decrease quickly when H_2 (ethanol or H_2S) is introduced onto their surface [116–118]. In air, oxygen molecules would spontaneously be adsorbed on the active sites of the surface of the ZnO to form chemisorbed oxygen molecules according to Equation (2). Then, the chemisorbed oxygen molecules can capture electrons from the conductive bands of the ZnO to become the oxygen species (O_2^-: <150 °C, O^-: 150 °C~400 °C and O^{2-}: >400 °C) based on Equation (3), which builds a depletion layer in the ZnO surface and a high resistance in air. When H_2 gas is introduced, the H_2 molecules will interact with the pre-adsorbed oxygen species to form H_2O based on Equation (4), releasing electrons back to ZnO. This response process increases the concentration of electrons and decreases the thickness of the depletion layer in ZnO, leading to a decrease in the resistance of the sensor based on ZnO-based materials.

$$O_2(g) + e^- \leftrightarrow O_2(ad) \quad (2)$$

$$O_2(ad) + e^- \leftrightarrow O_2^-(ad) \quad (3)$$

$$2H_2(g) + O_2^-\ (ad) = 2H_2O(g) + e^- \quad (4)$$

In contrast, composites made of p-n or p-p heterojunctions show typical p-type sensing performances towards reducing gases. As reported, the resistance of CuO nanowires increased when used to detect hydrogen gas (at working temperatures between 150 °C and 400 °C) [119]. When the CuO nanowires were placed in an air atmosphere, the oxygen molecule could also adsorb on the active sites in the surface of CuO to form adsorbed oxygen species (O^-) based on Equation (5), releasing holes to CuO and thus increasing the concentration of holes. This forms an accumulation layer in the sensing material, which acts as the conduction channel for carriers in CuO. In an H_2 atmosphere, hydrogen molecules interact with the adsorbed oxygen molecules according to Equation (6), reducing the concentration of carriers and the thickness of the accumulation layer and induces the formation of a depletion layer on the surface of CuO. Therefore, the resistance of a sensor based on CuO nanowires increases in reducing gas environments [108,119].

$$O_2(g) \leftrightarrow O_2^-(ad) + h^+ \leftrightarrow 2O^-(ad) + 2h^+ \quad (5)$$

$$2H_2(g) + 2O^-(ad) + 2h^+ = 2H_2O(g) \quad (6)$$

The synthesis of TiO_2 nanotubes decorated with SnO_2 nanoparticles and their H_2 sensing performance has been reported [89] and is selected to analyse the important role of the typical n-n heterojunction in improving the sensing performance of the sensor towards reducing gases. The results showed that the H_2 sensing property of the TiO_2-based composite was highly improved with the help of the heterojunction between TiO_2 and SnO_2. It was reported that the Fermi level of TiO_2 was higher than that of SnO_2, resulting in the electron transfer to SnO_2 from TiO_2 until achieving the equilibrium states of their Fermi levels. This would make a thick depletion layer formed at the interface between TiO_2 and SnO_2 and induce a high potential barrier built in air due to the adsorption of oxygen molecules. The potential barrier always acts as the obstacle to the transportation of electrons in the sensing materials, resulting in the high resistance of the composites. The accumulation of electrons in the SnO_2 side would induce more oxygen molecules adsorbed onto the surface of the composite. When hydrogen gas is introduced, the hydrogen gas interacts with the adsorbed oxygen species on the surfaces of TiO_2 and SnO_2 immediately and releases electrons back to the sensing materials. The released electrons would decrease the thickness of the depletion layers between TiO_2 and SnO_2, further resulting in the decrease in the height of the potential barrier. This process would increase the conductivity of the sensor and significantly enhance the H_2 sensing performance of the composite. The porous MoO_3/SnO_2 nanoflakes with n-n junctions was also reported to show an improved gas sensing property with a higher gas sensor response being 43.5 towards 10 ppm H_2S at 115 °C compared with that of the pure SnO_2, which could also be attributed to the reasons mentioned previously (see Figure 7a1,a2) [120]. Moreover, the improvement in the H_2S or xylene sensing performance of TiO_2-decorated α-Fe_2O_3 nanorods [121] or Fe_2O_3 nanoparticles-decorated MoO_3 nanobelts [122] could also be explained by the enhanced gas sensing mechanism above.

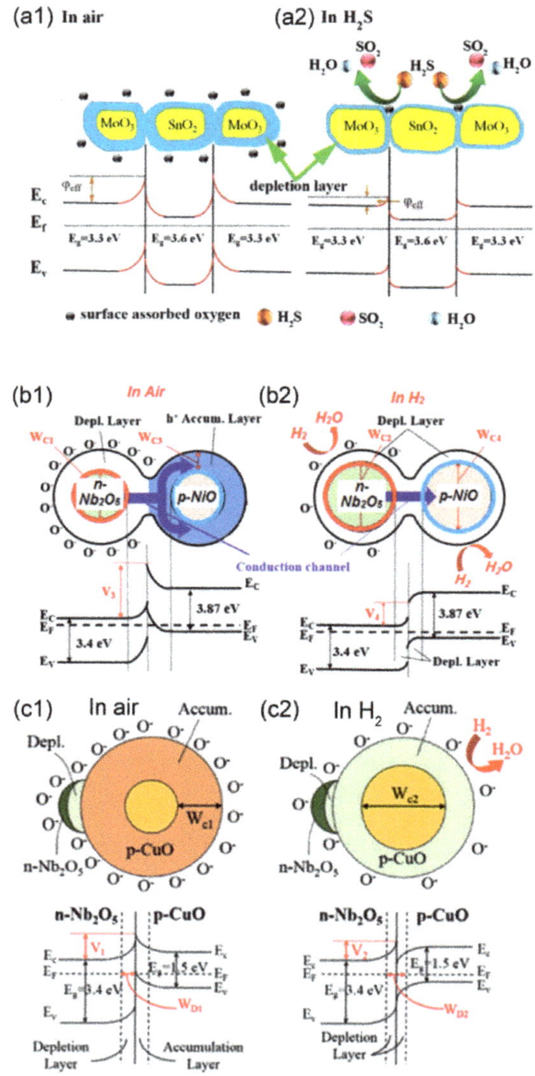

Figure 7. The enhanced gas sensing mechanisms of (**a1**,**a2**) MoO_3/SnO_2 nanoflakes (n-n heterojunction) (Copied with permission [120]. Copyright 2019, American Chemical Society), (**b1**,**b2**) NiO-Nb_2O_5 composite nanoparticles (n-p heterojunction) (Copied with permission [123]. Copyright 2017, Elsevier) and (**c1**,**c2**) Nb_2O_5 nanoparticle-decorated CuO nanorods (p-n heterojunction) (Copied with permission [119]. Copyright 2017, Springer Nature) to reducing gases.

In the case of sensors based on n-p heterojunctions, the NiO-decorated Nb_2O_5 nanocomposites have been reported to exhibit a significant improvement in the H_2 gas sensing performance compared with that of the pure Nb_2O_5 nanoparticles [123]. When the NiO nanoparticles are loaded onto the surface of the Nb_2O_5 nanoparticles, the electrons diffuse to the Nb_2O_5 and the holes move toward the NiO, causing the Fermi levels of the two different metal oxides to reach an equilibrium state. In air, the adsorption of the oxygen molecules on the surfaces of the NiO and the Nb_2O_5 also results in the formation of an accumulation layer of holes in the NiO side and a depletion layer in the Nb_2O_5 side. This causes the energy bands of NiO to bend upwards, increasing the potential barrier at the interfaces in the region of heterojunctions. When the NiO-decorated Nb_2O_5 nanocomposites is exposed to H_2, the interaction with H_2 and adsorbed oxygen species releases electrons to Nb_2O_5 but captures the holes in the NiO. This process induces the formation of a depletion layer between NiO and Nb_2O_5 and makes the energy bands of NiO bend downwards, dramatically decreasing the height of the potential barrier at the heterojunction (see Figure 7b1,b2). The NiO nanoparticles have also been reported to be an excellent catalyst to effectively oxidise H_2, causing reactions between the adsorbed H_2 and the adsorbed oxygen species to occur more sufficiently and smoothly. As a result, the NiO-decorated Nb_2O_5 nanocomposites exhibit an improved gas sensing property to H_2. The enhanced gas sensing performances of the Co_3O_4-decorated WO_3 nanowires [124], SnO_2-Co_3O_4 composite nanofibres [67], CuO-loaded In_2O_3 nanofibres [125], hierarchical SnO/SnO_2 nanocomposites [126], ZnO nanowire arrays/CuO nanospheres heterostructures [127] and p-NiS/n-In_2O_3 heterojunction nanocomposites [34] towards reducing gases can also be attributed to the reasons listed above.

For the p-n junction, Lee et al. reported the sensor based on Nb_2O_5 nanoparticles-decorated CuO nanorods to be more sensitive towards hydrogen molecules than the pure CuO nanorods [119]. The higher Fermi level of Nb_2O_5 makes the electrons diffuse to the CuO and the holes transfer in an opposite orientation, leading to the bending of energy bands. In air, the adsorption of oxygen molecules captures the electrons from Nb_2O_5 but releases the holes to the CuO, resulting in the formation of a thick depletion layer in Nb_2O_5 and a thick accumulation layer in the CuO. This leads to the high potential barrier in the composite in air. As reported, the accumulation layer in the CuO can act as a conduction channel for carriers in the sensing material. When exposed to H_2, the hydrogen molecule can interact with the adsorbed oxygen species on the Nb_2O_5 and the CuO, releasing the electrons back to the Nb_2O_5 but capturing the holes in the CuO. Effectively, this decreases the thickness of the depletion layer in Nb_2O_5 and significantly thins the accumulation layer in CuO with the possible formation of a depletion layer in the CuO, attributed to more oxygen molecules adsorbed on the surface of the CuO due to the formation of heterojunctions. This dramatic decrease in the thickness of the accumulation layer greatly narrows the conduction channel width for carriers, as shown Figure 7(c1,c2). As a result, the Nb_2O_5 nanoparticle-decorated CuO nanorods exhibited an improved p-type sensing performance to H_2. The improved sensing mechanism could also reasonably explain the enhancements in the gas sensing performances of the In_2O_3-decorated CuO nanowires [93], SnO_2-decorated NiO nanostructure [91], hierarchical α-Fe_2O_3/NiO composites [128], SnO_2-decorated NiO foam [129] and Cu_xO-modified ZnO composite [130] with a hollow structure towards H_2, the toluene and the acetone.

Similarly, in the case of the p-p heterojunction, p-NiO-decorated p-CuO microspheres were prepared through a hydrothermal process and studied the enhanced gas sensing performance of the obtained composite with p-p heterojunctions [131]. The Fermi level of the CuO was higher than that of the NiO, resulting in the transfer of holes from NiO to CuO and the diffusion of electrons to NiO from CuO. As such, accumulation and depletion layers of holes build on the CuO and the NiO sides, respectively. In air atmosphere, the adsorption of the oxygen molecules on the surfaces of CuO and NiO releases holes to the sensing material as previously mentioned. As a result, the thickness of the depletion layer of the holes in the NiO decreases, but the thickness of the accumulation layer of the holes in the CuO increases. The width of the conduction channel increased in the heterojunctions between CuO and NiO, causing a low resistance of the composite. When a reducing gas was introduced on the surface of the composite, the adsorbed oxygen ions reacted with the introduced gas molecules, capturing the holes from CuO and NiO. This process decreases the concentration of the carriers in the sensing materials, further increasing the thickness of the depletion layer of the holes in NiO and decreasing the thickness of the accumulation layer of holes in CuO. Therefore, the width of the conduction channel increased in the heterojunctions was significantly narrowed, causing an increase in the resistance of the composites. Therefore, the gas sensor response of the p-NiO-decorated p-CuO microspheres was highly improved towards reducing gases. The enhanced gas sensing mechanisms of p-NiO/p-$NiCr_2O_4$ nanocomposites [132] or the Cr_2O_3-Co_3O_4 nanofibres [133] to xylene or C_2H_5OH can also be explained by the above discussions.

Apart from H_2, there are also a number of references reporting similar improved sensing mechanisms of sensors based on metal oxide heterojunctions towards NH_3, another widely investigated reducing gas. The work conducted by Shi et al. showed that the sensor response of WO_3@$CoWO_4$ (n-n type) heterojunction nanofibres was over 10 times higher than that of the bare WO_3 at room temperature [134]. The authors pointed out the formation of a number of heterojunctions for WO_3 composited with $CoWO_4$ in the sensing material. The differences in the Fermi levels of WO_3 and $CoWO_4$ cause band bending and trigger the transfer of electrons and holes between them until an equilibrium in final Fermi levels is reached. In air, oxygen gas can be adsorbed on the surface of the two different sensing materials and capture electrons from their conductive bands to form chemisorbed oxygen ions (O_2^- at room temperature) according to Equation (3). This process results in a wider depletion layer and constructs a higher potential barrier near the surface of the heterojunction in the composite than those in the pure $CoWO_4$. NH_3 molecules could interact with the O_2^- according to the following equation: $4NH_3(ad) + 5O_2^-\ (ad) \rightarrow 4NO(g) + 6H_2O(g) + 5e^-$. Electrons were then released back to the sensing materials of the WO_3@$CoWO_4$ composite, reducing the thickness of the depletion layer and decreasing the height of the potential barrier at the heterojunction. As a result, the sensing performance of WO_3@$CoWO_4$ heterojunction nanofibres towards NH_3 could be significantly enhanced at room temperature. Additionally, the specific surface area of the composite was higher, allowing more electrons to be transferred from the shallow donor levels of the WO_3 nanoparticles to $CoWO_4$ nanoparticles in the composite, thus enabling the enhanced NH_3 sensing property of the sensor based on the metal oxide heterojunction. The study of Gong et al. also revealed that the enhanced NH_3 sensing performance of the flower-like n-ZnO decorated with p-NiO with hierarchical structure was mainly attributed to the formation of the depletion layer and the modulation of the potential barrier height at the surface of the heterojunction [135]. A similar improved sensing mechanism was also reported in the enhanced NH_3 sensing performance of the sensors based on other heterojunctions, including polyaniline/$SrGe_4O_9$ nanocomposite [136], polyaniline nanograin enchased TiO_2 fibres [137], SnO_2@polyaniline nanocomposites [138], V_2O_5/$CuWO_4$ heterojunctions [139], Fe_2O_3-ZnO nanocomposites [49], rGO/WO_3 nanowire nanocomposites [140], WO_3@SnO_2 core-shell nanosheets [141], PANI-CeO_2 nanocomposite thin films [142], CuPc-loaded ZnO nanorods [143], Co_3O_4 nanorod-decorated MoS_2 nanosheets [144], SnO_2/NiO composite nanowebs [145], bilayer SnO_2-WO_3 nanofilms [146], Cu_2O nanoparticles decorated with p-type MoS_2 nanosheets [147], TiO_2 and NiO nanostructured bilayer thin films [148] and mesoporous In_2O_3@CuO multijunction nanofibres [149].

3.2. *Improved Gas Sensing Mechanism towards Oxidising Gases*

In contrast to the sensing behaviour of n-n or n-p heterojunctions towards reducing gases, sensors based on n-n or the n-p heterojunctions were reported to exhibit a typical p-type sensing performance towards the oxidising gases. Many researchers have studied the oxidising gas (such as NO_2) sensing performance of heterojunctions based on n-type metal oxides at working temperatures within the range of 150 °C to 400 °C. The ZnO nanoparticles exhibited a typical p-type sensing performance towards 0.3–10 ppm NO_2 at the working temperature of 250 °C, with the resistance of the sensor increasing quickly when exposed to an NO_2 gas atmosphere [150]. In air, oxygen molecules can adsorb onto the active sites of the surface of the nanostructured ZnO according to Equation (7), capturing electrons from the conductive bands of ZnO. This process causes a decrease in carriers in ZnO and the formation of a depletion layer on the surface of ZnO. When exposed to an NO_2 atmosphere, the NO_2 molecules can interact with the ZnO directly and with adsorbed O^- on the ZnO according to Equations (8) and (9), respectively. Generally, NO_2 can be adsorbed onto the active sites of the ZnO surface based on Equation (8), capturing the electrons from the ZnO to form NO_2^-. The NO_2^- can then further react with adsorbed oxygen species following Equation (9), grabbing electrons from ZnO. These sensing processes decrease the concentration of the carriers in the ZnO and increase the thickness of the depletion layer in the surface of the ZnO, resulting in a significant increase in the resistance of ZnO and the p-type sensing performance to NO_2 gas.

$$1/2O_2(g) + e^- \leftrightarrow O^-(ad) \tag{7}$$

$$NO_2(g) + e^- \leftrightarrow NO_2^-(ad) \tag{8}$$

$$NO_2^-(ad) + O^-(ad) + 2e^- \leftrightarrow NO(g) + 2O_2^- \tag{9}$$

In addition, sensors based on p-n or p-p heterojunctions have been found to show typical n-type sensing performances towards oxidising gases. The resistances of the sensors assembled by heterojunctions based on p-type metal oxides decrease rapidly when exposed to oxidising gases at the working temperatures of approximately 200 °C. Taking the sensor based on Co_3O_4 as an example, oxygen molecules can be adsorbed onto the active sites on the surface of Co_3O_4 according to Equation (5), releasing holes to Co_3O_4 and resulting in the formation of chemisorbed oxygen species (mainly O^-). This process can also induce the establishment of an accumulation layer in the surface of Co_3O_4. In an NO_2 atmosphere, NO_2 molecules have also been reported to adsorb onto the active sites of a Co_3O_4 surface based on Equation (10), releasing holes to the sensing materials and forming NO_2^-. The adsorbed NO_2^- can then interact with the adsorbed oxygen species according to Equation (11), releasing more holes to Co_3O_4. These processes make the accumulating layer thicker on the surface of Co_3O_4 and cause the resistance of the sensor to decrease, leading to an n-type sensing performance of the Co_3O_4-based sensor towards NO_2 [64].

$$NO_2(g) \leftrightarrow NO_2^-(ad) + h^+ \tag{10}$$

$$NO_2^-(ad) + O^-(ad) \leftrightarrow NO(g) + 2O_2^-(ad) + 2h^+ \tag{11}$$

For the sensor based on an n-n heterojunction, the study of the gas sensing properties of ZnO-SnO_2 hollow nanofibres showed that the composites exhibit a much higher sensor response towards NO_2 than pure SnO_2 [151]. In the composite, the Fermi level of the SnO_2 is higher than that of the ZnO. The lower Fermi level of the ZnO can thus lead to the transfer of the holes from ZnO to SnO_2 and the diffusion of electrons to ZnO from SnO_2 until their Fermi levels reach an equilibrium state. This process can then form a thick accumulation layer on the ZnO side and a thin depletion layer on the SnO_2 side. In air atmosphere, oxygen molecules can adsorb onto the surface of SnO_2, which would capture electrons from SnO_2 and increase the thickness of the built-in depletion layer. The accumulation layer of the electrons in ZnO can cause more molecules to absorb onto its surface in air, capturing electrons and significantly decreasing the thickness of the established accumulation layer and even lead to the formation of a thin depletion layer (see Figure 8(a1)). In an NO_2 atmosphere, adsorbed NO_2 molecules on the surfaces of the metal oxides and the reaction between NO_2 and adsorbed oxygen molecules further capture electrons in ZnO and SnO_2, significantly increase in the depletion layer at the interfaces between ZnO and SnO_2 (see Figure 8(a2)). As a result, the height of the potential barrier increases greatly, making the ZnO-SnO_2 hollow nanofibres exhibit an improved NO_2 gas sensing property. The sensors based on ZnO nanorods decorated with TiO_2 nanoparticles [61], Bi_2O_3-branched SnO_2 nanowires [112], In_2O_3-composited SnO_2 nanorods [152] and SnO_2-core/ZnO-shell nanowires [153] also exhibited improved NO_2 gas sensing performances according to the mechanism discussed above.

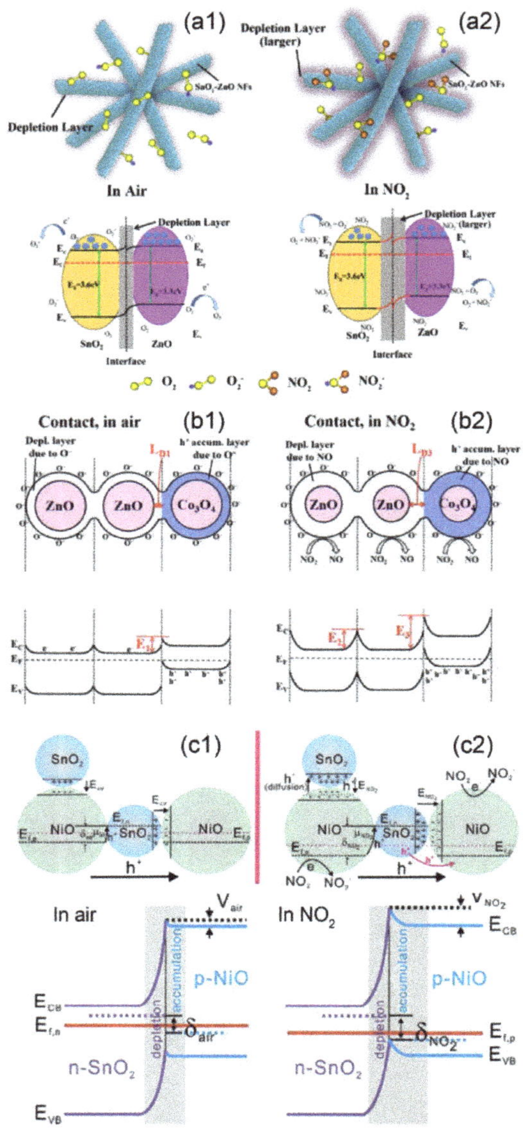

Figure 8. The enhanced gas sensing mechanisms of (**a1**,**a2**) the hollow nanofibres of ZnO-SnO_2 (n-n heterojunction) (Copied with permission [151]. Copyright 2018, Elsevier.), (**b1**,**b2**) the ZnO/Co_3O_4 composite nanoparticle (n-p heterojunction) (Copied with permission [64]. Copyright 2016, Elsevier.) and (**c1**,**c2**) NiO-SnO_2 nanocomposites (p-n heterojunction) (Copied with permission [154]. Copyright 2016, Royal Society of Chemistry.) towards oxidising gases.

In the case of the n-p junction, Co_3O_4-decorated ZnO nanoparticles have been established by Lee et al. and showed a significant enhancement in the NO_2 gas sensing performance [64]. The Fermi level of ZnO is higher than that of Co_3O_4, inducing the transfer of carriers between them and the formation of a depletion layer at the heterojunction. In air, the adsorption of the oxygen molecules on the surfaces of ZnO and Co_3O_4 capture the electrons from ZnO and release holes to Co_3O_4. This leads to the building of a depletion layer on the ZnO side and an accumulation layer on the Co_3O_4 side as well as a significant bending in their energy bands (see Figure 8(b1)). As a result, a potential barrier is formed at the interfaces between ZnO and Co_3O_4, resulting in a higher resistance of the sensor based on the composites than that of pure ZnO. In an NO_2 atmosphere, the adsorption of NO_2 molecules on the surfaces of ZnO and Co_3O_4 can lead to the capture of electrons from ZnO but the release of the holes to Co_3O_4. The variation in the carriers in ZnO and Co_3O_4 causes both the depletion layer in ZnO and the accumulation layer in Co_3O_4 to become thicker. This sensing process increases the height of the potential barrier in the heterojunctions and significantly increases the resistance of the Co_3O_4-decorated ZnO composite (see Figure 8(b1,b2)). Meanwhile, the catalytic property of Co_3O_4 to NO_2 also acts as a positive factor for the improved NO_2 gas sensing performance of the composite. Oxygen molecules are reported to be more easily adsorbed onto the surface of p-type metal oxides, which is another reason for the improved NO_2 gas sensing performance of the Co_3O_4-decorated ZnO nanoparticles. The improvements in the NO_2 gas sensing properties of the SnO-SnO_2 nanocomposites [90], CuO-decorated ZnO nanowires [155], TeO_2/SnO_2 brush- nanowires [156] and ultra-long ZnO@Bi_2O_3 heterojunction nanorods [157] can also be attributed to the reasons listed above.

Sensors based on the p-n heterojunctions have also been reported to be effective at detecting oxidising gases. For example, NiO-SnO_2 nanocomposites (p-n junctions) were found to exhibit an improved gas sensing performance towards NO_2 compared with pure NiO [154]. The Fermi level of the n-type SnO_2 was found to be higher than that of the p-type NiO. The electrons would be transferred from SnO_2 to the NiO, and the holes would diffuse from the NiO to the SnO_2. In an air atmosphere, the adsorption of oxygen molecules would capture electrons in the SnO_2 and release the holes to NiO, resulting the formation of a thin depletion layer in SnO_2 and an accumulation layer in NiO. A potential barrier is then established between NiO and SnO_2, and the carriers in the sensing materials are mainly transported through the accumulation layer. In an NO_2 atmosphere, the adsorption of NO_2 molecules on the surface of NiO results in more holes being released to NiO, further increasing the thickness of the accumulation layer in the NiO layer. The adsorption of NO_2 molecules on the surface of SnO_2 would allow more electrons to be grabbed from the SnO_2, further increasing the thickness of the depletion layer in SnO_2. Moreover, NO_2 can be adsorbed on SnO_2 more easily due to its higher electron concentration. There would be more NO_2 molecules adsorbed on the NiO-SnO_2 nanocomposites, further resulting in the great modulation in the accumulation layer of the heterojunction nanocomposites. The increase in the thickness of the accumulation layer in NiO would widen the conduction channel for the carriers, which would result in a significant decrease in the resistance of the sensing material (see Figure 8(c1,c2)). Therefore, the sensors based on NiO-SnO_2 nanocomposites exhibit a higher sensor response to NO_2 than that of bare NiO.

For the p-p heterojunction, the NO_2 gas sensing performance of sensors based on p-type NiO nanosheets could be successfully improved through modifying them with the p-type CuO nanoparticles [158]. In the CuO-decorated NiO nanosheets, the differences in Fermi levels of CuO and NiO lead to the transfer of carriers between the two, resulting in the formation of a hole depletion layer and a hole accumulation layer between their interfaces. In air, the adsorption of oxygen molecules on the surfaces of CuO and NiO can release holes to the sensing materials, leading to the increase in the thickness of the accumulation layer at the interfaces between CuO and NiO. When NO_2 is introduced and interacts with the sensing material, more holes are released to CuO and NiO. Moreover, more NO_2 molecules become adsorbed on the sensing material due to the accumulation of holes in the composite and its higher specific surface area. This sensing process would more effectively increase the carriers (holes) in the composites and widen the accumulation layer between CuO and NiO. As a result, the resistance of the p-p heterojunctions significantly decreased and the CuO-decorated NiO nanosheets presented an enhanced NO_2 gas sensing performance. The enhanced NO_2 sensing mechanism discussed above can also be applied to explain the improved NO_2 sensing properties of the sensors based on CuO-decorated TeO_2 nanorods [102] and vertically aligned $Cu_3Mo_2O_9$ micro/nanorods on a CuO layer ($Cu_3Mo_2O_9$@CuO nanorods) [159].

The discussions of the enhanced gas sensing mechanisms of the n-n, n-p, p-n and p-p heterojunctions reveal that modulations of the height of the potential barriers and the thickness of the accumulation layer in the heterojunctions are responsible for the improvements of the gas sensing performances of the nanocomposites. The different Fermi levels of the metal oxides induce band bending in the heterojunctions, leading to the formation of potential barriers and accumulation layers in n-type and p-type heterojunctions, respectively. The interactions between the targeted gases and the sensing materials cause variations in the height of the potential barriers in n-type heterojunctions (n-n or n-p heterojunctions) and the thickness of the accumulation layer in p-type heterojunctions (p-n or p-p heterojunctions), inducing enhancements of the sensing performance of the composites.

4. Conclusions

The gas sensing performances of metal oxides have been successfully improved through assembling heterojunctions in sensing materials. The heterojunctions are usually constructed via combined methods of electrospinning, thermal oxidation, ALD, PLD, hydrothermal process and CVD. The sensor response, response time or recovery time based on n-n, n-p, p-n or p-p heterojunctions can be effectively enhanced. Modulations in the built-in heterojunctions are mainly responsible for the enhanced gas sensing performances of the sensors based on n-n or n-p junctions. The improvement in the gas sensing behaviours of the sensors based on p-n or p-p heterojunctions can be attributed to variations in the thicknesses of the accumulation layers in the junctions. n-type or p-type nanostructured metal oxides with different morphologies can be selected to assemble heterojunctions and their concentrations can modified, indicating that more interesting gas sensors based on nanostructured metal oxide heterojunctions might be explored.

Supplementary Materials: The following are available online at https://www.mdpi.com/article/10.3390/nano11041026/s1, Table S1: The assembled strategies of n-n heterojunctions and their gas sensing performances. Table S2: The assembled strategies of n-p heterojunctions and their gas sensing performances. Table S3: The assembled strategies of p-n heterojunctions and their gas sensing performances. Table S4: The assembled strategies of p-p heterojunctions and their gas sensing performances.

Author Contributions: Writing—original draft preparation and editing, formal analysis and investigation, S.Y. and Z.W.; investigation and resources, G.L., Z.W. and Z.L.; writing—review, supervision and project administration, H.X. and H.G. All authors helped in drafting the final manuscript. All authors have read and agreed to the published version of the manuscript.

Funding: This research was funded by the National Natural Science Foundation of China (Grant No. 51802109 and 51972102), the Key Project of the Education Department of Hubei Province (Grant No. D20202903) and the ChuTian Scholars Program of Hubei Province.

Conflicts of Interest: The authors declare no conflict of interest.

References

1. Li, T.; Zeng, W.; Wang, Z. Quasi-one-dimensional metal-oxide-based heterostructural gas-sensing materials: A review. *Sens. Actuators B Chem.* **2015**, *221*, 1570–1585. [CrossRef]
2. Mirzaei, A.; Leonardi, S.G.; Neri, G. Detection of hazardous volatile organic compounds (VOCs) by metal oxide nanostructures-based gas sensors: A review. *Ceram. Int.* **2016**, *42*, 15119–15141. [CrossRef]
3. Zhang, J.; Qin, Z.; Zeng, D.; Xie, C. Metal-oxide-semiconductor based gas sensors: Screening, preparation, and integration. *Phys. Chem. Chem. Phys.* **2017**, *19*, 6313–6329. [CrossRef] [PubMed]
4. Sahay, P.P. Multifunctional metal oxide nanomaterials for chemical gas sensing. *Procedia Eng.* **2017**, *215*, 145–151. [CrossRef]
5. Seiyama, T.; Kato, A.; Fujiishi, K.; Nagatani, M. A new detector for gaseous components using semiconductive thin films. *Anal. Chem.* **1962**, *34*, 1502–1503. [CrossRef]
6. Hoa, N.D.; Duy, N.V.; El-Safty, S.A.; Hieu, N.V. Meso-/Nanoporous semiconducting metal oxides for gas sensor applications. *J. Nanomater.* **2015**, *16*, 186. [CrossRef]
7. Vuong, N.M.; Kim, D.; Kim, H. Surface gas sensing kinetics of a WO_3 nanowire sensor: Part 2—Reducing gases. *Sens. Actuators B Chem.* **2016**, *224*, 425–433. [CrossRef]

8. Vuong, N.M.; Kim, D.; Kim, H. Surface gas sensing kinetics of a WO_3 nanowire sensor: Part 1—Oxidizing gases. *Sens. Actuators B Chem.* **2015**, *220*, 932–941. [CrossRef]
9. Cao, P.-J.; Li, M.; Rao, C.N.; Han, S.; Xu, W.-Y.; Fang, M.; Liu, X.-K.; Zeng, Y.-X.; Liu, W.-J.; Zhu, D.-L.; et al. High sensitivity NO_2 gas sensor based on 3D WO_3 microflowers assembled by numerous nanoplates. *J. Nanosci. Nanotechnol.* **2020**, *20*, 1790–1798. [CrossRef] [PubMed]
10. Wang, C.; Yin, L.; Zhang, L.; Xiang, D.; Gao, R. Metal oxide gas sensors: Sensitivity and influencing factors. *Sensors* **2010**, *10*, 2088–2106. [CrossRef] [PubMed]
11. Xu, K.; Fu, C.; Gao, Z.; Wei, F.; Ying, Y.; Xu, C.; Fu, G. Nanomaterial-based gas sensors: A review. *Instrum. Sci. Technol.* **2018**, *46*, 115–145. [CrossRef]
12. Wang, Z.; Hu, Y.; Wang, W.; Zhang, X.; Wang, B.; Tian, H.; Wang, Y.; Guan, J.; Gu, H. Fast and highly-sensitive hydrogen sensing of Nb_2O_5 nanowires at room temperature. *Int. J. Hydrog. Energy* **2012**, *37*, 4526–4532. [CrossRef]
13. Yang, S.; Wang, Z.; Hu, Y.; Luo, X.; Lei, J.; Zhou, D.; Fei, L.; Wang, Y.; Gu, H. Highly responsive room-temperature hydrogen sensing of α-MoO_3 nanoribbon membranes. *ACS Appl. Mater. Interfaces* **2015**, *7*, 9247–9253. [CrossRef]
14. Lin, Y.; Wang, Y.; Wei, W.; Zhu, L.; Wen, S.; Ruan, S. Synergistically improved formaldehyde gas sensing properties of SnO_2 microspheres by indium and palladium co-doping. *Ceram. Int.* **2015**, *41*, 7329–7336. [CrossRef]
15. Ahn, M.-W.; Park, K.-S.; Heo, J.-H.; Park, J.-G.; Kim, D.-W.; Choi, K.J.; Lee, J.-H.; Hong, S.-H. Gas sensing properties of defect-controlled ZnO-nanowire gas sensor. *Appl. Phys. Lett.* **2008**, *93*, 263103. [CrossRef]
16. Choi, Y.J.; Hwang, I.S.; Park, J.G.; Choi, K.J.; Park, J.H.; Lee, J.H. Novel fabrication of an SnO_2 nanowire gas sensor with high sensitivity. *Nanotechnology* **2008**, *19*, 095508. [CrossRef] [PubMed]
17. Lin, S.; Li, D.; Wu, J.; Li, X.; Akbar, S.A. A selective room temperature formaldehyde gas sensor using TiO_2 nanotube arrays. *Sens. Actuators B Chem.* **2011**, *156*, 505–509. [CrossRef]
18. Ionescu, R.; Hoel, A.; Granqvist, C.G.; Llobet, E.; Heszler, P. Low-level detection of ethanol and H_2S with temperature-modulated WO_3 nanoparticle gas sensors. *Sens. Actuators B Chem.* **2005**, *104*, 132–139. [CrossRef]
19. Lim, S.K.; Hwang, S.-H.; Chang, D.; Kim, S. Preparation of mesoporous In_2O_3 nanofibers by electrospinning and their application as a CO gas sensor. *Sens. Actuators B Chem.* **2010**, *149*, 28–33. [CrossRef]
20. Miller, D.R.; Akbar, S.A.; Morris, P.A. Nanoscale metal oxide-based heterojunctions for gas sensing: A review. *Sens. Actuators B Chem.* **2014**, *204*, 250–272. [CrossRef]
21. Bao, H.-F.; Yue, T.-T.; Zhang, X.-X.; Dong, Z.; Yan, Y.; Feng, W. Enhanced ethanol-sensing properties based on modified NiO-ZnO p-n heterojunction nanotubes. *J. Nanosci. Nanotechnol.* **2020**, *20*, 731–740. [CrossRef]
22. Nakate, U.T.; Ahmad, R.; Patil, P.; Wang, Y.; Bhat, K.S.; Mahmoudi, T.; Yu, Y.T.; Suh, E.-K.; Hahn, Y.-B. Improved selectivity and low concentration hydrogen gas sensor application of Pd sensitized heterojunction n-ZnO/p-NiO nanostructures. *J. Alloys Compd.* **2019**, *797*, 456–464. [CrossRef]
23. Sun, Q.; Wang, J.; Hao, J.; Zheng, S.; Wan, P.; Wang, T.; Fang, H.; Wang, Y. SnS_2/SnS p–n heterojunctions with an accumulation layer for ultrasensitive room-temperature NO_2 detection. *Nanoscale* **2019**, *11*, 13741–13749. [CrossRef]
24. Sun, J.; Sun, L.; Han, N.; Pan, J.; Liu, W.; Bai, S.; Feng, Y.; Luo, R.; Li, D.; Chen, A. Ordered mesoporous WO_3/ZnO nanocomposites with isotype heterojunctions for sensitive detection of NO_2. *Sens. Actuators B Chem.* **2019**, *285*, 68–75. [CrossRef]
25. Jayababu, N.; Poloju, M.; Shruthi, J.; Reddy, M.V.R. NiO decorated CeO_2 nanostructures as room temperature isopropanol gas sensors. *RSC Adv.* **2019**, *9*, 13765–13775. [CrossRef]
26. Nguyen, D.D.; Do, D.T.; Vu, X.H.; Dang, D.V.; Nguyen, D.C. ZnO nanoplates surfaced-decorated by WO_3 nanorods for NH_3 gas sensing application. *Adv. Nat. Sci. Nanosci. Nanotechnol.* **2016**, *7*, 015004. [CrossRef]
27. Zhao, X.; Ji, H.; Jia, Q.; Wang, M. A nanoscale Co_3O_4–WO_3 p-n junction sensor with enhanced acetone responsivity. *J. Mater. Sci. Mater. Electron.* **2015**, *26*, 8217–8223. [CrossRef]
28. Park, S.; Kheel, H.; Sun, G.-J.; Ko, T.; Lee, W.I.; Lee, C. Acetone gas sensing properties of a multiple-networked Fe_2O_3-functionalized CuO nanorod sensor. *J. Nanomater.* **2015**, *2015*, 1–6. [CrossRef]
29. Wang, L.; Lou, Z.; Wang, R.; Fei, T.; Zhang, T. Ring-like PdO-decorated NiO with lamellar structures and their application in gas sensor. *Sens. Actuators B Chem.* **2012**, *171–172*, 1180–1185. [CrossRef]
30. Choi, K.J.; Jang, H.W. One-dimensional oxide nanostructures as gas-sensing materials: Review and issues *Sensors* **2010**, *10*, 4083–4099. [CrossRef]
31. Karnati, P.; Akbar, S.; Morris, P.A. Conduction mechanisms in one dimensional core-shell nanostructures for gas sensing: A review. *Sens. Actuators B Chem.* **2019**, *295*, 127–143. [CrossRef]
32. Meng, D.; Liu, D.; Wang, G.; Shen, Y.; San, X.; Li, M.; Meng, F. Low-temperature formaldehyde gas sensors based on NiO-SnO_2 heterojunction microflowers assembled by thin porous nanosheets. *Sens. Actuators B Chem.* **2018**, *273*, 418–428. [CrossRef]
33. Wang, Y.; Zhou, Y.; Meng, C.; Gao, Z.; Cao, X.; Li, X.; Xu, L.; Zhu, W.; Peng, X.; Zhang, B.; et al. A high-response ethanol gas sensor based on one-dimensional TiO_2/V_2O_5 branched nanoheterostructures. *Nanotechnology* **2016**, *27*, 425503. [CrossRef] [PubMed]
34. Huang, X.-Y.; Chi, Z.-T.; Liu, J.; Li, D.-H.; Sun, X.-J.; Yan, C.; Wang, Y.-C.; Li, H.; Wang, X.-D.; Xie, W.-F. Enhanced gas sensing performance based on p-NiS/n-In_2O_3 heterojunction nanocomposites. *Sens. Actuators B Chem.* **2020**, *304*, 127305. [CrossRef]
35. Giebelhaus, I.; Varechkina, E.; Fischer, T.; Rumyantseva, M.; Ivanov, V.; Gaskov, A.; Morante, J.R.; Arbiol, J.; Tyrra, W.; Mathur, S. One-dimensional CuO–SnO_2 p–n heterojunctions for enhanced detection of H_2S. *J. Mater. Chem. A* **2013**, *1*, 11261. [CrossRef]

36. Ju, D.; Xu, H.; Xu, Q.; Gong, H.; Qiu, Z.; Guo, J.; Zhang, J.; Cao, B. High triethylamine-sensing properties of NiO/SnO_2 hollow sphere P-N heterojunction sensors. *Sens. Actuators B Chem.* **2015**, *215*, 39–44. [CrossRef]
37. Chu, X.; Liang, S.; Chen, T.; Zhang, Q. Trimethylamine sensing properties of CdO–Fe_2O_3 nano-materials prepared using co-precipitation method in the presence of PEG400. *Mater. Chem. Phys.* **2010**, *123*, 396–400. [CrossRef]
38. Li, S.; Xie, L.; He, M.; Hu, X.; Luo, G.; Chen, C.; Zhu, Z. Metal-Organic frameworks-derived bamboo-like CuO/In_2O_3 Heterostructure for high-performance H_2S gas sensor with Low operating temperature. *Sens. Actuators B Chem.* **2020**, *310*, 127828. [CrossRef]
39. Liu, J.; Wang, T.; Wang, B.; Sun, P.; Yang, Q.; Liang, X.; Song, H.; Lu, G. Highly sensitive and low detection limit of ethanol gas sensor based on hollow ZnO/SnO_2 spheres composite material. *Sens. Actuators B Chem.* **2017**, *245*, 551–559. [CrossRef]
40. Zhao, S.; Wang, G.; Liao, J.; Lv, S.; Zhu, Z.; Li, Z. Vertically aligned MoS_2/ZnO nanowires nanostructures with highly enhanced NO_2 sensing activities. *Appl. Surf. Sci.* **2018**, *456*, 808–816. [CrossRef]
41. Bai, S.; Guo, W.; Sun, J.; Li, J.; Tian, Y.; Chen, A.; Luo, R.; Li, D. Synthesis of SnO_2-CuO heterojunction using electrospinning and application in detecting of CO. *Sens. Actuators B Chem.* **2016**, *226*, 96–103. [CrossRef]
42. Lou, Z.; Deng, J.; Wang, L.; Wang, L.; Fei, T.; Zhang, T. Toluene and ethanol sensing performances of pristine and PdO-decorated flower-like ZnO structures. *Sens. Actuators B Chem.* **2013**, *176*, 323–329. [CrossRef]
43. Choi, K.S.; Park, S.; Chang, S.-P. Enhanced ethanol sensing properties based on SnO_2 nanowires coated with Fe_2O_3 nanoparticles. *Sens. Actuators B Chem.* **2017**, *238*, 871–879. [CrossRef]
44. Liang, Y.Q.; Cui, Z.D.; Zhu, S.L.; Li, Z.Y.; Yang, X.J.; Chen, Y.J.; Ma, J.M. Design of a highly sensitive ethanol sensor using a nano-coaxial p-Co_3O_4/n-TiO_2 heterojunction synthesized at low temperature. *Nanoscale* **2013**, *5*, 10916–10926. [CrossRef] [PubMed]
45. Wang, L.; Li, J.; Wang, Y.; Yu, K.; Tang, X.; Zhang, Y.; Wang, S.; Wei, C. Construction of 1D SnO_2-coated ZnO nanowire heterojunction for their improved n-butylamine sensing performances. *Sci. Rep.* **2016**, *6*, 35079. [CrossRef]
46. Chowdhuri, A.; Gupta, V.; Sreenivas, K. Fast response H_2S gas sensing characteristics with ultra-thin CuO islands on sputtered SnO_2. *Sens. Actuators B Chem.* **2003**, *93*, 572–579. [CrossRef]
47. Park, S.; Park, S.; Jung, J.; Hong, T.; Lee, S.; Kim, H.W.; Lee, C. H_2S gas sensing properties of CuO-functionalized WO_3 nanowires. *Ceram. Int.* **2014**, *40*, 11051–11056. [CrossRef]
48. Han, T.-H.; Bak, S.-Y.; Kim, S.; Lee, S.H.; Han, Y.-J.; Yi, M. Decoration of CuO NWs Gas Sensor with ZnO NPs for Improving NO_2 Sensing Characteristics. *Sensors* **2021**, *21*, 2103. [CrossRef]
49. Tang, H.; Yan, M.; Zhang, H.; Li, S.; Ma, X.; Wang, M.; Yang, D. A selective NH_3 gas sensor based on Fe_2O_3-ZnO nanocomposites at room temperature. *Sens. Actuators B Chem.* **2006**, *114*, 910–915. [CrossRef]
50. Yoo, R.; Yoo, S.; Lee, D.; Kim, J.; Cho, S.; Lee, W. Highly selective detection of dimethyl methylphosphonate (DMMP) using CuO nanoparticles/ZnO flowers heterojunction. *Sens. Actuators B Chem.* **2017**, *240*, 1099–1105. [CrossRef]
51. Ling, C.; Xue, Q.; Han, Z.; Lu, H.; Xia, F.; Yan, Z.; Deng, L. Room temperature hydrogen sensor with ultrahigh-responsive characteristics based on Pd/SnO_2/SiO_2/Si heterojunctions. *Sens. Actuators B Chem.* **2016**, *227*, 438–447. [CrossRef]
52. Wang, L.; Lou, Z.; Zhang, R.; Zhou, T.; Deng, J.; Zhang, T. Hybrid Co_3O_4/SnO_2 core-shell nanospheres as real-time rapid-response sensors for ammonia gas. *ACS Appl. Mater. Interfaces* **2016**, *8*, 6539–6545. [CrossRef] [PubMed]
53. Hui, G.; Zhu, M.; Yang, X.; Liu, J.; Pan, G.; Wang, Z. Highly sensitive ethanol gas sensor based on CeO_2/ZnO binary heterojunction composite. *Mater. Lett.* **2020**, *278*, 128453. [CrossRef]
54. Choi, S.-W.; Katoch, A.; Sun, G.-J.; Kim, S.S. Synthesis and gas sensing performance of ZnO-SnO_2 nanofiber-nanowire stem-branch heterostructure. *Sens. Actuators B Chem.* **2013**, *181*, 787–794. [CrossRef]
55. Chen, Y.-J.; Xiao, G.; Wang, T.-S.; Zhang, F.; Ma, Y.; Gao, P.; Zhu, C.-L.; Zhang, E.; Xu, Z.; Li, Q.-H. α-MoO_3/TiO_2 core/shell nanorods: Controlled-synthesis and low-temperature gas sensing properties. *Sens. Actuators B Chem.* **2011**, *155*, 270–277. [CrossRef]
56. Huang, H.; Gong, H.; Chow, C.L.; Guo, J.; White, T.J.; Tse, M.S.; Tan, O.K. Low-temperature growth of SnO_2 nanorod arrays and tunable n-p-n sensing response of a ZnO/SnO_2 heterojunction for exclusive hydrogen sensors. *Adv. Funct. Mater.* **2011**, *21*, 2680–2686. [CrossRef]
57. Jin, C.; Park, S.; Kim, H.; Lee, C. Ultrasensitive multiple networked Ga_2O_3-core/ZnO-shell nanorod gas sensors. *Sens. Actuators B Chem.* **2012**, *161*, 223–228. [CrossRef]
58. Li, J.; Liu, H.; Fu, H.; Xu, L.; Jin, H.; Zhang, X.; Wang, L.; Yu, K. Synthesis of 1D α-MoO_3/0D ZnO heterostructure nanobelts with enhanced gas sensing properties. *J. Alloys Compd.* **2019**, *788*, 248–256. [CrossRef]
59. Galatsis, K.; Li, Y.X.; Wlodarski, W.; Kalantar-Zadeh, K. Sol-gel prepared MoO_3-WO_3 thin-films for O_2 gas sensing. *Sens. Actuators B Chem.* **2001**, *77*, 478–483. [CrossRef]
60. Jiang, D.; Wei, W.; Li, F.; Li, Y.; Liu, C.; Sun, D.; Feng, C.; Ruan, S. Xylene gas sensor based on α-MoO_3/α-Fe_2O_3 heterostructure with high response and low operating temperature. *RSC Adv.* **2015**, *5*, 39442–39448. [CrossRef]
61. Zou, C.W.; Wang, J.; Xie, W. Synthesis and enhanced NO_2 gas sensing properties of ZnO nanorods/TiO_2 nanoparticles heterojunction composites. *J. Colloid Interface Sci.* **2016**, *478*, 22–28. [CrossRef] [PubMed]
62. Hsu, K.-C.; Fang, T.-H.; Chen, S.-H.; Kuo, E.-Y. Gas sensitivity and sensing mechanism studies on ZnO/$La_{0.8}Sr_{0.2}Co_{0.5}Ni_{0.5}O_3$ heterojunction structure. *Ceram. Int.* **2019**, *45*, 8744–8749. [CrossRef]

63. Zhu, L.; Zeng, W.; Li, Y. A novel cactus-like WO_3-SnO_2 nanocomposite and its acetone gas sensing properties. *Mater. Lett.* **2018**, *231*, 5–7. [CrossRef]
64. Park, S.; Kim, S.; Kheel, H.; Lee, C. Oxidizing gas sensing properties of the n-ZnO/p-Co_3O_4 composite nanoparticle network sensor. *Sens. Actuators B Chem.* **2016**, *222*, 1193–1200. [CrossRef]
65. Poloju, M.; Jayababu, N.; Ramana Reddy, M.V. Improved gas sensing performance of Al doped ZnO/CuO nanocomposite based ammonia gas sensor. *Mater. Sci. Eng. B* **2018**, *227*, 61–67. [CrossRef]
66. Wang, T.-S.; Wang, Q.-S.; Zhu, C.-L.; Ouyang, Q.-Y.; Qi, L.-H.; Li, C.-Y.; Xiao, G.; Gao, P.; Chen, Y.-J. Synthesis and enhanced H_2S gas sensing properties of α-MoO_3/CuO p-n junction nanocomposite. *Sens. Actuators B Chem.* **2012**, *171–172*, 256–262. [CrossRef]
67. Kim, J.-H.; Lee, J.-H.; Mirzaei, A.; Kim, H.W.; Kim, S.S. Optimization and gas sensing mechanism of n-SnO_2-p-Co_3O_4 composite nanofibers. *Sens. Actuators B Chem.* **2017**, *248*, 500–511. [CrossRef]
68. Kabcum, S.; Channei, D.; Tuantranont, A.; Wisitsoraat, A.; Liewhiran, C.; Phanichphant, S. Ultra-responsive hydrogen gas sensors based on PdO nanoparticle-decorated WO_3 nanorods synthesized by precipitation and impregnation methods. *Sens. Actuators B Chem.* **2016**, *226*, 76–89. [CrossRef]
69. Xu, Q.; Ju, D.; Zhang, Z.; Yuan, S.; Zhang, J.; Xu, H.; Cao, B. Near room-temperature triethylamine sensor constructed with CuO/ZnO P-N heterostructural nanorods directly on flat electrode. *Sens. Actuators B Chem.* **2016**, *225*, 16–23. [CrossRef]
70. Sun, G.-J.; Choi, S.-W.; Katoch, A.; Wu, P.; Kim, S.S. Bi-functional mechanism of H_2S detection using CuO-SnO_2 nanowires. *J. Mater. Chem. C* **2013**, *1*, 5454. [CrossRef]
71. Huang, J.; Dai, Y.; Gu, C.; Sun, Y.; Liu, J. Preparation of porous flower-like CuO/ZnO nanostructures and analysis of their gas-sensing property. *J. Alloys Compd.* **2013**, *575*, 115–122. [CrossRef]
72. Zhang, Y.-B.; Yin, J.; Li, L.; Zhang, L.-X.; Bie, L.-J. Enhanced ethanol gas-sensing properties of flower-like p-CuO/n-ZnO heterojunction nanorods. *Sens. Actuators B Chem.* **2014**, *202*, 500–507. [CrossRef]
73. Xu, L.; Zheng, R.; Liu, S.; Song, J.; Chen, J.; Dong, B.; Song, H. NiO@ZnO heterostructured nanotubes: Coelectrospinning fabrication, characterization, and highly enhanced gas sensing properties. *Inorg. Chem.* **2012**, *51*, 7733–7740. [CrossRef]
74. Lokesh, K.; Kavitha, G.; Manikandan, E.; Mani, G.K.; Kaviyarasu, K.; Rayappan, J.B.B.; Ladchumananandasivam, R.; Aanand, J.S.; Jayachandran, M.; Maaza, M. Effective ammonia detection using n-ZnO/p-NiO heterostructured nanofibers. *IEEE Sens. J.* **2016**, *16*, 2477–2483. [CrossRef]
75. Wang, H.; Chen, M.; Rong, Q.; Zhang, Y.; Hu, J.; Zhang, D.; Zhou, S.; Zhao, X.; Zhang, J.; Zhu, Z. Ultrasensitive xylene gas sensor based on flower-like SnO_2/Co_3O_4 nanorods composites prepared by facile two-step synthesis method. *Nanotechnology* **2020**, *31*, 255501. [CrossRef]
76. Ng, S.; Kuberský, P.; Krbal, M.; Prikryl, J.; Gärtnerová, V.; Moravcová, D.; Sopha, H.; Zazpe, R.; Yam, F.K.; Jäger, A.; et al. ZnO coated anodic 1D TiO_2 nanotube layers: Efficient photo-electrochemical and gas sensing heterojunction. *Adv. Eng. Mater.* **2018**, *20*, 1700589. [CrossRef]
77. Yao, Y.; Yuan, J.; Chen, X.; Tan, L.; Gu, Q.; Zhao, W.; Chen, J. In situ construction and sensing mechanism of TiO_2-WO_3 composite coatings based on the semiconductor heterojunctions. *J. Mater. Res. Technol.* **2019**, *8*, 3580–3588. [CrossRef]
78. Zhao, P.X.; Tang, Y.; Mao, J.; Chen, Y.X.; Song, H.; Wang, J.W.; Song, Y.; Liang, Y.Q.; Zhang, X.M. One-dimensional MoS_2-decorated TiO_2 nanotube gas sensors for efficient alcohol sensing. *J. Alloy. Compd.* **2016**, *674*, 252–258. [CrossRef]
79. Sun, P.; Cai, Y.; Du, S.; Xu, X.; You, L.; Ma, J.; Liu, F.; Liang, X.; Sun, Y.; Lu, G. Hierarchical α-Fe_2O_3/SnO_2 semiconductor composites: Hydrothermal synthesis and gas sensing properties. *Sens. Actuators B Chem.* **2013**, *182*, 336–343. [CrossRef]
80. Zhang, S.; Song, P.; Wang, Q. Enhanced acetone sensing performance of an α-Fe_2O_3-In_2O_3 heterostructure nanocomposite sensor. *J. Phys. Chem. Solids* **2018**, *120*, 261–270. [CrossRef]
81. Feng, C.; Li, X.; Ma, J.; Sun, Y.; Wang, C.; Sun, P.; Zheng, J.; Lu, G. Facile synthesis and gas sensing properties of In_2O_3-WO_3 heterojunction nanofibers. *Sens. Actuators B Chem.* **2015**, *209*, 622–629. [CrossRef]
82. Chi, X.; Liu, C.; Liu, L.; Li, S.; Li, H.; Zhang, X.; Bo, X.; Shan, H. Enhanced formaldehyde-sensing properties of mixed Fe_2O_3-In_2O_3 nanotubes. *Mater. Sci. Semicond. Process.* **2014**, *18*, 160–164. [CrossRef]
83. Park, S. Acetone gas detection using TiO_2 nanoparticles functionalized In_2O_3 nanowires for diagnosis of diabetes. *J. Alloys Compd.* **2017**, *696*, 655–662. [CrossRef]
84. Du, H.; Wang, J.; Sun, Y.; Yao, P.; Li, X.; Yu, N. Investigation of gas sensing properties of SnO_2/In_2O_3 composite hetero-nanofibers treated by oxygen plasma. *Sens. Actuators B Chem.* **2015**, *206*, 753–763. [CrossRef]
85. Wang, X.; Li, S.; Xie, L.; Li, X.; Lin, D.; Zhu, Z. Low-temperature and highly sensitivity H_2S gas sensor based on ZnO/CuO composite derived from bimetal metal-organic frameworks. *Ceram. Int.* **2020**, *46*, 15858–15866. [CrossRef]
86. Chang, H.-K.; Ko, D.-S.; Cho, D.-H.; Kim, S.; Lee, H.-N.; Lee, H.S.; Kim, H.-J.; Park, T.J.; Park, Y.M. Enhanced response of the photoactive gas sensor on formaldehyde using porous SnO_2@TiO_2 heterostructure driven by gas-flow thermal evaporation and atomic layer deposition. *Ceram. Int.* **2021**, *47*, 5985–5992. [CrossRef]
87. Katoch, A.; Kim, J.H.; Kwon, Y.J.; Kim, H.W.; Kim, S.S. Bifunctional sensing mechanism of SnO_2-ZnO composite nanofibers for drastically enhancing the sensing behavior in H_2 gas. *ACS Appl. Mater. Interfaces* **2015**, *7*, 11351–11358. [CrossRef]
88. Ma, L.; Fan, H.; Tian, H.; Fang, J.; Qian, X. The n-ZnO/n-In_2O_3 heterojunction formed by a surface-modification and their potential barrier-control in methanal gas sensing. *Sens. Actuators B Chem.* **2016**, *222*, 508–516. [CrossRef]
89. Xun, H.; Zhang, Z.; Yu, A.; Yi, J. Remarkably enhanced hydrogen sensing of highly-ordered SnO_2-decorated TiO_2 nanotubes. *Sens. Actuators B Chem.* **2018**, *273*, 983–990. [CrossRef]

90. Yu, H.; Yang, T.; Wang, Z.; Li, Z.; Zhao, Q.; Zhang, M. p-N heterostructural sensor with SnO-SnO_2 for fast NO_2 sensing response properties at room temperature. *Sens. Actuators B Chem.* **2018**, *258*, 517–526. [CrossRef]
91. Gao, H.; Zhao, L.; Wang, L.; Sun, P.; Lu, H.; Liu, F.; Chuai, X.; Lu, G. Ultrasensitive and low detection limit of toluene gas sensor based on SnO_2-decorated NiO nanostructure. *Sens. Actuators B Chem.* **2018**, *255*, 3505–3515. [CrossRef]
92. Zhang, L.; Gao, Z.; Liu, C.; Zhang, Y.; Tu, Z.; Yang, X.; Yang, F.; Wen, Z.; Zhu, L.; Liu, R.; et al. Synthesis of TiO_2 decorated Co_3O_4 acicular nanowire arrays and their application as an ethanol sensor. *J. Mater. Chem. A* **2015**, *3*, 2794–2801. [CrossRef]
93. Li, X.; Li, X.; Chen, N.; Li, X.; Zhang, J.; Yu, J.; Wang, J.; Tang, Z. CuO-In_2O_3core-shell nanowire based chemical gas sensors. *J. Nanomater.* **2014**, *2014*, 973156. [CrossRef]
94. Wang, Z.; Zhang, K.; Fei, T.; Gu, F.; Han, D. α-Fe_2O_3/NiO heterojunction nanorods with enhanced gas sensing performance for acetone. *Sens. Actuators B Chem.* **2020**, *318*, 128191. [CrossRef]
95. Deng, J.; Wang, L.; Lou, Z.; Zhang, T. Design of CuO-TiO_2 heterostructure nanofibers and their sensing performance. *J. Mater. Chem. A* **2014**, *2*, 9030–9034. [CrossRef]
96. Shi, S.; Zhang, F.; Lin, H.; Wang, Q.; Shi, E.; Qu, F. Enhanced triethylamine-sensing properties of P-N heterojunction Co_3O_4/In_2O_3 hollow microtubes derived from metal-organic frameworks. *Sens. Actuators B Chem.* **2018**, *262*, 739–749. [CrossRef]
97. Mandal, S.; Rakibuddin, M.; Ananthakrishnan, R. Strategic synthesis of SiO_2-modified porous Co_3O_4 nano-octahedra through the nanocoordination polymer route for enhanced and selective sensing of H_2 gas over NOx. *ACS Omega* **2018**, *3*, 648–661. [CrossRef]
98. Zhang, B.; Li, M.; Song, Z.; Kan, H.; Yu, H.; Liu, Q.; Zhang, G.; Liu, H. Sensitive H_2S gas sensors employing colloidal zinc oxide quantum dots. *Sens. Actuators B Chem.* **2017**, *249*, 558–563. [CrossRef]
99. Zhang, L.; Dong, B.; Xu, L.; Zhang, X.; Chen, J.; Sun, X.; Xu, H.; Zhang, T.; Bai, X.; Zhang, S.; et al. Three-dimensional ordered ZnO-Fe_3O_4 inverse opal gas sensor toward trace concentration acetone detection. *Sens. Actuators B Chem.* **2017**, *252*, 367–374. [CrossRef]
100. Kim, J.-H.; Jeong, H.-M.; Na, C.W.; Yoon, J.-W.; Abdel-Hady, F.; Wazzan, A.A.; Lee, J.-H. Highly selective and sensitive xylene sensors using Cr_2O_3-$ZnCr_2O_4$ hetero-nanostructures prepared by galvanic replacement. *Sens. Actuators B Chem.* **2016**, *235*, 498–506. [CrossRef]
101. Koo, W.T.; Yu, S.; Choi, S.J.; Jang, J.S.; Cheong, J.Y.; Kim, I.D. Nanoscale PdO catalyst functionalized Co_3O_4 hollow nanocages using MOF templates for selective detection of acetone molecules in exhaled breath. *ACS Appl. Mater. Interfaces* **2017**, *9*, 8201–8210. [CrossRef] [PubMed]
102. Park, S.; Kim, S.; Sun, G.-J.; Lee, W.; Kim, K.K.; Lee, C. Fabrication and NO_2 gas sensing performance of TeO_2-core/CuO-shell heterostructure nanorod sensors. *Nanoscale Res. Lett.* **2014**, *9*, 638. [CrossRef]
103. Xu, Y.; Fan, Y.; Tian, X.; Liang, Q.; Liu, X.; Sun, Y. p-p heterojunction composite of $NiFe_2O_4$ nanoparticles-decorated NiO nanosheets for acetone gas detection. *Mater. Lett.* **2020**, *270*, 127728. [CrossRef]
104. Su, C.; Zhang, L.; Han, Y.; Ren, C.; Zeng, M.; Zhou, Z.; Su, Y.; Hu, N.; Wei, H.; Yang, Z. Controllable synthesis of heterostructured CuO-NiO nanotubes and their synergistic effect for glycol gas sensing. *Sens. Actuators B Chem.* **2020**, *304*, 127347. [CrossRef]
105. Minh Nguyet, Q.T.; Van Duy, N.; Phuong, N.T.; Trung, N.N.; Hung, C.M.; Hoa, N.D.; Van Hieu, N. Superior enhancement of NO_2 gas response using n-p-n transition of carbon nanotubes/SnO_2 nanowires heterojunctions. *Sens. Actuators B Chem.* **2017**, *238*, 1120–1127. [CrossRef]
106. Kwon, Y.J.; Kang, S.Y.; Mirzaei, A.; Choi, M.S.; Bang, J.H.; Kim, S.S.; Kim, H.W. Enhancement of gas sensing properties by the functionalization of ZnO-branched SnO $_2$ nanowires with Cr_2O_3 nanoparticles. *Sens. Actuators B Chem.* **2017**, *249*, 656–666. [CrossRef]
107. Pang, Z.; Nie, Q.; Lv, P.; Yu, J.; Huang, F.; Wei, Q. Design of flexible PANI-coated CuO-TiO_2-SiO_2 heterostructure nanofibers with high ammonia sensing response values. *Nanotechnology* **2017**, *28*, 225501. [CrossRef]
108. Volanti, D.P.; Felix, A.A.; Orlandi, M.O.; Whitfield, G.; Yang, D.-J.; Longo, E.; Tuller, H.L.; Varela, J.A. The role of hierarchical morphologies in the superior gas sensing performance of CuO-based chemiresistors. *Adv. Funct. Mater.* **2013**, *23*, 1759–1766. [CrossRef]
109. Yang, S.; Lei, G.; Lan, Z.; Xie, W.; Yang, B.; Xu, H.; Wang, Z.; Gu, H. Enhancement of the room-temperature hydrogen sensing performance of MoO_3 nanoribbons annealed in a reducing gas. *Int. J. Hydrog. Energy* **2019**, *44*, 7725–7733. [CrossRef]
110. Park, S.; Kim, S.; Sun, G.-J.; Choi, S.; Lee, S.; Lee, C. Ethanol sensing properties of networked In2O3 nanorods decorated with Cr_2O_3-nanoparticles. *Ceram. Int.* **2015**, *41*, 9823–9827. [CrossRef]
111. He, M.; Xie, L.; Zhao, X.; Hu, X.; Li, S.; Zhu, Z.-G. Highly sensitive and selective H_2S gas sensors based on flower-like WO_3/CuO composites operating at low/room temperature. *J. Alloys Compd.* **2019**, *788*, 36–43. [CrossRef]
112. Bang, J.H.; Choi, M.S.; Mirzaei, A.; Kwon, Y.J.; Kim, S.S.; Kim, T.W.; Kim, H.W. Selective NO_2 sensor based on Bi_2O_3 branched SnO_2 nanowires. *Sens. Actuators B Chem.* **2018**, *274*, 356–369. [CrossRef]
113. Tan, W.; Tan, J.; Fan, L.; Yu, Z.; Qian, J.; Huang, X. Fe_2O_3-loaded NiO nanosheets for fast response/recovery and high response gas sensor. *Sens. Actuators B Chem.* **2018**, *256*, 282–293. [CrossRef]
114. Ju, D.; Xu, H.; Qiu, Z.; Guo, J.; Zhang, J.; Cao, B. Highly sensitive and selective triethylamine-sensing properties of nanosheets directly grown on ceramic tube by forming NiO/ZnO PN heterojunction. *Sens. Actuators B Chem.* **2014**, *200*, 288–296. [CrossRef]
115. Yu, Q.; Zhu, J.; Xu, Z.; Huang, X. Facile synthesis of α-Fe_2O_3@SnO_2 core–shell heterostructure nanotubes for high performance gas sensors. *Sens. Actuators B Chem.* **2015**, *213*, 27–34. [CrossRef]

116. Rout, C.S.; Hari Krishna, S.; Vivekchand, S.R.C.; Govindaraj, A.; Rao, C.N.R. Hydrogen and ethanol sensors based on ZnO nanorods, nanowires and nanotubes. *Chem. Phys. Lett.* **2006**, *418*, 586–590. [CrossRef]
117. Feng, P.; Wan, Q.; Wang, T.H. Contact-controlled sensing properties of flowerlike ZnO nanostructures. *Appl. Phys. Lett.* **2005**, *87*, 213111. [CrossRef]
118. Yu, H.-L.; Li, L.; Gao, X.-M.; Zhang, Y.; Meng, F.; Wang, T.-S.; Xiao, G.; Chen, Y.-J.; Zhu, C.-L. Synthesis and H_2S gas sensing properties of cage-like α-MoO_3/ZnO composite. *Sens. Actuators B Chem.* **2012**, *171–172*, 679–685. [CrossRef]
119. Kheel, H.; Sun, G.-J.; Lee, J.K.; Mirzaei, A.; Choi, S.; Lee, C. Hydrogen gas detection of Nb_2O_5 nanoparticle-decorated CuO nanorod sensors. *Met. Mater. Int.* **2017**, *23*, 214–219. [CrossRef]
120. Gao, X.; Ouyang, Q.; Zhu, C.; Zhang, X.; Chen, Y. Porous MoO3/SnO2 nanoflakes with n-n junctions for sensing H_2S. *ACS Appl. Nano Mater.* **2019**, *2*, 2418–2425. [CrossRef]
121. Kheel, H.; Sun, G.-J.; Lee, J.K.; Lee, S.; Dwivedi, R.P.; Lee, C. Enhanced H_2S sensing performance of TiO2-decorated α-Fe_2O_3 nanorod sensors. *Ceram. Int.* **2016**, *42*, 18597–18604. [CrossRef]
122. Qu, F.; Zhou, X.; Zhang, B.; Zhang, S.; Jiang, C.; Ruan, S.; Yang, M. Fe_2O_3 nanoparticles-decorated MoO_3 nanobelts for enhanced chemiresistive gas sensing. *J. Alloys Compd.* **2019**, *782*, 672–678. [CrossRef]
123. Mirzaei, A.; Sun, G.-J.; Lee, J.K.; Lee, C.; Choi, S.; Kim, H.W. Hydrogen sensing properties and mechanism of NiO-Nb_2O_5 composite nanoparticle-based electrical gas sensors. *Ceram. Int.* **2017**, *43*, 5247–5254. [CrossRef]
124. Park, S.; Sun, G.-J.; Kheel, H.; Hyun, S.K.; Jin, C.; Lee, C. Hydrogen gas sensing of Co_3O_4-Decorated WO_3 nanowires. *Met. Mater. Int.* **2016**, *22*, 156–162. [CrossRef]
125. Liang, X.; Kim, T.-H.; Yoon, J.-W.; Kwak, C.-H.; Lee, J.-H. Ultrasensitive and ultraselective detection of H_2S using electrospun CuO-loaded In_2O_3 nanofiber sensors assisted by pulse heating. *Sens. Actuators B Chem.* **2015**, *209*, 934–942. [CrossRef]
126. Shanmugasundaram, A.; Basak, P.; Satyanarayana, L.; Manorama, S.V. Hierarchical SnO/SnO_2 nanocomposites: Formation of in situ p-n junctions and enhanced H_2 sensing. *Sens. Actuators B Chem.* **2013**, *185*, 265–273. [CrossRef]
127. Cai, L.; Li, H.; Zhang, H.; Fan, W.; Wang, J.; Wang, Y.; Wang, X.; Tang, Y.; Song, Y. Enhanced performance of the tangerines-like CuO-based gas sensor using ZnO nanowire arrays. *Mater. Sci. Semicond. Process.* **2020**, *118*, 105196. [CrossRef]
128. Wang, C.; Cheng, X.; Zhou, X.; Sun, P.; Hu, X.; Shimanoe, K.; Lu, G.; Yamazoe, N. Hierarchical alpha-Fe_2O_3/NiO composites with a hollow structure for a gas sensor. *ACS Appl. Mater. Interfaces* **2014**, *6*, 12031–12037. [CrossRef] [PubMed]
129. Liu, T.; Yu, Z.; Liu, Y.; Gao, J.; Wang, X.; Suo, H.; Yang, X.; Zhao, C.; Liu, F. Gas sensor based on Ni foam: SnO_2-decorated NiO for Toluene detection. *Sens. Actuators B Chem.* **2020**, *318*, 128167. [CrossRef]
130. Xue, X.-T.; Zhu, L.-Y.; Yuan, K.-P.; Zeng, C.; Li, X.-X.; Ma, H.-P.; Lu, H.-L.; Zhang, D.W. ZnO branched p-CuxO@n-ZnO heterojunction nanowires for improving acetone gas sensing performance. *Sens. Actuators B Chem.* **2020**, *324*, 128729. [CrossRef]
131. Wang, Y.; Qu, F.; Liu, J.; Wang, Y.; Zhou, J.; Ruan, S. Enhanced H_2S sensing characteristics of CuO-NiO core-shell microspheres sensors. *Sens. Actuators B Chem.* **2015**, *209*, 515–523. [CrossRef]
132. Gao, H.; Guo, J.; Li, Y.; Xie, C.; Li, X.; Liu, L.; Chen, Y.; Sun, P.; Liu, F.; Yan, X.; et al. Highly selective and sensitive xylene gas sensor fabricated from NiO/$NiCr_2O_4$ p-p nanoparticles. *Sens. Actuators B Chem.* **2019**, *284*, 305–315. [CrossRef]
133. Yoon, J.-W.; Kim, H.-J.; Jeong, H.-M.; Lee, J.-H. Gas sensing characteristics of p-type Cr_2O_3 and Co_3O_4 nanofibers depending on inter-particle connectivity. *Sens. Actuators B Chem.* **2014**, *202*, 263–271. [CrossRef]
134. Zhao, Y.; Ikram, M.; Wang, J.; Liu, Z.; Du, L.; Zhou, J.; Kan, K.; Zhang, W.; Li, L.; Shi, K. Ultrafast NH_3 sensing properties of WO_3@$CoWO_4$ heterojunction nanofibres at room temperature. *Aust. J. Chem.* **2018**, *71*, 87–94. [CrossRef]
135. Zhou, C.; Xu, L.; Song, J.; Xing, R.; Xu, S.; Liu, D.; Song, H. Ultrasensitive non-enzymatic glucose sensor based on three-dimensional network of ZnO-CuO hierarchical nanocomposites by electrospinning. *Sci. Rep.* **2014**, *4*, 7382. [CrossRef] [PubMed]
136. Zhang, Y.; Zhang, J.; Jiang, Y.; Duan, Z.; Liu, B.; Zhao, Q.; Wang, S.; Yuan, Z.; Tai, H. Ultrasensitive flexible NH_3 gas sensor based on polyaniline/$SrGe_4O_9$ nanocomposite with ppt-level detection ability at room temperature. *Sens. Actuators B Chem.* **2020**, *319*, 128293. [CrossRef]
137. Gong, J.; Li, Y.; Hu, Z.; Zhou, Z.; Deng, Y. Ultrasensitive NH_3 gas sensor from polyaniline nanograin enchased TiO_2 fibers. *J. Phys. Chem. C* **2010**, *114*, 9970–9974. [CrossRef]
138. Li, S.; Liu, A.; Yang, Z.; He, J.; Wang, J.; Liu, F.; Lu, H.; Yan, X.; Sun, P.; Liang, X.; et al. Room temperature gas sensor based on tin dioxide@polyaniline nanocomposite assembled on flexible substrate: Ppb-level detection of NH_3. *Sens. Actuators B Chem.* **2019**, *299*, 126970. [CrossRef]
139. Naderi, H.; Hajati, S.; Ghaedi, M.; Dashtian, K.; Sabzehmeidani, M.M. Sensitive, selective and rapid ammonia-sensing by gold nanoparticle-sensitized V_2O_5/$CuWO_4$ heterojunctions for exhaled breath analysis. *Appl. Surf. Sci.* **2020**, *501*, 144270. [CrossRef]
140. Hung, C.M.; Dat, D.Q.; Van Duy, N.; Van Quang, V.; Van Toan, N.; Van Hieu, N.; Hoa, N.D. Facile synthesis of ultrafine rGO/WO_3 nanowire nanocomposites for highly sensitive toxic NH_3 gas sensors. *Mater. Res. Bull.* **2020**, *125*, 110810. [CrossRef]
141. Yuan, K.-P.; Zhu, L.-Y.; Yang, J.-H.; Hang, C.-Z.; Tao, J.-J.; Ma, H.-P.; Jiang, A.-Q.; Zhang, D.W.; Lu, H.-L. Precise preparation of WO_3@SnO_2 core shell nanosheets for efficient NH_3 gas sensing. *J. Colloid Interface Sci.* **2020**, *568*, 81–88. [CrossRef]
142. Liu, C.; Tai, H.; Zhang, P.; Yuan, Z.; Du, X.; Xie, G.; Jiang, Y. A high-performance flexible gas sensor based on self-assembled PANI-CeO_2 nanocomposite thin film for trace-level NH_3 detection at room temperature. *Sens. Actuators B Chem.* **2018**, *261*, 587–597. [CrossRef]
143. Huang, J.; Jiang, D.; Zhou, J.; Ye, J.; Sun, Y.; Li, X.; Geng, Y.; Wang, J.; Du, Y.; Qian, Z. Visible light-activated room temperature NH_3 sensor base on CuPc-loaded ZnO nanorods. *Sens. Actuators B Chem.* **2021**, *327*, 128911. [CrossRef]

144. Zhang, D.; Jiang, C.; Li, P.; Sun, Y.E. Layer-by-layer self-assembly of Co_3O_4 nanorod-decorated MoS_2 nanosheet-based nanocomposite toward high-performance ammonia detection. *ACS Appl. Mater. Interfaces* **2017**, *9*, 6462–6471. [CrossRef]
145. Wang, Y.; Zhang, H.; Sun, X. Electrospun nanowebs of NiO/SnO_2 p-n heterojunctions for enhanced gas sensing. *Appl. Surf. Sci.* **2016**, *389*, 514–520. [CrossRef]
146. Van Toan, N.; Hung, C.M.; van Duy, N.; Hoa, N.D.; Le, D.T.; van Hieu, N. Bilayer SnO_2–WO_3 nanofilms for enhanced NH_3 gas sensing performance. *Mater. Sci. Eng. B* **2017**, *224*, 163–170. [CrossRef]
147. Ding, Y.; Guo, X.; Du, B.; Hu, X.; Yang, X.; He, Y.; Zhou, Y.; Zang, Z. Low-operating temperature ammonia sensor based on Cu_2O nanoparticles decorated with p-type MoS_2 nanosheets. *J. Mater. Chem. C* **2021**, *9*, 4838–4846. [CrossRef]
148. Kumar, A.; Sanger, A.; Kumar, A.; Chandra, R. Fast response ammonia sensors based on TiO_2 and NiO nanostructured bilayer thin films. *RSC Adv.* **2016**, *6*, 77636–77643. [CrossRef]
149. Zhou, J.; Ikram, M.; Rehman, A.U.; Wang, J.; Zhao, Y.; Kan, K.; Zhang, W.; Raziq, F.; Li, L.; Shi, K. Highly selective detection of NH_3 and H_2S using the pristine CuO and mesoporous In_2O_3@CuO multijunctions nanofibers at room temperature. *Sens. Actuators B Chem.* **2018**, *255*, 1819–1830. [CrossRef]
150. Kim, S.S.; Choi, S.-W.; Na, H.G.; Kwak, D.S.; Kwon, Y.J.; Kim, H.W. ZnO–SnO_2 branch–stem nanowires based on a two-step process: Synthesis and sensing capability. *Curr. Appl. Phys.* **2013**, *13*, 526–532. [CrossRef]
151. Bai, S.; Fu, H.; Zhao, Y.; Tian, K.; Luo, R.; Li, D.; Chen, A. On the construction of hollow nanofibers of ZnO-SnO_2 heterojunctions to enhance the NO_2 sensing properties. *Sens. Actuators B Chem.* **2018**, *266*, 692–702. [CrossRef]
152. Xu, S.; Gao, J.; Wang, L.; Kan, K.; Xie, Y.; Shen, P.; Li, L.; Shi, K. Role of the heterojunctions in In_2O_3-composite SnO_2 nanorod sensors and their remarkable gas-sensing performance for NOx at room temperature. *Nanoscale* **2015**, *7*, 14643–14651. [CrossRef] [PubMed]
153. Park, S.; An, S.; Mun, Y.; Lee, C. UV-enhanced NO_2 gas sensing properties of SnO_2-core/ZnO-shell nanowires at room temperature. *ACS Appl. Mater. Interfaces* **2013**, *5*, 4285–4292. [CrossRef]
154. Zhang, J.; Zeng, D.; Zhu, Q.; Wu, J.; Huang, Q.; Zhang, W.; Xie, C. Enhanced room temperature NO_2 response of NiO-SnO_2 nanocomposites induced by interface bonds at the p-n heterojunction. *Phys. Chem. Chem. Phys.* **2016**, *18*, 5386–5396. [CrossRef] [PubMed]
155. Hoa, L.T.; Hur, S.H. Highly sensitive NO_2 sensors based on local p-n heterojunctions composed of 0D CuO nanoparticles and 1D ZnO nanorods. *Phys. Status Solidi A* **2013**, *210*, 1213–1216. [CrossRef]
156. Yeh, B.Y.; Huang, P.F.; Tseng, W.J. Enhanced room-temperature NO_2 gas sensing with TeO_2/SnO_2 brush- and bead-like nanowire hybrid structures. *Nanotechnology* **2017**, *28*, 045501. [CrossRef] [PubMed]
157. Ramakrishnan, V.; Nair, K.G.; Dhakshinamoorthy, J.; Ravi, K.R.; Pullithadathil, B. Porous, n–p type ultra-long, ZnO@Bi_2O_3 heterojunction nanorods-based NO_2 gas sensor: New insights towards charge transport characteristics. *Phys. Chem. Chem. Phys.* **2020**, *22*, 7524–7536. [CrossRef] [PubMed]
158. Xu, H.; Zhang, J.; Rehman, A.U.; Gong, L.; Kan, K.; Li, L.; Shi, K. Synthesis of NiO@CuO nanocomposite as high-performance gas sensing material for NO_2 at room temperature. *Appl. Surf. Sci.* **2017**, *412*, 230–237. [CrossRef]
159. Adamu, B.I.; Falak, A.; Tian, Y.; Tan, X.; Meng, X.; Chen, P.; Wang, H.; Chu, W. p-p heterojunction sensors of p-$Cu_3Mo_2O_9$ micro/nanorods vertically grown on p-CuO layers for room-temperature ultrasensitive and fast recoverable detection of NO_2. *ACS Appl. Mater. Interfaces* **2020**, *12*, 8411–8421. [CrossRef]

nanomaterials

MDPI

Article

Multi-Angular Colorimetric Responses of Uni- and Omni-Directional Femtosecond Laser-Induced Periodic Surface Structures on Metals

Taek-Yong Hwang [1,*], Yong-dae Kim [1], Jongweon Cho [2], Hai-Joong Lee [3], Hyo-Soo Lee [3] and Byounghwak Lee [4,*]

1 Shape Manufacturing R&D Department, Korea Institute of Industrial Technology, Bucheon 14441, Korea; ydkim@kitech.re.kr
2 Department of Physics, Myongji University, Yongin 17058, Korea; jwcho@mju.ac.kr
3 Advanced Materials and Process R&D Department, Korea Institute of Industrial Technology, Incheon 21999, Korea; rookiehj@kitech.re.kr (H.-J.L.); todd3367@kitech.re.kr (H.-S.L.)
4 Department of Physics and Chemistry, Korea Military Academy, Seoul 01805, Korea
* Correspondence: taekyong@kitech.re.kr (T.-Y.H.); lebaiii@mnd.go.kr (B.L.)

Abstract: We investigated the colorimetric behaviors of metal surfaces with unidirectional low-spatial-frequency laser-induced periodic surface structures (UD-LSFLs) and omnidirectional LSFLs (OD-LSFLs) fabricated using femtosecond laser pulse irradiation. With the CIE standard illuminant D65, incident at −45°, we show that UD-LSFLs on metals transform polished metals to gonio-apparent materials with a unique behavior of colorimetric responses, depending on both the detection and rotation angles, whereas OD-LSFLs have nearly uniform gonio-apparent colors at each detection angle, regardless of their rotation. These colorimetric behaviors can be observed not only at the angles of diffraction but also near the angle of reflection, and we find that the power redistribution due to Rayleigh anomalies also plays an important role in the colorimetric responses of UD- and OD-LSFLs, in addition to diffraction.

Keywords: laser-induced periodic surface structures (LIPSS); laser ablation; structural color

Citation: Hwang, T.-Y.; Kim, Y.-d.; Cho, J.; Lee, H.-J.; Lee, H.-S.; Lee, B. Multi-Angular Colorimetric Responses of Uni- and Omni-Directional Femtosecond Laser-Induced Periodic Surface Structures on Metals. *Nanomaterials* **2021**, *11*, 2010. https://doi.org/10.3390/nano11082010

Academic Editor: Vincenzo Amendola

Received: 10 June 2021
Accepted: 4 August 2021
Published: 5 August 2021

1. Introduction

Nature creates a broad range of colors through periodic structures, pigments, and bioluminescence [1]. Periodic structures, when compared to the other two coloration mechanisms, present a unique capability to engineer colors through the modification of their periods and orientations as well as source and observer positions, since structural coloration by the periodic structures is rooted in diffraction [1,2].

In the past few decades, with femtosecond (fs) laser pulse irradiation, several types of quasi-periodic structures can be fabricated on metals [3–9]. Among these fs laser-induced periodic surface structures (LIPSSs), regarding structural coloration, low-spatial-frequency LIPSSs (LSFLs) have been popularly investigated because diffraction from LSFLs can be occurred in the entire range of visible wavelengths [4,5,9–15], and the viewing direction of the structural color can be adjusted easily by controlling the period and orientation of LSFLs with the incident angle and the polarization direction of fs laser pulses, respectively [7,8,13,15–17]. Moreover, LSFLs on metals with high hardness can be used to imprint themselves to the surface of soft metal such as Al [18], and this is promising for mass production. Accordingly, the structural coloration using LSFLs illustrates the broad applicability in industries.

Recently, we fabricated a new type of LSFL pattern, namely omnidirectional LSFLs (OD-LSFLs), expanding the viewing angle of structural colors by periodically ordering the orientation of LSFLs within the scanline [19]. Furthermore, compared to traditional LSFLs with a single orientation, unidirectional LSFLs (UD-LSFLs), OD-LSFLs uniformly distribute structural colors to all the azimuthal angles and show rotationally symmetric colorization under the normal illumination incidence [19].

The most notable applications for structural color materials are optical encryption and anti-counterfeiting due to their controllable optical properties [20]. As mentioned earlier, the structural colors from LSFLs can be easily altered by the polarization direction and wavelength of fs laser pulses. In particular, due to a quasi-periodic nature of LSFLs, their structural colors are unique in that it is exceedingly difficult to be duplicated through other classical means [4]. LSFLs are therefore uniquely suited for optical encryption and anti-counterfeiting [4,15]. To make use of UD- and OD-LSFLs for these applications, it is essential to understand how the colorimetric responses of UD- and OD-LSFLs change with their orientations and what mechanisms contribute to these colorimetric responses under specific positions of source and observer [4,15,20–23].

In this paper, we measure the modified colorimetric behavior of metal surfaces at multiple detection angles due to two types of LSFL patterns, UD-LSFLs and OD-LSFLs, produced by fs laser pulse irradiation with linear and periodically rotating polarizations. This work builds on our earlier study [19] and reports on angle-resolved distinctive characteristics of colors arising from quasi-periodic structural details at the nanometer scale. With the CIE standard illuminant D65, we show that both UD- and OD-LSFLs on metals can transform polished metals to gonio-apparent materials, and each has its unique behavior of colorimetric responses, depending on both the detection and rotation angles. In addition to diffraction, we find that the power redistribution due to Rayleigh anomalies also significantly affects the colorimetric responses of UD- and OD-LSFLs.

2. Materials and Methods

The laser employed in this experiment is an amplified Ti:sapphire fs laser system that generates 33.6-fs laser pulses with the maximum pulse energy of 1.2 mJ at a central wavelength and repetition rate of 800 nm and 5 kHz, respectively. At the fs laser output, the $1/e^2$ intensity radius is about 5 mm. We first prepare pure polycrystalline Ni (99.9%) with a thickness of 2 mm, and the surfaces are mechanically polished with 80-nm-grade colloidal silica. The average roughness of the polished Ni samples measured is 9.4 nm. To create UD-LSFLs on Ni, linearly polarized fs laser pulses are focused onto the samples with a 100 mm focal length plano-convex lens, as shown in Figure 1a. The surface of each sample is located at 1.5 mm before the focal plane. The $1/e^2$ intensity spot radius is about 80 μm at this defocused distance. To manipulate the polarization direction of fs laser pulses for the fabrication of OD-LSFLs, we insert a liquid crystal polymer patterned depolarizer (LCPPD) right before the focusing lens. A birefringent liquid crystal polymer film protected by two BK7 plates is used in the LCPPD, as shown in Figure 1b, and introduces an optical path difference of 380–430 nm between its fast and slow axes at our laser wavelength of 800 nm so that it nearly acts as a half-wave plate. Its fast axis also rotates 2° about the optical axis every 25 μm across the laser spot in the y-direction. Therefore, the polarization direction of fs laser pulses rotates periodically along the y-direction after LCPPD, and the period of the rotation by 180° is 1.125 mm, as visualized in Figure 1c. The $1/e^2$ intensity spot radius slightly increases to about 90 μm only in the y-direction at a defocused distance of 1.5 mm due to our LCPPD. Accordingly, the geometrical average of the spot radius, 85 μm, is used to calculate the laser fluence with our LCPPD. For colorimetric measurements with our spectrophotometer MA94 (X-Rite, Grand Rapids, MI, USA), UD- and OD-LSFLs are fabricated by raster scanning fs laser pulses with an area of 625 mm^2 (25 mm × 25 mm), sufficiently covering the circular area required for measurements. The light source for illumination in our spectrophotometer is a gas-filled tungsten lamp, and the colorimetric response due to the light source is calibrated so that the D65 colorimetric illuminant with the CIE 10° 1964 standard observer is used in our experiments. The colorimetric responses of the samples are measured at five different angles of detection, −65°, −30°, 0°, 20°, and 30°, with our spectrophotometer at an illumination incident angle of −45°, as shown in Figure 2. Each sample is mounted on the rotation stage and rotated about the z-axis, while the spectrophotometer itself stands still. The rotation angle of the sample about the z-axis is indicated by φ, defined as $0°$ when the x-axis is antiparallel to the j-axis, depicted in Figure 2.

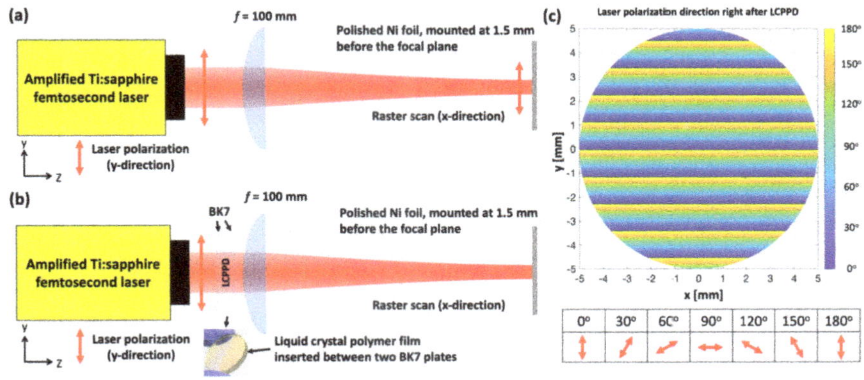

Figure 1. (**a**) Schematic of UD-LSFL fabrication in our experiments. The polarization direction of the fs laser pulses is in the y-direction. (**b**) Schematic of OD-LSFL fabrication in our experiments. The polarization direction of the fs laser pulses is manipulated by our liquid crystal polymer patterned depolarizer (LCPPD). (**c**) Calculated directions of polarization within the $1/e^2$ intensity radius of the fs laser pulses right after LCPPD. Angles denote the rotated angles of polarization direction due to LCPPD with respect to the y-direction.

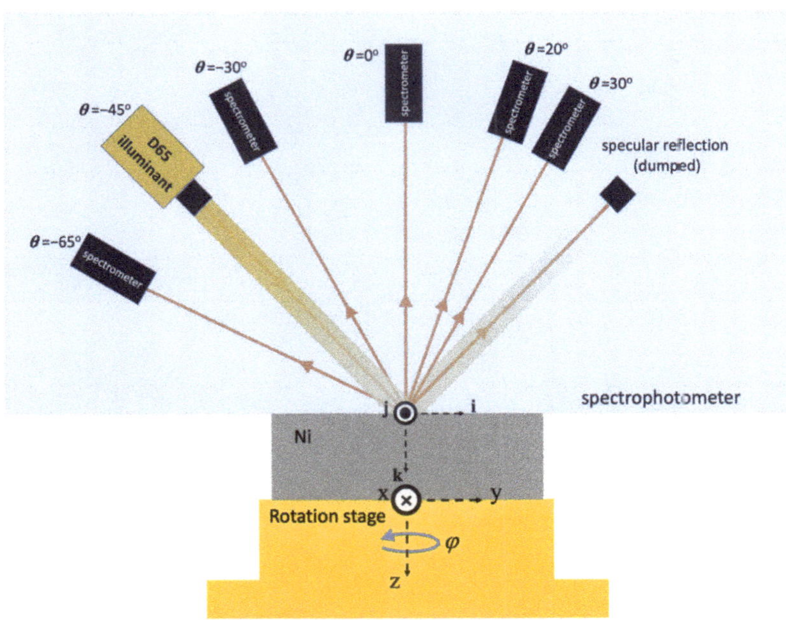

Figure 2. Configuration of the colorimetric measurements from the polished Ni, UD- and OD-LSFLs on Ni by using a spectrophotometer with detection angles of $30°$, $20°$, $0°$, $-30°$, and $-65°$. The *ijk* axes are the space-fixed frame with their origin located in the center of the D65 illuminant spot at the sample surface. The *xyz* axes are the body-fixed frame, rotating with the Ni sample, and the sample rotation about the z-axis is indicated by φ, defined as $0°$ when the x-direction is antiparallel to the j-direction.

3. Results

Figure 3 shows the scanning electron microscope (SEM) images of UD- and OD-LSFLs, fabricated at normal incidence with laser fluences of 0.16 J/cm^2 and 0.14 J/cm^2 by raster scanning fs laser pulses with scanning speeds of 8 mm/s and 4 mm/s, respectively. Under these experimental conditions, the period of UD- and OD-LSFLs groove estimated by using the Fourier transform of the SEM images is about 0.64 $\pm$ 0.03 μm, and the modulation depth of their grooves is about 0.42 $\pm$ 0.15 μm, measured by a confocal laser scanning microscope. As described by the double-headed red arrows in Figure 3, the grating vectors of UD- and OD-LSFLs are determined by the polarization direction of the incident fs pulses, since the formation of LSFLs is mainly attributed to the inhomogeneous energy deposition due to the interference of the incident fs pulse with surface plasmon polaritons, excited by the incident pulse [2]. The thickness of the scanlines for both UD- and OD-LSFLs is in the range of 68–74 μm, while the distance between the scanlines is 90 μm. Accordingly, there exists a gap of 16–22 μm between the scanlines. This gap is deliberately inserted to eliminate any unwanted effects on structural coloration, resulting from the decreases in the reflectance [2] and period [24] of LSFLs by means of extra pulses of irradiation due to the overlapping of scanlines. Within a single scanline of OD-LSFLs, the number and period of LSFL orientation rotations by 180° are about 4 and 18 μm, respectively, and this is consistent with our previous work where the same defocused distance was used [19].

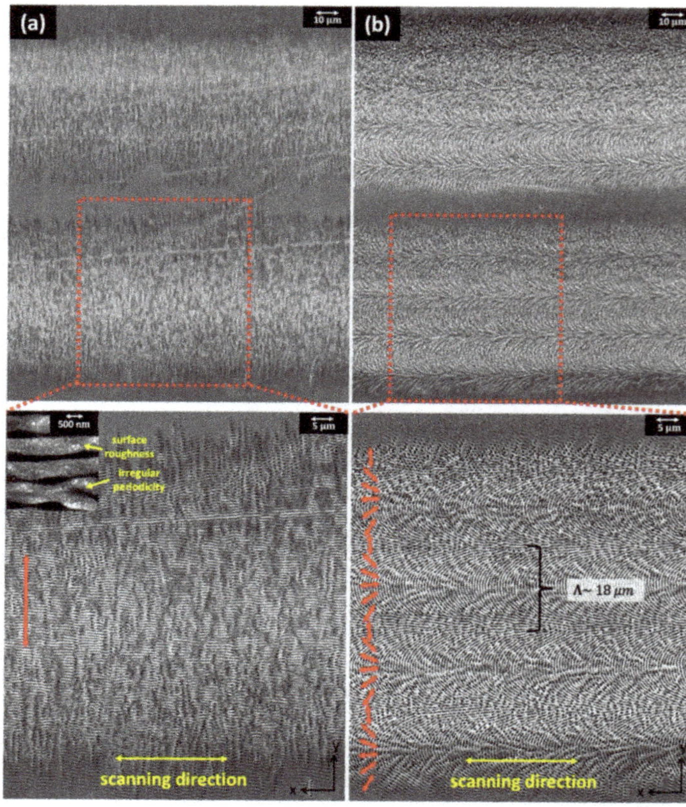

Figure 3. SEM images of (**a**) UD-LSFLs and (**b**) OD-LSFLs fabricated at a defocused distance of 1.5 mm. Double-headed red arrows indicate the grating vector (orientation) of UD- and OD-LSFLs and the polarization direction of fs laser pulses. The inset in (**a**) shows the surface roughness and irregular periodicity of UD-LSFLs on Ni.

Figure 4 shows the dependence of structural colors from the three samples on the rotation and detection angles. The CIELAB color space is employed because it is device-independent [25] and the color and its difference can be easily quantified by three coordinates, L^*, a^*, and b^* [26,27]. L^* stands for the lightness value, brightness in the color space, and a^* and b^* represent colors changing from green to red and blue to yellow, respectively, as their values increase from negative to positive. The neutral gray color appears when a^* and b^* equal zero. The difference in color between two points, (L_1^*, a_1^*, b_1^*) and (L_2^*, a_2^*, b_2^*) in the CIELAB color space, ΔE, is defined as the Euclidian distance between these points, $\Delta E = \sqrt{(L_1^* - L_2^*)^2 + (a_1^* - a_2^*)^2 + (b_1^* - b_2^*)^2}$ [26,27], and is perceptible when $\Delta E > 3.0$ [25]. By considering the structural symmetry of UD- and OD-LSFLs in terms of their orientations, L^*, a^*, and b^* values are measured in the rotation angle (φ) range of 180° at all detection angles (θ), and each data point is described by a small dot with actual true color, as shown in Figure 4.

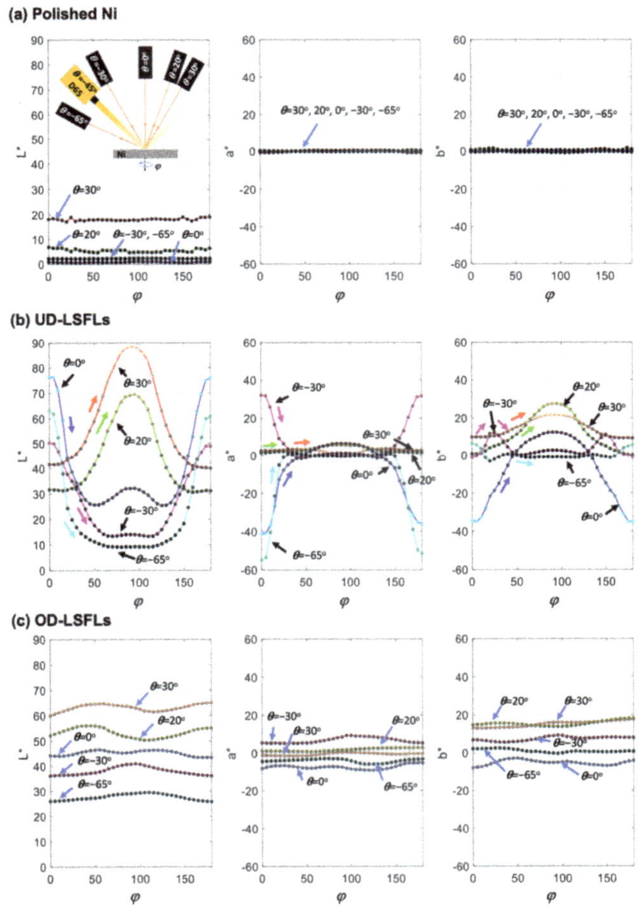

Figure 4. L^*, a^*, and b^* in the CIELAB color space are measured in a rotation angle (φ) range of 0° to 180° on (**a**) the polished Ni surface, (**b**) UD-LSFLs on Ni, and (**c**) OD-LSFLs on Ni. Red, green, blue, magenta, and cyan solid lines are used to represent our colorimetric measurements at detection angles (θ) of 30°, 20°, 0°, −30°, and −65°, respectively. Each data point is represented by a small dot color-coded with actual angle-resolved true color. In (**b**), changing directions of L^*, a^*, and b^* during an increase in φ from 0° to 90° are denoted by single-headed red, green, blue, magenta, and cyan arrows for detection angles of 30°, 20°, 0°, −30°, and −65°, respectively.

On the polished Ni surface without LSFLs, as shown in Figure 4a, a^* and b^* are almost independent of the detection (θ) and rotation angles (φ), and located near zero. This indicates that the color of the polished sample is very close to neutral gray in our experiments. During the rotation of the sample, the maximum difference in color, ΔE_{max}, within each detection angle is less than 3.0.

Figure 4b shows L^*, a^*, and b^* for UD-LSFLs on Ni, and the changing directions of L^*, a^*, and b^* during an increase in the rotation angle (φ) from 0° to 90° are represented by single-headed red, green, blue, magenta, and cyan arrows for detection angles of 30°, 20°, 0°, −30°, and −65°, respectively. Different from the polished Ni, a^* and b^* clearly change both with θ and φ, and the structural color can significantly deviate from neutral gray. For two backward detection angles of −30° and −65°, L^*, $|a^*|$, and $|b^*|$ tend to decrease as φ increases from 0° to 90°. Changes in b^* with φ are a little bit complicated at $\theta = -30°$, where b^* reaches its maximum around $\varphi \sim 25°$. In the case of two forward detection angles of 20° and 30°, both L^* and b^* monotonically elevate as φ changes from 0° to 90°. Compared to L^* and b^*, the change in a^* is relatively small at these forward detection angles. For $\theta = 0°$, an increase in φ from 0° to 45° decreases L^*, $|a^*|$ and $|b^*|$, and the structural color from UD-LSFLs becomes a neutral gray color around $\varphi = 45°$. Then, L^*, a^*, and b^* elevate all together when φ changes from 45° to 90°. During the rotation of the sample, the changes in color from UD-LSFLs are significantly large at all detection angles. The maximum values of measured color difference,ΔE_{max}, are 49.8, 47.0, 80.1, 48.7, and 77.1 at θ = 30°, 20°, 0°, −30°, and −65°, respectively, during the change in φ from 0° to 90°. Due to the structural symmetry of UD-LSFLs and our color measurement configuration, the changes in the structural color of UD-LSFLs while φ increases from 0° to 90° are equivalent to the changes in φ from 180° to 90° at all detection angles, as shown in Figure 4b.

For OD-LSFLs, as shown in Figure 4c, the structural color changes mostly with the detection angle (θ), and appears to have a weak dependence on the rotation angle (φ) as compared to that for UD-LSFLs: during the rotation of OD-LSFLs, L^*, a^*, and b^* are positioned within the variation range of those measured from UD-LSFLs within each detection angle. The maximum difference in color during the rotation of OD-LSFLs is less than 7.0 at each detection angle.

4. Discussion

As briefly mentioned earlier in Section 1, periodic structures can colorize material surfaces through diffraction. In fact, diffraction from periodic structures can be simply understood as the transfer of wavevector from the longitudinal direction to the transverse direction, where the amount of the transfer is the grating vector of periodic structures multiplied by integer numbers [28,29]. Assuming that our periodic structures are fabricated in the *ij* plane, as shown in Figure 2, the wavevector of diffracted light as functions of the wavelength of light (λ) and the rotation angle (φ) about the *z*-axis can be calculated by the following equation [29,30]:

$$\begin{aligned} \mathbf{k}_{\text{diff}}(\text{m}) = \hat{\mathbf{i}}\left(-k\sin\theta_i + m\left|\mathbf{K}_g\right|\cos\varphi\right) + \hat{\mathbf{j}}\,m\left|\mathbf{K}_g\right|\sin\varphi \\ -\hat{\mathbf{k}}\sqrt{k^2 - \left(-k\sin\theta_i + m\left|\mathbf{K}_g\right|\cos\varphi\right)^2 - \left(m\left|\mathbf{K}_g\right|\sin\varphi\right)^2}, \end{aligned} \quad (1)$$

where $\mathbf{k}_{\text{diff}}$ and $\mathbf{K}_g$ are the wavevectors of diffracted light and LSFLs, respectively, k is $2\pi/\lambda$, $|\mathbf{K}_g|$ is $2\pi/d$, d is the period of one-dimensional periodic structures, and θ_i is the incident angle of illumination, −45° in our color measurement conditions. As described in Figure 2, $\hat{\mathbf{i}}$, $\hat{\mathbf{j}}$, and $\hat{\mathbf{k}}$ are the unit vectors along the *ijk* axes of the space-fixed frame, and the origin of this system is located in the center of the D65 illuminant spot at the sample surface. In Equation (1), m represents the mth order of diffraction, available when the $\hat{\mathbf{k}}$ component of $\mathbf{k}_{\text{diff}}$ is real.

In the case of the polished Ni sample, no deliberately fabricated periodic structures are present at its surface, and local roughness variations largely due to structural defects and polishing-induced surface scratches can effectively diffuse the reflected light. Because roughness-induced diffuse light plays a dominant role in its coloration, this scenario can account for our observation that L^*, a^*, and b^* are nearly independent of φ, exhibiting neutral gray color at all detection angles, as shown in Figure 4a.

In the case of UD-LSFLs, however, diffraction comes into play on the colorimetric response of the surface due to a quasi-periodic nature of UD-LSFLs, and the surface roughness and quasi-periodicity of UD-LSFLs shown in the inset of Figure 3a diffuse the light near the diffraction and reflection angles.

With a UD-LSFL period range of 0.61 to 0.67 μm and the CIE 10° 1964 standard observer, the wavelength range of diffracted light can be obtained for $\varphi = 0$ at each detection angle with the help of Equation (1), as shown in Figure 5a. The −1st, −2nd, and −3rd orders of diffraction occur at our detection angles (θ) in the wavelength range of D65. However, any effects on color from the −3rd order of diffraction can be ignored, since the spectral luminous efficiency in photopic and scotopic vision is negligible at wavelengths below 380 nm [27].

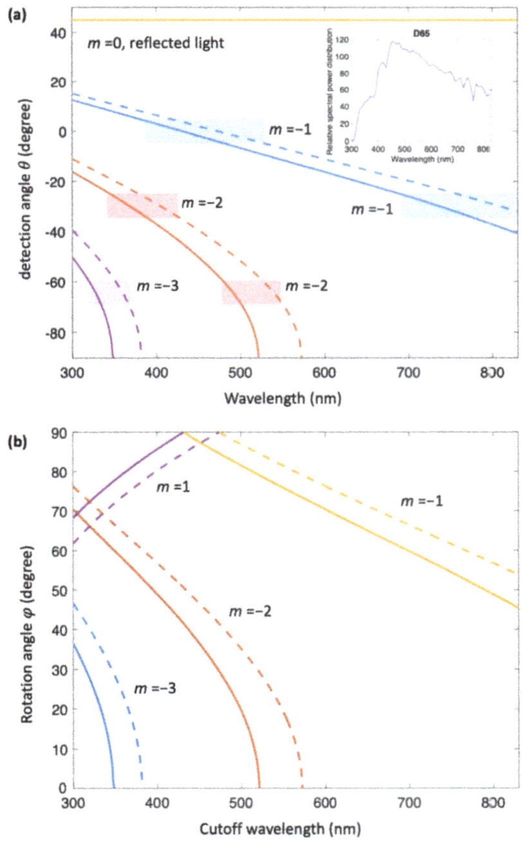

Figure 5. (**a**) Calculated mth order diffraction angles as a function of the wavelength of light. The ranges of the wavelength detected by three detection angles of 0°, −30°, and −65° are described by rectangular boxes filled with blue, red, and purple for $m = -1, -2$, and -3, respectively. For these boxed regions, the CIE 10° 1964 standard observer and a groove period range of 610–670 nm are considered. (**b**) Rotation angle φ versus the cutoff wavelengths for $m = -3, -2, -1$, and 1. Solid and dashed lines are used to describe the cases with groove periods of 610 and 670 nm, respectively.

Accordingly, for $\theta = -65°$, the structural color of UD-LSFLs can be mainly attributed to the −2nd order of diffraction at $\varphi = 0$, and is close to bluish-green, the mixture of blue and green colors, where a dominant wavelength range of 480–552 nm is considered to primarily determine the perceived color. For detection angles of −30°, the −2nd and −1st orders of diffraction corresponding to wavelength ranges of 345–429 nm and 689–859 nm, both contribute to the color of UD-LSFLs respectively, leading the perceived color to be similar to purplish-pink. By increasing φ from 0° to 90°, the propagation direction of diffraction starts to have the *j* component, as described in Equation (1). Since our color measurements are restricted near the *ik* plane shown in Figure 2, the effect on L^*, $|a^*|$, and $|b^*|$ due to diffraction at our spectrometers should decrease while $|\sin \varphi|$ increases. As shown in Figure 4b, on UD-LSFLs for $\theta = -65°$ and −30°, the structural color of UD-LSFLs shifts toward neutral gray at $\varphi = 90°$, indicating that diffraction by UD-LSFLs mostly contributes to their color.

For forward detection angles of $\theta = 20°$ and 30°, the structural color of UD-LSFLs deviates from neutral gray at most of φ measured in our experiments, and the changes in L^* and b^* at these forward detection angles are quite different from those at the backward detection angles. According to Equation (1) and Figure 5, there is no diffracted light propagating toward these detection angles, and therefore it is expected that the diffused light due to surface roughness and the irregular periodicity of UD-LSFLs mainly affects the coloration of UD-LSFLs.

To better understand our observations, we consider Rayleigh or threshold anomalies [28,29]. Rayleigh anomalies explain the abrupt power changes to the propagating orders of diffracted lights and the reflected light at a cutoff wavelength (λ_c) due to the power redistribution when a specific diffraction order disappears or appears with the changes in the wavelength and/or incident angle of light [28,29]. Considering no real part of the $\hat{\mathbf{k}}$ component in $\mathbf{k}_{diff}$ in Equation (1), the cutoff wavelength (λ_c) at all available diffraction orders (m) for the period range of UD-LSFLs, 0.61–0.67 μm, is calculated as a function of φ in Figure 5b. For $\lambda > \lambda_c$, no diffraction occurs because the $\hat{\mathbf{k}}$ component of $\mathbf{k}_{diff}$ is imaginary.

Based on our calculations, the cutoff wavelengths for $m = -2$ are in a wavelength range of 520–570 nm at $\varphi = 0$, and this range can be further broadened by the surface roughness and irregular periodicity of UD-LSFLs. Therefore, not all wavelength range of D65 illuminant in the visible spectral region is fully diffracted when $m = -2$, and some portion of power from the −2nd order of diffraction with the wavelengths longer than about 520 nm is redistributed into the diffused light, propagating toward detection angles (θ) of 20° and 30°. It follows that the structural color of UD-LSFLs becomes yellowish-pink, as a result of the contribution of red and some amount of yellow and green, whose corresponding wavelengths are mostly longer than about 550 nm in the visible spectral region, and the structural color deviates from the color of the polished sample, neutral gray. As φ increases until 45°, the cutoff wavelengths for $m = -2$ continuously decrease to about 416–458 nm shown in Figure 5b. More power from shorter wavelengths will be redistributed, and this leads to an increase in L^*, as shown in Figure 4b. Additionally, the structural color of UD-LSFLs at these forward detection angles shifts toward orange-yellow, because the contribution of additional green and yellow with a small amount of blue to yellowish pink becomes larger with the decrease in the cutoff wavelengths for $m = -2$. This is equivalent to an increase in b^* since the contribution of more yellow than blue elevates b^* in the CIELAB color space. When we further increase φ to 90°, the cutoff wavelength for $m = -2$ can reach the wavelength below 400 nm, and the −2nd order of diffraction does not affect the structural color of UD-LSFLs anymore. Instead, the −1st order of diffraction comes into play by disappearing from red to blue. Consequently, the disappeared color from the −1st order of diffraction is continuously transferred toward our forward detection angles and keeps increasing L^* and b^*, the same reason for $\varphi < 45°$.

At $\theta = 0°$, the effects from both diffraction and Rayleigh anomalies are clearly observed on UD-LFSLs. When $\varphi = 0°$, the −1st order of diffraction with a wavelength range of 378–

533 nm mostly determines the structural color of UD-LSFLs, greenish-blue. As φ increases, any effects due to the -1st order of diffraction reduce, due to our color measurement configuration limited near the plane of illumination incidence, and the structural color of UD-LSFLs shifts to neutral gray. However, once φ reaches 50° and further increases to 90°, Rayleigh anomalies clearly have an effect on the color of UD-LSFLs by disappearing the -1st order of diffraction as the cutoff wavelength decreases, and lead to the increases in L^* and b^*. More discussion about the angular-dependence of colorimetric responses shown in Figure 4b is available in Supplementary Materials.

In the case of OD-LSFLs, their structural color depends mainly on the detection angle, but does not change much on the rotation angle (φ). Moreover, it is not neutral gray, and varies within the colorimetric response variation of UD-LSFLs with φ at each detection angle, as shown in Figure 4c. This indicates that the structural color of OD-LSFLs is also subject to both diffraction and Rayleigh anomalies, and mostly is attributed to overall contribution from all the colorimetric responses of UD-LSFLs with the φ range of 0° to 180° since OD-LSFLs are UD-LSFLs with a periodic ordering of their orientations.

5. Conclusions

The colorimetric behaviors of the polished Ni, UD-LSFLs, and OD-LSFLs on Ni are investigated by using the CIE standard illuminant D65 at an illumination angle of $-45°$. We demonstrate that UD-LSFLs on metals can transform polished metals to gonio-apparent materials with their colorimetric responses that depend on the rotation angle of the sample, whereas the colorimetric response of OD-LSFLs changes only with the detection angle, and is nearly independent of the rotation angles. Furthermore, we find that these colorimetric responses of UD- and OD-LSFLs are attributed not only to diffraction but also to the wavelength-dependent power redistribution to the diffused light due to Rayleigh anomalies.

Supplementary Materials: The following are available online at https://www.mdpi.com/article/10.3390/nano11082010/s1, More discussion about the angular-dependence of colorimetric responses shown in Figure 4b is available.

Author Contributions: Conceptualization, T.-Y.H., Y.-d.K., B.L.; methodology, H.-J.L., H.-S.L.; software, T.-Y.H.; validation, T.-Y.H.; formal analysis, J.C.; investigation, J.C., H.-J.L., H.-S.L.; resources, H.-J.L., H.-S.L.; data curation, T.-Y.H.; writing—original draft preparation, T.-Y.H., B.L.; writing—review and editing, T.-Y.H., J.C.; visualization, T.-Y.H., J.C.; supervision, T.-Y.H., Y.-d.K.; project administration, Y.-d.K.; funding acquisition, Y.-d.K. All authors have read and agreed to the published version of the manuscript.

Funding: This study has been conducted with the support of the Korea Institute of Industrial Technology as "Development of root technology for multi-product flexible production (KITECH EO-21-0008)".

Institutional Review Board Statement: Not applicable.

Informed Consent Statement: Not applicable.

Data Availability Statement: The data presented in this study are available on request from the corresponding author.

Acknowledgments: We thank Heedeuk Shin at Pohang University of Science and Technology for discussion and technical assistance.

Conflicts of Interest: The authors declare no conflict of interests.

References

1. Sun, J.; Bhushan, B.; Tong, J. Structural coloration in nature. *RSC Adv.* **2013**, *3*, 14862–14889. [CrossRef]
2. Hecht, E. *Optic*, 4th ed.; Addison-Wesley: Reading, MA, USA, 2002.
3. Vorobyev, A.Y.; Guo, C. Direct femtosecond laser surface nano/microstructuring and its applications. *Laser Photonics Rev.* **2013**, *7*, 385–407. [CrossRef]

4. Bonse, J.; Höhm, S.; Kirner, S.V.; Rosenfeld, A.; Krüger, J. Laser-induced periodic surface structures—A scientific evergreen. *IEEE J. Sel. Top. Quantum Electron.* **2017**, *23*, 109–123. [CrossRef]
5. Buividas, R.; Mikutis, M.; Juodkazis, S. Surface and bulk structuring of materials by ripples with long and short laser pulses: Recent advances. *Prog. Quantum Electron.* **2014**, *38*, 119–156. [CrossRef]
6. Weck, A.; Crawford, T.H.R.; Wilkinson, D.S.; Haugen, H.K.; Preston, J.S. Ripple formation during deep hole drilling in copper with ultrashort laser pulses. *Appl. Phys. A Mater. Sci. Process.* **2007**, *89*, 1001–1003. [CrossRef]
7. Hwang, T.Y.; Guo, C. Angular effects of nanostructure-covered femtosecond laser induced periodic surface structures on metals. *J. Appl. Phys.* **2010**, *108*, 73523. [CrossRef]
8. Zuhlke, C.A.; Tsibidis, G.D.; Anderson, T.; Stratakis, E.; Gogos, G.; Dennis, R.; Zuhlke, C.A.; Tsibidis, G.D.; Anderson, T.; Stratakis, E.; et al. Investigation of femtosecond laser induced ripple formation on copper for varying incident angle. *AIP Adv.* **2018**, *8*, 15212. [CrossRef] [PubMed]
9. Garrelie, F.; Colombier, J.P.; Pigeon, F.; Tonchev, S.; Faure, N.; Bounhalli, M.; Reynaud, S.; Parriaux, O. Evidence of surface plasmon resonance in ultrafast laser-induced ripples. *Opt. Express* **2011**, *19*, 9035–9043. [CrossRef]
10. Vorobyev, A.Y.; Guo, C. Colorizing metals with femtosecond laser pulses. *Appl. Phys. Lett.* **2008**, *92*, 041914. [CrossRef]
11. Ionin, A.A.; Kudryashov, S.I.; Makarov, S.V.; Seleznev, L.V.; Sinitsyn, D.V.; Golosov, E.V.; Golosova, O.A.; Kolobov, Y.R.; Ligachev, A.E. Femtosecond laser color marking of metal and semiconductor surfaces. *Appl. Phys. A* **2012**, *107*, 301–305. [CrossRef]
12. Jwad, T.; Penchev, P.; Nasrollahi, V.; Dimov, S. Laser induced ripples' gratings with angular periodicity for fabrication of diffraction holograms. *Appl. Surf. Sci.* **2018**, *453*, 449–456. [CrossRef]
13. Livakas, N.; Skoulas, E.; Stratakis, E. Omnidirectional iridescence via cylindrically-polarized femtosecond laser processing. *Opto-Electronic Adv.* **2020**, *3*, 190035. [CrossRef]
14. Tamamura, Y.; Miyaji, G. Structural coloration of a stainless steel surface with homogeneous nanograting formed by femtosecond laser ablation. *Opt. Mater. Express* **2019**, *9*, 2902–2909. [CrossRef]
15. Dusser, B.; Sagan, Z.; Soder, H.; Faure, N.; Colombier, J.P.; Jourlin, M.; Audouard, E. Controlled nanostructrures formation by ultrafast laser pulses for color marking. *Opt. Express* **2010**, *18*, 2913–2924. [CrossRef]
16. Lim, H.U.; Kang, J.; Guo, C.; Hwang, T.Y. Manipulation of multiple periodic surface structures on metals induced by femtosecond lasers. *Appl. Surf. Sci.* **2018**, *454*, 327–333. [CrossRef]
17. Beresna, M.; Gecevičius, M.; Kazansky, P.G.; Gertus, T. Radially polarized optical vortex converter created by femtosecond laser nanostructuring of glass. *Appl. Phys. Lett.* **2011**, *98*, 201101. [CrossRef]
18. La, M.; Hwang, T.Y.; Choi, J.Y.; Hong, S.; Jea Park, S.; Kang, J.; Choi, D. Development of a metal-to-metal imprinting process: Transcription quality analysis and surface wettability characterization. *Appl. Surf. Sci.* **2020**, *527*, 146823. [CrossRef]
19. Hwang, T.Y.; Shin, H.; Lee, H.J.; Lee, H.S.; Guo, C.; Lee, B. Rotationally symmetric colorization of metal surfaces through omnidirectional femtosecond laser-induced periodic surface structures. *Opt. Lett.* **2020**, *45*, 3414–3417. [CrossRef]
20. Hong, W.; Yuan, Z.; Chen, X. Structural Color Materials for Optical Anticounterfeiting. *Small* **2020**, *16*, 1–25. [CrossRef]
21. Bonse, J.; Kirner, S.V.; Höhm, S.; Epperlein, N.; Spaltmann, D.; Rosenfeld, A.; Krüger, J. Applications of laser-induced periodic surface structures (LIPSS). In Proceedings of the Laser-Based Micro- and Nanoprocessing XI, San Francisco, CA, USA, 17 February 2017; Volume 10092, p. 100920N.
22. Ageev, E.I.; Veiko, V.P.; Vlasova, E.A.; Karlagina, Y.Y.; Krivonosov, A.; Moskvin, M.K.; Odintsova, G.V.; Pshenichnov, V.E.; Romanov, V.V.; Yatsuk, R.M. Controlled nanostructures formation on stainless steel by short laser pulses for products protection against falsification. *Opt. Express* **2018**, *26*, 2117. [CrossRef] [PubMed]
23. Qian, J.; Zhao, Q.Z. Anti-counterfeiting microstructures induced by ultrashort laser pulses. *Phys. Status Solidi Appl. Mater. Sci.* **2020**, *217*, 1–7. [CrossRef]
24. Vorobyev, A.Y.; Makin, V.S.; Guo, C. Periodic ordering of random surface nanostructures induced by femtosecond laser pulses on metals. *J. Appl. Phys.* **2007**, *101*, 34903. [CrossRef]
25. Yuan, J.; Christensen, P.R.; Wolf, M.O. Dynamic anti-counterfeiting security features using multicolor dianthryl sulfoxides. *Chem. Sci.* **2019**, *10*, 10113–10121. [CrossRef] [PubMed]
26. MacAdam, D.L. *Color Measurement. Theme and Variations*, 2nd ed.; Springer: Berlin/Heidelberg, Germany, 1985.
27. Ohta, N.; Robertson, A.R. *Colorimetry: Fundamentals and Applications*; John Wiley & Sons Ltd.: Chichester, UK, 2005.
28. Loewen, E.G.; Popov, E. *Diffraction Gratings and Applications*; Marcel Dekker: New York, NY, USA, 1997.
29. Palmer, C. *Diffraction Grating Handbook*, 6th ed.; Newport Corporation: Rochester, NY, USA, 2005.
30. Harvey, J.E.; Vernold, C.L. Description of Diffraction Grating Behavior in Direction Cosine Space. *Appl. Opt.* **1998**, *37*, 8158. [CrossRef]

nanomaterials

MDPI

Article

In-Sn-Zn Oxide Nanocomposite Films with Enhanced Electrical Properties Deposited by High-Power Impulse Magnetron Sputtering

Hui Sun [1,†], Zhi-Yue Li [2,3,4,†], Sheng-Chi Chen [5,6,*], Ming-Han Liao [7], Jian-Hong Gong [8], Zhamatuofu Bai [1] and Wan-Xia Wang [8]

1 School of Space Science and Physics, Shandong University, Weihai 264209, China; huisun@sdu.edu.cn (H.S.); jamtuub@foxmail.com (Z.B.)
2 Department of Physics, The University of Hong Kong, Hong Kong 999077, China; lizhiyue007@hotmail.com
3 Shenzhen Institute for Quantum Science and Engineering, Southern University of Science and Technology, Shenzhen 518055, China
4 Department of Physics, Southern University of Science and Technology, Shenzhen 518055, China
5 Department of Materials Engineering and Center for Plasma and Thin Film Technologies, Ming Chi University of Technology, Taipei 243, Taiwan
6 College of Engineering and Center for Green Technology, Chang Gung University, Taoyuan 333, Taiwan
7 Department of Mechanical Engineering, National Taiwan University, Taipei 106, Taiwan; mhliaoa@ntu.edu.tw
8 School of Mechanical, Electrical and Information Engineering, Shandong University, Weihai 264200, China; gongjh@sdu.edu.cn (J.-H.G.); wang_wanxia@sdu.edu.cn (W.-X.W.)
* Correspondence: chensc@mail.mcut.edu.tw; Tel.: +886-2-2908-9899 (ext. 4679)
† These authors contributed equally to this work.

Abstract: In-Sn-Zn oxide (ITZO) nanocomposite films have been investigated extensively as a potential material in thin-film transistors due to their good electrical properties. In this work, ITZO thin films were deposited on glass substrates by high-power impulse magnetron sputtering (HiPIMS) at room temperature. The influence of the duty cycle (pulse off-time) on the microstructures and electrical performance of the films was investigated. The results showed that ITZO thin films prepared by HiPIMS were dense and smooth compared to thin films prepared by direct-current magnetron sputtering (DCMS). With the pulse off-time increasing from 0 μs (DCMS) to 2000 μs, the films' crystallinity enhanced. When the pulse off-time was longer than 1000 μs, In_2O_3 structure could be detected in the films. The films' electrical resistivity reduced as the pulse off-time extended. Most notably, the optimal resistivity of as low as 4.07×10^{-3} Ω·cm could be achieved when the pulse off-time was 2000 μs. Its corresponding carrier mobility and carrier concentration were 12.88 $cm^2V^{-1}s^{-1}$ and 1.25×10^{20} cm^{-3}, respectively.

Keywords: ITZO film; high-power impulse magnetron sputtering; duty cycle; pulse off-time; electrical properties

Citation: Sun, H.; Li, Z.-Y.; Chen, S.-C.; Liao, M.-H.; Gong, J.-H.; Bai, Z.; Wang, W.-X. In-Sn-Zn Oxide Nanocomposite Films with Enhanced Electrical Properties Deposited by High-Power Impulse Magnetron Sputtering. *Nanomaterials* **2021**, *11*, 2016. https://doi.org/10.3390/nano11082016

Academic Editors: Jihoon Lee and Ming-Yu Li

Received: 28 June 2021
Accepted: 28 July 2021
Published: 6 August 2021

1. Introduction

Si-based thin-film transistors (TFTs) are widely used in liquid crystal displays, sensors, logic integrated circuits, etc. [1–3]. However, the high temperatures required for the formation of Si materials severely limits their applications in novel optoelectronic devices. For instance, flexible electrical devices and wearable devices are being developed rapidly nowadays but the flexible substrates used in these devices possess poor heat resistance, which makes Si-based TFTs unusable in these fields. Conversely, amorphous oxide semiconductors can be fabricated at room temperature [4–6]. Such materials combine good light transmittance and conductivity. Therefore, TFTs based on such materials are gradually replacing Si-based TFTs in certain areas. To date, the amorphous oxide semiconductor that has been most widely studied and has achieved commercial applications is In-Ga-Zn-O (IGZO) [7,8]. Its good uniformity and high carrier mobility (~10 cm^2/Vs) have led to it attracting much attention in recent years [9]. Unfortunately, during the traditional back-channel-etching process used to manufacture amorphous IGZO TFTs, IGZO reacts easily with weak acids [10,11]. Furthermore, the field-effect mobility of IGZO TFT is still inadequate to drive high-frame-rate displays [12]. Therefore, it is necessary to explore other amorphous oxide semiconductors.

In-Sn-Zn-O (ITZO) is a novel transparent conductive material that replaces Ga_2O_3 in IGZO with more chemically stable SnO_2, which helps to endow ITZO with better etching-resistance ability [13]. Meanwhile, the direct spatial overlap of the orbitals between Sn 5s orbital and In 5s can enhance the mobility of the electrons within the conduction band minimum, leading to a higher carrier mobility [14]. In addition, compared with the substitution of Zn^{2+} by trivalent Ga^{3+}, the substitution of Zn^{2+} by tetravalent Sn^{4+} will release more free electrons and improve the electrical properties of the films [15]. As a result, ITZO-based TFTs have high potential for the development of next generation displays due to their good etching-resistance during the back-channel-etching process.

Currently, magnetron sputtering and the sol-gel method are most commonly used to prepare ZnO-based thin films [16–18]. In particular, magnetron sputtering has attracted much attention due to its low deposition temperature, fast sputtering speed, uniform film formation, and good repeatability [19–21]. However, the films deposited by traditional magnetron sputtering method present loose structure with many defects, greatly affecting the films' performance [22]. The relatively recently developed high-power impulse magnetron sputtering (HiPIMS) technology has an important advantage in its high target ionization rate, which can improve the activity of the various species during the sputtering process [23–26]. In addition, due to the high instantaneous power density applied on the target, the energy of the incident species to the substrate is effectively increased, resulting in the formation of a denser and more uniform film, thereby reducing the carrier scattering and enhancing the carrier mobility [27,28]. To the best of our knowledge, no other groups have prepared ITZO films using HiPIMS technology. In the current work, the optoelectronic properties of ITZO films prepared by HiPIMS technology under different duty cycles were investigated.

2. Experimental Details

ITZO thin films with a thickness of 100 nm were deposited through HiPIMS technology on glass and silicon substrates from an ITZO target (99.9% purity, In_2O_3:SnO_2:ZnO = 30:35:35 at.%, Φ = 76.2 mm) at room temperature. The working pressure was 0.7 Pa with the Ar flow rate maintained at 20 sccm. The sputtering power of the HiPIMS power supply was 300 W, while the pulse on-time (t_{on}) remained at 50 μs. The pulse off-time (t_{off}) varied from 0 to 2000 μs during the deposition process. The duty cycle is defined as the ratio between the t_{on} and the sum of t_{on} plus t_{off}, and therefore reduces with an increase in t_{off}. The deposition parameters maintained during the deposition are summarized in Table 1.

Table 1. Sputtering parameters maintained during deposition of ITZO thin films.

Parameters	Value	Parameters	Value
Target	ITZO	Background pressure (Pa)	$<7 \times 10^{-4}$
Substrate	Glass/silicon	Working pressure (Pa)	0.7
t_{on} (μs)	50	Ar flow rate (sccm)	20
t_{off} (μs)	$0 \rightarrow 2000$	Deposition temp. (°C)	ambient
Power (W)	300	Film thickness (nm)	~100

The sputtering voltage and current variation of HiPIMS power output were monitored by oscilloscope (Rigol DS5202CA, Rigol Technologies. Inc., Beijing, China). The films' thickness was detected by step profiler (Kosaka Surfcoder, Kosaka Laboratory Ltd., Tokyo, Japan). The films' composition was characterized by electron probe *X*-ray microanalyzer (EPMA, JEOL JXA-8200, JEOL, Tokyo, Japan). The structural properties of ITZO films were analyzed through *X*-ray diffractometer (XRD, Rigaku Ultima IV, Tokyo, Japan). The surface roughness of the films was measured by atomic force microscope (AFM, DI-Dimension 3100, Digital instruments, Bresso, Italy). The microstructure of the specimens prepared by focused ion beam (FIB) milling was observed on cross-sections by high resolution transmission electron microscopy (HR-TEM, JEOL JEM-2100, JEOL, Tokyo, Japan). The films' electrical properties were obtained by the Hall effect measurement system (AHM-800B, Agilent Technologies, Santa Clare, CA, USA).

3. Results

Figure 1 shows the variation of the sputtering voltage and current on the target with the pulse off-time prolongation during the deposition process. Both of them increased with the extension of the pulse off-time. Consequently, the peak power density on the target also rises gradually. The variation of the duty cycle and the calculated peak power density as a function of t_{off} are given in Table 2. As the pulse off-time extended from 0 μs to 2000 μs, the target peak power density rises greatly from 6.42 to 531.97 W·cm^{-2}. However, the deposition rate monotonically decreased with the extension of pulse off-time (Figure 2). With increasing pulse off-time, the reduction in the effective sputtering period resulted in fewer target atoms being sputtered, which in turn reduced the deposition rate.

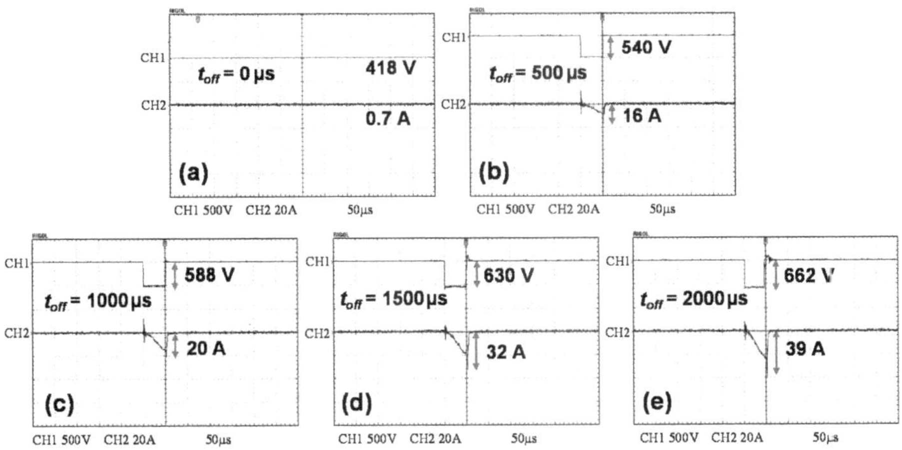

Figure 1. The sputtering voltage and current with different sputtering pulse off-times (t_{off}): (**a**) 0 μs, (**b**) 500 μs, (**c**) 1000 μs, (**d**) 1500 μs, and (**e**) 2000 μs.

Table 2. The variation of duty cycle and peak power density as a function of t_{off} used in each experiment (t_{on} = 50 μs).

t_{off} (μs)	0	500	1000	1500	2000
Duty cycle (%)	100	9.09	4.72	3.23	2.44
Peak power density (W/cm^2)	6.42	189.47	257.89	442.11	531.97

Table 3 shows the relationship between the pulse off-time and the film's composition. The content of In, Sn, Zn, and O in the film changed slightly as the pulse off-time was extended. When the pulse off-time was 0 μs, the sputtering mode was equivalent to conventional DCMS, where the target ionization rate is limited. Upon extension of the pulse off-time, the peak power density applied on the target surface increased markedly, and the instantaneous energy released on the target rises considerably, resulting in a significant increase in the ionization rate of the target species. The ionized target species possess higher activity and react more easily with the reactive O atoms. Therefore, O content in the films deposited using HiPIMS mode was higher than that in the films deposited by DCMS mode. Nevertheless, the films' composition remained almost unchanged and they were always oxygen-deficient, resulting in the formation of donor defects such as oxygen vacancies, thereby improving their conductivity.

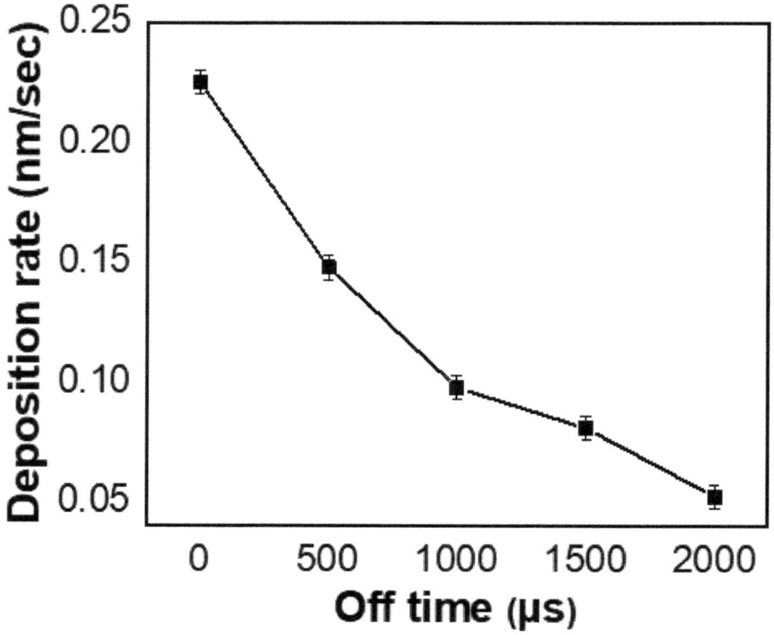

Figure 2. Deposition rate at different pulse off-times.

Table 3. Atomic concentration of ITZO films deposited at different pulse off-times.

t_{off} (μs)	In	Sn	Zn	O
0	54.69 (±0.1) at.%	3.49 (±0.1) at.%	1.19 (±0.1) at.%	40.63 (±0.1) at.%
500	52.95 (±0.1) at.%	3.47 (±0.1) at.%	1.03 (±0.1) at.%	42.56 (±0.1) at.%
1000	52.31 (±0.1) at.%	3.32 (±0.1) at.%	1.19 (±0.1) at.%	43.18 (±0.1) at.%
1500	52.91 (±0.1) at.%	3.56 (±0.1) at.%	1.03 (±0.1) at.%	42.50 (±0.1) at.%
2000	51.67 (±0.1) at.%	3.89 (±0.1) at.%	1.14 (±0.1) at.%	43.30 (±0.1) at.%

Figure 3 shows the XRD spectra of the ITZO thin films deposited with various pulse off-times. Amorphous-like structures were obtained when the pulse off-times were 0 μs and 500 μs. No obvious diffraction peak could be detected in either of these films. As the pulse off-time extended to 1000 μs and 1500 μs, ITZO films began to crystallize and an In_2O_3 (222) diffraction peak emerged. Upon further extending the pulse off-time to 2000 μs, the films' crystallinity increased significantly. Additional diffraction peaks of In_2O_3 (222), In_2O_3 (400), In_2O_3 (440), and In_2O_3 (622) were also identified. This behavior was related to higher instantaneous energy being bombarded on the target with the extension of pulse off-time; therefore, the sputtering species consequently possessed more kinetic energy. This promoted the nucleation and orderly growth of the ITZO films, thus improving the films' crystallinity. The films' crystalline features can be analyzed by TEM, as shown in Figure 4. The close-up lattice images in this figure are taken from the areas marked by red squares and produced through inverse Fourier transformation. In Figure 4a, many amorphous areas can be found in the ITZO film deposited with the pulse off-time of 0 μs. In contrast, In_2O_3 (222) planes with interplanar lattice spacing of about 2.9 Å can be clearly observed in Figure 4b, indicating that the films' crystallinity increased with the extension of pulse off-time.

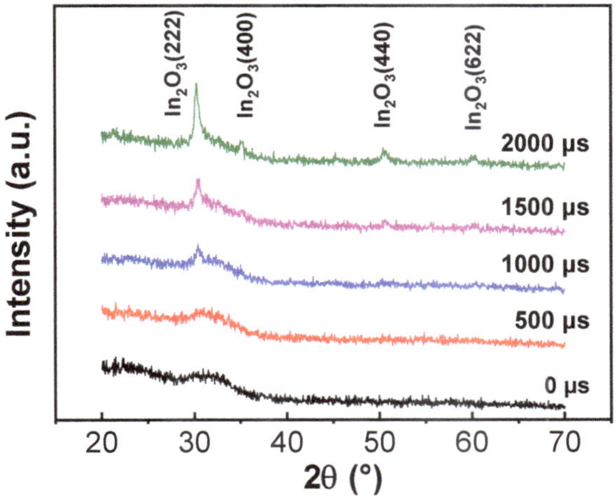

Figure 3. XRD patterns of ITZO films deposited at different pulse off-times.

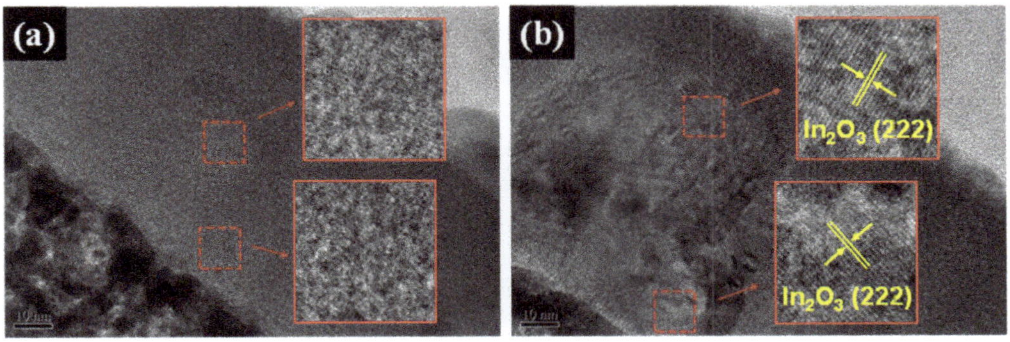

Figure 4. The cross-sectional TEM images as well as the corresponding enlarged images of ITZO films deposited with pulse off-times of (**a**) 0 μs and (**b**) 2000 μs.

The films' surface morphology was characterized through AFM analysis (Figure 5). The roughness of the films decreased from 2.17 nm to 0.85 nm and further to 0.70 nm, as the pulse off-time extended from 0 μs to 1000 μs and on to 2000 μs. Due to the high-energy sputtering species bombardment of the substrate during the deposition process, the films became much denser and smoother under extended pulse off-time.

The films' electrical properties as analyzed by Hall measurement are shown in Figure 6. The variation of the carrier concentration and the carrier mobility are summarized in Figure 6a. As the pulse off-time was raised from 0 to 500 μs, the deposition mode changed from DC mode to HiPIMS mode. Due to the higher kinetic energy of the sputtering species bombarding the substrate during the HiPIMS deposition mode, a denser film with a low amount of defects was obtained. As a result, the carrier concentration decreased from 3.92×10^{19} cm^{-3} to 9.11×10^{18} cm^{-3}; while the carrier scattering reduced, and the carrier mobility increased greatly from 3.99 cm^{2}V^{-1}s^{-1} to 31.25 cm^{2}V^{-1}s^{-1}. Upon further extending the pulse off-time, the films changed from an isotropic amorphous structure to an anisotropic polycrystalline structure, with more grain boundaries introduced. This increased the probability of grain boundary scattering and hindered the carrier migration, resulting in a decrease in the carrier mobility. Similar behavior has also been found in nitrogen-doped ITZO films [29]. In addition, the target ionization rate enhanced with the extension of the pulse off-time, thereby raising the activity of the doping species during the sputtering process. Thus, the substitution of In^{3+} ions by Sn^{4+} occurred more readily. This substitution leads to lattice distortion and aids in the formation of V_O (oxygen vacancies) and Sn_{In} (the substitution of In^{3+} by Sn^{4+}) donor defects, which both improve the carrier concentration.

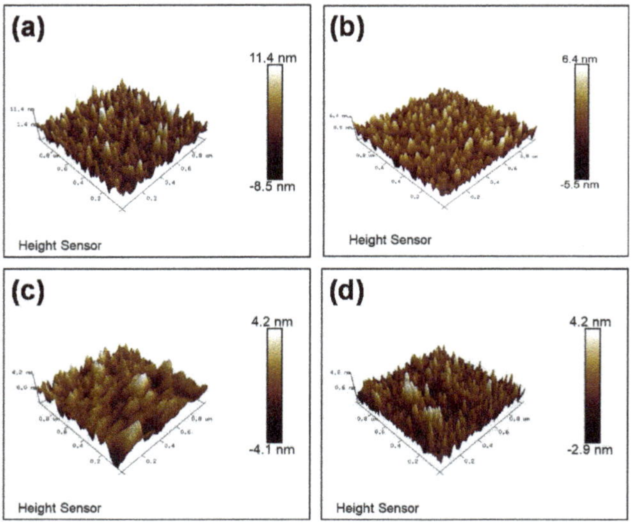

Figure 5. AFM images of ITZO films deposited with pulse off-times of (**a**) 0 μs, (**b**) 500 μs, (**c**) 1000 μs, and (**d**) 2000 μs.

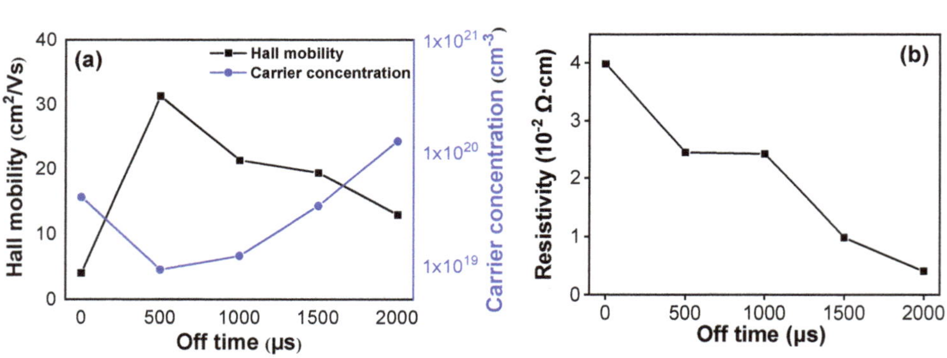

Figure 6. Electrical properties of ITZO films deposited at different pulse off-times: (**a**) the carrier mobility and carrier concentration and (**b**) the films' resistivity.

The film's resistivity is related to the carrier mobility and carrier concentration. Their relationship is calculated using the following equation [30]:

$$\rho = \frac{1}{(e \times \mu \times N)} \tag{1}$$

where ρ is the film's resistivity, e is the electron charge, μ is the carrier mobility, and N is the carrier concentration. Through the combined effects of carrier mobility and carrier concentration, the variation of the films' resistivity as a function of pulse off-time is shown in Figure 6b. It decreased from 3.98×10^{-2} Ω·cm to 4.07×10^{-3} Ω·cm as the pulse off time rises from 0 μs to 2000 μs.

4. Conclusions

In this work, ITZO thin films were deposited on glass substrates at room temperature through HiPIMS technology with various pulse off-times. The microstructures and electrical properties of the films were investigated. The results show that compared with the ITZO film deposited under DCMS mode, ITZO films deposited using HiPIMS mode are denser and possess smoother surface morphology. As the pulse off-time was extended, the crystallinity of ITZO films enhanced, and the film's resistivity effectively reduced. The optimal resistivity of about 4.07×10^{-3} Ω·cm was achieved when the pulse off-time was 2000 μs. This result indicates that through utilizing HiPIMS technology, ITZO films with controllable carrier concentration and carrier mobility in addition to controllable resistivity can be produced, which is desirable in the production of films for applications in various optoelectronic devices.

Author Contributions: Conceptualization, H.S. and S.-C.C.; methodology, Z.-Y.L.; investigation, M.-H.L.; writing—original draft preparation, W.-X.W.; writing—review and editing, J.-H.G. and Z.B.; supervision, H.S. and S.-C.C.; funding acquisition, S.-C.C. All authors have read and agreed to the published version of the manuscript.

Funding: We gratefully acknowledge the National Natural Science Foundation of China (No. 62004117), the Ministry of Science and Technology of Taiwan (No. 108-2221-E-131-014), and the Young Scholars Program of Shandong University, Weihai for their financial support.

Acknowledgments: We also thank the Physical–Chemical Materials Analytical & Testing Center of Shandong University at Weihai for their assistance with characterization and Hsin-Chih Lin and Chung-Yuan Kao of the Instrumentation Center, National Taiwan University for their assistance with EPMA experiments.

Conflicts of Interest: The authors declare no conflict of interest.

References

1. Lee, C.C.; Wang, C.W. Interfacial fracture investigation of patterned active matrix OLED driven by amorphous-Si TFTs under film-type packaging technology. *Appl. Surf. Sci.* **2020**, *510*, 145428. [CrossRef]
2. Hong, Y.; Wu, M.; Bae, J.H.; Hong, S.; Jeong, Y.; Jang, D.; Kim, J.S.; Hwang, C.S.; Park, B.-G.; Lee, J.H. A new sensing mechanism of Si FET-based gas sensor using pre-bias. *Sens. Actuators B Chem.* **2020**, *302*, 127147. [CrossRef]
3. Song, J.H.; Kim, K.S.; Mo, Y.G.; Choi, R.; Jeong, J.K. Achieving High Field-Effect Mobility Exceeding 50 cm^2/Vs in In-Zn-Sn-O Thin-Film Transistors. *IEEE Electron Device Lett.* **2014**, *35*, 853. [CrossRef]
4. Sun, H.; Chen, S.C.; Peng, W.C.; Wen, C.K.; Wang, X.; Chuang, T.H. The influence of oxygen flow ratio on the optoelectronic properties of p-Type $Ni_{1-x}O$ films deposited by ion beam assisted sputtering. *Coatings* **2018**, *8*, 168. [CrossRef]
5. Tiwari, B.; Bahubalindruni, P.G.; Santos, A.; Santa, A.; Figueiredo, C.; Pereira, M.; Martins, R.; Fortunato, E.; Barquinha, P. Low-Voltage High-Speed Ring Oscillator With a-InGaZnO TFTs. *IEEE J. Electron Devices Soc.* **2020**, *8*, 584–588. [CrossRef]
6. Marrani, A.G.; Bonomo, M.; Dini, D. Adsorption Dynamics of Redox Active Species onto Polarized Surfaces of Sensitized NiO. *ACS Omega* **2019**, *4*, 1690–1699. [CrossRef]
7. Kim, J.; Park, J.; Yoon, G.; Khushabu, A.; Kim, J.S.; Pae, S.; Cho, E.C.; Yi, J. Effect of IGZO thin films fabricated by Pulsed-DC and RF sputtering on TFT characteristics. *Mater. Sci. Semicond. Process.* **2020**, *120*, 105264. [CrossRef]
8. Sun, H.; Jen, S.U.; Chiang, H.P.; Chen, S.C.; Lin, M.H.; Chen, J.Y.; Wang, X. Investigation of optoelectronic performance in In, Ga co-doped ZnO thin films with various In and Ga levels. *Thin Solid Films* **2017**, *641*, 12–18. [CrossRef]
9. Chae, M.S.; Park, J.H.; Son, H.W.; Hwang, K.S.; Kim, T.G. IGZO-based electrolyte-gated field-effect transistor for in situ biological sensing platform. *Sens. Actuators B Chem.* **2018**, *262*, 876–883. [CrossRef]
10. Jia, J.; Torigoshi, Y.; Shigesato, Y. In situ analyses on negative ions in the indium-gallium-zinc oxide sputtering process. *Appl. Phys. Lett.* **2013**, *103*, 013501. [CrossRef]
11. Tomai, S.; Nishimura, M.; Itose, M.; Matuura, M.; Kasami, M.; Matsuzaki, S.; Kawashima, H.; Utsuno, F.; Uano, K. High-Performance Thin Film Transistor with Amorphous In_2O_3–SnO_2–ZnO Channel Layer. *Jpn. J. Appl. Phys.* **2012**, *51*, 03CB01. [CrossRef]
12. Jang, J.; Kim, D.G.; Kim, D.M.; Choi, S.J.; Lim, J.H.; Lee, J.H.; Kim, Y.S.; Ahn, B.D.; Kim, D.H. Investigation on the negative bias illumination stress-induced instability of amorphous indium-tin-zinc-oxide thin film transistors. *Appl. Phys. Lett.* **2014**, *105*, 152108. [CrossRef]
13. Jia, J.; Torigoshi, Y.; Kawashima, E.; Utsuno, F.; Yano, K.; Shigesato, Y. Amorphous indium-tin-zinc oxide films deposited by magnetron sputtering with various reactive gases: Spatial distribution of thin film transistor performance. *Appl. Phys. Lett.* **2015**, *106*, 023502. [CrossRef]

14. Noh, J.Y.; Kim, H.; Nahm, H.H.; Kim, Y.S.; Kim, D.H.; Ahn, B.D.; Lim, J.H.; Kim, G.H.; Lee, J.H.; Song, J. Cation composition effects on electronic structures of In-Sn-Zn-O amorphous semiconductors. *J. Appl. Phys.* **2013**, *113*, 183706. [CrossRef]
15. Li, Z.Y.; Chen, S.C.; Huo, Q.H.; Liao, M.H.; Dai, M.J.; Lin, S.S.; Yang, T.L.; Sun, H. Influence of sputtering power on the electrical properties of In-Sn-Zn oxide thin films deposited by high power impulse magnetron sputtering. *Coatings* **2019**, *9*, 715. [CrossRef]
16. Wen, L.; Sahu, B.B.; Kim, H.R.; Han, J.G. Study on the electrical, optical, structural, and morphological properties of highly transparent and conductive AZO thin films prepared near room temperature. *Appl. Surf. Sci.* **2019**, *473*, 649–656. [CrossRef]
17. Ayaz, S.; Mishra, P.K.; Sharma, R.K.; Kamal, S.; Sen, S. Structural, Optoelectronic, and Electrochemical Properties of $Zn_{1-x}(Ga_{0.5}Al_{0.5})_xO$ Nanoparticles for Supercapacitor Applications. *ACS Appl. Nano Mater.* **2020**, *3*, 4562–4573. [CrossRef]
18. Li, Z.Y.; Yang, H.Z.; Chen, S.C.; Lu, Y.B.; Xin, Y.Q.; Yang, T.L.; Sun, H. Impact of active layer thickness of nitrogen-doped In–Sn–Zn–O films on materials and thin film transistor performances. *J. Phys.* **2018**, *51*, 175101. [CrossRef]
19. Cui, Y.; Li, C.J.; Li, J.; Xiong, L.Y.; Liu, S. Characterization of FeCeAlY thin film deposited by magnetron sputtering and its corrosion resistance under high-temperature water vapor environment. *Surf. Technol.* **2020**, *49*, 72–78.
20. Song, S.; Sun, H.; Chen, S.C.; Dai, M.; Wang, K.; Zheng, X.; Lu, Y.B.; Yang, T.L.; Yue, Z.M. The adhesion strength and mechanical properties of SiC films deposited on SiAlON buffer layer by magnetron sputtering. *Surf. Coat. Technol.* **2019**, *360*, 116–120. [CrossRef]
21. Fan, X.; Huai, X.; Wang, J.; Jing, L.C.; Wang, T.; Liu, J.; Geng, H.Z. Low surface roughness graphene oxide film reduced with aluminum film deposited by magnetron sputtering. *Nanomaterials* **2021**, *11*, 1428. [CrossRef] [PubMed]
22. Chen, S.C.; Kuo, T.Y.; Lin, H.C.; Chen, R.Z.; Sun, H. Optoelectronic properties of p-type NiO films deposited by direct current magnetron sputtering versus high power impulse magnetron sputtering. *Appl. Surf. Sci.* **2020**, *508*, 145106. [CrossRef]
23. Lin, S.S.; Gao, D.; Su, Y.F.; Xu, W.; Guo, C.Q.; Li, H.; Shi, Q.; Wei, C.B.; Dai, M.J.; Yang, J.C. Effect of bias voltage on structure and properties of DLC films deposited by high power pulse magnetron sputtering. *Mater. Res. Appl.* **2020**, *14*, 1–8.
24. Sun, H.; Kuo, T.Y.; Chen, S.C.; Chen, Y.H.; Lin, H.C.; Yazdi, M.A.P.; Billard, A. Contribution of enhanced ionization to the optoelectronic properties of p-type NiO films deposited by high power impulse magnetron sputtering. *J. Eur. Ceram. Soc.* **2019**, *39*, 5285–5291. [CrossRef]
25. Ghailane, A.; Larhlimi, H.; Tamraoui, Y.; Makha, M.; Busch, H.; Fischer, C.B.; Alami, J. The effect of magnetic field configuration on structural and mechanical properties of TiN coatings deposited by HiPIMS and dcMS. *Surf. Coat. Technol.* **2020**, *404*, 126572. [CrossRef]
26. Chuang, T.H.; Wen, C.K.; Chen, S.C.; Liao, M.H.; Liu, F.; Sun, H. p-type semi-transparent conductive NiO films with high deposition rate produced by superimposed high power impulse magnetron sputtering. *Ceram. Int.* **2020**, *46*, 27695–27701. [CrossRef]
27. Wang, Z.; Li, Q.; Yuan, Y.; Yang, L.; Zhang, H.; Liu, Z.; Ouyang, J.T.; Chen, Q. The semi-conductor of ZnO deposited in reactive HiPIMS. *Appl. Surf. Sci.* **2019**, *494*, 384–390. [CrossRef]
28. Stranak, V.; Bogdanowicz, R.; Sezemsky, P.; Wulff, H.; Kruth, A.; Smietana, M.; Kratochvil, J.; Cada, M.; Hubicka, Z. Towards high quality ITO coatings: The impact of nitrogen admixture in HiPIMS discharges. *Surf. Coat. Technol.* **2018**, *335*, 126–133. [CrossRef]
29. Jia, J.; Torigoshi, Y.; Suko, A.; Nakamura, S.I.; Kawashima, E.; Utsuno, F.; Shigesato, Y. Effect of nitrogen addition on the structural, electrical, and optical properties of In-Sn-Zn oxide thin films. *Appl. Surf. Sci.* **2017**, *396*, 897–901. [CrossRef]
30. Chen, S.C.; Huang, S.Y.; Sakalley, S.; Paliwal, A.; Chen, Y.H.; Liao, M.H.; Sun, H.; Biring, S. Optoelectronic properties of Cu_3N thin films deposited by reactive magnetron sputtering and its diode rectification characteristics. *J. Alloys Compd.* **2019**, *789*, 428–434. [CrossRef]

nanomaterials

MDPI

Review

Metal-Based Nanomaterials: Work as Drugs and Carriers against Viral Infections

Junlei Yang [1,2], Lihuan Yue [3,4], Zhu Yang [1,2], Yuqing Miao [1], Ruizhuo Ouyang [1,*] and Yihong Hu [2,3,*]

1 Institute of Bismuth Science, University of Shanghai for Science and Technology, Shanghai 200093, China; fuyangjunlei@163.com (J.Y.); 203612304@st.usst.edu.cn (Z.Y.); yqmiao@usst.edu.cn (Y.M.)
2 Institut Pasteur of Shanghai, Chinese Academy of Sciences, University of Chinese Academy of Sciences, Yueyang Road 320, Shanghai 200031, China
3 CAS Key Laboratory of Molecular Virology & Immunology, Institutional Center for Shared Technologies and Facilities, Pathogen Discovery and Big Data Center, Institut Pasteur of Shanghai, Chinese Academy of Sciences, Yueyang Road 320, Shanghai 200031, China; lhyue@ips.ac.cn
4 Department of Bioengineering, School of Life Sciences and Biotechnology, Shanghai Jiao Tong University, Shanghai 200240, China
* Correspondence: ouyangrz@usst.edu.cn (R.O.); yhhu@ips.ac.cn (Y.H.); Tel.: +86-21-18616897485 (R.O.); +86-21-54923052 (Y.H.)

Abstract: Virus infection is one of the threats to the health of organisms, and finding suitable antiviral agents is one of the main tasks of current researchers. Metal ions participate in multiple key reaction stages of organisms and maintain the important homeostasis of organisms. The application of synthetic metal-based nanomaterials as an antiviral therapy is a promising new research direction. Based on the application of synthetic metal-based nanomaterials in antiviral therapy, we summarize the research progress of metal-based nanomaterials in recent years. This review analyzes the three inhibition pathways of metal nanomaterials as antiviral therapeutic materials against viral infections, including direct inactivation, inhibition of virus adsorption and entry, and intracellular virus suppression; it further classifies and summarizes them according to their inhibition mechanisms. In addition, the use of metal nanomaterials as antiviral drug carriers and vaccine adjuvants is summarized. The analysis clarifies the antiviral mechanism of metal nanomaterials and broadens the application in the field of antiviral therapy.

Keywords: metal-based nanomaterials; characteristics; antiviral therapy; mechanism; application

Citation: Yang, J.; Yue, L.; Yang, Z.; Miao, Y.; Ouyang, R.; Hu, Y. Metal-Based Nanomaterials: Work as Drugs and Carriers against Viral Infections. *Nanomaterials* **2021**, *11*, 2129. https://doi.org/10.3390/nano11082129

Academic Editors: Jihoon Lee and Ming-Yu Li

Received: 1 July 2021
Accepted: 12 August 2021
Published: 20 August 2021

1. Introduction

Virus infection has always been a threat to human and animal health. Typical viruses include hepatitis B virus [1], influenza virus [2], human immunodeficiency virus (HIV) [3], and coronavirus [4], etc., which can cause severe disease. Therefore, antiviral drug development is a major research direction for scientists. At present, the main treatment methods for viral infections include developing vaccines and screening antiviral drugs. However, the cycle for virus vaccine development and drug screening is currently too long. Additionally, the emergence of antiviral drug resistance has brought considerable challenges to successfully suppressing viral infections, so the development of new antiviral drugs is particularly important.

As one of the emerging fields in recent years, nanomaterials can be simply divided into one-dimensional, two-dimensional, and three-dimensional nanostructures according to the morphology of the material. There are many methods to synthesize nanomaterials, such as mechanical grinding synthesis [5], chemical vapor synthesis [6], chemical liquid reaction [7], physical vapor deposition [8], and phytosynthesis [9,10]. Among them, the phytosynthesis method is the most remarkable method of preparing nanoparticles now. It has the advantages of environmental protection and low energy consumption, but it also has the disadvantages of limited reaction yield and poor particle uniformity.

Since then, further research has made substantial progress on the preparation methods of nanomaterials by adjusting the ratio of raw materials and exploring suitable reaction conditions. Nanomaterials have the advantages of small size, high specific surface area, adjustable particle size, and easier surface functionalization, which make them widely used in sensing [11], catalysis [12], energy storage [13], and the medical treatment field [14]. Metal-based nanomaterials could be widely used in biomedical fields, as metal ions are essential in living organisms. Since the particle size of the virus particles ranges from tens to hundreds of nanometers, the surface activity of the metal material is enhanced after the metal is nanosized. It can be used in the inhibition of virus infection. In addition, according to the statistics on Web of Science, the number of articles published on the application of metal nanomaterials in the field of viruses has been increasing year-on-year (Figure 1). More importantly, some nanomaterials such as Ag [15], Au [16], ZnO [17], etc., which have an inhibitive effect on bacteria and viruses, highlight their potential for antiviral applications. Notably, the pandemic of severe acute respiratory syndrome coronavirus 2 (SARS-CoV-2) in 2019 poses a huge threat to human health. As such, progress on virus pathogenesis new antiviral drug development is a research priority [18].

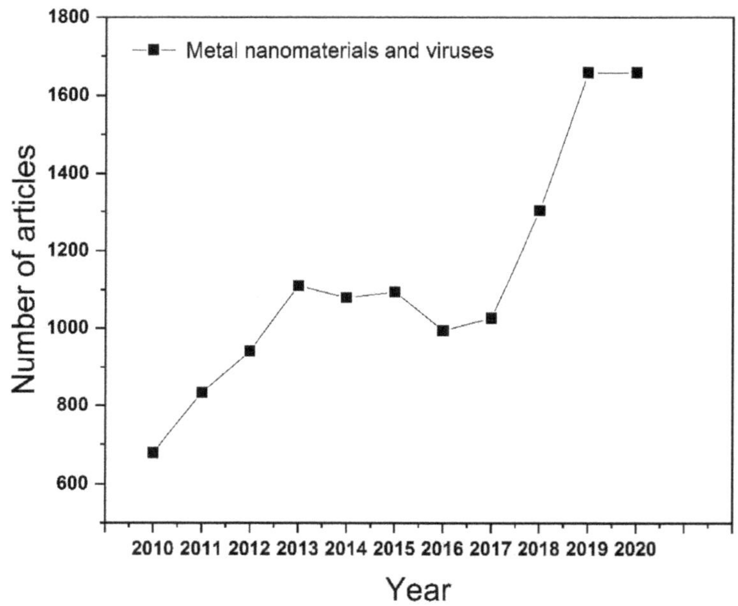

Figure 1. The number of articles published each year on the application of metallic nanomaterials in the field of viruses.

This review first provides a more comprehensive summary of the different stages of the role of metal nanomaterials in inhibiting viral infections in recent years. (1) Directly inactivating the virus; (2) inhibiting virus adsorption and entry; (3) intracellular virus suppression. Subsequently, it summarizes the use of metal-based nanomaterials as drugs and vaccine carriers. The application of metal nanomaterials in the field of antiviral research is introduced and its prospects are discussed. This review focuses on metal nanomaterials by comparing research progress in related fields in recent decades [19–21]. The inhibitory effect of metal nanomaterials on virus infection and their application as drugs and vaccine carriers are discussed and demonstrated.

2. Inhibition of Virus Infections

Virus infection of cells can be divided into three stages. The first is the early adsorption of the virus onto cell membrane, where the virus binds to cell surface receptors through its surface protein. The second is the process of uncoating, replication, and translation after the virus enters the cell. Finally, the virus repackages to release new progeny viruses. In these processes, if metal nanomaterials can inhibit the virus in the early stage and prevent the virus from invading cells, they can be used in medical protective equipment, which has great potential for preventing the virus spread (Figure 2).

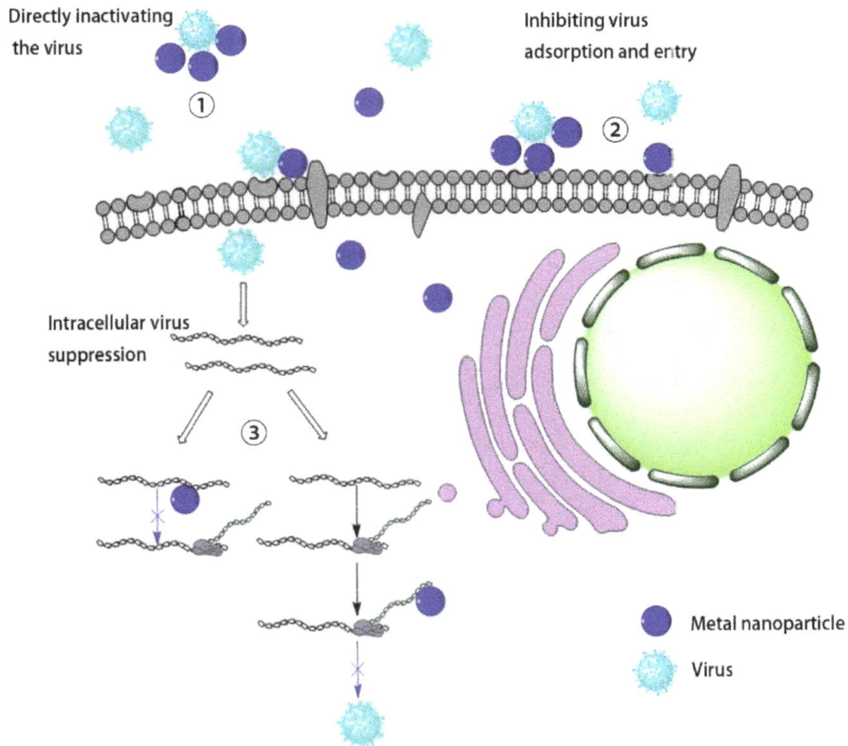

Figure 2. The mechanism of metallic nanomaterials inhibiting virus infection.

2.1. Direct Inactivation of the Virus

If metal-based nanomaterials inactivate viruses directly before viruses invade cells, this requires that nanomaterials possess excellent antiviral performance and good biological safety. Studies found that metal-based nanomaterials such as Au [22], Ag [23], ZnO [24], etc., have good antibacterial properties. Among them, TiO_2 nanomaterials with photocatalytic properties can generate active substances by using light or through their affinity with biological macromolecules. They can damage the biofilm and kill the bacteria. Since the surface of the virus also has a biofilm, these metal-based nanomaterials can be used for the direct inactivation of the virus. H. Cui and colleagues found that TiO_2 has good photocatalytic properties and can generate reactive oxygen species (ROS) under ultraviolet irradiation. ROS has oxidative properties. Therefore, using external interference in tumor therapy, the balance of ROS expression in tumor cells will be affected, which accelerates tumor cell apoptosis. At the same time, its excessive presence can also destroy the biofilm and cause it to lose its biological activity. Related genes in cells regulate the normal expression level of ROS. H. Cui's group explored the inhibitory effect of TiO_2 on the influenza virus under ultraviolet irradiation (Figure 3). Compared with normal cells infected with influenza virus (Figure 3b), the cell activity after TiO_2 treatment is greatly enhanced (Figure 3a), which intuitively shows the inhibitory effect of TiO_2 on the virus. Later studies showed that with the increase of TiO_2, virus culture time, and ultraviolet light time, the inhibitory effect of TiO_2 on the virus is increased. Considering that the material can generate ROS under ultraviolet light irradiation, it can inactivate viruses [25]. Similarly, N. A. Mazurkova explored the antiviral properties of TiO_2 under sunlight and ultraviolet rays. It was observed by scanning electron microscopy that the material adsorbed to the surface of the virus envelope in the early stage, and then destroyed it, thereby inactivating the virus. Later experimental results showed that the inhibitory effect was positively correlated with the concentration of TiO_2 and the incubation time [26]. Ahmad Tavakoli and colleagues also used CuO nanomaterials to generate ROS and explored its inhibitory effect on herpes simplex virus (HSV). During the study, CuO was added after the virus infected the cells. After 48 h of culture, the results showed that the inhibitory effect increased as the CuO concentration increased [27].

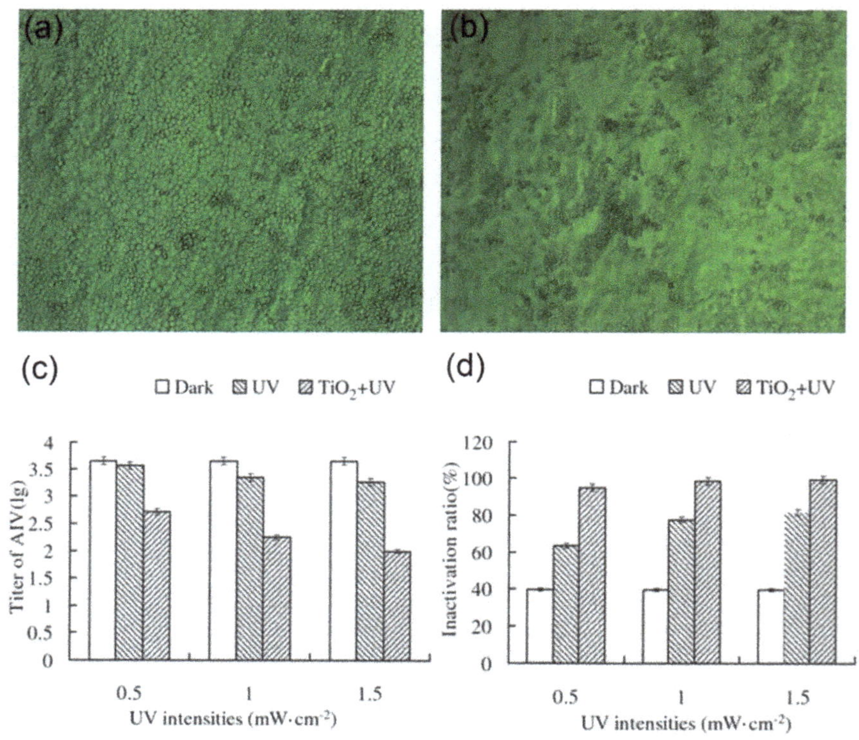

Figure 3. Cytopathic effect on Madin-Darby canine kidney cells infected with avian influenza virus eluted from nano-TiO_2 film (**a**) and control (**b**); changes in titer (**c**) and inactivation efficiency (**d**) of H9N2 vs. ultraviolet intensity. (Reproduced with permission from Ref. [25], Copyright 2010, Wiley).

Compared with some metal-based nanomaterials that use light to produce ROS to inactivate viruses, the mutual combination of metal ions and protein molecules can often change the protein conformation, causing irreversible damage to the effect of inhibiting virus infection. SungJun Park synthesized a magnetic hybrid colloid loaded with Ag nanoparticles of different sizes. Using the interaction between Ag and biological macromolecules, this nanomaterial's inhibitory effect on bacteriophages, norovirus, and adenovirus was explored. The results show that the virus binds to the sulfhydryl-containing protein on the surface of the virus through Ag nanoparticles, thereby destroying the virus envelope to inhibit the virus. Similarly, the use of the affinity of metal ions with proteins can bind to the outer surface proteins of virus particles and destroy the virus structure to inhibit the virus [28]. F. Pfaff tested its inhibitory effect on modified vaccinia virus Ankara (MVA), human adenovirus serotype 5 (HAdV-5), poliovirus type 1 (PV-1), and murine norovirus (MNV) by co-cultivating WC material and virus. The author shows that WC tends to reunite, which can encapsulate virus particles, thereby destroying the nucleic acid of the virus and inactivating the virus [29].

2.2. Inhibiting Virus Adsorption and Entry

The binding process of the virus and the cell is through the targeted binding of the envelope protein on the virus surface to the cell surface receptor. The expression of these virus surface proteins is stable and not prone to gene mutation. Therefore, nanomaterials can be used to adsorb viruses directly, simulate virus-connected receptors on the cell surface, and competitively bind virus particles before the virus invades the cell to prevent the virus from infecting the cell. The mechanism provides a new trend for the research of antiviral materials.

To prevent the virus from invading the cell, scientists target the combination of virus and cell. The most direct method can be used to bind the viral capsid protein to occupy the receptor of the virus-bound cell, thereby preventing the disease from invading the cell. Metal ions have a good affinity with biological macromolecules, and the combination of virus and metal nanomaterials inhibits virus infection on cells. Rishikesh Kumar's lab has synthesized 10 nm Fe_2O_3 nanoparticles with the external glycine reducing the biological toxicity of Fe_2O_3 nanomaterials, and this shows a specific inhibitory effect on influenza virus H1N1. The results show that, with smaller particle size and a higher specific surface area, the combination of Fe_2O_3 nanomaterials and viruses is easier. The Fe_2O_3 nanoparticles combine with viruses to produce a particular steric hindrance, inhibiting the virus from invading cells [30]. Yasmin Abo-Zeid explored the interaction of Fe_2O_3/Fe_3O_4 with hepatitis C glycoproteins E1, E2, and the spike protein receptor binding domain of COVID-19, respectively. She found that the ends of these viral proteins have richer amino acid residues. However, iron oxide nanoparticles can bind to them to occupy the virus's binding site and the cell, successfully interfering with the absorption of the virus into the cell (Figure 4) [31].

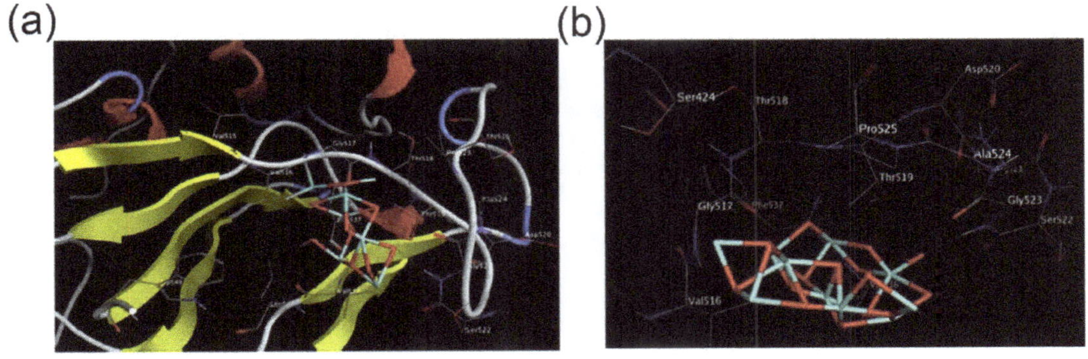

Figure 4. Three-dimensional interaction diagram showing (**a**) Fe_2O_3; and (**b**) Fe_3O_4 docking interactions with the key amino acids in the HCV E2 glycoprotein. Three-dimensional interaction diagram showing docking interactions with the key amino acids in the HCV E2 glycoprotein. (Reproduced with permission from Ref. [31], Copyright 2020, Elsevier)

Metal-based nanomaterials have a small particle size and a high specific surface area, convenient for material modification, and can be functionalized in combination with the characteristics of virus receptors on cell surface. Studies have shown that when HSV invades cells, glycoproteins on the envelope's surface bind to heparin sulfate (HS) on the cell membrane [32]. In this binding process, the sulfonate ion in the protein plays an important role. Therefore, the surface of metal-based nanomaterials can be functionalized to modify sulfonate groups and simulate HS on the cells' outer surface to prevent viruses from invading cells through competitive binding with viruses. Dana Baram-Pinto and colleagues used the affinity of the noble metals (Au and Ag) to sulfhydryl groups and external sulfonate ions as shown in Figure 5 to synthesize metal-based nanomaterials that simulate HS. Studies have shown that the material has successfully added sulfhydryl sulfonate, and the later anti-HSV experiments have shown that it has good virus inhibition characteristics. Influenza virus invades cells through the highly conserved protein hemagglutinin (HA), which contains six disulfide bonds on the surface of the virus, binding cell surface receptors and entering the cell. The study of metal nanoparticles focuses on binding HA and destroying its structure to prevent viruses from invading cells [22,23]. In addition, Jonathan Vonnemann explored the strength of the adsorption between Au and viruses by synthesizing Au with different particle sizes and then connecting sulfonate particles. The results show that when the diameter of Au nanoparticles is less than 52 nm, although the nanomaterials are adsorbed on the virus envelope, the Au nanoparticles' surface area is not enough to prevent the adsorption of the virus and cells, as shown in Figure 6. When Au nanoparticles are larger than 52 nm, the nanoparticles occupy the surface area of the virus. The steric hindrance is increased, which significantly reduces the contact between the virus and the cell, thereby reducing the probability of the virus infecting cells [33].

Au/Ag is known to have a good affinity with S. Jinyoung Kim synthesized Au and Ag with different morphologies and with or without holes. After co-cultivating with the virus for a certain period, the adsorbed and unadsorbed viruses were separated by centrifugation. The interaction between the nanoparticles and the virus was observed by transmission electron microscope, and the number of disulfide bonds in the viral hemagglutinin was quantified to determine the effective destruction of the disulfide bonds by the metal-based nanoparticles. The results show that, compared with non-porous Au and Ag nanomaterials, porous Au nanoparticles can achieve a more apparent antiviral effect by destroying the disulfide bonds in the hemagglutinin on the surface of the virus [34]. This provides a new application direction for porous Au nanoparticles to suppress viruses containing enveloped spike proteins.

The process of a virus invading cells involves specific binding, which provides a good starting point for preventing viruses from infecting cells. The primary inhibition method is to competitively bind to the virus by simulating cell surface virus receptors and inhibiting the virus from invading cells. However, this method can stop the virus from spreading further and does not inactivate the virus. There are certain drawbacks to the application. It is worth noting that these nanomaterials can be adsorbed on medical protective fabrics and concrete surfaces, thereby reducing the spread and infection of viruses.

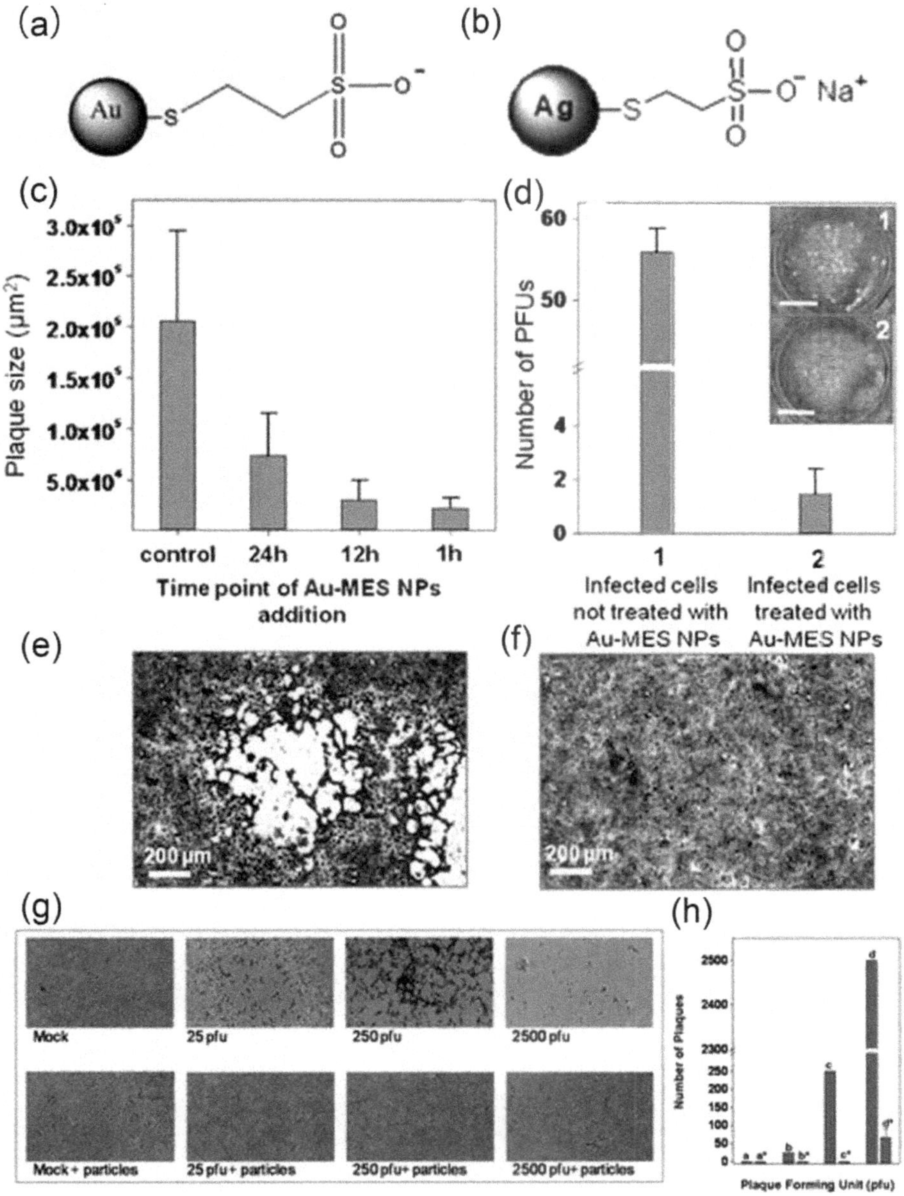

Figure 5. Characterization of (**a**) Au-MES (MES = mercaptoethane sulfonate) NPs; (**b**) Ag-MES NPs; (**c**) variation of plaque size as a function of increasing time lapse between primary infection and the administration of Au-MES NPs; (**d**) evaluation of inhibition of virus entry into host cells. Plaque numbers observed 48 h after infection (1) without and (2) with nanoparticles; infected cell culture in the absence of Au-MES NPs (**e**) and in the presence of Au-MES NPs (**f**). (Reproduced by permission from Ref. [22], Copyright 2010, Wiley.) (**g**) Optical microscopic image (×10 magnification) demonstrating the effect of Ag-MES nanoparticles on HSV-1 infectivity; (**h**) quantitative analysis of the Ag-MES inhibition efficiency (reproduced with permission from Ref. [23], Copyright 2009, American Chemical Society).

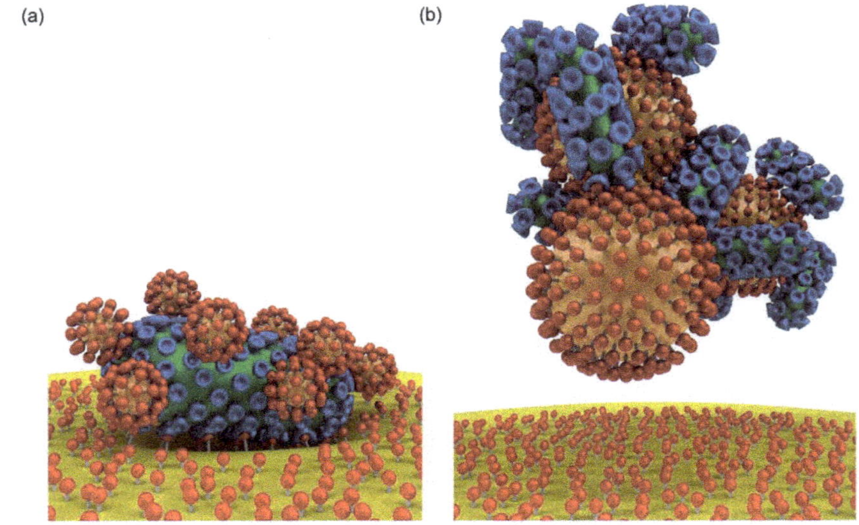

Figure 6. Schematic representation of the size-dependent virus inhibition by ligand functionalized gold nanoparticles according to the TEM data. (**a**) Although smaller sized gold nanoparticles decorate virions, the inhibition of virus-cell binding was shown to be inefficient; (**b**) larger virus-sized gold nanoparticles induce the formation of virus-inhibitor clusters, inhibiting the virus-cell binding more efficiently. (Reproduced with permission from Ref. [33], Copyright 2014, Royal Society of Chemistry).

2.3. Intracellular Virus Suppression

After the virus enters the cell, it uses the intracellular machinery to carry out protein replication and translation. Since the virus is uncoated to expose the genetic material DNA/RNA, this process can provide the possibility for external drugs to destroy the viral nucleic acid or interfere with the process of translation, replication, and release.

When the virus enters the cell, it begins the process of unpacking, replication, and translation. During this period, the RNA/DNA and protein of the virus are synthesized in large quantities; this happens to be the most effective moment for drugs to inactivate viruses. By importing the drug into the cell, it acts on the viral RNA/DNA and protein, destroying the replication process of the virus and inhibiting its reproduction. Most nanomaterials have good biocompatibility. The advantages of small particle size and high specific surface area can help nanomaterials enter cells efficiently. With the capability of binding metal ions with biological macromolecules, the metal nanomaterials electrostatically interact with viral nucleic acids or proteins to form chemical bonds, destroy their structures, or cause irreversible conformational changes of viral proteins, thereby achieving the purpose of inhibiting virus replication. Ting Du synthesized glutathione-encapsulated Ag_2S nanoparticles to explore their inhibitory effect on coronaviruses. The author explored the expression levels of the virus by adding Ag_2S materials at different times after virus infection. Since the particle size of Ag_2S is about 2.5 nm, it has a more profound cell penetration ability and can better interact with viruses. PEDV nucleocapsid (N) protein is a protein that binds to viral RNA, and its expression is tested to verify the inhibition of Ag_2S. The results show that Ag_2S inhibits the virus by inhibiting the negative-strand RNA synthesis and budding of PEDV [35].

The relatively high surface activity of metal nanomaterials also brings the disadvantage of high cytotoxicity. However, more amphiphilic polymers such as PEG and PVP have been studied at this stage [36]. They can be easily combined with metal nanoparticles to enhance the water solubility of materials, reduce biological toxicity, and increase the internalization rate of cells to materials. ZnO is the main component of many sunscreens [37] and antibacterial materials [38], and its application in the field of antibacterial products has been extensively studied. However, the effect of ZnO on viruses is less reported. Studies have shown that synthetic micron-sized filamentous foot-like ZnO prevents HSV-1 from entering the cell [17]. Later studies showed that naked ZnO and PEG-modified ZnO had a dramatic inactivating effect on HSV-1, which was mainly achieved by interfering with the expression of early viral proteins. The results showed that PEG-modified ZnO had a better virus inhibition effect [39]. Similarly, through the inhibition of naked ZnO and PEG-modified ZnO on H1N1 influenza virus, the virus inhibition stage is shown to be after the virus entry [24].

3. Loading Drug Synergy

The high infectivity and pathogenicity of viruses has caused people to pay more attention to antiviral drug research and development. The current research and development of viral drugs mainly include screening natural drugs and synthesizing organic small-molecule drugs. However, the screening of natural medicines requires a high workload, and the inhibitory effect of natural medicines on viruses is not satisfactory. The inhibition mechanism remains to be explored. The research and development of organic small-molecule drugs, finding the relevant derivatives of existing antiviral small-molecule drugs, and further exploring drugs with viral inhibitory effects have good application prospects. However, small organic molecules have disadvantages such as low water solubility, poor biocompatibility, and high toxicity. Moreover, the development and application of vaccines have also opened up new areas for virus therapy. Vaccines are preventive treatments that can significantly reduce the incidence of viral infections. However, vaccines for injection have biological activity, and instability in transportation and storage is still an open question. Hence, many scholars have developed numerous carrier materials, such as colloids [40], magnetic nanomaterials [41], inorganic nanomaterials [42], and metal organic framework materials [43], to increase the loading rate of drug molecules by high biocompatibility and high specific surface area of the carrier material. Thereby, the carrier nanomaterial increases the uptake rate of the cell to the drug, and enhances the stability of the vaccine and the immune response of the body.

The human body contains many kinds of metal elements, including Al, Ca, Mg, Fe, Na, Mn, Zn, K, Li, Cu, Se, etc. These metals are essential for participating in various life cycle processes in the human body. Therefore, small amounts of metal ions are less harmful to the human body, and the metal ions could be prepared as nanomaterials with a high specific surface area of nanomaterials, porosity of some materials, and good biological activity as a promising drug carrier.

3.1. Loading Drugs

The poor water solubility of antiviral drugs makes their bioavailability low, and higher doses are often required to achieve the desired therapeutic effect. However, higher doses will produce certain toxicity to organisms, so the emergence of drug carriers can improve the drugs' bioavailability and reduce the damage to organisms. When selecting metal nanoparticles as drug carriers, the toxicity of metal ions to organisms and the loading efficiency of materials should be considered. Using metal ions or inert ions, with higher content in human cells, can reduce the toxicity of materials and increase the materials' natural metabolism.

Human Acquired Immune Deficiency Syndrome (AIDS), caused by HIV, is a disease in which immune cells are the primary targets. These cells can accumulate viruses for a long time to be used as targeted sites for drug delivery. Studies have shown that Au nanoparticles can be more internalized by human macrophages and have less biological toxicity [44]. Hinojal Zazo and colleagues immobilized stavudine, an antiretroviral drug, on the surface of Au nanoparticles to deliver targeted drugs to macrophages. In addition, when Au is loaded with drugs, it can increase the expression of inflammatory genes in macrophages and induce a specific inflammatory response in the body [45].

Quantum dots (QD) are a type of low-dimensional semiconductor material and often have smaller sizes. Therefore, applying QD materials to the field of biological therapy can often improve the therapeutic effect. Ranjeet Dungdung's group used ZnS quantum dots as a drug carrier, loaded with mycophenolic acid (MPA), an immunosuppressant against dengue fever virus. It was found that cells have a higher internalization rate of ZnS-coupled MPA, and its inhibitory effect on dengue virus was significantly improved and the selectivity index was increased by two orders of magnitude [46]. The study shows that quantum dots can significantly increase the uptake rate and therapeutic effect of drugs, indicating that QD as drug carriers have great application prospects.

The application of antiviral drugs can significantly reduce the infection rate of the virus. However, the presence of the blood–brain barrier reduces the scope of action of the drug, which is also a shortcoming in the therapeutic application of most drugs. When Madhavan Nair and colleagues studied the antiretroviral application of HIV, they found that the antiretroviral drug azidothymidine 5′-triphosphate (AZTTP) has low efficiency in passing through the blood–brain barrier, which makes the treatment of the virus not very efficient. Therefore, it is crucial to find an excellent carrier to load AZTTP to achieve targeted drug release. Studies have shown that by synthesizing 30 nm magnetic $CoFe_2O_4@BaTiO_3$, AZTTP is bonded to nanomaterials' surface through ionic bonds. As shown in Figure 7, the delivery and trigger release of the drug are controlled by applying the strength and frequency of the magnetic field in different directions to achieve the purpose of release on demand. Later studies have shown that the load rate of AZTTP for three hours at 37 °C is about 24% through spectrophotometric detection. Depending on the magnetic field strength and frequency, they are adjusted to achieve quantitative release of AZTTP [47].

In the treatment of viral infections, in addition to traditional antiviral drugs, nucleic acid drugs such as oligonucleotides [48], siDNA [49], DNAzymes [50], etc., have also achieved good results. These nucleic acid biological macromolecules can often target the RNA fragments of the virus and interfere with the virus's replication process in the cell to achieve the purpose of inhibition. However, these nucleic acid drugs still have the disadvantages of low cell uptake rate and low bioavailability. Using these nucleic acid drugs, the methods of transporting them into the cell through the carrier can increase the uptake rate of nucleic acid drugs to the cell and improve the therapeutic effect. Soo-Ryoon Ryoo and colleagues selected a therapeutic oligonucleotide called DNAzyme that specifically targets hepatitis C virus (HCV) mRNA and splits specific genes N53, a helicase and ribozyme gene involved in hepatitis C virus replication. As shown in Figure 8a,b, iron oxide nanoparticles (MN) were conjugated to DNAzyme as a carrier, and then connected to the outer layer of cell-penetrating peptide (CPP) through disulfide bonds and modified fluorescein CyC5.5 (D2-MPAP-MN). The material group connected with DNAzyme showed a significant inhibitory effect (Figure 8c–e), by examining the gene expression of virus N53, compared with the carrier material (MN/MPAP-MN) [51].

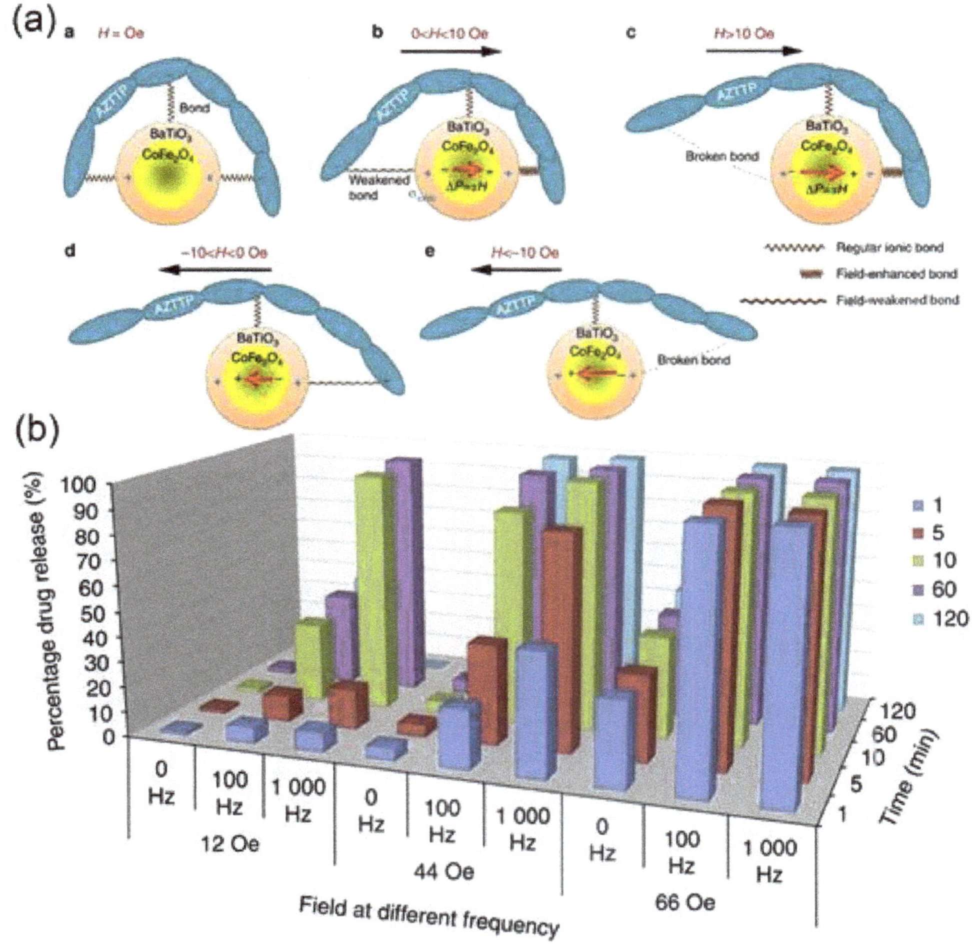

Figure 7. (**a**) Illustration of the underlying physics of the a.c.-field-triggered release; (**b**) pharmacokinetics study: three-dimensional chart representation of the drug release percentage at various combinations of the field strength, the frequency, and the treatment duration. (Reproduced with permission from Ref. [47], Copyright 2013, *Nature* Publishing Group).

Studies have shown that small interfering RNA 331 (siRNA331) can target the conserved region of hepatitis C virus RNA [50]. Inspired by this, Zhongliang Wang and colleagues used Au nanoparticles as a carrier platform to link endoribonucleotides and DNA oligonucleotides on the outer surface through non-covalent adsorption and Au-S bonds to form nanozymes. Here, endoribonucleotides are used for non-specific degradation of single-stranded RNA. The DNA oligonucleotide design contains siRNA331, which is used to improve the hybridization and cleavage efficiency of the targeted gene of nanozyme. The nanozyme specifically binds HCV RNA and cuts it to inhibit the replication process of HCV. Through specific experiments in the later period, it was found that nanozymes have a better ability to precisely cut HCV genes. The experimental results in cells and mice are consistent with expected results [52].

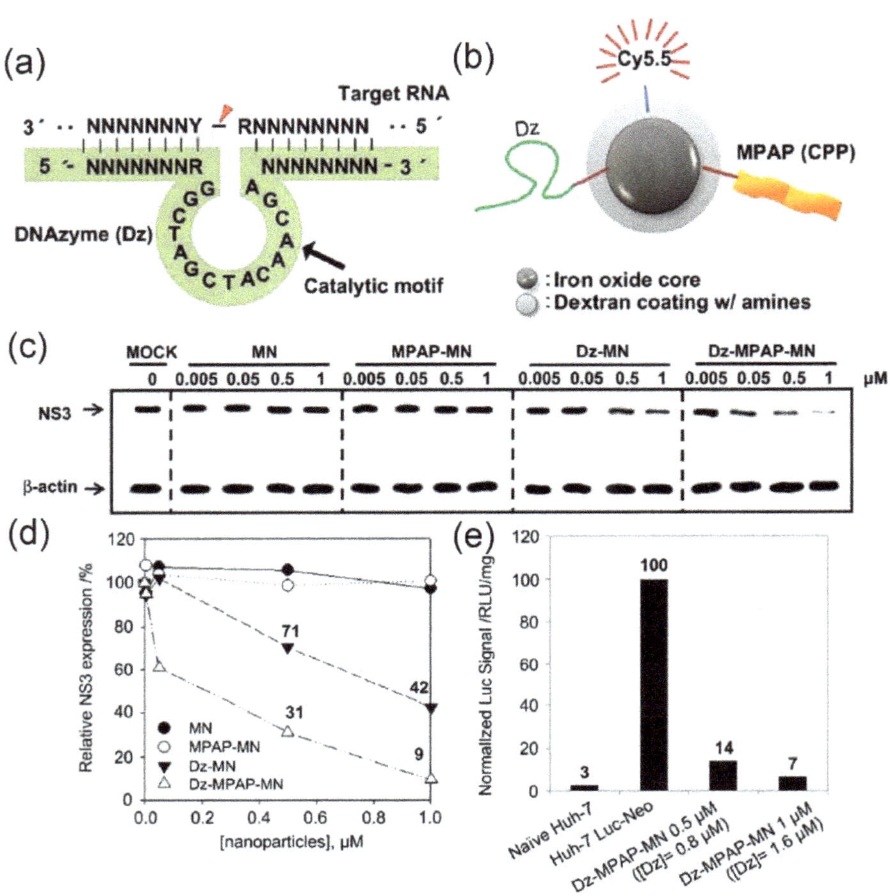

Figure 8. Material construction and performance testing. (**a**) Structure of the 10–23 synthetic DNAzyme (shaded in green color). A conserved 15-base unpaired motif serves as the catalytic core and is flanked by variable binding domains at its 50- and 30-ends. The point of scission within target RNA is indicated by a red color arrow; (**b**) multifunctional nanoparticle formulation for DNAzyme delivery. (Dz, DNAzyme; MPAP, myristoylated polyarginine peptide; Cy5.5, fluorescent dye; CPP, cell-penetrating peptide); (**c**) Western blot analysis of HCV NS3 expression in cultured Huh-7 Luc-Neo cells treated with Dz-conjugated multifunctional nanoparticles; (**d**) densitometric analysis of the data shown in (**c**). Density values for NS3 were normalized to the β-actin band density; (**e**) luciferase (reporter gene) assays of Huh-7 Luc-Neo cells treated with Dz-MPAP-MN indicate dose-dependent downregulation of the NS3 target gene. (Reproduced with permission from Ref. [51], Copyright 2012, Elsevier).

3.2. Load Vaccine

Vaccines are used to prevent infectious diseases by artificially attenuating and inactivating pathogenic microorganisms and their metabolites to produce an immune response in the body. After immunization, the organism will produce immune antibodies to protect itself, and the immune system will form a memory of pathogenic microorganisms. When it invades again, the immune system can react in time and produce corresponding antibodies to protect the organism. The treatment of viral infections is often more complicated. However, it is often a valid preventive method to cause the body to produce specific antibodies through the previous vaccine injection.

Vaccines can be divided into live-attenuated vaccines, inactivated vaccines, recombinant vaccines, and so on. Among them, live vaccines (adenovirus vaccines [53], measles vaccines [54], and polio vaccines [55]) can stimulate the body's comprehensive systemic immunity and long-lasting immune response with the disadvantages of antigen interference and enhanced virulence. Inactivated vaccine (influenza split vaccine [56], rabies vaccine [57], and hepatitis A vaccine [58]) are viruses that have been inactivated by chemical or physical methods, and still maintain the immunogenicity of their immune antigens. Killed vaccines have high safety and good stability. However, adjuvants are often needed to enhance the immune effect. Gene vaccines (hepatitis B vaccine [59], HIV vaccine [60]) are not infectious, convenient for mass production, and safe. They also need adjuvants to enhance the immune effect during usage. Therefore, adding vaccine adjuvants to non-specifically increase the body's immune response in traditional vaccine production is often a good preparation plan. These adjuvants often increase the immune response ability in organisms through their own physical and chemical properties or by changing antigens' physical properties.

For nanoparticles, by relying on the diversity of their structures and morphologies, they can be used as adjuvants and suitable carriers of antigens to enhance the immune response. This requires easy functional modification of nanomaterials' surface, and the vaccine adjuvants are physically or chemically connected. So, the adjuvant is released slowly to achieve the effect of prolonging the immune time. Meanwhile, some metal-based nanomaterials have adjuvant properties, and their use with vaccines can increase the body's immune effect. Besides, the affinity of metal ions and biological macromolecules can be used to combine the surface proteins of the virus in the vaccine to enhance its stability, thereby improving the immune effect.

On the other hand, the storage and transportation of vaccines are more challenging, and the biological activity needs to be guaranteed. The vaccine can be protected by adding nanomaterials to increase its stability during transportation and storage and reduce costs. They are widely used as carriers in the biological field [61]. Metal-organic framework materials are nanomaterials with porous structures. Their surface is rich in unsaturated metal sites and is easier to modify after synthesis. Therefore, it is widely used in drug transportation [62]. In the same way, vaccine reagents are biologically active and can use porous materials to be adsorbed in the pores through static electricity to achieve the effect of vaccine protection. In addition, the affinity of metal ions and biological proteins is used to coordinate and adsorb the metal nanomaterials and vaccine proteins to ensure the biological activity of the vaccine. This indicates extensive application prospects in the field of vaccine protection with metal nanomaterials.

ZIF-8 has mild synthesis conditions, and it is possible to encode ovalbumin (OVA), an antigen model, into ZIF-8 for co-transport of antigen and adjuvant by surface electrostatic adsorption of immune adjuvant cytosine-phosphate-guanine (CpG) [63]. According to the characteristics of ZIF-8 in which protonation in a weakly acidic environment leads to structural degradation, the release of antigens and adjuvants is achieved, and a robust immune response is caused. Similarly, Yong Yang [64] first synthesized MIL-101-Fe-NH_2, which is safe and degradable, and then modified disulfide bonds on the outer surface. The immune adjuvant OVA is bound to the outer surface through disulfide bonds, and then CpG is adsorbed on the outer surface of MIL-101-Fe-NH_2 through electrostatic adsorption. Since the glutathione (GSH) content in the cell is higher than that in the extracellular tissue, the synthetic material undergoes a redox reaction between GSH and the disulfide bond in the material to release OVA. Studies have shown that the material's better transportation performance increases the immune effect and the memory of the immune system. Furthermore, metal elemental nanomaterials have a small particle size and are easily taken up by cells. Therefore, the cellular uptake of antigens and adjuvants that they carried increases [65,66]. More importantly, studies using Ag as the metallic element combined with influenza-inactivated virus vaccines have shown that the Ag nanoparticles can stimulate the regeneration of bronchial-associated lymphoid tissues and the expression of antigen-specific IgA, acting as an immune adjuvant [67]. Moreover, calcium phosphate nanoparticles (CAP) have been proven to be a good vaccine adjuvant [68,69].

Due to the excellent physical and chemical properties of metal-based nanoparticles, their biological applications have been continuously expanded. Previous work has found that some metal-based materials themselves have adjuvant properties. During the early vaccination process, alum is a promising adjuvant widely used in vaccines. Later studies found that CAP can stimulate the organism's immune response, and its effect is better than that of alum [68]. On this basis, Marina A. Volkova and colleagues studied the immune response induced by CAP when combined with an inactivated Newcastle disease virus vaccine. It is found that CAP can induce effective mucosal immunity, thereby enhancing the effectiveness of the vaccine. CAP can also be used as a vaccine carrier. By synthesizing CAP nanoparticles, the immunologically active TLR9 ligand and influenza A antigen are loaded. Analogously, studies have shown that the loaded CAP nanoparticles can be effectively taken up by dendritic cells and can stimulate an effective immune response in the body [70].

Some adjuvants need to be added into inactivated virus vaccines to increase the immune effect when used in vivo. Here, adjuvants can stimulate the immune response, and protect the activity of the inactivated virus antigen. The inactivated foot-and-mouth-disease virus vaccine is not stable when used. The histidine residues on the surface protein are protonated, causing it to lose immune activity [71,72]. Metal ions have a good affinity with amino acid residues. Shuai Li and colleagues coordinated with histidine residues on the surface protein of the foot-and-mouth-disease virus through Zn^{2+} to prevent protonation in organisms and improve the stability of antigens [73]. Similarly, by taking advantage of the good affinity of metal ions and biological macromolecules, Ni coordination will bind to the virion and the target protein, activate the antigen-presenting cells in the body, and promote the efficiency of antigen cross-presentation [74]. In terms of vaccine transportation protection, Michael A. Luzuriaga selected the metal organic framework material ZIF-8 with milder synthesis conditions as the protection model. By carrying out surface biomineralization on the surface of the tobacco mosaic virus, the growth of the ZIF-8 shell achieves its purpose of protection under changing environmental conditions. After treatment with pressure, heating, methanol, guanidine hydrochloride, and ethyl acetate, compared to the protected virus, the degree of change in the surface protein of the virus after wrapping ZIF-8 is significantly smaller, which protects the virus from the external environment. After peeling off the outer layer of ZIF-8, it was found that the surface protein and RNA of the virus remained active. The subsequent results of mouse experiments also confirmed the above results [75].

We briefly summarize below the therapeutic mechanism of nanomaterials in viral infection inhibition. As shown in Table 1, according to different viral inhibition mechanisms, it is divided into direct inactivating, inhibition of absorption, intracellular inhibition, and as a carrier. The analysis shows that the metal nanometer that directly inactivates the virus mainly destroys the virus envelope by contacting the virus or releasing active substances to damage the virus. Inhibition of virus absorption is via competitive binding to cells, or binding the surface of the virus through metal nanomaterials, occupying the binding site of the virus and the cell to prevent the virus from invading the cell. Intracellular inhibition uses metal nanomaterials to enter the virus-infected cell, and metal ions damage the viral genetic material or interfere with its replication process to inhibit further virus replication. With high biocompatibility and high cellular uptake rate, the metal nanomaterial is used as a carrier to improve the therapeutic effect of antiviral drugs. In addition, metal nanomaterials are also used as vaccine adjuvants to improve immune response and enhance vaccine effects.

Table 1. Investigations on metal nanomaterials inhibiting viral infections.

Materials	Size/nm	Synthetic Method	Antiviral Mechanism	Detection Methods	Antiviral Types	Ref.
TiO_2	8	Sonochemistry	Direct inactivating	Hemagglutination assay (HA)	Newcastle virus	[76]
TiO_2	4–5	Hydrolysis, Calcination		TCID50	H3N2	[26]
TiO_2	50	Hydrothermal		SEM, Cytopathic effect (CPE), HA	H9N2	[25]
Fe_2O_3/Fe_3O_4	/	/		Protein conformation analysis	COVID-19, Hepatitis C virus (HCV)	[31]
Ag	10, 75, 110	Commercialization		TCID50	Feline Calicivirus (FCV)	[77]
Cu (II)-zeolite	/	Mechanochemistry		TCID50, HA/NA, RT-PCR, indirect fluorescent antibody (IFA)	H5N1, H5N3	[78]
CuI	100–400	Commercialization		Plaque assay	FCV	[79]
CuI	160	Commercialization		Plaque assay, SDS-PAGE	H1N1	[80]
Cu-doped TiO_2	20	Sol-gel		qRT-PCR	Human norovirus	[81]
Te/BSA	57 ± 7.6	Redox		TCID50, IFA	Porcine epidemic diarrhea virus	[82]
SnO_2	50–100	Flame transport synthesis		Plaque assay, fluorescence imaging	Herpes simplex virus-1	[83]
ZnO	100–500	Flame transport synthesis		Reporter Lysis assay	Herpes simplex virus-1	[17]
Ag	10	Commercialization		TCID50, RT-qPCR, IFA	SARS-CoV-2	[84]
Ga	/	/		Plaque assay, fluorescence imaging,	HIV	[85]
PEGylated ZnO-NPs	16–20	Commercialization	Intracellular inhibition (replication)	TCID50	H1N1	[24]
CuO	40	Commercialization		TCID50, IFA	Herpes simplex virus-1	[27]
Fe_3O_4	10–15	Hydrothermal		TCID50, RT-PCR	H1N1	[30]
Ga(III) Nanoparticles		Mechanochemistry		TCID50	HIV	[86]
Au@Ag		Seed-mediated growth		TCID50	Coronavirus	[87]
TiO_2	20–25	/		RT-PCR, TCID50, alkaline single-cell gel electrophoresis (comet)	Noroviruses	[88]
Ag_2S	3.7, 5.3	Sol-gel	Intracellular inhibition (replication and release)	TCID50	Porcine epidemic diarrhea virus	[35]
CdTe	3.2 ± 0.8	Hydrothermal		TCID50	Pseudorabies virus	[89]
Ag@OTV	2	Sonochemistry	Drug-loading	Caspase-3 activity, TEM	H1N1	[90]
TiO_2~DNA	~5	Hydrolysis synthesis		TCID50	Influenza A virus	[91]
Ag/TiO_2	12–32	Sol-gel		/	H1N1, enterovirus	[92]
Au	35	Seed-mediated growth		Immunohistochemistry	Neurotropic virus	[93]

4. Conclusions

The spread of viruses such as SARS-CoV and SARS-CoV-2 as well as influenza viruses poses a huge threat to human health. It is generally known that viruses are prone to mutate due to external influences to produce different virus subtypes, limiting the use of traditional antiviral drugs and vaccines. Therefore, novel drug discovery and vaccine development for the treatment of viral infectious diseases is very important.

Metal-based nanomaterials have the advantages of high specific surface area and small particle size, making them widely used in the biological field. In terms of antiviral activity, some gold-based nanomaterials such as Au, Ag, CuI, TiO_2, etc., have virus-inactivating ability and can damage their surface proteins or their genetic material by combining with viruses. In comparison, nanoparticles such as ZnO and Fe_3O_4 can prevent the virus from further infecting cells by interfering with the replication, translation, and release of the virus. Some other metal nanoparticles have a small particle size and good biocompatibility, are easily taken up by cells, and can be used as a suitable carrier for antiviral drugs. Moreover, utilizing the binding of metal ions and biological macromolecules, metal nanoparticles can be used as vaccine adjuvants or adjuvant carriers to promote the occurrence of immune response of the body. However, there are still some problems to be solved in the antiviral application of metal-based nanomaterials. First of all, the toxicity of metal ions in organisms is still a major obstacle to their application. The biological toxicity of metal ions has always been a major obstacle to the application of metal-related materials in biological treatments. While ensuring their therapeutic effect, reducing the in vivo toxicity of metal ions is one of the most important directions for researchers to explore. At this stage, there are mainly two methods; first, by reducing the concentration of metal materials used, and second, by optimizing the metabolism of metal ions. However, the effect is not yet satisfactory. Next, the antiviral mechanism of some metal ions has not yet been explored clearly. Finally, there are many metal nanomaterials, and the current research is limited to a few metal ions. The antiviral properties of other metals still need to be further studied. Based on their excellent properties, the application of metal-based nanomaterials in the antiviral field is a promising research field for expansion.

Funding: This research was funded by grants from the China National Mega-projects for Infectious Diseases, grant number 2017ZX10103009-002, Natural Science Foundation of Shanghai, grant number 19ZR1434800, and the Open Foundation of National Virus Resource Center, grant number NCRC-PY-01.

Institutional Review Board Statement: Not applicable.

Informed Consent Statement: Not applicable.

Data Availability Statement: Data sharing is not applicable for this article.

Acknowledgments: We thank Gary Wong for proofreading the article.

Conflicts of Interest: The authors declare no conflict of interest.

References

1. Tan, M.; Bhadoria, A.S.; Cui, F. Estimating the proportion of people with chronic hepatitis B virus infection eligible for hepatitis B antiviral treatment worldwide: A systematic review and meta-analysis. *Lancet Gastroenterol. Hepatol.* **2021**, *6*, 106–119. [CrossRef]
2. Beerens, N.; Germeraad, E.A.; Venema, S.; Verheij, E.; Pritz-Verschuren, S.B.E.; Gonzales, J.L. Comparative pathogenicity and environmental transmission of recent highly pathogenic avian influenza H5 viruses. *Emerg. Microbes Infect.* **2021**, *10*, 97–108. [CrossRef]
3. Cohen, M.S.; Chen, Y.Q.; McCauley, M.; Gamble, T.; Hosseinipour, M.C.; Kumarasamy, N.; Hakim, J.G.; Kumwenda, J.; Grinsztejn, B.; Pilotto, J.H.S.; et al. Prevention of HIV-1 infection with early antiretroviral therapy. *N. Engl. J. Med.* **2011**, *365*, 493–505. [CrossRef]
4. Zhou, P.; Yang, X.L.; Wang, X.G.; Hu, B.; Zhang, L.; Zhang, W.; Si, H.R.; Zhu, Y.; Li, B.; Huang, C.L.; et al. A pneumonia outbreak associated with a new coronavirus of probable bat origin. *Nature* **2020**, *579*, 270–273. [CrossRef]
5. Shen, L.; Bao, N.; Yanagisawa, K.; Domen, K.; Gupta, A.; Grimes, C.A. Direct synthesis of ZnO nanoparticles by a solution-free mechanochemical reaction. *Nanotechnology* **2006**, *17*, 5117–5123. [CrossRef]
6. Chen, Z.; Ren, W.; Gao, L.; Liu, B.; Pei, S.; Cheng, H.-M. Three-dimensional flexible and conductive interconnected graphene networks grown by chemical vapour deposition. *Nat. Mater.* **2011**, *10*, 424–428. [CrossRef]

7. Kołodziejczak-Radzimska, A.; Jesionowski, T. Zinc Oxide—From Synthesis to Application: A Review. *Materials* **2014**, *7*, 2833–2881. [CrossRef]
8. Pascu, A.; Stanciu, E.M.; Croitoru, C.; Roata, I.C.; Tierean, M.H. Carbon Nanoparticle-Supported Pd Obtained by Solar Physical Vapor Deposition. *Adv. Mater. Sci. Eng.* **2018**, *2018*, 1–7. [CrossRef]
9. Rai, M.; Yadav, A.; Gade, A. CRC 675—Current Trends in Phytosynthesis of Metal Nanoparticles. *Crit. Rev. Biotechnol.* **2008**, *28*, 277–284. [CrossRef]
10. Mohamed, H.; Afridi, S.; Khalil, A.T.; Zohra, T.; Alam, M.M.; Ikram, A.; Shinwari, Z.K.; Maaza, M. Phytosynthesis of $BiVO_4$ nanorods using Hyphaene thebaica for diverse biomedical applications. *AMB Express* **2019**, *9*, 1–14. [CrossRef]
11. Howes, P.D.; Chandrawati, R.; Stevens, M.M.; Stevens, M.M. Colloidal nanoparticles as advanced biological sensors. *Science* **2014**, *346*, 1247390. [CrossRef] [PubMed]
12. Tong, H.; Ouyang, S.; Bi, Y.; Umezawa, N.; Oshikiri, M.; Ye, J. Nano-photocatalytic materials: Possibilities and challenges. *Adv. Mater.* **2012**, *24*, 229–251. [CrossRef] [PubMed]
13. Tale, B.; Nemade, K.R.; Tekade, P.V. Graphene based nano-composites for efficient energy conversion and storage in Solar cells and Supercapacitors: A Review. *Polym. Plast. Technol. Mater.* **2021**, *60*, 784–797.
14. Liang, G.F.; Wang, H.J.; Shi, H.; Wang, H.T.; Zhu, M.X.; Jing, A.H.; Li, J.H.; Li, G.D. Recent progress in the development of upconversion nanomaterials in bioimaging and disease treatment. *J. Nanobiotechnol.* **2020**, *18*, 154. [CrossRef]
15. Morris, D.; Ansar, M.; Speshock, J.; Ivanciuc, T.; Qu, Y.; Casola, A.; Garofalo, R.P. Antiviral and Immunomodulatory Activity of Silver Nanoparticles in Experimental RSV Infection. *Viruses* **2019**, *11*, 732. [CrossRef]
16. Halder, A.; Das, S.; Ojha, D.; Chattopadhyay, D.; Mukherjee, A. Highly monodispersed gold nanoparticles synthesis and inhibition of herpes simplex virus infections. *Mater. Sci. Eng. C Mater. Biol. Appl.* **2018**, *89*, 413–421. [CrossRef]
17. Mishra, Y.K.; Adelung, R.; Röhl, C.; Shukla, D.; Spors, F.; Tiwari, V. Virostatic potential of micro–nano filopodia-like ZnO structures against herpes simplex virus-1. *Antivir. Res.* **2011**, *92*, 305–312. [CrossRef]
18. Guan, W.J.; Ni, Z.Y.; Hu, Y.; Liang, W.; Ou, C.; He, J.; Liu, L.; Shan, H.; Lei, C.; Hui, D.S.C.; et al. Clinical Characteristics of Coronavirus Disease 2019 in China. *N. Engl. J. Med.* **2020**, *382*, 1708–1720. [CrossRef]
19. Maduray, K.; Parboosing, R. Metal Nanoparticles: A Promising Treatment for Viral and Arboviral Infections *Biol. Trace Elem. Res.* **2021**, *199*, 3159–3176. [CrossRef] [PubMed]
20. Rai, M.; Deshmukh, S.D.; Ingle, A.; Gupta, I.; Galdiero, M.; Galdiero, S. Metal nanoparticles: The protective nanoshield against virus infection. *Crit. Rev. Microbiol.* **2016**, *42*, 46–56. [CrossRef]
21. Marija, M.; Aleksandar, A.; Jelena, M.; Tatjana, K.; Nebojsa, A. Antimicrobial Nanoarchitectonics | | Nanoparticles in Antiviral Therapy. *Antimicrob. Nanoarchitecton.* **2017**, *14*, 383–410.
22. Baram-Pinto, D.; Shukla, S.; Gedanken, A.; Sarid, R. Inhibition of HSV-1 attachment, entry, and cell-to-cell spread by functionalized multivalent gold nanoparticles. *Small* **2010**, *6*, 1044–1050. [CrossRef]
23. Baram-Pinto, D.; Shukla, S.; Perkas, N.; Gedanken, A.; Sarid, R. Inhibition of Herpes Simplex Virus Type 1 Infection by Silver Nanoparticles Capped with Mercaptoethane Sulfonate. *Bioconjug. Chem.* **2009**, *20*, 1497–1502. [CrossRef]
24. Ghaffari, H.; Tavakoli, A.; Moradi, A.; Tabarraei, A.; Bokharaei-Salim, F.; Zahmatkeshan, M.; Farahmand, M.; Javanmard, D.; Kiani, S.J.; Esghaei, M.; et al. Inhibition of H1N1 influenza virus infection by zinc oxide nanoparticles: Another emerging application of nanomedicine. *J. Biomed. Sci.* **2019**, *26*, 70. [CrossRef]
25. Cui, H.; Jiang, J.; Gu, W.; Sun, C.J.; Wu, D.L.; Yang, T.; Yang, G.C. Photocatalytic Inactivation Efficiency of Anatase Nano-TiO_2 Sol on the H9N2 Avian Influenza Virus. *Photochem. Photobiol.* **2010**, *86*, 1135–1139. [CrossRef]
26. Mazurkova, N.A.; Spitsyna, Y.E.; Shikina, N.V.; Ismagilov, Z.R.; Zagrebel'nyi, S.N.; Ryabchikova, E.I. Interaction of titanium dioxide nanoparticles with influenza virus. *Nanotechnol. Russ.* **2010**, *5*, 417–420. [CrossRef]
27. Tavakoli, A.; Hashemzadeh, M.S. Inhibition of herpes simplex virus type 1 by copper oxide nanoparticles. *J. Virol. Methods* **2020**, *275*, 113688. [CrossRef] [PubMed]
28. Park, S.; Park, H.H.; Kim, S.Y.; Kim, S.J.; Woo, K.; Ko, G. Antiviral Properties of Silver Nanoparticles on a Magnetic Hybrid Colloid. *Appl. Environ. Microbiol.* **2014**, *80*, 2343–2350. [CrossRef]
29. Pfaff, F.; Glück, B.; Hoyer, T.; Rohländer, D.; Sauerbrei, A.; Zell, R. Tungsten carbide nanoparticles show a broad spectrum virucidal activity against enveloped and nonenveloped model viruses using a guideline-standardized in vitro test. *Lett. Appl. Microbiol.* **2019**, *69*, 302–309. [CrossRef] [PubMed]
30. Kumar, R.; Nayak, M.; Sahoo, G.C.; Pandey, K.; Sarkar, M.C.; Ansari, Y.; Das, V.N.R.; Topno, R.K.; Bhawna; Madhukar, M.; et al. Iron oxide nanoparticles based antiviral activity of H1N1 influenza A virus. *J. Infect. Chemother.* **2019**, *25*, 325–329. [CrossRef] [PubMed]
31. Abo-Zeid, Y.; Ismail, N.S.M.; Mclean, G.R.; Hamdy, N.M. A molecular docking study repurposes FDA approved iron oxide nanoparticles to treat and control COVID-19 infection. *Eur. J. Pharm. Sci.* **2020**, *153*, 105465. [CrossRef]
32. Tyagi, M.; Rusnati, M.; Presta, M.; Giacca, M. Internalization of HIV-1 Tat Requires Cell Surface Heparan Sulfate Proteoglycans. *J. Biol. Chem.* **2001**, *276*, 3254–3261. [CrossRef]
33. Vonnemann, J.; Sieben, C.; Wolff, C.; Ludwig, K.; Bottcher, C.; Herrmann, A.; Haag, R. Virus inhibition induced by polyvalent nanoparticles of different sizes. *Nanoscale* **2014**, *6*, 2353–2360. [CrossRef] [PubMed]
34. Kim, J.; Yeom, M.; Lee, T.; Kim, H.O.; Na, W.; Kang, A.; Lim, J.W.; Park, G.; Park, C.; Song, D.; et al. Porous gold nanoparticles for attenuating infectivity of influenza A virus. *J. Nanobiotechnol.* **2020**, *18*, 54. [CrossRef] [PubMed]

35. Du, T.; Liang, J.; Dong, N.; Lu, J.; Fu, Y.Y.; Fang, L.R.; Xiao, S.B.; Han, H.Y. Glutathione-Capped Ag2S Nanoclusters Inhibit Coronavirus Proliferation through Blockage of Viral RNA Synthesis and Budding. *ACS Appl. Mater. Interfaces.* **2018**, *10*, 4369–4378. [CrossRef] [PubMed]
36. Pang, C.; Zhang, P.; Mu, Y.; Ren, J.Z.; Zhao, B. Transformation and Cytotoxicity of Surface-Modified Silver Nanoparticles Undergoing Long-Term Aging. *Nanomaterials* **2020**, *10*, 2255. [CrossRef]
37. Asok, A.; Kulkarni, A.R.; Gandhi, M.N. Defect-rich ZnO quantum dots as a potential multifunctional sunscreen and cosmetic active ingredient. *Pure Appl. Chem.* **2015**, *87*, 971–977. [CrossRef]
38. Siddiqi, K.S.; Rahman, A.; Husen, A. Properties of Zinc Oxide Nanoparticles and Their Activity against Microbes. *Nanoscale Res. Lett.* **2018**, *13*, 141.
39. Tavakoli, A.; Ataei-Pirkooh, A.; Sadeghi, G.M.M.; Bokharaei-Salim, F.; Sahrapour, P.; Kiani, S.J.; Moghoofei, M.; Farahmand, M.; Javanmard, D.; Monavari, S.H. Polyethylene glycol-coated zinc oxide nanoparticle an efficient nanoweapon to fight against herpes simplex virus type 1. *Nanomedicine* **2018**, *13*, 2675–2690. [CrossRef]
40. Saeedi, T.; Alotaibi, H.F.; Prokopovich, P. Polymer colloids as drug delivery systems for the treatment of arthritis. *Adv. Colloid Interface Sci.* **2020**, *285*, 102273. [CrossRef]
41. Patil-Sen, Y.; Torino, E.; De Sarno, F.; Ponsiglione, A.M.; Chhabria, V.; Ahmed, W.; Mercer, T. Biocompatible superparamagnetic core-shell nanoparticles for potential use in hyperthermia-enabled drug release and as an enhanced contrast agent. *Nanotechnology* **2020**, *31*, 375102. [CrossRef]
42. Dos Apostolos, R.C.R.; Cipreste, M.F.; De Sousa, R.G.; De Sousa, E.M.B. Multifunctional hybrid nanosystems based on mesoporous silica and hydroxyapatite nanoparticles applied as potential nanocarriers for theranostic applications. *J. Nanopart. Res.* **2020**, *22*, 368. [CrossRef]
43. Cai, W.; Chu, C.-C.; Liu, G.; Wang, Y.-X.J. Metal-Organic Framework-Based Nanomedicine Platforms for Drug Delivery and Molecular Imaging. *Small* **2015**, *11*, 4806–4822. [CrossRef]
44. Li, Z.; Huang, H.; Tang, S.; Li, Y.; Yu, X.F.; Wang, H.Y.; Li, P.H.; Sun, Z.B.; Zhang, H.; Liu, C.L.; et al. Small gold nanorods laden macrophages for enhanced tumor coverage in photothermal therapy. *Biomaterials* **2016**, *74*, 144–154. [CrossRef]
45. Zazo, H.; Colino, C.I.; Warzecha, K.T.; Hoss, M.; Gbureck, U.; Trautwein, C.; Tacke, F.; Lanao, J.M.; Bartneck, M. Gold Nanocarriers for Macrophage-Targeted Therapy of Human Immunodeficiency Virus. *Macromol. Biosci.* **2017**, *17*, 1600359. [CrossRef] [PubMed]
46. Dungdung, R.; Bayal, M.; Valliyott, L.; Unniyampurath, U.; Nair, S.S.; Pilankatta, R. A slow, efficient and safe nanoplatform of tailored ZnS QD-mycophenolic acid conjugates for in vitro drug delivery against dengue virus 2 genome replication. *Nanoscale Adv.* **2020**, *2*, 5777–5789. [CrossRef]
47. Nair, M.; Guduru, R.; Liang, P.; Hong, J.; Sagar, V.; Khizroev, S. Externally controlled on-demand release of anti-HIV drug using magneto-electric nanoparticles as carriers. *Nat. Commun.* **2013**, *4*, 1707. [CrossRef] [PubMed]
48. Sadewasser, A.; Dietzel, E.; Michel, S.; Kluver, M.; Helfer, M.; Thelemann, T.; Klar, R.; Eickmann, M.; Becker, S.; Jaschinski, F. Anti-Niemann Pick C1 Single-Stranded Oligonucleotides with Locked Nucleic Acids Potently Reduce Ebola Virus Infection In Vitro. *Mol. Ther. Nucleic Acids* **2019**, *16*, 686–697. [CrossRef]
49. Moelling, K.; Heinrich, J.; Matskevich, A.; Wittmer-Elzaouk, L.; Kwok, T. Silencing of viral RNAs by small double-stranded siDNA. *Retrovirology* **2009**, *6*, 58. [CrossRef]
50. Robaldo, L.; Berzal-Herranz, A.; Montserrat, J.M.; Iribarren, A.M. Activity of core-modified 10–23 DNAzymes against HCV. *Chem. Med. Chem.* **2014**, *9*, 2172–2177. [CrossRef]
51. Ryoo, S.R.; Jang, H.; Kim, K.S.; Lee, B.; Kim, K.B.; Kim, Y.K.; Yeo, W.S.; Lee, Y.; Kim, D.E.; Min, D.H. Functional delivery of DNAzyme with iron oxide nanoparticles for hepatitis C virus gene knockdown. *Biomaterials* **2012**, *33*, 2754–2761. [CrossRef] [PubMed]
52. Wang, Z.; Liu, H.; Yang, S.H.; Wang, T.; Liu, C.; Cao, Y.C. Nanoparticle-based artificial RNA silencing machinery for antiviral therapy. *Proc. Natl. Acad. Sci. USA* **2012**, *109*, 12387–12392. [CrossRef] [PubMed]
53. Kuschner, R.A.; Russell, K.L.; Abuja, M.; Bauer, K.M.; Faix, D.J.; Hait, H.; Henrick, J.; Jacobs, M.; Liss, A.; Lynch, J.A.; et al. A phase 3, randomized, double-blind, placebo-controlled study of the safety and efficacy of the live, oral adenovirus type 4 and type 7 vaccine, in U.S. military recruits. *Vaccine* **2013**, *31*, 2963–2971. [CrossRef] [PubMed]
54. Simon, J.K.; Pasetti, M.F.; Viret, J.-F.; Munoz, A.; Lagos, R.; Levine, M.M.; Campbell, J.D. A Clinical Study to Assess the Safety and Immunogenicity of Attenuated Measles Vaccine Administered Intranasally to Healthy Adults. *Hum. Vaccin.* **2014**, *3*, 54–58. [CrossRef]
55. Eswaran, S.P.; Praharaj, A.K.; Chander, Y.; Nagendra, A. Potency Titration of Oral Polio Vaccine by Estimation of Live Virus Content Using Tissue Culture Technique. *Med. J. Armed Forces India* **2003**, *59*, 105–107. [CrossRef]
56. Delore, V.; Salamand, C.; Marsh, G.; Arnoux, S.; Pepin, S.; Saliou, P. Long-term clinical trial safety experience with the inactivated split influenza vaccine, Vaxigrip®. *Vaccine* **2006**, *24*, 1586–1592. [CrossRef]
57. Finke, S.; Karger, A.; Freuling, C.; Muller, T. Assessment of inactivated human rabies vaccines: Biochemical characterization and genetic identification of virus strains. *Vaccine* **2012**, *30*, 3603–3609. [CrossRef]
58. Zhang, Z.; Zhu, X.; Hu, Y.; Liang, M.; Sun, J.; Song, Y.F.; Yang, Q.; Ji, H.Q.; Zeng, G.; Song, L.F.; et al. Five-year antibody persistence in children after one dose of inactivated or live attenuated hepatitis A vaccine. *Hum. Vaccin. Immunother.* **2017**, *13*, 1–6. [CrossRef]

59. Hernandez-Bernal, F.; Aguilar-Betancourt, A.; Aljovin, V.; Arias, G.; Valenzuela, C.; De Alejo, K.P.; Hernandez, K.; Oquendo, O.; Figueredo, N.; Figueroa, N.; et al. Comparison of four recombinant hepatitis B vaccines applied on an accelerated schedule in healthy adults. *Hum. Vaccin.* **2011**, *7*, 1026–1036. [CrossRef]
60. Ellenberger, D.; Li, B.; Smith, J.; Yi, H.; Folks, T.; Robinson, H.; Butera, S. Optimization of a multi-gene HIV-1 recombinant subtype CRF02_AG DNA vaccine for expression of multiple immunogenic forms. *Virology* **2004**, *319*, 118–130. [CrossRef]
61. Huda, S.; Alam, M.A.; Sharma, P.K. Smart nanocarriers-based drug delivery for cancer therapy: An innovative and developing strategy. *J. Drug Deliv. Sci. Technol.* **2020**, *60*, 102018. [CrossRef]
62. Wu, M.-X.; Yang, Y.-W. Metal-Organic Framework (MOF)-Based Drug/Cargo Delivery and Cancer Therapy. *Adv. Mater.* **2017**, *29*, 1606134. [CrossRef] [PubMed]
63. Zhang, Y.; Wang, F.; Ju, E.; Liu, Z.; Chen, Z.W.; Ren, J.S.; Qu, X.G. Metal-Organic-Framework-Based Vaccine Platforms for Enhanced Systemic Immune and Memory Response. *Adv. Funct. Mater.* **2016**, *26*, 6454–6461. [CrossRef]
64. Yang, Y.; Chen, Q.; Wu, J.P.; Kirk, T.B.; Xu, J.K.; Liu, Z.H.; Xue, W. Reduction-Responsive Codelivery System Based on a Metal-Organic Framework for Eliciting Potent Cellular Immune Response. *ACS Appl. Mater. Interfaces* **2018**, *10*, 12463–12473. [CrossRef] [PubMed]
65. Tao, W.; Hurst, B.L.; Shakya, A.K.; Uddin, M.J.; Ingrole, R.S.J.; Hernandez-Sanabria, M.; Arya, R.P.; Bimler, L.; Paust, S.; Tarbet, E.B.; et al. Consensus M2e peptide conjugated to gold nanoparticles confers protection against H1N1, H3N2 and H5N1 influenza A viruses. *Antivir. Res.* **2017**, *141*, 62–72. [CrossRef] [PubMed]
66. Wang, C.; Zhu, W.; Luo, Y.; Wang, B.Z. Gold nanoparticles conjugating recombinant influenza hemagglutinin trimers and flagellin enhanced mucosal cellular immunity. *Nanomedicine* **2018**, *14*, 1349–1360. [CrossRef] [PubMed]
67. Sanchez-Guzman, D.; Le Guen, P.; Villeret, B.; Sola, N.; Le Borgne, R.; Guyard, A.; Kemmel, A.; Crestani, B.; Sallenave, J.M.; Garcia-Verdugo, I. Silver nanoparticle-adjuvanted vaccine protects against lethal influenza infection through inducing BALT and IgA-mediated mucosal immunity. *Biomaterials* **2019**, *217*, 119308. [CrossRef] [PubMed]
68. He, Q.; Mitchell, A.; Morcol, T.; Bell, S.J.D. Calcium Phosphate Nanoparticles Induce Mucosal Immunity and Protection against Herpes Simplex Virus Type 2. *Clin. Diagn Lab. Immunol.* **2002**, *9*, 1021–1024. [CrossRef]
69. Volkova, M.A.; Irza, A.V.; Chvala, I.A.; Frolov, S.F.; Drygin, V.V.; Kapczynski, D.R. Adjuvant effects of chitosan and calcium phosphate particles in an inactivated Newcastle disease vaccine. *Avian Dis.* **2014**, *58*, 46–52. [CrossRef]
70. Knuschke, T.; Sokolova, V.; Rotan, O.; Wadwa, M.; Tenbusch, M.; Hansen, W.; Staeheli, P.; Epple, M.; Buer, J.; Westendorf, A.M. Immunization with biodegradable nanoparticles efficiently induces cellular immunity and protects against influenza virus infection. *J. Immunol.* **2013**, *190*, 6221–6229. [CrossRef]
71. Rincon, V.; Rodriguez-Huete, A.; Lopez-Arguello, S.; Ibarra-Molero, B.; Sanchez-Ruiz, J.M.; Harmsen, M.M.; Mateu, M.G. Identification of the structural basis of thermal lability of a virus provides a rationale for improved vaccines. *Structure.* **2014**, *22*, 1560–1570. [CrossRef] [PubMed]
72. Harmsen, M.M.; Fijten, H.P.; Westra, D.F.; Dekker, A. Stabilizing effects of excipients on dissociation of intact (146S) foot-and-mouth disease virions into 12S particles during storage as oil-emulsion vaccine. *Vaccine* **2015**, *33*, 2477–2484. [CrossRef] [PubMed]
73. Li, S.; Yang, Y.; Lin, X.; Li, Z.J.; Ma, G.H.; Su, Z.G.; Zhang, S.P. A Novel Particulate Delivery System Based on Antigen-Zn(2+) Coordination Interactions Enhances Stability and Cellular Immune Response of Inactivated Foot and Mouth Disease Virus. *Mol. Pharm.* **2020**, *17*, 2952–2963. [CrossRef] [PubMed]
74. Dong, W.; Bhide, Y.; Marsman, S.; Holtrop, M.; Meijerhof, T.; De Vries-Idema, J.; De Haan, A.; Huckriede, A. Monophosphoryl Lipid A-Adjuvanted Virosomes with Ni-Chelating Lipids for Attachment of Conserved Viral Proteins as Cross-Protective Influenza Vaccine. *Biotechnol. J.* **2018**, *13*, e1700645. [CrossRef]
75. Luzuriaga, M.A.; Welch, R.P.; Dharmarwardana, M.; Benjamin, C.E.; Li, S.B.; Dharmarwardana, M.; Benjamin, C.E.; Li, S.B.; Shahrivarkevishahi, A.; Popal, S.; et al. Enhanced Stability and Controlled Delivery of MOF-Encapsulated Vaccines and Their Immunogenic Response In Vivo. *ACS Appl. Mater. Interfaces* **2019**, *11*, 9740–9746. [CrossRef] [PubMed]
76. Akhtar, S.; Shahzad, K.; Mushtaq, S.; Ali, I.; Rafe, M.H.; Fazal-Ul-Karim, S.M. Antibacterial and antiviral potential of colloidal Titanium dioxide (TiO_2) nanoparticles suitable for biological applications. *Mater. Res. Express* **2019**, *6*, 105409. [CrossRef]
77. Bekele, A.Z.; Gokulan, K.; Williams, K.M.; Khare, S. Dose and Size-Dependent Antiviral Effects of Silver Nanoparticles on Feline Calicivirus, a Human Norovirus Surrogate. *Foodborne Pathog. Dis.* **2016**, *13*, 239–244. [CrossRef]
78. Imai, K.; Ogawa, H.; Bui, V.N.; Inoue, H.; Fukuda, J.; Ohba, M.; Yamamoto, Y.; Nakamura, K. Inactivation of high and low pathogenic avian influenza virus H5 subtypes by copper ions incorporated in zeolite-textile materials. *Antivir. Res.* **2012**, *93*, 225–233. [CrossRef]
79. Shionoiri, N.; Sato, T.; Fujimori, Y.; Nakayama, T.; Nemoto, M.; Matsunaga, T.; Tanaka, T. Investigation of the antiviral properties of copper iodide nanoparticles against feline calicivirus. *J. Biosci. Bioeng.* **2012**, *113*, 580–586. [CrossRef]
80. Fujimori, Y.; Sato, T.; Hayata, T.; Nagao, T.; Nakayama, M.; Nakayama, T.; Sugamata, R.; Suzuki, K. Novel Antiviral Characteristics of Nanosized Copper(I) Iodide Particles Showing Inactivation Activity against 2009 Pandemic H1N1 Influenza Virus. *Appl. Environ. Microbiol.* **2012**, *78*, 951–955. [CrossRef]
81. Moon, E.W.; Lee, H.W.; Rok, J.H.; Ha, J.H. Photocatalytic inactivation of viral particles of human norovirus by Cu-doped TiO_2 non-woven fabric under UVA-LED wavelengths. *Sci. Total Environ.* **2020**, *749*, 141574. [CrossRef]

82. Zhou, Y.; Jiang, X.; Tong, T.; Fang, L.R.; Wu, Y.; Liang, J.; Xiao, S. High antiviral activity of mercaptoethane sulfonate functionalized Te/BSA nanostars against arterivirus and coronavirus. *RSC Adv.* **2020**, *10*, 14161–14169. [CrossRef]
83. Trigilio, J.; Antoine, T.E.; Paulowicz, I.; Mishra, Y.K.; Adelung, R.; Shukla, D. Tin Oxide Nanowires Suppress Herpes Simplex Virus-1 Entry and Cell-to-Cell Membrane Fusion. *PLoS ONE* **2012**, *7*, e48147. [CrossRef]
84. Jeremiah, S.S.; Miyakawa, K.; Morita, T.; Yamaoka, Y.; Ryo, A. Potent antiviral effect of silver nanoparticles on SARS-CoV-2. *Biochem. Biophys. Res. Commun.* **2020**, *533*, 195–200. [CrossRef]
85. Soto, E.R.; O'connell, O.; Dikengil, F.; Peters, P.J.; Clapham, P.R.; Ostroff, G.R. Targeted Delivery of Glucan Particle Encapsulated Gallium Nanoparticles Inhibits HIV Growth in Human Macrophages. *J. Drug Deliv.* **2016**, *2016*, 8520629. [CrossRef] [PubMed]
86. Choi, S.-R.; Britigan, B.E.; Narayanasamy, P. Ga(III) Nanoparticles Inhibit Growth of both Mycobacterium tuberculosis and HIV and Release of Interleukin-6 (IL-6) and IL-8 in Coinfected Macrophages. *Antimicrob. Agents Chemother.* **2017**, *61*, e02505–e02516. [CrossRef] [PubMed]
87. Du, T.; Zhang, J.; Li, C.; Song, T.; Liu, J.F.; Du, X.J.; Wang, S. Gold/Silver Hybrid Nanoparticles with Enduring Inhibition of Coronavirus Multiplication through Multisite Mechanisms. *Bioconj. Chem.* **2020**, *31*, 2553–2563. [CrossRef]
88. Agnihothram, S.; Mullis, L.; Townsend, T.; Watanabe, F.; Mustafa, T.; Biris, A.; Manjanatha, M.; Azevedo, M. Titanium dioxide nanoparticles evoke proinflammatory response during Murine norovirus infection despite having minimal effects on virus replication. *Int. J. Nanotechnol. Med. Eng.* **2016**, *1*, 63–73.
89. Du, T.; Cai, K.; Han, H.; Fang, L.; Liang, J.; Xiao, S. Probing the interactions of CdTe quantum dots with pseudorabies virus. *Sci. Rep.* **2015**, *5*, 16403. [CrossRef]
90. Li, Y.; Lin, Z.; Zhao, M.; Xu, T.; Wang, C.; Hua, L.; Wang, H.; Xia, H.; Zhu, B. Silver Nanoparticle Based Codelivery of Oseltamivir to Inhibit the Activity of the H1N1 Influenza Virus through ROS-Mediated Signaling Pathways. *ACS Appl. Mater. Interfaces* **2016**, *8*, 24385–24393. [CrossRef]
91. Levina, A.S.; Repkova, M.N.; Zarytova, V.F.; Mazurkova, N.A. TiO_2~DNA nanocomposites as efficient site-specific antiviral agents against influenza A virus in cell culture. In Proceedings of the 2015 IEEE 15th International Conference on Nanotechnology (IEEE-NANO), Rome, Italy, 27–30 July 2015; pp. 1509–1512.
92. Moongraksathum, B.; Chien, M.Y.; Chen, Y.W. Antiviral and Antibacterial Effects of Silver-Doped TiO_2 Prepared by the Peroxo Sol-Gel Method. *J. Nanosci. Nanotechnol.* **2019**, *19*, 7356–7362. [CrossRef] [PubMed]
93. Ren, M.; Zhou, J.; Song, Z.; Mei, H.; Zhou, M.; Fu, Z.; Han, H.; Zhao, L. Aptamer and RVG functionalized gold nanorods for targeted photothermal therapy of neurotropic virus infection in the mouse brain. *Chem. Eng. J.* **2021**, *411*, 128557. [CrossRef]

nanomaterials

MDPI

Review

Recent Advances in Synthesis, Medical Applications and Challenges for Gold-Coated Iron Oxide: Comprehensive Study

Mohammed Ali Dheyab [1,2,*], Azlan Abdul Aziz [1,2,*], Mahmood S. Jameel [1,2] and Pegah Moradi Khaniabadi [3]

[1] Nano-Biotechnology Research and Innovation (NanoBRI), Institute for Research in Molecular Medicine (INFORMM), Universiti Sains Malaysia, Pulau Pinang 11800, Malaysia; mahmood@student.usm.my
[2] Nano-Optoelectronics Research and Technology Lab (NORLab), School of Physics, Universiti Sains Malaysia, Pulau Pinang 11800, Malaysia
[3] Department of Radiology and Molecular Imaging, College of Medicine and Health Science, Sultan Qaboos University, P.O. Box 35, Al Khod, Muscat 123, Oman; p.khaniabadi@squ.edu.om
* Correspondence: mdali@usm.my (M.A.D.); lan@usm.my (A.A.A.)

Abstract: Combining iron oxide nanoparticles (Fe_3O_4 NPs) and gold nanoparticles (Au NPs) in one nanostructure is a promising technique for various applications. Fe_3O_4 NPs have special supermagnetic attributes that allow them to be applied in different areas, and Au NPs stand out in biomaterials due to their oxidation resistance, chemical stability, and unique optical properties. Recent studies have generally defined the physicochemical properties of nanostructures without concentrating on a particular formation strategy. This detailed review provides a summary of the latest research on the formation strategy and applications of Fe_3O_4@Au. The diverse methods of synthesis of Fe_3O_4@Au NPs with different basic organic and inorganic improvements are introduced. The role and applicability of Au coating on the surface of Fe_3O_4 NPs schemes were explored. The 40 most relevant publications were identified and reviewed. The versatility of combining Fe_3O_4@Au NPs as an option for medical application is proven in catalysis, hyperthermia, biomedical imaging, drug delivery and protein separation.

Keywords: inorganic nanoparticles; chemical method; formation strategy; medical applications

Citation: Ali Dheyab, M.; Abdul Aziz, A.; Jameel, M.S.; Moradi Khaniabadi, P. Recent Advances in Synthesis, Medical Applications and Challenges for Gold-Coated Iron Oxide: Comprehensive Study. *Nanomaterials* **2021**, *11*, 2147. https://doi.org/10.3390/nano11082147

Academic Editors: Jihoon Lee and Ming-Yu Li

Received: 26 July 2021
Accepted: 13 August 2021
Published: 23 August 2021

1. Introduction

Coated nanoparticles, or core@shell nanoparticles, consist of two or more nanoparticles that contain a wide variety of organic as well as inorganic nanoparticles, where one serves as a core while the other is centered on the core and named the shell [1]. Knowledge of core@shell synthesis is a pioneering step of nanoscience, as the way to manipulate the nanoparticles' structure has enabled us to generate a variety of hybrid NPs [2,3]. Core@shell NPs, with the potential to be used as core or shell in a wide variety of materials, will reflect their satisfying distinctive properties and custom functions. Core or shell products can be chosen, depending on the intent of the study [4]. The core@shell property can be modified by causing changes to the components that make up the shell layer or core [5]. Characteristics and distinctive attributes such as optical, magnetic, biological, compatibility, chemical stability and physicochemical properties can be realized when different nanoparticles are incorporated, such as gold nanoparticles (Au NPs) on iron oxide nanoparticles (Fe_3O_4 NPs). In recent years, substantial attempts have probably been introduced to evaluate the biomedical applications of Fe_3O_4 NPs, including protein purification, immunoassays, hyperthermia, drug delivery, magnetic resonance imaging (MRI), and computed tomography (CT) [6]. Fe_3O_4 NPs are the most favored nanomaterials in medical applications because of their minimal toxicity features and excellent physicochemical characteristics such as stability, biocompatibility and supermagnetism [7]. The magnetic response stability of Fe_3O_4 is due to its low oxidation sensitivity [8]. In addition, size control, preventing aggregation via coating, precise dispersion and interaction, as well as the penetration of tissue and cell barriers all give Fe_3O_4 NPs an advantage over other metal nanoparticles. Fe_3O_4 NPs provide a forum for therapeutic uses where they can be utilized for their contrast agent characteristics in MRI diagnostics, as well as for therapeutics in the form of bio-catalysis, drug delivery and protein purification [9].

Various kinds of functional materials, including silica, polymers and Au have been formed on the Fe_3O_4 NPs surface to improve biocompatibility, chemical stability as well as processability for broader applications [10,11]. Au is considered to be the most desired coating material for the production of Fe_3O_4@Au NPs due to its surface functionality, catalytic activity and superior optical properties [12–15]. Because of the variety of physicochemical features and the ability to change the magnetic and optical property by modifying the charge, size, shape, surface modification and thickness of the Au shell, Fe_3O_4@Au NPs have been widely considered the most effective candidature for medical applications [16]. Several studies have been reported for the synthesis of Fe_3O_4@Au NPs. These studies generally described nanoparticles' physicochemical properties without focusing on a specific formation strategy [17–19].

For this reason, the current review will: (1) summarize the latest progress (2018–2020) in the design and synthesis of the Fe_3O_4@Au and elaborate upon the strategies involved in the formation Fe_3O_4@Au NPs core@shell, Fe_3O_4@Au HNPs, Fe_3O_4@Au core@satellite NPs as and as nanodumbbells, Fe_3O_4@Au DNPs; (2) explore the schemes of each manufacturing strategy for Au-coated Fe_3O_4; and (3) present the potency of Fe_3O_4@Au as a promising candidature for medical applications in areas of catalysis, hyperthermia, biomedical imaging, drug delivery and protein separation (2018–2020).

2. Synthesis of Fe_3O_4@Au

Fe_3O_4@Au NPs can be classified as Fe_3O_4@Au NPs core@shell, Fe_3O_4@Au HNPs, Fe_3O_4@Au core@satellite NPs as and as nanodumbbells, Fe_3O_4@Au DNPs structures. In this section, the synthesis of all structures will be introduced.

2.1. *Core@Shell Structure of Fe_3O_4@Au*

Core@shell nanoparticles have various properties, such as magnetism, metallicity and semiconductivity. These attributes come either through the core or shell materials, or both (Figure 1). In this review, we will discuss Fe_3O_4 NPs as a core and Au NP as a shell.

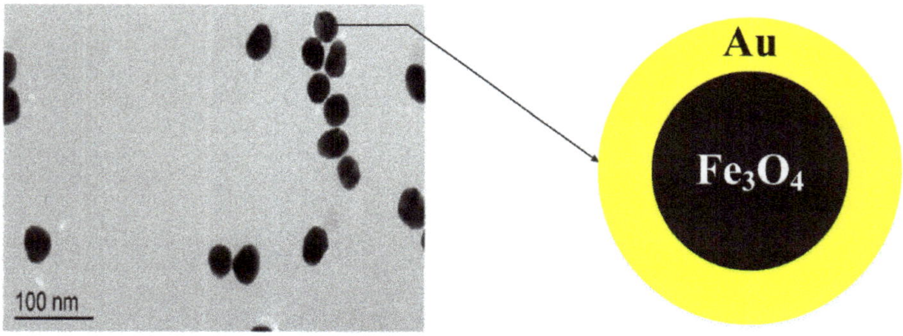

Figure 1. TEM image [20] and schematic diagram of Fe_3O_4@Au NPs core@shell construction.

Recently, Xie et al. [21] have synthesized novel core@shell NPs for use in fast, sensitive, convenient and good surface-enhanced Raman scattering. This method involved two steps for the formation of core@shell NPs: (i) the preparation of Fe_3O_4@silica NPs, using an ultrasound technique to deposit silica oxide on the surface of Fe_3O_4 for 5 min; and (ii) the preparation of the Fe_3O_4@SiO_2@Au seed, using a seed growth method. A similar study was submitted by He et al. [20]. Fe_3O_4@Au NPs were coated with glutathione to reduce the dose dependence of the anticancer medication, doxorubicin (DOX), by covering the glutathione shell on Fe_3O_4@Au NPs [22]. The former Fe_3O_4@Au NPs produce involved this process. Twenty milliliters (20 mL) of $HAuCl_4$ solution (0.1%) was added to 40 mL of distilled water in a 250 mL flask. The solution was refluxed and Fe_3O_4 was applied to the mixture, then the mixture was boiled for 45 min. A reduction (sodium citrate) was quickly applied under vigorous stirring to the boiling mixture. Citrate addition contributes to the color shift from a grey to a red solution. The solution was boiled for 15 min and then stirred until the solution reached room temperature. The use of glutathione conjugations thus acts as an effective vehicle of drug delivery. In addition to causing drug release with redox-activated glutathione, it requires significantly low levels of glutathione @Au@ Fe_3O_4 NPs for DOX releases. The sonochemical approach effectively synthesizes monodispersive and highly stable Fe_3O_4@Au NPs, with a size distribution of approximately 20 nm during 8 min [23]. Utilizing surface response (RSM) methodology, test runs of 14 dissimilar variations of gold ions, sonication frequency and sodium citrate (independent variables) have been conducted at two-center points to optimize testing procedures. Variance analysis (ANOVA) has been used to achieve optimal conditions for experimental results. The optimal zeta potential value of about −46.125 mV was reached under the ideal conditions of independent variables, which is compatible (at approximately 99.2%) with the real zeta potential value (−45.8 mV). The monodispersity and stability of the Fe_3O_4 NPs effectively coordinated a transition to the core@shell, as demonstrated by a rise in zeta potential value from −24 mV to −45 mV. To date, no work has been reported which produced core@shell NPs for large-scale production. The sonochemical method is widely considered to be one of the most promising methods for preparing, encapsulating and modifying nanoparticles due to its safe, rapid, low-cost and environmentally friendly characteristics [24]. Various types and shapes of nanomaterials have been prepared using a sonochemical method [25,26]. In addition, the sonochemical method ensures the uniformity, homogeneity and monodispersity of the nanoparticles produced [27]. From this point of view, all these advantages and properties of this method may have the potential to be more applicable to large-scale production.

Somayeh et al. [28] carried out a simple and eco-friendly green method for the preparation of spherical Fe_3O_4@Au with a size of 31 nm, utilizing the aqueous extract of the Carum carvi seed which plays three functions such as reduction, capping, and stabilizer agents during the Fe_3O_4@Au synthesis process (Table 1). The seeds of Carum carvi were thoroughly washed with distilled water, followed by drying at 25 °C for 2 days. In the end, the resultant was milled to produce a powder. In order to prepare the aqueous extract, the powder was steeped in 100 mL of distilled water for 10 h at a temperature of 25 °C and then purified using filter paper to acquire a clear solution. To synthesize Fe_3O_4@Au, 50 mg of Fe_3O_4 was dissolved in 100 mL of aqueous extract of Carum carvi and the mixture solution was stirred for around 10 min. Twenty milliliters (20 mL) $HAuCl_4$ solution (5 mM) was then applied to the mixture solution. Finally, the mixture solution was kept for 24 h and then dried overnight at 70 °C. The green, rapid and low-cost preparation of core@shell Fe_3O_4@Au NPs using natural honey as a reducing as well as stabilizing agent through hydrothermal method for 20 min was reported by Rasouli et al. [29] Fe_3O_4 NPs were dissolved in 50 mL ultrapure water and sonicated for 2 min, to which 25 mL of $HAuCl_4$ (0.005 M) was added and stirred for 15 min to achieve the full adsorption of gold ions on the surface of Fe_3O_4. Subsequently, 0.25 g of the natural honey was added to the mixture solution, held under the hydrothermal method at 120 °C for 20 min. Eventually, Fe_3O_4@Au NPs were separated from the excess result solution using a permeant magnet and washed three times through ultrapure water. TEM images revealed that the synthesis of Fe_3O_4@Au NPs has a diameter ranging between 3.49 and 4.11 nm. Tarhan et al. [30] announced that novel Fe_3O_4@Au NPs, functionalized via maltose, have been prepared as a favorable carrier matrix for easy and efficient L-asparaginase immobilization. The findings show that NPs are monodispersed to 9.0 emu/g magnetization with a size of 10 nm. Tarhan et al. [30] expect that flexible carriers will lead to new possibilities for applications in the fields of biomedicine, biotechnology and biochemistry on the basis of the success of the procedure and the promising findings achieved from their novel process.

Table 1. Summary of the recently published studies on the synthesis methods of Fe_3O_4@Au NPs.

No.	Nanoparticles Structure	Synthesis Method	Size/Shape	Applications	Ref
1	Core@shell	Growth method	5 nm/spherical	Food application	[21]
2	Core@shell	Sonochemical	~40 nm/flower	Food application	[20]
3	Core@shell	Green method	31/spherical	Antimicrobial activity	[28]
4	Core@shell	Green method	3.49–4.11 nm/semispherical	Drug delivery	[29]
5	Core@shell	Reduction	10 nm/amorphous	Enzyme immobilization	[30]
6	Core@shell	Sonochemical	20–50 nm/spherical	Cancer biomarkers	[31]
7	Core@shell	Seeds growth	9.49, 10.04 and 8.95 nm/flower	Catalytic reduction of RhB	[32]
8	Core@shell	Seeding technique	15–57 nm/		[33]
9	Core@shell	Nano-emulsion technique	11 nm/semispherical		[34]
10	Core@shell	Laser ablation	20 nm/spherical		[35]
11	Core@shell	Reduction	20–50 nm/semispherical	Cytotoxicity assay in MDCK cell line	[36]
12	Core@shell	Reduction	~100 nm/flower		[37]
13	HNPs	Reduction	10 nm/spherical	Hyperthermia	[38]
14	Au/PDA hybrid	In situ redox-oxidize polymerization	25 nm/spherical	Catalysis and adsorption	[39]
15	Fe_3O_4@Au@CeO_2 hybrid	Redox reaction	17 nm/nanofibers	Catalysis	[40]
16	HNPs	Thermal decomposition	25 nm/octahedral	Theranostics	[41]
17	HNPs	Seeds growth	90 nm/spiky	Multimodal in vivo imaging	[42]
18	HNPs	Chemical reduction	31 nm/spherical		[43]
19	Core@satellite	Seed-mediated growth	65 nm/cubic	Catalysis	[44]
20	Core@satellite	Hydrothermal treatment, and freeze-drying technologies	300–400 nm/spherical	Microbial fuel cells	[45]
21	Dumbbell NPs	Reduction	22 nm/spherical	Radiation therapy	[46]
22	Dumbbell NPs	Thermal decomposition	7 nm/spherical		[47]

Fe_3O_4@Au NPs have been produced as novel electrochemical immunosensors for the use of cancer biomarkers [31]. The morphology of Fe_3O_4@Au NPs was that of a spherical shape with an average size of approximately 20–50 nm. This novel strategy has shown simpler construction, easier operation and a wider linear range. The proposed approach and the use of a screen-printed carbon electrode provided for the development of a simple electrochemical immunosensor that could be disposable, portable and cheap without using additional labeling. For 15 min under sonication, the suspension of $HAuCl_4$ has been stirred with Fe_3O_4 solution. Subsequently, the reduction agent solution ($NaBH_4$) was quickly added to the cooled suspension, which was then sonicated for another 10 min. Kou et al. [32] reported the custom design of extremely effective catalysts for Fe_3O_4@Au NPs. Fe_3O_4 was formed with three different morphologies using engineered quantities of urea, and the probable mechanism was proposed. Therefore, by measuring the amount of Au seeds, they achieved Fe_3O_4@Au with different morphologies and tunable Au deposition. The catalytical ability of Fe_3O_4@Au with several structures was compared through the application to degrade 4-nitrophenol and catalytic rhodamine while systematically investigating the correlation of the Au seed amount to the turnover frequency and the catalytic capability of Fe_3O_4@Au. They observed that the flower-like Fe_3O_4@Au NPs with 20 mL of Au seeds applied had the highest degradation rate of 96.7%, and after recycling, their catalytic ability was almost unchanged. The formation of Fe_3O_4@Au NPs was accomplished by reducing the Au ions on the Fe_3O_4 surface using the seeding technique [33]. In a definite volume of glycerin, different concentrations of oxidized Fe_3O_4 or the Au-shell reaction were used. The reaction solution, including the reduction agent and Fe_3O_4 cores, was first sonicated for 15 min, then heated with vigorous stirring to approximately 150 °C. Once the reaction solution reached 150 °C, a drop-specific solution was added for $HAuCl_4$. Fifteen minutes after the addition of Au salts, the heating system was stopped but the stirrer proceeded while the mixture was refreshed at room temperature. The component ratio adaptation allowed the Fe_3O_4@Au NPs particle shell thickness to be tuned. The present route produces well-determined structures of the Fe_3O_4@Au NPs of various sizes between 15 and 57 nm, with the Au noble metal varying from Fe_3O_4 NPs. Bi-phase Fe_3O_4@Au NPs were provided using a nano-emulsion technique [34]. Characterization reveals that the Fe_3O_4@Au nanostructure produced a particle size and distribution of approximately 11 nm in size. The NPs are non-toxic, water-soluble and stable due to the capping agent covering the particles. Optical and magnetic data indicate that the NPs have a narrow-band surface absorption of plasmon and an increased susceptibility to the Au shell. As a result, the bi-phase Fe_3O_4@Au NPs are challenging for various applications such as magnetic separation, optical detection and photonic therapy. In a different process, Au and Fe_3O_4 representing magneto-plasmonic NPs were obtained in two successive steps in an aqueous environment by the laser ablation of the Au and Fe_3O_4 targets [35]. Au NPs are trapped in a Fe_3O_4 mucilaginous matrix, which was established by both microscopic and spectroscopic observation as magnetite. The plasmonic property of the colloids obtained was tested with surface-enhanced Raman scattering spectroscopy, as well as their adsorption capability. In addition to those inherent in Au NPs, the presence of Fe_3O_4 offers the bimetallic colloid new avenues of adsorption, particularly with respect to organic contaminants and heavy metals, allowing them to be extracted from the aqueous environment to promote a magnetic field. In addition, these NPs are low in toxicity, making them promising for biomedical applications. Fe_3O_4@Au in a size range of about 20–50 nm and significant magnetization saturation using a solvothermal one-pot process was recorded by Ángeles-Pascual et al. [36]. $NaBH_4$ gradually reduced the $HAuCl_4$ solution into 9 mL of the black NP solution to create a thin gold shell on the Fe_3O_4 NPs surface. The solution, under intense stirring, was heated up to 70 °C and allowed to naturally cool down to room temperature. Afterwards, Fe_3O_4@Au was separated using the neodymium magnet and rinsed to remove the excess of chemicals from the reagents. To examine the biocompatibility of NPs, a cytotoxicity assay was performed in the MDCK cell line. The tests for the Fe_3O_4@Au NPs exhibited higher cell viability, indicating their excellent biocompatibility and their potential for medical application. A novel and direct method for preparing Fe_3O_4@Au NPs comprising a Fe_3O_4 core coated with an Au shell was identified [37]. The synthesis incorporates ease of operation, minimal control and high reproducibility while at the same time being environmentally friendly. The shell of Au NPs with a controllable thickness of 30 nm was developed on the Fe_3O_4 core of 20 nm in size by reducing Au salt in the ultrasonic bath. Au shell thickness might be adjusted by means of varying the quantity of Au salt applied. Fe_3O_4@Au NPs of sizes ranging between 80 and 160 nm were prepared. The Fe_3O_4@Au NPs were studied for their magnetic and plasmonic behavior. Functionalization with polyethene glycol was conducted to explore its possible use in biomedical applications. Unlike Fe_3O_4@Au DNPs, core@shell was commonly utilized as a contrast agent in dual MR and CT imaging techniques.

2.2. The Hybrid Structure of Fe_3O_4@Au (HNPs)

The synthesis of hybrid Fe_3O_4@Au NPs (HNPs) with appropriate size, design and properties is difficult, and has gained considerable attention among researchers in material sciences. It is possible to tune the design of Fe_3O_4@Au HNPs by selecting the proper technique and controlling the processing parameters during the synthesis.

Fe_3O_4@Au HNPs have single-hybrid nanoparticles consisting of an entire-layer Au ion-reducing coating on the Fe_3O_4 surface. In addition to biocompatibility, the structure of Fe_3O_4@Au HNPs can also impart the NPs surface with appropriate biological and chemical interface activity [48]. A well-defined novel structure can easily be formed by the Au shell uniformly coated on the surface of Fe_3O_4 NPs with sulfur-based ligands. A considerable amount of work has been performed during the last two decades to develop Fe_3O_4@Au HNPs using various techniques, including co-precipitation, seed-mediated growth, direct coating and thermal decomposition methods. The most popular method for preparing Fe_3O_4@Au HNPs is the Au shell's direct coating on the Fe_3O_4 surface. In this approach, two strategies for forming the shell of Au on the Fe_3O_4 surface were observed. The first method is a one-pot process in which the Au ions extend to form the shell on the Fe_3O_4 NPs surface. For the second process, Au NPs are internally produced, then seeded into a suspension of Fe_3O_4 NPs to create Fe_3O_4@Au HNPs [18]. Sood et al. [49] observed that the Au shell's direct coating on the Fe_3O_4 NPs surface loaded with small ligands, including ascorbic acid and citric acid, may be more successful.

Park et al. [38] described the hyperthermic features of Fe_3O_4@Au HNPs within a 200 kHz and 1.5 kA m^{-1} biocompatible alternating magnetic field (AMF). In the air atmosphere, a 0.4 mL precursor of iron was added to a mixture of both oleic acid with octyl ether at 100 °C. The solution was stirred during 1.5 h before being cooled at room temperature. A mixture of oleylamine (0.5 mmol) and $HAuCl_4$ (1.3 mmol) in the 5 mL of chloroform was added two times at intervals of approximately 5 minutes with vigorous stirring with a solvent of oleylamine (2 mmol) and Fe_3O_4 NPs (0.1 mg) in chloroform (10 mL). HNPs were produced by growing Au NPs on the Fe_3O_4 NPs surface with an average size of 10 nm. Due to the decrease in the saturation value of the HNP solution relative to the Fe_3O_4 NPs, the initial heating rate was set to lower than the Fe_3O_4 NPs solution. The continued application of the AMF gradually increased the HNP solution temperature, while the solution of the Fe_3O_4 NPs achieved thermal equilibrium. A similar AMF condition was demonstrated with the heating efficiency of Au NPs combined with non-conductive and diamagnetic SiO_2 NPs, which demonstrates that sustained heat for HNPs may be due to the supplementary heating of the Au NPs in a radiation frequency solenoid belt (RF). A novel hollow nanosphere Fe_3O_4@Au/polydopamine (Au/PDA) capable of absorbing potentially toxic ions plus catalyzing the decrease in 4-nitrophenol has been published [39]. The hybrid shell has well encapsulated the hollow nanosphere Fe_3O_4(Au/PDA) to create the dual-functioning magnetics hollow nanocomposites utilizing an easy redox-oxidizing polymerization technique (Figure 2). Due to its uniform, hollow interior and usable PDA coating with a strong activity of the Au nanoshell, the eventual hollow nanosphere Fe_3O_4@Au/PDA has great potential for drug delivery and nanocatalysis. In brief, the multifunctional Fe_3O_4@Au/PDA nanosphere has wide application potential for coexisting toxic water contamination, green and simple synthesis and ease of manipulation, effective adsorption efficiency and strong catalytic activity. Au NPs play a crucial part in heterogeneous catalytic reactions. Nevertheless, Au NPs typically have low selectivity and complex recyclability. Fe_3O_4@Au@CeO_2 hybrid nanofibers were prepared in the presence of Fe_3O_4 nanofibers, through a simple one-pot redox reaction between $HAuCl_4$ and Ce $(NO_3)_3$ [40]. On the Fe_3O_4 nanofibers' surface, the CeO_2 shell was uniformly coated to form a unique hybrid structure, while the Au NPs were encapsulated within the CeO_2 shell. As a result of the CeO_2 shell formation, Fe_3O_4@Au@CeO_2 hybrid nanofibers are positively charged surfaces, allowing them to be excellent choices for the predominantly sensitive catalytic action against the degradation of negatively charged organic colors. The Fe_3O_4@Au@CeO_2 hybrid nanofibers have demonstrated magnetic properties, giving them good recyclable usability. This research provides a simple and efficient solution for preparing the hybrid nanomaterials of magnetic noble metal/metal oxide with a distinctive surface characteristic and chemical structure for offering applications in heterogeneous catalysis. A high temperature wet chemical method was used for the synthesis of Fe_3O_4@Au HNPs with a diameter of 25 nm [41]. Fe_3O_4@Au HNPs with Au seeds produced in situ were derived at high temperatures through the thermal decomposition of $HAuCl_4$ and $Fe(CO)_5$. Fe_3O_4@Au HNPs revealed the best features for application as hyperthermic and contrast agents for MRI. Due to the large saturation magnetization and octahedral shape of the magnetite particles, Fe_3O_4@Au HNPs obtained a particular loss power of approximately 617 W·gFe $^{-1}$ with an exceptionally high r_2-relaxivity of about 495 $mM^{-1}s^{-1}$.

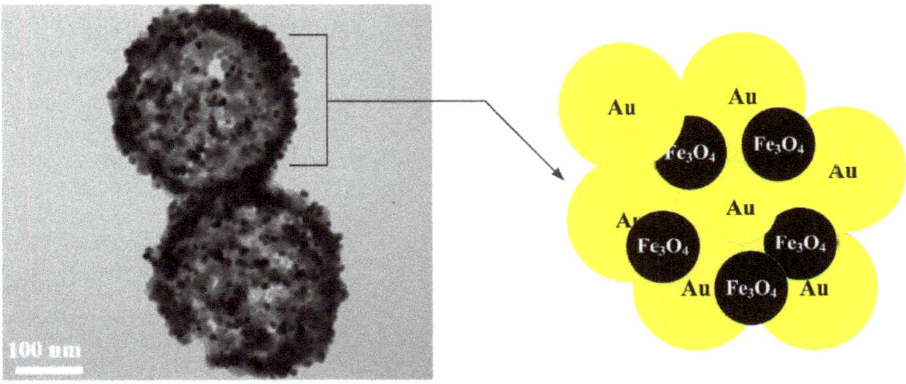

Figure 2. TEM image [39] and schematic illustration for the production of hybrid Fe_3O_4@Au HNP.

Wang et al. [42] stated that the novel structure of spiky Fe_3O_4@Au (SPs) is used for multi-modal imaging and phototherapy agents. The uniformly sized Fe_3O_4@Au SPs were synthesized in two steps. First, citrate-stabilized Fe_3O_4 NPs of the average size of 10 nm was synthesized, then the Au layer was coated on the Fe_3O_4 NPs surface to create Fe_3O_4@Au HNPs, which were used for the production of Fe_3O_4@Au SPs. The SPs exhibit great photodynamic effects and therapeutic photothermal, with a photothermal conversion efficiency of about 31%, and enable tumor-targeted imaging, such as MRI, photoacoustic and computed tomography. The SPs display good biocompatibility, in vivo as well as in vitro. Additionally, the SPs obliterated a tumor below 808 nm of radiation owing to its unique absorption in the near-infrared field. SPs represent a convenient product for application in clinical practice with their potential for deeply integrated multi-modal imaging as well as multiple therapeutic functions. Fe_3O_4@Au HNPs have been produced, characterized and presented as a new porous marker to increase micro-/nano-based pores found and quantified by SEM in the shale [43]. With the presynthesized Fe_3O_4 NPs in a solution, the Fe_3O_4@Au HNP shale has been synthesized using the chemical reduction technique. Because of the superparamagnetic properties, the nanomarker is easily operated via the external magnetic field to appoint in pores and provides a sharp contrast picture between the pores and shale matrix, making it much easier and more accurate to recognize micro/nano-sized pores in shales. Moreover, as Au NPs are particularly rare noble metals in the shale, Au's energy-dispersive X-ray mapping was used to accurately calculate area porosity in a shale. A precise and realistic technology is recommended to enable the characterization of micro/nano-pores in the shale in conjunction with the aforementioned merits of the nanomarker. The design and synthesis of hybrid NPs with distinct morphologies can draw the interest of scientists to hybrid biosynthesis NPs.

2.3. Core@Satellite Structures

One of the popular frameworks for Fe_3O_4@Au NPs is core@satellite (Cs). This structure has a single core of Fe_3O_4 with the binding by covalent bonds of numerous Au NPs similar to satellites. The CsFe_3O_4@Au NPs comprise a residually exposed core surface of Fe_3O_4 suitable for MR imaging and further functionalization. In addition, the Cs structure consists of many peripheral Au NPs with a large surface area of the satellite nanoparticle that is advantageous for imaging as well as photothermal capabilities [50]. CsFe_3O_4@Au NPs are drawn up using different methods. Liu et al. [44] announced that a seed deposition method was used to produce CsFe_3O_4@Au nanocubes (Figure 3). Ten milliliters (10 mL) of Au seeds were applied dropwise to obtain Fe_3O_4@PEI nanocubes dispersed in deionized water through ultrasonic treatment. The CsFe_3O_4@PEI@Au nanotubes were thoroughly washed with deionized water after 2 h of sonication. Recently, Song et al. [45] succeeded in developing CsFe_3O_4@Au NPs that combined three-dimensional microporous graphene foam was formed by an efficient approach which integrated in situ growth, hydrothermal treatment and freeze-drying methods. Ultrasonic treatment was required during the sample preparation to help form a stable mixed colloidal suspension of precursors. Nevertheless, it is notable for Au NPs to be removed from the CsFe_3O_4/Au NPs by using ultrasound. The binding force between the products of CsFe_3O_4/Au NPs must be powerful enough to solve this problem. As a result, the Cs Fe_3O_4/Au NPs used in this method were provided using an in situ growth technique, where Fe_3O_4 NPs coated with citric acid were utilized as seeds to reduce gold ions ($HAuCl_4$) with the asset of sodium citrate for the nucleation and growth of Au NPs on Fe_3O_4 NPs surfaces.

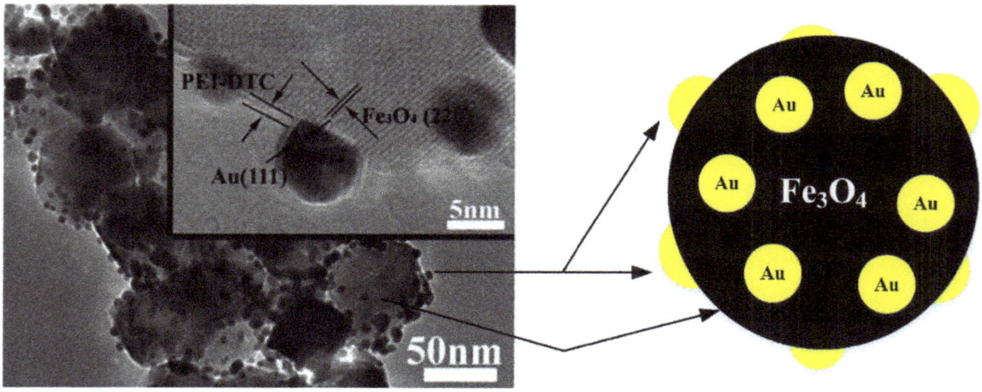

Figure 3. TEM image [44] and schematic diagram for the preparation of core@satellite CsFe_3O_4@Au NPs.

2.4. Fe_3O_4@Au Nanodumbbells

Dumbbell NPs (DNPs) consist of a tightly interacting heterostructure together with one NP at the other end (Figure 4). The separate NPs are dumbbell-like or resemble particles in near contact with each other. In contrast to Fe_3O_4@Au HNPs in which Au shields the Fe_3O_4 core, the Fe_3O_4@Au DNP's have a broad-based functional surface and active interface which improves their applications for diagnostics and therapy as theranostics [51]. Fe_3O_4@Au DNPs have unique features, including (1) the ability to allocate various functionalities to delivery applications and particular target imaging; (2) the magnetic detection and simultaneous optical abilities; and (3) the ability to customize optical and magnetic features by adjusting the size of Fe_3O_4@Au HNPs [52]. Fe_3O_4@Au DNPs can be regularly produced through the epitaxial growth of one NP to another form of NPs called NP seed. During the procedure, the nucleation should be properly regulated to generate heterogeneous nucleation on a particular crystalline phase around the seed NPs [53]. Klein et al. [46] developed a simple one-pot synthesis method for the preparation of Fe_3O_4@Au DNPs using a sonication process. In their analysis, Fe_3O_4@Au DNPs were achieved by the co-precipitation of Fe_3O_4 NPs in an aqueous solution of $HAuCl_4$. Subsequently, 3-mercaptopropionic acid was added to a mixture to stabilize Fe_3O_4@Au DNPs. The resulting DNPs were collected by permanent magnetism and washed three times with 20 mL of ultrapure water. Kostevsek et al. developed Fe_3O_4@Au DNPs coated with chitosan using a two-step process [47]. First, Fe_3O_4@Au DNPs were provided through the reduction of Au ions using the thermal decomposition of the Fe pentacarbonyl ($Fe(CO)_5$) with the existence of oleic acid, oleylamine and 1,2-hexadecanediol at the same time. An Au NP was observed to develop at first in the mixture, during the reaction because of a larger variance in the potential for reduction between Fe and Au. Afterwards, Au NPs were used to break down $Fe(CO)_5$ to produce Fe_3O_4@Au at higher temperatures. Second, the surface of Fe_3O_4@au presynthesized NPs was changed to produce highly biocompatible Fe_3O_4@Au DNPs coated with chitosan, utilizing hydrocaffeic acid and thioglycolic acid-conjugated chitosan. Fe_3O_4@Au DNPs were shown to be biocompatible within a certain range of concentrations that can be employed for optical and magnetic applications in biomedicine [54]. Despite the fact that much work has been expended in the synthesis of Fe_3O_4@Au CNPs for MR@CT imaging applications, the synthesis and development of these nanoparticle systems remain an open area with significant challenges. For example, Fe_3O_4@Au nanodumbbells have not been extensively used for dual-mode MR/CT imaging applications.

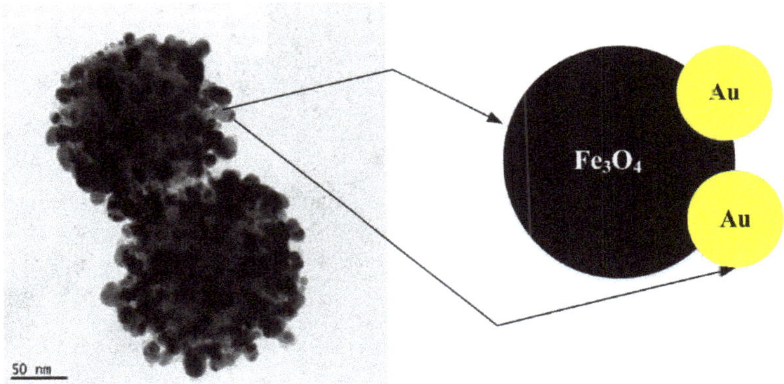

Figure 4. TEM image [55] and schematic drawing of Fe_3O_4@Au DNPs dumbbell preparation.

3. Medical Application of Fe_3O_4@Au NPs

Nanoscience currently ranks among the world's most desirable sciences due to its interdisciplinary research field, which can be used in many applications [55,56]. Fe_3O_4@Au with enhanced properties possesses a specific economic value relative to single NPs due to the current increase in performance, durability and a wide range of industrial, engineering and medical applications. Recently, Fe_3O_4@Au NPs have attracted many researchers due to their wide variety of features, potential, structures, easy control and simple production methods, as discussed above. Fe_3O_4@Au NPs were employed for a wide range of applications, including catalysis [57], hyperthermia [58], biomedical imaging [59], drug delivery [29] and protein separation [60]. Thereby, the most desirable applications will be discussed (Figure 5).

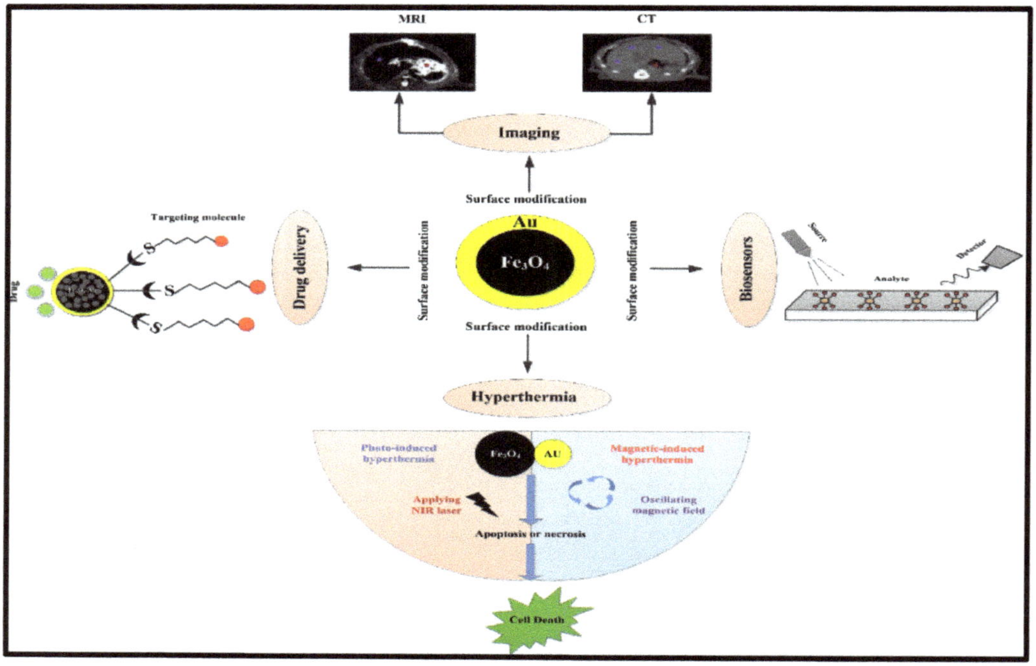

Figure 5. Schematic representation of the medical applications of Fe_3O_4@Au NPs [5].

Izadiyan et al. [61] documented the construction of Fe_3O_4@Au NPs using a modern two-step synthesis technique made of green husk extract from Juglans regia. Their analysis shows that Fe_3O_4/Au NPs' structure, physical and chemical properties exhibit Fe_3O_4 and Au's intrinsic features. The Fe_3O_4@Au NPs display 235 µg/mL of inhibitory concentration $(IC)_{50}$ against colorectal cancer cells (HT-29). Once measured against non-cancer cells, not even up to 500 µg/mL of IC_{50} was obtained. This result exhibited the promising properties of Fe_3O_4@Au NPs for cancer treatment and different biomedical applications. Au shell coating over Fe_3O_4 NPs provides an appropriate platform for adequate modification via therapeutic agents, which is one of the main challenges for the use of Fe_3O_4@Au NPs through cancer therapy. Cancer cells lack the necessary heat-shock reaction and therefore start dying before normal cells when the temperature of the tissue is above 42 °C, and the time necessary to achieve the therapeutic temperature was indeed faster for Fe_3O_4@Au NPs than for naked Fe_3O_4 NPs [36].

Zhao et al. [62] reported the production of Fe_3O_4@Au HNPs at room temperature, which concurrently improved X-ray attenuation as well as showed fluorescence and magnetic properties. Findings from the in vitro fluorescence experiment revealed that the NPs were extremely photostatic and could prevent endosome degradation in cells. Additionally, the in vivo study of normal mice showed 34.61 times more contrast under MR guidance 15 min after the administration of the Fe_3O_4@Au HNPs. The most elevated Hounsfield unit (HU) stood at 174 for 30 min after injections of Fe_3O_4@Au HNPs by CT. In vivo studies of Fe_3O_4@Au HNPs in rat models carrying three different viral infections were further evaluated. For fatty liver models, almost constant contrast improvement was observed without focus dysfunction or nodules under CT and MR (72 HU) and (the highest contrast ratio was 47.33). At the same time, the pronounced enhancement of HCC and cirrhotic liver under CT and MR guidance might be observed in liver parenchyma following Fe_3O_4@Au HNPs injection with highlighted lesions. In addition, the biochemical, hematological and pathological analysis revealed a lack of chronic and acute toxicity and demonstrated the biocompatibility of Fe_3O_4@Au HNPs in vivo applications. These Fe_3O_4@Au HNPs have shown great potential as a bio-image and multi-modality candidate. Recently, our team recorded Fe_3O_4@Au NPs developed through sonochemical production for MR and CT imaging [63]. The Fe_3O_4 NPs were produced by co-precipitation, followed by the reduction of a gold ion on the Fe_3O_4 surface utilizing a simple and rapid sonochemical process, in just 10 min. Viability testing for a human embryonic kidney cell line (HEK-293) with various doses (100 to 500 Fe μg/mL) for Fe_3O_4 and also Fe_3O_4@Au was performed for various incubation periods (24, 48 and 72 h). Significant reduction in the viability of HEK-293 cells could indeed be identified through an increase in the NPs dose. HEK-293 cells were cultivated with various concentrations of coating NPs (Fe_3O_4@Au), which were higher than that of bare Fe_3O_4 due to the biocompatibility properties of the Au shell. This result means that the Au shells could decrease the toxicity of Fe_3O_4 [64]. Fe_3O_4@Au NPs were first distributed as a control sample in various agar gel concentrations (0.1 to 0.5 mg) using a simple agar gel (Figure 6a). The brightness of the Fe_3O_4@Au NPs MRI images reduces if the dose increases, leading to a decrease in the MRI signal strength via the increasing Fe dose [65]. Transverse relaxivity (r_2) is typically used as a contrast agent to measure the effectiveness of Fe_3O_4. Illustration 6 (b) provides a relaxation rate (T_2) as a variable of the Fe_3O_4@Au NP dose in which T_2 linearly increases during the increase in the Fe dose with an r_2 slope value of about 222.28 mM^{-1} s^{-1} (Table 2). Fe_3O_4@Au NPs' r_2 value is high, probably due to the water protons that can be obtained at the Fe_3O_4 surface of the shell during the interstitial spaces of Au shells. The result of the sensitivity supports the possible use of Fe_3O_4@Au NPs in MRI applications as a T_2-shortening agent. The X-ray attenuation of various concentrations of Fe_3O_4@Au NPs has been studied, employing agar as a sample group to assess the potential of the use of Fe_3O_4@Au NPs as a contrast agent for CT (Figure 6c). The sensitivity of the CT picture improves with the concentration of Au. Illustration 6 (d) exhibits that the Fe_3O_4@Au NPs' CT value (HU) gradually increases with the concentration of Au shell (HU = 418) [66]. This report reveals that the attenuation rate of Fe_3O_4@Au under parallel concentrations of iodine is significantly higher than Omnipaque. This reduction was consistent with an earlier report [66]. Sun et al. stated that because of their higher surface-to-volume ratio, ultrafine Au shells demonstrate higher X-ray attenuation compared to their larger equivalents [67]. This function is imperative since Fe_3O_4@Au NPs' strong X-ray attenuation capability is a prerequisite for their future utilization as a CT contrast agent. In vitro results (r_2 and HU) support the efficacy of Fe_3O_4@Au in MR and CT imaging. In general, Fe_3O_4@Au NPs' MRI contrast influence depends on Fe_3O_4 concentration, whereas the Au shell serves an essential function via the X-ray attenuation of CT imaging.

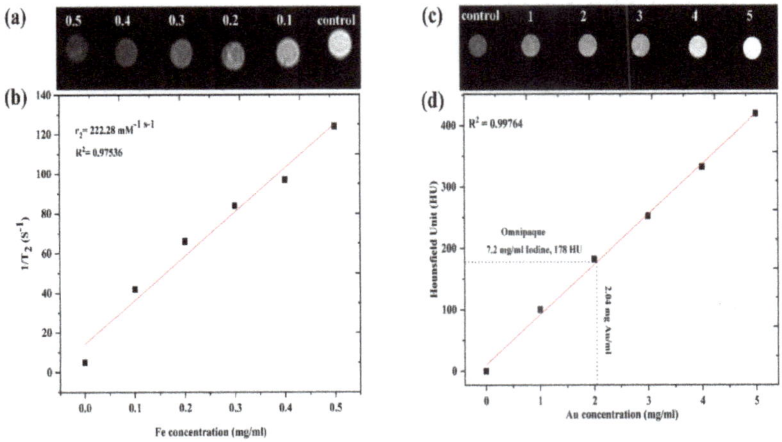

Figure 6. (**a**) MR images with various Fe doses; (**b**) T_2 linear fitting of Fe_3O_4@Au NPs; (**c**) CT images of various Au doses; and (**d**) the intensity of X-ray attenuation [63].

Table 2. Summary of the recently published studies on medical applications of Fe_3O_4@Au NPs.

No.	Nanoparticles Type	Application	Results	Ref
1	Core–shell Fe_3O_4/Au	Anticancer	The Fe_3O_4@Au NPs display 235 μg/mL of inhibitory concentration $(IC)_{50}$ against colorectal cancer cells (HT-29).	[61]
2	Fe_3O_4@Au HNPs	CT-MR dual-modality contrast agents	In vitro phantom studies revealed that these NPs provided superior contrast enhancement for CT and MR imaging.	[62]
3	Fe_3O_4@Au NPs	MRI and CT imaging	The in vitro findings (r_2 = 222.28 mM^{-1} s^{-1}, HU = 418) substantiate the effectiveness of Fe_3O_4@Au NPs in MRI and CT imaging.	[63]
4	Fe_3O_4@Au NPs	Photothermal therapy	The findings demonstrated that Fe_3O_4@Au NPs have the ability to be used as a phototherapeutic agent to enhance the eradication of breast cancer cells.	[68]
5	Fe_3O_4@Au core/shell	Biosensors	Fe_3O_4@Au NPs as new multiplex biosensors of real laboratory testing in complex matrices.	[69]
6	Spiky Fe_3O_4@Au NPs	Theranosticagents	The serum biochemistry results showed that the spiky Fe_3O_4@Au NPs had no discernible toxicity in vivo and could not accurately depict liver and kidney failure.	[70]
7	Fe_3O_4@Au NPs	Dual-modal imaging	The Fe_3O_4@Au NPs proved to be a successful candidate to image tumors for Vivo PA/MR through intravenous injection.	[71]
8	Fe_3O_4@Au NPs	Antibacterial study	Fe_3O_4@Au NPs revealed good antibacterial activity against Gram-positive and Gram-negative pathogens which are found in water.	[72]

Mohammed et al. [68] announced that the sonochemical method successfully synthesized Fe_3O_4@Au with a mean size of 20.8 nm. Fe_3O_4@Au NPs demonstrated slight toxicity to MCF-7 cell lines within 24 h, even with the maximum NPs concentration. The laser irradiation time, power, and wavelength used to treat both cells and NPs were 10 min, 200 mW and 808 nm, respectively. Cell viability decreased dramatically after treatment with 50 μg Fe/mL Fe_3O_4@Au NPs. The findings in this study conclude that Fe_3O_4@Au NPs have the ability to be used as a phototherapeutic agent to improve breast cancer treatment. Fe_3O_4@Au NPs were designed for a plasmon signal enhancement label for nucleotide and serum marker combined detection by Premaratne et al. [69]. The Fe_3O_4@Au NPs' integrated plasmon and magnetic enhancement features proved capable of quickly and magnetically separating the detection-attached sensors and magnifying the SPR signal's performance whilst reducing the non-particular signals of a serum matrix. Such features enhanced the assay's dynamics as well as its selectivity and sensitivity. With the recently developed emphasis on in vitro diagnostic imaging for painless/non-invasive disease and abnormality detection, results showed Fe_3O_4@Au NPs to be new multiplex biosensors of real laboratory testing in complex matrices. Spiky Fe_3O_4@Au NPs were proven to be efficacious theranostic agents in photothermal treatment, a drug-targeted delivery and genetic transmission system [70]. The clearance, biocompatibility and biodistribution of the spiky Fe_3O_4@Au were studied in mice. The organ distributions revealed that the intravenously administered spiky Fe_3O_4@Au NPs were mainly accumulated in the spleen and liver, and the size of the particles significantly affected their actions in vivo. The biochemistry and electron transmission microscopy serum of ultra-histologic structures revealed that spiky Fe_3O_4@Au NPs had no significant in vivo toxicity and did not present a potential risk of kidney and liver dysfunction. Such results lay the groundwork for the development of future theranostic agents. Kang et al. [71] studied the dual-mode imaging of Fe_3O_4@Au NPs as contrast agents for magnetic resonance (MR) and photoacoustic (PA) imaging. MR imaging offers a time-dependent location for the tumor, while PA imaging demonstrates the presence of high-resolution blood vessels within the tumor. The Fe_3O_4@Au NPs display a greater value of r_2—approximately 329 mM^{-1} s^{-1}. The Fe_3O_4@Au NPs were also added to the tumor-bearing mice of LNCaP as a successful candidate to image tumors for Vivo PA/MR through intravenous injection. MR/PA imagery results in the tumor area show a substantially improved MR/PA image. In multi-modal imaging, the prepared Fe_3O_4/Au NPs will be widely applied. Fe_3O_4@Au NPs were synthesized through the chemical reduction approach [72]. TEM analysis revealed the production of Fe_3O_4@Au with a mean size of approximately 18 nm. The decreased size allows these Fe_3O_4@Au NPs to effectively infiltrate the bacterial crust, resulting in membrane reliability failure. These Fe_3O_4@Au NPs indicate high antibacterial activity in the water against Gram-positive and Gram-negative pathogens. The result achieved showed that the Fe_3O_4@Au NPs were a strong antibacterial agent. Using invented Fe_3O_4@Au NPs in the medical industry is still challenging because the results of clinical trials have yet to be released.

4. Conclusions and Challenges

Fe_3O_4@Au NPs provide numerous possibilities for a powerful platform for medical applications due to their special optical and magnetic properties. Owing to advances in synthesis methods, various forms of Fe_3O_4@Au NPs such as core@shell NPs, core@shell HNPs, core@satellite NPs and dumbbell NPs have recently been explored. The physicochemical characteristics of Fe_3O_4@Au were controlled by manipulating each NP in terms of composition, size, shape and interparticle correlations according to their needs. Fe_3O_4@Au NPs have been commonly regarded as therapeutic agents for various uses due to various functional materials, including catalysis, hyperthermia, biomedical imaging, drug delivery, and protein separation. Nonetheless, the use of Fe_3O_4@Au NPs as a medical agent is still in its infancy and is faced with many doubts and challenges. It is very challenging to develop more effective, smart and secure Fe_3O_4@Au NPs for medical applications. While several Fe_3O_4@Au NPs have been established, translating these components into real clinical applications has not yet been carried out. To address these drawbacks, efforts should be made to produce Fe_3O_4@Au NPs, where each functionality performs in a combined way without affecting other features and functionality. In addition, these components should be precisely applied to long-term toxicity investigations, biodistribution evaluation and several other preclinical tests. Despite these challenges, medical applications based on Fe_3O_4@Au NPs will indeed find real-time applications due to their special features. Collective efforts from researchers from multidisciplinary backgrounds can enhance the success of using Fe_3O_4@Au HNPs as a medical agent.

Author Contributions: Conceptualization, M.A.D. and A.A.A.; methodology, M.A.D., M.S.J. and P.M.K.; software, M.A.D.; validation, M.A.D. and A.A.A.; formal analysis, M.A.D., M.S.J. and P.M.K.; investigation, M.A.D.; resources, M.A.D.; data curation, M.A.D., M.S.J. and P.M.K.; writing—original draft preparation, M.A.D.; writing—review and editing, M.A.D. and A.A.A.; visualization, M.A.D.; supervision, A.A.A.; project administration, A.A.A.; funding acquisition, A.A.A. All authors have read and agreed to the published version of the manuscript.

Funding: This work was funded by the Malaysian Ministry of Higher Education FRGS grant and the APC was funded by (203/PFIZIK/6711678).

Acknowledgments: We would like to thank the academics, authors and researchers at Universiti Sains Malaysia for their efforts which contributed to making this research work readable and the authors would also like to thank the School of Physics of Universiti Sains Malaysia for supporting this research work.

Conflicts of Interest: The authors declare no conflict of interest.

References

1. Dheyab, M.A.; Aziz, A.A.; Jameel, M.S.; Khaniabadi, P.M.; Mehrdel, B. Mechanisms of effective gold shell on Fe_3O_4 core nanoparticles formation using sonochemistry method. *Ultrason. Sonochem.* **2020**, *64*, 104865. [CrossRef]
2. Nasrabadi, H.T.; Abbasi, E.; Davaran, S.; Kouhi, M.; Akbarzadeh, A. Bimetallic nanoparticles: Preparation, properties, and biomedical applications. *Artif. Cell Nanomed. B.* **2016**, *44*, 376–380. [CrossRef]
3. Gawande, M.B.; Goswami, A.; Asefa, T.; Guo, H.; Biradar, A.V.; Peng, D.-L.; Zboril, R.; Varma, R.S. Core–shell nanoparticles: Synthesis and applications in catalysis and electrocatalysis. *Chem. Soc. Rev.* **2015**, *44*, 7540–7590. [CrossRef] [PubMed]
4. Khatami, M.; Alijani, H.Q.; Nejad, M.S.; Varma, R.S. Core@shell Nanoparticles: Greener Synthesis Using Natural Plant Products. *Appl. Sci.* **2018**, *8*, 411. [CrossRef]
5. Dheyab, M.A.; Aziz, A.A.; Jameel, M.S.; Abu Noqta, O.; Mehrdel, B. Synthesis and coating methods of biocompatible iron oxide/gold nanoparticle and nanocomposite for biomedical applications. *Chin. J. Phys.* **2020**, *64*, 305–325. [CrossRef]
6. Jiang, H.; Zeng, X.; He, N.; Deng, Y.; Lu, G.; Li, K. Preparation and biomedical applications of gold-coated magnetic nanocomposites. *J. Nanosci. Nanotechnol.* **2013**, *13*, 1617–1625. [CrossRef]
7. Coricovac, D.-E.; Moacă, E.-A.; Pinzaru, I.; Cîtu, C.; Soica, C.; Mihali, C.-V.; Păcurariu, C.; Tutelyan, V.A.; Tsatsakis, A.; Dehelean, C.-A. Biocompatible Colloidal Suspensions Based on Magnetic Iron Oxide Nanoparticles: Synthesis, Characterization and Toxicological Profile. *Front. Pharmacol.* **2017**, *8*, 154. [CrossRef] [PubMed]
8. Tran, N.; Webster, T.J. Magnetic nanoparticles: Biomedical applications and challenges. *J. Mater. Chem.* **2010**, *20*, 8760–8767. [CrossRef]
9. Shete, P.; Patil, R.; Tiwale, B.; Pawar, S. Water dispersible oleic acid-coated Fe_3O_4 nanoparticles for biomedical applications. *J. Magn. Magn. Mater.* **2015**, *377*, 406–410. [CrossRef]
10. Sousa, J.B.; Ramos-Jesus, J.; Silva, L.; Pereira, C.; De-Los-Santos-Álvarez, N.; Fonseca, R.A.; Miranda-Castro, R.; Delerue-Matos, C.; Júnior, J.R.S.; Barroso, M.F. Fe_3O_4@Au nanoparticles-based magnetoplatform for the HMGA maize endogenous gene electrochemical genosensing. *Talanta* **2020**, *206*, 120220. [CrossRef]
11. Smith, M.; McKeague, M.; DeRosa, M.C. Synthesis, transfer, and characterization of core-shell gold-coated magnetic nanoparticles. *MethodsX* **2019**, *6*, 333–354. [CrossRef]
12. Shiji, R.; Joseph, M.M.; Unnikrishnan, B.; Preethi, G.; Sreelekha, T. Fluorescent gold nanoclusters as a powerful tool for sensing applications in cancer management. In *Advances in Biomaterials for Biomedical Applications*; Springer: Singapore, 2017; pp. 385–428.
13. Rabeea, M.A.; Owaid, M.N.; Aziz, A.A.; Jameel, M.S.; Dheyab, M.A. Mycosynthesis of gold nanoparticles using the extract of Flammulina velutipes, Physalacriaceae, and their efficacy for decolorization of methylene blue. *J. Environ. Chem. Eng.* **2020**, *8*, 103841. [CrossRef]
14. Dheyab, M.A.; Owaid, M.N.; Rabeea, M.A.; Aziz, A.A.; Jameel, M.S. Mycosynthesis of gold nanoparticles by the Portabello mushroom extract, Agaricaceae, and their efficacy for decolorization of Azo dye. *Environ. Nanotechnol. Monit. Manag.* **2020**, *14*, 100312. [CrossRef]
15. Owaid, M.N.; Rabeea, M.A.; Aziz, A.A.; Jameel, M.S.; Dheyab, M.A. Mushroom-assisted synthesis of triangle gold nanoparticles using the aqueous extract of fresh Lentinula edodes (shiitake), Omphalotaceae. *Environ. Nanotechnol. Monit. Manag.* **2019**, *12*, 100270. [CrossRef]
16. Rajkumar, S.; Prabaharan, M. Theranostics Based on Iron Oxide and Gold Nanoparticles for Imaging-Guided Photothermal and Photodynamic Therapy of Cancer. *Curr. Top. Med. Chem.* **2017**, *17*, 1858–1871. [CrossRef]
17. Salihov, S.V.; Ivanenkov, Y.A.; Krechetov, S.P.; Veselov, M.; Sviridenkova, N.V.; Savchenko, A.G.; Klyachko, N.L.; Golovin, Y.I.; Chufarova, N.V.; Beloglazkina, E.K.; et al. Recent advances in the synthesis of Fe_3O_4@Au core/shell nanoparticles. *J. Magn. Magn. Mater.* **2015**, *394*, 173–178. [CrossRef]
18. Sabale, S.; Kandesar, P.; Jadhav, V.; Komorek, R.; Motkuri, R.K.; Yu, X.-Y. Recent developments in the synthesis, properties, and biomedical applications of core/shell superparamagnetic iron oxide nanoparticles with gold. *Biomater. Sci.* **2017**, *5*, 2212–2225. [CrossRef]
19. Sun, S.-N.; Wei, C.; Zhu, Z.-Z.; Hou, Y.-L.; Venkatraman, S.S.; Xu, Z.-C. Magnetic iron oxide nanoparticles: Synthesis and surface coating techniques for biomedical applications. *Chin. Phys. B* **2014**, *23*, 037503. [CrossRef]
20. He, H.; Sun, D.-W.; Pu, H.; Huang, L. Bridging Fe_3O_4@Au nanoflowers and Au@Ag nanospheres with aptamer for ultrasensitive SERS detection of aflatoxin B. *Food Chem.* **2020**, *324*, 126832. [CrossRef]
21. Xie, Y.; Chen, T.; Guo, Y.; Cheng, Y.; Qian, H.; Yao, W. Rapid SERS detection of acid orange II and brilliant blue in food by using Fe_3O_4@Au core–shell substrate. *Food Chem.* **2019**, *270*, 173–180. [CrossRef]
22. Singh, N.; Nayak, J.; Sahoo, S.K.; Kumar, R. Glutathione conjugated superparamagnetic Fe_3O_4-Au core shell nanoparticles for pH controlled release of DOX. *Mater. Sci. Eng. C* **2019**, *100*, 453–465. [CrossRef]
23. Dheyab, M.A.; Aziz, A.A.; Jameel, M.S. Synthesis and optimization of the sonochemical method for functionalizing gold shell on Fe_3O_4 core nanoparticles using response surface methodology. *Surf. Interfaces* **2020**, *21*, 100647. [CrossRef]
24. Dheyab, M.A.; Aziz, A.A.; Khaniabadi, P.M.; Jameel, M.S. Potential of a sonochemical approach to generate MRI-PPT theranostic agents for breast cancer. *Photodiagn. Photodyn. Ther.* **2021**, *33*, 102177. [CrossRef] [PubMed]
25. Xu, H.; Zeiger, B.; Suslick, K. Sonochemical synthesis of nanomaterials. *Chem. Soc. Rev.* **2013**, *42*, 2555–2567. [CrossRef] [PubMed]
26. Pokhrel, N.; Vabbina, P.K.; Pala, N. Sonochemistry: Science and Engineering. *Ultrason. Sonochem.* **2016**, *29*, 104–128. [CrossRef] [PubMed]

27. Dheyab, M.A.; Aziz, A.A.; Khaniabadi, P.M.; Jameel, M.S.; Ahmed, N.M.; Ali, A.T. Distinct advantages of using sonochemical over laser ablation methods for a rapid-high quality gold nanoparticles production. *Mater. Res. Express* **2021**, *8*, 015009. [CrossRef]
28. Mirsadeghi, S.; Zandavar, H.; Yousefi, M.; Rajabi, H.R.; Pourmortazavi, S.M. Green-photodegradation of model pharmaceutical contaminations over biogenic Fe_3O_4/Au nanocomposite and antimicrobial activity. *J. Environ. Manag.* **2020**, *270*, 110831. [CrossRef] [PubMed]
29. Rasouli, E.; Basirun, W.J.; Johan, M.R.; Rezayi, M.; Darroudi, M.; Shameli, K.; Shanavaz, Z.; Akbarzadeh, O.; Izadiyan, Z. Facile and greener hydrothermal honey-based synthesis of Fe_3O_4/Au core/shell nanoparticles for drug delivery applications. *J. Cell. Biochem.* **2019**, *120*, 6624–6631. [CrossRef] [PubMed]
30. Tarhan, T.; Ulu, A.; Sariçam, M.; Çulha, M.; Ates, B. Maltose functionalized magnetic core/shell Fe_3O_4@Au nanoparticles for an efficient L-asparaginase immobilization. *Int. J. Biol. Macromol.* **2020**, *142*, 443–451. [CrossRef] [PubMed]
31. Butmee, P.; Tumcharern, G.; Thouand, G.; Kalcher, K.; Samphao, A. An ultrasensitive immunosensor based on manganese dioxide-graphene nanoplatelets and core shell Fe_3O_4@Au nanoparticles for label-free detection of carcinoembryonic antigen. *Bioelectrochemistry* **2020**, *132*, 107452. [CrossRef]
32. Kou, Y.; Wu, T.; Xing, G.; Huang, X.; Han, D.; Yang, S.; Guo, C.; Gao, W.; Yang, J.; Liu, Y.; et al. Highly efficient and recyclable catalyst: Porous Fe_3O_4–Au magnetic nanocomposites with tailored synthesis. *Nanotechnology* **2020**, *31*, 225701. [CrossRef]
33. Al-Sherbini, A.-S.A.; El-Ghannam, G.; Yehya, H.; Nassef, O.A. Optical and Magnetic Studies of Fe_3O_4/Au Core/Shell Nanocomposites. *Int. J. Nanosci.* **2019**, *18*, 1850033. [CrossRef]
34. Ma, C.; Shao, H.; Zhan, S.; Hou, P.; Zhang, X.; Chai, Y.; Liu, H. Bi-phase dispersible Fe_3O_4@Au core–shell multifunctional nanoparticles: Synthesis, characterization and properties. *Compos. Interfaces* **2019**, *26*, 537–549. [CrossRef]
35. Muniz-Miranda, M.; Muniz-Miranda, F.; Giorgetti, E. Spectroscopic and Microscopic Analyses of Fe_3O_4/Au Nanoparticles Obtained by Laser Ablation in Water. *Nanomaterials* **2020**, *10*, 132. [CrossRef]
36. Ángeles-Pascual, A.; Piñón-Hernández, J.; Estevez-González, M.; Pal, U.; Velumani, S.; Pérez, R.; Esparza, R. Structure, magnetic and cytotoxic behaviour of solvothermally grown Fe_3O_4@Au core-shell nanoparticles. *Mater. Charact.* **2018**, *142*, 237–244. [CrossRef]
37. Billen, A.; de Cattelle, A.; Jochum, J.K.; Van Bael, M.J.; Billen, J.; Seo, J.W.; Brullot, W.; Koeckelberghs, G.; Verbiest, T. Novel synthesis of superparamagnetic plasmonic core-shell iron oxide-gold nanoparticles. *Phys. B Condens. Matter* **2019**, *560*, 85–90. [CrossRef]
38. Park, S.-I.; Chung, S.-H.; Kim, H.-C.; Lee, S.G.; Lee, S.J.; Kim, H.; Kim, H.; Jeong, S.W. Prolonged heating of Fe_3O_4–Au hybrid nanoparticles in a radiofrequency solenoid coil. *Colloids Surf. A* **2018**, *538*, 304–309. [CrossRef]
39. Xu, K.; Wu, J.; Fang, Q.; Bai, L.; Duan, J.; Li, J.; Xu, H.; Hui, A.; Hao, L.; Xuan, S. Magnetically separable h-Fe_3O_4@Au/polydopamine nanosphere with a hollow interior: A versatile candidate for nanocatalysis and metal ion adsorption. *Chem. Eng. J.* **2020**, *398*, 125571. [CrossRef]
40. Chen, S.; Qiu, S.; Zhong, M.; Tian, D.; Wang, C.; Lu, X. Constructing magnetic Fe_3O_4-Au@CeO_2 hybrid nanofibers for selective catalytic degradation of organic dyes. *Appl. Organomet. Chem.* **2019**, *33*, e5253. [CrossRef]
41. Efremova, M.V.; Nalench, Y.A.; Myrovali, E.; Garanina, A.; Grebennikov, I.S.; Gifer, P.K.; Abakumov, M.; Spasova, M.; Angelakeris, M.; Savchenko, A.G.; et al. Size-selected Fe_3O_4–Au hybrid nanoparticles for improved magnetism-based theranostics. *Beilstein J. Nanotechnol.* **2018**, *9*, 2684–2699. [CrossRef]
42. Wang, W.; Hao, C.; Sun, M.; Xu, L.; Xu, C.; Kuang, H. Spiky Fe_3O_4@Au Supraparticles for Multimodal In Vivo Imaging. *Adv. Funct. Mater.* **2018**, *28*, 1800310. [CrossRef]
43. Li, N.; Zhong, T.; Liu, J.-L.; Zheng, J.; Deng, H.-C.; Zhou, W.; Li, M.; Feng, M.-S.; Liu, Q.-J.; Li, C.-Y.; et al. Precise Identification and Analysis of Micro/Nano-Sized Pore Structure in Shale with Fe_3O_4/Au Hybrid Nanocomposite. *Anal. Chem.* **2018**, *90*, 12706–12713. [CrossRef]
44. Liu, Y.; Zhang, Y.; Kou, Q.; Wang, D.; Han, D.; Lu, Z.; Chen, Y.; Chen, L.; Wang, Y.; Yang, J.; et al. Fe_3O_4/Au binary nanocrystals: Facile synthesis with diverse structure evolution and highly efficient catalytic reduction with cyclability characteristics in 4-nitrophenol. *Powder Technol.* **2018**, *338*, 26–35. [CrossRef]
45. Song, R.-B.; Zhou, S.; Guo, D.; Li, P.; Jiang, L.-P.; Zhang, J.-R.; Wu, X.; Zhu, J.-J. Core/Satellite Structured Fe_3O_4/Au Nanocomposites Incorporated with Three-Dimensional Macroporous Graphene Foam as a High-Performance Anode for Microbial Fuel Cells. *ACS Sustain. Chem. Eng.* **2019**, *8*, 1311–1318. [CrossRef]
46. Klein, S.; Hübner, J.; Menter, C.; Distel, L.V.R.; Neuhuber, W.; Kryschi, C. A Facile One-Pot Synthesis of Water-Soluble, Patchy Fe_3O_4-Au Nanoparticles for Application in Radiation Therapy. *Appl. Sci.* **2019**, *9*, 15. [CrossRef]
47. Kostevsek, N.; Locatelli, E.; Garrovo, C.; Arena, F.; Monaco, I.; Nikolov, I.P.; Sturm, S.; Rozman, K.Z.; Lorusso, V.; Giustetto, P.; et al. The one-step synthesis and surface functionalization of dumbbell-like gold–iron oxide nanoparticles: A chitosan-based nanotheranostic system. *Chem. Commun.* **2016**, *52*, 378–381. [CrossRef] [PubMed]
48. Silva, S.M.; Tavallaie, R.; Sandiford, L.; Tilley, R.; Gooding, J.J. Gold coated magnetic nanoparticles: From preparation to surface modification for analytical and biomedical applications. *Chem. Commun.* **2016**, *52*, 7528–7540. [CrossRef]
49. Sood, A.; Arora, V.; Shah, J.; Kotnala, R.; Jain, T.K. Multifunctional gold coated iron oxide core-shell nanoparticles stabilized using thiolated sodium alginate for biomedical applications. *Mater. Sci. Eng. C* **2017**, *80*, 274–281. [CrossRef]
50. Leung, K.C.-F.; Xuan, S.; Zhu, X.; Wang, D.; Chak, C.-P.; Lee, S.-F.; Ho, W.K.-W.; Chung, B.C.-T. Gold and iron oxide hybrid nanocomposite materials. *Chem. Soc. Rev.* **2012**, *41*, 1911–1928. [CrossRef]

51. Huang, Q.; Li, W.; Lin, Q.; Pi, D.; Hu, C.; Shao, C.; Zhang, H. A review of significant factors in the synthesis of hetero-structured dumbbell-like nanoparticles. *Chin. J. Catal.* **2016**, *37*, 681–691. [CrossRef]
52. Liu, J.; Zhang, W.; Zhang, H.-L.; Yang, Z.; Li, T.; Wang, B.; Huo, X.; Wang, R.; Chen, H. A multifunctional nanoprobe based on Au–Fe_3O_4 nanoparticles for multimodal and ultrasensitive detection of cancer cells. *Chem. Commun.* **2013**, *49*, 4938. [CrossRef]
53. Wang, C.; Xu, C.; Zeng, H.; Sun, S. Recent Progress in Syntheses and Applications of Dumbbell-like Nanoparticles. *Adv. Mater.* **2009**, *21*, 3045–3052. [CrossRef]
54. Gupta, A.K.; Naregalkar, R.R.; Vaidya, V.D.; Gupta, M. Recent advances on surface engineering of magnetic iron oxide nanoparticles and their biomedical applications. *Nanomedicine* **2007**, *2*, 23–39. [CrossRef]
55. Zhao, X.; Zeng, L.; Hosmane, N.; Gong, Y.; Wu, A. Cancer cell detection and imaging: MRI-SERS bimodal splat-shaped Fe_3O_4/Au nanocomposites. *Chin. Chem. Lett.* **2019**, *30*, 87–89. [CrossRef]
56. Rajkumar, S.; Prabaharan, M. Multi-functional core-shell Fe_3O_4@Au nanoparticles for cancer diagnosis and therapy. *Colloids Surf. B* **2019**, *174*, 252–259.
57. Sarno, M.; Iuliano, M. Highly active and stable Fe_3O_4/Au nanoparticles supporting lipase catalyst for biodiesel production from waste tomato. *Appl. Surf. Sci.* **2019**, *474*, 135–146. [CrossRef]
58. Ge, Y.; Zhong, Y.; Ji, G.; Lu, Q.; Dai, X.; Guo, Z.; Zhang, P.; Peng, G.; Zhang, K.; Li, Y. Preparation and characterization of Fe_3O_4@Au-C225 composite targeted nanoparticles for MRI of human glioma. *PLoS ONE* **2018**, *13*, e0195703. [CrossRef]
59. Ding, L.; Wang, R.; Hu, Y.; Xu, F.; Zhang, N.; Cao, X.; Wang, X.; Shi, X.; Guo, R. Folic acid-modified Laponite®-stabilized Fe_3O_4 nanoparticles for targeted T2-weighted MR imaging of tumor. *Appl. Clay Sci.* **2020**, *186*, 105447. [CrossRef]
60. Karami, C.; Taher, M.A. A catechol biosensor based on immobilizing laccase to Fe_3O_4@Au core-shell nanoparticles. *Int. J. Biol. Macromol.* **2019**, *129*, 84–90. [CrossRef]
61. Izadiyan, Z.; Shameli, K.; Miyake, M.; Teow, S.-Y.; Peh, S.-C.; Mohamad, S.E.; Taib, S.H.M. Green fabrication of biologically active magnetic core-shell Fe_3O_4/Au nanoparticles and their potential anticancer effect. *Mater. Sci. Eng. C* **2019**, *96*, 51–57. [CrossRef]
62. Zhao, H.Y.; Liu, S.; He, J.; Pan, C.C.; Li, H.; Zhou, Z.Y.; Ding, Y.; Huo, D.; Hu, Y. Synthesis and application of strawberry-like Fe_3O_4-Au nanoparticles as CT-MR dual-modality contrast agents in accurate detection of the progressive liver disease. *Biomaterials* **2015**, *51*, 194–207. [CrossRef] [PubMed]
63. Dheyab, M.A.; Aziz, A.A.; Jameel, M.S.; Abu Noqta, O.; Khaniabadi, P.M.; Mehrdel, B. Excellent relaxivity and X-ray attenuation combo properties of Fe_3O_4@Au CSNPs produced via Rapid sonochemical synthesis for MRI and CT imaging. *Mater. Today Commun.* **2020**, *25*, 101368. [CrossRef]
64. Keshtkar, M.; Shahbazi-Gahrouei, D.; Mehrgardi, M.; Aghaei, M.; Khoshfetrat, S. Synthesis and Cytotoxicity Assessment of Gold-coated Magnetic Iron Oxide Nanoparticles. *J. Biomed. Phys. Eng.* **2018**, *8*, 357–364.
65. Nosrati, H.; Salehiabar, M.; Kheiri Manjili, H.; Davaran, S.; Danafar, H. Theranostic nanoparticles based on magnetic nanoparticles: Design, preparation, characterization, and evaluation as novel anticancer drug carrier and MRI contrast agent. *Drug Dev. Ind. Pharm.* **2018**, *44*, 1668–1678. [CrossRef] [PubMed]
66. Li, J.; Zheng, L.; Cai, H.; Sun, W.; Shen, M.; Zhang, G.; Shi, X. Facile One-Pot Synthesis of Fe_3O_4@Au Composite Nanoparticles for Dual-Mode MR/CT Imaging Applications. *ACS Appl. Mater. Interfaces* **2013**, *5*, 10357–10366. [CrossRef] [PubMed]
67. Xu, C.; Tung, G.A.; Sun, S. Size and Concentration Effect of Gold Nanoparticles on X-ray Attenuation As Measured on Computed Tomography. *Chem. Mater.* **2008**, *20*, 4167–4169. [CrossRef]
68. Dheyab, M.; Aziz, A.; Jameel, M.; Khaniabadi, P.; Mehrdel, B.; Khaniabadi, B. Gold-coated iron oxide nanoparticles as a potential photothermal therapy agent to enhance eradication of breast cancer cells. *J. Phys. Conf. Ser.* **2020**, *1497*, 012003. [CrossRef]
69. Premaratne, G.; Dharmaratne, A.C.; Al Mubarak, Z.H.; Mohammadparast, F.; Andiappan, M.; Krishnan, S. Multiplexed surface plasmon imaging of serum biomolecules: Fe_3O_4@Au Core/shell nanoparticles with plasmonic simulation insights. *Sens. Actuators B Chem.* **2019**, *299*, 126956. [CrossRef]
70. Zhou, H.; Oh, S.; Kim, J.E.; Zou, F.; Hwang, D.Y.; Lee, J. In Vivo Study of Spiky Fe_3O_4@Au Nanoparticles with Different Branch Lengths: Biodistribution, Clearance, and Biocompatibility in Mice. *ACS Appl. Bio Mater.* **2019**, *2*, 163–170. [CrossRef]
71. Kang, N.; Xu, D.; Han, Y.; Lv, X.; Chen, Z.; Zhou, T.; Ren, L.; Zhou, X. Magnetic targeting core/shell Fe_3O_4/Au nanoparticles for magnetic resonance/photoacoustic dual-modal imaging. *Mater. Sci. Eng. C* **2019**, *98*, 545–549. [CrossRef]
72. Jency, D.A.; Sathyavathi, K.; Umadevi, M.; Parimaladevi, R. Enhanced bioactivity of Fe_3O_4-Au nanocomposites—A comparative antibacterial study. *Mater. Lett.* **2020**, *258*, 126795. [CrossRef]

nanomaterials

MDPI

Article

Evaluating a Human Ear-Inspired Sound Pressure Amplification Structure with Fabry–Perot Acoustic Sensor Using Graphene Diaphragm

Cheng Li [1,2,*], Xi Xiao [1], Yang Liu [1] and Xuefeng Song [3]

[1] School of Instrumentation Science and Opto-Electronics Engineering, Beihang University, Beijing 100191, China; xiaoxi@buaa.edu.cn (X.X.); sy2017328@buaa.edu.cn (Y.L.)
[2] Research Institute of Beihang University in Shenzhen, Shenzhen 518055, China
[3] Shenzhen Institute for Quantum Science and Engineering, Southern University of Science and Technology, Shenzhen 518055, China; song@sustc.edu.cn
* Correspondence: licheng@buaa.edu.cn

Abstract: In order to enhance the sensitivity of a Fabry–Perot (F-P) acoustic sensor without the need of fabricating complicated structures of the acoustic-sensitive diaphragm, a mini-type external sound pressure amplification structure (SPAS) with double 10 μm thickness E-shaped diaphragms of different sizes interconnected with a 5 mm length tapered circular rod was developed based on the acoustic sensitive mechanism of the ossicular chain in the human middle ear. The influence of thickness and Young's modulus of the two diaphragms with the diameters of 15 mm and 3 mm, respectively, on the amplification ratio and frequency response were investigated via COMSOL acoustic field simulation, thereby confirming the dominated effect. Then, three kinds of dual-diaphragm schemes relating to steel and thermoplastic polyurethanes (TPU) materials were introduced to fabricate the corresponding SPASs. The acoustic test showed that the first scheme achieved a high resonant response frequency with lower acoustic amplification due to strong equivalent stiffness; in contrast, the second scheme offered a high acoustic amplification but reduced frequency range. As a result of sensitivity enhancement, adapted with the steel/TPU diaphragm structure, an optical fiber Fabry–Perot sensor using a multilayer graphene diaphragm with a diameter of 125 μm demonstrated a remarkable sensitivity of 565.3 mV/Pa @1.2 kHz due to the amplification ratio of up to ~29.9 in the range of 0.2–2.3 kHz, which can be further improved by miniaturizing structure dimension, along with the use of microstructure packaging technology.

Keywords: sound pressure amplification structure; double diaphragm scheme; Fabry–Perot sensor; graphene diaphragm; sensitivity enhancement

Citation: Li, C.; Xiao, X.; Liu, Y.; Song, X. Evaluating a Human Ear-Inspired Sound Pressure Amplification Structure with Fabry–Perot Acoustic Sensor Using Graphene Diaphragm. *Nanomaterials* **2021**, *11*, 2284. https://doi.org/10.3390/nano11092284

Academic Editors: Jihoon Lee and Ming-Yu Li

Received: 30 July 2021
Accepted: 31 August 2021
Published: 2 September 2021

1. Introduction

Acoustic pressure sensing plays an important role in applications such as environmental noise monitoring [1], photoacoustic spectroscopy [2] and human–machine interaction [3], etc. Due to the advantages including immunity to electromagnetic interference, the capability of performing remote sensing, very high resolution, fast response, and compact size [4], optical sensors have been attracting more research interests among various acoustic sensors. It is important to note that planar diaphragms are generally used to sense the sound pressure in optical sensors. To strengthen acoustic sensitivity, various diaphragm structures have been reported in recent years. The diaphragms with large areas and small thickness were generally adopted to increase the sensitivity [5]. Unfortunately, a larger ratio of diameter to thickness may lead to low resonance frequency and uneven frequency response [6–8]. Although a metal diaphragm-based sensor developed by Gaomi Wu [9] showed a sensitivity larger than 800 mV/Pa, the high sensitivity mainly resulted from the larger film area. When the film diameter was reduced to 125 μm, the diaphragm with the same area ratio would only have an extremely low sensitivity of 8 mV/Pa. Recently, the corrugated diaphragm was introduced to improve acoustic sensitivity [10–14]. For example, the average mechanical sensitivity of the corrugated silver diaphragm-based optical fiber microphone was 52 nm/Pa in the low frequency range of 63 Hz–1 kHz, which was twice that of the planar diaphragm with the same larger diameter of 2.5 mm and thickness of 200 nm [13]. In their latest work in 2020 [14], the mechanical sensitivity was increased to 82.65 nm/Pa by optimizing the depth of the corrugated diaphragm. However, fabricating a corrugated diaphragm is limited by the diaphragm material and complicated micro-nano fabrication technology, in addition to the need for a larger size diaphragm. Instead of the use of a peripherally clamped circular diaphragm, cantilever type sensors have been developed [15–18]. In Ref. [17], the mechanical sensitivity of 198.3 nm/Pa at 1 kHz was demonstrated by using a steel cantilever with the larger size of 1.8 mm $\times$ 1 mm and the thickness of 10 μm. Although the cantilever structure has the advantage of high sensitivity, the asymmetric feature makes the sensor mainly operate at close to the resonance frequency with an extremely narrow band and nonlinear response [19]. Another feasible way is to employ external acoustic amplification structures. In 2011, Eui Sung Jung [20] proposed a microphone with a spiral-type acoustic tube, which generated a resonance effect between the diaphragm and the acoustic transducer inside a case. The underwater frequency response of the microphone at 3–4 kHz was improved by approximately 20 dB. Nevertheless, this structure was specifically for sensing diaphragms with large areas, especially those with a diameter of over 10 mm. Then, in 2015, Pan Hu [21] designed a steel megaphone at the front of an Fabry–Perot (F-P) acoustic sensor, which showed the sensitivity of 56.99 mV/Pa at 4.5 kHz. However, the long megaphone was confined to short-distance sound detection. Recently, in 2018, Renxi Gao [22] proposed a Helmholtz resonator to encapsulate an optical fiber vibration sensor with multiple response peaks below 1 kHz; however, the bandwidth of each resonance peak was less than 100 Hz, meaning an extremely uneven frequency response. Therefore, an external sound pressure-enhanced structure suitable for miniaturized sensors, capable of long-distance detection in a wide frequency range, is of great significance to F-P acoustic sensors.

In reality, human organs have subtle sound sensing structures that have recently been used to improve the frequency response and sensitivity. As we know, the basilar membrane (BM) in the human cochlear is naturally designed with an asymmetric trapezoidal shape that has numerous resonance frequencies by varying the width and thickness of the BM along the cochlea spiral, thereby enabling the resonance-based sensing and frequency tuning capability [23]. For example, in 2016, Jongmoon Jang [24] reported a trapezoidal-shape triboelectric-based artificial basilar membrane (TEABM) using eight double-ended polymer beams fixed on the substrate. The beam length was in the range of 8.2–32 mm and its width was 6–8 mm. Compared with their previous artificial basement membrane with a cantilever array, this fabricated TEABM displayed multiple harmonic response spikes in the lower range of 294.8–2311 Hz, and the sensitivity to sound was improved from 0.35–1.67 mV/Pa to 1.74–13.1 mV/Pa. Then, in 2018, Jae Hyun Han [23] developed a curved-shape basilar membrane-inspired flexible piezoelectric acoustic sensor (f-PAS) using seven polymer beams with a length in the range of 0.5–1 mm on the plastic substrate. The curved-shape f-PAS exhibited a non-flat resonance response distribution composed of many discrete harmonic peaks in the range of 0.1–4 kHz. The acoustic test showed four to eight times higher sensitivity compared to the conventional condenser sensor. However, multi-channel simultaneous detection is also required for the beam array structure due to the limited effective bandwidth, therefore restricting the miniaturization and complicating the potential application because of the access of multiple acoustic probes and their matching conditioning devices. Furthermore, biological studies on the human ear have verified that external sound pressure is transmitted and then amplified from the tympanic membrane to the stapes footplate in the middle ear [25–27], wherein the middle-ear sound pressure gain can reach 23.5 dB around 1–2 kHz. As a result, the middle ear-inspired sound pressure amplification structure (SPAS) is a great option for enhancing acoustic sensitivity.

Hence, in this paper, a simple low-frequency SPAS primarily consisting of dual diaphragms of different diameters interconnected with a tapered rod was introduced to drive the adapting F-P acoustic sensor by imposing the amplified sound pressure onto the graphene diaphragm adhered on the endface of the sensor, which offers a feasible sensitivity-enhanced way without the need of modifying an F-P sensor by fabricating complicated diaphragm structures. Then, in virtue of the sound reinforcement principle of the human middle ear, the acoustic amplification performance of the developed structure was investigated for the determination of the structural parameters of SPAS by COMSOL Multiphysics simulation. Then, three types of SPASs were fabricated and further evaluated by acoustic test, which characterized the significant role of acoustic amplification for SPAS and a high acoustic sensitivity of 255.5 mV/Pa @ 1 kHz was achieved for an F-P acoustic sensor with SPAS. Compared with the external amplification structures previously reported in Refs. [20,22], the simple bionic SPAS can offer an enhanced sensitivity in a relatively wide frequency band ranging from 0.2 Hz to 2.3 kHz. It is necessary to note that, herein, the F-P acoustic sensor adapted with the SPAS employed a graphene diaphragm due to the small thickness and excellent mechanical strength of graphene [28]. Although recent research described a large-area graphene oxide (GO) diaphragm-based F-P sensor with a sensitivity of 25.84 mV/Pa [29], it is more appropriate for the SPAS to use a smaller cavity accommodating the F-P probe. Hence, the F-P acoustic sensor with graphene diaphragm was introduced in the manuscript to evaluate the SPAS performance.

2. The Model Adapted to SPAS

Figure 1 illustrates the schematic diagram of the SPAS. Since the lymphatic fluid that is full of the human cochlea would cause higher acoustic loss, an ossicular chain structure in the middle ear is capable of enhancing the input external acoustic pressure and then transfer into the inner ear for effectively matching sound passing from the low impedance of external air to the high impedance of cochlear fluid [25]. It is known that a human middle ear is made of a tympanic membrane ™, middle ear cavity, and ossicular chain which consists of the malleus, incus, and stapes [30]. Referring to the structural illustration of a human middle ear in Figure 1a, external air pressure is strengthened via large TM vibration and then lever amplification offered by malleus and incus. In this case, the amplified acoustic pressure is passed to the following inner ear by the use of the stapes. According to the acoustic signal enhancement scheme including the large area ratio between input TM and output stapes and the lever magnifying mechanism formed by malleus and incus, a mini-type SPAS of simple structure with a large area ratio of the outer film (diaphragm 1) to inner film (diaphragm 2) is proposed, as shown in Figure 1b, wherein the two films of different sizes are equivalent as the TM and stapes in Figure 1a for acoustic signal amplification. Then, a conical rod, serving as a pressure transmission medium, is fixedly connected with the two diaphragms at its upper and lower endfaces, respectively. The diaphragms attached with the rod have adhered to the supporting outer casing that also offers a guide tube for assembling an F-P probe, thereby forming two sealed air cavities (cavity 1 and cavity 2). The compressed air in cavity 2 caused by the SPAS would increase the pressure imposed on the sensitive diaphragm such as graphene suspended on the ferrule endface. As a consequence, acoustic sensitivity can be improved with the aid of the SPAS for a diaphragm-typed F-P acoustic sensor.

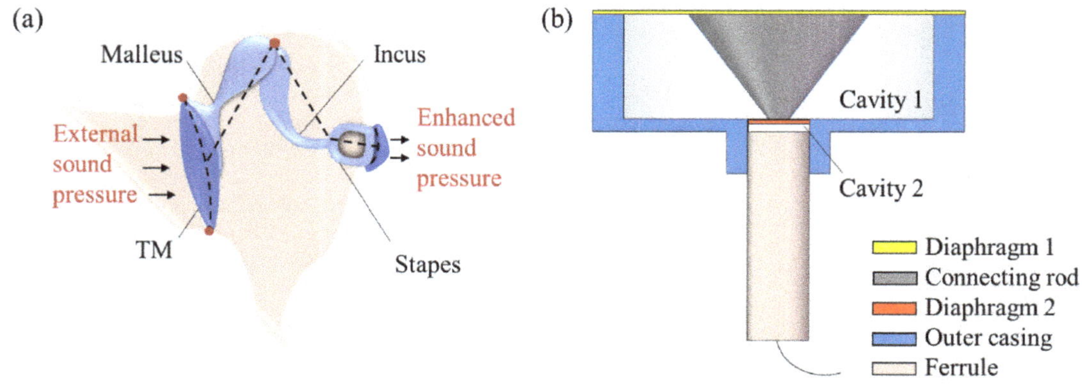

Figure 1. Schematic illustrations of (**a**) a human middle ear structure and (**b**) the presented SPAS with an F-P probe.

For the SPAS displayed in Figure 1b, an equivalent schematic diagram of the force model is depicted in Figure 2. For the sensitivity increase, the radius R_1 of the diaphragm 1 is five times larger than the radius R_2 of the diaphragm 2. In this case, considering the smaller deflection deformation of diaphragm 1 resulted from the applied lower external acoustic pressure ranging from 20 μPa to 20 Pa, the relative volume change in cavity 1 is by far lower than that in cavity 2, thus neglecting the effect of pressure change in cavity 1. For the purpose of modeling the load-deflection behavior, in view of the rigid connecting rod, the effective working part of each diaphragm is an annular region, which can be regarded as an E-shaped diaphragm under the uniform load on one side and the concentrated load on the other side. According to the small deformation theory of the E-shaped plates [31], the normal displacement $\omega_1(r)$ about the distance r from the center of diaphragm 1 under the uniform dynamic load p_1 and the central load p_{c1} can be approximated as:

$$\begin{cases} \omega_1(r) = \frac{12(1-\mu_1{}^2)}{E_1 H_1{}^3}\left[\frac{r^4}{64}p_1 - \frac{p_{c1}r_1{}^2 r^2}{8}\left(\ln\frac{r}{R_1}-1\right)\right. \\ \qquad \left. + \frac{C_1}{4}r^2 + C_2\ln\frac{r}{R_1} + C_3\right] \\ C_1 = -\frac{p_{c1}r_1{}^4\ln\frac{r_1}{R_1}}{2(R_1{}^2-r_1{}^2)} - \frac{p_{c1}r_1{}^2}{4} - \frac{p_1}{8}\left(R_1{}^2+r_1{}^2\right) \\ C_2 = \frac{p_1}{16}R_1{}^2 r_1{}^2 + \frac{p_{c1}R_1{}^2 r_1{}^4}{4(R_1{}^2-r_1{}^2)}\ln\frac{r_1}{R_1} \\ C_3 = \frac{p_1}{64}R_1{}^2\left(R_1{}^2+2r_1{}^2\right) + \frac{p_{c1}R_1{}^2 r_1{}^4}{8(R_1{}^2-r_1{}^2)}\ln\frac{r_1}{R_1} - \frac{p_{c1}R_1{}^2 r_1{}^2}{16} \end{cases} \quad (1)$$

where R_1 and H_1 are the radius and the thickness of diaphragm 1, respectively; r_1 is the radius of the upper endface of the connecting rod; E_1 and μ_1 are Young's modulus and Poisson's ratio of the diaphragm material. Similarly, the deflection $\omega_2(r)$ of diaphragm 2 under the uniform dynamic load p_2 and the central load p_{c2} can be obtained by:

$$\begin{cases} \omega_2(r) = -\frac{12(1-\mu_2{}^2)}{E_2 H_2{}^3}\left[\frac{r^4}{64}p_2 - \frac{p_{c2}r_2{}^2 r^2}{8}\left(\ln\frac{r}{R_2}-1\right)\right. \\ \qquad \left. + \frac{C_4}{4}r^2 + C_5\ln\frac{r}{R_2} + C_6\right] \\ C_4 = -\frac{p_{c2}r_2{}^4\ln\frac{r_2}{R_2}}{2(R_2{}^2-r_2{}^2)} - \frac{p_{c2}r_2{}^2}{4} - \frac{p_2}{8}\left(R_2{}^2+r_2{}^2\right) \\ C_5 = \frac{p_2}{16}R_2{}^2 r_2{}^2 + \frac{p_{c2}R_2{}^2 r_2{}^4}{4(R_2{}^2-r_2{}^2)}\ln\frac{r_2}{R_2} \\ C_6 = \frac{p_2}{64}R_2{}^2\left(R_2{}^2+2r_2{}^2\right) + \frac{p_{c2}R_2{}^2 r_2{}^4}{8(R_2{}^2-r_2{}^2)}\ln\frac{r_2}{R_2} - \frac{p_{c2}R_2{}^2 r_2{}^2}{16} \end{cases} \quad (2)$$

where R_2 and H_2 are the radius and the thickness of diaphragm 1, respectively; r_2 is the radius of the lower endface of the connecting rod; E_2 and μ_2 are Young's modulus and Poisson's ratio of the diaphragm material.

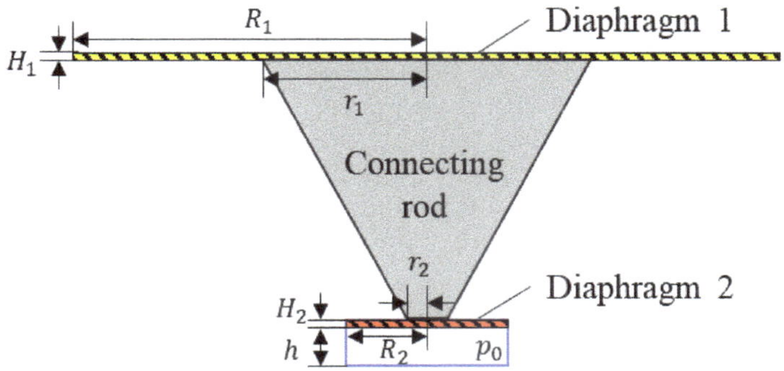

Figure 2. An equivalent schematic diagram of the force model for the presented SPAS.

Due to the rigid connection of the tapered rod with the diaphragms 1 and 2, the center deflection for the two diaphragms can be approximated as:

$$\omega_1(r_1) = \omega_2(r_2) \tag{3}$$

Then, based on the force conduction in the rod, the following relationship can be obtained by:

$$\pi {r_1}^2 p_{c1} = \pi {r_2}^2 p_{c2} \tag{4}$$

In terms of the ideal gas equation of state [32], the air pressure (p_0 + p_2) in cavity 2 at a constant room temperature can be given by:

$$p_0 + p_2 = \frac{\pi {R_2}^2 h p_0}{\pi {R_2}^2 h - \Delta V} \tag{5}$$

where p_0 and ΔV are the initial pressure (1×10^5 Pa) and the variation of volume in the cavity 2, respectively; h is the height of cavity 2. ΔV can be determined by:

$$\Delta V = \pi {r_2}^2 \omega_2(r_2) + \int_{r_2}^{R_2} 2\pi r \omega_2(r) dr \tag{6}$$

In this case, the generated dynamic acoustic load p_2 in cavity 2 can be further written by:

$$p_2 = \left[\frac{\pi {R_2}^2 h}{\pi {R_2}^2 h - \left(\pi {r_2}^2 \omega_2(r_2) + \int_{r_2}^{R_2} 2\pi r \omega_2(r) dr \right)} - 1 \right] p_0 \tag{7}$$

For a specific SPAS with known structural parameters, due to a set input load p_1, the pressure values p_2, p_{c1} and p_{c2} can be calculated by the aforementioned Equations (1)–(4) and (7). Therefore, the acoustic amplification factor K of the SPAS can be confirmed by:

$$K = \frac{p_2}{p_1} \tag{8}$$

In this way, the structural parameters of SPAS could be optimized by the mechanical amplification responsivity on basis of Equation (8).

3. Simulation on Acoustic Amplification Effect

3.1. Effect of SPAS Structural Parameters on Amplification Ratio

An investigation of the radius and thickness of diaphragm and rod as dominant structural parameters was introduced to evaluate the amplification ratio of SPAS. For simplified analysis, considering the reduced size and ease of fabrication for the SPAS, the material and thickness of the two diaphragms are identical by defining H_1 = H_2 = H. Taking a steel diaphragm as an example, $E_1 = E_2 = E = 2.07 \times 10^{11}$ Pa and $\mu_1 = \mu_2 = \mu = 0.29$. The related simulation parameters for the design of SPAS are listed in Table 1. It should be added that since the fabrication of SPAS and the connection of the amplification structure with an F-P sensor were performed at room temperature and pressure, the initial pressure in cavity 2 was assumed to be about 1×10^5 Pa in Table 1. Then, based on the established model mentioned above, Figure 3 shows the influence of structural parameters on K.

Table 1. Simulation parameters for the design of SPAS.

Structural Parameter	Value/mm	Material Parameter	Value
R_1, R_2	7.5, 1.5	E	2.07×10^{11} Pa
r_1, r_2	4.5, 0.5	μ	0.29
H	0.01	**Pressure in cavity 2**	**Value**
h	0.1	p_0	1.01×10^5 Pa

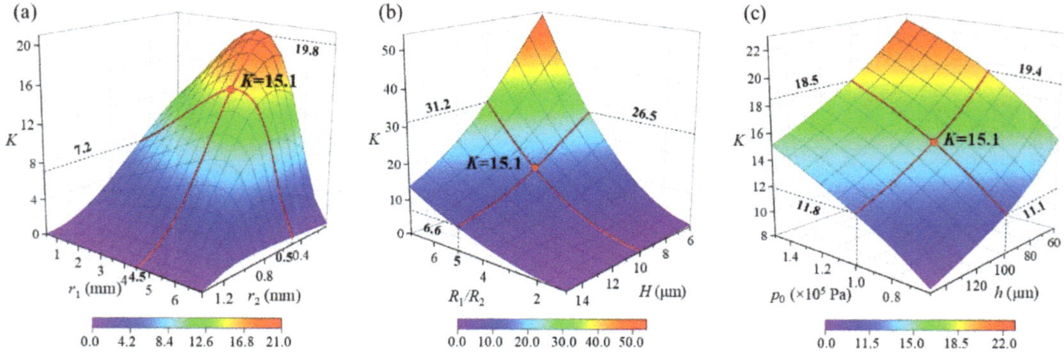

Figure 3. Simulation on (**a**) the effect of the radius r_1 and r_2 of the connecting rod on K, (**b**) the effect of the ratio of radius R_1/R_2 and the thickness H of two diaphragms, and (**c**) the effect of the height h and the initial pressure p_0 of cavity 2 on K.

Referring to Figure 3a, although as a whole the increased r_1 or reduced r_2 contributes to an increase in K, the radius at both the upper and lower surfaces of the pressure conduction rod produces the non-monotonic effect on K. It can be seen from the red line in Figure 3 that for a certain r_2 value (0.5 mm), r_1 is confirmed as 4.5 mm instead of a maximum value to achieve an extreme value of 15.1 for K. In contrast, when r_1 is chosen as 4.5 mm, r_2 can be further reduced to obtain a higher K; however, a smaller r_2 is limited by the fixation technology of the connecting rod and the supporting diaphragm at its lower endface. In other words, r_2 followed by r_1 is confirmed by the model simulation. In Figure 3b, K is directly positively proportional to R_1/R_2 and $1/H$. Moreover, the enhancement effect of R_1/R_2 is more obvious than that of H. For example, for a specific diaphragm thickness of 10 μm, K equals to 1 when $R_1/R_2 = 1$; nevertheless, it correspondingly rises from 15.1 to 31.2 with an amplification factor of ~2.1 when R_1/R_2 changes from 5 to 7. Hence, the preference for dual thin diaphragms with a large area ratio should be given on the basis of the thinning process of diaphragm material and miniaturization of SPAS. In addition to the aforementioned main structural parameters, the reduced height (h) of cavity 2 and the increased initial pressure (p_0) in the cavity also demonstrate clearly consistent effects on K as shown in Figure 3c. According to the red line in Figure 3c, for a cavity height of 100 μm at an approximate median point in the range of 60–150 μm, K is increased to 18.5 at 1.5×10^5 Pa from 11.1 at 0.6×10^5 Pa with a 66.7% rate of increase. By contrast, a 64.4% rate of increase for K is obtained when the cavity height (h) decreases from 150 μm to 60 μm at 1×10^5 Pa. Although the magnitude of the impact with little difference, whether raising p_0 or reducing h closely depends upon the sealing outer casing and miniaturization process of SPAS.

3.2. Effect of Diaphragm Parameters on Acoustic Amplification

According to the above simulation on K, the structural parameters of dual diaphragms are major influence factors, wherein the diaphragm thickness is more critical regarding the limitation in the physical dimension of SPAS. Hence, acoustic amplification performance by COMSOL multiphysics simulation using the initial parameters in Table 1 mentioned above is performed. By imposing a 1-Pa dynamic acoustic pressure whose frequency ranges from 200 Hz to 10 kHz on diaphragm 1, the acoustic pressure response is investigated in cavity 2 to estimate the dynamic amplification performance. Assuming the excitation frequency is 20 Hz. Figure 4 illustrates the dynamic acoustic response (f and K) verse diaphragm thickness (H) and Young's modulus (E) of the diaphragm, respectively, wherein f is the resonance frequency at the first vibration mode of the diaphragm. It can be clearly found in Figure 4a that f and K represent the opposite response to H or E. To achieve a wider frequency response for acoustic amplification, the thicker diaphragm is much better; however, the expense of lower K needs to be paid. Hence, a trade-off between f and K should be considered. It should be pointed out that the simulated sound pressures are, respectively, 15.18374 Pa and 15.18375 Pa at the edge and center of the diaphragm, with a negligible pressure difference of 10 μPa. That is to say, the acoustic pressure in cavity 2 can be regarded to be uniform. Similarly, in Figure 4b, the preferable response to E primarily lies in the region of 20–1000 GPa, as indicated in the orange area. In view of acoustic sensitivity evaluation at 1 kHz with a higher K value, the points closing to the intersection in Figure 4a are chosen so as to obtain K = 15.2 and f = 953 Hz, thereby confirming the corresponding film thickness of 10 μm and Young's modulus of ~207 GPa. Note that the K value equals 15.2 by COMSOL simulation in Figure 4 agrees well with the one that is calculated as 15.1 in Figure 3, thus favorably signifying the availability of the established theoretical model for the SPAS.

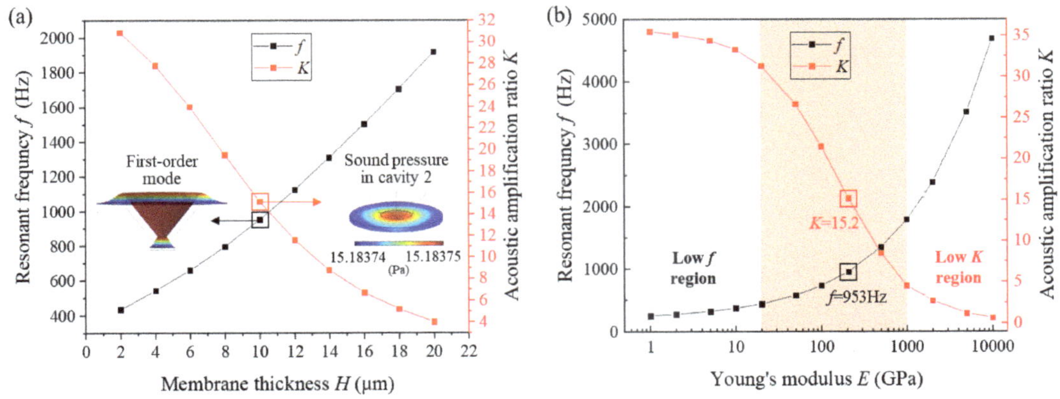

Figure 4. Acoustic simulation on (**a**) the effect of diaphragm thickness H on K and f and (**b**) the effect of Young's modulus E of the diaphragm on K and f.

4. Experiment and Result Analysis

4.1. SPSA Fabrication

Figure 5 presents the fabrication process of the SPAS. As shown in Figure 5a, the diaphragm 2 was clamped between two 100 μm thickness steel gaskets with a 3 mm diameter inner hole by the use of epoxy glue. Then, as shown in Figure 5b, based on the same adhesive bonding method, the diaphragm 1 was glued onto the endface of a machined steel casing with an inner diameter of 15 mm and a height of 5 mm. Then, a PLA tapered rod with upper and lower diameters of 9 mm and 1 mm was 3D printed, whose upper surface was adhered to the diaphragm 1. After that, referring to Figure 5c, both the casing with the tapered rod and the gaskets holding the diaphragm 2 were glued together via a location base whose inner diameter is the same as the outer diameter of the casing. The following process is to insert an F-P probe into a steel location base that would be held together with the casing in the subsequent procedure. As indicated in Figure 5d, an F-P probe with a 13-layer graphene diaphragm with a thickness of ~4.3 nm was inserted into a location base with an inner diameter of 2.5 mm until it contacted the surface of a glass base. In this case, the epoxy glue was dipped into the joint between the F-P probe and the location plate in order to stabilize the two components. Next, in a similar manner, the steel plate with an F-P probe was rigidly attached to the steel casing by glue as given in Figure 5e. Note that the detailed process of transferring a graphene diaphragm onto the endface of a ferrule with an inner diameter of 125 μm in the inset in Figure 5f can refer to the Ref. [33], wherein a wetting transfer method was adopted. In short, firstly, the polymethyl methacrylate (PMMA) substrate was etched off by immersing the divided graphene sample into acetone solution for about 1 h. Secondly, the ferrule was moved down slowly toward the floating graphene diaphragm until it touched the graphene diaphragm. Finally, the graphene diaphragm-covered fiber-capillary tip assembly was then left to dry in a cabinet for about half an hour. In this way, the preparation of an F-P probe coated with the graphene diaphragm was completed. As a result, the SPAS assembled with the F-P probe was displayed in Figure 5f, wherein the SPAS shows the physical dimension of Φ18 mm × 9 mm. Additionally, the inset in Figure 5f shows that certain wrinkles occurred on the surface of the diaphragm, which was to a certain extent caused by the uneven stress in the diaphragm. The uneven film stress would affect the flatness of dynamic frequency response of the F-P acoustic sensor and the upper limiting frequency, which could be improved by optimizing a high-quality transfer method of graphene diaphragm in the future study. In addition, it should be added that during acoustic test, diaphragm 2 made of steel or thermoplastic polyurethanes (TPU) material was introduced successively to fabricate different SPAS devices for comparison of acoustic amplification behaviors, in combination with the use of reflective coating on diaphragm 2.

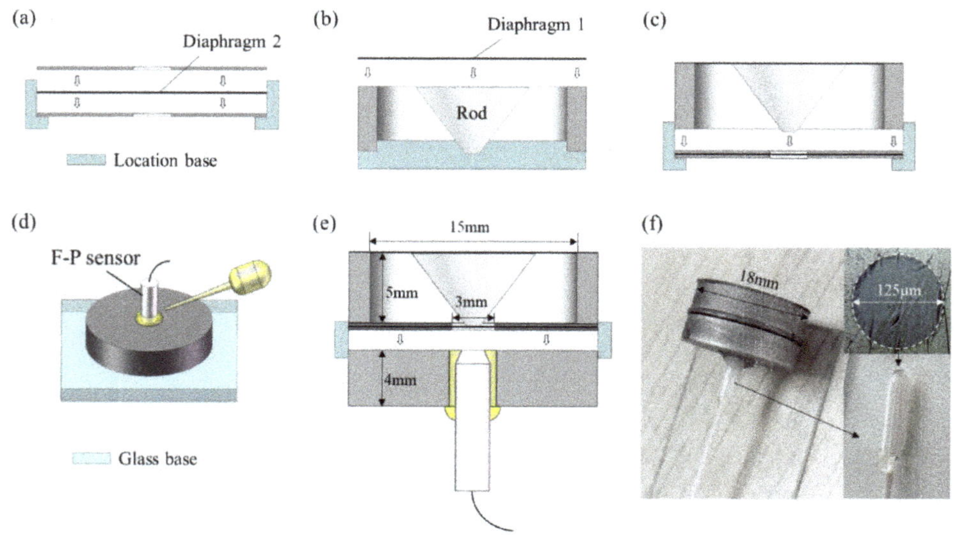

Figure 5. Fabrication process of the SPAS. (**a**) Clamping diaphragm 2 between two steel gaskets. (**b**) Bonding diaphragm 1 onto the endface of a steel casing. (**c**) Fixing the parts that make up the SPAS. (**d**) Inserting an F-P probe into a location base. (**e**) Assembling the casing with the locating base installed with an F-P probe by epoxy glue. (**f**) Picture of the assembled SPAS.

4.2. Acoustic Measurement Setup

To examine the sensitivity-enhancing effect induced by the introduction of SPAS, an acoustic pressure measurement setup was established as shown in Figure 6. A signal generator (DG5102, Rigol, Beijing, China) offered a dynamic acoustic signal in the range of 0.2–10 kHz via a conventional loudspeaker. The generated acoustic signal was detected simultaneously by the developed F-P sensor and a reference microphone (BK4189, BK, Nærum, Denmark) with a sensitivity of 50 mV/Pa, which were placed at the two symmetry positions closer to the loudspeaker along the central axis of the loudspeaker in a self-made soundproof box. The F-P sensor was driven by 1550 nm incident light with an optical power of −16.3 dB sent by a tunable laser (AP3350A, APEX, France) for the optimal voltage output. Then the resulting interference signal through a three-port circulator (6015-3-APC, Thorlabs, Newton, NJ, USA) was fed into an oscilloscope (DPO3054, Tektronix, Beaverton, OR, USA) for optical intensity demodulation through a low noise photodetector (PR-200K4177, Conquer, Beijing, China) with a 200 kHz bandwidth. In this way, the voltage output from BK4189 was amplified by a conditioning amplifier (BK1708, BK, Nærum, Denmark) and then external acoustic pressure applied on the diaphragm 1 could be determined by the amplified voltage and the reference microphone output. Hence, the sensitivity verse acoustic pressure can be obtained for the developed F-P sensor with or without SPAS. Then, the amplification ratio K in different acoustic frequencies can further be solved by the ratio of the two measured acoustic responses when the SPAS was used or not.

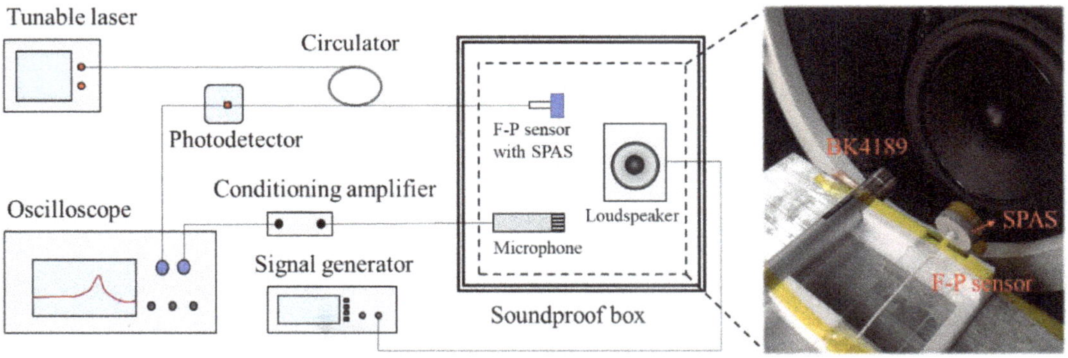

Figure 6. Experimental setup for the acoustic test.

4.3. Results and Discussion

In order to optimize the SPAS for acoustic amplification, three types of SPASs, named as SPAS1, SPAS2, and SPAS3, were fabricated by employing a steel diaphragm, TPU diaphragm with nano-carbon power coating, and TPU diaphragm with black paint coating as diaphragm 2, respectively, as depicted in Figure 7a. Through the experimental setup in Figure 6, it can be concluded from the acoustic frequency response in Figure 7b that the F-P sensors with SPAS could remarkably achieve much higher output voltage; moreover, their responses near resonant frequencies are even superior to that (~500 mV) from the reference microphone. In addition, SPAS2 and SPAS3 can offer relatively stronger sound pressure signals than SPAS1, with the sacrifice of the slightly declining resonant frequency from 1.7 kHz to 1.3 kHz and then 1.2 kHz. It is further worth pointing out that in comparison with SPAS2, SPAS3 provided more preferable acoustic frequency output with a double peak amplitude in a wider band of 1.8 kHz, where the latter demonstrated a more sensitive response to sound information than BK4189. The corresponding K values were calculated by dividing the response from the F-P sensor using SPAS by the one without SPAS, as indicated in Figure 7c. Although the three types of SPASs obtained roughly identical bands corresponding to the region of K greater than 1 (0.2–2.5 kHz, 0.2–2 kHz and 0.2–2.3 kHz), which agrees well with the simulation data as labelled by dashed line with shadow area, SPAS3 exhibited the optimal amplification performance conforming to the sound-enhanced trend in Figure 7b due to higher K values in a comparatively broad frequency domain. The maximum K values corresponding to the resonant frequencies 1.7 kHz, 1.3 kHz, and 1.2 kHz for SPAS1, SPAS2, and SPAS3, respectively, were calculated as 7.6, 11.1, and 29.9. The varying K or amplitude response verse frequency is dependent upon the resonant amplification effect of the presented SPAS. Therefore, further research on a SPAS with a greater harmonic response with a wide and flat low-pass band is needed to improve the aforementioned frequency response fluctuations.

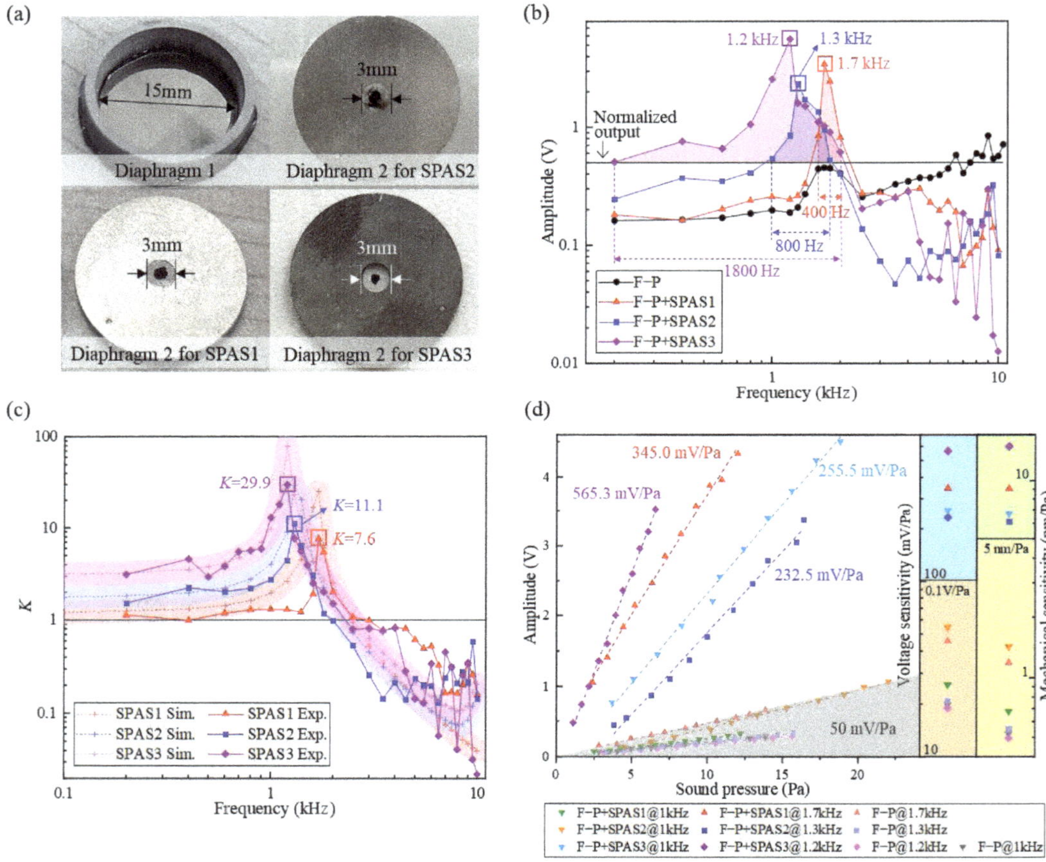

Figure 7. (**a**) Diaphragms 1 and 2 for the acoustic test. (**b**) Measured frequency response with or without SPAS, (**c**) K value and (**d**) acoustic sensitivity for the F-P sensors with SPAS1, SPAS2, and SPAS3. Note that the black horizontal line in (**b**) represents the normalized output from the reference microphone.

The acoustic sensitivities of these graphene-based F-P sensors with the SPAS under different acoustic pressure levels at the representative frequencies were measured as shown in Figure 7d. Compared with the F-P sensors without SPAS, those F-P probes with SPAS behaved more outstanding acoustic pressure characteristics. Take the SPAS3 among the three SPAS as an example. The F-P sensor with the SPAS3 exhibited the acoustic sensitivities of 255.5 mV/Pa @ 1 kHz and 565.3 mV/Pa @ 1.2 kHz, respectively. By comparison, the F-P sensor without SPAS only showed a sensitivity of ~20 mV/Pa. Hence, there is no doubt about the effective sound-enhanced effect induced by the SPAS. Additionally, although the enhanced sensitivity for the F-P sensor using the SPAS1 or SPAS2 was limited at typical 1 kHz, the voltage sensitivities of the F-P sensor were estimated to be 345.0 mV/Pa @1.7 kHz and 232.5 mV/Pa @1.3 kHz by using a least square fitting method with a fitting coefficient of 99.6% and 99.8%, respectively.

Note that the measured acoustic sensitivity mentioned above is the voltage sensitivity, which is also related to the signal conditioning unit in addition to the sensor performance itself. Thus, the mechanical sensitivity, closely concerned with the intrinsic response of the sensor, should also be evaluated on the basis of the mechanical response characteristic of the pressure-sensitive diaphragm. In this case, it can be inferred from the large deflection behaviors of circular graphene diaphragm [34] that the diaphragm deflection responses for the F-P sensor with the SPAS3 under acoustic pressure were derived as ~6.7 nm/Pa and ~14.8 nm/Pa, respectively, corresponding to 255.5 mV/Pa @1 kHz and 565.3 mV/Pa @1.2 kHz. The calculated mechanical sensitivity by far higher than 6 nm/Pa is obviously preceded over conventional diaphragm-type F-P acoustic sensors with the measured optimal sensitivity of ~2.38 nm/Pa in Ref. [34] and ~1.1 nm/Pa in Ref. [35]. Referring to Figure 7b, the F-P sensor with the SPAS3 also revealed a sensitivity-enhanced frequency band of 0.2–2.3 kHz (bandwidth: 2.1 kHz), which was obviously greater than the sound field enhancement bandwidth of ~1 kHz generated by a spiral-type acoustic tube in Ref. [20] and ~100 Hz confined by a single resonant peak excited by a Helmholtz resonator [22]. This phenomenon verified the advantage and the applicability of the presented external SPAS. Additionally, it further reveals that a higher resonant frequency as the cut-off frequency with a flat low-frequency band is oriented to structural improvement in future research.

5. Conclusions

In order to enhance the acoustic sensitivity for an F-P acoustic sensor via an amplification structure, a human ear-inspired cylindrical SPAS, whose outer diameter and height were 18 mm 9 mm, was designed by means of two circular diaphragms interconnected with a conical round rod. Then, the COMSOL-based sound field simulation of the amplification structure demonstrated the dominating influence of structural parameters of the diaphragm on the amplification ratio and resonant frequency response. According to the COMSOL simulation analysis, the radius ratio of the two circular diaphragms in SPAS was set as ~5, along with a radius ratio of 0.6 for the upper endfaces of the connecting round rod. Furthermore, steel and TPU materials were used in combination so as to fabricate three types of SPAS parts (SPAS1, SPAS2, and SPAS3). Subsequently, an F-P acoustic sensor with a multilayer graphene diaphragm suspended onto a ferrule with a diameter of 125 μm was inserted into the SPAS for acoustic test. The measured results showed that the SPAS3 achieved the optimal acoustic harmonic response that is higher than the reference microphone within the range of 0.2–2 kHz. Moreover, the maximum gain factor of ~29.9 for the SPAS3 was obtained at 1.2 kHz, which is obviously greater than 7.6 and 11.1 for the other two SPASs with a narrower bandwidth of 400–800 Hz. In this case, the enhanced acoustic sensitivity of 255.5 mV/Pa @ 1 kHz was achieved for the F-P sensor with the SPAS3, which is significantly superior to the conventional F-P acoustic sensor reported previously and the reference electric microphone (~50 mV/Pa @ 1 kHz).

Author Contributions: C.L. conceived the idea for the research; X.X. performed the simulation; X.X. and C.L. wrote the paper; Y.L. and X.X. performed the experiments; X.S. analyzed the simulation and proofread the manuscript. All authors have read and agreed to the published version of the manuscript.

Funding: This work was supported by National Natural Science Foundation (62173021), Beijing Natural Science Foundation (4212039), Science Technology and Innovation Commission of Shenzhen Municipality (JCYJ20180504165721952, JCYJ20170817111857745), and Aviation Science Foundation of China (2020Z073051002).

Institutional Review Board Statement: Not applicable.

Informed Consent Statement: Not applicable.

Conflicts of Interest: The authors declare no conflict of interest.

References

1. Kanjo, E. Noisespy: A real-time mobile phone platform for urban noise monitoring and mapping. *Mob. Netw. Appl.* **2010**, *15*, 562–574. [CrossRef]
2. Cao, Y.; Jin, W.; Ho, H.L.; Ma, J. Miniature fiber-tip photoacoustic spectrometer for trace gas detection. *Opt. Lett.* **2013**, *38*, 434–436. [CrossRef]
3. Akhbari, S.; Sammoura, F.; Eovino, B.; Yang, C.; Lin, L. Bimorph piezoelectric micromachined ultrasonic transducers. *J. Microelectromech. Syst.* **2016**, *25*, 326–336. [CrossRef]
4. Leal, A.G.; Marques, C. Diaphragm-embedded optical fiber sensors: A review and tutorial. *IEEE Sens. J.* **2021**, *21*, 12719–12733. [CrossRef]
5. Liu, B.; Lin, J.; Liu, H.; Jin, A.; Jin, P. Extrinsic Fabry-Perot fiber acoustic pressure sensor based on large-area silver diaphragm. *Microelectron. Eng.* **2016**, *166*, 50–54. [CrossRef]
6. Xu, F.; Shi, J.; Gong, K.; Li, H.; Hui, R.; Yu, B. Fiber-optic acoustic pressure sensor based on large-area nanolayer silver di-aghragm. *Opt. Lett.* **2014**, *39*, 2838–2840. [CrossRef]
7. Gong, Z.; Chen, K.; Zhou, X.; Yang, Y.; Zhao, Z.; Zou, H.; Yu, Q. High-sensitivity Fabry-Perot interferometric acoustic sensor for low-frequency acoustic pressure detections. *J. Light. Technol.* **2017**, *35*, 5276–5279. [CrossRef]
8. Fan, P.; Yan, W.; Lu, P.; Zhang, W.; Zhang, W.; Fu, X.; Zhang, J. High sensitivity fiber-optic Michelson interferometric low-frequency acoustic sensor based on a gold diaphragm. *Opt. Express* **2020**, *28*, 25238. [CrossRef]
9. Wu, G.; Xiong, L.; Dong, Z.; Liu, X.; Cai, C.; Qi, Z.-M. Development of highly sensitive fiber-optic acoustic sensor and its preliminary application for sound source localization. *J. Appl. Phys.* **2021**, *129*, 164504. [CrossRef]
10. Li, X.; Lin, R.; Kek, H.; Miao, J.; Zou, B. Sensitivity-improved silicon condenser microphone with a novel single deeply corru-gated diaphragm. *Sens. Actuators A Phys.* **2001**, *92*, 257–262. [CrossRef]
11. Yang, W.; Jin, L.; Liang, Y.; Ma, J.; Guan, B. Corrugated-diaphragm based fiber laser hydrophone with Sub-100 μPa/Hz1/2 resolution. *Sensors* **2017**, *17*, 1219. [CrossRef]
12. Zhu, J.; Wang, M.; Chen, L.; Ni, X.; Ni, H. An optical fiber Fabry–Perot pressure sensor using corrugated diaphragm and angle polished fiber. *Opt. Fiber Technol.* **2017**, *34*, 42–46. [CrossRef]
13. Liu, B.; Zhou, H.; Liu, L.; Wang, X.; Shan, M.; Jin, P.; Zhong, Z. An optical fiber Fabry-Perot microphone based on corrugated silver diaphragm. *IEEE Trans. Instrum. Meas.* **2018**, *67*, 1994–2000. [CrossRef]
14. Liu, B.; Zheng, G.; Wang, A.; Gui, C.; Yu, H.; Shan, M.; Jin, P.; Zhong, Z. Optical fiber Fabry–Perot acoustic sensors based on corrugated silver diaphragms. *IEEE Trans. Instrum. Meas.* **2020**, *69*, 3874–3881. [CrossRef]
15. Koskinen, V.; Fonsen, J.; Roth, K.; Kauppinen, J. Cantilever enhanced photoacoustic detection of carbon dioxide using a tunable diode laser source. *Appl. Phys. A* **2007**, *86*, 451–454. [CrossRef]
16. Liu, J.; Yuan, L.; Lei, J.; Zhu, W.; Cheng, B.; Zhang, Q.; Song, Y.; Chen, C.; Xiao, H. Micro-cantilever-based fiber optic hy-drophone fabricated by a femtosecond laser. *Opt. Lett.* **2017**, *42*, 2459–2462. [CrossRef]
17. Chen, K.; Yang, B.; Deng, H.; Guo, M.; Zhang, B.; Yang, Y.; Liu, S.; Zhao, Y.; Peng, W.; Yu, Q. Simultaneous measurement of acoustic pressure and temperature using a Fabry-Perot interferometric fiber-optic cantilever sensor. *Opt. Express* **2020**, *28*, 15050–15061. [CrossRef]
18. Li, W.; Wang, Z.; Feng, C.; Li, Q.; Yu, H. High sensitivity all-optical acoustic pressure sensor based on resonant micro-opto-mechanical cantilever with integrated rib waveguide. *Sens. Actuators A Phys.* **2019**, *285*, 300–307. [CrossRef]
19. Qi, X.; Wang, S.; Jiang, J.; Liu, K.; Zhang, P.; Li, R.; Liu, T. Flywheel-like diaphragm-based fiber-optic Fabry–Perot frequency tailored acoustic sensor. *J. Phys. D Appl. Phys.* **2020**, *53*, 415102. [CrossRef]
20. Jung, E.S.; Seong, K.W.; Lim, H.G.; Lee, J.H.; Cho, J.H. Implantable microphone with acoustic tube for fully implantable hearing devices. *IEICE Electron. Express* **2011**, *8*, 215–219. [CrossRef]
21. Hu, P.; Tong, X.; Zhao, M.; Deng, C.; Guo, Q.; Mao, Y.; Wang, K. Study on high temperature Fabry–Perot fiber acoustic sensor with temperature self-compensation. *Opt. Eng.* **2015**, *54*, 097104. [CrossRef]
22. Gao, R.; Wang, H.; Zhu, D.; Fan, G.; Jiao, W.; Liang, C.; Liu, Y.; Wang, Y.; Liu, W.; He, X.; et al. Sensitivity enhancement of optical fiber vibration sensor through encapsulation of acoustic Helmholtz resonator. *Optik* **2018**, *169*, 33–40. [CrossRef]
23. Han, J.H.; Kwak, J.-H.; Joe, D.J.; Hong, S.K.; Wang, H.S.; Park, J.H.; Hur, S.; Lee, K.J. Basilar membrane-inspired self-powered acoustic sensor enabled by highly sensitive multi tunable frequency band. *Nano Energy* **2018**, *53*, 198–205. [CrossRef]
24. Jang, J.; Lee, J.; Jang, H.; Choi, H. A triboelectric-based artificial basilar membrane to mimic cochlear tonotopy. *Adv. Healthc. Mater.* **2016**, *5*, 2481–2487. [CrossRef]
25. Kurokawa, H.; Goode, R.L. Sound pressure gain produced by the human middle ear. *Otolaryngol. Head Neck Surg.* **1995**, *113*, 349–355. [CrossRef]
26. Aibara, R.; Welsh, J.T.; Puria, S.; Goode, R.L. Human middle-ear sound transfer function and cochlear input impedance. *Hear. Res.* **2001**, *152*, 100–109. [CrossRef]
27. Ferris, P.; Prendergast, P. Middle-ear dynamics before and after ossicular replacement. *J. Biomech.* **2000**, *33*, 581–590. [CrossRef]
28. Cui, Q.; Thakur, P.; Rablau, C.; Avrutsky, I.; Cheng, M.M.-C. Miniature optical fiber pressure sensor with exfoliated graphene diaphragm. *IEEE Sens. J.* **2019**, *19*, 5621–5631. [CrossRef]

29. Wang, S.; Chen, W. A large-area and nanoscale graphene oxide diaphragm-based extrinsic fiber-optic Fabry–Perot acoustic sensor applied for partial discharge detection in air. *Nanomaterials* **2020**, *10*, 2312. [CrossRef]
30. Sun, Q.; Gan, R.Z.; Chang, K.-H.; Dormer, K.J. Computer-integrated finite element modeling of human middle ear. *Biomech. Model. Mechanobiol.* **2002**, *1*, 109–122. [CrossRef]
31. Timoshenko, S. *Theory of Plates and Shells*; Hill Book Limited: New York, NY, USA, 1959.
32. Borgnakke, C.; Sonntag, R. *Fundamentals of Thermodynamics*, 8th ed.; John Wiley & Sons: New Jersey, NJ, USA, 2013.
33. Li, C.; Xiao, J.; Guo, T.; Fan, S.; Jin, W. Interference characteristics in a Fabry–Perot cavity with graphene membrane for optical fiber pressure sensors. *Microsyst. Technol.* **2014**, *21*, 2297–2306. [CrossRef]
34. Li, C.; Gao, X.; Guo, T.; Xiao, J.; Fan, S.; Jin, W. Analyzing the applicability of miniature ultra-high sensitivity Fabry–Perot acoustic sensor using a nanothick graphene diaphragm. *Meas. Sci. Technol.* **2015**, *26*, 085101. [CrossRef]
35. Ma, J.; Xuan, H.; Ho, H.L.; Jin, W.; Yang, Y.; Fan, S. Fiber-optic Fabry–Pérot acoustic sensor with multilayer graphene dia-phragm. *IEEE Photon. Technol. Lett.* **2013**, *25*, 932–935. [CrossRef]

nanomaterials

MDPI

Communication

Photocatalytic Degradation of Tobacco Tar Using CsPbBr3 Quantum Dots Modified Bi2WO6 Composite Photocatalyst

Runda Huang [1], Menglong Zhang [2,*], Zhaoqiang Zheng [1], Kunqiang Wang [3], Xiao Liu [2,*], Qizan Chen [1] and Dongxiang Luo [1,3,*]

1 School of Materials and Energy, Guangdong University of Technology, Guangzhou 510006, China; 2111902118@mail2.gdut.edu.cn (R.H.); zhengzhq5@mail2.sysu.edu.cn (Z.Z.); qizanchen@gdut.edu.cn (Q.C.)
2 Institute of Semiconductors, South China Normal University, Guangzhou 510631, China
3 School of Chemistry and Chemical Engineering, Institute of Clean Energy and Materials, Guangzhou Key Laboratory for Clean Energy and Materials, Huangpu Hydrogen Innovation Center, Guangzhou University, Guangzhou 510006, China; kunqiangwang@m.scun.edu.cn
* Correspondence: mlzhang@m.scnu.edu.cn (M.Z.); liuxiao@m.scnu.edu.cn (X.L.); luodx@gdut.edu.cn (D.L.)

Abstract: Polycyclic aromatic hydrocarbons (PAHs) in tobacco tar are regarded as a significant threat to human health. PAHs are formed due to the incomplete combustion of organics in tobacco and cigarette paper. Herein, for the first time, we extended the application of $CsPbBr_3$ quantum dots ($CsPbBr_3$) to the photocatalytic degradation of tobacco tar, which was collected from used cigarette filters. To optimize the photoactivity, $CsPbBr_3$ was coupled with Bi_2WO_6 for the construction of a type-II photocatalyst. The photocatalytic performance of the $CsPbBr_3/Bi_2WO_6$ composite was evaluated by the degradation rate of PAHs from tobacco tar under simulated solar irradiation. The results revealed that $CsPbBr_3/Bi_2WO_6$ possesses a large specific surface area, outstanding absorption ability, good light absorption and rapid charge separation. As a result, in addition to good stability, the composite photocatalyst performed remarkably well in degrading PAHs (over 96% were removed in 50 mins of irradiation by AM 1.5 G). This study sheds light on promising novel applications of halide perovskite.

Keywords: nanocomposites; photocatalytic degradation; perovskite; tobacco tar

Citation: Huang, R.; Zhang, M.; Zheng, Z.; Wang, K.; Liu, X.; Chen, Q.; Luo, D. Photocatalytic Degradation of Tobacco Tar Using CsPbBr3 Quantum Dots Modified Bi2WO6 Composite Photocatalyst. *Nanomaterials* **2021**, *11*, 2422. https://doi.org/10.3390/nano11092422

Academic Editors: Jihoon Lee, Ming-Yu Li and Antonino Gulino

Received: 20 July 2021
Accepted: 14 September 2021
Published: 17 September 2021

1. Introduction

The development of economies can create problems such as energy shortages and environmental pollution, which are regarded as critical challenges to global sustainable development. In recent years, industrial and environmental pollution caused by organic pollutants has attracted considerable public attention [1,2]. In terms of human health, the main concerns include benzene compounds [3], formaldehyde [4], bacterial infections [5], and tobacco tar [6]. In particular, it is well known that tobacco tar is formed due to the incomplete combustion of organics in tobacco and cigarette paper. The substrates of tobacco are very complicated, and among these organic substrates, polycyclic aromatic hydrocarbons (PAHs) are considered as the major pollutants. This can be attributed to their bio-accumulative, toxic, carcinogenic, and recalcitrant features [7]. Thus, minimizing the toxicity of tobacco tar is desirable for human health and the environment. However, until now, not much research has been carried out on the degradation of tobacco tar via photocatalysis.

Among the techniques used for organic pollutant degradation, photocatalytic oxidation uses low-cost semiconductive materials and is considered as an environmentally-friendly path for the purification or transformation of organic pollutants [8–11]. Semiconductor-based photocatalytic reactions have already attracted extensive attention because they are a simple and environmentally friendly way to utilize solar energy. Therefore, numerous types of photocatalysts have been extensively investigated, such as TiO_2, ZnO, SnO_2, $SrTiO_3$,

$LaFeO_3$, $BaTiO_3$, and so on [12–17]. However, these semiconductors still have some common shortcomings, such as a narrow light absorption range and a high recombination ratio of photogenerated charges [18]. In this case, there is an urgent need to develop visible-light driven and highly active photocatalysts.

Bi_2WO_6 is currently regarded as a promising photocatalyst for its outstanding photo-redox ability, nontoxicity, and thermal and chemical stability [19]. Its potential applications include solar cells, photoelectrodes and photocatalysis [20–22]. However, the rapid recombination of exitonic pairs, ineffective charge separation and unsatisfactory visible-light absorption have limited the photocatalytic performance of pure Bi_2WO_6 [23]. Hence, different strategies have been applied to improve the photocatalytic properties of Bi_2WO_6, such as coupling with other semiconductors, morphological control, ion doping and surface modification [24–28]. One of the most accessible approaches is the deposition of quantum dots (QDs) onto the semiconductor surface, which is beneficial due to their noteworthy properties, including high charge separation efficiency and quantum confinement effects [29,30]. Thus, modification with QDs is regarded as an effective method to improve the photocatalytic performance of semiconductor materials.

Among the various QDs materials, $CsPbBr_3$ QDs have received increasing interest due to their high luminescent quantum yields [31] in solid states in the green range (band gap energy = 2.4 eV) and superior environmental stability in comparison to other perovskites with organic cations [32]. Perovskite materials possess unique advantages in regard to effective charge separation and electron hole diffusion, which makes them suitable not only for the photovoltaic field, but also for several photocatalytic reaction types [33,34]. In addition, $CsPbBr_3$ is very attractive due to its high thermal-stability and photo-resistance [35]. Considering these advantages, $CsPbBr_3$ was used to decorate Bi_2WO_6 via a hydrothermal method in this work and the photoactivity was evaluated through the photocatalytic degradation of PAHs in tobacco tar under simulated solar light. Additionally, the $CsPbBr_3/Bi_2WO_6$ heterogeneous photocatalyst exhibited increased specific surface area and improved light absorption in the visible range, which effectively promoted the photocatalytic activity [36]. Compared with pure Bi_2WO_6, the $CsPbBr_3/Bi_2WO_6$ composites exhibit superior photocatalytic activity and stability. This work presents a promising strategy for designing a more efficient composite photocatalyst-based semiconductor photocatalyst for application to pollutant purification. In addition, to the best of our knowledge, this work is the first to utilize perovskite for the degradation of PAHs.

2. Experimental Section

2.1. Reagents and Materials

Bismuth nitrate pentahydrate ($Bi(NO_3)_3{\cdot}5H_2O$, 99%), ethanol (CH_3CH_2OH, 99.7%), nitric acid (HNO_3, 68%), sodium tungstate dehydrates ($Na_2WO_4{\cdot}2H_2O$, 99.5%) were purchased from Aladdin Industrial Co, Ltd. (Shanghai, China) and $CsPbBr_3$ perovskite QDs was purchased from Nanjing MKNANO, China and used as received. The used cigarette filters were collected randomly. The deionized water used throughout all experiments was purified through a Millipore system (Millipore, Billerica, MA, USA).

2.2. Synthesis of $CsPbBr_3/Bi_2WO_6$

Pure Bi_2WO_6 was synthesized via a hydrothermal method. First, $Bi(NO_3)_3{\cdot}5H_2O$ (1 mmol) was dissolved in an aqueous solvent mixture of 30 mL HNO_3 (pH = 3) and 20 mL $Na_2WO_4{\cdot}2H_2O$ (0.5 mmol). The mixed solution was then vigorously stirred until clear, under room temperature. The solution was subsequently heat-treated in a muffle furnace at 120 °C for 2 h. The resultant precipitate was dried to obtain the pure Bi_2WO_6 photocatalyst (light-yellow powder). Subsequently, 1 mL of $CsPbBr_3$ suspension (10 mg/mL, purchased from Nanjing MKNANO, Nanjing, China) was ultrasonicated for 30 min to generate $CsPbBr_3$ QDs dispersion. Then, a certain amount of Bi_2WO_6 was added into the above $CsPbBr_3$ suspension and stirred for 20 h. For example, to prepared 10 wt% $CsPbBr_3/Bi_2WO_6$, 90 mg of Bi_2WO_6 was added to the 1 mL suspension. Subsequently, the mixture was evaporated at 40 °C for 10 h and the composite photocatalyst was obtained after drying.

2.3. Photocatalytic Experiments

The tobacco tar was obtained by immersing 10 randomly collected used cigarette filters in 50 mL trifluorotoluene, followed by 10 min ultra-sonification to give a brown-yellow tobacco tar solution. The cigarette filters were then removed from the solution. The concentrations of the substrates in the tobacco tar solution were confirmed using gas chromatography (GC). The as-prepared photocatalyst was uniformly dispersed in 2 mL of tobacco tar solution with 20 min ultrasonic treatment. Then, the samples were illuminated by Xe lamp with an AM 1.5 G filter as the simulated solar source. The photoactivity was confirmed by monitoring the concentrations of PAHs under different illumination time.

2.4. Characterization

The crystal structure of the samples was studied by X-ray diffractometry (XRD; Bruker-AXS D8 Advance, Karlsruhe, Germany). Field emission scanning electron microscopy (SEM) (Hitachi S-4800, Tokyo, Japan) and transmission electron microscopy (TEM) (JEOL JEM-2100, Tokyo, Japan) were performed for the morphological and microstructural analyses, respectively, of the samples. The UV-vis absorption spectroscopy of the samples was performed by a UV-vis near-infrared spectrophotometer (Agilent Cary 5000, Palo Alto, CA, USA). The XPS energy spectrum was determined by an X-ray photoelectron spectrometer (Thermo Fisher/Escalab 250Xi, Waltham, MA, USA). Measuring the surface area and pore size distribution of samples was performed using a specific surface and porosity analyzer (Micromeritics ASAP2460, Atlanta, GA, USA). TGA analysis was done by using the Mettler Toledo TGA/DSC 3+ instrument (Zurich, Switzerland). Total organic carbon (TOC) content in solvent was assessed by a TOC analyzer (Shimadzu, TOC-L-VCPN, Tokyo, Japan). The photocatalytic result was characterized by single quadrupole gas chromatography-mass spectrometry (Thermo Fisher/Trace1300 ISQ QD, Waltham, MA, USA).

3. Results and Discussion

3.1. Properties of the Composite Photocatalysts

Powder X-ray diffraction (PXRD) (Figure 1a) was initially exploited to evaluate the phase and crystallinity of the as-prepared $CsPbBr_3$ QDs-decorated Bi_2WO_6 ($CsPbBr_3/Bi_2WO_6$). A p-FTO glass slide was used as a substrate to support the samples for characterization. The main peaks at 2θ of 28.3°, 32.8°, 47.0°, 55.8° and 58.5° can be indexed to an orthorhombic Bi_2WO_6 (JCPDS#39-0256), and corresponded to the indices of (131), (200), (260), (331) and (262) planes, respectively. On the other hand, the diffraction patterns of cubic $CsPbBr_3$ corresponded to JCPDS#54-0752 and a previous report [32,37]. The transmission electron microscopy (TEM, Figure 1b) revealed that the $CsPbBr_3$ QDs of a few nanometers were dispersed on Bi_2WO_6 with a quasi-rectangle morphology. The two materials were subsequently confirmed by high-resolution transmission electron microscopy (HR-TEM), which showed the lattice spacing of $CsPbBr_3$ ([220] = 0.20 nm) and Bi_2WO_6 ([200] = 0.28) (Figure 1d), which can be compared to the diffraction results from PXRD. In addition, as illustrated in Figure 1c, the connected boundary presented as a semi-coherent interface involving several unit cells. The lattice mismatch between the crystal planes of these two components can be defined as $\Delta d/d_{Bi2WO6}$ = 28%. Furthermore, as shown in Figure 1c,d, $CsPbBr_3$ and Bi_2WO_6 exhibited periodically coincided lattice (7d ($CsPbBr_3$) = 5d (Bi_2WO_6)), suggesting that intimate energetic and chemical interactions were formed at the semi-coherent interface of the two components [38].

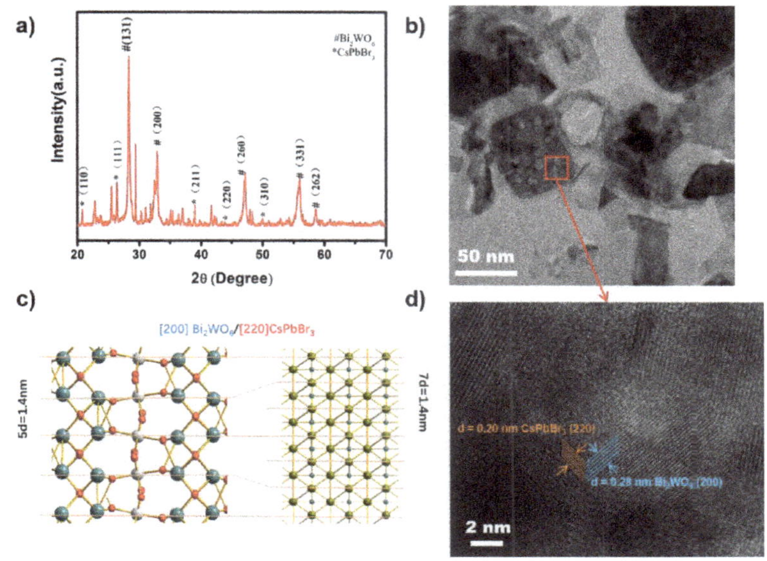

Figure 1. (**a**) PXRD patterns and (**b**) TEM image of the as-prepared 15 wt% $CsPbBr_3/Bi_2WO_6$ photocatalyst; (**c**) schematic diagram of semi-coherent interface of the two components; (**d**) HR-TEM image of the as-prepared 15 wt% $CsPbBr_3/Bi_2WO_6$ photocatalyst.

The elemental composition of the as-prepared pure Bi_2WO_6 and $CsPbBr_3/Bi_2WO_6$ were confirmed by X-ray photoelectron spectroscopy (XPS), which demonstrated that the $CsPbBr_3/Bi_2WO_6$ sample contained no other elements apart from O, Bi, Cs, Pb, Br, C 1 and W (Figure 2a). The surface atomic content was calculated by the fitting peak area and sensitivity factor, as shown in Table 1 (the carbon atomic content comes from the XPS instrument). Besides, the altered chemical state of the coupled materials was recorded by XPS as well; as shown in Figure 2b–d, the characteristic peaks of $W4f_{5/2}$ and $W4f_{7/2}$ in pristine $CsPbBr_3/Bi_2WO_6$ were located at 37.8 and 35.7 eV (Figure 2b), the characteristic peaks of $Bi4f_{5/2}$ and $Bi4f_{7/2}$ were located at 164.8 and 159.3 eV (Figure 2c), and the characteristic peaks of O1s were located at 533.3, 532.0 and 530.4 eV (Figure 2d). In the Bi_2WO_6 sample, the characteristic peaks of $W4f_{5/2}$ and $W4f_{7/2}$ were located at 37.4 and 35.3 eV (Figure 2b), the characteristic peaks of $Bi4f_{5/2}$ and $Bi4f_{7/2}$ were located at 164.4 and 158.9 eV (Figure 2c), and the characteristic peak of O1s was 533.4, 532.1 and 530.2 eV (Figure 2d). This showed that the valence state of Bi, W and O in the composite sample was +3, +6 and −2, respectively. In comparison to pristine Bi_2WO_6, the binding energy of Bi4f, W4f in the $CsPbBr_3/Bi_2WO_6$ samples shifted ca. 0.7 and 0.4 towards the higher energy, respectively. This could be attributed to the regional environment and electron density change of Bi_2WO_6 and $CsPbBr_3$ in the composites due to strong heterojunction formation [39]. In detail, the Bi_2WO_6 displayed increased electron accepting capacity and decreased electron density. On the other hand, this is accompanied by an increase in the electron density and an improvement in the electron-donating capacity on the $CsPbBr_3$ side. Consequently, the internal electric field was established, leading to a tendency for electron transfer from $CsPbBr_3$ to Bi_2WO_6, and as a result, a typical type-II band structure was formed.

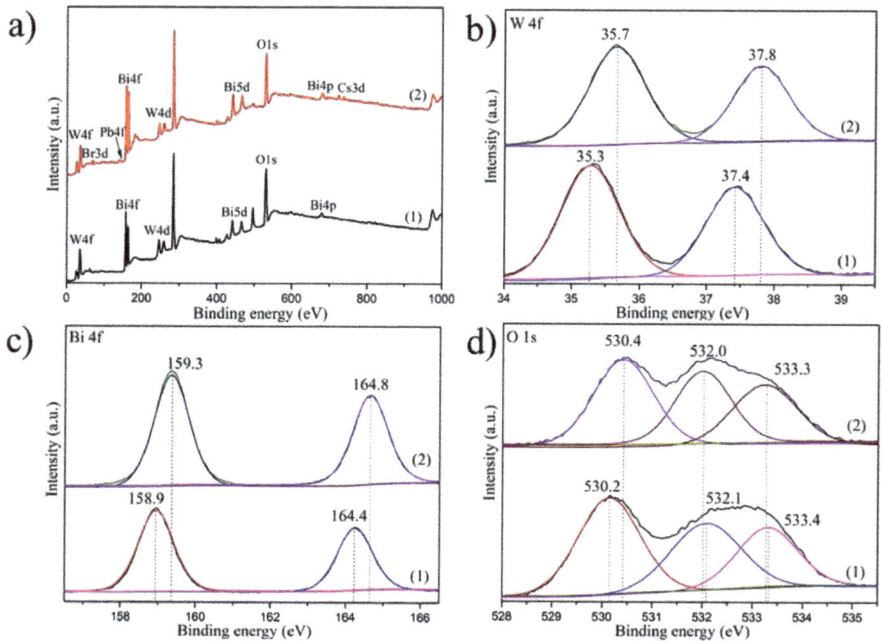

Figure 2. XPS spectra of (1) pure Bi_2WO_6 and (2) 15 wt% $CsPbBr_3/Bi_2WO_6$ samples. (**a**) Survey scan; (**b**) W4f; (**c**) Bi4f; (**d**) O1s.

Table 1. Relative atomic content of pure Bi_2WO_6 and 15 wt% $CsPbBr_3/Bi_2WO_6$ samples.

Sample	Bi4f/%	W4f/%	O1s/%	C1s/%	Br3d/%	Pb4f/%	Cs4d/%
Bi_2WO_6	1.62	2.53	24.04	71.8	0	0	0
$CsPbBr_3/Bi_2WO_6$	2.21	2.15	21.7	72.74	0.96	0.09	0.16

In order to explore the electronic features of $CsPbBr_3/Bi_2WO_6$, Hall effect measurement was employed to monitor the charge carrier density and the Hall mobility. Hall effect measurement is an important approach for investigating the charge transport mechanisms in Bi_2WO_6 with or without the coupling of $CsPbBr_3$ QDs. Here, as shown in Table 2, the Bi_2WO_6 sample exhibited lower charge carrier density (2.738×10^{11} cm^{-3}) than that of $CsPbBr_3/Bi_2WO_6$ (5.154×10^{14} cm^{-3}). In contrast to the reduced charge density in the Z-scheme due to the recombination of electrons and holes between the two component materials, this result was consistent with the type-II feature of the composite materials, which possessed accumulated charges. Meanwhile, greater mobility was observed from pristine Bi_2WO_6 in comparison to that of $CsPbBr_3/Bi_2WO_6$; this can be attributed to the effective mass of the carriers. According to the equation: $\mu = \frac{\tau e}{m}$, the mobility (μ) is proportional to the elemental charge (e) and carrier mean free time (τ). In contrast, the mobility is inversely proportional to the effective mass of the carriers (m), and a greater m typically indicates larger carrier density [40].

Table 2. Hall effect measurement of pure Bi_2WO_6 and 15 wt% $CsPbBr_3/Bi_2WO_6$ samples.

Sample	Carrier Density (cm^{-3})	Hall Mobility (cm^{2}V^{-1}s^{-1})	p/n Type
Bi_2WO_6	2.738×10^{11}	25.316	n
$CsPbBr_3/Bi_2WO_6$	5.154×10^{14}	4.672	p

The morphologies of the as-prepared samples were confirmed by scanning electron microscopy (SEM). The Bi_2WO_6 displayed particles of a few hundred nanometers with exposed flat and smooth facets as shown in Figure 3a. In comparison to the pure Bi_2WO_6, the as-prepared $CsPbBr_3/Bi_2WO_6$ composite exhibited a rougher morphology due to the decoration of $CsPbBr_3$ dots on the facets of Bi_2WO_6 (Figure 3b). These SEM results were consistent with the cubic-shaped particles on the micro-square observed by TEM under low magnification (Figure 1b). The rough morphologies of the composite material lead to efficient light absorption due to their light trapping effect [41]. Besides, the surface area could be increased due to the presence of $CsPbBr_3$ quantum dots on Bi_2WO_6, which was confirmed by N_2 adsorption-desorption isotherms. The samples presented typical Type-IV adsorption isotherms with a H_3 hysteresis loop (Figure S1a), illustrating the porous nature of the catalysts [42]. Furthermore, the specific surface area of the Bi_2WO_6 and $CsPbBr_3/Bi_2WO_6$ were compared and $CsPbBr_3/Bi_2WO_6$ exhibited a more than 4-fold larger surface area than that of Bi_2WO_6. The details of the isotherm studies can be found in Table 3. The high surface area of $CsPbBr_3/Bi_2WO_6$ suggested an improvement in the active sites and adsorption capacity, which allows rapid interaction with the reactants. Next, the pure Bi_2WO_6 and $CsPbBr_3/Bi_2WO_6$ were subjected to thermal gravimetric analysis (TGA) using -temperatures ranging from room temperature to 600 °C with a heat rate of 5 °C/min in fluid He (20 mL/min). As illustrated in Figure S1b, $CsPbBr_3/Bi_2WO_6$ exhibited improved thermal stability in comparison to the pure Bi_2WO_6, especially from 250 °C to 600 °C. This could be attributed to the strong interaction between $CsPbBr_3$ and Bi_2WO_6.

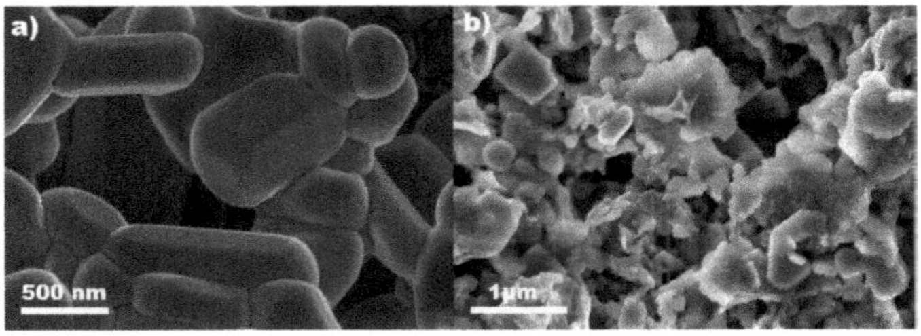

Figure 3. SEM images of (**a**) Bi_2WO_6 and (**b**) $CsPbBr_3/Bi_2WO_6$ photocatalysts.

Table 3. The specific surface of Bi_2WO_6 and $CsPbBr_3/Bi_2WO_6$.

Catalyst	BET Surface Area (m^2/g)	t-Plot External Surface Area (m^2/g)
$CsPbBr_3/Bi_2WO_6$	9.88	9.70
Bi_2WO_6	5.04	5.34

In Figure 4a, the UV-vis absorption spectra of pure Bi_2WO_6 and the $CsPbBr_3/Bi_2WO_6$ composite are illustrated. The pure Bi_2WO_6 presented an absorption edge at ca. 445 nm, corresponding to its typical band gap energy of 2.78 eV. After loading with $CsPbBr_3$ QDs, a dramatically enhanced absorption below ca. 510 nm was observed, which indicated the optimized light harvesting capacity of the composite photocatalyst [38,43–45]. In addition, as presented in Figure 4b, a decreased intensity and slightly blue-shift in the photoluminescence (PL) peak was observed when coupling the $CsPbBr_3$ with Bi_2WO_6, which indicated that the recombination rate of electron-hole pairs were suppressed, and thus they promote the photocatalytic efficiency [46].

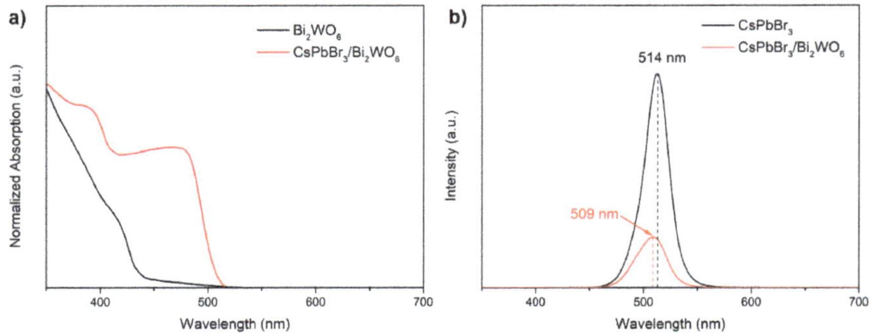

Figure 4. (**a**) UV-vis absorption spectra and (**b**) steady-state PL spectra of $CsPbBr_3$ and 15 wt% $CsPbBr_3/Bi_2WO_6$.

3.2. *Activity for Photodegradation of PAHs from Tobacco Tar*

Trifluorotoluene was employed as the solvent to dissolve the tobacco tar due to its apolar nature, good solubility, and superior chemical stability comparing with toluene. A list of the main substrates in the tobacco tar solution identified by gas chromatography-mass spectrometry (GC-MS) is given in Table S1. A Xe lamp (100 mW cm^{-2}) coupled with AM 1.5 G filters were utilized as the simulated solar source for illumination. GC-MS and UV-visible absorption spectra (UVAS) were exploited to evaluate the products and degradation rate of the photocatalysts. A series of tests was carried out under room temperature, and the photocatalytic reactor was saturated with different matter (air, H_2O, and O_2) to investigate the key radicals for promoting the degradation rate. To reach adsorption equilibrium, the as-prepared photocatalysts were initially mixed with tobacco tar under stirring (1200 rpm) in the dark for 30 min prior to the photocatalytic reactions.

As reported previously [47,48], the highly reactive hydroxyl radicals ($\cdot OH$) and superoxide radicals ($\cdot O_2^-$) formed in the photocatalytic process are desirable for degrading PAHs into less toxic products. Thus, the samples were saturated with air to supply the source H_2O and O_2 needed to accelerate the generation of hydroxyl and superoxide radicals. As shown in Figure 5, approximately 32% of the PAHs was degraded with the pure Bi_2WO_6 photocatalyst in 50 min. This can be compared to the $CsPbBr_3/Bi_2WO_6$ composite, which presented a significantly improved degradation rate. This can be attributed to the efficient charge separation of the type-II band structure and the enhanced light absorption of the $CsPbBr_3$. In addition, the degradation rate was systematically enhanced with the increasing weight ratio of the $CsPbBr_3$ in the composites (Figure S2), and reached its maximum at 96% when loaded with 15 wt% $CsPbBr_3$ (Figure 5a,b). The total organic carbon (TOC) content in the solvent was assessed by a TOC analyzer to further reflect the degree of PAHs degradation. As shown in Table S2, the TOC content in the reaction system decreased from 5.36 mg/L before the reaction to 2.3 mg/L after the reaction, indicating that the $CsPbBr_3/Bi_2WO_6$ composite photocatalyst is ideal for mineralizing PAHs. The products mainly consisted of monocyclic aromatic hydrocarbons with low toxicity and carbon oxide, as listed in Table S3. Although increasing the ratio of $CsPbBr_3$ in the composites enhances the light absorption due to the outstanding light harvesting capacity of perovskite, overloading leads to aggregations resulting in a longer migration pathway for photogenerated charges. In addition, the overloading might also reduce the exposed surface area of Bi_2WO_6 to reactants. Consequently, increasing the perovskite ratio further is not necessary to optimize the degradation rate. To the best of our knowledge, there is no previous research on the photocatalytic degradation of PAHs in tobacco tar. Thus, we compared our optimal sample (over 96% degradation in 50 min) with that for PAH removal in water treatment, such as TiO_2 (over 99% in 24 h) [49], ZnO/TiO_2 (98% in 2 h) [50], TiO_2 (over 66% in 39 min) [51], Evonik P25 (70% in 8 h) [52] and Pr/TiO_2 (over 99% in 3 h) [53].

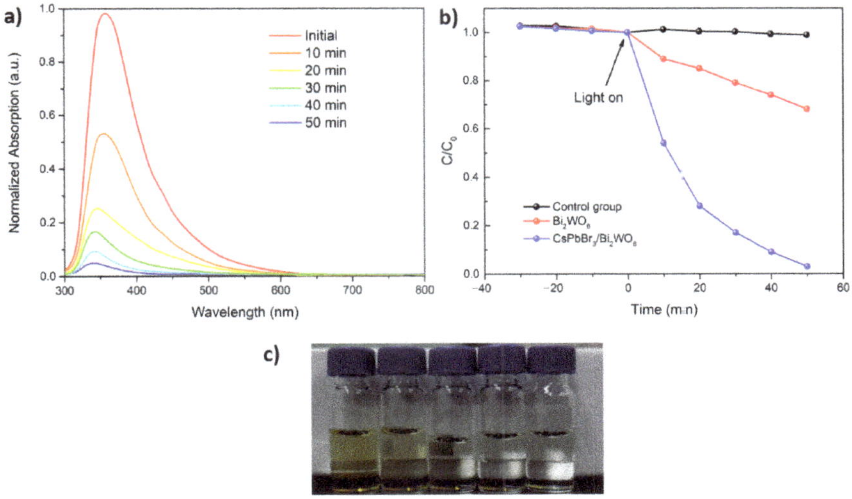

Figure 5. (**a**) The absorption spectra of the samples using 15 wt% $CsPbBr_3/Bi_2WO_6$ as a photocatalyst for different times; (**b**) photocatalytic degradation of tobacco tar using pure Bi_2WO_6, 15 wt% $CsPbBr_3/Bi_2WO_6$ and control group (without photocatalyst); (**c**) digital photographs of the sample using 15 wt% $CsPbBr_3/Bi_2WO_6$ as a photocatalyst for 10 to 50 min (from left to right). Reaction condition: 50 mg photocatalyst, saturated with air, stirred in the dark for 30 min before experiments; then, the reactions were conducted under AM 1.5 G simulated light irradiation for 50 min.

To investigate the photodegradation process, electron paramagnetic resonance (EPR) spectra were employed to confirm the generated reactive radicals under simulated solar light. The EPR signals of hydroxyl and superoxide radicals (Figure 6a) were recorded [38,54,55] after illumination for 5 min as these two radicals are typically considered responsible for the high degradation rate of PAHs. Control experiments were carried out to conf irm the respective influence of these radicals on the photodegradation. To this end, one sample was degassed by Ar and then saturated with O_2 to eliminate the source of hydroxyl generation, another sample was degassed with Ar and then 20 μL H_2O was added to inhibit the superoxide generation. Compared to the sample saturated with air, the degradation rate dramatically reduced in the absence of a source for hydroxyl generation (Figure 6b). In contrast, the sample showed a dramatically decreased degradation rate when the source of superoxide radicals was excluded. These results suggested that the hydroxyl radicals with oxidation reactivity were predominant in the degradation of PHAs, and degradation benefits from the presence of water and oxygen. Aside from photocatalytic activity, the stability of a photocatalyst is an important criterion. As shown in Figure S3, the photocatalytic activity only decreased by 7% after six consecutive cycles, suggesting that $CsPbBr_3/Bi_2WO_6$ photocatalyst has high stability and reusability. The XRD spectra in Figure S4 shows that the diffraction peaks of the material's crystal structure did not change significantly after six cycles. In addition, no trace of Pd ions was found after the reaction, which can be attributed to the stable nature of $CsPbBr_3$ in apolar solvent.

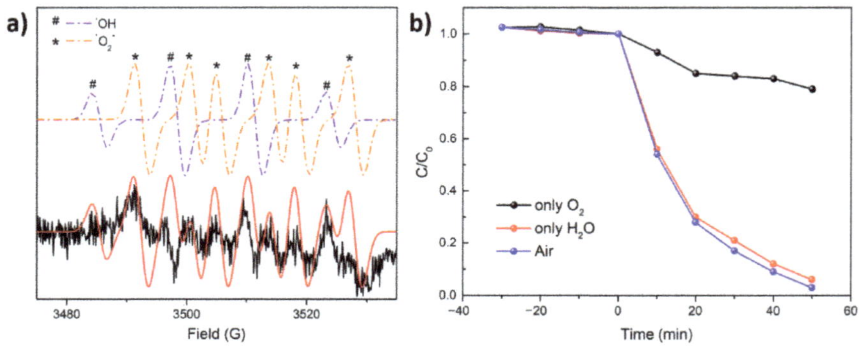

Figure 6. (**a**) EPR spectra of $CsPbBr_3/Bi_2WO_6$ sample saturated with air under illumination for 5 min and 5,5-dimethyl-pyrroline N-oxide (DMPO) was used as trap agent; (**b**) photocatalytic degradation rate of 15 wt% $CsPbBr_3/Bi_2WO_6$ sample when saturated with different gases.

4. Photocatalytic Mechanism

Based on the Tauc plots shown in Figure S5a,b, the band gap energy (E_g) of the Bi_2WO_6 and $CsPbBr_3$ samples were measured to be 2.71 eV, and 2.39 eV from the linear part of $(\alpha hv)^2 - (hv)$ curves. The VB values (E_{VB}) and CB values (E_{CB}) for a semiconductor can be calculated according to the following equations:

$$E_{CB} = E_{VB} - E_g \tag{1}$$

$$E_{VB} = \chi - E_e + 0.5E_g \tag{2}$$

The electronegativity (χ) of Bi_2WO_6 and $CsPbBr_3$ is 6.36 eV and 4.38 eV, respectively, according to previous reports [56,57]. E_e was 4.5 eV, which is the energy of free electrons on a hydrogen atom. Thus, we can conclude that the E_{VB} of Bi_2WO_6 and $CsPbBr_3$ was 3.25 and 1.075 eV while their associated E_{CB} was 0.505 eV and −1.315 eV, respectively, versus normal hydrogen electrode (NHE). On the basis of the above results, a possible photocatalytic mechanism for degradation of PAHs was proposed (Figure 7). The type II heterojunction was formed at the interfaces between $CsPbBr_3$ and Bi_2WO_6. Under illumination, both $CsPbBr_3$ and Bi_2WO_6 were excited to produce photogenerated electrons and holes. Then, to achieve Fermi level balance, the conduction electrons of $CsPbBr_3$ were transferred to the conduction band of Bi_2WO_6 under the potential difference of the band energy. At the same time, the valence holes of Bi_2WO_6 were transferred to the VB of $CsPbBr_3$. Finally, both $CsPbBr_3$ and Bi_2WO_6 have improved charge separation efficiency, and hence increased activity for oxidation and reduction, respectively. The reserved holes can oxide water molecules or hydroxyl ions to produce hydroxyl radicals. Meanwhile, the reserved electrons can reduce O_2 to yield superoxide radicals. Finally, according to the EPR spectra results, PAHs were degraded into octopamine, carbon dioxide, toluene and other substances under the joint action of superoxide and hydroxyl radicals. Of the two active radicals, hydroxyl radicals were predominant in the photocatalytic reaction.

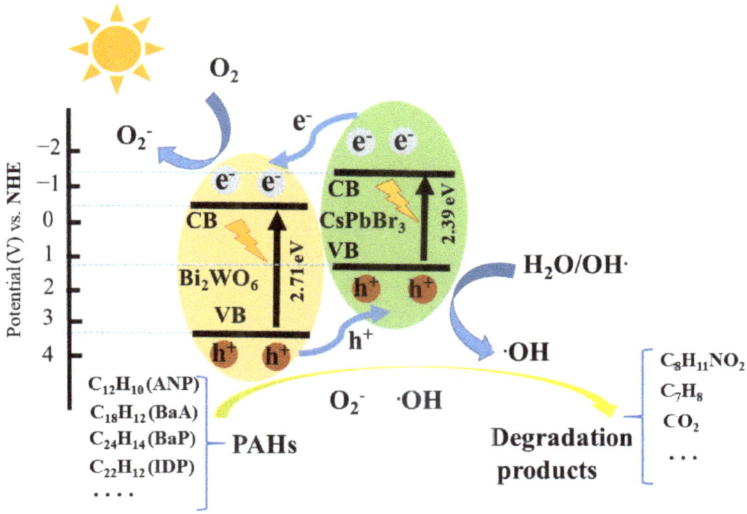

Figure 7. Schematic photocatalytic reaction process and charge transfer of the $CsPbBr_3/Bi_2WO_6$ composite catalyst under solar light irradiation.

5. Conclusions

In this research, $CsPbBr_3$ QDs decorated Bi_2WO_6 was successfully prepared by using a hydrothermal method. The sample was used as a photocatalyst to investigate its novel application for photodegradation of tobacco tar under visible radiation. The $CsPbBr_3/Bi_2WO_6$ presented outstanding activity compared to classic metal oxides catalysts, and the degradation products were mainly low toxic compounds. The findings of this work may open a promising avenue for producing high-efficiency photocatalysts with special heterostructures for removing organic contaminants from used tobacco tar, and could potentially be applied to novel cigarette filters in the future.

Supplementary Materials: The following are available online at https://www.mdpi.com/article/10.3390/nano11092422/s1, Figure S1. (a) adsorption/desorption isotherms of pure Bi_2WO_6 and $CsPbBr_3/Bi_2WO_6$; (b) the TGA curves of pure Bi_2WO_6 and $CsPbBr_3/Bi_2WO_6$; Figure S2. Degradation rate of PAHs using 15 wt% $CsPbBr_3/Bi_2WO_6$ composite photocatalyst with different $CsPbBr_3$ weight ratios; Figure S3. Degradation rate of PAHs using 15 wt% CsPbBr3/Bi2WO6 composite photocatalyst with air in different reaction cycles; Figure S4. XRD patterns of 15 wt% $CsPbBr_3/Bi_2WO_6$ before reaction and after six reactions; Figure S5. Tauc plots of the Bi_2WO_6 (a) and $CsPbBr_3$ (b); Table S1. The total organic carbon (TOC) content of PAHs with 15 wt% $CsPbBr_3/Bi_2WO_6$ nanocomposite; Table S2. The main substrates from tobacco tar identified by GC-MS; Table S3. The main products from the photodegradation of tobacco tar. Noted that there are side-products with low concentration.

Author Contributions: Conceptualization, D.L. and M.Z.; methodology, R.H. and Q.C.; software, K.W.; validation, X.L.; formal analysis, R.H.; investigation, X.L.; resources, D.L.; data curation, Z.Z.; writing—original draft preparation, R.H. and M.Z.; writing—review and editing, M.Z. and D.L.; visualization, R.H.; supervision, D.L.; project administration, M.Z.; funding acquisition, D.L. All authors have read and agreed to the published version of the manuscript.

Funding: The authors are grateful to the Science and Technology Program of Guangzhou (202103030001), Guangdong Basic and Applied Basic Research Foundation (Grant No. 2020B1515020032), the National Natural Science Foundation of China (Grant No. 62074060), the Guangdong Basic and Applied Basic Research Foundation (Grant No. 2020B1515120022) and the Guangdong Science and Technology Plan (Grant No. 2019B040403003).

Institutional Review Board Statement: Not applicable.

Informed Consent Statement: Not applicable.

Data Availability Statement: Not applicable.

Acknowledgments: The authors thank Weizhe Wang for proceeding the EPR data.

Conflicts of Interest: The authors declare no conflict of interest.

References

1. Singh, J.; Kumar, V.; Jolly, S.S.; Kim, K.-H.; Rawat, M.; Kukkar, D.; Tsang, Y.F. Biogenic synthesis of silver nanoparticles and its photocatalytic applications for removal of organic pollutants in water. *J. Ind. Eng. Chem.* **2019**, *80*, 247–257. [CrossRef]
2. Soltan, W.B.; Ammar, S.; Olivier, C.; Toupance, T. Influence of zinc doping on the photocatalytic activity of nanocrystalline SnO2 particles synthesized by the polyol method for enhanced degradation of organic dyes. *J. Alloys Compd.* **2017**, *729*, 638–647. [CrossRef]
3. Andreoli, R.; Spatari, G.; Pigini, D.; Poli, D.; Banda, I.; Goldoni, M.; Riccelli, M.G.; Petyx, M.; Protano, C.; Vitali, M.; et al. Urinary biomarkers of exposure and of oxidative damage in children exposed to low airborne concentrations of benzene. *Environ. Res.* **2015**, *142*, 264–272. [CrossRef]
4. Yan, Z.; Xu, Z.; Yu, J.; Jaroniec, M. Enhanced formaldehyde oxidation on CeO2/AlOOH-supported Pt catalyst at room temperature. *Appl. Catal. B Environ.* **2016**, *199*, 458–465. [CrossRef]
5. Husman, T. Health effects of indoor-air microorganisms. *Scand. J. Work Environ. Health* **1996**, *22*, 5–13. [CrossRef]
6. Alexandrov, L.B.; Ju, Y.S.; Haase, K.; Van Loo, P.; Martincorena, I.; Nik-Zainal, S.; Totoki, Y.; Fujimoto, A.; Nakagawa, H.; Shibata, T.; et al. Mutational signatures associated with tobacco smoking in human cancer. *Science* **2016**, *354*, 618–622. [CrossRef]
7. Enguita, F.J.; Leitao, A.L. Hydroquinone: Environmental pollution, toxicity, and microbial answers. *BioMed Res. Int.* **2013**, *2013*, 542168. [CrossRef]
8. Chen, X.; Sun, H.; Zelekew, O.A.; Zhang, J.; Guo, Y.; Zeng, A.; Kuo, D.-H.; Lin, J. Biological renewable hemicellulose-template for synthesis of visible light responsive sulfur-doped TiO2 for photocatalytic oxidation of toxic organic and As(III) pollutants. *Appl. Surf. Sci.* **2020**, *525*, 146531. [CrossRef]
9. Nagajyothi, P.; Vattikuti, S.; Devarayapalli, K.; Yoo, K.; Shim, J.; Sreekanth, T. Green synthesis: Photocatalytic degradation of textile dyes using metal and metal oxide nanoparticles-latest trends and advancements. *Crit. Rev. Environ. Sci. Technol.* **2019**, *50*, 2617–2723. [CrossRef]
10. Liu, Y.; Li, M.; Zhang, Q.; Qin, P.; Wang, X.; He, G.; Li, L. One-step synthesis of a WO_3-CuS nanosheet heterojunction with enhanced photocatalytic performance for methylene blue degradation and Cr(VI) reduction. *J. Chem. Technol. Biotechnol.* **2019**, *95*, 665–674. [CrossRef]
11. Thirukumaran, P.; Atchudan, R.; Parveen, A.S.; Kalaiarasan, K.; Lee, Y.R.; Kim, S.-C. Fabrication of ZnO nanoparticles adorned nitrogen-doped carbon balls and their application in photodegradation of organic dyes. *Sci. Rep.* **2019**, *9*, 19509. [CrossRef] [PubMed]
12. He, F.; Muliane, U.; Weon, S.; Choi, W. Substrate-specific mineralization and deactivation behaviors of TiO2 as an air-cleaning photocatalyst. *Appl. Catal. B Environ.* **2020**, *275*, 119145. [CrossRef]
13. Wang, H.; Qiu, X.; Liu, W.; Yang, D. Facile preparation of well-combined lignin-based carbon/ZnO hybrid composite with excellent photocatalytic activity. *Appl. Surf. Sci.* **2017**, *426*, 206–216. [CrossRef]
14. Ullah, I.; Munir, A.; Muhammad, S.; Ali, S.; Khalid, N.; Zubair, M.; Sirajuddin, M.; Hussain, S.Z.; Ahmed, S.; Khan, Y.; et al. Influence of W-doping on the optical and electrical properties of SnO2 towards photocatalytic detoxification and electrocatalytic water splitting. *J. Alloys Compd.* **2020**, *827*, 154247. [CrossRef]
15. Parida, K.M.; Reddy, K.H.; Martha, S.; Das, D.P.; Biswal, N. Fabrication of nanocrystalline LaFeO3: An efficient sol–gel auto-combustion assisted visible light responsive photocatalyst for water decomposition. *Int. J. Hydrogen Energy* **2010**, *35*, 12161–12168. [CrossRef]
16. Yadav, A.A.; Hunge, Y.M.; Mathe, V.L.; Kulkarni, S.B. Photocatalytic degradation of salicylic acid using BaTiO3 photocatalyst under ultraviolet light illumination. *J. Mater. Sci. Mater. Electron.* **2018**, *29*, 15069–15073. [CrossRef]
17. Zhang, G.; Liu, G.; Wang, L.; Irvine, J.T.S. Inorganic perovskite photocatalysts for solar energy utilization. *Chem. Soc. Rev.* **2016**, *45*, 5951–5984. [CrossRef]
18. Xie, T.; Liu, Y.; Wang, H.; Wu, Z. Synthesis of α-Fe2O3/Bi2WO6 layered heterojunctions by in situ growth strategy with enhanced visible-light photocatalytic activity. *Sci. Rep.* **2019**, *9*, 7551. [CrossRef]
19. Dumrongrojthanath, P.; Phuruangrat, A.; Thongtem, S.; Thongtem, T. Hydrothermal preparation of visible-light-driven Br-doped Bi2WO6 photocatalyst. *Mater. Lett.* **2017**, *209*, 501–504. [CrossRef]
20. Zhang, P.; Teng, X.; Feng, X.; Ding, S.; Zhang, G. Preparation of Bi2WO6 photocatalyst by high-energy ball milled Bi2O3-WO3 mixture. *Ceram. Int.* **2016**, *42*, 16749–16757. [CrossRef]
21. Kaur, A.; Kansal, S.K. Bi2WO6 nanocuboids: An efficient visible light active photocatalyst for the degradation of levofloxacin drug in aqueous phase. *Chem. Eng. J.* **2016**, *302*, 194–203. [CrossRef]
22. Zhuo, Y.; Huang, J.; Cao, L.; Ouyang, H.; Wu, J. Photocatalytic activity of snow-like Bi2WO6 microcrystalline for decomposition of Rhodamine B under natural sunlight irradiation. *Mater. Lett.* **2013**, *90*, 107–110. [CrossRef]
23. Liu, M.; Xue, X.; Yu, S.; Wang, X.; Hu, X.; Tian, H.; Chen, H.; Zheng, W. Improving Photocatalytic Performance from Bi2WO6@MoS2/graphene Hybrids via Gradual Charge Transferred Pathway. *Sci. Rep.* **2017**, *7*, 3637. [CrossRef]

24. Zhao, S.; Jin, H.; Yang, Y.; Cui, J. Magnetically retrievable Bi2WO6/Fe3O4/Na-MMT composite: Fabrication and photocatalytic activity. *Res. Chem. Intermed.* **2020**, *46*, 4579–4593. [CrossRef]
25. Shen, Z.; Li, H.; Hao, H.; Chen, Z.; Hou, H.; Zhang, G.; Bi, J.; Yan, S.; Liu, G.; Gao, W. Novel Tm3+ and Yb3+ co-doped bismuth tungstate up-conversion photocatalyst with greatly improved photocatalytic properties. *J. Photochem. Photobiol. A Chem.* **2019**, *380*, 111864. [CrossRef]
26. Liu, Z.; Liu, X.; Wei, L.; Yu, C.; Yi, J.; Ji, H. Regulate the crystal and optoelectronic properties of Bi2WO6 nanosheet crystals by Sm3+ doping for superior visible-light-driven photocatalytic performance. *Appl. Surf. Sci.* **2020**, *508*, 145309. [CrossRef]
27. Hou, X.; Liu, J.; Guo, W.; Li, S.; Guo, Y.; Shi, Y.; Zhang, C. A novel 3D-structured flower-like bismuth tungstate/mag-graphene nanoplates composite with excellent visible-light photocatalytic activity for ciprofloxacin degradation. *Catal. Commun.* **2019**, *121*, 27–31. [CrossRef]
28. Kadeer, K.; Tursun, Y.; Dilinuer, T.; Okitsu, K.; Abulizi, A. Sonochemical preparation and photocatalytic properties of CdS QDs/Bi2WO6 3D heterojunction. *Ceram. Int.* **2018**, *44*, 13797–13805. [CrossRef]
29. Parnicka, P.; Mikolajczyk, A.; Pinto, H.P.; Lisowski, W.; Klimczuk, T.; Trykowski, G.; Bajorowicz, B.; Zaleska-Medynska, A. Experimental and DFT insights into an eco-friendly photocatalytic system toward environmental remediation and hydrogen generation based on AgInS2 quantum dots embedded on Bi2WO6. *Appl. Surf. Sci.* **2020**, *525*, 146596. [CrossRef]
30. Ge, L.; Liu, J. Efficient visible light-induced photocatalytic degradation of methyl orange by QDs sensitized CdS-Bi2WO6. *Appl. Catal. B Environ.* **2011**, *105*, 289–297. [CrossRef]
31. Schlaus, A.P.; Spencer, M.S.; Miyata, K.; Liu, F.; Wang, X.; Datta, I.; Lipson, M.; Pan, A.; Zhu, X.Y. How lasing happens in CsPbBr3 perovskite nanowires. *Nat. Commun.* **2019**, *10*, 265. [CrossRef]
32. Liu, B.; Feng, S.; Yang, L.; Li, C.; Luo, Z.; Wang, T.; Gong, J. Bifacial passivation of n-silicon metal–insulator–semiconductor photoelectrodes for efficient oxygen and hydrogen evolution reactions. *Energy Environ. Sci.* **2020**, *13*, 221–228. [CrossRef]
33. Corti, M.; Bonomi, S.; Chiara, R.; Romani, L.; Quadrelli, P.; Malavasi, L. Application of Metal Halide Perovskites as Photocatalysts in Organic Reactions. *Inorganics* **2021**, *9*, 56. [CrossRef]
34. Zhu, X.; Lin, Y.; San Martin, J.; Sun, Y.; Zhu, D.; Yan, Y. Lead halide perovskites for photocatalytic organic synthesis. *Nat. Commun.* **2019**, *10*, 2843. [CrossRef] [PubMed]
35. Cho, S.; Yun, S.H. Structure and optical properties of perovskite-embedded dual-phase microcrystals synthesized by sonochemistry. *Commun. Chem.* **2020**, *3*, 15. [CrossRef]
36. Mano, G.; Harinee, S.; Sridhar, S.; Ashok, M.; Viswanathan, A. Microwave assisted synthesis of ZnO-PbS heterojuction for degradation of organic pollutants under visible light. *Sci. Rep.* **2020**, *10*, 2224. [CrossRef]
37. Hu, J.; Yang, M.; Ke, X.; Yang, S.; Wang, K.; Huang, H.; Wang, W.; Luo, D.; Zheng, Z.; Huang, L.; et al. Cubic-cubic perovskite quantum dots/PbS mixed dimensional materials for highly efficient CO2 reduction. *J. Power Sources* **2021**, *481*, 228838. [CrossRef]
38. Huang, H.; Zhao, J.; Du, Y.; Zhou, C.; Zhang, M.; Wang, Z.; Weng, Y.; Long, J.; Hofkens, J.; Steele, J.A.; et al. Direct Z-Scheme Heterojunction of Semicoherent FAPbBr3/Bi2WO6 Interface for Photoredox Reaction with Large Driving Force. *ACS Nano* **2020**, *14*, 16689–16697. [CrossRef]
39. Choi, H.; Stathatos, E.; Dionysiou, D.D. Photocatalytic TiO2 films and membranes for the development of efficient wastewater treatment and reuse systems. *Desalination* **2007**, *202*, 199–206. [CrossRef]
40. Lin, R.; Wan, J.; Xiong, Y.; Wu, K.; Cheong, W.C.; Zhou, G.; Wang, D.; Peng, Q.; Chen, C.; Li, Y. Quantitative Study of Charge Carrier Dynamics in Well-Defined WO3 Nanowires and Nanosheets: Insight into the Crystal Facet Effect in Photocatalysis. *J. Am. Chem. Soc.* **2018**, *140*, 9078–9082. [CrossRef]
41. Tekin, D.; Tekin, T.; Kiziltas, H. Photocatalytic degradation kinetics of Orange G dye over ZnO and Ag/ZnO thin film catalysts. *Sci. Rep.* **2019**, *9*, 17544. [CrossRef]
42. Kasinathan, M.; Thiripuranthagan, S.; Sivakumar, A. A facile fabrication of Br-modified g-C3N4/rGO composite catalyst for enhanced visible photocatalytic activity towards the degradation of harmful dyes. *Mater. Res. Bull.* **2020**, *130*, 110870. [CrossRef]
43. Zhang, J.; Fu, X.; Hao, H.; Gan, W. Facile synthesis 3D flower-like Ag@WO3 nanostructures and applications in solar-light photocatalysis. *J. Alloys Compd.* **2018**, *757*, 134–141. [CrossRef]
44. Pang, H.-F.; Xiang, X.; Li, Z.-J.; Fu, Y.-Q.; Zu, X.-T. Hydrothermal synthesis and optical properties of hexagonal tungsten oxide nanocrystals assisted by ammonium tartrate. *Phys. Status Solidi* **2012**, *209*, 537–544. [CrossRef]
45. Bhagyalakshmi, S.B.; Ghimire, S.; Takahashi, K.; Yuyama, K.I.; Takano, Y.; Nakamura, T.; Biju, V. Nonradiative Energy Transfer through Distributed Bands in Piezochemically Synthesized Cesium and Formamidinium Lead Halide Perovskites. *Chemistry* **2020**, *26*, 2133–2137. [CrossRef]
46. Thorat, A.A.; Dalvi, S.V. Liquid antisolvent precipitation and stabilization of nanoparticles of poorly water soluble drugs in aqueous suspensions: Recent developments and future perspective. *Chem. Eng. J.* **2012**, *181–182*, 1–34. [CrossRef]
47. Chen, S.; Lu, W.; Shen, H.; Xu, S.; Chen, X.; Xu, T.; Wang, Y.; Chen, Y.; Gu, Y.; Wang, C.; et al. The development of new pigments: Colorful g-C3N4-based catalysts for nicotine removal. *Appl. Catal. B Environ.* **2019**, *254*, 500–509. [CrossRef]
48. Liu, B.; Chen, B.; Zhang, B.; Song, X.; Zeng, G.; Lee, K. Photocatalytic ozonation of offshore produced water by TiO2 nanotube arrays coupled with UV-LED irradiation. *J. Hazard. Mater.* **2021**, *402*, 123456. [CrossRef]
49. Woo, O.T.; Chung, W.K.; Wong, K.H.; Chow, A.T.; Wong, P.K. Photocatalytic oxidation of polycyclic aromatic hydrocarbons: Intermediates identification and toxicity testing. *J. Hazard. Mater.* **2009**, *168*, 1192–1199. [CrossRef] [PubMed]

50. Vela, N.; Martínez-Menchón, M.; Navarro, G.; Pérez-Lucas, G.; Navarro, S. Removal of polycyclic aromatic hydrocarbons (PAHs) from groundwater by heterogeneous photocatalysis under natural sunlight. *J. Photochem. Photobiol. A Chem.* **2012**, *232*, 32–40. [CrossRef]
51. Lair, A.; Ferronato, C.; Chovelon, J.-M.; Herrmann, J.-M. Naphthalene degradation in water by heterogeneous photocatalysis: An investigation of the influence of inorganic anions. *J. Photochem. Photobiol. A Chem.* **2008**, *193*, 193–203. [CrossRef]
52. Benkhennouche-Bouchene, H.; Mahy, J.G.; Wolfs, C.; Vertruyen, B.; Poelman, D.; Eloy, P.; Hermans, S.; Bouhali, M.; Souici, A.; Bourouina-Bacha, S.; et al. Green Synthesis of N/Zr Co-Doped TiO2 for Photocatalytic Degradation of p-Nitrophenol in Wastewater. *Catalysts* **2021**, *11*, 235. [CrossRef]
53. Wen, S.; Zhao, J.; Sheng, G.; Fu, J.; Peng, P. Photocatalytic reactions of phenanthrene at TiO2/water interfaces. *Chemosphere* **2002**, *46*, 871–877. [CrossRef]
54. Luo, H.; Zeng, Y.; He, D.; Pan, X. Application of iron-based materials in heterogeneous advanced oxidation processes for wastewater treatment: A review. *Chem. Eng. J.* **2021**, *407*, 127191. [CrossRef]
55. Su, F.; Mathew, S.C.; Lipner, G.; Fu, X.; Antonietti, M.; Blechert, S.; Wang, X. mpg-C3N4-Catalyzed Selective Oxidation of Alcohols Using O-2 and Visible Light. *J. Am. Chem. Soc.* **2010**, *132*, 16299–16301. [CrossRef] [PubMed]
56. Wu, Z.; Jin, C.; Li, Z.; Yang, D.; Xie, Y.; Wang, M. MoS_2 and g-C_3N_4 nanosheet co-modified Bi_2WO_6 ternary heterostructure catalysts coupling with H_2O_2 for improved visible photocatalytic activity. *Mater. Chem. Phys.* **2019**, *272*, 124982. [CrossRef]
57. Gong, Y.; Shen, J.; Zhu, Y.; Yan, W.; Zhu, J.; Hou, L.; Xie, D.; Li, C. Tailoring charge transfer in perovskite quantum dots/black phosphorus nanosheets photocatalyst via aromatic molecules. *Appl. Surf. Sci.* **2021**, *545*, 149012. [CrossRef]

nanomaterials

Article

Nitrogen-Containing Gas Sensing Properties of 2-D Ti_2N and Its Derivative Nanosheets: Electronic Structures Insight

Hongni Zhang [1,2], Wenzheng Du [1], Jianjun Zhang [1], Rajeev Ahuja [3,4] and Zhao Qian [1,*]

1 Key Laboratory of Liquid-Solid Structural Evolution and Processing of Materials (Ministry of Education), Shandong University, Jinan 250061, China; nini810@126.com (H.Z.); 201813740@mail.sdu.edu.cn (W.D.); 15163625981@163.com (J.Z.)
2 College of Industry and Commerce, Shandong Management University, Jinan 250357, China
3 Condensed Matter Theory, Department of Physics and Astronomy, Ångström Laboratory, Uppsala University, 75120 Uppsala, Sweden; rajeev.ahuja@physics.uu.se
4 Department of Materials Science and Engineering, KTH Royal Institute of Technology, 10044 Stockholm, Sweden
* Correspondence: qianzhao@sdu.edu.cn

Abstract: In this work, the potentials of two-dimensional Ti_2N and its derivative nanosheets Ti_2NT_2(T=O, F, OH) for some harmful nitrogen-containing gas (NCG) adsorption and sensing applications have been unveiled based on the quantum-mechanical Density Functional Theory calculations. It is found that the interactions between pure Ti_2N and NCGs (including NO, NO_2, and NH_3 in this study) are very strong, in which NO and NO_2 can even be dissociated, and this would poison the substrate of Ti_2N monolayer and affect the stability of the sensing material. For the monolayer of Ti_2NT_2(T=O, F, OH) that is terminated by functional groups on surface, the adsorption energies of NCGs are greatly reduced, and a large amount of charges are transferred to the functional group, which is beneficial to the reversibility of the sensing material. The significant changes in work function imply the good sensitivity of the above mentioned materials. In addition, the fast response time further consolidates the prospect of two-dimensional Ti_2NT_2 as efficient NCGs' sensing materials. This theoretical study would supply physical insight into the NCGs' sensing mechanism of Ti_2N based nanosheets and help experimentalists to design better 2-D materials for gas adsorption or sensing applications.

Keywords: atomic structure; electronic structure; Density Functional Theory (DFT); MXenes; two-dimensional materials

Citation: Zhang, H.; Du, W.; Zhang, J.; Ahuja, R.; Qian, Z. Nitrogen-Containing Gas Sensing Properties of 2-D Ti_2N and Its Derivative Nanosheets: Electronic Structures Insight. *Nanomaterials* **2021**, *11*, 2459. https://doi.org/10.3390/nano11092459

Academic Editor: Jihoon Lee

Received: 12 August 2021
Accepted: 14 September 2021
Published: 21 September 2021

1. Introduction

Industrial pollutant emissions account for a large share of environmental pollution, and one of the most harmful is nitrogen-containing gas (NCGs), such as NO, NO_2, and NH_3. At the same time, the nitrogen-containing gases themselves are generally toxic and can damage the respiratory cells of humans and animals [1–4]. Therefore, there is an urgent need to explore efficient and alternative materials for the sensing or capturing of these gases. Although some current solid-state sensors can also detect toxic gases, there is still much room for improvement due to the increasingly high demand for sensitivity, selectivity, stability, and other factors [5,6]. Two-dimensional (2-D) materials have large contact areas with gas and special electronic properties; thus, their interactions with gas molecules have been widely studied, especially in gas sensing and capturing [7]. For example, Zhang et al. studied the adsorption behavior between four modified graphene substrates and small gas molecules (CO, NO, NO_2, and NH_3) and significantly improved the performance of the gas sensor by introducing doping or defects [8]. Liu et al. found that the GeSe monolayer had potential for NH_3 gas capture and storage, as well as for the detection and catalysis of SO_2 and NO_2. The optical properties of this GeSe monolayer can also be tuned as a candidate material for optical sensors by adsorbing different small molecules [9]. Li et al. compared the structures and electronic properties of pure and two vacancy-defected MoS_2 nanosheets with NO adsorption. Defects can effectively improve the NO-sensing performance [10].

Transition metal carbide or nitride MXenes are also one family of promising gas-sensing materials [11,12], where M is a transition metal and X is carbon or nitrogen. The general formula of MXenes is written as M_nX_{n-1}. During the preparation process in an aqueous solution containing fluoride ions, there may be some O, F, and OH functional groups formed [13–15]. This type of MXene with functional groups is written as $M_nX_{n-1}T_n$(T=O, F, OH). MXenes are also widely reported in energy storage, electrode materials, ion batteries, and the spintronic and optical devices, due to their unique structures and properties [16–24].

After our review of previous literature, we found that some carbide-based MXenes have the potential to be used as sensors for molecules, such as amino acids and nucleobase [25,26]. With less exfoliation energy compared with other types, the synthesizability of the nitride MXenes is good [27], while there are fewer studies on their use in sensing NCGs [28–32]. In order to obtain a theoretical understanding of the nitride MXenes in terms of their sensing or capturing performance towards nitrogen-containing gas (NCGs), in this work, we conducted a detailed computational study on the adsorption behaviors of NCGs (including NO, NO_2, and NH_3) of Ti_2N monolayer with bare surface and terminated with O, F, and OH functional groups. Density Functional Theory–based first-principles calculations were performed to unveil the adsorption structures, charge transfer, band structures, density of states, and work functions to stimulate and inspire the deep mechanism investigations.

2. Methods

DFT calculations were employed by using the projector augmented wave (PAW) [33] pseudo-potentials in a Vienna *Ab initio* Simulation Package (VASP) [34]. For the exchange-correlation functional, we used the Perdew-Burke-Ernzerhof (PBE) [35] form of the generalized gradient approximation (GGA). Grimme's DFT-D2 method was used to describe the van der Waals interactions [36]. This method adds semi-empirical dispersion on the basis of conventional Kohn–Sham DFT energy. In calculations, Ti_2N and Ti_2NT_2(T=O, F, OH) were constructed as $4 \times 4 \times 1$ supercells, and a 20 Å vacuum layer was added in *Z*-axis direction to eliminate the interlayer interaction. The energy cutoff of 520 eV was used for wave functions' expansion. The Brillouin zone integration was sampled with 5×5 **k**-grid mesh for geometry optimizations and 15×15 **k**-grid mesh for electronic properties calculations to achieve high accuracy. Geometry optimizations were performed by using the conjugated gradient method, and the convergence criterion was set to be 0.02 eV/Å on force and 1×10^{-5} eV/atom on energy. The adsorption energy (E_{ads}) of nitrogen-containing gas (NCGs) on Ti_2N or Ti_2NT_2(T=O, F, OH) is defined as follows:

$$E_{ads} = E_{MXene + NCGs} - E_{NCGs} - E_{MXene} \quad (1)$$

where E_{ads}, $E_{MXene + NCGs}$, E_{MXene}, and E_{NCGs} stand for the adsorption energy, total energies of optimized Ti_2N or Ti_2NT_2(T=O, F, OH) MXene with adsorbed NCGs molecules, energies of pure MXene, and isolated NCGs (NO, NO_2, and NH_3) molecules respectively.

3. Results and Discussion

First of all, the Ti_2N MXene has a hexagonal structure with three layers of atoms: the middle is N-atoms layer; and the upper (a) and lower (b) layers are Ti atoms, as is shown in Figure 1a. After structural optimizations, the Ti-N bond length is 2.07 Å, the bond angle of the same layer Ti and N (Ti–N–Ti) and N–Ti–N are both 92.4°, and the bond angle of different-layer Ti and N (Ti–N–Ti) is 87.6°. The lattice parameter of the 4×4 supercell is a = b = 11.98 Å. These parameters are very consistent with the results of Ti_2N studied in other works in the literature [37,38], indicating the reliability of the calculation results. The band structure and PDOS of the material are shown in Figure 1b; it can be seen that there are many electronic states available for occupation at the Fermi level, and Ti_2N exhibits metallic properties. The states near the Fermi level are occupied by Ti electrons, where N's contribution is relatively small, meaning that low-energy carriers mainly originate from Ti.

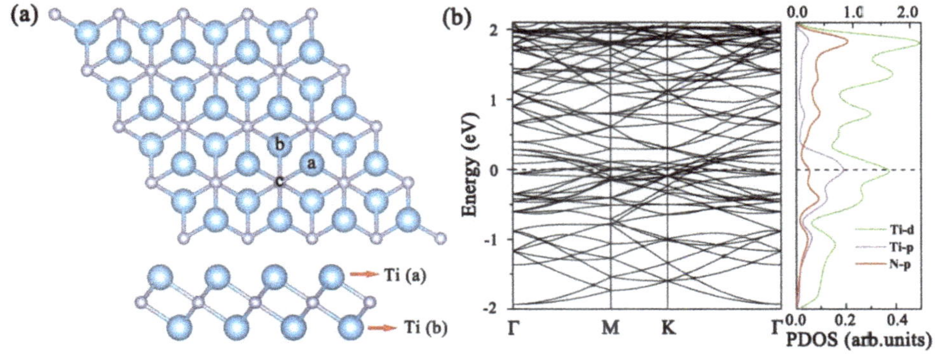

Figure 1. (**a**) Top and side view of the optimized structures for Ti_2N monolayer (blue and silver balls represent Ti and N atoms, respectively). (**b**) Band structure and projected density of states of Ti_2N monolayer (Ti*d*-orbital uses top abscissa and other orbitals use bottom coordinate).

In order to study the most stable adsorption configurations of NCGs on Ti_2N MXene, the NO, NO_2, and NH_3 molecules were respectively placed at three different adsorption sites, namely on top of the Ti(a) atom shown in Figure 1a (site "a"), on top of the hollow Ti(b) atom (site "b"), or on top of the N atom (site "c"). The screened most stable configurations after adsorbing NCGs are shown in Figure 2. Both NO and NO_2 show dissociative adsorption. The adsorption energy is far greater than those of other two-dimensional materials [39–44]. When NO is adsorbed on Ti_2N, N and O atoms are adsorbed at the adjacent site "b" respectively. The distance between N and O is 3.07 Å, which is much larger than 1.15 Å of free NO gas molecule, indicating that the N–O bond is broken and NO dissociates. The dissociated N and O atoms form new bonds with Ti atoms in the upper layer of Ti_2N, with bond lengths of 1.93 and 1.97 Å, respectively. Ti atoms protrude slightly upwards from its binding, and the Ti–N bond length on the surrounding matrix increases by about 0.14 Å. It can be seen from Figure 3a that, after dissociation, the N 2*p* and O 2*p* in NO and the nearby Ti 3*d*-orbitals have similar peak positions and obvious *p-d* hybridization in PDOS diagram. There is a strong bonding between NO and Ti_2N, and the adsorption energy is as high as −9.809 eV. At the same time, the band structure shows that the dissociative adsorption of NO on Ti_2N does not change the original metallic properties.

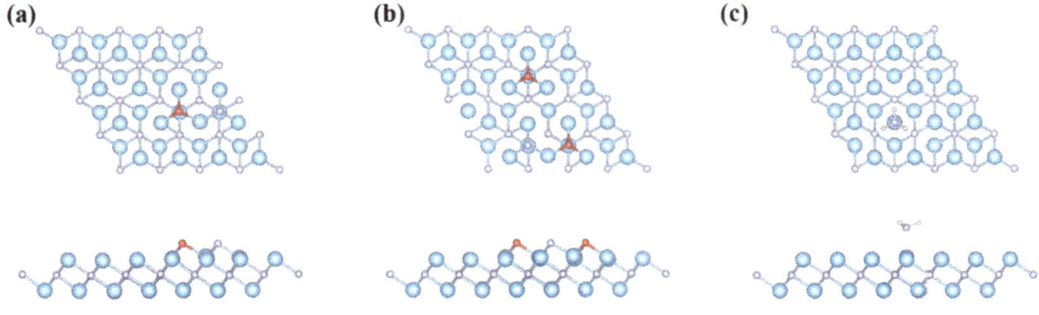

Figure 2. Top and side view of the optimized atomic structures for Ti_2N monolayer after adsorbing (**a**) NO. (**b**) NO_2, and (**c**) NH_3 (blue, silver, red, and pink balls represent Ti, N, O, and H atoms, respectively).

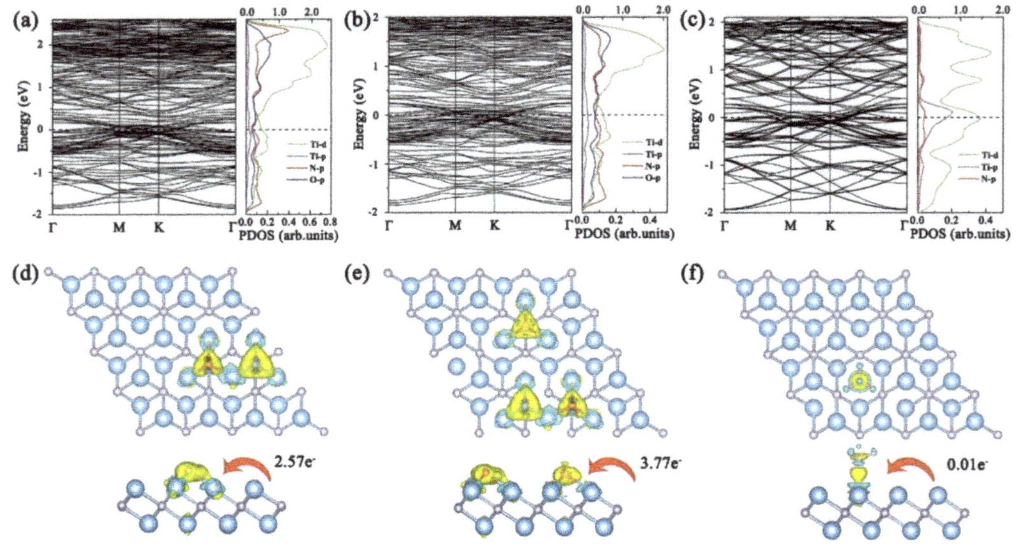

Figure 3. Band structures, PDOS and differential charge densities of Ti_2N after adsorbing (**a**,**d**) NO, (**b**,**e**) NO_2, and (**c**,**f**) NH_3 (Ti*d*-orbital uses top abscissa and other orbitals use bottom coordinate; isovalue of NO and NO_2 is set to 0.01 e/Å^3, NH_3 is set to 0.005 e/Å^3).

When NO_2 interacts with Ti_2N, it is interesting that NO_2will also dissociate similar to NO adsorption, and all the N and O atoms are adsorbed at the site "b". One O-atom keeps the same shape adsorbed at adjacent position, while the another O-atom is adsorbed away from it. The N-O bond length in $NO_2$is increased to 3.07 Å. The dissociative adsorption energy is up to −14.336 eV. As can be seen from the electronic structure in Figure 3b, the *p-d* hybridization of N 2*p*-, O 2*p*-, and Ti 3*d*-orbitals corresponds with the strong bonding of Ti-N and Ti-O, respectively. The metallic behavior of Ti_2N remains after NO_2 dissociative adsorption. From the differential charge density in Figure 3d–f, it can be seen that the dissociation of NO and NO_2 is caused by the excessive charge transfer from Ti_2N MXene to the gas molecule. The Bader charge analysis shows that 2.57 e^- and 3.77 e^- charges were transferred from Ti_2N to NO and NO_2, respectively. The bulk of these transferred charges comes from the surrounding Ti atoms, and this is also the fundamental reason why the N-O bond is broken.

The adsorption of NH_3 is completely different from those of NO and NO_2. It is bound by associative adsorption with the binding distance of 2.2 Å and the E_{ads} value of−1.286 eV. As shown in Figure 3c, NH_3 tends to be adsorbed at site "a" with N-atom below and three H-atoms evenly distributed above. The distance between the Ti atom at site "a" and the surrounding N atom is elongated to 2.14 Å, and the N–Ti–N bond angle is reduced to 88.3°. To figure out the binding and charge transfer mechanism between NH_3 molecule and Ti_2N monolayer, the band structure and PDOS were analyzed. It is difficult to find a pronounced overlap between N *p* and Ti*d* close to the Fermi level in Figure 3c. The amount of charge transfer is only 0.01 e^-, which is far less than those of NO and NO_2. The bonding between NH_3 and Ti_2N is very weak, and the band structure that is almost the same as the bare Ti_2N monolayer also confirms this.

For ideal gas sensing application, excessive adsorption energy is obviously not feasible. The dissociative adsorption of NO and NO_2 on Ti_2N monolayer will obviously poison the substrate and make it fail to work cyclically. The dissociation of gas molecules results from the strong interaction between the bare Ti_2N surface and O atoms. Thus, the pure Ti_2N monolayer can be used as a kind of gas-trapping material instead of highly efficient sensor for NO and NO_2. Compared with those two gases, the adsorption effect of NH_3

is satisfactory. The associative adsorption with lower energy will not cause chemical poisoning, which is beneficial to Ti_2N as a promising candidate material for reversible NH_3 sensors. In actual preparation process, some functional groups are usually introduced onto the Ti_2N monolayer, and these functional groups may have significant impacts on adsorption properties of the material. Considering this reality, in the following part, we recount how we studied the adsorption behaviors of the Ti_2N monolayer terminated by O, F, and OH towards NCGs.

We introduce O, F, and OH functional groups at three adsorption positions (site "a", "b", and "c") of Ti_2N, of which OH only considers the vertical direction (O atom is close to the matrix). All the functional groups tend to adsorb at site "b" with the lowest energy, as shown in Figure 4a–c. After adsorbing different functional groups, the Ti–N bonds show different degrees of extension: 2.164, 2.052, and 2.065 Å for Ti_2NO_2,Ti_2NF_2, and $Ti_2N(OH)_2$, respectively, which is due to the bonding between Ti and functional groups. Gouveia et al. reported possible atomic-layer-stacking-phase transitions of some MXenes [41]; thus, we also carefully checked this point in our research. It is found that the material system maintains phase stability and stacking way after introduction of those functional groups. The Bader charge analysis shows that 1.1, 0.8, and 0.8 e^- are respectively transferred to O, F, and OH from Ti_2N. It can also be observed in Figure 4d–f that there is a strong overlap between the Ti*d*-orbital and the T (O, F, or OH)*d*-orbitals.

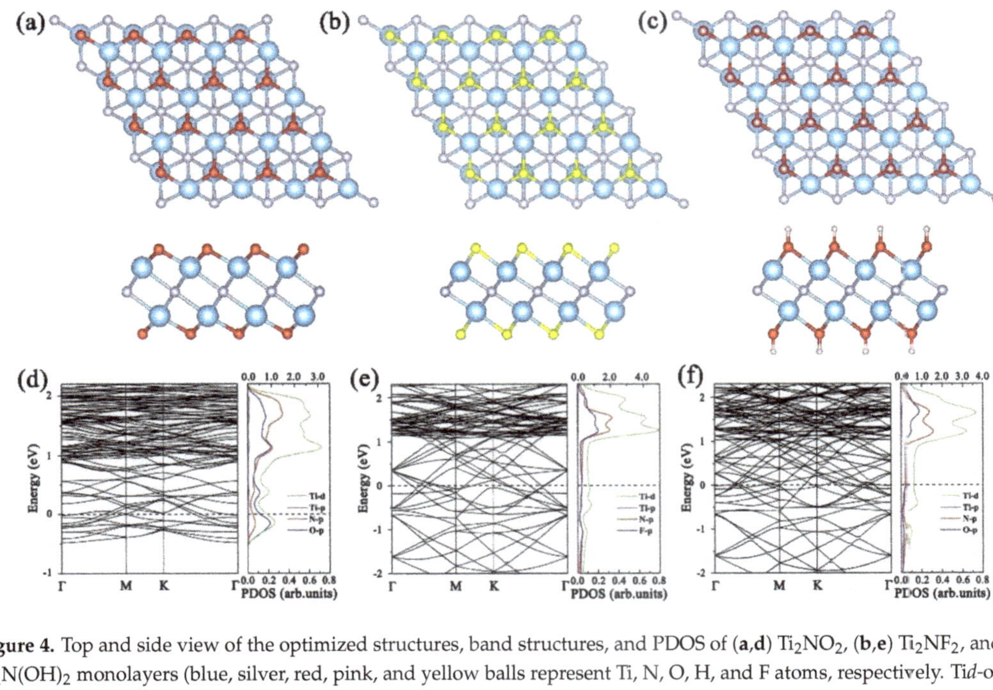

Figure 4. Top and side view of the optimized structures, band structures, and PDOS of (**a**,**d**) Ti_2NO_2, (**b**,**e**) Ti_2NF_2, and (**c**,**f**) $Ti_2N(OH)_2$ monolayers (blue, silver, red, pink, and yellow balls represent Ti, N, O, H, and F atoms, respectively. Ti*d*-orbital uses top abscissa and other orbitals use bottom coordinate).

After obtaining the stable configurations of Ti_2NT_2(T=O, F, OH), we placed various NCGs molecules at different adsorption sites (site "a", "b", and "c"), such as those in pure Ti_2N. It can be seen from the most stable adsorption structures after optimizations in Figure 5 that all NCGs have no dissociative adsorption on Ti_2NT_2(T=O, F, OH). The binding distance is defined as the shortest direct distance between the NCGs molecule and the matrix; hereby, the binding distances of NO, NO_2, and NH_3 on Ti_2NO_2 are 2.082, 1.824, and 2.172 Å, respectively. The bond lengths of NO and NH_3 after adsorption are the same as those of free gas molecules, while one of the N–O bonds of NO_2 is stretched to 1.530 Å.

The 2-D Ti_2NO_2 does not change significantly under NO environment, while the Ti atoms exposed to the adsorption positions of NO_2 and NH_3 are obviously dragged out. For NCGs adsorption on the surface of Ti_2NF_2, the molecular bond lengths and the matrix are almost unchanged with the adsorption distances of 2.951, 2.800, and 2.947 Å, respectively. It is surprising when NO is adsorbed on $Ti_2N(OH)_2$. The H atoms on the substrate bind with NO, where three H atoms combine with the N atom and one H atom combines with the O atom to form a hydroxyl lammonium-like product. The adsorption distances of NO_2 and $NH_3$are 1.625 and 1.702 Å, which are measured from the nearest H atom; thus, they are shorter than those of other Ti_2NT_2(T=O, F). While, the adsorption is not strong, especially the NH_3 molecule has basically no change compared with the free molecule.

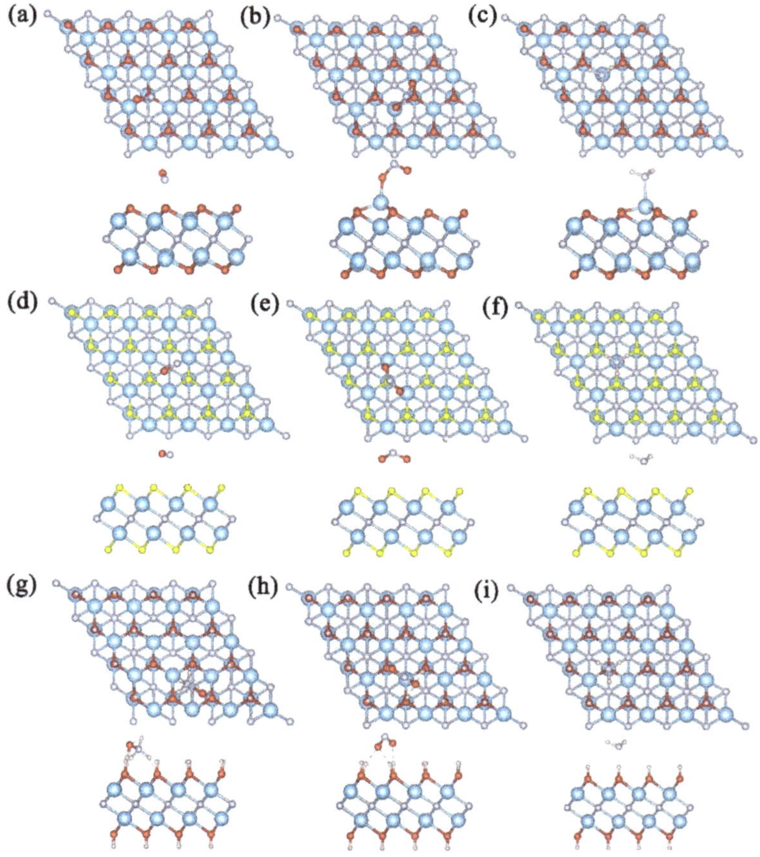

Figure 5. Top and side view of the optimized structures for Ti_2NO_2 after adsorbing (**a**) NO, (**b**) NO_2, and (**c**) NH_3; Ti_2NF_2 after adsorbing (**d**) NO, (**e**) NO_2, and (**f**) NH_3;and $Ti_2N(OH)_2$ after adsorbing (**g**) NO, (**h**) NO_2, and (**i**) NH_3 (blue, silver, red, pink, and yellow balls represent Ti, N, O, H, and F atoms, respectively).

We can clearly see that the absolute values of adsorption energy of NCGs on Ti_2NT_2(T=O, F, OH) decrease sharply compared with the pure substrate in Table 1, especially on the surface of Ti_2NF_2. Since the H atoms of the OH functional groups fall off the $Ti_2N(OH)_2$ substrate and bind with NO, their decrease is not that large. On the pure Ti_2N monolayer, a large amount of charge is transferred from Ti_2N to NO and NO_2. After the introduction of the functional group O, F, or OH, most of the charge transfer from Ti_2N is captured by these functional groups, which leaves a very small charge amount for exchange with NCGs.

In addition, one can see that the difference in adsorption energy of each NCG on substrate is obvious, indicating that the selectivity is good. As mentioned above, NO and NO_2 are dissociatively adsorbed on pure Ti_2N, which results in the poisoning of the Ti_2N surface. Similarly, NO binds with H atoms of $Ti_2N(OH)_2$ and destroys the surface integrity of the substrate, which is detrimental to the sensing material. Thus, $Ti_2N(OH)_2$ is not suitable for NO sensing. From the perspective of adsorption energy, those of NO on Ti_2NO_2, NO_2on Ti_2NF_2, and NH_3on $Ti_2N(OH)_2$ are between physisorption and chemisorption, which are important to make sensors work reversibly and increase their efficiency [45].

Table 1. Adsorption energy (E_{ads}) and charge-transfer profiles of NCGs adsorption on 2-D Ti_2N and Ti_2NT_2(T=O, F, OH).

NCGs	Adsorption Energy (eV)				Charge Transfer (e^-)			
	Ti_2N	Ti_2NO_2	Ti_2NF_2	$Ti_2N(OH)_2$	Ti_2N	Ti_2NO_2	Ti_2NF_2	$Ti_2N(OH)_2$
NO	−9.81	−0.68	−0.15	−5.55	2.57	−0.24	−0.07	0.50
NO_2	−14.34	−2.38	−0.36	−3.78	3.77	0.49	0.29	1.78
NH_3	−1.29	−1.16	−0.13	−0.57	0.01	−0.18	−0.03	0.15

To better understand the effects of NCGs molecules on Ti_2NT_2(T=O, F, OH), the band structure and PDOS of all the adsorption systems are calculated and shown in Figure 6. Regardless of the type of NCGs adsorption, the system retains metallic properties. Only when Ti_2NO_2 is exposed to NO_2, the Ti*d*-, O*p*-,and N *p*-orbitals show a certain degree of d-p hybridization, while the *p*- and *d*-orbitals in other PDOS do not overlap. When Ti_2NO_2 adsorbs NO, a flat band appears near the Fermi energy and indicates that the electrons are delocalized, which further proves that the interaction between NO and Ti_2NO_2 is not strong. This phenomenon is more obvious when NCGs are adsorbed on Ti_2NF_2 shown in Figure 6d–f. These results are consistent with the above discussion on energetics and charge-transfer profiles.

An ideal sensing material not only needs to possess excellent gas-binding capability but also has a requirement for good sensitivity. The conductivity of materials, especially two-dimensional materials, is closely related with work function. During adsorption, if the electron affinity of gas molecule surpasses the work function of substrate materials, the molecule will grab electrons from the surface of materials and the molecule would be charged negatively; in contrast, the adsorbed molecule would be charged positively. Both two cases would change the electronic structure, as well as conductivity of materials. The work function [46] is defined as the least energy required to free an electron from the surface of a system, which can be calculated by the following equation:

$$\varphi = V_\infty - E_f \quad (2)$$

where φ, V_∞, and E_f are the work function, electrostatic potential at the vacuum level, and the Fermi energy, respectively. Figure 7 shows the change of work function of Ti_2N and Ti_2NT_2(T=O, F, OH) before and after adsorption of various NCGs. "Bare" represents the bare system (before adsorption), and the names of NCGs in abscissa represent various systems after adsorption of different gases. The work functions of Ti_2NO_2 and $Ti_2N(OH)_2$ decrease obviously after adsorbing NO, while that of Ti_2NF_2 increases significantly after adsorbing NO_2. After adsorbing NH_3, the work functions of Ti_2N, Ti_2NF_2, and Ti_2NO_2 all decrease, and the change of Ti_2NO_2 is the most obvious. Because all the systems are metallic, even small changes in work function can cause large variation in conductivity [43]. Comprehensively considering the above results on adsorption structure and energy, Ti_2NO_2 can be used as a superior sensing candidate material for NO and NH_3, and Ti_2NF_2 can be utilized as that for NO_2. In addition, the potential of sensing properties of Ti_2N, Ti_2NF_2, and $Ti_2N(OH)_2$ towards NH_3 is also good.

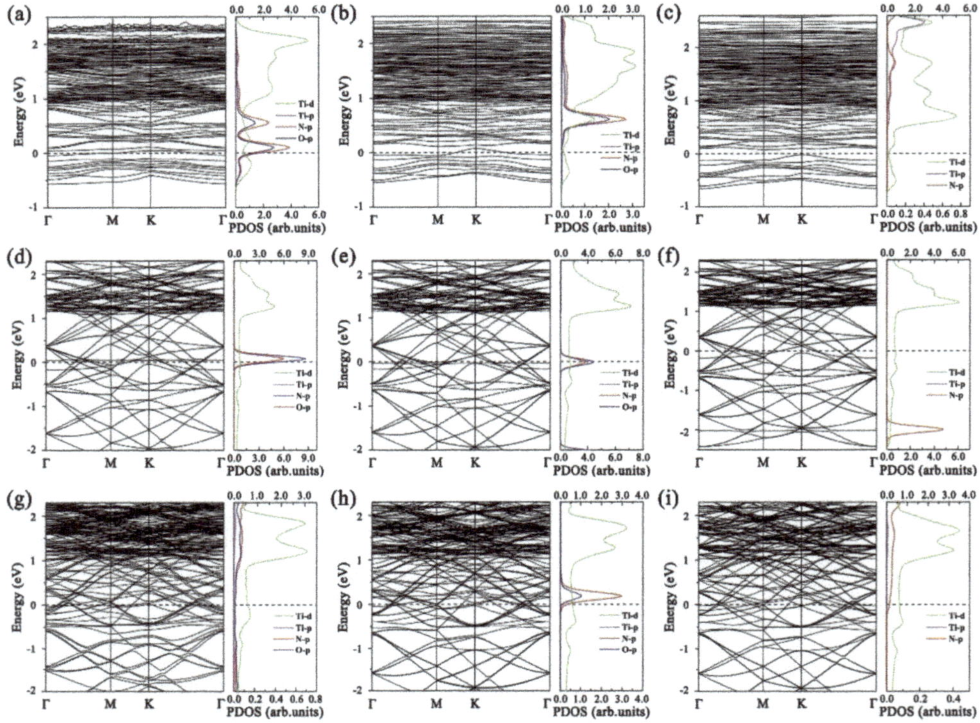

Figure 6. Band structures and PDOS of Ti_2NO_2 after adsorbing (**a**) NO, (**b**) NO_2, and (**c**) NH_3; Ti_2NF_2 after adsorbing (**d**) NO, (**e**) NO_2, and (**f**) NH_3; $Ti_2N(OH)_2$ after adsorbing (**g**) NO, (**h**) NO_2, and (**i**) NH_3 (Ti*d*-orbital uses top abscissa and other orbitals use bottom coordinate).

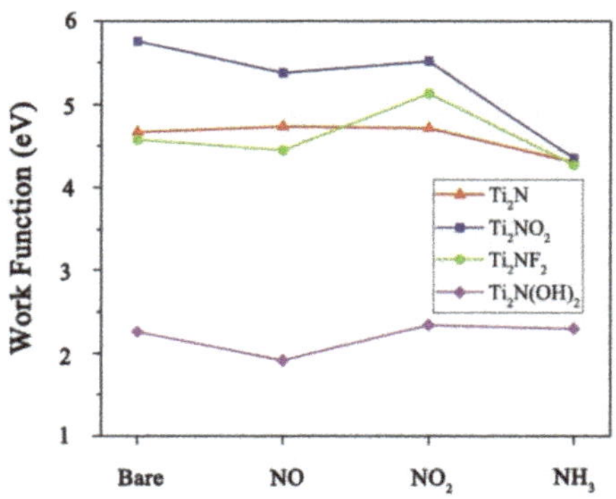

Figure 7. Work function change of Ti_2N and Ti_2NT_2(T=O, F, OH) before and after adsorption of various NCGs.

Finally, we discuss the recovery time (τ). An ideal gas-sensing material needs appropriate adsorption energy, and a shorter recovery time is better. To make a gas sensor work efficiently, a short recovery time is required. It can be calculated through transition state theory, namely by the following relationship [47,48]:

$$\tau = \upsilon^{-1}\exp(-E_{ads}/kT) \tag{3}$$

where υ, E_{ads}, k, and T are the attempt frequency, adsorption energy, Boltzmann's constant (8.62×10^{-5} eV K^{-1}), and operational temperature, respectively. Under the same conditions, the increase in adsorption energy will lead to an exponential increase in recovery time; then the higher operating temperature compensation is required, which is detrimental to the sensor performance. In order to get a relatively reasonable value, we set the attempt frequency here to $10^{12}s^{-1}$ based on traditional transition state theory [49–51]. The recovery time of NO on Ti_2NO_2 is calculated to be 2.16×10^5 us, and that of NO_2/Ti_2NF_2 and NH_3/Ti_2NF_2 is 1.29 and 1.71 $\times 10^{-4}$ us, respectively. Here it can be seen in Figure 8 that the adsorption energy difference of less than 1 eV can be enlarged to several orders of magnitude in recovery time. Although NO on Ti_2NO_2 has the longest recovery time, the value is only 0.216 s. For the system of NH_3/Ti_2NF_2, it has the shortest recovery time, which shows that the MXene system in this research can be promising. Therefore, when other conditions are met, the candidate materials with appropriate gas adsorption energy and short response time should be selected.

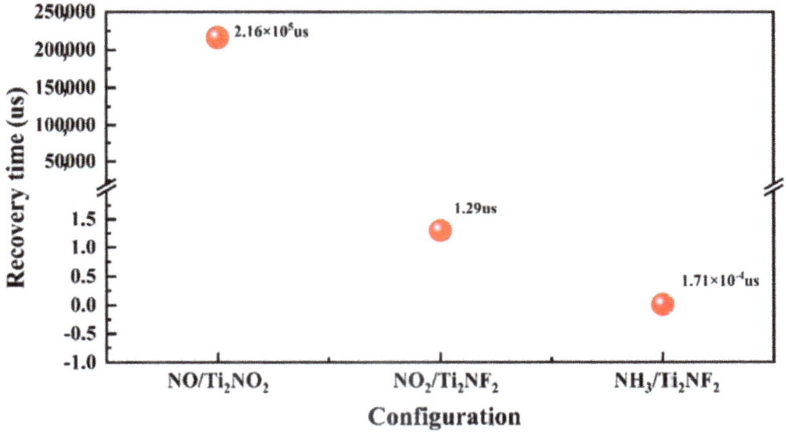

Figure 8. Recovery time of different sensing systems.

4. Summary and Outlook

In this work, we used the first-principles DFT calculations to investigate the 2-D Ti_2N nanosheet and its Ti_2NT_2(T=O, F, OH) derivative materials for harmful nitrogen-containing gas (NGCs) adsorption and sensing applications. When NCGs are exposed on the surface of Ti_2N, the dissociative adsorption of NO and NO_2 occurs and destroys the stability of the nanosheet, which would make it hard to work reversibly, while NH_3 does not have this effect on Ti_2N. When studying the interaction mechanisms between NCGs and Ti_2NT_2(T=O, F, OH), it can be found that a large amount of charges accumulates at functional groups, thus greatly reducing the adsorption energy of NCGs in absolute value. The obvious change in work function, coupled with the metallic nature of the systems, would improve gas sensitivity. Appropriate gas adsorption energy can directly determine fast recovery time to ensure the efficiency and reversibility of sensors. This research gives the electronic structures origin and application prospect of Ti_2NT_2(T=O, F, OH) as efficient NCGs sensing materials and would inspire experimentalists to explore better 2-D candidates in the field.

Author Contributions: Formal analysis, H.Z., W.D., J.Z., R.A., and Z.Q.; project administration, Z.Q.; supervision, Z.Q.; writing—original draft, H.Z., W.D., and Z.Q.; writing—review and editing, H.Z. and Z.Q.; funding acquisition, Z.Q. All authors have read and agreed to the published version of the manuscript.

Funding: We are grateful for the support from the Natural Science Foundation of China (51801113 and 52171182), the Sino-German Center for Research Promotion and NSFC (GZ1673), the Natural

Science Foundation of Shandong Province (ZR2018MEM001), the Young Scholars Program of Shandong University (YSPSDU), the Education Program of Shandong University (2020Y299), and the Fundamental Research Funds for the Central Universities.

Data Availability Statement: Data are contained within the article.

Acknowledgments: The National Supercomputer Centre (NSC) and the HPC Cloud Platform of Shandong University are acknowledged.

Conflicts of Interest: The authors declare no conflict of interest.

References

1. Ramanathan, V.; Feng, Y. Air pollution, greenhouse gases and climate change: Global and regional perspectives. *Atmos. Environ.* **2009**, *43*, 37–50. [CrossRef]
2. Koolen, C.D.; Rothenberg, G. Air pollution in Europe. *ChemSusChem* **2018**, *12*, 164–172. [CrossRef]
3. Hasselblad, V.; Eddy, D.M.; Kotchmar, D.J. Synthesis of environmental evidence: Nitrogen dioxide epidemiology studies. *J. Air Waste Manag. Assoc.* **1992**, *42*, 662–671. [CrossRef]
4. Garrett, M.H.; Hooper, M.A.; Hooper, B.M.; Abramson, M.J. Respiratory symptoms in children and indoor exposure to nitrogen dioxide and gas stoves. *Am. J. Respir. Crit. Care Med.* **1998**, *158*, 891–895. [CrossRef] [PubMed]
5. Ouyang, T.; Qian, Z.; Ahuja, R.; Liu, X. First-principles investigation of CO adsorption on pristine, C-doped and N-vacancy defected hexagonal AlN nanosheets. *Appl. Surf. Sci.* **2018**, *439*, 196–201. [CrossRef]
6. Liu, Y.; Parisi, J.; Sun, X.; Lei, Y. Solid-state gas sensors for high temperature applications—A review. *J. Mater. Chem. A* **2014**, *2*, 9919–9943. [CrossRef]
7. Tang, X.; Du, A.; Kou, L. Gas sensing and capturing based on two-dimensional layered materials: Overview from theoretical perspective. *Wiley Interdiscip. Rev. Comput. Mol. Sci.* **2018**, *8*, e1361. [CrossRef]
8. Zhang, Y.-H.; Chen, Y.-B.; Zhou, K.-G.; Liu, C.; Zeng, J.; Zhang, H.-L.; Peng, Y. Improving gas sensing properties of graphene by introducing dopants and defects: A first-principles study. *Nanotechnology* **2009**, *20*, 185504. [CrossRef]
9. Liu, L.; Yang, Q.; Wang, Z.; Ye, H.; Chen, X.; Fan, X.; Zhang, G. High Selective Gas Detection for small molecules based on Germanium selenide monolayer. *Appl. Surf. Sci.* **2018**, *433*, 575–581. [CrossRef]
10. Li, F.; Shi, C. NO-sensing performance of vacancy defective monolayer MoS_2 predicted by density function theory. *Appl. Surf. Sci.* **2018**, *434*, 294–306. [CrossRef]
11. Yu, X.-F.; Li, Y.-C.; Cheng, J.-B.; Liu, Z.-B.; Li, Q.-Z.; Li, W.-Z.; Yang, X.; Xiao, B. Monolayer Ti_2CO_2: A Promising Candidate for NH_3 Sensor or Capturer with High Sensitivity and Selectivity. *ACS Appl. Mater. Interfaces* **2015**, *7*, 13707–13713. [CrossRef]
12. Xiao, B.; Li, Y.-C.; Yu, X.-F.; Cheng, J.-B. MXenes: Reusable materials for NH_3 sensor or capturer by controlling the charge injection. *Sens. Actuators B Chem.* **2016**, *235*, 103–109. [CrossRef]
13. Naguib, M.; Kurtoglu, M.; Presser, V.; Lu, J.; Niu, J.; Heon, M.; Hultman, L.; Gogotsi, Y.; Barsoum, M.W. Two-dimensional nanocrystals produced by exfoliation of Ti_3AlC_2. *Adv. Mater.* **2011**, *23*, 4248–4253. [CrossRef] [PubMed]
14. Naguib, M.; Mashtalir, O.; Carle, J.; Presser, V.; Lu, J.; Hultman, L.; Gogotsi, Y.; Barsoum, M.W. Two-dimensional transition metal carbides. *ACS Nano* **2012**, *6*, 1322–1331. [CrossRef]
15. Feng, A.; Yu, Y.; Wang, Y.; Jiang, F.; Yu, Y.; Mi, L.; Song, L. Two-dimensional MXene Ti_3C_2 produced by exfoliation of Ti_3AlC_2. *Mater. Des.* **2017**, *114*, 161–166. [CrossRef]
16. Lashgari, H.; Abolhassani, M.; Boochani, A.; Elahi, S.; Khodadadi, J. Electronic and optical properties of 2D graphene-like compounds titanium carbides and nitrides: DFT calculations. *Solid State Commun.* **2014**, *195*, 61–69. [CrossRef]
17. Weng, H.; Ranjbar, A.; Liang, Y.; Song, Z.; Khazaei, M.; Yunoki, S.; Arai, M.; Kawazoe, Y.; Fang, Z.; Dai, X. Large-gap two-dimensional topological insulator in oxygen functionalized MXene. *Phys. Rev. B* **2015**, *92*, 075436–0754367. [CrossRef]
18. Pan, H. Electronic properties and lithium storage capacities of two-dimensional transition-metal nitride monolayers. *J. Mater. Chem. A* **2015**, *3*, 21486–21493. [CrossRef]
19. Wang, D.; Liu, Y.; Meng, X.; Wei, Y.; Zhao, Y.; Pang, Q.; Chen, G. Two-dimensional VS_2 monolayers as potential anode materials for lithium-ion batteries and beyond: First-principles calculations. *J. Mater. Chem. A* **2017**, *5*, 21370–21377. [CrossRef]
20. Kumar, H.; Frey, N.C.; Dong, L.; Anasori, B.; Gogotsi, Y.; Shenoy, V.B. Tunable magnetism and transport properties in nitride MXenes. *ACS Nano* **2017**, *11*, 7648–7655. [CrossRef]
21. Yang, Z.; Zheng, Y.; Li, W.; Zhang, J. Investigation of two-dimensional hf-based MXenes as the anode materials for li/na-ion batteries: A DFT study. *J. Comput. Chem.* **2019**, *40*, 1352–1359. [CrossRef] [PubMed]
22. Gao, G.; Ding, G.; Li, J.; Yao, K.; Wu, M.; Guangqian, D. Monolayer MXenes: Promising half-metals and spin gapless semiconductors. *Nanoscale* **2016**, *8*, 8986–8994. [CrossRef] [PubMed]
23. Pang, J.; Mendes, R.G.; Bachmatiuk, A.; Zhao, L.; Ta, H.Q.; Gemming, T.; Liu, H.; Liu, Z.; Rummeli, M.H. Applications of 2D MXenes in energy conversion and storage systems. *Chem. Soc. Rev.* **2018**, *48*, 72–133. [CrossRef] [PubMed]
24. Shukla, V.; Jena, N.K.; Naqvi, S.R.; Luo, W.; Ahuja, R. Modelling high-performing batteries with Mxenes: The case of S-functionalized two-dimensional nitride Mxene electrode. *Nano Energy* **2019**, *58*, 877–885. [CrossRef]

25. Gouveia, J.; Novell-Leruth, G.; Vines, F.; Illas, F.; Gomes, J. The Ti_2CO_2MXene as a nucleobase 2D sensor: A first-principles study. *Appl. Surf. Sci.* **2021**, *544*, 148946. [CrossRef]
26. Gouveia, J.; Novell-Leruth, G.; Reis, P.M.L.S.; Viñes, F.; Illas, F.; Gomes, J. First-principles calculations on the adsorption behavior of amino acids on a titanium carbide MXene. *ACS Appl. Biol. Mater.* **2020**, *3*, 5913–5921. [CrossRef]
27. Dolz, D.; Morales-García, A.; Viñes, F.; Illas, F. Exfoliation energy as a descriptor of MXenes synthesizability and surface chemical activity. *Nanomaterials* **2021**, *11*, 127. [CrossRef] [PubMed]
28. Soundiraraju, B.; George, B.K. Two-dimensional Titanium Nitride (Ti2N) MXene: Synthesis, characterization, and potential application as surface-enhanced ramanscattering substrate. *ACS Nano* **2017**, *11*, 8892–8900. [CrossRef]
29. Li, Y.; Guo, Y.; Chen, W.; Jiao, Z.; Ma, S. Reversible hydrogen storage behaviors of Ti_2NMXenes predicted by first-principles calculations. *J. Mater. Sci.* **2018**, *54*, 493–505. [CrossRef]
30. Lin, H.; Yang, D.-D.; Lou, N.; Zhu, S.-G.; Li, H.-Z. Functionalized titanium nitride-based MXenes as promising host materials for lithium-sulfur batteries: A first principles study. *Ceram. Int.* **2018**, *45*, 1588–1594. [CrossRef]
31. Urbankowski, P.; Anasori, B.; Makaryan, T.; Er, D.; Kota, S.; Walsh, P.L.; Zhao, M.; Shenoy, V.B.; Barsoum, M.W.; Gogotsi, Y. Synthesis of two-dimensional titanium nitride Ti_4N_3(MXene). *Nanoscale* **2016**, *8*, 11385–11391. [CrossRef]
32. Xiao, X.; Yu, H.; Jin, H.; Wu, M.; Fang, Y.; Sun, J.; Hu, Z.; Li, T.; Wu, J.; Huang, L.; et al. Salt-templated synthesis of 2D metallic MoN and other nitrides. *ACS Nano* **2017**, *11*, 2180–2186. [CrossRef]
33. Blöchl, P.E. Projector augmented-wave method. *Phys. Rev. B* **1994**, *50*, 17953–17979. [CrossRef]
34. Kresse, G.; Joubert, D. From ultrasoft pseudopotentials to the projector augmented-wave method. *Phys. Rev. B* **1999**, *59*, 1758–1775. [CrossRef]
35. Perdew, J.P.; Burke, K.; Ernzerhof, M. Generalized gradient approximation made simple. *Phys. Rev. Lett.* **1996**, *77*, 3865–3868. [CrossRef]
36. Grimme, S. Semiempirical GGA-type density functional constructed with a long-range dispersion correction. *J. Comput. Chem.* **2006**, *27*, 1787–1799. [CrossRef]
37. Wang, D.; Gao, Y.; Liu, Y.; Jin, D.; Gogotsi, Y.; Meng, X.; Du, F.; Chen, G.; Wei, Y. First-principles calculations of Ti_2N and Ti_2NT_2 (T=O, F, OH) monolayers as potential anode materials for lithium-ion batteries and beyond. *J. Phys. Chem. C* **2017**, *121*, 13025–13034. [CrossRef]
38. Xie, Y.; Kent, P.R.C. Hybrid density functional study of structural and electronic properties of functionalized $Ti_{n+1}X_n$ (X=C, N) monolayers. *Phys. Rev. B* **2013**, *87*, 235441. [CrossRef]
39. Prasongkit, J.; Amorim, R.; Chakraborty, S.; Ahuja, R.; Scheicher, R.; Amornkitbamrung, V. Highly sensitive and selective gas detection based on silicene. *J. Phys. Chem. C* **2015**, *119*, 16934–16940. [CrossRef]
40. Leenaerts, O.; Partoens, B.; Peeters, F.M. Adsorption of H_2O, NH_3, CO, NO_2, and NO on graphene: A first-principles study. *Phys. Rev. B* **2008**, *77*, 125416. [CrossRef]
41. Gouveia, J.D.; Viñes, F.; Illas, F.; Gomes, J.R.B. MXenes atomic layer stacking phase transitions and their chemical activity consequences. *Phys. Rev. Mater.* **2020**, *4*, 054003. [CrossRef]
42. Ouyang, T.; Qian, Z.; Hao, X.; Ahuja, R.; Liu, X. Effect of defects on adsorption characteristics of AlN monolayer towards SO_2 and NO_2: Ab initio exposure. *Appl. Surf. Sci.* **2018**, *462*, 615–622. [CrossRef]
43. Kou, L.; Frauenheim, T.; Chen, C. Phosphorene as a superior gas sensor: Selective adsorption and distinct *I–V* response. *J. Phys. Chem. Lett.* **2014**, *5*, 2675–2681. [CrossRef] [PubMed]
44. Yue, Q.; Shao, Z.; Chang, S.; Li, J. Adsorption of gas molecules on monolayer MoS_2 and effect of applied electric field. *Nanoscale Res. Lett.* **2013**, *8*, 425. [CrossRef] [PubMed]
45. Hussain, T.; Singh, D.; Gupta, S.K.; Karton, A.; Sonvane, Y.; Ahuja, R. Efficient and selective sensing of nitrogen-containing gases by Si_2BN nanosheets under pristine and pre-oxidized conditions. *Appl. Surf. Sci.* **2018**, *469*, 775–780. [CrossRef]
46. Gholizadeh, R.; Yu, Y.-X. Work functions of pristine and heteroatom-doped graphenes under different external electric fields: An ab initio DFT study. *J. Phys. Chem. C* **2014**, *118*, 28274–28282. [CrossRef]
47. Barsan, N.; Koziej, D.; Weimar, U. Metal oxide-based gas sensor research: How to? *Sens. Actuators B Chem.* **2007**, *121*, 18–35. [CrossRef]
48. Nagarajan, V.; Chandiramouli, R. Investigation of NH_3 adsorption behavior on graphdiyne nanosheet and nanotubes: A first-principles study. *J. Mol. Liq.* **2018**, *249*, 24–32. [CrossRef]
49. Eyring, H. The activated complex in chemical reactions. *J. Chem. Phys.* **1935**, *3*, 107–115. [CrossRef]
50. Hadipour, N.L.; Peyghan, A.A.; Soleymanabadi, H. Theoretical study on the Al-doped ZnO nanoclusters for CO chemical sensors. *J. Phys. Chem. C* **2015**, *119*, 6398–6404. [CrossRef]
51. Ahmadi, A.; Hadipour, N.L.; Kamfiroozi, M.; Bagheri, Z. Theoretical study of aluminum nitride nanotubes for chemical sensing of formaldehyde. *Sens. Actuators B Chem.* **2012**, *161*, 1025–1029. [CrossRef]

nanomaterials

MDPI

Article

Extensive Broadband Near-Infrared Emissions from Ge_xSi_{1-x} Alloys on Micro-Hole Patterned Si(001) Substrates

Kun Peng, Ningning Zhang, Jiarui Zhang, Peizong Chen, Jia Yan, Changlin Zheng, Zuimin Jiang and Zhenyang Zhong *

State Key Laboratory of Surface Physics, Department of Physics, Fudan University, Shanghai 200438, China; 15110190021@fudan.edu.cn (K.P.); 16110190037@fudan.edu.cn (N.Z.); 17110190029@fudan.edu.cn (J.Z.); 16110190016@fudan.edu.cn (P.C.); 19110190054@fudan.edu.cn (J.Y.); zcl@fudan.edu.cn (C.Z.); zmjiang@fudan.edu.cn (Z.J.)
* Correspondence: zhenyangz@fudan.edu.cn

Abstract: Broadband near-infrared (NIR) luminescent materials have been continuously pursued as promising candidates for optoelectronic devices crucial for wide applications in night vision, environment monitoring, biological imaging, etc. Here, graded Ge_xSi_{1-x} (x = 0.1–0.3) alloys are grown on micro-hole patterned Si(001) substrates. Barn-like islands and branch-like nanostructures appear at regions in-between micro-holes and the sidewalls of micro-holes, respectively. The former is driven by the efficient strain relation. The latter is induced by the dislocations originating from defects at sidewalls after etching. An extensive broadband photoluminescence (PL) spectrum is observed in the NIR wavelength range of 1200–2200 nm. Moreover, the integrated intensity of the PL can be enhanced by over six times in comparison with that from the reference sample on a flat substrate. Such an extensively broad and strong PL spectrum is attributed to the coupling between the emissions of GeSi alloys and the guided resonant modes in ordered micro-holes and the strain-enhanced decomposition of alloys during growth on the micro-hole patterned substrate. These results demonstrate that the graded Ge_xSi_{1-x} alloys on micro-hole pattered Si substrates may have great potential for the development of innovative broadband NIR optoelectronic devices, particularly to realize entire systems on a Si chip.

Keywords: broadband near-infrared; ordered micro-holes; graded alloy; photoluminescence; guided resonant mode

Citation: Peng, K.; Zhang, N.; Zhang, J.; Chen, P.; Yan, J.; Zheng, C.; Jiang, Z.; Zhong, Z. Extensive Broadband Near-Infrared Emissions from Ge_xSi_{1-x} Alloys on Micro-Hole Patterned Si(001) Substrates. *Nanomaterials* **2021**, *11*, 2545. https://doi.org/10.3390/nano11102545

Academic Editors: Jihoon Lee and Ming-Yu Li

Received: 18 August 2021
Accepted: 24 September 2021
Published: 28 September 2021

1. Introduction

The near-infrared (NIR) spectroscopy is becoming an indispensable technique for a wide range of applications, such as night vision [1,2], surveillance [3], food analyses and environmental monitoring [4,5], biological imaging [6,7], and medical diagnostics [8,9]. The dominant principle of these applications is based on the unique absorption or scattering features related to the fundamental carbon-hydrogen (C-H) and oxygen-hydrogen (O-H) vibration modes, which generally cover a broad spectra range of ~1000–2000 nm in the NIR region (or so-called short-wavelength infrared) [4]. Accordingly, the broadband NIR light source and detector are highly in demand for simultaneous identification of various substances via the same photonic circuit. The traditional broadband NIR light sources are halogen lamps, high-pressure discharge lamps and super continuum lasers. They have disadvantages in spectra-stability, size, energy consumption and cost [9]. Many efforts have been devoted to realizing innovative broadband NIR light sources, such as light-emitting diodes based on unique phosphors [2,10] and PbS quantum dots in layered perovskite [11]. Considering the potential of monolithic optical-electronic integrated circuits on Si substrates to realize entire 'systems on a chip' [12], Si-based broadband NIR light sources and detectors are of great interest. In general, normal Si cannot emit or absorb light of wavelength longer than ~1.1 μm due to the limitation associated with the bandgap energy of ~1.12 eV. There have been several attempts to overcome this fundamental limitation. A hybrid structure of metal nanoantenna/Si has been proposed for the NIR detector by taking advantage of the direct light-harvesting and the energetic or "hot" electron generation during the plasma decay of nanoantenna [13]. The deep levels in Si formed via an ion implantation have been exploited to fabricate a waveguide-integrated infrared avalanche photodiode [14]. Dual Inversion Layers and Fowler–Nordheim Tunneling in a p-i-n junction (p-Si/AlOx/n-ZnO nanowire) have been employed to realize broadband photodetection [15]. Tensile-strained Si nanomembrane has been applied in photodetectors with an absorption limit of up to 1550 nm [16]. For longer wavelengths in mid-infrared range, GeSn alloy films on silicon substrates have also been studied due to the possible smaller direct bandgap [17]. More frequently, given the small bandgap of ~0.66 eV and the compatibility with the Si integration technology of Ge, Ge_xSi_{1-x} ($0 \leq x \leq 1$) alloy promises to extend the operation wavelength in the NIR region [18–23]. Particularly, a stack of GeSi alloy quantum-wells (QWs) has been exploited to realize broadband NIR photodetectors in terms of the miniband formation due to the coupling of neighboring QWs or the composition of different energy transitions in QWs with different thicknesses and/or Ge contents [23]. Whereas the large lattice mismatch between Si and Ge restricts the number, the thickness and the Ge content of QWs to avoid strain-induced defects [24], their design and growth are quite complicated. In addition, charge tunneling to the smaller bandgap QWs can take place and, in turn, prevents equally efficient emission from all QWs. There is a well-known tradeoff between the wide spectral bandwidth and the degradation of performance of NIR optoelectronic devices based on GeSi alloys with higher Ge contents. So far, the broadband GeSi NIR light source or detector is still a significant challenge.

In this report, Ge_xSi_{1-x} alloys with a graded increase of Ge content (x = 0.1, 0.15, 0.2, 0.25, 0.3) are grown on micro-hole patterned Si(001) substrates. The growth characteristics are disclosed. A strong and extensively broad photoluminescence (PL) spectrum is observed in the NIR wavelength range from 1200 nm to 2200 nm. This exhibits the broadest emissions from Ge_xSi_{1-x} alloys and covers the widest NIR spectrum for practical applications. Detailed analyses of power- and temperature-dependent PL spectra, as well as the three-dimensional finite-difference time-domain (FDTD) simulations, provide an insight into the broad and strong PL spectrum. Our results demonstrate an innovative strategy of alloy growth on micro-hole patterned substrates to realize extraordinary broadband emissions. Given the rather broad and strong NIR emission and the compatibility with the Si integration technology, the graded Ge_xSi_{1-x} alloys on micro-hole patterned Si substrates promise to be the superior candidate for a broadband NIR light source, particularly to realize entire systems on a Si chip.

2. Materials and Methods

The samples are grown on micro-hole patterned Si(001) substrates by solid source molecular beam epitaxy(MBE) in a Riber Eva-32 system. The ordered micro-holes in a hexagonal lattice on Si(001) (p-type, Boron doping) substrates are fabricated by a nanosphere lithograph [25]. The templates are cleaned by the RCA method with a subsequent HF dip to obtain a hydrogen-terminated surface. After a thermal desorption, a Si buffer layer of 100 nm with Boron (B) doping of ~3×10^{18} cm^{-3} is grown at a rate of 1Å s^{-1} at 500 °C. Then an intrinsic Si layer of 1 μm is grown. Subsequently, a stack of five Ge_xSi_{1-x} (x = 0.1, 0.15, 0.2, 0.25 and 0.3) alloy films is deposited at 480 °C. The thickness of each alloy film is 80 nm. The nominal compositions and the thicknesses of alloy films are determined by the growth rates of Ge and Si. The graded increase of Ge content with the largest value of 0.3 can effectively reduce misfit-induced defects. In addition, a $Ge_{0.3}Si_{0.7}$ alloy layer of 100 nm with phosphorus (P) doping of ~6×10^{18} cm^{-3} is grown at 430 °C. Finally, a Si capping layer of 40 nm with P-doping of ~1.3×10^{19} cm^{-3} is grown at 430 °C. The overall layer structure is schematically shown in Figure 1a. The same layer structure is also grown on a flat Si(001) substrate as a reference sample. Figure 1b schematically illustrates the corresponding energy band diagram of the reference sample. The B doping and the P doping naturally create a built-in electric field related to the p-i-n structure, which tilts the band alignment. This band alignment facilitates the accumulation of electrons and holes in regions E (around the interface between the n-GeSi and the n-Si) and H (intrinsic $Ge_{0.3}Si_{0.7}$ alloy layer), respectively, as shown in Figure 1b. The p-i-n structure also facilitates the realization of optoelectronic devices.

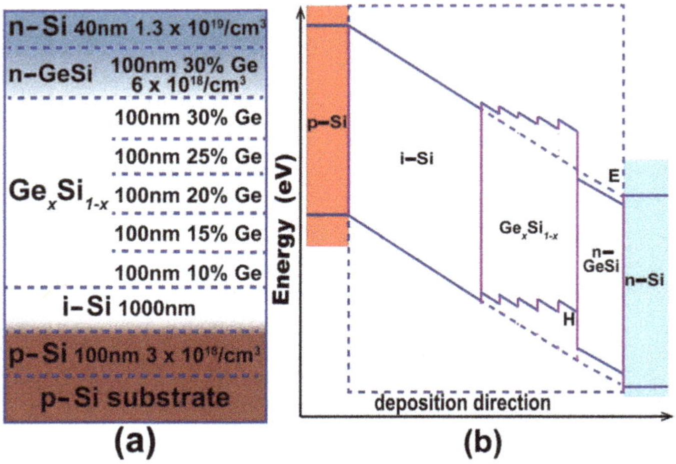

Figure 1. Schematic layer structure and energy band diagram. (**a**) The layer structure grown on the micro-hole patterned and the flat Si(001) substrates, (**b**) the energy band diagram of the sample.

The surface morphology of the sample is characterized by scanning electron microscopy (SEM) (Zeiss Sigma) and atomic force microscopy (AFM) (Veeco DI Multimode V SPM) in tapping mode. The structure and composition of the sample are also investigated from a cross-sectional view using a field emission transmission electron microscope (TEM) (Thermo Scientific Talos F200i) operating at 200 kV. The electron-transparent TEM foil is prepared by focused ion beam (FIB) (Thermo Scientific Helios G4 CX). The Ge and Si distributions in the sidewalls of micro-holes are analyzed by X-ray energy dispersive spectroscopy (EDS) (Bruker Xflash 6T-30) attached to the microscope. PL measurements are performed in a closed-cycle helium cryostat with a temperature range of 17 to 300 K. The excitation source is a semiconductor laser (473 nm). The luminescence is analyzed by a monochromator (Omni-λ500, Zolix Instruments Co.) and detected with an extended InGaAs photodetector using the standard lock-in technique. The simulations of the emission spectra of Ge_xSi_{1-x} alloys grown on micro-hole patterned Si (001) substrates are also carried out by the FDTD method. The boundary conditions of the perfectly matched layer are imposed in the z axis (out of plane). Periodic boundary conditions are employed in the x and y axes (in-plane). The emission intensity is obtained by integrating pointing vectors at the input port (x-y plane at z = 4.5 μm) in the time domain.

3. Results and Discussion

3.1. Morphologies and Structure Properties

Figure 2a shows the typical top-view SEM image of ordered micro-holes on a Si(001) substrate. The micro-holes are arranged in a hexagonal lattice. The period, the diameter and the depth of micro-holes are about 1.0, 0.7 and 4.0 μm, respectively. These parameters can be readily modulated in the nanosphere lithograph [25]. Figure 2b shows the corresponding top-view SEM image of the micro-hole array after the growth of Si and graded Ge_xSi_{1-x} alloys. The micro-holes remain with slightly reduced diameter of ~0.6 μm, which results from partial growth at their sidewalls. Moreover, barn-like islands with a base radius of ~260 nm and height of ~180 nm appear at the center regions in between the neighboring three micro-holes, as denoted by yellow dotted-circles in Figure 2a,b. This is consistent with the previous report on the low mismatch Ge_xSi_{1-x} alloy islands on Si(100) substrates [26]. This barn-like island, with facets {1, 1, 1}, {1, 1, 3}, {15, 3, 23} and {1, 0, 5}, originates from the low surface tension and efficient strain relaxation of these facets [26]. The partial strain relaxation at the edges of the micro-holes facilitates the location of the barn-like island at the center region in-between the micro-holes, where misfit strain can be relaxed by the island formation. These microstructures are well ordered in a large area, as demonstrated in Figure 2c. Although some domain boundaries and point defects appear to degrade the long-range ordering of micro-holes, they can be considerably reduced by optimizing the nanosphere lithograph [25].

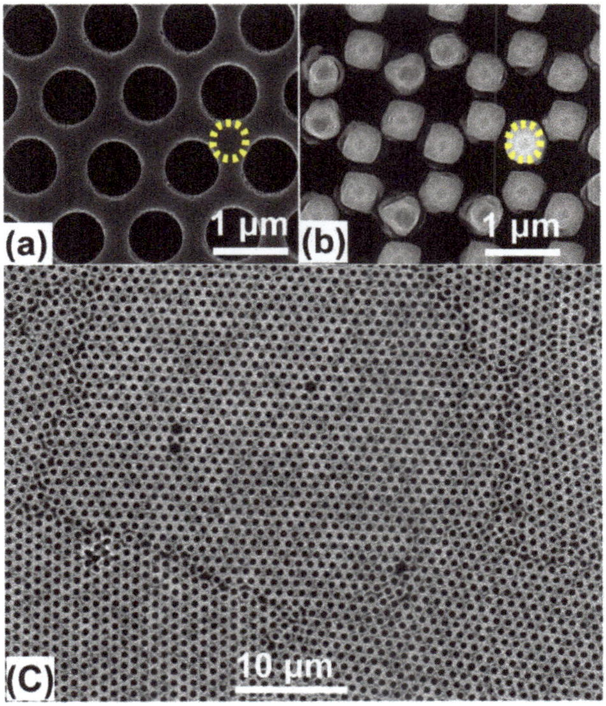

Figure 2. Top-view SEM images. (**a**) Ordered micro-holes on a Si(001) substrate, the micro-hole array after the growth of Si and graded Ge_xSi_{1-x} alloy via MBE in (**b**) a small area, (**c**) a large area. The yellow dotted circles in (**a**,**b**) denote the regions in-between three micro-holes.

Figure 3a shows the scanning transmission electron microscopy (STEM)-EDS mapping of Ge and Si in the sidewalls of micro-holes. Obviously, most Ge appears on the top of the sidewalls. Few Ge can be found at the bottom-half of the sidewalls. Figure 3b shows the corresponding high-angle annular dark field (HAADF) STEM image across the nearest neighboring micro-holes. Some branch-like nanostructures appear at the middle of the sidewalls of the micro-holes, considerably away from the growth onset of Ge_xSi_{1-x} alloys denoted by dotted lines in Figure 3a,b. By careful inspection of Figure 3a,b, three main features can be seen. (i) The Si and GeSi alloys essentially grow on the upper-half of the sidewalls of micro-holes. Given the general incident flux of Si and Ge atoms tilted away from the axis direction of [001] of the micro-holes, the incident flux of Si and Ge atoms can hardly attach to the bottom part of the deep micro-holes, particularly after the formation of branch-like nanostructures at the sidewalls. (ii) The Ge-rich edge can be distinguished near the top of the sidewall. This demonstrates the segregation of Ge to the edge of the micro-holes during GeSi alloy growth. (iii) The branch-like nanostructures initiate during the Si growth at the sidewalls, since their root is mainly Si. These are induced by dislocations in the Si sidewalls, as shown in the enlarged STEM-HAADF image of Figure 3c. In the upper Si sidewalls with little dislocation, no branch-like nanostructure appears. We argue that these dislocations originate from the defects introduced during the long-time (over 20 min) plasma etching to obtain micro-holes with a depth of over 4 μm. These dislocations are mainly stacking faults, as demonstrated by the atomic resolution STEM-HAADF image (along the [110] zone axis) and the extra spots in the corresponding Fourier transformation (FT) patterns in Figure 3d. To learn about the evolution of those dislocations, further systematic studies are necessary. Considering the reduction of defects by the shorter etching time and the growth of GeSi alloy all over the sidewalls of the micro-holes, the depth of the micro-hole may be decreased. Figure 3e shows the AFM image of the surface morphology of the reference sample after the growth of Si and Ge_xSi_{1-x} alloys on a flat substrate. The surface is quite smooth without a cross-hatch pattern [24], as it demonstrates few dislocations in the GeSi alloy layers. This result indicates that the graded increase of Ge content in GeSi alloy layers can efficiently suppress the formation of dislocations.

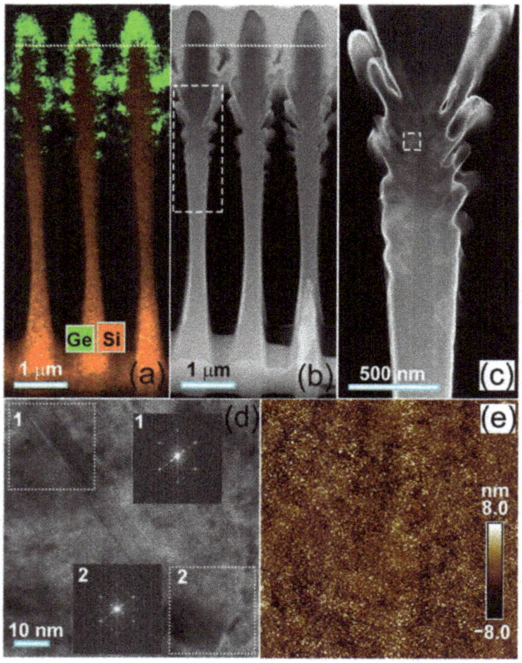

Figure 3. STEM and AFM images. (**a**) STEM-EDS mapping of Ge and Si in sidewalls, (**b**) STEM-HAADF image across the nearest neighboring micro-holes, (**c**) enlarged STEM-HAADF image of dash-boxed region in (**b**), (**d**) high resolution STEM image of dash-boxed region in (**c**), (**e**) AFM image (5×5 μm^2) of the reference sample after growth on a flat substrate. The dotted line in (**a**,**b**) denotes the growth onset of Ge_xSi_{1-x} alloys. The insets (1, 2) in (**d**) are the FT patterns of STEM images in the dot-boxes (1, 2).

3.2. *Power Dependent PL Spectra*

Figure 4 shows the PL spectra of graded Ge_xSi_{1-x} alloys grown on the micro-hole patterned Si(001) substrate and the flat Si(001) substrate under the same growth conditions. Obviously, the PL spectrum of graded GeSi alloys on the micro-hole patterned substrate is substantially different from that of the reference sample on the flat substrate. Although there are five Ge_xSi_{1-x} alloy layers with different Ge compositions (x = 0.1, 0.15, 0.2, 0.25, 0.3), the PL spectrum of the reference sample on the flat substrate is essentially composed of two peaks in the wavelength range of 1300–1600 nm, whereas the spectrum of the

Ge_xSi_{1-x} alloy on the micro-hole patterned substrate covers a rather broad wavelength range of 1200–2200 nm. Moreover, it can be separated into three regions. (I) In the range of 1200–1500 nm, the spectrum is similar to that of the reference sample on the flat substrate. (II) In the range of 1500–1850 nm, five satellite peaks can be distinguished. (III) In the range of 1850–2200 nm, a peak shoulder appears in the spectrum. Considering the longer wavelength and the unique peak features, the spectrum in regions (II) and (III) cannot be obtained directly from the graded Ge_xSi_{1-x} alloys with x = 0.1–0.3.

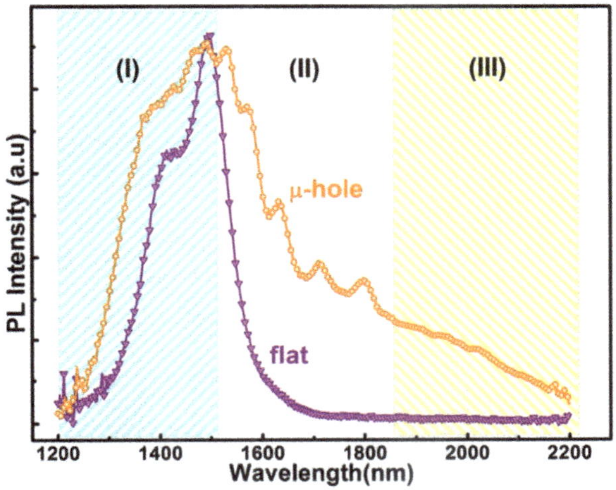

Figure 4. PL spectra. PL of graded Ge_xSi_{1-x} alloys on the micro-hole patterned and the flat Si(001) substrates, denoted by 'μ-hole' and 'flat', respectively, at 17 K with the excitation power of 800 mW. (I), (II) and (III) denote three ranges of wavelength.

To clarify the origin of the extensive broadband spectrum, power-dependent PL spectra of the graded Ge_xSi_{1-x} alloy layers on the flat and the micro-hole patterned substrate are obtained, as shown in Figure 5a,b, respectively. The spectra of the reference sample can be fitted by two Gaussian peaks around the wavelengths of 1408 nm and 1501 nm, as shown in the inset of Figure 5a. These two peaks are both blue-shifted with the excitation power, as shown in Figure 5c. This is the typical behavior of the type-II band alignment in GeSi/Si heterostructure due to band-filling and/or band-bending effects [27], whereas their energy difference of ~55 meV is nearly power-independent. These results demonstrate that these two PL peaks of the reference sample can be assigned as the non-phonon(NP) emission and its transverse-optical(TO) phonon replica of the GeSi alloy [27]. In addition, they mainly originate from the recombination of electrons and holes in regions E (around the interface between n-type $Ge_{0.3}Si_{0.7}$ and Si) and H (intrinsic $Ge_{0.3}Si_{0.7}$), as denoted in Figure 1b, which are the energetically favorable positions for electrons and holes, respectively. This means that the photon-generated carriers can efficiently drift to the energetically favorable positions around the Ge_xSi_{1-x} alloy layer with the largest x value of 0.3 before their recombination. Emissions of the shorter wavelength from Ge_xSi_{1-x} alloy layers with x < 0.3 can hardly be observed. This is consistent with the narrow electroluminescence spectrum of a quantum dot (QD) ensemble by charge tunneling to the smaller bandgap QDs [4]. This also suppresses the broadband emission. For PL spectra of the graded Ge_xSi_{1-x} alloys on the micro-hole patterned substrate; the spectra in regions (I) of 1200–1500 nm and (III) of 1850–2200 nm are also slightly blue-shifted with the excitation power. Accordingly, they are dominated by the emissions of Ge_xSi_{1-x} alloys but with x value different from 0.3, whereas, the positions of the satellite peaks in region (II) of 1500–1850 nm denoted by Pi (i = 1–5) are power-independent, as demonstrated in Figure 5b. This is distinctly different from the power-dependence of PL peaks of GeSi alloys [27]. Considering the ordered micro-holes in a hexagonal lattice, these satellite peaks are attributed to the coupling between the emissions of the graded Ge_xSi_{1-x} alloys and the guided resonance modes in the ordered micro-holes [25]. Such a coupling not only effectively broadens the PL spectrum but also enhances the PL intensity. We can obtain the enhancement factor (EF) from $EF = I_{\mu\text{-hole}}/I_{flat}$, where $I_{\mu\text{-hole}}$ and I_{flat} are the integrated intensity of PL peaks of graded GeSi alloys on the micro-hole patterned and the flat substrates, respectively. This is larger than one and remarkably increased with the excitation power, as shown in Figure 5d. The enhancement of PL spectra is generally attributed to the increased extraction efficiency of PL from the micro-structured surface and/or the reduced lifetime of excitation for the radiative recombination due to the coupling into the guided resonant modes of the ordered micro-holes. The former is essentially determined by the microstructure on the sample surface, which is power-independent. The present power-dependent EF demonstrates that the latter play the dominant role in the enhancement of PL spectra of the sample on a micro-hole patterned substrate. This result is consistent with the previous report on the enhanced PL spectra of SiGe coaxial quantum wells on ordered Si nanopillars [28].

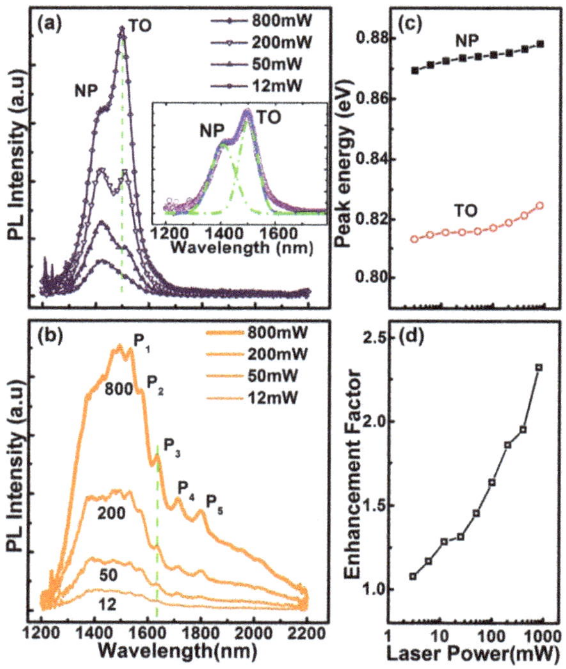

Figure 5. Power-dependent PL spectra. PL spectra of graded Ge_xSi_{1-x} alloys (**a**) on the flat, (**b**) on the micro-hole patterned Si(001) substrate, (**c**) energies of peaks NP and TO in (**a**) as a function of excitation power, (**d**) the enhancement factor vs. the excitation power. The inset in (**b**) shows the fitting of the PL spectrum at 800 mW by two Gaussian peaks. The dashed green line in (**a**,**b**) is for eye guidance.

3.3. FDTD Simulations

The origin of the peaks Pi (i = 1–5) in region (II) of 1500–1850nm can be confirmed from the FDTD simulation of the emission spectrum of graded Ge_xSi_{1-x} (x = 0.1–0.3) alloys grown on micro-hole patterned Si (001) substrates, as shown in Figure 6a. The geometrical profiles employed in the simulation are schematically shown in Figure 6b. Based on the SEM and the TEM images and the discussions above, the diameters at the top, the middle and the bottom of micro-holes are simplified to be 600 nm, 730 nm and 550 nm, respectively, as schematically demonstrated in the bottom-panel of Figure 6b. The barn-like morphologies at the regions in-between neighboring three micro-holes are also considered in the simulation. The simulated spectrum clearly demonstrates that there are guided resonant modes with the wavelengths well-matched with the experimental PL peaks Pi (i = 1–5) in region (II) of 1500–1850nm, whereas two additional peaks in the simulated spectrum with the shorter wavelengths (<1500 nm) are not as distinguishable in the experimental one. This may be attributed to the simple geometrical model used in the FDTD simulation, particularly without the consideration of the roughness at the sidewalls of the micro-holes.

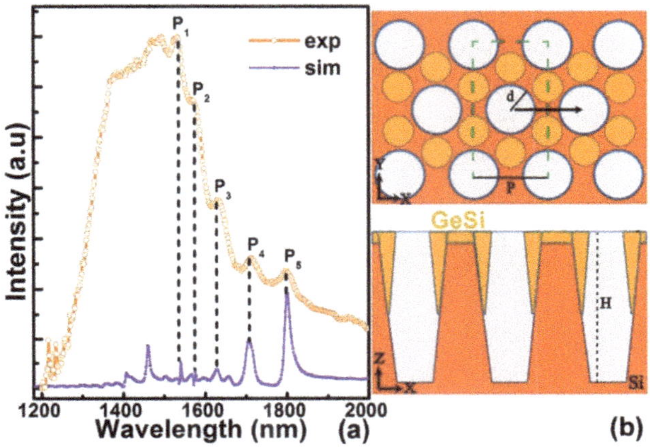

Figure 6. FDTD simulations. (**a**) Simulated spectrum of guided resonant modes of ordered micro-holes after graded Ge_xSi_{1-x} (x = 0.1–0.3) alloys growth, as well as the experimental PL spectrum, (**b**) schematic illustration of the top view (up-panel) and the cross-sectional (along the black arrow in the up-panel) side-view (down-panel) of ordered micro-holes. The green dashed lines in the up-panel of (**b**) denotes the simulation cell of 1.02 μm × 1.77 μm × 11 μm. The mesh size in the FDTD simulation is 10 nm × 8.6 nm × 5 nm. The depth H of micro-hole is 5.8 μm. The period P is 1.02 μm.

3.4. Temperature Dependent PL Spectra

Figure 7a,b show the temperature-dependent PL spectra of graded Ge_xSi_{1-x} alloy layers on the flat and the micro-hole patterned substrate, respectively. In both cases, a red-shift of PL peak with temperature is distinguished since the band gaps of both GeSi alloy and Si decrease with temperature. Even the peaks related to the guided resonant modes of the ordered micro-holes red-shift slightly with temperature. Such a red-shift is attributed to the variation of the refractive index of material with temperature. Surprisingly, along with the increase in temperature, a distinguished peak evolves from the former shoulder in region (III) of 1850–2200 nm in the PL spectra of the sample on the micro-hole patterned substrate, as demonstrated in Figure 7b. This is beyond the emission even of pure Ge. We argue that this peak mainly originates from Ge-rich domains in the graded Ge_xSi_{1-x} alloys on the micro-hole patterned substrate. It has been found that segregation and/or spinodal decomposition are enhanced by strain during the step-flow growth of an alloy on a vicinal surface [29,30]. In addition, nonuniform surface chemical potential originated from the strain field and the vicinal surface energy biases the incorporations of the alloy components in different regions around the patterned surface. The vicinal surface and the nonuniform strain distribution naturally exist around the edge of the micro-hole. Accordingly, some Ge-rich domains can self-assemble around the edge of the micro-hole during the growth of graded Ge_xSi_{1-x} alloys, particularly for x = 0.3, due to the accumulation of strain energy and the relative high Ge content [30]. This is confirmed by the TEM images in Figure 3. The rest in turn is $Ge_{0.3-\delta}Si_{0.7+\delta}$ alloy with a small δ that denotes the percentage of Ge separated into the Ge-rich domains. To obtain more details about the Ge-rich domain, further studies are necessary. Such a strain-dependent decomposition is also consistent with the previous report on non-uniform composition in GeSi islands [31]. The emissions of these Ge-rich domains dominate the spectrum in the long wavelength region (III) of 1850–2200 nm, particularly at high temperatures, since the higher Ge content provides deeper potential for holes. Similar emissions with smaller energy than the band gap of unstrained bulk Ge has been observed from Ge hut clusters on Si substrates due to the type-II band alignment and the misfit strain [32]. Given the limited number of Ge-rich domains, some photon-generated carriers can also radiatively recombine around $Ge_{0.3-\delta}Si_{0.7+\delta}$ alloys to dominate the spectra in the short wavelength region (I) of 1200–1500 nm. This scenario accounts for the slight blue-shift of the spectrum in region (I) of 1200–1500 nm in comparison with the PL peaks of the reference sample due to the small difference of Ge composition, as shown in Figure 4.

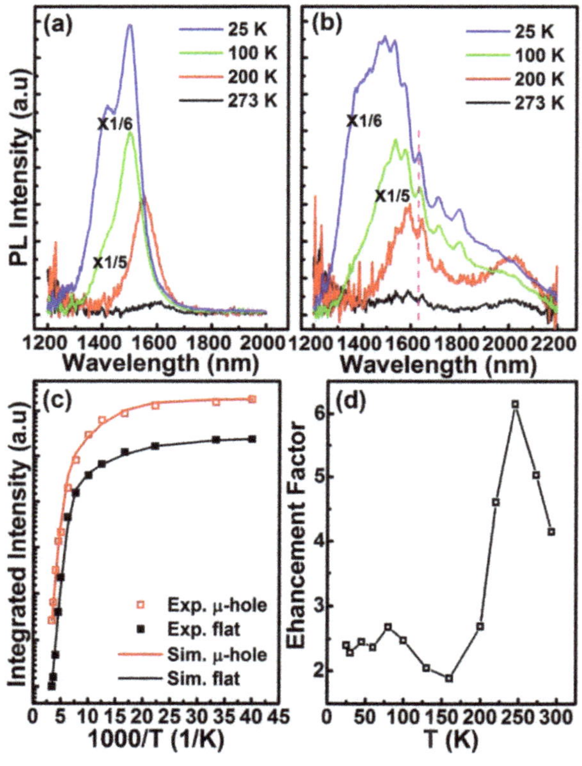

Figure 7. Temperature-dependent PL spectra. PL spectra of graded Ge_xSi_{1-x} alloys (**a**) on the flat Si(001) substrate, (**b**) on the micro-hole patterned Si(001) substrate, (**c**) the integrated PL intensity vs. temperature, (**d**) the EF vs. temperature. For comparisons, the PL spectra at 25 K and 100 K in (**a**,**b**) are reduced by multiplying 1/6 and 1/5, as is denoted by 'x1/6' and 'x1/5', respectively. 'Exp.', 'Sim.', 'μ-hole' and 'flat' in (**c**) denote experiment, simulation, micro-hole patterned substrate and flat substrate, respectively. The dashed pink line is for eye-guide.

The formation of Ge-rich domains in the Ge_xSi_{1-x} alloy grown on the micro-hole patterned substrate is also confirmed by the temperature-dependence of the integrated PL intensity, as shown in Figure 7c. Similar to the GeSi QDs, the integrated PL intensities of GeSi alloys as a function of temperature can be fitted by the following Equation (1) [33,34],

$$I(T) = \frac{I(0)}{1 + c_1 \exp(-E_1/kT) + c_2 \exp(-E_2/kT)} \quad (1)$$

where *I(0)* is approximated to be the integrated PL intensity at 17 K, c_1 and c_2 are the fitting parameters, E_1 is the activation energy for the carriers to escape from the potential confinement of GeSi alloys, and E_2 is the binding energy related to exciton dissociation. By fitting the experimental data in Figure 7c, the energies (E_1, E_2) for the samples on the flat and the micro–hole patterned substrates are (123 meV, 12.5 meV) and (132 meV, 19.8 meV), respectively. As shown in Figure 7b, the emissions in the long wavelength region (III) of 1850–2200 nm become more pronounced with the increase of temperature. The larger E_1 demonstrates the deeper potential confinement, which corresponds to the larger Ge content, i.e., the Ge-rich domain. The larger E_2 indicates the relatively small size of the Ge-rich domain [35]. Figure 7d shows the EF of PL intensity as a function of temperature. The light field distributions of guided resonant modes around the ordered micro-holes are not uniform and depend on the frequencies [25]. The strong coupling between the exciton emissions and the cavity modes of ordered micro-holes occurs mainly for excitons localizing at the light field maxima [36]. The variation of spatial distributions of excitons in the Ge_xSi_{1-x} alloys on the micro-hole patterned substrate with the increase of temperature may affect the coupling between the emissions of Ge_xSi_{1-x} alloys and the guided resonant modes. Accordingly, the EF is not monotonically changed with temperature, as shown in Figure 7d. The maximum EF is over 6 at 245 K. It is worth mentioning that the PL intensity of Ge_xSi_{1-x} alloys on the micro-hole patterned substrate is reduced by defects in the sidewalls of micro-holes. By optimizing plasma etching processes to suppress defect formation in the Si sidewalls of micro-holes, the EF can be further increased. In addition, the cutoff wavelength of emissions will be shifted by applying an external voltage in the p-i-n structure due to a quantum confine Stark effect. However, such a shift is not so obvious for the general bias voltages [37]. By optimizing the parameters of micro-holes and the growth conditions (e.g., reducing Ge content), as well as taking into account the contributions of Si, the emission band can be extended further into the shorter NIR wavelengths.

4. Conclusions

In conclusion, a p-i-n structure with graded Ge_xSi_{1-x} (x = 0.1, 0.15, 0.2, 0.25, 0.3) alloys is grown on micro-hole patterned Si(001) substrates. To relax misfit strain, barn-like islands self-assemble at the center regions in-between three neighboring micro-holes. Branch-like nanostructures appears at the sidewalls with many dislocations, which originate from the defects introduced during plasma etching. An extensive broadband PL spectrum is observed in the NIR range of 1200–2200 nm. It is attributed to the coupling between the emissions of Ge_xSi_{1-x} alloy and the guided resonant modes in the ordered micro-holes, as well as the strain-enhanced decomposition of Ge_xSi_{1-x} alloys during their growth on micro-hole patterned substrates. Moreover, the intensity of the PL spectra is remarkably enhanced, which can be over six times of that from the reference sample on the flat substrate at 245 K. Our results demonstrate an innovative strategy of growth of alloy on micro-hole patterned substrate to realize strong broadband spectrum. Considering the strong and rather broad wavelength range (1200–2200 nm) of emissions and the compatibility with the Si integration technology, the graded Ge_xSi_{1-x} alloy layers on micro-hole patterned Si substrates will have a great potential in the broadband NIR light source particularly to realize entire systems on a Si chip.

Author Contributions: Conceptualization, K.P. and Z.Z.; methodology, K.P., N.Z., J.Z. and C.Z.; software, K.P. and N.Z.; validation, P.C., J.Y. and Z.J.; formal analysis, K.P., N.Z. and J.Z.; investigation, K.P., J.Z. and J.Y.; resources, K.P. and J.Z.; data curation, K.P. and N.Z.; writing—original draft preparation, K.P.; writing—review and editing, Z.Z.; visualization, C.Z. and Z.J.; supervision, Z.Z.; project administration, Z.Z.; funding acquisition, Z.Z. All authors have read and agreed to the published version of the manuscript.

Funding: This research was funded by National Natural Science Foundation of China, grant number 11774062, 12074077, 61871134 and Shanghai Municipal Science and Technology Commission (18JC1410300).

Acknowledgments: Part of the characterization was performed at the Fudan Nano-Fabrication Laboratory.

Conflicts of Interest: The authors declare no conflict of interest.

References

1. Sun, L.; Choi, J.J.; Stachnik, D.; Bartnik, A.C.; Hyun, B.; Malliaras, G.G.; Hanrath, T.; Wise, F.W. Bright infrared quantum-dot light-emitting diodes through inter-dot spacing control. *Nat. Nanotechnol.* **2012**, *7*, 369–373. [CrossRef] [PubMed]
2. Qiao, J.; Zhou, G.; Zhou, Y.; Zhang, Q.; Xia, Z. Divalent europium-doped near-infrared-emitting phosphor for light-emitting diodes. *Nat. Commun.* **2019**, *10*, 5267. [CrossRef] [PubMed]
3. Goossens, S.; Navickaite, G.; Monasterio, C.; Gupta, S.; Piqueras, J.J.; Pérez, R.; Burwell, G.; Nikitskiy, I.; Lasanta, T.; Galán, T.; et al. Broadband image sensor array based on graphene–CMOS integration. *Nat. Photonics* **2017**, *11*, 366–371. [CrossRef]
4. Pradhan, S.; Dalmases, M.; Konstantatos, G. Solid-State Thin-Film Broadband Short-Wave Infrared Light Emitters. *Adv. Mater.* **2020**, *32*, 2003830. [CrossRef]
5. He, S.; Zhang, L.; Wu, H.; Wu, H.; Pan, G.; Hao, Z.; Zhang, X.; Zhang, L.; Zhang, H.; Zhang, J. Efficient Super Broadband NIR Ca2LuZr2Al3O12:Cr3+,Yb3+ Garnet Phosphor for pc-LED Light Source toward NIR Spectroscopy Applications. *Adv. Optical Mater.* **2020**, *8*, 1901684. [CrossRef]
6. Medintz, I.L.; Uyeda, H.T.; Goldman, E.R.; Mattoussi, H. Quantum dot bioconjugates for imaging, labelling and sensing. *Nat. Mater.* **2005**, *4*, 435–446. [CrossRef]
7. Thimsen, E.; Sadtler, B.; Berezin, M.Y. Shortwave-infrared (SWIR) emitters for biological imaging: A review of challenges and opportunities. *Nanophotonics* **2017**, *6*, 1043–1054. [CrossRef]
8. Eggebrecht, A.T.; Ferradal, S.L.; Robichaux-Viehoever, A.; Hassanpour, M.S.; Dehghani, H.; Snyder, A.Z.; Hershey, T.; Culver, J.P. Mapping distributed brain function and networks with diffuse optical tomography. *Nat. Photonics* **2014**, *8*, 448–454. [CrossRef]
9. Hayashi, D.; van Dongen, A.M.; Boerekamp, J.; Spoor, S.; Lucassen, G.; Schleipen, J. A broadband LED source in visible to short-wave-infrared wavelengths for spectral tumor diagnostics. *Appl. Phys. Lett.* **2017**, *110*, 233701. [CrossRef]
10. Basore, E.T.; Xiao, W.; Liu, X.; Wu, J.; Qiu, J. Broadband Near-Infrared Garnet Phosphors with Near-Unity Internal Quantum Efficiency. *Adv. Optical Mater.* **2020**, *8*, 2000296. [CrossRef]
11. Gao, L.; Quan, L.; García de Arquer, F.P.; Zhao, Y.; Munir, R.; Proppe, A.; Quintero-Bermudez, R.; Zou, C.; Yang, Z.; Saidaminov, M.I.; et al. Efficient near-infrared light-emitting diodes based on quantum dots in layered perovskite. *Nat. Photonics* **2020**, *14*, 227. [CrossRef]
12. Atabaki, A.H.; Moazeni, S.; Pavanello, F.; Gevorgyan, H.; Notaros, J.; Alloatti, L.; Wade, M.T.; Sun, C.; Kruger, S.A.; Meng, H.; et al. Integrating photonics with silicon nanoelectronics for the next generation of systems on a chip. *Nature* **2018**, *556*, 349. [CrossRef]
13. Knight, M.W.; Sobhani, H.; Nordlander, P.; Halas, N.J. Photodetection with Active Optical Antennas. *Science* **2011**, *332*, 702. [CrossRef] [PubMed]
14. Ackert, J.J.; Karar, A.S.; Paez, D.J.; Jessop, P.E.; Cartledge, J.C.; Knights, A.P. 10 Gbps silicon waveguide-integrated infrared avalanche photodiode. *Opt. Express* **2013**, *21*, 19530. [CrossRef]
15. Zou, H.; Li, X.; Dai, G.; Peng, W.; Ding, Y.; Zhang, Y.; Wang, A.C.; Zhang, S.L.; Xu, C.; Zhang, S.; et al. Dramatically Enhanced Broadband Photodetection by Dual Inversion Layers and Fowler-Nordheim Tunneling. *ACS Nano* **2019**, *13*, 2289–2297. [CrossRef] [PubMed]
16. Katiyar, A.K.; Thai, K.Y.; Yun, W.S.; Lee, J.; Ahn, J. Breaking the absorption limit of Si toward SWIR wavelength range via strain engineering. *Sci. Adv.* **2020**, *6*, eabb0576. [CrossRef] [PubMed]
17. Atalla, M.R.M.; Assali, S.; Attiaoui, A.; Lemieux-Leduc, C.; Kumar, A.; Abdi, S.; Moutanabbir, O. All-Group IV Transferable Membrane Mid-Infrared Photodetectors. *Adv. Funct. Mater.* **2020**, *31*, 2006329. [CrossRef]
18. Kruck, P.; Helm, M.; Fromherz, T.; Bauer, G.; Nutzel, J.F.; Abstreiter, G. Mediumwavelength, normalincidence, ptype Si/SiGe quantum well infrared photodetector with background limited performance up to 85 K. *Appl. Phys. Lett.* **1996**, *69*, 3372–3374. [CrossRef]
19. Krapf, D.; Adoram, B.; Shappir, J.; Sa'ar, A.; Thomas, S.G.; Liu, J.L.; Wang, K.L. Infrared multispectral detection using Si/Si_x Ge $_{1-x}$ quantum well infrared Photodetectors. *Appl. Phys. Lett.* **2001**, *78*, 495–497. [CrossRef]
20. Rauter, P.; Fromherz, T.; Falub, C.; Grutzmacher, D.; Bauer, G. SiGe quantum well infrared photodetectors on pseudosubstrate. *Appl. Phys. Lett.* **2009**, *94*, 081115. [CrossRef]
21. Michel, J.; Liu, J.; Kimerling, L.C. High-performance Ge-on-Si photodetectors. *Nat. Photonics* **2010**, *4*, 527–534. [CrossRef]
22. Rauter, P.; Mussler, G.; Grutzmacher, D.; Fromherz, T. Tensile strained SiGe quantum well infrared photodetectors based on a light-hole ground state. *Appl. Phys. Lett.* **2011**, *98*, 211106. [CrossRef]
23. Bandara, S.V.; Gunapala, S.D.; Liu, J.K.; Rafol, S.B.; Hill, C.J.; Ting, D.Z.-Y.; Mumolo, J.M.; Trinh, T.Q.; Fastenau, J.M.; Liu, A.W.K. Tuning and tailoring of broadband quantum-well infrared photodetector responsivity Spectrum. *Appl. Phys. Lett.* **2005**, *86*, 151104. [CrossRef]
24. Gallas, B.; Hartmann, J.M.; Berbezier, I.; Abdallah, M.; Zhang, J.; Harris, J.J.; Joyce, B.A. Influence of misfit and threading dislocations on the surface morphology of SiGe graded-layers. *J. Cryst. Growth* **1999**, *201*, 547–550. [CrossRef]
25. Peng, K.; Zhang, N.; Chen, P.; Zhang, L.; Jiang, Z.; Zhong, Z. Manipulations of light by ordered micro-holes in silicon substrates. *Opt. Express* **2020**, *28*, 16151–16162. [CrossRef] [PubMed]

26. Sutter, E.; Sutter, P.; Bernard, J.E. Extended shape evolution of low mismatch $Si_{1-x}Gex$ alloy islands on Si(100). *Appl. Phys. Lett.* **2004**, *84*, 2262–2264. [CrossRef]
27. Wan, J.; Jin, G.; Jiang, Z.; Luo, Y.; Liu, J.; Wang, K.L. Band alignments and photon-induced carrier transfer from wetting layers to Ge islands grown on Si(001). *Appl. Phys. Lett.* **2001**, *78*, 1763–1765. [CrossRef]
28. Wu, Z.; Lei, H.; Zhou, T.; Fan, Y.; Zhong, Z. Fabrication and characterization of SiGe coaxial quantum wells on ordered Si nanopillars. *Nanotechnology* **2014**, *25*, 055204. [CrossRef] [PubMed]
29. Venezuela, P.; Tersoff, J.; Floro, J.A.; Chason, E.; Follstaedt, D.M.; Liu, F.; Lagally, M.G. Self-organized growth of alloy superlattices. *Nature* **1999**, *397*, 678–681. [CrossRef]
30. Aqua, J.-N.; Berbezier, I.; Favre, L.; Frisch, T.; Ronda, A. Growth and self-organization of SiGe nanostructures. *Phys. Rep.* **2013**, *522*, 59–189. [CrossRef]
31. Gatti, R.; Uhlík, F.; Montalenti, F. Intermixing in heteroepitaxial islands: Fast, self-consistent calculation of the concentration profile minimizing the elastic energy. *New J. Phys.* **2008**, *10*, 083039. [CrossRef]
32. Denker, U.; Stoffel, M.; Schmidt, O.G.; Sigg, H. Ge hut cluster luminescence below bulk Ge band gap. *Appl. Phys. Lett.* **2003**, *82*, 454–456. [CrossRef]
33. Kamenev, B.V.; Lee, E.-K.; Chang, H.-Y.; Han, H.; Grebel, H.; Tsybeskov, L.; Kamins, T.I. Excitation-dependent photoluminescence in Ge Si Stranski-Krastanov nanostructures. *Appl. Phys. Lett.* **2006**, *89*, 153106. [CrossRef]
34. Wang, Z.; Wang, S.; Yin, Y.; Liu, T.; Lin, D.; Li, D.; Yang, X.; Jiang, Z.; Zhong, Z. Promising features of low-temperature grown Ge nanostructures on Si(001) Substrates. *Nanotechnology* **2017**, *28*, 115701. [CrossRef] [PubMed]
35. Takagahara, T.; Takeda, K. Theory of the quantum confinement effect on excitons in quantum dots of indirect-gap materials. *Phys. Rev. B* **1992**, *46*, 15578–15581. [CrossRef] [PubMed]
36. Schatzl, M.; Hackl, F.; Glaser, M.; Rauter, P.; Brehm, M.; Spindlberger, L.; Simbula, A.; Galli, M.; Fromherz, T.; Schaffler, F. Enhanced telecom emission from single group-iv quantum dots by precise cmos-compatible positioning in photonic crystal cavities. *ACS Photonics* **2017**, *4*, 665–673. [CrossRef] [PubMed]
37. Tong, S.; Liu, J.L.; Wan, J.; Wang, K.L. Normal-incidence Ge quantum-dot photodetectors at 1.5 μm based on Si substrate. *Appl. Phys. Lett.* **2002**, *80*, 1189–1191. [CrossRef]

nanomaterials

Article

Simple Self-Assembly Strategy of Nanospheres on 3D Substrate and Its Application for Enhanced Textured Silicon Solar Cell

Dan Su [1,†], Lei Lv [2,†], Yi Yang [2], Huan-Li Zhou [2], Sami Iqbal [2] and Tong Zhang [1,2,3,*]

1 Key Laboratory of Micro-Inertial Instrument and Advanced Navigation Technology, Ministry of Education, School of Instrument Science and Engineering, Southeast University, Nanjing 210096, China; jssysls@163.com
2 Joint International Research Laboratory of Information Display and Visualization, School of Electronic Science and Engineering, Southeast University, Nanjing 210096, China; lvlei@seu.edu.cn (L.L.); yyang19x@163.com (Y.Y.); huanli_zhou@163.com (H.-L.Z.); 101300112@seu.edu.cn (S.I.)
3 Suzhou Key Laboratory of Metal Nano-Optoelectronic Technology, Southeast University Suzhou Campus, Suzhou 215123, China
* Correspondence: tzhang@seu.edu.cn
† Equally contributed.

Abstract: Nanomaterials and nanostructures provide new opportunities to achieve high-performance optical and optoelectronic devices. Three-dimensional (3D) surfaces commonly exist in those devices (such as light-trapping structures or intrinsic grains), and here, we propose requests for nanoscale control over nanostructures on 3D substrates. In this paper, a simple self-assembly strategy of nanospheres for 3D substrates is demonstrated, featuring controllable density (from sparse to close-packed) and controllable layer (from a monolayer to multi-layers). Taking the assembly of wavelength-scale SiO_2 nanospheres as an example, it has been found that textured 3D substrate promotes close-packed SiO_2 spheres compared to the planar substrate. Distribution density and layers of SiO_2 coating can be well controlled by tuning the assembly time and repeating the assembly process. With such a versatile strategy, the enhancement effects of SiO_2 coating on textured silicon solar cells were systematically examined by varying assembly conditions. It was found that the close-packed SiO_2 monolayer yielded a maximum relative efficiency enhancement of 9.35%. Combining simulation and macro/micro optical measurements, we attributed the enhancement to the nanosphere-induced concentration and anti-reflection of incident light. The proposed self-assembly strategy provides a facile and cost-effective approach for engineering nanomaterials at 3D interfaces.

Keywords: self-assembly; 3D substrate; solar cell

Citation: Su, D.; Lv, L.; Yang, Y.; Zhou, H.-L.; Iqbal, S.; Zhang, T. Simple Self-Assembly Strategy of Nanospheres on 3D Substrate and Its Application for Enhanced Textured Silicon Solar Cell. *Nanomaterials* **2021**, *11*, 2581. https://doi.org/10.3390/nano11102581

Academic Editors: Jihoon Lee and Ming-Yu Li

Received: 10 August 2021
Accepted: 27 September 2021
Published: 30 September 2021

1. Introduction

The development of nanomaterials and nanostructures provides new opportunities for performance boosting of optical and optoelectronic devices [1–5]. For example, dielectric nanostructures are used to enhance the transmission and brightness of different transparent windows or display screens [6–9], and plasmonic or dielectric nanocoatings are widely proposed and applied to enhance the efficiency of photovoltaic devices or sensitivity of photodetectors [1,10]. In many optoelectronic devices, there are complex 3D surfaces [11,12]. For example, micron and nanoscale heterogeneities exist in various photovoltaic devices [13], such as the intrinsic grains in polycrystalline solar cells [14–16] and textured surfaces in the commercial silicon solar cells for light-trapping [17].

In this context, there has been a drive for designing and fabricating nanostructures on 3D surfaces with desired control. On planar substrates or interfaces, both top-down and bottom-up self-assembly strategies enable precise control over distribution density and geometrical shapes [8,18–23]. Although substantial advances have been made, control over nanofabrication on complex 3D surfaces is still very challenging [7]. Top-down strategies, such as electron beam lithography and focused ion beam, show some difficulties in the fabrication on 3D substrates limited by the inherent process characteristics of photoresist coating or beam focus [24]. Self-assembly methods, using colloidal nanoparticles as building blocks, show great potential in assembling nanostructures on 3D substrates [18]. Previous reports have demonstrated the self-assembly results on 3D fabrics and textured wafers [25–27]. However, the nanostructures usually accumulate and form multi-layers at the bottom of valleys [25,26], while precise control over the distribution density and layer numbers on the 3D substrate has remained elusive.

We emphasize the importance of controllable nanosphere assembly for understanding the mechanism of nanosphere-enhanced textured photovoltaics. For photovoltaic devices application, wavelength scale dielectric nanospheres have attracted much attention. Nanosphere array supports Whispering Gallery Modes (WGM), which can be coupled with the guided modes in photovoltaic thin-film to improve the light absorption of the devices [10,28]. Some studies have shown that dielectric microspheres can also enhance planar bulk material cells [29,30], mainly due to the anti-reflection effect of close-packed SiO_2 spheres. Besides, SiO_2 arrays enable colorful photovoltaic devices without sacrificing device performance [31]. In 2020, Bek et al. proposed that in planar cells, coexisting anti-reflection and light concentration effects are the reason for the nanosphere enhancement [32]. The former contributes to an enhanced photocurrent, and the latter contributes to an enhanced fill factor of the devices. When it comes to textured solar cells, it is highly desirable to control nanosphere distribution density with specific layer numbers to explore the optimized parameters and discern the underlying mechanism.

In this work, we propose a simple assembly strategy on 3D substrates featuring controllable distribution density and coating layers. The influence of gravitational sedimentation on assembly behavior was analyzed by comparing substrates with different surface features (planar and textured) and under different orientations (upward and inverted) in the SiO_2 colloid. The effect of SiO_2 microsphere coating on the device performance under different assembly conditions was further explored. For the first time, the nanosphere anti-reflection and light concentration mechanisms were analyzed on the textured solar cells with numerical simulation, macro and micro-region optical characterization.

2. Materials and Methods

2.1. Materials

Tetraethyl orthosilicate (TEOS, 98%) and ammonium hydroxide solution (NH_4OH, 28.0–30.0%) were bought from Sigma-Aldrich (Saint Louis, MO, USA). Absolute ethanol ($\geq$95%) was purchased from Sinopharm Chemical Reagent Co., Ltd. (Ningbo, China). Poly-L-lysine solution has a concentration of 0.1 mg/mL and degree of polymerization from 144 to 335, purchased from Xiya Regent (Shanghai, China). Ultrapure water used in the experiments has a resistivity larger than 18.2 MΩ·cm (Milli-Q). All chemicals were used without any purification. Shanghai Shenzhou New Energy Co., Ltd. (Shanghai, China) supplied six-inch polycrystalline silicon solar cells. The solar cells were cut into 4 cm^2 ones.

2.2. Synthesis of SiO_2 Nanosphere

The ~550 nm diameter SiO_2 nanosphere was synthesized according to previous work [33]. Briefly, 25 mL ultrapure water, 62 mL ethanol, and 9 mL NH_4OH were mixed and stirred at a moderate speed for 30 min. Then, 4.5 mL TEOS was added into the mixed solution and reacted for 3 h at 30 °C. The resulting products were centrifuged three times using absolute ethanol and dried in an oven at 200 °C for an hour to obtain SiO_2 nanosphere powders for future use. For the sake of a fast self-assembly process, SiO_2 nanospheres were redispersed in absolute ethanol with a concentration of 50 mg/mL, and we call it SiO_2 assembly solution.

2.3. Self-Assembly Process of SiO_2 Nanosphere

The four-step self-assembly processes are outlined schematically in Figure 1. We first immersed the substrate into the poly-L-lysine solution (pH value about 6.2) for 4 min.

The ionic form of the ε-amino group in poly-L-lysine solution depends primarily on the pH [34,35]. In the acid environment, the ε-amino group is in a positive charge state ($-NH_3^+$), widely used in biology and chemistry [34,36]. In such a dip process, the absorption of poly-L-lysine on the textured silicon substrate yields a positively charged surface. Then we used ultra-pure water to remove the redundant poly-L-lysine. Following this, the substrates were dip-coated onto the solution of SiO_2 assembly solution for different time durations. It has been reported that the solution synthesized SiO_2 nanospheres with an isoelectric point (IEP = 2.0) [37]. At a pH value higher than 2, the SiO_2 nanospheres are abundant with OH^- and negatively charged. The negatively charged SiO_2 surface will promote the electrostatic attraction to the positive textured surface while resisting the absorption of multi-layer SiO_2. At last, the silicon solar cell was rinsed with ultra-pure water a second time to remove the unabsorbed SiO_2 particles, which are not adhesive to the devices. It should be noticed that if the first two steps are not adopted, the SiO_2 nanospheres could not be maintained on the surface after washing with ultra-pure water.

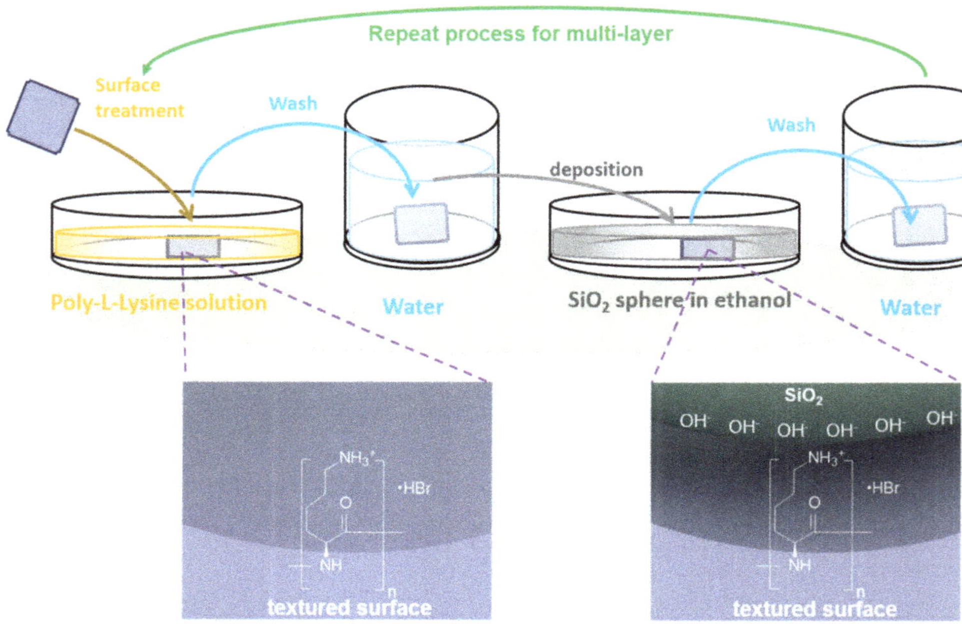

Figure 1. Schematic illustration and self-assembly process. The four-step self-assembly procedure of SiO_2 nanosphere onto the textured 3D surface, and the inset figures show the electrostatic force-directed assembly mechanism.

When the assembly process is repeated, another layer of poly-L-lysine will be absorbed upon the top surface of the SiO_2 nanocoating assembled on the textured surface, forming a positive surface and enabling a layer-by-layer assembly process.

2.4. Characterization of Textured Silicon Solar Cells

Brightfield reflection and darkfield scattering optical images are taken from a Nikon (Ti-U) inverted microscope under the illumination of a halogen lamp. The grayscale transformation and statical analysis of the optical micro-region images were conducted using a MATLAB code. A long working distance 100×/0.80 NA objective lens (Nikon Plan Fluor ELWD 100×) was used to collect bright and darkfield images. The reflection spectra were taken from a fiber spectrometer (Nova, Ideaoptics Co., Ltd., Shanghai, China) coupled with an integrating sphere (IS-30-6-R). Electrical properties of solar cells are measured with a high-precision source-meter-unit (SMU, Keithley 2651) under the illumination of a 1000 W xenon lamp equipped with an AM1.5 filter (Crowntech Inc., Indianapolis, IN, USA).

2.5. Modeling of Silicon Solar Cell with SiO_2 Nanospheres

The three-dimensional finite element method (COMSOL software) was used to simulate the light-concentrating performance of silica nanospheres, and the simulation process is similar to previously reported work [32]. In our simulation model, a single silica nanosphere or a vertically aligned dimmer is placed on the surface of the silicon substrate. The simulation width of the device is 2000 nm, the size of the nanoparticle is 550 nm, and the thickness of the air layer is 1500 nm. The linear-polarization plane wave is used as the incident excitation, and its amplitude is 1 V/m. In addition, to eliminate the unwanted reflection of the interface, the boundary of the device region was selected and was delimited by the perfectly matched layer (PML). The optical constants of SiO_2 and Si are obtained by linear interpolation in the optical manual [38].

3. Results and Discussion

3.1. Self-Assembled SiO_2 Nanocoating on Planar and Textured 3D Substrates

In order to fully demonstrate the assembly characteristics of the SiO_2 nanosphere coating on the 3D textured surfaces, we compared the assembly results on the flat surface and the texturing Si wafer surface under the same assembly time. As shown in Figure 2a, on the surface of the flat substrate, the assembly of SiO_2 nanospheres has two characteristics. Firstly, some of the SiO_2 nanospheres become aggregated during the assembly process. Similar aggregates, also found in previously reported gold or dielectric nanocoatings via electrostatic assembly [39,40], can be attributed to the capillary force [41] (originating from water bridges formed between particles when the substrate is taken out of the solution). Secondly, the covered area of nanospheres is estimated at less than 35% according to SEM analysis (using ImageJ software packages). This situation has also been widely studied in previous electrostatic assemblies [39,42]. In the absence of external force, the assembly process of spherical nanoparticles usually conforms to the random sequential adsorption (RSA) model [43], where particles are supposed to be fixed on the substrate and cannot be moved after adsorption. Under this model, the surface coverage of particles could not exceed the jamming limit (54.7%) [39]. Furthermore, the repulsion between charged particles will further reduce surface coverage. Therefore, in the self-assembly coating on a planar substrate without external forces, the surface coverage of nanospheres usually does not exceed 40% [39].

Under the same assembly time, we observed that the particle coverage of silica nanospheres on the 3D substrate is significantly larger than that of a two-dimensional substrate, as shown in Figure 2b. Except for the sharp features, a single layer of densely arranged SiO_2 nanospheres is almost formed on the entire substrate. Note that this assembly result is significantly different from the multi-layer accumulation only located at the bottom of the valley obtained by spin-coating or dip-coating [25,26]. In addition, our dense arrangement result is similar to the coating effect formed by the Langmuir-budget liquid-liquid assembly with an external force and the following transfer method [31]. However, our assembly method is simple and does not require the introduction of other external forces. We infer that the dense nanosphere assembly originates due to gravitational sedimentation [44]. For ~550 nm particles, the gravitational potential energy of the particle is comparable to the thermal energy accounting for thermal motion [45]. Therefore, the sedimentation of nanospheres in micron-scale valleys may lead to dense assembly behavior. Previously, the literature also proposed that even if there is repulsion between particles, nanoparticles (100–1000 nm) will still undergo sedimentation in the solvent [45].

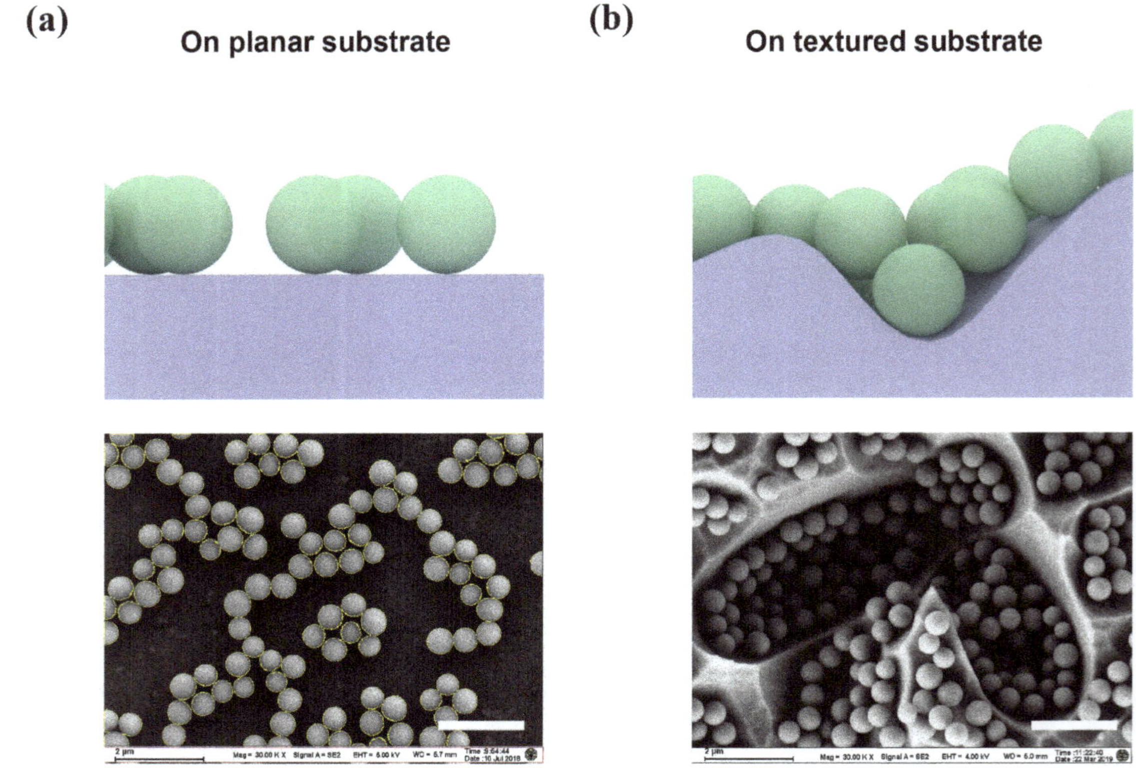

Figure 2. Schematic illustration and SEM graphs of self-assembled SiO_2 nanocoating on the planar and textured substrates: (**a**) Assembly on a planar substrate, schematic representation (**top**), and the SEM graphs (**down**). The yellow circles around nanospheres were used to calculate the surface coverage of nanospheres with the ImageJ software; (**b**) Assembly on a textured substrate, schematic representation (**top**), and the SEM graphs (**down**). The dipping time in SiO_2 assembly solution is the same (10 min) for the planar and textured substrate. The scale bar is 2 µm.

To verify the influence of gravitational sedimentation in the assembly process, the extured substrates were set at different orientations in the SiO_2 colloid. As shown in Figure 3, the assembly procedures with the substrate upward (Figure 3a) and inverted (Figure 3d) were performed under the same assembly time. The resultant assembly characteristics under the different orientations of substrate changed dramatically. A relatively dense assembly is achieved by an upward substrate (shown in Figure 3b,c). However, when the substrate was placed invertedly in the SiO_2 colloid, the resultant surface coverage of SiO_2 nanospheres was obviously reduced (shown in Figure 3d,e). The assembly time was increased to 30 min for further investigation of the assembly characteristics. As shown in Figure S1, for substrates with different assembly orientations, the surface coverages of SiO_2 nanospheres were similar compared to that of 10 min assembly time. It can be concluded that the gravitational sedimentation contributes to a dense-packed nanocoating in the textured solar cell when the substrate is placed upward. Notably, gravitational sedimentation is a natural phenomenon, and therefore the dense-packed assembly can be achieved without other external forces such as electric field force or pressure [8,19,31].

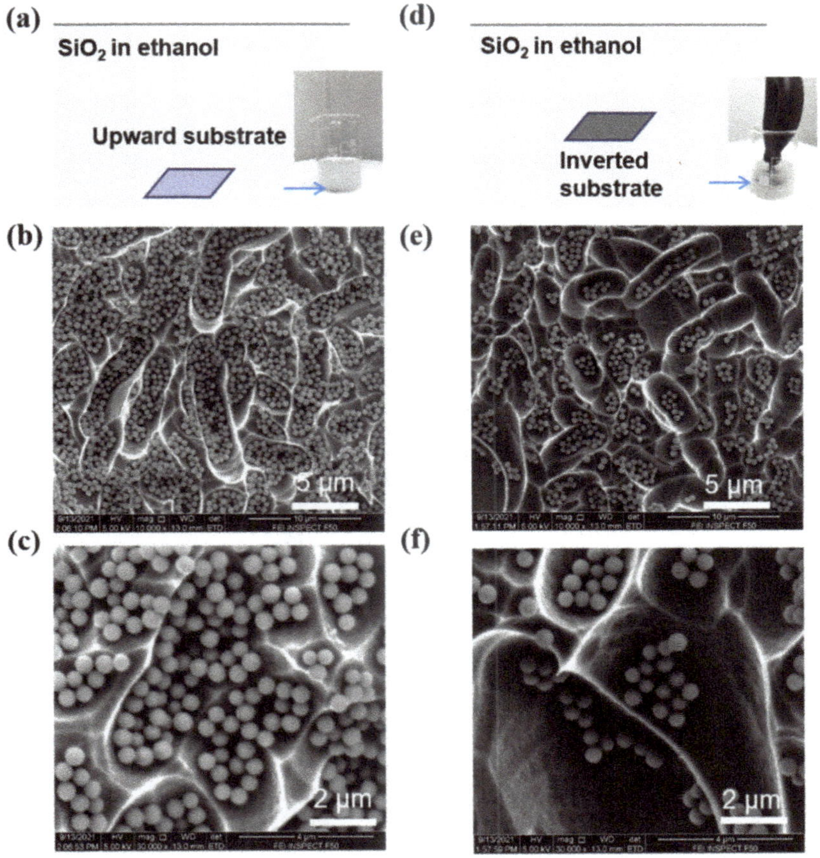

Figure 3. Self-assembly nanocoating under different substrate orientations in the SiO_2 assembly colloid: (**a–c**) Self-assembly with an upward substrate; (**d–f**) Self-assembly with an inverted substrate. The assembly time was fixed to 10 min for all the assembly processes, and the inset figures in (**a**) and (**d**) are the images of assembly in colloidal SiO_2. The blue arrows indicate the location of the devices. A vacuum suction ball was used to keep the substrate upside-down during the assembly process in the inverted condition.

The assembly behavior of nanoparticles on 3D texturing substrates (textured Si solar cells) under different timelines (10 s, 30 s, 10 min) and different assembly rounds was further investigated. The samples for different assembly times were named sample 1 (S1), sample 2 (S2), and sample 3 (S3) for 10 s, 30 s, 10 min, respectively. Figure 4a–i is one round of self-assembly SEM images and schematic diagrams of different assembly times. It can be found that, in the initial stage of assembly (10 s), the particles first deposit at the valley bottom, which indicates that the gravity of the nanoparticles is greater than the particle-substrate electrostatic attraction in this stage. As the assembly process proceeds, the silica nanospheres gradually assemble along the sidewalls of the valley. Subsequently, we repeated the assembling process twice, and the SEM images are shown in Figure 4j,k. As the schematic diagram Figure 4l shows, multiple layers appeared on the surface of the textured substrate, especially inside the valley. The sample for two round assembly was named sample 4 (S4).

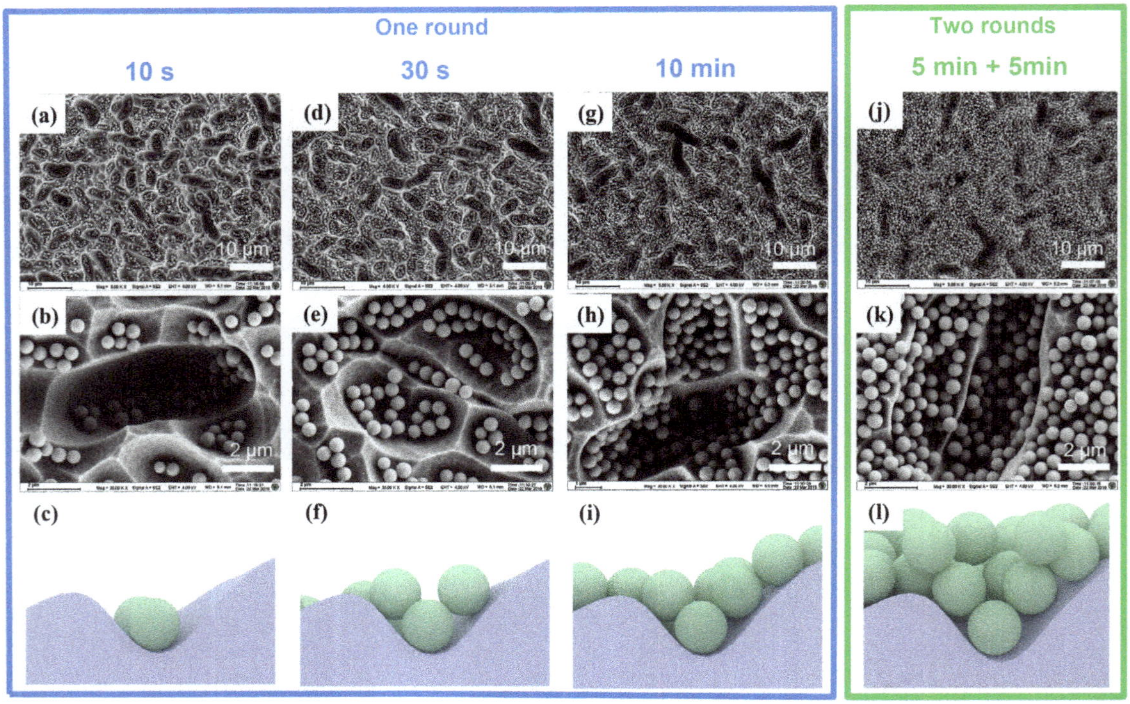

Figure 4. Self-assembly nanocoating under different timelines and repeated assembly process. SEM graphs under different magnification and the schematic diagram: (**a–c**) One round with an assembly time of 10 s; (**d–f**) One round with an assembly time of 30 s; (**g–i**) One round with an assembly time of 10 min; (**j–l**) Two rounds with assembly times of 5 min + 5 min.

3.2. The Performance Analysis of Textured Si Solar Cells with SiO_2 Nanosphere Coatings

We compared the effects of different nanosphere coatings (different distribution densities and number of layers) on the electrical performance of the textured Si solar cells to obtain optimized parameters, as shown in Table 1 and Figure 5. Figure 5a shows a structural diagram of the textured Si solar cell with nanocoatings. The polycrystalline silicon solar cell thickness is ~180 μm, and the junction depth is about 500 nm. The SiO_2 nanosphere coating is assembled on the textured upper surface of the device. As shown in Table 1, we tested the electrical properties of the devices under a solar simulator with an irradiance of 1000 W/m^2. In the case of a single-layer SiO_2 coating, as the distribution density increases, the relative enhancement of device efficiency gradually increases. When the nanospheres are in close packing, the maximum efficiency enhancement reaches 9.35%.

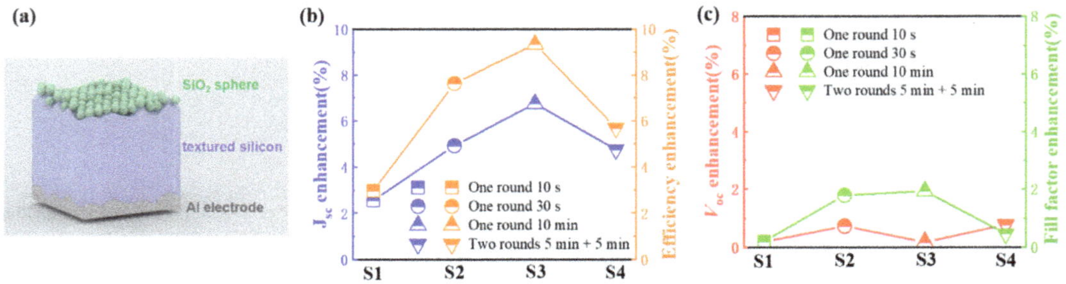

Figure 5. Comparison of electrical parameters enhancement with different SiO_2 nanocoatings. (**a**) Schematic illustration of the textured Si solar cell with SiO_2 nanosphere coating; (**b**) Relative enhancement of J_{sc} and power conversion efficiency of the textured Si solar cells; (**c**) Relative enhancement of V_{oc} and fill factor of the textured Si solar cells.

Table 1. Relative enhancement of electrical properties of samples before and after SiO_2 nanosphere assembly.

Description	V_{oc} (V)	J_{sc} (mA/cm^2)	Fill Factor (%)	Efficiency (%)
S1	0.537 ± 0.001	24.98 ± 0.32	50.2 ± 0.1	6.73 ± 0.10
S1–One round 10 s	0.538 ± 0.001	25.62 ± 0.29	50.3 ± 0.2	6.93 ± 0.12
Relative enhancement (%)	0.19	2.55	0.20	2.96
S2	0.546 ± 0.001	24.93 ± 0.13	61.4 ± 0.1	8.36 ± 0.04
S2–One round 30 s	0.550 ± 0.001	26.16 ± 0.17	62.5 ± 0.1	8.99 ± 0.07
Relative enhancement (%)	0.73	4.93	1.79	7.63
S3	0.567 ± 0.001	27.06 ± 0.24	61.3 ± 0.1	9.40 ± 0.10
S3–One round 10 min	0.569 ± 0.001	28.89 ± 0.19	62.5 ± 0.1	10.28 ± 0.09
Relative enhancement (%)	0.18	6.76	1.96	9.35
S4	0.550 ± 0.001	27.13 ± 0.26	64.8 ± 0.2	9.67 ± 0.07
S4–Two rounds 5 min + 5 min	0.553 ± 0.001	28.42 ± 0.26	65.1 ± 0.1	10.22 ± 0.10
Relative enhancement (%)	0.55	4.76	0.46	5.69

With a two-layer SiO_2 coating (S4), the improvement is lower than that of the single-layer close-packed situation. So, we achieve an optimized self-assembly distribution for SiO_2 coating on textured Si solar cells. Further optimizations in particle size or dielectric constant of nanospheres may obtain a higher efficiency enhancement, as discussed in planar devices [10].

Further, we hope to discuss the possible enhancement mechanism via analyzing the changing trends of electrical properties for different nanocoating, as shown in Figure 5b,c. We first noticed that the relative enhancement of short-circuit current density (J_{sc}) follows the trend of enhancement of the solar cell efficiency, and the maximum relative enhancement is achieved under the single-layer close-packed situation. More importantly, it is noteworthy that efficiency enhancement is more significant than the J_{sc} enhancement. The results are different from the previous results on thin-film solar cells, where the efficiency enhancements were almost the same compared to J_{sc} enhancement [28]. Therefore, we further compared the open-circuit voltage (V_{oc}) and fill factor of the devices, shown in Figure 5c. It is also found that the V_{oc} has a slight increase after coating with SiO_2 nanospheres due to the logarithmic relation between V_{oc} and J_{sc}.

Next, we focused on the interesting changing trend of fill factor after adding nanosphere coatings. When the density increased under single-layer conditions (S1, S2, and S3), the fill factor of the device gradually increased, which is similar to the changing trend of the current. However, although the J_{sc} enhancement of S4 (with two layers of SiO_2 coating) is very close to that of S2 (single-layer, non-close-packed), their fill factor enhancement is significantly different (0.46% compared to 1.79%). This phenomenon indicates that multiple mechanisms exist accounting for enhancing the J_{sc} and fill factor of the textured solar cells. In the applications of dielectric nanospheres for enhanced photovoltaics, J_{sc} enhancement relates to more light being coupled into the photovoltaic device, increasing the generation rate of the photo-generated carriers [28–30]. Moreover, the increase in the fill factor was recently discovered and elucidated in planar solar cells [32]. It was attributed to the nanosphere concentrating effects. Here, we also found direct evidence for fill factor enhancement on textured solar cells, indicating the nanosphere's light concentration could also enhance the efficiency of textured solar cells. Our results suggest that the nanosphere's concentrating effect enables further efficiency enhancement, even under similar light absorption enhancement (with similar J_{sc} enhancement).

In order to clarify the light concentration effects for S2 and S4, we simplify the model to a single nanostructure (a single SiO_2 sphere or two vertically arranged SiO_2 dimmers) on a flat Si substrate. In the model, the transmission direction of incident light is along the direction of $-Z$, and the polarization direction is along the x-axis. From Figure 6a, it can be observed that a single SiO_2 nanosphere focuses the incident electromagnetic energy into the solar cell region. As the incident light is transmitted into the device, the electric field gradually diverges, similar to the focusing phenomenon of a single macroscopic lens. As shown in Figure 6b, when two vertically arranged nanospheres are located on the surface of the device, they do not show apparent focusing and divergence effects. We further calculated the electric field distribution on different Z planes, as shown in Figure 6c,d. It is found that the electric field distribution caused by a single nanosphere at a distance of 15 nm from the surface in a silicon cell is similar to that of the vertically arranged dimmer. By contrast, a single nanosphere causes a more intensive electric field near the junction area. Therefore, the single-layer ~550 nm SiO_2 nanospheres will concentrate more light energy into the junction region of the device where the extraction efficiency of photo-generated carriers is largest, thereby effectively increasing the output power.

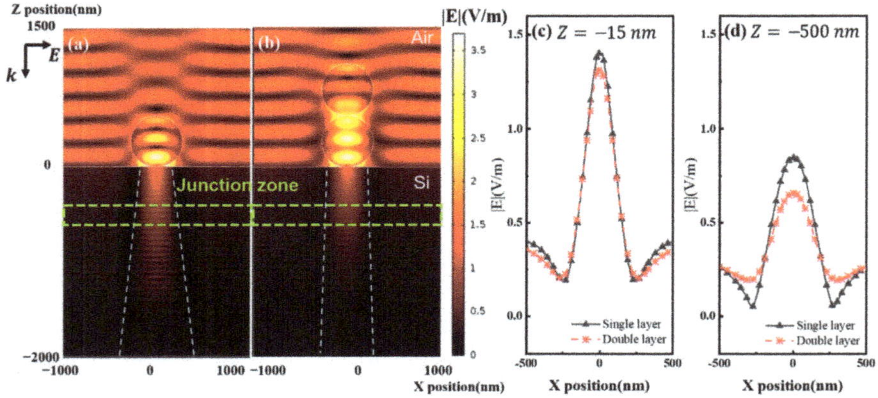

Figure 6. Electric field simulation of single-layer and double-layer nanospheres on Si solar cells: (**a**) Electric field distribution of single SiO_2 nanosphere atop Si solar cell; (**b**) Electric field distribution of vertically aligned SiO_2 nanosphere dimmer atop Si solar cell; (**c**) Electric field intensity distribution beneath the Si surface at Z = −15 nm; (**d**) Electric field intensity distribution near the junction region of Si solar cell at Z = −500 nm.

3.3. The Influence of Poly-L-Lysine on the Performance of Solar Cells

Another question we want to discuss is the influence of poly-L-lysine on solar cells. Both optical and electrical properties are discussed. The reflection spectra and the *J-V* characteristics of the original S3, S3 treated by the poly-L-lysine, and S3 with 10 min SiO_2 nanosphere coating are shown in Figure 7a,b. Both the reflectance spectrum and *J-V* characteristic cure of S3 are overlapped with those of S3 treated by poly-L-lysine. Besides, S3 with 10 min SiO_2 nanosphere coating shows a noticeable reduction in reflection and increased current density compared to that of the original S3. That is to say, the influence of the poly-L-lysine could be neglected on the performances of solar cells.

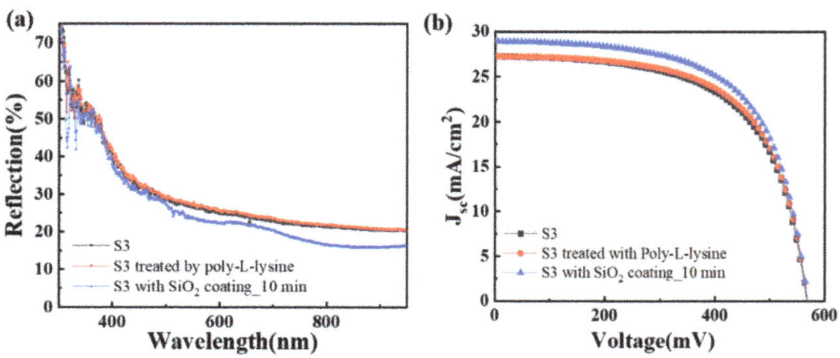

Figure 7. The reflection spectra and current density-voltage characteristics of the solar cells were treated only with poly-L-lysine and after deposition of SiO_2 nanospheres: (**a**) Reflection spectra and (**b**) Current density-voltage characteristics of the original solar cell (S3, black): S3 after being treated by poly-L-lysine (red), and S3 with 10 min self-assembly of SiO_2 nanosphere (blue).

3.4. Optical Analysis of Textured Si Solar Cells with SiO_2 Nanosphere Coatings

In order to analyze the influence of the introduction of SiO_2 nanocoating on the optical properties of surface-textured Si solar cells, we systematically analyzed the optical properties of the devices.

First, we test the macro-reflectance spectra of uncoated and coated solar cells. As shown in Figure 8a, when there is only one single layer of SiO_2 coating on the surface of the device (one round with assembly time of 10 s, 30 s, 10 min), the reflectivity of the device is reduced in a broad spectrum. Moreover, the larger the particle distribution density, the more significant the decrease in reflectivity of the device. When the assembly time is 10 min, that is, when the close-packed condition is reached, the reflectance of the device is the lowest. On the contrary, the reflection will increase (especially in the 700 nm to 850 nm region) when two-layer multi-layer nanocoating is applied (also shown in Figure S2). The change in reflection spectra is well-matched with the variation of photocurrent of the device. We noticed no narrow-band peaks in the reflection spectrum (the feature of WGM) in all textured devices. On the textured surfaces, the height of the nanospheres is different, so the WGM effect caused by the planar close-packed photonic crystals no longer exists [28–30]. To further analyze the reflectance variation at different wavelengths, we normalized the reflectance of the device with SiO_2 nanosphere coating to that of the uncoated device. As shown in Figure 8b, there are two dips in the normalized reflectance spectra. The first dip is in the 550–600 nm band, and the second is in the 800 to 850 nm band. Moreover, as the particle distribution density increases, both dips have a redshift. Macroscopically, the equivalent refractive index can be used for analysis. When the particle density increases, the equivalent refractive index of the nanosphere coating increases. Similar to the anti-reflective coating in the device, the increase in refractive index causes a redshift of the reflection spectrum [29].

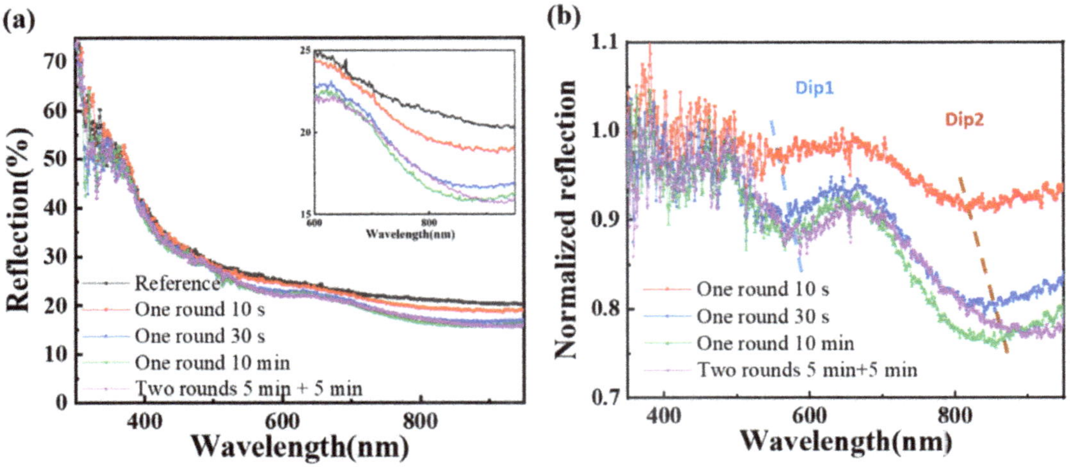

Figure 8. The reflection spectra analysis: (**a**) Reflection spectra from the textured solar cells without or with SiO_2 nanocoating. The inset graph is the magnified reflection spectra over the wavelength range from 600 nm to 950 nm; (**b**) Normalized reflection spectra of the textured solar cells with SiO_2 nanocoating.

Then, brightfield and darkfield optical microscopy imaging techniques are used to analyze the influence of nanocoating with different assembly times and rounds on the optical reflection and scattering of the textured solar cells. A wide-spectrum halogen tungsten lamp was used as the white light source. Besides, a scientific-grade color camera is used to collect micro-region brightfield (Figure 9a–e) and dark-field graphs (Figure 9k–o). The color graphs were transformed into grayscale images by a grayscale transformation code (MATLAB), shown in Figure S3. The gray image divides the intensity into 256 levels. We analyzed the number of pixels on different gray levels in the form of histograms, shown in Figure 9f–j, for brightfield images, and Figure 9p–t for darkfield images. In order to analyze the intensity and contrast of the image, we further calculated the average and standard deviation (std. in Table 2) of the intensity of the brightfield grayscale image and the darkfield grayscale image, as shown in Table 2.

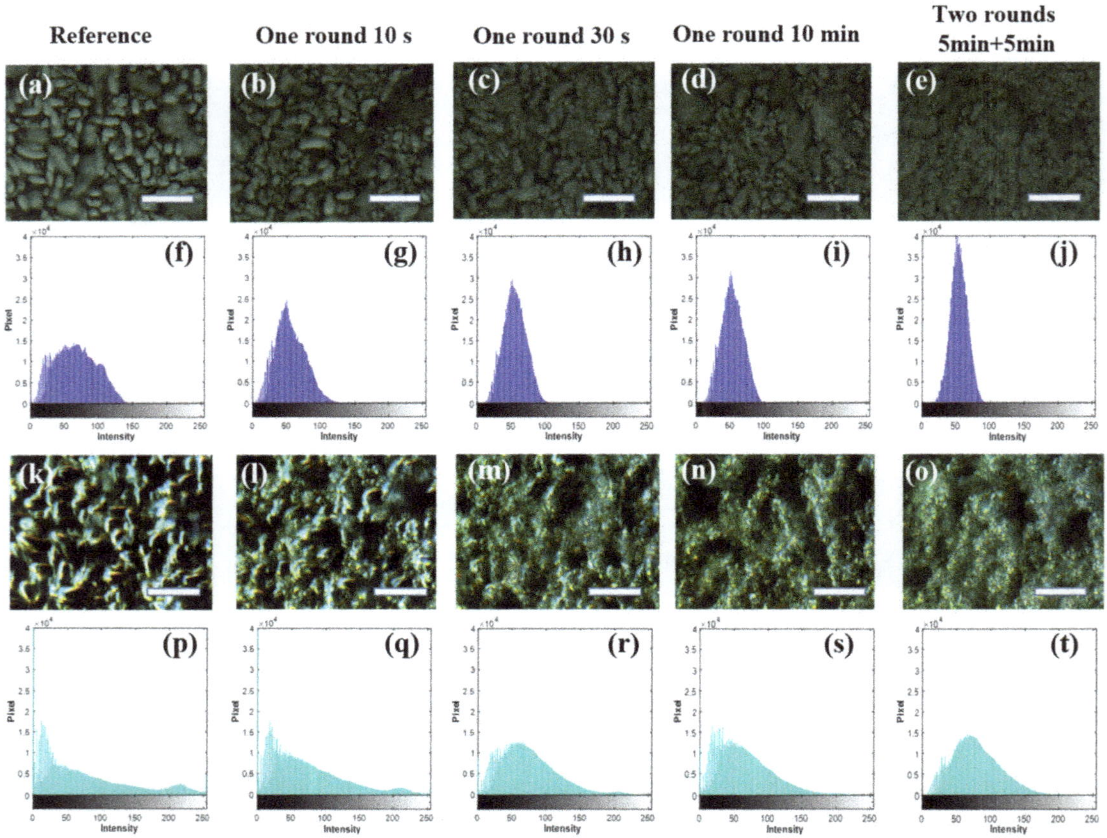

Figure 9. The optical micrographs of the solar cells with different deposition times of SiO_2 nanosphere coatings: (**a–e**) The brightfield images and (**f–j**) the corresponding grayscale image histograms; (**k–o**) Darkfield images and (**p–t**) the corresponding grayscale image histograms. The scale bar is 10 μm.

Table 2. Statistical analysis of brightfield (BF) and darkfield (DF) grayscale images.

Description	Reference	One Round 10 s	One Round 30 s	One Round 10 min	Two Rounds 5 min + 5 min
BF-average	57.76	45.61	46.09	43.59	45.43
DF-average	57.47	61.7	62	53.18	67.48
BF-std.	36.67	29.91	25.92	25.69	23.22
DF-std.	73.42	63.47	51.61	48.54	45.83

Brightfield images are usually used to reflect the specular reflectance of the devices [46,47]. By adding a single-layer SiO_2 coating, the specular reflectance of the textured solar cell is obviously reduced. When the assembly is carried out for two rounds, the average reflection intensity is increased. Therefore, one-round assembly for 10 min yields the lowest average reflection intensity value. Darkfield images reflect the large-angle scattering ability for the textured solar cells [48]. It can be found that the textured surface of polycrystalline Si solar cell exhibits strong backward light scattering, especially in the sharp regions. By adding SiO_2 nanospheres, the backscattering of these sharp features can be weakened. At the same time, the SiO_2 nanospheres could also enhance the scattering of the original silicon devices in areas with extremely weak scattering. Statistics data show that the average backscattering on the device surface was the weakest when 10 min one-round assembly was conducted. Fewer particles or multiple layers of particles will increase the backscattering. The brightfield and darkfield analysis verify that the specular reflection and backscattering of the textured solar cell can be simultaneously suppressed when coated with a single-layer close-packed SiO_2 nanocoating. In addition, the standard deviations of the brightfield and darkfield scattering intensities are gradually reduced with increased SiO_2 nanosphere distribution density. The trend still holds after a five-round assembly (Figure S4). That is to say, except for reduced reflection and scattering, SiO_2 nanocoating also makes reflection and scattering intensity more uniform from the textured Si surface.

4. Conclusions

In this work, a simple assembly scheme on a 3D substrate with controllable distribution is demonstrated. By employing this strategy, for the first time, we realize the controllable assembly of SiO_2 nanospheres on the surface of textured silicon. The distribution density can be varied from sparse to dense, and the number of layers can be varied from single-layer to multi-layer. This assembly method allows us to study the effect of SiO_2 nanosphere arrangement on the performance parameters of textured Si solar cells. The optimized enhancement was achieved by close-packed single-layer nanospheres. The efficiency increased by 9.35%, the current increased by 6.76%, and the fill factor increased by 1.96%.

In comparison with the planar substrate, it is found that 3D substrate promotes dense packing due to gravitational sedimentation effects. The quantitative mechanism analysis still needs further study, including morphology of 3D substrate, the combined effect of gravitational sedimentation, the attractive and repulsive force between particles, and that of the substrate with particles. The quantitative mechanism analysis still needs further study, including morphology of 3D substrate, the combined effect of gravity, the attractive and repulsive force between particles, and that of the substrate with particles. The resulted enhanced fill factor of textured Si solar cells is evidence of the nanosphere concentration effect. Reduced surface reflection and backward scattering play a vital role in photocurrent enhancement. Morphology engineering of the nanosphere may further increase the concentration effects. This research provides heuristic guidelines for nanofabrication and photon management on complex 3D surfaces at the nanoscale.

Supplementary Materials: The following are available online at https://www.mdpi.com/article/10.3390/nano11102581/s1, Figure S1: Self-assembly nanocoating under different substrate orientations in the SiO_2 assembly colloid with an assembly time of 30 min. (a–c) Schematic illustration and SEM graphs of the assembly with an upward substrate, (d–f) Schematic illustration and SEM graphs of the assembly with an inverted substrate. Figure S2: Optical reflection measurement of the textured silicon solar before (black line) and after (green line) five-round SiO_2 repeated assembly process. Figure S3: Grayscale image transformation of the micro-region optical images. Bright filed grayscale image for (a) reference solar cell, (b) after one round 10 s assembly, (c) after one round 30 s assembly, (d) after one round 10 min assembly, and (e) after two round 5 min + 5 min assembly. Dark filed grayscale image for (f) reference solar cell, (g) after one round 10 s assembly, (h) after one round 30 s assembly, (i) after one round 10 min assembly, and (j) after two round 5 min + 5 min assembly. Figure S4: Micro-region optical image analysis of solar cell after five-round SiO_2 repeated assembly process. (a) Original bright field image, (b) Bright field gray image after grayscale transformation, (c) Intensity distribution histogram of bright field gray image, (d) Original dark field image, (e) Dark field gray image after grayscale transformation, (f) Intensity distribution histogram of dark field gray image.

Author Contributions: Conceptualization, T.Z.; methodology, D.S., Y.Y. and H.-L.Z.; formal analysis, D.S. and L.L.; investigation, D.S.; writing—original draft preparation, D.S.; writing—review and editing, L.L., Y.Y., H.-L.Z. and S.I.; supervision, T.Z.; funding acquisition, T.Z. All authors have read and agreed to the published version of the manuscript.

Funding: This research was funded by NSFC under grant numbers 61875241, and MOST under grant number 2017YFA0205800.

Data Availability Statement: Data are contained within the article.

Acknowledgments: We thank Shanghai Shenzhou New Energy Co., Ltd., for providing the silicon solar cells. Some early results were reported in our previous conference proceedings paper: Dan Su, Ning Zhao, Shan-Jiang Wang, Huan-li Zhou, Yi Yang, Meng Xiong, Tong Zhang, Lysine-assisted self-assembly of silica microspheres on textured polycrystalline silicon solar cells for light trapping. *Photonics & Electromagnetics Research Symposium-Spring* **2019**, 1252, 7.

Conflicts of Interest: The authors declare no conflict of interest.

References

1. Garnett, E.C.; Ehrler, B.; Polman, A.; Alarcon-Llado, E. Photonics for Photovoltaics: Advances and Opportunities. *ACS Photonics* **2021**, *8*, 61–70. [CrossRef] [PubMed]
2. Oliveira, R.; Mouquinho, A.; Centeno, P.; Alexandre, M.; Haque, S.; Martins, R.; Fortunato, E.; Águas, H.; Mendes, M. Colloidal Lithography for Photovoltaics: An Attractive Route for Light Management. *Nanomaterials* **2021**, *11*, 1665. [CrossRef] [PubMed]
3. Wu, J.-Y.; Chun, Y.T.; Li, S.; Zhang, T.; Wang, J.; Shrestha, P.K.; Chu, D. Broadband MoS2 Field-Effect Phototransistors: Ultrasensitive Visible-Light Photoresponse and Negative Infrared Photoresponse. *Adv. Mater.* **2018**, *30*, 1705880. [CrossRef] [PubMed]
4. Zhang, T.; Su, D.; Li, R.-Z.; Wang, S.-J.; Shan, F.; Xu, J.-J.; Zhang, X.-Y. Plasmonic nanostructures for electronic designs of photovoltaic devices: Plasmonic hot-carrier photovoltaic architectures and plasmonic electrode structures. *J. Photonics Energy* **2016**, *6*, 042504. [CrossRef]
5. Wang, L.; Liu, F.; Zhao, B.; Ning, Y.; Zhang, L.; Bradley, R.; Wu, W. Carbon Nanobowls Filled with MoS2 Nanosheets as Electrode Materials for Supercapacitors. *ACS Appl. Nano Mater.* **2020**, *3*, 6448–6459. [CrossRef]
6. Choi, Y.J.; Pyun, S.B.; Cho, E.C. Slippery Colloidal Crystal Monolayers for Sustainable Enhancement of Commercial Solar Cell Performance. *ACS Appl. Energy Mater.* **2021**, *4*, 303–311. [CrossRef]
7. Hou, Y.; Wang, Z.; Cai, C.; Hao, X.; Li, N.; Zhao, N.; Zhao, Y.; Chen, L.; Ma, H.; Xu, J. Conformal Nanocoatings with Uniform and Controllable Thickness on Microstructured Surfaces: A General Assembly Route. *Adv. Mater.* **2018**, *30*, 1704131. [CrossRef]
8. Montelongo, Y.; Sikdar, D.; Ma, Y.; McIntosh, A.J.S.; Velleman, L.; Kucernak, A.R.; Edel, J.B.; Kornyshev, A.A. Electrotunable nanoplasmonic liquid mirror. *Nat. Mater.* **2017**, *16*, 1127–1135. [CrossRef]
9. Li, S.; Xu, Q.; Zhou, Z.; Zhao, W.; Li, X.; Wen, Z.; Yao, Y.; Xu, H.; Deng, H.; Dai, N.; et al. Flexible Transparent Heat Mirror for Thermal Applications. *Nanomaterials* **2020**, *10*, 2479. [CrossRef]
10. Wang, B.; Leu, P.W. High index of refraction nanosphere coatings for light trapping in crystalline silicon thin film solar cells. *Nano Energy* **2015**, *13*, 226–232. [CrossRef]
11. Li, R.-Z.; Peng, R.; Kihm, K.D.; Bai, S.; Bridges, D.; Tumuluri, U.; Wu, Z.; Zhang, T.; Compagnini, G.; Feng, Z.; et al. High-rate in-plane micro-supercapacitors scribed onto photo paper using in situ femtolaser-reduced graphene oxide/Au nanoparticle microelectrodes. *Energy Environ. Sci.* **2016**, *9*, 1458–1467. [CrossRef]
12. Li, R.-Z.; Hu, A.; Bridges, D.; Zhang, T.; Oakes, K.D.; Peng, R.; Tumuluri, U.; Wu, Z.; Feng, Z. Robust Ag nanoplate ink for flexible electronics packaging. *Nanoscale* **2015**, *7*, 7368–7377. [CrossRef]
13. Zhang, T.; Iqbal, S.; Zhang, X.-Y.; Wu, W.; Su, D.; Zhou, H.-L. Recent advances in highly efficient organic-silicon hybrid solar cells. *Sol. Energy Mater. Sol. Cells* **2020**, *204*, 110245. [CrossRef]
14. Hao, J.; Hao, H.; Li, J.; Shi, L.; Zhong, T.; Zhang, C.; Dong, J.; Xing, J.; Liu, H.; Zhang, Z. Light Trapping Effect in Perovskite Solar Cells by the Addition of Ag Nanoparticles, Using Textured Substrates. *Nanomaterials* **2018**, *8*, 815. [CrossRef]
15. Tennyson, E.M.; Doherty, T.A.S.; Stranks, S.D. Heterogeneity at multiple length scales in halide perovskite semiconductors. *Nat. Rev. Mater.* **2019**, *4*, 573–587. [CrossRef]
16. Tennyson, E.M.; Howard, J.M.; Leite, M.S. Mesoscale Functional Imaging of Materials for Photovoltaics. *ACS Energy Lett.* **2017**, *2*, 1825–1834. [CrossRef]
17. Su, D.; Zhao, N.; Wang, S.J.; Zhou, H.L.; Yang, Y.; Xiong, M.; Zhang, T. Lysine Assisted Self-assembly of Silica Micro-spheres on Textured Polycrystalline Silicon Solar Cells for Light Trapping. In Proceedings of the 2019 PhotonIcs & Electromagnetics Research Symposium—Spring (PIERS-Spring), Rome, Italy, 17–20 June 2019; pp. 1252–1257.
18. Grzelczak, M.; Vermant, J.; Furst, E.; Liz-Marzán, L.M. Directed Self-Assembly of Nanoparticles. *ACS Nano* **2010**, *4*, 3591–3605. [CrossRef] [PubMed]
19. Lu, X.; Huang, Y.; Liu, B.; Zhang, L.; Song, L.; Zhang, J.; Zhang, A.; Chen, T. Light-Controlled Shrinkage of Large-Area Gold Nanoparticle Monolayer Film for Tunable SERS Activity. *Chem. Mater.* **2018**, *30*, 1989–1997. [CrossRef]
20. Salih, A.E.; Elsherif, M.; Alam, F.; Yetisen, A.K.; Butt, H. Gold Nanocomposite Contact Lenses for Color Blindness Management. *ACS Nano* **2021**, *15*, 4870–4880. [CrossRef]

21. Shin, Y.; Song, J.; Kim, D.; Kang, T. Facile Preparation of Ultrasmall Void Metallic Nanogap from Self-Assembled Gold-Silica Core-Shell Nanoparticles Monolayer via Kinetic Control. *Adv. Mater.* **2015**, *27*, 4344–4350. [CrossRef]
22. Zhang, X.-Y.; Hu, A.; Zhang, T.; Lei, W.; Xue, X.-J.; Zhou, Y.; Duley, W.W. Self-Assembly of Large-Scale and Ultrathin Silver Nanoplate Films with Tunable Plasmon Resonance Properties. *ACS Nano* **2011**, *5*, 9082–9092. [CrossRef]
23. Zhang, X.-Y.; Shan, F.; Zhou, H.-L.; Su, D.; Xue, X.-M.; Wu, J.-Y.; Chen, Y.-Z.; Zhao, N.; Zhang, T. Silver nanoplate aggregation based multifunctional black metal absorbers for localization, photothermic harnessing enhancement and omnidirectional light antireflection. *J. Mater. Chem. C* **2017**, *6*, 989–999. [CrossRef]
24. Cai, H.; Meng, Q.; Ding, H.; Zhang, K.; Lin, Y.; Ren, W.; Yu, X.; Wu, Y.; Zhang, G.; Li, M.; et al. Utilization of Resist Stencil Lithography for Multidimensional Fabrication on a Curved Surface. *ACS Nano* **2018**, *12*, 9626–9632. [CrossRef] [PubMed]
25. Gao, J.; Pei, K.; Sun, T.; Wang, Y.; Zhang, L.; Peng, W.; Lin, Q.; Giersig, M.; Kempa, K.; Ren, Z.; et al. Transparent Nanowire Network Electrode for Textured Semiconductors. *Small* **2012**, *9*, 733–737. [CrossRef] [PubMed]
26. Niesen, B.; Blondiaux, N.; Boccard, M.; Stuckelberger, M.; Pugin, R.; Scolan, E.; Meillaud, F.; Haug, F.-J.; Hessler-Wyser, A.; Ballif, C. Self-Patterned Nanoparticle Layers for Vertical Interconnects: Application in Tandem Solar Cells. *Nano Lett.* **2014**, *14*, 5085–5091. [CrossRef]
27. Wu, M.; Ma, B.; Pan, T.; Chen, S.; Sun, J. Silver-Nanoparticle-Colored Cotton Fabrics with Tunable Colors and Durable Antibacterial and Self-Healing Superhydrophobic Properties. *Adv. Funct. Mater.* **2016**, *26*, 569–576. [CrossRef]
28. Grandidier, J.; Callahan, D.M.; Munday, J.N.; Atwater, H.A. Light Absorption Enhancement in Thin-Film Solar Cells Using Whispering Gallery Modes in Dielectric Nanospheres. *Adv. Mater.* **2011**, *23*, 1272–1276. [CrossRef]
29. Ha, D.; Gong, C.; Leite, M.; Munday, J.N. Demonstration of Resonance Coupling in Scalable Dielectric Microresonator Coatings for Photovoltaics. *ACS Appl. Mater. Interfaces* **2016**, *8*, 24536–24542. [CrossRef]
30. Ha, D.; Yoon, Y.; Zhitenev, N.B. Nanoscale imaging of photocurrent enhancement by resonator array photovoltaic coatings. *Nanotechnology* **2018**, *29*, 145401. [CrossRef]
31. Zhang, Y.; Chen, S.; Hu, D.; Xu, Y.; Wang, S.; Qin, F.; Cao, Y.; Guan, B.-O.; Miroshnichenko, A.; Gu, M.; et al. Coloring solar cells with simultaneously high efficiency by low-index dielectric nanoparticles. *Nano Energy* **2019**, *62*, 682–690. [CrossRef]
32. Esmaeilzad, N.S.; Demir, A.K.; Hajivandi, J.; Ciftpinar, H.; Turan, R.; Kurt, H.; Bek, A. Nanosphere Concentrated Photovoltaics with Shape Control. *Adv. Opt. Mater.* **2021**, *9*, 2000943. [CrossRef]
33. Dabirian, A.; Byranvand, M.M.; Naqavi, A.; Kharat, A.N.; Taghavinia, N. Self-Assembled Monolayer of Wavelength-Scale Core–Shell Particles for Low-Loss Plasmonic and Broadband Light Trapping in Solar Cells. *ACS Appl. Mater. Interfaces* **2016**, *8*, 247–255. [CrossRef] [PubMed]
34. Jiang, Y.; Liu, C.; Wang, X.; Wang, T.; Jiang, J. Fluorescent Phthalocyanine Assembly Distinguishes Chiral Isomers of Different Types of Amino Acids and Sugars. *Langmuir* **2017**, *33*, 7239–7247. [CrossRef] [PubMed]
35. Li, W.-C.; Lee, S.-W. The characteristics of lysine-mediated self-assembly of gold nanoparticles on the ITO glass. *Curr. Appl. Phys.* **2012**, *12*, 1361–1365. [CrossRef]
36. Hartman, R.; Schwaner, R.C.; Hermans, J. Beta poly(l-lysine): A model system for biological self-assembly. *J. Mol. Biol.* **1974**, *90*, 415–429. [CrossRef]
37. Wu, S.-H.; Mou, C.-Y.; Lin, H.-P. Synthesis of mesoporous silica nanoparticles. *Chem. Soc. Rev.* **2013**, *42*, 3862–3875. [CrossRef] [PubMed]
38. Palik, E.D. Preface. In *Handbook of Optical Constants of Solids*; Palik, E.D., Ed.; Academic Press: Cambridge, MA, USA, 1985.
39. Huwiler, C.; Kunzler, T.P.; Textor, M.; Vörös, J.; Spencer, N.D. Functionalizable Nanomorphology Gradients via Colloidal Self-Assembly. *Langmuir* **2007**, *23*, 5929–5935. [CrossRef]
40. Zhu, S.-Q.; Zhang, T.; Guo, X.-L.; Zhang, X.-Y. Self-assembly of large-scale gold nanoparticle arrays and their application in SERS. *Nanoscale Res. Lett.* **2014**, *9*, 114. [CrossRef]
41. Aizenberg, J.; Braun, P.V.; Wiltzius, P. Patterned Colloidal Deposition Controlled by Electrostatic and Capillary Forces. *Phys. Rev. Lett.* **2000**, *84*, 2997–3000. [CrossRef]
42. Malakhovsky, P.; Reznikov, I.; Aniskevich, Y.; Murausky, D.; Artemyev, M. Electrostatic Deposition Kinetics of Colloidal Silver Nanoplates onto Optically and E-Beam Transparent Water-Insoluble Polycationic Films. *J. Phys. Chem. C* **2021**, *125*, 17870–17880. [CrossRef]
43. Evans, J.W. Random and cooperative sequential adsorption. *Rev. Mod. Phys.* **1993**, *65*, 1281–1329. [CrossRef]
44. Luo, D.; Yan, C.; Wang, T. Interparticle Forces Underlying Nanoparticle Self-Assemblies. *Small* **2015**, *11*, 5984–6008. [CrossRef] [PubMed]
45. Boles, M.A.; Engel, M.; Talapin, D.V. Self-Assembly of Colloidal Nanocrystals: From Intricate Structures to Functional Materials. *Chem. Rev.* **2016**, *116*, 11220–11289. [CrossRef] [PubMed]
46. Nishijima, Y.; Komatsu, R.; Ota, S.; Seniutinas, G.; Balčytis, A.; Juodkazis, S. Anti-reflective surfaces: Cascading nano/microstructuring. *APL Photonics* **2016**, *1*, 076104. [CrossRef]
47. Sasikumar, H.; Varma, M. Metal-enhanced bright-field microscopy. *Appl. Opt.* **2020**, *59*, 9971–9977. [CrossRef] [PubMed]
48. Sönnichsen, C.; Alivisatos, P. Gold Nanorods as Novel Nonbleaching Plasmon-Based Orientation Sensors for Polarized Single-Particle Microscopy. *Nano Lett.* **2005**, *5*, 301–304. [CrossRef] [PubMed]

nanomaterials

MDPI

Article

The Deactivation Mechanism of the Mo-Ce/Zr-PILC Catalyst Induced by Pb for the Selective Catalytic Reduction of NO with NH_3

Chenxi Li [1], Jin Cheng [1], Qing Ye [1,*], Fanwei Meng [1], Xinpeng Wang [1] and Hongxing Dai [2,*]

[1] Key Laboratory of Beijing on Regional Air Pollution Control, Department of Environmental Science, School of Environmental and Chemical Engineering, Faculty of Environment and Life, Beijing University of Technology, Beijing 100124, China; lcx@emails.bjut.edu.cn (C.L.); chengjin3708@163.com (J.C.); MengFanwei@emails.bjut.edu.cn (F.M.); 17121679@bjtu.edu.cn (X.W.)

[2] Beijing Key Laboratory for Green Catalysis and Separation, Key Laboratory of Beijing on Regional Air Pollution Control, Key Laboratory of Advanced Functional Materials, Education Ministry of China, and Laboratory of Catalysis Chemistry and Nanoscience, Department of Environmental Chemical Engineering, School of Environmental and Chemical Engineering, Faculty of Environment and Life, Beijing University of Technology, Beijing 100124, China

* Correspondence: yeqing@bjut.edu.cn (Q.Y.); hxdai@bjut.edu.cn (H.D.)

Abstract: As a heavy metal, Pb is one component in coal-fired flue gas and is widely considered to have a strong negative effect on catalyst activity in the selective catalytic reduction of NO_x by NH_3 (NH_3-SCR). In this paper, we investigated the deactivation mechanism of the Mo-Ce/Zr-PILC catalyst induced by Pb in detail. We found that NO conversion over the 3Mo4Ce/Zr-PILC catalyst decreased greatly after the addition of Pb. The more severe deactivation induced by Pb was attributed to low surface area, lower amounts of chemisorbed oxygen species and surface Ce^{3+}, and lower redox ability and surface acidity (especially a low number of Brønsted acid sites). Furthermore, the addition of Pb inhibited the formation of highly active intermediate nitrate species generated on the surface of the catalyst, hence decreasing the NH_3-SCR activity.

Keywords: Pb poisoning; Mo-Ce/Zr-PILC catalyst; NH_3-SCR; surface acidity

Citation: Li, C.; Cheng, J.; Ye, Q.; Meng, F.; Wang, X.; Dai, H. The Deactivation Mechanism of the Mo-Ce/Zr-PILC Catalyst Induced by Pb for the Selective Catalytic Reduction of NO with NH_3. *Nanomaterials* **2021**, *11*, 2641. https://doi.org/10.3390/nano11102641

Academic Editors: Jihoon Lee and Ming-Yu Li

Received: 12 August 2021
Accepted: 30 September 2021
Published: 7 October 2021

1. Introduction

Nitrogen oxides (NO_x) are not only harmful to human health but are also important precursors of secondary pollution (e.g., photochemical smog, fine particles, and acid deposition) [1–3]. NH_3-SCR is an efficient and economical technology that has recently been used in the elimination of NO_x from the exhausts of stationary and mobile sources. In addition, the $V_2O_5-WO_3$ (MoO_3)/TiO_2 catalysts are the most widely used catalyst systems in stationary sources [4]. However, these systems still have some inevitable practical defects, such as low N_2 selectivity in the high-temperature range, high SO_2 to SO_3 conversion rates, and the strong biological toxicity of vanadium species [5]. Therefore, it is urgent and necessary to develop a new SCR catalyst with low toxicity and high activity. Among the possible substitutes, Ce-based oxides have attracted extensive attention due to their excellent oxygen storage/release ability and redox performance [6,7]. It is reported that Mo modification could significantly promote the adsorption and activation of NH_3 species, which is beneficial for improvement in NH_3-SCR activity [8]. Therefore, Mo-Ce mixed-oxide catalysts have been extensively investigated. It was found that these materials were the most promising SCR catalysts [9], among which the Mo-Ce/Zr-PILC catalysts exhibited outstanding NH_3-SCR activity.

Deactivation induced by the poisoning of K, Na, Ca, Pb, or As is the main reason for the poor performance of the SCR catalysts. A number of experiments have been focused on the poisoning effects of alkali or alkaline earth metals on the NH_3-SCR catalyst [10,11].

However, very few studies have been done regarding the effects of heavy metals on the activity of SCR catalysts. Pb is a typical heavy metal contained in the flue gas derived from coal-fired and municipal solid waste incineration power plants. Tokarz et al. [12] observed a strong accumulation effect of Pb on the NH_3-SCR catalyst, where the Pb concentration reached 3350 ppm after 1908 h operation of a municipal waste incinerator. This finding revealed that lead dust exhibited a strong accumulative poisoning effect on the NH_3-SCR catalyst. However, there have been few reports in the literature on the poisoning mechanisms of heavy metals (e.g., Pb) on NH_3-SCR catalysts. Hence, it is necessary to investigate the effects of Pb compounds on the performance of NH_3-SCR catalysts and the involved deactivation mechanisms.

To elucidate the inhibitory effect of Pb on Mo-Ce/Zr-PILC in the NH_3-SCR reaction, we prepared 3Mo4Ce/Zr-PILC (Mo and Ce contents were 3 and 4 wt%, respectively) and Pb-doped 3Mo4Ce/Zr-PILC catalysts. The physicochemical properties of these samples were characterized using the BET, XRD, H_2-TPR, XPS, NH_3-TPD, and in situ DRIFTS techniques. The main purpose of the present work is to clarify the effect of Pb doping on the catalytic performance of Ce-based catalysts, and to elucidate the involved poisoning mechanisms.

2. Experimental

2.1. Catalyst Preparation

The zirconium pillared clay was prepared using the pillaring method. An appropriate amount of zirconium oxychloride solution was added dropwise to the montmorillonite suspension, stirred for 12 h, aged at ambient room conditions for 24 h, dried at 100 °C for 12 h, and calcined at 400 °C for 2 h (named as Zr-PILC). The 3Mo/Zr-PILC catalyst with a Mo loading of 3 wt% was prepared by impregnating Zr-PILC with a proper amount of ammonium molybdate under stirring for 1 h, and then was put into a rotary evaporator to dry and calcine in air at 400 °C for 2 h. The 3Mo4Ce/Zr-PILC (Ce content was 4 wt%) catalyst was prepared by impregnating the 3Mo/Zr-PILC powders with a $Ce(NO_3)_3$ aqueous solution, and its preparation method followed the steps above. The catalysts poisoned by Pb were prepared using the impregnating method, by mixing the 3Mo4Ce/Zr-PILC catalyst with $Pb(CH_3COO)_2{\cdot}3H_2O$ at different molar Pb/Ce ratios, and the subsequent procedures were the same as the steps above.

2.2. Catalyst Characterization

The crystal structures of the catalysts were determined using a Bruker D8 advance diffractometer (Bruker, Karlsruhe, Germany) equipped with a Cu Kα detector. N_2 adsorption–desorption isotherms were measured at −196 °C on a W-BK132F apparatus (JWGB, Beijing, China). X-ray photoelectron spectroscopy (XPS, ESCALAB 250Xi, Thermo Fisher, Waltham, MA, USA) with Al Kα radiation (1486.6 eV) was used to analyze the surface compositions, metal oxidation states, and oxygen species of the catalysts. The sample powders were vacuumed, heated, and glued to the tape, and then the test was carried out immediately. The hydrogen temperature-programmed reduction (H_2-TPR) and ammonia temperature-programmed desorption (NH_3-TPD) experiments were carried out in a quartz U-tube reactor on a PCA-1200 analyzer (Beijing Builder Electronic Technology, Beijing, China) equipped with a thermal conductivity detector (TCD). Before H_2-TPR measurement, the catalyst was pretreated in 5 vol% O_2/N_2 flow at 400 °C for 1 h and then cooled to room temperature (RT). The H_2-TPR profiles were recorded in a 5 vol% H_2/N_2 flow of 30 mL/min from RT to 950 °C at a heating rate of 10 °C/min. Prior to NH_3-TPD measurement, the catalyst was pretreated in a He flow of 30 mL/min at 400 °C for 1 h, followed by NH_3 adsorption at 100 °C. Subsequently, the catalyst was purged in a He flow of 30 mL/min for 30 min to remove the physically adsorbed NH_3. Finally, the NH_3-TPD profiles were recorded in a He flow of 30 mL/min from 30 to 850 °C at a heating rate of 10 °C/min. The in situ diffuse reflectance Fourier transform infrared spectroscopy (in situ DRIFTS) experiments were performed on a Bruker TENSOR II spectrometer (Bruker, Karlsruhe,

Germany). The catalyst was pretreated at 400 °C for 1 h to remove any adsorbed moisture and impurities. The background spectra were recorded by exposing the catalysts to a N_2 flow of 100 mL/min at 200 °C. The absorption spectra of NH_3 and (NO + O_2) were recorded in a 100 mL/min flow of (1100 ppm NH_3 + 1000 ppm NO + 4 vol% O_2 + N_2 (balance) and (1000 ppm NO + 4 vol% O_2 + N_2 (balance)), respectively.

2.3. NH_3-SCR Activity Evaluation

The activity measurements were performed in a fixed-bed quartz reactor using 0.3 g of the catalyst (40–60 mesh) mixed with 0.3 g of quartz sand (40–60 mesh). The simulated exhaust gas was composed of 1100 ppm NH_3 + 1000 ppm NO + 4 vol% O_2 + 5 vol% H_2O + N_2 (balance). The total flow rate of the feed gas was 500 mL/min, and the space velocity (SV) was 100,000 mL/(g h). NO concentrations were measured using a MODEL1080 analyser (Beijing SDL Technology, Beijing, China). The NO conversion was calculated according to the following equation:

$$\text{NO conversion}(\%) = \frac{[\text{NO}]_{\text{in}} - [\text{NO}]_{\text{out}}}{[\text{NO}]_{\text{in}}} \times 100\%$$

3. Results and Discussion

3.1. NH_3-SCR Activity

In the present study, we investigated the poisoning effects that Pb had on the catalytic performance of 3Mo4Ce/Zr-PILC. Figure 1 shows the NO conversion as a function of temperature (150–450 °C) over the 3Mo4Ce/Zr-PILC and Pb-poisoned 3Mo4Ce/Zr-PILC catalysts. Obviously, 3Mo4Ce/Zr-PILC exhibited excellent catalytic activity, with a NO conversion of 90% in the 300–450 °C range. However, with a rise in the doped Pb amount of 3Mo4Ce/Zr-PILC, NO conversion decreased considerably. When the Pb/Ce molar ratio was 0.2, catalytic efficiency in the whole temperature range was decreased by 5–25%. When the Pb/Ce molar ratio rose to 1, the activity of the catalyst was less than 50% at 150–450 °C. Obviously, the deNO$_x$ performance of the catalyst was greatly inhibited after Pb doping, and this inhibitive effect was enhanced with increasing Pb/Ce molar ratio.

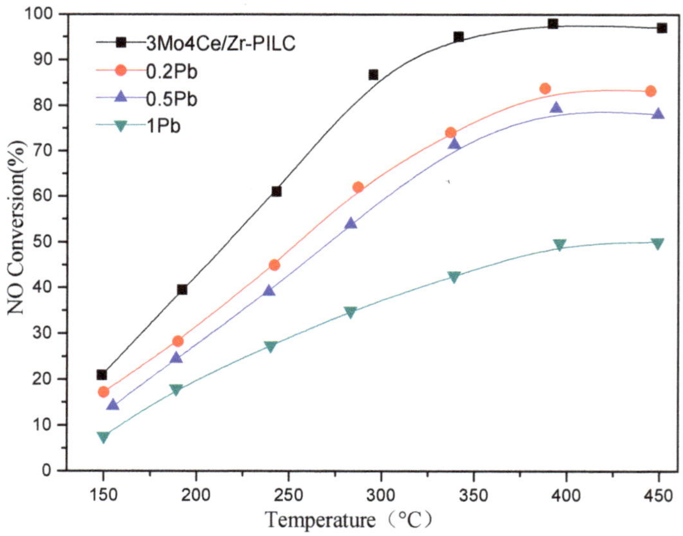

Figure 1. Catalytic activity for the NH_3-SCR reaction over the 3Mo4Ce/Zr-PILC and Pb-poisoned 3Mo4Ce/Zr-PILC catalysts under the reaction conditions of (1000 ppm NO + 1100 ppm NH_3 + 4 vol% O_2 + 5 vol% H_2O + N_2 (balance)) and SV = 100,000 mL/(g h).

3.2. Crystal Phase and Textural Characteristics

The crystal phase compositions of the 3Mo4Ce/Zr-PILC and Pb-poisoned 3Mo4Ce/Zr-PILC catalysts were measured using XRD. Figure 2 shows their diffraction patterns. According to our previous research [13], the XRD pattern of the PILC sample displayed a high-intensity basement (001) reflection peak at 2θ = 9.3°, indicating that the montmorillonite clay possessed a regular and ordered interlayer structure. The (001) reflection peak of the 3Mo4Ce/Zr-PILC and Pb-poisoned 3Mo4Ce/Zr-PILC catalysts was shifted to a lower angle and its intensity decreased, indicating an increase in interlayer distance and a disorder in the structure of the clay, which was caused by the effect of the strong delamination when the metal oxide entered the interlayer of clay. For the 3Mo4Ce/Zr-PILC catalyst, two-dimensional *hk* reflections were present at 2θ = 19.8° and 34.9°, in which the former was a summation of the *hk* indices of (02) and (11), while the latter was a summation of the *hk* indices of (13) and (20) [14]. The peaks at 2θ = 26.6° and 28° were reflections of the quartz and cristobalite impurities in the clay [15]. No diffraction peaks assignable to the crystalline cubic CeO_2 and MoO_3 were detected, suggesting that the cerium and molybdenum species were widely dispersed on the Zr-PILC surface, or were present in an amorphous state. As shown in Figure 2, after Pb doping, some obvious diffraction signals assignable to the $PbMoO_4$ phase were recorded at 2θ = 27.5°, 29.5°, 32.9°, 37.8°, 44.9°, 47.3°, 50.9°, 55.7°, and 56.7°, and the intensity of the diffraction peaks ascribable to the $PbMoO_4$ phase increased gradually with a rise in Pb doping, indicating partial accumulation of the $PbMoO_4$ phase. The accumulation of the $PbMoO_4$ species would result in the partial deactivation of the 3Mo4Ce/Zr-PILC catalyst, which was consistent with its NH_3-SCR activity.

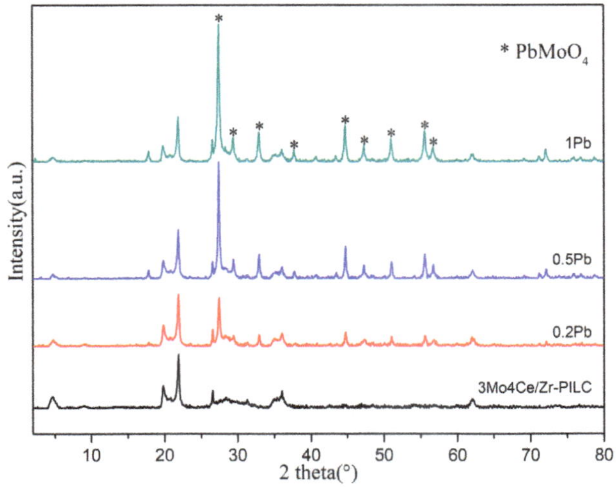

Figure 2. XRD patterns of the 3Mo4Ce/Zr-PILC and Pb-poisoned 3Mo4Ce/Zr-PILC catalysts.

N_2 adsorption–desorption isotherms of the 3Mo4Ce/Zr-PILC and Pb-poisoned 3Mo4Ce/Zr-PILC catalysts are shown in Figure 3. It can be seen that the isotherms of the catalysts were obviously in accordance with type I in the low relative pressure range, according to the IUPAC classification, which conformed to the characteristics of microporous materials [16]. The adsorbed volumes of the catalysts increased significantly in the p/p_0 (<0.01) range, which was due to the formation of a large number of micropores during the pillaring process. Additionally, the hysteresis loops of all of the catalysts corresponded to type H3, which appeared in the higher relative pressure range, indicating the presence of mesopores.

The textural parameters of the 3Mo4Ce/Zr-PILC and Pb-poisoned catalysts are listed in Table 1. As seen in Table 1, 3Mo4Ce/Zr-PILC displayed a surface area and a pore volume of 270 m^2/g and 0.164 cm^3/g, respectively. Additionally, it was noticeable that the doping of Pb resulted in significant decreases in surface area and pore volume, which might be attributable to the partial blocking of the micropores in the catalysts, the agglomeration of the particles, or both.

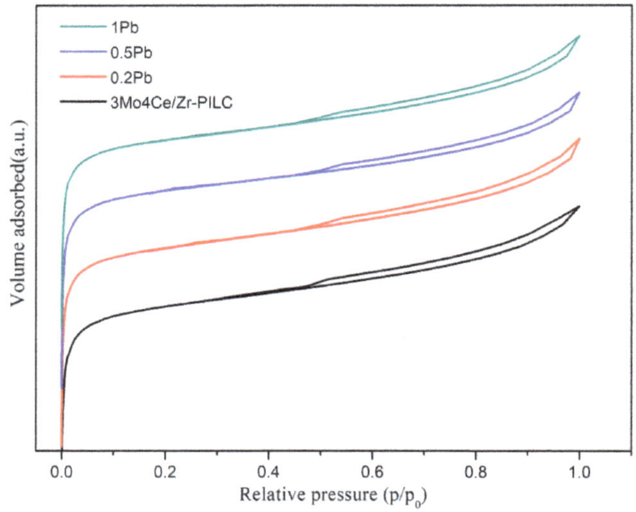

Figure 3. N_2 adsorption-desorption isotherms of the 3Mo4Ce/Zr-PILC and Pb-poisoned 3Mo4Ce/Zr-PILC catalysts.

The SEM images of the 3Mo4Ce/Zr-PILC and Pb-poisoned 3Mo4Ce/Zr-PILC catalysts are shown in Figure 4. It can be observed that the 3Mo4Ce/Zr-PILC catalyst displayed a microporous structure, while particles appeared on the surface of the Pb-poisoned catalysts. Moreover, for the Pb-poisoned 3Mo4Ce/Zr-PILC catalysts with higher quantities of doped Pb, the number of particles increased but the pore size decreased, which was consistent with the BET results.

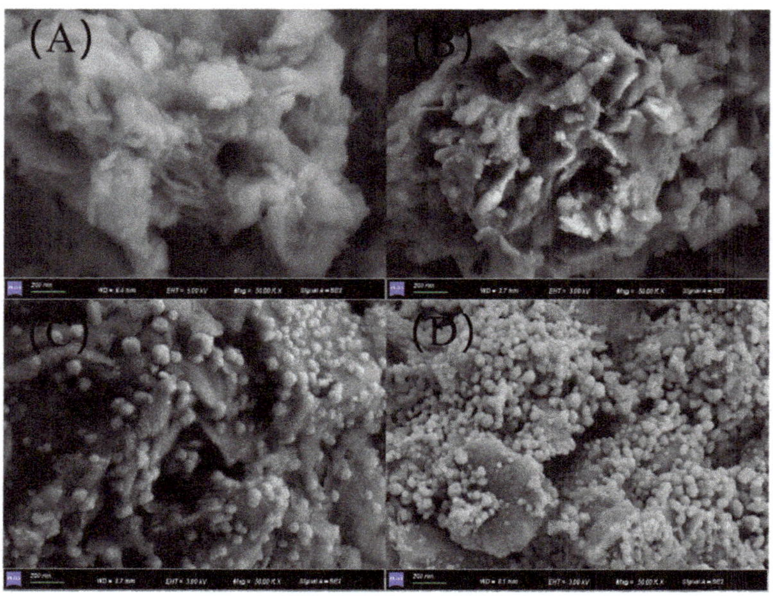

Figure 4. SEM images of (**A**) 3Mo4Ce/Zr-PILC, (**B**) 0.2Pb-Mo4Ce/Zr-PILC, (**C**) 0.5Pb-Mo4Ce/Zr-PILC, and (**D**) 1Pb-Mo4Ce/Zr-PILC.

Table 1. BET surface areas and pore volumes of the 3Mo4Ce/Zr-PILC and Pb-poisoned 3Mo4Ce/Zr-PILC catalysts.

Catalyst	BET Surface Area (m^2/g)	Pore Volume (cm^3/g)
3Mo4Ce/Zr-PILC	270	0.164
0.2Pb	259	0.161
0.5Pb	256	0.154
1Pb	243	0.152

3.3. Redox Properties

The redox property of a catalyst is an important factor in NH_3-SCR reactions. The reducibility of Ce and Mo species in Pb-free and Pb-doped catalysts was investigated using the H_2-TPR technique, so that the effect of Pb doping on the catalytic activity of 3Mo4Ce/Zr-PILC could be clarified. Meanwhile, after quantitative analysis of the H_2-TPR peaks, the H_2 consumption and reduction peak positions are summarized in Table 2. H_2-TPR profile of 3Mo4Ce/Zr-PILC displayed three main reduction peaks at 463, 527, and 723 °C (Figure 5), which could be interpreted as the reduction of the surface Ce^{4+} to Ce^{3+} species, co-reduction of the octahedrally structured well-dispersed Mo and iron species, and the co-reduction of the tetrahedrally structured Mo species and bulk CeO_2 species [17–19], respectively. In profiles of the Pb-poisoned catalysts, the first peak due to the reduction of the Ce species was shifted to a higher temperature, indicating that doping of Pb stabilized the Ce species and made it less reducible. The temperature of the low-temperature reduction peak increased in the order 3Mo4Ce/Zr-PILC (463 °C) > 0.2Pb-3Mo4Ce/Zr-PILC (470 °C) > 0.5Pb-3Mo4Ce/Zr-PILC (478 °C) > 1Pb-3Mo4Ce/Zr-PILC (483 °C), which was in accordance with their changing trend in SCR performance. In addition, two new reduction peaks at 575 and 790 °C appeared in the Pb-poisoned catalysts, which might be attributable to the reduction of the surface or bulk oxygen coordinated with the Pb atoms [20,21].

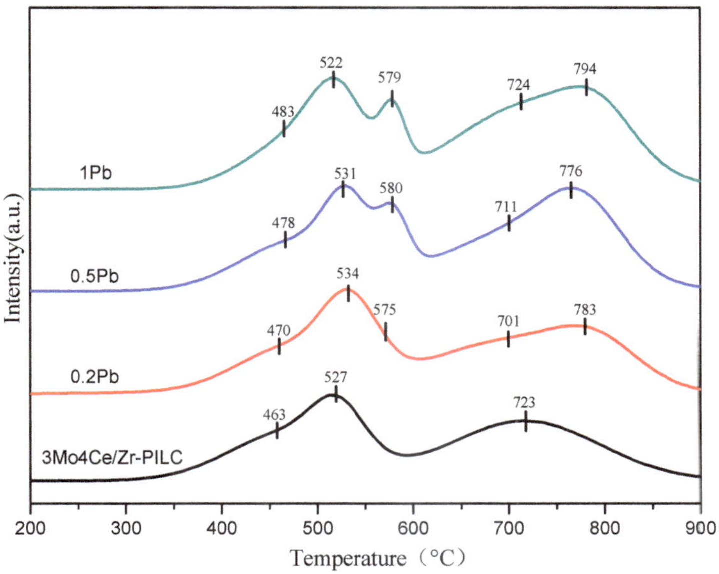

Figure 5. H_2-TPR profiles of the 3Mo4Ce/Zr-PILC and Pb-poisoned 3Mo4Ce/Zr-PILC catalysts.

Table 2. Reduction peak temperatures and H_2 consumption of the 3Mo4Ce/Zr-PILC and Pb-poisoned 3Mo4Ce/Zr-PILC catalysts.

Catalyst	Reduction Peak Temperature (°C)					H_2 Consumption (mmol/g)
	Peak 1	Peak 2	Peak 3	Peak 4	Peak 5	
3Mo4Ce/Zr-PILC	463	527	-	723	-	0.79
0.2Pb	470	534	575	701	783	0.90
0.5Pb	478	531	580	711	776	1.01
1Pb	483	522	579	724	794	1.10

3.4. Surface Property

To gain better insights into the surface compositions and chemical states of the catalysts, we conducted XPS characterization, and the results are shown in Figure 6 and Table 3.

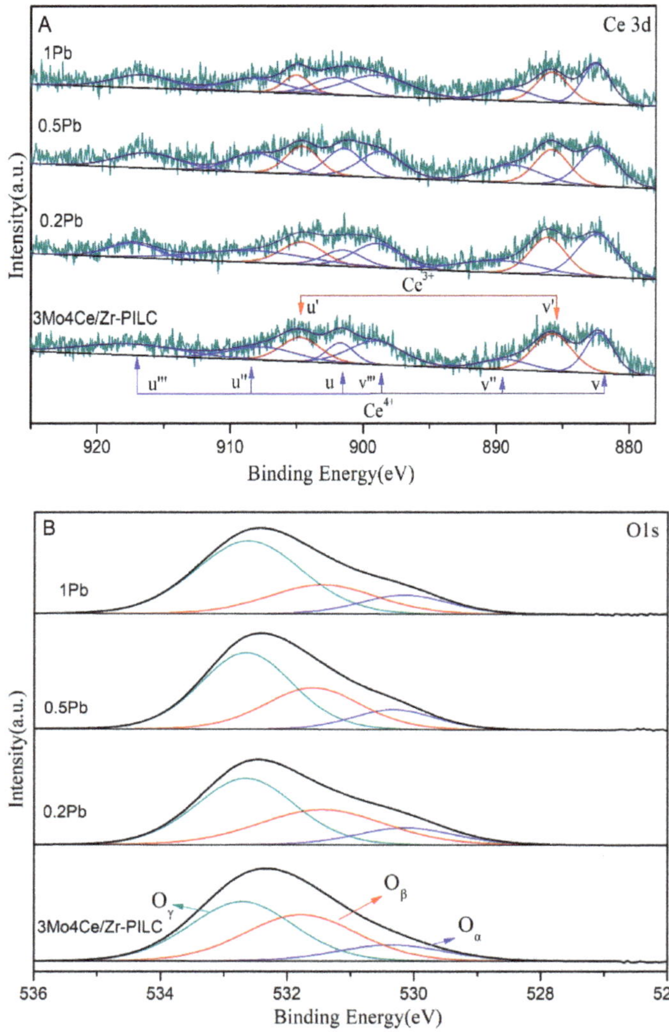

Figure 6. (**A**) Ce 3d and (**B**) O 1s spectra of the 3Mo4Ce/Zr-PILC and Pb-poisoned 3Mo4Ce/Zr-PILC catalysts.

Table 3. Surface element compositions of the 3Mo4Ce/Zr-PILC and Pb-poisoned 3Mo4Ce/Zr-PILC catalysts.

Catalyst	Composition of Cerium Species (at%)			Composition of Oxygen Species (mol%)		
	Ce^{3+}	Ce^{4+}	Ce^{3+}/Ce^{4+} Atomic Ratio	O_α	O_β	O_γ
3Mo4Ce/Zr-PILC	30.3	69.7	0.43	13.4	39.4	47.2
0.2Pb	26.2	73.8	0.36	13.1	34.7	52.2
0.5Pb	24.3	75.7	0.32	13.1	31.4	55.5
1Pb	21.1	78.9	0.27	13.4	24.8	61.8

The Ce 3d XPS spectra of the 3Mo4Ce/Zr-PILC and Pb-poisoned 3Mo4Ce/Zr-PILC catalysts are presented in Figure 6A. The peaks labeled as v and u corresponded to Ce $3d_{5/2}$ and Ce $3d_{3/2}$ spin-orbit peaks [22,23], respectively, which could be deconvoluted into eight components: the u^I and v^I components were assigned to the surface Ce^{3+} species corresponding to the $3d^{10}4f^1$ initial electronic state, while the other components were due to the surface Ce^{4+} species corresponding to the $3d^{10}4f^0$ initial electronic state. The Ce^{3+}/Ce^{4+} atomic ratios were calculated according to the integral areas of the corresponding peaks. With an increase in Pb doping, the Ce^{3+}/Ce^{4+} atomic ratio decreased in the order 3Mo4Ce/Zr-PILC (0.43) > 0.2Pb-3Mo4Ce/Zr-PILC (0.36) > 0.5Pb-3Mo4Ce/Zr-PILC (0.32) > 1Pb-3Mo4Ce/Zr-PILC (0.27), which was the same trend seen in the catalytic activity. The Ce^{3+} species exerted a positive effect on catalytic activity owing to its strong ability to create charge imbalance and form oxygen vacancies and unsaturated chemical bonds on the catalyst surface, which could increase the amount of surface chemisorbed oxygen species [24]. Some oxygen vacancies were occupied due to Pb introduction, which inhibited the transformation of Ce^{4+} to Ce^{3+}. Therefore, the decrease in the amount of the surface Ce^{3+} species on 1Pb-3Mo4Ce/Zr-PILC was an important parameter that gave rise to this system's worse catalytic activity.

O 1s XPS spectra of the catalysts are illustrated in Figure 6B, which could be decomposed into three components at BE = 532.61–532.70, 531.48–531.78, and 530.16–531.36 eV, assignable to the surface lattice oxygen of the Si–O bond (O_γ), chemisorbed oxygen (O_β), and lattice oxygen (O_α) [25–27], respectively. Chemisorbed oxygen was mainly the oxygen species adsorbed at the oxygen vacancies, as well as the defect oxides and hydroxyl-like groups. Based on that curve-fitting approach, the O_β concentrations on the surface of the catalysts were calculated, and the results are listed in Table 3. A distinct decrease in the amount of O_β species was observed after the Pb doping. The $O_\beta/(O_\alpha + O_\beta + O_\gamma)$ molar ratio decreased in the order 3Mo4Ce/Zr-PILC (0.394) > 0.2Pb-3Mo4Ce/Zr-PILC (0.347) > 0.5Pb-3Mo4Ce/Zr-PILC (0.314) > 1Pb-3Mo4Ce/Zr-PILC (0.248). These results were consistent with the order of NH_3-SCR activity, indicating that the lower amounts of O_β species on the Pb-poisoned catalysts were also responsible for their lower catalytic activities. It has been previously demonstrated that O_β species are the most effective oxygen species in the SCR reaction [27]. Moreover, the oxygen in the gas phase and the oxygen adsorbed on the catalyst surface could be easily exchanged with the O_β species, and the active transportation of the O_β species could also facilitate the SCR reaction [28].

3.5. Surface Acidity

The NH_3 adsorption capacity has been widely reported to influence the SCR reaction, and it has also been strongly associated with the surface acidity of a catalyst [29]. Therefore, NH_3-TPD experiments on the 3Mo4Ce/Zr-PILC and Pb-poisoned 3Mo4Ce/Zr-PILC catalysts were carried out to examine the influence of Pb doping on surface acidity, and the results are illustrated in Figure 7.

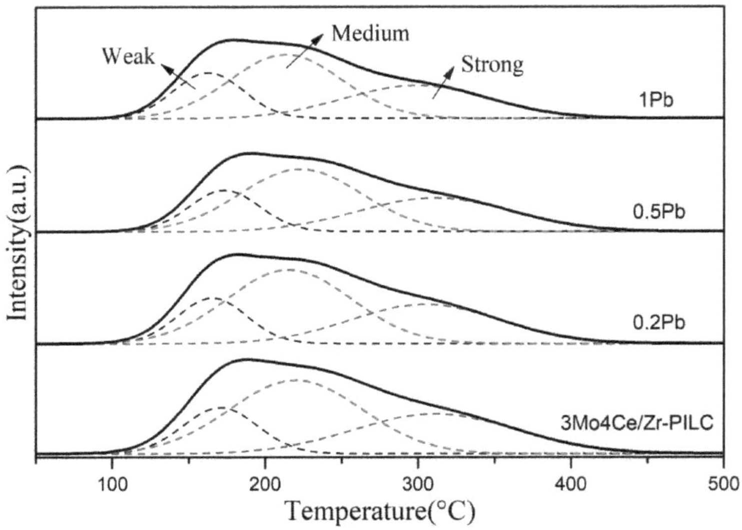

Figure 7. NH_3-TPD profiles of the 3Mo4Ce/Zr-PILC and Pb-poisoned 3Mo4Ce/Zr-PILC catalysts.

There were three desorption peaks in the NH_3-TPD profile of each sample (Figure 7). The first peak was located at 163–173 °C, and signified the desorption of the physically adsorbed NH_3 species and some NH_4^+ species at the weak Brønsted acid sites (i.e., weak acid sites); the second peak at 214–223 °C was attributed to the NH_4^+ species desorbed from the strong Brønsted acid sites (i.e., medium acid sites); and the third peak at 301–311 °C was assigned to the desorption of the coordinated NH_3 bound to the Lewis acid sites (i.e., strong acid sites) [29,30]. Acid site numbers were quantitatively analyzed according to the NH_3-TPD profiles, as listed in Table 4. The amount of Brønsted acid sites was in the order: 3Mo4Ce/Zr-PILC (0.131 mmol/g) > 0.2Pb-3Mo4Ce/Zr-PILC (0.125 mmol/g) > 0.5Pb-3Mo4Ce/Zr-PILC (0.103 mmol/g) > 1Pb-3Mo4Ce/Zr-PILC (0.096 mmol/g). The total acid amount decreased in the sequence 3Mo4Ce/Zr-PILC (0.264 mmol/g) > 0.2Pb-3Mo4Ce/Zr-PILC (0.247 mmol/g) > 0.5Pb-3Mo4Ce/Zr-PILC (0.213 mmol/g) > 1Pb-3Mo4Ce/Zr-PILC (0.202 mmol/g). It is noteworthy that a substantial decrease in the desorbed ammonia amount was observed when the amount of doped Pb increased. One reason for this phenomenon was probably due to the agglomeration of Pb species on the catalyst surface, partially blocking the micropores and covering some of the acid sites in the catalyst. In addition, it can be seen from the above Ce 3d XPS results that the concentration of the surface Ce^{3+} species on the sample decreased after Pb doping. According to the literature [31], the decrease in Ce^{3+} atomic concentration leads to a decrease in amount of oxide defects or hydroxyl-like groups, resulting in a decrease in the number of Brønsted acid sites. Therefore, we can conclude that the reduced number of acid sites (especially Brønsted acid sites) in the Pb-poisoned catalysts was one of the main reasons for the decrease in catalytic activity.

Table 4. NH_3-TPD temperatures and desorption amounts of the 3Mo4Ce/Zr-PILC and Pb-poisoned 3Mo4Ce/Zr-PILC catalysts.

Catalyst	Temperature (°C)			Acidity ($mmol_{NH3}$/g)			Total Desorption Amount ($mmol_{NH3}$/g) (mmol/g)
	Weak Peak	Medium Peak	Strong Peak	Weak Peak	Medium Peak	Strong Peak	
3Mo4Ce/Zr-PILC	171	220	311	0.046	0.131	0.087	0.264
0.2Pb	166	215	305	0.043	0.125	0.079	0.247
0.5Pb	173	223	310	0.040	0.103	0.070	0.213
1Pb	163	214	301	0.042	0.096	0.064	0.202

3.6. Switching Feed Gas from $NO + O_2 + N_2 + NH_3$ to $NO + O_2 + N_2$ on Various Catalysts

In order to investigate activity of the NO_x species and their role in the NH_3-SCR reaction, transient experiments were performed over the 3Mo4Ce/Zr-PILC catalyst. Figure 8 shows the DRIFTS spectra at 200 °C recorded after switching from 1100 ppm NH_3 + 1000 ppm NO + 4% O_2 + N_2 (balance) to a flow containing 1000 ppm NO + 4% O_2 + N_2 (balance) over the 3Mo4Ce/Zr-PILC catalyst.

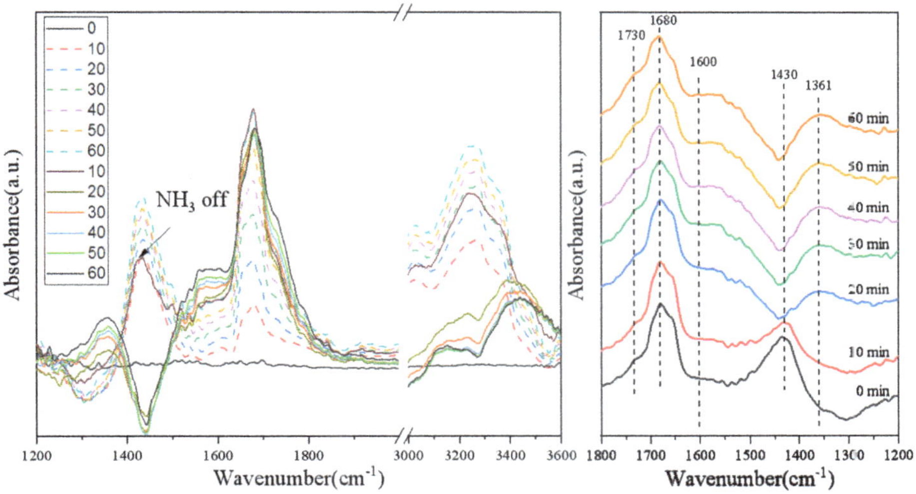

Figure 8. In situ DRIFTS spectra of the 3Mo4Ce/Zr-PILC catalyst exposed to (1100 ppm NH_3 + 1000 ppm NO + 4% O_2 + N_2 (balance)) for 1 h and then switched to (1000 ppm NO + 4% O_2 + N_2 (balance)) for 1 h at 200 °C.

From the transient experimental results of the 3Mo4Ce/Zr-PILC catalyst, it can be seen that bands attributable to the nitrate species and NH_3 species appeared on the surface of 3Mo4Ce/Zr-PILC after the flow containing $NH_3 + NO + O_2 + N_2$ was introduced. Several bands were observed at 1430, 1600, 1680, and 3200–3400 cm^{-1}, in which the bands at 1430 and 1680 cm^{-1} were attributed to the symmetric bending vibrations of NH_4^+ that was chemisorbed at the Brønsted acid sites, and the band at 3200–3400 cm^{-1} was assigned to the NH_3 species adsorbed at the Lewis acid sites [32]. In addition, the band at 1600 cm^{-1} could be ascribed to the bridged nitrate, and the band attributable to the adsorbed nitrogen oxide species was at 1680 cm^{-1} [31,33]. It is well known that the adsorbed NO_2 species can participate in the "fast NH_3-SCR" reaction, which promotes the NH_3-SCR performance [34]. When the feed gas was switched to $NO + O_2 + N_2$, the bands at 1430 and 3200–3400 cm^{-1} disappeared rapidly with respect to time, indicating that the NH_3 adsorbed at the Lewis and Brønsted acid sites reacted with the nitrate species, and the NH_3 species were gradually consumed during the reaction. Simultaneously, the intensity of the bands at 1331 and 1600 cm^{-1} was attributable to monodentate nitrate, and bridge nitrate also increased rapidly, indicating that the monodentate nitrates and bridge nitrates were highly active NO_x species and played an important role in the NH_3-SCR reaction.

Figure 9 shows the DRIFTS spectra of the 1Pb-3Mo4Ce/Zr-PILC catalyst. The band intensity of the NH_3 species at 1430 and 3200–3400 cm^{-1} was lower than that of the 3Mo4Ce/Zr-PILC catalyst after exposure to a flow of $NH_3 + NO + O_2 + N_2$. The intensity of the bands at 1365 and 1604 cm^{-1} was attributable to the monodentate nitrate, and bridge nitrate also increased slightly after cutting off NH_3 introduction. These results indicate that the doping of Pb inhibited the adsorption of the NH_3 species on the 1Pb-3Mo4Ce/Zr-PILC catalyst, and less of the main active species monodentate nitrate and bridge nitrate were generated, which would influence NH_3-SCR activity of the catalyst.

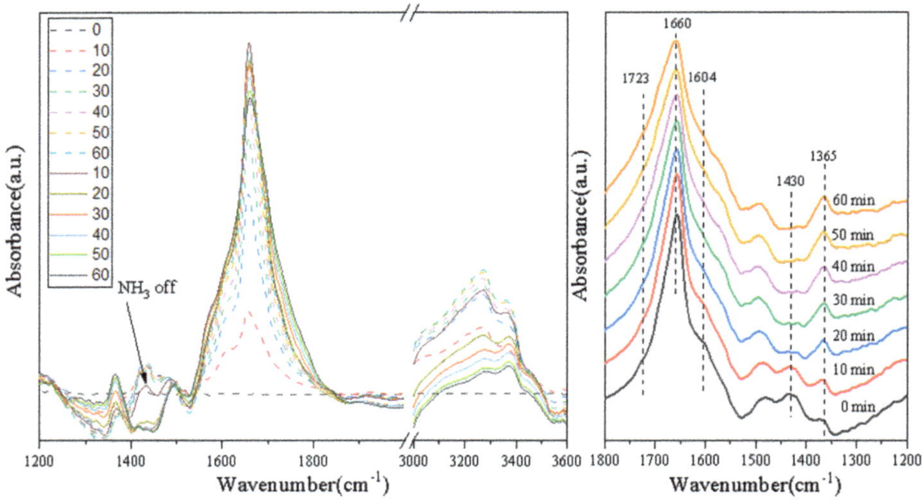

Figure 9. In situ DRIFTS spectra of the 1Pb-3Mo4Ce/Zr-PILC catalyst exposed to (1100 ppm NH_3 + 1000 ppm NO + 4% O_2 + N_2 (balance)) for 1 h and then switched to (1000 ppm NO + 4% O_2 + N_2 (balance)) for 1 h at 200 °C.

Therefore, it can be concluded that the monodentate nitrate, bridge nitrate, and adsorbed NO_2 species were the highly active intermediate species which could promote the NH_3-SCR reaction. Meanwhile, the doping of Pb inhibited the formation of intermediate nitrate species on the surface of the catalyst, thereby inhibiting the NH_3-SCR activity.

3.7. Pb Poisoning Mechanism

Pb is an ingredient found in coal-fired flue gas and inhibits SCR catalysts. The $DeNO_x$ performance of the catalyst was greatly suppressed after the Pb doping, and this inhibition effect was enhanced with increasing Pb/Ce molar ratio. XRD patterns (Figure 2) and SEM images (Figure 4) demonstrate that there was obvious agglomeration of the metal oxides after the Pb doping, and some $PbMoO_4$ species were generated on the surface of the catalysts. In combination with the Ce 3d XPS results (Figure 6A), this suggests that Pb might occupy the oxygen vacancies in ceria, which could weaken the $Ce^{3+} \leftrightarrow Ce^{4+}$ redox cycle and suppress the surface $Ce^{4+} \rightarrow Ce^{3+}$ transformation. In addition, based on the characterization results of NH_3-TPD (Figure 7), the number of both Brønsted and Lewis acid sites decreased significantly (especially the number of Brønsted acid sites) after the Pb doping. Since NH_3 was more readily adsorbed on the catalyst surface than NO in the SCR reaction, the adsorbed ammonia species were critical in the SCR reaction [35]. Furthermore, Pb inhibited the formation of the intermediate nitrate species (i.e., monodentate nitrates and bridge nitrates) on the surface of catalysts in the SCR reaction. The decrease in ammonia absorbance at the acid sites, the decrease of the highly active intermediate species, and the vanishing of active ammonia species were the most important influencing factors in the Pb-induced deactivation of the 3Mo4Ce/Zr-PILC catalyst.

4. Conclusions

Pb doping caused a serious deactivation of the 3Mo4Ce/Zr-PILC catalyst for the NH_3-SCR reaction in this study. NO conversion decreased from 98% to 49% over the 3Mo4Ce/Zr-PILC catalyst at 450 °C when the Pb/Ce molar ratio was 1. The characterization results revealed that as the amount of surface Ce^{3+} and chemisorbed oxygen species increased, the interaction between PbO and 3Mo4Ce/Zr-PILC increased as well, with more $PbMoO_4$ species formed on the catalyst surface, less intermediate nitrate species, and lower redox ability and surface acidity.

Author Contributions: Conceptualization, C.L. and J.C.; Methodology, C.L. and J.C.; Formal analysis, C.L.; Investigation, C.L. and J.C.; Software, F.M. and X.W.; Validation, Q.Y.; Writing—original draft, C.L. and J.C.; Writing—review& editing, C.L., Q.Y. and H.D.; Supervision, Q.Y. and H.D. All authors have read and agreed to the published version of the manuscript.

Funding: This research received no external funding.

Data Availability Statement: The data presented in this study are available on request from the corresponding author.

Acknowledgments: This work was supported by the National Natural Science Foundation of China (Grant Nos. 21277008 and 20777005) and the National Key Research and Development Program of China (Grant No. 2017YFC0209905).

Conflicts of Interest: The authors declare no conflict of interest.

References

1. Shan, W.; Yu, Y.; Zhang, Y.; He, G.; Peng, Y.; Li, J.; He, H. Theory and practice of metal oxide catalyst design for the selective catalytic reduction of NO with NH_3. *Catal. Today* **2021**, *376*, 292–301. [CrossRef]
2. Zhang, X.; Diao, Q.; Hu, X.; Wu, X.; Xiao, K.; Wang, J. Modification of V_2O_5-WO_3/TiO_2 catalyst by loading of MnO_x for enhanced low-temperature NH_3-SCR performance. *Nanomaterials* **2020**, *10*, 1900. [CrossRef] [PubMed]
3. Xu, W.; Zhang, G.; Chen, H.; Zhang, G.; Han, Y.; Chang, Y.; Gong, P. Mn/beta and Mn/ZSM-5 for the low-temperature selective catalytic reduction of NO with ammonia: Effect of manganese precursors. *Chin. J. Catal.* **2018**, *39*, 118–127. [CrossRef]
4. Zhang, N.; He, H.; Wang, D.; Li, Y. Challenges and opportunities for manganese oxides in low-temperature selective catalytic reduction of NO_x with NH_3:H_2O resistance ability. *J. Solid State Chem.* **2020**, *289*. [CrossRef]
5. Chen, L.; Li, J.; Ge, M. Promotional effect of Ce-doped V_2O_5-WO_3/TiO_2 with low vanadium loadings for selective catalytic reduction of NO_x by NH_3. *J. Phys. Chem. C* **2009**, *113*, 21177–21184. [CrossRef]
6. Hao, Z.; Jiao, Y.; Shi, Q.; Zhang, H.; Zhan, S. Improvement of NH_3-SCR performance and SO_2 resistance over Sn modified $CeMoO_x$ electrospun fibers at low temperature. *Catal. Today* **2019**, *327*, 37–46. [CrossRef]
7. Xiong, Y.; Tang, C.; Yao, X.; Zhang, L.; Li, L.; Wang, X.; Deng, Y.; Gao, F.; Dong, L. Effect of metal ions doping (M = Ti^{4+}, Sn^{4+}) on the catalytic performance of MnO_x/CeO_2 catalyst for low temperature selective catalytic reduction of NO with NH_3. *Appl. Catal. A Gen.* **2015**, *495*, 206–216. [CrossRef]
8. Zhan, S.; Shi, Q.; Zhang, Y.; Li, Y.; Tian, Y. Preparation of novel $CeMo_x$ hollow microspheres for low-temperature SCR removal of NO_x with NH_3. *RSC Adv.* **2016**, *6*, 59185–59194. [CrossRef]
9. Tang, X.; Shi, Y.; Gao, F.; Zhao, S.; Yi, H.; Xie, Z. Promotional role of Mo on $Ce_{0.3}FeO_x$ catalyst towards enhanced NH_3-SCR catalytic performance and SO_2 resistance. *Chem. Eng. J.* **2020**, *398*. [CrossRef]
10. Li, L.; Chen, L.; Kong, M.; Liu, Q.; Ren, S. New insights into the deactivation mechanism of V_2O_5-WO_3/TiO_2 catalyst during selective catalytic reduction of NO with NH_3: Synergies between arsenic and potassium species. *RSC Adv.* **2019**, *9*, 37724–37732. [CrossRef]
11. Wang, H.; Chen, X.; Gao, S.; Wu, Z.; Liu, Y.; Weng, X. Deactivation mechanism of Ce/TiO_2 selective catalytic reduction catalysts by the loading of sodium and calcium salts. *Catal. Sci. Technol.* **2013**, *3*, 715–722. [CrossRef]
12. Tokarz, M.; Jarars, S.; Persson, B. Poisoning of De-NO_x SCR catalyst by flue gases from a waste incineration plant. In *Catalyst Deactivation 1991, Proceedings of the 5th International Symposium, Studies in Surface Science & Catalysis*; Elsevier: Amsterdam, The Netherlands, 1991; Volume 68, pp. 523–530.
13. Cheng, J.; Ye, Q.; Li, C.; Cheng, S.; Kang, T.; Dai, H. Ceria-modified Al-Mn-pillared clay catalysts for the selective catalytic reduction of NO with NH_3 at low temperatures. *Asia Pac. J. Chem. Eng.* **2019**, *15*.
14. Yang, S.; Fu, Y.; Liao, Y.; Xiong, S.; Qu, Z.; Yan, N.; Li, J. Competition of selective catalytic reduction and non-selective catalytic reduction over MnO_x/TiO_2 for NO removal: The relationship between gaseous NO concentration and N_2O selectivity. *Catal. Sci. Technol.* **2014**, *4*, 224–232. [CrossRef]
15. Zang, S.; Zhang, G.; Qiu, W.; Song, L.; Zhang, R.; He, H. Resistance to SO_2 poisoning of V_2O_5/TiO_2-PILC catalyst for the selective catalytic reduction of NO by NH_3. *Chin. J. Catal.* **2016**, *37*, 888–897. [CrossRef]
16. Li, C.; Cheng, J.; Ye, Q.; Meng, F.; Wang, X.; Dai, H. Poisoning effects of alkali and alkaline earth metal doping on selective catalytic reduction of NO with NH_3 over the Nb-Ce/Zr-PILC catalysts. *Catalysts* **2021**, *11*, 329. [CrossRef]
17. Gao, G.; Shi, J.-W.; Liu, C.; Gao, C.; Fan, Z.; Niu, C. Mn/CeO_2 catalysts for SCR of NO_x with NH_3: Comparative study on the effect of supports on low-temperature catalytic activity. *Appl. Surf. Sci.* **2017**, *411*, 338–346. [CrossRef]
18. Ding, S.; Liu, F.; Shi, X.; Liu, K.; Lian, Z.; Xie, L.; He, H. Significant promotion effect of Mo additive on a novel Ce-Zr mixed oxide catalyst for the selective catalytic reduction of NO_x with NH_3. *ACS Appl. Mater. Interfaces* **2015**, *7*, 9497–9506. [CrossRef]
19. Rajkumar, T.; Sapi, A.; Abel, M.; Kiss, J.; Szenti, I.; Baan, K.; Gomez-Perez, J.F.; Kukovecz, A.; Konya, Z. Surface engineering of CeO_2 catalysts: Differences between solid solution based and interfacially designed $Ce_{1-x}M_xO_2$ and MO/CeO_2 (M = Zn, Mn) in CO_2 hydrogenation reaction. *Catal. Lett.* **2021**. [CrossRef]

20. Zhou, L.; Li, C.; Zhao, L.; Zeng, G.; Gao, L.; Wang, Y.; Yu, M. The poisoning effect of PbO on Mn-Ce/TiO_2 catalyst for selective catalytic reduction of NO with NH_3 at low temperature. *Appl. Surf. Sci.* **2016**, *389*, 532–539. [CrossRef]
21. Gao, X.; Du, X.; Fu, Y.; Mao, J.; Luo, Z.; Ni, M.; Cen, K. Theoretical and experimental study on the deactivation of V_2O_5 based catalyst by lead for selective catalytic reduction of nitric oxides. *Catal. Today* **2011**, *175*, 625–630. [CrossRef]
22. Wang, P.; Chen, S.; Gao, S.; Zhang, J.; Wang, H.; Wu, Z. Niobium oxide confined by ceria nanotubes as a novel SCR catalyst with excellent resistance to potassium, phosphorus, and lead. *Appl. Catal. B Environ.* **2018**, *231*, 299–309. [CrossRef]
23. Fan, J.; Ning, P.; Song, Z.; Liu, X.; Wang, L.; Wang, J.; Wang, H.; Long, K.; Zhang, Q. Mechanistic aspects of NH_3-SCR reaction over $CeO_2/TiO_2-ZrO_2-SO_4^{2-}$ catalyst: In situ DRIFTS investigation. *Chem. Eng. J.* **2018**, *334*, 855–863. [CrossRef]
24. Guo, R.; Lu, C.; Pan, W.; Zhen, W.; Wang, Q.; Chen, Q.; Ding, H.; Yang, N. A comparative study of the poisoning effect of Zn and Pb on Ce/TiO_2 catalyst for low temperature selective catalytic reduction of NO with NH_3. *Catal. Commun.* **2015**, *59*, 136–139. [CrossRef]
25. Lei, Z.; Hao, S.; Yang, J.; Zhang, L.; Fang, B.; Wei, K.; Qi, L.; Jin, S.; Wei, C. Study on denitration and sulfur removal performance of Mn-Ce supported fly ash catalyst. *Chemosphere* **2021**, *270*, 128646. [CrossRef]
26. Kim, M.; Park, G.; Shin, M.; Kim, H.-D.; Cho, D.-S.; Lee, H. Electron redistribution and acid site of tungsten-doped $CeTiO_x$ with oxygen vacancies. *Adv. Appl. Ceram.* **2021**, *120*, 187–191. [CrossRef]
27. Wang, J.; Dong, X.; Wang, Y.; Li, Y. Effect of the calcination temperature on the performance of a $CeMoO_x$ catalyst in the selective catalytic reduction of NO_x with ammonia. *Catal. Today* **2015**, *245*, 10–15. [CrossRef]
28. Liu, J.; Shi, X.; Shan, Y.; Yan, Z.; Shan, W.; Yu, Y.; He, H. Hydrothermal stability of CeO_2-WO_3-ZrO_2 mixed oxides for selective catalytic reduction of NO_x by NH_3. *Environ. Sci. Technol.* **2018**, *52*, 11769–11777. [PubMed]
29. Xu, H.; Wang, Y.; Cao, Y.; Fang, Z.; Lin, T.; Gong, M.; Chen, Y. Catalytic performance of acidic zirconium-based composite oxides monolithic catalyst on selective catalytic reduction of NO_x with NH_3. *Chem. Eng. J.* **2014**, *240*, 62–73. [CrossRef]
30. Yang, N.; Guo, R.; Pan, W.; Chen, Q.; Wang, Q.; Lu, C.; Wang, S. The deactivation mechanism of Cl on Ce/TiO_2 catalyst for selective catalytic reduction of NO with NH_3. *Appl. Surf. Sci.* **2016**, *378*, 513–518. [CrossRef]
31. Li, L.; Tan, W.; Wei, X.; Fan, Z.; Liu, A.; Guo, K.; Ma, K.; Yu, S.; Ge, C.; Tang, C.; et al. Mo doping as an effective strategy to boost low temperature NH_3-SCR performance of CeO_2/TiO_2 catalysts. *Catal. Commun.* **2018**, *114*, 10–14. [CrossRef]
32. Li, X.; Li, Y. Molybdenum modified $CeAlO_x$ catalyst for the selective catalytic reduction of NO with NH_3. *J. Mol. Catal. A Chem.* **2014**, *386*, 69–77. [CrossRef]
33. Jiang, Y.; Yang, L.; Liang, G.; Liu, S.; Gao, W.; Yang, Z.; Wang, X.; Lin, R.; Zhu, X. The poisoning effect of PbO on CeO_2-MoO_3/TiO_2 catalyst for selective catalytic reduction of NO with NH_3. *Mol. Catal.* **2020**, 486. [CrossRef]
34. Yan, Z.; Shi, X.; Yu, Y.; He, H. Alkali resistance promotion of Ce-doped vanadium-titanic-based NH_3-SCR catalysts. *J. Environ. Sci.* **2018**, *73*, 155–161. [CrossRef] [PubMed]
35. Liu, Z.; Zhang, S.; Li, J.; Ma, L. Promoting effect of MoO_3 on the NO_x reduction by NH_3 over CeO_2/TiO_2 catalyst studied with in situ DRIFTS. *Appl. Catal. B Environ.* **2014**, *144*, 90–95. [CrossRef]

nanomaterials

Article

Mesoporous Zn/MgO Hexagonal Nano-Plates as a Catalyst for *Camelina* Oil Biodiesel Synthesis

Lai-Fan Man [1], Tsz-Lung Kwong [2], Wing-Tak Wong [3,*] and Ka-Fu Yung [3,*]

1 Department of Chemistry, The University of Hong Kong, Pokfulam Road, Hong Kong, China; h0992119@hku.hk
2 Institute of Textiles and Clothing, The Hong Kong Polytechnic University, Hung Hom, Kowloon, Hong Kong, China; tlsamuel.kwong@polyu.edu.hk
3 Department of Applied Biology and Chemical Technology, The Hong Kong Polytechnic University, Hung Hom, Kowloon, Hong Kong, China
* Correspondence: w.t.wong@polyu.edu.hk (W.-T.W.); bckfyung@polyu.edu.hk (K.-F.Y.)

Abstract: A novel mesoporous Zn/MgO hexagonal-nano-plate catalyst was synthesized by a simple template-free hydrothermal method and applied in the base-catalyzed transesterification of *Camelina* oil for biodiesel synthesis. The Zn/MgO catalyst calcinated at 873 K exhibited the highest catalytic activity with a yield of 88.7%. This catalytic reaction was performed using 3% *w*/*w* of the catalyst with a methanol-to-oil molar ratio of 24:1 at 393 K in 8 h. The excellent catalytic performance is possibly attributed to its favorable textural features with relatively high surface area (69.1 $m^2\ g^{-1}$) and appropriate size of the mesopores (10.4 nm). In addition, the as-synthesized catalyst demonstrated a greater basic sites density than single mesoporous MgO, which might have been promoted by the addition of Zn, leading to a synergetic interaction that enhanced its catalytic activity. This catalytic system demonstrated high stability for five catalytic runs and catalytic activity with over 84% yield.

Keywords: Zn/MgO; nano-plates; mesoporous; heterogeneous base catalyst; biodiesel; transesterification; *Camelina* oil

Citation: Man, L.-F.; Kwong, T.-L.; Wong, W.-T.; Yung, K.-F. Mesoporous Zn/MgO Hexagonal Nano-Plates as a Catalyst for *Camelina* Oil Biodiesel Synthesis. *Nanomaterials* **2021**, *11*, 2690. https://doi.org/10.3390/nano11102690

Academic Editors: Jihoon Lee and Ming-Yu Li

Received: 23 September 2021
Accepted: 11 October 2021
Published: 13 October 2021

1. Introduction

Biomass was one of the major sources of energy and fuel supply in the late 1800s; however, lower diesel fuel production costs have retarded its development. The recent concerns about the finite reserves and environmental impact of diesel fuel have driven scientists to explore a cheap and energy-efficient way to produce fuel from biomass, the only natural and renewable carbon resource abundant enough to substitute fossil fuel. Several reports have indicated that the combination of biodiesel with various catalysts improves the combustion behavior of diesel engines and produces great environmental benefits because of the closed carbon cycle and lower exhaust emissions of unburnt hydrocarbons, particulate matter, sulfur oxides, carbon monoxide, etc. [1–6]. Moreover, biodiesel possesses similar combustion properties as commercial diesel fuel and thus can be adapted to regular diesel engines without modification of the latter [7–9].

Vegetable oils, animal fat, low-valued non-edible oil, and waste lipid can be used as renewable feedstock sources. However, they cannot be used directly as fuel because of their high viscosity and low volatility that cause poor atomization and deposition in the injector in the diesel engine. *Camelina* oil was selected as a potential feedstock for biodiesel synthesis in this study. *Camelina sativa* is an underexploited and low-input crop species of tremendous economic potential. Its seed contains 30 to 40% of oil that is rich in Omega-3 fatty acids [10]. Cold-pressed filtered *Camelina* seed oil produces a maximum power output of 43.35 kW, while that of mineral fuel is only 38.50 kW [11]. It exhibits a positive energy balance, with a net energy ratio of 1.47 in biodiesel production [12]. Moreover, its substantially lower production cost makes it an attractive candidate over other oil crops such as soybean, rapeseed, corn, etc.

Transesterification is a typical organic reaction employed for biodiesel synthesis, in which triglycerides are transformed into mono-alkyl esters through the interchange of the alkoxy moiety with the alcohol molecule in the presence of a catalyst [13]. Alkyl ester products exhibit reduced viscosity and high volatility and can readily be combusted as a fuel. Conventionally, homogeneous acids and bases such as concentrated sulphuric acid (H_2SO_4) and sodium hydroxide (NaOH) are employed for biodiesel synthesis. However, feedstock containing high levels of free fatty acids cause saponification and emulsification of biodiesel products.

These problems can be eliminated by switching to the use of heterogeneous catalysts. Calcium oxide (CaO) and magnesium oxide (MgO) were reported in many pioneering works as heterogeneous base catalysts for the transesterification of vegetable oil for biodiesel synthesis [14,15]. However, few studies have been performed because of the relatively low catalytic activity of alkaline earth metal oxides owing to their weak basic strength and high solubility in methanol [16,17]. The modification of MgO chemistry by adding other catalyst species such as metallic ions and oxides to obtain a binary metal oxide system and enhance the catalyst activity and stability has been widely reported [18–20]. Extensive studies have revealed that metal-doped MgO demonstrated excellent catalytic activity in various base-catalyzed organic reactions such as propane oxidation [21] and aldol condensation [22]. Metal-doped MgO has also been employed in transesterification for biodiesel synthesis. Dahlquist et al. [23] have successfully synthesized a Li-doped MgO catalyst that gave a high biodiesel yield of 93.9% at 333 K in 2 h. However, significant metal leaching of active species was found, causing catalyst deactivation. Cobalt-doped MgO supplied a high biodiesel yield of 90% in reaction conditions including a methanol-to-oil molar ratio of 9:1, using 5.00 wt.% catalyst at 423 K for 2 h. However, this catalyst demonstrated low catalytic stability, and the biodiesel yield dropped by 50%. It was reported that catalyst deactivation is attributed to catalyst poisoning by organic molecules in the reactant mixture and catalyst leaching of the active species [24]. Solid MgO–ZnO mixed metal oxides were also employed in biodiesel synthesis. It was found that the bi-functional system of Mg and Zn exhibited a synergetic interaction that is effective for enhancing both catalytic activity and stability towards transesterification. Different synthetic methods such as the sol–gel method [25], pulsed-laser deposition [26], and thermal decomposition [27] have been employed for the synthesis of MgO–ZnO mixed metal oxides. However, they usually involve complicated synthetic procedures, comparatively high temperature, and expensive equipment.

In order to extend the lifetime of the MgO catalyst, modification of the chemistry of the MgO catalytic system is needed to decrease its solubility in alcohols and increase its surface basicity. Reinoso and co-workers reported the application of zinc oleate as a catalyst for biodiesel synthesis. A biodiesel yield of 95% was obtained [28]. Our previous studies also reported the development of zinc oxide (ZnO) nano-stars and zinc glycerolates (ZnGly) nano-plates as effective catalysts in simultaneous esterification and transesterification, with excellent biodiesel yield over 95% [29,30]. These results indicate that zinc is a promising active transition metal for catalyzing oil transesterification with different alcohols. Therefore, we considered a bi-functional concept consisting in incorporating Zn^{2+} dopant into the crystal lattice of MgO without changing its original size and morphology. The synergistic effect of a bi-functional system would enhance its catalytic activity and the reusability towards transesterification.

Moreover, solid catalysts encounter mass-transfer limitations to the transesterification system, and their low active sites available for catalytic reaction have limited their application. The development of nano-catalysts with mesoporous structures has improved the situation by increasing the ratio between surface area and volume. Mesoporous structures also increase the bulk particle size, allowing a more effective separation by filtration than nano-sized catalysts.

In the present study, a simple template-free hydrothermal synthesis of a mesoporous Zn/MgO hexagonal nano-plates catalytic system was developed. This catalytic system exhibited a hierarchical combination of interconnected channels, thus minimized the diffusional limitations of the reactants and products. The as-synthesized Zn/MgO catalyst was found highly stable and active towards transesterification with methanol for biodiesel synthesis from *Camelina* oil.

2. Materials and Methods

2.1. Synthesis of the Catalysts

Mesoporous Zn/MgO was synthesized through an alkali hydrothermal approach, as shown in Scheme 1. A precursor solution containing $Mg(NO_3)_2{\cdot}6H_2O$ (513 mg, 2 mmol) and $Zn(NO_3)_2{\cdot}6H_2O$ (595 mg, 2 mmol) was mixed in an alkaline medium of NaOH(aq) (20 mL, 10 M) and ethanol (20 mL) in a 100 mL Teflon cup inside a stainless steel autoclave. The reaction mixture was heated at 423 K for 24 h under static conditions. The as-synthesized catalyst was filtered out, washed, and then dried in an oven at 383 K for 12 h. The catalyst was then calcinated in the air at designated temperatures of 773 K, 873 K, 973 K, and 1073 K for 3 h. Single mesoporous MgO was also prepared through the same synthetic protocol with the addition of the $Mg(NO_3)_2{\cdot}6H_2O$ precursor only. No agglomeration was observed during the synthesis of the mesoporous MgO and Zn/MgO catalysts.

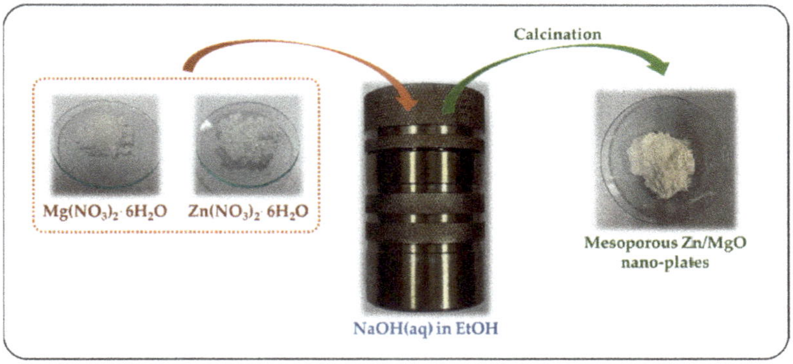

Scheme 1. Scheme diagram showing the hydrothermal synthesis of mesoporous Zn/MgO nanoplates catalyst.

2.2. Characterization of the Catalysts

The crystal phases of the as-synthesized catalysts were identified by powder X-ray diffraction (XRD) using a Rigaku SmartLab with a CuKα (λ= 1.541862Å) radiation, operated at 45 kV and 200 mA, with a scan rate of 2-theta, ranging from 20° to 80°. The morphology of the as-synthesized catalysts was characterized by a scanning electron microscope (SEM) (Hitachi S4800 FEG SEM system), equipped with energy-dispersive X-ray spectroscopy (EDS) (Horiba EMAX EDS) detectors for elemental analysis, and a transmission electron microscope (TEM) (Philips Tecnai G2 20 S-TWIN TEM system) in conjunction with INCAx-sight EDS detectors for elemental analysis.

Nitrogen adsorption–desorption isotherms of the catalysts were measured on a Micromeritics ASAP 2020 instrument. The catalysts were pre-treated by outgassing under a high vacuum at 350 °C for 6 h before the measurement. The specific surface areas were calculated by the Brunauer–Emmett–Teller (BET) method, while pore size distribution (average pore diameter and pore volume) was calculated according to the Barrett–Joyner–Halenda (BJH) method.

The basic strengths of the samples were determined by the Hammett indicator method. The solid catalyst sample (20 mg) was suspended in a methanol solution of Hammett indicator (1 mL, 0.02 mol/L) and left for 2 h to achieve equilibration. A change from acidic color to basic color was observed if the catalyst possessed higher basic strength than the indicator.

Hammett indicators–benzoic acid titration is a common method for determining the basic strength and the amount of basic sites of a catalyst. Basicity (basic site distribution) of these catalysts was evaluated by the Hammett indicator–benzene carboxylic acid titration method. The catalyst (20 mg) was suspended in methanolic Hammett indicator solution (1 mL, 0.02 mol/L) and left for 2 h. The above mixture was then transferred to a 50 mL Erlenmeyer flask containing methanol (10 mL), followed by titration with benzoic acid in methanol (0.01 M). The end point was noted as the point at which the basic color of the indicator disappeared. Triplicated measurements were done to minimize the end-point color determination error. The basicity is expressed as mmol g^{-1} calculated from the benzoic acid in methanol titrant (0.01 M) needed for the specific amount of catalyst used.

The Hammett indicators employed to study basic strength and basicity were as follows: neutral red ($H\_ = 6.8$), bromothymol blue ($H\_ = 7.2$), phenolphthalein ($H\_ = 9.3$), Nile blue ($H\_ = 10.1$), Tropaeolin O ($H\_ = 11$), and 2,4-dinitroaniline ($H\_ = 15$).

X-ray photoelectron spectroscopy (XPS) studies were also performed using a Kratos Axis Ultra XPS system equipped with monochromatic Al−Kα radiation of 1486.6 eV and with an electron take-off angle of 90°. The pressure of the sample chamber was kept at 10^{-8} Torr during the measurements. The spectrum was recorded in the binding energy (B.E.) range of 0.00 to 1400.00 eV with a step size of 1.00 eV. Energy calibration was performed with the C 1s peak of carbon at 285.0 eV.

2.3. Catalytic Reactions and Biodiesel Determination

Catalytic transesterification experiments were carried out in a stirred batch reactor containing the catalyst (66 mg, 3% *w/w*, with respect to the weight of oil), oil (2.4 mL), and methanol (2.4 mL, 24:1 MeOH-to-oil molar ratio), and the reaction was heated at the temperature of 393 K for 8 h with a stirring rate of 1000 rpm. After the reaction, the liquid product was isolated from the solid catalyst. The upper layer containing the biodiesel was extracted from glycerol with *n*-hexane. Residue methanol and *n*-hexane were then removed under vacuum. The biodiesel product was re-dissolved in $CDCl_3$ for ^{1}H NMR analysis with an AV400 Bruker FT-NMR spectrometer. The biodiesel yield was calculated using the signal integral of H of the methoxy group (–OCH_3) of FAME and that of the α-methylene group (α-CH_2) of both FAME and oil with the equation below [31].

$$\text{Biodiesel Yield } (\%) = \frac{2 \times \text{integral of H of -OCH}_3}{3 \times \text{integral of H of } \alpha\text{-CH}_2} \times 100\% \quad (1)$$

Reusability studies were carried out with solid catalysts recovered by centrifugation, followed by washing with *n*-hexane and then methanol. The catalyst was dried at 393 K for 5 h and evaluated under the same reaction conditions.

3. Results and Discussions

3.1. Characterization of the Catalysts

The powder X-ray diffraction (XRD) patterns of mesoporous Zn/MgO calcinated at different temperatures are shown in Figure 1. The XRD patterns of the Zn/MgO catalyst calcinated at 773 K, 873 K, 973 K, and 1073 K showed characteristic diffraction peaks of the single cubic periclase phases (JCPDS 01-076-6597) as the single mesoporous MgO. The absence of characteristic diffraction peaks of ZnO revealed that Zn^{2+} had in art replaced Mg^{2+} in the host lattice, giving a homogeneous solid solution. The close ionic radii of Zn^{2+} (0.74 Å, coordination number CN = 6) and Mg^{2+} (0.72 Å, coordination number CN = 6) favor the easy incorporation of Zn^{2+} into the MgO crystal lattice [32–35].

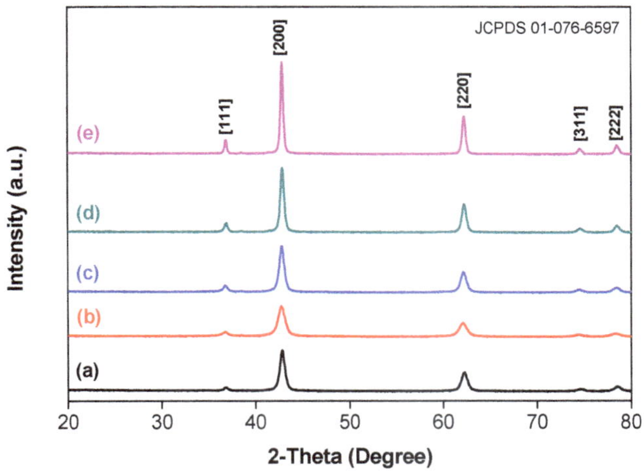

Figure 1. XRD patterns of (**a**) mesoporous MgO calcinated at 873 K and mesoporous Zn/MgO calcinated at (**b**) 773 K, (**c**) 873 K, (**d**) 973 K, and (**e**) 1073 K.

The surface morphology of mesoporous Zn/MgO calcinated at various temperatures was examined by SEM, and the micrographs are displayed in Figure 2. No agglomeration was observed during the synthesis of the mesoporous MgO and Zn/MgO catalysts. Zn/MgO-773 K (Figure 2b) comprised layers of hexagonal plates with lateral length between 100 to 500 nm and thickness in the range of 50 to 200 nm, with uniform mesopores evenly distributed on the plate surfaces. The formation of such mesoporous Zn/MgO nano-plate is the result of the solvothermal annealing process. The phenomenon was reported for single mesoporous MgO. The hexagonal $Mg(OH)_2$ prisms change into 3D wormhole-like mesoporous prisms with the preservation of the hexagonal morphology of $Mg(OH)_2$ due to the thermal decomposition process. The solvothermal process leads to the formation of octahedron $Mg(OH)_2$, with Mg ions layers separated by two adjacent hydroxyl ions layers. The subsequent calcination process leads to the partial dissociation of the hydrogens bonds associated with the two adjacent hydroxyl ions layers, releasing water molecules and generating the mesopores [36]. The morphology of Zn/MgO was retained when the calcination temperature was raised. However, it could be observed that the pore size increased as the calcination temperature of the samples increased. The pores of Zn/MgO-1073 K (Figure 2e) appeared to be the largest in size when comparing all catalysts, the number of pores was lower, and the pore distribution appeared less orderly on the plate surface. Interestingly, single mesoporous MgO (Figure 2a) also exhibited a similar hexagonal morphology with similar mesoporosity. Therefore, it is likely that Zn ions were incorporated into the MgO crystal lattice without changing the original morphology of MgO.

Representative TEM micrographs and the corresponding SAED pattern of Zn/MgO-873 K are displayed in Figure 3. It was observed that the catalyst comprised layers of hexagonal plates with uniform mesopores evenly distributed on the plate surfaces. The SAED pattern (Figure 3c) confirmed the existence of the cubic periclase phase. High-resolution transmission electron microscopy (HR-TEM) and the corresponding fast Fourier transform (FFT) image (Figure 3e) identified the surface termination as exposing polar (111) facets as the main surface on the hexagonal side (Figure 3f), with the generation of low coordination defect sites, which are reported to be highly basic, leading to high catalytic activity [35,37,38].

The average atomic percentage of Zn in Zn/MgO was found to be 7.8% by EDX analysis, which agreed well with the results obtained by inductively coupled plasma mass spectrometry (ICP–MS), with 8.4% of Zn. Furthermore, ICP–MS confirmed the absence of sodium (Na) in the catalysts.

Figure 2. SEM micrographs of mesoporous (**a**) single mesoporous MgO-873 K, (**b**) Zn/MgO-773 K, (**c**) Zn/MgO-873 K, (**d**) Zn/MgO-973 K, and (**e**) Zn/MgO-1073 K.

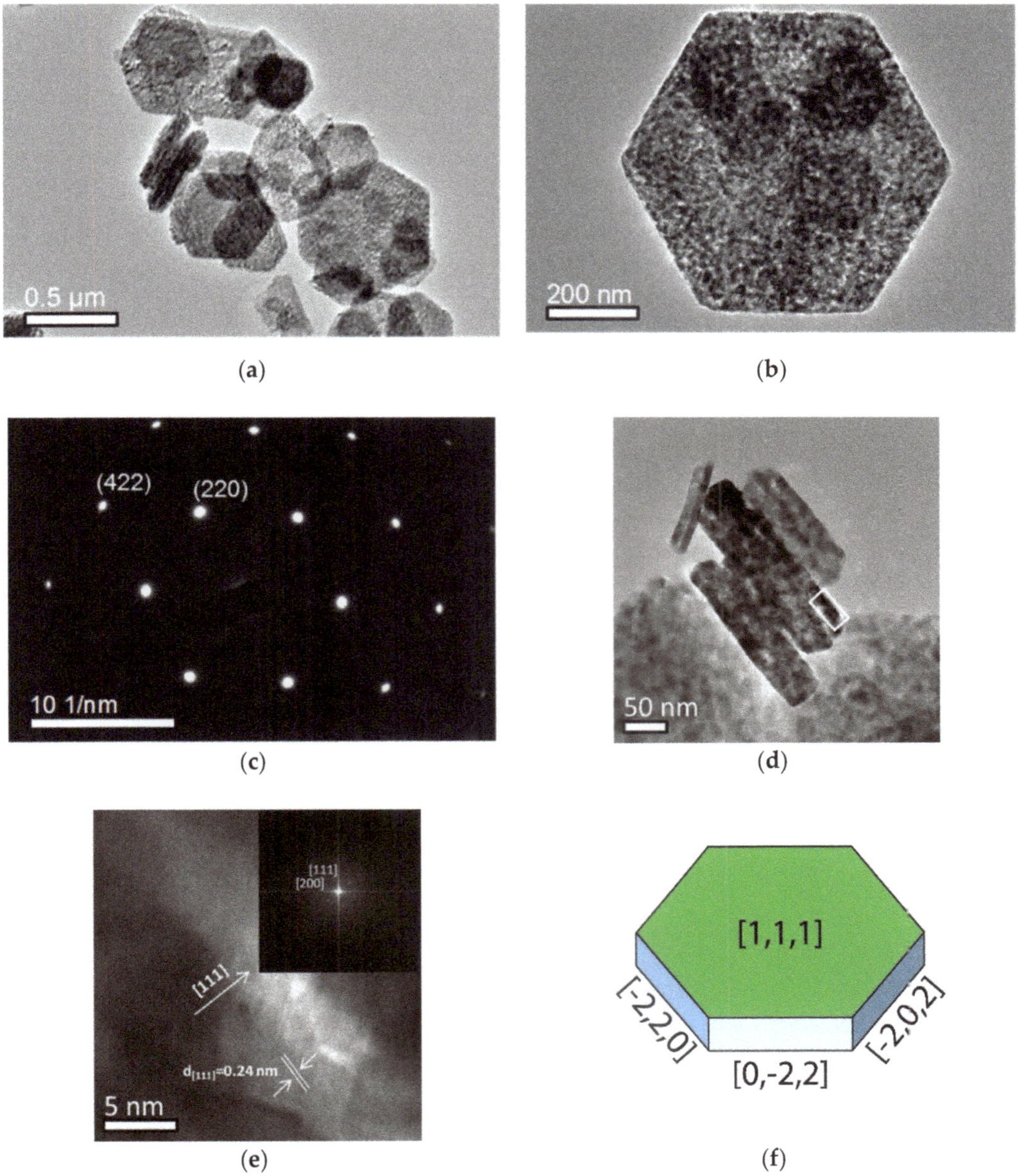

Figure 3. (**a**) Overall TEM micrograph of Zn/MgO-873 K, (**b**) TEM micrograph of the front side of the hexagonal nano-plate and (**c**) corresponding SAED pattern, (**d**) TEM micrograph of the lateral side of the hexagonal nano-plate and (e) corresponding HR-TEM image (insert: FFT image), and (**f**) schematic diagram showing the different facets of the Zn/MgO-873 K nano-plate.

The physical properties, such as crystallite size, BET surface area, total pore volume, and pore diameter of all Zn/MgO nano-plates calcinated at different temperatures are summarized in Table 1. The crystallite sizes of Zn/MgO were calculated from the XRD pattern using Scherrer's equation, and it was found that the crystallite size increased with the increase of the calcination temperature because of the sintering effect [39]. The results were aligned to the results of the BET analysis. The surface areas decreased with the increase in crystallite size. The nitrogen adsorption–desorption isotherms and the pore size distribution of all Zn/MgO catalysts calcinated at a temperature varying from 773 K to 1073 K are displayed in Figure 4a,b respectively. Zn/MgO calcinated at 773 K, 873 K, and 973 K exhibited a type IV isotherm, which is a characteristic of material possessing mesoporosity. Dual hysteresis loops were observed in the samples with an H1 hysteresis loop and an H3 hysteresis loop, which indicated the presence of both mesopores (2 to 50 nm) and macropores (>50 nm) in the samples [40,41]. The hysteresis loop type H1 in the P/P_0 range between 0.5 and 0.9 indicated the presence of approximately even and uniform mesopores in a reasonably regular array, accompanied by a narrow distribution of pore size. On the other hand, the H3 hysteresis loop in a P/P_0 range from 0.9 to 1.0 did not clearly show any adsorption isotherm plateau in the P/P_0 range close to unity, indicating the presence of slit-shaped pores with the Barrett–Joyner–Halenda (BJH) pore size distribution extending to the macropore range. It likely resulted from the aggregation of plate-like particles [40]. Zn/MgO-1073 K showed a type IV isotherm that is a characteristic of material with nonporous or potential macroporous structure and with high energy of adsorption. In addition, Zn/MgO calcinated at 773 K, 873 K, and 973 K possessed a relatively narrow distribution of pore size (Figure 4b), with average pore diameters of 5.4 nm, 10.4 nm, and 18.8 nm, respectively (Table 1). However, the Zn/MgO-1073 K catalyst exhibited a broader pore size distribution with an average pore diameter of 22.7 nm. Interestingly, Zn/MgO-873 K possessed comparable textural properties to those of mesoporous MgO-873 K, as shown in Table 1.

Table 1. Physical properties of mesoporous MgO and Zn/MgO nano-plates calcinated at different temperatures.

Catalyst	Crystallite Size [1] (nm)	S_{BET} [2] ($m^2\ g^{-1}$)	V [3] ($cm^3\ g^{-1}$)	D_{aver} [4] (nm)
Meso. MgO-873 K	12.6	73.1	0.267	12.0
Zn/MgO-773 K	9.7	151.1	0.269	5.4
Zn/MgO-873 K	12.7	69.1	0.233	10.4
Zn/MgO-973 K	17.8	45.3	0.263	18.8
Zn/MgO-1073 K	24.1	22.6	0.140	22.7

[1] Calculated from XRD patterns by employing Scherrer's equation; [2] BET surface area; [3] Total pore volume; [4] Average pore width.

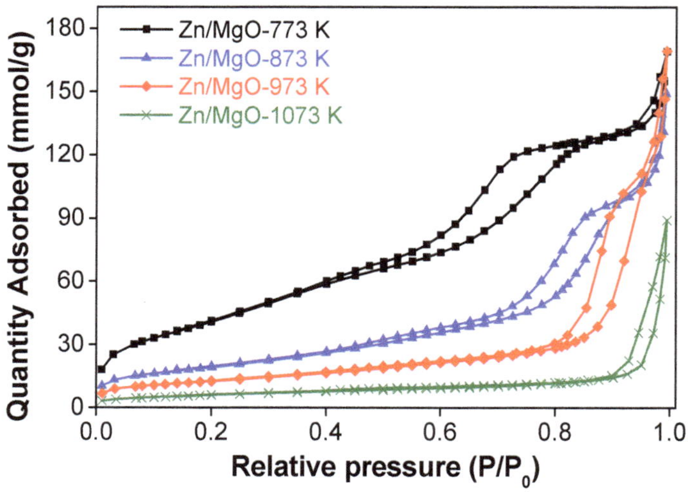

(a)

Figure 4. *Cont.*

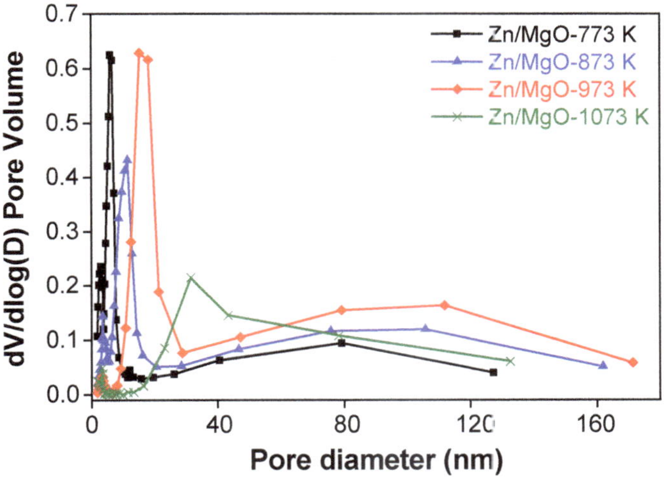

(b)

Figure 4. (**a**) N_2 adsorption–desorption isotherms and (**b**) BJH pore size distribution of mesoporous MgO and Zn/MgO nano-plates calcinated at different temperatures.

The results of the comparison of basic strength, basic site distribution, and total basicity between all Zn/MgO catalysts and single mesoporous MgO are summarized in Table 2. All Zn/MgO catalysts possessed a basic strength of $9.3 < H\_ < 10.1$ and superior total basicity as compared to single mesoporous MgO. Among all Zn/MgO catalysts, Zn/MgO-873 K revealed the greatest total basicity (1.52 mmol g^{-1}), with 0.90 mmol g^{-1}, 0.15 mmol g^{-1}, and 0.47 mmol g^{-1} basic sites of $6.8 < H\_$, $6.8 < H\_ < 7.2$ and $7.2 < H\_ < 9.3$, respectively. It suggests that at least three types of basic sites exist on the catalyst surface. The remarkably larger number of basic sites in 7.2 < H_ < 9.3 of all Zn/MgO catalysts indicates a higher density of stronger basic sites in comparison with single mesoporous MgO. These findings are well complemented by Olutoye's studies which have found that the basicity of a catalyst was enhanced by the synergetic interaction between Mg and Zn.

Table 2. Surface basicity of mesoporous MgO and Zn/MgO nano-plates catalysts calcinated at different temperatures.

Catalyst	Basic Strength by Hammett Indicator	Basicity by Benzoic Titration (mmol g^{-1})			
		$6.8 < H\_$	$6.8 < H\_ < 7.2$	$7.2 < H\_ < 9.3$	Total Basicity
MgO-873 K	$9.3 < H\_ < 10.1$	0.79	0.25	0.01	1.05
Zn/MgO-773 K	$9.3 < H\_ < 10.1$	0.60	0.15	0.32	1.07
Zn/MgO-873 K	$9.3 < H\_ < 10.1$	0.90	0.15	0.47	1.52
Zn/MgO-973 K	$9.3 < H\_ < 10.1$	0.67	0.30	0.45	1.42
Zn/MgO-1073 K	$9.3 < H\_ < 10.1$	0.59	0.34	0.54	1.47

The XPS survey data of Mg, Zn, and O elements for Zn/MgO-873 K catalyst compared to those of single MgO and ZnO are depicted in Table 3. The binding energy of 49.3 eV, 1021.1 eV, 529.6 eV, and 531.6 eV corresponds to Mg 2p, $Zn2p_{3/2}$, O_{lat}, and O_{hyd}, respectively. The slight shift in the Auger parameter of Zn was attributed to the incorporation of Zn^{2+} into the MgO crystal lattice as Zn–O–Mg [42]. The asymmetric O 1s peak of MgO, ZnO, and Zn/MgO was further resolved into the lattice oxygen (O_{lat}) and the surface hydroxyl group (O_{hyd}) [43]. The peak at lower binding energy was ascribed to lattice oxygen, while the one at higher binding energy was attributed to the hydroxyl group. A lower O 1s binding energy indicates a higher electron pair donation ability and, therefore, a stronger basic strength [44]. Zn/MgO exhibited a slightly higher O_{lat} binding energy than MgO, which might be attributed to the incorporation of Zn^{2+} into the MgO crystal lattice [42]. Lattice oxygen on the catalyst surface serves as a Lewis basic site for transesterification [45]. Figure 5 shows the respective resolved O 1s peaks of mesoporous MgO and Zn/MgO; the surface percentage of lattice oxygen was calculated as summarized in Table 4. It is noted that the Zn/MgO catalyst contained a higher percentage of surface lattice oxygen than MgO. A higher percentage of surface lattice O^{2-} indicates a higher number of active Lewis basic sites. The increase in the percentage of surface lattice O^{2-} contrasts with the insignificant increase in the binding energy of O_{lat} (decrease in basic strength). The XPS results aligned well with the data of basicity determination. Therefore, Zn/MgO possesses a comparable basic strength but a substantially higher amount of active surface basic sites than single mesoporous MgO. Based on the results of benzoic titration and XPS analysis, it is interesting to note that there is a direct correlation between the catalytic transesterification activities and the surface O_{lat} concentration of the catalyst.

Table 3. Binding energy of Mg^{2+}, Zn^{2+}, O_{lat}, and O_{hyd} of mesoporous MgO, ZnO, and Zn/MgO calcinated at 873 K.

Catalyst	Binding Energy (eV)						
	Mg 2p	Zn $2p_{1/2}$	Zn $2p_{3/2}$	Zn LMM	Auger Parameter [1]	O_{lat}	O_{hyd}
MgO	49.3	-	-	-	-	529.5	531.2
ZnO	-	1045.2	1022.1	498.9	2009.8	531.0	532.5
Zn/MgO	49.3	1044.2	1021.1	497.7	2010.0	529.6	531.6

[1] Auger parameter = K.E. (Zn KLMM) + B.E. (Zn $2p_{3/2}$).

Table 4. Surface percentage of O_{lat} of single mesoporous MgO and Zn/MgO calcinated at 873 K.

Catalyst	Surface Percentage of O_{lat} (at.%)
Mesoporous MgO	23.5
Zn/MgO	32.8

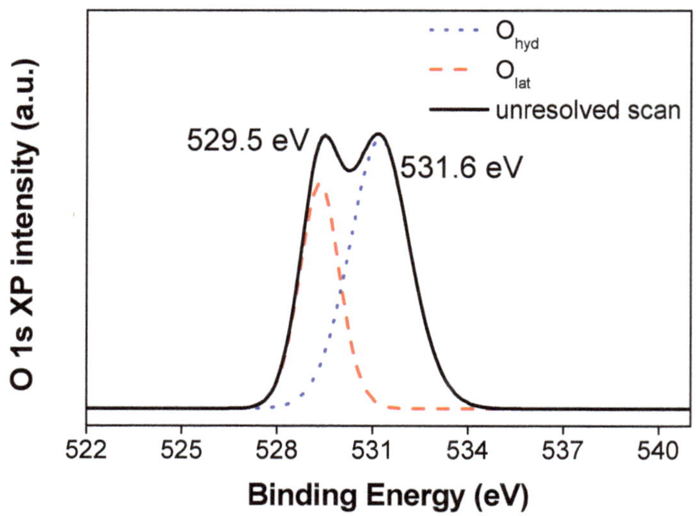

(a)

Figure 5. *Cont.*

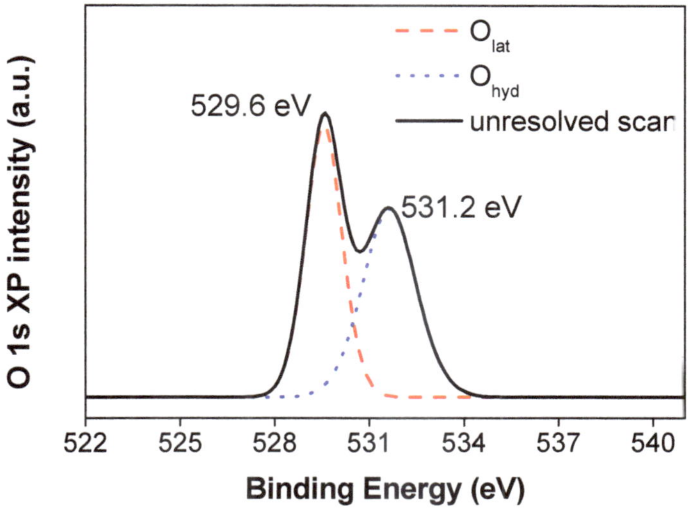

(b)

Figure 5. O1s XPS peaks of (**a**) mesoporous MgO and (**b**) Zn/MgO nano-plate catalyst calcinated at 873 K.

3.2. Catalytic Study

Catalytic activity and reusability of Zn/MgO nano-plates catalyst were investigated and compared with those of mesoporous MgO, commercial MgO, and ZnO, as depicted in Figure 6. Among all catalysts, Zn/MgO demonstrated the highest catalytic activity and stability, with high biodiesel yields of 88.7% and 86.7% for the first and second run, respectively. Commercial ZnO gave a 78.0% biodiesel yield in the first run, but its catalytic activity decreased drastically to 59.3% in the second reaction cycle. Catalyst deactivation might be due to the leaching of the active species into the reaction medium. Therefore, the high biodiesel yield of ZnO in the first cycle is likely to be due to a homogeneous pathway. Commercial and mesoporous MgO are quite stable in transesterification. Commercial MgO possessed a low BET surface area of 9.0 m^2 g^{-1}, with a relatively low number of active sites available for the catalytic reaction. Mesoporous MgO showed a higher BET surface area (73.1 m^2 g^{-1}) and possesses a mesoporous nature that may favor mass transfer and hence the catalytic transesterification process, leading to a higher biodiesel yield than commercial MgO. Zn/MgO was found to possess a comparable basic strength but a substantially higher number of active surface basic sites than single mesoporous MgO, as indicated by its higher total basicity determined from benzoic titration and an increased percentage of lattice O^{2-}, as shown in the XPS analysis. The enhanced catalytic activity of Zn/MgO compared to single mesoporous MgO is attributed to the substitution of Zn^{2+} into the MgO crystal lattice, causing lattice distortion with defects that might occur at corners, edges, and steps, which creates additional sites with ion pairs of low coordination numbers and provides a large active surface for substrates to bind. In addition, threefold-Mg^{2+}–threefold-O^{2-} ($Mg^{2+}{}_{3c}$–$O^{2-}{}_{3c}$) was reported to be the most reactive, as it is the most coordinatively unsaturated [38]. In addition, such defects could trap electrons on the surface of crystallites [46]. These might enhance the basicity and activity of the Zn/MgO catalyst. As discussed previously, the surface termination on the lateral side of hexagonal plates of Zn/MgO predominately exposes (111) and (200) facets, with the generation of low coordination defect sites [35,37,38]. The high surface energy polar (111) crystal plane comprises alternating monolayers of cations and anions, which leads to the creation of a strong electrostatic field perpendicular to the polar plane [29] and might enhance the interaction between reactant molecules with the surface. It is concluded that the results demonstrated a high biodiesel yield compared to single mesoporous MgO and commercial MgO and ZnO, proving that the successful incorporation of Zn^{2+} dopant into MgO crystal lattice creates a synergistic effect, attributed to the high catalytic activity of the Zn/MgO catalytic system.

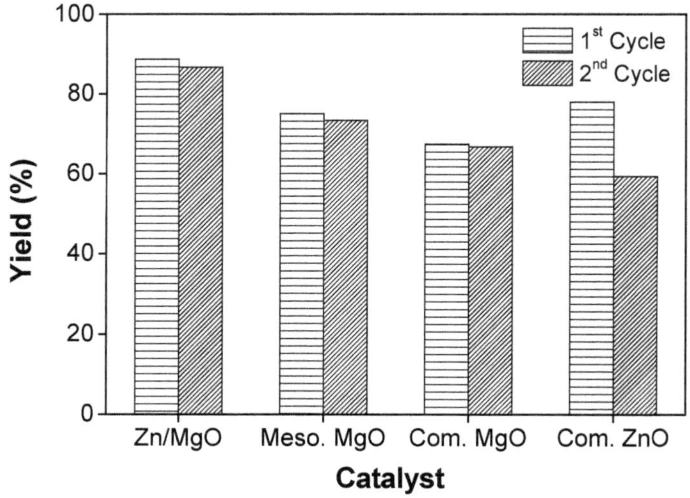

Figure 6. A comparison of catalytic activity and reusability of Zn/MgO with mesoporous MgO, commercial MgO, and ZnO (all calcinated at 873 K). Reaction conditions: catalyst (3% *w*/*w*), MeOH-to-oil molar ratio (24:1), reaction temperature (393 K), and reaction time (8 h).

The effect of the calcination temperature on biodiesel conversion by Zn/MgO was studied, and the results are depicted in Figure 7. Zn/MgO-773 K gave the lowest biodiesel yield of 79.3%. Though the catalyst exhibited the largest BET surface area of 151.1 $m^2\ g^{-1}$ as compared to those annealed at higher calcination temperature, its comparatively low average pore diameter (5.4 nm) restricted the accommodation of the bulky triglyceride moiety that possesses a diameter of approximately 5.8 nm [47]. Zn/MgO-873 K possessed a BET surface area with a value smaller than half of Zn/MgO-773 K; however, its larger pore diameter (10.4 nm) favored the accessibility of the triglyceride substrate to the active sites, resulting in an increased biodiesel yield from 79.3% to 88.7%. We observed a slight drop in biodiesel yield with a further increase in the calcination temperature because the decrease in BET surface area counteracted the effect of the increase in pore size. Apart from the textural properties, the surface basicity of the catalysts also significantly contributed to the catalytic activity. All the Zn/MgO catalysts calcinated at various temperatures exhibited an equal surface basic strength of $9.3 < H\_ < 10.1$; however, the total basicity was different, with the values of 1.07, 1.52, 1.42, and 1.47 mmol g^{-1} for Zn/MgO calcinated at 773 K, 873 K, 973 K, and 1073 K, respectively. These results align quite well with the trend of the respective catalytic performances in transesterification.

Catalyst reusability is one of the essential features determining the economic viability for the commercialization of an industrial process. Zn/MgO-873 K, which possessed the highest catalytic activity, was therefore subjected to reusability studies. It was found that the catalytic system had high stability in transesterification, with the biodiesel yield remaining over 84.0% in five catalytic runs, as shown in Figure 8. This high catalytic stability is attributed to the defects induced by the incorporation of Zn^{2+} into the MgO crystal lattice [47]. The reusability of the as-synthesized catalyst makes it advantageous over the homogenous counterpart, as it could lower the overall production cost. To the best of our knowledge, several studies reported the use of MgO–ZnO catalysts for biodiesel synthesis; however, harsh reaction conditions were applied. Olutoye et al. [47] employed $Mg_{1-x}Zn_{1+x}O_2$ in the transesterification of used vegetable cooking oil, with the highest

FAME yield of 78% achieved at a high reaction temperature of 461 K. Lee et al. [48] studied the activity of a series of Mg–Zn mixed metal oxides in the transesterification of *Jatropha* oil. The highest biodiesel yield of 83% was achieved for MgO–ZnO at a Mg/Zn atomic ratio of 8 (MZ8), with 3% catalyst dosage and a 24:1 methanol-to-oil molar ratio at 393 K for 3 h. MZ8 exhibited a consistent decrease in activity from 83% to 63% in five consecutive runs of reaction, with an activity retention of 76%. In comparison, our catalysts demonstrated a higher catalytic activity and stability towards transesterification, with the biodiesel yield decreased slightly from 89.7% to 84.0% for Zn/MgO-873 K. An activity retention of 94% was obtained after five successive catalytic runs for Zn/MgO-873 K.

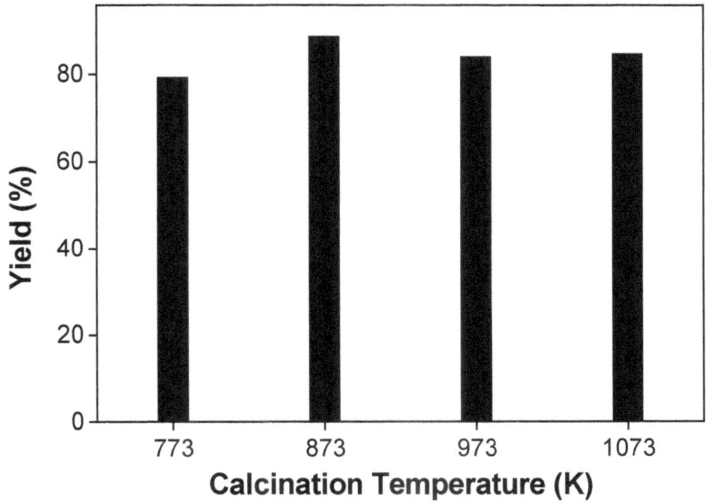

Figure 7. Biodiesel yield of Zn/MgO as a function of the calcination temperature. Reaction conditions: catalyst (3% *w*/*w*), MeOH-to-oil molar ratio (24: 1), reaction temperature (393 K), and reaction time (8 h).

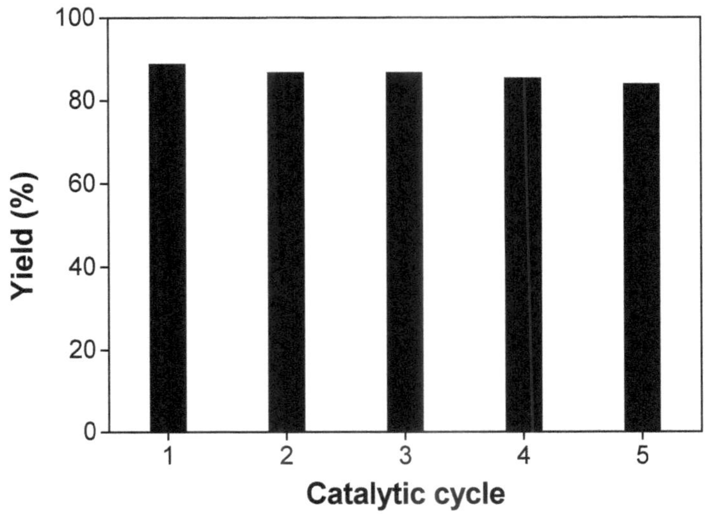

Figure 8. Reusability of Zn/MgO-600 °C. Reaction conditions: catalyst (3% *w*/*w*), MeOH-to-oil molar ratio (24:1), reaction temperature (393 K), and reaction time (8 h).

A study of the reaction extent for the as-synthesized Zn/MgO-873 K revealed a slow transesterification with low biodiesel yield at the beginning, which is proposed to be due to the low mass transfer limit within the three phases system. The transesterification proceeded much faster afterwards, reaching a maximum yield at 6 to 8 h. However, the biodiesel yield remained almost the same by extending the reaction time. Due to the excess of methanol used in this transesterification, the overall reaction most likely followed a pseudo-first-order reaction kinetics model, comparable to that of a similar catalytic system we reported [49].

Further development and investigation of this as-synthesized catalytic system during oil transesterification, including massive production of catalyst and large-scale biodiesel synthesis or even pilot-scale continuous flow reactions toward biodiesel synthesis, will be the subject of our future investigations. For large-scale biodiesel production, optimization and kinetic studies are important to analyze the role of each parameter (methanol-to-oil molar ratio, reaction temperature, catalyst loading, and reaction time) and pinpoint the most prominent parameter. This information will help design and adapt our synthesized catalytic system for large-scale biodiesel production. Moreover, the fuel properties of the final biodiesel product produced using the as-synthesized catalytic system in combustion engines are also important to understand whether the biodiesel fuel is associated with a lower emission of toxic gases. These experiments are now ongoing and will be discussed and published in the future.

4. Conclusions

A novel mesoporous Zn/MgO hexagonal-nano-plate catalyst was synthesized and demonstrated transesterification of *Camelina* oil, with biodiesel yield of 88.7%. This was achieved at 393 K in 8 h using 3% w/w of the catalyst with a MeOH-to-oil ratio of 24:1. Its excellent catalytic performance was mainly associated with a relatively high surface area (69.1 $m^2\ g^{-1}$), a pore size of 10.4 nm, and a high total basicity of 1.52 mmol g^{-1}, all these values being better than those measured for single MgO. The incorporation of Zn^{2+} into the MgO crystal lattice induced defects that exhibited high catalytic activity and stability for at least five catalytic cycles, with biodiesel yield above 84.0%.

Author Contributions: Conceptualization, T.-L.K. and K.-F.Y.; methodology, L.-F.M. and T.-L.K.; formal analysis, L.-F.M. and T.-L.K.; investigation, L.-F.M.; resources, W.-T.W. and K.-F.Y.; writing—original draft preparation, L.-F.M.; writing—review and editing, T.-L.K. and K.-F.Y.; visualization, T.-L.K.; supervision, W.-T.W. and K.-F.Y.; project administration, T.-L.K.; funding acquisition, K.-F.Y. All authors have read and agreed to the published version of the manuscript.

Funding: This research was funded by The Hong Kong Polytechnic University with grant numbers of PolyU 153053/17P and PolyU P0032339.

Acknowledgments: The authors are also very grateful to the University Research Facility in Chemical and Environmental Analysis (UCEA) of The Hong Kong Polytechnic University for supporting the instruments used in this research investigation.

Conflicts of Interest: The authors declare no conflict of interest.

References

1. Rahmanian, B.; Safaei, M.R.; Kazi, S.N.; Ahmadi, G.; Oztop, H.F.; Vafai, K. Investigation of pollutant reduction by simulation of turbulent non-premixed pulverized coal combustion. *Appl. Therm. Eng.* **2014**, *73*, 1222–1235. [CrossRef]
2. Sateesh, K.; Yaliwal, V.; Soudagar, M.E.M.; Banapurmath, N.; Fayaz, H.; Safaei, M.R.; Elfasakhany, A.; EL-Seesy, A.I. Utilization of biodiesel/Al_2O_3 nanoparticles for combustion behavior enhancement of a diesel engine operated on dual fuel mode. *J. Therm. Anal. Calorim.* **2021**, 1–15. [CrossRef]
3. Akkoli, K.; Banapurmath, N.; Shivashimpi, M.; Soudagar, M.E.M.; Badruddin, I.A.; Alazwari, M.A.; Yaliwal, V.; Mujtaba, M.; Akram, N.; Goodarzi, M. Effect of injection parameters and producer gas derived from redgram stalk on the performance and emission characteristics of a diesel engine. *Alex. Eng. J.* **2021**, *60*, 3133–3142. [CrossRef]
4. Khan, H.; Soudagar, M.E.M.; Kumar, R.H.; Safaei, M.R.; Farooq, M.; Khidmatgar, A.; Banapurmath, N.R.; Farade, R.A.; Abbas, M.M.; Afzal, A. Effect of nano-graphene oxide and n-butanol fuel additives blended with diesel—Nigella sativa biodiesel fuel emulsion on diesel engine characteristics. *Symmetry* **2020**, *12*, 961. [CrossRef]

5. Gavhane, R.S.; Kate, A.M.; Pawar, A.; Safaei, M.R.; Soudagar, M.E.; Mujtaba Abbas, M.; Muhammad Ali, H.; Banapurmath, N.R.; Goodarzi, M.; Badruddin, I.A. Effect of zinc oxide nano-additives and soybean biodiesel at varying loads and compression ratios on VCR diesel engine characteristics. *Symmetry* **2020**, *12*, 1042. [CrossRef]
6. Sadeghinezhad, E.; Kazi, S.; Sadeghinejad, F.; Badarudin, A.; Mehrali, M.; Sadri, R.; Safaei, M.R. A comprehensive literature review of bio-fuel performance in internal combustion engine and relevant costs involvement. *Renew. Sustain. Energy Rev.* **2014**, *30*, 29–44. [CrossRef]
7. Hajjari, M.; Tabatabaei, M.; Aghbashlo, M.; Ghanavati, H. A review on the prospects of sustainable biodiesel production: A global scenario with an emphasis on waste-oil biodiesel utilization. *Renew. Sustain Energy Rev.* **2017**, *72*, 445–464. [CrossRef]
8. Wategave, S.; Banapurmath, N.; Sawant, M.; Soudagar, M.E.M.; Mujtaba, M.; Afzal, A.; Basha, J.S.; Alazwari, M.A.; Safaei, M.R.; Elfasakhany, A. Clean combustion and emissions strategy using reactivity controlled compression ignition (RCCI) mode engine powered with CNG-Karanja biodiesel. *J. Taiwan Inst. Chem. Eng.* **2021**, *124*, 116–131. [CrossRef]
9. Banapurmath, N.; Chandrashekar, T.; Soudagar, M.E.M.; Anqi, A.E.; Mujtaba, M.; Goodarzi, M.; Elfasakhany, A.; Siddiqui, M.I.H.; Ali, M.A. Effect of Parameters Behavior of Simarouba Methyl Ester Operated Diesel Engine. *Energies* **2021**, *14*, 4973.
10. Bansal, S.; Durrett, T.P. Camelina sativa: An ideal platform for the metabolic engineering and field production of industrial lipids. *Biochimie* **2016**, *120*, 9–16. [CrossRef]
11. Bernardo, A.; Howard-Hildige, R.; O'Connell, A.; Nichol, R.; Ryan, J.; Rice, B.; Roche, E.; Leahy, J.J. Camelina oil as a fuel for diesel transport engines. *Ind. Crops Prod.* **2003**, *17*, 191–197. [CrossRef]
12. Patil, P.D.; Gude, V.G.; Deng, S. Biodiesel Production from Jatropha Curcas, Waste Cooking, and Camelina Sativa Oils. *Ind. Eng. Chem. Res.* **2009**, *48*, 10850–10856. [CrossRef]
13. Thangaraj, B.; Solomon, P.R.; Muniyandi, B.; Ranganathan, S.; Lin, L. Catalysis in biodiesel production—A review. *Clean Energy* **2019**, *3*, 2–23. [CrossRef]
14. Almerindo, G.I.; Buratto, S.C.; Wanderlind, E.H.; Nicolazi, L.M.; Sangaletti, P.; Medeiros, M.; Schneider, F.S.; Caramori, G.F.; Parreira, R.L.; Micke, G.A. Kinetics and adsorption calculations: Insights into the MgO-catalyzed detoxification of simulants of organophosphorus biocides. *J. Mater. Chem. A* **2020**, *8*, 19011–19021. [CrossRef]
15. Montero, J.; Isaacs, M.; Lee, A.; Lynam, J.; Wilson, K. The surface chemistry of nanocrystalline MgO catalysts for FAME production: An in situ XPS study of H_2O, CH_3OH and CH_3OAc adsorption. *Surf. Sci.* **2016**, *646*, 170–178. [CrossRef]
16. Poosumas, J.; Ngaosuwan, K.; Quitain, A.T.; Assabumrungrat, S. Role of ultrasonic irradiation on transesterification of palm oil using calcium oxide as a solid base catalyst. *Energy Convers. Manag.* **2016**, *120*, 62–70. [CrossRef]
17. Qu, T.; Niu, S.; Gong, Z.; Han, K.; Wang, Y.; Lu, C. Wollastonite decorated with calcium oxide as heterogeneous transesterification catalyst for biodiesel production: Optimized by response surface methodology. *Renew. Energy* **2020**, *159*, 873–884. [CrossRef]
18. Hu, S.; Zhou, Y.; Yuan, C.; Wang, W.; Hu, J.; Li, Q.; He, J. Surface-modification effect of MgO nanoparticles on the electrical properties of polypropylene nanocomposite. *High Volt.* **2020**, *5*, 249–255. [CrossRef]
19. Qasim, M.K. Modified nanostructure MgO superbasicity with CaO in heterogeneous transesterification of sunflower oil. *Egypt. J. Chem.* **2019**, *62*, 475–485. [CrossRef]
20. Foroutan, R.; Peighambardoust, S.J.; Mohammadi, R.; Ramavandi, B.; Boffito, D.C. One-pot transesterification of non-edible Moringa oleifera oil over a $MgO/K_2CO_3/HAp$ catalyst derived from poultry skeletal waste. *Environ. Technol. Innov.* **2021**, *21*, 101250. [CrossRef]
21. Barzegari, F.; Farhadi, F.; Rezaei, M.; Kazemeini, M.; Keshavarz, A. Influence of metal loading and reduction temperature on the performance of mesoporous NiO–MgO–SiO_2 catalyst in propane steam reforming. *J. Energy Inst.* **2021**, *96*, 38–51. [CrossRef]
22. Manríquez-Ramírez, M.E.; Elizalde, I.; Ortiz-Islas, E. Synthesis of MgO and MgO–CeO_2 by co-precipitation for the catalytic conversion of acetone by aldol condensation. *React. Kinet. Mech. Catal.* **2020**, *131*, 769–780. [CrossRef]
23. Wen, Z.; Yu, X.; Tu, S.-T.; Yan, J.; Dahlquist, E. Synthesis of biodiesel from vegetable oil with methanol catalyzed by Li-doped magnesium oxide catalysts. *Appl. Energy* **2010**, *87*, 743–748. [CrossRef]
24. Ab Rahman, N.A.; Olutoye, M.A.; Hameed, B.H. Synthesis of methyl esters from palm (Elaeis guineensis) oil using cobalt doped MgO as solid oxide catalyst. *Bioresour. Technol.* **2011**, *102*, 9749–9754. [CrossRef] [PubMed]
25. Mahraz, Z.A.S.; Sahar, M.; Ghoshal, S.; Saad, A.P.M.; Syahrom, A. Sol–gel grown MgO-ZnO-tricalcium-phosphate nanobioceramics: Evaluation of mechanical and degradation attributes. *Corros. Sci.* **2018**, *138*, 179–188. [CrossRef]
26. Negrea, R.; Busuioc, C.; Constantinoiu, I.; Miu, D.; Enache, C.; Iordache, F.; Jinga, S.-I. Akermanite-based coatings grown by pulsed laser deposition for metallic implants employed in orthopaedics. *Surf. Coat. Technol.* **2019**, *357*, 1015–1026. [CrossRef]
27. Evstropiev, S.; Soshnikov, I.; Kolobkova, E.; Evstropyev, K.; Nikonorov, N.; Khrebtov, A.; Dukelskii, K.; Kotlyar, K.; Oreshkina, K.; Nashekin, A. Polymer-salt synthesis and characterization of MgO-ZnO ceramic coatings with the high transparency in UV spectral range. *Opt. Mater.* **2018**, *82*, 81–87. [CrossRef]
28. Reinoso, D.M.; Damiani, D.E.; Tonetto, G.M. Efficient production of biodiesel from low-cost feedstock using zinc oleate as catalyst. *Fuel Process. Technol.* **2015**, *134*, 26–31. [CrossRef]
29. Kwong, T.-L.; Yung, K.-F. One-step production of biodiesel through simultaneous esterification and transesterification from highly acidic unrefined feedstock over efficient and recyclable ZnO nanostar catalyst. *Renew. Energy* **2016**, *90*, 450–457. [CrossRef]
30. Lau, P.-C.; Kwong, T.-L.; Yung, K.-F. Effective heterogeneous transition metal glycerolates catalysts for one-step biodiesel production from low grade non-refined Jatropha oil and crude aqueous bioethanol. *Sci. Rep.* **2016**, *6*, 23822. [CrossRef]

31. Yadav, M.; Singh, V.; Sharma, Y.C. Methyl transesterification of waste cooking oil using a laboratory synthesized reusable heterogeneous base catalyst: Process optimization and homogeneity study of catalyst. *Energy Convers. Manag.* **2017**, *148*, 1438–1452. [CrossRef]
32. Shannon, R. Revised effective ionic radii and systematic studies of interatomic distances in halides and chalcogenides. *Acta Crystallogr. Sect. A* **1976**, *32*, 751–767. [CrossRef]
33. Ohtomo, A.; Kawasaki, M.; Koida, T.; Masubuchi, K.; Koinuma, H.; Sakurai, Y.; Yoshida, Y.; Yasuda, T.; Segawa, Y. $Mg_xZn_{1-x}O$ as a II–VI widegap semiconductor alloy. *Appl. Phys. Lett.* **1998**, *72*, 2466–2468. [CrossRef]
34. Park, W.I.; Yi, G.-C.; Jang, H.M. Metalorganic vapor-phase epitaxial growth and photoluminescent properties of $Zn_{1-x}Mg_xO$ ($0 \leq x \leq 0.49$) thin films. *Appl. Phys. Lett.* **2001**, *79*, 2022–2024. [CrossRef]
35. Berger, T.; Schuh, J.; Sterrer, M.; Diwald, O.; Knözinger, E. Lithium ion induced surface reactivity changes on MgO nanoparticles. *J. Catal.* **2007**, *247*, 61–67. [CrossRef]
36. Ling, Z.; Zheng, M.; Du, Q.; Wang, Y.; Song, J.; Dai, W.; Zhang, L.; Ji, G.; Cao, J. Synthesis of mesoporous MgO nanoplate by an easy solvothermal–annealing method. *Solid State Sci.* **2011**, *13*, 2073–2079. [CrossRef]
37. Montero, J.M.; Brown, D.R.; Gai, P.L.; Lee, A.F.; Wilson, K. In situ studies of structure–reactivity relations in biodiesel synthesis over nanocrystalline MgO. *Chem. Eng. J.* **2010**, *161*, 332–339. [CrossRef]
38. Montero, J.M.; Gai, P.; Wilson, K.; Lee, A.F. Structure-sensitive biodiesel synthesis over MgO nanocrystals. *Green Chem.* **2009**, *11*, 265–268. [CrossRef]
39. Yu, X.; Wen, Z.; Li, H.; Tu, S.-T.; Yan, J. Transesterification of Pistacia chinensis oil for biodiesel catalyzed by CaO–CeO_2 mixed oxides. *Fuel* **2011**, *90*, 1868–1874. [CrossRef]
40. Li, W.-C.; Lu, A.-H.; Weidenthaler, C.; Schüth, F. Hard-Templating Pathway To Create Mesoporous Magnesium Oxide. *Chem. Mater.* **2004**, *16*, 5676–5681. [CrossRef]
41. Wang, G.; Zhang, L.; Dai, H.; Deng, J.; Liu, C.; He, H.; Au, C.T. P123-Assisted Hydrothermal Synthesis and Characterization of Rectangular Parallelepiped and Hexagonal Prism Single-Crystalline MgO with Three-Dimensional Wormholelike Mesopores. *Inorg. Chem.* **2008**, *47*, 4015–4022. [CrossRef]
42. Etacheri, V.; Roshan, R.; Kumar, V. Mg-Doped ZnO Nanoparticles for Efficient Sunlight-Driven Photocatalysis. *ACS Appl. Mater. Interfaces* **2012**, *4*, 2717–2725. [CrossRef]
43. Hattori, H. Heterogeneous Basic Catalysis. *Chem. Rev.* **1995**, *95*, 537–558. [CrossRef]
44. Rheinheimer, V.; Unluer, C.; Liu, J.; Ruan, S.; Pan, J.; Monteiro, P.J. XPS study on the stability and transformation of hydrate and carbonate phases within MgO systems. *Materials* **2017**, *10*, 75. [CrossRef] [PubMed]
45. Gerpen, J.V. Biodiesel processing and production. *Fuel Process. Technol.* **2005**, *86*, 1097–1107. [CrossRef]
46. Chiesa, M.; Paganini, M.C.; Giamello, E.; Di Valentin, C.; Pacchioni, G. First Evidence of a Single-Ion Electron Trap at the Surface of an Ionic Oxide. *Angew. Chem. Int. Ed.* **2003**, *42*, 1759–1761. [CrossRef] [PubMed]
47. Olutoye, M.; Hameed, B. Synthesis of fatty acid methyl ester from used vegetable cooking oil by solid reusable $Mg_{1-x}Zn_{1+x}O_2$ catalyst. *Bioresour. Technol.* **2011**, *102*, 3819–3826. [CrossRef] [PubMed]
48. Lee, H.; Taufiq-Yap, Y.; Hussein, M.; Yunus, R. Transesterification of jatropha oil with methanol over Mg–Zn mixed metal oxide catalysts. *Energy* **2013**, *49*, 12–18. [CrossRef]
49. Kwong, T.-L.; Yung, K.-F. Heterogeneous alkaline earth metal–transition metal bimetallic catalysts for synthesis of biodiesel from low grade unrefined feedstock. *RSC Adv.* **2015**, *5*, 83748–83756. [CrossRef]

nanomaterials

Article

Enhanced Hole Injection Characteristics of a Top Emission Organic Light-Emitting Diode with Pure Aluminum Anode

Chan Young Park and Byoungdeog Choi *

Department of Electrical and Computer Engineering, Sungkyunkwan University, 2066 Seobu-Ro, Jangan-Gu, Suwon 16419, Korea; parkcy@skku.edu
* Correspondence: bdchoi@skku.edu

Abstract: A top emitting organic light-emitting diode (OLED) device with pure aluminum (Al) anode for high-resolution microdisplays was proposed and fabricated. The low work function of the Al anode, even with a native oxide formed on the Al anode surface, increases the energy barrier of the interface between the anode and hole injection layer and has poor hole-injection properties, which causes the low efficiency of the device. To enhance the hole-injection characteristics of the Al anode, we applied hexaazatriphenylene hexacarbonitrile (HATCN) as the hole-injection layer material. The proposed OLED device with a pure Al anode and native oxide on the anode surface improved efficiency by up to 35 cd/A at 1000 nit, which is 78% of the level of normal OLEDs with indium tin oxide (ITO) anode.

Keywords: OLED; top emitting; hole injection; Al anode; ITO anode; HATCN; efficiency; native oxide; microdisplay; Si substrate

Citation: Park, C.Y.; Choi, B. Enhanced Hole Injection Characteristics of a Top Emission Organic Light-Emitting Diode with Pure Aluminum Anode. *Nanomaterials* **2021**, *11*, 2869. https://doi.org/10.3390/nano11112869

Academic Editor: Baoquan Sun

Received: 30 September 2021
Accepted: 24 October 2021
Published: 27 October 2021

1. Introduction

Organic light-emitting diodes (OLEDs) are currently applied to display panels for smart phones, tablets, and TVs. The characteristics of OLEDs, such as their being self-emitting, and having a thin form factor, vivid color, and fast response time, attract attention for use in display panels and various other applications. However, OLEDs have a limited resolution caused by the fabrication method of pixel-patterning using a fine metal mask (FMM), which is a shadow mask with tiny holes for patterning pixels from evaporated organic materials. The resolution of an OLED display panel using FMM is approximately 800 ppi at present [1]. For a microdisplay, which has high resolution of over 2000 ppi, applied to Virtual Reality (VR) devices and Augmented Reality (AR) devices, the FMM fabrication method is unavailable because the hole size of FMM is smaller than 10 um. Another approach was used to fabricate OLED microdisplays, involving a separated backplane and OLED panel [2,3]. The backplane driving each pixel is made of a single crystal silicon wafer substrate with a complementary metal-oxide-semiconductor (CMOS) process at the semiconductor foundry. OLEDs are fabricated on the wafer backplane using the evaporation process. Instead of separated RGB pixels, white OLED-stacked RGB organic layers, and RGB color filters corresponding to each pixel, are used for color patterning OLED microdisplays [3].

Since the wafer backplane is not transparent, as opposed to the glass substrate normally used in the display device process, a top emitting OLED structure has to be applied to OLED microdisplays [4]. In top emitting OLEDs, the anode materials should be highly reflective in the visible range to achieve a high luminance efficiency, and should have a high work function for the efficient injection of holes. Many previous studies have been carried out on the anode for top emitting OLEDs. Silver (Ag) has a high reflectivity of 94% at wavelengths of 520 nm and a low electrical resistivity of 1.47 μΩ cm at a temperature of 298 K [5]. An Ag/indium tin oxide (ITO) composite bi-layer structure combined the advantages of both the high reflectivity of Ag and the high work function of ITO [6,7].

Other materials have been studied for use as the anode in top emitting OLEDs such as aluminum (Al)/ITO [8], Ag/Al [9], gold (Au) [10], molybdenum (Mo) [11], platinum (Pt) [12], and vanadium (V) [13]. However, despite their good electrical and optical characteristics, the materials mentioned above are not supported by the current general CMOS foundry. Al, titanium (Ti), and Tungsten (W) are available as the anode for top emitting OLEDs at the general CMOS foundry. Some approaches have been studied that use available materials as the anode of top emitting OLEDs, supported by a general CMOS foundry, such as TiN, Al/TiN bi-layer [14–16]. However, top emitting OLEDs with Ti and TiN anodes have limited efficiency due to the low reflectivity of the anode. Other approaches have been studied, applying Al as the anode of top emitting OLEDs using oxide buffer materials [17] or inverted OLED structures [18,19]. Al has a high reflectance that leads to highly efficient top emitting OLEDs. However, Al has a low work function of 4.2 eV compared to ITO, with a work function of 4.7 eV, which is normally used as the OLEDs anode. The energy barrier between the low-work-function anode and hole-injection layer (HIL) in OLEDs is higher, so poor hole-injection characteristics decrease the efficiency of OLED devices. In addition, the native oxide (Al_2O_3) formed on the Al anode surface, decreases efficiency.

In this study, we proposed a top emitting OLED device structure with an Al anode. To enhance the hole-injection characteristics of the Al anode, we applied hexaazatriphenylene hexacarbonitrile (HATCN) as the HIL material. The fabricated top emitting OLED devices, with enhanced hole-injection characteristics, improved the efficiency of normal OLEDs with an ITO anode by 78%.

2. Materials and Methods

Figure 1 presents the schematic structure of the device. In the OLED microdisplay, the backplane fabrication process and OLED fabrication process are separated. Therefore, the backplane was fabricated at the general CMOS foundry until the Al anode layer formed. The OLED was fabricated at a different site, with the Al anode formed backplane, from HIL to capping layer (CPL) using the thermal evaporation method. During the transport from the foundry to OLED evaporation site, the Al anode had to be exposed to air, causing native oxide on the surface of Al anode. We used 8-inch (200 mm) silicon wafers as substrate at DB HiTek (Bucheon, Korea). Al of 150 nm thickness was deposited on the substrate using the sputter method for a reflective anode. We stored the Al-deposited substrate in a clean room for one day to purposefully form native oxide, assuming an extreme case. We then deposited a HIL, hole-transport layer (HTL), green emission layer (EML), electron-transport layer (ETL), cathode, and CPL in sequence using the thermal vacuum evaporation method. We used hexaazatriphenylene hexacarbonitrile (HATCN) of 30 nm thickness as the HIL, N′-bis(phenyl)benzidine (NPB) of 20 nm thickness as the HTL. GGH1 (Gracel, Korea) of 16 nm thickness was used as green EML and hydroxyquinolatolithium (Liq) of 20 nm of thickness was used as ETL. Mg:Ag of 13 nm thickness was used as the semi-transparent cathode. Most materials used in the devices were provided by Sigma-Aldrich (St. Louis, MO, USA) as commercial-grade powder sources, except GGH1, from Cracel (Seoul, Korea). The total thickness of the OLED device between the reflective anode and semi-transparent cathode was decided as 86 nm as in (1) to improve the efficiency using a micro-cavity effect [20]. The thickness of the layers of the devices was determined in situ by the crystal monitor in the vacuum chamber. We also fabricated an ITO anode OLED device for reference to compare the characteristics of OLEDs between Al anode and ITO anode

$$I_{\mathrm{cav}}(\lambda) = \frac{(1-R_2)\left[1+R_1+2(R_1)^{0.5}\cos\left(\frac{4\pi d}{\lambda}\right)\right]}{1+R_1R_2-2(R_1R_2)^{0.5}\cos\left(\frac{4\pi L}{\lambda}\right)} \times I_{\mathrm{nc}}(\lambda) \tag{1}$$

where I_{cav} is the spectrum intensity, R_1 and R_2 are the reflectivities of the anode and cathode mirrors, respectively, d is the effective distance of the emitting layer from the anode, L is the total optical thickness of the cavity, and I_{nc} is the free space electroluminescence intensity at wavelength λ.

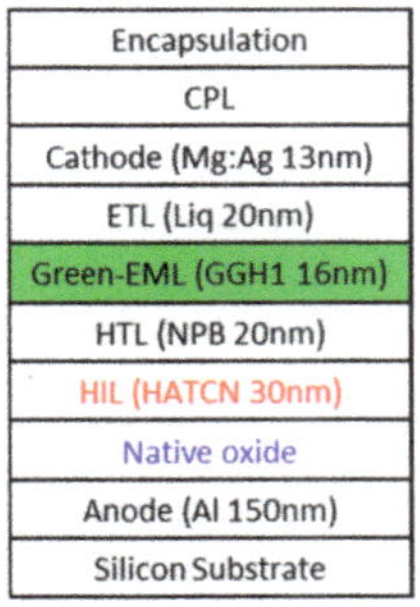

Figure 1. Schematic diagram of proposed structure of green top emitting OLED device.

Measurements of OLED properties were performed by recording current–voltage characteristics as well as electro-luminescence (EL) spectra. The current–voltage was acquired using a Keithley 236 voltage source unit, while the EL intensity, luminance and spectral characteristics of the devices were measured with a calibrated silicon photodiode (Hamamatsu Photonics, Hamamatsu, Japan, S5227-1010BQ), a photomultiplier tube, and a spectroradiometer (Minolta, Osaka, Japan, CS-1000).

3. Results and Discussion

The characteristics of the OLED device with an Al anode are presented in Table 1. The current efficiency of the OLED device is 35 cd/A at 1000 nit, which is 78% of the conventional OLED device with an ITO anode. Usually, the efficiency of OLEDs with an Al anode is below 50% of OLEDs with an ITO anode [21]. The proposed OLED device with an Al anode has high efficiency, even though native oxide was formed on the Al anode. This means that hole injection characteristics of the proposed OLED device are enhanced by Al anode and HATCN HIL, even though an energy barrier exists between the Al anode and HATCN HIL.

Table 1. Measured characteristics of green top emitting OLED devices with Al anode and ITO anode.

Parameters	Device with Al Anode	Device with ITO Anode	Notes
Current efficiency (cd/A)	35	45	@1000 nit
Driving voltage (V)	4.2	3.8	@1000 nit
Peak wavelength (nm)	525	525	
CIE color coordination	(0.22, 0.71)	(0.23, 0.71)	

Figure 2 shows the optical and electrical characteristics of the OLED devices with an Al anode. In the spectrum of OLED devices, the main peak in the spectrum, at 525 nm, was not changed by the anode materials. The electroluminescence spectrum of the OLED devices is determined by the thickness of the OLED device. Therefore, the spectrum is not related to the hole-injection characteristics of the OLED device. The driving voltage of the OLED device with an Al anode was 4.2 V, an increase of 0.4 V compared to the OLED device with an ITO anode at 10 mA/cm^2. The increase in voltage is caused by native oxide in the Al anode, which increases the resistance of the interface between the anode and HIL. The power consumption of the OLED device with an Al anode can be increased slightly, but this does not matter when operating an OLED microdisplay panel with an Al anode. The current efficiency of the OLED device with an Al anode was 35 cd/A at 1000 nit, which is 78% of the current efficiency of the OLED device with an ITO anode at 44 cd/A.

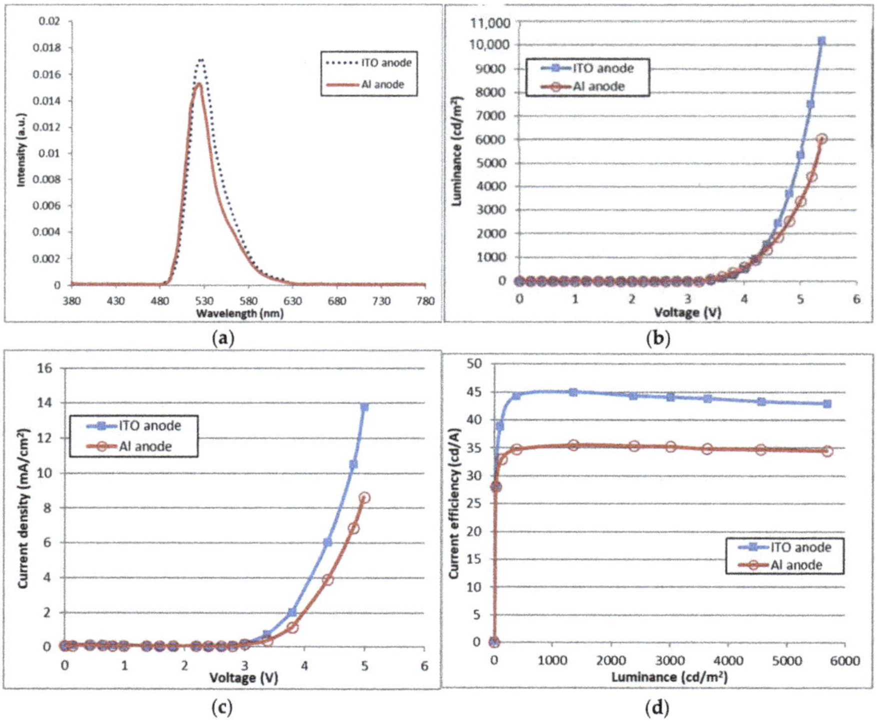

Figure 2. Characteristics of OLED devices with Al anode and HATCN HIL: (**a**) Electroluminescence spectrum of the OLED devices; (**b**) Characteristics of luminance for different voltages; (**c**) Characteristics of current density for different voltages; and (**d**) Characteristics of current efficiency for difference luminance.

Figure 3 shows the energy band diagram of each of the layers comprising the proposed OLED device. As shown in Figure 3, there is a high energy barrier between the Al anode and HIL, so it is hard to inject holes from the Al anode, with a low work function of 4.2 eV, to the HATCN HIL at 9.5 eV, which is the highest occupied molecular orbital (HOMO). However, the proposed OLED device shows high efficiency with the low-work-function Al anode, even with native oxide on the surface of the Al anode. This can be explained by the charge generation at the HATCN/NPB interface and the high electron-withdrawing properties of HATCN to the anode [22–24]. HATCN has good electron-transport characteristics. At the interface of HATCN and NPB, the energy level difference between the lowest occupied molecular orbital (LUMO) of HATCN and HOMO of NPB is small enough for electrons in the HOMO of NPB to move to the LUMO of HATCN. Moreover, while the electron-holed pairs are generated at the interface of HATCN and NPB, HATCN constantly transports electrons to the Al anode. Since the electrons are effectively transported from HATCN to the Al anode, the hole-injection property is enhanced. Therefore, HATCN improved hole-injection characteristics from the Al anode to HIL in the proposed OLED structure.

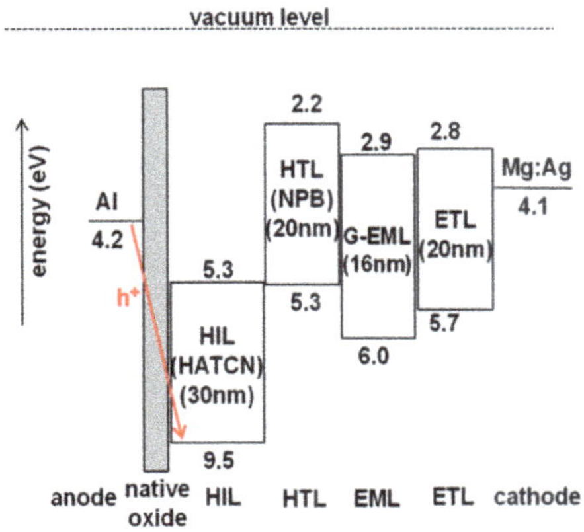

Figure 3. Energy band diagram of the proposed OLED device with Al anode and HATCN HIL.

4. Conclusions

We investigated the characteristics of OLED devices with a reflective Al anode and fabricated a high-efficiency, green top emitting OLED device with an Al anode by improving the hole-injection characteristics. We obtained an efficiency of 35 cd/A at 1000 nit from the OLED with an Al anode, which is 78% of a normal OLED with an ITO anode. Using HATCN as the HIL improved the hole-injection characteristics of the low-work-function Al anode, even though native oxide had formed on the Al anode surface, which increased the energy barrier of the interface between the Al anode and HIL. This result indicates that a high-efficiency OLED microdisplay using an Al anode for high-resolution patterning can be fabricated using the enhanced hole-injection structure of an OLED device. Thus, we believe that this study can provide a simple, practical, and low-cost method for improving the performance of OLED microdisplay products in the current foundry infrastructure.

Author Contributions: Conceptualization, C.Y.P.; B.C. methodology, C.Y.P. investigation, C.Y.P. writing, C.Y.P. supervision, B.C. funding acquisition, B.C. All authors have read and agreed to the published version of the manuscript.

Funding: This research was funded and conducted under 「the Competency Development Program for Industry Specialists of the Korean Ministry 」 of Trade, Industry and Energy (MOTIE), operated by Korea Institute for Advancement of Technology (KIAT). (No. P0012453, Next-generation Display Expert Training Project for Innovation Process and Equipment, Materials Engineers).

Conflicts of Interest: The authors declare no conflict of interest.

References

1. Kim, C.; Kim, K.; Kwon, O.; Jung, J.; Park, J.K.; Kim, D.H.; Jung, K. Fine metal mask material and manufacturing process for high-resolution active-matrix organic light-emitting diode displays. *J. Soc. Inf. Disp.* **2020**, *28*, 668–679. [CrossRef]
2. Vogel, U.; Kreye, D.; Richter, B.; Bunk, G.; Reckziegel, S.; Herold, R.; Scholles, M.; Törker, M.; Grillberger, C.; Amelung, J.; et al. 8.2: Bi-Directional OLED Microdisplay for Interactive HMD. *SID Symp. Dig. Tech. Pap.* **2008**, *39*, 81–84. [CrossRef]
3. Levy, G.; Evans, W.; Ebner, J.; Farrell, P.; Hufford, M.; Allison, B.; Wheeler, D.; Lin, H.; Prache, O.; Naviasky, E. An 852×600 pixel OLED-on-silicon color microdisplay using CMOS subthreshold-voltage-scaling current drivers. *IEEE J. Solid State Circuits* **2002**, *37*, 1879–1889. [CrossRef]
4. Chen, S.; Deng, L.; Xie, J.; Peng, L.; Xie, L.; Fan, Q.; Huang, W. Recent Developments in Top-Emitting Organic Light-Emitting Diodes. *Adv. Mater.* **2010**, *22*, 5227–5239. [CrossRef]
5. Peng, H.J.; Zhu, X.L.; Yu, X.M.; Wong, M.; Sun, J.X.; Kwok, H.S. Efficiency improvement of phosphorescent organic light-emitting diodes using semitransparent Ag as anode. *Appl. Phys. Lett.* **2006**, *88*, 033509. [CrossRef]
6. Lu, M.-H.; Weaver, M.S.; Zhou, T.X.; Rothman, M.A.; Kwong, R.C.; Hack, M.; Brown, J.J. High-Efficiency top-emitting organic light-emitting devices. *Appl. Phys. Lett.* **2002**, *81*, 3921–3923. [CrossRef]
7. Park, M.J.; Kim, S.K.; Pode, R.; Kwon, J.H. Low absorption semi-transparent cathode for micro-cavity top-emitting organic light emitting diodes. *Org. Electron.* **2018**, *52*, 153–158. [CrossRef]

8. Jean, F.; Mulot, J.-Y.; Geffroy, B.; Denis, C.; Cambon, P. Microcavity organic light-emitting diodes on silicon. *Appl. Phys. Lett.* **2002**, *81*, 1717–1719. [CrossRef]
9. Gohri, V.; Hofmann, S.; Reineke, S.; Rosenow, T.; Thomschke, M.; Levichkova, M.; Lüssem, B.; Leo, K. White top-emitting organic light-emitting diodes employing a heterostructure of down-conversion layers. *Org. Electron.* **2011**, *12*, 2126–2130. [CrossRef]
10. Hatton, R.A.; Willis, M.R.; Chesters, M.A.; Rutten, F.J.M.; Briggs, D. Enhanced hole injection in organic light-emitting diodes using a SAM-derivatised ultra-thin gold anode supported on ITO glass. *J. Mater. Chem.* **2002**, *13*, 38–43. [CrossRef]
11. Kanai, K.; Koizumi, K.; Ouchi, S.; Tsukamoto, Y.; Sakanoue, K.; Ouchi, Y.; Seki, K. Electronic structure of anode interface with molybdenum oxide buffer layer. *Org. Electron.* **2010**, *11*, 188–194. [CrossRef]
12. Qiu, C.F.; Peng, H.J.; Chen, H.Y.; Xie, Z.L.; Wong, M.; Kwok, H.S. Top emitting OLED using praseodymium oxide coated platinum as hole injectors. *IEEE Trans. Electron. Devices* **2004**, *51*, 1207–1210. [CrossRef]
13. Wu, J.; Hou, J.; Cheng, Y.; Xie, Z.; Wang, L. Efficient top-emitting organic light-emitting diodes with a V_2O_5 modified silver anode. *Semicond. Sci. Technol.* **2007**, *22*, 824–826. [CrossRef]
14. Ventsch, F.; Gather, M.C.; Meerholz, K. Towards organic light-emitting diode microdisplays with sub-pixel patterning. *Org. Electron.* **2010**, *11*, 57–61. [CrossRef]
15. Ji, Y.; Ran, F.; Xu, H.; Shen, W.; Zhang, J. Improved performance and low cost OLED microdisplay with titanium nitride anode. *Org. Electron.* **2014**, *15*, 3137–3143. [CrossRef]
16. Lee, H.; Cho, H.; Byun, C.-W.; Kang, C.-M.; Han, J.-H.; Lee, J.-I.; Kim, H.; Kim, M.; Cho, N.S. Device Characteristics of Top-Emitting Organic Light-Emitting Diodes Depending on Anode Materials for CMOS-Based OLED Microdisplays. *IEEE Photonics J.* **2018**, *10*, 1–9. [CrossRef]
17. Lee, J.; Li, P.; Kung, H.; Lu, Z. Highly efficient top-emission organic light-emitting diode on an oxidized aluminum anode. *J. Appl. Phys.* **2019**, *125*, 145501. [CrossRef]
18. Zhou, X.; Pfeiffer, M.; Huang, J.S.; Blochwitz-Nimoth, J.; Qin, D.S.; Werner, A.; Drechsel, J.; Maennig, B.; Leo, K. Low-Voltage inverted transparent vacuum deposited organic light-emitting diodes using electrical doping. *Appl. Phys. Lett.* **2002**, *81*, 922–924. [CrossRef]
19. Dobbertin, T.; Kroeger, M.; Heithecker, D.; Schneider, D.J.; Metzdorf, D.; Neuner, H.; Becker, E.; Johannes, H.-H.; Kowalsky, W. Inverted top-emitting organic light-emitting diodes using sputter-deposited anodes. *Appl. Phys. Lett.* **2003**, *82*, 284–286. [CrossRef]
20. Dodabalapur, A.; Rothberg, L.J.; Jordan, R.H.; Miller, T.M.; Slusher, R.E.; Phillips, J.M. Physics and applications of organic microcavity light emitting diodes. *J. Appl. Phys.* **1996**, *80*, 6954–6964. [CrossRef]
21. Ali, T.A.; Khayrullin, I.I.; Vazan, F.; Ziesmer, S.A.; Prache, O.; Jones, G.W.; Ghosh, A.P. 59.2: High Performance Top Emitting OLED Devices. *SID Symp. Dig. Tech. Pap.* **2007**, *38*, 1691–1694. [CrossRef]
22. Kim, Y.-K.; Kim, J.W.; Park, Y. Energy level alignment at a charge generation interface between NPB and HATCN. *Appl. Phys. Lett.* **2009**, *94*, 43.
23. Lee, J.-H.; Kim, J.-J. Interfacial doping for efficient charge injection in organic semiconductors. *Phys. Status Solidi A* **2012**, *209*, 1399–1413. [CrossRef]
24. Yoo, S.J.; Chang, J.-H.; Lee, J.-H.; Moon, C.-K.; Wu, C.-I.; Kim, J.-J. Formation of perfect ohmic contact at indium tin oxide/N,N′-di(naphthalene-1-yl)-N,N′-diphenyl-benzidine interface using ReO_3. *Sci. Rep.* **2014**, *4*, 3902. [CrossRef]

nanomaterials

MDPI

Article

Extended Graphite Supported Flower-like MnO_2 as Bifunctional Materials for Supercapacitors and Glucose Sensing

Han-Wei Chang [1,2,*], Chung-Li Dong [3], Yan-Hua Chen [1], Yuan-Zhang Xu [1], Tzu-Chi Huang [1], Song-Chi Chen [1], Feng-Jiin Liu [1,2], Yin-Hung Lai [1,2,4,*] and Yu-Chen Tsai [5,*]

1 Department of Chemical Engineering, National United University, Miaoli 360302, Taiwan; U0714022@gm.nuu.edu.tw (Y.-H.C.); U0714139@gm.nuu.edu.tw (Y.-Z.X.); j927528@gmail.com (T.-C.H.); chen31423@yahoo.com.tw (S.-C.C.); liu@nuu.edu.tw (F.-J.L.)

2 Pesticide Analysis Center, National United University, Miaoli 360302, Taiwan

3 Department of Physics, Tamkang University, Tamsui, New Taipei City 25137, Taiwan; cldong@mail.tku.edu.tw

4 Institute of Food Safety and Health Risk Assessment, National Yang Ming Chiao Tung University, Taipei 11221, Taiwan

5 Department of Chemical Engineering, National Chung Hsing University, Taichung 40227, Taiwan

* Correspondence: hwchang@nuu.edu.tw (H.-W.C.); laiyh@nuu.edu.tw (Y.-H.L.); yctsai@dragon.nchu.edu.tw (Y.-C.T.); Tel.: +886-37-382216 (H.-W.C.); +886-37-382205 (Y.-H.L.); +886-4-22857257 (Y.-C.T.); Fax: +886-37-382189 (H.-W.C.); +886-37-382189 (Y.-H.L.); +886-4-22854734 (Y.-C.T.)

Citation: Chang, H.-W.; Dong, C.-L.; Chen, Y.-H.; Xu, Y.-Z.; Huang, T.-C.; Chen, S.-C.; Liu, F.-J.; Lai, Y.-H.; Tsai, Y.-C. Extended Graphite Supported Flower-like MnO_2 as Bifunctional Materials for Supercapacitors and Glucose Sensing. *Nanomaterials* **2021**, *11*, 2881. https://doi.org/10.3390/nano11112881

Academic Editors: Jihoon Lee and Ming-Yu Li

Received: 3 September 2021
Accepted: 24 October 2021
Published: 28 October 2021

Abstract: A simple, efficient, and cost-effective extended graphite as a supporting platform further supported the MnO_2 growth for the construction of hierarchical flower-like MnO_2/extended graphite. MnO_2/extended graphite exhibited an increase in sp^2 carbon bonds in comparison with that of extended graphite. It can be expected to display better electrical conductivity and further promote electron/ion transport kinetics for boosting the electrochemical performance in supercapacitors and glucose sensing. In supercapacitors, MnO_2/extended graphite delivered an areal capacitance value of 20.4 mF cm^{-2} at 0.25 mA cm^{-2} current densities and great cycling stability (capacitance retention of 83% after 1000 cycles). In glucose sensing, MnO_2/extended graphite exhibited a good linear relationship in glucose concentration up to about 5 mM, sensitivity of 43 μA $mM^{-1}cm^{-2}$, and the limit of detection of 0.081 mM. It is further concluded that MnO_2/extended graphite could be a good candidate for the future design of synergistic multifunctional materials in electrochemical techniques.

Keywords: MnO_2/extended graphite; supercapacitors; glucose sensing

1. Introduction

Over the years, fossil fuel consumption and public health problems have posed serious issues, including global warming and substantial human diseases, that have caused significant negative economic and human health impacts. The relationship expressed above indicated that key technological changes and suitable implementation strategies might be necessary for promoting opportunities to reduce fossil fuel dependency, and, simultaneously, implement health promotion/disease prevention. Supercapacitors serve as intermediate energy storage systems between traditional capacitors and batteries, due to their simple construction, short charging time, safety, high power density, and excellent lifetime, which could replace fossil fuel use to meet energy requirements for reducing the fossil fuel energy consumption and further relaxing environmental pollution.

World-class public health is another globally important issue and challenge. The International Diabetes Federation (IDF) estimated that the number of people worldwide with diabetes (aged 20–79 years) would increase to 700 million in 2045 [1]. In Taiwan, diabetes is the fifth leading cause of death (data obtained from Ministry of Health and Welfare in Taiwan) [2]. Clearly, a fast, simple, efficient, and cost-effective approach needs to be developed in order to monitor blood glucose, allowing for improving glycemic control. Previous studies demonstrated that electrochemical techniques with moderate cost, instrumental simplicity, high portability, and suitable performance showed promise in a diverse range of applications such as solar cells, batteries, sensors, and supercapacitors, and hence provided more realistic solutions to address both the energy and public health challenges.

The choice of suitable electrode materials, which will be used to build novel high-performance electrodes in future electrochemical applications, is critical in the modeling and optimization of electrochemical devices. MnO_2 nanostructures with well-controlled morphology and crystal phase are expected to exhibit improved electrochemical performance when being used in supercapacitors and electrochemical sensing. More advantages for the use of MnO_2 as the electrode material were explained in previous reports. Yang et al. reported that MnO_2 had great advantages in terms of high theoretical specific capacity (up to 1370 F g^{-1}), low cost, natural abundance, and multiple crystallographic phases to be one of the most promising candidates for supercapacitors and glucose sensing. The structural transformation of multiple MnO_2 polymorphic forms composed of MnO_6 octahedrons by edge and/or corner sharing made differences in the intrinsic channel dimensions of MnO_2 to possess different electrochemical activities. MnO_2 based composites for supercapacitor applications with enhanced capacitive behaviors were attributed to electrochemically induced polymorphic phase and structural transformations of MnO_2, thus affecting the surface area and interfacial resistance [3]. Ponnusamy et al. also reported that MnO_2 based materials as a result of electrocatalytic effect of Mn^{3+}/Mn^{4+} redox couples had excellent electrocatalytic oxidation ability for glucose in the alkaline environment. The result implied that the MnO_2 based materials could be used as electrochemical electrodes, and they exhibited great excellent performance in electrochemical glucose sensing [4].

However, MnO_2 based materials possessed inherently low conductivity, which might restrict their practical use in electrochemical techniques [5]. The conductive carbon materials (such as graphite, carbon nanotubes, and graphene, etc.) serving as a conductive component could possibly form the conducting interconnection of carbon materials/MnO_2 nanocomposites, allowing fast ions and electrons transporting inside interior space of MnO_2 pore/channel structure, thus dramatically enhancing the electrocatalytic activity [6,7]. Many efforts described that carbon materials served mainly as a conductive component that provided a platform to integrate MnO_2 pore/channel structure in constructing hierarchical carbon materials/MnO_2 nanocomposites with three-dimensional (3D) interconnections. It could be expected that MnO_2 integrated within carbon materials generated a synergistic effect, which would improve electrocatalytic activity for electrochemical multifunctional applications. Several groups of researchers reported that the synthesis of carbon materials/MnO_2 nanocomposite with enhanced electrochemical activity showed effective characterization in electrochemical applications (supercapacitors and sensors). The improvement in electrochemical performance could be mainly attributed to the synergistic effect between carbon materials and MnO_2 that tightly connected to obtain dense hierarchical interconnected architectures in shortening the ions diffusion path and facilitating the charge transport [8–10].

Among carbon materials, simple, efficient, cost effective, and ecofriendly graphite made it an ideal substrate to grow MnO_2 for generating carbon materials/MnO_2 interconnected architectures, and had its own advantages to increase a wide range of industrial-scale production in electrochemical applications. Despite these advantages, pristine graphite had a lower specific surface area, thus causing a lower electrochemical performance. The surface modifications of activated graphite could be obtained by various activation techniques to provide larger specific surface areas at the optimum activation parameters for future enhancement of electrochemical performance. Kabir et al. also reported an electrochemically induced activation in aqueous electrolyte to exfoliate graphite to make surface functionalization. Activation could facilitate the formation of activated regions to improve the specific surface area that ensured excellent surface functional group coverage on graphite surfaces [11]. Chang et al. reported that graphite was efficiently activated by CO_2

treatment to change the surface area and surface functional groups, leading to low internal resistance and improved charge collection ability, suggesting the enhanced electrochemical behavior of graphite electrodes [12]. Some previous studies also demonstrated that surface modifications of carbon materials led to an increase in polarity of the surface layer, which facilitated the strong interactions to form the strong covalent interfacial bonding with selected substances for the formation of carbon materials/metal nanocomposites, and further highlight its capability to induce the application of a novel electrochemical-based techniques [13–16].

In this work, the formation of extended graphite through electrochemically activated treatment of graphite as a supporting platform further supports the formation of MnO_2 for the construction of hierarchical flower-like MnO_2/extended graphite through spontaneous redox reaction in the $KMnO_4$ solution. MnO_2/extended graphite are anticipated to serve as bifunctional materials for supercapacitors and glucose sensing. This work extends the understanding of electrocatalytic reaction mechanisms between the synergistic effect of MnO_2 and extended graphite for further electrochemical multifunctional applications in supercapacitors and glucose sensing.

2. Materials and Methods

2.1. Reagents

Potassium permanganate ($KMnO_4$), sodium hydroxide (NaOH), sodium sulfate (Na_2SO_4), glucose, carbon black graphite powder, and terpineol were purchased from Sigma-Aldrich (St. Louis, MO, USA). All aqueous solutions were prepared using deionized water (DI water) through a Milli-Q water purification system (Millipore, Burlington, MA, USA). All chemicals were used without further purification.

2.2. Preparation of MnO_2/Extended Graphite Nanocomposite

To obtain a series of extended graphite with different activation degrees, the graphite paste electrodes (3 mm diameter) were immersed in 0.2 M HNO_3 solution, and potential scanning was obtained at a scan rate of 50 mV s^{-1} within the applied voltage range of 1.0 to 2.0 V vs. Ag/AgCl for different numbers of electrochemical activation cycles (5, 10, and 20 cycles). The extended graphite with different activation degrees were conducted by using cyclic voltammetry (CV) cycling processes, which can be expected to induce extended defect on the surface of graphite, in this text referred to as extended graphite, providing a new set of growth platforms to form hierarchical flower-like MnO_2. The obtained extended graphite with three different degree activations (low, medium, and high degree activation) were denoted as gra-L, gra-M, and gra-H, respectively. The obtained extended graphite was then washed with DI water to remove the remaining reagents and collected for subsequent MnO_2 synthesis. MnO_2 was deposited on the extended graphite by a spontaneous redox reaction between transition metal ions and carbon materials, which involve simultaneous oxidation of the extended graphite and reduction of $KMnO_4$. The increased structure defects of the carbon materials may optimally tune the oxidation potential of carbon materials and thus more easily promote $KMnO_4$ reduction to MnO_2 [17]. The synthesis procedures in detail were as follows: gra-L, gra-M, and gra-H were immersed in 50 mL aqueous solutions of 0.001 M $KMnO_4$ with stirring at 80 °C. Subsequently, these obtained products (MnO_2/extended graphite) were rinsed three times by DI water and dried at 60 °C for 3 h (denoted as MnO_2/gra-L, MnO_2/gra-M, and MnO_2/gra-H, respectively). The resulted products were then collected for the following characterization.

2.3. Apparatus

The morphology was characterized by using field emission scanning electron microscopy (FESEM, JSM-7410F, JEOL, Akishima, Japan) and field emission transmission electron microscopy (FETEM, JEM-2100F, JEOL, Akishima, Japan). The chemical structure and composition were determined by X-ray photoelectron spectroscopy (XPS, PHI-5000 Versaprobe, ULVAC-PHI, Tokyo, Japan). Synchrotron X-ray absorption spectroscopy (XAS) at C K-, O K- and Mn L-edges were performed at BL20A of the National Synchrotron Radiation Research Center (NSRRC), Hsinchu, Taiwan. Electrochemical measurements were performed using a three-electrode system comprised of as-prepared sample-graphite paste electrode as current collector (mixing graphite powder and carbon black in 4:1 weight ratio dispersed terpineol) by casting on plastic substrate with an active material mass loading of 0.4 mg cm^{-2}, a platinum wire counter electrode; and an Ag/AgCl (3 M KCl) reference electrode by an electrochemical analyzer (Autolab, model PGSTAT30, Eco Chemie, Utrecht, The Netherlands). The electrochemical testing in supercapacitors and electrochemical glucose sensing was performed in the two different aqueous electrolyte solutions as supporting electrolyte (using 1 M Na_2SO_4 for supercapacitors and 0.1 M NaOH for electrochemical glucose sensing). In the supercapacitor, the capacitive performance was evaluated thorough cyclic voltammetry (CV) and galvanostatic charge–discharge (GCD) in 1 M Na_2SO_4 within the applied voltage range of 0.0 to 1.0 V vs. Ag/AgCl. In electrochemical glucose sensing, the sensing performance of electrochemical sensor was evaluated using CV and amperometry in the 0.1 M NaOH in the absence and presence of glucose. LCMS/MS analysis was conducted using a Q-TOF mass spectrometer (LCMS-9030, Shimadzu Corp., Kyoto, Japan) and a triple quadrupole mass spectrometer (LCMS-8045, Shimadzu Corp., Kyoto, Japan). A Unison UK-Amino column (UKA24U, Imtakt USA, Portland, OR, USA) was utilized for the chromatographic separation.

3. Results

The morphologies of extended graphite and MnO_2/extended graphite are characterized by FETEM and FESEM. FETEM images in Figure 1a display the general morphologies and detailed microstructures of extended graphite. It clearly indicates that the surfaces of extended graphite are rough. This result implies that the defect densities in graphite can be fine-tuned through electrochemically activated treatment of graphite in the HNO_3 solution that effectively expose more edge and basal plane defect sites in the 2D structure graphite layers. This work presents a facile synthesis strategy to synthesize extended defect structures in graphite layers and further introduces functional groups at defect sites along the extended graphite. Extended graphite attached to these defect sites offers an ideal platform for the growth of MnO_2 in the $KMnO_4$ solution by spontaneous redox reaction. Figure 1b,c shows that MnO_2/extended graphite has a 3D hierarchical flower-like MnO_2 construction consisting of many assembled ultrathin 2D nanosheets, and it develops an interconnected network within extended graphite, confirming the successful synthesis of the MnO_2/extended graphite. The detailed structural morphologies of 3D hierarchical flower-like MnO_2 over the extended graphite with the three different degrees of electrochemically activated treatment (MnO_2/extended graphite) are represented by EFSEM. Figure 2 displays the EFSEM images of extended graphite with different activation degrees (gra-L, gra-M, and gra-H) and MnO_2/extended graphite (MnO_2/gra-L, MnO_2/gra-M, and MnO_2/gra-H). It can be clearly seen in Figure 2a–c that the surface morphologies of extended graphite show no obvious differences when the activation degrees are controlled by electrochemically treatment. With extended graphite as an ideal platform for the growth of MnO_2, the 3D hierarchical flower-like MnO_2 is grown at the defect structures of extended graphene through in situ nucleation and growth processes (see Figure 2d–f). The extended graphite via electrochemically treatment exposes numerous edge and basal plane defect sites that promote the subsequent nucleation and growth of 3D hierarchical flower-like MnO_2, especially in view of MnO_2/gra-H. The FESEM image of MnO_2/gra-H in Figure 2f shows the conductive gra-H interconnected with the 3D hierarchical flower-like MnO_2 owing to clear and distinct crumpled layer structure consisting of many assembled ultrathin 2D nanosheets, which is consistent with the TEM images in the Figure 1b. The 3D hierarchical structure of MnO_2/gra-H constructed by the formation of the interconnected conducting networks between defect sites of extended graphite and ultrathin MnO_2 nanosheets ensures good electrical contact and also provides a large electrochemically active surface area, which further not only contributes to a continuous electron pathway, but also facilitates ion transport by shortening diffusion pathways. The above results clearly reveal that the 3D hierarchical MnO_2/extended graphite could serve as functional materials in electrochemical techniques.

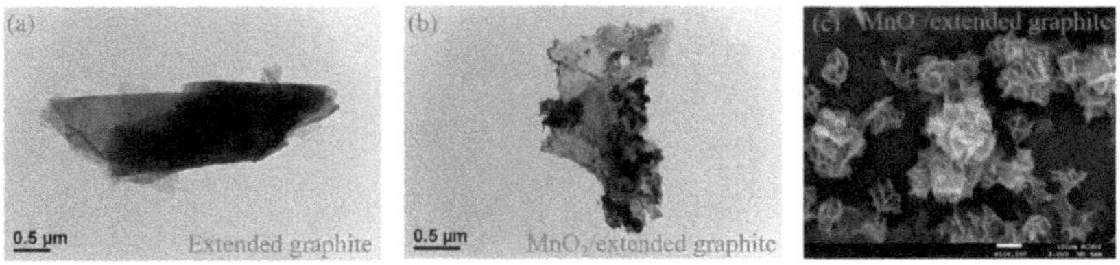

Figure 1. FETEM of (**a**) extended graphite and (**b**) MnO_2/extended graphite. (**c**) FESEM of MnO_2/extended graphite.

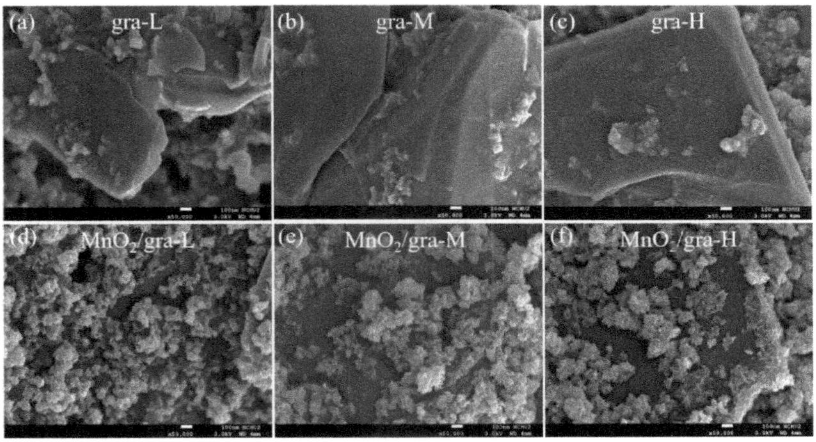

Figure 2. FESEM of (**a**) gra-L, (**b**) gra-M, (**c**) gra-H, (**d**) MnO_2/gra-L, (**e**) MnO_2/gra-M, and (**f**) MnO_2/gra-H.

The surface chemical and electronic states of extended graphite and MnO_2/extended graphite are characterized by XPS and XAS. Figure 3a presents the XPS full survey spectrum of extended graphite and MnO_2/extended graphite. The results indicate the coexistence of C and O elements in the extended graphite with different activation degrees (gra-L, gra-M, and gra-H). After further growth of 3D hierarchical flower-like MnO_2 within extended graphite, the full survey spectrum shows signals from C, O, and Mn elements, suggesting the obtained MnO_2/extended graphite was successfully synthesized. The high-resolution C 1s XPS spectrum of gra-L, gra-M, and gra-H is shown in Figure 3b. The high-resolution C 1s spectrum of gra-L, gra-M, and gra-H (Figure 3b) shows four different peaks located at 284.5, 285.0, 286.4, and 288.9 eV, which correspond to C–C (sp^2), C–C (sp^3), C-OH (hydroxyl), and O-C-O/O-C=O (epoxy and carboxyl) functional groups, respectively [18]. The XPS quantitative analysis results are summarized in Table 1, Table 2, Table 3. From Table 1, it is clear that the content of sp^3 carbon bonds, hydroxyl, epoxy, and carboxyl functional groups within extended graphite are increased due to the destruction of the sp^2 carbon bonds and the formation of defective sites with increasing activation degree, which proves that extended graphite exists from an electrochemically activated transformation and reflects the relative sp^3/sp^2 carbon bonds ratio [19,20]. It is well known that the functional groups are mainly located at the basal and edges planes of graphite, which not only facilitates the exfoliation process of graphite sheets, but also modifies surface hydrophilicity of exfoliate graphite [21–24]. The extended graphite by electrochemically activated treatment leads to the formation of sp^3 carbon bonds and creates defect sites, providing numerous nucleation sites that facilitate the uniform nucleation and growth of the selected substances [25,26].

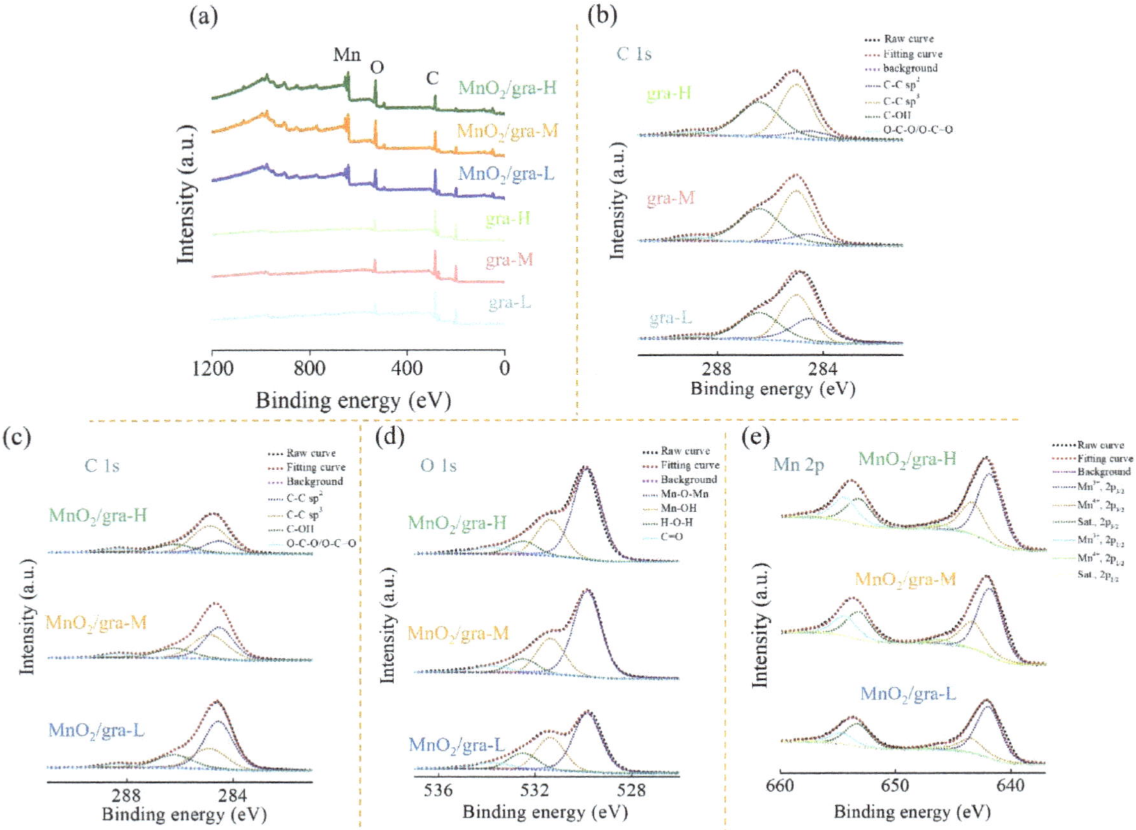

Figure 3. (**a**) Full scan (**b**,**c**) C 1s, (**d**) O 1s, and (**e**) Mn 2p XPS spectra of extended graphite and MnO_2/extended graphite.

Table 1. Fitted results of C 1s XPS spectra.

Samples	Fitted Results of C 1s XPS Spectra			
	C-C sp^2 (%)	C-C sp^3 (%)	C-OH (%)	O-C-O/O-C=O (%)
gra-L	28.3	37.0	32.5	2.2
gra-M	12.6	42.4	41.4	3.6
gra-H	10.1	43.9	42.3	3.7
MnO_2/gra-L	53.5	23.5	19.3	3.7
MnO_2/gra-M	42.9	35.3	17.0	4.8
MnO_2/gra-H	22.5	52.3	18.8	6.4

Table 2. Fitted results of O 1s XPS spectra.

Samples	Fitted Results of O 1s XPS Spectra			
	Mn-O-Mn (%)	Mn-OH (%)	H-O-H (%)	C=O (%)
MnO_2/gra-L	51.1	26.0	13.5	9.4
MnO_2/gra-M	59.5	22.9	9.5	8.1
MnO_2/gra-H	61.5	22.2	8.5	7.8

Table 3. Fitted results of Mn 2p XPS spectra.

Samples	Fitted Results of Mn 2p XPS Spectra						
	Mn^{3+} $2p_{3/2}$ (%)	Mn^{3+} $2p_{1/2}$ (%)	Mn^{4+} $2p_{3/2}$ (%)	Mn^{4+} $2p_{1/2}$ (%)	Sat. $2p_{3/2}$ (%)	Sat. $2p_{1/2}$ (%)	Mn^{3+}/Mn^{4+} Ratio
MnO_2/gra-L	47.1	20.2	17.2	10.3	3.1	2.1	2.4
MnO_2/gra-M	41.9	18.2	20.3	14.2	3.2	2.0	1.7
MnO_2/gra-H	39.4	16.2	22.7	16.4	3.1	2.2	1.4

Figure 3c–e show the high-resolution C 1s, O 1s, and Mn 2p XPS spectrum of MnO_2/extended graphite. The high-resolution O 1s XPS spectrum of the MnO_2/extended graphite can be deconvoluted into four peaks corresponding to Mn-O-Mn, Mn-O-H, H-O-H, and C=O located at 529.8, 531.4, 532.5, and 533.8 eV. The Mn 2p XPS spectrum of the MnO_2-coated different carbon materials in Figure 3f consist of two doublet peaks of Mn $2p_{3/2}$ and $2p_{1/2}$ located at 642 and 653 eV, with a typical spin energy separation (~11.7 eV) of the Mn 2p doublet peaks. Furthermore, the overlapped Mn $2p_{3/2}$ ($2p_{1/2}$) spectra can be fitted into three peaks at 641.9 (653.3), 643.4 (654.5), and 645.3 (657.5) eV, which are characteristics of Mn^{3+}, Mn^{4+}, and shake-up satellite (denoted as Sat.), respectively, further confirming the presence of mixed-valance manganese oxide with oxidation state Mn^{3+} and Mn^{4+} [27,28]. A deeper analysis into the XPS quantitative analysis in the MnO_2/extended graphite (Table 3) shows that the Mn^{3+}/Mn^{4+} in the MnO_2/extended graphite decreases with MnO_2 growth on the extended graphite by gradual increase of activation degrees. This suggests that higher content of Mn^{4+} is more pronounced for MnO_2/gra-H. Second, MnO_2/extended graphite exhibits an increase in sp^2 carbon bonds and the decrease in hydroxyl in comparison with that of electrochemically activated extended graphite (see Figure 3c and Tables 1 and 2), which indicates that the introduction of MnO_2 within electrochemically activated extended graphite can greatly increase the relative content of sp^2 carbon bonds, further improving the microcrystalline structure and graphitization degree. It can be expected to display better electrical conductivity and further promote a rapid and efficient charge transfer at the active electrode material/electrolyte interfaces, thus leading to a significantly enhanced electrochemical performance. Third, the decrease in the sp^2 carbon bonds causes the gradual increase in Mn-O-Mn with the MnO_2 growth over the three different activation degrees of the extended graphite, especially when MnO_2 grows on the gra-H due to the more severe activation degree, which exposes abundant defective sites and favors the nucleation and growth of MnO_2 [29–31].

Figure 4 presents the C K-edge, O K-edge, and Mn L3,2-edge XAS of extended graphite and MnO_2/extended graphite. XAS is known to be a powerful tool sensitive to understanding the information on local atomic and electronic structure details caused by electronic transition from a core level to various unoccupied states. Figure 4a shows the C K-edge XAS of extended graphite and MnO_2/extended graphite. Features A_4, B_4, and C_4 are associated with the electronic transition from C 1s to the unoccupied C–C π^* (ring) orbitals, the chemically functionalized carbon atoms and/or defects, and unoccupied C–C σ^* (ring) orbitals, respectively. With MnO_2 growth on the extended graphite, the pronounced C–C π^* (A_4) resulting from significant charge-transfer interactions between extended graphite and Mn atoms reveals that MnO_2/extended graphite has the perfect graphitic crystalline structure of graphite sheets and demonstrates excellent electrical conductivity. Additionally, these charge-transfer interactions between MnO_2 and electrochemically activated extended graphite induce dramatic changes in Feature B_4, which pertains to the construction of new Mn–O–C bonds. It is reasonable to believe that the oxygen functional groups on extended graphite can act as anchoring or nucleation sites for the growth of MnO_2. The results support the fact that the electrochemically activated extended graphite provides an interconnected conducting network, suggesting that these ultrathin MnO_2 nanosheets are tightly anchored within extended graphite. The features (denoted as *) at 298.6 and 301.0 eV are associated with the K L3- and L2-edge that can be explained by the presence of potassium ions and water molecules generally intercalated within the interlayer space of MnO_2 structure via spontaneous redox reaction in the $KMnO_4$ solution [32], especially in view of MnO_2/gra-H.

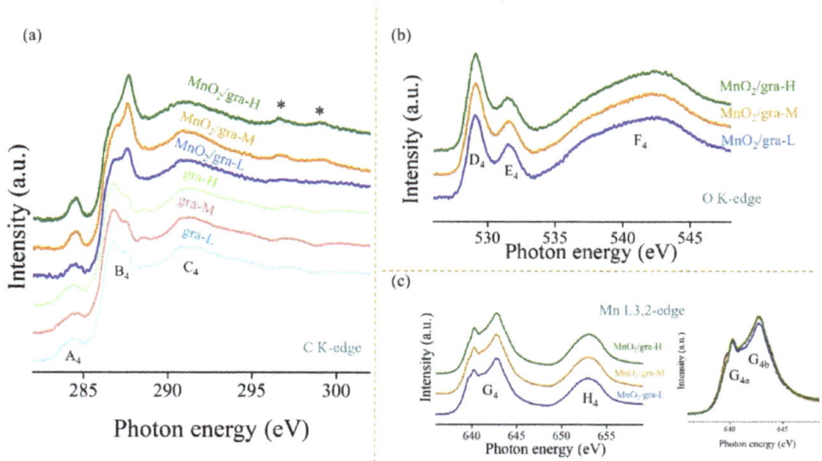

Figure 4. (**a**) C K-edge, (**b**) O K-edge, and (**c**) Mn L3, 2-edge XAS spectra of extended graphite and MnO_2/extended graphite. Inset in (**c**) is the region of Mn L3-edge XAS spectra. The features (denoted as *) at 298.6 and 301.0 eV are potassium ions and water molecules within the interlayer space of MnO_2 structure via spontaneous redox reaction in the $KMnO_4$ solution.

Figure 4b,c show the O K-edge and Mn L3,2-edge XAS of MnO_2/extended graphite. These O K-edge XAS spectra present similar features (labeled as D_4, E_4, and F_4) from the MnO_2/extended graphite (MnO_2/gra-L, MnO_2/gra-M, and MnO_2/gra-H). Feature F_4 can be assigned to electronic transition from O 1s to Mn 4sp transition. Features D_4 and E_4 are related to the electronic transition from hybridization of O 2p and metal Mn 3d t_{2g} and 3d e_g orbitals. Variation in the three features represents that electrochemically activated extended graphite can possess the Mn-O bonding and the surface configuration, favoring the electronic interactions between MnO_2 and electrochemically activated extended graphite. The ratio of the intensity of Features D_4 and E_4 increases with the MnO_2 growth within more severe activation degree of extended graphite, which suggests that the surface modification (structural defects and chemical functionalization) of extended graphite affects the nucleation process. It is believed that the surface sufficient functional groups within extended graphite can provide reactive sites for the nucleation and growth of MnO_2. The Mn L3,2-edge XAS of MnO_2/extended graphite display doublet features (labeled as G_4 and H_4) that arise from the Mn $2p_{3/2}$ and $2p_{1/2}$ core level states to the unoccupied Mn 3d states electronic transition, which can be correlated with the valence state of Mn atoms in the MnO_2 [33]. Notably, Feature G_4 can be split into two peaks (G_{4a} and G_{4b}) (inset in Figure 4c), and the mixed valence states of Mn^{3+} and Mn^{4+} and their ratios within MnO_2/extended graphite can be estimated from the ratio of G_{4a}/G_{4b}. A decrease in the G_{4a}/G_{4b} ratio with the MnO_2 growth within more severe activation degrees of extended graphite may give evidence that the MnO_2/gra-H contains Mn in a higher oxidation state [34,35]. These XAS results are consistent with the XPS results in Figure 3.

Hence, the overall analysis of the XPS and XAS results demonstrate that extended graphite by electrochemically activated treatment can create defect sites and provide abundant nucleation sites for the growth of MnO_2; in the meantime, MnO_2/extended graphite favorably attain the sp^2 carbon bonds character in graphite sheet, resulting in better electrical conductivity. It is concluded that MnO_2/extended graphite could be a good candidate for the future design of synergistic multifunctional materials in electrochemical techniques. In this study, MnO_2/extended graphite are synthesized by spontaneous redox reaction and used as bifunctional electrode materials for supercapacitors and glucose sensing.

For the potential use of MnO_2/extended graphite in supercapacitors, Figure 5 shows the electrochemical capacitive performance of extended graphite and MnO_2/extended graphite thorough cyclic voltammetry (CV) and galvanostatic charge–discharge (GCD) in 1 M Na_2SO_4. In comparison to the CV results of extended graphite and MnO_2/extended graphite in Figure 5a,b, these results clearly reveal that extended graphite (gra-L, gra-M, and gra-H) have greater electrical double-layer capacitive behavior with more severe activation degree, which might be attributed to the functionalized and defective extended graphite owing to these strong interfacial interactions. Thus, it could be expected that increasing the interfacial area and interfacial interactions within extended graphite helps build an intimate interface to facilitate interactions between extended graphite and $KMnO_4$ interfaces for MnO_2 growth by spontaneous redox reaction. Figure 5b shows that the CV curve of MnO_2/extended graphite gives a quasi-rectangular shape with two pairs of well-defined redox peaks corresponding to both the functionalized extended graphite and the Mn^{3+}/Mn^{4+} redox reaction. Figure 4a reveals that controlled growth of 3D hierarchical flower-like MnO_2 within extended graphite results in more perfect crystalline structure of the graphitic carbon. MnO_2/extended graphite, because of its perfect graphitic crystalline structure, should possess good conductivity to effectively ensure electron transport, and hence the redox reaction can proceed more effectively and reversibly; there most likely exists two distinct pseudocapacitance charge-storage mechanisms. It demonstrates that the synergistic effect of extended graphite and MnO_2 results in the coexistence of both the electrical double-layer capacitance of extended graphite and pseudocapacitance of extended graphite and MnO_2. Hence, the CV result in Figure 5b indicates that MnO_2/gra-H displays better capacitive performance compared to MnO_2/gra-M and MnO_2/gra-L. Thanks to the excellent surface modification of gra-H, there is a perfect graphitic crystalline structure present in MnO_2/gra-H, which provides a continuous conductive network to optimize electron/ion transport kinetics for boosting the electrochemical performance. Figure 5c displays the CV curves of MnO_2/gra-H measured at different scan rates of 10, 20, 50, and 100 mV s^{-1}. The CV curves maintain their shape and reversibility even at relatively high scan rates, suggesting excellent capacitive performance of MnO_2/gra-H. The GCD measurements are used for further discussing the enhanced capacitive performance of MnO_2/gra-H (Figure 5d). MnO_2/gra-H displays quasi-triangular symmetrical shape by GCD measurements at different current densities of 0.25, 0.5, 1, 2, and 4 mA cm^{-2}, and indicates the coexistence of both electrical double-layer capacitance and pseudocapacitance, which are consistent with the CV curves. The areal capacitance values of MnO_2/gra-H can be calculated according to the following equation: $C = (I\ \Delta t)/(\Delta V\ A)$, where C is areal capacitance (mF cm^{-2}), I is the charge/discharge current (mA), Δt is the discharge time (s), ΔV is the voltage change during discharge (V), and A is the area of the electrode (cm^2). The areal capacitance (gravimetric capacitance) values at 0.25, 0.5, 1, 2, and 4 mA cm^{-2} current density are calculated to be 20.4, 16.8, 16.0, 15.8, and 15,7 mF cm^{-2} (51, 42.0, 40.0, 39.5, and 39.3 F g^{-1}), respectively. The capacitive performance of MnO_2/gra-H is comparable to other previously reported results in transition metal-based electrode materials (Table 4) [36–40]. In order to further research the long-term cycling stability of the MnO_2/gra-H, MnO_2/gra-H is performed over 1000 consecutive galvanostatic charge–discharge cycles at a constant current density of 4 mA cm^{-2} (Figure 5e). After 1000 cycles, the capacitance is equivalent to 83% of initial value, demonstrating a good cycling performance. These results once again demonstrated that the MnO_2/extended graphite consisting of the interconnected conducting networks between 3D hierarchical flower-like MnO_2 and extended graphite provide tight contact and stable interface, which facilitates an enhanced capacitive performance for supercapacitors.

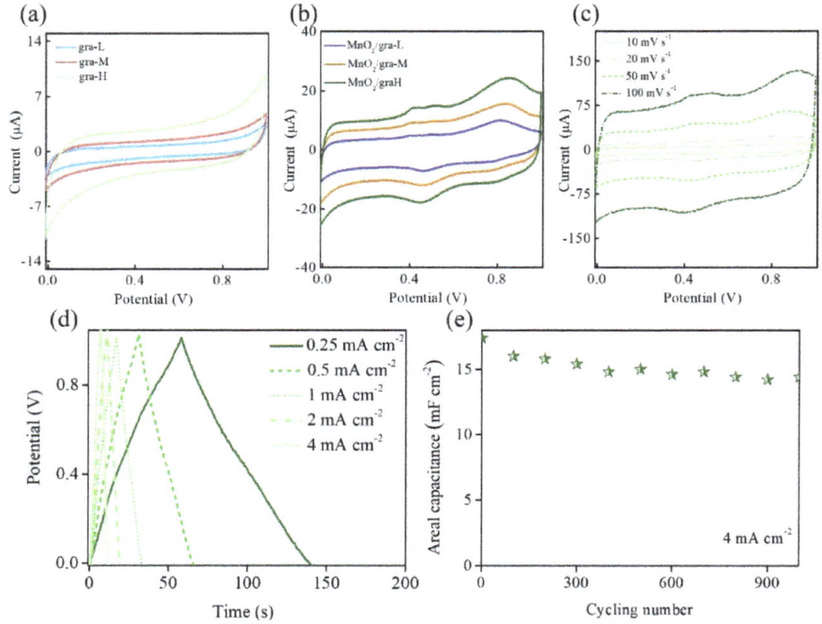

Figure 5. CV curves of (**a**) gra-L, gra-M, and gra-H, and (**b**) MnO_2/gra-L, MnO_2/gra-M, and MnO_2/gra-H in 1 M Na_2SO_4 at a scan rate of 20 mV s^{-1}. (**c**) CV curves of MnO_2/gra-H in 1 M Na_2SO_4 at different scan rates. (**d**) Galvanostatic charge/discharge curves of MnO_2/gra-H in 1 M Na_2SO_4 at different current densities. (**e**) Cyclic stability of MnO_2/gra-H in 1 M Na_2SO_4 is recorded every 100 cycles during cycling (denoted by green star).

Table 4. Comparison of the capacitive performance with transition metal-based electrode materials.

Electrode Materials	Electrolyte	Current Density	Areal Capacitance (mF cm^{-2})	Reference
Mn_3N_2	Na_2SO_4	1.00 mA cm^{-2}	74.0	[36]
MnO_x	Na_2SO_4	0.25 mA cm^{-2}	19.3	[37]
TiO_2	Na_2SO_4	2.00 mV s^{-1}	23.2	[38]
$NiCo_2O_4$	KOH	0.25 mA cm^{-2}	28.0	[39]
$ZnCo_2O_4$	KOH	0.01 mA cm^{-2}	16.1	[40]
MnO_2/gra-H	Na_2SO_4	0.25 mA cm^{-2}	20.4	This work

It is anticipated that the findings of this work could open another opportunity to explore its possible applications in non-enzymatic glucose sensing. Figure 6 displays the electrochemical sensing performance of gra-H and MnO_2/gra-H using cyclic voltammetry (CV) and amperometry towards glucose electrooxidation in alkaline solution. In Figure 6a, gra-H and MnO_2/gra-H are examined using CV under the scan rate of 50 mV s^{-1} in 0.1 M NaOH in the absence and presence of 10 mM glucose. The successive addition of glucose results in an increase in the anodic (oxidation) current at 0.4~0.8 V. Specifically, the anodic (oxidation) peak current of glucose remarkably increases at MnO_2/gra-H. This result reveals that the MnO_2/gra-H has great potential to electrocatalyze glucose oxidation reactions effectively involving the redox-active Mn mediated electrocatalytic reaction mechanisms [41]. The mechanisms can be illustrated as follows.

$$Mn^{3+} + OH^- \leftrightarrow Mn^{4+} + H_2O + e^- \quad (1)$$

$$Mn^{4+} + \text{glucose} + e^{-} \rightarrow Mn^{3+} + \text{gluconolactone} \quad (2)$$

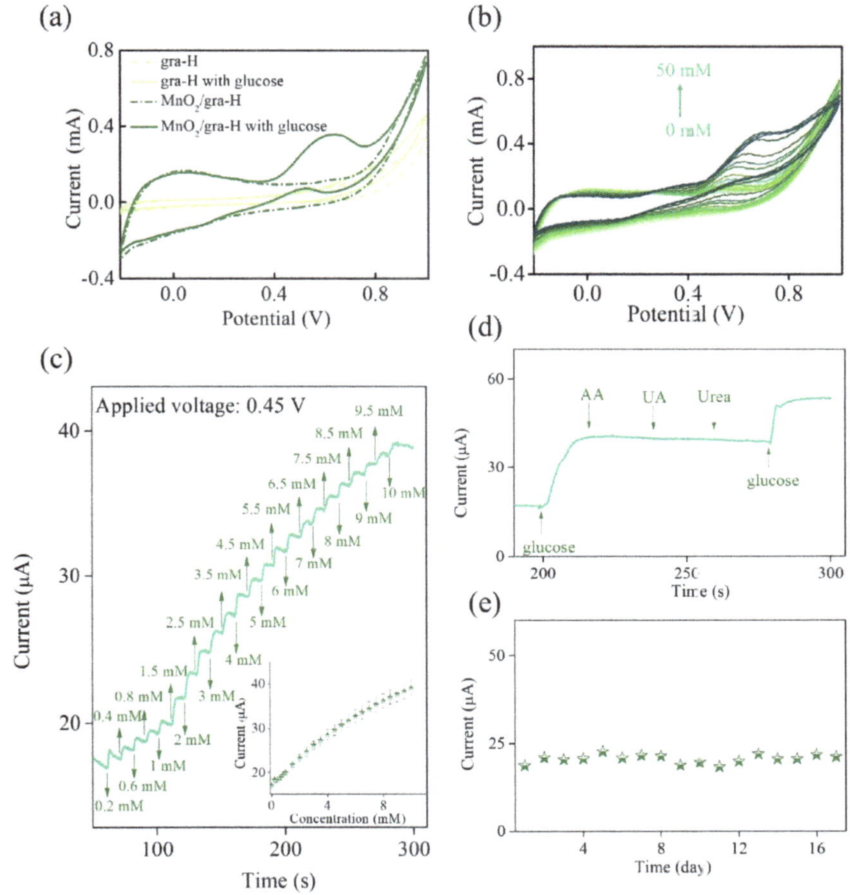

Figure 6. (**a**) CV curves of gra-H and MnO_2/gra-H in 0.1 M NaOH in the absence (dashed lines) and presence (solid lines) of 10 mM glucose at a scan rate of 50 mV s^{-1}. (**b**) CV curves of MnO_2/gra-H in 0.1 M NaOH with successive addition of various glucose concentrations. (**c**) Amperometry of MnO_2/gra-H in 0.1 M NaOH with successive addition of various glucose concentrations at applying voltage +0.45 V. In inset of (**c**) is the corresponding calibration plot of the response current versus glucose concentration. (**d**) Interference tests and (**e**) long-term stability of MnO_2/gra-H in 0.1 M NaOH at applying voltage +0.45 V (denoted by green star).

When increasing the glucose concentration in the range from 0 to 50 mM, MnO_2/gra-H exhibits a gradual increase anodic (oxidation) peak current, demonstrating that MnO_2/gra-H provides cooperative effects in promoting the electrocatalytic activity toward glucose electrooxidation. Furthermore, there is continuously a progressive positive shift in anodic (oxidation) peak potential at a voltage region of 0.4~0.8 V, which is caused by a kinetic limitation in the glucose electrooxidation between redox sites within MnO_2/gra-H (Figure 6b) [42]. Figure 6c shows the amperometry of MnO_2/gra-H towards glucose electrooxidation in 0.1 M NaOH at 0.45 V, with the gradual additions of glucose concentrations up to 10 mM. The inset of Figure 6c shows the corresponding calibration plot of the response current (I) versus glucose concentration ($Conc._{glu}$) (error bars represent the standard deviations for the 3-times repeated measurements). Response current increases linearly with increasing glucose concentration up to about 5 mM. The corresponding calibration equation is I (mA) = 0.003 $Conc._{glu}$ (mM) + 0.017, having the resulting correlation coefficient value (R^2) 0.994, sensitivity 43 μA $mM^{-1}cm^{-2}$, and the limit of detection 0.081 mM based on a signal-to-noise ratio of 3 (S/N = 3), indicating the significantly improved electrocatalytic activity of MnO_2/gra-H for glucose electrooxidation. The present MnO_2/gra-H exhibits better or comparable performance with some previously reported results with similar electrode materials [43–46]. The comparison results are summarized in Table 5. The interference test is conducted by amperometry in 0.1 M NaOH at 0.45 V with the gradual additions of 4.83 mM glucose, 0.05 mM ascorbic acid (AA), 0.5 mM uric acid (UA), 5 mM urea, and 4.83 mM glucose, which does not cause any observable interference signal to glucose detection (Figure 6d). This means that the proposed MnO_2/gra-H possesses excellent selectivity. It is well known that the long-term stability is also an important index to evaluate the sensing performance. Hence, the long-term stability of the proposed MnO_2/gra-H is evaluated by measuring its amperometric current response to 1 mM glucose every day for 2 weeks. Figure 6e reveals that the amperometric current response stabilizes to approximately 80~90% of its initial value, implying an excellent stability during long-term storage.

Table 5. Comparison of the non-enzymatic glucose sensing performance.

Electrode Materials	Applied Potential (V vs. Ag/AgCl)	Linear Range (mM)	Sensitivity (μA $mM^{-1}cm^{-2}$)	Detection Limit (mM)	Reference
MnCo–carbon nanofibers/Nafion	0.60	0.5–7.0	36	0.050	[43]
PtAu–MnO_2	0.00	0.1–30.0	59	0.020	[44]
MnCu/MWCNT/GO	−0.05	1.0–32.0	59	0.001	[45]
PDDA-RGO/MnO_2/AuNPs	0.60	0.02–0.85	84	0.002	[46]
MnO_2/gra-H	0.45	Up to 5.0	43	0.080	This work

To further demonstrate the potential of the proposed MnO_2/gra-H in practical applications, the recovery tests of glucose are carried out by adding a known concentration (1 mM) to samples, which is then evaluated by analyzing the amperometric response to glucose by 3-time repeated measurements. The results of the recovery tests obtained by the proposed amperometric method are in good agreement with the results obtained by the reference method (mass spectrometry), and the comparison results are given in Table 6. The recovery test demonstrates that the proposed MnO_2/gra-H makes satisfactory recovery of the added glucose concentrations.

Table 6. The results of the recovery tests obtained by the electrochemical sensing and the mass spectrometric methods for glucose.

Added (mM)	Found by Electrochemical Sensing (mM)	Recovery (%)	RSD (%)	Found by Mass Spectrometry (mM)	Recovery (%)	RSD (%)
1	1.02	102	5.2	0.94	94	0.9

4. Conclusions

In this study, MnO_2/extended graphite displayed high bifunctional electro-activity performances in electrochemical multifunctional applications (including supercapacitors and glucose sensing). Thanks to the excellent surface modification of gra-H through electrochemically activated treatment of graphite inducing extended defects on the surface of graphite, providing a new set of growth platforms to form hierarchical flower-like MnO_2, and thus a perfect graphitic crystalline structure present in MnO_2/gra-H, which provide a continuous conductive network to optimize electron/ion transport kinetics for boosting the electrochemical performances of supercapacitors (areal capacitance value of 20.4 mF cm^{-2} at 0.25 mA cm^{-2} and capacitance retention of 83% after 1000 cycles) and glucose sensing (linear range up to about 5 mM, sensitivity of 43 μA $mM^{-1}cm^{-2}$, and the limit of detection of 0.081 mM). Because of its electrochemical synergistic effects, MnO_2/extended graphite is a promising candidate for the development of electrochemical multifunctional applications.

Author Contributions: Conceptualization, H.-W.C.; methodology, H.-W.C.; software, Y.-H.C., Y.-Z.X., T.-C.H., S.-C.C. and F.-J.L.; validation, H.-W.C.; formal analysis, H.-W.C., F.-J.L. and Y.-H.L.; investigation, H.-W.C., Y.-H.C., Y.-Z.X., T.-C.H., S.-C.C. and Y.-H.L.; resources, C.-L.D., F.-J.L., Y.-H.L. and Y.-C.T.; data curation, H.-W.C., Y.-H.C., Y.-Z.X., T.-C.H. and S.-C.C.; writing—original draft preparation, H.-W.C.; writing—review and editing, H.-W.C. and Y.-C.T.; visualization, H.-W.C., C.-L.D., Y.-H.L. and Y.-C.T.; supervision, H.-W.C., C.-L.D. and Y.-C.T.; project administration, H.-W.C. and Y.-C.T. All authors have read and agreed to the published version of the manuscript.

Funding: This study acknowledges financial support from the Ministry of Science and Technology, and National United University, Taiwan (MOST 110-2221-E-239-001-, MOST 110-2622-E-005-013-, MOST 110-2221-E-005-009-, and 1100002).

Acknowledgments: This study acknowledges financial support from the Ministry of Science and Technology, and National United University, Taiwan.

Conflicts of Interest: The authors declare no conflict of interest.

References

1. International Diabetes Federation. *IDF Diabetes Atlas*, 9th ed.; International Diabetes Federation: Brussels, Belgium, 2019.
2. Moha, W. *Taiwan Health and Welfare Report*; Ministry of Health and Welfare: Taipei, Taiwan, 2018.
3. Yang, X.; Qiao, Z.; Liu, F.; Yang, S.; Zhang, L.; Cao, B. In-depth study of electrochemical capacitor performance of MnO_2 during phase transition from Ramsdellite-MnO_2 to Birnessite-MnO_2. *Electrochimica Acta* **2018**, *280*, 77–85. [CrossRef]
4. Ponnusamy, R.; Venkatesan, R.; Kandasamy, M.; Chakraborty, B.; Rout, C.S. MnO_2 polymorph selection for non-enzymatic glucose detection: An integrated experimental and density functional theory investigation. *Appl. Surf. Sci.* **2019**, *487*, 1033–1042. [CrossRef]
5. Jeon, H.; Jeong, J.; Hong, S.B.; Yang, M.; Park, J.; Kim, D.H.; Hwang, S.Y.; Choi, B.G. Facile and fast microwave-assisted fabrication of activated and porous carbon cloth composites with graphene and MnO_2 for flexible asymmetric supercapacitors. *Electrochim. Acta* **2018**, *280*, 9–16. [CrossRef]
6. Wang, Y.; Hao, L.; Zeng, Y.; Cao, X.; Huang, H.; Liu, J.; Chen, X.; Wei, S.; Gan, L.; Yang, P. Three-dimensional hierarchical porous carbon derived from resorcinol formaldehyde-zinc tatrate/poly (styrene-maleic anhydride) for high performance su-percapacitor electrode. *J. Alloys Compd.* **2021**, *886*, 161176. [CrossRef]
7. Liu, M.; Shi, M.; Lu, W.; Zhu, D.; Li, L.; Gan, L. Core–shell reduced graphene oxide/MnOx@ carbon hollow nanospheres for high performance supercapacitor electrodes. *Chem. Eng. J.* **2017**, *313*, 518–526. [CrossRef]
8. Xiong, C.; Lin, X.; Liu, H.; Li, M.; Li, B.; Jiao, S.; Zhao, W.; Duan, C.; Dai, L.; Ni, Y. Fabrication of 3D Expanded Graphite-Based (MnO_2 Nanowalls and PANI Nanofibers) Hybrid as Bifunctional Material for High-Performance Supercapacitor and Sensor. *J. Electrochem. Soc.* **2019**, *166*, A3965–A3971. [CrossRef]
9. Xiong, C.; Li, M.; Zhao, W.; Duan, C.; Ni, Y. Flexible N-doped reduced graphene oxide/carbon nanotube-MnO_2 film as a multifunctional material for high-performance supercapacitors, catalysts and sensors. *J. Mater.* **2020**, *6*, 523–531. [CrossRef]
10. Saraf, M.; Natarajan, K.; Mobin, S.M. Robust Nanocomposite of Nitrogen-Doped Reduced Graphene Oxide and MnO2 Nanorods for High-Performance Supercapacitors and Nonenzymatic Peroxide Sensors. *ACS Sustain. Chem. Eng.* **2018**, *6*, 10489–10504. [CrossRef]
11. Kabir, H.; Gyan, I.O.; Cheng, I.F. Electrochemical modification of a pyrolytic graphite sheet for improved negative electrode performance in the vanadium redox flow battery. *J. Power Sources* **2017**, *342*, 31–37. [CrossRef]
12. Chang, Y.-C.; Chen, J.-Y.; Kabtamu, D.M.; Lin, G.-Y.; Hsu, N.-Y.; Chou, Y.-S.; Wei, H.-J.; Wang, C.-H. High efficiency of CO_2-activated graphite felt as electrode for vanadium redox flow battery application. *J. Power Sources* **2017**, *364*. 1–8. [CrossRef]
13. Panda, S.; Rout, T.K.; Prusty, A.D.; Ajayan, P.M.; Nayak, S.; Panda, S. Electron Transfer Directed Antibacterial Properties of Graphene Oxide on Metals. *Adv. Mater.* **2018**, *30*, 1702149. [CrossRef] [PubMed]
14. Li, M.; Bai, L.; Wen, X.; Guan, J. Ultrafine Mn_3O_4 nanoparticles supported on functionalized-graphite towards efficient photochemical and electrochemical water oxidation catalysis. *Int. J. Hydrog. Energy* **2018**, *43*, 15807–15814. [CrossRef]

15. Huang, Y.; Sun, L.; Yu, Z.; Jiang, R.; Huang, J.; Hou, Y.; Yang, F.; Zhang, B.; Zhang, R.; Zhang, Y. Adjustable anchoring of Ni/Co cations by oxygen-containing functional groups on functionalized graphite paper and accelerated mass/electron transfer for overall water splitting. *Catal. Sci. Technol.* **2020**, *10*, 2627–2643. [CrossRef]
16. Yue, L.; Jia, D.; Tang, J.; Zhang, A.; Liu, F.; Chen, T.; Barrow, C.; Yang, W.; Liu, J. Improving the rate capability of ultrathin NiCo-LDH nanoflakes and FeOOH nanosheets on surface electrochemically modified graphite fibers for flexible asymmetric supercapacitors. *J. Colloid Interface Sci.* **2020**, *560*, 237–246. [CrossRef]
17. Chang, H.-W.; Lu, Y.-R.; Chen, J.-L.; Chen, C.-L.; Lee, J.-F.; Chen, J.-M.; Tsai, Y.-C.; Chang, C.-M.; Yeh, P.-H.; Chou, W.-C.; et al. Nanoflaky MnO_2/functionalized carbon nanotubes for supercapacitors: An in situ X-ray absorption spectroscopic investigation. *Nanoscale* **2015**, *7*, 1725–1735. [CrossRef] [PubMed]
18. Luo, D.; Zhang, X. The effect of oxygen–containing functional groups on the H2 adsorption of graphene–based nanomaterials: Experiment and theory. *Int. J. Hydrogen Energy* **2018**, *43*, 5668–5679. [CrossRef]
19. Kavinkumar, T.; Sastikumar, D.; Manivannan, S. Effect of functional groups on dielectric, optical gas sensing properties of graphene oxide and reduced graphene oxide at room temperature. *RSC Adv.* **2015**, *5*, 10816–10825. [CrossRef]
20. Gong, L.; Yin, B.; Li, L.-P.; Yang, M.-B. Nylon-6/Graphene composites modified through polymeric modification of graphene. *Compos. Part B Eng.* **2015**, *73*, 49–56. [CrossRef]
21. Jung, J.Y.; Hong, Y.L.; Kim, J.-G.; Kim, M.J.; Kim, Y.-K.; Kim, N.D. New insight of tailor-made graphene oxide for the formation of atomic Co-N sites toward hydrogen evolution reaction. *Appl. Surf. Sci.* **2021**, *563*, 150254. [CrossRef]
22. Radovic, L.R.; Mora-Vilches, C.V.; Salgado-Casanova, A.J.; Buljan, A. Graphene functionalization: Mechanism of carboxyl group formation. *Carbon* **2018**, *130*, 340–349. [CrossRef]
23. Ahmad, H.; Fan, M.; Hui, D. Graphene oxide incorporated functional materials: A review. *Compos. Part B: Eng.* **2018**, *145*, 270–280. [CrossRef]
24. Yang, D.; Yu, Q.; Gao, L.; Mao, L.; Yang, J.-H. The additive effect of graphene in nickel phosphate/graphene composite and enhanced activity for electrochemical oxidation of methanol. *Appl. Surf. Sci.* **2017**, *416*, 503–510. [CrossRef]
25. Zhang, W.; Liu, F.; Li, Q.; Shou, Q.; Cheng, J.; Zhang, L.; Nelson, B.J.; Zhang, X. Transition metal oxide and graphene nanocomposites for high-performance electrochemical capacitors. *Phys. Chem. Chem. Phys.* **2012**, *14*, 16331–16337. [CrossRef] [PubMed]
26. Christy, M.; Jang, H.; Nahm, K.S. Cobaltite oxide nanosheets anchored graphene nanocomposite as an efficient oxygen re-duction reaction (ORR) catalyst for the application of lithium-air batteries. *J. Power Sources* **2015**, *288*, 451–460.
27. Makgopa, K.; Ejikeme, P.M.; Jafta, C.J.; Raju, K.; Zeiger, M.; Presser, V.; Ozoemena, K.I. A high-rate aqueous symmetric pseudocapacitor based on highly graphitized onion-like carbon/birnessite-type manganese oxide nanohybrids. *J. Mater. Chem. A* **2015**, *3*, 3480–3490. [CrossRef]
28. Chen, H.; Wang, M.Q.; Yu, Y.; Liu, H.; Lu, S.; Bao, S.-J.; Xu, M. Assembling Hollow Cobalt Sulfide Nanocages Array on Graphene-like Manganese Dioxide Nanosheets for Superior Electrochemical Capacitors. *ACS Appl. Mater. Interfaces* **2017**, *9*, 35040–35047. [CrossRef] [PubMed]
29. Song, Z.; Ma, Y.-L.; Li, C.-E. The residual tetracycline in pharmaceutical wastewater was effectively removed by using MnO_2/graphene nanocomposite. *Sci. Total. Environ.* **2019**, *651*, 580–590. [CrossRef]
30. Song, L.; Li, C.; Chen, W.; Liu, B.; Zhao, Y. Highly efficient MnO2/reduced graphene oxide hydrogel motors for organic pol-lutants removal. *J. Mater. Sci.* **2020**, *55*, 1984–1995. [CrossRef]
31. Wang, D.; Wang, K.; Sun, L.; Wu, H.; Wang, J.; Zhao, Y.; Yan, L.; Luo, Y.; Jiang, K.; Li, Q.; et al. MnO_2 nanoparticles anchored on carbon nanotubes with hybrid supercapacitor-battery behavior for ultrafast lithium storage. *Carbon* **2018**, *139*, 145–155. [CrossRef]
32. Lee, S.-W.; Bak, S.-M.; Lee, C.-W.; Jaye, C.; Fischer, D.A.; Kim, B.-K.; Yang, X.-Q.; Nam, K.-W.; Kim, K.-B. Structural Changes in Reduced Graphene Oxide upon MnO2 Deposition by the Redox Reaction between Carbon and Permanganate Ions. *J. Phys. Chem. C* **2014**, *118*, 2834–2843. [CrossRef]
33. Wu, L.; Klie, R.; Zhu, Y.; Jooss, C. Experimental confirmation of Zener-polaron-type charge and orbital ordering in Pr1− xCax MnO_3. *Phys. Rev. B* **2007**, *76*, 174210. [CrossRef]
34. Céspedes, E.; Laguna-Marco, M.; Jiménez-Villacorta, F.; Chaboy, J.; Boada, R.; Guglieri, C.; Andrés, A.d.; Prieto, C. On the origin of the magnetism of Mn–Zn–O systems: Structural, electronic, and magnetic study of exotic $MnO_{2-\delta}$/ZnO thin films. *J. Phys. Chem. C* **2011**, *115*, 24092–24101. [CrossRef]
35. Chang, H.-W.; Huang, Y.-C.; Chen, J.-L.; Chen, C.-L.; Chen, J.-M.; Wei, D.-H.; Chou, W.-C.; Dong, C.-L.; Tsai, Y.-C. Soft X-ray absorption spectroscopic investigation of MnO_2/graphene nanocomposites used in supercapacitor. *Catal. Today* **2021**, *45*, 17120–17127. [CrossRef]
36. Durai, G.; Kuppusami, P.; Maiyalagan, T.; Ahila, M.; Kumar, P.V. Supercapacitive properties of manganese nitride thin film electrodes prepared by reactive magnetron sputtering: Effect of different electrolytes. *Ceram. Int.* **2019**, *45*, 17120–17127. [CrossRef]
37. Achour, A.; Guerra, A.; Moulaï, F.; Islam, M.; Hadjersi, T.; Ahmad, I.; Parvez, S.; Boukherroub, R.; Pireaux, J.-J. MnOx thin film based electrodes: Role of surface point defects and structure towards extreme enhancement in specific capacitance. *Mater. Chem. Phys.* **2019**, *242*, 122487. [CrossRef]
38. Zhang, J.; Wang, Y.; Wu, J.; Shu, X.; Yu, C.; Cui, J.; Qin, Y.; Zhang, Y.; Ajayan, P.M.; Wu, Y. Remarkable supercapacitive performance of TiO_2 nanotube arrays by introduction of oxygen vacancies. *Chem. Eng. J.* **2017**, *313*, 1071–1081. [CrossRef]

39. Liu, Y.; Wang, N.; Yang, C.; Hu, W. Sol–gel synthesis of nanoporous $NiCo_2O_4$ thin films on ITO glass as high-performance supercapacitor electrodes. *Ceram. Int.* **2016**, *42*, 11411–11416. [CrossRef]
40. Prasad, K.; Rajasekhara Reddy, G.; Rajesh, M.; Babu, P.R.; Shanmugam, G.; Sushma, N.J.; Pratap Reddy, M.S.; Deva Prasad Raju, B.; Mallikarjuna, K. Electrochemical performance of 2D-hierarchical sheet-Like $ZnCo_2O_4$ microstructures for superca-pacitor applications. *Crystals* **2020**, *10*, 566. [CrossRef]
41. Zhu, H.; Li, L.; Zhou, W.; Shao, Z.; Chen, X. Advances in non-enzymatic glucose sensors based on metal oxides. *J. Mater. Chem. B* **2016**, *4*, 7333–7349. [CrossRef]
42. Mirzaei, H.; Nasiri, A.A.; Mohamadee, R.; Yaghoobi, H.; Khatami, M.; Azizi, O.; Zaimy, M.A.; Azizi, H. Direct growth of ternary copper nickel cobalt oxide nanowires as binder-free electrode on carbon cloth for nonenzymatic glucose sensing. *Microchem. J.* **2018**, *142*, 343–351. [CrossRef]
43. Li, M.; Liu, L.; Xiong, Y.; Liu, X.; Nsabimana, A.; Bo, X.; Guo, L. Bimetallic MCo (M=Cu, Fe, Ni, and Mn) nanoparticles doped-carbon nanofibers synthetized by electrospinning for nonenzymatic glucose detection. *Sens. Actuators B Chem.* **2015**, *207*, 614–622. [CrossRef]
44. Xiao, F.; Li, Y.; Gao, H.; Ge, S.; Duan, H. Growth of coral-like PtAu–MnO_2 binary nanocomposites on free-standing graphene paper for flexible nonenzymatic glucose sensors. *Biosens. Bioelectron.* **2013**, *41*, 417–423. [CrossRef] [PubMed]
45. Lin, K.-C.; Huang, L.-H.; Chen, S.-M. Electrochemical synthesis of mixed-valence manganese/copper hybrid composite using graphene oxide and multi-walled carbon nanotubes for nonenzymatic glucose sensor. *J. Electroanal. Chem.* **2014**, *735*, 36–42. [CrossRef]
46. Zhang, C.; Zhang, Y.; Miao, Z.; Ma, M.; Du, X.; Lin, J.; Han, B.; Takahashi, S.; Anzai, J.-I.; Chen, Q. Dual-function amperometric sensors based on poly(diallydimethylammoniun chloride)-functionalized reduced graphene oxide/manganese dioxide/gold nanoparticles nanocomposite. *Sens. Actuators B Chem.* **2016**, *222*, 663–673. [CrossRef]

nanomaterials

Article

Vertically Aligned Binder-Free TiO_2 Nanotube Arrays Doped with Fe, S and Fe-S for Li-ion Batteries

Suriyakumar Dasarathan [1,2,†], Mukarram Ali [2,3,†], Tai-Jong Jung [2,3], Junghwan Sung [1,2], Yoon-Cheol Ha [3], Jun-Woo Park [3] and Doohun Kim [1,2,*]

1 Nano Hybrid Technology Research Center, Electrical Materials Research Division, Korea Electrotechnology Research Institute, Changwon 51543, Korea; suriyakumar@keri.re.kr (S.D.); swg9311@keri.re.kr (J.S.)
2 Department of Electro-Functionality Materials Engineering, University of Science and Technology (UST), Daejeon 305-333, Korea; mali@keri.re.kr (M.A.); tjjung@keri.re.kr (T.-J.J.)
3 Next Generation Battery Research Center, Electrical Materials Research Division, Korea Electrotechnology Research Institute, Changwon 51543, Korea; ycha@keri.re.kr (Y.-C.H.); parkjw@keri.re.kr (J.-W.P.)
* Correspondence: kdh0121@keri.re.kr
† S.D. and M.A. contributed equally to this work.

Abstract: Vertically aligned Fe, S, and Fe-S doped anatase TiO_2 nanotube arrays are prepared by an electrochemical anodization process using an organic electrolyte in which lactic acid is added as an additive. In the electrolyte, highly ordered TiO_2 nanotube layers with greater thickness of 12 μm, inner diameter of approx. 90 nm and outer diameter of approx. 170 nm are successfully obtained. Doping of Fe, S, and Fe-S via simple wet impregnation method substituted Ti and O sites with Fe and S, which leads to enhance the rate performance at high discharge C-rates. Discharge capacities of TiO_2 tubes increased from 0.13 mAh cm^{-2}(bare) to 0.28 mAh cm^{-2} for Fe-S doped TiO_2 at 0.5 C after 100 cycles with exceptional capacity retention of 85 % after 100 cycles. Owing to the enhancement of thermodynamic and kinetic properties by doping of Fe-S, Li-diffusion increased resulting in remarkable discharge capacities of 0.27 mAh cm^{-2} and 0.16 mAh cm^{-2} at 10 C, and 30 C, respectively.

Keywords: Li-ion batteries; binder-free electrodes; TiO_2 nanotube arrays; electrochemical anodization; elemental doping

Citation: Dasarathan, S.; Ali, M.; Jung, T.-J.; Sung, J.; Ha, Y.-C.; Park, J.-W.; Kim, D. Vertically Aligned Binder-Free TiO_2 Nanotube Arrays Doped with Fe, S and Fe-S for Li-ion Batteries. *Nanomaterials* **2021**, *11*, 2924. https://doi.org/10.3390/nano11112924

Academic Editor: Christophe Detavernier

Received: 17 September 2021
Accepted: 25 October 2021
Published: 31 October 2021

1. Introduction

Titanium based oxides have drawn great attention in the lithium ion battery (LIB) world because of their superior thermal stability compared with the conventional graphite anode. Moreover, this class of active material shows other interesting features such as low cost, non-toxicity, and small volume change process (2–3%) during the lithium insertion and extraction, along with an excellent cycling life [1]. In general, bulk TiO_2 shows a low theoretical capacity of 175–360 mAh g^{-1} and a low electrical conductivity. The electrochemical performance and the reversible capacity of titanium-based oxides mainly depend on their microscopic structure, morphology, and particle size [2]. Interestingly, the nanostructured titanium oxide leads to a superior capacity, longer cycling life, and higher rate capacity than bulk TiO_2 [2,3].

TiO_2 shows excellent safety and stability characteristics at the operation potential of 1.5 V vs. Li/Li^+. Moreover, TiO_2 has high electro-activity, strong oxidation capability, good chemical stability, high abundance, and structural diversity [4–6]. Where TiO_2 based materials possess boosting the performance of battery, still they have limitations such as poor electrical conductivity and low Li-ion diffusivity, which result in poor electrochemical performance, thus hindering their practical application [4–7]. So far, many attempts have been made to compensate for this problem by means of using low-dimension (1 D, 2 D) TiO_2 nanostructures composites [8,9].

The reduction of $Ti^{4+} \rightarrow Ti^{3+}$ is accompanied by Li ion insertion/extraction into the oxide structure. Electrochemical fading in the crystal structure due to pulverizing of electrodes during volume expansion and reduction. In the advantage of taking forward TiO_2 nanotubes (TNTs) as anodes, (i) the well-ordered electrode geometry reduces the ion diffusion path (ii) mechanical strain arising during Li ion insertion/extraction cycling can be accommodated, and (iii) therefore, the structural stability is maintained [10].

On the other hand, introducing the heteroatoms such as C, Nb, N, B, W, Sn, and Fe into TiO_2 nanostructures is a promising way to stabilize these structures and improve the electron flow to accelerate the kinetics during electrochemical processes [11–15]. This will enhance the rate performance, cycling behavior, and specific capacity of TNTs in Li-ion battery application. However, the single substitutional doping in TiO_2 mostly has a low thermodynamic solubility [16–19]. Therefore, a concept of co-doping idea is attempted in various studies as noted (including carbon, nitrogen; fluorine, nitrogen; chromium, nitrogen; and sulfur, nitrogen) in TiO_2 nanoparticles as an anode material for Li-ion batteries [2,8,9,12,15,17–19]. In the previous studies, it has been demonstrated that the Fe-S co-dopant pairs can substantially narrow band gap and effectively modify the electronic structure of TiO_2 [9,19]. However, so far only limited studies on doping of metal atoms in TNTs as anode for LIB are available [1,18–20].

In this study, vertically aligned self-organized TiO_2 nanotubes are prepared by an electrochemical anodizing technique [21–24]. In conventional LIBs, up to 10% of "dead weights" loss occur when the additives such as polymeric binder and carbon conductor are used. However, these anodic TiO_2 nanotube arrays directly formed on Ti can be used as an anode in LIB, so-called binder-free electrode. Furthermore, in order to enhance the electrochemical performance of TNTs, Fe-S co-doping was done by a wet immersion technique for boosting the electronic pathways and lithium ion diffusion coefficient, which closure with the result of greater storage performance.

2. Materials and Methods

2.1. Preparation of TNTs

Pre-cleaned Ti-foils (0.1 mm thick, 99.99% purity, Nilaco) were anodized in an electrolyte composed of 1.5 M lactic acid, 0.1 M ammonium fluoride and 5 wt. % deionized water in ethylene glycol [25]. The anodization was carried out in a two-electrode cell configuration: a Pt mesh was used as the counter electrode and the Ti foils were used as the working electrode. The anodization was conducted by using a high-voltage potentiostat (OPS-22101, ODA, Incheon, Korea) at a DC voltage of 120 V for 300 s, 600 s, and 800 s with the electrolyte temperature at 60 °C as shown in Figure 1. The obtained samples were rinsed in ethanol and dried in an oven at room temperature.

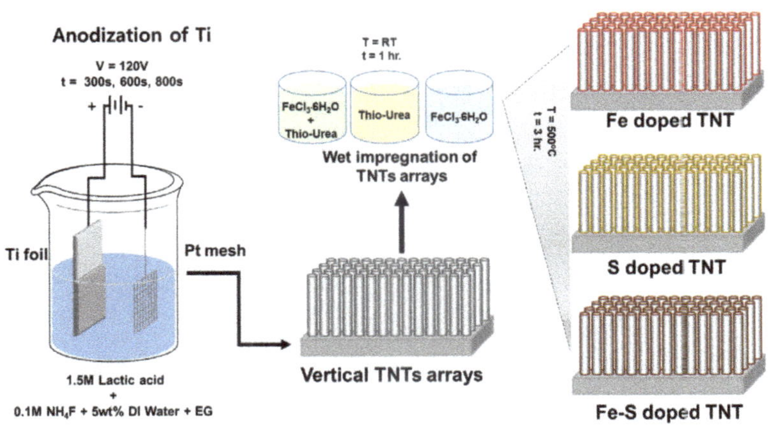

Figure 1. A schematic illustration of the preparation steps for Fe, S, and Fe-S doped TNTs.

2.2. Preparation of Fe-S co-doped TNTs

Fe, S, and Fe-S co-doped samples were prepared by using simple wet impregnation method as mentioned in a previous report [20]. For the solution preparation, 0.48 g of $FeCl_3 \cdot 6H_2O$, 0.274 g of thiourea, mixture of 0.48 g of $FeCl_3 \cdot 6H_2O$ and 0.274 g of thiourea were dissolved in 20 mL of absolute ethanol under vigorous stirring until the mixture turned into clear. Subsequently, the prepared TNT samples were soaked in the solutions for 1 hr and then kept for drying at room temperature. In order to obtain the anatase crystalline TNTs, the dried samples were annealed at 500 °C for 3 h. using a tube furnace (XY-1400S, Hantech, Ulsan, Korea).

2.3. Materials Characterization

The morphological study of the synthesized materials was performed using a field-emission microscope (Hitachi FE-SEM S4800, Chiyoda City, Tokyo, Japan) equipped with energy dispersive spectroscopy as well. The structure and crystalline phase of the samples were characterized by X-ray powder diffraction (XRD, Philips, X-pert PRO MPD, Amsterdam, Netherlands) with Cu Kα ($\lambda = 0.15406$ nm). The electronic states of elements were characterized by X-ray photoelectron spectroscopy (XPS, K-Alpha + XPS System, Thermo Scientific, Loughborough, UK). XPS was conducted with a monochromatic Al Kα source ($h\upsilon$= 1486.6 eV) with a spot size of 400 μm.

2.4. Electrochemical Characterization

The vertically aligned TNTs arrays doped with Fe, S, and Fe-S grown on Ti foils was cut into disk (diameter of 14 mm) and used as the anode for the electrochemical tests. The weight of the Ti substrate and the active material (bare, Fe, S, and Fe-S doped TNT layers) are shown in Table S2. A coin half-cell with a polypropylene membrane separator (Celgard 2325, Celgard Inc., Charlotte, NC, USA) and a Li-metal foil (thickness = 500 μm, purity 99.9%) as the counter-electrode was used to evaluate the electrochemical performance. The fabricated TNT disks were directly used as a working electrode in the electrochemical cell without adding any conductive carbon or binder. The electrolyte was 1 M $LiPF_6$ dissolved in 1:1:1, *v/v/v* mixture of ethylene carbonate, dimethyl carbonate, and ethyl methyl carbonate (EC: DMC: EMC) with 5% fluoroethylene carbonate (FEC). The amount of the used electrolyte was approximately 15 mL (g) and the ratios of electrolyte/active material are provided in Table S2. The assembled cells were galvanostatically cycled at different C-rates ranging from 0.2 C to 30 C in a potential range of 0.5–3 V using a multi-channel battery tester (MACCOR). Cyclic voltammetry (CV) was performed by using an electrochemical workstation (VMP3, Bio-Logic, Claix, France) with the same coin cell in the scan range of 0.5–3.0 V at a scan rate of either 1 mV s^{-1} or 0.5 mV s^{-1}. Electrochemical impedance spectroscopy (EIS) test was performed by using an electrochemical workstation (VSP-300, Bio-Logic, Claix, France) in the frequency range of 10^{-2}–10^{+5} Hz.

3. Results and Discussion

3.1. Morphology and Crystal Structure

Self-organized vertically aligned TiO_2 nanotube layers were prepared via electrochemical anodization of Ti foils in the lactic acid added F^- ion containing electrolyte with 120 V at 60 °C for 300 s, 600 s, and 800 s to obtain layer thickness of 12 μm, 36 μm, and 60 μm, respectively. The length of the nanotubes mainly depends on the anodizing time, applied voltage, electrolyte temperature, etc [26]. Lactic acid added electrolyte was used to obtain a high layer thickness (i.e., a long length) in a relatively short time. This additive stabilized the TNTs formation and allowed the application of the high working voltage of 120 V and also the high temperature of 60 °C in the anodization process [25]. Figure 2a–c shows the FE-SEM results for the prepared TNTs with different anodizing time. The inset Figure 2a(1) shows the top view image which represents highly ordered TNT array having a uniform porosity, while Figure 2a(2) shows a cross-sectional view image which indicates the vertically aligned TNT array, and Figure 2a(3) shows the bottom view image of the hexagonally close packed TNT structure [27]. Figure 2d shows the SEM image of EDS mapping and elemental distribution result of Fe-S doped TNTs, where Fe, Ti, O peaks clearly shows the even distribution of elements.

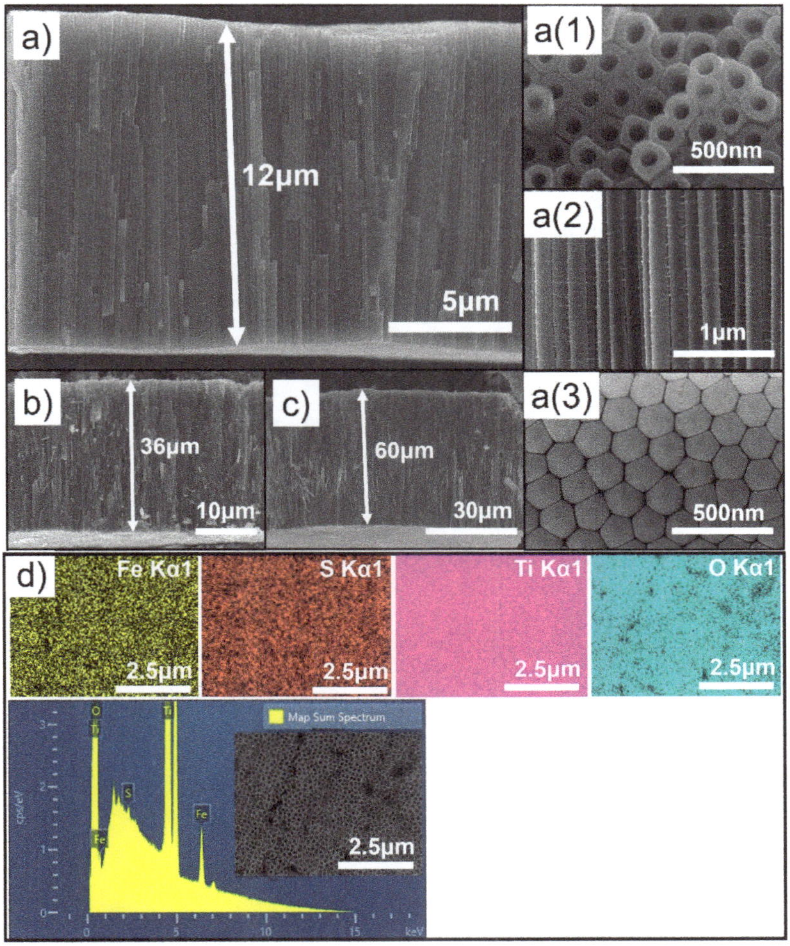

Figure 2. (**a**–**c**) shows FE-SEM images of as prepared TNTs. 12 μm, 35 μm, and 60 μm thick TNTs were prepared by anodization process at 120 V–60 °C for 300, 600, and 800 s, respectively. a(1), a(2), and a(3) shows the top, cross-section, and bottom view of 12 μm thick TNTs. (**d**) EDS results with elemental mapping of Fe-S doped TNTs. The inset SEM image shows the measured site.

Figure 3a shows the XRD patterns for bare, S, Fe, and Fe-S doped TNTs. It reveals the existence of the anatase TiO_2 for all the doped samples after annealing at 500 °C for 3 hrs with characteristic peaks at 24.025° (ICSD# 98-000-9852). A high peak intensity of the (004) orientation can be observed in all of the TNTs, which indicates a high percentage of (101) orientation in the growth direction of the TNTs (Figure S1a). Moreover, doping of different atoms did not affect the orientation except the intensity difference in (101) peak [25]. In Figure 3c, the diffraction peaks (101) of Fe and Fe-S doped TNTs shift to both lower angle and higher angle (Figure S6) are analyzed and the lattice parameters d_{011} are increased slightly from 0.450 nm for bare to 0.454 nm and 0.455 nm for Fe and Fe-S doped TNTs, respectively. This occurs due to the incorporation of Fe^{3+} (0.650 Å) having larger radius than Ti^{4+} (0.606 Å), suggesting that the Fe atoms have been successfully incorporated into the crystal structure of TiO_2 [12,25]. However, in case of S doped TNTs, the peak shift is towards a higher angle. This phenomenon can be attributed towards the larger radius of Ti^{4+} (0.606 Å) than S^{4+} (0.370 Å) and S^{6+} (0.290 Å) [28]. However, the doping amount of S is kept very small in both Fe-S, S doped TNTs due to its high reactivity. Hence the effect is minimal in the lattice parameter change (Table S1). Similarly, the effect of the doping on crystallite size can also be observed. The crystallite size decreases significantly as a result of S, Fe-S doping, while in case of Fe doping the decrease in the crystallite size is not as drastic as the former ones. The average crystallite size, calculated from Scherrer equation of Fe-S, Fe, S doped, and bare TNTs are approximately 32.38 nm, 39.64 nm, 30.31 nm, and 45.93 nm, respectively (Table S1).

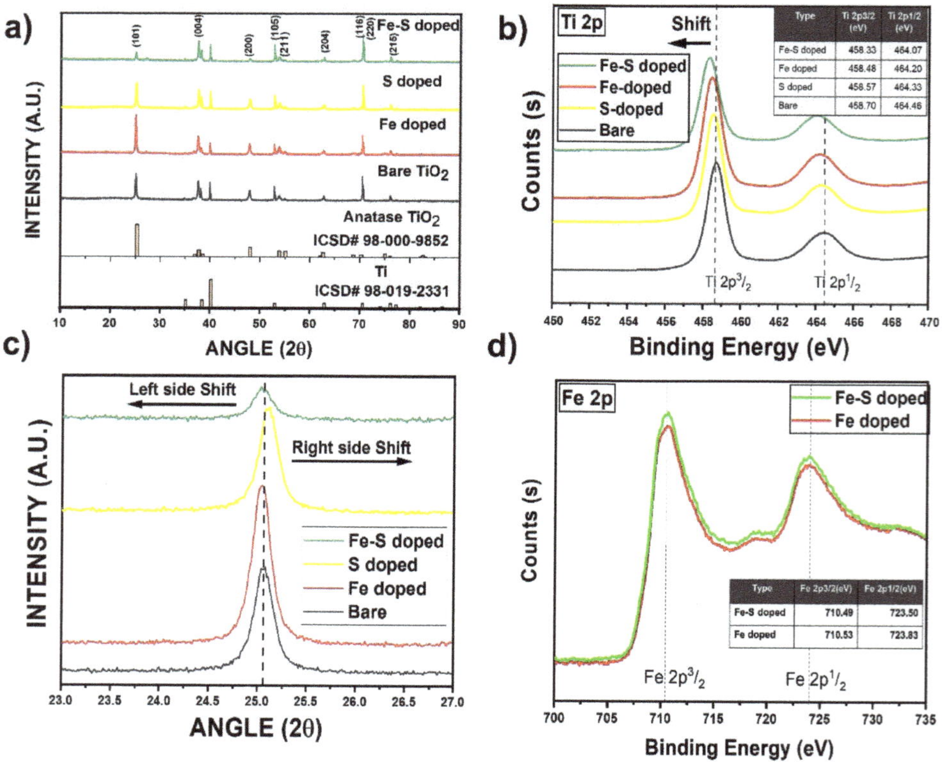

Figure 3. (**a**) XRD results for bare, Fe, S, and Fe-S doped TNTs. (**b**) Ti 2p XPS results for bare, Fe, S, and Fe-S doped TNTs showing a peak shift towards lower energies due to the presence of doped elements in the TNTs framework. (**c**) A zoomed in view of (101) XRD peak showing clear peaks shift for Fe, S, and Fe-S doped TNTs. (**d**) Fe 2p XPS results for Fe, Fe-S doped TNTs showing the incorporation of Fe in the TiO_2 framework.

The electronic states of the dopants and the parent atoms in TNTs were analyzed by (XPS). As illustrated in Figure 3b, the most intense peaks at 458.70 eV and 464.46 eV correspond to Ti 2p3/2 and Ti 2p1/2 spin–orbit splitting peaks, respectively [29,30]. In the case of Fe, S, and Fe-S doped TNTs, the peaks shift towards lower energy. This confirms the presence of dopants which have replaced Ti and O atoms due to the difference in the ionization energy decreases [31,32]. In case of Fe-S doped TNTs, not only S^{+2} and S^{+6} replaced Ti but there is a small peak observed at 163.01 eV (Figure S1c) which shows that there is also S^{-2} replacing O^{-2} as well [33].

Figure 3d shows the XPS spectrum of Fe 2p, where the binding energies at 710.53 eV, 723.83 eV, and 710.49 eV, 723.50 eV corresponding to Fe 2p3/2 and Fe 2p1/2 can be easily observed in Fe doped and Fe-S doped samples. This indicates that the doped Fe is mainly in 3^+ oxidation state. Due to the similarities of the radius between Fe^{3+} (0.650 Å) and Ti^{4+} (0.606 Å), Fe^{3+} can be incorporated into the lattice of TiO_2 to form Ti–O–Fe bonds [12]. In the case of S, Fe-S doped TNTs, the doping amounts for S are extremely small as discussed previously and accordingly S2p spectra is not observed in case of S however, in case of Fe-S doped TNTs S peaks were observed. In Figure S1b, the O 1s XPS spectra of Fe-S doped TNTs are split as two peaks. The energy of the peak located at 529.56 eV is equal to the O 1s electron binding energy for TiO_2. The other peak at 531.13 eV is ascribed to S–O–S bond, which confirms that the sulfur atoms replace a part of Ti sites [34] and in Figure S1c peak located at 163.01 eV [33]. This shows that sulfur is present in the form of S^{2-} by replacing O^{2-} which contradicts the previous report using thio-urea as a dopant precursor for sulfur [35]. This remarkable structural stability is expected to be conducive to reversible lithium storage with excellent cycle performances.

3.2. Electrochemical Performance

In order to investigate the effect of different dopants in TNTs on the electrochemical performances, cyclic voltammetry experiments for bare, Fe, S, and Fe-S doped TNTs electrodes were conducted at 0.5 mV s^{-1} as shown in Figure 4a. Moreover, CV for TNTs with different thickness were also done 1 mV s^{-1} to investigate the effect of higher thickness on lithiation/delithiation of TNTs (Figure 4b). In principle, the reaction equation can be used to express the lithium insertion and extraction in anatase TNTs electrode:

$$TiO_2 + nLi^+ + ne^- \leftrightarrow Li_nTiO_2 \quad (1)$$

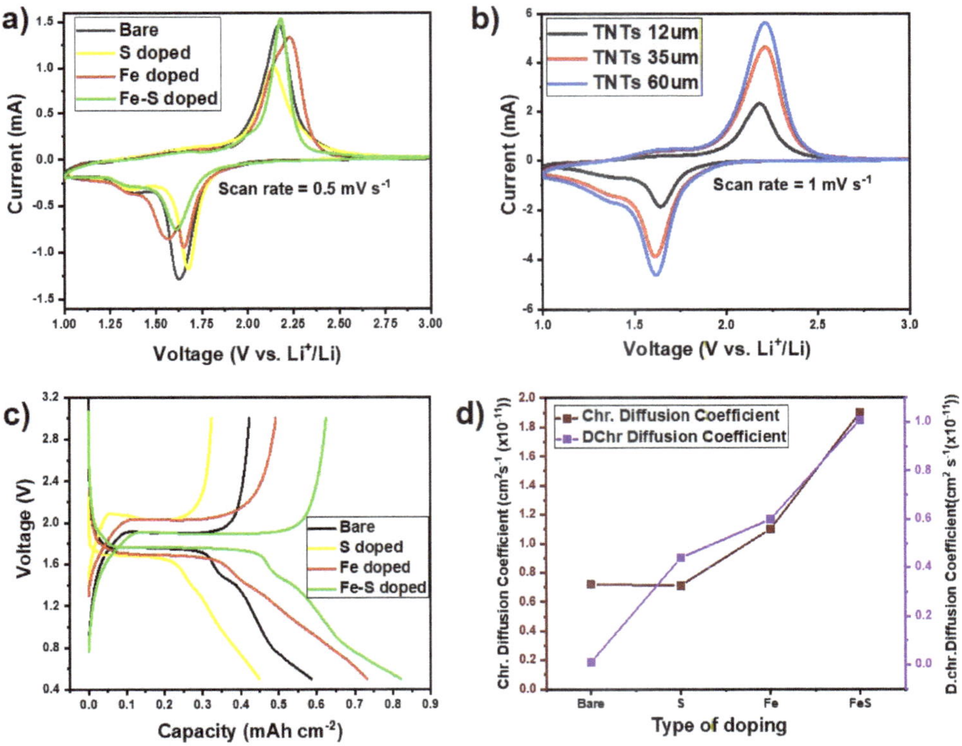

Figure 4. (**a**) Cyclic voltammograms of bare, Fe, S, and Fe-S doped TNTs with respect to Li metal at a scan rate of 0.5 mV s^{-1}. (**b**) Cyclic voltammograms of TNTs with the different thickness of 12 μm, 36 μm, and 60 μm. (**c**) 1st charge-discharge profiles of bare, Fe, S, and Fe-S doped 12 μm thick TNTs at 0.5 C. (**d**) Diffusion coefficient values calculated from Randles–Sevcik equation for bare, Fe, S, and Fe-S doped TNTs.

All the sample electrodes (Figure 4a) exhibit peak couples at 2.14/1.63 V (bare TNTs), 2.17/1.67 V (S doped TNTs), 2.19/1.64 V (Fe doped TNTs), and 2.20/1.66 V (Fe-S doped TNTs) corresponding to the transition of Li poor α-Li_xTiO_2 ($0.01 < x \leq 0.21$) with anatase structure to the orthorhombic β Li_xTiO_2(x~0.55) phase, their positions are in good agreement with those reported in the literature [36]. In case Fe-doped TiO_2 in reduction peaks, there is an additional reaction happens due to the over potential of phase transformation from TiO_2 to Li_xTiO_2. In order to reduce the polarization effect happens between the electrode and electrolyte interface, the Fe-doping in TiO_2 will reduce this effect and enhance the electronic conductivity, which results that further enhancement in electrochemical storage [37]. Moreover, there is a small peak pair visible at 1.62 V (lithiation) and 1.4 V (delithiation), which corresponds to a second phase change to fully lithiated $LiTiO_2$ [36]. This feature is more prominent in the case of higher-length tubes as shown in Figure 4b. Moreover, it can be seen that upon delithiation peak broadening is observed which is a characteristic for self-oriented TNTs. In contrast to bare, Fe, S, and Fe-S TNTs show a prominent second phase transition upon delithiation.

The diffusion coefficient of lithium ion during the Li^+ intercalation/de-intercalation processes can be calculated according to Randles–Sevcik equation [38]. Figures 4d and S4b shows the calculated diffusion coefficient values for doped TNTs and elongated TNTs. The Li^+ diffusion increases by doping of Fe, S, Fe-S. In case of doping Li^+ ion diffusion increased from 0.75×10^{-11} cm^2 s^{-1}(bare TNTs) to 0.13×10^{-10} cm^2 s^{-1} (Fe-S doped TNTs) for de-intercalation and 0.2×10^{-11} cm^2 s^{-1} (bare TNTs) to 0.12×10^{-10} cm^2 s^{-1} (Fe-S doped TNTs) for intercalation. It should be noted that Li^+ diffusion is greatly improved by co-doping with both Fe and S in anodic TNT's framework.

Moreover, preferentially oriented tubes are derived by anodization process and even directly use of these TNTs in LIB increases the charge/discharge capacities [39]. However, these performances are considerably enhanced by co-doping with Fe-S in the TNT structures. In addition, the CV curves at a scan rate of 0.5 mV s^{-1} as shown in Figure S2 are stable with almost overlaps from the second cycle, which indicate well posited dopant in the structures and excellent stability for Fe, S, and Fe-S doped TNT electrodes.

Galvanostatic charging/discharging was carried out at C-rate of 0.5 C in the range of 0.5–3.0 V vs. Li^+/Li^-, Initial discharge capacities for bare, S, Fe, and Fe-S doped TNTs anodes came out to be around 0.6 mAh cm^{-2}, 0.45 mAh cm^{-2}, 0.75 mAh cm^{-2}, and 0.84 mAh cm^{-2} (Figure 4c). It can be seen that the discharge capacities for S doped TNTs are lower when compared to others. This is because of the reaction of sulfur with Li to form Li_xS which in result decreases the discharge capacity [40].

However, further in this study it will be shown that this does not affect the cycling behavior of TNTs but increase cyclability of S doped TNTs. It can also be observed that the charge (1.9 V) and discharge (1.65 V) plateaus for Fe-S doped TNTs show lowest and highest among the other electrodes, respectively. This indicates the lowest electrochemical polarization and the best charge/discharge energy density of Fe, S doped TNTs anode. Moreover, the specific capacity is proportional to time interval (i.e., scan rate) of charge/discharge [11,40]. The sloping region for the Fe-S doped TNTs electrode below the plateau corresponds to the pseudo-capacitive lithium storage in the surface area [41].

$$C = i/(dV/dT) \tag{2}$$

Hence the sloping region will indicate the pseudo-capacitive lithium storage in the structures of TNTs. This shows that the shallowest slope (dV/dc) observed with the Fe-S doped TNTs electrode represents the highest capacitance values as compared to the steep slopes from the other three kinds of electrodes. The calculated capacitance values of bare, Fe, S, Fe-S doped TNTs are 46.5%, 52%, 50%, and 54%. The corresponding values shows the Fe-S doped TNTs have the high pseudo-capacitance behavior [42], which leads to increase in the capacity of TNTs from 0.6 mAh cm^{-2} for bare TNTs to 0.81 mAh cm^{-2} for Fe-S doped TNTs. It is noted that the capacity values are reported with areal unit (mAh cm^{-2}) due to the reason that the mass loading in anodized TiO_2 nanotubes (as an active materials) directly grown on Ti substrates (as a current collector) cannot be precisely measurable with gravimetric unit measurement since the electrode itself together with the TiO_2 active materials and the Ti current collector is one solid body and, moreover, with the areal capacity we can further discuss with our previous study of anodic TiO_2 in LIBs [43–47]. However, in Figure S3, we provide both areal and "estimated" gravimetric capacities for bare and Fe-S doped TNTs. It is clear that Fe-S doped TNTs show a prominent cycling as compared to bare TNTs. In Figure S5, we show the high rate of gravimetric capacity for different C-rates.

Moreover, in case of Fe, S, and Fe-S after 60 cycles (Figure 5a), they show higher discharge capacities of 0.24 mAh cm^{-2}, 0.19 mAh cm^{-2}, and 0.28 mAh cm^{-2} as compared to 0.13 mAh cm^{-2} of bare TNTs. Fe-S doped and bare TNTs were further cycled to 100. In the cycling results, it was found that Fe-S retained 85% discharge capacity (taken after 3rd cycle) as compared to bare TNTs with 65% (i.e., doping of Fe, S together increases the capacity retention and increased capacitive properties). Moreover, columbic efficiency in the 1st cycle is 76% for Fe-S doped TNTs and 60% for bare TNTs. These high irreversible capacities can be because of the higher length of the tubes and side reactions with sulfur and absorbed moisture due to high specific surface area [48]. However, this is redeemed in the 2nd cycle with 90.9% for Fe-S and 93.8% for bare TNTs. The decrease in columbic efficiency for Fe-S as compared to bare TNTs can be because of reaction of Li with S to form Li_xS, however, this product is redox active [40]. Hence, this will help in further cycling and increase the electrochemical stability with the high capacity retention as shown in Figure 5a.

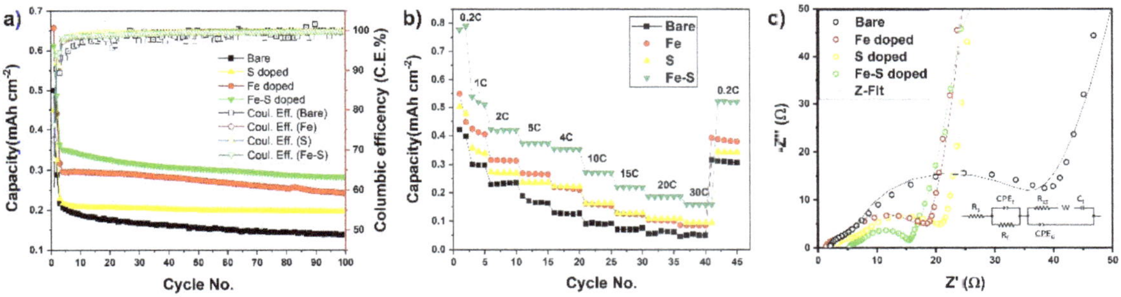

Figure 5. (**a**) Cycle vs. discharge capacity curves for bare, Fe, S, and Fe-S doped TNTs cycled at 0.5 C. (**b**) Rate capablity of bare, Fe, S, and Fe-S doped TNTs. (**c**) EIS curves of bare, Fe, S, and Fe-S doped TNTs after 100 cycles.

Figure 5b shows rate capability cycling results for bare, Fe, S, and Fe-S doped TNTs. The cells were charged at 0.2 C and discharged at different C-rate. It is evident that Fe-S doped TNTs shows the best rate capabilities at various C-rates compared to charge-discharge study is applied from 0.2 C to 30 C. All the doped TNTs showed exceptional rate capabilities at higher rates as compared to bare TNTs. Fe-S doped TNTs show 0.27 mAh cm^{-2} at 10 C and 0.14 mAh cm^{-2} at 30 C for two different C-rates as compared to 0.07 mAh cm^{-2} at 10 C and 0.03 mAh cm^{-2} at 30 C of bare TNTs. It is important to note that Fe-S doped TNTs retained the reversible specific capacities of 0.51 mAh cm^{-2} at 0.2 C as well when cycled again. These results clearly validate that the Fe-S doped TNTs electrodes exhibit superior lithium storage properties with prolonged cycle life and great rate capability for the fast discharge process.

To further understand the origin of the superior electrochemical properties of doped TNTs and to study differentiate kinetic and thermodynamic properties of the TNTs electrodes upon lithiation and delithiation, electrochemical impedance spectroscopy (EIS) was employed in the frequency range of 10^{-2}–10^{+5} Hz. This allows for differentiation of processes taking place at different time scales during Li insertion and extraction. Figure 5c shows us the EIS spectrum and the fitting circuit of bare, Fe, S, and Fe-S doped TNTs after 100 cycles. In all of the observed EIS spectrums, at the highest frequencies, a depressed semicircle can be observed which corresponds to the parallel combination of surface film resistance R_f and surface film capacitance C_f [48]. These thin surface films originate from the decomposition of compounds from the carbonate-based (EC, DMC) $LiPF_6$ electrolytes [49]. The second medium-frequency semi-circle relates to the interfacial charge transfer, represented by the charge transfer resistance, R_{ct}, and the double-layer capacitance, C_{dl}, in parallel insertion [7]. It can be seen that, after 100 cycles R_{ct} values decreased when compared to bare TNT electrodes. R_{ct} values of 20.78 ohm, 18.49 ohm, and 15.54 ohm were achieved for Fe, S, and Fe-S doped TNTs as compared to 38.47 ohm for bare TNTs. However, in the nanotube system, the charge transfer may not be the only rate-determining step; therefore, it is also necessary to consider solid state diffusion in the bulk material [50]. This shows that both the increase in Li^+ diffusion and charge transfer resistances of doped TNTs increased the cycling and rate cycling properties. In order to observe the difference between before and after cycling effect in Fe-S doped TNTs, EIS was done before cycling of Fe-S doped TNTs samples as well. Figure S4a shows the EIS spectrum and fitting of before and after 100 cycles. It can be observed that R_{ct} values for Fe-S doped TNTs decreased from 269 ohm to 15.54 ohm after 100 cycles. During lithiation, in case of TiO_2, fully lithiated Li_1TiO_2 forms with higher charge resistance, moreover, the decomposition of $LiPF_6$ electrolyte to LiF and Li_2CO_3 also increase the charge transfer resistance between electrode/electrolyte interfaces. However, by the introduction of Fe and S as co dopants the fully lithiated phase transition is improved by decreasing the charge transfer resistance and as a result higher reversibility during lithiation and de-lithiation is achieved [51]. This incredible increase in electronic conductivity and charge transfer can be attributed towards the modified electronic structure resulting in exceptional properties and uniform pathways for Li^+ intercalation/de-intercalation. Anodic TNTs improved the structural as well as the directional properties compared to wet synthesized TNTs. TiO_2 Nanoparticle of 0D material, which are smaller in size shortens the electronic pathways and increased the Li^+ diffusion in the lattice. Similarly, the 2D nanotube shows relative high surface area which provides more active surface sites compared to TiO_2 nanocrystals. It further enhances the fast lithium ion transfer between electrode and electrolyte, which leads to closure in deducing the charge transfer as shown in EIS results. Moreover, the increase in tubular size to 12 μm as compared to conventional small sized tubes helped in increasing the discharge capacities [20]. Doping of Fe-S, Fe, and S in TNTs increased the interlayer spacing, thus favoring Li^+ intercalation/de-intercalation as well. Fe-S doped TNTs showed best results with exceptional increase in cycling and rate cycling discharge capacities at exceptionally higher rates.

4. Conclusions

As a binder free anode, vertically aligned TNT layers with different dopants are prepared and performed as LIBs anodes. Co-doping of Fe and S proved to be fruitful, and amplified the storage performance to a comparable level to bare TNTs with notable high rates and cycling stability. This was attributed to boosting the Li^+ diffusion (D) and deducing the charge transfer resistance (R_{ct}). Moreover, the enhanced performance is due to the well-ordered geometry with a high aspect ratio and improved crystallinity of the TNT anodes.

Supplementary Materials: The following are available online at https://www.mdpi.com/article/10.3390/nano11112924/s1, Figure S1: (a) XRD results of amorphous and crystalline TNTs. The diffraction peak of TiO_2 amorphous annotate Ti metal. (b) XPS spectra for O 1s of Fe-S doped TNTs. (c) XPS spectra for S+4and S-2 of Fe-S doped TNTs, Figure S2: (a–d) Cyclic voltammetry curves at a scan rate 0.5mVs-1 for Fe-S, Fe, S doped TNTs and bare 12μm TNTs, respectively, Figure S3: Gravimetric and areal capacity retention with cycling of Fe-S doped and bare TNT anodes discharged at C-rate of 0.5 C, Figure S4: (a) EIS spectrum of before and after cycling for Fe-S doped TNTs. (b) Diffusion coefficient values as calculated using Randles–Sevcik equation for doped and elongated TNTs, respectively, Figure S5: High rate gravimetric capacities for bare, Fe, S, and Fe-S doped TNT's discharged at different C-rates, Figure S6: A zoomed in view of (105) and (211) XRD peak showing clear peaks shift for Fe, S, and Fe-S doped TNTs, Table S1: Calculated lattice parameters and crystallite sizes of TNTs, Table S2: Weight of Ti substrate and active material of TiO_2 nanotube layers and ratio of electrolyte/active material

Author Contributions: Conceptualization, S.D. and D.K.; methodology, S.D., T.-J.J. and J.S.; software, J.S.; validation, T.-J.J., M.A. and J.S.; formal analysis, M.A., J.S., Y.-C.H. and J.-W.P.; investigation, D.K., Y.-C.H. and J.-W.P.; resources, D.K.; data curation, S.D. and M.A.; writing—original draft preparation, S.D. and M.A.; writing—review and editing, D.K. and Y.-C.H.; visualization, J.-W.P.; supervision, D.K., Y.-C.H. and J.-W.P.; project administration, D.K., Y.-C.H. and J.-W.P.; funding acquisition, D.K. All authors have read and agreed to the published version of the manuscript.

Funding: This research was funded by the Primary Research Program(21A01042) of the Korea Electro-Technology Research Institute and by the Technology Innovation Program of the Ministry of Trade, Industry & Energy (MOTIE, Korea), grant number (20014581) and The APC was funded by 20014581.

Institutional Review Board Statement: Not applicable.

Informed Consent Statement: Not applicable.

Acknowledgments: This work was supported by the Primary Research Program (21A01042) of the Korea Electro- Technology Research Institute and by the Technology Innovation Program (20014581) of the Ministry of Trade, Industry & Energy (MOTIE, Korea).

Conflicts of Interest: The authors declare no conflict of interest.

References

1. Goriparti, S.; Miele, E.; De Angelis, F.; Di Fabrizio, E.; Proietti Zaccaria, R.; Capiglia, C. Review on recent progress of nanostructured anode materials for Li-ion batteries. *J. Power Sources* **2014**, *257*, 421–443. [CrossRef]
2. Hwang, H.; Kim, H.; Cho, J. MoS_2 Nanoplates consisting of disordered graphene-like layers for high rate lithium battery anode materials. *Nano Lett.* **2011**, *11*, 4826–4830. [CrossRef] [PubMed]
3. Moretti, A.; Kim, G.T.; Bresser, D.; Renger, K.; Paillard, E.; Marassi, R.; Winter, M.; Passerini, S. Investigation of different binding agents for nanocrystalline anatase TiO_2 anodes and its application in a novel, green lithium-ion battery. *J. Power Sources* **2013**, *221*, 419–426. [CrossRef]
4. Chen, Z.; Belharouak, I.; Sun, Y.K.; Amine, K. Titanium-based anode materials for safe lithium-ion batteries. *Adv. Funct. Mater.* **2013**, *23*, 959–969. [CrossRef]
5. Wagemaker, M.; Mulder, F.M. Properties and promises of nanosized insertion materials for li-ion batteries. *Acc. Chem. Res.* **2013**, *46*, 1206–1215. [CrossRef]
6. Meng, X.; Banis, M.N.; Geng, D.; Li, X.; Zhang, Y.; Li, R.; Abou-Rachid, H.; Sun, X. Controllable atomic layer deposition of one-dimensional nanotubular TiO_2. *Appl. Surf. Sci.* **2013**, *266*, 132–140. [CrossRef]
7. Acevedo-Peña, P.; Haro, M.; Rincón, M.E.; Bisquert, J.; Garcia-Belmonte, G. Facile kinetics of Li-ion intake causes superior rate capability in multiwalled carbon nanotube@TiO_2 nanocomposite battery anodes. *J. Power Sources* **2014**, *268*, 397–403. [CrossRef]
8. Guo, Y.G.; Hu, Y.S.; Sigle, W.; Maier, J. Superior electrode performance of nanostructured mesoporousTiO_2 (Anatase) through efficient hierarchical mixed conducting networks. *Adv. Mater.* **2007**, *19*, 2087–2091. [CrossRef]
9. Yang, Y.; Ji, X.; Jing, M.; Hou, H.; Zhu, Y.; Fang, L.; Yang, X.; Chen, Q.; Banks, C.E. Carbon dots supported upon N-doped TiO_2 nano rods applied into sodium and lithium ion batteries. *J. Mater. Chem. A* **2015**, *3*, 5648–5655. [CrossRef]
10. Lee, K.; Mazare, A.; Schmuki, P. One-Dimensional titanium dioxide nanomaterials: Nanotubes. *Chem. Rev.* **2014**, *114*, 9385–9454. [CrossRef]
11. Xu, J.; Wang, Y.; Li, Z.; Zhang, W.F. Preparation and electrochemical properties of carbon-doped TiO_2 nanotubes as an anode material for lithium-ion batteries. *J. Power Sources* **2008**, *175*, 903–908. [CrossRef]
12. Andriamiadamanana, C.; Laberty-Robert, C.; Sougrati, M.T.; Casale, S.; Davoisne, C.; Patra, S.; Sauvage, F. Room-Temperature synthesis of iron-doped anatase TiO_2 for lithium-ion batteries and photocatalysis. *Inorg. chem.* **2014**, *53*, 10129–10139. [CrossRef]
13. Lübke, M.; Johnson, I.; Makwana, N.M.; Brett, D.; Shearing, P.; Liu, Z.; Darr, J.A. High power TiO_2 and high capacity Sn-doped TiO_2 nanomaterial anode for lithium-ion batteries. *J. Power Sources* **2015**, *294*, 94–102. [CrossRef]
14. Zhang, Y.; Du, F.; Yan, X.; Jin, Y.; Zhu, K.; Wang, X.; Li, H.; Chen, G.; Wang, C.; Wei, Y. Improvements in the electrochemical kinetic properties and rate capability of anatase titanium dioxide nanoparticles by nitrogen doping. *ACS Appl. Mater. Interfaces* **2014**, *6*, 4458–4465. [CrossRef] [PubMed]
15. Lübke, M.; Shin, J.; Marchand, P.; Brett, D.; Shearing, P.; Liu, Z.; Darr, J.A. Highly pseudo capacitive Nb-doped TiO_2 high power anodes for lithium-ion batteries. *J. Mater. Chem. A* **2015**, *3*, 22908–22914. [CrossRef]
16. Tan, Y.N.; Wong, C.L.; Mohamed, A.R. An Overview on the photocatalytic activity of nano-doped-TiO_2 in the degradation of organic pollutants. *ISRN Mater. Sci.* **2011**, *2011*, 1–18. [CrossRef]

17. Yu, L.; Wang, Z.; Zhang, L.; Wu, H.; Bin, X.; Lou, W. TiO_2 nanotube arrays grafted with Fe_2O_3 hollow nanorods as integrated electrodes for lithium-ion batteries. *J. Mater. Chem. A* **2013**, *1*, 122–127. [CrossRef]
18. Chamanzadeh, Z.; Noormohammadi, M. Self-organized and uniform TiO_2 nanotube arrays with optimized NH_4F concentration in electrolyte by high voltage electrochemical anodization. *Mater. Res. Express* **2018**, *5*, 055025. [CrossRef]
19. Sabaghi, D.; Madian, M.; Omar, A.; Oswald, S.; Uhlemann, M.; Maghrebi, M.; Baniadam, M.; Mikhailova, D. Directly anodized sulfur-doped TiO_2 nanotubes as improved anodes for Li-ion batteries. *Batteries* **2020**, *6*, 51. [CrossRef]
20. Fang, J.; Liu, W.; Yu, F.; Qin, F.; Wang, M.; Zhang, K.; Lai, Y. Fe, S co-doped anatase TiO_2 nanotubes as anodes with improved electrochemical performance for lithium ion batteries. *RSC Adv.* **2016**, *6*, 70133–70140. [CrossRef]
21. Macák, J.M.; Tsuchiya, H.; Schmuki, P. High-Aspect-Ratio TiO_2 nanotubes by anodization of titanium. *Angew. Chem Int. Ed.* **2005**, *44*, 2100–2102. [CrossRef]
22. Roy, P.; Berger, S.; Schmuki, P. TiO_2 nanotubes: Synthesis and Applications. *Angew. Chem. Int. Ed.* **2011**, *50*, 2904–2939. [CrossRef]
23. Albu, S.P.; Ghicov, A.; Macak, J.M.; Schmuki, P. 250 μm long anodic TiO_2 nanotubes with hexagonal self-ordering. *Physic. Stat. Solid.* **2007**, *1*, R65–R67. [CrossRef]
24. Macák, J.M.; Tsuchiya, H.; Ghicov, A.; Yasuda, K.; Hahn, R.; Bauer, S.; Schmuki, P. TiO_2 nanotubes: Self-organized electrochemical formation, properties and applications. *Curr. Opin. Solid State Mater. Sci.* **2007**, *11*, 3–18. [CrossRef]
25. So, S.; Lee, K.; Schmuki, P. Ultrafast growth of highly ordered anodic TiO_2 nanotubes in lactic acid electrolytes. *J. Am. Chem. Soc.* **2012**, *134*, 11316–11318. [CrossRef]
26. Kowalski, D.; Kim, D.; Schmuki, P. TiO_2 nanotubes, nanochannels and mesosponge: Self-organized formation and applications. *Nano Today* **2013**, *8*, 235–264. [CrossRef]
27. Macák, J.M.; Tsuchiya, H.; Taveira, L.; Aldabergerova, S.; Schmuki, P. Smooth Anodic TiO_2 nanotubes. *Angew. Chem. Int. Ed.* **2005**, *44*, 7463–7465. [CrossRef]
28. Bayati, M.R.; Moshfegh, A.Z.; Golestani-Fard, F. Micro-arc oxidized S-TiO_2 nanoporous layers: Cationic or anionic doping? *Mater. Lett.* **2010**, *64*, 2215–2218. [CrossRef]
29. Zhou, M.; Yu, J.; Cheng, B. Effects of Fe-doping on the photocatalytic activity of mesoporous TiO_2 powders prepared by an ultrasonic method. *J. Hazard. Mater.* **2006**, *137*, 1838–1847. [CrossRef] [PubMed]
30. Hiroyuki, U.; Sho, Y.; Kuniaki, W.; Masahiro, S.; Hiroki, S. Nb-doped rutile TiO_2: A potential anode material for Na-ion battery. *ACS Appl. Mater. Interfaces* **2015**, *7*, 6567–6573.
31. Yan, D.; Yu, C.; Bai, Y.; Zhang, W.; Chen, T.; Hu, B.; Sun, Z.; Pan, L. Sn-doped TiO_2 nanotubes as superior anode materials for sodium ion batteries. *Chem. Commun.* **2015**, *51*, 8261–8264. [CrossRef] [PubMed]
32. Ni, J.; Fu, S.; Wu, C.; Maier, J.; Yu, Y.; Li, L. Self-Supported nanotube arrays of sulfur-doped TiO_2 enabling ultrastable and robust sodium storage. *Adv. Mater.* **2016**, *28*, 2259–2265. [CrossRef] [PubMed]
33. Piatkowska, A.; Janus, M.; Szymanski, K.; Mozia, S. C-, N- and S-doped TiO_2 photo catalysts: A review. *Catalysts* **2021**, *11*, 144. [CrossRef]
34. Umebayashi, T.; Yamaki, T.; Itoh, H.; Asai, K. Band gap narrowing of titanium dioxide by sulfur doping. *Appl. Phys. Lett.* **2002**, *81*, 454–456. [CrossRef]
35. Ohno, T.; Akiyoshi, M.; Umebayashi, T.; Asai, K.; Mitsui, T.; Matsumura, M. Preparation of S-doped TiO_2 photocatalysts and their photocatalytic activities under visible light. *Appl. Catal. A Gen.* **2004**, *265*, 115–121. [CrossRef]
36. Brumbarov, J.; Vivek, J.P.; Leonardi, S.; Valero-Vidal, C.; Portenkirchner, E.; Kunze-Liebhäuser, J. Oxygen deficient, carbon coated self-organized TiO_2 nanotubes as anode material for Li-ion intercalation. *J. Mater. Chem. A* **2015**, *3*, 16469–16477. [CrossRef]
37. Guan, D.; Wang, Y. Electrodeposition of Ag nanoparticles onto bamboo-type TiO_2 nanotube arrays to improve their lithium-ion intercalation performance. *Ionics* **2013**, *19*, 879–885. [CrossRef]
38. Aljabali, A.A.A.; Barclay, J.E.; Butt, J.N.; Lomonossoff, G.P.; Evans, D. Redox-active ferrocene-modified cowpea mosaic virus nanoparticles. *J. Dalt. Trans.* **2010**, *39*, 7569–7574. [CrossRef]
39. Arrouvel, C.; Parker, S.C.; Saiful Islam, M. Lithium insertion and transport in the TiO_2-B anode Material: A Computational Study. *Chem. Mater.* **2009**, *21*, 4778–4783. [CrossRef]
40. Lee, C.W.; Pang, Q.; Ha, S.; Cheng, L.; Han, S.D.; Zavadil, K.R.; Gallagher, K.G.; Nazar, L.F.; Balasubramanian, M. Directing the lithium−sulfur reaction pathway via sparingly solvating electrolytes for high energy density batteries. *ACS Cent. Sci.* **2017**, *3*, 605–613. [CrossRef] [PubMed]
41. Zhu, K.; Wang, Q.; Kim, J.; Pesaran, A.; Frank, A.J. Pseudocapacitive lithium-ion storage in oriented anatase TiO_2 nanotube arrays. *J. Phys. Chem. C* **2012**, *116*, 11895–11899. [CrossRef]
42. Liu, Y.; Jiang, S.P.; Shao, Z. Intercalation pseudocapacitance in electrochemical energy storage: Recent advances in fundamental understanding and materials development. *Mater. Today Adv.* **2020**, *7*, 100072. [CrossRef]
43. Pervez, S.; Kim, D.; Doh, C.H.; Farooq, U.; Choi, H.Y.; Choi, J.H. Anodic WO_3 mesosponge @ carbon: A novel binder-less electrode for advanced energy storage devices. *ACS Appl. Mater. Interface* **2015**, *7*, 7635–7643. [CrossRef] [PubMed]
44. Pervez, S.; Kim, D.; Farooq, U.; Yaqub, A.; Choi, J.H.; Lee, Y.J.; Doh, C.-H. Comparative electrochemical analysis of crystalline and amorphous anodized iron oxide nanotube layers as negative electrode for LIB. *ACS Appl. Mater. Interfaces* **2014**, *6*, 11219–11224. [CrossRef] [PubMed]
45. Pervez, S.; Farooq, U.; Yaqub, A.; Doh, C.-H.; Kim, D.; Sim, S.; Hwang, M.; Choi, J.H.; Lee, Y.J. Improved performance of Ag-nanoparticle-decorated TiO_2 nanotube Arrays in li-ion batteries. *J. Korean Phys. Soc.* **2013**, *63*, 1809–1814. [CrossRef]

46. Pervez, S.; Kim, D.; Doh, C.H.; Farooq, U.; Yaqub, A.; Choi, J.H.; Lee, Y.J.; Saleem, M. High areal capacity for battery anode using rapidly growing self-ordered TiO_2 nanotubes with a high aspect ratio. *Mater. Lett.* **2014**, *137*, 347–350. [CrossRef]
47. Pervez, S.; Kim, D.; Farooq, U.; Yaqub, A.; Choi, J.H.; Lee, Y.J.; Muhammad, S.; Doh, C.-H. Crystalline iron oxide nanotube arrays with high aspect ratio as binder free anode for li-ion batteries. *Phys. Status Solidi* **2014**. *211*, 1889–1894. [CrossRef]
48. Zhang, S.; Zheng, M.; Lin, Z.; Li, N.; Liu, Y.; Zhao, B.; Pang, H.; Cao, J.; He, P.; Shi, Y. Activated carbon with ultrahigh specific surface area synthesized from natural plant material for lithium–sulfur batteries. *J. Mater. Chem. A* **2014**, *2*, 15889–15896. [CrossRef]
49. Gauthier, M.; Carney, T.J.; Grimaud, A.; Giordano, L.; Pour, N.; Chang, H.H.; Fenning, D.P.; Lux, S.F.; Paschos, O.; Bauer, C.; et al. The electrode-electrolyte interface in li-ion batteries: Current understanding and new Insights. *J. Phys. Chem. Lett.* **2015**, *6*, 4653–4672. [CrossRef]
50. Wagemaker, M.; Borghols, W.J.H.; Mulder, F.M. Large impact of particle size on insertion reactions. A Case for anatase Li_xTiO_2. *J. Am. Chem. Soc.* **2007**, *129*, 4323–4327. [CrossRef]
51. Auer, A.; Portenkirchner, E.; Götsch, T.; Valero-Vidal, C.; Penner, S.; Kunze Liebhäuser, H. Preferentially oriented TiO_2 nanotubes as anode material for li-ion batteries: Insight into li-ion storage and lithiation kinetics. *ACS Appl. Mater. Interfaces* **2017**, *9*, 36828–36836. [CrossRef] [PubMed]

nanomaterials

Article

Effect of Point Defects on Electronic Structure of Monolayer GeS

Hyeong-Kyu Choi, Janghwan Cha, Chang-Gyu Choi, Junghwan Kim and Suklyun Hong *

Department of Physics, Graphene Research Institute, and GRI-TPC International Research Center, Sejong University, Seoul 05006, Korea; chk75541@gmail.com (H.-K.C.); heatpoint@nate.com (J.C.); power_ccq@naver.com (C.-G.C.); uprightshine0415@naver.com (J.K.)
* Correspondence: hong@sejong.ac.kr

Abstract: Using density functional theory calculations, atomic and electronic structure of defects in monolayer GeS were investigated by focusing on the effects of vacancies and substitutional atoms. We chose group IV or chalcogen elements as substitutional ones, which substitute for Ge or S in GeS. It was found that the bandgap of GeS with substitutional atoms is close to that of pristine GeS, while the bandgap of GeS with Ge or S vacancies was smaller than that of pristine GeS. In terms of formation energy, monolayer GeS with Ge vacancies is more stable than that with S vacancies, and notably GeS with Ge substituted with Sn is most favorable within the range of chemical potential considered. Defects affect the piezoelectric properties depending on vacancies or substitutional atoms. Especially, GeS with substitutional atoms has almost the same piezoelectric stress coefficients e_{ij} as pristine GeS while having lower piezoelectric strain coefficients d_{ij} but still much higher than other 2D materials. It is therefore concluded that Sn can effectively heal Ge vacancy in GeS, keeping high piezoelectric strain coefficients.

Keywords: germanium monosulfide (GeS); defect; formation energy; vacancy healing; electronic structure; piezoelectric coefficients; DFT calculations; two-dimensional materials

Citation: Choi, H.-K.; Cha, J.; Choi, C.-G.; Kim, J.; Hong, S. Effect of Point Defects on Electronic Structure of Monolayer GeS. *Nanomaterials* **2021**, *11*, 2960. https://doi.org/10.3390/nano11112960

Academic Editors: Jihoon Lee and Ming-Yu Li

Received: 30 September 2021
Accepted: 23 October 2021
Published: 4 November 2021

1. Introduction

Since theoretically proposed graphene [1] was experimentally separated from graphite [2], the research on graphene has been actively conducted because of its fast charge mobility [3,4], good thermal conductivity [3,5], and so on. Such interest has expanded to a variety of two-dimensional materials with sizable bandgap and outstanding electronic properties, such as transition metal dichalcogenides (TMDs) [6–16] and black phosphorus (BP) [17–20]. Since BP has been considered as a promising material for applications, group IV monochalcogenides with a similar structure to BP, such as GeS, GeSe, SnS, and SnSe, have also attracted attention. Interestingly, group IV monochalcogenides are more resistant to oxidation than BP [21]. These materials, such as group IV monochalcogenides, can be used in photovoltaics [22,23] and thermoelectrics [24–31]. In addition, the problem of efficient energy storage [32–36] and very strong piezoelectric effect [37–39] have received much attention. Especially, multi-layered GeS photodetectors exhibit device robustness, photoswitching stability, and long-term durability [40]. Among MX (M = Ge, Sn; X = S, Se) sheets, GeS sheet has the lowest activation barriers and holds Li atoms strongly with the highest adsorption energies; at the same time, low, open-circuit voltage and higher capacity can make it the best choice for Li-ion batteries [32,41]. The extraordinary sensitivity and selectivity of GeS monolayer for NO_2 molecules suggest that GeS monolayer is a potential sensing material for NO_2 gas [42]. Moreover, GeS monolayers are proposed as efficient photocatalysts for water splitting [43–45].

In general, defects play an important role in determining electronic properties. In 2D materials, defects, such as edges, grain boundaries, strain, vacancies, substitutional atoms, and dopants, can lead to localized electronic properties, which is different from the case of pristine 2D materials. These properties are sometimes beneficial ones or sometimes harmful ones that should be removed. For example, mechanically isolated MoS_2 shows better mobility than chemically deposited MoS_2 [46,47]. Furthermore, MoS_2 samples obtained using chemical vapor deposition, physical vapor deposition, and mechanical delamination have about ten times lower mobility than theoretical expectations because of the presence of grain boundaries or point defects [48]. Such defects can cause optical properties to change. In MoS2, the shift of the first-order Raman mode is affected by the concentration of S vacancies [49]. Moreover, intensive calculations related to defects and modification of 2D materials have been reported [50–56].

On the other hand, the defects of group IV monochalcogenides have been actively studied. For example, group IV monochalcogenides are more resistant to oxidation than BP, which have the same structure with group IV monochalcogenides [21]. It was predicted by simulation studies that the original bandgap can be recovered by substituting chalcogen vacancies with oxygen. In terms of formation energy, Ge vacancies are more likely to form than S vacancies, but there has been little discussion on how to heal Ge vacancies [21]. Therefore, further discussion is needed on how to effectively reduce and heal vacancies in group IV monochalcogenides.

In this study, we investigate structural and electronic properties related to point defects, such as Ge and S vacancies, in GeS that is one of group IV monochalcogenides. Compared to pristine GeS, we focus on the change caused by vacancies and substitutional atoms. Group IV elements, such C, Si, and Sn, are chosen as the substitutional atom of Ge, while chalcogen elements, such as Se and Te, are chosen as that of S. Specifically, the focus is on the case where the properties of defective GeS are restored by substitutional atoms.

2. Calculation Method

To understand the structural and electronic properties of group IV monochalcogenides, we employed a first-principles approach based on the spin-polarized density functional theory [57,58], as implemented in the Vienna ab-initio simulation package (VASP) [59,60], which is used for calculations of density of states (DOS) and elastic and piezoelectric coefficients. Note that spin-polarized calculations lead to no spin polarization for all the systems considered. The core and valence electrons were treated with the projector-augmented wave (PAW) [61] method. The exchange correlation energy is described by the generalized gradient approximation (GGA) using the Perdew-Burke-Ernzerhof (PBE) [62] functional. The Kohn-Sham orbitals are expanded in a plane-wave basis with a cutoff energy of 400 eV. To model the pristine and defective GeS, the 3×3 supercells are periodic in the monolayer plane, and large vacuum regions (12 Å) are included to impose periodic boundary conditions in the perpendicular direction. The Brillouin zone (BZ) is sampled using a Γ-centered $14 \times 10 \times 1$ grid for the monolayer, following the scheme proposed by Monkhorst-Pack [63]. The convergence criteria for electronic and ionic relaxations are 10^{-6} eV and 10^{-3} eV/Å, respectively. The charge transfers are calculated with decomposition of charge density into atomic contributions by using the Bader charge analysis method [64]. To obtain the piezoelectric properties, the elastic coefficient C_{ij} is calculated by using strain–stress relationship (SSR) [65], and the piezoelectric stress coefficients e_{ij} are calculated by the density functional perturbation theory (DFPT) method [66,67] by the VASP code.

3. Results and Discussion

3.1. Crystal Structure and Energetics

The pristine GeS is one of group IV monochalcogenide that has an orthorhombic structure. Bulk GeS forms a $Pnma - D_{2h}^{16}$ space group and shows a lower symmetry than bulk BP ($Pnma - D_{2h}^{18}$) because it is composed of two kinds of atoms, unlike BP. If the dimension is reduced to monolayer, the space group changes to $Pmn2_1 - C_{2v}^{7}$. Such difference in symmetry affects the piezoelectric coefficient, which will be discussed below in Section 3.3. The in-plane lattice constants a = 4.54 Å, b = 3.63 Å are used as shown in Figure 1. In this study, we use 3×3 supercell including about 12 Å vacuum.

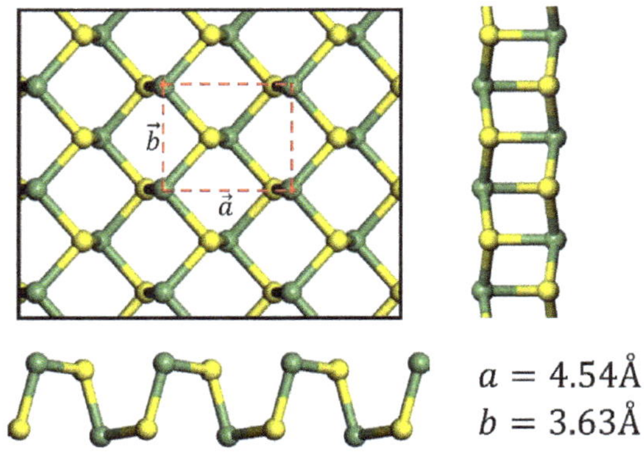

Figure 1. Structure and lattice parameters of the pristine GeS layer. Germanium and sulfur atoms are represented by green and yellow balls, respectively. The 1 × 1 unit cell is marked with the red box.

Vacancies or substitutional atoms are introduced as defects, of which structures are shown in Figure 2. Vacancies can occur at either S or Ge position. For S vacancy (V_S), the Ge atoms nearest to the S vacancy site gather inside the GeS layer, while for Ge vacancy (V_{Ge}), the S atoms nearest to the Ge vacancy site move away. When foreign atoms are introduced into vacancy sites, the structural changes occur: the substituting atoms for the vacancy positions lie in the GeS plane or protrude from the plane, depending on their sizes. If C atoms substitute for Ge atoms (C_{Ge}), the C atoms lie in the plane of GeS layer, while if Sn atoms replaces Ge atoms (Sn_{Ge}), the Sn atoms are placed slightly above the GeS plane. For the same reason, if Te is placed in the S vacancy site (Te_S), it is placed slightly above the GeS plane. Note that in the case of C_{Ge}, the C atom binds to the surrounding S atoms, and thus, the S atoms move too far towards C, breaking its bonds with Ge atoms.

To find out the stability of monolayer GeS with defects, we consider the formation energy of the system with defects. The formation energy $E_{formation}$ is defined as follows:

$$E_{formation} = E_{defect} - (E_{pristine} + \sum n_i \mu_i) \tag{1}$$

where $E_{pristine}$ is the total energy of pristine GeS, and E_{defect} is that of monolayer GeS with defects (vacancies or substitutional atoms) at the Ge or S position. Here, n_i is the number of the i element that has been removed or introduced for substitution, and μ_i is the corresponding chemical potential. Since only one element is removed or substitutes in the 3 × 3 supercell, the value of n_i is −1 or +1. To measure the chemical potential, we used the relation that $\mu_{GeS} \cong \mu_{Ge} + \mu_S$, where μ_{GeS} means the total energy per formula unit in pristine GeS. Since the systems under study are assumed to be in chemical and thermal equilibrium with bulk, one may be able to use the bulk energy, i.e., the total energy per atom of the specific bulk crystal, as the chemical potential [68–70]. That is, the chemical potentials μ_i for substitutional atoms are calculated from their bulk structures. In the Ge-rich environment, μ_{Ge} is the total energy per atom in the diamond-structured solid Ge, which enables one to calculate μ_S as $\mu_{GeS} - \mu_{Ge}$. On the other hand, in an S-rich environment, μ_S is the total energy per atom in α-S8 crystal, and thus, μ_{Ge} is calculated as $\mu_{GeS} - \mu_S$. The chemical potentials of chalcogen elements, such as Te and Se, are calculated from the crystal structures of $P3_121$ Te [71] and hexagonal Se [72–74], respectively.

The results of the formation energies as a function of chemical potential are represented in Figure 3. Specifically, the formation energies are calculated at two points, such as Ge-rich and S-rich environments for each line. For the case of vacancies, it is natural that V_S has difficulty in creating in S-rich environment, while V_{Ge} has difficulty making a Ge-rich environment. The formation energy of V_{Ge} is smaller than that of V_S except for the chemical potential range around Ge-rich environment, so it is considered that V_{Ge} is more likely to be found than V_S in relatively wider range of chemical potential considered. Indeed, it is experimentally expected that V_{Ge} is more frequently observed than V_S. A remarkable point is that the formation energy of Sn_{Ge} is negative for any value of chemical potential between in S-rich and Ge-rich environments, which implies that Sn_{Ge} is energetically favorable and forms spontaneously in the presence of Ge vacancies and Sn atoms. Such Sn substitution may play an important role in effective vacancy healing. In addition, Si_{Ge} has negative formation energy in an S-rich environment. In contrast, the formation energies of C_{Ge} are relatively large, positive values in the chemical potential range considered, leading to its instability.

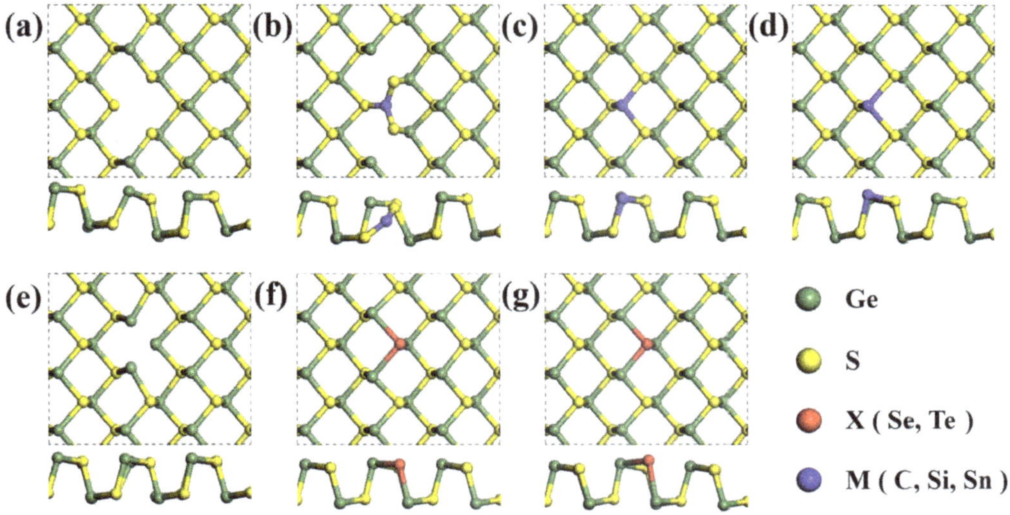

Figure 2. Structures of (**a**) Ge vacancy (V_{Ge}) and (**b**) substitutional C (C_{Ge}), (**c**) substitutional Si (Si_{Ge}), and (**d**) substitutional Sn (Sn_{Ge}) for Ge vacancy; structures of (**e**) S vacancy (V_S) and (**f**) substitutional Se (Se_S) and (**g**) substitutional Te (Te_S) for S vacancy. For perspective views of atomic structures of pristine and defective GeS are shown in Figure S1.

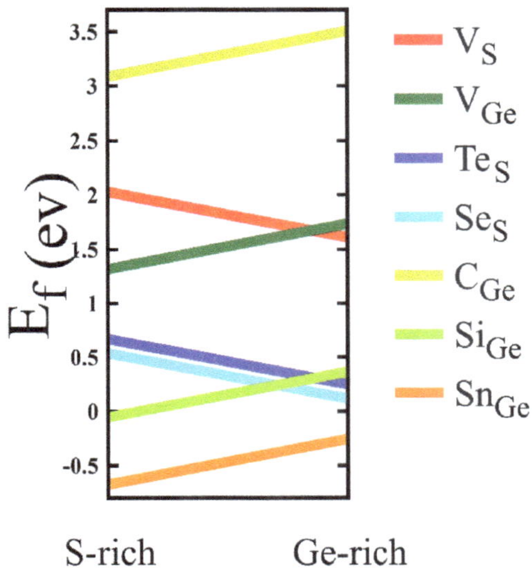

Figure 3. Formation energies of defected structures as a function of chemical potential. Especially, the formation energy of Sn_{Ge} is negative for S-rich and Ge-rich conditions, which implies that Sn_{Ge} is energetically favorable.

3.2. Electronic Properties

To study the electronic structure of defective GeS, we calculated density of states (DOS) of GeS. Figure 4 shows total DOS and partial DOS (PDOS) for (a) pristine GeS and defective GeS with (b) V_S; (c) Se_S and Te_S; (d) V_{Ge}; and (e) C_{Ge}, Si_{Ge}, and Sn_{Ge}. Here, the black line represents total DOS, while colored lines represent PDOS corresponding to the same-colored atoms drawn in dark in the inset. The bandgap of pristine GeS is 1.76 eV, which is listed in Table 1 along with the bandgaps of defective GeS. When vacancies are introduced, the bandgaps become smaller for the cases of both V_{Ge} and V_S, of which bandgaps are 0.04 eV and 1.04 eV, respectively. Defects such as V_{Ge} and V_S present vacancy states inside the forbidden band region of pristine GeS, leading to the doping effect. This agrees well with previous research results [21]. Refer to Figure S2 for band structures of pristine and defective GeS.

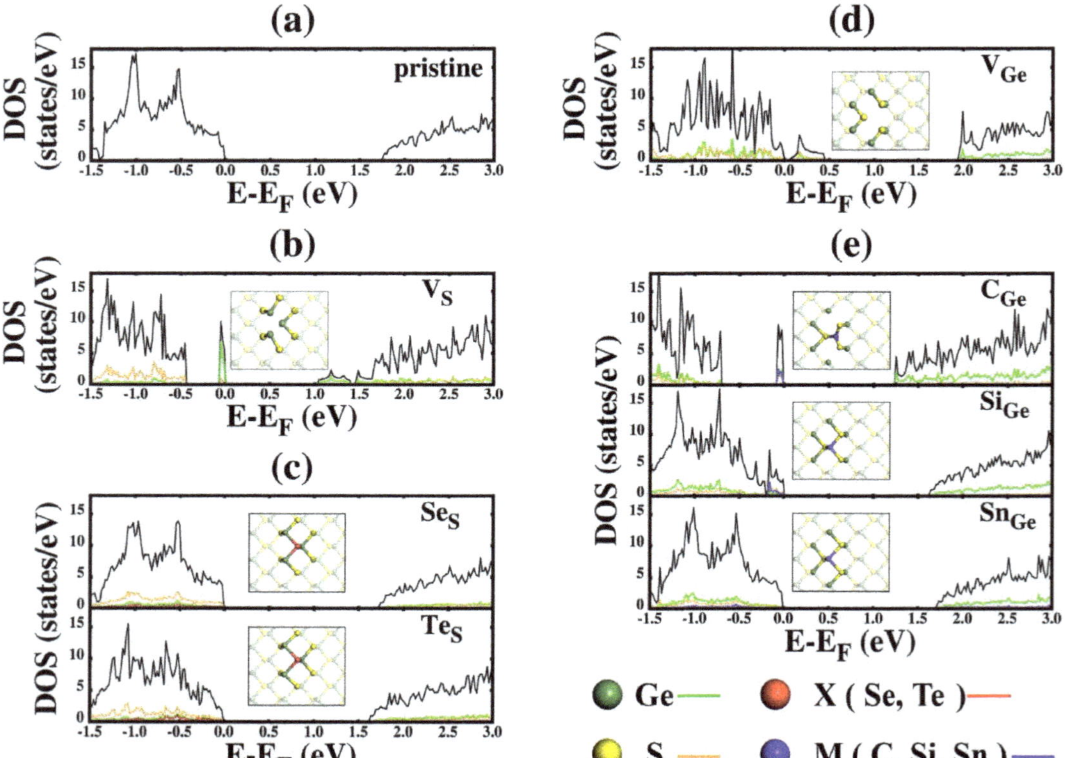

Figure 4. Total DOS and PDOS of monolayer pristine GeS and defective GeS: (**a**) pristine GeS, (**b**) GeS with vacancy at S site (V_S), (**c**) GeS with substitutional atoms at S site (Se_S and Te_S), (**d**) GeS with vacancy at Ge site (V_{Ge}), and (**e**) GeS with substitutional atoms at Ge site (C_{Ge}, Si_{Ge}, Sn_{Ge}). The black line represents total DOS, while the colored PDOS lines represents contributions by the atoms shown in the same color in the inset.

Table 1. Bandgaps obtained from DFT calculations for GeS without and with defect.

	Pristine GeS	V_{Ge}	V_S	C_{Ge}	Si_{Ge}	Sn_{Ge}	Se_S	Te_S	Pristine GeS DFT [75]/Exp [76,77]
E_G (eV)	1.76	0.04	1.04	1.26	1.60	1.71	1.71	1.63	1.65/1.65, 1.70–1.96

For the case of V_S, the vacancy state in PDOS of V_S is located just below the Fermi energy, which makes the bandgap become narrower and act as a donor state. The vacancy state is created by adjacent Ge atoms surrounding S vacancy. The contribution to the vacancy state by the Ge atoms amounts to the most portion of vacancy state. Structurally, when V_S forms, three surrounding Ges gather to be close each other and thus contribute to the formation of the vacancy state, while atoms farther away from S vacancy do not contribute the vacancy state. In contrast, in the case of V_{Ge}, a relatively wider vacancy state appears just above the Fermi energy acting as an acceptor state. In addition, it is difficult to say that the states just below the Fermi energy are made only by the contributions of adjacent atoms.

When the substitutional atom is introduced for either Ge site or S site, the bandgap becomes larger than those of both cases of V_{Ge} and V_S and has a value close (within ~10%) to that of pristine GeS, as shown in Table 1. When S is substituted with Se or Te, the structural properties are almost the same as that of pristine GeS because S, Se, and Te belong to the same group, which makes the electronic properties, such as bandgap value, almost similar. Similarly, when Ge is substituted with Si and Sn, it is likewise the case. However, in case of C_{Ge}, as mentioned earlier in the structural description, the S atoms around the substituted C atom become closer to the C atom and further away from outer Ge atoms, so C_{Ge} may have an electronic structure similar to V_S. Especially, considering the formation energy discussed above, Sn_{Ge} is energetically most stable among the substitutional cases, so when Ge is substituted with Sn to replace Ge in GeS, the original electronic properties of pristine GeS are almost recovered.

It is worthwhile to ask whether the band gaps of pristine and defective GeS are direct or indirect. Pristine monolayer GeS has an indirect band gap, which has been already reported in many previous studies [75–77]. Comparison of PDOS profiles between pristine and substitutional GeS except C_{Ge} shows similarities to each other, which makes one expect that substitutional GeS except C_{Ge} have indirect band gaps like the pristine case. This can be confirmed by the band structure plots shown in Figure S2. For V_{Ge}, V_S and C_{Ge}, the PDOS profiles are much changed from that of the pristine case. For detailed band structures, including the locations of the valence band minimum (VBM) and conduction band minimum (CBM), see Figure S2: the band gaps of V_{Ge}, V_S and C_{Ge} are indirect.

Based on the Bader charge analysis, we found charge transfers between the substitutional atom and neighboring atoms. For Ge-substituted cases, Si loses $0.36e$ to neighboring S atoms, while C and Sn gain $2.55e$ and $0.11e$ from them, respectively. Note that a relatively large gain occurs for C due to large deformation around C. For S-substituted cases, Se and Te lose $1.28e$ and $1.62e$ to neighboring Ge atoms, respectively. In the cases of Ge or S vacancies, the S atoms nearest to the Ge vacancy site lose $0.25e$ per S, while the Ge atoms nearest to S vacancy site gain $0.62e$ per Ge compared to the pristine case.

3.3. Piezoelectric Properties

Since the piezoelectric property is a ground-state one, DFT is a useful tool for predicting those properties, and a great deal of previous studies on GaN [78] and MoS2 [79,80] show high agreement between DFT and experiments. As mentioned earlier, monolayer GeS belongs to monolayer group IV monochalcogenides, which belongs to $Pmn2_1 - C_{2v}^7$ space group. Because of the prediction that this group has strong piezoelectric properties using DFT [37,38], the group has received a great deal of attention and has been followed by related subsequent studies. Specifically, there have been attempts to maximize the piezoelectric effects by slightly modifying the structure of group IV monochalcogenides, but the piezoelectric effect has not been greatly increased [39]. Here, we investigated how defects influence the piezoelectric properties of GeS depending on vacancies or substitutional atoms. In particular, based on the formation energy results, the more stable cases (Sn_{Ge}, Se_S) among the substitutions of Ge or S with foreign atoms in GeS are considered.

To know the piezoelectric properties, we calculated the elastic coefficients C_{ij} and further piezoelectric coefficients by using VASP [65–67]. The elastic energy Δu is defined by

$$\Delta u \sum_{i=1}^{6}\sum_{j=1}^{6}\frac{1}{2}C_{ij}\varepsilon_i\varepsilon_j \tag{2}$$

where C_{ij} is the elastic coefficient, and ε_i is the strain in the i direction in Voigt notation [81]. Since we dealt with 2D monolayer GeS, the elastic energy Δu is given as energy per unit area. The number of 36 components in Δu can be reduced by considering the structural symmetry. Orthorhombic structure needs just nine independent elements [82]: C_{11}, C_{22}, C_{33}, C_{44}, C_{55}, C_{66}, C_{12}, C_{13}, and C_{23}. For monolayer GeS, only in-plane directions are considered with the restriction that angles between lattice vectors do not change. Then, only C_{11}, C_{22}, and C_{12} need to be considered among nine elastic coefficients. The elastic energy expression is now reduced to

$$\Delta u = \frac{1}{2}C_{11}\varepsilon_1^2 + \frac{1}{2}C_{22}\varepsilon_2^2 + C_{12}\varepsilon_1\varepsilon_2, \tag{3}$$

The calculation results for elastic coefficients are listed in Table 2, where only the results for the relaxed-ion case that can be compared with the experiment are given along with the values from previous studies for pristine GeS and MoS_2 for comparison. It was found that the present results agree well with the previous theoretical calculations.

Table 2. Elastic coefficients for pristine GeS, GeS with vacancies (V_{Ge}, V_S), and GeS with substitutional atoms (Sn_{Ge}, Se_S) compared with previous studies for pristine GeS and MoS_2.

	Pristine GeS	V_{Ge}	V_S	Sn_{Ge}	Se_S	**Pristine GeS [33]/[34]**	MoS_2
$C_{11}(N/m)$	13.28	11.01	11.84	14.15	13.95	15.24/20.87	129.94/130 [53]
$C_{22}(N/m)$	44.28	42.74	43.76	48.87	49.16	45.83/53.40	130.57/130 [53]
$C_{12}(N/m)$	18.71	12.13	16.11	19.09	19.26	21.62/22.22	32.03/32 [53]

Next, we calculated piezoelectric stress coefficients e_{ij}, showing how polarization changes with strain. The piezoelectric stress coefficients e_{ij} are given by

$$e_{ij} = \frac{\partial P_i}{\partial \varepsilon_j} \tag{4}$$

The calculation results are listed in Table 3. Note that V_{Ge} and V_S have different piezoelectric stress coefficients from each other. V_{Ge} has piezoelectric stress coefficients e_{11} and e_{12} larger by 36% and 81% than those of pristine GeS, respectively. V_S has e_{11} close to (slightly smaller than) that of pristine GeS while having smaller e_{12} by 20%. However, Sn_{Ge} and Se_S have e_{11} and e_{12} close to those of pristine GeS, which means that they will experience similar polarization changes for similar distortions.

Table 3. Piezoelectric stress coefficients e_{ij} for pristine GeS, GeS with vacancies (V_{Ge}, V_S), and GeS with substitutional atoms (Sn_{Ge}, Se_S) compared with previous studies for pristine GeS and MoS_2.

	Pristine GeS	V_{Ge}	V_S	Sn_{Ge}	Se_S	**Pristine GeS [33]/[34]**	MoS_2
$e_{11}\left(C/Å\right)$	5.84	7.92	5.72	5.94	5.83	7.28/4.6	3.66/3.64 [53]
$e_{12}\left(C/Å\right)$	−4.59	−8.29	−3.67	−4.36	−4.51	−4.97/−10.1	−3.67/−3.64 [53]

On the other hand, the relationship between d_{ij}, C_{ij}, and e_{ij} is given by

$$e_{ij} = \sum d_{ik}C_{kj} \tag{5}$$

where e_{ij} or d_{ij} are piezoelectric (stress or strain) coefficients that represent how polarization changes when strain or stress is changed, respectively. Note that both piezoelectric coefficients e_{ij} and d_{ij} are connected by elastic constants C_{ij}, showing how stress is induced with strain. Since it is sometimes convenient to use d_{ij} (change of polarization depending on stress), we found d_{ij} from Equation (5). The results for piezoelectric strain coefficients d_{ij} are listed in Table 4. V_{Ge} has piezoelectric strain coefficient d_{11} close (within 6%) to pristine GeS, and V_S, Sn_{Ge}, and Se_S have smaller d_{11} by 17~21%, while all defective GeS's have smaller d_{12} by 19~27%. Based on these results, it was found that V_{Ge} has larger e_{ij} and d_{ij} values than V_S.

Table 4. Piezoelectric strain coefficients d_{ij} for pristine GeS, GeS with vacancies (V_{Ge}, V_S), and GeS with substitutional atoms (Sn_{Ge}, Se_S) compared with previous studies for pristine GeS and MoS_2.

	Pristine GeS	V_{Ge}	V_S	Sn_{Ge}	Se_S	Pristine GeS [33]/[34]	MoS_2
$d_{11}(pm/V)$	144.71	135.75	119.66	114.19	118.62	190.92/75.43	3.74/3.73 [53]
$d_{12}(pm/V)$	−71.51	−57.92	−52.44	−53.53	−55.64	−100.91/−50.42	−3.73/−3.73 [53]

From the results listed in Tables 2–4, it is worthwhile to emphasize the relation between d_{ij}, C_{ij}, and e_{ij}: Sn_{Ge} and Se_S have larger elastic constants than those of pristine GeS. Especially, for Sn_{Ge} and Se_S, the larger value of C_{22} by ~10% makes them have smaller piezoelectric strain coefficients d_{11} and d_{12} even though they have piezoelectric stress coefficients e_{11} and e_{12} close to those of pristine GeS. Note that even defective GeS have larger piezoelectric strain coefficients d_{ij} by one or two orders of magnitude than other bulk piezoelectric materials, which are frequently used ones, such as α-quartz [83] and wurtzite AlN [84], and recently emerging 2D materials, such as MoS_2 [79] and GaSe [85]. Such larger piezoelectric coefficients can be used for devices applications. For better applications of piezoelectric properties of defective materials with vacancies or substitutional atoms, detailed experimental studies elucidating the relation between d_{ij}, C_{ij}, and e_{ij} are needed.

4. Conclusions

We have studied the effects of defects on atomic and electronic structure in monolayer GeS. Vacancies in GeS create doping states, which are located just above or below the Fermi energy. Such vacancy states disappear, and the original bandgap of pristine GeS is recovered when the vacancies are substituted with foreign atoms. In terms of the formation energy, GeS with Ge replaced by Sn is found to be energetically most favorable in both S-rich and Ge-rich environments among all the defective GeS cases considered. Defective GeS has smaller piezoelectric strain coefficients d_{ij} by 20–30% than pristine GeS except for d_{11} of V_{Ge}, but these values (d_{ij}) are much larger than those of other piezoelectric materials. Moreover, piezoelectric stress coefficients e_{ij} of GeS with substitutional atoms are almost the same as those of pristine GeS. Therefore, it is concluded that substitutional atoms act as vacancy healers to help restore the properties of pristine GeS; Especially, the Sn substitution would make GeS with Ge vacancies stable and allow it to be utilized for applications in sensitive devices.

Supplementary Materials: The following are available online at https://www.mdpi.com/article/10.3390/nano11112960/s1, Figure S1: Perspective views of atomic structures of pristine and defective GeS; Figure S2: Band structures of pristine and defective GeS.

Author Contributions: Conceptualization, S.H.; data curation, formal analysis, and investigation, H.-K.C., J.C., C.-G.C., J.K. and S.H.; writing, H.-K.C., J.K. and S.H.; funding acquisition, S.H.; supervision, S.H. All authors have read and agreed to the published version of the manuscript.

Funding: This work was supported by Global Research and Development Center Program (2018K1A4A3A01064272) and Basic Science Research Program (2021R1A4A1031900) through the NRF grant funded by the Korea government (MSIT).

Conflicts of Interest: The authors declare no conflict of interest.

References

1. Wallace, P.R. The Band Theory of Graphite. *Phys. Rev.* **1947**, *71*, 622. [CrossRef]
2. Novoselov, K.S.; Geim, A.K.; Morozov, S.V.; Jiang, D.; Zhang, Y.; Dubonos, S.V.; Grigorieva, I.V.; Firsov, A.A. Electric Field Effect in Atomically Thin Carbon Films. *Science* **2004**, *306*, 666. [CrossRef]
3. Neto, A.H.C.; Guinea, F.; Peres, N.M.R.; Novoselov, K.S.; Geim, A.K. The electronic properties of graphene. *Rev. Mod. Phys.* **2009**, *81*, 109–162. [CrossRef]
4. Banszerus, L.; Schmitz, M.; Engels, S.; Dauber, J.; Oellers, M.; Haupt, F.; Watanabe, K.; Taniguchi, T.; Beschoten, B.; Stampfer, C. Ultrahigh-mobility graphene devices from chemical vapor deposition on reusable copper. *Sci. Adv.* **2015**, *1*, e1500222. [CrossRef]
5. Pop, E.; Varshney, V.; Roy, A.K. Thermal properties of graphene: Fundamentals and applications. *MRS Bull.* **2012**, *37*, 1273–1281. [CrossRef]
6. Chhowalla, M.; Shin, H.S.; Eda, G.; Li, L.-J.; Loh, K.; Zhang, H. The chemistry of two-dimensional layered transition metal dichalcogenide nanosheets. *Nat. Chem.* **2013**, *5*, 263–275. [CrossRef] [PubMed]
7. Wang, Q.H.; Kalantar-Zadeh, K.; Kis, A.; Coleman, J.; Strano, M.S. Electronics and optoelectronics of two-dimensional transition metal dichalcogenides. *Nat. Nanotechnol.* **2012**, *7*, 699–712. [CrossRef]
8. Lv, R.; Robinson, J.A.; Schaak, R.E.; Sun, D.; Sun, Y.; Mallouk, T.E.; Terrones, M. Transition Metal Dichalcogenides and Beyond: Synthesis, Properties, and Applications of Single- and Few-Layer Nanosheets. *Accounts Chem. Res.* **2015**, *48*, 56–64. [CrossRef] [PubMed]
9. Mak, K.F.; Shan, K.F.M.J. Photonics and optoelectronics of 2D semiconductor transition metal dichalcogenides. *Nat. Photonics* **2016**, *10*, 216–226. [CrossRef]
10. Cha, J.; Min, K.-A.; Sung, D.; Hong, S. Ab initio study of adsorption behaviors of molecular adsorbates on the surface and at the edge of MoS 2. *Curr. Appl. Phys.* **2018**, *18*, 1013–1019. [CrossRef]
11. Kim, J.; Choi, C.-G.; Min, K.-A.; Cho, K.; Hong, S. Effect of atomic passivation at Ni-MoS2 interfaces on contact behaviors. *Curr. Appl. Phys.* **2020**, *20*, 132–136. [CrossRef]
12. Min, K.-A.; Park, J.; Wallace, R.; Cho, K.; Hong, S. Reduction of Fermi level pinning at Au–MoS 2 interfaces by atomic passivation on Au surface. *2D Mater.* **2016**, *4*, 015019. [CrossRef]
13. Zhang, Z.Y.; Si, M.S.; Wang, Y.H.; Gao, X.P.; Sung, D.; Hong, S.; He, J. Indirect-direct band gap transition through electric tuning in bilayer MoS2. *J. Chem. Phys.* **2014**, *140*, 174707. [CrossRef] [PubMed]
14. Santosh, K.C.; Zhang, C.; Hong, S.; Wallace, R.M.; Cho, K. Phase stability of transition metal dichalcogenide by competing ligand field stabilization and charge density wave. *2D Mater.* **2015**, *2*, 035019. [CrossRef]
15. Zhang, C.; Gong, C.; Nie, Y.; Min, K.-A.; Liang, C.; Oh, Y.J.; Zhang, H.; Wang, W.; Hong, S.; Colombo, L.; et al. Systematic study of electronic structure and band alignment of monolayer transition metal dichalcogenides in Van der Waals heterostructures. *2D Mater.* **2016**, *4*, 015026. [CrossRef]
16. Zhang, C.; Kc, S.; Nie, Y.; Liang, C.; Vandenberghe, W.G.; Longo, R.C.; Zheng, Y.; Kong, F.; Hong, S.; Wallace, R.M.; et al. Charge Mediated Reversible Metal–Insulator Transition in Monolayer MoTe2 and WxMo1–xTe2 Alloy. *ACS Nano* **2016**, *10*, 7370–7375. [CrossRef]
17. Fei, R.; Yang, L. Strain-Engineering the Anisotropic Electrical Conductance of Few-Layer Black Phosphorus. *Nano Lett.* **2014**, *14*, 2884–2889. [CrossRef]
18. Qiao, J.; Kong, X.; Hu, Z.-X.; Yang, F.; Ji, W. High-mobility transport anisotropy and linear dichroism in few-layer black phosphorus. *Nat. Commun.* **2014**, *5*, 4475. [CrossRef]
19. Li, L.; Yu, Y.; Ye, G.J.; Ge, Q.; Ou, X.; Wu, H.; Feng, D.; Chen, X.H.; Zhang, Y. Black phosphorus field-effect transistors. *Nat. Nanotechnol.* **2014**, *9*, 372–377. [CrossRef]
20. Liu, Q.; Zhang, X.; Abdalla, L.B.; Fazzio, A.; Zunger, A. Switching a Normal Insulator into a Topological Insulator via Electric Field with Application to Phosphorene. *Nano Lett.* **2015**, *15*, 1222–1228. [CrossRef]
21. Gomes, L.C.; Carvalho, A.; Neto, A.H.C. Vacancies and oxidation of two-dimensional group-IV monochalcogenides. *Phys. Rev. B* **2016**, *94*, 054103. [CrossRef]
22. Haq, B.U.; AlFaify, S.; Ahmed, R.; Butt, F.K.; Laref, A.; Shkir, M. Exploring single-layered SnSe honeycomb polymorphs for optoelectronic and photovoltaic applications. *Phys. Rev. B* **2018**, *97*, 075438. [CrossRef]
23. Xia, C.; Du, J.; Xiong, W.; Jia, Y.; Wei, Z.; Li, J. A type-II GeSe/SnS heterobilayer with a suitable direct gap, superior optical absorption and broad spectrum for photovoltaic applications. *J. Mater. Chem. A* **2017**, *5*, 13400–13410. [CrossRef]
24. Hu, Z.-Y.; Li, K.-Y.; Lu, Y.; Huang, Y.; Shao, X.-H. High thermoelectric performances of monolayer SnSe allotropes. *Nanoscale* **2017**, *9*, 16093–16100. [CrossRef]
25. Zhao, L.-D.; Lo, S.-H.; Zhang, Y.S.; Sun, H.; Tan, G.H.; Uher, C.; Wolverton, C.M.; Dravid, V.P.; Kanatzidis, M.G. Ultralow thermal conductivity and high thermoelectric figure of merit in SnSe crystals. *Nat. Cell Biol.* **2014**, *508*, 373–377. [CrossRef]
26. Zhao, L.D.; Tan, G.; Hao, S.; He, J.; Pei, Y.; Chi, H.; Wang, H.; Gong, S.; Xu, H.; Dravid, V.P.; et al. Ultrahigh power factor and thermoelectric performance in hole-doped single-crystal SnSe. *Science* **2016**, *351*, 141–144. [CrossRef]
27. Shi, G.; Kioupakis, E. Quasiparticle band structures and thermoelectric transport properties of p-type SnSe. *J. Appl. Phys.* **2015**, *117*, 065103. [CrossRef]

28. Solanki, G.K.; Deshpande, M.P.; Agarwal, M.K.; Patel, P.D.; Vaidya, S.N. Thermoelectric power factor measurements in GeSe single crystals grown using different transporting agents. *J. Mater. Sci. Lett.* **2003**, *22*, 985–987. [CrossRef]
29. Chen, C.-L.; Wang, H.; Chen, Y.-Y.; Day, T.; Snyder, G.J. Thermoelectric properties of p-type polycrystalline SnSe doped with Ag. *J. Mater. Chem. A* **2014**, *2*, 11171–11176. [CrossRef]
30. Tan, Q.; Zhao, L.-D.; Li, J.-F.; Wu, C.-F.; Wei, T.-R.; Xing, Z.-B.; Kanatzidis, M.G. Thermoelectrics with earth abundant elements: Low thermal conductivity and high thermopower in doped SnS. *J. Mater. Chem. A* **2014**, *2*, 17302–17306. [CrossRef]
31. Zhu, H.; Sun, W.H.; Armiento, R.; Lazic, P.; Ceder, G. Band structure engineering through orbital interaction for enhanced thermoelectric power factor. *Appl. Phys. Lett.* **2014**, *104*, 082107. [CrossRef]
32. Ji, Y.; Dong, H.; Yang, M.; Hou, T.; Li, Y. Monolayer germanium monochalcogenides (GeS/GeSe) as cathode catalysts in nonaqueous Li–O2batteries. *Phys. Chem. Chem. Phys.* **2017**, *19*, 20457–20462. [CrossRef] [PubMed]
33. Kim, H.; Son, Y.; Lee, J.; Lee, M.; Park, S.; Cho, J.; Choi, H.C. Nanocomb Architecture Design Using Germanium Selenide as High-Performance Lithium Storage Material. *Chem. Mater.* **2016**, *28*, 6146–6151. [CrossRef]
34. Lee, D.-H.; Park, C.-M. Tin Selenides with Layered Crystal Structures for Li-Ion Batteries: Interesting Phase Change Mechanisms and Outstanding Electrochemical Behaviors. *ACS Appl. Mater. Interfaces* **2017**, *9*, 15439–15448. [CrossRef]
35. Guo, Y.; Wei, Y.; Li, H.; Zhai, T. Layer Structured Materials for Advanced Energy Storage and Conversion. *Small* **2017**, *13*. [CrossRef] [PubMed]
36. Li, W.; Sun, X.; Yu, Y. Si-, Ge-, Sn-Based Anode Materials for Lithium-Ion Batteries: From Structure Design to Electrochemical Performance. *Small Methods* **2017**, *1*. [CrossRef]
37. Gomes, L.C.; Carvalho, A.; Neto, A.H.C. Enhanced piezoelectricity and modified dielectric screening of two-dimensional group-IV monochalcogenides. *Phys. Rev. B* **2015**, *92*, 214103. [CrossRef]
38. Fei, R.; Li, W.; Li, J.; Yang, L. Giant piezoelectricity of monolayer group IV monochalcogenides: SnSe, SnS, GeSe, and GeS. *Appl. Phys. Lett.* **2015**, *107*, 173104. [CrossRef]
39. Hu, T.; Dong, J. Two new phases of monolayer group-IV monochalcogenides and their piezoelectric properties. *Phys. Chem. Chem. Phys.* **2016**, *18*, 32514–32520. [CrossRef]
40. Ulaganathan, R.K.; Lu, Y.-Y.; Kuo, C.-J.; Tamalampudi, S.R.; Sankar, R.; Boopathi, K.M.; Anand, A.; Yadav, K.; Mathew, R.J.; Liu, C.-R.; et al. High photosensitivity and broad spectral response of multi-layered germanium sulfide transistors. *Nanoscale* **2016**, *8*, 2284–2292. [CrossRef]
41. Zhou, Y. MX (M = Ge, Sn; X = S, Se) sheets: Theoretical prediction of new promising electrode materials for Li ion batteries. *J. Mater. Chem. A* **2016**, *4*, 10906–10913. [CrossRef]
42. Wang, S.-G.; Tan, C.-J.; Yang, Q.; Xu, Y.-X.; Li, S.-L.; Chen, X.-P. A Novel Ultra-Sensitive Nitrogen Dioxide Sensor Based on Germanium Monosulfide Monolayer. *IEEE Electron. Device Lett.* **2017**, *38*, 1590–1593. [CrossRef]
43. Ju, L.; Dai, Y.; Wei, W.; Li, M.; Huang, B. DFT investigation on two-dimensional GeS/WS2 van der Waals heterostructure for direct Z-scheme photocatalytic overall water splitting. *Appl. Surf. Sci.* **2018**, *434*, 365–374. [CrossRef]
44. Chowdhury, C.; Karmakar, S.; Datta, A. Monolayer Group IV–VI Monochalcogenides: Low-Dimensional Materials for Photocatalytic Water Splitting. *J. Phys. Chem. C* **2017**, *121*, 7615–7624. [CrossRef]
45. Lv, X.; Wei, W.; Sun, Q.; Li, F.; Huang, B.; Dai, Y. Two-dimensional germanium monochalcogenides for photocatalytic water splitting with high carrier mobility. *Appl. Catal. B Environ.* **2017**, *217*, 275–284. [CrossRef]
46. Liu, H.; Si, M.; Najmaei, S.; Neal, A.; Du, Y.; Ajayan, P.M.; Lou, J.; Ye, P.D. Statistical Study of Deep Submicron Dual-Gated Field-Effect Transistors on Monolayer Chemical Vapor Deposition Molybdenum Disulfide Films. *Nano Lett.* **2013**, *13*, 2640–2646. [CrossRef] [PubMed]
47. Yu, Z.; Pan, Y.; Shen, Y.; Wang, Z.; Ong, Z.-Y.; Xu, T.; Xin, R.; Pan, L.; Wang, B.; Sun, L.; et al. Towards intrinsic charge transport in monolayer molybdenum disulfide by defect and interface engineering. *Nat. Commun.* **2014**, *5*, 5290. [CrossRef] [PubMed]
48. Kaasbjerg, K.; Thygesen, K.S.; Jacobsen, K.W. Phonon-limited mobility inn-type single-layer MoS2from first principles. *Phys. Rev. B* **2012**, *85*. [CrossRef]
49. Parkin, W.; Balan, A.; Liang, L.; Das, P.M.; Lamparski, M.; Naylor, C.H.; Rodríguez-Manzo, J.A.; Johnson, A.T.C.; Meunier, V.; Drndić, M. Raman Shifts in Electron-Irradiated Monolayer MoS2. *ACS Nano* **2016**, *10*, 4134–4142. [CrossRef]
50. Bafekry, A.; Faraji, M.; Fadlallah, M.M.; Khatibani, A.B.; Ziabari, A.A.; Ghergherehchi, M.; Nedaei, S.; Shayesteh, S.F.; Gogova, D. Tunable electronic and magnetic properties of MoSi2N4 monolayer via vacancy defects, atomic adsorption and atomic doping. *Appl. Surf. Sci.* **2021**, *559*, 149862. [CrossRef]
51. Bafekry, A.; Faraji, M.; Fadlallah, M.M.; Mortazavi, B.; Ziabari, A.A.; Khatibani, A.B.; Nguyen, C.V.; Ghergherehchi, M.; Gogova, D. Point Defects in a Two-Dimensional ZnSnN2 Nanosheet: A First-Principles Study on the Electronic and Magnetic Properties. *J. Phys. Chem. C* **2021**, *125*, 13067–13075. [CrossRef]
52. Faraji, M.; Bafekry, A.; Fadlallah, M.; Molaei, F.; Hieu, N.N.; Qian, P.; Ghergherehchi, M.; Gogova, D. Surface modification of titanium carbide MXene monolayers (Ti2C and Ti3C2) via chalcogenide and halogenide atoms. *Phys. Chem. Chem. Phys.* **2021**, *23*, 15319–15328. [CrossRef] [PubMed]
53. Bafekry, A.; Karbasizadeh, S.; Faraji, M.; Khatibani, A.B.; Sarsari, I.A.; Gogova, D.; Ghergherehchi, M. Van der Waals heterostructure of graphene and germanane: Tuning the ohmic contact by electrostatic gating and mechanical strain. *Phys. Chem. Chem. Phys.* **2021**, *23*, 21196–21206. [CrossRef] [PubMed]

54. Bafekry, A.; Stampfl, C.; Naseri, M.; Fadlallah, M.M.; Faraji, M.; Ghergherehchi, M.; Gogova, D.; Feghhi, S.A.H. Effect of electric field and vertical strain on the electro-optical properties of the MoSi2N4 bilayer: A first-principles calculation. *J. Appl. Phys.* **2021**, *129*, 155103. [CrossRef]
55. Bafekry, A.; Karbasizadeh, S.; Stampfl, C.; Faraji, M.; Hoat, D.M.; Sarsari, A.S.; Feghhi, S.A.H.; Ghergherehchi, M. Two-dimensional Janus semiconductor BiTeCl and BiTeBr monolayers: A first-principles study on their tunable electronic properties via an electric field and mechanical strain. *Phys. Chem. Chem. Phys.* **2021**, *23*, 15216–15223. [CrossRef] [PubMed]
56. Bafekry, A.; Mortazavi, B.; Faraji, M.; Shahrokhi, M.; Shafique, A.; Jappor, H.R.; Nguyen, C.; Ghergherehchi, M.; Feghhi, S.A.H. Ab initio prediction of semiconductivity in a novel two-dimensional Sb2X3 (X = S, Se, Te) monolayers with orthorhombic structure. *Sci. Rep.* **2021**, *11*, 10366. [CrossRef]
57. Kohn, W.; Sham, L.J. Self-Consistent Equations Including Exchange and Correlation Effects. *Phys. Rev.* **1965**, *140*, A1133–A1138. [CrossRef]
58. Hohenberg, P.; Kohn, W. Inhomogeneous Electron Gas. *Phys. Rev.* **1964**, *136*, B864. [CrossRef]
59. Kresse, G.; Furthmller, J. Efficiency of ab-initio total energy calculations for metals and semiconductors using a plane-wave basis set. *Comput. Mater. Sci.* **1996**, *6*, 15. [CrossRef]
60. Kresse, G.; Furthmuller, J. Efficient iterative schemes for ab initio total-energy calculations using a plane-wave basis set. *Phys. Rev. B* **1996**, *54*, 11169. [CrossRef]
61. Blochl, P.E. Projector augmented-wave method. *Phys. Rev. B* **1994**, *50*, 17953. [CrossRef] [PubMed]
62. Perdew, J.P.; Burke, K.; Ernzerhof, M. Generalized Gradient Approximation Made Simple. *Phys. Rev. Lett.* **1996**, *77*, 3865. [CrossRef] [PubMed]
63. Monkhorst, H.J.; Pack, J.D. Special points for Brillonin-zone integrations. *Phys. Rev. B* **1976**, *13*, 5188. [CrossRef]
64. Henkelman, G.; Arnaldsson, A.; Jónsson, H. A fast and robust algorithm for Bader decomposition of charge density. *Comput. Mater. Sci.* **2006**, *36*, 354–360. [CrossRef]
65. Le Page, Y.; Saxe, P. Symmetry-general least-squares extraction of elastic data for strained materials from ab initio calculations of stress. *Phys. Rev. B* **2002**, *65*, 104104. [CrossRef]
66. Gajdoš, M.; Hummer, K.; Kresse, G.; Furthmüller, J.; Bechstedt, F. Linear optical properties in the projector-augmented wave methodology. *Phys. Rev. B* **2006**, *73*, 045112. [CrossRef]
67. Baroni, S.; Resta, R. Ab initio calculation of the macroscopic dielectric constant in silicon. *Phys. Rev. B* **1986**, *33*, 7017. [CrossRef]
68. Northrup, J.E. Structure of Si(100)H: Dependence on the H chemical potential. *Phys. Rev. B* **1991**, *44*, 1419–1422. [CrossRef]
69. Hong, S.; Chou, M.Y. Theoretical study of hydrogen-covered diamond (100) surfaces: A chemical-potential analysis. *Phys. Rev. B* **1997**, *55*, 9975–9982. [CrossRef]
70. Hong, S. Surface energy anisotropy of iron surfaces by carbon adsorption. *Curr. Appl. Phys.* **2003**, *3*, 457–460. [CrossRef]
71. Liu, Y.; Hu, S.; Caputo, R.; Sun, K.; Li, Y.; Zhao, G.; Ren, W. Allotropes of tellurium from first-principles crystal structure prediction calculations under pressure. *RSC Adv.* **2018**, *8*, 39650–39656. [CrossRef]
72. Abdullaev, G.B.; Asadov, Y.G.; Mamedov, K.P. The Growth of α- and β-red Monoclinic Selenium Crystals and an Investigation of Some of Their Physical Properties. In *The Physics of Selenium and Tellurium*; Cooper, W.C., Ed.; Pergamon Press: New York, NY, USA, 1969; p. 179. [CrossRef]
73. Unger, P.; Cherin, P. Coordination and Thermal Motion in Crystalline Selenium and Tellurium. In *The Physics of Selenium and Tellurium*; Cooper, W.C., Ed.; Pergamon Press: New York, NY, USA, 1969; p. 223. [CrossRef]
74. Burbank, R.D. The crystal structure of α-monoclinic selenium. *Acta Crystallogr.* **1951**, *4*, 140. [CrossRef]
75. Gomes, L.C.; Carvalho, A. Phosphorene analogues: Isoelectronic two-dimensional group-IV monochalcogenides with orthorhombic structure. *Phys. Rev. B* **2015**, *92*, 085406. [CrossRef]
76. Eymard, R.; Otto, A. Optical and electron-energy-loss spectroscopy of GeS, GeSe, SnS, and SnSe single crystals. *Phys. Rev. B* **1977**, *16*, 1616–1623. [CrossRef]
77. Malone, B.D.; Kaxiras, E. Quasiparticle band structures and interface physics of SnS and GeS. *Phys. Rev. B* **2013**, *87*, 245312. [CrossRef]
78. Shimada, K. First-Principles Determination of Piezoelectric Stress and Strain Constants of Wurtzite III-V Nitrides. *Jpn. J. Appl. Phys.* **2006**, *45*, L358–L360. [CrossRef]
79. Duerloo, K.-A.N.; Ong, M.T.; Reed, E.J. Intrinsic Piezoelectricity in Two-Dimensional Materials. *J. Phys. Chem. Lett.* **2012**, *3*, 2871–2876. [CrossRef]
80. Zhu, H.; Wang, Y.; Xiao, J.; Liu, M.; Xiong, S.; Wong, Z.J.; Ye, Z.; Ye, Y.; Yin, X.; Zhang, X. Observation of piezoelectricity in free-standing monolayer MoS2. *Nat. Nanotechnol.* **2015**, *10*, 151–155. [CrossRef]
81. Nye, J.F.; Lindsay, R.B. Physical Properties of Crystals: Their Representation by Tensors and Matrices. *Phys. Today* **1957**, *10*, 26. [CrossRef]
82. Levy, M. Introduction to fundamentals of elastic constants. In *Experimental Methods in the Physical Sciences Vol. 39*; Academic Press: Cambridge, MA, USA, 2001; pp. 1–35. [CrossRef]
83. Bechmann, R. Elastic and Piezoelectric Constants of Alpha-Quartz. *Phys. Rev.* **1958**, *110*, 1060. [CrossRef]
84. Lueng, C.M.; Chan, H.L.W.; Surya, C.; Choy, C.L. Piezoelectric coefficient of aluminum nitride and gallium nitride. *J. Appl. Phys.* **2000**, *88*, 5360–5363. [CrossRef]
85. Li, W.; Li, J. Piezoelectricity in two-dimensional group-III monochalcogenides. *Nano Res.* **2015**, *8*, 3796–3802. [CrossRef]

nanomaterials

MDPI

Article

Degradation of Acid Red 1 Catalyzed by Peroxidase Activity of Iron Oxide Nanoparticles and Detected by SERS

Edna Vázquez-Vélez *, Horacio Martínez * and Fermín Castillo

Instituto de Ciencias Físicas, Universidad Nacional Autónoma de México. Av. Universidad 1000, Col. Chamilpa, 62210 Cuernvaca, Morelos, Mexico; ciro@icf.unam.mx

* Correspondence: velez.edna@icf.unam.mx (E.V.-V.); hm@icf.unam.mx (H.M.)

Abstract: Magnetic iron oxide nanoparticles (MIONPs) were synthesized using tannic acid and characterized by Raman, FTIR, UV, and DRX spectroscopy. In a heterogeneous Fenton-like reaction, the catalytic peroxidase-like activity of MIONPs in the degradation of Acid Red 1 (AR 1) dye was investigated. TEM/STEM was used to determine the quasi-spherical morphology and particle size (3.2 nm) of the synthesized MIONPs. The XRD powder patterns were indexed according to the reverse spinel structure of magnetite, and SEM-EDS analysis confirmed their chemical composition. At pH = 3.5, the decomposition of H_2O_2 in hydroxyl radicals by MIONPs results in high AR 1 degradation (99%). This behavior was attributed to the size and surface properties of the MIONPs. Finally, the Surface Enhanced Raman Spectroscopy (SERS) technique detected intermediary compounds in the degradation process.

Keywords: magnetite; peroxidase activity; SERS

Citation: Vázquez-Vélez, E.; Martínez, H.; Castillo, F. Degradation of Acid Red 1 Catalyzed by Peroxidase Activity of Iron Oxide Nanoparticles and Detected by SERS. *Nanomaterials* **2021**, *11*, 3044. https://doi.org/10.3390/nano11113044

Academic Editors: Jihoon Lee and Ming-Yu Li

Received: 1 October 2021
Accepted: 3 November 2021
Published: 12 November 2021

1. Introduction

The textile industry represents two-thirds of the use of dyes resulting in one of the main sources of contamination. Due to the high concentration of dye in the wastewater from these installations, it is not easy to treat it satisfactorily. It is estimated that 2% of the colorants are discharged directly into the aqueous effluent [1]. Azo-type dyes are the most commonly used polluting compounds, accounting for 70% of the total. Notably, the acid red 1 (AR 1) dye is among the 11 non-biodegradable azo compounds listed [2]. Nowadays, the importance of this type of dye increases in the field of biomedicine, electronics, and energy because dyes can absorb visible electromagnetic radiation with high efficiency. As a result, removing the non-biodegradable dye is a significant environmental issue. These compounds can be found in the aquatic environment, and they are poisonous to both marine and human organisms. Besides, they are mutagenic and carcinogenic [3]. For decades, there has been a growing interest in removing these difficult-to-remove compounds [4,5]. However, new technology is required to complete the mineralization of various dyes while minimizing their impact on the environment and humans.

Nanotechnology is one of the fastest-growing and emerging research areas today. Problems related to water quality can be solved or improved through it. The synthesis of nanoparticles is an economical, effective, efficient, and sustainable alternative. Its use makes treatment processes less polluting than traditional methods [6,7]. In recent decades, magnetic iron oxide nanoparticles (MIONPs) have sparked interest in catalytic wet peroxide oxidation (CWPO) or Fenton heterogeneous oxidation. The potential of these materials stems from their more remarkable ability to degrade recalcitrant contaminants, which involves the generation of hydroxyl radicals in sufficient quantities to allow oxidation [8,9]. The hydroxyl radical constitutes one of the most powerful oxidants (E0 = 2.73 V), and is much stronger than other conventional oxidizing species such as hydrogen peroxide (E0 = 1.31 V) or ozone (E0 = 1.52 V) [10]. MIONPs represent a promising alternative to the conventional catalysts used in CWPO due to their higher activity, easy recovery, and further reusability. In this context, one of the most critical challenges in CWPO is the development of more active and stable catalysts. Several studies on the degradation of recalcitrant dyes using MIONPs have been published [10]. Iron oxide nanoparticles have also been supported on various porous materials to improve their efficiency in heterogeneous Fenton-like catalysis [11]. Other studies have used ultrasonic radiation to improve the degradation efficiency of dyes [12]. The peroxidase-like activity of magnetite nanoparticles to generate hydroxyl radicals, in particular, has been extensively studied in the degradation of molecules [13]. However, its study on the degradation of recalcitrant contaminants continues to be of great interest due to being one of the most cost-effective CWPO.

This work describes an accessible synthesis of magnetic iron oxide nanoparticles (MIONPs) utilizing tannic acid, a water-soluble polyphenol frequently found in herbaceous and woody plants [14]. Furthermore, the peroxidase-like activity of synthesized MIONPs in the oxidative degradation of acid red 1 dye was investigated. The quantification of hydrogen peroxide generation confirmed that AR 1 was degraded by hydroxyl radicals. Finally, the SERS technique was used to characterize the intermediate compounds in the degradation process.

2. Materials and Methods

2.1. Materials

All chemical reagents and solvents were purchased from Merck KGaA, Darmstadt, Germany. Deionized water Millipore was used for NPs formation, and all degradation experiments were performed with ultrapure water. The acid red 1 solution was prepared fresh before each experiment to ensure the accuracy of the experimental data. The pH of the solution was adjusted using NaOH or HCl solution (5%).

2.2. Synthesis of MIONPs

Magnetic Iron oxide nanoparticles (MIONPs) were synthesized by dissolving 0.8 g (3 mmol) of iron (III) chloride hexahydrate ($FeCl_3{\cdot}6H_2O$) in 20 mL of deionized water to which the capping agent was added: 0.25 g of tannic acid (0.14 mmol) dissolved in 20 mL of deionized water. The solution immediately produces a black-colored solution, which was mechanically stirred at 250 rpm. The pH of the solution was then adjusted to 7.6 ± 0.2 by adding NaOH solution drop by drop. The result was the formation of a magnetic black precipitate. The solid was washed three times with deionized water (40 mL) to remove the salts, once with acetone to remove residual water, and three times with ethanol to eliminate the organic matter. All washes were accomplished with the assistance of an external magnetic field to separate the liquid from the solid. Finally, the black solid was dried at 80 °C for two hours and used immediately for degradation tests.

2.3. Characterization Techniques

Morphological and size study of the nanoparticles was analyzed by field emission transmission electron microscopy (TEM/STEM, JEOL JEM-2010 FEG, Akishima, Tokyo, Japan) operating at 200 kV. A drop of the solution (NPs dispersed in EtOH) was placed on a carbon-coated copper grid to prepare the sample. Scanning electron microscopy (SEM) and energy-dispersive X-ray spectroscopy (EDS) was performed on MIONPs using a TESCAN MIRA 3 GMU microscope from Brno—Kohoutovice, Czech Republic; coupled with energy dispersive X-ray analysis (Brucker). A drop of the solution of NPs dispersed was placed on carbon tape to create the sample. The X-ray diffraction (XRD) patterns of the synthesized nanoparticles were recorded using a Rigaku Miniflex DMAX 2200 X-ray diffractometer, Austin, TX, USA. The solid was subjected to Cu Kα radiation (1.54 Å) with graphite monochromator in the 2θ range of 5–80°. For the characterization by UV-vis spectrophotometry, the solid was dispersed in ethanol with the help of ultrasonic vibrations. The solution was analyzed using an Ocean View UV-vis spectrophotometer, Orlando, FL, USA. FTIR spectroscopy was used to characterize the nanoparticles using a Bruker ATR-FTIR spectrometer Alpha II, Ettlingen, Germany. Raman spectroscopy using a confocal microscope Raman Bruker SENTERRA II from Ettlingen, Germany, confirmed the composition and crystallographic phase of Iron NPs. A low laser power (1 mW) was used to avoid sample degradation due to laser heating. The accumulation time was 10,000 ms. Bleaching of 1 mW and 100 ms was performed to attenuate the fluorescence of the sample. A power of 25 mW and an integration time of 20,000 ms were used for the samples analyzed by the SERS technique.

2.4. Degradation Procedure

The pH effect on dye degradation was studied by adjusting the starting pH of the AR 1 solution at a concentration of 50 mg/L. AR 1 solution was carried to a pH = 3.5, 4.5, 6.5, 8.5, and 10.5 using either solution of HCl or NaOH (5%). MIONPs (2 g/L) were dispersed in AR 1 solution (50 mL) using ultrasonic vibrations for 5 s. A 2 mM concentration of H_2O_2 was then added to begin heterogeneous Fenton-like catalysis. This amount was found to be the bare minimum required to initiate the peroxidase-like activity of MIONPs. The oxidation process was carried out in the absence of light. The absorbance was measured at 505.06 nm after 60 min of treatment. On the other hand, the degradation of AR 1 was investigated according to the treatment time. For this, 50 mL of a 200 mg/L solution was prepared. This concentration required a minimum of 4 g/L of MIONPs to start the catalytic oxidation. At the end of the process, MNIONPs were removed from the solution using a magnet. After the first degradation cycle, the solid was dried at 90 °C and immediately reused for the second cycle of degradation under the same conditions.

2.5. UV-Vis Spectroscopic Analysis

The concentration of AR 1 was measured at 505.06 nm by UV-vis spectrophotometry, using an Ocean View UV-Vis spectrophotometer. Besides, the concentration of H_2O_2 during the degradation process was quantified using titanium sulfate spectrophotometry [15]. The absorbance of this solution was measured at 400.13 nm. Previously, a calibration plot based on Beer–law Lambert's equation was established by relating the absorption to AR 1 and the hydrogen peroxide concentration. AR 1 dye degradation efficiency (%) was calculated using equation 1, where Ci is the initial concentration, and Cf is the final concentration of solution after degradation:

$$\text{Degradation } (\%) = \frac{(\text{Ci} - \text{Cf})}{\text{Ci}} \cdot 100 \tag{1}$$

3. Results and Discussion

3.1. Characterization of MIONPs

The morphology and size of the synthesized MIONPs were analyzed by TEM/STEM images (Figure 1). Figure 1a,b shows TEM images of nanoparticles with a quasi-spherical morphology and a size of about 3 nm. MIONPs can aggregate and form large particles due to their strong anisotropy dipolar [16], so certain needle forms are observed in Figure 1c. Using the Digital Micrograph (DM) software 3.7 Gatan, Pleasanton, CA, USA, one large aggregate (Figure 1d) was analyzed and digitally processed to obtain fast Fourier transform (FFT). The bright circular rings indicate the polycrystalline phase in the selected area electron diffraction (SAED) patterns. The spaced-resolved lattice fringes with an interplanar distance of 0.25 nm agree well with the lattice spacing of Fe_3O_4 (311) planes [11]. Figure 1e,f shows the STEM analysis of NPs; these images revealed a better quasi-spherical morphology. The size of NPs in Figure 1d was analyzed to perform a histogram employing the DM software, which confirms an average size of 3.2 nm.

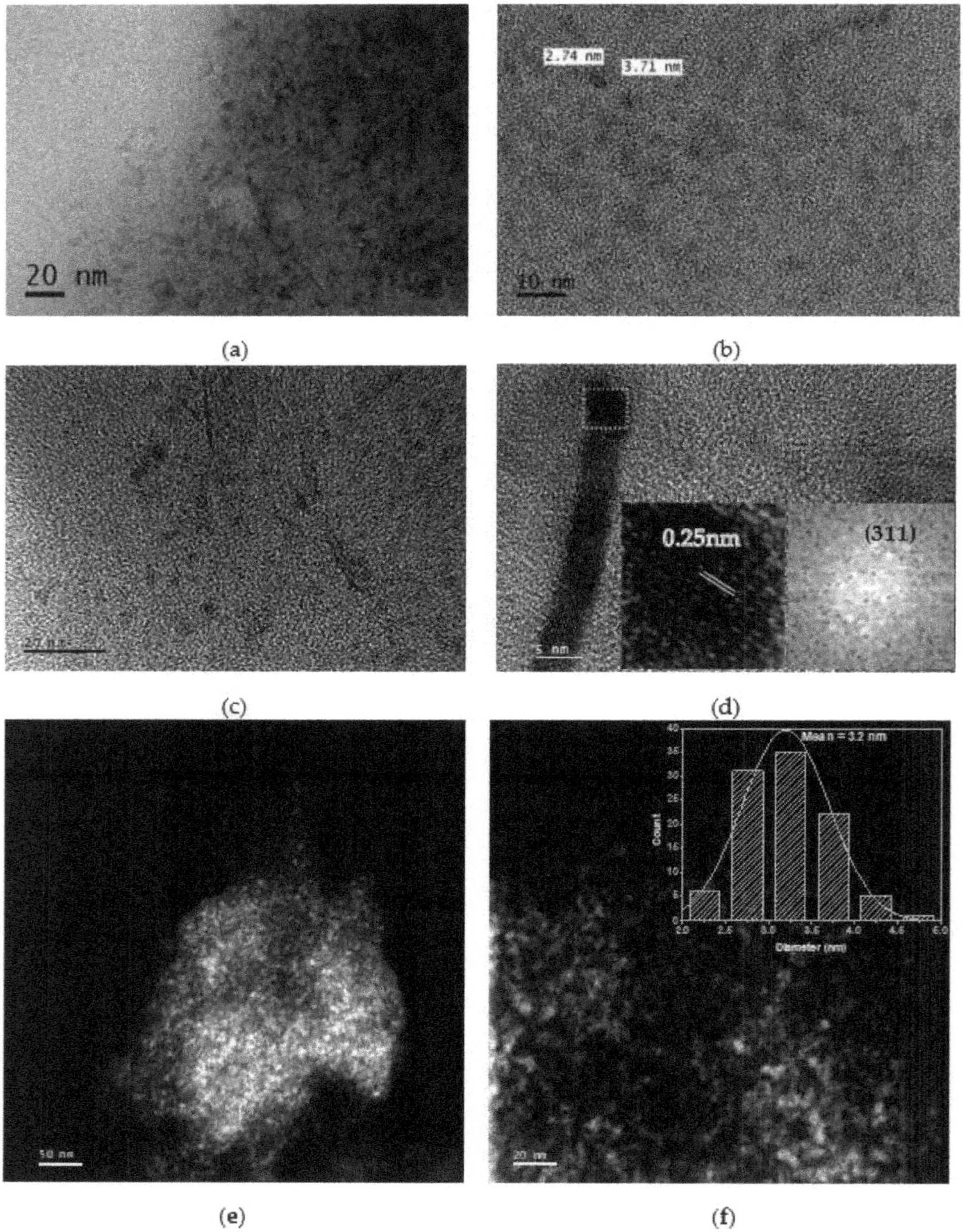

Figure 1. (**a**–**d**) TEM images and (**e**–**f**) STEM images of synthesized MIONPs.

The SEM image is seen in Figure 2a with its point EDS analysis applied in the box marked. The EDS spectrum recorded for the nanoparticles is shown in Figure 2b, where a strong signal for elemental iron is observed at 6.2 keV and another small signal at 7.1 keV. The presence of an oxygen peak at 0.6 keV indicates that iron oxides were formed. The mass relationship corresponds to Fe_3O_4. Similar results were obtained using Ridge gourd peel extract [17] to synthesize magnetite nanoparticles.

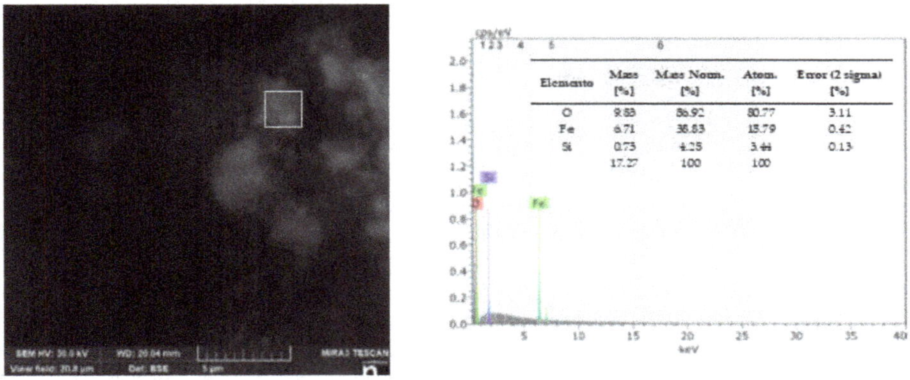

Elemento	Mass [%]	Mass Norm. [%]	Atom. [%]	Error (2 sigma) [%]
O	9.83	56.92	80.77	3.11
Fe	6.71	38.83	15.79	0.42
Si	0.73	4.25	3.44	0.13
	17.27	100	100	

(a) (b)

Figure 2. (**a**) SEM image and (**b**) EDS spectrum of MIONPs.

Figure 3a shows the XRD spectrum of synthesized MIONPs. The broad diffraction peaks confirm the formation of an amorphous solid. Here, the width in the middle of the maximum is related to the smallest particle size [18]. The diffractogram depicts the initial crystallizing phase, which corresponds primarily to the magnetite phase [19]. The diffraction patterns are according to the inverse spinel structure of magnetite shown in down of figure. The diffraction pattern card from magnetite powder (19–0629) was obtained from Columbian Carbon Co., in New York, USA. The five characteristic peaks at 2 = 30.50, 35.87, 43.65, 57.54, and 63.28° are assigned to (220), (311), (400), (511), and (440) crystalline planes, respectively. However, it is well established that maghemite (γ-Fe_2O_3) and magnetite (Fe_3O_4) reveal similar XRD profiles [20]. Bibi et al. [21] published a diffractogram of γ-Fe_2O_3 Nps as well as its absorbance spectrum, which showed a peak at 371.71 nm. Compared to this study, MIONPs revealed a different diffractogram in their intensities. Figure 2b depicts the absorbance spectrum showing a peak at 228 nm, consistent with previous reports for the magnetite phase [22].

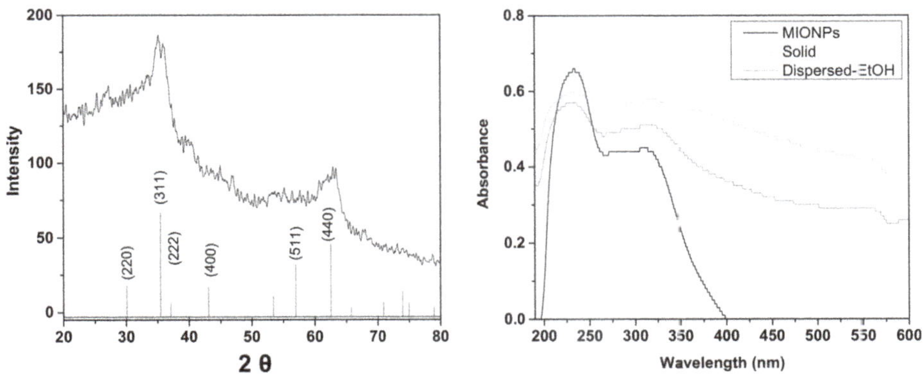

(a) (b)

Figure 3. Spectra of MIONPs: (**a**) XRD and (**b**) UV-visible.

UV-visible spectroscopy was performed to analyze the stability of magnetite nanoparticles. Figure 3b depicts the absorbance spectrum of MIONPs in a solid-state after two days of conservation in a closed container without an inert atmosphere. The peak at 228 nm decreases, while a broad peak of 350–600 increases, indicating a change to the magnetite phase. However, the same figure also shows the spectrum of MIONPs conserved in an anhydrous ethanol dispersion after fifteen days. The peak at 228 nm is still visible, but the broad peak at 350–600 nm appears. The peak at about 350 nm is assigned to octahedral Fe^{3+} in small oligomeric FeO_x clusters, and the bands at 450−600 nm are characteristic of the Fe_2O_3 aggregates [23]. These results indicate that MIONPs can be stable in the dispersion of anhydrous ethanol during a particular time.

ATR-FTIR analysis was performed to observe the surface purity of the MIONPs, see Figure 4a. According to the literature, the magnetite FTIR spectrum has two strong absorption bands at 570 and 390 cm^{-1}, which can be attributed to the Fe-O stretching mode of the tetrahedral and octahedral sites, respectively [24]. In this study, the Fe-O vibration signal of MIONPs was observed at 540 and 494 cm^{-1}. This shift can be attributed to the small size of the nanoparticles [25]. However, a weak shoulder was observed at 594 cm^{-1}, which could be due to the beginning of the γ-Fe_2O_3 formation, with absorption bands at 630, 590, and 430 cm^{-1} [24]. Nevertheless, none of the vibration band was observed around 3400 cm^{-1}, indicating the vibration stretching of the O-H group from tannic acid. So, the MIONPs used for the degradation test was free of organic matter.

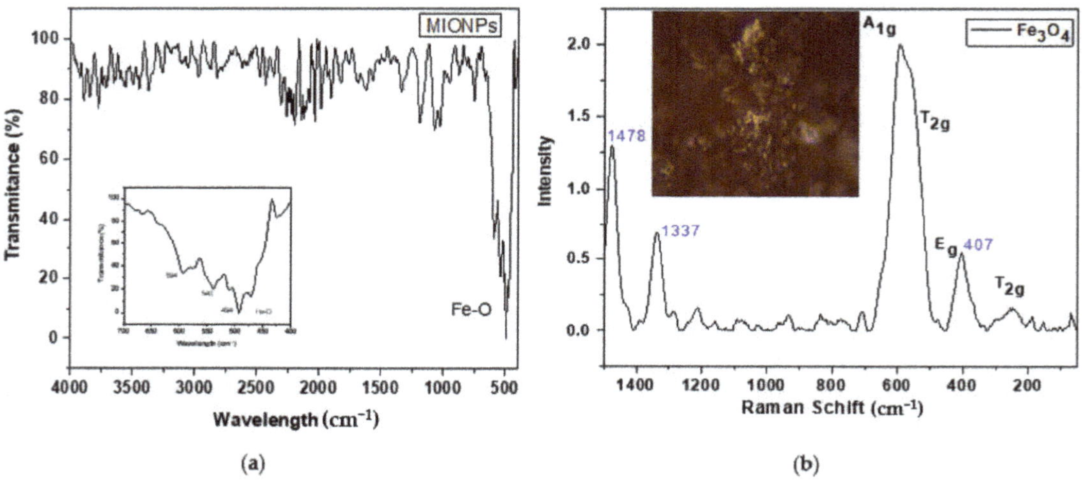

Figure 4. (**a**) FTIR and (**b**) Raman spectra with optical imagen at 50× magnification, of MIONPs.

On the other hand, in Raman spectroscopy, magnetite has a spinel structure, and five phonon bands have been theoretically predicted: one from A_{1g}, a second from E_g, and three from T_{2g} [26]. Zhang et al. reported the Fe_3O_4 nanoparticles Raman spectrum. The T_{2g} modes were observed at 305 and another at 534 cm^{-1}, the E_g mode at 513 cm^{-1}, and the A_{1g} mode at 660 cm^{-1} [20]. The Raman spectrum of synthesized MIONPs is shown in Figure 3b. The band assigned to the A_{1g} mode was shifted to 602 cm^{-1} due to particle size. However, Profile Breit Wigner Fano (BWF) [18] can better describe this shift to low frequency. Nevertheless, the phonon mode (A_{1g}) of γ-Fe_2O_3 was not observed at 710 cm^{-1}, implying no phases coexist on the surface of MIONPs [19]. On the other hand, the peaks at 407, 1337, and 1478 cm^{-1} indicate the start of the oxidation to hematite by laser action. This fact is due to the increased temperature of laser-heated spots results in a softening of phonon frequency [27].

MIONPs were successfully synthesized by the coprecipitation method using tannic acid at pH < 8. Synthesis of magnetite nanoparticles at pH = 8 has been reported using a capping agent synthesized from gallic acid [28]. However, the formation γ-Fe_2O_3 from the synthesis with tannic acid has also been described at pH = 12 [29]. Then, the pH control in the magnetite synthesis plays an essential role in controlling the oxidation of Fe^{2+} to Fe^{3+} [30].

3.2. Degradation Catalyzed by MIONPs

Figure 5a shows the absorbance value of the AR 1 solutions treated at different pH and their corresponding percent degradation in heterogeneous Fenton-like catalysis. The most representative value was the absorbance of 0.01 for the solution at pH = 3.5 with 99.12% of degradation efficiency. These findings are consistent with the Fenton reaction, which shows that the highest degradation efficiency occurs at pH = ~4. The degradation efficiency at pH > 4 is around 85%, except at pH = 6.5, which is 92%. The oxidation process produces inactive ferric oxyhydroxides at pH > 4, which reduces degradation efficiency. However, a different behavior was observed when the pH = 6.5. Here, the inherent pH of the solution prevented the beginning formation of oxyhydroxides. Furthermore, the adsorption of the dye on Fe^{3+} probably occurs through a ligand with the non-protonated carbonyl groups of AR 1. The geometry of the molecule adsorbed on MIONPs surface through a complex plays an essential role in the activation of reactive species, like it has been reported for EDTA complex, which generates hydroxyl and oxygen radicals [31].

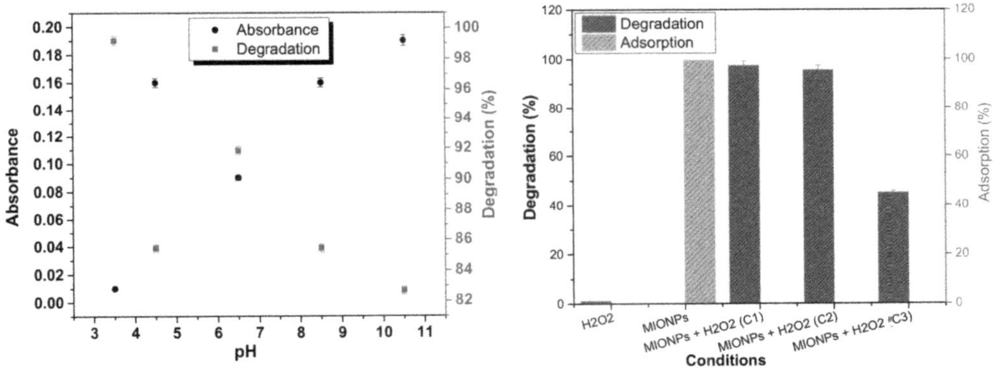

Figure 5. Graphic of (**a**) absorbance and degradation (%) at different pH and (**b**) degradation and adsorption at different conditions of the reaction.

Figure 5b shows the degradation percentage of AR 1 in an aqueous solution at 200 mg/L under different conditions after 60 min of treatment. The adsorption of AR 1 by MIONPs was 99.5% after 10 min of their addition. The addition of only H_2O_2 did not achieve the AR 1 degradation. The dilution of the solution by adding H_2O_2 caused just an instantaneous decrease in the AR 1 concentration. However, the oxidation process immediately occurred when H_2O_2 was introduced into the system to initiate the catalytic reaction following AR 1 adsorption on MIONPs. The degradation of AR 1 reached 97.4%. MIONPs stability and reusability were studied because of the essential use for their application. In the second degradation cycle (C2), the efficiency decreased to 96% and 45% in the third cycle (C3). Figure 6 shows the Raman spectrum of MIONPs before and after of AR 1 degradation process. The spectrum after degradation corresponds mainly to the magnetite phase, but the maghemite phase begins to be observed at 770 cm^{-1}. In the optical images of MIONPs, a change in porosity, shape, and aggregation state was detected after the treatment process, which corroborates that the MIONPs act as adsorbents and exhibit catalytic activity for decomposition H_2O_2.

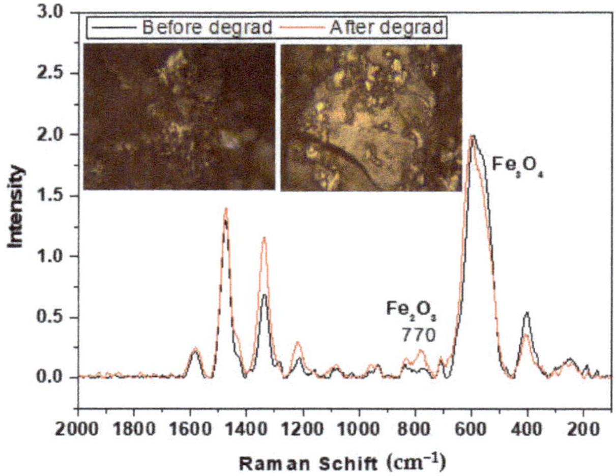

Figure 6. Raman spectra of MIONPs before and after AR 1 degradation.

3.3. *Peroxidase-like Activity of MIONPs*

The degradation of AR 1 as a function of time is shown in Figure 7a for a concentration of 200 mg/L. This concentration is mainly reported for effluents that need to be treated [1]. The double peak with a maximum at λ = 531 and 505.5 nm is the characteristic band of the conjugated electron structure (chromophore group) corresponding to the π-π^* transition bond of N=N, while the second peak at about 325 nm corresponding to the π-π^* bond of N-H. The 240 nm and 280 nm are the bands due to benzene and naphthalene rings [32]. The solution's absorbance was measured at 505 nm. Figure 7a depicts the discoloration image of solutions treated each 10 min, with their absorbance spectra. The graph shows that the double absorption peak at 505 nm and 531 nm decreases with increasing time, indicating that AR 1 degradation is effective. The intensity of this band significantly reduced after the addition of H_2O_2 (initial). The highest percentage of degradation (84%) occurs within the first 10 min, and degradation was 97.4% completed up to 40 min. UV-visible spectra indicate the AR 1 degradation to small molecules from the chromophore groups.

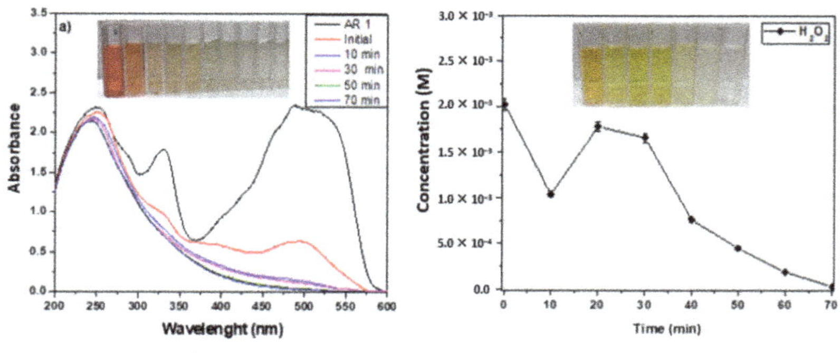

(a) (b)

Figure 7. (**a**) UV-Visible spectra of AR 1 and (**b**) concentration of H_2O_2 in the degradation process.

Figure 7b shows the concentration of H_2O_2 quantified during the heterogeneous Fenton-like catalysis depicted in Figure 7a. The titanium sulfate spectrophotometry method was used to determine the concentration of peroxide. It was observed that the H_2O_2 decreased in the first ten minutes, which corresponded to the highest degradation seen in the first minutes after initiation with hydrogen peroxide. After, H_2O_2 generation catalyzed by MIONPs is followed in the next 10 min to achieve the maximum dye degradation. Finally, the peroxide consumption is observed slowly and concludes in 50 min more.

3.4. Mechanism of AR 1 Degradation

Peroxidases are a group of enzymes capable of catalyzing the oxidation of hydrogen peroxide into hydroxyl radicals. These radicals further participate in electron exchange with substrates producing color on oxidation. This activity is performed by the HEMO group (a Fe ion coordinated with a porphyrin that acts as a tetradentate ligand), which is found in peroxidase-type enzymes. Gao et al. [33] reported on the peroxidase-like activity of magnetite nanoparticles in the presence of peroxide. However, it has been stated that Fe_3O_4 MNPs are not as effective in the treatment of hazardous organic pollutants. It is necessary to increase the H_2O_2 activation capacity of Fe_3O_4 MNPs, with the assistance of ultrasonic irradiation [13]. In this study, we investigate the peroxidase-like activity of MIONPs in the oxidation of a recalcitrant dye in the absence of assisted irradiation. The reaction mechanism of MIONPs in the generation of OH radicals is due to their actuation as redox enzymes by the electron exchange escalated by their superficial atoms. Besides, when the organic molecule (AR 1) is adsorbed on the surface of MIONPs through electronic interaction at the molecule-metal interface to form a monolayer, these electrostatic and steric interactions promote the catalytic activity on its degradation. Then, the redox reaction occurs on the surface of MIONPs, but is related to the heterogeneous Fenton reaction in solution through the following equations [31]:

$$Fe^{2+} + H_2O_2 \longrightarrow Fe^{3+} + {}^{\bullet}OH + OH^- \quad (2)$$

$$Fe^{3+} + H_2O_2 \longrightarrow Fe^{2+} + {}^{\bullet}OOH + H^+ \quad (3)$$

$$HO_2^{\bullet} + HO_2^{\bullet} \longrightarrow H_2O_2 + O_2 \quad (4)$$

$${}^{\bullet}OH + AR\ 1 \leftrightarrow \text{intermediates} \leftrightarrow CO_2 + H_2O \quad (5)$$

AR 1 oxidation begins when the bare minimum of H_2O_2 is added to the dye solution to begin the catalytic activity of MIONPs. This relationship is not stoichiometric, unlike the conventional Fenton reaction. Then, the degradation of AR 1 occurs mainly at the solid-liquid interfaces of MIONPs, where the formation of hydroxyl radicals is due to the catalyzed decomposition of hydrogen peroxide by the active sites (Fe^{2+} and Fe^{3+}) of MIONPs.

Figure 8a shows the Raman spectrum of AR 1 (down), where a broad and strong peak was observed at 1580 cm^{-1} corresponding to the vibrational signal C=C of the aromatic rings. The carbonyl signal of the amide group coexists in this peak because the electric field from the aromatic ring causes a shift at a low frequency. The azo group N=N signal was detected at 1353 cm^{-1}, and the aromatic ring C-N bond vibration signal was detected at 1330 cm^{-1}. The sulfate group S=O and the vibration signal from the aromatic ring C-S bond were observed at 1156 cm^{-1} and 656 cm^{-1}. The AR 1 spectrum (up) adsorbed on the surface of MIONPs showed the signal of C=O from the amide group displaced at 1748 cm^{-1} because of electrostatic interaction with MIONPs. The vibrational signal from the aromatic ring occurs at C=C at 1580 cm^{-1}, but another peak corresponding to this signal appears at 880 cm^{-1}. This unusual signal is caused by a change in the symmetry of the molecule that interacts with the MIONPs surface. Finally, two vibrational signals appear that correspond to the C-O and N-O bonds, most likely as a result of AR 1 forming a coordination ligand with Fe^{2+} through the amide and azo groups. The strong signals at 253 and 375 cm^{-1} correspond to γ-FeOOH [26].

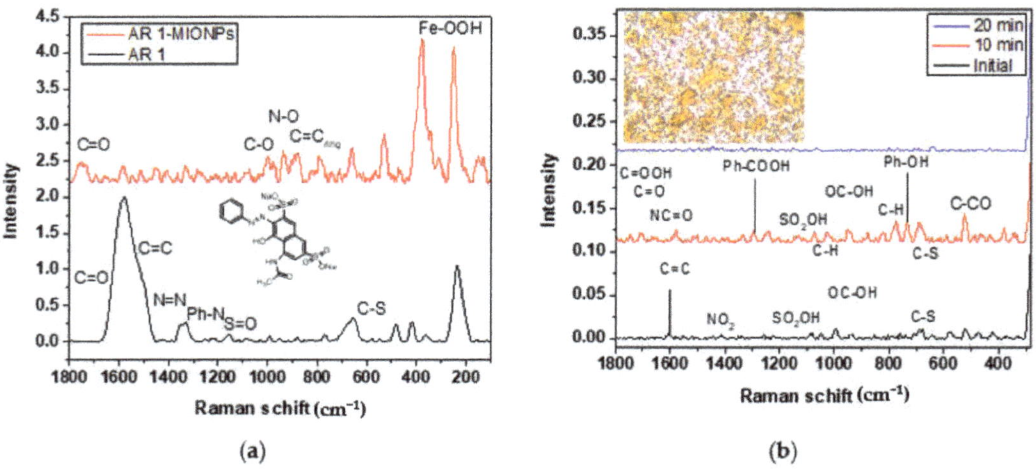

Figure 8. Raman spectra of (**a**) AR 1 (down) and AR 1 adsorbed on MIONPs (up); (**b**) intermediaries' compounds, and optical image at 50× magnification of the SERS substrate (initial).

3.5. Compounds Detected by SERS in AR 1 Degradation

SERS analysis was performed on solutions obtained from Figure 7a. To detect AR 1, the initial and treated solutions were deposited on Ag nanoflowers [34] (on Si wafer substrate). Figure 8b shows the spectra of the initial, at 10 min, and 20 min solutions from the degradation process. After this time, no signal was detected for the following solutions. The spectra present the signals corresponding to different functional groups linked to intermediaries' compounds reported and detected by mass spectroscopy during the H_2O_2-photolysis degradation of AR 1 [32].

Figure 9 shows the AR 1 degradation mechanism through oxidation by radicals •OH. In the initial spectrum (immediate analysis after adding peroxide), the signal from azo group N=N at 1353 cm^{-1} in compound 1 was not observed. Instead, the vibrational signal for the NO_2 group was detected at 1412 cm^{-1} because of N=N bond cleavage to obtain compound 2 and nitrophenol ($C_6H_5NO_2$). Besides, the signal of the SO_2OH group in intermediaries' compounds was detected at 1084 cm^{-1}. The spectrum corresponding to ten minutes of treatment revealed more signals with higher intensities from the functional groups in compounds 4, 5, 6, and 7. The signal corresponding to the C=O from carboxylic acid in compounds 5, 6, and 7 was detected at 1747 cm^{-1}; the vibrational stretching of the C=O bond in compound 4 was observed at 1706 cm^{-1}, and the signal of C=O from the amide group in compounds 3, 4, and 6 was seen at 1660 cm^{-1}. At 1580 cm^{-1}, the vibrational signal was observed at C=C from the aromatic rings. At 1291 cm^{-1} and 731 cm^{-1}, the benzoic acid and phenol peaks were detected, respectively. Then, the main intermediaries' compounds in the degradation of AR 1 can be observed after ten minutes of catalytic activity of MIONPs. However, signals in the spectrum of the solution after 20 min of treatment decrease considerably.

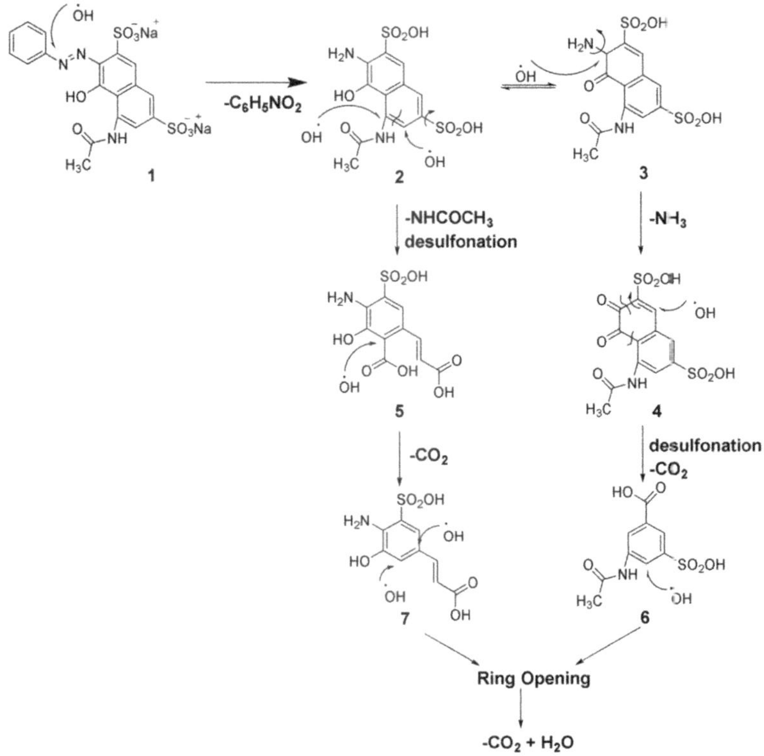

Figure 9. AR 1 degradation mechanism through oxidation by radicals •OH.

The results obtained demonstrate that the application of MIONPs in AR 1 degradation suggests the superiority of the conventional Fenton-like catalysts through a dye highly recalcitrant. In this context, several studies on the degradation of azo-type organic dyes using iron oxide nanoparticles in heterogeneous Fenton-like catalysis reactions have been published. Table 1 depicts a comparison of those studies and this work. The peroxidase-like activity of MIONPs was reported for the degradation of Rhodamine (RhB) using ultrasonic irradiation (UI) [13,35]. The efficiency of degradation (90%) is due to the synergy of the ultrasonic and Fenton processes. However, this technology has the disadvantage of ultrasonic systems of high power consumption. Alternatively, magnetite nanoparticles supported on graphene oxide (GO) presented a good efficiency in degrading Acid orange 7 (AO 7) in normal conditions without irradiation. The Magnetite NPs without support of GO showed the same efficiency after 45 min of the process achieving 80%, and those supported in GO reached 98% of AO 7 degradation after three hours [11]. Nanocomposites of stable hematite nanoparticles with SiO_2 have also been reported for heterogeneous Fenton-like catalysis. The degradation efficiency was excellent, but the treatment times were too long [23,36]. Compared to the other reported results, the synthesized MIONPs in this work present an excellent peroxidase-like activity in degrading a recalcitrant contaminant at a high concentration. The high efficiency of degradation under normal temperature conditions without irradiation is one of the study's advantages. Besides, the particle size (3.2 nm) of MIONPs and the adsorption of AR 1 via—interactions from the aromatic ring structure on its basal plane and NPs surface contribute to the efficiency of this work. This dye adsorption increases the local concentration of AR 1 within the vicinity of the active sites to be further oxidized by the generated hydroxyl radicals.

Table 1. Degradation studies of azo dyes using different nanomaterials and process conditions.

Organic Dyes [Concentration]	Catalyst	Conditions	Degradation and Time	Reference
RhB (10 mg/L)	Fe_3O_4 NPs (10–15 nm)	Cat = 0.5 g/L, H_2O_2 40 mM pH = 5, T = 25 °C, UI = 20 kHz	90% 1 h	[13]
RhB (10 mg/L)	Ag_3PO_4–Fe_3O_4@AB (10 nm)	Cat = 0.1 g/L, H_2O_2 = 1.5 mM pH = 3.5, T = 25 °C, UI = 40–970 kHz	99.3% 1 h	[35]
AO 7 (35 mg/L)	GO–Fe_3O_4 NPs Fe_3O_4 NPs (10 nm)	Cat = 0.2 g/L, H_2O_2 = 22 mM pH = 3, T = 25 °C	98%, 3 h ~80%	[11]
MB (120–300 mg/L)	α-Fe_2O_3/SiO_2	Cat = 1 g/L, H_2O_2 = 2 M pH = 3, T = 25 °C	99% 60 h	[23]
MB (50 mg/L) RhB (25 mg/L))	α-Fe_2O_3@Ti-tmSiO_2	Cat = 0.25 g/L, H_2O_2 = 53 mM pH = 4, T = 50 °C	100%, 5.5 h 100%, 9h	[36]
AR 1 (200 mg/L)	Fe_3O_4 NPs (3.2 nm)	Cat = 4 g/L, H_2O_2 = 2 mM pH = 3.5, T = 25 °C	97.4% 1 h	This work

4. Conclusions

Synthesis MIONPs with tannic acid is a good and efficient method in the obtention of magnetite nanoparticles. The nanoparticles were characterized and found to be 3.2 nm in average size, with a quasi-spherical morphology. The various spectroscopies and microscopies showed their chemical composition and magnetite phase. The peroxidase-like activity of the synthesized nanoparticles was verified by a heterogeneous Fenton-like catalysis reaction involving the degradation of AR 1. A high concentration of AR 1 was mineralized with an efficiency of 97.4% in one hour. MIONPs activity is strongly dependent on their size, shape, and surface structure; these characteristics can be appropriately tuned during their synthesis. Furthermore, the SERS technique identified the intermediary compounds in the degradation process. In the future, these nanoparticles could be supported in other nanomaterials to investigate their efficiency in heterogeneous Fenton-like catalysis, which would address the disadvantage of NPs instability and aggregation.

Author Contributions: Conceptualization, H.M. and E.V.-V.; Methodology, E.V.-V.; Formal Analysis, H.M. and E.V.-V.; Investigation, E.V.-V.; Resources, H.M. and F.C.; Data Curation, E.V.-V.; Writing—Original Draft Preparation, E.V.-V.; Writing—Review & Editing, H.M. and F.C.; Visualization, H.M. and F.C.; Supervision, H.M.; Project Administration, H.M.; Funding Acquisition, H.M. All authors have read and agreed to the published version of the manuscript.

Funding: This research received no external funding.

Institutional Review Board Statement: Not applicable.

Informed Consent Statement: Not applicable.

Data Availability Statement: Not applicable.

Acknowledgments: The authors thank Guadalupe Zavala from the Institute of Biotechnology, UNAM, and Samuel Tehuacanero-Cuapa from Physic Institute, UNAM; for their microscopy analysis support. To María Luisa Ramon-García from the Renewable Energy Institute, UNAM, for his X-ray diffraction analysis support. To Osvaldo Flores-Cedillo, and Ing. Hector H. Hinojosa-Galvan from ICF, UNAM for their technical support. Finally, to Cecilia A. Reyes-Aranda and Reynel Hernández-Martínez from UAEM for their support in the experiments.

Conflicts of Interest: The authors declare no conflict of interest.

References

1. Singh, K.; Arora, S. Removal of synthetic textile dyes from wastewaters: A critical review on present treatment technologies. *Crit. Rev. Environ. Sci. Technol.* **2011**, *41*, 807–878. [CrossRef]
2. Shaul, G.M.; Holdsworth, T.J.; Dempsey, C.R.; Dostal, K.A. Fate of water-soluble azo dyes in the activated sludge process. *Chemosphere* **1991**, *22*, 107–119. [CrossRef]

3. UNESCO. *The United Nations World Water Development Report 4: Managing Water Report under Uncertainty and Risk*; United Nations Educational, Scientific and Cultural Organization: Paris, France, 2012; Volume 1.
4. Brillas, E.; Martínez-Huitle, C.A. Decontamination of wastewaters containing synthetic organic dyes by electrochemical methods. An up dated review. *Appl. Catal. B Environ.* **2015**, *166*, 603–643. [CrossRef]
5. Anjaneyulu, Y.; Sreedhara Chary, N.; Suman Raj, D.S. Decolorization of industrial effluents–available methods and emerging technologies—A review. *Rev. Environ. Sci. Biotechnol.* **2005**, *4*, 245–273. [CrossRef]
6. Florenza, X.; Sales Solano, A.M.; Centellas, F.; Martínez-Huitle, C.A.; Brillasa, E.; Garcia-Segura, S. Degradation of the azo dye Acid Red 1 by anodic oxidation and indirectelectrochemical processes based on Fenton's reaction chemistry. Relationship between decolorization, mineralization and products. *Electrochim. Acta* **2014**, *142*, 276–288. [CrossRef]
7. Do Vale-Júnior, E.; Da Silva, D.R.; Fajardo, A.S.; Martínez-Huitle, C.A. Treatment of an azo dye effluent by peroxi-coagulation and its comparison to traditional electrochemical advanced processes. *Chemosphere* **2018**, *204*, 548–555. [CrossRef]
8. Javaid, R.; Qazi, U.Y. Catalytic Oxidation Process for the Degradation of Synthetic Dyes: An Overview. *Int. J. Environ. Res. Public Health* **2019**, *16*, 2066. [CrossRef]
9. Hodges, B.C.; Cates, E.L.; Kim, J.-H. Challenges and prospects of advanced oxidation water treatment processes using catalytic nanomaterials. *Nat. Nanotechn.* **2018**, *13*, 642–650. [CrossRef] [PubMed]
10. Munoz, M.; de Pedro, Z.M.; Casas, J.A.; Rodriguez, J.J. Preparation of magnetite-based catalysts and their application in heterogeneous Fenton oxidation–A review. *Appl. Catalys. B Environ.* **2015**, *176*, 249–265. [CrossRef]
11. Zubir, N.A.; Yacou, C.; Motuzas, J.; Zhang, X.; Diniz da Costa, J.C. Structural and functional investigation of graphene oxide–Fe3O4 nanocomposites for the heterogeneous Fenton-like reaction. *Sci. Rep.* **2014**, *4*, 4594. [CrossRef]
12. Maharjan, A.; Dikshit, P.K.; Gupta, A.; Kim, B.S. Catalytic activity of magnetic iron oxide nanoparticles for hydrogen peroxide decomposition: Optimization and characterization. *J. Chem. Technol. Biotechnol.* **2020**, *95*, 2495–2508. [CrossRef]
13. Wang, N.; Zhu, L.; Wang, M.; Wang, D.; Tang, H. Sono-enhanced degradation of dye pollutants with the use of H2O2 activated by Fe3O4 magnetic nanoparticles as peroxidase mimetic. *Ultrason. Sonochem.* **2010**, *17*, 78–83. [CrossRef] [PubMed]
14. Robles, H. Tannic Acid. In *Encyclopedia of Toxicology*, 3rd ed.; Elsevier: Amsterdam, The Netherlands, 2014; pp. 474–475. [CrossRef]
15. Du, X.; Xu, Y.; Qin, L.; Lu, X.; Liu, Q.; Bai, Y. Simple and Rapid Spectrophotometric Determination of Titanium on Etched Aluminum Foils. *Am. J. Anal. Chem.* **2014**, *5*, 149–156. [CrossRef]
16. Deng, J.; Wen, X.; Wang, Q. Solvothermal in situ synthesis of Fe3O4-multiwalled carbon nanotubes with enhanced heterogeneous Fenton-like activity. *Mater. Res. Bull.* **2012**, *47*, 3369–3376. [CrossRef]
17. Cheera, P.; Karlapudi, S.; Sellola, G.; Ponneri, V. A facile green synthesis of spherical Fe3O4 magnetic nanoparticles and their effect on degradation of methylene blue in aqueous solution. *J. Mol. Liq.* **2016**, *221*, 993–998. [CrossRef]
18. Bhattacharjee, S.; Mazumder, N.; Mondal, S.; Panigrahi, K.; Banerjee, A.; Das, D.; Sarkar, S.; Roy, D.; Kumar Chattopadhyay, K. Size-modulation of functionalized Fe3O4: Nanoscopic customization to devise resolute piezoelectric nanocomposites. *Dalton Trans.* **2020**, *49*, 7872. [CrossRef]
19. Ali, S.; Khan, S.A.; Yamani, Z.H.; Qamar, M.T.; Morsy, M.A.; Sarfraz, S. Shape and size-controlled superparamagnetic iron oxidenanoparticles using various reducing agents and their relaxometric properties by Xigo acorn area. *Appl. Nanosci.* **2019**, *9*, 479–489. [CrossRef]
20. Zhang, X.; Niu, Y.; Meng, X.; Li, Y.; Zhao, J. Structural evolution and characteristics of the phase transformations between α-Fe2O3, Fe3O4 and γ-Fe2O3 nanoparticles under reducing and oxidizing atmospheres. *Cryst. Eng. Comm.* **2013**, *15*, 8166–8172. [CrossRef]
21. Bibi, I.; Nazar, N.; Ata, S.; Sultan, M.; Ali, A.; Abbas, A.; Jilani, K.; Kamal, S.; Sarim, F.M.; Khan, M.I.; et al. Green synthesis of iron oxide nanoparticles using pomegranate seeds extract and photocatalytic activity evaluation for the degradation of textile dye. *J. Mater. Res. Technol.* **2019**, *8*, 6115–6124. [CrossRef]
22. Razack, S.A.; Suresh, A.; Sriram, S.; Ramakrishnan, G.; ·Sadanandham, S.; Veerasamy, M.; Nagalamadaka, R.B.; ·Sahadevan, R. Green synthesis of iron oxide nanoparticles using Hibiscus rosa-sinensis for fortifying wheat biscuits. *SN Appl. Sci.* **2020**, *2*, 898. [CrossRef]
23. Wu, Z.; Zhu, W.; Zhang, M.; Lin, Y.; Xu, N.; Chen, F.; Wang, D.; Chen, Z. Adsorption and Synergetic Fenton-like Degradation of Methylene Blue by a Novel Mesoporous α-Fe2O3/SiO2 at Neutral pH. *Ind. Eng. Chem. Res.* **2018**, *57*, 5539–5549. [CrossRef]
24. Stoia, M.; Istratie, R.; Păcurariu, C. Investigation of magnetite nanoparticles stability in air by thermal analysis and FTIR spectroscopy. *J. Therm. Anal. Calorim.* **2016**, *125*, 1185–1198. [CrossRef]
25. Chamritski, I.; Burns, G. Infrared- and Raman-Active Phonons of Magnetite, Maghemite, and Hematite: A Computer Simulation and Spectroscopic Study. *J. Phys. Chem. B* **2005**, *109*, 11. [CrossRef]
26. De Faria, D.L.A.; Venaüncio Silva, S.; De Oliveira, M.T. Raman Microspectroscopy of Some Iron Oxides and Oxyhydroxides. *J. Raman Spectrosc.* **1997**, *28*, 873–878. [CrossRef]
27. Shebanova, O.N.; Lazor, P. Raman spectroscopic study of magnetite (FeFe2O4): A new assignment for the vibrational spectrum. *J. Sol. Stat. Chem.* **2003**, *174*, 424–430. [CrossRef]
28. Guin, D.; Manorama, S.V. Room temperature synthesis of monodispersed iron oxide nanoparticles. *Mater. Lett.* **2008**, *62*, 3139–3142. [CrossRef]
29. Herrera-Becerra, R.; Rius, J.L.; Zorrilla, C. Tannin biosynthesis of iron oxide nanoparticles. *Appl. Phy. A* **2010**, *100*, 453–459. [CrossRef]

30. Morgan, B.; Lahav, O. The effect of pH on the kinetics of spontaneous Fe(II) oxidation by O2 in aqueous solution–basic principles and a simple heuristic description. *Chemosphere* **2007**, *68*, 2080–2084. [CrossRef] [PubMed]
31. Pignatello, J.J.; Oliveros, E.; MacKay, A. Advanced Oxidation Processes for Organic Contaminant Destruction Based on the Fenton Reaction and Related Chemistry. *Crit. Rev. Environ. Sci. Technol.* **2006**, *36*, 1–84. [CrossRef]
32. Shoniya, T.; Sreekanth, R.; Sijumon, V.A.; Usha, K.A.; Aravinda, K.C.T. Oxidative degradation of Acid Red 1 in aqueous medium. *Chem. Eng. J.* **2014**, *244*, 473–482. [CrossRef]
33. Gao, L.; Zhuang, J.; Nie, L.; Zhang, J.; Zhang, Y.; Gu, N.; Wang, T.; Feng, J.; Yang, D.; Perrett, S.; et al. Intrinsic peroxidase-like activity of ferromagnetic nanoparticles. *Nat. Nanotechnol.* **2007**, *2*, 577–583. [CrossRef] [PubMed]
34. Tong, J.; Xu, Z.; Bian, Y.; Niu, Y.; Zhang, Y.; Wang, Z. Flexible and smart fibers decorated with Ag nanoflowers for highly active surface-enhanced Raman scattering detection. *J. Raman Spectrosc.* **2019**, *50*, 1468–1476. [CrossRef]
35. Jun, B.-M.; Kim, Y.; Yoon, Y.; Yea, Y.; Park, C.M. Enhanced sonocatalytic degradation of recalcitrant organic contaminants using a magnetically recoverable Ag/Fe-loaded activated biochar composite. *Ceram. Intern.* **2020**, *46*, 22521–22531. [CrossRef]
36. Lv, Q.; Li, G.; Sun, H.; Kong, L.; Lu, H.; Gao, X. Preparation of magnetic core/shell structured c-Fe2O3@Ti-tmSiO2 and its application for the adsorption and degradation of dyes. *Microporous Mesoporous Mater.* **2014**, *186*, 7–13. [CrossRef]

nanomaterials

Article

$Ti_3Si_{0.75}Al_{0.25}C_2$ Nanosheets as Promising Anode Material for Li-Ion Batteries

Jianguang Xu *, Qiang Wang, Boman Li, Wei Yao and Meng He *

School of Materials Science and Engineering, Yancheng Institute of Technology, Yancheng 224051, China; wangqiang970808@sina.com (Q.W.); liboman0919@sina.com (B.L.); weiyao@ycit.edu.cn (W.Y.)
* Correspondence: xujg@ycit.cn or jgxu@163.com (J.X.); hemeng666@ycit.edu.cn or hemeng315@163.com (M.H.)

Abstract: Herein we report that novel two-dimensional (2D) $Ti_3Si_{0.75}Al_{0.25}C_2$ (TSAC) nanosheets, obtained by sonically exfoliating their bulk counterpart in alcohol, performs promising electrochemical activities in a reversible lithiation and delithiation procedure. The as-exfoliated 2D TSAC nanosheets show significantly enhanced lithium-ion uptake capability in comparison with their bulk counterpart, with a high capacity of ≈350 mAh g^{-1} at 200 mA g^{-1}, high cycling stability and excellent rate performance (150 mAh g^{-1} after 200 cycles at 8000 mA g^{-1}). The enhanced electrochemical performance of TSAC nanosheets is mainly a result of their fast Li-ion transport, large surface area and small charge transfer resistance. The discovery in this work highlights the uniqueness of a family of 2D layered MAX materials, such as Ti_3GeC_2, Ti_3SnC_2 and Ti_2SC, which will likely be the promising choices as anode materials for lithium-ion batteries (LIBs).

Keywords: MAX phases; ultrathin nanosheets; lithium-ion battery; electrochemical performance; anode materials

Citation: Xu, J.; Wang, Q.; Li, B.; Yao, W.; He, M. $Ti_3Si_{0.75}Al_{0.25}C_2$ Nanosheets as Promising Anode Material for Li-Ion Batteries. *Nanomaterials* **2021**, *11*, 3449. https://doi.org/10.3390/nano11123449

Academic Editors: Jihoon Lee and Ming-Yu Li

Received: 27 November 2021
Accepted: 17 December 2021
Published: 20 December 2021

1. Introduction

Since they were first reported, lamellar ternary carbides and nitrides have been named "MAX phases" or "$M_{n+1}AX_n$ phases (n = 1, 2 or 3)", where M represents an early transition metal, A is an element of IIIA to VIA groups and X is carbon or nitrogen. They have attracted great attention because of their special combination of metallic and ceramic properties [1]. For example, due to their inherent layered structure with alternately arranged MX and A layers, the MAX phases display a superior resistance to oxidation, thermal energy and corrosion, very good electrical conductivity, high strength and elastic modulus and excellent machinability [2–6]. Owing to their lamellar structure and excellent conductivity, MAX phases show great potential in lithium-ions storage for a Li-ion battery (LIB) or capacitor [7–14]. However, the reported capacities of MAX phases are relatively low, particularly in the initial few charge–discharge cycles, which may restrict their real application in LIB. Thus, it is important to optimize the lithium-ion uptake property of MAX phases. It has been well accepted that the nanoscale materials, particularly the ultrathin two-dimensional (2D) nanosheets, have improved properties compared to their corresponding bulk counterparts [15–18]. For instance, the reversible capacity of free-standing graphene nanosheet (GNS) was found to be 540 mAh g^{-1} [19], and that of N-GNS even reached a high value of 684 mAh g^{-1} [20], both of which are over the theoretical reversible capacity of graphite. In that case, it is anticipated that MAX nanosheets can exhibit an enhanced lithium-ion storage property compared to their bulk materials.

Unfortunately, unlike the inorganic graphene analog (IGA) with weak van der Waals force between its layers, the MAX phases have relatively robust connections between the MX and A layers, and it seems difficult to exfoliate the bulk MAX materials into ultrathin nanosheets by a facile sonic exfoliation process [21]. Particularly, most of the Ti_3SiC_2 and Ti_2SC particles were broken into small species instead of being delaminated to ultrathin nanosheets by increasing the power of sonication [10]. To overcome this limitation, we developed an available substitutional-solid-solution-based exfoliation process for the large-scale fabrication of ultrathin nanosheets of A-layer-activated MAX phases. As a result, ultrathin $Ti_3Si_{0.75}Al_{0.25}C_2$ (TSAC) nanosheets with a very thin thickness of 4 nm were prepared based on this strategy, which can be used as a promising filler for polymer composites [22,23].

Herein, we extend the application of these TSAC nanosheets to the electrode for LIBs because the size and morphology of MAX phases promise improvements on their electrochemical performance for LIBs [9–11]. The $Ti_3Si_{0.75}Al_{0.25}C_2$ nanosheets have very thin thickness, so it can be expected that Li ions can easily be intercalated into the TSAC nanosheets layers. In addition, the electrical conductivity of a MAX phase is normally higher than its MXene counterpart, which usually has excellent conductivity [24], because the A-group layer increases the metallic properties of the material for the improvement of their overall electrochemical properties [1]. Hence, per the diagram in Figure 1, combining the large surface area and excellent conductivity with $Ti_3Si_{0.75}Al_{0.25}C_2$ nanosheets, high performance can be anticipated for the $Ti_3Si_{0.75}Al_{0.25}C_2$ nanosheets electrode.

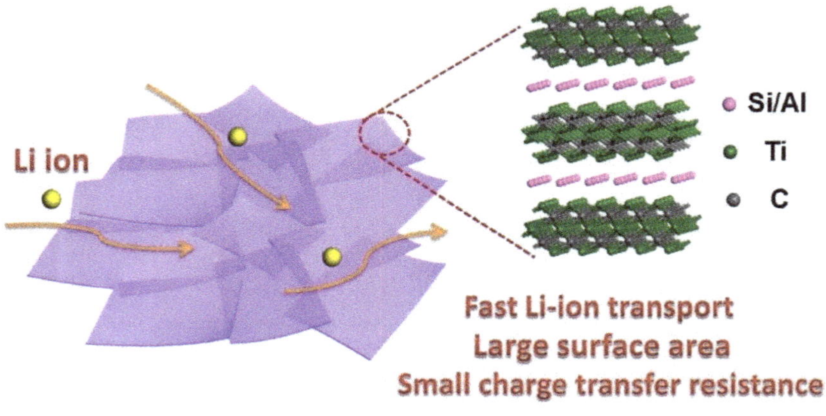

Figure 1. Schematic illustration of the advantages by using TSAC nanosheets-based electrode for Li-ion battery.

2. Experimental Section

2.1. Materials

Titanium powder (300 mesh, 99.9 wt.%) was purchased from Guangzhou metallurgy (Guangzhou, China). Silicon powder (200 mesh, 99.0 wt.%), aluminum (Al) powder (200 mesh, 99.0 wt.%), graphite powder (30 μm, 99.85 wt.%) and absolute alcohol (AR) were purchased from Sinopharm (Shanghai, China).

2.2. Preparation of $Ti_3Si_{0.75}Al_{0.25}C_2$ Powder

The $Ti_3Si_{0.75}Al_{0.25}C_2$ powder (bulk TSAC) was synthesized via a self-propagation high-temperature synthesis (SHS) process. In detail, 14.3 g Ti powder, 4.21 g Si powder, 0.68 g Al powder and 2.4 g graphite powder were blended using a QM-BP Ball Mill at 300 rpm for 2 h, in which the molar ratio of Ti: Si: Al: C was about 3:1.5:0.25:2. The as-received mixture was then put into a self-propagation high-temperature reactor, and then ignited by a tungsten filament under the protection of pure Ar gas. After combustion, a gray product was collected for further processing. The atomic ratio of the Ti, Si, Al and C of the as-received powder is about 3:0.75:0.25:2, which was determined by X-ray fluorescence spectroscopy (XRF) in our previous work [22].

2.3. Preparation of Ultrathin $Ti_3Si_{0.75}Al_{0.25}C_2$ Nanosheets

The $Ti_3Si_{0.75}Al_{0.25}C_2$ nanosheets (TSAC nanosheets) were obtained by the liquid exfoliation of TSAC powder in absolute alcohol via sonication. In detail, 2 g bulk TSAC powder was dispersed in 0.2 L absolute alcohol, and then sonicated for 24 h. After sonication, the resulting dispersion liquid was then centrifuged at 2000 rpm for 20 min to remove most residual large-size particles. Finally, TSAC nanosheets were dried and collected for the following tests through vacuum filtration of the resulting supernatant.

2.4. Preparation of Electrodes

The electrochemical behaviors of TSAC nanosheets and bulk TSAC were studied using CR-2032 coin-cell with lithium metal as the counter electrode and reference electrode. The batteries were based on Li metal (−) | | TSAC (+) with liquid electrolyte (1M solution of $LiPF_6$ in ethyl carbonate (EC)-dimethyl carbonate (DMC)-ethyl methyl carbonate (EMC) (1:1:1, $v/v/v$)). Microporous polypropylene membrane (Celgard2500) was used as a separator. Next, 80 wt.% TSAC nanosheets or bulk TSAC, 10 wt.% acetylene black and 10 wt.% polyvinylidene fluoride (PVDF) were dispersed in N-methylpyrrolidone (NMP) and uniformly mixed into a viscous slurry. Then, the slurry was deposited on a copper foil current collector. The electrodes were then vacuum dried for 12 h at 120 °C, followed by electrochemical evaluation. The loading of active material was about 0.65~0.85 mg cm^{-2}. Finally, the cells were assembled in a glove box filled with 99.99 wt.% Ar gas.

2.5. Characterization

The TSAC powder was characterized by X-ray powder diffraction (XRD) with a Japan Rigaku Dmax X-ray diffractometer (Tokyo, Japan) equipped with graphite monochromatized high-intensity Cu-Kα radiation (λ = 1.54178 Å). The field emission scanning electron microscopy (FE-SEM) images were performed using a FEI Nova NanoSEM 450 scanning electron microscope (Hillsboro, OR, USA). The transmission electron microscopy (TEM) images were taken on a JEM-2100F field emission electron microscope (Tokyo, Japan) with X-MaxN 80T IE250 Energy Disperse Spectroscopy (UK). The Brunauer–Emmett–Teller (BET) specific surface areas of the samples were determined using a Micromeritics TriStar II 3020 system (Norcross, GA, USA). X-ray photoelectron spectroscopy (XPS, Thermo Fisher ESCALAB 250Xi, Waltham, MA, USA) was utilized to explore the electron-binding energy of TSAC nanosheets. The Raman spectra of the samples were recorded by a Renishaw inVia Reflex system (Gloucestershire, UK) equipped with an argon ion Laser with a wavelength of 633 nm. The cells were charged and discharged galvanostatically in a fixed voltage window from 0.001 to 3 V on a Shenzhen Neware battery cycler (Shenzhen, China) at room temperature. All the gravimetric capacity data related to as-prepared samples were based on the mass of TSAC nanosheets or bulk TSAC. Cyclic voltammetry and electrochemical impedance spectroscopy (EIS, frequency range: 0.1–10^5 Hz, amplitude: 5 mV) analysis were carried out by a Zahner-Zennium electrochemical workstation (Kronach, Germany).

3. Results and Discussion

The feasibility of using bulk TSAC and exfoliated TSAC nanosheets as electrodes for LIBs was investigated. The XRD pattern and SEM image of a bulk TSAC are shown in Figure 2. The primary phase of this product is very close to the pattern of Ti_3SiC_2 (JCPDS card No. 89-1356). A small blue shift also appears (Figure 2a), indicating the formation of a solid solution $Ti_3Si_{0.75}Al_{0.25}C$, as discussed in our previous work [22]. In addition, a small amount of TiC is also found in the XRD pattern of bulk TSAC, in most situations which is generated with the formation of Ti_3SiC_2 [25]. The micrograph of the product (Figure 2b) shows a distinct lamellar structure, which is in agreement with the crystal structure of Ti_3SiC_2. As shown in Figure 1, the crystal structure of Ti_3SiC_2 is composed of Ti_3C_2 layer and Si layer along c-direction. Because the connection between the Ti_3C_2 layer and the Si layer is relatively low and can be exfoliated, Ti_3SiC_2 and other MAX phase compounds generally display a lamellar appearance.

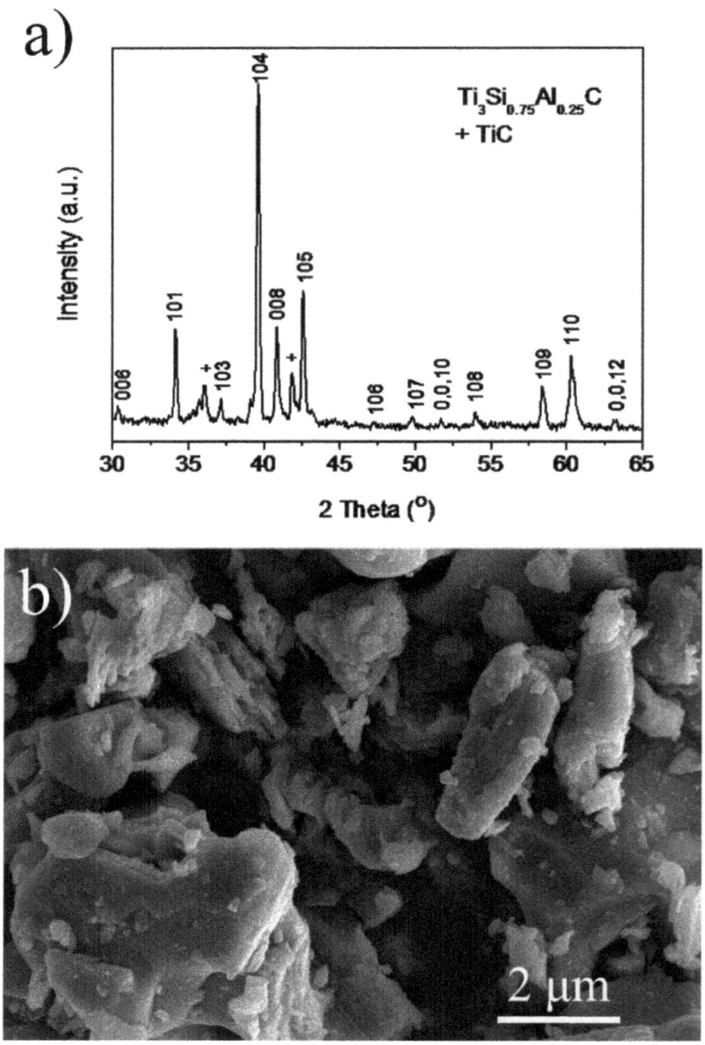

Figure 2. (**a**) XRD pattern and (**b**) SEM micrograph of the product synthesized by SHS.

The microstructure of the TSAC nanosheets is shown in Figure 3. The sample exhibits a sheet-like morphology with a size from around 100 to 1000 nm. According to a TEM image of the nanosheets (Figure 3b), most of them exhibit ultrathin sheet-like structures, which render them transparent or semitransparent. Moreover, it could be detected from a magnified nanosheet in Figure 3c that TSAC nanosheets are composed of only a few layers, indicating that bulk TSAC has been delaminated successfully and the crystal structure of $Ti_3Si_{0.75}Al_{0.25}C_2$ (inserted in Figure 3c) is well maintained during sonication. The BET results further confirmed the exfoliated sheet-like structure. Compared with the bulk TSAC, the specific surface area (SSA) of the TSAC nanosheets increased from 4.25 m^2 g^{-1} to 11.68 m^2 g^{-1}. In addition, on the basis of EDX analysis in Figure 3d, the atomic ratios of Ti:(Si + Al):C and Si:Al are close to 3:1:2 and 3:1, respectively, suggesting that the composition of $Ti_3Si_{0.75}Al_{0.25}C_2$ particles has no obvious change after delamination.

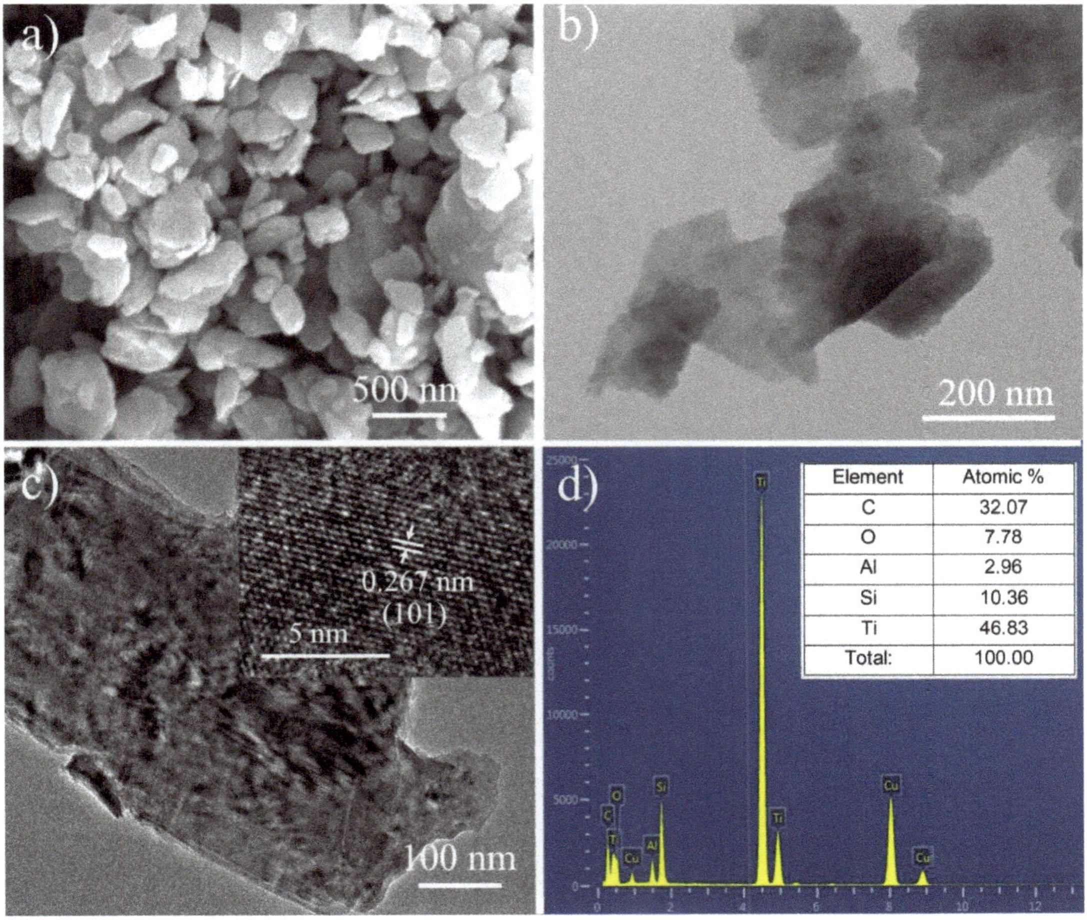

Figure 3. (**a**) SEM and (**b**,**c**) TEM images of the exfoliated TSAC nanosheets. (**d**) Energy-dispersive X-ray spectroscopy (EDX) analysis of TSAC nanosheets, performed on the center of the nanosheet in Figure 3c.

Figure 4a–e show the high resolution XPS spectra of TSAC nanosheets. The XPS spectra of Ti 2p and C 1s are similar to those spectra of Ti_3C_2 MXene [24,26,27], which also includes Ti–C, Ti–O, TiO_2, C–C, C–O and O–C=O bonds, indicating both Ti_3C_2 and TSAC nanosheets have similar intralayer and surface structure of MX layer. In addition, based on the existence of Ti–O, C–O, Si–O and Al–O bonds [28], it can be deduced there is an oxide layer covered on the surface of TSAC nanosheets, which is in agreement with previous reports of MAX phases [28]. In the spectra of Si 2p and Al 2p in Figure 4b,c, Si metal and Al metal bonds are detected [29], suggesting weak connections between the MX layer and the A layer in TSAC MAX phase nanosheets. Thus, bulk TSAC particles have the possibility to be exfoliated into thin nanosheets. The Raman spectra of bulk TSAC and TSAC nanosheets in Figure 4f further proved the layered structure of TSAC MAX phase, because both samples show distinct Raman peaks of Ti_3SiC_2 as in previous reports [30,31]. After delamination, these peaks of the received TSAC nanosheets exhibit small red shifts, which may be caused by the expanding crystal structure of the MAX phase [10]. Hence, by combining the results of SEM, TEM, XPS and Raman of TSAC nanosheets, it can be concluded that the exfoliated products are mainly composed of $Ti_3Si_{0.75}Al_{0.25}C_2$ phase and the layered structure of $Ti_3Si_{0.75}Al_{0.25}C_2$ phase has been well maintained during sonication.

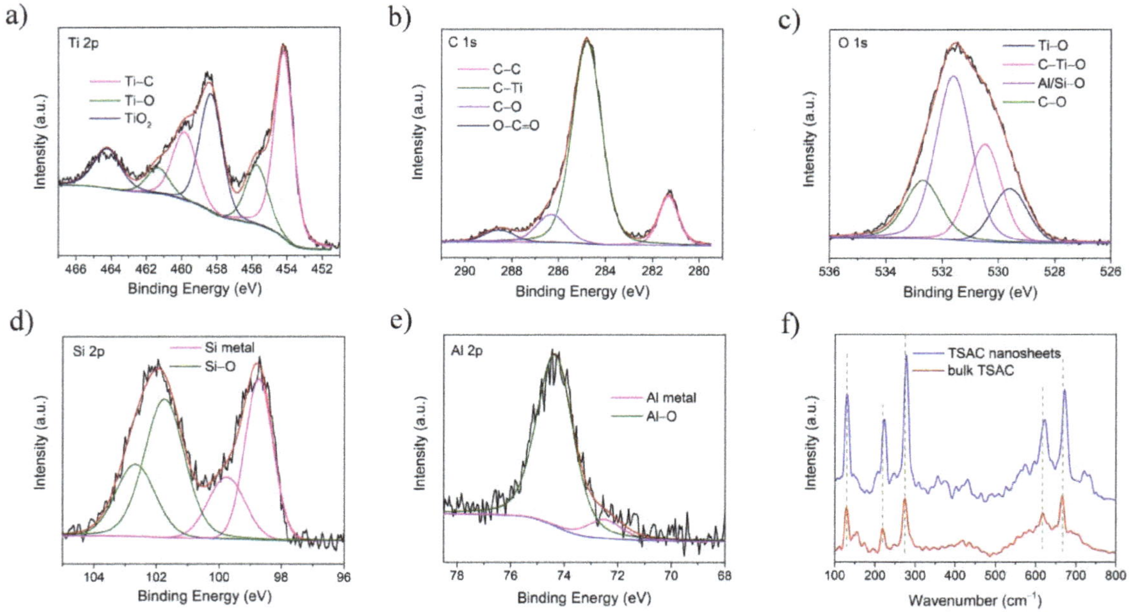

Figure 4. High resolution XPS spectra of TSAC nanosheets: (**a**) Ti 2p; (**b**) C 1s; (**c**) O1s; (**d**) Si 2p; (**e**) Al 2p; (**f**) Raman spectra of bulk TSAC and TSAC nanosheets.

The cyclic voltammetry (CV) and galvanostatic charge–discharge cycling (GV) of bulk TSAC and TSAC nanosheets were tested in this work. In Figure 5a,b, both TSAC nanosheets and bulk TSAC exhibit similar CV behaviors, except there is an additional cathodic peak of TSAC nanosheets around 2.4 V, showing that the exfoliated TSAC nanosheets perform a special Li^+ deintercalation process. During the first reduction process of both electrodes, there is an obvious peak located at about 1.2 V associated with the generation of a solid electrolyte interphase (SEI) film [32]. This peak then practically disappears during the following reduction process, which can be ascribed to the isolation between the TSAC anode and electrolyte by the dense SEI film formed on the surface of the TSAC anode in the first reduction process. Then, a peak around 0.75 V is detected and kept at the following cycles, where the reaction of Li^+ with TSAC nanosheets such as Ti_3C_2 probably occurs [32]. In the following cycles, two broad redox reversible peaks at 1.8 and 2.4 V corresponding to the reaction between Li^+ and titanium oxide on the surface of TSAC nanosheets appear, because similar pairs were also observed in Ti_2C and Ti_3C_2 nanosheets [32–35]. Interestingly, during the second and third cycles of TSAC nanosheets, a small cathodic peak at 0.1 V and two anodic peaks at 0.15 and 0.5 V can be observed, probably corresponding to lithiation process of Si and delithiation of Li_xSi [36]. Meanwhile, for bulk TSAC, these redox peaks are not obvious because there are less exposed Si atoms in bulk TSAC particles. This alloying procedure of exposed "A" layer of MAX phases is also found in those Sn-containing MAX phases, such as Ti_2SnC, Nb_2SnC and V_2SnC [7–9]. In addition, the differences in the CV profiles between these initial two cycles and the similarity in the second and the third cycles indicates that the irreversible capacity losses of TSAC nanosheets and bulk TSAC mainly take place in the first cycle.

The CV profiles of TSAC nanosheets in Figure 5a also show no obvious charge and discharge capacity at potential higher than 3 V vs. Li/Li^+. Thus, the lithium-ion storage tests for TSAC were performed from 0.001 to 3 V. The voltage profile for TSAC nanosheets at 80 mA g^{-1} (Figure 6a) delivers a beginning charge capacity (lithiation) of 862 mAh g^{-1}. This capacity is much bigger than that of bulk TSAC with only 215 mAh g^{-1} (Figure 6b). It can also be detected from the charge profile in Figure 6a that more than three-quarters of the capacity are under 1.5 V, indicating that TSAC nanosheets could function better as anode materials for LIBs. Figure 6c shows the lithium-ion storage behavior of bulk TSAC and TSAC nanosheets with charge/discharge cycles at 200 mA g^{-1}. The coulomb efficiency of TSAC nanosheets at the first cycle is 47.2%, which then increases rapidly and reaches 93.1% at the fifth cycle. The reason for the irreversibility in the initial cycles could be attributed to the formation of solid electrolyte interphase (SEI) or because of some irreversible reactions of Li ions with the surface groups and/or water molecules in TSAC nanosheets. On the whole, this irreversibility could be minimized by tailoring the surface structure of TSAC nanosheets or by prelithiating the electrode material as mentioned in other materials [37]. Then the reversible performance becomes stable after the first few lithiation/delithiation cycles. A stable cycle capacity of TSAC nanosheets is around 350 mAh g^{-1} at a current density of 200 mA g^{-1} and higher than the maximum theoretical capacity of f-Ti_3C_2 predicted by Tang et al. [38]. In addition, TSAC nanosheets show very good reversibility and stability, and a reversible capacity of 350 mAh g^{-1} is still kept after 100 cycles while it is only around 70 mAh g^{-1} for bulk TSAC.

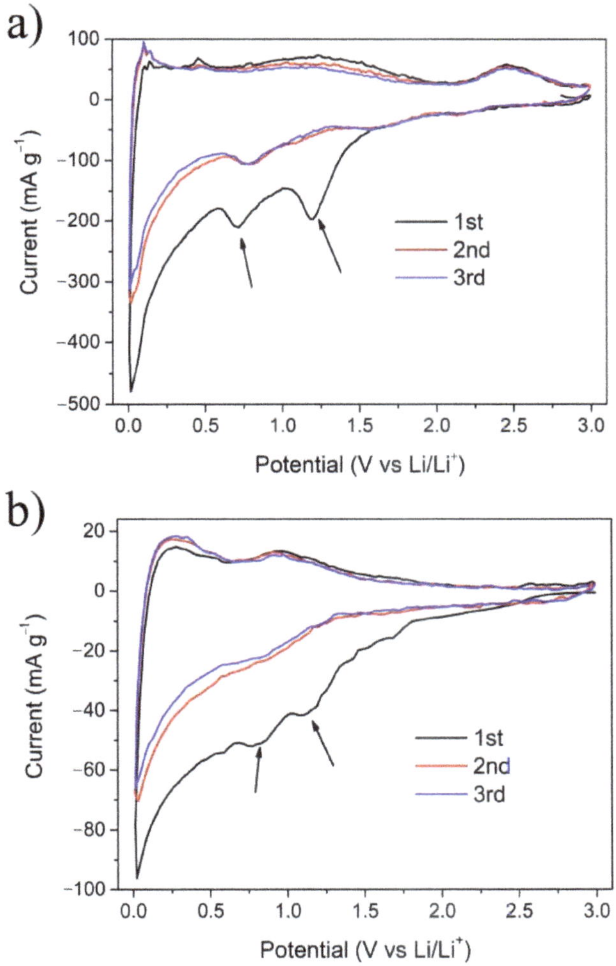

Figure 5. CV curves of (**a**) TSAC nanosheets and (**b**) bulk TSAC at a scan rate of 0.1 mV s^{-1}.

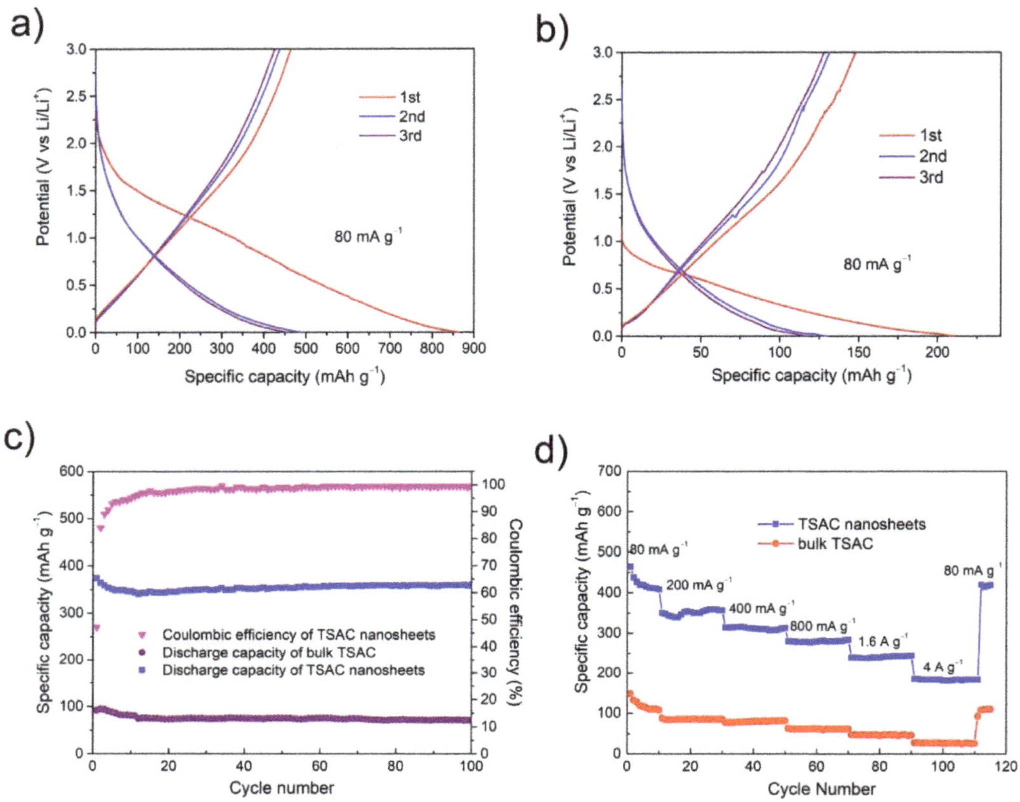

Figure 6. (**a**) The galvanostatic charge/discharge lithiation/delithiation curves of TSAC nanosheets; (**b**) the galvanostatic lithiation/delithiation curves of bulk TSAC; (**c**) specific delithiation (discharge) capacities of TSAC nanosheets and bulk TSAC vs. cycle number. The tests were carried at a current density of 200 mA g^{-1}; (**d**) specific delithiation (discharge) capacities of TSAC nanosheets and bulk TSAC electrodes cycled at various current densities.

Rate-capability tests were also performed to further assess the electrochemical activities of TSAC nanosheets. As shown in Figure 6d, both TSAC nanosheets and bulk TSAC anodes manifest excellent cycling stability at various current densities. After 10 cycles at 80, 200, 400, 800, 1600 and 4000 mA g^{-1}, TSAC nanosheets deliver delithiation capacities of 410, 353, 312, 279, 240 and 181 mAh g^{-1}, while that of bulk TSAC is only 108, 85, 80, 62, 46 and 26 mAh g^{-1}, respectively, suggesting a better rate performance for TSAC nanosheets at various rates. Even at a high rate of 8000 mA g^{-1}, the TSAC nanosheets anode is capable of maintaining a discharge capacity of 150 mAh g^{-1} after 200 cycles (Figure 7a), which equals to that of the third cycle. In addition, the coulombic efficiency of TSAC nanosheets at 8000 mA g^{-1} is between 98–100% after the first five cycles, indicating a relatively stable SEI formation and negligible side reactions of the electrode. The superior rate performances suggest that TSAC nanosheets can be promising anode materials for LIBs, particularly in high-power applications. For instance, lithium titanate is well known due to its capability to handle high cycling rates, even at 10 C, and the capacity of $Li_4Ti_5O_{12}$ is around 108 mAh g^{-1} [39]. Moreover, after hybridizing with graphene and Ag, high capacities of 133 and 156 mAh g^{-1} at 10 C are achieved for the resulting LTO/graphene and LTO/Ag, respectively [40,41]. In addition, MXenes with similar laminar structure are also capable of handling high cycling rates, such as V_2CT_x (110 mAh g^{-1} at 10 C), Nb_2CT_x (125 mAh g^{-1} at 10 C), annealed Nb_2CT_x (342 mAh g^{-1} at 2 A g^{-1}), f-T_3C_2 (110 mAh g^{-1} at 36 C) and porous $T_3C_2T_x$ foam (101 mAh g^{-1} at 18 A g^{-1}) [42–45].

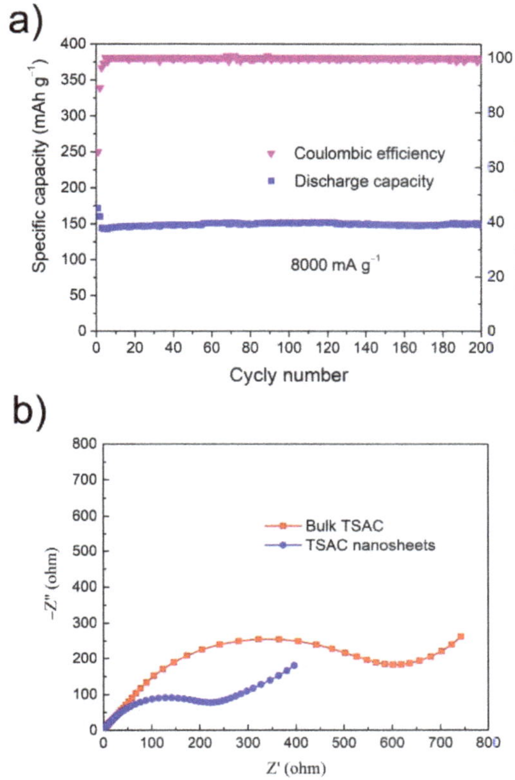

Figure 7. (**a**) Specific discharge capacity of TSAC nanosheets vs. cycle number at a current density of 8000 mA g^{-1}; (**b**) Nyquist plots of bulk TSAC and TSAC nanosheets.

Through Figures 6 and 7, compared with bulk TSAC, TSAC nanosheets exhibited enhanced electrochemical properties as anode materials for LIB. According to the studies of lithium-ion uptake of other MAX phases, both MX and A layers show a possible redox reaction capability with lithium-ion [7–9]. In addition, after etching A layer from MAX phases, the resulting MXenes also show promising lithium-ion storage ability [44,46–48]. Compared to bulk TSAC, TSAC nanosheets can provide larger active surface area and more exposed Si atoms to promote the redox reactions of Ti_3C_2-Li and Si-Li; thus, the specific capacity of TSAC nanosheets is superior to that of bulk TSAC. The specific surface area (SSA) calculated using the BET equation for the TSAC nanosheets is 11.68 m^2 g^{-1}. This value is about three times higher than the bulk TSAC powders measured at around 4.25 m^2 g^{-1}. Moreover, TSAC nanosheets electrodes have better conductivity from their 2D graphene-like nanostructure, which can effectively promote the electron transfer and shorten the Li^+ ions diffusion distance and the polarization, resulting in the improvement of their electrochemical performance. It is obvious that the semicircular arc of the electrochemical impedance spectroscopy (EIS) of TSAC nanosheets is smaller than that of bulk TSAC (Figure 7b), indicating that it has smaller charge transfer resistance. Specially, it can be anticipated that the electrochemical performance of TSAC nanosheets would be enhanced by optimizing and engineering the materials' surfaces, structures and compositions, and/or by introducing additives as reported for other 2D materials, such as graphene [41,49–51], MoS_2 [52–54] et al.

4. Conclusions

In this work, novel $Ti_3Si_{0.75}Al_{0.25}C_2$ ultrathin nanosheets as promising anode material for LIB are successfully developed by facile sonic exfoliating in alcohol. The nanosheets have a high capacity of $\approx$350 mAh g^{-1} at 200 mA g^{-1}, high cycling stability and excellent rate performance (150 mAh g^{-1} after 200 cycles at 8000 mA g^{-1}), which enhances the lithium-ion uptake capability in comparison with their bulk counterparts. It is noted that the reversible capacity of the nanosheets is about six times higher than the pristine bulk $Ti_3Si_{0.75}Al_{0.25}C_2$. In addition, more than 100 compounds have been found in the MAX family to date, and they all have similar potential due to their superior electrical conductivity, lamellar structure, very good stability even under severe environments and activated "A" layers for using as LIBs electrodes. Thus, our findings in this work are opening the door for the study on 2D MAX compounds as valuable LIBs electrodes, particularly for high power applications.

Author Contributions: Conceptualization, J.X. and M.H.; formal analysis, Q.W. and W.Y.; investigation, B.L.; writing—original draft preparation, J.X.; writing—review and editing, J.X. and M.H.; supervision, J.X. and M.H.; project administration, J.X. and M.H.; funding acquisition, J.X. and W.Y. All authors have read and agreed to the published version of the manuscript.

Funding: This research was funded by National Natural Science Foundation of China, grant number 21671167 and 51602277. This research was also supported by Qinglan Project of Jiangsu Province.

Institutional Review Board Statement: Not applicable.

Informed Consent Statement: Not applicable.

Data Availability Statement: No new data were created or analyzed in this study. Data sharing is not applicable to this article.

Acknowledgments: The authors would like to thank Shiyanjia Lab (www.shiyanjia.com, accessed on 15 December 2021) for the support of Raman analysis.

Conflicts of Interest: The authors declare no conflict of interest.

References

1. Eklund, P.; Beckers, M.; Jansson, U.; Högberg, H.; Hultman, L. The $M_{n+1}AX_n$ phases: Materials science and thin-film processing. *Thin Solid Films* **2010**, *518*, 1851–1878. [CrossRef]
2. Wang, X.H.; Zhou, Y.C. Layered Machinable and Electrically Conductive Ti_2AlC and Ti_3AlC_2 Ceramics: A Review. *J. Mater. Sci. Technol.* **2010**, *26*, 385–416. [CrossRef]
3. Fu, L.; Xia, W. MAX Phases as Nanolaminate Materials: Chemical Composition, Microstructure, Synthesis, Properties, and Applications. *Adv. Eng. Mater.* **2021**, *23*, 2001191. [CrossRef]
4. Sokol, M.; Natu, V.; Kota, S.; Barsoum, M.W. On the Chemical Diversity of the MAX Phases. *Trends Chem.* **2019**, *1*, 210–223. [CrossRef]
5. Gupta, S.; Barsoum, M.W. On the tribology of the MAX phases and their composites during dry sliding: A review. *Wear* **2011**, *271*, 1878–1894. [CrossRef]
6. Sun, Z.M. Progress in research and development on MAX phases: A family of layered ternary compounds. *Int. Mater. Rev.* **2013**, *56*, 143–166. [CrossRef]
7. Wu, H.; Zhu, J.; Liu, L.; Cao, K.; Yang, D.; Gong, C.; Lei, H.; Hang, H.; Yao, W.; Xu, J. Intercalation and delamination of Ti_2SnC with high lithium ion storage capacity. *Nanoscale* **2021**, *13*, 7355–7361. [CrossRef] [PubMed]
8. Zhao, S.; Agnese, Y.D.; Chu, X.; Zhao, X.; Gogotsi, Y.; Gao, Y. Electrochemical Interaction of Sn-Containing MAX Phase (Nb_2SnC) with Li-Ions. *ACS Energy Lett.* **2019**, *4*, 2452–2457. [CrossRef]
9. Li, Y.; Ma, G.; Shao, H.; Xiao, P.; Lu, J.; Xu, J.; Hou, J.; Chen, K.; Zhang, X.; Li, M.; et al. Electrochemical Lithium Storage Performance of Molten Salt Derived V_2SnC MAX Phase. *Nano-Micro Lett.* **2021**, *13*, 158. [CrossRef]
10. Xu, J.; Zhao, M.-Q.; Wang, Y.; Yao, W.; Chen, C.; Anasori, B.; Sarycheva, A.; Ren, C.E.; Mathis, T.; Gomes, L.; et al. Demonstration of Li-Ion Capacity of MAX Phases. *ACS Energy Lett.* **2016**, *1*, 1094–1099. [CrossRef]
11. Sengupta, A.; Rao, B.V.B.; Sharma, N.; Parmar, S.; Chavan, V.; Singh, S.K.; Kale, S.; Ogale, S. Comparative evaluation of MAX, MXene, NanoMAX, and NanoMAX-derived-MXene for microwave absorption and Li ion battery anode applications. *Nanoscale* **2020**, *12*, 8466–8476. [CrossRef] [PubMed]
12. Luan, S.; Zhou, J.; Xi, Y.; Han, M.; Wang, N.; Gao, J.; Hou, L.; Gao, F. High Lithium-Ion Storage Performance of Ti_3SiC_2 MAX by Oxygen Doping. *ChemistrySelect* **2019**, *4*, 5319–5321. [CrossRef]

13. Chen, X.; Zhu, Y.; Zhu, X.; Peng, W.; Li, Y.; Zhang, G.; Zhang, F.; Fan, X. Partially Etched Ti_3AlC_2 as a Promising High-Capacity Lithium-Ion Battery Anode. *ChemSusChem* **2018**, *11*, 2677–2680. [CrossRef] [PubMed]
14. Zhu, J.; Chroneos, A.; Wang, L.; Rao, F.; Schwingenschlögl, U. Stress-enhanced lithiation in MAX compounds for battery applications. *Appl. Mater. Today* **2017**, *9*, 192–195. [CrossRef]
15. Sun, Y.; Cheng, H.; Gao, S.; Sun, Z.; Liu, Q.; Liu, Q.; Lei, F.; Yao, T.; He, J.; Wei, S.; et al. Freestanding Tin Disulfide Single-Layers Realizing Efficient Visible-Light Water Splitting. *Angew. Chem. Int. Ed.* **2012**, *51*, 8727–8731. [CrossRef] [PubMed]
16. Stark, M.S.; Kuntz, K.L.; Martens, S.J.; Warren, S.C. Intercalation of Layered Materials from Bulk to 2D. *Adv. Mater.* **2019**, *31*, 1808213. [CrossRef]
17. Li, Z.; Zhang, X.; Cheng, H.; Liu, J.; Shao, M.; Wei, M.; Evans, D.G.; Zhang, H.; Duan, X. Confined Synthesis of 2D Nanostructured Materials toward Electrocatalysis. *Adv. Energy Mater.* **2019**, *10*, 1900486. [CrossRef]
18. Tao, P.; Yao, S.; Liu, F.; Wang, B.; Huang, F.; Wang, M. Recent advances in exfoliation techniques of layered and non-layered materials for energy conversion and storage. *J. Mater. Chem. A* **2019**, *7*, 23512–23536. [CrossRef]
19. Yoo, E.; Kim, J.; Hosono, E.; Zhou, H.-S.; Kudo, T.; Honma, I. Large Reversible Li Storage of Graphene Nanosheet Families for Use in Rechargeable Lithium Ion Batteries. *Nano Lett.* **2008**, *8*, 2277–2282. [CrossRef]
20. Li, X.; Geng, D.; Zhang, Y.; Meng, X.; Li, R.; Sun, X. Superior cycle stability of nitrogen-doped graphene nanosheets as anodes for lithium ion batteries. *Electrochem. Commun.* **2011**, *13*, 822–825. [CrossRef]
21. Naguib, M.; Mashtalir, O.; Carle, J.; Presser, V.; Lu, J.; Hultman, L.; Gogotsi, Y.; Barsoum, M.W. Two-Dimensional Transition Metal Carbides. *ACS Nano* **2012**, *6*, 1322–1331. [CrossRef]
22. Zhang, X.; Xu, J.; Wang, H.; Zhang, J.; Yan, H.; Pan, B.; Zhou, J.; Xie, Y. Ultrathin Nanosheets of MAX Phases with Enhanced Thermal and Mechanical Properties in Polymeric Compositions: $Ti_3Si_{0.75}Al_{0.25}C_2$. *Angew. Chem. Int. Ed.* **2013**, *52*, 4361–4365. [CrossRef]
23. Xu, J.; Yan, H.; Gu, D. Friction and wear behavior of polytetrafluoroethene composites filled with Ti_3SiC_2. *Mater. Des.* **2014**, *61*, 270–274. [CrossRef]
24. Xu, J.; Zhu, J.; Gong, C.; Guan, Z.; Yang, D.; Shen, Z.; Yao, W.; Wu, H. Achieving high yield of $Ti_3C_2T_x$ MXene few-layer flakes with enhanced pseudocapacior performance by decreasing precursor size. *Chin. Chem. Lett.* **2020**, *31*, 1039–1043. [CrossRef]
25. Fan, X.; Yin, X.; Wang, L.; Greil, P.; Travitzky, N. Synthesis of Ti3SiC2-based materials by reactive melt infiltration. *Int. J. Refract. Met. Hard Mater.* **2014**, *45*, 1–7. [CrossRef]
26. Schultz, T.; Frey, N.C.; Hantanasirisakul, K.; Park, S.; May, S.J.; Shenoy, V.B.; Gogotsi, Y.; Koch, N. Surface Termination Dependent Work Function and Electronic Properties of $Ti_3C_2T_x$ MXene. *Chem. Mater.* **2019**, *31*, 6590–6597. [CrossRef]
27. Han, F.; Luo, S.; Xie, L.; Zhu, J.; Wei, W.; Chen, X.; Liu, F.; Chen, W.; Zhao, J.; Dong, L.; et al. Boosting the Yield of MXene 2D Sheets via a Facile Hydrothermal-Assisted Intercalation. *ACS Appl. Mater. Interfaces* **2019**, *11*, 8443–8452. [CrossRef]
28. Chen, K.; Qiu, N.; Deng, Q.; Kang, M.; Yang, H.; Baek, J.; Koh, Y.; Du, S.; Huang, Q.; Kim, H. Cytocompatibility of Ti_3AlC_2, Ti_3SiC_2, and Ti_2AlN: In Vitro Tests and First-Principles Calculations. *ACS Biomater. Sci. Eng.* **2017**, *3*, 2293–2301. [CrossRef]
29. Guo, W.; Posadas, A.B.; Demkov, A.A. Deal–Grove-like thermal oxidation of Si (001) buried under a thin layer of $SrTiO_3$. *J. Appl. Phys.* **2020**, *127*, 055302. [CrossRef]
30. Spanier, J.E.; Gupta, S.; Amer, M.; Barsoum, M.W. Vibrational behavior of the $M_{n+1}AX_n$ phases from first-order raman scattering (M = Ti, V, Cr, A = Si, X = C, N). *Phys. Rev. B* **2005**, *71*, 012103.1–012103.4. [CrossRef]
31. Presser, V.; Naguib, M.; Chaput, L.; Togo, A.; Hug, G.; Barsoum, M.W. First-order Raman scattering of the MAX phases: Ti_2AlN, $Ti_2AlC_{0.5}N_{0.5}$, Ti_2AlC, $(Ti_{0.5}V_{0.5})_2AlC$, V_2AlC, Ti_3AlC_2, and Ti_3GeC_2. *J. Raman Spectrosc.* **2012**, *43*, 168–172. [CrossRef]
32. Liu, Y.; Wang, W.; Ying, Y.; Wang, Y.; Peng, X. Binder-free layered Ti_3C_2/CNTs nanocomposite anodes with enhanced capacity and long-cycle life for lithium-ion batteries. *Dalton Trans.* **2015**, *44*, 7123–7126. [CrossRef] [PubMed]
33. Pourali, Z.; Sovizi, M.; Yaftian, M. Two-Dimensional $Ti_3C_2T_X$/CMK-5 nanocomposite as high performance anodes for lithium batteries. *J. Alloy. Compd.* **2018**, *738*, 130–137. [CrossRef]
34. Lin, Z.; Sun, D.; Huang, Q.; Yang, J.; Barsoum, M.W.; Yan, X. Carbon nanofiber bridged two-dimensional titanium carbide as a superior anode for lithium-ion batteries. *J. Mater. Chem. A* **2015**, *3*, 14096–14100. [CrossRef]
35. Naguib, M.; Come, J.; Dyatkin, B.; Presser, V.; Taberna, P.-L.; Simon, P.; Barsoum, M.W.; Gogotsi, Y. MXene: A promising transition metal carbide anode for lithium-ion batteries. *Electrochem. Commun.* **2012**, *16*, 61–64. [CrossRef]
36. Shao, R.; Niu, J.; Zhu, F.; Dou, M.; Zhang, Z.; Wang, F. A facile and versatile strategy towards high-performance Si anodes for Li-ion capacitors: Concomitant conductive network construction and dual-interfacial engineerin. *Nano Energy* **2019**, *63*, 103824. [CrossRef]
37. Wang, F.; Wang, B.; Li, J.; Wang, B.; Zhou, Y.; Wang, D.; Liu, H.; Dou, S. Prelithiation: A Crucial Strategy for Boosting the Practical Application of Next-Generation Lithium Ion Battery. *ACS Nano* **2021**, *15*, 2197–2218. [CrossRef]
38. Tang, Q.; Zhou, Z.; Shen, P. Are MXenes Promising Anode Materials for Li Ion Batteries? Computational Studies on Electronic Properties and Li Storage Capability of Ti_3C_2 and $Ti_3C_2X_2$ (X = F, OH) Monolayer. *J. Am. Chem. Soc.* **2012**, *134*, 16909–16916. [CrossRef]
39. Michalska, M.; Krajewski, M.; Ziolkowska, D.; Hamankiewicz, B.; Andrzejczuk, M.; Lipinska, L.; Korona, K.P.; Czerwinski, A. Influence of milling time in solid-state synthesis on structure, morphology and electrochemical properties of $Li_4Ti_5O_{12}$ of spinel structure. *Powder Technol.* **2014**, *266*, 372–377. [CrossRef]

40. Michalska, M.; Krajewski, M.; Hamankiewicz, B.; Ziolkowska, D.; Korona, K.P.; Jasinski, J.B.; Kaminska, M.; Lipinska, L.; Czerwinski, A. $Li_4Ti_5O_{12}$ modified with Ag nanoparticles as an advanced anode material in lithium-ion batteries. *Powder Technol.* **2014**, *266*, 372–377. [CrossRef]
41. Yan, H.; Yao, W.; Fan, R.; Zhang, Y.; Luo, J.; Xu, J. Mesoporous Hierarchical Structure of $Li_4Ti_5O_{12}$/Graphene with High Electrochemical Performance in Lithium-Ion Batteries. *ACS Sustain. Chem. Eng.* **2018**, *6*, 11360–11366. [CrossRef]
42. Mashtalir, O.; Naguib, M.; Mochalin, V.N.; Agnese, Y.D.; Heon, M.; Barsoum, M.W.; Gogotsi, Y. Intercalation and delamination of layered carbides and carbonitrides. *Nat. Commun.* **2013**, *4*, 1716. [CrossRef] [PubMed]
43. Naguib, M.; Halim, J.; Lu, J.; Cook, K.M.; Hultman, L.; Gogotsi, Y.; Barsoum, M.W. New Two-Dimensional Niobium and Vanadium Carbides as Promising Materials for Li-Ion Batteries. *J. Am. Chem. Soc.* **2013**, *135*, 15966–15969. [CrossRef]
44. Zhao, J.; Wen, J.; Xiao, J.; Ma, X.; Gao, J.; Bai, L.; Gao, H.; Zhang, X.; Zhang, Z. Nb_2CT_x MXene: High capacity and ultra-long cycle capability for lithium-ion battery by regulation of functional groups. *J. Energy Chem.* **2021**, *53*, 387–395. [CrossRef]
45. Zhao, Q.; Zhu, Q.; Miao, J.; Zhang, P.; Wan, P.; He, L.; Xu, B. Flexible 3D Porous MXene Foam for High-Performance Lithium-Ion Batteries. *Small* **2019**, *15*, 1904293. [CrossRef]
46. Come, J.; Naguib, M.; Rozier, P.; Barsoum, M.W.; Gogotsi, Y.; Taberna, P.-L.; Morcrette, M.; Simon, P. A Non-Aqueous Asymmetric Cell with a Ti_2C-Based Two-Dimensional Negative Electrode. *J. Electrochem. Soc.* **2012**, *159*, A1368–A1373. [CrossRef]
47. Hui, X.; Zhao, D.; Wang, P.; Di, H.; Ge, X.; Zhang, P.; Yin, L. Oxide Nanoclusters on Ti_3C_2 MXenes to Deactivate Defects for Enhanced Lithium Ion Storage Performance. *Small* **2021**, 2104439. [CrossRef] [PubMed]
48. Liu, F.; Zhou, J.; Wang, S.; Wang, B.; Shen, C.; Wang, L.; Hu, Q.; Huang, Q.; Zhou, A. Preparation of High-Purity V2C MXene and Electrochemical Properties as Li-Ion Batteries. *J. Electrochem. Soc.* **2017**, *164*, A709–A713. [CrossRef]
49. Li, Y.; Ou, C.; Zhu, J.; Liu, Z.; Yu, J.; Li, W.; Zhang, H.; Zhang, Q.; Guo, Z. Ultrahigh and Durable Volumetric Lithium/Sodium Storage Enabled by a Highly Dense Graphene-Encapsulated Nitrogen-Doped Carbon@Sn Compact Monolith. *Nano Lett.* **2020**, *20*, 2034–2046. [CrossRef] [PubMed]
50. Xu, X.; Zeng, H.; Han, D.; Qiao, K.; Xing, W.; Rood, M.J.; Yan, Z. Nitrogen and Sulfur Co-Doped Graphene Nanosheets to Improve Anode Materials for Sodium-Ion Batteries. *ACS Appl. Mater. Interfaces* **2018**, *10*, 37172–37180. [CrossRef]
51. Zhao, J.; Zhang, Y.; Zhang, F.; Liang, H.; Ming, F.; Alshareef, H.N.; Gao, Z. Partially Reduced Holey Graphene Oxide as High Performance Anode for Sodium-Ion Batteries. *Adv. Energy Mater.* **2018**, *9*, 1803215. [CrossRef]
52. Wang, J.; Luo, C.; Gao, T.; Langrock, A.; Mignerey, A.C. An Advanced MoS_2/Carbon Anode for High-Performance Sodium-Ion Batteries. *Small* **2015**, *11*, 473–481. [CrossRef] [PubMed]
53. Jiang, H.; Ren, D.; Wang, H.; Hu, Y.; Guo, S.; Yuan, H.; Hu, P.; Zhang, L.; Li, C. 2D Monolayer MoS_2-Carbon Interoverlapped Superstructure: Engineering Ideal Atomic Interface for Lithium Ion Storage. *Adv. Mater.* **2015**, *27*, 3687–3695. [CrossRef] [PubMed]
54. Xu, X.; Fan, Z.; Yu, X.; Ding, S.; Yu, D.; Lou, X.W.D. A Nanosheets-on-Channel Architecture Constructed from MoS_2 and CMK-3 for High-Capacity and Long-Cycle-Life Lithium Storage. *Adv. Energy Mater.* **2014**, *4*, 1400902. [CrossRef]

nanomaterials

Article

Process Optimization for Manufacturing Functional Nanosurfaces by Roll-to-Roll Nanoimprint Lithography

Usama Tahir [1,2,†], Jin Il Kim [2,†], Shama Javeed [3,†], Amin Khaliq [1], Jun-Hyun Kim [1], Doo-In Kim [2] and Myung Yung Jeong [1,2,*]

1 Department of Cogno-Mechatronics Engineering, Pusan National University, Busan 46241, Korea; usama@pusan.ac.kr (U.T.); aaminkhaliq@pusan.ac.kr (A.K.); oaksa03@pusan.ac.kr (J.-H.K.)
2 Department of Opto-Mechatronics Engineering, Pusan National University, Busan 46241, Korea; real_one@pusan.ac.kr (J.I.K.); dooin.kim@pusan.ac.kr (D.-I.K.)
3 Department of Mathematics, Pusan National University, Busan 46241, Korea; shamajaveed@pusan.ac.kr
* Correspondence: myjeong@pusan.ac.kr
† These authors contributed equally to this work.

Abstract: Roll-to-roll nanoimprint lithography (RTR-NIL) is a low-cost and continuous fabrication process for large-area functional films. However, the partial ultraviolet (UV) resin filling obstructs the ongoing production process. This study incorporates UV resin filling process into the nanopillars and nanopores by using RTR-NIL. A multiphase numerical model with a sliding mesh method is proposed in this study to show the actual phenomena of imprint mold rotation and feeding of UV resin on the polyethylene terephthalate (PET) substrate. The implementation of UV resin filling under environmental conditions was performed by utilizing the open-channel (OC) boundary conditions. The numerical model was solved by using the explicit volume of fluid (VOF) scheme to compute the filling on each node of the computational domain. The effects of different processing parameters were investigated through the proposed numerical model such as imprinting speed (IS), contact angles (CAs), viscosity, initial thickness of the PET, and supporting roll diameter. A good agreement was found between numerical simulations and experimental results. The proposed numerical model gives better insights of the filling process for the mass production of functional surfaces with nanopillars and nanopores patterns for different applications on an industrial scale.

Keywords: nanopillars/pores; roll-to-roll imprinting; multiphase flow; imprinting-speed; functional surfaces; sliding mesh method

Citation: Tahir, U.; Kim, J.I.; Javeed, S.; Khaliq, A.; Kim, J.-H.; Kim, D.-I.; Jeong, M.Y. Process Optimization for Manufacturing Functional Nanosurfaces by Roll-to-Roll Nanoimprint Lithography. *Nanomaterials* **2022**, *12*, 480. https://doi.org/10.3390/nano12030480

Academic Editor: Paola Tiberto

Received: 24 November 2021
Accepted: 28 January 2022
Published: 29 January 2022

1. Introduction

Mass production of nanopatterns by using roll-to-roll UV nanoimprint lithography (RTR-UV-NIL) drew the attention of several researchers because of its practical significance in the micro/nanofabrication industry [1]. The smooth fabrication process has various defects because of the incomplete filling of the UV resin into the nanopatterns from the left/right (based on the filling direction) end of the imprint mold. Several studies [2–9] were conducted in the literature to optimize the filling processes. Pietarinen et al. [10] introduced the micro-optics replication process with the solvent-assisted filling into the micro-pyramid structures to avoid the defects due to air entrapment. Hiroshi et al. [11] used pentafluoropropane gas to replace air during the UV imprinting process. But these methods are difficult to implement on large scale due to experimental limitations. The UV resin filling process in RTR-UV-NIL is hardly explored by the scientific community due to system complexity.

The mathematical models are the most thorough techniques to investigate the RTR-UV-NIL process. Numerous studies [2,3,5–9,12] analyzed the RTR system by using numerical modeling for the UV resin filling process and proposed various strategies to attain high fidelity for the continuous fabrication process. Song et al. [13] investigated the influence of backing film together with UV resin flow into the cavities by utilizing coupled Eulerian–Lagrangian technique. They performed numerical simulations without accounting for air in the model. On an industrial scale, the role of air in RTR-UV-NIL cannot be overlooked. Peng et al. [12] employed multiphase model and investigated the UV resin filling into the pyramid shape cavity by accounting air in their numerical model. Zhou et al. [9] employed multiphase modeling technique and discussed UV resin accumulation as an important parameters during the filling process. A two-body system with an interface between the imprint mold and the outside region was used in their numerical model. In our previous work [5], we utilized multiphase numerical modeling with open-channel (OC) boundary conditions to analyze the filling behavior into the microcavities. The numerical model is based upon a single zone and optimize the filling behavior on the prior and succeeding ends of the imprint mold. However, the numerical modeling of the UV resin filling process into the nanocavities for mass production was not discussed in the literature until now.

In this study, we proposed a theoretical model to optimize the filling of UV resin into the nanocavities for the mass production of functional films. Experiments were carried out utilizing the optimized parameters, resulting in successfully imprinted nanopillars and nanopores patterns without defects. The numerical formulation of the sliding mesh method, geometric model, and explicit finite volume method for the solution to the multiphase problem are presented in Section 2. The numerical simulations' results are discussed in Section 3. The fabrication of soft molds, roll-to-roll experimental setup, and experimental results based upon numerical simulations are presented in Section 4. Finally, Section 5 concludes this study.

2. Numerical Model

The RTR-UV imprinting technique involves the rotation of the imprint mold, as well as the translational movement of the polyethylene terephthalate (PET) substrate, can be calculated by $u = r\theta$. Here, u and θ represents the translational and rotation speed of PET-Web and imprinting mold. r is the radius of imprint mold and supporting roller. A thoroughly constructed model that includes the entire physics of the problem is required to examine the filling process and optimization of the imprinting speed (IS). The majority of the above studies utilized different numerical models but their models have several limitations, such as the flow of UV resin between two parallel plates and the modeling of a tiny part of the imprinting process without taking air into account, as illustrated in Figure 1a.

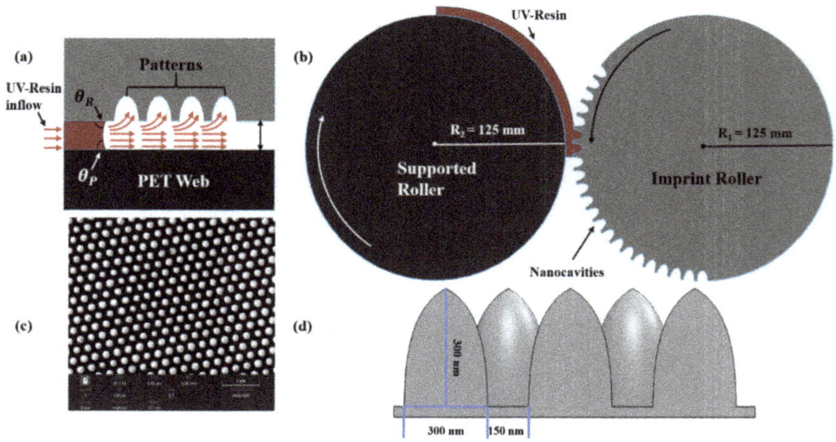

Figure 1. Schematic diagram of ultraviolet (UV) resin filling for elliptical nanocavities. (**a**) Between two parallel plates; (**b**) mass fabrication of nanocavities; (**c**) soft-mold; (**d**) 3D view of nanocavities.

To simulate UV resin flow into nanopillars and nanopores under environmental conditions, we used a 2D multiphase volume of fluid model (VOF) model [14,15] with moving walls and OC boundary conditions. The mathematical formulation of multiphase VOF model is explained in [5]. The fixed Eulerian mesh method was used to define the two phases in multiphase VOF model. Since the flow fields of the two phases have the same velocity, a single momentum equation is therefore needed to solve throughout the computational domain in VOF model. Furthermore, according to the specification of current VOF model, the phases cannot overlap or interpenetrate, and their volume fraction is specified between 0 and 1. In addition to the previous study, we employed the sliding mesh algorithm for the rotation of imprint mold. The sliding mesh algorithm is a robust approach for simulating the rotating flows or dynamic motion of one part to another, for example, rotation of blade in centrifugal pump [16]. Although its computational cost is high, it correctly describes the transient behavior of the flow in any problem. In this technique, each cell zone slide is relative to the other cell zone and is connected through an interface between them. The interfaces between two cell zones should be of same shape to employ sliding mesh method accurately; for example, if we slide one cell zone of different types of micro/nanopatterns (V-shape, oval, rectangular) into some area of difference between them, relative to other cell zones, the common interface will give some leeway due to tolerance and a diverging solution to the problem. More precisely we will be unable to solve the problem explicitly in which we kept the Courant–Friedrichs–Lewy (CFL) < 1 to compute the flow at each node of the computational domain. By using backward difference scheme, the general form of fluid transport equation for dynamic sliding mesh for any arbitrary volume U can be written as,

$$\frac{d}{dt}\int_{U}\rho\phi dU = \frac{(\rho\phi U)^{n+1} - (\rho\phi U)^{n}}{\Delta t} \tag{1}$$

Here, $\frac{dU}{dt}$ represents the control volume derivative. The conservation equation to satisfy the mesh can be written as,

$$\frac{dU}{dt} = \int_{dU} \vec{u}_g.d\vec{A} = \sum_{j}^{n_f} \vec{u}_{g,j}.\vec{A}_j \tag{2}$$

where n_f, $\vec{A}_j$, and $\vec{u}_{g,j}.\vec{A}_j$ represent the total number faces of control volume, face area vector, and dot product of each control volume, respectively. Equations (1) and (2) update the flow on each node of mesh of the sliding zone with respect to time. The numerical simulations are carried out using the computationally fluid dynamic software ANSYS-Fluent [17] to provide a full overview of the UV filling process into nanopillars and nanopores.

2.1. Geometric Model and Open-Channel Boundary Conditions

A 2D model was built to explore the filling behavior of elliptical nanopillars and nanopores. Figure 1b depicts the RTR-UV-NIL schematic diagram, which incorporates the movement of the PET-Web from left to right with the assistance of a supporting roller, as well as the rotation of the imprint mold. To assure mechanical precision, the angular velocity of rollers is set in the same way as the linear velocity of the PET-Web. Figure 2a,b shows the detailed simulation model of elliptical nanopillars and nanopores.

UV resin above the PET-Web is driven towards the imprint roller from the left (inlet) side. Above the supporting roller, the PET-Web is treated as a stretched and moving wall boundary condition. The computational domain based on sliding mesh algorithm slides the rotating zone relative to the fixed zone. The rotating zone consist of nanopillars and nanopores and a mesh interface. This mesh interface is connected to the fixed zone interface of same shape, as shown in the magnified view of Figure 2. More precisely, these two cell zones are geometrically separated but numerically connected. The fixed cell zone consists of other parts and boundary conditions of computational domain, such as UV resin supply and OC boundary conditions. The OC-VOF model's computational domain also includes the initial thickness and velocity of UV resin on the PET substrate. Further to investigate the UV resin filling process under environmental circumstances, the inflow, outflow, and pressure conditions are set as atmospheric above the PET-Web and imprint roller. The movement of fixed and rotating zones are updated with respect to time in the computational domain.

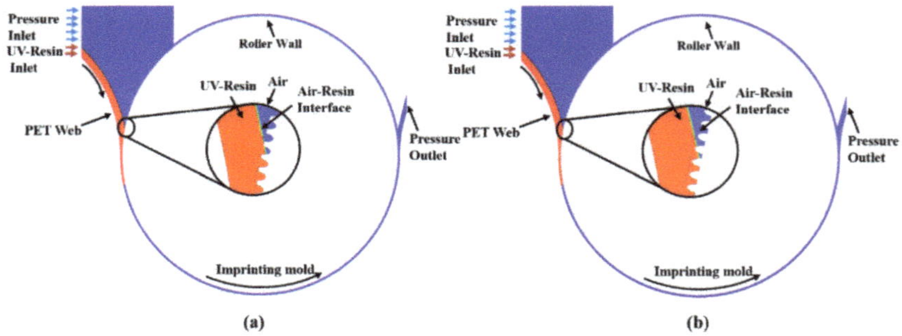

Figure 2. Computational multiphase model scheme with sliding mesh method based upon stationary and fixed zones with open-channel (OC) boundary conditions for (**a**) nanopillars and (**b**) nanopores patterns.

Table 1 lists the simulation parameters such as height and width of nanopillars/pores, density, viscosity, initial thickness, and CAs for the UV resin filling process. θ_P represents the angle between PET-web and UV resin, and θ_R represents the angle between the imprint roller and UV resin.

Table 1. Processing parameters for numerical simulations.

Parameters	Values
Nanopillars/pores height(H)	300 nm
Nanopillars/pores width (W)	300 nm
Density of UV resin	1196 kg/m^3
Viscosity of UV resin	150–200 cp
CA σ_R	7.2°
CA σ_P	16.9°
PET-Web Speed	10–50 mm/s
UV resin initial thickness	1–5 micrometer

2.2. Explicit Finite Volume Method

The simulations for both cases (nanopillars and nanopores UV resin filling) were performed using an explicit scheme for finite volume method, because this method is supremum to implicit scheme regarding numerical accuracy when we need to calculate the solution on each node of the computational domain. The transient problem was solved by employing coupled algorithm for pressure–velocity coupling in a pressure-based solver. In spatial discretization, the gradient calculations were performed using the least-squares cell-based method because it provides good accuracy in comparison with that of computationally expensive methods. To reduce the numerical diffusion to solve the momentum equations, a second-order up-wind scheme is utilized by keeping the pressure at PREssure Staggering Option (PRESTO). Further geometric reconstruction scheme was utilized in spatial discretization for face fluxes interpolation.

For temporal discretization, the first-order implicit scheme was used to see the transient behavior of the multiphase problem. The time step was chosen by keeping the Courant number in the acceptable range so that UV resin flow can be computed on each node of the computational domain. The variable time step was set between 2×10^{-5} s to 5×10^{-5} s to keep the Courant number under 1. The residual during the simulations shows the convergence of the solution up to 10^{-6} s.

A mesh independence study was also implemented for the current study by utilizing the mesh grids of sizes 10 × 10 nm, 25 × 25 nm, and 30 × 30 nm to see the UV resin filling more accurately into the bubble cavity. By employing the above mesh sizes, the difference of bubble elimination into the bubble cavity was less than 1%. It is also investigated by employing the adaptive mesh refinement technique, which also does not affect the filling behavior into the nanocavities. We use four levels of adaptive mesh refinement for each time step during our simulations. The mesh utilized in the simulations is 30 × 30 nm, which was chosen as a compromise between computing time and accuracy [17–19].

3. Simulation Results

3.1. UV Resin Filling Process into the Nanopillars

The filling in elliptical nanopillars at different intervals is shown in Figure 3. Figure 3a depicts the air above the UV resin at time t = 0 s. When the imprint mold begins to rotate, the PET-Web delivers UV resin into the nanopillars. The initial thickness of UV resin, viscosity, and PET-Web speed was kept at 1.5 µm, 200 cp, and 18.75 mm/s, respectively. At the same time a little amount of UV resin backflow at the prior end of the imprint roller and successfully fills the nanopillars, except cavity *A* as shown in Figure 3b–d. In cavity *A*, the air is locked above the UV resin and has no way to escape throughout the filling process.

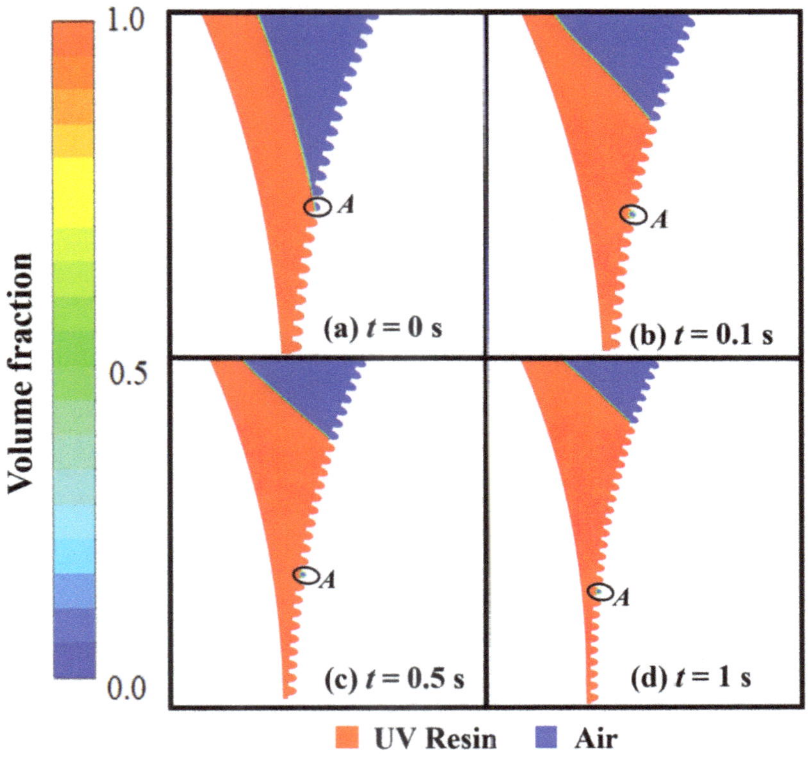

Figure 3. UV resin incomplete filling into elliptical nanopillars with bubble defects: (**a**) t = 0 s, (**b**) t = 0.1 s, (**c**) t = 0.5 s, and (**d**) t = 1 s, at a PET-Web speed of 18.75 mm/s with viscosity 200 cp ($\theta_R = 7.2°, \theta_P = 16.9°$).

This kind of bubble entrapment happens due to the contact UV resin initially has with the nanocavities on the imprint mold as shown in Figure 4a. In our numerical model, the imprint mold changes its reference position with time due to the employment of the sliding mesh method, so cavity *A* moves towards the narrow channel where this bubble is impossible to remove during the filling process (see Supplementary Materials Video S1a).

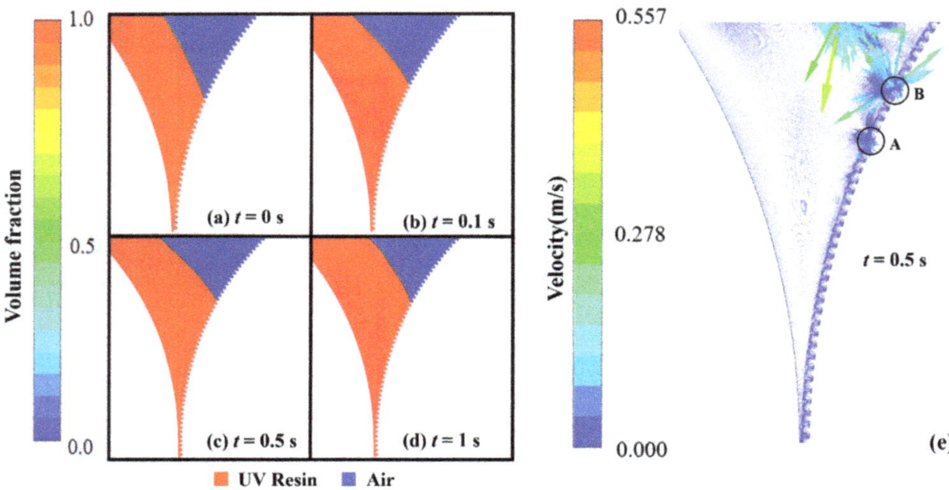

Figure 4. UV resin complete filling into elliptical nanopillars without bubble defects: (**a**) t = 0 s, (**b**) t = 0.1 s, (**c**) t = 0.5 s, and (**d**) t = 1 s, at a polyethylene terephthalate (PET)-Web speed of 18.75 mm/s with viscosity 180 cp. (**e**) Velocity vectors of moving PET and rotating roller mold at t = 0.5 s.

We increase the initial thickness of UV resin above the PET substrate to change the initial contact of UV resin with nanocavities on the imprinting mold (see Supplementary Materials Video S1b). The viscosity and PET-Web speed was kept at 200 cp and 18.75 mm/s during the filling process.

The UV resin successfully fills the cavities at the preceding end of the imprint mold, but a small amount of air is still let into cavity *A*. This problem arises due to the viscosity of UV resin. To resolve this problem, we decrease the viscosity of UV resin to 180 cp during the filling process. Figure 4b–d represents the UV resin filling at different time steps. Nanopillars are successfully filled with UV resin (see Supplementary Materials Video S1c). The UV resin successfully fills the nanopillars at the prior and succeeding ends of the imprint mold during simulations. The processing parameters such as initial thickness, CAs, and PET-Web speed were kept the same as in previous simulation. For the filling process, $\theta_R = 7.2°$ and $\theta_P = 16.2°$ are the appropriate CAs measured after running several simulations. In numerical simulations, the CA θ_R values vary depending on the pattern shape. As a result, we kept them in a feasible range for successful filling of nanocavities. Figure 4e shows the contours for the velocity vector distribution of UV resin filling and rotational motion of imprinting mold at t = 0.5 s.

The vector distribution of the velocity of UV resin in cavities *A* and B depicts that UV resin enters the cavity and slows down to form vortices. The velocity distribution of vectors is more in cavity *B* than cavity *A* because the UV resin starts to fill in cavity *B* at t = 0.5 s. At high speeds, the imprint mold has less time to fill nanocavities with UV resin. Therefore, bubble entrapment occurs at a high speed. Figure 5 depicts the incomplete filling of UV resin at t = 0.5 s with IS at 50 mm/s. Continuous bubble entrapment can be seen in different nanopillars at high IS. Figure 6 shows the UV resin filling process at different ISs. It is also evident that UV resin leaves the nanocavities without filling at high IS.

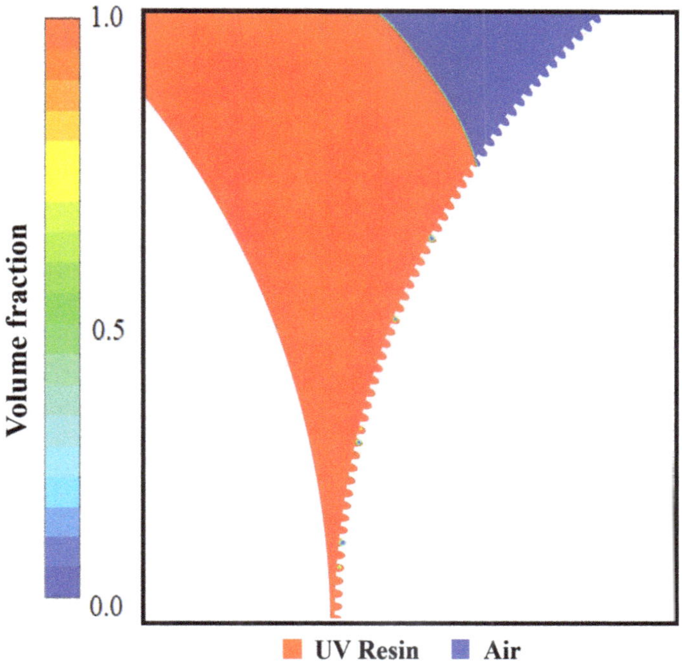

Figure 5. UV resin incomplete filling into elliptical nanopillars with continuous bubble defects at Web speed of 50 mm/s with viscosity 150 cp ($\theta_R = 7.2°$, $\theta_P = 16.9°$).

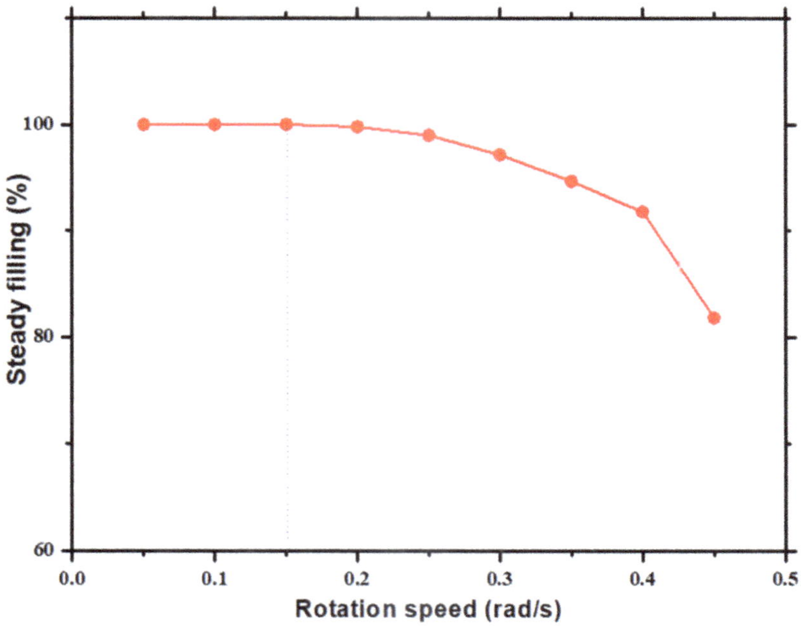

Figure 6. UV resin filling varies from uniform to nonuniform at varying impression mold rotation speeds.

3.2. UV Resin Filling Process into the Nanopores

UV resin filling process into the nanopores is different than the nanopillars due to the reverse filling direction. For simple understanding in the case of hydrophobic surfaces, if the gap between two consecutive nanocavities is less, then fabricated nanosurfaces will show more hydrophobicity for antiwetting phenomena [20]. It is also observed during the numerical simulation that the UV resin easily fills into nanopillars with zero-gap during RTR imprinting, but it is impossible to fill into the nanopores with zero-gap for RTR imprinting because in the reverse direction the filling is impossible between the edges of two consecutive nanocavities.

Therefore, bubble entrapment continuously occurs during the filling process between the consecutive edges of the nanopores. Figure 7 shows the incomplete filling with bubble entrapment between two consecutive nanocavities with a gap of 100 nm at $t = 0.5$ s. We performed numerical simulation by increasing the gap between imprinting mold and supporting roll and also decreasing the supporting roll diameter for the easy escape of air, as shown in Figure 7.

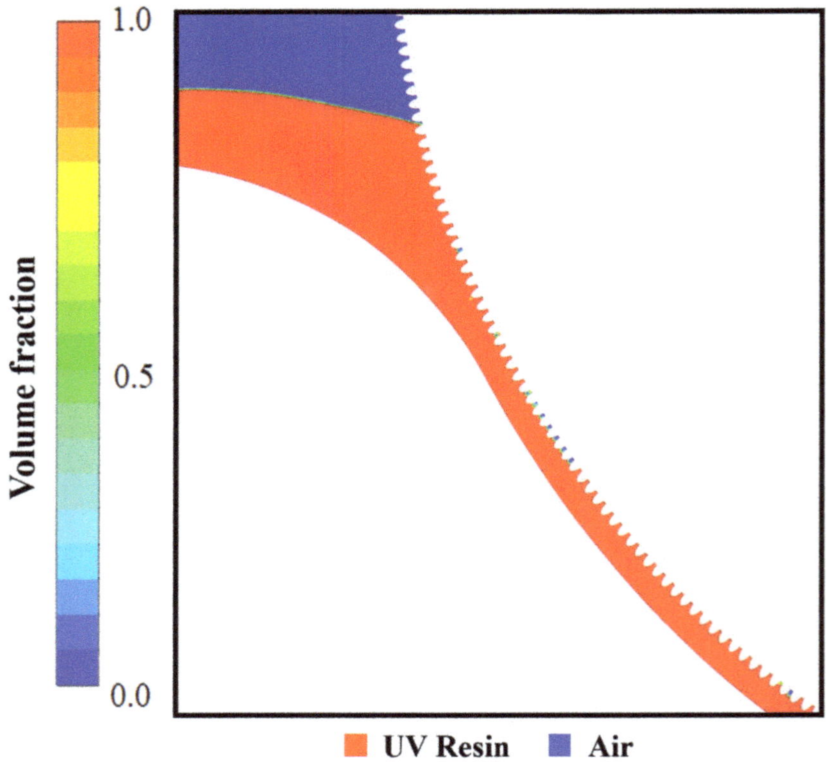

Figure 7. UV resin continuous bubble defects into nanopores with 100 nm gap between edges of two consecutive nanopores at Web speed of 18.75 mm/s with viscosity 180 cp ($\theta_R = 7.2°$, $\theta_P = 16.9°$).

The UV resin fails to fill into the nanocavities because the gap between the edges is very narrow during the filling process. The Ni mold was fabricated based on these investigations and further used in our experiments for mass production of nanopatterns.

Different simulations were performed to optimize the filling process with a suitable gap of 150 nm between two consecutive nanocavities.

Same as nanopillars, bubble entrapment also occurred in the cavities when we kept the UV resin initial thickness of 1.5 μm on the PET substrate (see Supplementary Materials Video S1d).

To counter this bubble entrapment, we change the UV resin initial contact with the imprinting mold by increasing the initial thickness of the UV resin above the PET substrate. Figure 8a–d represents the UV resin filling into the nanopores at different time steps. The viscosity and PET-Web speed was kept at 200 cp and 18.75 mm/s during the filling process. Figure 8b–d shows that bubble entrapment significantly decreases by changing the UV resin initial contact with the imprinting mold. However, there are still small bubbles that exist during the filling process due to the viscosity of UV resin (see Supplementary Materials Video S1e). Figure 9 represents the complete filling process of UV resin into nanopores. The UV resin filling process is shown at different time steps in Figure 9a–d.

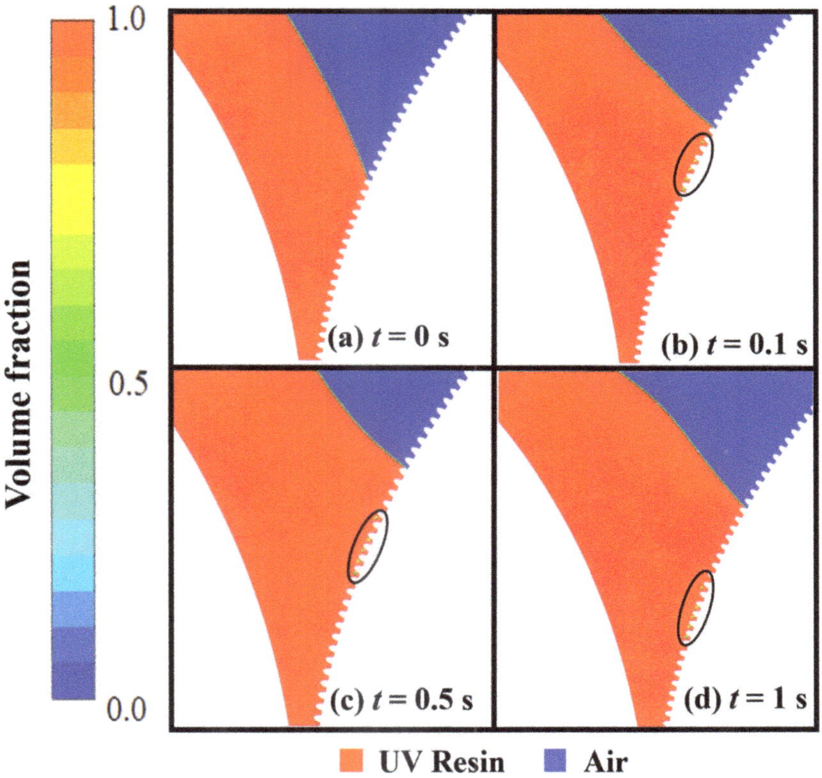

Figure 8. UV resin incomplete filling into nanopores with bubble defects: (**a**) t = 0 s, (**b**) t = 0.1 s, (**c**) t = 0.5 s, and (**d**) t = 1 s, at a PET-Web speed of 18.75 mm/s with viscosity 200 cp ($\theta_R = 7.2°$, $\theta_P = 16.9°$).

To get the successful filling at both ends of the imprinting mold without bubble entrapment, we decrease the viscosity of the UV resin up to 180 cp by keeping other parameters the same as in the previous simulation (see Supplementary Materials Video S1f). Figure 9e shows the velocity vectors distribution of UV resin filling and rotation of the imprinting mold at t = 0.5 s. Due to surface tension of UV resin and adhesive force from the imprinting mold, the velocity vectors distribution significantly increases in cavity A and replaces the air with UV resin.

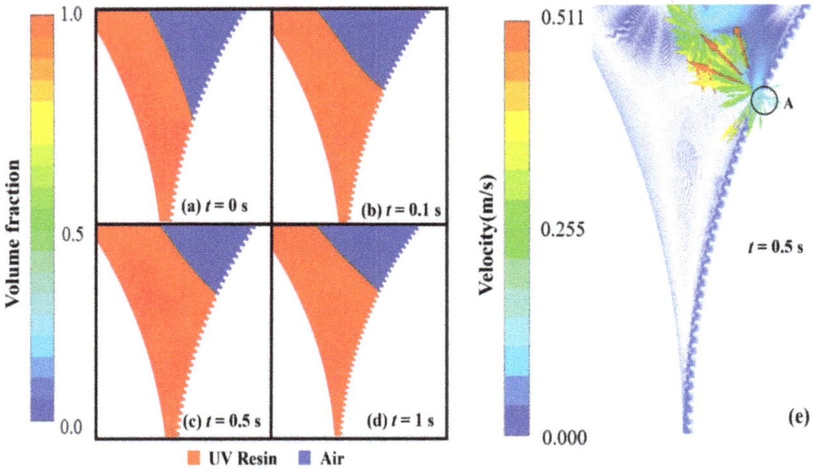

Figure 9. UV resin complete filling into nanopores without bubble defects: (**a**) $t = 0$ s, (**b**) $t = 0.1$ s, (**c**) $t = 0.5$ s, and (**d**) $t = 1$ s, with initial thickness of five micrometer at a PET-Web speed of 18.75 mm/s with viscosity 180 cp. (**e**) Velocity vectors of moving PET and rotating roller mold at $t = 0.5$ s.

3.3. Effect of Supporting Roller Diameter on UV Resin Filling Process

In this study, we also investigated the effect of supporting roll diameter during the filling process because it has a direct impact on the initial contact of UV resin with the imprinting mold. The imprint mold and supporting roller of the same diameter are shown in Figure 10a.

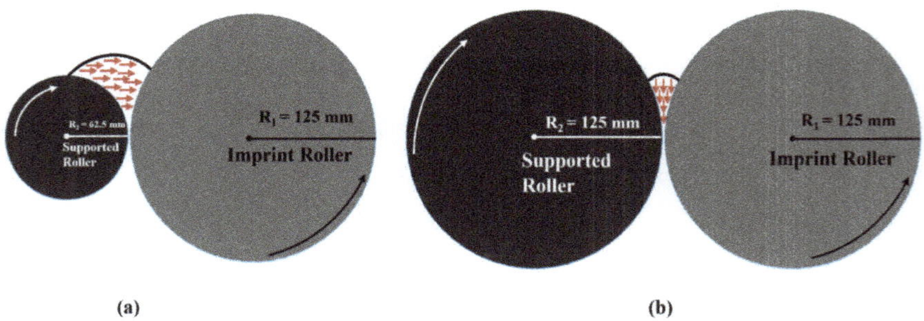

Figure 10. (**a**) Schematic diagram of filling direction of UV resin with supported roller with half of diameter of imprinting mold. (**b**) Schematic diagram of filling direction of UV resin with supported roller and impression mold of same diameter.

In this scenario, the UV resin fills vertically into the nanocavities and leaves the bubble entrapment alone if we reduce the initial thickness of UV resin above the PET substrate. If the supporting roll diameter is half of the diameter of the imprinting mold diameter, then UV resin easily goes into the nanocavities during the imprinting process as shown in Figure 10b. Figure 11a–d shows the complete filling of UV resin into the nanocavities at different time steps. Reducing the diameter of supporting roll shows that the UV resin fills the nanocavities with a minimum initial thickness of 1.5 µm on the PET substrate.

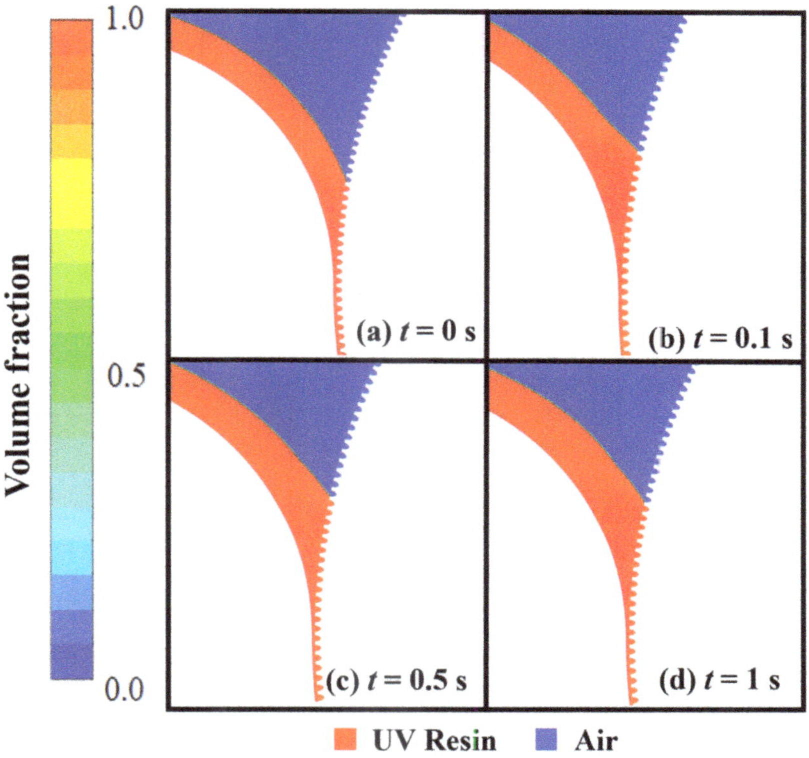

Figure 11. UV resin complete filling into nanopillars without bubble defects with supporting roll diameter of half of imprinting mold: (**a**) t = 0 s, (**b**) t = 0.1 s, (**c**) t = 0.5 s, and (**d**) t = 1 s, at a PET-Web speed of 18.75 mm/s with viscosity 180 cp ($\theta_R = 7.2°$, $\theta_P = 16.9°$).

4. Experimental Section

4.1. Fabrication of Nanopillars and Nanopores Soft Molds

The fabrication process of the soft mold with elliptical nanopillars and nanopores patterns consists of three steps: the fabrication of nanopores soft mold from nanopillars nickel (Ni) master mold, surface treatment for smooth release of soft mold, and the fabrication of nanopillar soft mold from treated soft mold. Figure 1c represents the Ni mold or master used in this study, which was obtained by electroplating process [21] that has elliptical nanopillars patterns. The schematic diagram of elliptical patterns with dimensions and the gap between two consecutive nanocavities are shown in Figure 1d. In the first step of the soft mold fabrication process, a uniform coating of 50 µm of UV resin was obtained by using the spin coating method on the Ni master mold. Then, we covered the PET substrate over to well coat UV-curing resin layer and baked it under UV exposure with an intensity of 800 W/cm^2 for 4 min. The baked UV-curing resin was exposed under the LICHTZEN HG UV lamp system and the liquid state resin transformation to a nontacky-solid state by the photoinitiator. After releasing the PET substate from the Ni mold, the reverse (nanopores) patterns transform onto the PET substrate. In the second step, a surface treatment process was used to reduce the peel-off force during the RTR imprinting process. In this step, the reverse soft mold was dipped into a primer agent for 20 min to develop crosslinking over the patterned surface. The dipped soft mold was kept under room temperature for 1 h to chemically stabilize the cross-linking agent on the surface of the soft mold. Further, the soft mold was again dipped into releasing agent with a fluorinated additive for 20 min to make a low surface energy layer in the soft mold. After forming the low surface energy layer onto the soft mold, it was kept for 24 h under room temperature for chemical stabilization.

This process provides more excellent releasing properties during the imprinting process. In the third step above, the process was repeated again by using negative soft mold for the fabrication of soft mold with positive elliptical nanocavities.

4.2. Roll-to-Roll Imprinting Setup

Apart from the imprinting module, the experimental system consists of other modules named as unwinding module, EPC module for PET alignment, slot-die module, dispensing module, demolding module, and rewinding module to support continuous imprinting process. The uncoated PET film is delivered by the unwinding module.

The PET film employed as a substrate was 125 μm thick, with an average transmittance of 87% under visible radiation (Kolon polyester film, Daegu, Korea). The tension module generates tension on the PET film's ends so that it may correctly align during the imprinting process. The Edge Position Control (EPC) module is responsible for the alignment of PET substrate from unwinding to rewinding module. In this study, we also used a PET substrate treatment module as shown in Figure 12b. This treatment module uses the corona effect to increase the adhesion of UV resin with PET substrate when the coating is applied on the PET substrate. The new slot-die dispensing module is shown in Figure 12c. This module consists of a UV resin injection reservoir that can contain 30 mL of UV resin. This reservoir supplies the UV resin to slot-die with the help of pipe as shown above of the slot die in Figure 12c. Then, slot-die dispenses the UV resin onto the PET substrate by using the meniscus effect. The gap between the slot-die and the roller underneath the slot-die is responsible for the thickness height of UV resin above the PET substrate. In our experiments, we kept this gap up to 5 μm based on our simulations. Figure 12d represents the uniform thickness of 5 μm of UV resin above the PET substrate. The main module of the RTR system is the imprinting module that consists of a supporting roller and imprint-roller of the same diameter of 250 mm as shown in Figure 12a. The imprint mold contains the elliptical nanopillars and nanopores patterns of height and width of 300 nm (AR = 1). To cure the UV resin after filling a UV light is placed underneath the imprint mold.

After curing, in the demolding module, a roll was used above the imprinting roll to detach the imprinted patterns smoothly. The imprinted pattern was dragged back by the rewinding module. Scanning electron microscopy ((SEM) S8000G, TESCAN ORSAY HOLDING, a.s. Brno, Kohoutovice, Brno, Czech Republic) with an electron energy of 5 kV was used to examine the patterns. A layer of Au (40 mA, 4 min) was sputtered on the samples to increase the conductivity for better resolution of SEM images. The UV lamp (UV-LED, UV LED-SPOT-100-HP-IC-365, Dr Hönle AG, Lochhamer Schlag 1 D-82166 Gräfelfing, Munich, Germany) emits the UV light with 1500 mW/cm^2 intensity on the focus point and cured the UV resin at a wavelength of 460 ± 5 nm. Minuta Technology Co., Ltd. (MINS-311R, Osan, Korea; viscosity: 200 cp at 25 °C) provided the UV resin used in this research.

4.3. Experimental Results

In our experimental system, we used a treatment module with a corona effect and slot-die module for the uniform coating of UV resin above the PET substrate. These two modules are quite effective in comparison to coating modules used in previous studies.

The experiments were performed by utilizing the optimized parameters through the numerical simulations such as viscosity, IS, gap between two consecutive nanocavities, and initial thickness of UV resin above the PET substrate. Figure 13a shows the hexagonal arrangement of elliptical nanopillars. Figure 13b,c represents the top and enlarged view of fabricated nanopatterns by using soft mold through the RTR imprinting process. The diameter, depth, and width of these patterns were 300 nm. Figure 13d shows the SEM images of hexagonal elliptical nanopores of depth 300 nm. Figure 13e,f shows the enlarged view of elliptical nanopores. These SEM images represent the successful imprinting of nanopatterns by using RTR imprinting.

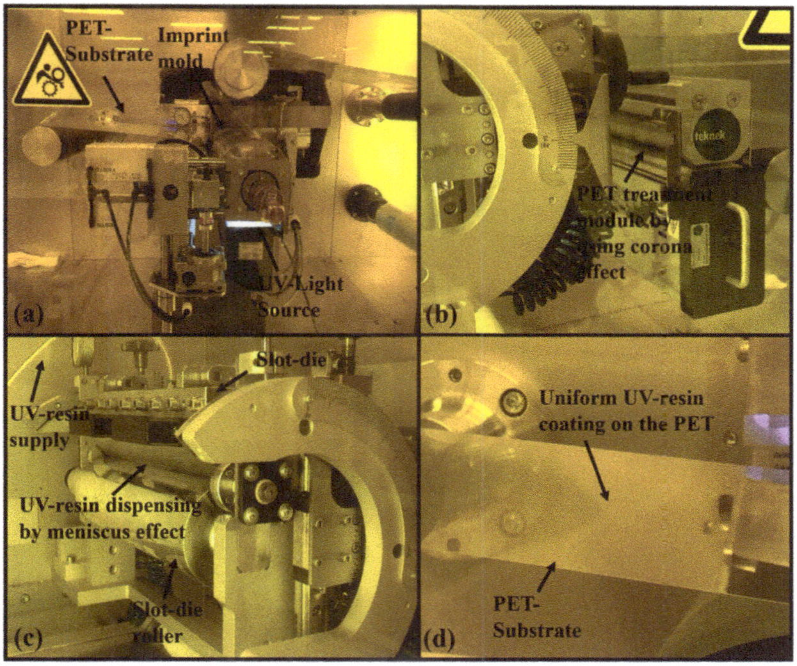

Figure 12. (**a**) Imprinting unit of roll-to-roll (RTR) system with PET-Web of width 250 mm, capable of imprinting the patterns with a speed up to 10 m/min. (**b**) PET substrate treatment unit that incorporates with corona effect to increase adhesion of UV resin with PET substrate. (**c**) Slot-die unit with a dispensing source and dispensing of UV resin on PET substrate by using meniscus effect. (**d**) Uniform coating of height 5 µm of UV resin on PET substrate.

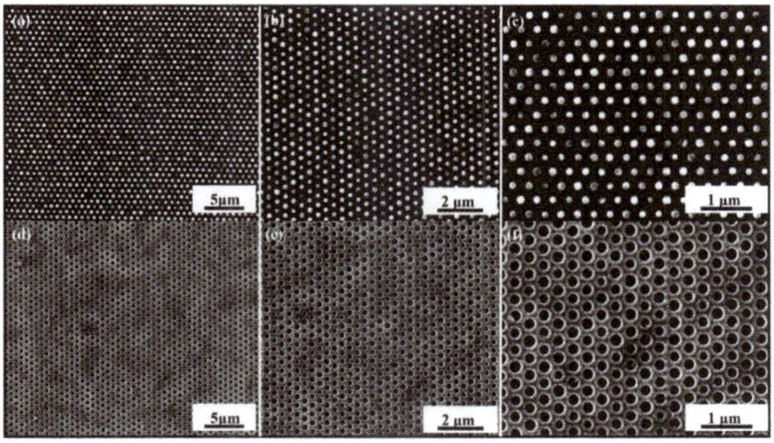

Figure 13. Scanning electron microscopy (SEM) images of fabricated patterns: (**a**) low-magnificent image of elliptical nanopillars via RTR imprinting at Web speed of 18 mm/s; (**b**,**c**) closeup image of nanopillars; (**d**) low-magnificent images of successful imprinting of nanopores; (**e**,**f**) closeup images of nanopores without defects with smooth edges at Web speed of 18 mm/s.

The parameters during the imprinting were kept the same as in the numerical simulations as shown in Figures 4 and 9, respectively. The Web speed for the fabrication of elliptical nanopillars and nanopores was kept 18 mm/min during the experiments. Figure 14a,b shows the top and enlarged view of SEM images for the fabrication of nanopores at an IS of 30 mm/s. At this IS, the fabrication shows the incomplete filling and the edges of nanopores are broken, leading to defects in the quality of the final product. Figure 14c–f shows the SEM images of nanosurfaces at IS 60 mm/s, 100 mm/s, 130 mm/s, and 160 mm/s, respectively. These SEM images show the severe defects and bad arrangement of fabricated patterns with broken edges at high IS. We also carried out experiments on RTR system with different types of nanosurfaces on different IS and found that the suitable speed for RTR nanoimprinting is in the range of 10–20 mm/s.

Figure 14. SEM images of fabricated elliptical nanopores with broken edges and defects of incomplete filling at higher IS. (**a**) Low-magnificent image at 30 mm/s; (**b**) closeup image at 30 mm/s; (**c**) 60 mm/s; (**d**) 100 mm/s; (**e**) 130 mm/s, and with maximum speed of the system; and (**f**) 160 mm/s, respectively.

5. Conclusions

The roll-to-roll nanoimprint lithography (RTR-NIL) offers a better solution for the fabrication of nanosurfaces because it is a high-speed and high-resolution manufacturing process. The following findings were reached in this study:

- A multiphase numerical model with open-channel (OC) boundary conditions was utilized in this study together with the sliding mesh method technique. The numerical model optimizes the ultraviolet (UV) resin filling into the nanopillars/pores. The explicit scheme was utilized to calculate the solution on each node of the computational domain. The processing parameters such as IS, viscosity effects, initial thickness on PET substrate, and the effects of supporting roll diameter on filling behavior were optimized by using the proposed numerical model. We also investigated the filling defects at high imprinting speed (IS), and the gap between two consecutive nanopores for UV resin filling was also optimized in this study. The optimal Web speed for the complete filling was noted at 18.75 mm/min. The proposed numerical model is more accurate than the existing models and applicable to any system with variable dimensions of the imprinting molds, and filling behavior with different pattern shapes can also be investigated through this numerical model, such as V-shape, square shape, and nano-lenses shape.
- The 3D hexagonal arrangement of elliptical nanopillar and nanopore soft molds were fabricated by using the soft lithography technique, and these flexible arrays were

directly obtained by using the RTR imprinting process under the optimized parameters by using numerical modeling. This is an economical approach that provides a solution for the low-cost fabrication of nanosurfaces. It was also found in the experiments that at high imprinting speed, the fabrication process of nanosurfaces shows continuous defects and broken edges. It was also determined that highly ordered soft molds could be prepared with the established method. In addition, we tested the reproducibility of these soft mold fabrication techniques by running the RTR-NIL process for up to 50 roll revolutions (785.4 m) with a Web speed of 18 mm/s. In summary, these results provide better significance for the economical and continuous mass production of nanosurfaces without defects.

Supplementary Materials: The following are available online at https://www.mdpi.com/article/10.3390/nano12030480/s1, Video S1: UV resin filling in to the nanopillars and nanopores.

Author Contributions: Conceptualization, U.T., J.I.K., S.J. and A.K.; formal analysis, U.T., S.J. and J.I.K.; methodology, U.T., J.I.K. and S.J.; project administration, M.Y.J.; resources, J.I.K., J.-H.K., D.-I.K. and M.Y.J.; software, U.T., S.J. and A.K.; supervision, D.-I.K. and M.Y.J.; validation, U.T. and S.J.; writing—original draft, U.T. and S.J.; writing—review and editing, M.Y.J. All authors have read and agreed to the published version of the manuscript.

Funding: This work was supported by the Technology Innovation Program (P0008763) funded by the Ministry of Trade, Industry, and Energy (MOTIE, Korea) and by the National Research Foundation of Korea (NRF) grant funded by the Korea government (MSIT) (No.2017R1A2B2006999).

Conflicts of Interest: The authors declare no conflict of interest.

References

1. Tahir, U.; Shim, Y.B.; Kamran, M.A.; Kim, D.-I.; Jeong, M.Y. Nanofabrication Techniques: Challenges and Future Prospects. *J. Nanosci. Nanotechnol.* **2021**, *21*, 4981–5013. [CrossRef] [PubMed]
2. Gomez-Constante, J.P.; Pagilla, P.R.; Rajagopal, K.R. A thermomechanical description of the mold filling process in roll-to-roll nanoimprinting lithography. *Appl. Eng. Sci.* **2020**, *1*, 100001. [CrossRef]
3. Jain, A.; Bonnecaze, R.T. Fluid management in roll-to-roll nanoimprint lithography. *J. Appl. Phys.* **2013**, *113*, 234511. [CrossRef]
4. Tahir, U.; Kamran, M.A.; Jang, M.H.; Jeong, M.Y. Thin-film coating on cylinder for fabrication of cylindrical mold: Roll-to-roll nano-imprint lithography. *Microelectron. Eng.* **2019**, *211*, 5–12. [CrossRef]
5. Tahir, U.; Kamran, M.A.; Jeong, M.Y. Numerical study on the optimization of roll-to-roll ultraviolet imprint lithography. *Coatings* **2019**, *9*, 573. [CrossRef]
6. Ye, H.; Shen, L.; Li, M.; Zhang, Q. Bubble defect control in low-cost roll-to-roll ultraviolet imprint lithography. *Micro Nano Lett.* **2014**, *9*, 28–30. [CrossRef]
7. Ye, H.; Shen, L.-G.; Tao, B.; Li, M.-J. A study of the bubbles in UV micro roll-to-roll imprinting. *Int. Polym. Process.* **2014**, *29*, 602–606. [CrossRef]
8. Yoo, D.; Seok, J.-M.; Kim, N.W. Three-dimensional numerical investigations of air bubble defects during antireflective pattern fabrication via ultraviolet nanoimprint lithography. *J. Vac. Sci. Technol. B Nanotechnol. Microelectron. Mater. Process. Meas. Phenom.* **2014**, *32*, 61602. [CrossRef]
9. Zhou, Y.; Li, M.; Shen, L.; Ye, H.; Wang, J.; Huang, S. Effect of resin accumulation on filling process in roll-to-roll UV imprint lithography. *J. Vac. Sci. Technol. B Nanotechnol. Microelectron. Mater. Process. Meas. Phenom.* **2017**, *35*, 31602. [CrossRef]
10. Pietarinen, J.; Kalima, V.; Pakkanen, T.T.; Kuittinen, M. Improvement of UV-moulding accuracy by heat and solvent assisted process. *Microelectron. Eng.* **2008**, *85*, 263–270. [CrossRef]
11. Hiroshima, H.; Komuro, M. UV-nanoimprint with the assistance of gas condensation at atmospheric environmental pressure. *J. Vac. Sci. Technol. B Microelectron. Nanometer Struct. Process. Meas. Phenom.* **2007**, *25*, 2333–2336. [CrossRef]
12. Peng, L.; Wu, H.; Yi, P.; Lai, X. Study on bubble defects in roll-to-roll UV imprinting process for micropyramid arrays II: Numerical study. *J. Vac. Sci. Technol. B Nanotechnol. Microelectron. Mater. Process. Meas. Phenom.* **2016**, *34*, 51203. [CrossRef]
13. Song, X.; Shan, X.; Chow, S.; Deng, X.; Teo, W. Numerical and experimental study of the filling stage of roll-to-roll UV embossing process with micro features. *Microsyst. Technol.* **2015**, *21*, 1729–1738. [CrossRef]
14. Hirt, C.W.; Nichols, B.D. Volume of fluid (VOF) method for the dynamics of free boundaries. *J. Comput. Phys.* **1981**, *39*, 201–225. [CrossRef]
15. Versteeg, H.K.; Malalasekera, W. *An Introduction to Computational Fluid Dynamics: The Finite Volume Method*; Pearson Education: London, UK, 2007.
16. Lewis, R.; Cross, M.; Ludlow, D. The influence of rotating wheels on the external aerodynamic performance of a vehicle. In *Proceedings of the The International Vehicle Aerodynamics Conference*; SAE International: Warrendale, PA, USA, 2014; pp. 161–173.

17. ANSYS. *ANSYS Fluent Theory Guide*; Release 16.0; ANSYS: Canonsburg, PA, USA, 2006.
18. Jose, J.T.; Dunne, J.J.F. Numerical simulation of single-droplet dynamics, vaporization, and heat transfer from impingement onto static and vibrating surfaces. *Fluids* **2020**, *5*, 188. [CrossRef]
19. Massey, B.S.; Ward-Smith, J. *Mechanics of Fluids*; CRC Press: Boca Raton, FL, USA, 2018.
20. Hauschwitz, P.; Bičštová, R.; Brodsky, A.; Kaplan, N.; Cimrman, M.; Huynh, J.; Brajer, J.; Rostohar, D.; Kopeček, J.; Smrž, M.J.N. Towards Rapid Fabrication of Superhydrophobic Surfaces by Multi-Beam Nanostructuring with 40,401 Beams. *Nanomaterials* **2021**, *11*, 1987. [CrossRef] [PubMed]
21. Peng, L.; Zhang, C.; Wu, H.; Yi, P.; Lai, X.; Ni, J. Continuous fabrication of multiscale compound eyes arrays with antireflection and hydrophobic properties. *IEEE Trans. Nanotechnol.* **2016**, *15*, 971–976. [CrossRef]

nanomaterials

MDPI

Article

Reduced Electron Temperature in Silicon Multi-Quantum-Dot Single-Electron Tunneling Devices

Youngmin Lee [1], So Hyun Lee [2], Hyo Seok Son [2] and Sejoon Lee [1,2,*]

[1] Quantum-Functional Semiconductor Research Center, Dongguk University-Seoul, Seoul 04620, Korea; ymlee@dongguk.edu
[2] Department of Semiconductor Science, Dongguk University-Seoul, Seoul 04620, Korea; thgus1731@naver.com (S.H.L.); shs_0213@naver.com (H.S.S.)
* Correspondence: sejoon@dongguk.edu

Abstract: The high-performance room-temperature-operating Si single-electron transistors (SETs) were devised in the form of the multiple quantum-dot (MQD) multiple tunnel junction (MTJ) system. The key device architecture of the Si MQD MTJ system was self-formed along the volumetrically undulated [110] Si nanowire that was fabricated by isotropic wet etching and subsequent oxidation of the e-beam-lithographically patterned [110] Si nanowire. The strong subband modulation in the volumetrically undulated [110] Si nanowire could create both the large quantum level spacings and the high tunnel barriers in the Si MQD MTJ system. Such a device scheme can not only decrease the cotunneling effect, but also reduce the effective electron temperature. These eventually led to the energetic stability for both the Coulomb blockade and the negative differential conductance characteristics at room temperature. The results suggest that the present device scheme (i.e., [110] Si MQD MTJ) holds great promise for the room-temperature demonstration of the high-performance Si SETs.

Keywords: quantum dot; single-electron transistor; Coulomb blockade; charge stability; effective electron temperature

Citation: Lee, Y.; Lee, S.H.; Son, H.S.; Lee, S. Reduced Electron Temperature in Silicon Multi-Quantum-Dot Single-Electron Tunneling Devices. *Nanomaterials* **2022**, *12*, 603. https://doi.org/10.3390/nano12040603

Academic Editors: Seiichi Miyazaki and Efrat Lifshitz

Received: 6 January 2022
Accepted: 9 February 2022
Published: 11 February 2022

1. Introduction

The semiconductor single-electron transistors (SETs), which comprise either the double-barrier tunnel junction (DTJ) with a single quantum dot (QD) or the multiple tunnel junction (MTJ) with multiple quantum dots (MQDs), allow single-electron transport through the discrete quantum energy states of the semiconductor QDs [1–5]. In short, the electron can transfer one-by-one through the quantum states via the Coulomb blockade effect with its corresponding quantum–mechanical single-electron tunneling events. This leads to the unique transfer and output characteristics, such as Coulomb blockade oscillation (CBO) and negative differential conductance (NDC), respectively [6–15]. For example, the precise control of the single-electron (or even single-spin) transport characteristics was demonstrated on various types of semiconductor QD-based DTJ and MTJ device schemes (e.g., room temperature observation of multiple CBO peaks from multiple quantum states in a Si single-QD device [10], simultaneous observation of both sharp CBO and NDC peaks from a Si-QD DTJ device [9–14], bias voltage-controlled precise modulation of energetic Coulomb blockade conditions in a Si single-QD transistor [8], high-fidelity q-bit processing in Si MQD [16–19] and GaAs MQD [20–22] devices). Such an extremely high precision of the single-charge manipulation could enable us to extend the SET application toward the broad area of the sensing metrology. Namely, when the external stimuli transfer from the sensing object to the SET, it would significantly influence the electronic charging energy of the QD; hence, the tunneling conductance of the SET could be tuned via changing in the energetic Coulomb blockade condition by the stimuli from the sensing target. In this circumstance, the SET-based sensors could also reveal the higher sensitivity than that of complementary metal–oxide–semiconductor (CMOS)-based sensors, because in the SET, the conductance values at the CBO peaks and valleys (i.e., on- and off-tunneling states) are also precisely controllable by changing the gate and/or the drain bias voltages [23]. Owing to such an astonishing physical mechanism, the electric current standard device [24], the thermometers [25,26], the charge sensors [16–19], the photon detectors [27,28], the ion sensors [29,30], and the mechanical displacement detectors [1–3] were conceived and reported as feasible applications of the SET-based sensors.

In typical SETs, however, the thermal fluctuation and the thermally activated carrier conduction may cause the malfunction of the sensors because those give rise to both the thermal quenching and thermal broadening of CBO and NDC. At the elevated temperature, therefore, the SET will eventually result in the ambiguous operation of the sensors. According to previous literature, thermal quenching and thermal broadening of CBO and NDC are closely relevant to the cotunneling phenomena at the Coulomb blockade state [31,32]. Cotunneling events can be categorized into two different types, i.e., one is elastic cotunneling that occurs via additional electron tunneling through the intermediate virtual quantum levels in the QD, and the other is inelastic cotunneling that takes place via the in- and out-tunneling of other electrons through other quantum levels [33]. These may in turn increase the leakage current at the Coulomb blockade state (i.e., valley current of CBO); hence, the SET will lead to the impairable operation of the SET-based sensors. To increase the efficiency of the SET-based sensors, therefore, suppressing the cotunneling effect is vital. Furthermore, since electron cotunneling strongly relies on the effective electron-temperature in the QD device, the low operation temperature of the conventional SETs makes the thermal fluctuation and the leakage current issues more critical.

All these above backgrounds prompt us to investigate the fabrication and the characterization of the high-performance room-temperature-operating SETs, in which the cotunneling behaviors should be effectively suppressed. Herein, to take full advantage of the scientific and technical knowledge on the Si nanoelectronic devices, we fabricated and characterized the CMOS-compatible Si SETs that could steadily operate at room temperature. The devices were devised in the form of the gate-all-around (GAA) Si nanowire-channel metal–oxide–semiconductor field-effect transistor (MOSFET), where the MTJs were self-formed through isotropic wet etching of the undulated [110] Si nanowire that had been created by *e*-beam lithography. The transport characteristics of the fabricated SETs are thoroughly examined, and their effective electron temperatures are analyzed and discussed by means of the cotunneling current characterization.

2. Experimental Section

Figure 1a displays the schematic illustration of the Si SET, which comprises a device scheme of the CMOS-compatible Si nanowire-channel GAA MOSFET. To construct such a device architecture, as a primary task, the [110] Si nanowire-channel (length $\approx$ 200 nm, width $\approx$ 40 nm) was patterned on the ~10 nm-thick silicon-on-insulator substrate by using *e*-beam lithography. Next, to shrink the volumetric nanowire size, isotropic wet etching was carried out by using the SC-1 solution (NH_4OH:H_2O_2:H_2O = 1:1:6). Then, the size of the etched nanowire ($\approx$15 nm) became narrower than the initial size ($\approx$40 nm) of the *e*-beam-lithographically patterned Si nanowire (Figure 1b). To configure the GAA structure, subsequently, a part of the buried oxide underneath the Si nanowire-channel was etched out by dipping the sample into the dilute hydrogen fluoride acid solution (HF:H_2O = 1:10). Then, the Si nanowire-channel could be suspended from the buried oxide because of the large supporting areas of source (S) and drain (D). Thereafter, the surface of the suspended Si nanowire was oxidized by dry oxidation at 900 °C to form the gate oxide layer. During this step, the final diameter size of the Si nanowire was further shrunken down to <5 nm [7–11]. Through the sequential deposition of additional SiO_2 ($\approx$30 nm) and n^+ poly-Si ($\approx$250 nm) gate (G), finally, the formation of the GAA stacks was finalized. The rest processes for forming the n^+-S ($\approx 10^{20}$ cm^{-3}) and n^+-D ($\approx 10^{20}$ cm^{-3}) reservoirs were followed by the CMOS-compatible process steps with the P^+ ion implantation and the thermal activation of the dopants at 950 °C.

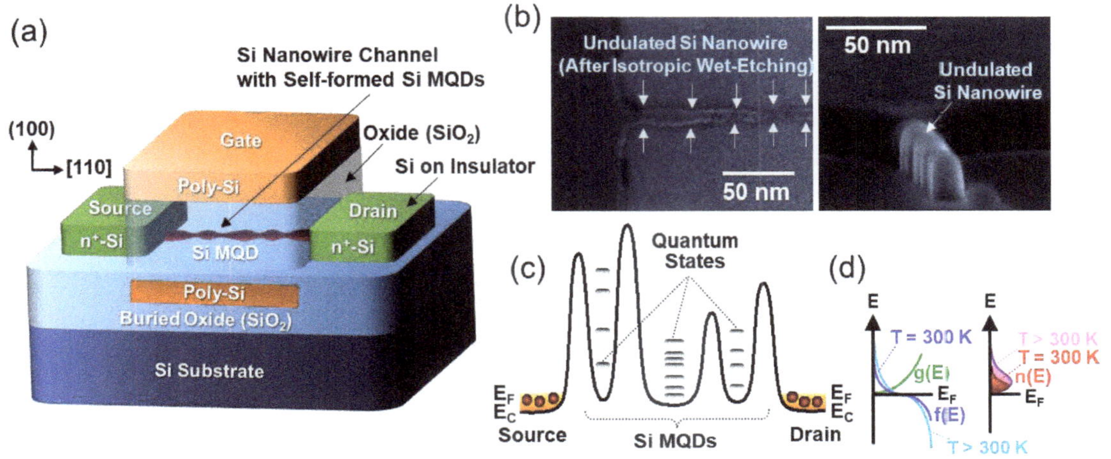

Figure 1. (**a**) Schematic of the fabricated SHT device, (**b**) scanning electron microscopy images (left, top view; right, tilted view) of the undulated Si nanowire channel obtained from dummy samples, (**c**) expected energy band diagram of the undulated Si nanowire-channel at the conduction band region, and (**d**) Fermi–Dirac distribution function, f(E) at T = 300 K and T > 300 K; density of state function, g(E); and electron distribution function, n(E) at T = 300 K and T > 300 K in the source and the drain reservoirs. In (**c**,**d**), E_c and E_F denote the conduction band and the Fermi level, respectively.

Here, one needs to remind that, during the volumetric shrinkage of the Si nanowire by isotropic wet etching, the diameter sizes of the nanowire were volumetrically undulated along the direction normal to the channel axis (Figure 1b) and such a volumetric undulation would become significant during the thermal oxidation of the Si nanowire surface. Since the diameter size of the Si nanowire became narrow (≈5 nm), the volumetrically undulated areas would be much narrower than 5 nm. Due to the strong quantum-mechanical sub-band modulation in [110] Si nanowires [34,35], in the volumetrically undulated Si nanowire, the MTJ system could be self-created along the nanowire–channel direction. In short, some parts of the Si nanowire would be squeezed (<<5 nm), and the rest of the parts (≈5 nm) would be connected in series along with the squeezed regions. According to Refs. [36,37], the sub-band modulation becomes significant as the diameter of the [110] Si nanowire decreases. For example, in the [110] Si nanowire with the smaller diameter of <2 nm, the ground state could locate at 500 meV above the conduction band (E_C) of bulk Si [36,37]. This would eventually create the energy band fluctuation at E_C along the Si nanowire (Figure 1c). Accordingly, the squeezed regions (<<5 nm) and the unsqueezed areas (≈5 nm) may act as the tunneling barriers and QDs, respectively. Hence, the MTJ system could be formed along the Si nanowire for the fabricated device to operate as a MQD Si SET. In this circumstance, the single-electron tunneling transport would strongly depend on both the quantum level spacings of the QDs and the carrier distribution functions of the electron reservoirs (Figure 1d). Therefore, the Coulomb blockade characteristics would strongly rely on the effective electron temperature and its corresponding cotunneling effect.

3. Results and Discussion

Figure 2 shows the transfer characteristics of three different SETs that were fabricated through the identical process procedures described above. For convenience only, we simply refer to the three SETs as SET-A, SET-B, and SET-C, respectively. Figure 2a–c displays the drain current vs. gate voltage (I_D–V_G) curves of SET-A, SET-B, and SET-C under the drain voltage (V_D) of 1 mV at room temperature, respectively. The SETs reveal the typical transfer characteristics of the MQD SET. Namely, the devices exhibit the clear CBO peaks together with the multiple humps, arising from the stochastic tunneling events in the MTJ system [38,39]. Since the peak and valley of CBO correspond to the on- and off-resonance states for single-electron tunneling via the Coulomb blockade event, the large magnitude of the maximum peak-to-valley current ratio (>20) depicts the large quantum level spacings to exist in the QDs. As aforementioned, the QDs were self-formed along the volumetrically undulated Si nanowire-channels. In such a geometrical structure, the large sub-band modulation at the squeezed nanowire regions (<<5 nm) could create the large potential barriers, which are big enough to energetically separate the unsqueezed nanowire areas (≈5 nm). According to Refs. [36,37], as the diameter of the [110] Si nanowire decreases, the quantum level spacings could increase up to 75 meV because of the decreased effective mass ($m_e \sim 0.11m_0$ [36]) at the two-fold Γ valley. One can therefore conjecture the unsqueezed nanowire areas to act as the QDs, possessing the large quantum level spacings. Accordingly, the fabricated devices could operate as room-temperature-operating MQD SETs.

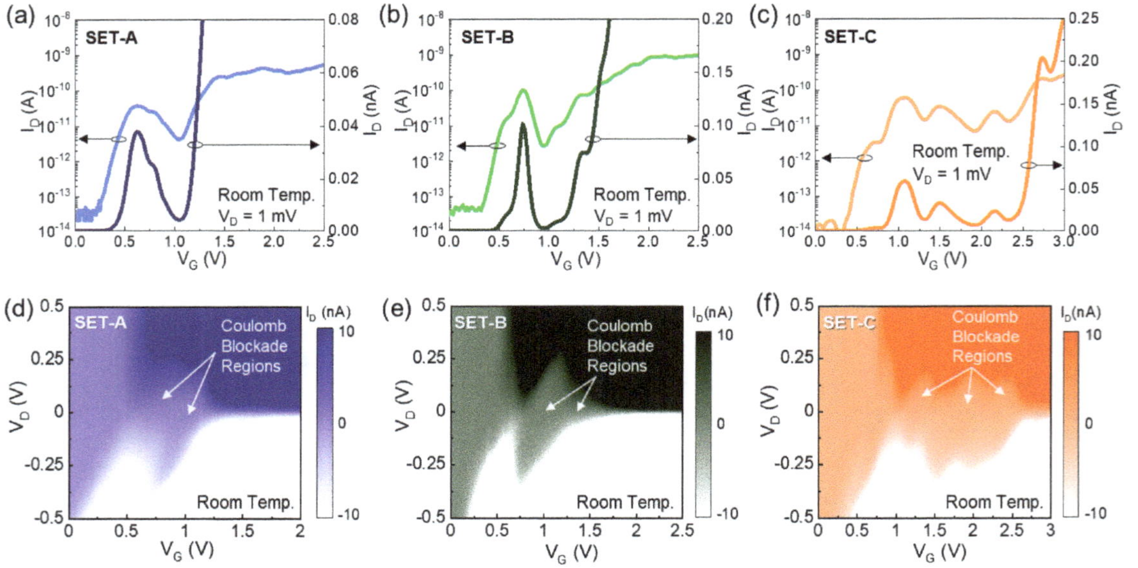

Figure 2. Transfer characteristic curves (i.e., I_D–V_G at V_D = 1 mV) at room temperature of (**a**) SET-A, (**b**) SET-B, and (**c**) SET-C; and contour plots of I_D as functions of V_G and V_D at room temperature for (**d**) SET-A, (**e**) SET-B, and (**f**) SET-C.

The MQD behavior of the fabricated SETs can also be traced from the charge stability diagram (i.e., Coulomb diagram). Figure 2d–f shows the contour plots of I_D as functions of V_G and V_D for SET-A, SET-B, and SET-C, respectively. The SETs clearly display the typical shape of the rhombus Coulomb blockade regions, indicative of the single-electron tunneling transport characteristics via the Coulomb blockade effect. Here, it should be noted that some parts of the Coulomb blockade regions are overlapped by their adjacent Coulomb blockade regions. Such an overlapped blockade feature can be interpreted by the irregular MQD system. As can be expected from Figure 1a–c, both the tunnel barrier heights and the quantum-dot sizes would be inhomogeneous in the present devices because those were self-created through the volumetric shrinkage of the undulated Si nanowire. In this case, the shapes and sizes of the QDs as well as the heights and curvatures of the tunnel barriers would be irregular so that the QDs would have different quantum level spacings. Since such an inhomogeneity causes the imbalance of the energetic Coulomb blockade conditions for every dot in MQDs, the present type of the MQD system would show the stochastic tunneling characteristics, resulting in the appearance of the overlapped Coulomb blockade regions in the Coulomb diagram.

Here, we also note that the three SETs reveal quite different Coulomb blockade features, even though those were fabricated in a same bath with the identical fabrication process. As mentioned earlier, the energy band profile of the present device scheme strongly depends on the degree of volumetric undulation along the nanowire channel direction. In this case, the energy band profile would alter device by device because the strong and weak sub-band modulations at the squeezed and unsqueezed areas are responsible for the self-formation of both the tunnel barriers and the isolated dots. In brief, the inhomogeneity of volumetric undulation leads to the randomness of the MQD MTJ profile with the different number of QDs. Accordingly, the SETs fabricated in a single chip showed different CBO features (Figure S1, Supplementary Materials). In the application point of view, such inhomogeneous device characteristics may restrict the reliability of the circuit integration. Hence, the key issue could become the fabrication of the device structure with a clear and regular succession of quantum dots at regular distances. To release this issue, therefore, the advanced sub-5 nm patterning techniques can be suggested as feasible ways to improve the device homogeneity. For example, recent advances in nanofabrication technology, such as scanning probe lithography [40], heavy ion lithography [41], extreme ultraviolet lithography [42,43], block copolymer self-assembly [44], may allow the precise undulation of the Si nanowire because these methods enable us to control both the fine size and the exact site of the sub-5 nm patterns.

In the MQD system, the cotunneling effect can be effectively suppressed because of the following reason. According to the single-electron tunneling transport model [32,33], the magnitude of I_D is proportional to the multiplication factor;

$$\left(g_T e^2/h\right)^{N+1} \tag{1}$$

where g_T is the tunnel conductance of the single tunnel barrier, e is the unit charge, h is the Planck constant, and N is the number of QDs. Since the magnitude of $(g_T e^2/h)$ is much smaller than 1 at the Coulomb blockade state (i.e., very low g_T at the off-tunneling state), the multiplication factor $\left(g_T e^2/h\right)^{N+1}$ would drastically decrease with increasing N in the MQD system. To briefly sum up, the cotunneling current (i.e., valley current (I_{valley})) could be effectively decreased as one increases the number of QDs. Based upon the above model, for the MQD system with N QDs, the value of I_{valley} can be described by [45]

$$\begin{aligned} I_{valley} &\propto \left(g_T e^2/h\right)^{N+1}\left\{(eV_D)^2 + (2\pi k_B T_{eff})^2\right\}^N V_D \\ &\equiv G_b^{N+1}\left\{(eV_D)^2 + (2\pi k_B T_{eff})^2\right\}^N V_D \end{aligned} \tag{2}$$

where G_b^{N+1} is the multiplication of the tunnel barrier conductance, k_B is the Boltzmann constant, and T_{eff} is the effective electron temperature. For example, the I_{valley} values for the single (N = 1), double (N = 2), and triple (N = 3) QD systems can be derived by Equations (3)–(5), respectively [9,10]:

$$I_{valley(N=1)} = \alpha G_S G_D\left\{e^2 V_D^3 - (2\pi k_B T_{eff})^2 V_D\right\} \tag{3}$$

$$I_{valley(N=2)} = \beta G_S G_i G_D\left\{e^4 V_D^5 + 2e^2(2\pi k_B T_{eff})^2 V_D^3 + (2\pi k_B T_{eff})^4 V_D\right\} \tag{4}$$

$$I_{valley(N=3)} = \gamma G_S G_{i1} G_{i2} G_D\left\{e^6 V_D^7 + 3e^4(2\pi k_B T_{eff})^2 V_D^5 + 3e^2(2\pi k_B T_{eff})^4 V_D^3 + (2\pi k_B T_{eff})^6 V_D\right\} \tag{5}$$

where α, β, and γ are the proportional factors, and G_S, G_i, and G_D are the source, intermediate, and drain conductance values, respectively.

To assess the cotunneling characteristics of the present devices, we examined the V_D dependence of the CBO evolution (Figure 3a–c) and plotted the values of I_{valley} as a function of V_D (Figure 3d–f). As can be seen from Figure 3a–c, the devices exhibit the clear valley states even at higher V_D up to 0.5 V. In general, the cotunneling events would become significant at the higher bias voltages because the higher external electric field from the over-driving voltage gives rise to the increase in the excess energy in the QD system [15,46,47]. Therefore, the clear valley states at higher V_D depict the present devices to hold a weak cotunneling effect. Nevertheless, the magnitude of I_{valley} goes out of the single tendency when V_D exceeds 0.3–0.35 V. In the present type of the SETs, the tunnel barriers are created by the sub-band modulation at the squeezed Si nanowire regions but not the material barriers, such as SiO_2. In this case, the tunnel barriers would be lowered with increasing V_D, particularly at the drain region, because the tunnel barrier is capacitively coupled in between the dot and the electrode. Thus, the stochastic tunneling events would alter and/or be broken at the higher V_D region so that the I_{valley} values become irrespective of the above cotunneling model. For data fitting to the above equations, we therefore chose only the V_D region, in which I_D follows the I_{valley} vs. V_D functions in Equations (3)–(5).

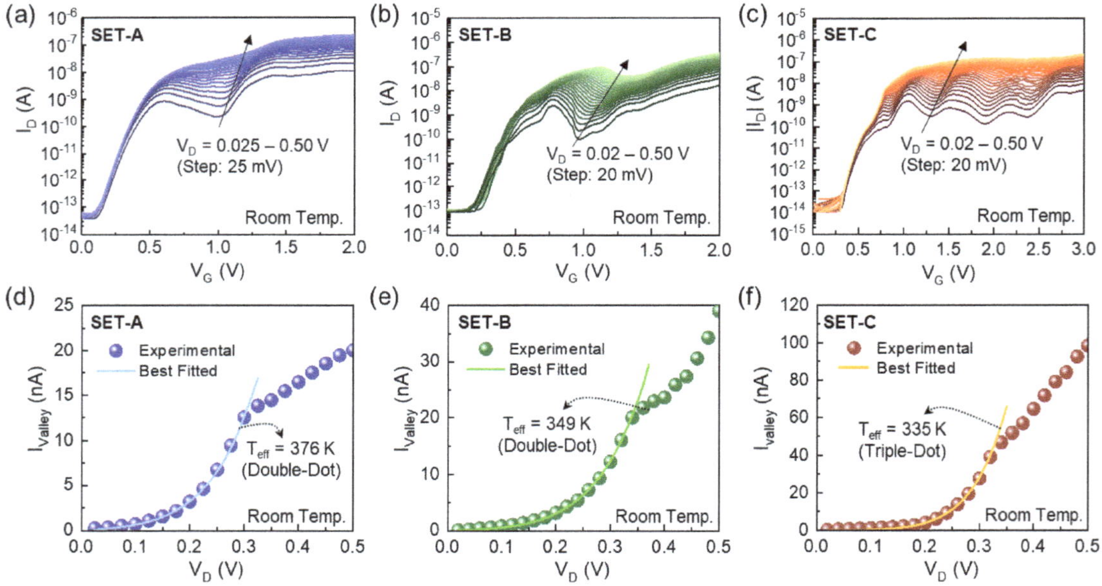

Figure 3. Evolution of the CBO peaks at the positive V_D region (i.e., I_D–V_G curves at V_D = 0.02 – 0.5 V) for (**a**) SET-A, (**b**) SET-B, and (**c**) SET-C; and the I_{valley} as a function of V_D for (**d**) SET-A, (**e**) SET-B, and (**f**) SET-C.

By fitting the measured I_{valley} values to the above equations, we found that the SET-A, SET-B, and SET-C devices were composed of the MQD systems with N = 2 (double), 2 (double), and 3 (triple), respectively. Namely, the I_{valley} data could be well fitted only to Equation (4) for SET-A and SET-B and to Equation (5) for SET-C. From the fitting curves, the T_{eff} values were determined to be 376, 349, and 335 K for SET-A, SET-B, and SET-C, respectively. Accordingly, the excess energy ($E_{exc} = E_{e\text{-}f} - E_{env}$, where E_{env} is the thermal energy at the environmental system) could be deduced to be 6.5, 4.2, and 3.0 meV for SET-A, SET-B, and SET-C, respectively. In addition, the other SETs (N = 2 or 3) fabricated in a single chip were confirmed to have similar values to the above (Figure S1, Supplementary Materials). These values are much smaller than those of other single-dot SETs/SHTs and are comparable to those of the state-of-the-art single-dot SETs that comprised the ellipsoidal

QDs produced by sophisticate fabrication processes (Table 1). As a result, forming the MQD system would effectively lead to the decrease in the cotunneling effect; hence, the Coulomb blockade state (i.e., valley state) could be stabilized even at higher V_D.

Table 1. Comparison of T_{eff} and E_{exc} for various CMOS-compatible Si SETs with different device configurations.

Number of QD	Device Type	Si Nanowire Direction	T_{eff} (K)	E_{exc} (meV)	Ref.
Single	Si SHT	[100]	1260	82.7	[15]
	Si SHT	[100]	870	49.1	[15]
	Si SHT	[100]	415	9.9	[15]
	Si SET	[110]	312	1.0	[10]
	Si SET	[100]	338	3.3	[9]
Double	Si SET (A)	[110]	376	6.5	This Work
	Si SET (B)	[110]	349	4.2	
	Si SET (S1)	[110]	384	7.2	
	Si SET (S2)	[110]	389	7.7	
	Si SET (S3)	[110]	397	8.4	
Triple	Si SET (C)	[110]	335	3.0	
	Si SET (S4)	[110]	342	3.4	

As mentioned earlier, T_{eff} affects not only the cotunneling characteristics at the Coulomb blockade states, but also the thermally activated carrier conduction (i.e., thermal fluctuation of the quantum states). To verify the energetic stability of the quantum states, we examined the NDC characteristics via observing the V_D-dependent single-electron tunneling current at V_G near the Coulomb blockade state. Figure 4a–c displays the room-temperature I_D–V_D characteristic curves at various V_G conditions near the Coulomb blockade regions for SET-A, SET-B, and SET-C, respectively. All the devices clearly exhibit the I_D humps or knees, as indicated by the arrows. For example, in the case of SET-A (Figure 4a), the I_D hump begins to appear at V_G = 0.7 V, and the position of the I_D hump gradually moves to the higher V_D and higher I_D region as V_G increases.

A similar feature can be also observable from SET-B (Figure 4b) and SET-C (Figure 4c). Namely, SET-B and SET-C show the I_D knees in their I_D–V_D characteristic curves. As shown in Figure 4d–f, the I_D humps and knees can be confirmed to originate from the NDC characteristics. These are attributable to the sudden drop of the drain conductance due to the off resonance at the forbidden energy gaps [10,12]. In other words, the tunneling processes could be prohibited at specific V_D bias voltages because of the large quantum level spacings in the ultra-small Si QDs. Based upon all the above results, therefore, it can be concluded that both the cotunneling effects and the thermal fluctuation behaviors could be effectively reduced by forming the MQD system. Furthermore, the Si MQD system formed along the [110] Si nanowire can be suggested as a commendable strategy to reduce the T_{eff} value for the room-temperature application of the CMOS-compatible Si SETs.

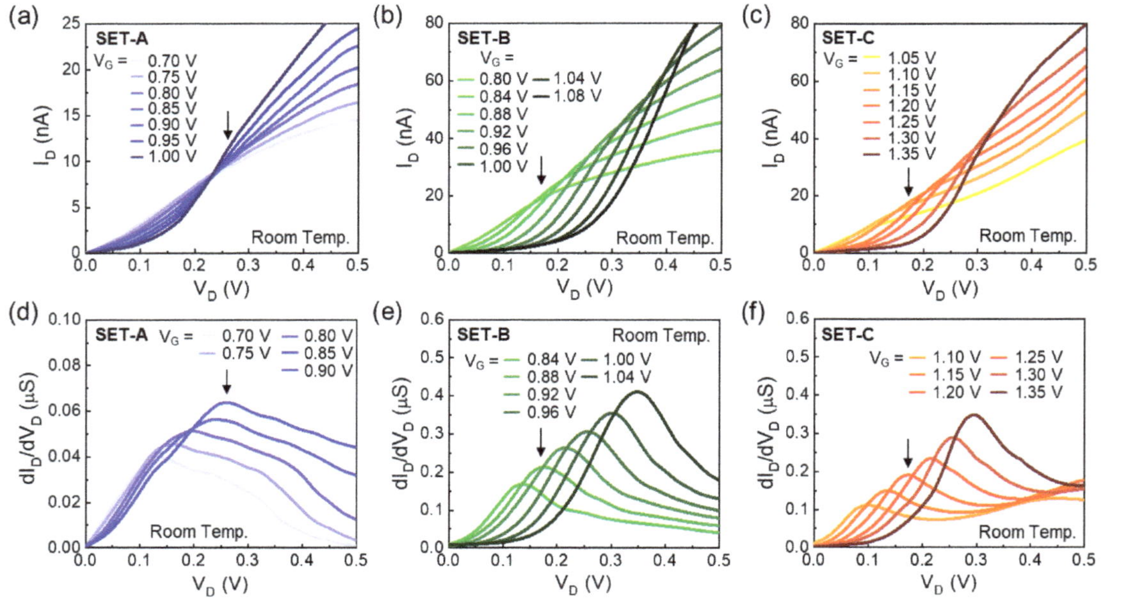

Figure 4. I_D–V_D output characteristic curves at various V_G near the Coulomb blockade regions for (**a**) SET-A, (**b**) SET-B, and (**c**) SET-C, and dI_D/dV_D–V_D curves of (**d**) SIA, (**e**) SET-B, and (**f**) SET-C.

4. Summary and Conclusions

The CMOS-compatible Si MQD SETs were fabricated in the form of the Si nanowire-channel MOSFETs, in which the multiple Si QDs were self-formed through isotropic wet etching of the *e*-beam-lithographically patterned [110] Si nanowires. Owing to the large sub-band modulation in the volumetrically undulated [110] Si nanowire, the Si MQD MTJ system with large quantum level spacings could be achieved. Although the volumetrically undulation method (i.e., self-formation of the Si MQD MTJ system) did not fully guarantee the identical Coulomb blockade characteristics for all the devices in a single chip, in terms of the theoretical fitting model, we found that the MQD MTJ system could allow us to effectively reduce both the cotunneling current and the effective electron temperature. These eventually led to the room-temperature manipulation of clear CBO and NDC peaks at wide bias voltage ranges. Consequently, the formation of the [110] Si MQD MTJ system could be an effective strategy to fabricate the high-performance CMOS-compatible Si SETs.

Supplementary Materials: The following supporting information can be downloaded at: https://www.mdpi.com/article/10.3390/nano12040603/s1, Figure S1: Coulomb blockade characteristics of (a) SET-S1, (b) SET-S2, (c) SET-S3, and (d) SET-S4 that had been fabricated in a single chip studied in the present work.

Author Contributions: Y.L.: data curation, formal analysis, investigation, methodology, and writing—original draft. S.H.L.: investigation and methodology. H.S.S.: investigation and methodology. S.L.: conceptualization, supervision, writing—review and editing. All authors have read and agreed to the published version of the manuscript.

Funding: This research was supported by the National Research Foundation (NRF) of Korea through the Basic Science Research Programs ((2016R1A6A1A03012877; 2019R1A2C1085448; 2021R1I1A1A01049638) funded by the Korean Government.

Institutional Review Board Statement: Not applicable.

Informed Consent Statement: Not applicable.

Data Availability Statement: Not applicable.

Conflicts of Interest: The authors declare no conflict of interest.

References

1. Knobel, R.G.; Cleland, A.N. Nanometre-scale displacement sensing using a single electron transistor. *Nature* **2003**, *424*, 291–293. [CrossRef] [PubMed]
2. Knobel, R.; Cleland, A.N. Piezoelectric displacement sensing with a single-electron transistor. *Appl. Phys. Lett.* **2002**, *81*, 2258–2260. [CrossRef]
3. Mahboob, I.; Nishiguchi, K.; Fujiwara, A.; Yamaguchi, H. Room temperature piezoelectric displacement detection via a silicon field effect transistor. *Appl. Phys. Lett.* **2009**, *95*, 233102. [CrossRef]
4. van der Wiel, W.G.; De Franceschi, S.; Elzerman, J.M.; Fujisawa, T.; Tarucha, S.; Kouwenhoven, L.P. Electron transport through double quantum dots. *Rev. Mod. Phys.* **2002**, *75*, 1–22. [CrossRef]
5. Oda, S.; Ferry, D.K. *Silicon Nanoelectronics*; Taylor & Francis: New York, NY, USA, 2006.
6. Jing, Y.; Huang, S.; Wu, J.; Meng, M.; Li, X.; Zhou, Y.; Peng, H.; Xu, H. A Single-Electron Transistor Made of a 3D Topological Insulator Nanoplate. *Adv. Mater.* **2019**, *31*, 1903686. [CrossRef]
7. Lee, S.; Lee, Y.; Song, E.B.; Wang, K.L.; Hiramoto, T. Gate-Tunable Selective Operation of Single Electron/Hole Transistor Modes in a Silicon Single Quantum Dot at Room Temperature. *Appl. Phys. Lett.* **2013**, *102*, 083504. [CrossRef]
8. Lee, S.; Lee, Y.; Song, E.B.; Hiramoto, T. Modulation of Peak-to-Valley Current Ratio of Coulomb Blockade Oscillations in Si Single Hole Transistors. *Appl. Phys. Lett.* **2013**, *103*, 103502. [CrossRef]
9. Lee, Y.; Lee, J.W.; Lee, S.; Hiramoto, T.; Wang, K.L. Reconfigurable Multivalue Logic Functions of a Silicon Ellipsoidal Quantum-Dot Transistor Operating at Room Temperature. *ACS Nano* **2021**, *15*, 18483–18493. [CrossRef]
10. Lee, S.; Lee, Y.; Song, E.B.; Hiramoto, T. Observation of Single Electron Transport *via* Multiple Quantum States of a Silicon Quantum Dot at Room Temperature. *Nano Lett.* **2014**, *14*, 71–77. [CrossRef]
11. Lee, S.; Lee, Y.; Song, E.B.; Hiramoto, T. The Characteristic of Elongated Coulomb-Blockade Regions in a Si Quantum-Dot Device Coupled *via* Asymmetric Tunnel Barriers. *J. Appl. Phys.* **2013**, *114*, 164513. [CrossRef]
12. Lee, Y.; Lee, S.; Hiramoto, T. Transport Behaviors and Mechanisms in Cuspidal Blockade Region for Silicon Single-Hole Transistor. *Curr. Appl. Phys.* **2014**, *14*, 428–432. [CrossRef]
13. Lee, S.; Miyaji, K.; Kobayashi, M.; Hiramoto, T. Extremely High Flexibilities of Coulomb Blockade and Negative Differential Conductance Oscillations in Room-Temperature-Operating Silicon Single Hole Transistor. *Appl. Phys. Lett.* **2008**, *92*, 073502. [CrossRef]
14. Lee, Y.; Lee, S.; Im, H.; Hiramoto, T. Multiple Logic Functions from Extended Blockade Region in a Silicon Quantum-Dot Transistor. *J. Appl. Phys.* **2015**, *117*, 064501. [CrossRef]
15. Lee, S.; Hiramoto, T. Strong Dependence of Tunneling Transport Properties on Overdriving Voltage for Room-Temperature-Operating Single Electron/Hole Transistors Formed with Ultranarrow [100] Silicon Nanowire Channel. *Appl. Phys. Lett.* **2008**, *93*, 043508. [CrossRef]
16. Eenink, H.G.J.; Petit, L.; Lawrie, W.I.L.; Clarke, J.S.; Vandersypen, L.M.K.; Veldhorst, M. Tunable Coupling and Isolation of Single Electrons in Silicon Metal-Oxide-Semiconductor Quantum Dots. *Nano Lett.* **2019**, *19*, 8653–8657. [CrossRef]
17. Yuan, M.; Yang, Z.; Savage, D.E.; Lagally, M.G.; Eriksson, M.A.; Rimberg, A.J. Charge sensing in a Si/SiGe quantum dot with a radio frequency superconducting single-electron transistor. *Appl. Phys. Lett.* **2012**, *101*, 142103. [CrossRef]
18. Stuyck, N.D.; Li, R.; Kubicek, S.; Mohiyaddin, F.A.; Jussot, J.; Chan, B.T.; Simion, G.; Govoreanu, B.; Heyns, M.; Radu, I. An Integrated Silicon MOS Single-Electron Transistor Charge Sensor for Spin-Based Quantum Information Processing. *IEEE Electron. Device Lett.* **2020**, *41*, 1253–1256. [CrossRef]
19. Chanrion, E.; Niegemann, D.J.; Bertrand, B.; Spence, C.; Jadot, B.; Li, J.; Mortemousque, P.-A.; Hutin, L.; Maurand, R.; Jehl, X.; et al. Charge Detection in an Array of CMOS Quantum Dots. *Phys. Rev. Appl.* **2020**, *14*, 024066. [CrossRef]
20. Hensgens, T.; Fujita, T.; Janssen, L.; Li, X.; Van Diepen, C.J.; Reichl, C.; Wegscheider, W.; Das Sarma, S.; Vandersypen, L.M.K. Quantum simulation of a Fermi–Hubbard model using a semiconductor quantum dot array. *Nature* **2017**, *548*, 70–73. [CrossRef]
21. Reed, M.D.; Maune, B.M.; Andrews, R.W.; Borselli, M.G.; Eng, K.; Jura, M.P.; Kiselev, A.A.; Ladd, T.D.; Merkel, S.T.; Milosavljevic, I.; et al. Reduced Sensitivity to Charge Noise in Semiconductor Spin Qubits via Symmetric Operation. *Phys. Rev. Lett.* **2016**, *116*, 110402. [CrossRef]
22. Martins, F.; Malinowski, F.K.; Nissen, P.D.; Barnes, E.; Fallahi, S.; Gardner, G.C.; Manfra, M.J.; Marcus, C.M.; Kuemmeth, F. Noise Suppression Using Symmetric Exchange Gates in Spin Qubits. *Phys. Rev. Lett.* **2016**, *116*, 116801. [CrossRef] [PubMed]
23. Nakajima, A. Application of Single-Electron Transistor to Biomolecule and Ion Sensors. *Appl. Sci.* **2016**, *6*, 94. [CrossRef]
24. Brun-Picard, J.; Djordjevic, S.; Leprat, D.; Schopfer, F.; Poirier, W. Practical Quantum Realization of the Ampere from the Elementary Charge. *Phys. Rev. X* **2016**, *6*, 041051. [CrossRef]
25. Chen, I.H.; Lai, W.T.; Li, P.W. Realization of solid-state nanothermometer using Ge quantum-dot single-hole transistor in few-hole regime. *Appl. Phys. Lett.* **2014**, *104*, 243506. [CrossRef]
26. Meschke, M.; Kemppinen, A.; Pekola, J.P. Accurate Coulomb blockade thermometry up to 60 kelvin. *Philos. Trans. A Math. Phys. Eng. Sci.* **2016**, *374*, 20150052. [CrossRef]
27. Ghirri, A.; Cornia, S.; Affronte, M. Microwave Photon Detectors Based on Semiconducting Double Quantum Dots. *Sensors* **2020**, *20*, 4010. [CrossRef]

28. Nishiguchi, K.; Ono, Y.; Fujiwara, A.; Yamaguchi, H.; Inokawa, H.; Takahashi, Y. Infrared detection with silicon nano-field-effect transistors. *Appl. Phys. Lett.* **2007**, *90*, 223108. [CrossRef]
29. Nishiguchi, K.; Clement, N.; Yamaguchi, T.; Fujiwara, A. Si nanowire ion-sensitive field-effect transistors with a shared floating gate. *Appl. Phys. Lett.* **2009**, *94*, 163106. [CrossRef]
30. Clément, N.; Nishiguchi, K.; Dufreche, J.F.; Guerin, D.; Fujiwara, A.; Vuillaume, D. A silicon nanowire ion-sensitive field-effect transistor with elementary charge sensitivity. *Appl. Phys. Lett.* **2011**, *98*, 014104. [CrossRef]
31. Schupp, F.J. Single-electron devices in silicon. *Mater. Sci. Technol.* **2017**, *33*, 944–962. [CrossRef]
32. Averin, D.V.; Odintsov, A.A.; Vyshenskii, S.V. Ultimate accuracy of single-electron dc current standards. *J. Appl. Phys.* **1993**, *73*, 1297–1308. [CrossRef]
33. Ohkura, K.; Kitade, T.; Nakajima, A. Cotunneling Current in Si Single-Electron Transistor Based on Multiple Islands. *Appl. Phys. Lett.* **2006**, *89*, 183520. [CrossRef]
34. Kobayashi, M.; Hiramoto, T. Experimental Study on Quantum Confinement Effects in Silicon Nanowire Metal-Oxide-Semiconductor Field-Effect Transistors and Single-Electron Transistors. *J. Appl. Phys.* **2008**, *103*, 053709. [CrossRef]
35. Kobayashi, M.; Hiramoto, T. Large Coulomb-Blockade Oscillations and Negative Differential Conductance in Silicon Single-Electron Transistors with [100]- and [110]-Directed Channels at Room Temperature. *Jpn. J. Appl. Phys.* **2007**, *46*, 24–27. [CrossRef]
36. Neophytou, N.; Paul, A.; Lundstrom, M.S.; Klimeck, G. Bandstructure Effects in Silicon Nanowire Electron Transport. *IEEE Trans. Electron Dev.* **2008**, *55*, 1286–1297. [CrossRef]
37. Yi, K.S.; Trivedi, K.; Floresca, H.C.; Yuk, H.; Hu, W.; Kim, M.J. Room-Temperature Quantum Confinement Effects in Transport Properties of Ultrathin Si Nanowire Field-Effect Transistors. *Nano Lett.* **2011**, *11*, 5465–5470. [CrossRef]
38. Kobayashi, M.; Saitoh, M.; Hiramoto, T. Large Temperature Dependence of Coulomb Blockade Oscillations in Room-Temperature-Operating Silicon Single-Hole Transistor. *Jpn. J. Appl. Phys.* **2006**, *45*, 6157–6161. [CrossRef]
39. Basu, T.S.; Wakayama, Y.; Hayakawa, R. Theoretical Insight into Quantum Transport Via Molecular Dots in a Vertical Tunnel Transistor. *ACS Appl. Electron. Mater.* **2021**, *3*, 973–978. [CrossRef]
40. Gotszalk, T.; Jóźwiak, G.; Radojewski, J.; Fröhlich, T.; Füssl, R.; Manske, E.; Holz, M.; Ivanov, T.; Ahmad, A.; Rangelow, I.W. Tip-Based Nano-Manufacturing and -Metrology. *J. Vac. Sci. Technol. B* **2019**, *37*, 030803. [CrossRef]
41. Liu, Q.; Zhao, J.; Guo, J.; Wu, R.; Liu, W.; Chen, Y.; Du, G.; Duan, H. Sub-5 nm Lithography with Single GeV Heavy Ions Using Inorganic Resist. *Nano Lett.* **2021**, *21*, 2390–2396. [CrossRef]
42. Li, L.; Liu, X.; Pal, S.; Wang, S.; Ober, C.K.; Giannelis, E.P. Extreme Ultraviolet Resist Materials for Sub-7 nm Patterning. *Chem. Soc. Rev.* **2017**, *46*, 4855–4866. [CrossRef]
43. Hiramoto, T. Five Nanometre CMOS Technology. *Nat. Electron.* **2019**, *2*, 557–558. [CrossRef]
44. Kwak, J.; Mishra, A.K.; Lee, J.; Lee, K.S.; Choi, C.; Maiti, S.; Kim, M.; Kim, J.K. Fabrication of Sub-3 nm Feature Size Based on Block Copolymer Self-Assembly for Next-Generation Nanolithography. *Macromolecules* **2017**, *50*, 6813–6818. [CrossRef]
45. Averin, D.V.; Nazarov, Y.V. Virtual Electron Diffusion During Quantum Tunneling of the Electric Charge. *Phys. Rev. Lett.* **1990**, *65*, 2446–2449. [CrossRef] [PubMed]
46. Takahashi, Y.; Horiguchi, S.; Fujiwara, A.; Murase, K. Co-Tunneling Current in Very Small Si Single-Electron Transistors. *Physica B* **1996**, *227*, 105–108. [CrossRef]
47. De Franceschi, S.; Sasaki, S.; Elzerman, J.M.; van der Wiel, W.G.; Tarucha, S.; Kouwenhoven, L.P. Electron Cotunneling in a Semiconductor Quantum Dot. *Phys. Rev. Lett.* **2001**, *86*, 878–881. [CrossRef]

nanomaterials

MDPI

Article

Deformation Mechanism of Depositing Amorphous Cu-Ta Alloy Film via Nanoindentation Test

Weibing Li [1,†], Xiao Wang [2,†], Xiaobin Feng [3,4,5,†], Yao Du [6], Xu Zhang [7], Yong Xie [8], Xiaoming Chen [9], Yang Lu [3,4,*] and Weidong Wang [2,3,*]

1 ZNDY of Ministerial Key Laboratory, Nanjing University of Science and Technology, Nanjing 210094, China; njustlwb@163.com
2 School of Mechano-Electronic Engineering, Xidian University, Xi'an 710071, China; wangxiao_9626@stu.xidian.edu.cn
3 CityU-Xidian Joint Laboratory of Micro/Nano-Manufacturing, Shenzhen 518057, China; xbfeng2-c@my.cityu.edu.hk
4 Department of Mechanical Engineering, City University of Hong Kong, Kowloon 999077, Hong Kong SAR, China
5 Hubei Key Laboratory of Theory and Application of Advanced Materials Mechanics, Wuhan University of Technology, Wuhan 430070, China
6 Department of Materials Science and Engineering, Northwestern University, Evanston, IL 60208, USA; yaodu2015@u.northwestern.edu
7 Department of Mechanical Engineering, Northwestern University, Evanston, IL 60208, USA; xuzhang2017@u.northwestern.edu
8 School of Advanced Materials and Nanotechnology, Xidian University, Xi'an 710071, China; yxie@xidian.edu.cn
9 Micro- and Nanotechnology Research Center, State Key Laboratory for Manufacturing Systems Engineering, Xi'an Jiaotong University, Xi'an 710049, China; xiaomingchen@xjtu.edu.cn
* Correspondence: yanglu@cityu.edu.hk (Y.L.); wangwd@mail.xidian.edu.cn (W.W.)
† These authors contributed equally to this work.

Abstract: As a representative of immiscible alloy systems, the Cu-Ta system was the research topic because of its potential application in industry, military and defense fields. In this study, an amorphous Cu-Ta alloy film was manufactured through magnetron sputter deposition, which was characterized by scanning electron microscopy (SEM) and transmission electron microscopy (TEM). Mechanical properties of Cu-Ta film were detected by the nanoindentation method, which show that the elastic modulus of Cu3.5Ta96.5 is 156.7 GPa, and the hardness is 14.4 GPa. The nanoindentation process was also simulated by molecular dynamic simulation to indicate the deformation mechanism during the load-unload stage. The simulation results show that the structure <0,2,8,4> and <0,2,8,5> Voronoi cells decreased by 0.1% at 50 Ps and then remained at this value during the nanoindentation process. In addition, the number of dislocations vary rapidly with the depth between indenter and surface. Based on the experimental and simulation results, the Voronoi structural changes and dislocation motions are the key reasons for the crystallization of amorphous alloys when loads are applied.

Keywords: Cu-Ta alloy; deform mechanism; magnetron sputtering; nanoindentation; molecular dynamics (MD)

Citation: Li, W.; Wang, X.; Feng, X.; Du, Y.; Zhang, X.; Xie, Y.; Chen, X.; Lu, Y.; Wang, W. Deformation Mechanism of Depositing Amorphous Cu-Ta Alloy Film via Nanoindentation Test. *Nanomaterials* **2022**, *12*, 1022. https://doi.org/10.3390/nano12061022

Academic Editors: Eva Pellicer and Jordi Sort

Received: 21 February 2022
Accepted: 16 March 2022
Published: 21 March 2022

1. Introduction

An important player in the alloy family, metallic glasses (MG) were first prepared by Duwenz et al. [1] in 1960. They obtained the first amorphous alloy chips in the Au-Si system by means of the liquid spray technique. Subsequently, Cohen et al. [2–4] proposed criteria for the formation of metallic glasses and verified the presence of glass transport in amorphous alloys in terms of specific heat theory. Since then, more amorphous alloys based on different metals have been prepared [5–8]. In addition, these amorphous alloy systems have excellent mechanical strength and high elastic deformation.

Copper-based amorphous alloys, a branch of MG, have excellent strength and elasticity capabilities. Much attention has been paid by many researchers to its preparation and mechanical characteristics. As early as 1989, Subrananlan et al. [9] introduced the phase diagram of the Cu-Ta binary alloy, elucidating that Ta is difficult to miscible in Cu. Several preparation methods have been used to produce Cu-Ta alloys, including ball milling, ion beam mixing, physical vapor deposition and magnetron sputtering [10–14]. Depending on the concentration of Ta, Cu-Ta systems can form crystal clusters or amorphous alloys. Ta films can act as diffusion barriers for Cu-based semiconductor substrates [15–17]. Francis et al. [18] investigated the growth of Ta films in (011) body-centered cubic (BCC) Ta and the deposition of CuTa films on (111) face-centered cubic (FCC) Cu. Frolov et al. [19] investigated the effect of Ta concentration on grain growth and mechanical strength in nanocrystalline (NC) Cu-Ta alloys. Li et al. [20] evaluated the axial tensile strength and elastic modulus of NC Cu-Ta alloys by molecular dynamics methods, and from their results it appears that the strength of the alloy varies with the grain size in the Hall-Petch effect. In particular, many researchers have given more attention to the interface between the amorphous state and NC. Gu et al. [21] prepared CuTa/Cu multilayers with different nanoscale Cu layer thicknesses and investigated the structure-related size effects by nanoindentation tests. Capenter et al. [22] prepared CuNi multilayer films by magnetron sputtering and in which the deformation mechanisms of the CLS (Constrained Layer Slip) model and grain boundary excitation were explained. According to their conclusions, increasing the concentration of nickel contributes to the contribution of CLS and slip to hardness.

Meanwhile, nanoindentation plays a key role in nanoscale mechanisms as an important characterization method. Ruestes et al. [23] used three different interatomic potentials for indentation simulations of tantalum disulfide and investigated the effects of indentation tip and penetration velocity by molecular dynamics (MD) methods. Zeng et al. [24] examined by nanoindentation methods Ta content of 25.6 to 96 at% of the mechanical properties of Cu-Ta amorphous films. They found that the incremental elastic modulus per Ta atom was 1.35 GPa and the incremental hardness per Ta atom was 0.205 GPa. Fang et al. [25] investigated the deformation behavior of high entropy bodies by nanoindentation MD simulations. In addition, many researchers have studied the mechanical properties of different materials by nanoindentation experiments and MD methods [26–35].

Although the mechanical properties at the interface between MG and NC in Cu-Ta alloys have been investigated in the above studies, the details of the deformation have not been precisely explained. In the present study, amorphous Cu-Ta alloy films were first prepared by magnetron sputtering. The deposition and nanoindentation processes were then simulated by MD methods. It was found that the Voronoi cell has a strong influence on the amorphization of the metal. Finally, the deformation behavior and the motivation of the interfacial structure are discussed.

2. Experimental

2.1. Magnetron Sputter Deposition

Cu-Ta films were synthesized by direct current (DC) magnetron sputter-deposition in an argon atmosphere (99.99%) using Ta (99.5% purity) and Cu (99.5% purity) targets. We used 1.8 cm × 1.8 cm silicon substrates. The substrates were ultrasonically cleaned in acetone and methanol bath for 10 min sequentially. The base pressure of the deposition chamber was about 1.3×10^{-5} Pa. Deposition was performed at an Ar pressure of 0.4 Pa, substrate bias of −100 V with the substrate held at ambient temperature. The target power was set at 100 W for Ta and 50 W for Cu, respectively. The Cu concentration in the film was controlled by adjusting the duty cycle of the shutter in front of the Cu target. The thickness of all films was about 1 μm, as determined by optical system.

2.2. Nanoindentation

Cu-Ta alloy films were indented using Hysitron 950 Triboindenter with a Berkovich tip (about 50 nm radius). The samples were cut to fit the microscope stubs and combine them using Ag paste. 2 × 3 array was selected on each sample. The maximum depth was about 100 nm less than 10% of films thickness to avoid substrate influence. The load resolution of the instrument is 30 nN and the displacement resolution is less than 0.2 nm. After the test, the area function was corrected on a standard sample of fused silicon.

2.3. Microscopy

A FEI QUANTA 450 SEM was used in ultrahigh resolution mode to investigate surface morphology and element distribution. The SEM was operated at 10 kV in high vacuum mode and the sample concentrations were detected by energy-dispersive spectroscopy (EDS). The lamellae were produced with a FEI Helios 600i dual-beam FIB/SEM, operated at 30 kV of Ga+ source and 10 kV of electron source. A JEM-2100 F TEM operating at 100 kV acceleration voltage was used to determine orientation relationship and microstructure of the amorphous Cu-Ta alloy film. As the Figure 1 shows, there are fewer particles on the surface of the Ta-rich film. Figure 1c shows more clearly that the thickness of the Cu-Ta alloy film is 1.087 μm.

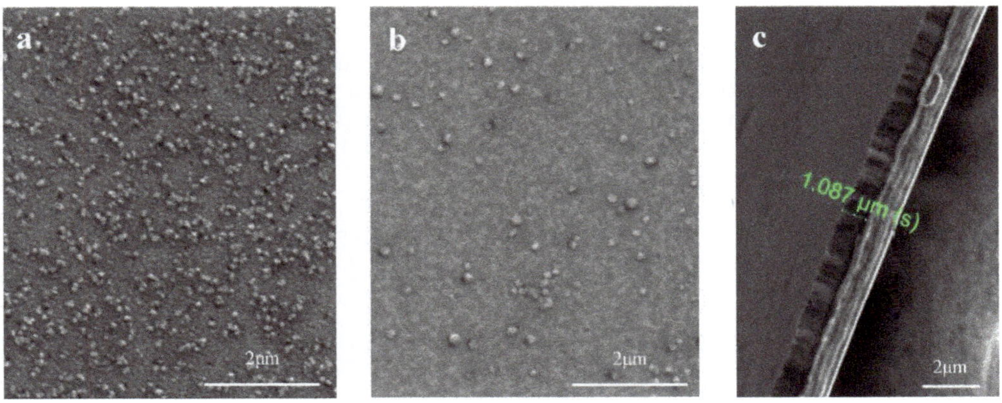

Figure 1. Surface morphology of Cu-Ta alloy films. (**a**) The surface of Cu84Ta16 (where 84 represents the atom concentration of Cu is 84%); (**b**) The surface of Cu3.5Ta96.5; (**c**) The thickness of Cu-Ta alloy film (where the sample is Cu84Ta16).

2.4. Theroy and Modeling

2.4.1. Elastic Modulus and Hardness

We can obtain the elastic modulus and hardness from the indentation test. The most popular method calculating modulus is O&P method [36]. This method fits the unload curve of indentation range from the bottom of unload points to 40% of unload curve. Details are shown below.

$$P_u = a(h - h_r)^m \quad (1)$$

where a is the fitting parameter, h_r is the residual depth, and the index m is the shape parameter of the indenter. Stiffness S can be determined by derivative load P_u as Equation (2) shows:

$$S = \left.\frac{dP_u}{dh}\right|_{h=h_{max}} = ma(h_{max} - h_r)^{m-1} \quad (2)$$

The reduced elastic modulus and true elastic modulus can be obtained from the next 2 equations, and A represent the contact area.

$$E_r = \frac{\sqrt{\pi}}{2} \cdot \frac{S}{\sqrt{A}} \quad (3)$$

$$\frac{1}{E_r} = \frac{1-v_s^2}{E_s} + \frac{1-v_i^2}{E_i} \quad (4)$$

where E_r is reduced elastic modulus, E_s and E_i represent elastic modulus of material and indenter, respectively. v_s and v_i are passion ratio of material and indenter. It is worth mentioning that some other authors [37,38] recommend using a model for all penetration depth/thickness ratio. These models are mostly based on weight function, depending on coating and substrate modulus. Thus, composite modulus is close to both initial moduli.

The O&P method is very popular in the indentation experiment results. However, in the atomic scale, this method may be failure [39]. In MD simulations of nanoindentation, most researchers derive the modulus of elasticity and nanohardness from Hertz theory [40]. According to Hertz, the starting point of the load-depth curve is the absolute elastic phase. In this phase, the load and depth satisfy the following relationship. It should be noted that the derivation of the following equations is only for molecular dynamics simulations of nanoindentation experiments in this work.

$$P_s = \frac{4}{3} E^* R^{\frac{1}{2}} h_t^{\frac{3}{2}} \quad (5)$$

Here, R is the radius of virtual sphere indenter, h_t is the total indent depth. E^* is the reduced elastic modulus (To distinguish from the real experimental reduced modulus E_r). Since in our model the radius of indenter can't far exceed the indent depth, according to Pathak et al. [41,42] the last equation should be modified as bellow:

$$P_s = \frac{4}{3} E^* \frac{\left(Rh_t - h_t^2/4\right)^{\frac{3}{2}}}{R} \quad (6)$$

According to the Equation (5), the reduced elastic modulus can be obtained by fitting the curve range from the start to the first dislocation emission.

Again, as mentioned above, the load-depth curve can be converted into a strain-stress curve according to Hertz theory. Details are given below.

$$\sigma = \frac{P}{\pi\left(Rh_t - h_t^2/4\right)} \quad (7)$$

$$\varepsilon = \frac{\left(Rh_t - h_t^2/4\right)^{\frac{1}{2}}}{R} \cdot \frac{4}{3\pi} \quad (8)$$

From the two equations above, we can easily obtain the elastic modulus in the linear phase.

2.4.2. Simulation Model

In order to explore the deformation mechanism of Cu-Ta alloy film, the simulated sputter depositions and nanoindentation were executed by the Large-scale Atomistic/Molecular Massively Parallel Simulation (LAMMPS) [43]. The embedded atom model (EAM) developed by Foiles [44] and Ravelo [45] are used to describe the force on Cu-Cu pair and Ta-Ta pair, respectively. The angular-dependent interatomic potential (ADP) [46] was set to describe the function of Cu-Ta pair.

As with the sputter deposition experiments, a copper substrate was set up at the bottom of the system. The substrate is divided into three zones, including a fixed zone, a thermally controlled zone and a free zone. In the thermally controlled area a Langevin thermostat is set to maintain the temperature at 300 K. The injected atoms leave the incidence plane and reach the free zone. Two important sections are set up at the top of the simulation box. The incident layer is located one lattice distance from the entire top of the box. As shown in Figure 2, two virtual walls are set up to prevent the emitted atoms from leaving the system. The frequency of the emitted atoms was controlled so that the

Ta content approximately met that of the experimentally prepared amorphous film. The boundaries are set to be periodic in the x and y directions and contracted in the z direction.

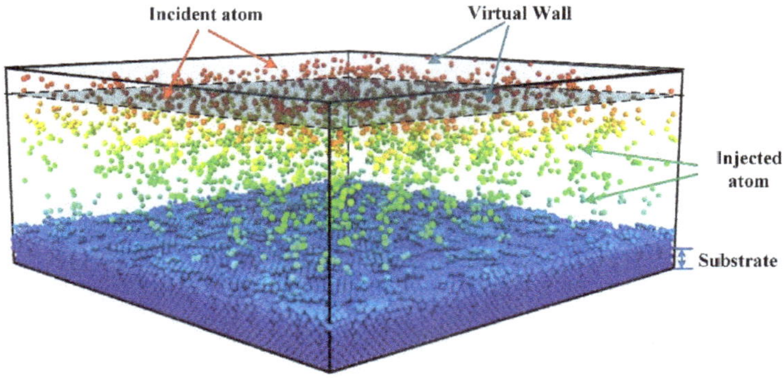

Figure 2. Physical model of sputter deposition. The bottom blue atoms represent Cu atoms. The region red atoms located is the birthplace of incident atom. The atoms in the middle of the simulation box are the emitted atoms before they reach the free surface.

Next, the output of the deposition simulation can be used as the initial model for the indentation. The indentation model is shown in Figure 3. A virtual sphere indenter with a radius of 3 nm is set at the top center of the deposition output model. The distance between the indenter edge and the top of the deposition output model is 1 Å. The speed of the indenter downwards is set to 2 Å/s. At the start of the simulation, energy minimization and relaxation were performed to bring the system to an equilibrium state with minimum energy. During the indentation test, the model boundary was set to be periodic in the X and Y directions and contracted in the Z direction.

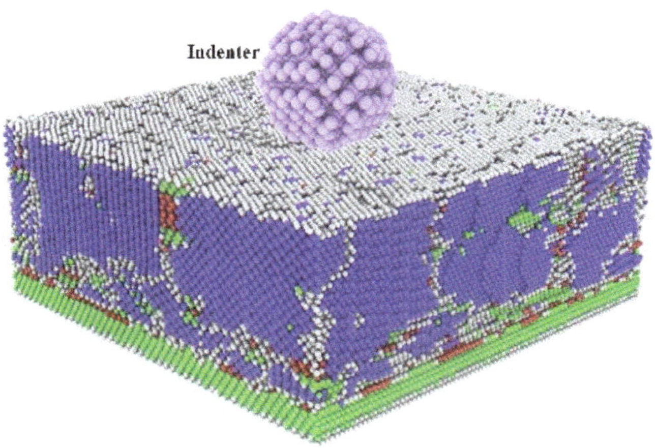

Figure 3. Physical model of nanoindentation. The top purple sphere represents the virtual indenter. The bottom block is output model of sputter deposition, which color coding with common neighbor analysis (CNA) (Reproduced with permission from [47]. Journal of Physical Chemistry, 1987). The green and blue atoms represent FCC and BCC structure type, respectively.

3. Results and Discussion

In the results and discussion, it is divided into two parts: experiments and simulations. The experimental part shows the structure of our prepared amorphous Ta-rich Cu-Ta films and the results of nanoindentation experiments. In addition, the focus is on the analysis of the results of the atomic simulations to obtain the deformation mechanism of the deposited films during the indentation process.

3.1. Experimental Results

The internal structure of the prepared Cu3.5Ta96.5 films was characterized using a JEM-2100 F TEM. First, a 1 μm slice was cut out at the interface between the film and the substrate using Focused Ion beam (FIB), and the thickness of both the substrate and the film was about 500 nm. As can be seen in Figure 4a,b, there is no apparent array arrangement inside the film, which is in a distinctly amorphous state. To further verify this, the red circle in (a) was characterized by selected area electron diffraction and the image was found to be a distinct aperture as shown in Figure 4d, which further confirms that the prepared film is amorphous. Figure 4c shows the EDS spectrum of the atomic composition of the film, confirming the Ta content.

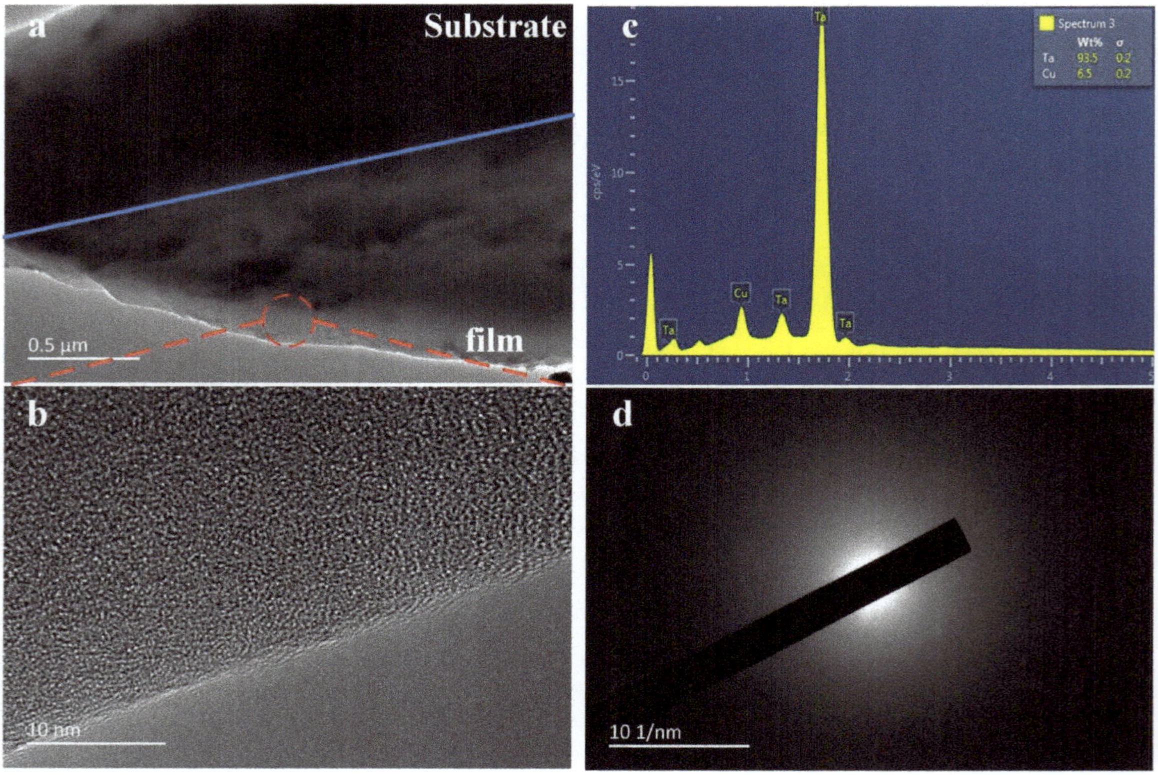

Figure 4. TEM characterization of Cu3.5Ta96.5 film. (**a**) The inner structure of Cu-Ta alloy film; (**b**) A partial enlargement of red circle in (**a**), obviously amorphous state; (**c**) The EDS spectrum of Cu3.5Ta96.5; (**d**) Selected area diffraction patterns sampling Cu3.5Ta96.5.

The results of the nanoindentation experiments are shown in Figure 5, which can be illustrated by the complete load-displacement curve. The hardness and elastic modulus are 14.4 GPa and 156.7 GPa, respectively.

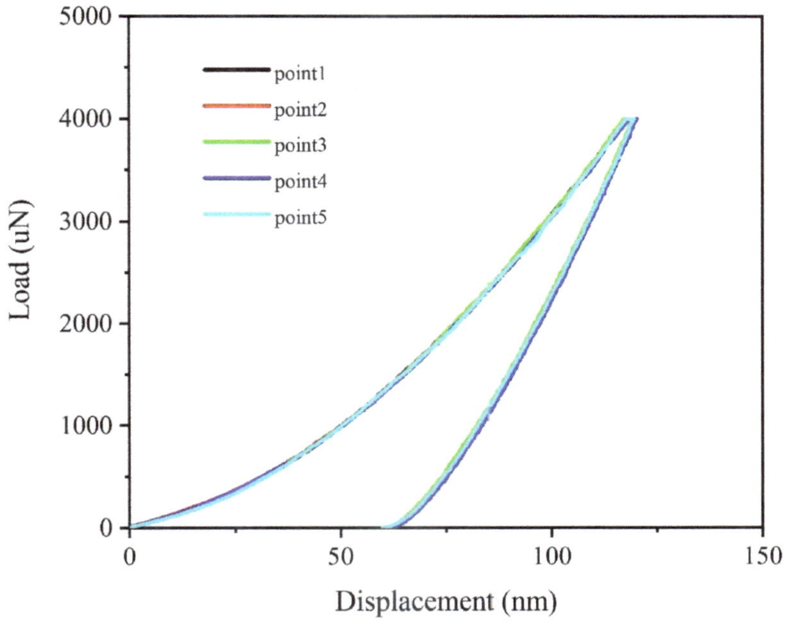

Figure 5. Load-displacement curves of nanoindentation experiment for Ta-rich film. Five points in sample were detected and the elastic modulus and hardness were obtained from these curves.

3.2. Atomic Simulation Results

In the atomic simulation results, we discuss the results of the simulated deposition and analyze the changes of the Voronoi cell during the deposition process. In addition, we simulated the nanoindentation process of the deposited film, and as a complement to the experimental results dynamically analyzed the motion behavior of the internal atoms.

3.2.1. Deposition Properties

In order to observe the activity of atoms and the changes in internal structure during deposition simulation, we used the Voronoi index and radial distribution function (RDF) to characterize the changes in the crystal structure of the model during deposition. At the same time, we characterized the internal structure and crystal properties of the prepared Cu-Ta films by High Resolution Transmission Electron Microscope (HRTEM). As shown in Figure 4, it can be seen that the Cu3.5Ta96.5 films obtained by sputtering have an amorphous structure. This is also well supported by the results of the atomic simulations. Figure 6 shows the statistics of the change in the number of the first five Voronoi structures in the whole model at different sputtering times. It can be seen that the proportion of <0,12,0,0> cells keep decreasing as the deposited atoms are continuously injected. This is because the <0,12,0,0> units correspond to FCC structures centered on copper atoms. Before deposition starts, single-crystal copper is used as a substrate and <0,0,12,0> also occupies the largest proportion. In addition, <0,5,4,0> is a common defect structure to be observed in single crystal copper. At the end of deposition, the proportion of <0,12,0,0> reaches a minimum of 6.3%. In addition, the proportion of <0,6,0,8> centered on tantalum atoms increases from 0 to 55.8% with the implantation of atoms, making it the largest Voronoi unit in the whole model. It was shown that the degree of amorphization during metal vitrification is associated with Voronoi polyhedral with a particular structure [48]. In our simulations, the variation of <0,6,0,8> polyhedral clearly affects the vitrification process of Cu-Ta alloy films.

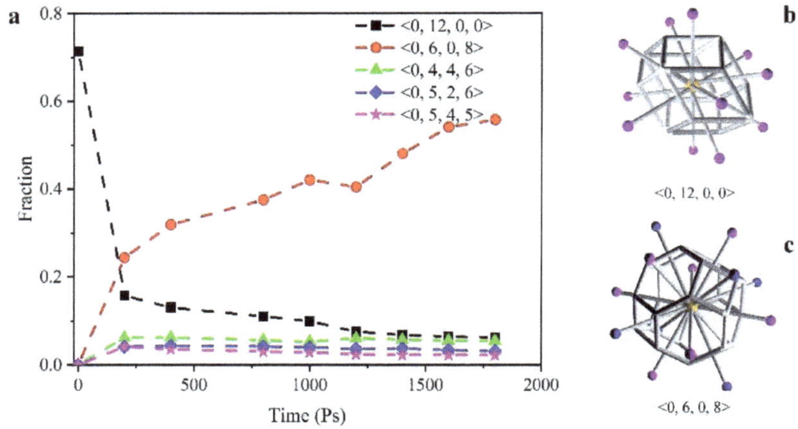

Figure 6. The change of Voronoi polyhedron with the increase of deposited atoms. (**a**) <0,12,0,0>, <0,6,0,8>, <0,4,4,6>, <0,5,2,6> and <0,5,4,5> fractions with the deposit time; (**b**) the structure of <0,12,0,0> Voronoi cell; (**c**) the structure of <0,6,0,8> Voronoi cell. (For structural clarity, the central atom is indicated in yellow and the other atoms in purple or blue).

To demonstrate the amorphous structure of the films from macroscopic statistics of the atomic arrangement. We use RDF to describe the atomic distribution inside the film after different deposition times. Combined with the distribution of Voronoi cells at the corresponding time in Figure 7, the atomic composition inside the film at this time can be analyzed. Figure 7d shows the atomic distribution at 1800 Ps deposition. The smaller probability of finding a Cu atom at the same distance through the RDF of the Cu-Cu pair corresponds to the smallest proportion of Voronoi cells centered on Cu atoms. g(r) fluctuations between Cu-Ta and Ta-Ta are smaller and the amorphous character is more pronounced.

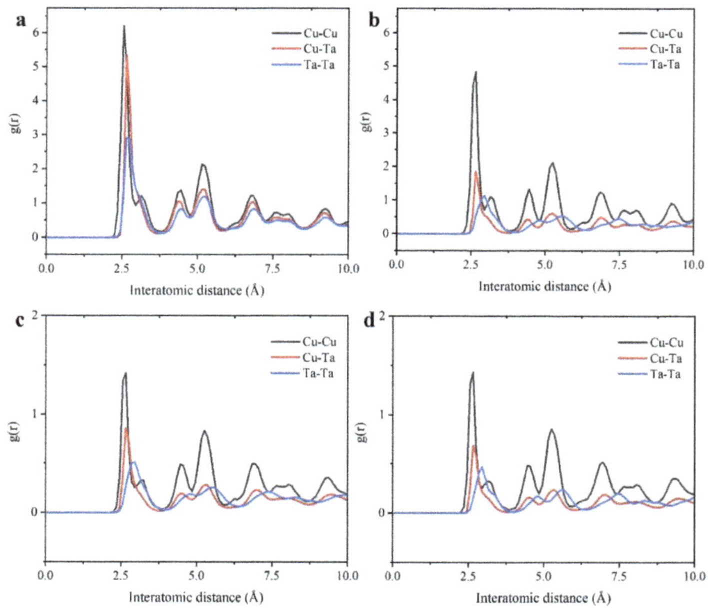

Figure 7. The RDF profile of deposit Cu-Ta model at different time. (**a**) 200 Ps; (**b**) 800 Ps; (**c**) 1200 Ps; (**d**) 1800 Ps.

3.2.2. Indentation Properties

The elastic modulus and nanohardness in the atomic simulations were obtained by Hertz theory. Figure 8a shows the load-displacement curve for the entire indentation process with a maximum indentation load of 0.51 μN. the maximum indentation depth is 2.8 nm. the residual depth is 2 nm. a short distance at the beginning of the loading phase is linearly increasing, indicating that the film is in the fully elastic deformation phase. The elastic modulus can be obtained by fitting a curve for this stage. Figure 8c shows the results of the Hertz fit for this phase, with a reduced elastic modulus of 111 GPa. In addition, by using Equations (8) and (9), the load-displacement curve can be converted to a stress-strain curve as show in Figure 8d, and then the curve for the linear phase can be fitted to obtain the reduced elastic modulus, which gives a result of 109 GPa, in agreement with (c). However, it differs from the experimental value (156.7 GPa) by 30%. After analysis, this difference may be caused by the amplification of the "substrate effect". During the indentation process, the simulated nanoscale film system has only a very small thickness (8 nm). If we use h (depth of indentation)/t (film thickness) as the normalized evaluation criterion, the experimental value is less than 10%, while the simulated value reaches 30%. The "substrate effect" at the atomic level would have a more pronounced effect. Therefore, the modulus values obtained by scaling up the simulated system will be accurate. Optimizing the simulated deposition times will be closer to the experimental values. When the strain is less than 0.1, the film is in the fully elastic deformation phase, where the strain also corresponds to the initial emission position of the dislocation. Figure 8b shows the load variation corresponding to different loading times. The three sub-images correspond to the depth of the indentation at different times and are color coding with indentation depth.

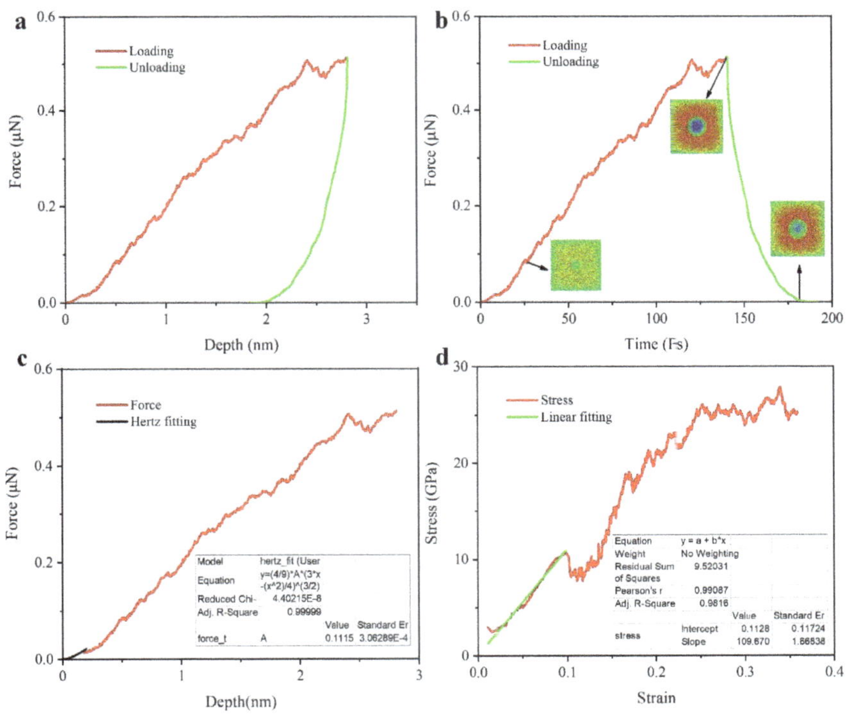

Figure 8. Indentation load curve. (**a**) Load-depth curve of simulated Ta-rich deposition output model; (**b**) Time-force of nanoindentation, three sub-images relate to the indentation depth at that time; (**c**) Load-depth curve of loading stage and Hertz fitting; (**d**) Strain-stress curve transfer from load-depth curve.

The nanohardness is calculated using the following formula.

$$H = \frac{P}{A} \tag{9}$$

where P is the max load, A is project area.

The nanohardness of the Cu3.5Ta96.5 film obtained was 12.6 GPa, a difference of 12% from the experimental value of 14.4 GPa. The H/E (hardness-to-elastic modulus ratio) is a key indicator of the film's wear resistance. The lower the value, the lower the abrasion resistance and the higher the wear resistance. The simulated Cu-Ta alloy film had an H/E of 0.114, while the prepared film had an H/E of 0.092. The results of the simulated system were also relatively close to the experimental value, which could reflect the high wear resistance of the Ta-rich film.

3.2.3. Deformation Mechanism

Figure 9 shows the change in depth of the film in the Z-direction during dynamic loading and unloading. The color of the atoms is determined by the depth in the indentation direction. The indentation depth increases with increasing load and reaches a maximum at an indentation depth of 2.8 nm. The corresponding picture is shown in (e). The force unloading process is then carried out. As the load decreases, the film bounces back. However, due to the effect of plastic deformation, the indentation depth no longer decreases but remains at this value when the indentation depth is 2.0 nm. The indentation depth at this point is also the residual indentation depth h_r.

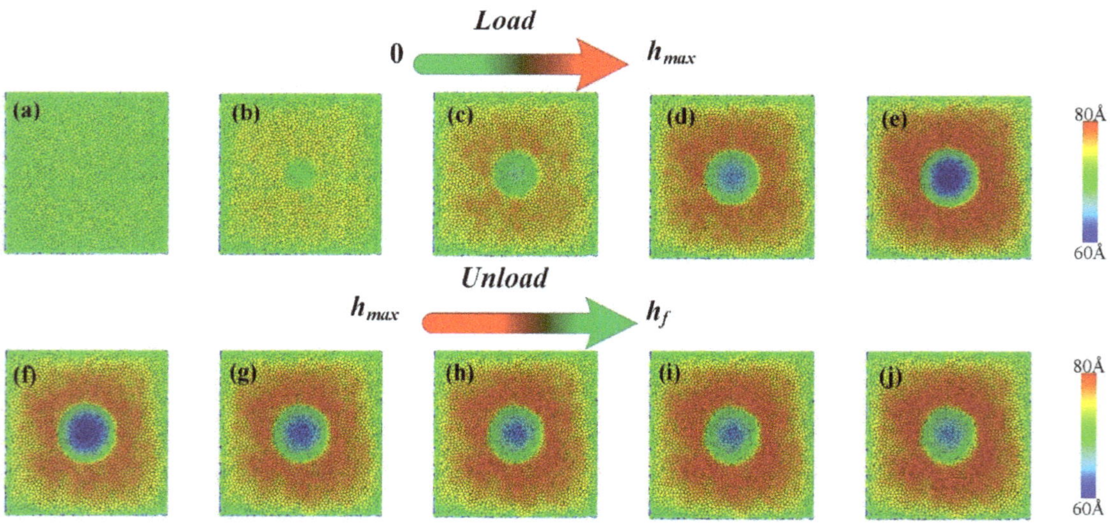

Figure 9. Dynamic loading and unloading process. (**a**–**e**) represent the process of load increase, (**f**–**j**) represent the process of unloading. The atoms are colored according to the depth in the loading direction (Z direction).

To show more clearly the displacement of atoms inside the film during the indentation process. Figure 10 shows the displacement of atoms in the XY-plane for different indentation depths using the slice method. The color of the atoms is determined by the size of the displacement. Figure 10c corresponds to a maximum indentation depth of 2.8 nm and (d) to the state of the atoms inside the model after unloading. At the same time, there is a clear "indentation" during loading, as shown by the red outline in Figure 10c. As the indenter penetrates into the film, the surrounding surface of the film bulges, forming a small convex surface. This is the "Pile-up" phenomenon, where it is known that the

Oliver-pharr method is no longer applicable, which is why we use the Hertz model in simulation. The distribution of atoms during the indentation process shows that the atoms on the original free surface have two trends of movement: one is the part of the atoms that moves downward with the indenter, i.e., the red part, and the other is the green atoms that occur in the Pile-up region. The green atoms are bulged by the joint action of the indenter and the surrounding atoms to form a convex package. After the unloading is completed, it can be seen from Figure 10d that some of the red atoms diffuse into green or yellow atoms due to the elastic effect. There are also light blue atoms in the lower part of the entire hemisphere due to plastic deformation, and according to their distribution, it can be approximated that if the indentation load continues to increase, these areas will become sprouts for crack expansion.

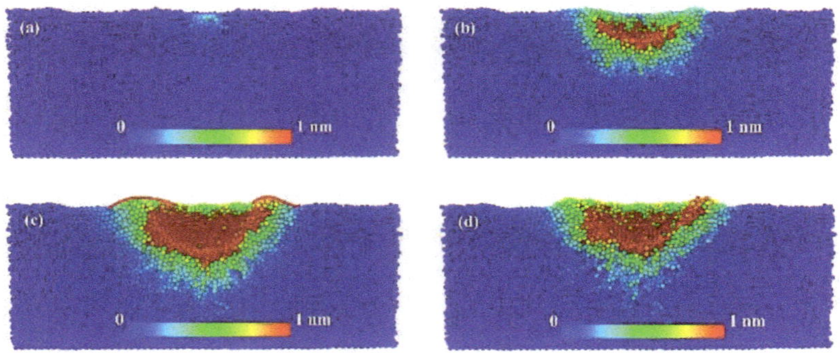

Figure 10. The atomic displacement of the Cu-Ta alloy film during nanoindentation process are the cross-section snapshots of XY plane at the indentation depths of (**a**) 0.2, (**b**) 1.2, (**c**) 2.8 and (**d**) 2 nm, respectively.

Figure 11 shows the displacement of the atoms during the indentation process, we then used the dislocation analysis (DXA) method [49] to calculate the dislocations in the XY plane at a distance of 1.5 nm from the indentation surface. At the same time, the variation of the Voronoi cells was accounted for. In general, the variation of the Voronoi cells during the indentation process is not very significant and most of the structures remain at a stable value. Only the Voronoi cells of <0,2,8,4> and <0,2,8,5> decreased by 0.1% at 50 Ps and then remained at this value. The change in the number of dislocations allows a better analysis of the changes in the internal structure of the film. It can be seen that there are still many defects in the initial state model. As the indenter penetrates deeper into the substrate, the number of in-plane dislocations changes significantly. From 80 Ps onwards, the number of dislocations increases significantly, which also corresponds to the end of the fitted curve in Figure 8c. In our calculations, we consider this moment as the time of initial dislocation emission. Subsequently, the number of dislocations at 160 Ps decreases rapidly. At this moment, the indenter starts to leave the substrate surface and the internal dislocations decrease. However, a few dislocations still occur during the recovery of the film. The number of dislocations remains constant until the indenter leaves the substrate completely.

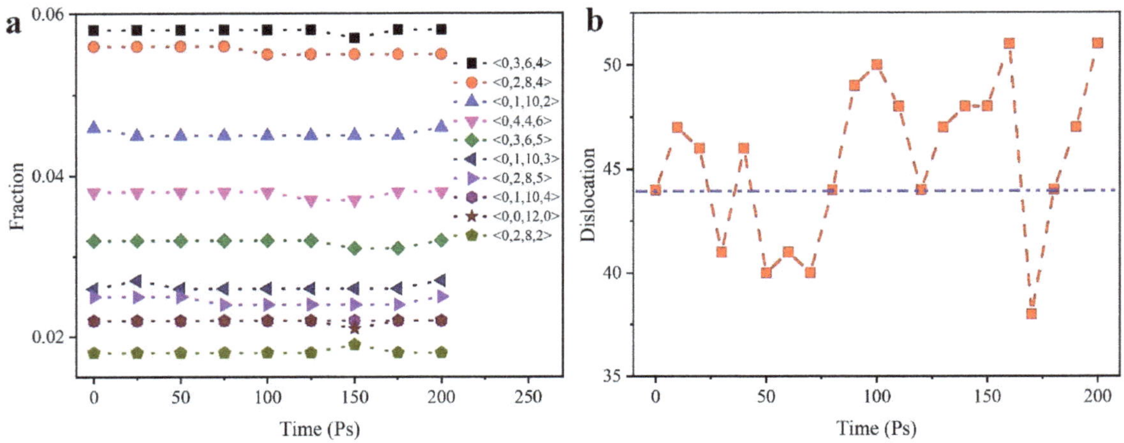

Figure 11. Changes in the internal structure of the model during indentation. (**a**) Voronoi cell statistics at different indent time; (**b**) Dislocation count in XY plane at different indent time.

4. Conclusions

Firstly, amorphous Cu-Ta alloy films, namely Cu3.5Ta96.5, were prepared by magnetron sputtering. The surface morphology and internal atomic structure of the films were characterized by SEM and TEM. The TEM results showed that the prepared Cu-Ta alloy films were amorphous, but some nano-clusters were present in them. The elastic modulus and hardness of the film were obtained by nanoindentation. the hardness and modulus of Cu3.5Ta96.5 were 14.4 GPa and 156.7 GPa, respectively. it is evident from the indentation experimental results that the elastic modulus and hardness increase with increasing Ta concentration. Then, both atomic deposition and nanoindentation were simulated using the MD method. The results of the atomic deposition simulations were compared with the TEM characterization results to further demonstrate the amorphous structure of the films. By analyzing the variation of Voronoi units, the main structure affecting the amorphization of the Cu-Ta system is probably the <0,6,0,8> structure centered on the Ta atom. The elastic modulus and nanohardness of the simulated nanoindentation were calculated using a Hertz fitting method. The results obtained were compared with the experimental results and the numerical differences were found to be within 30%. This may be due to the "substrate effect" of the single crystal copper substrate during the simulation and the size of the simulated system. Finally, the specific deformation behavior of the Cu-Ta alloy films was shown by analyzing the changes in atomic displacements during nanoindentation and the number of dislocations in the XY plane. The results show that the Voronoi unit and the number of dislocations per unit area dominates the deformation behavior of the material together.

Author Contributions: Writing—original draft preparation, W.L. and X.W.; writing—review and editing, W.W. and X.W.; methodology, X.F., Y.D. and X.Z.; software, X.W.; formal analysis, X.W.; investigation, X.C. and Y.X.; data curation, X.W.; supervision, W.W. and Y.L.; funding acquisition, W.W. All authors have read and agreed to the published version of the manuscript.

Funding: This research was funded by the Key Research and Development Program of Shaanxi (Program No. 2021GY-277 and 2022NY-211), and the Science and Technology on Vacuum Technology and Physics (No. HTKJ2019KL510007).

Data Availability Statement: Not applicable.

Conflicts of Interest: The authors declare no conflict of interest.

References

1. Jun, W.K.; Willens, R.H.; Duwez, P. Non-crystalline structure in solidfied gold-silicon alloys. *Nature* **1960**, *187*, 869–870.
2. Turnbull, C. Composition requirements for glass formation in metallic and ionic systems. *Nature* **1961**, *189*, 131–132.
3. Chen, H.S.; Turnbull, D. Thermal evidence of a glass transition in gold-silicon-germanium alloy. *Appl. Phys. Lett.* **1967**, *10*, 284–286. [CrossRef]
4. Drehman, A.J.; Greer, A.L.; Turnbull, D. Bulk formation of a metallic glass: $Pd_{40}Ni_{40}P_{20}$. *Appl. Phys. Lett.* **1982**, *41*, 716–717. [CrossRef]
5. Wang, A.; Zhao, C.; He, A.; Men, H.; Chang, C.; Wang, X. Composition design of high B-s Fe-based amorphous alloys with good amorphous-forming ability. *J. Alloy. Compd.* **2016**, *656*, 729–734. [CrossRef]
6. Zhao, K.; Jiao, W.; Ma, J.; Gao, X.Q.; Wang, W.H. Formation and properties of Strontium-based bulk metallic glasses with ultralow glass transition temperature. *J. Mater. Res.* **2012**, *27*, 2593–2600. [CrossRef]
7. Wang, Z.; Yu, H.-B.; Wen, P.; Bai, H.Y.; Wang, W.H. Pronounced slow β-relaxation in La-based bulk metallic glasses. *J. Phys. Condens. Matter.* **2011**, *23*, 142202. [CrossRef]
8. Takcuchi, A.; Inoue, A. Development of metallic glasses by semi-empirical calculation method. *J. Metastable Nanocrystalline Mater.* **2005**, *24–25*, 283–286. [CrossRef]
9. Subrananlan, P.R.; Laughlin, D.E. The Cu-Ta (copper-tantalum) system. *Bull. Alloy. Phase Diagr.* **1989**, *10*, 652–686. [CrossRef]
10. Hashibon, A.; Lozovoi, A.Y.; Mishin, Y.; Elsässer, C.; Gumbsch, P. Interatomic potential for the Cu-Ta system and its application to surface wetting and dewetting. *Phys. Rev. B* **2008**, *77*, 4131. [CrossRef]
11. Zhu, X.; Li, Y.; Zhang, G.; Zhu, S. On strain-localized damage in nanoscale Cu-Ta multilayers on a flexible substrate. *Mater. Sci. Eng. A* **2010**, *527*, 3279–3283. [CrossRef]
12. Gong, H.R.; Kong, L.T.; Lai, W.S.; Liu, B.X. Atomistic modeling of solid-state amorphization in an immiscible Cu-Ta system. *Phys. Rev. B* **2002**, *66*, 4204. [CrossRef]
13. Gong, H.R.; Liu, B.X. Interface stability and solid-state amorphization in an immiscible Cu-Ta system. *Appl. Phys. Lett.* **2003**, *83*, 4515. [CrossRef]
14. Darling, K.A.; Roberts, A.J.; Mishin, Y.; Mathaudhu, S.N.; Kecskes, L.J. Grain size stabilization of nanocrystalline copper at high temperatures by alloying with tantalum. *J. Alloy. Compd.* **2013**, *573*, 142–150. [CrossRef]
15. Ono, H.; Nakano, T.; Ohta, T. Diffusion barrier effects of transition metals for Cu/M/Si multilayers (M=Cr, Ti, Nb, Mo, Ta, W). *Appl. Phys. Lett.* **1994**, *64*, 1511–1513. [CrossRef]
16. Latt, K.M.; Lee, Y.K.; Osipowicz, T.; Park, H.S. Interfacial reactions and failure mechanism of Cu/Ta/$SaiC_2$/Si multilayer structure in thermal annealing. *Mater. Sci. Eng. B* **2002**, *94*, 111–120. [CrossRef]
17. Li, S.; Dong, Z.L.; Latt, K.M.; Park, H.S.; White, T. Formation of Cu diffusion channels in Ta layer of a Cu/Ta/SiO_2/Si structure. *Appl. Phys. Lett.* **2002**, *80*, 2296–2298.
18. Francis, M.F.; Neurock, M.N.; Zhou, X.W.; Quan, J.J.; Wadley HN, G.; Webb, E.B., III. Atomic assembly of Cu/Ta multilayers: Surface roughness, grain structure, misfit dislocation and amorphization. *J. Appl. Phys.* **2008**, *104*, 4310. [CrossRef]
19. Frolov, T.; Darling, K.A.; Kecskes, L.J.; Mishin, Y. Stablization and strengthening of nanocrystalline copper by alloying with tantalum. *Acta Mater.* **2012**, *60*, 2158–2168. [CrossRef]
20. Li, W.B.; Wang, X.; Gao, L.B.; Lu, Y.; Wang, W.D. Atomic Study on Tension Behaviors of Sub-10 nm Nano-polycrystalline Cu-Ta Alloy. *Materials* **2019**, *12*, 3913. [CrossRef]
21. Gu, C.; Wang, F.; Huang, P.; Xu, W.K.; Lu, T.J. Structure-dependent size effects in CuTa/Cu nanolaminates. *Mater. Sci. Eng. A* **2016**, *658*, 381–388. [CrossRef]
22. Carpenter, J.S.; Misra, A.; Anderson, P.M. Achieving maximum hardness in semi-coherent multilayer thin films with unequal layer thickness. *Acta Mater.* **2012**, *60*, 2625–2636. [CrossRef]
23. Ruestes, C.J.; Stukowski, A.; Tang, Y.; Tramontina, D.R.; Erhart, P.; Remington, B.A.; Bringa, E.M. Atomistic simulation of tantalum nanoindentation: Effects of indenter diameter, penetration velocity, and interatomic potentials on defect mechanisms and evolution. *Mater. Sci. Eng. A* **2014**, *613*, 390–403. [CrossRef]
24. Zeng, F.; Gao, Y.; Li, L.; Li, D.; Pan, F. Elastic modulus and hardness of Cu-Ta amorphous films. *J Alloy. Compd.* **2005**, *389*, 75–79. [CrossRef]
25. Fang, Q.; Yi, M.; Li, J.; Liu, B.; Huang, Z. Deformation behaviors of $Cu_{29}Zr_2Ti_{15}Al_5Ni_{19}$ high entropy bulk metallic glass during nanoindentation. *Appl. Surf. Sci.* **2018**, *443*, 122–130. [CrossRef]
26. Lee, Y.; Park, J.Y.; Kim, S.Y.; Jun, S.; Im, S. Atomistic simulations of incipient plasticity under Al(111) nanoindentation. *Mech. Mater.* **2005**, *37*, 1035–1048. [CrossRef]
27. Lilleodden, E.T.; Zimmerman, J.A.; Foiles, S.M.; Nix, W.D. Atomistic simulations of elastic deformation and dislocation nucleation during nanoindentation. *J. Mech. Phys. Solids* **2003**, *51*, 901–920. [CrossRef]
28. Mann, A.B.; D van Heerden Pethica, J.B.; Bowes, P.; Weihs, T.P. Contact resistance and phase transformations during nanoindentation of silicon. *Philos. Mag. A* **2002**, *82*, 1921–1929. [CrossRef]
29. Komvopoulos, K.; Yang, J. Dynamic analysis of single and cyclic indentation of an elastic–plastic multi-layered medium by a rigid fractal surface. *J. Mech. Phys. Solids* **2006**, *54*, 927–950. [CrossRef]
30. Lu, L.; Huang, C.; Pi, W.; Xiang, H.; Gao, F.; Fu, T.; Peng, X. Molecular dynamics simulation of effects of interface imperfections and modulation periods on Cu/Ta multilayers. *Comput. Mater. Sci.* **2018**, *143*, 63–70. [CrossRef]

31. Zhang, H.; Song, H.; Kang, Y.; Huang, G.; Qu, C. Experimental Analysis on Deformation Evolution and Crack Propagation of Rock Under Cyclic Indentation. *Rock Mech. Rock Eng.* **2013**, *46*, 1053–1059. [CrossRef]
32. Gerberich, W.W.; Stauffer, D.D.; Beaber, A.R.; Mook, W.M. Anomalous Plastic Deformation at Surfaces: Nanoindentation of Gold Single Crystals. *Phys. Rev. B* **1997**, *55*, 16057–16060.
33. Zarudi, I.; Zhang, L.C.; Swain, M.V. Behavior of monocrystalline silicon under cyclic microindentations with a spherical indenter. *Appl. Phys. Lett.* **2003**, *82*, 1027–1029. [CrossRef]
34. Cheng, Y.Q.; Ma, E. Atomic-level structure and structure–property relationship in metallic glasses. *Prog. Mater. Sci.* **2011**, *56*, 379–473. [CrossRef]
35. Maruschak, P.O.; Okipnyi, I.B.; Poberezhnyi, L.Y.; Maruschak, E.V. Study of heat-resistant steel strain hardening by indentation. *Metallurgist* **2013**, *56*, 946–951. [CrossRef]
36. Oliver, W.C.; Pharr, G.M. An improved technique for determining hardness and elastic modulus using load and displacement sensing indentation experiments. *J. Mater. Res.* **1992**, *7*, 1564–1583. [CrossRef]
37. Saha, R.; Nix, W.D. Effects of the substrate on the determination of thin film mechanical properties by nanoindentation. *Acta Mater.* **2002**, *50*, 23–38. [CrossRef]
38. Antunes, J.; Fernandes, J.; Sakharova, N.; Oliveira, M.; Menezes, L. On the determination of the Young's modulus of thin films using indentation tests. *Int. J. Solids Struct.* **2007**, *44*, 8313–8334. [CrossRef]
39. Jeng, Y.R.; Tan, C.M. Investigation into the nanoindentation size effect using static atomistic simulations. *Appl. Phys. Lett.* **2006**, *89*, 251901. [CrossRef]
40. Hertz, H. On the contact of elastic solids. Z. Reine Angew. *Mathematik* **1881**, *92*, 156–171.
41. Kontomaris, S.-V. The Hertz model in AFM nanoindentation experiments: Applications in biological samples and biomaterials. *Micro Nanosyst.* **2018**, *10*, 11–22. [CrossRef]
42. Pathak, S.; Kalidindi, S.R. Spherical nanoindentation stress–strain curves. *Mater. Sci. Eng. R Rep.* **2015**, *91*, 1–36. [CrossRef]
43. Plimpton, S. Fast Parallel Algorithms for Short-Range Molecular Dynamics. *J. Comp. Phys.* **1995**, *117*, 1–19. [CrossRef]
44. Foiles, S.M.; Baskes, M.I.; Daw, M.S. Embedded-atom-method functions for the fcc metals Cu, Ag, Au, Ni, Pd, Pt and their alloys. *Phys. Rev. B* **1986**, *33*, 7983–7991. [CrossRef]
45. Ravelo, R.; Germann, T.C.; Guerrero, O.; An, Q.; Holian, B.L. Shock-induced plasticity in tantalum single crystals: Interatomic potentials and large-scale molecular-dynamics simulations. *Phys. Rev. B* **2013**, *88*, 134101. [CrossRef]
46. Pun, G.P.; Darling, K.; Kecskes, L.; Mishin, Y. Angular-dependent interatomic potential for the Cu-Ta system and its application to structural stability of nano-crystalline alloys. *Acta Mater.* **2015**, *100*, 377–391.
47. Honeycutt, J.D.; Andersen, H.C. Molecular dynamics study of melting and freezing of small Lennard-Jones clusters. *J. Phys. Chem.* **1987**, *91*, 4950–4963. [CrossRef]
48. Sheng, H.W.; Luo, W.K.; Alamgir, F.; Bai, J.M.; Ma, E. Atomic packing and short-to-medium-range order in metallic glasses. *Nature* **2006**, *439*, 419–425. [CrossRef]
49. Stukowski, A.; Bulatov, V.V.; Arsenlis, A. Automated identification and indexing of dislocations in crystal interfaces. *Model. Simul. Mater. Sci. Eng.* **2012**, *20*, 085007. [CrossRef]

nanomaterials

MDPI

Article

Synchronous Defect and Interface Engineering of $NiMoO_4$ Nanowire Arrays for High-Performance Supercapacitors

Pengcheng Wang [1], Xinying Ding [1], Rongjie Zhe [1], Ting Zhu [1,2], Chen Qing [1,2,*], Yingkai Liu [1] and Hong-En Wang [1,2,*]

1 Yunnan Key Laboratory of Optoelectronic Information Technology, College of Physics and Electronics Information, Yunnan Normal University, Kunming 650500, China; pcwang0312@163.com (P.W.); xinyingding1998@163.com (X.D.); rongjiezhe@163.com (R.Z.); zhut0002@ynnu.edu.cn (T.Z.); ykliu@ynnu.edu.cn (Y.L.)

2 Key Laboratory of Advanced Technique & Preparation for Renewable Energy Materials, Ministry of Education, Yunnan Normal University, Kunming 650500, China

* Correspondence: qingchen1@126.com (C.Q.); hongenwang@whut.edu.cn (H.-E.W.)

Abstract: Developing high-performance electrode materials is in high demand for the development of supercapacitors. Herein, defect and interface engineering has been simultaneously realized in $NiMoO_4$ nanowire arrays (NWAs) using a simple sucrose coating followed by an annealing process. The resultant hierarchical oxygen-deficient $NiMoO_4$@C NWAs (denoted as "$NiMoO_{4-x}$@C") are grown directly on conductive ferronickel foam substrates. This composite affords direct electrical contact with the substrates and directional electron transport, as well as short ionic diffusion pathways. Furthermore, the coating of the amorphous carbon shell and the introduction of oxygen vacancies effectively enhance the electrical conductivity of $NiMoO_4$. In addition, the coated carbon layer improves the structural stability of the $NiMoO_4$ in the whole charging and discharging process, significantly enhancing the cycling stability of the electrode. Consequently, the $NiMoO_{4-x}$@C electrode delivers a high areal capacitance of 2.24 F cm^{-2} (1720 F g^{-1}) at a current density of 1 mA cm^{-2} and superior cycling stability of 84.5% retention after 6000 cycles at 20 mA cm^{-2}. Furthermore, an asymmetric super-capacitor device (ASC) has been constructed with $NiMoO_{4-x}$@C as the positive electrode and activated carbon (AC) as the negative electrode. The as-assembled ASC device shows excellent electrochemical performance with a high energy density of 51.6 W h kg^{-1} at a power density of 203.95 W kg^{-1}. Moreover, the $NiMoO_4$//AC ASC device manifests remarkable cyclability with 84.5% of capacitance retention over 6000 cycles. The results demonstrate that the $NiMoO_{4-x}$@C composite is a promising material for electrochemical energy storage. This work can give new insights on the design and development of novel functional electrode materials via defect and interface engineering through simple yet effective chemical routes.

Keywords: $NiMoO_4$ nanowire arrays; oxygen vacancies; core-shell electrode structure; asymmetric supercapacitors

Citation: Wang, P.; Ding, X.; Zhe, R.; Zhu, T.; Qing, C.; Liu, Y.; Wang, H.-E. Synchronous Defect and Interface Engineering of $NiMoO_4$ Nanowire Arrays for High-Performance Supercapacitors. *Nanomaterials* **2022**, *12*, 1094. https://doi.org/10.3390/nano12071094

Academic Editors: Jihoon Lee and Ming-Yu Li

Received: 1 March 2022
Accepted: 24 March 2022
Published: 26 March 2022

1. Introduction

Several alternative energy technologies have been under development globally in a great effort to mitigate the energy and environmental challenges faced and in accordance with the current "carbon neutral" policies. Supercapacitors (SCs), also known as electrochemical capacitors (ECs), have been considered as one of the most promising energy storage devices due to their unique characteristics of high power density (>10 kW/kg), fast charging and discharging capability (within a few seconds), long lifespan (over 100,000 cycles), and good operational safety. SCs have been widely applied in some important fields, including smart electric grids, memory back-ups, (hybrid) electric vehicles, and aerospace crafts. Although SCs have the advantages of high power density and very long calendar lives, their further application is still hindered by their limited energy density.

Therefore, it is crucial to develop advanced high-performance SCs with higher energy densities without severely compromising the power density and cyclability.

Recently, asymmetric supercapacitors (ASCs) have been regarded as promising due to their enhanced energy density. ASCs commonly combine pseudocapacitive materials (as positive electrode) and electric double-layer capacitive materials (as negative electrode), making use of the much higher specific capacitance derived from the pseudocapacitive electrode materials and a wider potential window during operation. In this sense, the electrochemical performances of ASCs are largely influenced by the structures and properties of the pseudocapacitive electrode materials. Thus, the development of high energy density ASCs heavily relies on the rational selection and design, as well as delicate fabrication of advanced electrode materials.

The use of transition metal oxides (TMOs) [1], hydroxides [2], sulfides [3], selenides [4], carbides [5], and their composites with conductive carbon and polymers [6], as possible electrode materials for ASCs, has been attempted. Particularly, TMOs, such as NiO [7], MoO_3 [8], Co_3O_4 [9], and MnO_2 [10], have received considerable attention due to their low cost, convenience of synthesis, environmental friendliness, and relatively high capacitance. Compared to binary TMOs, ternary TMOs materials containing two different metal cations, such as $NiCo_2O_4$ [11], $ZnCo_2O_4$ [12,13], $CoMoO_4$ [14], $MnMoO_4$ [15], and $NiMoO_4$ [16], as well as some solid solutions [17], exhibit higher electrochemical activity due to the rich redox reactions stemming from the multiple oxidation states of the transition-metal components. It has been proposed that $NiMoO_4$ has a good reversible capacitance and electrochemical characteristics for SCs/ASCs applications because of the electrochemical active Ni cation and improved electronic conductivity from Mo cation [16,18].

However, the practical application of $NiMoO_4$ electrodes in SCs and ASCs is still hampered by their poor electronic conductivity, insufficient ionic transport and diffusion, and structural instability during long-term cycling [19]. Therefore, the controllable fabrication of $NiMoO_4$ with desired nano- and microstructures and rational structural engineering is highly desired but remains challenging.

Various nanostructured $NiMoO_4$ materials, such as nanosheets and nanorods arrays [20], nanotubes [21], hollow nanorods [22], mesoporous nanospheres [23], nanoparticles, and quantum dots [24], have been designed to boost the electrochemical performances of the $NiMoO_4$ electrodes via increased exposed surface for ion adsorption and insertion, shortened path distances for ion transport and diffusion, and improved electrolyte impregnation and permeation. Specifically, various low-dimensional $NiMoO_4$ nanostructures directly grown on conductive substrates (e.g., Ni/Cu foams [25,26], graphene [27], and carbon substrates [28,29]) are particularly preferred for directional electron transport with reduced charge carrier scattering at grain boundaries and easy integration into flexible devices with some specific applications.

To overcome the poor electronic conductivity of pristine $NiMoO_4$, various $NiMoO_4$/carbon composites have been synthesized by hybridizing $NiMoO_4$ nanostructures with graphene [30,31], carbon nanotubes [32], conducting polymers [33], and porous carbon architectures [34,35] Alternatively, intentional doping of $NiMoO_4$ with several kinds of heteroatoms such as Mn [36,37], P [38], Zn [39], Ce [40], or the creation of oxygen vacancies [14,41–43] in the lattice have recently been reported. In addition, $NiMoO_4$ has also been coupled with other metal oxides [44–49] or sulfides [50–52] to form heterostructure electrodes for supercapacitors with improved electrochemical performances.

However, monotonous strategy sometimes has a limited contribution for the overall electrochemical performance improvement of $NiMoO_4$ materials. In addition, some reported approaches for hybridization or doping of $NiMoO_4$ involve multiple and complex chemical and physical processes that are not economically or environmentally friendly. Thus, the rational design and the design of a $NiMoO_4$-based composite electrode for high-performance supercapacitors remains a challenge.

In this work, we report the simultaneous defect and interface engineering of $NiMoO_4$ nanowires arrays (NWAs) using a simple and effective sucrose coating followed by a thermal treatment approach. In this process, an amorphous carbon shell was uniformly coated on the $NiMoO_4$ surface, effectively improving the electronic transport and structural integrity of the $NiMoO_4$ during electrochemical cycling. Additionally, oxygen-vacancy defects were incorporated into the $NiMoO_4$ during the carbonization process, further enhancing the electronic conductivity of $NiMoO_4$ and redox activity in the $NiMoO_4$ electrode surface. As expected, the resultant $NiMoO_{4-x}$@C composite exhibited a higher specific capacitance than that of the pristine $NiMoO_4$ NWAs. Furthermore, an asymmetric supercapacitor (ASC) was assembled with the $NiMoO_{4-x}$@C as positive electrode and activated carbon (AC) as negative electrode, delivering a remarkably high energy density of 51.6 W h kg^{-1} at a power density of 203 W kg^{-1} and an excellent cycling stability with a retention of 84.5% after 6000 cycles under a high current density of 10 A g^{-1}.

2. Results and Discussion

The synthesis route of oxygen-deficient $NiMoO_4$@carbon nanowire arrays ($NiMoO_{4-x}$@C) is schematically shown in Figure 1. The preparation process mainly involved three critical steps. Firstly, a $NiMoO_4$ nanowire arrays (NWAs) precursor ($NiMoO_4{\cdot}xH_2O$ NWAs, light green) was directly deposited on a ferronickel foam by a hydrothermal reaction process (Step 1, Figure 1). Secondly, the $NiMoO_4$ NWAs precursor was transferred into $NiMoO_4$ NWAs by annealing in Ar to remove crystal H_2O and improve crystallinity (Step 2, Figure 1). Finally, the as-obtained $NiMoO_4$ NWAs were immersed in a sucrose solution, followed by drying and annealing in an Ar atmosphere to fabricate oxygen-deficient $NiMoO_4$@carbon NWAs ($NiMoO_{4-x}$@C) (Step 3, Figure 1).

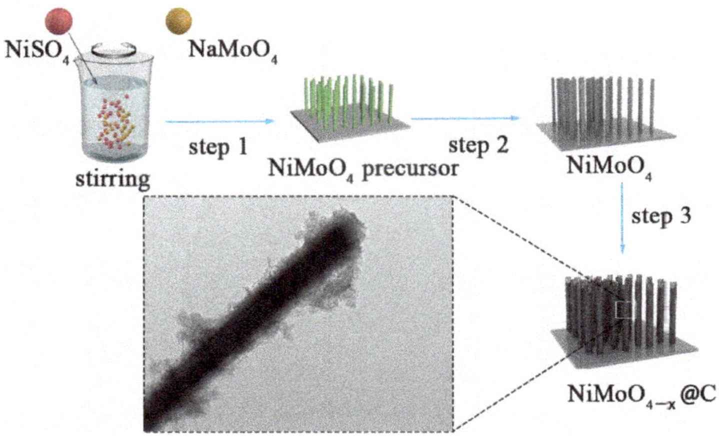

Figure 1. Schematic illustration of the synthesis process of oxygen-deficient $NiMoO_4$@carbon nanowire arrays (NWAs) (denoted as "$NiMoO_{4-x}$@C"). Step 1, growth of the $NiMoO_4$ NWAs precursor directly on a ferronickel foam substrate using a hydrothermal process; Step 2, conversion of the $NiMoO_4$ NWAs precursor into the $NiMoO_4$ NWAs via annealing in Ar; Step 3, fabrication of $NiMoO_{4-x}$@C composite by sucrose coating followed by annealing in Ar.

The crystal structures of $NiMoO_{4-x}$@C and neat $NiMoO_4$ NWAs samples were characterized by X-ray diffraction (XRD) analysis as depicted in Figure 2. The two strongest diffraction peaks, located at ca. 45° and 52°, were from the ferronickel foam substrate. The $NiMoO_{4-x}$@C and $NiMoO_4$ NWAs samples showed similar diffraction peak shapes and locations. The diffraction peaks located at 14.3°, 24.0°, 25.4°, 28.9°, 32.6°, 37.1°, 38.7°, 41.3°, and 47.4° corresponded to the (110), (021), (−112), (220), (022), (−113), (−132), (040), and (−204) crystal planes of orthorhombic $NiMoO_4$ (JCPDS card No. 86-0361) [37]. Compared to the $NiMoO_4$ NWAs samples, the $NiMoO_{4-x}$@C sample exhibited a slightly lower diffraction peak intensity possibly due to the covering of carbon on the $NiMoO_4$ surface as well as the reduced crystallinity of $NiMoO_4$ with increased structural defects. In addition, no characteristic peaks for carbon phases were noted, suggesting the amorphous nature of the carbon species in the $NiMoO_{4-x}$@C sample. The formation of amorphous carbon could be partially ascribed to the lower low annealing temperature (400 °C herein).

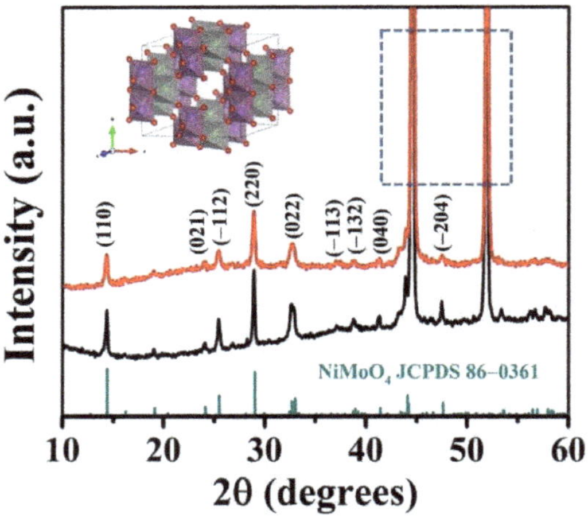

Figure 2. XRD patterns of $NiMoO_4$ NWAs (black) and $NiMoO_{4-x}$@C composite (red). The inset (top left) shows the structural model of $NiMoO_4$ crystal, whereas the green, blue, and red balls represent the Ni, Mo, and O atoms, respectively. The two strongest diffraction peaks in the square regions of dotted line are from the ferronickel foam substrate.

Raman spectra of pristine $NiMoO_4$ NWAs and $NiMoO_{4-x}$@C samples are illustrated in Figure 3. The bands at 961 cm^{-1} and 913 cm^{-1} corresponded to the symmetric and asymmetric stretching modes of Mo=O bonds, while the band at 706 cm^{-1} could be ascribed to the stretching mode of Ni/Mo-O bonds of the orthorhombic α-$NiMoO_4$ phase [53]. In addition, two bands ascribed to the presence of carbon species were identified. The band at around 1360 cm^{-1} could be attributed to the D band from defects and disorders in the amorphous carbon layers, while the other band at around 1590 cm^{-1} was related to the G band related to the vibration of sp^2-bonded carbon atoms [54]. This result implied the successful deposition of amorphous carbon layer on the surface of the $NiMoO_4$ NWAs.

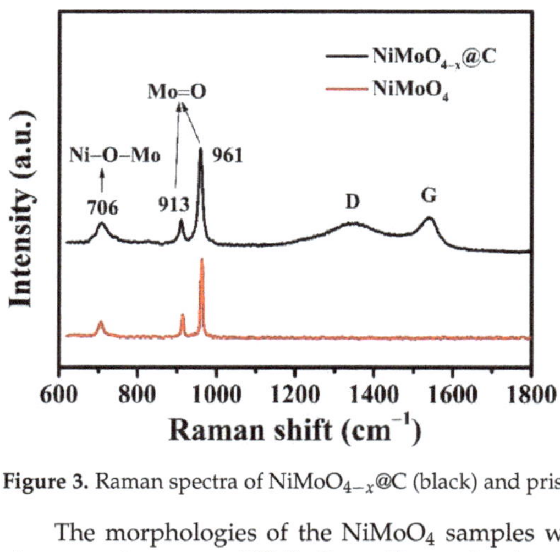

Figure 3. Raman spectra of $NiMoO_{4-x}$@C (black) and pristine $NiMoO_4$ NWAs (red) samples.

The morphologies of the $NiMoO_4$ samples were firstly observed using scanning electron microscopy (SEM). From Figure 4a, the pristine $NiMoO_4$ NWAs sample was composed of oriented nanowires (NWs) with a smooth surface. In addition, the $NiMoO_4$ NWAs have relatively uniform diameters of ~300 nm, on average, and lengths of several micrometers. After the coating of the carbon, the surface of the $NiMoO_{4-x}$@C sample became obviously coarse as shown in Figure 4b. The element composition analyses using energy-dispersive X-ray spectra (EDS) analysis indicated the existence of Ni, Mo, O, C, Fe, and Al elements in the $NiMoO_{4-x}$@C sample (Figure S1, Supporting Information). Note that the Fe and Al signals mainly stemmed from the ferronickel foam substrate and the sample holder, respectively.

Figure 4. SEM images of (**a**) neat $NiMoO_4$ NWAs and (**b**) $NiMoO_{4-x}$@C samples.

The microstructures of the pure $NiMoO_4$ NWAs and $NiMoO_{4-x}$@C samples were further investigated by transmission electron microscope (TEM) and high-resolution TEM (HRTEM) as shown in Figure 5. The TEM image (Figure 5a) revealed that the pure $NiMoO_4$ nanowire had a smooth surface with a diameter of about 200 nm. A selected-area electron diffraction (SAED) pattern (inset of Figure 5a) taken from this nanowire depicted a clear two-dimensional dot pattern, suggesting its single-crystalline structure in nature. Two diffraction spots, as marked by white circles, could be indexed to the (220) and (−222) crystal facets of orthorhombic $NiMoO_4$. From Figure 5b, the crystal plane with a lattice spacing of 2.73 Å in the HRTEM micrograph corresponded to the (−222) planes of $NiMoO_4$ [49,52]. In contrast, the TEM image in Figure 5c indicated that a layer of amorphous carbon film with a thickness of ca. 20~50 nm had been coated on the $NiMoO_4$ nanowire's surface, confirming the core-shell structure of the $NiMoO_{4-x}$@C composite sample with different brightness contrasts of $NiMoO_4$ and carbon. The deposition of amorphous carbon on the surface of $NiMoO_4$ can be further confirmed by HRTEM micrograph as shown in Figure 5d.

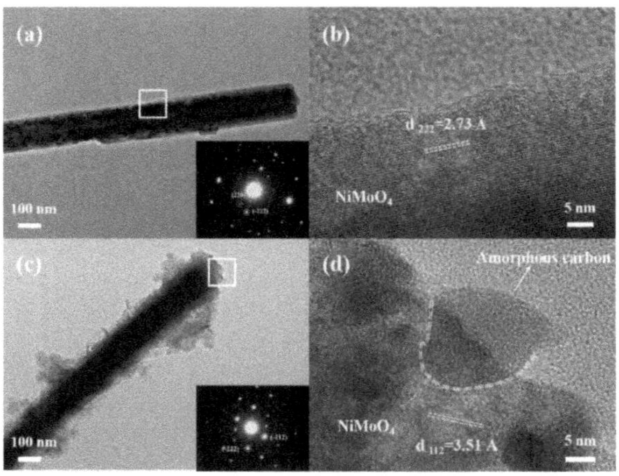

Figure 5. TEM images (**a**,**c**) and HRTEM micrographs (**b**,**d**) of pure $NiMoO_4$ NWAs (**a**,**b**) and $NiMoO_{4-x}$@C (**c**,**d**) samples. The insets in (**a**,**c**) are corresponding SAED patterns taken from a single nanowire.

Next, the chemical composition and valence states of element on the surface of $NiMoO_4$ NWAs and $NiMoO_{4-x}$@C samples were identified by X-ray photoelectron spectroscopy (XPS, Figure 6). In the high-resolution Ni 2p spectrum of the pristine $NiMoO_4$ NWAs sample (Figure 6a), two main peaks were observed at binding energies (BEs) of 873.4 eV and 856.3 eV with a spin-orbital splitting energy of 17.1 eV, corresponding to the Ni $2p_{1/2}$ and Ni $2p_{3/2}$ of Ni^{2+} in $NiMoO_4$ lattice [55]. In addition, two satellite peaks with BEs of 877.3 eV and 860.6 eV were noted for Ni^{2+}. In Figure 6b, similar peak locations and separations can also be observed in the Ni 2p spectra of $NiMoO_{4-x}$@C sample, indicating the coating of carbon had little effect on the chemical valence states of the Ni component in $NiMoO_4$. The Mo 6d spectrum of pure $NiMoO_4$ and $NiMoO_{4-x}$@C samples are shown in Figure 6c,d. Evidently both samples had two strong bands with BEs located at 236.0 eV and 232.9 eV, which could be assigned to Mo $3d_{3/2}$ and Mo $3d_{5/2}$ of Mo^{6+} cations in the $NiMoO_4$ lattice [56]. In addition, another pair of doublets was noted for the $NiMoO_{4-x}$@C sample, verifying the existence of Mo^{4+} in the $NiMoO_{4-x}$@C composite [41,42,57] possibly produced during the amorphous carbon coating process. From Figure 6e, the O 1s spectrum of the $NiMoO_4$ NWs sample was deconvoluted into three bands. The band centered at 530.1 eV was assigned to the lattice oxygen with O-Ni/O-Mo bonds, while the bands located at 531.3 eV and 532.9 eV correspond to the O-C and O=C bonds from moisture adsorbed on surface [58]. For the $NiMoO_{4-x}$@C sample, another band could be noted at 532 eV, suggesting the presence of oxygen vacancies (V_o) [59] at the $NiMoO_4$ surface (Figure 6f). The formation of and Mo^{4+} and V_o can tune the electronic structures and electrochemical properties of the $NiMoO_{4-x}$@C composite sample.

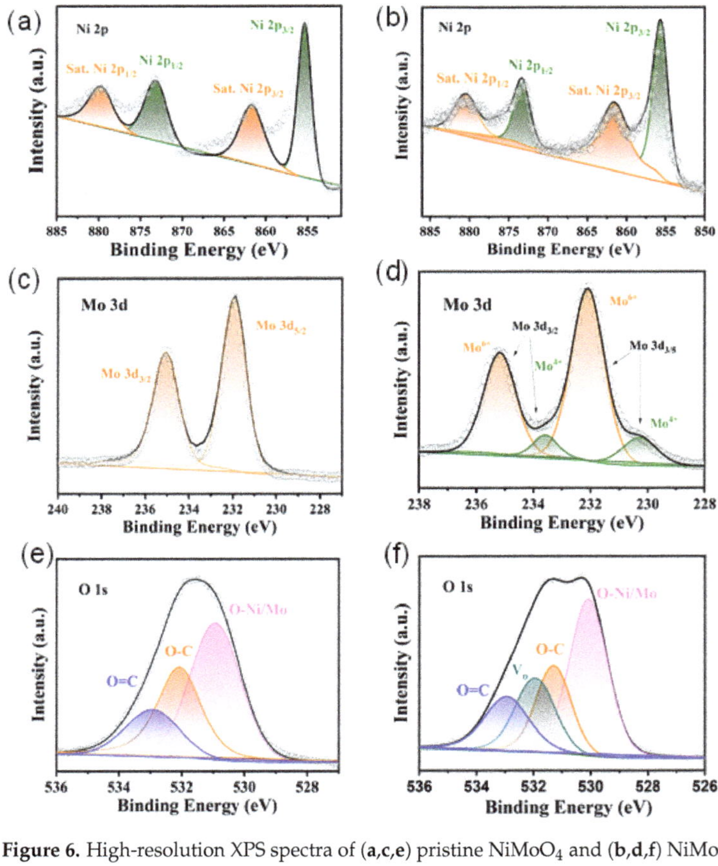

Figure 6. High-resolution XPS spectra of (**a**,**c**,**e**) pristine $NiMoO_4$ and (**b**,**d**,**f**) $NiMoO_{4-x}$@C samples; (**a**,**b**) Ni 2p, (**c**,**d**) Mo 3d, (**e**,**f**) O 1s.

Then, the effects of pyrolysis temperatures (from 200~800 °C) during the carbon coating of the morphologies and microstructures of the $NiMoO_4$/carbon composites were investigated. It is noted that some aggregates of residual sucrose were observed after annealing at 200 °C (Figure S2a), suggesting the carbonization of sucrose was incomplete under a lower temperature. This SEM result also coincides well with the thermogravimetric (TGA) and the differential scanning calorimetry (DSC) analyses (Figure S3), where the thermal decomposition process of sucrose mainly occurs between 223 and 389 °C. With the increase of annealing temperature, sucrose was decomposed, and the carbonization process occurred accompanied by the release of some gases (e.g., CO, CO_2). At a higher temperature, the generated reductive gases (e.g., CO) reacted with $NiMoO_4$ and generated some oxygen vacancies on the $NiMoO_4$ surface via abstracting some surface oxygen atoms. In contrast, well-defined nanowires were obtained for the samples prepared after annealing at 400 and 600 °C, respectively (Figure S2b,c). However, the nanowire structure was destroyed when the pyrolysis temperature was increased to 800 °C (Figure S2d), which might have been caused by the large inner stain in the $NiMoO_4$ NWAs or at the $NiMoO_{4-x}$@C interface. Thus, the standard annealing temperature was chosen as 400 °C.

To evaluate the electrochemical performance of $NiMoO_4$ NWAs and $NiMoO_{4-x}$@C samples, electrochemical measurements were tested by a three-electrode system with 2 M KOH electrolyte (Figure 7). Figure 7a shows the CV curves of $NiMoO_4$ NWAs and $NiMoO_{4-x}$@C samples at a scan rate of 20 mV s^{-1} with a potential window of 0 to 0.7 V. Overall, the $NiMoO_{4-x}$@C sample had a larger integral area than that of the $NiMoO_4$ NWAs sample, indicating a significant increase of capacitance after carbon deposition and introduction of oxygen vacancies. Meanwhile, the CV curves of the two samples exhibited typical oxidation peaks, demonstrating typical pseudocapacitive charge storage characteristics. In addition, the CV curves of four different pyrolysis temperatures of $NiMoO_{4-x}$@C samples are revealed in Figure S4. The sample collected at 400 °C shows the highest capacitance which is consistent with the result of SEM in Figure S2. The sample collected at 800 °C exhibited an unsatisfactory performance due to its collapsed morphology. Figure 7b shows the GCD curves of the two samples. It revealed that the discharge time of the $NiMoO_{4-x}$@C sample was almost twice as much as that of the $NiMoO_4$ NWAs sample at a current density of 1 A cm^{-2}. Figure 7c shows the electrochemical impedance spectroscopy (EIS) of $NiMoO_4$ NWAs and $NiMoO_{4-x}$@C. The direct impedance and charge transfer resistance of the $NiMoO_{4-x}$@C sample was significantly lower than that of the $NiMoO_4$ NWAs sample. The remarkably reduced size of the semicircle for the $NiMoO_{4-x}$@C indicated an improved charge transfer kinetics due to enhanced electrical conductivity provided by the carbon shell and oxygen vacancy defects. In addition, the $NiMoO_{4-x}$@C exhibits the steepest slope in the low-frequency region, clearly indicating the lowest Warburg impedance and, hence, the highest K-ion diffusion capability at the interface between the electrode and electrolyte. Through the AC EIS, we added the corresponding equivalent circuit diagram in Figure 7c.

The true impedance of capacitor can be estimated using the following Equation (1):

$$Z_{real} = R_{\Omega} + \frac{-j(wC)\left(R_{ct} + W \cdot jw^{1/2}\right)}{R_{ct} + W \cdot jw^{1/2} - j(wC)} \quad (1)$$

Figure 7d shows the capacitance of $NiMoO_4$ NWAs and $NiMoO_{4-x}$@C samples calculated from different current densities. After coating the carbon layer, the capacitance of the $NiMoO_{4-x}$@C sample was greatly increased. The areal capacitance can be calculated as high as 2.24 F cm^{-2} (1720 F g^{-1}) at a current density of 1 mA cm^{-2}. In contrast, the $NiMoO_4$ NWAs electrode only demonstrated a specific capacitance of 1.206 F cm^{-2} (927 F g^{-1}) at the same current density. The cycling performances of $NiMoO_4$ NWAs and $NiMoO_{4-x}$@C samples are presented in Figure 7e. The capacitance retention of $NiMoO_{4-x}$@C is 84.5% at 20 mA cm^{-2} after 6000 cycles, which is considerably better than that of the $NiMoO_4$ NWAs sample (63.1% after 6000 cycles). To illustrate the difference of the cycling process between $NiMoO_4$ NWAs and $NiMoO_{4-x}$@C samples, we also obtained the SEM results after cycling as shown in Figure S5. It is evident that the $NiMoO_{4-x}$@C sample still held some nanorod structures under the protection of amorphous carbon shell. Instead, $NiMoO_4$ NWAs were aggregated after 10,000 cycles, with unsatisfactory cycle abilities. In addition, we made a comparison of the C_s and cycling stability of this work with some previously reported $NiMoO_4$-based electrodes materials as summarized in Table S1.

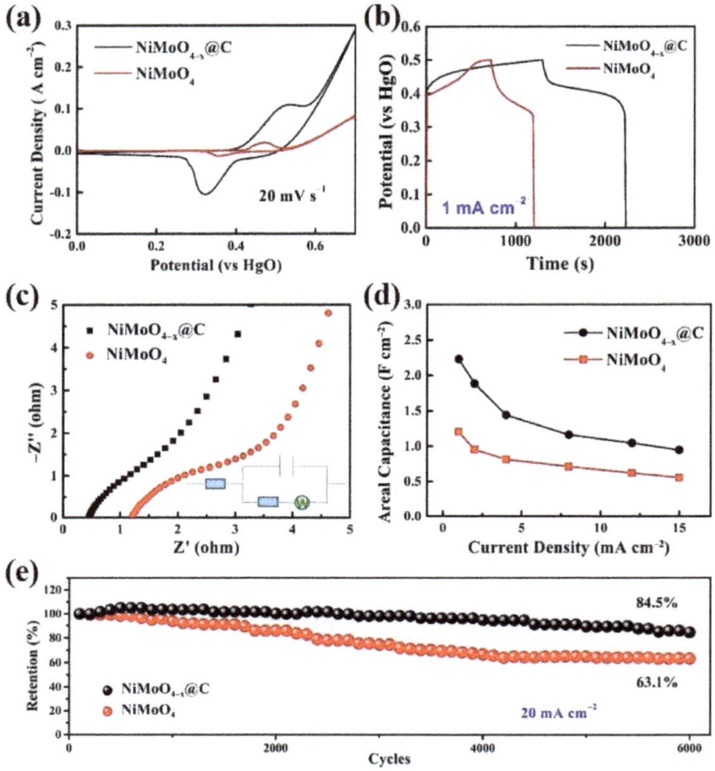

Figure 7. (**a**) CV curves of the $NiMoO_4$ NWAs and $NiMoO_{4-x}$@C at a scan rate of 20 mV s^{-1}. (**b**) GCD curves and (**c**) EIS spectra of $NiMoO_4$ NWAs and $NiMoO_{4-x}$@C and its corresponding equivalent fitting circuit. (**d**) GCD test of $NiMoO_{4-x}$@C at different current densities. (**e**) Cycle test of $NiMoO_4$ NWAs and $NiMoO_{4-x}$@C at 20 mA cm^{-2} over 6000 cycles.

First-principles density functional theory (DFT) simulations were next adopted to further probe the structure–performance relationship of the $NiMoO_{4-x}$@C composite electrode in supercapacitors. The optimized geometry configurations of pristine and oxygen-deficient $NiMoO_4$ (110) surface slabs are shown in Figure S6. The pristine $NiMoO_4$ (110) surface was flat and composed of fivefold Ni and Mo atoms and twofold O atoms (Figure S6a). The defective $NiMoO_4$ (110) surface can be produced after eliminating one surface O atom, leaving one threefold Ni and Mo atoms nearby (Figure S6b). The resultant oxygen-deficient $NiMoO_4$ (110) plane retains flat. Next, the adsorption behavior of OH group on the pristine $NiMoO_4$ (110) surface was first investigated. As shown in Figure 8, the OH can be adsorbed on the top of either the surface of the Mo atom (Figure 8a,b) or the Ni atom (Figure 8c,d), yielding an adsorption energy (E_{ads}) of -0.50 and -5.97 eV, respectively. Evidently, the adsorption of OH on the Ni site was much stronger than that on the Mo site, which is consistent with the fact that the Ni in $NiMoO_4$ is electrochemically active for pseudocapacitive charge storage process based on Faradic reactions. The chemisorption of OH on Ni and Mo sites of the $NiMoO_4$ (110) surface was further verified by the interfacial charge transfer from the charge density difference contours (Figure S7).

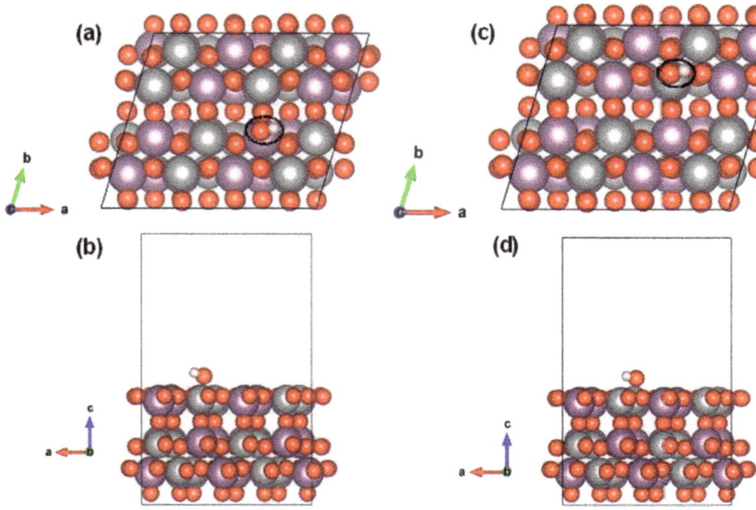

Figure 8. The optimized adsorption configurations of one HO molecule anchored on exposed (**a**,**b**) Mo- and (**c**,**d**) Ni- atoms of pristine $NiMoO_4$ (110) surfaces, leading to an E_{ads} of −0.50 and −5.97 eV, respectively. The black dotted ellipsoids in (**a**,**c**) mark the positions of the adsorbed HO molecule on the $NiMoO_4$ (110) surface.

In the following, the adsorption of OH adsorbed on the oxygen-deficient $NiMoO_4$ (110) surface was evaluated. Specifically, the adsorption on the Ni and Mo sites with lower coordination due to the removal of surface O was considered. Interestingly, it is noted that the OH group was preferred to be adsorbed at the vicinity of the oxygen-vacancy position (Figure 9a), leading to an E_{ads} of −3.63 eV and a charge transfer at the OH/$NiMoO_4$ interface (Figure 9b). This result suggests that the presence of surface oxygen vacancies offers more active sites for OH adsorption, concentration, and subsequent redox reactions for enhanced pseudocapacitive charge storage.

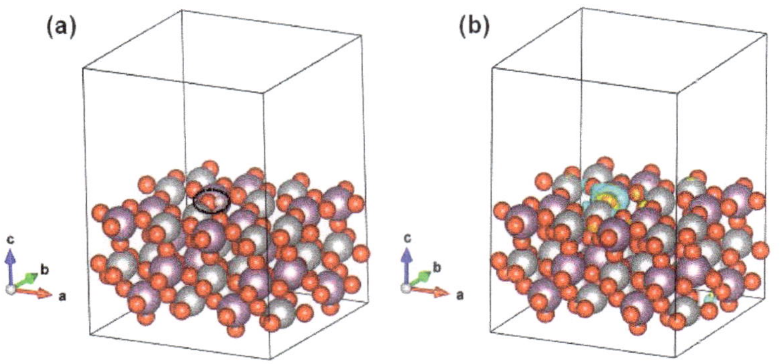

Figure 9. The optimized adsorption configurations of one HO molecule anchored on $NiMoO_4$ (110) surfaces with one surface oxygen vacancy (V_o), leading to an E_{ads} of −3.63 eV. The black dotted ellipsoid in (**a**) labels the position of the adsorbed HO molecule, which is at the vicinity of the V_o. The yellow and cyan colors in the charge density difference contour (**b**) represent the electron accumulation and depletion, respectively. The isosurface level is 0.001 e bohr^{-3}.

Based on the above experimental data and theoretical simulations, the significantly improved electrochemical performances of the $NiMoO_{4-x}$@C composite can be mainly attributed to the following points: (i) the deposition of amorphous carbon shell effectively enhances the electron transport of $NiMoO_4$ nanowires and charge transfer at $NiMoO_4$/C heterointerface; (ii) the deposited carbon layer also improves the structural integrity of the $NiMoO_4$ nanowire arrays during long-term electrochemical cycling; (iii) the creation of oxygen vacancies in $NiMoO_4$ accompanied by the coating of the carbon further enhances the electronic conductivity of the $NiMoO_4$ electrode and creates more active sites for pseudocapacitive charge storage. Therefore, the synergy of defect and interface engineering of $NiMoO_4$ NWAs realized by carbon deposition effectively improves the overall electrochemical performance of the resultant $NiMoO_{4-x}$@C composite electrode in supercapacitors.

To further assess the practical application potential of a $NiMoO_{4-x}$@C sample, an asymmetric supercapacitor device ($NiMoO_{4-x}$@C//AC) was assembled with the $NiMoO_{4-x}$@C as a positive electrode and activated carbon (AC) as a negative electrode. Before testing the ASC device, we performed the CV measurements of $NiMoO_{4-x}$@C and AC electrodes in a three-electrode system at a scan rate of 5 mV s^{-1} to estimate the suitable operating voltage range (Figure 10a). The maximum operating voltage of $NiMoO_{4-x}$@C//AC ASC was determined to be 1.6 V. The CV curves of $NiMoO_{4-x}$@C//AC ASC at different scan rates are shown in Figure 10b. All the curves display obvious redox peaks, indicating the main contribution from pseudocapacitance. An increased separation of redox peak position can be noted along with the increase of scan rates due to the increased polarization. Figure 10c shows the GCD curves of the $NiMoO_{4-x}$@C//AC ASC at different current densities while Figure 10d shows the specific capacitance calculated from different current densities. The overall capacitance of the ASC was calculated to be 1.01 F cm^{-2} (156.25 F g^{-1}) at 1 mA cm^{-2}. Furthermore, the $NiMoO_{4-x}$@C//AC ASC device has demonstrated a good capacitance retention of 83.6% after 6000 cycles at 20 mA cm^{-2} (Figure 10e). As a result, the ASC device can power a yellow LED (inset of Figure 10e), showing its potential in practical applications. In addition, the electrochemical performances of our $NiMoO_{4-x}$@C//AC ASC device are also superior or comparable to some recently reported $NiMoO_4$-based electrode materials for ASCs as summarized in Table S1.

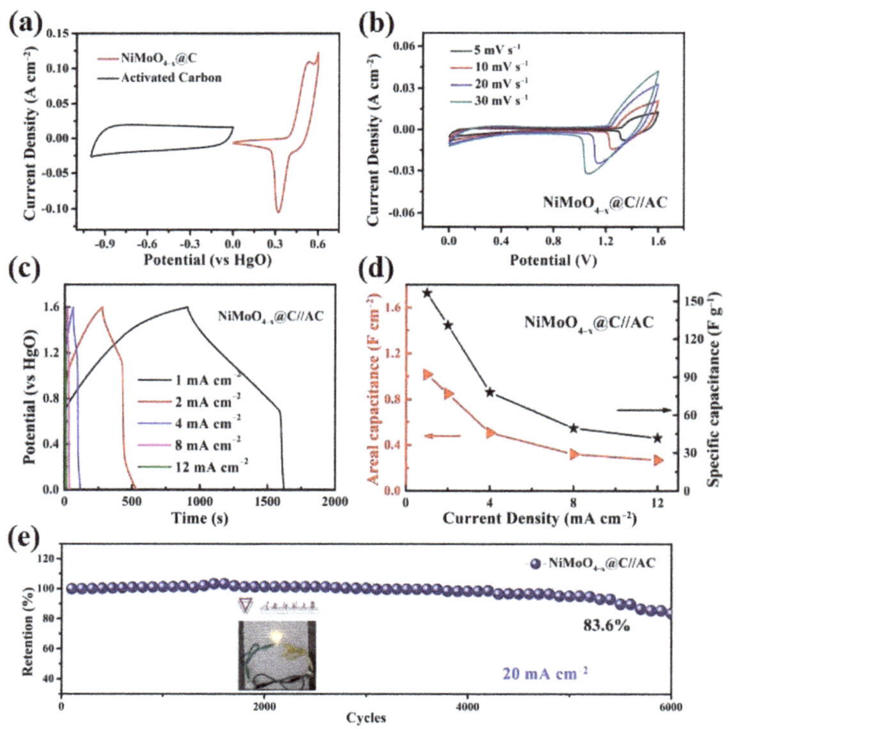

Figure 10. Electrochemical performance of the asymmetric $NiMoO_{4-x}$@C//AC supercapacitor. (**a**) CV curves of activated carbon (AC) and $NiMoO_{4-x}$@C electrodes from −1.0~0 V and 0~0.6 V, respectively

at 5 m V s^{-1}; (**b**) CV curves; (**c**) GCD curves; (**d**) areal capacitance and specific capacitance; (**e**) cyclic stability of $NiMoO_{4-x}$@C//AC.

3. Conclusions

In summary, synchronous defect and interface engineering was implemented in $NiMoO_4$ material via the formation of oxygen vacancies and the coating of the carbon on $NiMoO_4$ nanowire arrays through a simple hydrothermal method paired with sucrose pyrolysis. During this process, an amorphous carbon layer was homogeneously deposited on the surface of $NiMoO_4$ nanowires and oxygen vacancies were created on the $NiMoO_4$ surface during the carbonization of sucrose. The deposited carbon layer and formed oxygen vacancies in $NiMoO_4$ boosted the electronic conductivity of $NiMoO_4$ nanowires. In addition, the coated carbon layer also improved the structural integrity of the $NiMoO_4$ electrode during long-term operation in supercapacitors. Consequently, the resultant $NiMoO_{4-x}$@C heterostructure electrode achieved a high specific capacitance of 2.24 F cm^{-2} (1720 F g^{-1}) and maintained a good capacitance retention of about 83.6% after 6000 cycles at 20 mA cm^{-2}. In addition, the as-assembled $NiMoO_{4-x}$@C//activated carbon asymmetric supercapacitor device manifested a high energy density of 51.6 W h kg^{-1} at a high power density of 203.95 W kg^{-1}, indicating that $NiMoO_{4-x}$@C composite is a suitable electrode material for supercapacitor applications. The proposed synergistic defect and interface engineering strategy herein can be extended for the design and development of other novel composite electrode materials for applications in electrochemical energy storage and conversion.

Supplementary Materials: The following supporting information can be downloaded at https://www.mdpi.com/article/10.3390/nano12071094/s1, Figure S1: The EDS analysis of $NiMoO_{4-x}$@C composite and $NiMoO_4$ NWAs, Figure S2: SEM images of the $NiMoO_{4-x}$@C composite prepared under different annealing temperature: (a) 200 °C, (b) 400 °C, (c) 600 °C and (d) 800 °C, Figure S3: Thermogravimetric (TG) and differential scanning calorimetry (DSC) analysis of sucrose in Ar atmosphere at a ramping rate of 10 °C min^{-1}, Figure S4: (a) CV and (b) GCD of $NiMoO_{4-x}$@C samples prepared under different annealing temperatures, Figure S5: SEM images of the (a) $NiMoO_{4-x}$@C and (b) $NiMoO_4$ NWAs electrodes after 10,000 cycles at 20 mA cm^{-2}, Figure S6: Optimized geometry structures of the (a) pristine $NiMoO_4$ (110) surface and (b) oxygen-deficient $NiMoO_4$ (110) surface obtained by removing one surface O atom (green). The grey, light pink and red balls represent the Ni, Mo and O atoms, respectively [60–63], Figure S7: Charge density difference contours of after adsorption of one HO molecule on (a) Mo- and (b) Ni-side of $NiMoO_4$ (110) surface, respectively. The yellow and cyan colors denote the electron gain and loss, respectively. The isosurface level is 0.001 e $bohr^{-3}$, Table S1: Brief comparison of electrochemical performance of current work with recently relevant literature [20,22,30,55,64–72].

Author Contributions: Conceptualization, C.Q. and P.W.; methodology, P.W. and X.D.; software, R.Z. and H.-E.W.; validation, P.W., X.D. and R.Z.; formal analysis, P.W., X.D. and C.Q.; investigation, P.W. and X.D.; resources, C.Q., H.-E.W. and Y.L.; data curation, P.W., C.Q. and H.-E.W.; writing—original draft preparation, P.W.; writing—review and editing, C.Q., T.Z. and H.-E.W.; visualization, P.W. and R.Z.; supervision, C.Q. and H.-E.W.; project administration, C.Q. and Y.L.; funding acquisition, C.Q. and H.-E.W. All authors have read and agreed to the published version of the manuscript.

Funding: This work is supported by the National Natural Science Foundation of China (Grant No. 52162037) and Yunnan Fundamental Research Projects (grant No. 2019FD113 & 202101AT070120).

Data Availability Statement: All data included in this study are available upon request by contact with the corresponding author.

Conflicts of Interest: The authors declare no conflict of interest.

References

1. Augustyn, V.; Simon, P.; Dunn, B. Pseudocapacitive oxide materials for high-rate electrochemical energy storage. *Energy Environ. Sci.* **2014**, *7*, 1597–1614. [CrossRef]
2. Ma, H.N.; He, J.; Xiong, D.B.; Wu, J.S.; Li, Q.Q.; Dravid, V.; Zhao, Y.F. Nickel Cobalt Hydroxide Reduced Graphene Oxide Hybrid Nanolayers for High Performance Asymmetric Supercapacitors with Remarkable Cycling Stability. *ACS Appl. Mater. Interfaces* **2016**, *8*, 1992–2000. [CrossRef] [PubMed]
3. Yun, Q.B.; Lu, Q.P.; Zhang, X.; Tan, C.L.; Zhang, H. Three-Dimensional Architectures Constructed from Transition-Metal Dichalcogenide Nanomaterials for Electrochemical Energy Storage and Conversion. *Angew. Chem. Int. Ed.* **2018**, *57*, 626–646. [CrossRef] [PubMed]
4. Miao, C.X.; Zhou, C.L.; Wang, H.E.; Zhu, K.; Ye, K.; Wang, Q.; Yan, J.; Cao, D.X.; Li, N.; Wang, G.L. Hollow Co-Mo-Se nanosheet arrays derived from metal-organic framework for high-performance supercapacitors. *J. Power Sources* **2021**, *490*, 229532. [CrossRef]
5. Lukatskaya, M.R.; Kota, S.; Lin, Z.F.; Zhao, M.Q.; Shpigel, N.; Levi, M.D.; Halim, J.; Taberna, P.L.; Barsoum, M.; Simon, P.; et al. Ultra-high-rate pseudocapacitive energy storage in two-dimensional transition metal carbides. *Nat. Energy* **2017**, *2*, 17105. [CrossRef]
6. Ren, J.; Shen, M.; Li, Z.L.; Yang, C.M.; Liang, Y.; Wang, H.E.; Li, J.H.; Li, N.; Qian, D. Towards high-performance all-solid-state asymmetric supercapacitors: A hierarchical doughnut-like Ni_3S_2 PPy core-shell heterostructure on nickel foam electrode and density functional theory calculations. *J. Power Sources* **2021**, *501*, 230003. [CrossRef]
7. Sun, L.; Sun, Y.D.; Fu, Q.; Pan, C.X. Facile preparation of NiO nanoparticles anchored on N/P-codoped 3D carbon nanofibers network for high-performance asymmetric supercapacitors. *J. Alloys Compd.* **2021**, *888*, 161488. [CrossRef]
8. Liu, Y.; Wang, Y.F.; Meng, Y.; Plamthottam, R.; Tjiu, W.W.; Zhang, C.; Liu, T.X. Ultrathin Polypyrrole Layers Boosting MoO_3 as Both Cathode and Anode Materials for a 2.0 V High-Voltage Aqueous Supercapacitor. *ACS Appl. Mater. Interfaces* **2022**, *14*, 4490–4499. [CrossRef]
9. Shu, T.; Wang, H.; Li, Q.; Feng, Z.P.; Wei, F.X.; Yao, K.X.; Sun, Z.; Qi, J.Q.; Sui, Y.W. Highly stable Co_3O_4 nanoparticles/carbon nanosheets array derived from flake-like ZIF-67 as an advanced electrode for supercapacacitor. *Chem. Eng. J.* **2021**, *419*, 129631. [CrossRef]
10. Pandit, B.; Goda, E.S.; Abu Elella, M.H.; Rehman, A.U.; Hong, S.E.; Rondiya, S.R.; Barkataki, P.; Shaikh, S.F.; Al-Enizi, A.M.; El-Bahy, S.M.; et al. One-pot hydrothermal preparation of hierarchical manganese oxide nanorods for high-performance symmetric supercapacitors. *J. Energy Chem.* **2022**, *65*, 116–126. [CrossRef]
11. Wei, S.; Wan, C.C.; Zhang, L.Y.; Liu, X.Y.; Tian, W.Y.; Su, J.H.; Cheng, W.J.; Wu, Y.Q. N-doped and oxygen vacancy-rich $NiCo_2O_4$ nanograss for supercapacitor electrode. *Chem. Eng. J.* **2022**, *429*, 132242. [CrossRef]
12. Liu, C.X.; Zhao, Z.; Liu, Y.; Lu, Q.S. Carbon dots decorated zinc cobaltite nanowires-assembled hierarchical arrays supported on nickel foam as binder-free electrodes for high performance supercapacitors. *J. Power Sources* **2022**, *519*, 230780. [CrossRef]
13. Liu, H.Q.; Dai, M.Z.; Zhao, D.P.; Wu, X.; Wang, B. Realizing Superior Electrochemical Performance of Asymmetric Capacitors through Tailoring Electrode Architectures. *ACS Appl. Energy Mater.* **2020**, *3*, 7004–7010. [CrossRef]
14. Li, P.X.; Ruan, C.H.; Xu, J.; Xie, Y.B. Supercapacitive performance of $CoMoO_4$ with oxygen vacancy porous nanosheet. *Electrochim. Acta* **2020**, *330*, 135334. [CrossRef]
15. Zhu, Z.; Sun, Y.; Li, C.S.; Yang, C.; Li, L.; Zhu, J.H.; Chou, S.L.; Wang, M.M.; Wang, D.D.; Li, Y.L. Mini-review: Progress on micro/nanoscale $MnMoO_4$ as an electrode material for advanced supercapacitor applications. *Mater. Chem. Front.* **2021**, *5*, 7403–7418. [CrossRef]
16. Guo, D.; Zhang, P.; Zhang, H.M.; Yu, X.Z.; Zhu, J.; Li, Q.H.; Wang, T.H. $NiMoO_4$ nanowires supported on Ni foam as novel advanced electrodes for supercapacitors. *J. Mater. Chem. A* **2013**, *1*, 9024–9027. [CrossRef]
17. Denis, D.K.; Sun, X.; Zhang, J.Y.; Wang, Y.Y.; Hou, L.R.; Li, J.; Yuan, C.Z. Solid Solution Engineering of Co-Ni-Based Ternary Molybdate Nanorods toward Hybrid Supercapacitors and Lithium-Ion Batteries as High-Performance Electrodes. *ACS Appl. Energy Mater.* **2020**, *3*, 3955–3965. [CrossRef]
18. Guo, D.; Luo, Y.Z.; Yu, X.Z.; Li, Q.H.; Wang, T.H. High performance $NiMoO_4$ nanowires supported on carbon cloth as advanced electrodes for symmetric supercapacitors. *Nano Energy* **2014**, *8*, 174–182. [CrossRef]
19. Hong, J.; Lee, Y.W.; Hou, B.; Ko, W.; Lee, J.; Pak, S.; Hong, J.; Morris, S.M.; Cha, S.; Sohn, J.I.; et al. Solubility-Dependent $NiMoO_4$ Nanoarchitectures: Direct Correlation between Rationally Designed Structure and Electrochemical Pseudokinetics. *ACS Appl. Mater. Interfaces* **2016**, *8*, 35227–35234. [CrossRef]
20. Peng, S.J.; Li, L.L.; Wu, H.B.; Madhavi, S.; Lou, X.W. Controlled Growth of $NiMoO_4$ Nanosheet and Nanorod Arrays on Various Conductive Substrates as Advanced Electrodes for Asymmetric Supercapacitors. *Adv. Energy Mater.* **2015**, *5*, 1401172. [CrossRef]
21. Yin, Z.X.; Zhang, S.; Chen, Y.J.; Gao, P.; Zhu, C.L.; Yang, P.P.; Qi, L.H. Hierarchical nanosheet-based $NiMoO_4$ nanotubes: Synthesis and high supercapacitor performance. *J. Mater. Chem. A* **2015**, *3*, 739–745. [CrossRef]
22. Lin, L.Y.; Liu, T.M.; Liu, J.L.; Sun, R.; Hao, J.H.; Ji, K.M.; Wang, Z.C. Facile synthesis of groove-like $NiMoO_4$ hollow nanorods for high-performance supercapacitors. *Appl. Surf. Sci.* **2016**, *360*, 234–239. [CrossRef]
23. Yao, M.M.; Hu, Z.H.; Liu, Y.F.; Liu, P.P. A novel synthesis of size-controllable mesoporous $NiMoO_4$ nanospheres for supercapacitor applications. *Ionics* **2016**, *22*, 701–709. [CrossRef]
24. Jinlong, L.; Meng, Y.; Tongxiang, L. Enhanced performance of $NiMoO_4$ nanoparticles and quantum dots and reduced nanohole graphene oxide hybrid for supercapacitor applications. *Appl. Surf. Sci.* **2017**, *419*, 624–630. [CrossRef]

25. Huang, L.; Xiang, J.W.; Zhang, W.; Chen, C.J.; Xu, H.H.; Huang, Y.H. 3D interconnected porous $NiMoO_4$ nanoplate arrays on Ni foam as high-performance binder-free electrode for supercapacitors. *J. Mater. Chem. A* **2015**, *3*, 22081–22087. [CrossRef]
26. Wang, Z.J.; Wei, G.J.; Du, K.; Zhao, X.X.; Liu, M.; Wang, S.T.; Zhou, Y.; An, C.H.; Zhang, J. Ni Foam-Supported Carbon-Sheathed $NiMoO_4$ Nanowires as Integrated Electrode for High-Performance Hybrid Supercapacitors. *ACS Sustain. Chem. Eng.* **2017**, *5*, 5964–5971. [CrossRef]
27. Xu, Y.K.; Xuan, H.C.; Gao, J.H.; Liang, T.; Han, X.K.; Yang, J.; Zhang, Y.Q.; Li, H.; Han, P.D.; Du, Y.W. Hierarchical three-dimensional $NiMoO_4$-anchored rGO/Ni foam as advanced electrode material with improved supercapacitor performance. *J. Mater. Sci.* **2018**, *53*, 8483–8498. [CrossRef]
28. Huang, Z.Y.; Zhang, Z.; Qi, X.; Ren, X.H.; Xu, G.H.; Wan, P.B.; Sun, X.M.; Zhang, H. Wall-like hierarchical metal oxide nanosheet arrays grown on carbon cloth for excellent supercapacitor electrodes. *Nanoscale* **2016**, *8*, 13273–13279. [CrossRef]
29. Huang, Y.P.; Cui, F.; Zhao, Y.; Lian, J.B.; Bao, J.; Liu, T.X.; Li, H.M. $NiMoO_4$ nanorod deposited carbon sponges with ant-nest-like interior channels for high-performance pseudocapacitors. *Inorg. Chem. Front.* **2018**, *5*, 1594–1601. [CrossRef]
30. Murugan, E.; Govindaraju, S.; Santhoshkumar, S. Hydrothermal synthesis, characterization and electrochemical behavior of $NiMoO_4$ nanoflower and $NiMoO_4$/rGO nanocomposite for high-performance supercapacitors. *Electrochim. Acta* **2021**, *392*, 138973. [CrossRef]
31. Huang, B.J.; Yao, D.C.; Yuan, J.J.; Tao, Y.R.; Yin, Y.X.; He, G.Y.; Chen, H.Q. Hydrangea-like $NiMoO_4$-Ag/rGO as Battery-type electrode for hybrid supercapacitors with superior stability. *J. Colloid Interf. Sci.* **2022**, *606*, 1652–1661. [CrossRef] [PubMed]
32. Yao, P.P.; Li, C.Y.; Yu, J.L.; Zhang, S.; Zhang, M.; Liu, H.C.; Ji, M.W.; Cong, G.T.; Zhang, T.; Zhu, C.Z.; et al. High performance flexible energy storage device based on copper foam supported $NiMoO_4$ nanosheets-CNTs-CuO nanowires composites with core-shell holey nanostructure. *J. Mater. Sci. Technol.* **2021**, *85*, 87–94. [CrossRef]
33. Zhu, D.; Sun, X.; Yu, J.; Liu, Q.; Liu, J.Y.; Chen, R.R.; Zhang, H.S.; Song, D.L.; Li, R.M.; Wang, J. Three-dimensional heterostructured polypyrrole/nickel molybdate anchored on carbon cloth for high-performance flexible supercapacitors. *J. Colloid Interf. Sci.* **2020**, *574*, 355–363. [CrossRef] [PubMed]
34. Hao, Y.Y.; Huang, H.; Wang, Q.Q.; Wang, Q.; Zhou, G.W. Nitrogen-doped carbon/$NiMoO_4$ nanospheres assembled by nanosheets and ultrasmall nanoparticles for supercapacitors. *Chem. Phys. Lett.* **2019**, *728*, 215–223. [CrossRef]
35. Tong, B.L.; Wei, W.T.; Chen, X.L.; Wang, J.; Ye, W.Y.; Cui, S.Z.; Chen, W.H.; Mi, L.W. Designed synthesis of porous $NiMoO_4$/C composite nanorods for asymmetric supercapacitors. *Cryst. Eng. Comm.* **2019**, *21*, 5492–5499. [CrossRef]
36. Li, Y.; Zhang, S.M.; Ma, M.Y.; Mu, X.M.; Zhang, Y.X.; Du, J.W.; Hu, Q.; Huang, B.Y.; Hua, X.H.; Liu, G.; et al. Manganese-doped nickel molybdate nanostructures for high-performance asymmetric supercapacitors. *Chem. Eng. J.* **2019**, *372*, 452–461. [CrossRef]
37. Yuan, J.J.; Yao, D.C.; Jiang, L.; Tao, Y.R.; Che, J.F.; He, G.Y.; Chen, H.Q. Mn-Doped $NiMoO_4$ Mesoporous Nanorods/Reduced Graphene Oxide Composite for High-Performance All-Solid-State Supercapacitor. *ACS Appl. Energy Mater.* **2020**, *3*, 1794–1803. [CrossRef]
38. Wang, F.F.; Ma, K.; Tian, W.; Dong, J.C.; Han, H.; Wang, H.P.; Deng, K.; Yue, H.R.; Zhang, Y.X.; Jiang, W.; et al. P-Doped $NiMoO_4$ parallel arrays anchored on cobalt carbonate hydroxide with oxygen vacancies and mass transfer channels for supercapacitors and oxygen evolution. *J. Mater. Chem. A* **2019**, *7*, 19589–19596. [CrossRef]
39. Sharma, P.; Sundaram, M.M.; Watcharatharapong, T.; Laird, D.; Euchner, H.; Ahuja, R. Zn Metal Atom Doping on the Surface Plane of One-Dimesional $NiMoO_4$ Nanorods with Improved Redox Chemistry. *ACS Appl. Mater. Interfaces* **2020**, *12*, 44815–44829. [CrossRef]
40. Cui, S.Z.; Wang, F.Q.; Sun, K.J.; Wang, X.; Hu, Q.Z.; Peng, H.; Ma, G.F.; Lei, Z.Q. High-Performance Hybric Supercapacitors Based on Ce-Doped $NiMoO_4$ Nanosheets and Fe_3O_4@Bi_2O_3 Nanoarrays. *J. Phys. Chem. C* **2021**, *125*, 18129–18140. [CrossRef]
41. Qing, C.; Yang, C.X.; Chen, M.Y.; Li, W.H.; Wang, S.Y.; Tang, Y.W. Design of oxygen-deficient $NiMoO_4$ nanoflake and nanorod arrays with enhanced supercapacitive performance. *Chem. Eng. J.* **2018**, *354*, 182–190. [CrossRef]
42. Sivakumar, P.; Raj, C.J.; Park, J.; Jung, H. Synergistic effects of nanoarchitecture and oxygen vacancy in nickel molybdate hollow sphere towards a high-performance hybrid supercapacitor. *Int. J. Energy Res.* **2021**, *45*, 21516–21526. [CrossRef]
43. Zhu, S.; Le, J.Y.; Mao, Y.J.; Chen, S.X.; Han, X.X.; Zeng, Z.L.; Wang, J.; Deng, S.G. Synergistic engineering of fluorine doping and oxygen vacancies towards high-energy and long-lifespan flexible solid-state asymmetric supercapacitor. *Ionics* **2021**, *27*, 2649–2658. [CrossRef]
44. Zhang, X.Y.; Wei, L.; Guo, X. Ultrathin mesoporous $NiMoO_4$-modified MoO_3 core/shell nanostructures: Enhanced capacitive storage and cycling performance for supercapacitors. *Chem. Eng. J.* **2018**, *353*, 615–625. [CrossRef]
45. Shen, J.W.; Wang, Q.G.; Zhang, K.; Wang, S.M.; Li, L.; Dong, S.B.; Zhao, S.T.; Chen, J.; Sun, R.S.; Wang, Y.; et al. Flexible carbon cloth based solid-state supercapacitor from hierarchical holothurian-morphological $NiCo_2O_4$@$NiMoO_4$/PANI. *Electrochim. Acta* **2019**, *320*, 134578. [CrossRef]
46. Xu, R.; Lin, J.M.; Wu, J.H.; Huang, M.L.; Fan, L.Q.; Xu, Z.D.; Song, Z.Y. A high-performance pseudocapacitive electrode material for supercapacitors based on the unique $NiMoO_4$/NiO nanoflowers. *Appl. Surf. Sci.* **2019**, *463*, 721–731. [CrossRef]
47. Yu, D.Y.; Zhang, Z.Q.; Teng, Y.F.; Meng, Y.N.; Wu, Y.P.; Liu, X.L.; Hua, Y.J.; Zhao, X.D.; Liu, X.Y. Fabrication of CuO@$NiMoO_4$ core-shell nanowire arrays on copper foam and their application in high-performance all-solid-state asymmetric supercapacitors. *J. Power Sources* **2019**, *440*, 227164. [CrossRef]

48. Liu, Y.; Ma, Z.L.; Xin, N.; Ying, Y.L.; Shi, W.D. High-performance supercapacitor based on highly active P-doped one-dimension/two-dimension hierarchical $NiCo_2O_4/NiMoO_4$ for efficient energy storage. *J. Colloid Interf. Sci.* **2021**, *601*, 793–802. [CrossRef]
49. Zeng, Y.; Liao, J.Z.; Wei, B.B.; Huang, Z.; Zhu, W.J.; Zheng, J.X.; Liang, H.F.; Zhang, Y.Z.; Wang, Z.C. Tuning the electronic structure of $NiMoO_4$ by coupling with SnO2 for high-performance hybrid supercapacitors. *Chem. Eng. J.* **2021**, *409*, 128297. [CrossRef]
50. Chen, C.; Yan, D.; Luo, X.; Gao, W.J.; Huang, G.J.; Han, Z.W.; Zeng, Y.; Zhu, Z.H. Construction of Core-Shell $NiMoO_4$@Ni-Co-S Nanorods as Advanced Electrodes for High-Performance Asymmetric Supercapacitors. *ACS Appl. Mater. Interfaces* **2018**, *10*, 4662–4671. [CrossRef]
51. Chen, F.S.; Ji, S.; Liu, Q.B.; Wang, H.; Liu, H.; Brett, D.J.L.; Wang, G.X.; Wang, R.F. Rational Design of Hierarchically Core-Shell Structured Ni_3S_2@$NiMoO_4$ Nanowires for Electrochemical Energy Storage. *Small* **2018**, *14*, 1800791. [CrossRef] [PubMed]
52. Acharya, J.; Ojha, G.P.; Kim, B.S.; Pant, B.; Park, M. Modish Designation of Hollow-Tubular rGO-$NiMoO_4$@Ni-Co-S Hybrid Core-shell Electrodes with Multichannel Superconductive Pathways for High-Performance Asymmetric Supercapacitors. *ACS Appl. Mater. Interfaces* **2021**, *13*, 17487–17500. [CrossRef] [PubMed]
53. Hu, K.L.; Jeong, S.; Elumalai, G.; Kukunuri, S.; Fujita, J.; Ito, Y. Phase-Dependent Reactivity of Nickel Molybdates for Electrocatalytic Urea Oxidation. *ACS Appl. Energy Mater.* **2020**, *3*, 7535–7542. [CrossRef]
54. Wang, Y.; Liu, J.; Chen, X.; Kang, B.; Wang, H.-E.; Xiong, P.; Chen, Q.; Wei, M.; Li, N.; Qian, Q.; et al. Structural engineering of tin sulfides anchored on nitrogen/phosphorus dual-doped carbon nanofibres in sodium/potassium-ion batteries. *Carbon* **2022**, *189*, 46–56. [CrossRef]
55. Nti, F.; Anang, D.A.; Han, J.I. Facilely synthesized $NiMoO_4/CoMoO_4$ nanorods as electrode material for high performance supercapacitor. *J. Alloys Compd.* **2018**, *742*, 342–350. [CrossRef]
56. Zhang, Z.; Liu, Y.; Huang, Z.; Ren, L.; Qi, X.; Wei, X.; Zhong, J. Facile hydrothermal synthesis of $NiMoO_4$@$CoMoO_4$ hierarchical nanospheres for supercapacitor applications. *Phys. Chem. Chem. Phys.* **2015**, *17*, 20795–20804. [CrossRef]
57. Zhao, X.; Wang, H.E.; Chen, X.X.; Cao, J.; Zhao, Y.D.; Neale, Z.G.; Cai, W.; Sui, J.H.; Cao, G.Z. Tubular MoO_2 organized by 2D assemblies for fast and durable alkali-ion storage. *Energy Storage Mater.* **2018**, *11*, 161–169. [CrossRef]
58. Wang, H.E.; Zhao, X.; Yin, K.L.; Li, Y.; Chen, L.H.; Yang, X.Y.; Zhang, W.J.; Su, B.L.; Cao, G.Z. Superior Pseudocapacitive Lithium-Ion Storage in Porous Vanadium Oxides C Heterostructure Composite. *ACS Appl. Mater. Interfaces* **2017**, *9*, 43665–43673. [CrossRef]
59. Wang, H.E.; Yin, K.L.; Qin, N.; Zhao, X.; Xia, F.J.; Hu, Z.Y.; Guo, G.L.; Cao, G.Z.; Zhang, W.J. Oxygen-deficient titanium dioxide as a functional host for lithium-sulfur batteries. *J. Mater. Chem. A* **2019**, *7*, 10346–10353. [CrossRef]
60. Perdew, J.P.; Burke, K.; Ernzerhof, M. Generalized Gradient Approximation Made Simple. *Phys. Rev. Lett.* **1996**, *77*, 3865–3868. [CrossRef]
61. Kresse, G.; Joubert, D. From ultrasoft pseudopotentials to the projector augmented-wave method. *Phys. Rev. B* **1999**, *59*, 1758–1775. [CrossRef]
62. Momma, K.; Izumi, F. VESTA 3 for three-dimensional visualization of crystal, volumetric and morphology data. *J. Appl. Cryst.* **2011**, *44*, 1272–1276. [CrossRef]
63. Wang, V.; Xu, N.; Liu, J.C.; Tang, G.; Geng, W.T. VASPKIT: A user-friendly interface facilitating high-throughput computing and analysis using VASP code. *Comp. Phys. Comm.* **2021**, *267*, 108033. [CrossRef]
64. Zhao, Y.; Zhang, P.; Fu, W.; Ma, X.; Zhou, J.; Zhang, X.; Li, J.; Xie, E.; Pan, X. Understanding the role of Co_3O_4 on stability between active hierarchies and scaffolds: An insight into $NiMoO_4$ composites for supercapacitors. *Appl. Surf. Sci.* **2017**, *416*, 160–167. [CrossRef]
65. Xuan, H.; Xu, Y.; Zhang, Y.; Li, H.; Han, P.; Du, Y. One-step combustion synthesis of porous CNTs/C/$NiMoO_4$ composites for high-performance asymmetric supercapacitors. *J. Alloys Compd.* **2018**, *745*, 135–146. [CrossRef]
66. Huang, Y.; Cui, F.; Zhao, Y.; Lian, J.; Bao, J.; Li, H. Controlled growth of ultrathin $NiMoO_4$ nanosheets on carbon nanofiber membrane as advanced electrodes for asymmetric supercapacitors. *J. Alloys Compd.* **2018**, *753*, 176–185. [CrossRef]
67. Hussain, S.; Javed, M.S.; Asim, S.; Shaheen, A.; Khan, A.J.; Abbas, Y.; Ullah, N.; Iqbal, A.; Wang, M.; Qiao, G. Novel gravel-like $NiMoO_4$ nanoparticles on carbon cloth for outstanding supercapacitor applications. *Ceram. Int.* **2020**, *46*, 6406–6412. [CrossRef]
68. Feng, X.; Ning, J.; Wang, D.; Zhang, J.; Xia, M.; Wang, Y.; Hao, Y. Heterostructure arrays of $NiMoO_4$ nanoflakes on N-doping of graphene for high-performance asymmetric supercapacitors. *J. Alloys Compd.* **2020**, *816*, 152625. [CrossRef]
69. Zhong, Y.; Liu, T.; Zhang, A.; Cui, L.; Liu, X.; Zheng, R.; Liu, J. Controllable synthesis of $Ni_{1-x}Co_xMoO_4$ with tunable morphologies for high-performance asymmetric supercapacitors. *J. Alloys Compd.* **2021**, *850*, 156734. [CrossRef]
70. Chen, L.; Deng, W.; Chen, Z.; Wang, X. Hetero-architectured core–shell $NiMoO_4$@Ni_9S_8/MoS_2 nanorods enabling high-performance supercapacitors. *J. Mater. Res.* **2022**, *37*, 284–293. [CrossRef]
71. Muthu, D.; Vargheese, S.; Haldorai, Y.; Kumar, R.T.R. $NiMoO_4$/reduced graphene oxide composite as an electrode material for hybrid supercapacitor. *Mater. Sci. Semicond. Proc.* **2021**, *135*, 106078. [CrossRef]
72. Yi, T.-F.; Qiu, L.-Y.; Mei, J.; Qi, S.-Y.; Cui, P.; Luo, S.; Zhu, Y.-R.; Xie, Y.; He, Y.-B. Porous spherical NiO@$NiMoO_4$@PPy nanoarchitectures as advanced electrochemical pseudocapacitor materials. *Sci. Bull.* **2020**, *65*, 546–556. [CrossRef]

nanomaterials

MDPI

Review

Research Progress on Sound Absorption of Electrospun Fibrous Composite Materials

Xiuhong Li [1], Yujie Peng [1], Youqi He [1], Chupeng Zhang [1], Daode Zhang [1,*] and Yong Liu [2,*]

[1] School of Mechanical Engineering, Hubei University of Technology, Wuhan 430068, China; 20200005@hbut.edu.cn (X.L.); pengyj1024@163.com (Y.P.); heyouqi1995@163.com (Y.H.); zcp1988123@126.com (C.Z.)

[2] Beijing Key Laboratory of Advanced Functional Polymer Composites, College of Materials Science and Engineering, Beijing University of Chemical Technology, Beijing 100029, China

* Correspondence: hgzdd@126.com (D.Z.); yongliu@mail buct.edu.cn (Y.L.)

Abstract: Noise is considered severe environmental pollutant that affects human health. Using sound absorption materials to reduce noise is a way to decrease the hazards of noise pollution. Micro/nanofibers have advantages in sound absorption due to their properties such as small diameter, large specific surface area, and high porosity. Electrospinning is a technology for producing micro/nanofibers, and this technology has attracted interest in the field of sound absorption. To broaden the applications of electrospun micro/nanofibers in acoustics, the present study of electrospun micro/nano fibrous materials for sound absorption is summarized. First, the factors affecting the micro/nanofibers' sound absorption properties in the process of electrospinning are presented. Through changing the materials, process parameters, and duration of electrospinning, the properties, morphologies, and thicknesses of electrospun micro/nanofibers can be controlled. Hence, the sound absorption characteristics of electrospun micro/nanofibers will be affected. Second, the studies on porous sound absorbers, combined with electrospun micro/nanofibers, are introduced. Then, the studies of electrospun micro/nanofibers in resonant sound absorption are concluded. Finally, the shortcomings of electrospun micro/nano fibrous sound absorption materials are discussed, and the future research is forecasted.

Keywords: electrospun micro/nanofibers; sound absorption; porous materials; resonance

Citation: Li, X.; Peng, Y.; He, Y.; Zhang, C.; Zhang, D.; Liu, Y. Research Progress on Sound Absorption of Electrospun Fibrous Composite Materials. *Nanomaterials* **2022**, *12*, 1123. https://doi.org/10.3390/nano12071123

Academic Editors: Jihoon Lee and Ming-Yu Li

Received: 2 March 2022
Accepted: 22 March 2022
Published: 29 March 2022

1. Introduction

With the development of industrial production and urban construction, noise pollution is one of the four significant pollutants currently becoming an increasingly severe issue, highly endangering human health and restricting social development [1–3]. Therefore, controlling noise is essential, and using sound absorption materials is an efficient way to control noise pollution [4]. According to the sound absorption mechanisms, sound absorption materials can be classified as either porous or resonant sound absorption materials [5]. Porous sound absorption materials possess a large number of internal pores. When sound waves occur on their surfaces, friction between the materials and the air in the pores is generated. As a result, the sound energy will be converted into thermal energy and consumed due to the viscous and thermal effects. Resonant sound absorption materials are efficient for absorbing sound waves at low frequency, which are the equivalent of Helmholtz resonator, and consume sound energy by the resonance effect [6,7]. Porous sound absorption materials exhibit a good sound performance at high frequencies but perform poorly in the low and medium frequencies. However, resonant sound absorption materials show an opposite sound performance compared with the former. In addition, these materials also show disadvantages such as narrow sound absorption bands and bulky structures that need to be compounded with other materials to obtain a broader sound absorption band and a better sound absorption performance at high frequencies [8].

Low-frequency sound waves can easily bypass obstacles and are difficult to be absorbed by traditional sound absorption materials [9]. However, nanofibers with a high specific surface area can efficiently absorb the low-frequency sound waves. The large specific surface area of electrospun micro/nanofibers promotes the collision between the fibers and sound waves at low and medium frequencies, while the internal interconnected pores allow the diffusion of sound waves through the material, both of which boost the consumption of acoustic energy. Electrospinning is a technique for producing fibers with a diameter ranging from micron to nano scales, and it has been recently used to produce sound absorption materials [10–12].

Electrospun micro/nanofibers possess many advantages, such as small diameter, light weight, high specific surface area, and porosity. The sound absorption materials made of electrospun micro/nanofibers are promising for noise reduction, especially in traffic [13–15]. The sound absorption mechanism of electrospun micro/nanofibers is complex, and their sound absorption performance is mainly affected by several factors: material, morphology, thickness, spatial structure, and cavity depth. The type of materials is related to the properties of electrospun micro/nanofibers, such as piezoelectricity [16–18] and viscoelasticity [19], which influence the acoustic loss. Electrospun fiber properties such as fiber diameter [20], surface density [21,22], and microstructure [23,24] play essential roles in the process of sound absorption. The increase in the thickness of electrospun micro/nanofibers will not only lengthen the pore channels inside the material, but also enlarge the contact area between the material and acoustic waves, making it easier to consume acoustic energy [25]. Compared with 2D electrospun fiber membranes, three-dimensional (3D) electrospun fibers possess a larger surface area, manageable pore size, and mesoporous channels [26]. These features promote the consumption of sound energy, resulting in better sound absorption capacity of the 3D electrospun fibers [27,28]. Furthermore, setting a cavity behind the electrospun micro/nanofibers can form a Helmholtz resonator structure, improving their resonant sound absorption capacities [29,30].

Fibers fabricated by electrospinning have micro and nanoscale diameters, which provide high specific surface area and porosity. These features of electrospun micro/nanofibers lead to good sound absorption performance in low and medium frequencies [31]. In this review, we provide a focus on factors related to the sound absorption capacities of electrospun micro/nanofibers and their applications. Figure 1 presents an outline of this paper. First, the principles of how the materials and properties of electrospun micro/nanofibers affect the sound absorption performance are discussed. Second, recent research on electrospun micro/nanofiber sound absorption materials, which are classified as porous or resonant, is summarized. Finally, we outline the existing disadvantages of sound absorption materials produced by electrospinning and the prospects for their development.

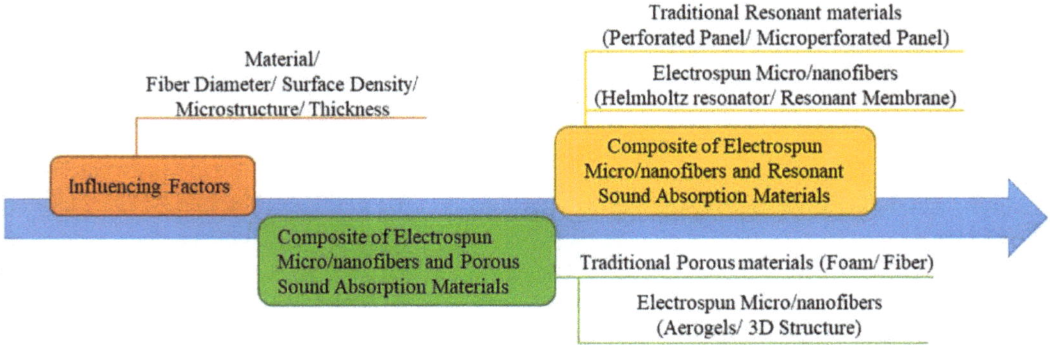

Figure 1. The outline map of this review.

2. Factors Influencing Sound Absorption Performance of Electrospun Micro/Nanofibers

Micro/nanofibers possess characteristics which are beneficial to the absorption of sound waves, namely small diameter, large specific surface area, and high porosity. However, micro/nanofibers usually have thin thicknesses, resulting in poor sound absorption. Micro/nanofibers are commonly combined with traditional sound absorption materials to achieve a better sound absorption effect. Electrospinning, one of the main methods of preparing micro/nanofibers, is divided into two types: solution and melt electrospinning [32]. There are significant limitations to melt electrospinning, such as a larger fiber diameter, the relative complexity of the design, and complicated devices [33]. Therefore, most researchers choose to fabricate micro/nanofibers by solution electrospinning.

The factors that influence the sound absorption performance of electrospun micro/nanofibers involve parameters such as materials, fiber diameter, surface density, microstructure, and thickness. Materials determine the properties of electrospun micro/nanofibers, such as piezoelectric and elastic properties. The fiber diameter, surface density, and microstructure of the fibers are related to the solution properties, process parameters, and environmental parameters of the electrospinning technology [34]. The thickness of electrospun micro/nanofiber membranes can be controlled by adjusting the electrospinning duration. Table 1 represents the recent research of the sound absorption coefficient (SAC) and properties of electrospun micro/nanofibers. The sound absorption capacity of the materials can be reflected by SAC, which is the ratio between the energy absorbed by the material and the total energy of the incident sound wave.

Table 1. Summary of the properties and sound absorption performance of electrospun micro/nanofibers.

Materials	Fiber Diameter and Surface Density	Microstructure	Thickness of Nanofibrous Structures	SACs	Reference
Polyvinylidene fluoride (PVDF)/carbon nanotubes (CNTs) nanofibers and foam	138 ± 21 nm 45.5 g/m^2	-	-	above 0.9 (1000 Hz)	[16]
PVDF/graphene (GP) nanofibers and acoustic nonwoven	169 ± 21 nm 45.5 g/m^2	-	-	0.87 (1000 Hz) 0.95 (4000 Hz)	[17]
Nylon-6 nanofibers and Polyurethane (PU) foam PU nanofibers and PU foam	180 ± 10 nm 300 ± 10 nm	-	-	0.81 (600 Hz) 0.59 (1700 Hz)	[35]
PU nanofibers and polyethylene terephthalate (PET) nonwovens	509.9 nm	-	1.548 mm	0.9 (1800 Hz)	[36]
Polyacrylonitrile (PAN) nanofiber and spacer-knitted fabrics	110 ± 7 nm 17 g/m^2	-	50 μm	0.7 (100–3200 Hz)	[37]
PAN/cellulose nanocrystals (CNC) aerogels	-	maze-like structure	50 mm	above 0.9 (600 Hz)	[38]
Poly (vinyl alcohol) (PVA) nanofibers	268 nm	Miura-ori structure	130 ± 5 μm	max value is 1.0	[39]
PU nanofibers and nonwovens	5–40 nm	nano-cobweb structure	1.2 mm	average value is 0.57	[40]
Polyvinylpyrrolidone (PVP) nonwoven mats of stacked nanofibrous layers	1.6/2.8 ± 0.5 μm 0.89 kg/m^2	-	2.54 cm	max value is above 0.9	[41]
PAN nanofiber and perforated panel	333 ± 58 nm	-	205 ± 4 μm	max value is 0.93	[42]
Nylon-6 nanofibers with back cavity (30 mm)	150–200 nm	-	10 μm	0.6 (2000 Hz)	[43]

Acoustic models can be applied to theoretically evaluate the sound absorption properties of the materials, which are useful for directing the design of sound absorption materials. Recently, several acoustic models used for investigating the SAC of sound absorption materials have been published [39,44]. Previously, there were two types of acoustic models commonly employed [4]. The first is the empirical model with few parameters, which is relatively easy to establish. The most representative empirical model is the Delany–Bazley model, related to the airflow resistivity of the materials, but it can only be employed for predicting the acoustic behaviors of porous materials in the frequency range of 250–4000 Hz [45,46]. In order to obtain more accurate predictions, some studies made specific corrections to the Delany–Bazley model and developed several new models. For example, Miki proposed a modified expression, known as the Delany–Bazley–Miki model, based on the Delany–Bazley model [47,48]. The second acoustic model is the phenomenological model. Compared to the empirical models, it involves non-acoustical physical parameters and can improve the prediction accuracy. In terms of phenomenological model, the Johnson–Champoux–Allard (JCA) model is the most-used method for describing the sound propagation in sound absorption porous materials [49,50].

2.1. Materials

The sorts of materials determine the properties of electrospun micro/nanofibers. Modifying the materials by altering the material or adding other matter can change the physical and chemical properties of the fibers, thus affecting the sound absorption performance of the materials.

Polyvinylidene fluoride (PVDF) is a material with the piezoelectric property that can convert acoustic energy into electric energy, thus facilitating the absorption of acoustic energy [51]. Wu et al. [16–18] conducted a series of studies on electrospun PVDF nanofibers. They added carbon nanotubes (CNTs), GP, and silver nanoparticles (AgNPs) into PVDF solutions and fabricated them into nanofiber membranes by electrospinning. The results show that the addition of CNTs and GP increases the surface area and the β-phase crystallinity of PVDF nanofiber membranes, enhancing the contact between materials and sound waves and improving piezoelectricity, thereby promoting the absorption of sound waves at low frequency. PVDF/AgNPs nanofiber membranes have excellent piezoelectricity and acoustoelectric conversion characteristics. The composite nanofiber membranes can convert acoustic energy into other forms of energy and show great potential for sound absorption applications. Another group reported that electrospun polyacrylonitrile (PAN) nanofiber membranes display greater piezoelectric conversion capability compared to PVDF nanofibers [52]. For this purpose, Shao et al. [53] prepared an acoustoelectric device by interposing an electrospun PAN fibrous membrane to two metal-coated polyethylene terephthalate (PET) films. The electrospun PAN nanofiber membranes can convert noise into electric power at low and medium frequencies. Furthermore, they prepared a single-layer nanofiber membrane made of a PAN/PVDF polymer blend, which possesses an acoustoelectric energy conversion efficiency as high as 25.6%. Figure 2 shows the noise harvester structure [54].

The elasticity of the materials may influence the resonant process between electrospun micro/nanofibers and sound waves. Good elasticity of materials may promote the consumption of acoustic energy by the resonance effect, thus improving the sound absorption performance of the materials. Park et al. [35] prepared polyurethane (PU) and nylon 6 composite nanofibers by electrospinning and then laminated them with polyurethane foam. They found that PU composite nanofibers perform better in sound absorption than nylon 6 composite nanofibers. Furthermore, the PU nanofibers exhibit superior air permeability and elasticity, and the sound waves can easily propagate into the material, leading to the increased consumption of acoustic energy by vibration. On the contrary, some researchers believe that electrospun micro/nanofibers with less elasticity have more friction with sound waves, promoting efficient sound absorption. Rabbi et al. [55] demonstrated the effects of applying PU and PAN nanofibers within polyester and wool nonwovens. Owing to the higher air permeability and elasticity of PU compared with PAN nanofiber layers, the composite with PAN nanofibers possesses a higher sound transmission loss.

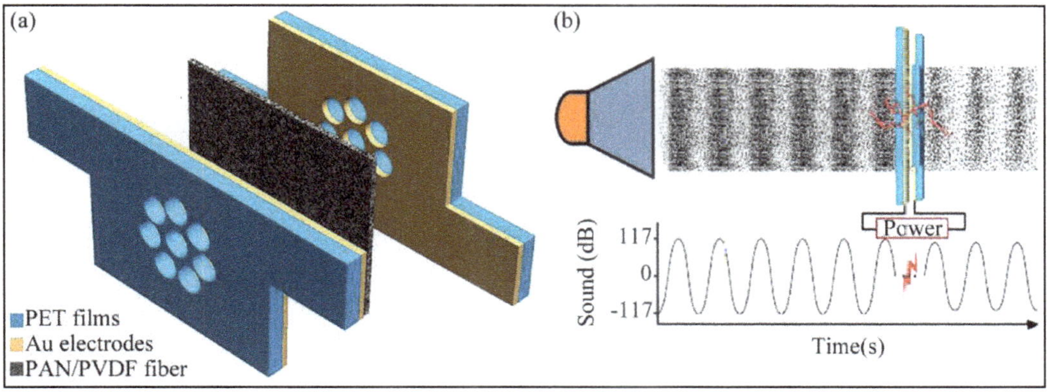

Figure 2. (**a**) Schematic illustration of the PAN/PVDF noise harvester structure; (**b**) Principle of sound energy harvest. Reprinted with permission from ref. [54]. Copyright 2021 Elsevier, Amsterdam, The Netherlands.

2.2. Diameter and Surface Density

Diameter and surface density are the critical factors affecting the flow resistivity and natural resonant frequency of the materials. Diameter impacts the specific surface area and porosity of electrospun micro/nanofibers, while the surface density determines the fibers' distribution. The sound absorption capability of materials can be represented by flow resistivity, which reflects the extent of curvature on the path of sound waves traveling through the materials. Materials with a low natural resonant frequency usually exhibit good sound absorption capacity at low frequency. Thus, suitable diameter and surface density for electrospun micro/nanofibers enable good sound absorption. For instance, Akasaka et al. [44] prepared nonwoven sheets composed of thin silica fibers with diameters ranging from 0.72 to 3.44 μm by electrospinning. They found that the sound absorption performance of the silica fibers is related to a critical value (approximately 3 μm) of the fiber diameter. When the fiber diameter is above the critical value, the sound absorption coefficient of the composite increases with the decrease in the fiber diameter in a wide frequency range. The diameter and surface density of electrospun micro/nanofibers can be controlled by varying the parameters of the electrospinning process. Shou [20] evaluated the effects of the processing parameters such as solution concentration, voltage, collecting distance, the diameter of the sprayer, and environmental temperature on the diameter of the electrospun fibers. The diameter of electrospun fibers can be changed by adjusting the processing parameters mentioned above, thus altering the flow resistivity and sound absorption efficiency of the materials. Du [56] adopted a nano spider spinning device to fabricate PAN nanofiber membranes and controlled the diameter and surface density of the fiber membranes by adjusting the voltage and transmission speed of the receiving device. The sound absorption performance of the composites composed of electrospun nanofiber membranes and nonwoven fabrics was tested. The results demonstrate that the magnitude of fiber diameter and surface density is inversely proportional to voltage, within a certain range. The increment of the transmission speed of the collector decreases the surface density of the composite and increases its porosity. When the sound waves' frequency is higher than 500 Hz, increasing the surface density of the nanofiber membrane or decreasing its diameter can improve the sound absorption coefficient of the composites. Kalinová [57] studied the effects of feeding rate and collecting distance on surface density and fiber diameter of electrospun nanofiber membranes. It was found that electrospun nanofiber membranes' surface density decreases with the increase in feed rate, and the larger the receiving distance, the smaller diameter. In addition, the resonant frequency of the nanofiber membrane is inversely proportional to both the average fiber diameter and the surface density.

2.3. Microstructure

The microstructures of the electrospun micro/nanofibers influence the sound absorption performance of the materials. For example, the bead structure will increase the porosity, surface area, and bulkiness of the fiber, causing irregular changes in the fiber diameter. Moreover, the core-shell or hollow structure may influence the density and natural resonant frequency of the electrospun micro/nanofibers, therefore modifying their response to sound waves at different frequencies. Therefore, electrospun micro/nanofibers with different microstructures can be fabricated by adjusting the process parameters and devices of electrospinning, and their sound absorption performance will be influenced. Zhang [23] discussed the relationship between poly(lactic acid) (PLA) solution concentration and the morphology of the electrospun micro/nanofibers and observed that the number of bead structures decreases with the increase in solution concentration. When the PLA solution concentration is 13% and 14%, the electrospun nanofibers with the fine diameter and a few bead structures perform best in sound absorption. Similarly, Gao [58] fabricated polyvinyl alcohol (PVA) electrospun fiber mats with different bead morphologies by varying the solution concentration and investigated the sound absorption performance of the composites with nonwoven fabrics. The results indicate that the bead structures appear in the electrospun fiber mats with 3 wt.% to 7 wt.% concentration of PVA solution. The diameter and the number of the beads will increase as the concentration of PVA solution decreases. Yoon et al. [59] prepared a micro-glass bead/PLA porous fiber composite by electrospinning. However, the sound absorption of the porous fibers is lower than normal fibers because the porous structures do not change the void ratio of the composite, but have increased its natural frequency. The electrospun micro/nanofibers with unique microstructures can also be produced using specially designed electrospinning devices. For instance, the core-shell or hollow structure may influence the fibers' density and resonance frequency, therefore modifying their response to different frequencies of sound waves. Bertocchi et al. [24] fabricated a core-shell fiber with polycaprolactone (PCL) as the surface layer and polyethylene glycol (PEG) as the core layer by using a coaxial electrospinning device that they installed themselves. Compared with single-phase fibers, the core-shell fiber structure exhibits a higher absorption capacity for all noises. The sound absorption ability of the core-shell fibers is proportional to the fluid viscosity of the core layer, and its sound absorption frequency range can be tuned by controlling the core layer's fluid viscosity.

2.4. Thickness

The thickness of electrospun micro/nanofibers is a critical factor for their acoustic performance, affecting the propagation distance of sound waves within the fibers. Thickness variation of electrospun micro/nanofibers can be achieved by adjusting the electrospinning duration. Zou [60] reported that the sound absorption performance of PU and PVDF nanofiber mats prepared by electrospinning for 2 h is similar to traditional foam. As the electrospinning time increased to 4 h, the sound absorption performance of PU and PVDF nanofiber mats was significantly enhanced. The sound absorption coefficient of the samples is above 0.5 when the frequency is nearly 1000 Hz, reflecting the excellent sound absorption performance of electrospun fibers at low and medium frequencies. Salehi et al. [61] reported a study applying multi-layered PET nonwoven structures integrated with PAN nanofibers. They obtained different deposition amounts of composite fibers using three electrospinning times of 15, 60, and 180 min. They suggested that the sample with the electrospinning time of 60 min shows the best sound absorption performance. The increase in nanofiber deposition amount can improve the sound absorption coefficient of nanofibers, but the best sound absorption performance could not be achieved by increasing deposition time continuously. Similarly, Özkal et al. [36] produced new sound absorption materials by incorporating PU nanofibers of several electrospinning durations (5, 20, 60, and 120 min) with recycled PET bottle waste nonwovens. It is found that the resonant frequency of the composite decreases with the increase in the spinning duration. Ding et al. [40] prepared multi-layer PU nano-membranes with a nano-cobweb structure using a needleless electrospinning method. The nano-cobweb multi-layer material with an average sound absorption coefficient of 0.63, prepared by electrospinning for 240 min, dramatically improves the sound absorption performance over the multi-layer. Avossa et al. [41] obtained polyvinylpyrrolidone (PVP) nonwoven mats of stacked nanofibrous layers by electrospinning that yielded reduced thickness and excellent sound absorption properties in the low and medium frequency range. The PVP mats' sound absorption coefficient is much higher than 0.9 at 450 Hz, and its sound absorption performance can be continuously tuned by changing the mass. Ji et al. [62] designed a sound absorber made of electrospun poly (vinylidene fluoride-co-hexafluoropropylene) (PVDF-HFP) fibrous membrane and melamine foam. The result shows that the sound absorption performance of the sound absorber is significantly affected by the thickness of the two materials.

3. Electrospun Micro/Nanofiber-Based Porous Sound Absorption Materials

Porous sound absorption materials have many internal pores, which are beneficial for sound waves to enter the interior of the materials and consume energy by friction. The sound absorption principle of porous sound absorption materials is shown in Figure 3 [63]. Porous sound absorption materials perform well in sound absorption and possess advantages of low cost, easy formation, and light weight. Porous acoustic materials are ideal noise reduction materials that can reduce noise in construction and transportation fields [4]. Foams and fibers are commonly used as porous sound absorption materials. Both possess high sound absorption coefficients at the high-frequency range, but show poor sound absorption at the low and medium frequencies [44]. The poor sound absorption performance of porous sound absorption materials in the low and medium frequencies limits their applications. To broaden the applications of porous sound absorption materials, researchers combine them with electrospun micro/nanofibers, which improves the sound absorption performance by expanding the contact area of the material with acoustic waves and enhances the material properties such as water resistance, high-temperature resistance, and mechanical strength. Furthermore, the addition of lightweight nanofibers will not vastly increase the weight and size of the sound absorption materials [43]. For instance, combining electrospun micro/nanofibers with traditional sound absorption foams and fibers can increase the contact area between the materials and sound waves, thus improving the sound absorption performance at low and medium frequencies. Moreover, electrospinning micro/nanofibers with special structures can be prepared by the refinement of electrospinning devices and material modifications, which shows promising application prospects with a good sound absorption performance at low and medium frequencies.

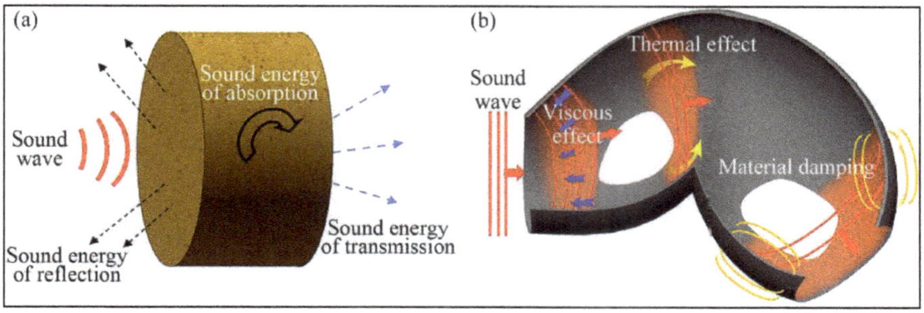

Figure 3. (**a**) Schematic illustration of the sound absorption process for porous materials; (**b**) Process of the sound energy consumption in the porous sound absorption materials. Reprinted with permission from ref. [63]. Copyright 2017 American Institute of Physics, College Park, MD, USA.

3.1. Composite of Sound Absorption Foam and Electrospun Micro/Nanofibers

Foams are typically porous materials with good sound absorption properties, and they have been widely used for sound absorption in the field of transportation [64]. The cellular morphology of foams includes a cavity with three pores (open, partially open, and closed) [65]. Open-cell foams have many interconnected pores, while the pores of the closed-cell foams are isolated from each other. The type of pores affects the sound absorption properties of the foams; the open-cell foams are more efficient than closed-cell foams in sound absorption [66]. The sound absorption principle of partially open-cell foams is similar to that of open-cell foams, but their sound absorption capability is inferior. Therefore, open-cell foams are the primary sound absorption foam materials.

Traditional foam sound absorption materials have large pore sizes and low porosity. The sound waves cannot vibrate sufficiently inside the pores, leading to the poor sound absorption performance of the foams at low frequency. Increasing the thickness of the material or setting a cavity behind the foam are common methods used to promote their sound absorption performance at low frequency. However, these methods will cause the waste of material and space. Laminating electrospun micro/nanofibers with foam materials can enhance the sound absorption performance of foams at all frequencies, especially at low frequencies [67]. Zou et al. [60] made three composite nanofiber mats by solution electrospinning of PU, PVDF, and PU/PVDF blends on the surface of porous foams and examined their sound absorption performance. The sound absorption coefficients of the composites are higher than the typical acoustic foam in the frequency range of 800–1600 Hz. The composite made of foam and PU/PVDF nanofibers has the highest absorption coefficient of nearly 0.7. Tomáš [68] found that the 10 mm-thick foam treated with electrospun nanofibers and the untreated 20 mm-thick foam sheets exhibit similarly shaped absorption coefficient curves in a wide frequency range. The foam combined with nanofiber membrane shows a higher absorption coefficient in the 1500–4000 Hz frequency range. The above studies suggest that composite performances of electrospun micro/nanofibers and foam acoustic materials are better than those of the typical foam acoustic materials.

The sponge is a type of acoustic foam with many applications in noise reduction. Still, its substantial water absorption property in wet environments leads to a poor sound absorption performance. Using electrospun micro/nanofibers to modify the sponge can improve its sound absorption performance and reduce the influence of the environment on sound absorption [35]. Xiang et al. [69] covered the sponge surface with electrospun polymethylsilsesquioxane (PMSQ). The addition of PMSQ enhances the sponge's sound absorption performance, and the resulting composite shows a strong water resistance property. As a result, the composite can be used for noise reduction in aqueous environments. Cao [70] fabricated ultralight polystyrene (PS) nanofiber sponges with excellent sound absorption properties and hydrophobicity using the humidity-assisted electrospinning method. The PS sponges significantly improve the sound absorption coefficients when compared to those of nonwovens and melamine foams in the frequency range of 500–1000 Hz. Furthermore, the PS sponges exhibit durable sound absorption performance at different humidity levels.

3.2. Composite of Sound Absorption Fibers and Electrospun Micro/Nanofibers

Fibrous structures can absorb, reflect, and transmit the incident sound waves simultaneously [71]. According to their compositions, acoustic fiber materials can be classified into natural, synthetic, and metal fibers. Natural and synthetic fibers are widely used due to their low price and superior sound absorption performance at high frequency. Their sound absorption performance at the low and medium frequencies can be effectively improved by compounding with electrospun fibers [72,73].

Composites of electrospun micro/nanofibers and natural fibers have improved sound absorption performance while retaining the environment-friendly and biodegradable features [74]. For instance, Selvaraj et al. [75] formed a composite by covering modified PVA electrospun nanofibers with coir fibers. Figure 4 shows the preparation process and the sound absorption performance of the samples (coir, PVA nanofibers/coir, and Fleshing hydrolysate (FH)/PVA nanofibers/coir). The composite shows a higher sound absorption coefficient in the frequency range of 400–1000 Hz and a lower frequency of sound absorption peaks compared with coir fibers. Na et al. [76] laminated melt electrospun nylon microfibers with wool fabric. The composite performs better than knitted wool fabric in sound absorption at frequencies between 1000 and 4000 Hz, with the highest sound absorption coefficient of 0.85 at 4000 Hz. Ozturk et al. [77] coated electrospun PAN nanofiber membranes on the surface of wool and jute fiber felts, respectively, and inspected their acoustic properties. The results suggest that the addition of nanofiber membranes would change the permeability of the fiber mats, further affecting the sound absorption performance. The sound absorption performance of wool felt is improved by four times after being composited with the PAN nanofibers. The sound absorption coefficient of jute felts could be improved at least 4 to 8 times, and its maximum could reach 0.4 in the frequency between 500 and 1500 Hz by coating nanofibers. In another work, an alternative sound absorption material was designed [78]. The material is composed of nanofiber resonant membranes together with wool felts covering polypropylene (PP) nonwovens, which displays better sound absorption properties than wool felts alone and is suitable for sound absorption on white goods.

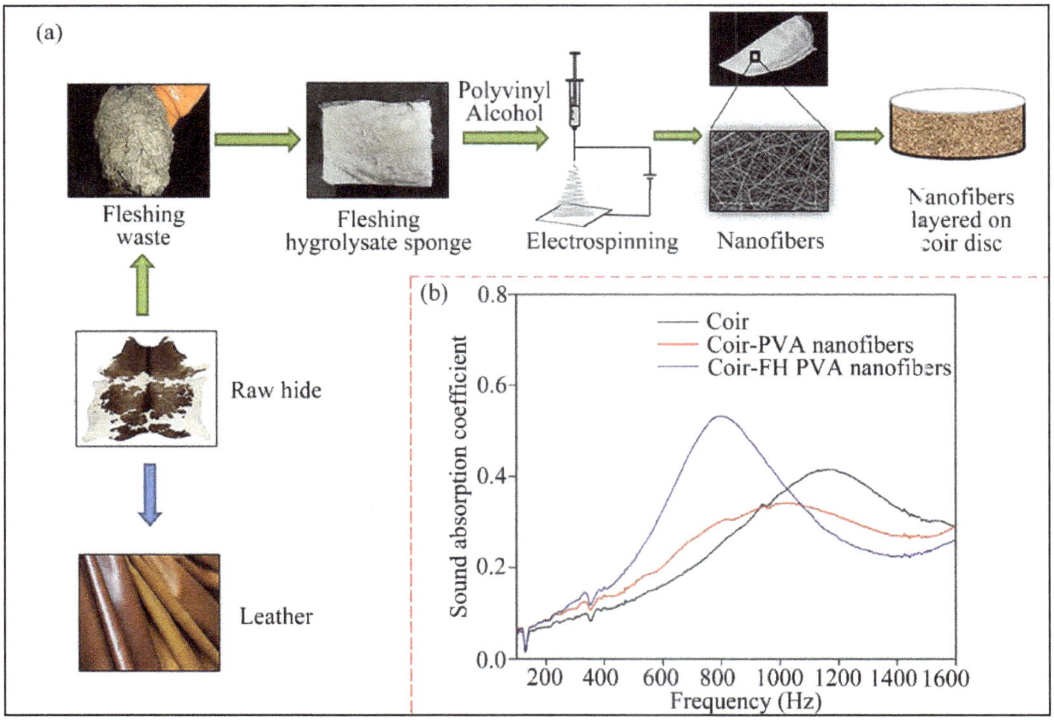

Figure 4. (**a**) The preparation process of nanofibers and coir composites by electrospinning; (**b**) The sound absorption coefficients of the samples. Reprinted with permission from ref. [75]. Copyright 2019 Elsevier.

Synthetic fibers are extensively applied in sound absorption fields due to their high performance and economical price [1,79]. The sound absorption performance of synthetic fibers at low frequency can be improved by combining them with electrospun micro/nanofibers [9,30]. For instance, electrospun nanofibers laminated with nonwoven fibers can effectively improve the sound absorption coefficient of the materials, and the composites display higher absorption coefficients when the nanofiber side is facing the sound source [23,56]. Yang et al. [25] prepared reduced graphene oxide (RGO)/PAN nanofiber nonwovens by applying electrospinning and bi-component melt-blown nonwovens technology. In the range of 500 to 6300 Hz, the sound absorption performance of the composites increased with the addition of RGO, and the sound absorption coefficient can increase by 24%. Liu et al. [80] prepared PVA/polyethylene oxide (PEO) electrospun nanofiber membranes with graphene oxide (GO) of various concentrations and laminated them with nonwoven fabrics. The addition of nanofibers significantly improves the sound absorption performance of the materials at a low frequency. Ozturk et al. [37] fabricated a nanofibrous coated composite consisting of electrospun PAN nanofibers and polyester spacer warp-knitted fabrics. By adjusting the deposition amount and surface coating arrangement, the sound absorption coefficient of the composite can reach 0.7 at low and medium frequencies. Özkal et al. [81] reported a sound absorption material made of electrospun PU nanofibers and polyester needle-punched nonwovens. This composite shows an excellent improvement in sound absorption compared with commercial sound absorption materials. Karaca et al. [82] covered thermoplastic polyurethane (TPU) and TPU/PS sub-micro fiber webs on the surface of rigid glass fiber fabric reinforced epoxy composites (GFEC) and flexible PP spun bond nonwovens. The sound absorption coefficient of GFEC is improved from 0.1 to 0.4 by combining them with sub-micro web layers with a thickness of less than 1 mm, and the maximum sound absorption coefficient of the PP nonwovens moves to a lower frequency after being covered by sub-micro fiber webs.

3.3. Electrospun Micro/Nanofiber Based Sound Absorption Materials

The high porosity and specific surface area of electrospun micro/nanofibers increase the contact between the material and sound waves. However, electrospun micro/nanofibers are usually deposited as thin mats or membranes, making them perform poorly in sound absorption. Thus, they cannot be used for noise reduction individually. For this purpose, researchers have investigated the development of new sound absorption materials made of electrospun micro/nanofibers.

Electrospun micro/nanofibers can be made into excellent sound absorption materials by changing their microscopic or spatial structure. Chang et al. [27] electrospun a nanofiber with a 3D lattice structure by using two syringes with opposite polarities, oppositely placed. The sound absorption performance of electrospun three-dimensional nanofiber is higher than that of the commercial acoustic cotton in the range of 400 to 900 Hz, and its sound absorption coefficient is up to 0.9. Selvaraj et al. [83] prepared a nanofiber layer using electrospun PVA and collagen hydrolysate extracted from waste leather trimmings. The layer is sandwiched between PAN nanofibrous layers. The composite structure is porous and shows good sound absorption performance in the 800–2500 Hz frequency range. Zong et al. [84] reported a fibrous sound absorption sponge with an interlocked dual-network-induced stable fluffy-stacked structure formed by PSU microfiber and PVDF nanofiber networks. The structure gives the fibrous sponge an excellent sound absorption property at low frequency, and its sound absorption coefficient could reach 0.93 at 1000 Hz. Similarly, Feng et al. [85] produced gradient structured fibrous sponges (GSFSs) with high sound absorption capacity in a wide frequency range by combining humidity-assisted multi-step electrospinning and physical/chemical double cross-linking techniques. Its fabrication process is shown in Figure 5a, and its sound absorption is illustrated in Figure 5b.

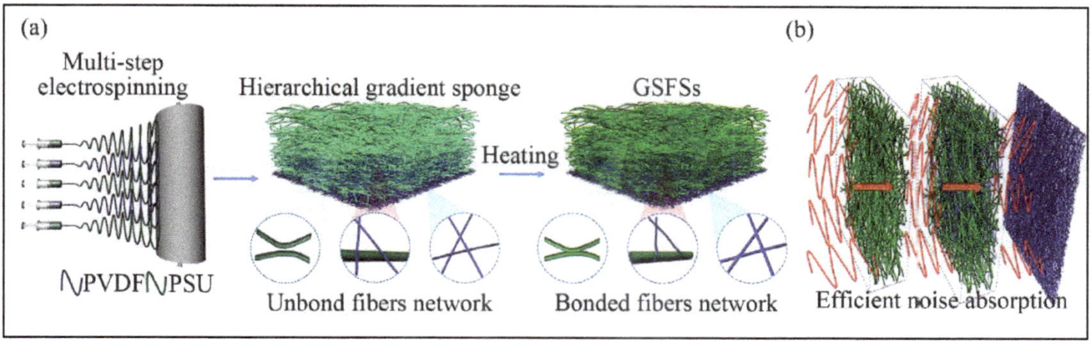

Figure 5. Fabrication of GSFSs (**a**) Fabrication process of GSFSs; (**b**) Sound absorption process of GSFSs. Reprinted with permission from ref. [85]. Copyright 2021 Elsevier.

The sound absorption performance of electrospun micro/nanofibers can also be influenced by adding fillers in the polymer solutions. Elkasaby et al. [86] reported that the sound absorption coefficient of electrospun polyvinyl chloride (PVC) nanofibers is influenced by adding CNTs. The addition of 5 wt% CNTs in PVC nanofibers increases the sound absorption by 62% compared with PVC electrospun nanofibers, and the sound absorption coefficient of the composite reaches the maximum value of 0.4 at 800 Hz. Sorbo et al. [87] found that the addition of graphene can affect the sound absorption properties of electrospun PVP mats in a non-monotonous manner.

Recently, sound absorption materials made of aerogels have been reported. Aerogels are mesoporous and open-cell solids with high porosity and low density [88,89]. The open-cell structure of aerogels is beneficial to increase the friction and collision between sound waves and materials in the transfer process of sound energy, endowing the aerogels with good sound absorption performance [28]. Ding and his colleagues have produced a series of nanofibrous aerogels by electrospinning. Si et al. [90] produced fibrous, isotropically bonded elastic reconstructed (FIBER) nanofibrous aerogels (NFAs) with 3D structure that are highly compressible and resilient using electrospinning and fibrous freeze-shaping techniques. The sound absorption coefficient of this material is higher than that of commercial non-woven materials at the frequency range from 100 to 6300 Hz (shown in Figure 6a–c). Cao et al. [91] designed nanofibrous aerogels with bamboo lashing-like structures through a freeze-drying method (shown in Figure 6d–f). Benefiting the internal hierarchical porous structures and hydrophobicity, the nanofibrous aerogels exhibit excellent moisture resistance, are ultralight, and show efficient sound absorption capability with a noise reduction coefficient (NFC) of 0.41. Cao et al. [38] fabricated composite nanofiber aerogels with a hierarchical maze-like microstructure by weaving cellulose nanocrystal lamellas with electrospun PAN nanofibers through the freeze-casting technique (shown in Figure 6g–i). The composite aerogels exhibit a remarkable sound absorption ability due to the maze-like structure that increases the contact area between the sound waves and the composite. Compared to the fiber aerogels with a network structure, the designed maze-like structure significantly enhances sound absorption capacity in the low-frequency range, and its NFC can reach 0.58.

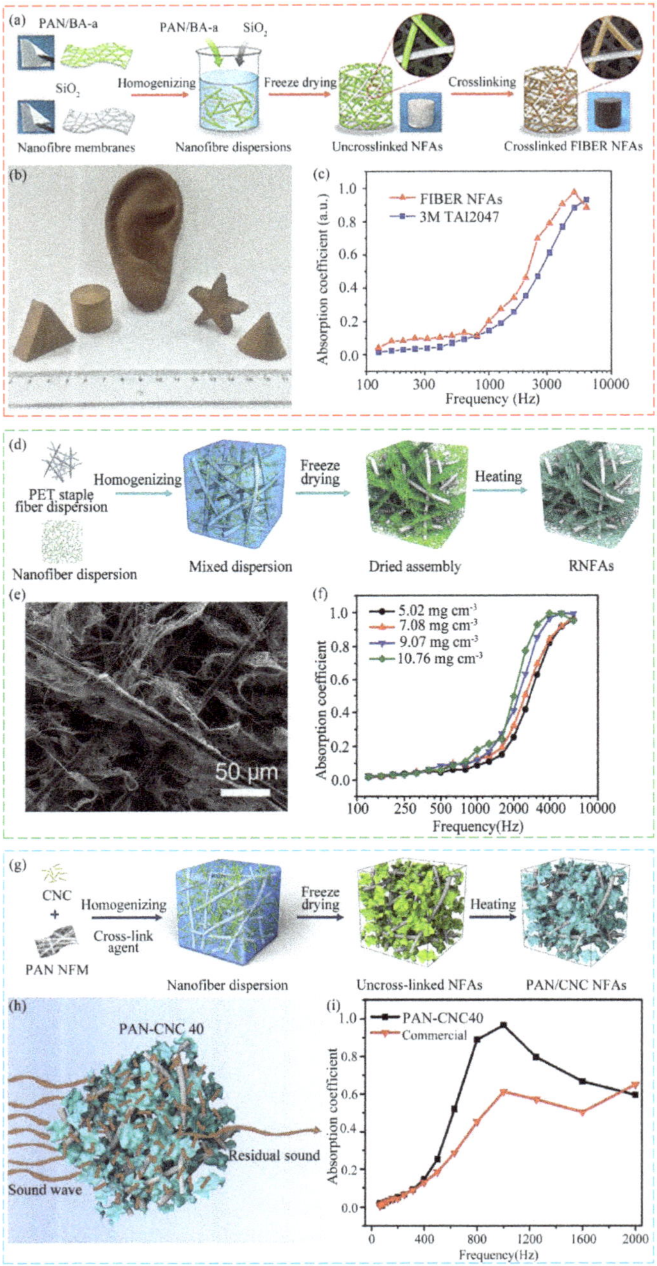

Figure 6. Several nanofiber aerogels for sound absorption. (**a**) Synthetic steps of FIBER NFAs; (**b**) Photographs of FIBER NFAs with several shapes; (**c**) Sound absorption coefficients of FIBER NFAs and commercial 3M TAI2047 nonwovens. Reprinted with permission from ref. [90]. Copyright 2014 Springer Nature. (**d**) Fabrication process of RNFAs; (**e**) Microstructure of RNFAs; (**f**) Sound absorption coefficient of RNFAs with different densities. Reprinted with permission from ref. [91]. Copyright 2019 Royal Society of Chemistry. (**g**) Fabrication of NFAs; (**h**) Sound absorption mechanism of PAN-CNC40; (**i**) Sound absorption coefficients of the PAN-CNC40 and the commercial fibrous sound absorption felt. Reprinted with permission from ref. [38]. Copyright 2021 Elsevier.

4. Electrospun Micro/Nanofiber Based Resonant Sound Absorption Materials

Resonant acoustic materials have a high absorption coefficient at low frequencies, making up for the lack of sound absorption capacity of porous acoustic materials at low and medium frequencies [92]. However, the sound absorption frequency range of traditional resonant acoustic materials, such as perforated panels and microperforated panels, where the thin plate is narrow, exhibits poor individual sound absorption effects. To achieve a better sound absorption effect, they are usually combined with other acoustic materials. Figure 7 shows a perforated porous material [93]. Electrospun micro/nanofibers yield both porous and resonant sound absorption capability. They can improve the sound absorption performance of traditional resonant sound absorption materials and can be used as a resonant acoustic material to absorb sound waves at low frequencies.

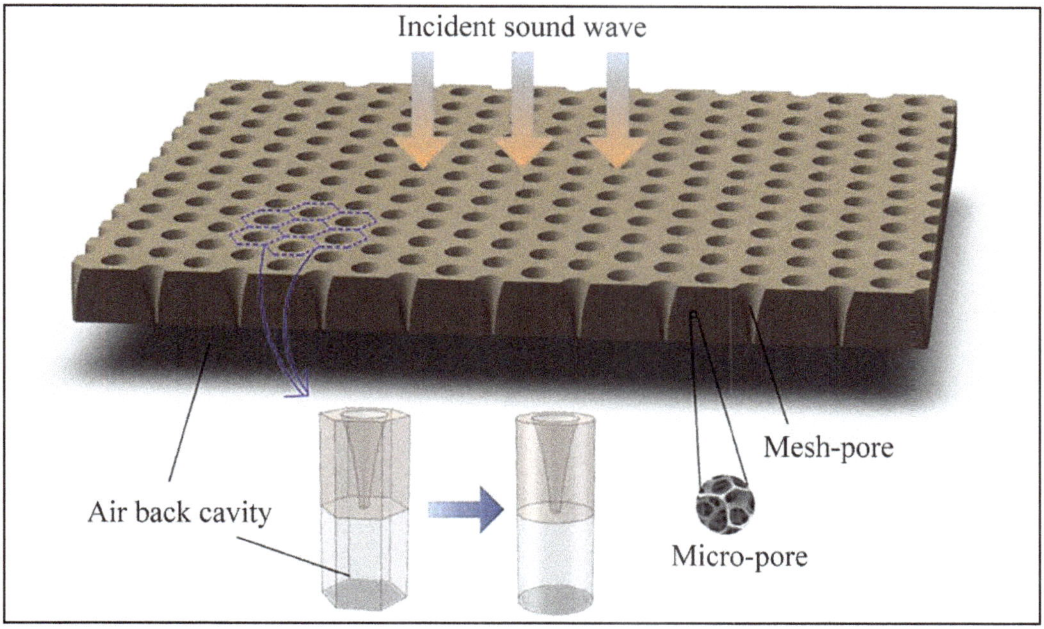

Figure 7. Schematic illustration of a gradually perforated porous material backed with a Helmholtz resonator. Reprinted with permission from ref. [93]. Copyright 2021 Elsevier.

Traditional resonant acoustic materials show narrow sound absorption bands. However, decorating them with electrospun micro/nanofibers can effectively broaden the absorption frequency range and improve their sound absorption coefficients. Guo et al. [94] prepared PVP nanofiber membranes using the electrospinning technique with a microperforated panel as a collector to deposit the fibers. The microperforated panel and obtained nanofibers form a composite sound absorption structure. The composite establishes a wider sound absorption frequency range than the single-layer microperforated panel, and its absorption coefficient is higher at the frequency range of 1500 to 2500 Hz. However, the sound absorption coefficient of the composite decreases from 2500 to 6000 Hz, and its sound absorption frequency range becomes narrow when the thickness of the nanofiber surpasses a certain value. Therefore, the thickness of the nanofiber membranes needs to be controlled within a certain range to attain an excellent sound absorption performance at low frequency. Tomáš et al. [68] adopted electrospun PVA nanofiber membranes to modify the perforated panel and investigated the effect of the back cavity depth on the sound absorption performance of the composite. The results reveal that the sound absorption peak of the composite shifts to a low frequency with the increase in the back cavity depth. When the depth of the back cavity is 34 mm, and the surface density of the nanofiber membrane is 0.6 gsm, the sound absorption coefficient of the composite material can reach 0.7 in the 300 to 400 Hz frequency range. Xiang et al. [42] fabricated three electrospun PAN nanofiber membranes with different thicknesses and porosities. They studied the sound absorption properties of the composites made of different nanofiber membranes and perforated panels, respectively. The results illustrate that the addition of a nanofiber membrane substantially improves the sound absorption performance of the perforated panel. When the depth of the back cavity is 10 mm, the highest sound absorption coefficient of the composite can reach 0.93, which is 3 times the absorption coefficient of a single perforated panel.

Although composites made of electrospun micro/nanofibers and traditional resonant acoustic materials have a good sound absorption performance, they still have problems such as the large size and bulky structure, causing the waste of space and limiting their applications. Using electrospun micro/nanofibers as resonant acoustic materials individually is an effective solution to save space, expanding the applications of resonant acoustic materials. The Helmholtz resonator is a structure that absorbs sound waves by resonance, and electrospun micro/nanofiber membranes can form this structure by incorporating a cavity. Therefore, the sound absorption performance of the electrospun fiber membranes at the low and medium frequencies can be promoted by setting an air layer between the fiber membrane and the rigid wall [30]. Liu et al. [29] compared the effects of different cavity depths on the sound absorption performance of three electrospun nanofibers of PAN, TPU, and thermoplastic polyester elastomer (TPEE). It is shown that the sound absorption coefficients of the nanofiber membranes increase in the frequency range from 100 to 1000 Hz as the cavity depths increase. Wang et al. [39] fabricated a PVA nanofibrous membrane with a Miura-ori sandwich structure by integrating electrospinning and paper folding techniques, and its fabrication process is shown in Figure 8. The sound absorption coefficient of Miura-ori sandwich metamaterial is improved to 1.0, and the first resonance frequency shifts to the lower frequency range compared with flat nanofiber membranes.

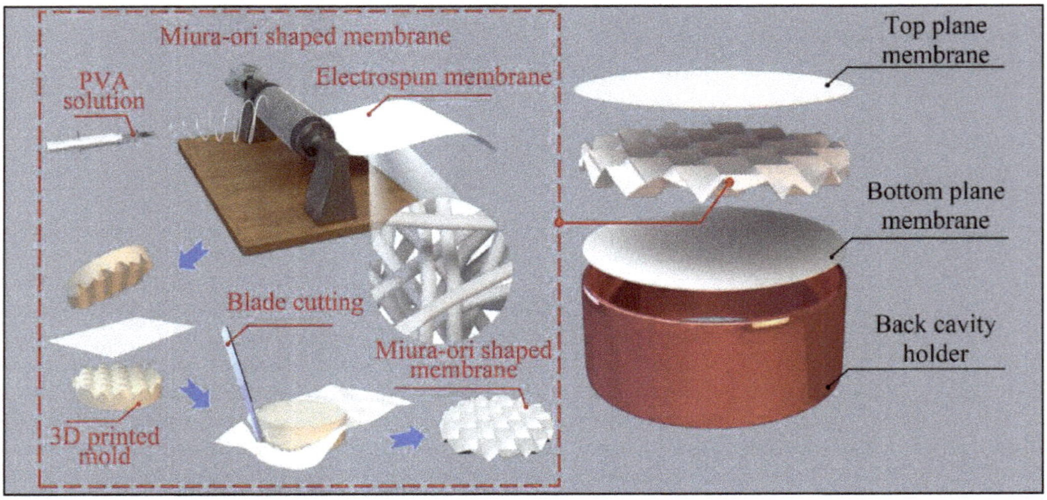

Figure 8. Fabrication of the Miura-ori sandwich structure. Reprinted with permission from ref. [39]. Copyright 2021 Wiley-VCH Verlag GmbH & Co. KGaA, Weinheim, Germany.

Furthermore, nanofiber membranes can vibrate at a low frequency when forced by sound waves, and the energy of sound waves will be absorbed by the principle of membrane resonance [95–97]. Kalinová et al. [98] designed a resonant structure with a nanofiber membrane, grid frame, and cavity based on the resonance principle. The nanofiber membrane converts the acoustic energy into kinetic energy, and the metal frame promotes energy dissipation. The size of the grid frame determines the resonant frequency of the structure, and the absorption coefficient is related to the depth of the cavity. When the grid size is 4.6 × 3.8 mm, and the cavity depth is 50 mm, the structure exhibits the best absorption performance, with the highest absorption coefficient of 1 at the frequency range between 600 and 700 Hz.

The optimization of the electrospun micro/nanofibers' structure is a way to enhance their resonant sound absorption ability, which can turn them into a sort of resonant sound absorption material. Xu et al. [99] produced nanoporous microspheres by electrospinning. These microspheres can be deemed an assembly of several thousand Helmholtz resonators, absorbing low-frequency sound waves by resonance. Shen et al. [100] reported a flexible PVA nanofiber microperforated structure prepared using electrospinning and a punching process (shown in Figure 9). The microperforated membrane possesses both porous and resonant sound absorption effects, and it can be used individually.

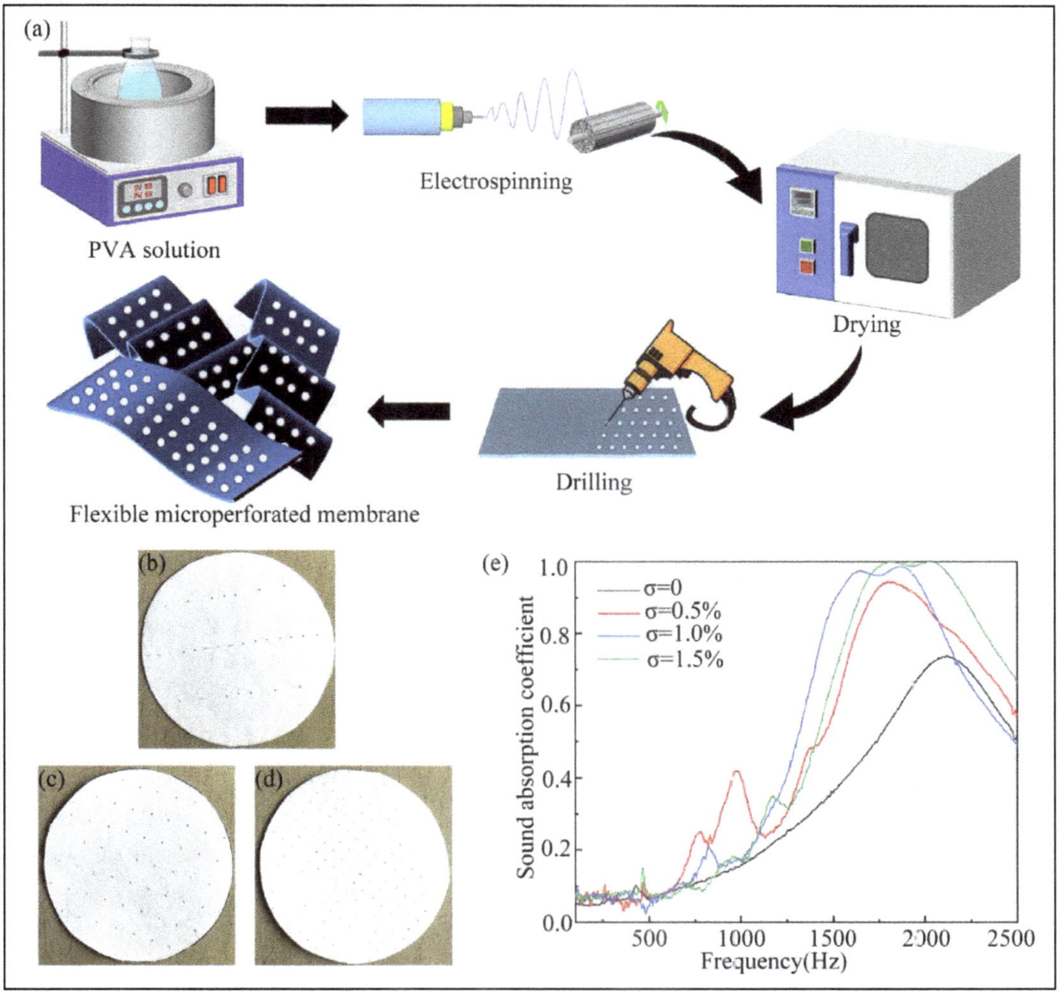

Figure 9. (**a**) The fabrication process of the flexible microperforated membrane; (**b–d**) Photographs of the microperforated samples with different perforation rates; (**e**) Sound absorption coefficient of the microperforated membranes with different perforation rates. Reprinted with permission from ref. [100]. Copyright 2022 Elsevier.

The sound absorption frequency range of traditional resonant acoustic materials can be broadened by compounding them with electrospun micro/nanofibers. Yet, their sizes are still too large for commercial applications. Resonant acoustic materials manufactured with electrospun micro/nanofibers possess less volume, and their acoustic properties can be controlled by adjusting electrospinning parameters. Therefore, it is necessary to study the resonant sound absorption materials prepared by electrospinning.

5. Conclusions

This review introduces the recent research of electrospun micro/nanofibers for sound absorption. Electrospun micro/nanofibers have advantages in sound absorption owing to their small fiber diameter, high specific surface area, and high porosity. Their sound absorption properties are determined by factors such as material, diameter, surface density, microstructure, and thickness, which can be controlled by changing the materials, parameters, and duration of electrospinning. Electrospun micro/nanofibers have various forms of sound absorption. The composite of electrospun micro/nanofibers and porous materials are effective in sound absorption at medium and high frequencies. Decorating resonant sound absorption materials with electrospun micro/nanofibers can enhance their sound absorption properties at a low frequency. In addition, electrospun micro/nanofibers with a 3D or Helmholtz resonator structure can also be deemed as sound absorption materials by themselves.

In the future, electrospun micro/nanofibers will surely be promising materials for the application of sound absorption in many fields. However, the application of electrospun micro/nanofibers for sound absorption faces several challenges. First, the sound absorption properties of electrospun micro/nanofibers and their composites at low frequency require additional improvement. Theoretical research can reduce the cost and pollution caused by the experimental process, and it can study the factors affecting the sound absorption performance of electrospun micro/nanofibers. The sound absorption performance of materials can be predicted by establishing a sound absorption model. The accuracy of the prediction can be increased by modifying the model according to the material and structure of the electrospun micro/nanofibers. Second, traditional electrospun micro/nanofibers have many limitations in its application under several different conditions, such as moist and high temperature environments, owing to their poor thermal stability, water resistance, and mechanical properties. Therefore, the multifunctional modification of the fibers is necessary. Electrospun micro/nanofibers can obtain properties such as good water resistance and thermal stability by modifying the materials or preparing them using an innovative process, which is beneficial for broadening their application for sound absorption. Finally, most electrospun micro/nanofibers are fabricated by solution electrospinning, which exhibits the problem of low efficiency, and the use of solvents is usually corrosive and toxic. These characteristics of solution electrospinning are not beneficial to the industrial production of electrospun micro/nanofibers and may lead to environmental concerns. Hence, it is urgent to find ways to solve the problems inherent to solution electrospinning. Using water-soluble polymers in solution electrospinning or choosing eco-friendly melt electrospinning to replace the former in the future are suggestions for solving these problems. In conclusion, further exploration of electrospun micro/nanofibers will promote their sound absorption applications in many fields.

Author Contributions: Conceptualization, X.L. and Y.P.; writing—original draft preparation, X.L., Y.P. and Y.H.; review and editing, C.Z., D.Z. and Y.L.; supervision, D.Z. and Y.L.; funding acquisition, X.L. All authors have read and agreed to the published version of the manuscript.

Funding: This work was supported by the National Natural Science Foundation of China (Grant No. 52003078), the Doctoral Scientific Research Foundation of Hubei University of Technology (Grant No. BSQD2020002), and the Hubei Key Laboratory of Modern Manufacturing Quality Engineering Foundation (Grant No. KFJJ-2020005).

Data Availability Statement: Not Applicable.

Conflicts of Interest: The authors declare no conflict of interest.

References

1. Yang, T.; Hu, L.; Xiong, X.; Petrů, M.; Noman, M.T.; Mishra, R.; Militký, J. Sound Absorption Properties of Natural Fibers: A Review. *Sustainability* **2020**, *12*, 8477. [CrossRef]
2. Gao, B.; Zuo, L.; Zuo, B. Sound absorption properties of spiral vane electrospun PVA/nano particle nanofiber membrane and non-woven composite material. *Fibers Polym.* **2016**, *17*, 1090–1096. [CrossRef]
3. Liu, H.; Zuo, B. Structure and Sound Absorption Properties of Spiral Vane Electrospun PVA/PEO Nanofiber Membranes. *Appl. Sci.* **2018**, *8*, 296. [CrossRef]
4. Cao, L.; Fu, Q.; Si, Y.; Ding, B.; Yu, J. Porous materials for sound absorption. *Compos. Commun.* **2018**, *10*, 25–35. [CrossRef]
5. Tang, X.; Yan, X. Acoustic energy absorption properties of fibrous materials: A review. *Compos. Part A Appl. Sci. Manuf.* **2017**, *101*, 360–380. [CrossRef]
6. Gai, X.-L.; Xing, T.; Li, X.-H.; Zhang, B.; Wang, W.-J. Sound absorption of microperforated panel mounted with helmholtz resonators. *Appl. Acoust.* **2016**, *114*, 260–265. [CrossRef]
7. Sakagami, K.; Morimoto, M.; Yairi, M. A note on the relationship between the sound absorption by microperforated panels and panel/membrane-type absorbers. *Appl. Acoust.* **2009**, *70*, 1131–1136. [CrossRef]
8. Ao, Q.; Wang, J.; Li, Y.; Ma, J.; Wu, C. Development of Lower Frequency Sound Absorption Materials. *J. Funct. Mater.* **2020**, *51*, 12045–12050. [CrossRef]
9. Li, H.; Xu, H.; Chen, M.; Yang, T.; Chen, X.; Yan, H.; Yang, W. Research Progress of Noise Reduction by Nanofibers. *J. Textile Res.* **2020**, *41*, 168–173. [CrossRef]
10. Passaro, J.; Russo, P.; Bifulco, A.; De Martino, M.T.; Granata, V.; Vitolo, B.; Iannace, G.; Vecchione, A.; Marulo, F.; Branda, F. Water Resistant Self-Extinguishing Low Frequency Soundproofing Polyvinylpyrrolidone Based Electrospun Blankets. *Polymers* **2019**, *11*, 1205. [CrossRef]
11. Ciaburro, G.; Iannace, G.; Passaro, J.; Bifulco, A.; Marano, A.D.; Guida, M.; Marulo, F.; Branda, F. Artificial neural network-based models for predicting the sound absorption coefficient of electrospun poly(vinyl pyrrolidone)/silica composite. *Appl. Acoust.* **2020**, *169*, 107472. [CrossRef]
12. Mishra, R.K.; Mishra, P.; Verma, K.; Mondal, A.; Chaudhary, R.G.; Abolhasani, M.M.; Loganathan, S. Electrospinning production of nanofibrous membranes. *Environ. Chem. Lett.* **2018**, *17*, 767–800. [CrossRef]
13. Giannella, V.; Branda, F.; Passaro, J.; Petrone, G.; Barbarino, M.; Citarella, R. Acoustic Improvements of Aircraft Headrests Based on Electrospun Mats Evaluated Through Boundary Element Method. *Appl. Sci.* **2020**, *10*, 5712. [CrossRef]
14. Giannella, V.; Colangeli, C.; Cuenca, J.; Citarella, R.; Barbarino, M. Experimental/Numerical Acoustic Assessment of Aircraft Seat Headrests Based on Electrospun Mats. *Appl. Sci.* **2021**, *11*, 6400. [CrossRef]
15. Petrone, G.; Branda, F.; Marulo, F. New Polyvinylpyrrolidone (PVP) Based Soundproofing Materials through Electrospinning. In Proceedings of the INTER-NOISE and NOISE-CON Congress, Hamburg, Germany, 21–24 August 2016; pp. 5374–5384.
16. Wu, C.M.; Chou, M.H. Polymorphism, piezoelectricity and sound absorption of electrospun PVDF membranes with and without carbon nanotubes. *Compos. Sci. Technol.* **2016**, *127*, 127–133. [CrossRef]
17. Wu, C.M.; Chou, M.H. Sound absorption of electrospun polyvinylidene fluoride/graphene membranes. *Eur. Polym. J.* **2016**, *82*, 35–45. [CrossRef]
18. Wu, C.M.; Chou, M.H. Acoustic–electric conversion and piezoelectric properties of electrospun polyvinylidene fluoride/silver nanofibrous membranes. *Express Polym. Lett.* **2020**, *14*, 103–114. [CrossRef]
19. Khan, W.S.; Asmatulu, R.; Yildirim, M.B. Acoustical Properties of Electrospun Fibers for Aircraft Interior Noise Reduction. *J. Aerosp. Eng.* **2012**, *25*, 376–382. [CrossRef]
20. Shou, D.H. Preparation and Applications of Electrospun Fibers. Master's Thesis, Donghua University, Shanghai, China, December 2008.
21. Rabbi, A.; Bahrambeygi, H.; Nasouri, K.; Shoushtari, A.M.; Babaei, M.R. Manufacturing of PAN or PU Nanofiber Layers/PET Nonwoven Composite as Highly Effective Sound Absorbers. *Adv. Polym. Technol.* **2014**, 33. [CrossRef]
22. Jia, W.; Qin, X. Acoustic Absorption Property of Electrospun PU Nanofibers Membrane. *J. Donghua Univ. (Nat. Sci.)* **2014**, *40*, 509–514.
23. Zhang, W.T. Absorption Coefficient and Structure of High Performance Sound Absorption Materials. Master's Thesis, South China University of Technology, Guangzhou, China, January 2017.
24. Bertocchi, M.J.; Vang, P.; Balow, R.B.; Wynne, J.H.; Lundin, J.G. Enhanced Mechanical Damping in Electrospun Polymer Fibers with Liquid Cores: Applications to Sound Damping. *ACS Appl. Polym. Mater.* **2019**, *1*, 2068–2076. [CrossRef]
25. Yang, Q.; Zhang, R. Preparation and properties of automotive sound absorbing nonwovens. *Shanghai Textile Sci. Technol.* **2020**, *48*, 29–32. [CrossRef]
26. Chen, Y.; Shafiq, M.; Liu, M.; Morsi, Y.; Mo, X. Advanced fabrication for electrospun three-dimensional nanofiber aerogels and scaffolds. *Bioact. Mater.* **2020**, *5*, 963–979. [CrossRef]
27. Chang, G.; Zhu, X.; Li, A.; Kan, W.; Warren, R.; Zhao, R.; Wang, X.; Xue, G.; Shen, J.; Lin, L. Formation and self-assembly of 3D nanofibrous networks based on oppositely charged jets. *Mater. Des.* **2016**, *97*, 126–130. [CrossRef]
28. Ding, B. Functional Polymeric Micro/Nano-fibrous Materials. *Acta Polym. Sin.* **2019**, *50*, 764–774. [CrossRef]

29. Liu, H.; Wang, D.; Zhao, N.; Ma, J.; Gong, J.; Yang, S.; Xu, J. Application of electrospinning fibres on sound absorption in low and medium frequency range. *Mater. Res. Innov.* **2014**, *18*, S4-888–S4-891. [CrossRef]
30. Ma, S.; Liu, X.; Xie, C.; Su, X. Sound Absorption Properties of Polyacrylonitrile Electrospinning Nanofiber Membrane and Its Laminated Materials. *J. Silk* **2020**, *57*, 13–19. [CrossRef]
31. Utkarsh, U. Development of Polymeric Nano-Composite Fiber Based Structures for Acoustics and Filtration Applications. Ph.D. Thesis, University of Ontario Institute of Technology, Oshawa, ON, Canada, August 2021.
32. Koenig, K.; Beukenberg, K.; Langensiepen, F.; Seide, G. A new prototype melt-electrospinning device for the production of biobased thermoplastic sub-microfibers and nanofibers. *Biomater. Res.* **2019**, *23*, 10. [CrossRef] [PubMed]
33. Ibrahim, Y.S.; Hussein, E.A.; Zagho, M.M.; Abdo, G.G.; Elzatahry, A.A. Melt Electrospinning Designs for Nanofiber Fabrication for Different Applications. *Int. J. Mol. Sci.* **2019**, *20*, 2455. [CrossRef]
34. Utkarsh, U.; Hegab, H.; Tariq, M.; Syed, N.A.; Rizvi, G.; Pop-Iliev, R. Towards Analysis and Optimization of Electrospun PVP (Polyvinylpyrrolidone) Nanofibers. *Adv. Polym. Technol.* **2020**, *2020*, 1–9. [CrossRef]
35. Park, M.; Park, H.K.; Shin, H.K.; Kang, D.; Pant, B.; Kim, H.; Song, J.-K.; Kim, H.Y. Sound Absorption and Insulation Properties of a Polyurethane Foam Mixed with Electrospun Nylon-6 and Polyurethane Nanofibre Mats. *J. Nanosci. Nanotechnol.* **2019**, *19*, 3558–3563. [CrossRef] [PubMed]
36. Özkal, A.; Çallıoğlu, F.C. Effect of nanofiber spinning duration on the sound absorption capacity of nonwovens produced from recycled polyethylene terephthalate fibers. *Appl. Acoust.* **2020**, *169*, 107468. [CrossRef]
37. Öztürk, M.K.; Ozden-Yenigun, E.; Nergis, B.; Candan, C. Nanofiber-enhanced lightweight composite textiles for acoustic applications. *J. Ind. Text.* **2016**, *46*, 1498–1510. [CrossRef]
38. Cao, L.; Yu, X.; Yin, X.; Si, Y.; Yu, J.; Ding, B. Hierarchically maze-like structured nanofiber aerogels for effective low-frequency sound absorption. *J. Colloid Interface Sci.* **2021**, *597*, 21–28. [CrossRef]
39. Wang, Y.; Guo, J.; Fang, Y.; Zhang, X.; Yu, H. Ultralight Metamaterial for Sound Absorption Based on Miura-Ori Tessellation Structures. *Adv. Eng. Mater.* **2021**, *23*, 2100563. [CrossRef]
40. Zhao-Xuan, D.; Ying, W. Optimization and Characterization of Polyurethane Electro-Spun Nano-Membranes Used for the Surfaces of Sound Absorbent Multi-Layer Sheets. *J. Macromol. Sci. Part B* **2021**, 1–16. [CrossRef]
41. Avossa, J.; Branda, F.; Marulo, F.; Petrone, G.; Guido, S.; Tomaiuolo, G.; Costantini, A. Light Electrospun Polyvinylpyrrolidone Blanket for Low Frequencies Sound Absorption. *Chin. J. Polym. Sci.* **2018**, *36*, 1368–1374. [CrossRef]
42. Xiang, H.-F.; Tan, S.-X.; Yu, X.-L.; Long, Y.-H.; Zhang, X.-L.; Zhao, N.; Xu, J. Sound absorption behavior of electrospun polyacrylonitrile nanofibrous membranes. *Chin. J. Polym. Sci.* **2011**, *29*, 650–657. [CrossRef]
43. Iannace, G. Acoustic Properties of Nanofibers. *Noise Vib. Worldw.* **2014**, *45*, 29–33. [CrossRef]
44. Akasaka, S.; Kato, T.; Azuma, K.; Konosu, Y.; Matsumoto, H.; Asai, S. Structure-sound absorption property relationships of electrospun thin silica fiber sheets: Quantitative analysis based on acoustic models. *Appl. Acoust.* **2019**, *152*, 13–20. [CrossRef]
45. Delany, M.; Bazley, E. Acoustical properties of fibrous absorbent materials. *Appl. Acoust.* **1970**, 3, 105–116. [CrossRef]
46. Liang, M.; Wu, H.; Liu, J.; Shen, Y.; Wu, G. Improved sound absorption performance of synthetic fiber materials for industrial noise reduction: A review. *J. Porous Mater.* **2022**, 1–24. [CrossRef]
47. Miki, Y. Acoustical properties of porous materials. Modifications of Delany-Bazley models. *J. Acoust. Soc. Jpn. (E)* **1990**, *11*, 19–24. [CrossRef]
48. Miki, Y. Acoustical properties of porous materials. Generalizations of empirical models. *J. Acoust. Soc. Jpn. (E)* **1990**, *11*, 25–28. [CrossRef]
49. Allard, J.-F.; Champoux, Y. New empirical equations for sound propagation in rigid frame fibrous materials. *J. Acoust. Soc. Am.* **1992**, *91*, 3346–3353. [CrossRef]
50. Chevillotte, F.; Perrot, C. Effect of the three-dimensional microstructure on the sound absorption of foams: A parametric study. *J. Acoust. Soc. Am.* **2017**, *142*, 1130–1140. [CrossRef]
51. Hwang, Y.J.; Choi, S.; Kim, H.S. Highly flexible all-nonwoven piezoelectric generators based on electrospun poly(vinylidene fluoride). *Sens. Actuators A Phys.* **2019**, *300*, 111672. [CrossRef]
52. Wang, W.; Zheng, Y.; Jin, X.; Sun, Y.; Lu, B.; Wang, H.; Fang, J.; Shao, H.; Lin, T. Unexpectedly high piezoelectricity of electrospun polyacrylonitrile nanofiber membranes. *Nano Energy* **2018**, *56*, 588–594. [CrossRef]
53. Shao, H.; Wang, H.; Cao, Y.; Ding, X.; Fang, J.; Niu, H.; Wang, W.; Lang, C.; Lin, T. Efficient conversion of sound noise into electric energy using electrospun polyacrylonitrile membranes. *Nano Energy* **2020**, *75*, 104956. [CrossRef]
54. Shao, H.; Wang, H.; Cao, Y.; Ding, X.; Bai, R.; Chang, H.; Fang, J.; Jin, X.; Wang, W.; Lin, T. Single-layer piezoelectric nanofiber membrane with substantially enhanced noise-to-electricity conversion from endogenous triboelectricity. *Nano Energy* **2021**, *89*, 106427. [CrossRef]
55. Rabbi, A.; Bahrambeygi, H.; Shoushtari, A.M.; Nasouri, K. Incorporation of Nanofiber Layers in Nonwoven Materials for Improving Their Acoustic Properties. *J. Eng. Fibers Fabr.* **2013**, *8*. [CrossRef]
56. Du, X.Y. Study on the Composite Sound Absorption and Thermal Insulation Nonwovens with Micro/Nano Fibers. Master's Thesis, Donghua University, Shanghai, China, May 2018.
57. Kalinová, K. Nanofibrous Resonant Membrane for Acoustic Applications. *J. Nanomater.* **2011**, *2011*, 1–6. [CrossRef]
58. Gao, B. Sound Absorption Properties of Electrospun PVA Nano Fiber Membrane. Master's Thesis, Soochow University, Suzhou, China, March 2017.

59. Yoon, C.K.; Park, B.K.; Lee, W.I. Characteristics of micro-glass bead/PLA porous composite prepared by electrospinning. *Adv. Compos. Mater.* **2017**, *27*, 183–193. [CrossRef]
60. Zou, Y.; Shi, L.; Zhou, Y.; Yao, L. Preparation of Sound Absorption on Nano-fiber Composite Mats and Its Absorption Property. *Tech. Text.* **2014**, *32*, 22–26.
61. Salehi, F.; Avanaki, M.J.; Nouri, M. An investigation into the designing of fibrous sound absorption materials incorporated with nanofibers: A case study of a multi-layered material composed of PET nonwoven and PAN nanofibers. *J. Text. Inst.* **2021**, 1–10. [CrossRef]
62. Ji, G.; Cui, J.; Fang, Y.; Yao, S.; Zhou, J.; Kim, J.-K. Nano-fibrous composite sound absorbers inspired by owl feather surfaces. *Appl. Acoust.* **2019**, *156*, 151–157. [CrossRef]
63. Rahimabady, M.; Statharas, E.C.; Yao, K.; Mirshekarloo, M.S.; Chen, S.; Tay, F.E.H. Hybrid local piezoelectric and conductive functions for high performance airborne sound absorption. *Appl. Phys. Lett.* **2017**, *111*, 241601. [CrossRef]
64. Sung, G.; Kim, S.K.; Kim, J.W.; Kim, J.H. Effect of isocyanate molecular structures in fabricating flexible polyurethane foams on sound absorption behavior. *Polym. Test.* **2016**, *53*, 156–164. [CrossRef]
65. Choe, H.; Sung, G.; Kim, J.H. Chemical treatment of wood fibers to enhance the sound absorption coefficient of flexible polyurethane composite foams. *Compos. Sci. Technol.* **2018**, *156*, 19–27. [CrossRef]
66. Rastegar, N.; Ershad-Langroudi, A.; Parsimehr, H.; Moradi, G. Sound-absorbing porous materials: A review on polyurethane-based foams. *Iran. Polym. J.* **2022**, *31*, 83–105. [CrossRef]
67. Bahrambeygi, H.; Sabetzadeh, N.; Rabbi, A.; Nasouri, K.; Shoushtari, A.M.; Babaei, M.R. Nanofibers (PU and PAN) and nanoparticles (Nanoclay and MWNTs) simultaneous effects on polyurethane foam sound absorption. *J. Polym. Res.* **2013**, *20*, 1–10. [CrossRef]
68. Ulrich, T.; Arenas, J.P. Sound Absorption of Sustainable Polymer Nanofibrous Thin Membranes Bonded to a Bulk Porous Material. *Sustainability* **2020**, *12*, 2361. [CrossRef]
69. Xiang, H.; Zhang, L.; Wang, Z.; Yu, X.; Long, Y.; Zhang, X.; Zhao, N.; Xu, J. Multifunctional polymethylsilsesquioxane (PMSQ) surfaces prepared by electrospinning at the sol–gel transition: Superhydrophobicity, excellent solvent resistance, thermal stability and enhanced sound absorption property. *J. Colloid Interface Sci.* **2011**, *359*, 296–303. [CrossRef] [PubMed]
70. Cao, L.; Si, Y.; Yin, X.; Yu, J.; Ding, B. Ultralight and Resilient Electrospun Fiber Sponge with a Lamellar Corrugated Microstructure for Effective Low-Frequency Sound Absorption. *ACS Appl. Mater. Interfaces* **2019**, *11*, 35333–35342. [CrossRef] [PubMed]
71. Mazrouei-Sebdani, Z.; Khoddami, A.; Hadadzadeh, H.; Zarrebini, M. Synthesis and performance evaluation of the aerogel-filled PET nanofiber assemblies prepared by electro-spinning. *RSC Adv.* **2015**, *5*, 12830–12842. [CrossRef]
72. Farahani, M.D.; Avanaki, M.J.; Jeddi, A.A. Sound absorption of warp knitted spacer fabrics based on knit structure and nanofiber enhancement. *J. Ind. Text.* **2020**. [CrossRef]
73. Tang, X.; Tang, X.; Yan, X.; Yan, X. Multi-layer fibrous structures for noise reduction. *J. Text. Inst.* **2017**, *108*, 2096–2106. [CrossRef]
74. Peng, M.; Zhao, X.M. Advances in the fiber-based sound-absorbing materials. *Mater. Rep.* **2019**, *33*, 3669–3677.
75. Selvaraj, S.; Jeevan, V.; Jonnalagadda, R.R.; Fathima, N.N. Conversion of tannery solid waste to sound absorbing nanofibrous materials: A road to sustainability. *J. Clean. Prod.* **2019**, *213*, 375–383. [CrossRef]
76. Na, Y.; Agnhage, T.; Cho, G. Sound absorption of multiple layers of nanofiber webs and the comparison of measuring methods for sound absorption coefficients. *Fibers Polym.* **2012**, *13*, 1348–1352. [CrossRef]
77. Öztürk, M.K.; Nergis, F.B.; Candan, C. Design of electrospun polyacrylonitrile nanofiber-coated nonwoven structure for sound absorption. *Polym. Adv. Technol.* **2017**, *29*, 1255–1260. [CrossRef]
78. Kucukali-Ozturk, M.; Nergis, B.; Candan, C. Design of layered structure with nanofibrous resonant membrane for acoustic applications. *J. Ind. Text.* **2017**, *47*, 1739–1756. [CrossRef]
79. Velayutham, T.; Manickam, R.K.; Sundararajan, P.; Chung, I.-M.; Prabakaran, M. A Study on the Effect of Natural Regenerated and Synthetic Non-woven Fabric Properties on Acoustic Applications. *J. Nat. Fibers* **2021**, 1–11. [CrossRef]
80. Liu, H.; Zuo, B. Sound absorption property of PVA/PEO/GO nanofiber membrane and non-woven composite material. *J. Ind. Text.* **2019**, *50*, 512–525. [CrossRef]
81. Özkal, A.; Çallıoğlu, F.C.; Akduman, C. Development of a new nanofibrous composite material from recycled nonwovens to improve sound absorption ability. *J. Text. Inst.* **2019**, *111*, 189–201. [CrossRef]
82. Karaca, N.; Yüksek, I.; Uçar, N.; Önen, A.; Baydoğan, M.; Kirbaş, C. Experimental analysis of sound absorption effect of specially developed elastomeric thin thermoplastic polyurethane sub-micron fibre web layers placed on rigid layer and flexible layers. *Plast. Rubber Compos.* **2021**, *50*, 507–515. [CrossRef]
83. Selvaraj, S.; Ramalingam, S.; Parida, S.; Rao, J.R.; Nishter, N.F. Chromium containing leather trimmings valorization: Sustainable sound absorber from collagen hydrolysate intercalated electrospun nanofibers. *J. Hazard. Mater.* **2021**, *405*, 124231. [CrossRef]
84. Zong, D.; Cao, L.; Li, Y.; Yin, X.; Si, Y.; Yu, J.; Ding, B. Interlocked Dual-Network and Superelastic Electrospun Fibrous Sponges for Efficient Low-Frequency Noise Absorption. *Small Struct.* **2020**, *1*. [CrossRef]
85. Feng, Y.; Zong, D.; Hou, Y.; Yin, X.; Zhang, S.; Duan, L.; Si, Y.; Jia, Y.; Ding, B. Gradient structured micro/nanofibrous sponges with superior compressibility and stretchability for broadband sound absorption. *J. Colloid Interface Sci.* **2021**, *593*, 59–66. [CrossRef] [PubMed]
86. Elkasaby, M.A.; Utkarsh, U.; Syed, N.A.; Rizvi, G.; Mohany, A.; Pop-Iliev, R. Evaluation of electro-spun polymeric nanofibers for sound absorption applications. *AIP Conf. Proc.* **2020**, *2205*, 020042. [CrossRef]

87. Del Sorbo, G.R.; Truda, G.; Bifulco, A.; Passaro, J.; Petrone, G.; Vitolo, B.; Ausanio, G.; Vergara, A.; Marulo, F.; Branda, F. Non Monotonous Effects of Noncovalently Functionalized Graphene Addition on the Structure and Sound Absorption Properties of Polyvinylpyrrolidone (1300 kDa) Electrospun Mats. *Materials* **2018**, *12*, 108. [CrossRef]
88. Mazrouei-Sebdani, Z.; Begum, H.; Schoenwald, S.; Horoshenkov, K.V.; Malfait, W.J. A review on silica aerogel-based materials for acoustic applications. *J. Non-Cryst. Solids* **2021**, *562*, 120770. [CrossRef]
89. Venkataraman, M.; Mishra, R.; Militky, J.; Xiong, X.; Marek, J.; Yao, J.; Zhu, G. Electrospun nanofibrous membranes embedded with aerogel for advanced thermal and transport properties. *Polym. Adv. Technol.* **2018**, *29*, 2583–2592. [CrossRef]
90. Si, Y.; Yu, J.; Tang, X.; Ge, J.; Ding, B. Ultralight nanofibre-assembled cellular aerogels with superelasticity and multifunctionality. *Nat. Commun.* **2014**, *5*, 5802. [CrossRef]
91. Cao, L.; Si, Y.; Wu, Y.; Wang, X.; Yu, J.; Ding, B. Ultralight, superelastic and bendable lashing-structured nanofibrous aerogels for effective sound absorption. *Nanoscale* **2019**, *11*, 2289–2298. [CrossRef] [PubMed]
92. Ding, L. Characteristics and Application of Resonant Sound Absorbing Materials in Audio Engineering. *Audio Eng.* **2019**, *43*, 12–18. [CrossRef]
93. Liu, X.; Yu, C.; Xin, F. Gradually perforated porous materials backed with Helmholtz resonant cavity for broadband low-frequency sound absorption. *Compos. Struct.* **2021**, *263*, 113647. [CrossRef]
94. Guo, L.J.; Zhao, Z.; Zhou, X.Y.; Fan, X. Research of Sound Absorption Characteristics for Micro-Perforated Panel Decorated by PVP Nanofiber. *Mater. Prot.* **2016**, *49*, 40–41. [CrossRef]
95. Mohrova, J.; Kalinova, K. Different Structures of PVA Nanofibrous Membrane for Sound Absorption Application. *J. Nanomater.* **2012**, *2012*, 1–4. [CrossRef]
96. Ozturk, M.K.; Kalinova, K.; Nergis, B.; Candan, C. Comparison of resonance frequency of a nanofibrous membrane and a homogeneous membrane structure. *Text. Res. J.* **2013**, *83*, 2204–2210. [CrossRef]
97. Ulrich, T.; Arenas, J.P. Role of porosity on nanofibrous membrane sound absorption properties. In Proceedings of the 26th International Congress on Sound and Vibration (ICSV 2019), Montreal, QC, Canada, 7–11 July 2019.
98. Kalinova, K. A sound absorptive element comprising an acoustic resonance nanofibrous membrane. *Recent Pat. Nanotechnol.* **2015**, *9*, 61–69. [CrossRef]
99. Xu, L.; Liu, F.; Faraz, N. Theoretical model for the electrospinning nanoporous materials process. *Comput. Math. Appl.* **2012**, *64*, 1017–1021. [CrossRef]
100. Shen, J.; Lee, H.P.; Yan, X. Sound absorption performance and mechanism of flexible PVA microperforated membrane. *Appl. Acoust.* **2021**, *185*, 108420. [CrossRef]

nanomaterials

MDPI

Article

Silkworm Protein-Derived Nitrogen-Doped Carbon-Coated $Li[Ni_{0.8}Co_{0.15}Al_{0.05}]O_2$ for Lithium-Ion Batteries

Gyu Sang Sim [1,†], Nitheesha Shaji [1,†], P. Santhoshkumar [2], Jae Woo Park [1], Chang Won Ho [1], Murugan Nanthagopal [1], Hong Ki Kim [1] and Chang Woo Lee [1,2,*]

1 Department of Chemical Engineering (Integrated Engineering), College of Engineering, Kyung Hee University, 1732 Deogyeong-daero, Giheung, Yongin 17104, Gyeonggi, Korea; simgyusang0215@khu.ac.kr (G.S.S.); nitheesha@khu.ac.kr (N.S.); jwpark82@khu.ac.kr (J.W.P.); ghckddnjs@khu.ac.kr (C.W.H.); nanthamurugan@khu.ac.kr (M.N.); hkkim95@khu.ac.kr (H.K.K.)

2 Center for the SMART Energy Platform, College of Engineering, Kyung Hee University, 1732 Deogyeong-daero, Giheung, Yongin 17104, Gyeonggi, Korea; santhoshkumar.palanisamy@gmail.com

* Correspondence: cwlee@khu.ac.kr; Tel.: +82-31-201-3825; Fax: +82-31-204-8114

† These authors contributed equally to this work.

Abstract: $Li[Ni_{0.8}Co_{0.15}Al_{0.05}]O_2$ (NCA) is a cathode material for lithium-ion batteries and has high power density and capacity. However, this material has disadvantages such as structural instability and short lifespan. To address these issues, herein, we explore the impact of N-doped carbon wrapping on NCA. Sericin, an easily obtained carbon- and nitrogen-rich component of silk cocoons, is utilized as the precursor material. The electrochemical performance evaluation of N-doped carbon-coated NCA shows that the capacity retention of 0.3 NC@NCA at 1C current density is 69.83% after 200 cycles, which is about 19% higher than the 50.65% capacity retention of bare NCA. The results reveal that the sericin-resultant N-doped carbon surface wrapping improves the cycling stability of NC@NCA.

Keywords: $Li[Ni_{0.8}Co_{0.15}Al_{0.05}]O_2$; sericin; N-doped carbon; lithium-ion batteries; surface wrapping

Citation: Sim, G.S.; Shaji, N.; Santhoshkumar, P.; Park, J.W.; Ho, C.W.; Nanthagopal, M.; Kim, H.K.; Lee, C.W. Silkworm Protein-Derived Nitrogen-Doped Carbon-Coated $Li[Ni_{0.8}Co_{0.15}Al_{0.05}]O_2$ for Lithium-Ion Batteries. *Nanomaterials* **2022**, *12*, 1166. https://doi.org/10.3390/nano12071166

Academic Editor: Carlos Miguel Costa

Received: 11 March 2022
Accepted: 28 March 2022
Published: 31 March 2022

1. Introduction

With the expansion of various devices, there has been a growing demand for various energy storage devices that may be utilized as energy sources [1,2]. Lithium-ion batteries (LIBs) have become the most frequently utilized secondary batteries among energy storage systems due to their attractive properties such as good energy density, operating potential, and extended cycle life [3,4]. The cathode is a crucial part of a battery that influences energy density, power density, longevity, and potential of the battery; it also accounts for 40% of the overall battery price [5,6]. The cathode materials of LIBs including $LiCoO_2$ (LCO), $LiMn_2O_4$ (LMO), and $LiFePO_4$ (LFP) are those with layered, spinel, and olivine structures, respectively [7–9]. Recently, layer structured, high nickel content cathodes, such as $Li[Ni_{0.8}Co_{0.15}Al_{0.05}]O_2$ (NCA) and $Li[Ni_{0.8}Co_{0.1}Mn_{0.1}]O_2$ (NCM811), have obtained widespread consideration because of their high capacity, relatively inexpensive price, and eco-friendly nature [10,11]. Among them, NCA has been popularized in the battery industry for its low price, high capacity, and power density [11,12]. However, NCA has many problems including poor lifespan, and low thermal and structural stability. Additionally, the degradation of NCA caused by side reactions with the electrolyte is also a major drawback [13,14].

There have been various attempts to address the drawbacks of NCA. Among these attempts, the active material surface wrapping can form a barrier to effectively protect the active material from an unstable structure by suppressing the surface degradation of the material and side reactions with the electrolyte; this improves the structural stability of the material [15–18]. Depending on the coating material used, the surface wrapping approach can also improve the overall conductivity of the electrode material [19,20].

Carbonaceous materials with wide availability, low cost, high conductivity, and diverse structure have gained significant research interest. Among them, N-doped carbon is an effective material for surface modification of various active materials because it efficiently suppresses surface degradation, prevents by-products from unnecessary side reactions with electrolytes, and improves the electronic conductivity of the host material [21–23]. Recently, Feng et al. reported on polyacrylonitrile (PAN)-induced conductive carbon coating on NCA. The PAN-induced carbon coated NCA, with a 5 nm thick coating layer, offered faster electron movement and hence provided an enhanced cycling performance [24]. Ion-conductive polymer Nafion-coating on NCA was investigated by Yigitalp et al. to obtain superior electrochemical performance. They concluded that Nafion-coating could suppress the passive layer formation and provide improved cycling stability [25]. Park et al. explored the conformal graphene coating on NCA by a scalable Pickering emulsion method. The conformal graphene coating reduces the surface degradation of NCA and hence delivered an excellent life cycle [26]. In this study, we utilized sericin to coat N-doped carbon on active materials; it is a biomaterial that is a component of silk cocoons. Sericin, a gummy component that surrounds fibroin and maintains the cohesiveness of the silk cocoon, accounts for about 20–30% of the total mass of a cocoon [27,28]. Sericin is a hydrophilic protein composed of various amino acids such as glycine, alanine, and arginine. It contains a nitrogen content of approximately 17%, rendering it appropriate as a surface modification agent for electrode materials [28,29].

Herein, we propose a simple solid-state technique to prepare N-doped carbon wrapping on the NCA surface using sericin powder extracted by the degumming process. To optimize the suitable weight percentages of the N-doped carbon for effective surface wrapping, different weight percentages of sericin were used. The as-prepared N-doped carbon-coated NCA material was then physically and electrochemically characterized to identify the effect of N-doped carbon surface wrapping to overcome the disadvantages associated with NCA.

2. Materials and Methods

2.1. Sericin Synthesis

Sericin was extracted through the degumming of the silkworm cocoon. Initially, 0.02 M Na_2CO_3 was dissolved in 100 mL distilled water and allowed to boil on a hot plate at 100 °C. Then, the cocoon was added to it and boiled for 2–3 h to extract the sericin content from the cocoon [30,31]. After extraction of the sericin content, the pale yellowish solution was filtered to eliminate the remaining silk fiber after the sericin was dissolved. Then, the solution was dried in a convection oven at 80 °C for 12 h, and the sericin powder was obtained after drying.

2.2. Synthesis of Nitrogen-Doped Carbon Coated NCA (NC@NCA)

Using a simple solid-state method, NC@NCAs were synthesized. Approximately 2 g of commercial NCA (EcoPro BM, Ltd., Cheongju, Korea) was ground with sericin powder for approximately 1 h for thorough mixing. Subsequently, the mixed powder was heated at 500 °C for 3h in the presence of argon gas to get NC@NCA. For comparison, different amounts of sericin were used, such as 0.2, 0.3, and 0.4 weight percent, to prepare NC@NCA and designated as 0.2 NC@NCA, 0.3 NC@NCA, and 0.4 NC@NCA, respectively.

2.3. Physical Measurements

The crystal structure and phase of the as-synthesized samples were analyzed with an X-ray diffractometer (XRD, Bruker D8 Advance, Billerica, MA, USA, Cu-Kα radiation, λ = 1.5406 Å) in the 2θ range of 10°–70°. X-ray photoelectron spectroscopy (XPS, K-Alpha; Thermo Fisher, Waltham, MA, USA) was conducted to determine the constituent elements on the surface of the material. We also carried out high-resolution field-emission scanning electron microscopy (HR-FE-SEM, MERLIN-LEO SUPRA 55, Carl Zeiss, Jena, Germany) embedded with energy-dispersive X-ray spectroscopy (EDX, Ox-ford Instruments, Abingdon,

UK) to determine the morphological characteristics and to conduct elemental analysis and elemental mapping.

2.4. Electrochemical Measurements

The obtained active materials were combined with Denka black as a conductive agent and polyvinylidene fluoride (PVDF) as a binder at a weight ratio of 80:10:10 to make the working electrode. Then, as a dispersion agent, an appropriate amount of N-methyl-2-pyrrolidone (NMP) was added to make a slurry. The prepared slurry was laminated on aluminum foil followed by room temperature drying for a day. The electrode was then dried in a convection oven at 120 °C for 5 h, lab-pressed, and punched into 14 mm diameter disks using a punching tool. Finally, the electrode was dried for 5 h at 100 °C in a vacuum oven. The as-produced electrode's electrochemical performance was assessed in 2032-type coin cells, with lithium foil as the reference electrode, Celgard® 2320 membrane as the separator, and 1 M $LiPF_6$ in ethylene carbonate (EC) and diethyl carbonate (DEC) (1:1 vol. percent) as the electrolyte. The average mass loading of the cathode was 6 mg cm^{-2}. The CR2032-type half-cells used to evaluate electrochemical performance were made in a glove box filled with Ar gas and aged for several hours. The electrochemical performance of the prepared cells was evaluated within a voltage range of 3.0–4.3 V using a cycler (ETH cycler, Hwaseong, Korea). Electrochemical impedance spectroscopy (EIS) was also conducted; the NCA cells completed for 200 cycles at a current density of 1 C were used and measured by applying an amplitude of 10 mV and a frequency range of 0.5 mHz–100 kHz. EIS measurements were carried out using an electrochemical workstation (Iviumstat, Ivium Technologies, Eindhoven, The Netherlands).

3. Results and Discussion

3.1. Physico-Chemical Characterizations

The XRD patterns of bare NCA, 0.2 NC@NCA, 0.3 NC@NCA, and 0.4 NC@NCA are shown in Figure 1, and their equivalent lattice parameters are exhibited in Table 1; these diffraction patterns and lattice parameters are used to identify differences in crystal structure and crystallinity. In Figure 1, the XRD patterns of bare NCA, 0.2 NC@NCA, 0.3 NC@NCA, and 0.4 NC@NCA exhibited no differences, and all the peaks are well matched to JCPDS NO. 87-1562, which is indexed to α-$NaFeO_2$ structure with $R\bar{3}m$ space group [32]. Furthermore, it is confirmed that the change of the lattice parameter did not appear significantly before and after coating, as indicated in Table 1. From these points, the N-doped carbon derived from sericin have no effect on the crystal structure of the bare NCA [33]. However, in terms of c/a values, there is little difference between the materials. The higher the ratio of c/a values, the higher the crystallinity and stability of NCA having a layered structure [34,35]. In this regard, it may be evident that 0.3 NC@NCA and 0.4 NC@NCA have higher crystallinity and stability than bare NCA. However, 0.2 NC@NCA has a lower c/a value than that of bare NCA, which means lower crystallinity and stability.

Table 1. Lattice parameters of bare NCA, 0.2 NC@NCA, 0.3 NC@NCA, and 0.4 NC@NCA.

	Bare NCA	0.2 NC@NCA	0.3 NC@NCA	0.4 NC@NCA
a (Å)	2.8636	2.8636	2.8625	2.8614
c (Å)	14.1532	14.1502	14.1643	14.1456
c/a	4.9425	4.9414	4.9482	4.9436

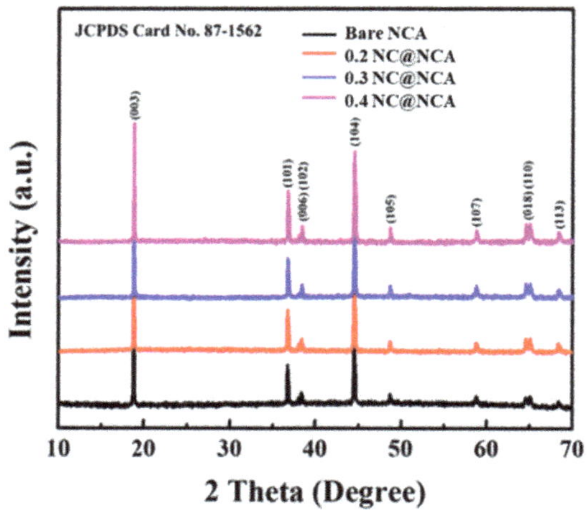

Figure 1. XRD patterns of bare NCA, 0.2 NC@NCA, 0.3 NC@NCA, and 0.4 NC@NCA.

XPS was conducted to study the chemical valence states and elemental composition of 0.3 NC@NCA. As shown in Figure 2a, the 0.3 NC@NCA XPS spectra exhibited the peaks of Ni, Co, Al, O, C, and N. The Ni 2p, Co 2p, Al 2p, and O 1s peaks fit well with the previous reports of the NCA [36–38]. Moreover, the C 1s and N 1s peaks are also observed, indicated by the N-doped carbon. Figure 2f displays the C 1s spectrum, which is fitted with peaks of C–C, C–O–C, and O–C=O bonds at binding energies of 284.0, 286.2, and 288.3 eV [39,40]. In Figure 2g, the N 1s spectra are fitted into two peaks for pyridinic N and pyrrolic N at binding energies of 397.8 and 399.8 eV [21,41]. The pyridinic N and pyrrolic N observed in 0.3 NC@NCA generate numerous active sites, which facilitate the rapid diffusion of lithium ions [42,43]. Consequently, the N-doped carbon layer contains pyridinic N and pyrrolic N, leading to improved electrochemical performance without disturbing the chemical states of NCA.

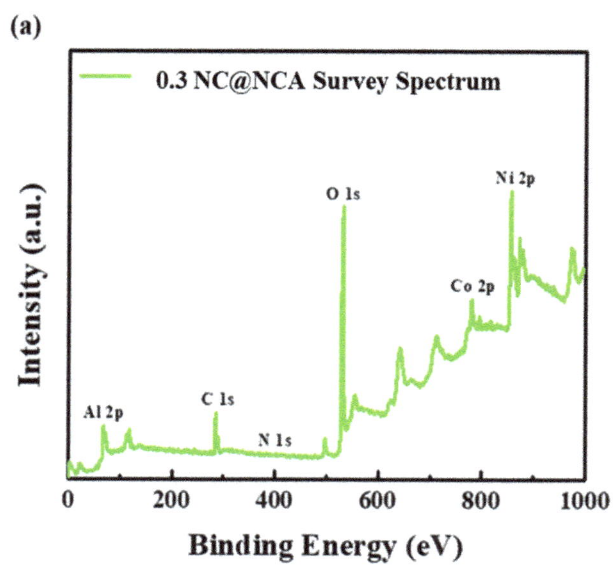

Figure 2. *Cont.*

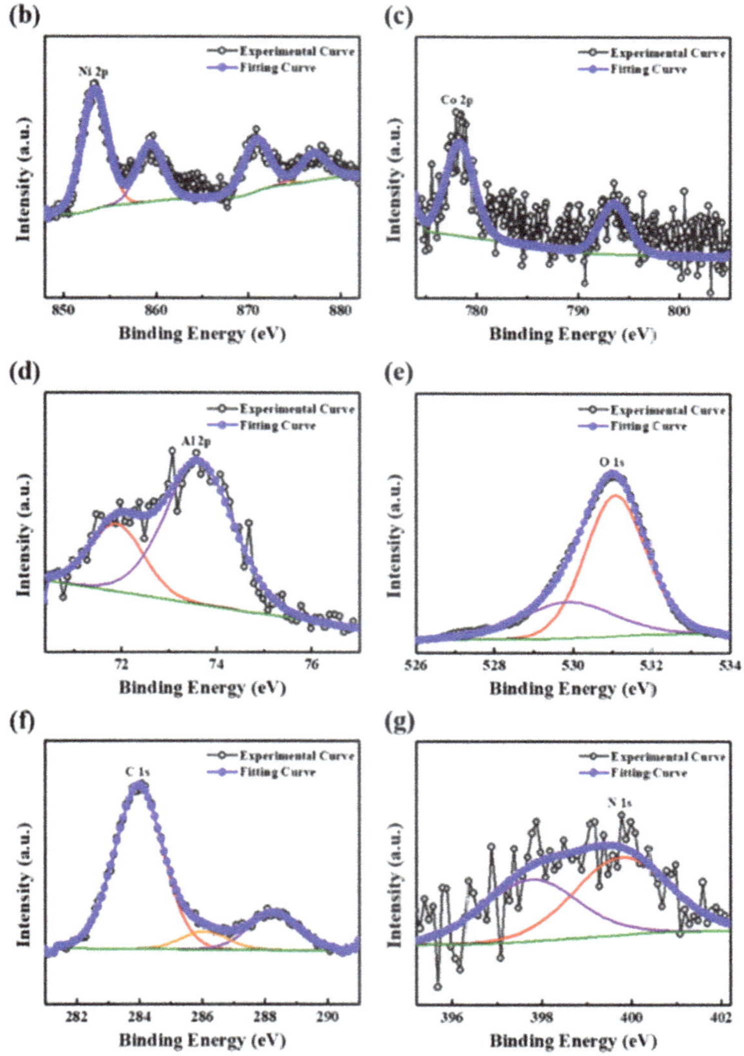

Figure 2. XPS spectra for 0.3 NC@NCA: (**a**) survey; (**b**) Ni 2p; (**c**) Co 2p; (**d**) Al 2p; (**e**) O 1s; (**f**) C 1s; and (**g**) N 1s.

Figure 3a–i show HR-FE-SEM images of bare NCA, 0.2 NC@NCA, 0.3 NC@NCA, and 0.4 NC@NCA. The surface of bare NCA, investigated by the HR-FE-SEM image, is observed as a rough surface. After the addition of sericin as the coating agent, the original shape of the materials is preserved for NC@NCA. Additionally, N-doped carbon on the surface of NC@NCA is clearly evident in the high-resolution images. In particular, in 0.3 NC@NCA, N-doped carbon particles are visible on the surface, and in 0.4 NC@NCA, more particles are noticeable on NCA. This indicates that sericin changes the surface of the NCA but does not affect the original spherical shape of the NCA.

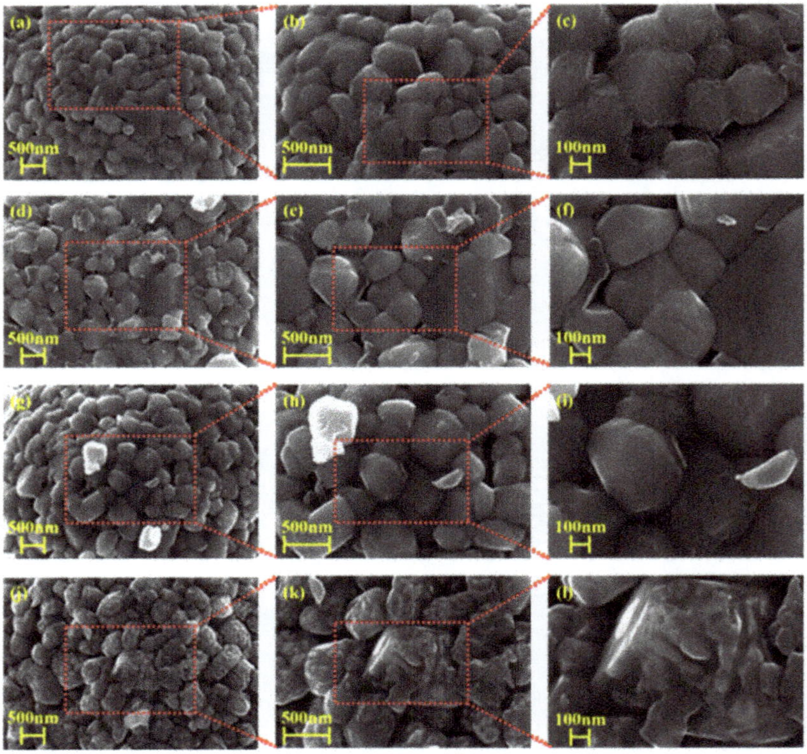

Figure 3. HR-FE-SEM images of (**a–c**) bare NCA; (**d–f**) 0.2 NC@NCA; (**g–i**) 0.3 NC@NCA; and (**j–l**) 0.4 NC@NCA.

Figure 4a–d show the EDS elemental mapping and spectra mapping to further investigate the existence of various elements of bare NCA, 0.2 NC@NCA, 0.3 NC@NCA, and 0.4 NC@NCA. It is confirmed that Ni, Co, and Al are evenly and densely spread on the surface of them through each elemental mapping image. For 0.2 NC@NCA, 0.3 NC@NCA, and 0.4 NC@NCA, it is observed that nitrogen and carbon are also evenly distributed; an increase in the nitrogen content is noticeable as the weight percent of sericin increased, and the weight and atomic ratio of nitrogen for the materials are given in Table 2. The EDS mapping results confirmed that NCA is coated with N-doped carbon; as the weight percent of sericin increased, the thickness of the wrapping layer also grew.

Table 2. Nitrogen weight ratio and nitrogen atomic ratio of bare NCA, 0.2 NC@NCA, 0.3 NC@NCA, and 0.4 NC@NCA.

	Bare NCA	0.2 NC@NCA	0.3 NC@NCA	0.4 NC@NCA
Nitrogen Weight Ratio (wt.%)	0	0.61	0.75	1.04
Nitrogen Atomic Ratio (%)	0	1.13	1.30	1.94

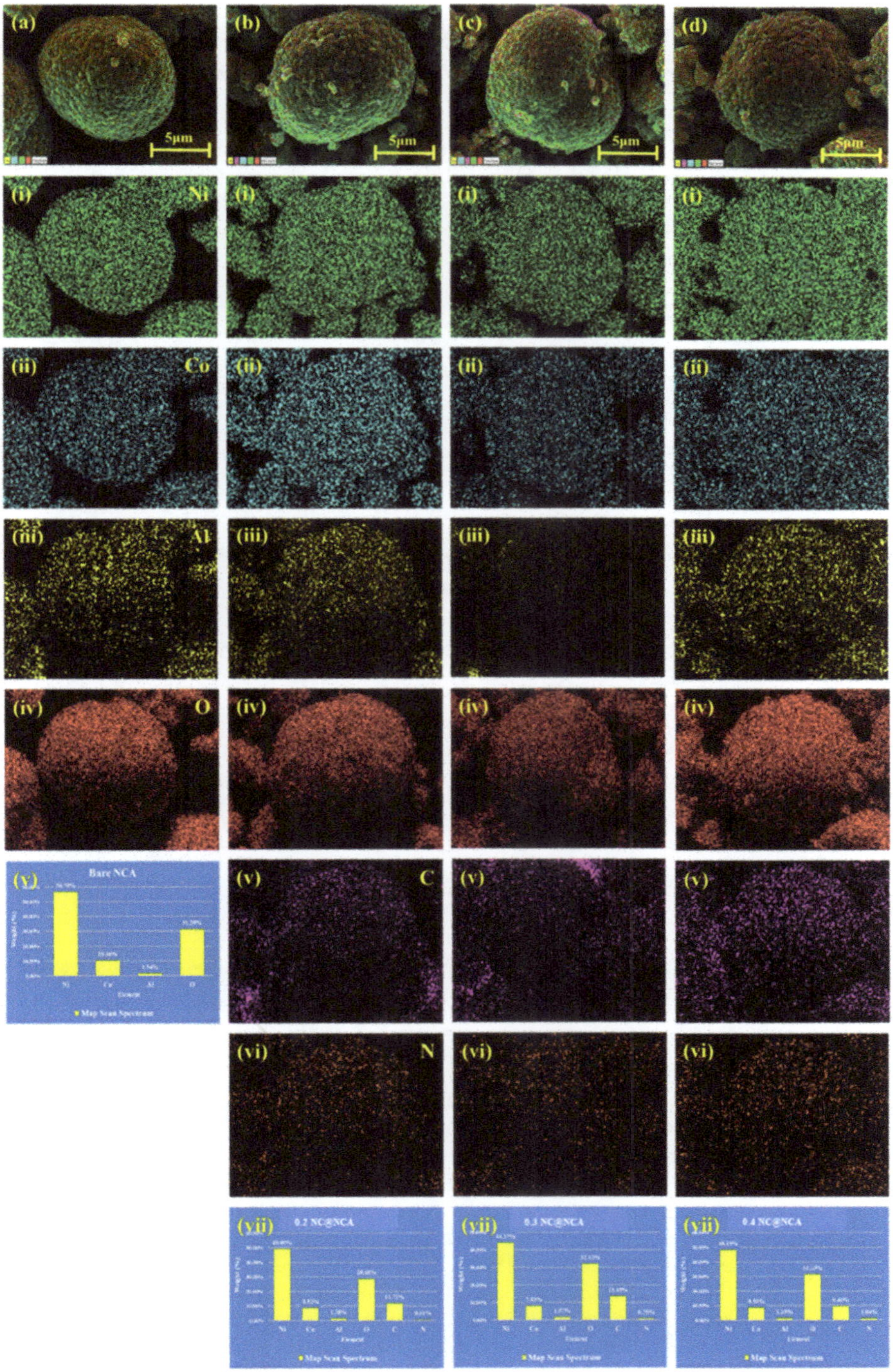

Figure 4. HR-FE-SEM images, elemental mapping spectra, and map scan spectra of (**a** (**i–v**)) bare NCA; (**b** (**i–vii**)) 0.2 NC@NCA; (**c** (**i–vii**)) 0.3 NC@NCA; and (**d** (**i–vii**)) 0.4 NC@NCA.

3.2. Electrochemical Characterizations

The galvanostatic charge/discharge behavior analysis of bare NCA, 0.2 NC@NCA, 0.3 NC@NCA, and 0.4 NC@NCA was conducted using a cycler, as shown in Figure 5a–d. The evaluation process was conducted within a voltage range of 3.0–4.3 V with a current density of 0.2 C. Figure 5a exhibits that the bare NCA delivers the highest initial specific capacity among its counterparts. It is considered that the N-doped carbon wrapping would have impeded the ability of lithium ions to penetrate the NCA structure. For 50 cycles, it is verified that the decrease in specific capacity of the bare NCA is greater than its counterparts. Therefore, it is considered that the N-doped carbon wrapping minimizes side reactions with the electrolyte and polarization.

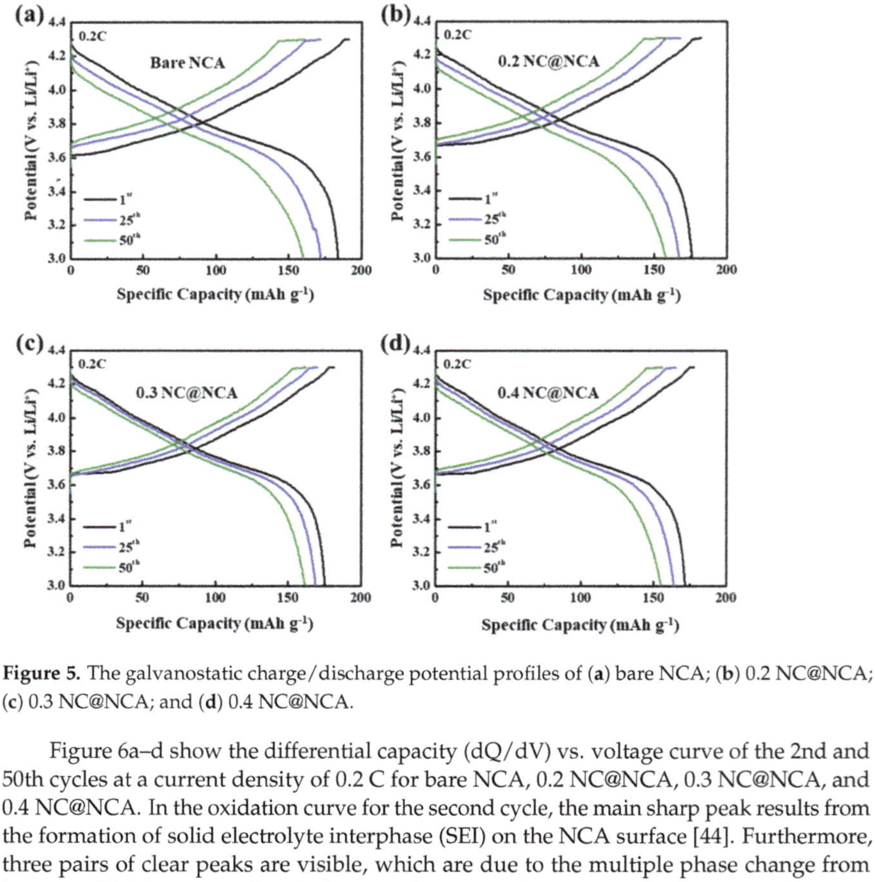

Figure 5. The galvanostatic charge/discharge potential profiles of (**a**) bare NCA; (**b**) 0.2 NC@NCA; (**c**) 0.3 NC@NCA; and (**d**) 0.4 NC@NCA.

Figure 6a–d show the differential capacity (dQ/dV) vs. voltage curve of the 2nd and 50th cycles at a current density of 0.2 C for bare NCA, 0.2 NC@NCA, 0.3 NC@NCA, and 0.4 NC@NCA. In the oxidation curve for the second cycle, the main sharp peak results from the formation of solid electrolyte interphase (SEI) on the NCA surface [44]. Furthermore, three pairs of clear peaks are visible, which are due to the multiple phase change from hexagonal (H1) to monoclinic (M), monoclinic (M) to hexagonal (H2) and hexagonal (H2) to hexagonal (H3) during cycling [45,46]. In the second cycle, the main oxidation peaks are 3.6431, 3.6679, 3.6677, and 3.6635 V, whereas, in the 50th cycle, the main oxidation peaks are shifted to 3.7595, 3.7205, 3.697, and 3.7118 V, respectively. For the bare NCA, the main oxidation peak shifted by 0.1164 V, and for the NC@NCA, the shifts in the main oxidation peaks are 0.0526, 0.0293, and 0.0483 V. Additionally, 0.2 NC@NCA, 0.3 NC@NCA, and 0.4 NC@NCA showed smaller shifts in oxidation peak compared with bare NCA; in particular, 0.3 NC@NCA experienced the smallest shift. The smaller shift in the oxidation peak means that the degree of reversibility during the cycling process improved [47]. That is, 0.2 NC@NCA, 0.3 NC@NCA, and 0.4 NC@NCA had better cycling reversibility than bare NCA; 0.3 NC@NCA, in particular, exhibited the best cycling reversibility.

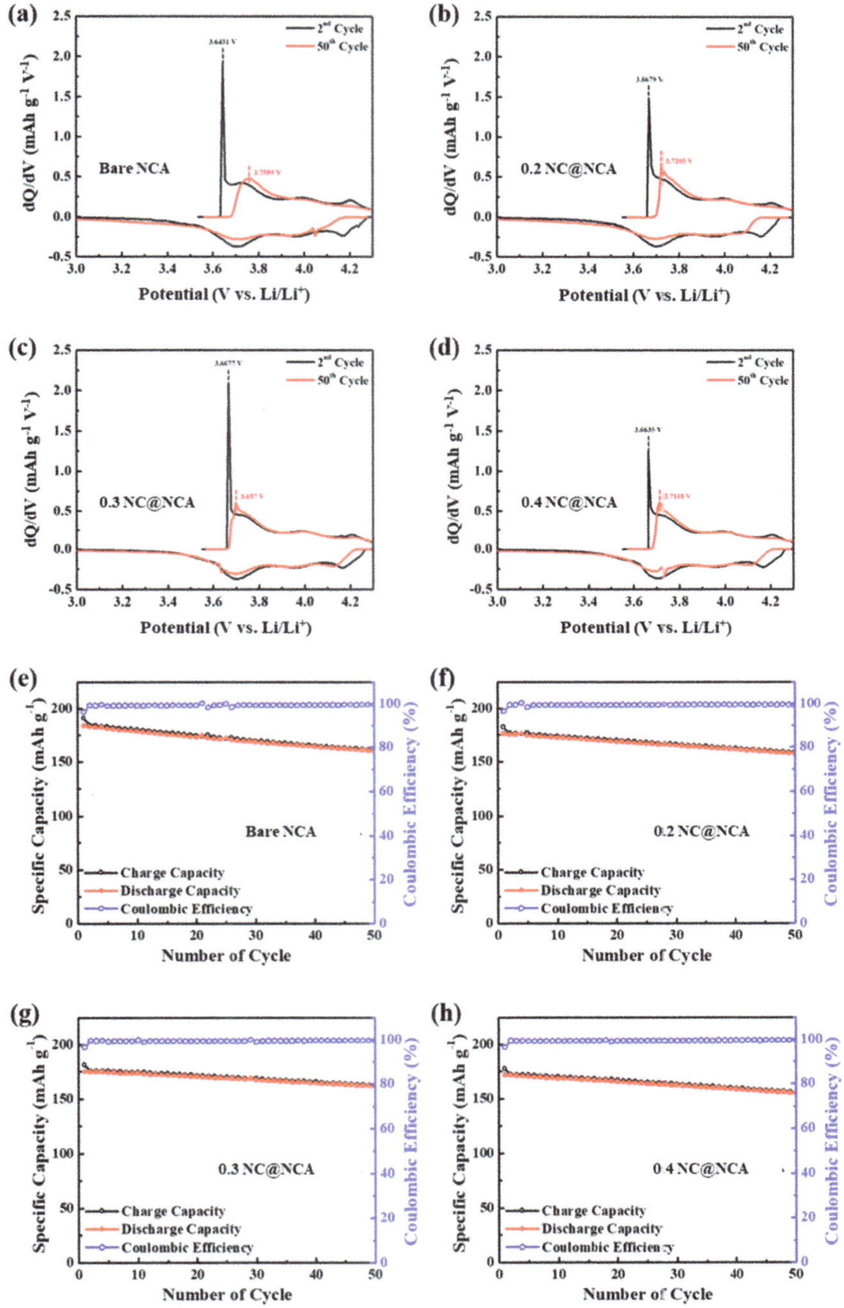

Figure 6. Differential capacity for (**a**) bare NCA, (**b**) 0.2 NC@NCA, (**c**) 0.3 NC@NCA, and (**d**) 0.4 NC@NCA; cycling performance profiles showing specific capacity and its coulombic efficiency at a current density of 0.2 C for (**e**) bare NCA, (**f**) 0.2 NC@NCA, (**g**) 0.3 NC@NCA, and (**h**) 0.4 NC@NCA.

The improved cycling reversibility was verified in Figure 6e–h; these figures exhibited the cycling performance of bare and N-doped carbon-coated NCA within a 3.0–4.3 V voltage range with a current density of 0.2 C. In the first cycle, bare NCA, 0.2 NC@NCA, 0.3 NC@NCA, and 0.4 NC@NCA exhibited specific discharge capacities of 183.95, 176.02, 175.31, and 171.78 mAh g^{-1}, respectively. However, after the 50 cycles, the specific discharge capacity of bare NCA is 160.27 mAh g^{-1}, whereas those of 0.2 NC@NCA, 0.3 NC@NCA, and 0.4 NC@NCA are 157.76, 161.63, and 155.10 mAh g^{-1}, respectively. When investigated by capacity retention, bare NCA, 0.2 NC@NCA, 0.3 NC@NCA, and 0.4 NC@NCA are 87.13, 89.62, 92.20, and 90.29%, respectively; 0.2 NC@NCA, 0.3 NC@NCA, and 0.4 NC@NCA exhibited higher capacity retention compared with bare NCA. These outcomes imply that the N-doped carbon wrapping minimizes side reactions between the NCA and the electrolyte, improving the structural stability and cycling reversibility of NCA.

Figure 7a shows the specific discharge capacity when 200 cycles of bare NCA, 0.2 NC@NCA, 0.3 NC@NCA, and 0.4 NC@NCA were conducted at a high current density of 1 C. In the initial cycle, the specific discharge capacity of bare NCA is higher than 0.2 NC@NCA, 0.3 NC@NCA, and 0.4 NC@NCA; however, after 200 cycles, the capacity retentions of 0.3 NC@NCA and 0.4 NC@NCA are 69.83% and 62.01%, respectively. This is higher than 0.2 NC@NCA with a capacity retention of 40.86% and a bare NCA of 50.65%. This low cycling stability in 0.2 wt.% N-C@NCA is attributed to the lower amount of sericin utilized. Therefore, it is considered that using a low amount of the coating agent could not form a stable coating layer to provide good cycling at a high current density, behaving instead as a minor variation. In addition, it is also considered that the low crystallinity and stability of 0.2 NC@NCA also contributed to the low-capacity retention at 1.0 C. For 0.4 NC@NCA, the capacity retention is higher than that of bare NCA; however, compared with 0.3 NC@NCA, the thicker coating layer impeded the movement of lithium ions and resulted in poor capacity retention.

Figure 7b shows the rate capability of bare NCA and 0.2 NC@NCA, 0.3 NC@NCA, and 0.4 NC@NCA. The process was conducted at various current densities from 0.2 C to 2.0 C. Bare NCA at a current density of 0.2 C exhibited the highest capacity, while at 0.5 C, it exhibited almost the same capacity as 0.3 NC@NCA. At current densities of 1.0 and 2.0 C, 0.4 NC@NCA exhibited a lower capacity than 0.3 NC@NCA. These results indicate that the 0.3 weight percent N-doped carbon wrapping is an effective coating layer capable of sufficiently protecting NCA during cycling at a high current rate.

To further analyze the effect of N-doped carbon on electrochemical performance, EIS was carried out for bare NCA, 0.2 NC@NCA, 0.3 NC@NCA, and 0.4 NC@NCA after 200 cycles at a 1 C current rate. Figure 7c shows that the semicircle in the high-frequency range, displayed in the Nyquist plot, is related to the charge transfer resistance (Rct), and the straight line in the low-frequency region is related to the Warburg resistance [48]. There is a large difference in charge transfer resistance, when comparing the bare NCA, 0.2 NC@NCA, 0.3 NC@NCA, and 0.4 NC@NCA. In numerical terms, the charge transfers resistances of the bare NCA, 0.2 NC@NCA, 0.3 NC@NCA, and 0.4 NC@NCA are 2091, 3049, 411, and 604 Ω, respectively. The charge transfer resistance is lowest for 0.3 NC@NCA, implying that this material showed the greatest improvement in ion and electron movement among them. This means that if 0.3 weight percent of sericin is used as the coating agent, the coating layer may serve as a protective layer for NCA and improve the movement of ions and electrons [49]. A table of comparison of previously reported similar studies with current work is presented in Table 3.

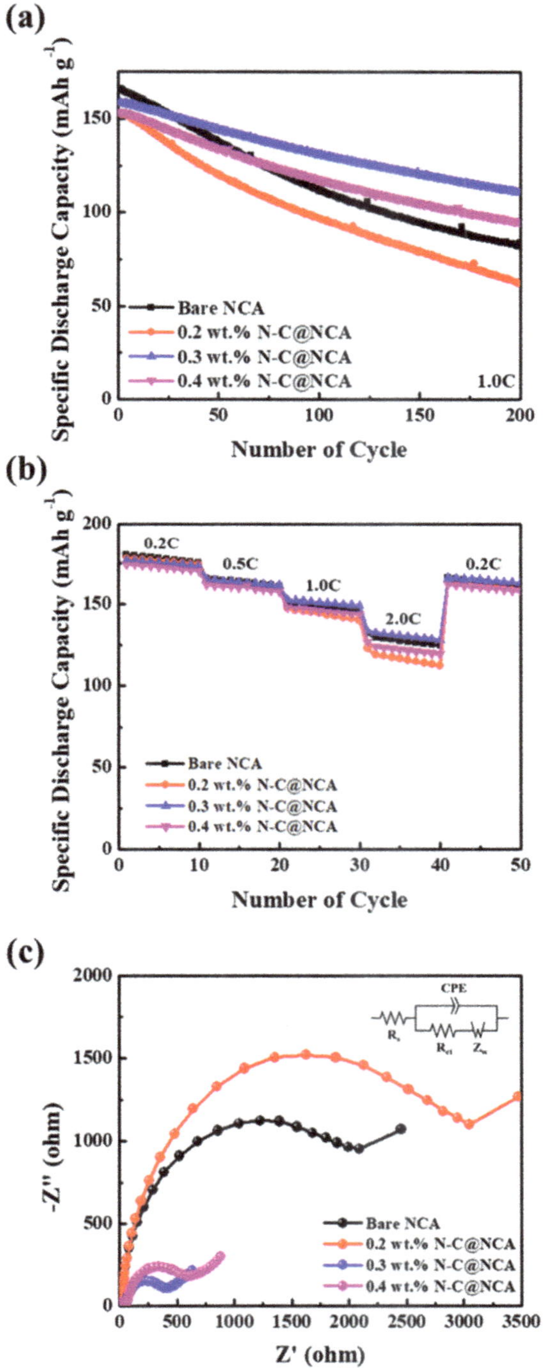

Figure 7. (**a**) Cycling performance at a current density of 1.0 C; (**b**) rate capability; (**c**) EIS analysis after 200 cycles at a current density of 1.0 C for bare NCA, 0.2 NC@NCA, 0.3 NC@NCA, and 0.4 NC@NCA.

Table 3. Comparison of previously reported similar studies with current work.

Material	Carbon Source	Discharge Capacity (mAh g^{-1})	Capacity Retention (%)	Ref.
Carbon nanotube coating on NCA (CNT-NCA)	Carbon nanotubes	205 (at 0.1 C)	91 (at 0.1 C)	[50]
Carbon-coated LNCAO (LNCAO/C)	Sodium dodecyl sulfate	183 (at 0.1 C)	93 (at 0.1 C)	[51]
NCA-graphene	Graphene nanoplatelets	180 (at 0.5 C)	97 (at 0.5 C)	[52]
Nanoscale carbon coating on NCA (C@NCA)	Glucose	260 (at 1 C)	88 (at 1.0 C)	[53]
N-doped carbon-coated NCA (NC@NCA)	Sericin	161 (at 0.2 C)	92 (at 0.2 C)	*This work

4. Conclusions

The N-doped carbon derived from sericin was successfully prepared using the solid-state method, and it was used to provide a coating layer for the NCA cathode material of LIBs. The influence of surface modification on the electrochemical characteristics of the NCA was characterized, and it was confirmed that this coating material is effective in improving electrochemical performance, particularly cycling stability. Specifically, 0.3 NC@NCA demonstrated the greatest improvements in cycling stability at current rates of 0.2 and 1.0 C as well as in rate capability. The measurement of charge transfer resistance using EIS confirmed that 0.3 NC@NCA showed the lowest charge transfer resistance. Based on these results, when 0.3 weight percent of sericin is used for the coating on NCA, it effectively suppresses side reactions with the electrolyte and surface degradation. Therefore, it could improve the structural stability of NCA, resulting in enhanced electrochemical performance, particularly in terms of cycling stability and rate capability.

Author Contributions: Conceptualization, methodology, validation, investigation, writing—original draft preparation, G.S.S. and N.S.; validation, formal analysis, P.S.; formal analysis, J.W.P., C.W.H., M.N. and H.K.K.; writing—review and editing, supervision, project administration, funding acquisition, C.W.L. All authors have read and agreed to the published version of the manuscript.

Funding: This research was supported by the Next Generation Engineering Researcher Program of National Research Foundation of Korea (NRF) funded by the Ministry of Science and ICT (No. 2017H1D8A2031138), and by the National Research Foundation of Korea (NRF) funded by the Ministry of Science and ICT (No. 2020K1A3A1A48111073). It was also supported by the Korea Institute for Advancement of Technology (KIAT) funded by the Ministry of Trade, Industry & Energy (MOTIE) of Korea (No. P0017363).

Institutional Review Board Statement: Not applicable.

Informed Consent Statement: Not applicable.

Data Availability Statement: Not applicable.

Conflicts of Interest: The authors declare no conflict of interest.

References

1. AL Shaqsi, A.Z.; Sopian, K.; Al-Hinai, A. Review of energy storage services, applications, limitations, and benefits. *Energy Rep.* **2020**, *6*, 288–306. [CrossRef]
2. Manthiram, A. A reflection on lithium-ion battery cathode chemistry. *Nat. Commun.* **2020**, *11*, 1550. [CrossRef] [PubMed]
3. Nitta, N.; Wu, F.; Lee, J.T.; Yushin, G. Li-ion battery materials: Present and future. *Mater. Today* **2015**, *18*, 252–264. [CrossRef]
4. Pender, J.P.; Jha, G.; Youn, D.H.; Ziegler, J.M.; Andoni, I.; Choi, E.J.; Heller, A.; Dunn, B.S.; Weiss, P.S.; Penner, R.M.; et al. Electrode Degradation in Lithium-Ion Batteries. *ACS Nano* **2020**, *14*, 1243–1295. [CrossRef] [PubMed]
5. Mohamed, N.; Allam, N.K. Recent advances in the design of cathode materials for Li-ion batteries. *RSC Adv.* **2020**, *10*, 21662–21685. [CrossRef]
6. Xi, Z.; Wang, Z.; Peng, W.; Guo, H.; Wang, J. Effect of copper and iron substitution on the structures and electrochemical properties of $LiNi_{0.8}Co_{0.15}Al_{0.05}O_2$ cathode materials. *Energy Sci. Eng.* **2020**, *8*, 1868–1879. [CrossRef]
7. Manthiram, A.; Kim, J. Low Temperature Synthesis of Insertion Oxides for Lithium Batteries. *Chem. Mater.* **1998**, *10*, 2895–2909. [CrossRef]
8. Chen, Z.; Zhang, W.; Yang, Z. A review on cathode materials for advanced lithium ion batteries: Microstructure designs and performance regulations. *Nanotechnology* **2020**, *31*, 012001. [CrossRef]
9. Guan, P.; Zhou, L.; Yu, Z.; Sun, Y.; Liu, Y.; Wu, F.; Jiang, Y.; Chu, D. Recent progress of surface coating on cathode materials for high-performance lithium-ion batteries. *J. Energy Chem.* **2020**, *43*, 220–235. [CrossRef]
10. Do, S.J.; Santhoshkumar, P.; Kang, S.H.; Prasanna, K.; Jo, Y.N.; Lee, C.W. Al-Doped $Li[Ni_{0.78}Co_{0.1}Mn_{0.1}Al_{0.02}]O_2$ for High Performance of Lithium Ion Batteries. *Ceram. Int.* **2019**, *45*, 6972–6977. [CrossRef]
11. Purwanto, A.; Yudha, C.S.; Ubaidillah, U.; Widiyandari, H.; Ogi, T.; Haerudin, H. NCA cathode material: Synthesis methods and performance enhancement efforts. *Mater. Res. Express* **2018**, *5*, 122001. [CrossRef]
12. Wang, D.; Liu, W.; Zhang, X.; Huang, Y.; Xu, M.; Xiao, W. Review of Modified Nickel-Cobalt Lithium Aluminate Cathode Materials for Lithium-Ion Batteries. *Int. J. Photoenergy* **2019**, *2019*, 2730849. [CrossRef]
13. Xu, J.; Lin, F.; Doeff, M.M.; Tong, W. A review of Ni-based layered oxides for rechargeable Li-ion batteries. *J. Mater. Chem. A* **2017**, *5*, 874–901. [CrossRef]
14. Xia, Y.; Zheng, J.; Wang, C.; Gu, M. Designing principle for Ni-rich cathode materials with high energy density for practical applications. *Nano Energy* **2018**, *49*, 434–452. [CrossRef]
15. Yang, S.; Zhou, X.; Zhang, J.; Liu, Z. Morphology-controlled solvothermal synthesis of $LiFePO_4$ as a cathode material for lithium-ion batteries. *J. Mater. Chem.* **2010**, *20*, 8086–8091. [CrossRef]
16. Chen, J.; Wang, X.L.; Jin, E.M.; Moon, S.G.; Jeong, S.M. Optimization of B_2O_3 coating process for NCA cathodes to achieve long-term stability for application in lithium ion batteries. *Energy* **2021**, *222*, 119913. [CrossRef]
17. Senthil, C.; Vediappan, K.; Nanthagopal, M.; Seop Kang, H.; Santhoshkumar, P.; Gnanamuthu, R.; Lee, C.W. Thermochemical conversion of eggshell as biological waste and its application as a functional material for lithium-ion batteries. *Chem. Eng. J.* **2019**, *372*, 765–773. [CrossRef]
18. Sim, G.S.; Santhoshkumar, P.; Park, J.W.; Ho, C.W.; Shaji, N.; Kim, H.K.; Nanthagopal, M.; Lee, C.W. Chitosan-derived nitrogen-doped carbon on $Li_2ZnTi_3O_8/TiO_2$ composite as an anode material for lithium-ion batteries. *Ceram. Int.* **2021**, *47*, 33554–33562. [CrossRef]
19. Park, K.S.; Son, J.T.; Chung, H.T.; Kim, S.J.; Lee, C.H.; Kang, K.T.; Kim, H.G. Surface modification by silver coating for improving electrochemical properties of $LiFePO_4$. *Solid State Commun.* **2004**, *129*, 311–314. [CrossRef]
20. Moskon, J.; Dominko, R.; Cerc-Korosec, R.; Gaberscek, M.; Jamnik, J. Morphology and electrical properties of conductive carbon coatings for cathode materials. *J. Power Sources* **2007**, *174*, 683–688. [CrossRef]
21. Nanthagopal, M.; Santhoshkumar, P.; Shaji, N.; Praveen, S.; Kang, H.S.; Senthil, C.; Lee, C.W. Nitrogen-doped carbon-coated $Li[Ni_{0.8}Co_{0.1}Mn_{0.1}]O_2$ cathode material for enhanced lithium-ion storage. *Appl. Surf. Sci.* **2019**, *492*, 871–878. [CrossRef]
22. Lee, D.J.; Ryou, M.H.; Lee, J.N.; Kim, B.G.; Lee, Y.M.; Kim, H.W.; Kong, B.S.; Park, J.K.; Choi, J.W. Nitrogen-doped carbon coating for a high-performance SiO anode in lithium-ion batteries. *Electrochem. Commun.* **2013**, *34*, 98–101. [CrossRef]
23. Zhao, L.; Hu, Y.S.; Li, H.; Wang, Z.; Chen, L. Porous $Li_4Ti_5O_{12}$ coated with N-doped carbon from ionic liquids for Li-ion batteries. *Adv. Mater.* **2011**, *23*, 1385–1388. [CrossRef] [PubMed]
24. Feng, D.; Liu, Q.; Hu, T.; Chen, Y.; Zeng, T. Boosting cyclability performance of the $LiNi_{0.8}Co_{0.15}Al_{0.05}O_2$ cathode by a polyacrylonitrile-induced conductive carbon surface coating. *Ceram. Int.* **2021**, *47*, 12706–12715. [CrossRef]
25. Yiğitalp, A.; Taşdemir, A.; Alkan Gürsel, S.; Yürüm, A. Nafion-coated $LiNi_{0.80}Co_{0.15}Al_{0.05}O_2$ (NCA) cathode preparation and its influence on the Li-ion battery cycle performance. *Energy Storage* **2020**, *2*, e154. [CrossRef]
26. Park, K.Y.; Lim, J.M.; Luu, N.S.; Downing, J.R.; Wallace, S.G.; Chaney, L.E.; Yoo, H.; Hyun, W.J.; Kim, H.U.; Hersam, M.C. Concurrently Approaching Volumetric and Specific Capacity Limits of Lithium Battery Cathodes via Conformal Pickering Emulsion Graphene Coatings. *Adv. Energy Mater.* **2020**, *10*, 2001216. [CrossRef]
27. Wang, H.Y.; Wang, Y.J.; Zhou, L.X.; Zhu, L.; Zhang, Y.Q. Isolation and bioactivities of a non-sericin component from cocoon shell silk sericin of the silkworm Bombyx mori. *Food Funct.* **2012**, *3*, 150–158. [CrossRef]
28. Kunz, R.I.; Brancalhão, R.M.C.; Ribeiro, L.D.F.C.; Natali, M.R.M. Silkworm Sericin: Properties and Biomedical Applications. *BioMed Res. Int.* **2016**, *2016*, 8175701. [CrossRef]

29. Das, G.; Shin, H.S.; Campos, E.V.R.; Fraceto, L.F.; del Pilar Rodriguez-Torres, M.; Mariano, K.C.F.; de Araujo, D.R.; Fernández-Luqueño, F.; Grillo, R.; Patra, J.K. Sericin based nanoformulations: A comprehensive review on molecular mechanisms of interaction with organisms to biological applications. *J. Nanobiotechnol.* **2021**, *19*, 30. [CrossRef]
30. Abdolmaleki, A.; Mahmoudian, M. Use of biomass sericin as matrices in functionalized graphene/sericin nanocomposites for the removal of phenolic compounds. *Heliyon* **2020**, *6*, e04955. [CrossRef]
31. Sim, G.S.; Nanthagopal, M.; Santhoshkumar, P.; Park, J.W.; Ho, C.W.; Shaji, N.; Kim, H.K.; Lee, C.W. Biomass-derived nitrogen-doped carbon on $LiFePO_4$ material for energy storage applications. *J. Alloys Compd.* **2022**, *902*, 163720. [CrossRef]
32. Park, K.J.; Hwang, J.Y.; Ryu, H.H.; Maglia, F.; Kim, S.J.; Lamp, P.; Yoon, C.S.; Sun, Y.K. Degradation Mechanism of Ni-Enriched NCA Cathode for Lithium Batteries: Are Microcracks Really Critical? *ACS Energy Lett.* **2019**, *4*, 1394–1400. [CrossRef]
33. Shan, W.; Huang, S.; Zhang, H.; Hou, X. Surface coating for high-nickel cathode materials to achieve excellent cycle performance at elevated temperatures. *J. Alloy. Compd.* **2021**, *862*, 158022. [CrossRef]
34. Wu, F.; Wang, M.; Su, Y.; Chen, S.; Xu, B. Effect of TiO_2-coating on the electrochemical performances of $LiCo_{1/3}Ni_{1/3}Mn_{1/3}O_2$. *J. Power Sources* **2009**, *191*, 628–632. [CrossRef]
35. Song, C.; Wang, W.; Peng, H.; Wang, Y.; Zhao, C.; Zhang, H.; Tang, Q.; Lv, J.; Du, X.; Dou, Y. Improving the electrochemical performance of $LiNi_{0.80}Co_{0.15}Al_{0.05}O_2$ in lithium ion batteries by $LiAlO_2$ surface modification. *Appl. Sci.* **2018**, *8*, 378. [CrossRef]
36. Li, P.; Jiao, Y.; Yao, S.; Wang, L.; Chen, G. Dual role of nickel foam in NiCoAl-LDH ensuring high-performance for asymmetric supercapacitors. *New J. Chem.* **2019**, *43*, 3139–3145. [CrossRef]
37. Gao, P.; Jiang, Y.; Zhu, Y.; Hu, H. Improved cycle performance of nitrogen and phosphorus co-doped carbon coatings on lithium nickel cobalt aluminum oxide battery material. *J. Mater. Sci.* **2018**, *53*, 9662–9673. [CrossRef]
38. Jamil, S.; Ran, Q.; Yang, L.; Huang, Y.; Cao, S.; Yang, X.; Wang, X. Improved high-voltage performance of $LiNi_{0.87}Co_{0.1}Al_{0.03}O_2$ by Li+-conductor coating. *Chem. Eng. J.* **2021**, *407*, 126442. [CrossRef]
39. Jabeen, M.; Ishaq, M.; Song, W.; Xu, L.; Deng, Q. Synthesis of Ni/Co/Al-layered triple hydroxide@brominated graphene hybrid on nickel foam as electrode material for high-performance supercapacitors. *RSC Adv.* **2017**, *7*, 46553–46565. [CrossRef]
40. Genieser, R.; Ferrari, S.; Loveridge, M.; Beattie, S.D.; Beanland, R.; Amari, H.; West, G.; Bhagat, R. Lithium ion batteries (NMC/graphite) cycling at 80 °C: Different electrolytes and related degradation mechanism. *J. Power Sources* **2018**, *373*, 172–183. [CrossRef]
41. Chulliyote, R.; Hareendrakrishnakumar, H.; Raja, M.; Gladis, J.M.; Stephan, A.M. Sulfur-Immobilized Nitrogen and Oxygen Co–Doped Hierarchically Porous Biomass Carbon for Lithium-Sulfur Batteries: Influence of Sulfur Content and Distribution on Its Performance. *ChemistrySelect* **2017**, *2*, 10484–10495. [CrossRef]
42. Xing, Z.; Ju, Z.; Zhao, Y.; Wan, J.; Zhu, Y.; Qiang, Y.; Qian, Y. One-pot hydrothermal synthesis of Nitrogen-doped graphene as high-performance anode materials for lithium ion batteries. *Sci. Rep.* **2016**, *6*, 26146. [CrossRef] [PubMed]
43. Zhu, H.; Deng, W.; Chen, L.; Zhang, S. Nitrogen doped carbon layer of Li_2MnSiO_4 with enhanced electrochemical performance for lithium ion batteries. *Electrochim. Acta* **2019**, *295*, 956–965. [CrossRef]
44. Zhou, P.; Zhang, Z.; Meng, H.; Lu, Y.; Cao, J.; Cheng, F.; Tao, Z.; Chen, J. SiO_2-coated $LiNi_{0.915}Co_{0.075}Al_{0.01}O_2$ cathode material for rechargeable Li-ion batteries. *Nanoscale* **2016**, *8*, 19263–19269. [CrossRef]
45. Han, C.J.; Yoon, J.H.; Cho, W., II; Jang, H. Electrochemical properties of $LiNi_{0.8}Co_{0.2-x}Al_xO_2$ prepared by a sol-gel method. *J. Power Sources* **2004**, *136*, 132–138. [CrossRef]
46. Huang, B.; Li, X.; Wang, Z.; Guo, H.; Xiong, X. Synthesis of Mg-doped $LiNi_{0.8}Co_{0.15}Al_{0.05}O_2$ oxide and its electrochemical behavior in high-voltage lithium-ion batteries. *Ceram. Int.* **2014**, *40*, 13223–13230. [CrossRef]
47. Chen, T.; Li, X.; Wang, H.; Yan, X.; Wang, L.; Deng, B.; Ge, W.; Qu, M. The effect of gradient boracic polyanion-doping on structure, morphology, and cycling performance of Ni-rich $LiNi_{0.8}Co_{0.15}Al_{0.05}O_2$ cathode material. *J. Power Sources* **2018**, *374*, 1–11. [CrossRef]
48. Santhoshkumar, P.; Prasanna, K.; Jo, Y.N.; Sivagami, I.N.; Kang, S.H.; Lee, C.W. A facile and highly efficient short-time homogenization hydrothermal approach for the smart production of high-quality α-Fe_2O_3 for rechargeable lithium batteries. *J. Mater. Chem. A* **2017**, *5*, 16712–16721. [CrossRef]
49. Nanthagopal, M.; Santhoshkumar, P.; Shaji, N.; Sim, G.S.; Park, J.W.; Senthil, C.; Lee, C.W. An encapsulation of nitrogen and sulphur dual-doped carbon over $Li[Ni_{0.8}Co_{0.1}Mn_{0.1}]O_2$ for lithium-ion battery applications. *Appl. Surf. Sci.* **2020**, *511*, 145580. [CrossRef]
50. Yu, J.; Li, H.; Zhang, G.; Li, X.; Huang, J.; Li, C.; Wei, C.; Xiao, C. Carbon nanotubes coating on $LiNi_{0.8}Co_{0.15}Al_{0.05}O_2$ as cathode materials for lithium battery. *Int. J. Electrochem. Sci.* **2017**, *12*, 11892–11903. [CrossRef]
51. Chung, Y.; Ryu, S.H.; Ju, J.H.; Bak, Y.R.; Hwang, M.J.; Kim, K.W.; Cho, K.K.; Ryu, K.S. A surfactant-based method for carbon coating of $LiNi_{0.8}Co_{0.15}Al_{0.05}O_2$ cathode in Li ion batteries. *Bull. Korean Chem. Soc.* **2010**, *31*, 2304–2308. [CrossRef]
52. Yoon, S.; Jung, K.N.; Yeon, S.H.; Jin, C.S.; Shin, K.H. Electrochemical properties of $LiNi_{0.8}Co_{0.15}Al_{0.05}O_2$-graphene composite as cathode materials for lithium-ion batteries. *J. Electroanal. Chem.* **2012**, *683*, 88–93. [CrossRef]
53. Liu, Z.; Wang, Z.; Lu, T.; Dai, P.; Gao, P.; Zhu, Y. Modification of $LiNi_{0.8}Co_{0.15}Al_{0.05}O_2$ using nanoscale carbon coating. *J. Alloy. Compd.* **2018**, *763*, 701–710. [CrossRef]

nanomaterials

Article

Optically Controlled Terahertz Dynamic Beam Splitter with Adjustable Split Ratio

Shan Yin [1], Dehui Zeng [1], Yuting Chen [1], Wei Huang [1,*], Cheng Zhang [2], Wentao Zhang [1,*] and Yiwen E [3]

1 Guangxi Key Laboratory of Optoelectronic Information Processing, School of Optoelectronic Engineering, Guilin University of Electronic Technology, Guilin 541004, China; syin@guet.edu.cn (S.Y.); mailzengdh@163.com (D.Z.); cyt5124@163.com (Y.C.)
2 Hubei Engineering Research Center of RF-Microwave Technology and Application, School of Science, Wuhan University of Technology, Wuhan 430070, China; czhang2020@whut.edu.cn
3 The Institute of Optics, University of Rochester, Rochester, NY 14627, USA; ye2@ur.rochester.edu
* Correspondence: weihuang@guet.edu.cn (W.H.); zhangwentao@guet.edu.cn (W.Z.)

Abstract: The beam splitter is an important functional device due to its ability to steer the propagation of electromagnetic waves. The split-ratio-variable splitter is of significance for optical, terahertz and microwave systems. Here, we are the first (to our knowledge) to propose an optically controlled dynamic beam splitter with adjustable split ratio in the terahertz region. Based on the metasurface containing two sets of reversed phase-gradient supercells, we split the terahertz wave into two symmetrical beams. Associated with the reconfigurable pump laser pattern programmed with the spatial light modulator, dynamic modulation of the split ratio varying from 1:1 to 15:1 is achieved. Meanwhile, the beam splitter works at a split angle of 36° for each beam. Additionally, we obtain an exponential relationship between the split ratio and the illumination proportion, which can be used as theoretical guidance for beam splitting with an arbitrary split ratio. Our novel beam splitter shows an outstanding level of performance in terms of the adjustable split ratio and stable split angles and can be used as an advanced method to develop active functional devices applied to terahertz systems and communications.

Keywords: beam splitter; metasurface; terahertz; dynamic

Citation: Yin, S.; Zeng, D.; Chen, Y.; Huang, W.; Zhang, C.; Zhang, W.; E, Y. Optically Controlled Terahertz Dynamic Beam Splitter with Adjustable Split Ratio. *Nanomaterials* **2022**, *12*, 1169. https://doi.org/10.3390/nano12071169

Academic Editor: Julian Maria Gonzalez Estevez

Received: 9 March 2022
Accepted: 30 March 2022
Published: 31 March 2022

1. Introduction

The manipulation of electromagnetic (EM) waves is significant for optoelectrical devices applied in communications [1], imaging [2] and detection [3]. Metasurfaces, emerging artificial microstructures that can manipulate EM waves, have attracted significant attention. With well-designed structures, metasurfaces can modulate the amplitude [4–6], phase [7], polarization [8], and chirality [9]. Owing to their excellent ability of control the wavefront with a phase-gradient configuration, metasurfaces have been developed for various functional devices like the meta-lens [10,11], beam splitter [12,13], and vortex generator [14,15].

The beam splitter is an important component of optical, microwave, and terahertz systems, because it can deflect and divide EM waves to expected angles [16]. Metasurface-based beam splitters were recently reported on. Niu et al. proposed a terahertz reflectarray as a polarizing beam splitter [17]. Ni et al. demonstrated a broadband non-polarizing terahertz beam splitter based on an all-dielectric metasurface [18]. Further, active devices that dynamically manipulate beam splitting have been proposed. The active metasurface terahertz deflector was presented and used to modulate the intensity of the deflected waves via the bias voltage [19]. Graphene-based metasurfaces were shown to dynamically steer the deflection angle by changing the Fermi energy [20]. A coding scheme was introduced to generate more beams by arranging different coding sequences [21,22]. Zhang et al. employed the Pancharatnam–Berry (PB) coding metasurface to form spin-controlled multiple beams with different polarizations [23]. Li et al. presented electronic controlled tunable dielectric metasurfaces modulated by liquid crystal with a beam deflection angle of 11° [24]. Previous work mainly focused on varying the beam splitting number, switching the deflection angles of different beams, or simultaneously modulating the intensity of beams, but active beam splitters with an adjustable split ratio have rarely been reported. In 2020, Kocer et al. presented a dynamic beam splitter employing an elastic substrate based all-dielectric metasurface to undergo splitting in three ways with different split angles and power ratios [25]. However, a dynamic beam splitter that can change the split ratio and maintain the same split angles has not yet been developed.

In this paper, we propose, for the first time (to our knowledge), a photo-induced terahertz beam splitter with an adjustable split ratio at certain split angles. The beam splitter is based on a metasurface composed of metallic structures deposited on the substrate of silicon on sapphire (SOS). The metallic structures consist of two groups of phase-gradient C-shaped split-ring resonators (SRRs). With the inverse phase gradient, the two groups of SRRs diffract the incident waves at two different angles. Combined with the pump laser pattern programmed by digital micromirror devices (DMD), we can control the illuminated area of the metasurface and, therefore, excite photoinduced carriers of the silicon layer. The structures located in the illuminated part will be opaque due to the metallized silicon, which reduces the intensity of the corresponding split beam. Accordingly, the split ratio can be arbitrarily adjusted by choosing the illumination ratio via the patterned pump light. We demonstrate that the split ratio of two beams can dynamically vary from 1:1 to 15:1 with an unchanged deflection angle of 36° for each beam. Implemented by the phase-gradient metasurface combined with the reconfigurable pump laser pattern, our novel beam splitter exhibits an excellent performance regarding the dynamic terahertz beam steering concept, and this optically controlled scheme provides an advanced method to develop more active functional devices that can be applied to terahertz systems and communications.

2. Design and Simulation

Our proposed terahertz beam splitter combined with the optically controlled scheme is presented in Figure 1a, and a detailed view of the beam splitter is shown in Figure 1b. C-shaped metallic split-ring resonators (SRRs) are deposited on the top surface, and 200 nm thick aluminum with a conductivity level of 3.56×10^7 S/m is used as a metallic material [26]. The substrate is a 1.5 μm thick silicon layer grown on a 500 μm thick sapphire wafer, and the permittivities of silicon and sapphire are 11.7 and 10.5, respectively [27]. For all simulations, the x-polarized terahertz plane wave was normally incident to the metasurface, and the y-polarized electric field was detected. The red stripes denote the reconfigurable pump laser pattern illuminating the SRRs and the silicon layer, which can be programmed by the digital micromirror devices (DMD) via a computer.

For each SRR whose structure is presented in Figure 1c,d, the width of the ring is w = 5 μm, the angle of the gap is $2a$, the external radius of the ring is r, and the axis of symmetry is rotated by α relative to the x-axis. The SRRs are aligned in the x and y directions with a period of P = 80 μm. Due to the asymmetry of the SRR under x-polarized wave incidence, the polarization can be partially converted to the orthogonal direction in a specific frequency range, and when $\alpha = \pm 45°$, the polarization conversion ratio is maximized [13]. Therefore, the phase and amplitude of the orthogonally polarized wave can be controlled simultaneously by varying the radius (r) and angle of the gap (a) of the SRRs.

In our previous work [11], the amplitude and phase of C-shaped SRRs were investigated. We chose eight pairs of specific parameters (a and r) of SRRs, as listed in Table 1, and their transmission phases and amplitudes at 0.8 THz are shown in Figure 2. Obviously, the phase of the eight SRRs successively increased by $\pi/4$, and the amplitude remained unchanged. Accordingly, we configured the eight SRRs as a positive phase-gradient supercell A whose phase uniformly ranged from 0 to 2π. Additionally, we inversely ordered the eight SRRs to construct another negative phase-gradient supercell B, whose phase uniformly decreased by $\pi/4$. Hence, we assembled the two supercells with inverse phase gradients as a basic unit (A + B) of the array, as shown in the inset of Figure 1a. It was possible to generate the two opposite deflected wavefronts, namely two splitting beams, simultaneously. The deflection angle depends on the phase gradient value of metasurfaces, and the relationship between the two obeys the generalized Snell's law of refraction defined as [7]:

$$n_t \sin\theta_t - n_i \sin\theta_i = \frac{\lambda_0}{2\pi}\frac{\mathrm{d}\Phi}{\mathrm{d}x} \tag{1}$$

where θ_t and θ_i represent the refractive and incident angles, respectively; n_t and n_i are the refractive indexes of the vacuum and sapphire, respectively; λ_0 indicates the wavelength in a vacuum; and $\mathrm{d}\Phi/\mathrm{d}x$ is the phase gradient of the metasurface. In our designed metasurface, the incident angle is $\theta_i = 0$, $n_t = 1$, and $n_i = 3.24$. The wavelength λ_0 corresponds to a frequency of 0.8 THz. $\mathrm{d}\Phi/\mathrm{d}x = 2\pi/D$ (D = 640 μm is the period of the supercell in x-direction). Therefore, by containing the positive and negative phase gradients simultaneously, the deflection angles of the two splitting beams were both calculated as $\pm 35.87°$ for our beam splitter.

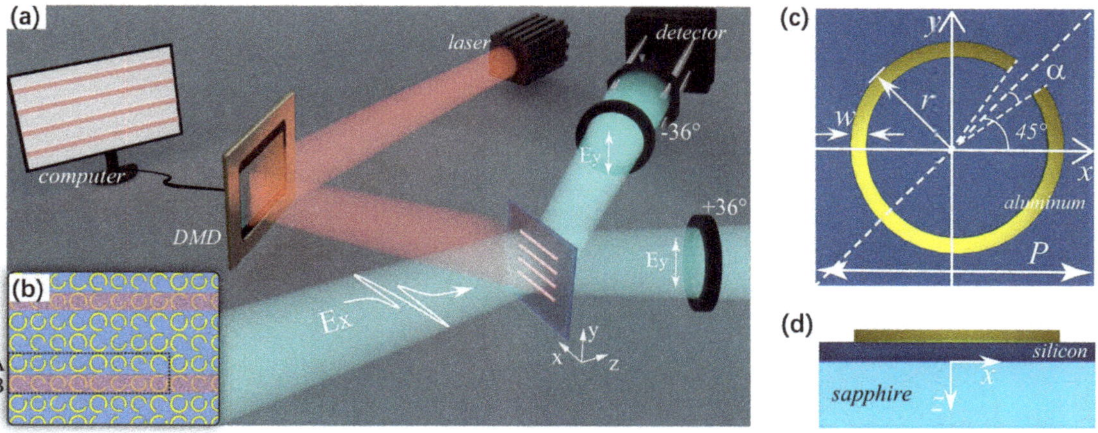

Figure 1. (**a**) Schematic of the proposed beam splitter combined with the optically controlled scheme. (**b**) Partial detailed view of the beam splitter. The dashed frame denotes the basic unit (A + B). (**c**) The geometrical parameters of an SRR. (**d**) Schematic of the multilayered structure.

Table 1. Geometrical parameters of the eight SRRs forming the supercell A.

Number	1	2	3	4	5	6	7	8
r (μm)	35	31.5	32	34.55	35	31.5	32	34.55
a (°)	10.5	15	43	73.5	−10.5	−15	−43	−73.5

To modulate the intensity of each splitting beam, we simulated the transmission amplitude of the SRRs excited by an external pump laser. The conductivity of photoconductive silicon σ_s is variable with the optical pump power [13,28]. When the metasurface is illuminated by the pump laser, the photoinduced carriers of the silicon layer will be excited, and σ_s = 8000 S/m was used to model the high conductivity of silicon. Due to the increased reflection and absorption of the terahertz wave in silicon under the pump, the transmission amplitude of the terahertz wave dropped to 0.015, as shown in Figure 2. Without the pump, the silicon layer returned to the static state (the conductivity σ_s is 0 S/m), and the transmission amplitude remained around 0.47. Hence, combined with the pump laser pattern programmed by the DMD, we were able to control the illuminated area of the metasurface, as shown in Figure 1a,b with the red reconfigurable stripe pattern, and therefore, we could modify the intensities of the splitting beams.

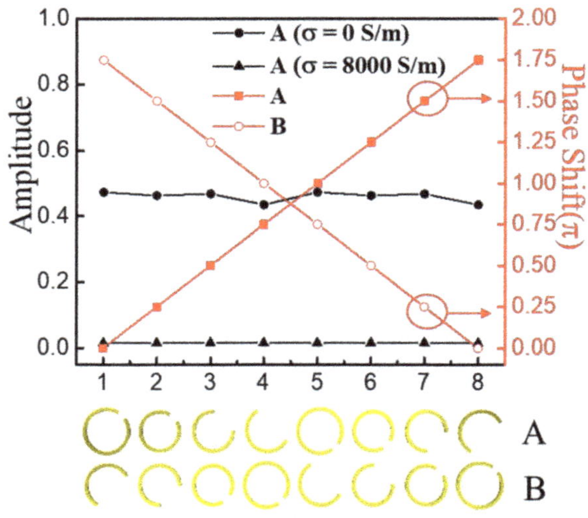

Figure 2. Transmission amplitudes and phases of eight SRRs in supercells A and B at 0.8 THz with the x-polarized terahertz incidence and y-polarized detection.

3. Results and Discussion

Since the beam splitter is formed by an array of two supercells, we were able to selectively illuminate the area of the metasurface, and the proportion of each supercell under illumination could be determined. The silicon located in the illuminated part turned opaque and reduced the intensity of the corresponding split beam. As a result, by controlling the illumination pattern, we were able to adjust the intensity of each splitting beam and, therefore, realize the variable split ratio (SR).

The function of the beam splitter is displayed in Figure 3. We simulated the electric field distributions of the deflected waves deflected by the beam splitter under four situations with different illuminated patterns at 0.8 THz. Figure 3a shows the situation without a pump laser, and Figure 3e demonstrates the electric field distributions. Beyond the near-field zone of coherent superposition (about 4.1 mm $\approx 10.9\lambda$), two deflected beams were symmetrically split when the terahertz wave was normally incident to the top metasurface. Generally, the split ratio (SR) can be defined as the ratio of the two splitting beams [29]:

$$\mathrm{SR} = \frac{I_{left}}{I_{right}} \tag{2}$$

where I_{left} and I_{right} represent the electric field intensities of the outgoing left and right beams, respectively. Here, the left and right beams were deflected by supercells A and B, respectively. Consequently, I_{left} and I_{right} were inversely correlated to the illuminated areas of supercells A and B, respectively. In this case, there was no illuminated area, so the intensities of the two splitting beams were equal (the numerical value was about 0.3 V/m), and the SR of our beam splitter was 1:1.

In Figure 3b, the horizontal stripe patterned pump light (denoted with red stripes) is used to illuminate the area of supercell B, where the width of each stripe is $d_1 = P = 80$ μm and is designed to overlap supercell B. With a period of $d = 4P$ covering 2 basic units, the stripe pattern illuminated half of the supercell B area; hence, the intensity of the right beam attenuated to about half the value of the original intensity (see Figure 3f), while the intensity of the left beam remained at the original level because supercell A was not pumped. As a result, the SR became 1.8:1 (0.30:0.17) in this situation. Similarly, by designing the pump laser pattern well and selectively illuminating fractional supercell B, we were able to obtain different split ratios. In Figure 3c, the period of the stripe illuminating pattern is shown to be $d = 10P$, which covers 5 basic units, and every period contains four stripes. Hence, 80% of supercell B was illuminated, and the right beam was evidently weakened, as shown in Figure 3g, so that the SR changed to 5:1 (0.30:0.06). This further increased the proportion of supercell B to 100%, corresponding to the pattern shown in Figure 3d, and the period of the stripe shrunk to $d = 2P$ while supercell B was fully selectively illuminated. The right beam nearly vanished while the left beam maintained the high intensity, leading to the the highest SR of 15:1 (0.30:0.02), as shown in Figure 3h.

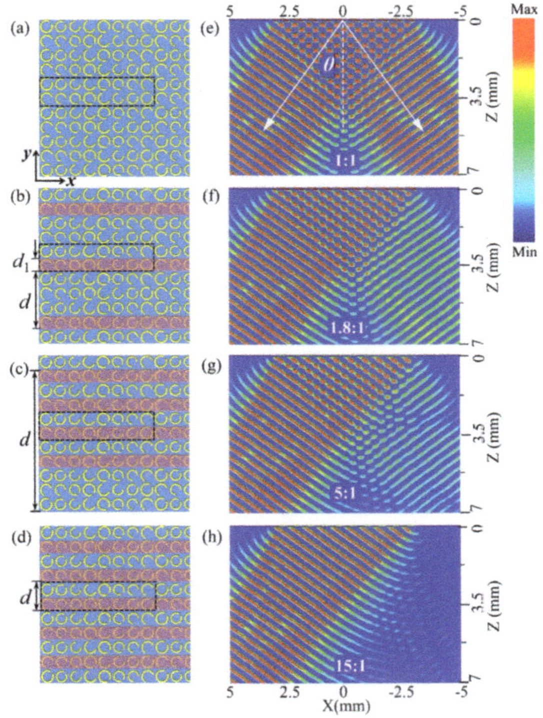

Figure 3. (**a–d**) Partial sketches of the designed metasurface with the red stripe patterned illumination. The dashed frames denote the basic unit cells. (**e–h**) Simulated electric field distributions of beam splitting in the corresponding illuminated patterns at 0 8 THz.

To clearly present the adjustable property of our beam splitter, we plotted the electric field distributions of the reflected waves versus the deflection angle, as shown in Figure 4. We output the electric field intensity of the splitting beams at a distance of 5 mm on the *xz*-plane. Obviously, the transmitted wave was symmetrically divided in two main directions, and the left split beam maintained the maximal value, while the right one attenuated, corresponding to the cases presented in Figure 3a–d. Note that the deflection angles of the two split beams were consistent in all situations, and the maximum value appeared at $\pm 36°$ (denoted with the vertical dashed in Figure 4), which is in excellent agreement with Equation (1). Hence, the separation angle between two split beams can reach 72°. Additionally, the split beams were highly concentrated, and the angle of divergence was less than 10°. These results indicate the outstanding performance of our beam splitter in terms of the adjustable split ratio and stable split angles. Compared with previous work [29,30] in which the split ratio or angle were changed by designing different structures, our beam splitter has the advantage of having an active metasurface that can realize multifunctionalization with one sample. This is beneficial for the development of integrated devices in terahertz systems.

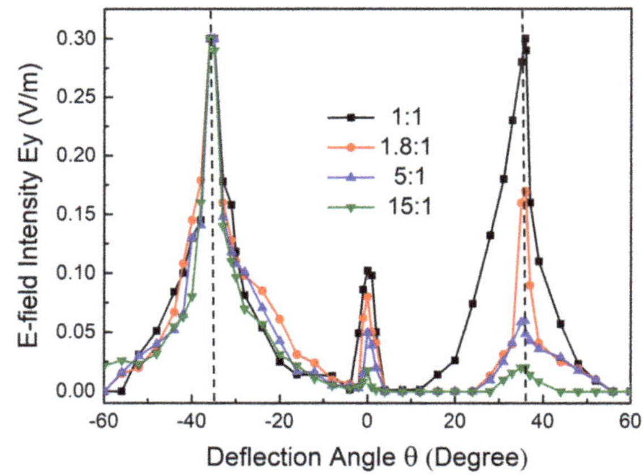

Figure 4. Electric field intensity of reflected waves. The vertical dashed lines denote the maximum values at ±36°.

Additionally, we extracted the relationship between the split ratio (SR) and the proportion of the pump illumination area for supercell B (η), and the results are plotted in Figure 5 with scattered dots. It was concluded that the split ratio increased with the enlargement of the illumination area, showing an exponentially increasing trend. Since the starting SR was 1:1, the relationship was fitted with the exponential formula $SR = 1 + 0.028 \exp(0.062\eta)$. As shown in Figure 5, the fitted curve (red line) agrees well with the simulated data (dots). Based on the SR-η formula, theoretically, the arbitrary intensity of the split beam can be manipulated through the optically controlled system by editing the pump pattern via DMD in real-time. For experimental feasibility, reconfigurable terahertz grating was demonstrated with a silicon wafer using DMD [31], and it was shown that the spatial profile at the silicon wafer can be controlled at a resolution of 10 μm [32], so it is likely that the flexible modulation of the split ratio in our beam splitter could be achieved through association with the reconfigurable pump laser pattern. Until now, this optically controlled scheme has not been applied to the dynamic beam splitter, and our proposal can provide a flexible means to manipulate the terahertz wave.

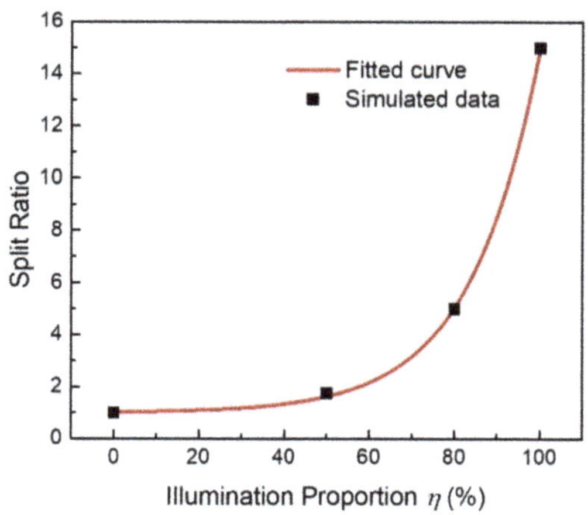

Figure 5. Relationship between the beam split ratio and the illumination proportion of supercell B.

4. Conclusions

A novel beam splitter with an arbitrary controllable split ratio in the terahertz region was proposed. Based on the metasurface containing two sets of reversed phase-gradient supercells, we split the terahertz wave into two symmetrical beams. Through an association with the reconfigurable pump laser pattern programmed with the spatial light modulation scheme, dynamic modulation of the split ratio varying from 1:1 to 15:1 was achieved. Meanwhile, the beam splitter was shown to work at a deflection angle of 36° for each beam, and the separation angle between two split beams could exceed 70°. We obtained an exponential relationship between the split ratio and the illumination proportion, which could be used as theoretical guidance for beam splitting with an arbitrary split ratio. To the best of our knowledge, this is the first proposal of a terahertz beam splitter with a dynamically controllable split ratio at stable split angles. We believe that this design represents an advanced way to develop more active function devices and to promote their practical application in optical, microwave, and terahertz systems.

5. Patent

Shan Yin, Dehui Zeng, Wei Huang, Ling Guo, Wentao Zhang, Xianming Xiong. A dynamic adjustable terahertz beam splitter based on composite metasurface. 202210034731.2 (China).

Author Contributions: Conceptualization, S.Y.; Data curation, D.Z.; Formal analysis, Y.C.; Funding acquisition, W.Z.; Investigation, Y.E.; Methodology, W.H.; Resources, C.Z.; Supervision, W.Z.; Writing—original draft, D.Z.; Writing—review & editing, S.Y. and W.H. All authors have read and agreed to the published version of the manuscript.

Funding: This work was supported by the National Natural Science Foundation of China (62005059, 61965005), the Innovation Project of GUET Graduate Education (2021YCXS127), funding from the Guangxi Oversea 100 Talent Project (W.H.), and funding from the Guangxi distinguished expert project (W.Z.).

Institutional Review Board Statement: Not applicable.

Informed Consent Statement: Not applicable.

Data Availability Statement: The raw data are available upon request from the authors.

Conflicts of Interest: The authors declare no conflict of interest.

References

1. Nagatsuma, T.; Ducournau, G.; Renaud, C.C. Advances in terahertz communications accelerated by photonics. *Nat. Photonics* **2016**, *10*, 371–379. [CrossRef]
2. Guerboukha, H.; Nallappan, K.; Skorobogatiy, M. Toward real-time terahertz imaging. *Adv. Opt. Photonics* **2018**, *10*, 843. [CrossRef]
3. Rogalski, A.; Sizov, F. Terahertz detectors and focal plane arrays. *Opto Electron. Rev.* **2011**, *19*, 346–404. [CrossRef]
4. Wang, L.; Zhang, Y.; Guo, X.; Chen, T.; Yang, Z. A review of thz modulators with dynamic tunable metasurfaces. *Nanomaterials* **2019**, *9*, 965. [CrossRef]
5. Dong, B.; Zhang, C.; Guo, G.; Zhang, X.; Wang, Y.; Huang, L.; Ma, H.; Cheng, Q. Bst-silicon hybrid terahertz meta-modulator for dual-stimuli-triggered opposite transmission amplitude control. *Nanophotonics* **2022**. [CrossRef]
6. Lu, X.; Dong, B.; Zhu, H.; Shi, Q.; Tang, L.; Su, Y.; Zhang, C.; Huang, W.; Cheng, Q. Two-channel vo2 memory meta-device for terahertz waves. *Nanomaterials* **2021**, *11*, 3409. [CrossRef]
7. Yu, N.; Genevet, P.; Kats, M.A.; Aieta, F.; Tetienne, J.P.; Capasso, F.; Gaburro, Z. Light propagation with phase discontinuities: Generalized laws of reflection and refraction. *Science* **2011**, *334*, 333–337. [CrossRef]
8. Zhang, H.; Zhang, X.; Xu, Q.; Tian, C.; Wang, Q.; Xu, Y.; Li, Y.; Gu, J.; Tian, Z.; Ouyang, C.; et al. High-efficiency dielectric metasurfaces for polarization-dependent terahertz wavefront manipulation. *Adv. Opt. Mater.* **2018**, *6*, 1700773. [CrossRef]
9. Wang, Q.; Plum, E.; Yang, Q.; Zhang, X.; Xu, Q.; Xu, Y.; Han, J.; Zhang, W. Reflective chiral meta-holography: Multiplexing holograms for circularly polarized waves. *Light Sci. Appl.* **2018**, *7*, 25. [CrossRef]
10. Khorasaninejad, M.; Chen, W.T.; Devlin, R.C.; Oh, J.; Zhu, A.Y.; Capasso, F. Metalenses at visible wavelengths: Diffraction-limited focusing and subwavelength resolution imaging. *Science* **2016**, *352*, 1190–1194. [CrossRef]
11. Yin, S.; Liang, Y.; Zeng, D.; Tian, Y.; Zhong, P.; Guo, L.; Huang, W.; Zhang, W. Dynamic switching of coaxial focus based on terahertz meta-lens. *Appl. Opt.* **2021**, *60*, 3629–3633. [CrossRef] [PubMed]

12. Fu, X.; Yang, F.; Liu, C.; Wu, X.; Cui, T.J. Terahertz beam steering technologies: From phased arrays to field-programmable metasurfaces. *Adv. Opt. Mater.* **2020**, *8*, 1900628. [CrossRef]
13. Cong, L.; Srivastava, Y.K.; Zhang, H.; Zhang, X.; Han, J.; Singh, R. All-optical active thz metasurfaces for ultrafast polarization switching and dynamic beam splitting. *Light Sci. Appl.* **2018**, *7*, 28. [CrossRef] [PubMed]
14. Yang, Y.; Wang, W.; Moitra, P.; Kravchenko, I.I.; Briggs, D.P.; Valentine, J. Dielectric meta-reflectarray for broadband linear polarization conversion and optical vortex generation. *Nano Lett.* **2014**, *14*, 1394–1399. [CrossRef]
15. Bao, Y.; Ni, J.; Qiu, C. A minimalist single-layer metasurface for arbitrary and full control of vector vortex beams. *Adv. Mater.* **2020**, *32*, 1905659. [CrossRef]
16. Liu, S.; Cui, T.J.; Zhang, L.; Xu, Q.; Wang, Q.; Wan, X.; Gu, J.Q.; Tang, W.X.; Qing Qi, M.; Han, J.G.; et al. Convolution operations on coding metasurface to reach flexible and continuous controls of terahertz beams. *Adv. Sci.* **2016**, *3*, 1600156. [CrossRef]
17. Niu, T.; Withayachumnankul, W.; Upadhyay, A.; Gutruf, P.; Abbott, D.; Bhaskaran, M.; Sriram, S.; Fumeaux, C. Terahertz reflectarray as a polarizing beam splitter. *Opt. Express* **2014**, *22*, 16148–16160. [CrossRef]
18. Ni, H.; Fang, X.; Hou, Z.; Li, Y.; Assouar, B. High-efficiency anomalous splitter by acoustic meta-grating. *Phys. Rev. B* **2019**, *100*, 104104. [CrossRef]
19. Su, X.; Ouyang, C.; Xu, N.; Cao, W.; Wei, X.; Song, G.; Gu, J.; Tian, Z.; O'Hara, J.F.; Han, J.; et al. Active metasurface terahertz deflector with phase discontinuities. *Opt. Express* **2015**, *23*, 27152–27158. [CrossRef]
20. Chen, D.; Yang, J.; Huang, J.; Bai, W.; Xie, W. The novel graphene metasurfaces based on split-ring resonators for tunable polarization switching and beam steering at terahertz frequencies. *Carbon* **2019**, *154*, 350–356. [CrossRef]
21. Cui, T.J.; Qi, M.Q.; Wan, X.; Zhao, J.; Cheng, Q. Coding metamaterials, digital metamaterials and programmable metamaterials. *Light Sci. Appl.* **2014**, *3*, e218. [CrossRef]
22. Xing, X.; Li, Y.; Lu, Y.; Zhang, W.; Zhang, X.; Han, J.; Zhang, W. Terahertz metamaterial beam splitters based on untraditional coding scheme. *Opt. Express* **2019**, *27*, A1627–A1635. [CrossRef] [PubMed]
23. Zhang, L.; Liu, S.; Li, L.; Cui, T.J. Spin-controlled multiple pencil beams and vortex beams with different polarizations generated by pancharatnam-berry coding metasurfaces. *ACS Appl. Mat. Interfaces* **2017**, *9*, 36447–36455. [CrossRef] [PubMed]
24. Li, S.-Q.; Xu, X.; Veetil, R.M.; Valuckas, V.; Paniagua-Domínguez, R.; Kuznetsov, A.I. Phase-only transmissive spatial light modulator based on tunable dielectric metasurface. *Science* **2019**, *364*, 1087–1090. [CrossRef]
25. Kocer, H.; Durna, Y.; Kurt, H.; Ozbay, E. Dynamic beam splitter employing an all-dielectric metasurface based on an elastic substrate. *Opt. Lett.* **2020**, *45*, 3521–3524. [CrossRef]
26. Srivastava, Y.K.; Manjappa, M.; Cong, L.; Cao, W.; Al-Naib, I.; Zhang, W.; Singh, R. Ultrahigh-q fano resonances in terahertz metasurfaces: Strong influence of metallic conductivity at extremely low asymmetry. *Adv. Opt. Mater.* **2015**, *4*, 457–463. [CrossRef]
27. Manjappa, M.; Srivastava, Y.K.; Cong, L.; Al-Naib, I.; Singh, R. Active photoswitching of sharp fano resonances in thz metadevices. *Adv. Mater.* **2017**, *29*, 1603355. [CrossRef]
28. Shen, N.H.; Massaouti, M.; Gokkavas, M.; Manceau, J.M.; Ozbay, E.; Kafesaki, M.; Koschny, T.; Tzortzakis, S.; Soukoulis, C.M. Optically implemented broadband blueshift switch in the terahertz regime. *Phys. Rev. Lett.* **2011**, *106*, 037403. [CrossRef]
29. Wei, M.; Xu, Q.; Wang, Q.; Zhang, X.; Li, Y.; Gu, J.; Tian, Z.; Zhang, X.; Han, J.; Zhang, W. Broadband non-polarizing terahertz beam splitters with variable split ratio. *Appl. Phys. Lett.* **2017**, *111*, 071101. [CrossRef]
30. He, Q.; Shen, Z. Polarization-insensitive beam splitter with variable split angles and ratios based on phase gradient metasurfaces. *Nanomaterials* **2021**, *12*, 113. [CrossRef]
31. He, J.W.; Wang, X.K.; Xie, Z.W.; Xue, Y.Z.; Wang, S.; Zhang, Y. Reconfigurable terahertz grating with enhanced transmission of te polarized light. *APL Photonics* **2017**, *2*, 076102. [CrossRef]
32. Kanda, N.; Konishi, K.; Kuwata-Gonokami, M. All-photoinduced terahertz optical activity. *Opt. Lett.* **2014**, *39*, 3274–3277. [CrossRef] [PubMed]

MDPI
St. Alban-Anlage 66
4052 Basel
Switzerland
www.mdpi.com

Nanomaterials Editorial Office
E-mail: nanomaterials@mdpi.com
www.mdpi.com/journal/nanomaterials

www.ingramcontent.com/pod-product-compliance
Lightning Source LLC
Chambersburg PA
CBHW061456220926
51850CB00004B/106
* 9 7 8 3 7 2 5 8 0 8 7 9 3 *